Organic Reactions

Organic Reactions

VOLUME 104

Published by John Wiley & Sons, Inc., Hoboken, New Jersey
Published simultaneously in Canada

For general information on our other products and services or for technical support, please contact our
Customer Care Department within the United States at (800) 762-2974, outside the United States at
(317) 572-3993 or fax (317) 572-4002.

Wiley also publishes its books in a variety of electronic formats. Some content that appears in print may
not be available in electronic formats. For more information about Wiley products, visit our web site at
www.wiley.com.

Library of Congress Cataloging-in-Publication Data:

ISBN: 978-1-119-65150-5

Printed in the United States of America

SKY10022349_110920

INTRODUCTION TO THE SERIES
BY ROGER ADAMS, 1942

In the course of nearly every program of research in organic chemistry, the investigator finds it necessary to use several of the better-known synthetic reactions. To discover the optimum conditions for the application of even the most familiar one to a compound not previously subjected to the reaction often requires an extensive search of the literature; even then a series of experiments may be necessary. When the results of the investigation are published, the synthesis, which may have required months of work, is usually described without comment. The background of knowledge and experience gained in the literature search and experimentation is thus lost to those who subsequently have occasion to apply the general method. The student of preparative organic chemistry faces similar difficulties. The textbooks and laboratory manuals furnish numerous examples of the application of various syntheses, but only rarely do they convey an accurate conception of the scope and usefulness of the processes.

For many years American organic chemists have discussed these problems. The plan of compiling critical discussions of the more important reactions thus was evolved. The volumes of *Organic Reactions* are collections of chapters each devoted to a single reaction, or a definite phase of a reaction, of wide applicability. The authors have had experience with the processes surveyed. The subjects are presented from the preparative viewpoint, and particular attention is given to limitations, interfering influences, effects of structure, and the selection of experimental techniques. Each chapter includes several detailed procedures illustrating the significant modifications of the method. Most of these procedures have been found satisfactory by the author or one of the editors, but unlike those in *Organic Syntheses*, they have not been subjected to careful testing in two or more laboratories. Each chapter contains tables that include all the examples of the reaction under consideration that the author has been able to find. It is inevitable, however, that in the search of the literature some examples will be missed, especially when the reaction is used as one step in an extended synthesis. Nevertheless, the investigator will be able to use the tables and their accompanying bibliographies in place of most or all of the literature search so often required. Because of the systematic arrangement of the material in the chapters and the entries in the tables, users of the books will be able to find information desired by reference to the table of contents of the appropriate chapter. In the interest of economy, the entries in the indices have been kept to a minimum, and, in particular, the compounds listed in the tables are not repeated in the indices.

The success of this publication, which will appear periodically, depends upon the cooperation of organic chemists and their willingness to devote time and effort to the preparation of the chapters. They have manifested their interest already by the almost unanimous acceptance of invitations to contribute to the work. The editors will welcome their continued interest and their suggestions for improvements in *Organic Reactions*.

INTRODUCTION TO THE SERIES
BY SCOTT E. DENMARK, 2008

In the intervening years since "The Chief" wrote this introduction to the second of his publishing creations, much in the world of chemistry has changed. In particular, the last decade has witnessed a revolution in the generation, dissemination, and availability of the chemical literature with the advent of electronic publication and abstracting services. Although the exponential growth in the chemical literature was one of the motivations for the creation of *Organic Reactions*, Adams could never have anticipated the impact of electronic access to the literature. Yet, as often happens with visionary advances, the value of this critical resource is now even greater than at its inception.

From 1942 to the 1980's the challenge that *Organic Reactions* successfully addressed was the difficulty in compiling an authoritative summary of a preparatively useful organic reaction from the primary literature. Practitioners interested in executing such a reaction (or simply learning about the features, advantages, and limitations of this process) would have a valuable resource to guide their experimentation. As abstracting services, in particular *Chemical Abstracts* and later *Beilstein*, entered the electronic age, the challenge for the practitioner was no longer to locate all of the literature on the subject. However, *Organic Reactions* chapters are much more than a surfeit of primary references; they constitute a distillation of this avalanche of information into the knowledge needed to correctly implement a reaction. It is in this capacity, namely to provide focused, scholarly, and comprehensive overviews of a given transformation, that *Organic Reactions* takes on even greater significance for the practice of chemical experimentation in the 21st century.

Adams' description of the content of the intended chapters is still remarkably relevant today. The development of new chemical reactions over the past decades has greatly accelerated and has embraced more sophisticated reagents derived from elements representing all reaches of the Periodic Table. Accordingly, the successful implementation of these transformations requires more stringent adherence to important experimental details and conditions. The suitability of a given reaction for an unknown application is best judged from the informed vantage point provided by precedent and guidelines offered by a knowledgeable author.

As Adams clearly understood, the ultimate success of the enterprise depends on the willingness of organic chemists to devote their time and efforts to the preparation of chapters. The fact that, at the dawn of the 21st century, the series continues to thrive is fitting testimony to those chemists whose contributions serve as the foundation of this edifice. Chemists who are considering the preparation of a manuscript for submission to *Organic Reactions* are urged to contact the Editor-in-Chief.

PREFACE TO VOLUME 104

The universe is asymmetric and I am persuaded that life, as it is known to us, is a direct result of the asymmetry of the universe or of its indirect consequences.

Louis Pasteur

The term *chirality* was originally coined by Lord Kelvin, and this concept now plays a central role in nearly every aspect of modern-day life. This phenomenon's impact on biological systems is immense and arguably, the most vital force for sustaining life on the planet. Louis Pasteur appreciated the implications of chirality after he inadvertently discovered molecular chirality in the spontaneous resolution of an aqueous solution of racemic sodium ammonium tartrate tetrahydrate in 1848. Although enantiomers primarily differ in their ability to rotate plane-polarized light, this definition is a gross oversimplification of the importance of homochirality. For example, Nature produces amino acids as single enantiomers, which provide the building blocks for proteins that recognize and differentiate between molecules with complementary shape and chirality. The origin of this preference for one-handedness remains a subject of significant debate and speculation. Pasteur also described the first chiral resolution, which involved the addition of the chiral base, cinchonine, to *rac*-tartaric acid to form diastereoisomers and thus established the basis for the classical chiral resolution process that is still widely employed today, particularly in the pharmaceutical industry. Based on these important discoveries, the idea that enantiomerically pure chiral molecules can only be formed in the presence of a chiral influence was formulated, which now forms the very basis of modern asymmetric catalysis. The following three chapters delineate the historical development of three entirely different transformations that, to varying degrees, incorporate the principles of chiral resolution and induction. Hence, the first chapter outlines non-enzymatic resolution reactions, while the second two chapters provide examples of challenging enantioselective and desymmetrization reactions.

The first chapter by Aileen B. Frost, Mark D. Greenhalgh, Elizabeth S. Munday, Stefania F. Musolino, James E. Taylor, and Andrew D. Smith provides an outstanding treatise on the desymmetrization and kinetic resolution of alcohols and amines by non-enzymatic enantioselective acylation reactions. The chapter aligns beautifully with the notion of efficiently separating enantiomers, which remains a stalwart approach in organic synthesis. Notably, the chapter describes the evolution of small molecules that emulate the efficiency and selectivity exhibited by enzymes. The discussion is organized in the context of stoichiometric and catalytic processes for the desymmetrization and kinetic resolution reactions of alcohols and amines in the context of mechanism, selectivity, scope and limitations, which illustrate the transition from the stoichiometric to the catalytic reaction manifold. The Mechanism and Stereochemistry section further subdivides the catalytic processes into the type of acylating agent and catalyst employed for a specific resolution. The Scope and Limitations component is categorized in the context of the substrate, namely, diols, alcohols, amines, diamines, amides, etc., which permits the reader to appreciate the expansive scope of this approach. The Applications to Synthesis illustrates how these methods have been implemented in the construction of some important pharmaceuticals and natural products, and the Comparison with Other Methods section provides a direct comparison with acylative and hydrolytic enzymatic kinetic resolution methods. The Tabular Survey summarizes the types of stoichiometric acylating agents and the various catalysts that have been employed to date, including oxidants and additives. The tables systematically provide examples of the types of substrates in the context of the associated approach and the organization mirrors the Scope and Limitations to permit the identification of suitable reaction conditions for a specific substrate. Overall, this is an excellent chapter on a particularly important and useful process, which will be an invaluable resource to anyone wishing to facilitate either a desymmetrization or kinetic resolution reaction of alcohol and amine derivatives.

The second chapter by Lucile Marin, Emmanuelle Schulz, David Lebœuf, and Vincent Gandon provides a scholarly account of the Piancatelli reaction or rearrangement, which is a useful process for the construction of 4-hydroxy-5-substituted-cyclopent-2-enones from 2-furylcarbinols. Piancatelli and coworkers reported this process in the course of studying acid-mediated reactions with heterocyclic steroid analogs in 1976. Notably, the rearrangement represents a rare example of a reaction that directly transforms a heterocycle into a carbocycle. The transformation is envisioned to proceed via an electrocyclic ring closure in a similar manner to the related Nazarov cyclization. Hence, while the preferred mechanism is a conrotatory 4π-electrocyclization of a transient pentadienyl carbocation, the Mechanism and Stereochemistry section also outlines some other possibilities, namely, ionic stepwise and aldol-type condensations. The Scope and Limitations portion is organized by the three variations of this process, namely, the oxa-, aza-, and carba-Piancatelli reactions, which each include sections on cascade processes. Interestingly, the enantio-determining step in this process, namely, a 4π-electrocyclization of a transient pentadienyl carbocation, makes the asymmetric version challenging. Nevertheless, the ability to employ chiral phosphoric acids to generate enantiomerically enriched substituted cyclopentenones (albeit limited to the aza-Piancatelli variant using anilines) represents a significant breakthrough for this process. The Applications to Synthesis describes the applications of this methodology to prostaglandin synthesis and some related natural products, and the Comparison with Other Methods section provides a relatively comprehensive assessment of other methods commonly deployed for the construction of this structural motif. The Tabular Survey incorporates reactions reported up to December 2019. The tables are uniquely organized based on the product framework with different substitutions to permit the identification of a suitable product. Overall, this is an important chapter on a remarkably useful reaction that has not been fully exploited in comparison with some of its related counterparts.

The third chapter by Constanze N. Neumann and Tobias Ritter outlines transition-metal-mediated and metal-catalyzed carbon-fluorine bond formation. The exponential growth in the development of methods that permit a late-stage fluorination can be ascribed to the unique physical properties that fluorine bestows on functional organic molecules, such as pharmaceuticals, agrochemicals, and materials. For instance, fluorine forms the strongest bond to carbon, which results in a highly polarized bond that has significant ionic character. Hence the large dipole moment provides a weak hydrogen bond acceptor that infers unique conformational behavior. The Mechanism and Stereochemistry component of this chapter categorizes the fluorination process in the context of nucleophilic and electrophilic fluorine sources, which are subdivided into the type of catalyst employed. Notably, the authors have devised an excellent classification system highlighting the knowledge gaps in this important and rapidly developing area that should stimulate further work in this field. The mechanistic classifications are then used throughout the remainder of the chapter to make cross-referencing a specific type of mechanism effortless for the reader. The Scope and Limitations part is organized by the substrate type, namely aryl, alkyl, and aliphatic substrates in the context of the aforementioned mechanistic variations, which permit one to identify the optimal approach for a particular substrate class. The substrates also address the critical challenges with site-selectivity (aryl) and stereocontrol (alkenyl and aliphatic) that are encountered with these substrate classes. A key and striking feature is the realization that the C-F bond can be introduced in a chemo-, regio-, and stereoselective manner. Consequently, several chiral catalysts have been developed that permit the asymmetric construction of carbon-fluorine bonds through desymmetrization and enantioselective reactions, which have proven particularly important in medicinal chemistry. The Applications to Synthesis section delineates the incorporation of fluorine into unnatural functionalized molecules, given the relatively few natural products that contain this motif. Fluorine in natural molecules is rare because of the difficulties that a haloperoxidase has to oxidize fluorine anion compared with other halide ions. Hence, this section outlines several successful applications to fluorine-18 positron-emission tomography (^{18}F-PET) tracer synthesis, an important and challenging aspect of late-stage fluorination given the relatively short half-life of the ^{18}F isotope. The Comparison with Other Methods portion describes some of the more classical fluorination methods, including nucleophilic aromatic substitution and displacement reactions with both nucleophilic and electrophilic fluorine sources. The Tabular Survey parallels the Scope and Limitations part in the context of aryl, alkenyl, and aliphatic fluorination reactions using both electrophilic and nucleophilic reaction conditions. Overall, this chapter provides the reader with an outstanding perspective on the recent developments of this important transformation, and represents a very important resource for the community.

I would be remiss if I did not acknowledge the entire *Organic Reactions* Editorial Board for their collective efforts in steering this chapter through the various stages of the editorial process. I would like particularly to thank Gary A. Molander (Chapter 1) and Steven M. Weinreb (Chapter 2), who each served as the Responsible Editor for the first two chapters and I was responsible for marshalling Chapter 3 through the various phases of development. I am also deeply indebted to Dr. Danielle Soenen for her heroic efforts as the Editorial Coordinator; her knowledge of *Organic Reactions* is critical to maintaining consistency in the series. Dr. Dena Lindsay (Secretary to the Editorial Board) is thanked for coordinating the contributions of the authors, editors, and publishers. In addition, the *Organic Reactions* enterprise could not maintain the quality of production without the efforts of Dr. Steven Weinreb (Executive Editor), Dr. Engelbert Ciganek (Editorial Advisor), Dr. Landy Blasdel (Processing Editor), and Dr. Debra Dolliver (Processing Editor). I would also like to acknowledge Dr. Barry Snider (Secretary) and Dr. Jeffery Press (Treasurer) for their efforts to keep everyone on task and make sure that we are fiscally solvent!

I am indebted to all the individuals that are dedicated to ensuring the quality of *Organic Reactions*. The unique format of the chapters, in conjunction with the collated tables of examples, make this series of reviews both unique and exceptionally valuable to the practicing synthetic organic chemist.

P. Andrew Evans
Kingston
Ontario, Canada

CONTENTS

CHAPTER 1

KINETIC RESOLUTION AND DESYMMETRIZATION OF ALCOHOLS AND AMINES BY NONENZYMATIC, ENANTIOSELECTIVE ACYLATION

Aileen B. Frost, Elizabeth S. Munday, Stefania F. Musolino, and Andrew D. Smith

University of St Andrews, School of Chemistry, North Haugh, St Andrews, Fife, KY16 9ST, U.K.

Mark D. Greenhalgh

University of Warwick, Department of Chemistry, Coventry, CV4 7AL, U.K.

James E. Taylor

University of Bath, Department of Chemistry, Claverton Down, Bath, BA2 7AY, U.K.

Edited by Gary A. Molander

CONTENTS

jet21@bath.ac.uk; ads10@st-andrews.ac.uk
How to cite: Frost A. B.; Munday E. S.; Musolino S. F.; Smith A. D.; Greenhalgh M. D.; Taylor J. E. Kinetic Resolution and Desymmetrization of Alcohols and Amines by Nonenzymatic, Enantioselective Acylation. *Org. React.* **2020**, *104*, 1–498.

ACKNOWLEDGMENTS

We are extremely grateful for editorial advice and proofreading from Prof. Gary Molander, Dr. Rebecca L. Grange, Prof. Tom Rovis, and Prof. Scott Denmark. We would also like to thank Dr. Linda Press, Dr. Danielle Soenen, and Dr. Dena Lindsay for their invaluable advice and expertise throughout the preparation of this chapter. We would also like to thank Prof. Vladimir Birman for his initial work on this chapter.

INTRODUCTION

Enantiomerically pure alcohols and amines are ubiquitous throughout Nature and are found within numerous biologically active compounds. Alcohol and amine functional groups are also synthetically versatile and can be incorporated within a diverse array of synthetic strategies. Consequently, significant efforts have been made toward the development of new methods that permit the preparation of enantiomerically pure alcohols and amines. In this regard, resolution methods in which the two enantiomers of a racemic mixture are separated are still widely used to obtain enantiomerically enriched alcohols and amines.

Kinetic resolution (KR) is a process by which enantiomeric enrichment of a racemic mixture is achieved through manufacturing a difference in the rate of reaction of the two enantiomers (Scheme 1). This is inherently challenging given that in the absence of a chiral environment the rate of reaction of two enantiomers is identical. Since the pioneering studies of Pasteur in the 19th century relating to resolution and stereochemistry,[1] the field of KR progressed at a modest rate until the early 1980s when landmark discoveries in enantioselective catalysis provided the platform for the development of new KR methods. A wide range of KR processes have subsequently been reported that employ a variety of transformations with different functional groups to facilitate the resolution and thereby enantiomerically enrich a racemic starting material. A number of comprehensive reviews and books are available that detail progress in the many different aspects of KR.[2−8]

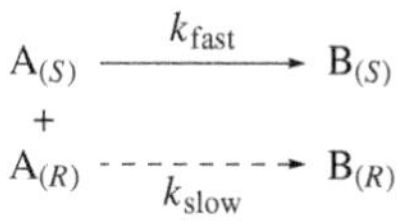

$$A_{(S)} \xrightarrow{\ k_{fast}\ } B_{(S)}$$

$$+$$

$$A_{(R)} \dashrightarrow[k_{slow}]{} B_{(R)}$$

Scheme 1

The efficiency of a KR process is often characterized by its selectivity factor (s), which is defined as the ratio of the rate constant for the fast-reacting enantiomer (k_{fast}) to the rate constant for the slow-reacting enantiomer (k_{slow}) (Eq. 1).

$$s = k_{fast}/k_{slow} \qquad\qquad (Eq.\ 1)$$

In practice, the selectivity factor cannot be easily obtained by directly measuring the individual rate constants. Consequently, Kagan developed an equation (Eq. 2)[1,9,10] based on the theoretical aspects of KR processes, which links the reaction conversion (C) to the enantiomeric excess of the recovered substrate (ee_A), both of which are easily measured. This equation is valid for a set of homocompetitive reactions in which the reaction is first-order with respect to the substrate. Alternative equations have also been derived for reactions using either scalemic catalysts and/or nonracemic substrates.[11] In some cases, the selectivity factor of a given KR process may vary with reaction conversion because of nonlinear effects associated with the kinetic partitioning of catalytic species.[12]

$$s = \ln[(1 - C)(1 - ee_A)]/\ln[(1 - C)(1 + ee_A)] \qquad\qquad (Eq.\ 2)$$

For a completely selective KR, in which only one enantiomer of a racemate reacts, the maximum theoretical yield of the recovered substrate is 50% (ca. $s > 500$). Nevertheless, reactions with lower selectivity can also be used to obtain enantiomerically pure recovered substrate by increasing the reaction conversion beyond 50%. For example, a reaction must proceed to 70% conversion with $s = 10$ to recover the unreacted substrate in 99:1 er, while a reaction with $s = 20$ requires 60% conversion to achieve the same result. Consequently, KR processes with $s > 10$ are considered synthetically useful, while reactions with $s > 50$ allow the isolation of highly enantiomerically enriched substrate (and product, if applicable) at 50% conversion.

Although various strategies for the KR of alcohols and amines are available, this chapter focuses on the use of nonenzymatic, acylative KR methods. In this case, one enantiomer of the racemic alcohol or amine selectively reacts with a suitable acylating agent to form the corresponding ester or amide, respectively (Scheme 2). Throughout this review, stoichiometric KRs are defined as those in which the enantioselectivity of the acylation is controlled using a chiral acylating agent, while catalytic KRs generally employ achiral acylating agents, and stereocontrol originates from the chiral catalyst.

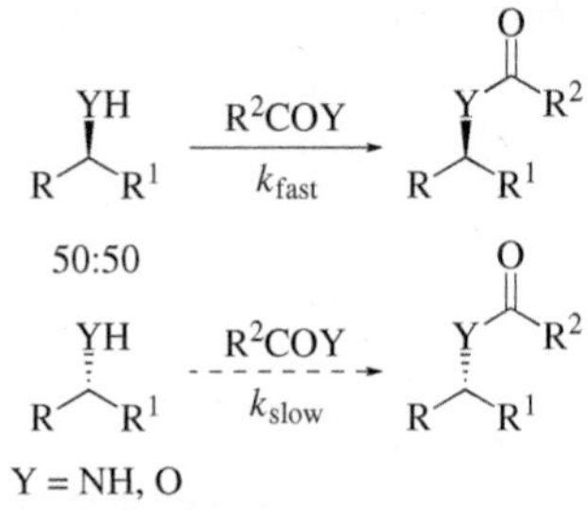

Scheme 2

Enantioselective acylation is a particularly attractive strategy for the KR of both alcohols and amines and has several advantages compared with other techniques. For example, the ester or amide products are often readily separable from the unreacted substrates, allowing the purification of the desired enantiomer. Furthermore, acylative KR allows both enantiomers of the substrate to be recovered, unlike some alternative methods in which one enantiomer is destroyed to perform the resolution. Once isolated from the initial KR, the product ester or amide can often be hydrolyzed to its parent alcohol or amine, giving access to both enantiomers of the substrate from a single KR process. Finally, the acylation of alcohols and amines is a well-studied field, and as such an array of acylating agents, catalysts, and conditions is available, that can act as a starting point for the development of a specific KR process.

Despite their similarities, the acylative KR of alcohols and amines presents distinctly different challenges. For example, the uncatalyzed acylation of amines with common reagents such as acid chlorides or anhydrides is often extremely rapid in comparison with the corresponding background acylation process for alcohols. Therefore, it is more challenging to develop a selective catalytic acylative KR for amines compared with alcohols. This difference is reflected in the literature to date, with many more methods and a broader substrate scope accessible for the acylative KR of alcohols than for amines, although recent advances in the latter suggest that further developments in this important process will be possible. Advances in acylative KR can also impact other areas of enantioselective synthesis. For example, the development of highly selective acyl transfer catalysts for the KR of alcohols preceded the renaissance in organocatalysis, and many of the catalysts explored in the context of KR have found broader utility in other areas.

Acylative desymmetrization processes of prochiral diols and diamines are also possible using either stoichiometric chiral acylating agents, or achiral acylating agents in combination with a chiral catalyst. Selective monoacylation of the prochiral starting material results in desymmetrization, with quantitative conversion into a single enantiomer of product theoretically possible. In many cases, the enantiomeric excess of the monoacylated desymmetrization product can be enhanced by a further in situ KR process to form the corresponding *meso* bisacylated product (Scheme 3).

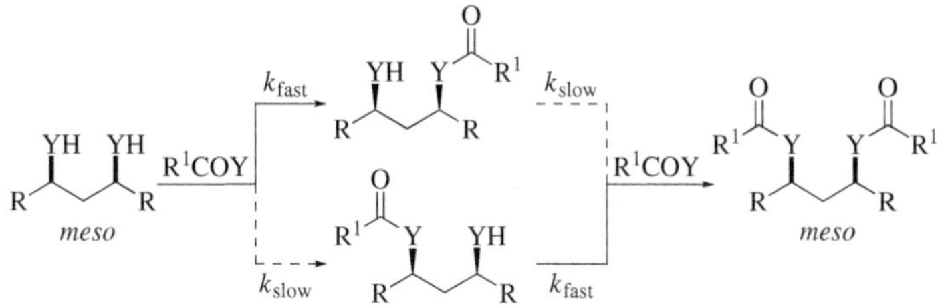

Scheme 3

This review aims to provide a comprehensive overview of the range of stoichiometric and catalytic methods available for the acylative KR of both alcohols and amines, highlighting the scope of substrates applicable to each process. The review covers relevant literature up until the end of 2019. Previous literature reviews on various aspects of acylative KR are also available.[1-8] Methods for the acylative desymmetrization of prochiral diols and diamines are also discussed and have been previously reviewed.[13,14] Although many enzyme-catalyzed, acylative KRs have been developed, these are outside the scope of this review and are covered elsewhere.[15-17] Related methods such as dynamic KR and parallel KR that often rely on enantioselective acylation are also not discussed but have been previously reviewed.[18-24]

MECHANISM AND STEREOCHEMISTRY

General Considerations

Many methods have been developed for KR and desymmetrization that vary in terms of reaction mechanism and the origin of stereoselectivity. For an effective KR, the two enantiomers of a starting material must be differentiated, with reliance upon diastereomeric interactions with either a chiral reagent or a chiral catalytic species, allowing reactions at different rates. For desymmetrization, a similar distinction must be made between either side of the mirror plane within the substrate to achieve enantiodiscrimination.

The choice of chiral reagent or chiral catalyst is key to the success of a given KR or desymmetrization process. Further details of the key interactions most commonly employed to achieve enantiodiscrimination in such processes are outlined in more detail in the following sections. The acylating agent selection is also crucial in many instances as it can have a dramatic effect on both the stereoselectivity and overall reactivity. The nature of KR dictates that a substoichiometric amount of acylating agent relative to the racemic substrate is often employed (0.5–1 equiv), whereas desymmetrization may require an excess of the reagent to reach completion. Many methods also require the addition of base to facilitate acylation and to sequester acidic byproducts, which typically involves organic tertiary amine bases. The reaction solvent can also influence the stereoselectivity of KR and desymmetrization reactions, albeit such effects are catalyst- and substrate-dependent and can be difficult to predict. Although there are many examples of KR and desymmetrization that work well at room temperature, lower temperatures are also commonly used to improve reaction stereoselectivity.

Stoichiometric KR and Desymmetrization

A stoichiometric acylative KR requires a chiral acylating agent to provide selectivity in the reaction. Stoichiometric acylative KRs have been reported in which the stereochemical control element is present within either the leaving group or the acyl group (Scheme 4), although the latter leads to diastereomeric products that have different physical properties and could, therefore, be considered a classical resolution.

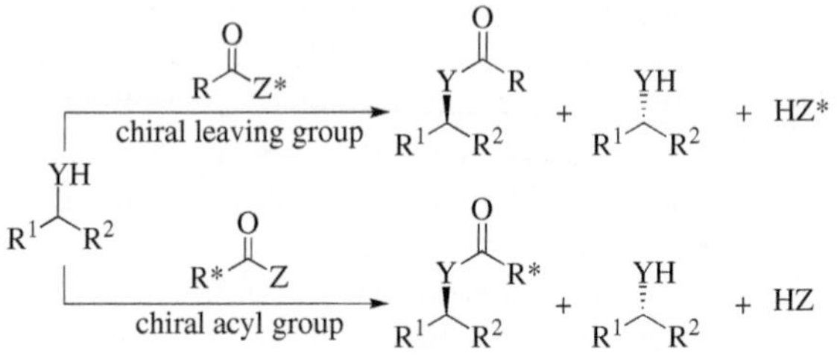

Scheme 4

For KRs in which the stereochemical control element resides in the leaving group, two potential steps could be stereodifferentiating (Scheme 5). The initial addition of the racemic starting material into the acylating agent may proceed at different rates ($k_1 \neq k_3$). Alternatively, the collapse of the diastereomeric, tetrahedral intermediate may be the stereodetermining step ($k_2 \neq k_4$), albeit this process would need to be reversible for the mismatched diastereoisomer.

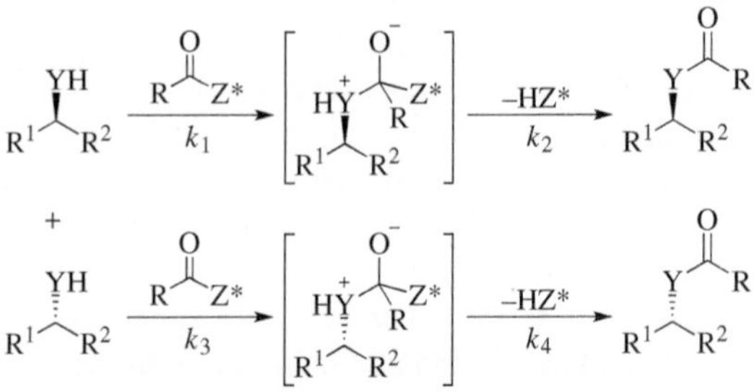

Scheme 5

There are relatively few examples of stoichiometric acylative KR and desymmetrization of alcohols, which may be ascribed to the many efficient catalytic methods that exist for these processes. In contrast, the stoichiometric acylative KR and desymmetrization of amines have been more widely studied, and several effective acylating agents for these substrates have been reported (Fig. 1).

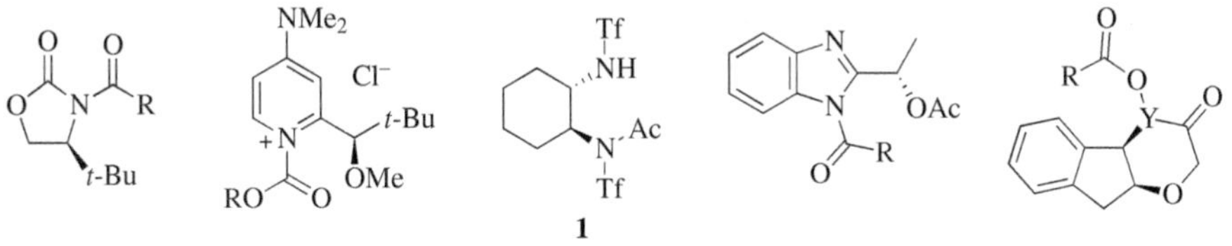

Figure 1. Examples of stoichiometric acylating agents used for KR.

Notably, the stereodetermining step is often the initial acylation of the racemic substrate, which in some cases is governed by the conformation of the acylating agent[25,26] and/or by noncovalent interactions such as H-bonding[27] and π-cation stacking[28] with the substrate. The stoichiometric, acylative KR of secondary benzylic amines such as 1-phenylethylamine (**2**) using chiral diamine **1** undergoes an interesting switch in stereoselectivity depending on the solvent used, with basic solvents such as DMF, DMPU, and ionic liquids giving amide (*S*)-**3**, whereas more acidic solvents such as CH_2Cl_2 furnish amide (*R*)-**3** selectively (Scheme 6).[29,30] The switch is rationalized by a change in the stereodetermining step in different solvents. In acidic solvents, reversible hemiaminal formation is nonselective, with subsequent collapse of the diastereomeric, tetrahedral intermediates determining the reaction selectivity. In more basic solvents, this initial equilibrium is proposed to be disfavored, and the selectivity determined by the initial enantioselective acylation of the racemic substrate.

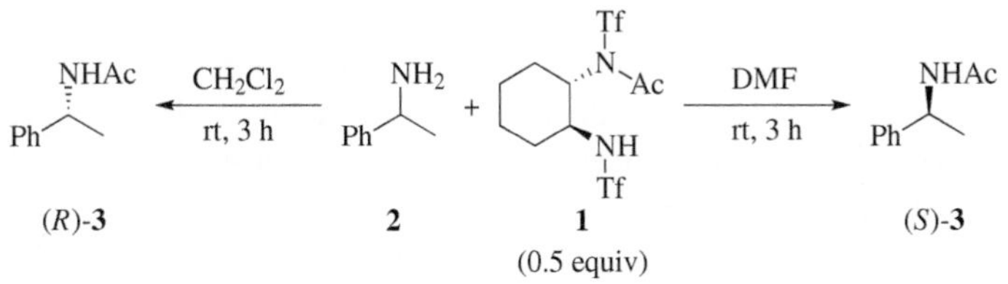

Scheme 6

Catalytic KR and Desymmetrization

A wide range of catalytic KR and desymmetrization reactions has been reported that permit the resolution of many classes of substrates. These processes employ achiral acylating agents and often require the addition of base to facilitate the acylation. The reaction solvent is highly dependent on the catalytic system, with reactions typically conducted at room temperature or below to achieve optimal selectivity.

Lewis acid and Lewis base catalysts have been the most widely explored for acylative KR and desymmetrization processes, with more details on the different modes
of activation and origins of enantiodiscrimination provided below.

Acylating Agents. A key consideration in catalytic, acylative KR and desymmetrization processes is the choice of achiral acylating agent, with the steric demand
of the acyl group often having an impact on the reaction selectivity. The rate of the
noncatalyzed reaction between the acylating agent and the racemic substrate must
be significantly slower than the catalytic, enantioselective process. For the KR and
desymmetrization of alcohols, such noncatalyzed reactions are rarely problematic,
and consequently, a wide range of acylating agents can be employed (Fig. 2). Readily
available acid chlorides and symmetrical anhydrides are by far the most commonly
used O-acylating agents and are compatible with many different catalysts under a
variety of conditions. In particular, aryl-substituted acid chlorides (e.g., benzoyl chloride) and short-chain alkyl acid anhydrides (e.g., acetic, propanoic, and isobutyric
anhydrides) are the most widely reported.

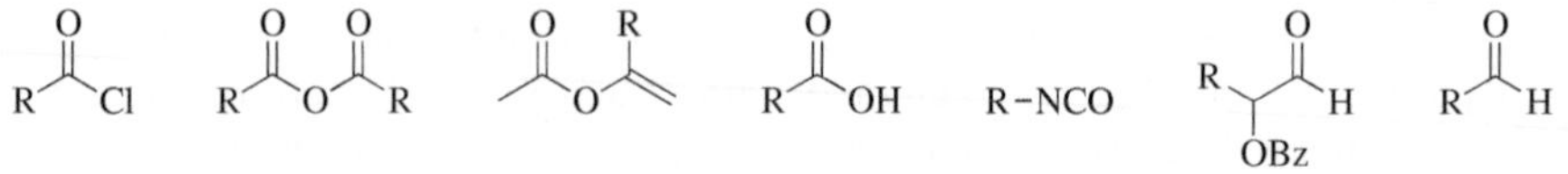

Figure 2. Examples of acylating agents used for the KR and desymmetrization of alcohols.

The use of vinyl esters as O-acylating agents for KR has also been reported,
and carboxylic acids require the in situ formation of a reactive mixed anhydride
using a sterically demanding anhydride (e.g., pivaloyl anhydride) that is unreactive to
the catalytic O-acylation. Isocyanates have also been used to form carbamate products instead of esters in both KR and desymmetrization processes. Acylating agents
that are not at the carboxylic acid oxidation level can also be used in conjunction
with an N-heterocyclic carbene (NHC) catalyst. For example, α-benzoyloxy aldehydes undergo a redox reaction in the presence of a suitable NHC to form an active
O-acylating agent in situ, and unsubstituted aldehydes can also be used in the presence
of an NHC with external oxidant.

The catalytic KR of amines is significantly more challenging compared with the
KR of alcohols because of the increased nucleophilicity of amines, which results
in faster rates of noncatalyzed N-acylation using common reagents such as acid
chlorides and acid anhydrides, even at low temperature. Consequently, choosing
an appropriate N-acylating agent that avoids unwanted background reactivity is
crucial for success (Fig. 3). Benzoic anhydride (**4**) is unreactive toward secondary
amines at low temperatures in toluene and has been used for the KR and desymmetrization of amines using a suitable chiral H-bonding catalyst in the presence of
4-(dimethylamino)pyridine (DMAP).[31,32] The KR of amines has also been reported
using O-acyl azlactone **5**, which preferentially reacts with the planar-chiral DMAP
catalyst rather than the racemic amine.[33] An alternative approach is to employ a

masked acylating agent that can only be revealed in the presence of a catalyst. For example, α'-hydroxyenone **6** undergoes a retro-benzoin reaction in the presence of an NHC catalyst, releasing acetone as the byproduct.[34] The resulting unsaturated Breslow intermediate undergoes protonation followed by tautomerization to form an acyl azolium ion, which undergoes preferential transesterification with a chiral hydroxylamine catalyst to form the key chiral *N*-acylating agent. Importantly, the intermediate acyl azolium ion reacts slowly with secondary amines, and therefore the undesired racemic *N*-acylation is minimized.

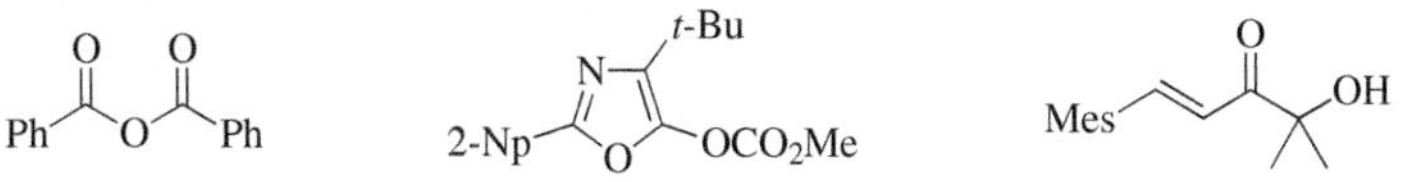

Figure 3. Examples of acylating agents used for the KR of amines.

Lewis Base Catalysis. Chiral Lewis bases are the most commonly used class of catalysts for the acylative KR and desymmetrization of alcohols and amines. In these cases, the Lewis base first reacts with the achiral acylating agent to form a more reactive chiral acylating agent (Scheme 7). Stereoselective acylation of one enantiomer of the substrate generates the product, with the associated counterion (Z^-) often crucial for concomitant proton transfer during this selectivity-determining step. In many protocols, an additional base is necessary to aid the dissociation of the resulting Lewis base-acid adduct to regenerate the catalyst. Many different classes of Lewis bases have been used for acylative KRs and desymmetrizations, but within each class, there are some common interactions and/or modes of recognition that allow the catalyst to discriminate between the two substrate enantiomers.

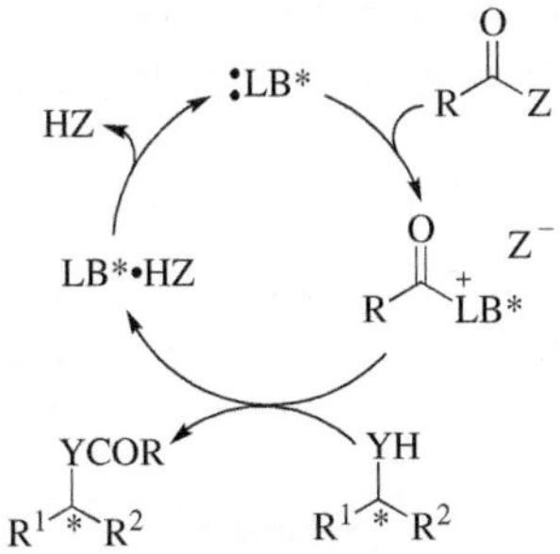

Scheme 7

Pyridine Analogues. Pyridine and analogues such as 4-dimethylaminopyridine (DMAP) are widely used as acyl transfer agents in many different reactions, including the acylation of both alcohols and amines.[35] As such, the investigation of chiral pyridine analogues was a logical step for the development of enantioselective acyl transfer agents for KR, and consequently, this class of catalyst has been the most widely explored to date. Mechanistic studies on the catalytic cycle in such processes are predominantly focused on the origins of stereoselectivity; however, details regarding the other steps in the catalytic cycle can be inferred from investigations into related (achiral) acyl transfer reactions.

The acylation of pyridine analogues to form the key N-acylpyridium intermediates is highly dependent upon the catalyst structure, the achiral acylating agent, and the solvent. For example, 4-pyrrolidinopyridine (PPY) readily reacts with acetyl chloride to form the corresponding N-acetylpyridium chloride quantitatively, whereas the analogous reaction with acetic anhydride results in <10% of the N-acetylpyridinium acetate at equilibrium.[35,36] Despite this difference, the rate of acylation of a tertiary alcohol is three times faster using acetic anhydride compared with acetyl chloride. These experiments highlight the importance of the N-acylpyridinium counterion,[37] in which computational studies suggest that the acetate counterion assists proton transfer in the turnover-limiting acylation step (Fig. 4).[38]

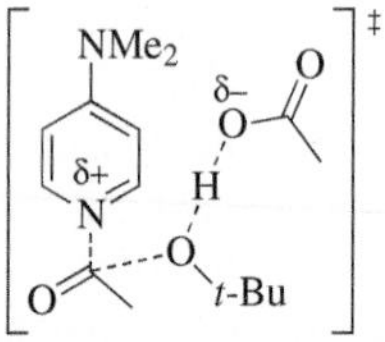

Figure 4. Computationally predicted role of acetate counterion in DMAP-catalyzed O-acylation using acetic anhydride.

Kinetic analysis of the DMAP-catalyzed acylation of cyclohexanol (7) with acetic anhydride and triethylamine shows that the reaction is first-order in the substrate, DMAP, and the anhydride, but zero-order in triethylamine (Scheme 8).[38] This provides evidence that substrate acylation is turnover-limiting and that the role of added base is to sequester acidic byproducts formed during this step.

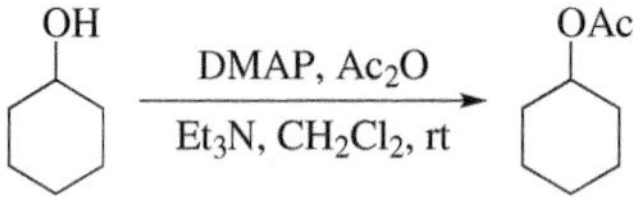

Component	Order
7	first
DMAP	first
Ac$_2$O	first
Et$_3$N	zero

Scheme 8

Various approaches for generating chiral analogues of the planar DMAP core have been explored in attempts to make efficient catalysts for acylative KR and desymmetrization (Fig. 5). The enantioselective acylation requires differentiation of the two faces of the intermediate N-acylpyridinium, and the two enantiomers of the racemic substrate must interact differently with the accessible face. One strategy is to introduce a substituent that contains a stereocenter onto the pyridine ring in either the

2- or 3-position; however, substitution in the 2-position generally leads to a significant decrease in reactivity because of the increased steric demand around the reactive nitrogen atom, and therefore such catalysts are often unsuitable for acyl transfer reactions.[39] Conversely, substitution in the 3-position retains the catalytic activity, but any stereocenter present is now further away from the reactive site of the N-acylpyridinium intermediate. A tactic for improving selectivity in these cases is to introduce substituents that may have stacking interactions with the N-acylpyridinium core.[40] For example, the N-acylpyridinium intermediates formed from 3-substituted DMAP analogues **8** and **9** are proposed to be stabilized by π–π and n_S–π interactions, respectively.[39,41] Such interactions restrict the conformation of the intermediate, providing enhanced differentiation between the two faces of the pyridinium core and enhancing the selectivity of the acylation step. Despite these advances in acylative KRs and desymmetrizations, catalysts that employ point chirality to facilitate enantiodiscrimination generally afford lower selectivity in contrast to those that use other forms of chirality.

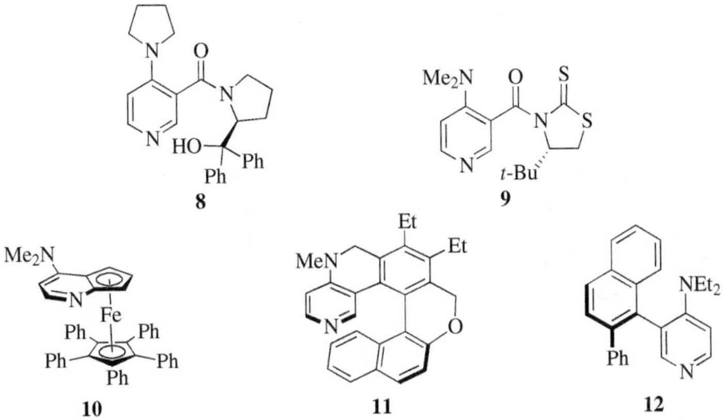

Figure 5. Examples of chiral DMAP-derived catalysts.

Consequently, the DMAP analogues **10** and **11** that possess planar chirality and helical chirality, respectively, provide high selectivity in acylative KR.[42,43] In these cases, the two faces of the intermediate N-acylpyridinium are clearly differentiated. These catalysts are particularly effective for the KR of benzylic secondary alcohols, as the two enantiomers of the alcohol have different stacking interactions with the N-acylpyridinium intermediate, leading to effective enantiodiscrimination. For example, π–π stacking in the proposed acylation transition state **14** between the aryl ring of the fast-reacting enantiomer of 1-phenylethanol (**13**) and the pyridinium ring, as well as minimization of steric interactions between the remaining substituents and the catalyst, provide selectivity (Scheme 9).[43] A similar model for enantiodiscrimination is proposed for helical catalyst **11** in the corresponding KR of secondary benzylic alcohols.[42]

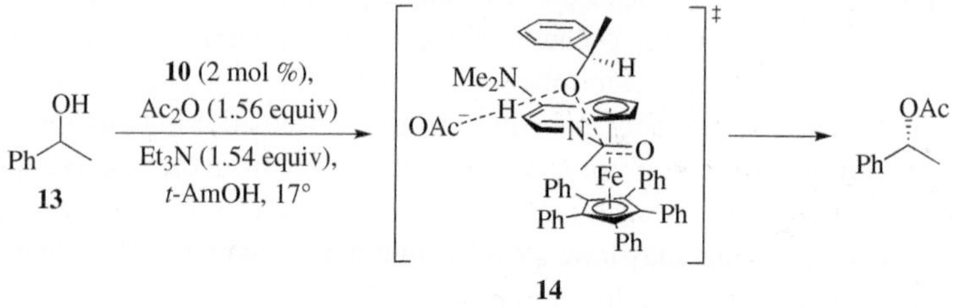

Scheme 9

Reaction progress kinetic analysis of the KR of 1-phenylethanol (**13**) using planar-chiral DMAP catalyst **10** and acetic anhydride suggests the process is first-order with respect to catalyst **10** but has fractional orders with respect to acetic anhydride and the racemic alcohol.[43] The fractional orders are attributed to the equilibrium between the free catalyst **10** and its *N*-acetylpyridinium ion. In contrast, the KR of secondary benzylic amines using a related planar-chiral DMAP analogue and *O*-acyl azlactone **5** is first-order in both catalyst and amine and zero-order in acylating agent, which is consistent with the corresponding *N*-acylpyridinium being the catalyst resting state in this case.[33]

The mechanism of KR using enantiomerically pure, axially chiral, atropisomeric DMAP analogues such as **12** has also been studied computationally.[44,45] The Lewis base mode of reaction is again favored, with the steric interactions between the substrate and the catalyst minimized during nucleophilic attack of the fast-reacting enantiomer on the *N*-acylpyridinium intermediate (Scheme 9a).

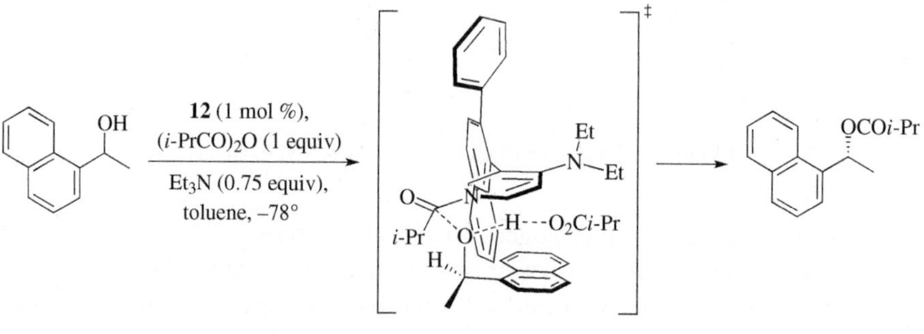

Scheme 9a

Amidines, Guanidines, and Isothioureas.   Several Lewis basic amidines, guanidines, and isothioureas have been explored as catalysts for the KR of various alcohols, typically using acid anhydrides as the acylating agents.[46] The catalysts usually contain a stereocenter adjacent to the nucleophilic nitrogen atom that controls the facial selectivity of the acylation step. The catalysts often contain an extended aromatic π-system within the backbone that may provide additional stabilization through stacking interactions in the acylation transition state of the fast-reacting enantiomer of certain substrate classes.

Enantiomerically pure amidines such as **15–17** (Fig. 6) provide high selectivity for the KR of a wide range of alcohol substrates. Secondary alcohols bearing one sp^2-hybridized substituent on the stereogenic carbinol carbon tend to afford the highest selectivities because of the presence of cation-π interactions within the favored transition state of the fast-reacting enantiomer.

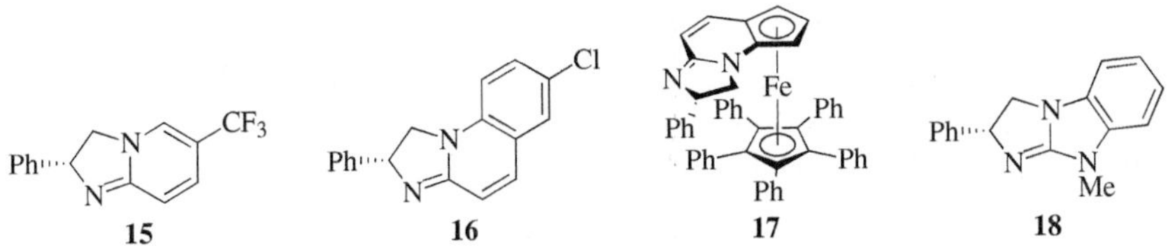

Figure 6. Examples of amidine and guanidine catalysts.

For example, computational modeling of the key acylation step in the KR of secondary benzylic alcohol **13** using **15** and propionic anhydride highlights the importance of a cation-π interaction between the pyridinium nitrogen atom and the phenyl substituent within the substrate transition state **19** for the acylation of the fast-reacting (R)-enantiomer (Scheme 10).[47,48] A similar stereochemical model is proposed for KRs of secondary alcohols using planar-chiral amidine catalyst **17**[49] and guanidine **18**.[50]

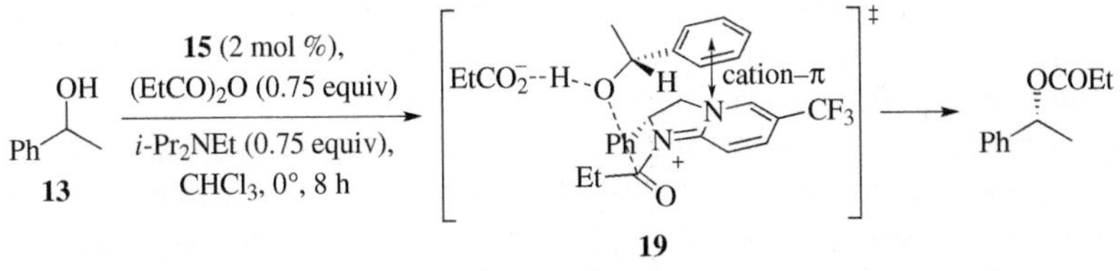

Scheme 10

For the KR of secondary alcohols bearing a cinnamyl substituent, the presence of an extended π-system in the catalyst leads to higher selectivity. For example, the selectivity in the KR of **20** using amidine **16** is proposed to be enhanced by both cation-π and π-π interactions between the substrate and catalyst in transition state **21** ($s = 27$) for the fast-reacting (R)-enantiomer (Scheme 11).[51] Computational modeling of the KR of aryl-substituted secondary amides using amidine **16** also suggests that the cation-π interaction is essential for enantiodiscrimination,[52] whereas alternative functional groups such as enones can also provide the π-system required for selective KR using an amidine catalyst.[53]

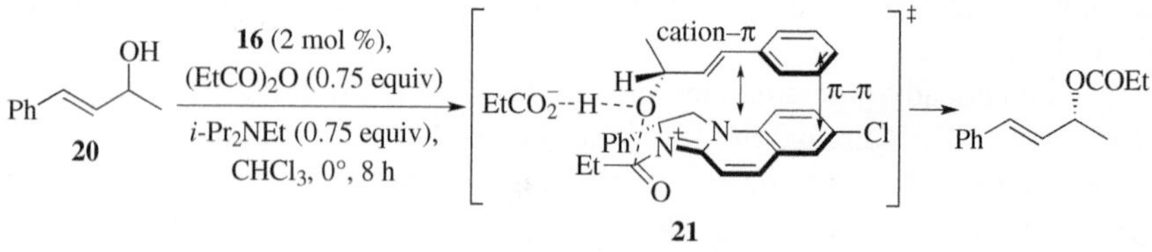

Scheme 11

Isothiourea-based catalysts **22–25** (Fig. 7) are effective for the acylative KR and desymmetrization of a wide range of substrates. The procedures that use these catalysts generally employ anhydrides as acylating agents. Kinetic analysis of the KR of a cyclic, secondary alcohol using isothiourea **24** suggests that the reaction is first-order with respect to the catalyst, alcohol, and anhydride.[54] KRs using isothiourea catalysts have also been reported using mixed anhydrides generated in situ from a carboxylic acid and a sacrificial anhydride, such as pivaloyl anhydride, which itself does not react with the racemic substrate.[55]

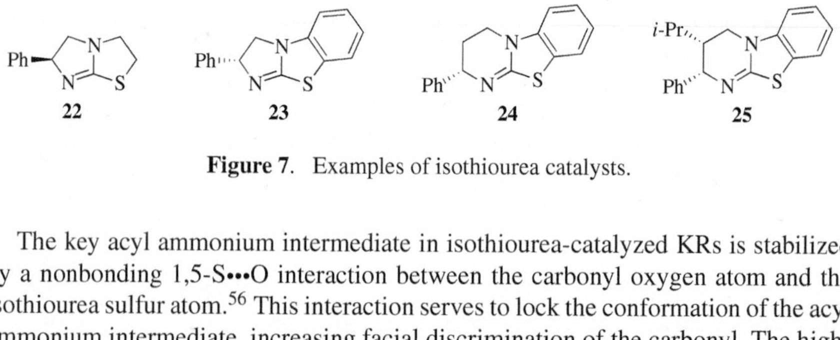

Figure 7.   Examples of isothiourea catalysts.

The key acyl ammonium intermediate in isothiourea-catalyzed KRs is stabilized by a nonbonding 1,5-S•••O interaction between the carbonyl oxygen atom and the isothiourea sulfur atom.[56] This interaction serves to lock the conformation of the acyl ammonium intermediate, increasing facial discrimination of the carbonyl. The highest levels of selectivity in isothiourea-catalyzed KRs are obtained when the racemic substrate bears either an sp^2- or sp-hybridized substituent on the stereogenic carbinol center. In these cases, the transition state for the acylation of the fast-reacting enantiomer is stabilized by cation-π interactions, while minimizing steric interactions of the other substituents with the acyl ammonium core.[57–59] For example, the KR of secondary benzylic alcohols using the isothiourea **25** is proposed to proceed through transition state **26** for the fast-reacting enantiomer (Scheme 12).[60] The acyl ammonium intermediate promotes *si* face attack onto the carbonyl opposite to the stereodirecting substituents on the isothiourea core, which is conformationally locked by the 1,5-S•••O interaction. Stabilizing cation-π interactions between the aryl ring and the isothiouronium core provides selectivity during the acylation step. In accord with the amidine-based catalysts, substituents that can participate in π-π interactions with the benzenoid ring of the isothiourea result in higher selectivity. Computational studies suggest that a favorable cation-π interaction is also responsible for the observed selectivity in the KRs of α-hydroxy esters,[61,62] α-hydroxy lactones,[63] and α-hydroxy phosphonate derivatives catalyzed by the isothiourea **23**.[64]

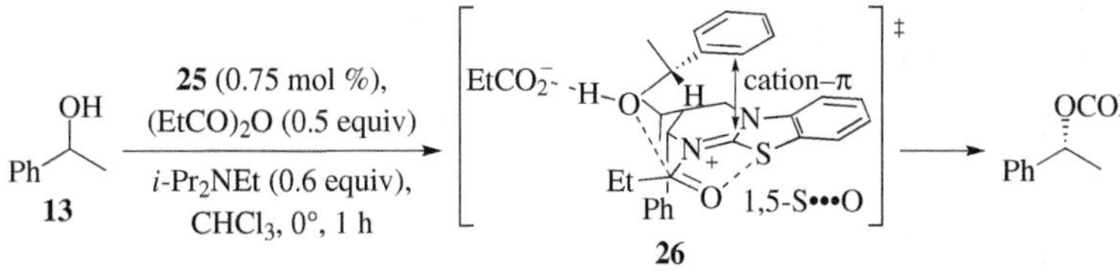

Scheme 12

Oligopeptide Analogues. Given the efficiency of enzymes as catalysts for acyla-tive KR, a significant amount of research has been directed to the use of short, syn-thetic peptides (e.g., **27, 29, 30**) as enzyme mimics in such processes (Fig. 8). To this end, these catalysts make use of a Lewis basic nitrogen within an *N*-methyl histidine residue as the acyl transfer moiety. Peptide-based catalysts typically adopt a folded conformation in solution, which is held in place by multiple H-bonding interactions that create a unique chiral environment for the enantiodiscrimination in the KR.

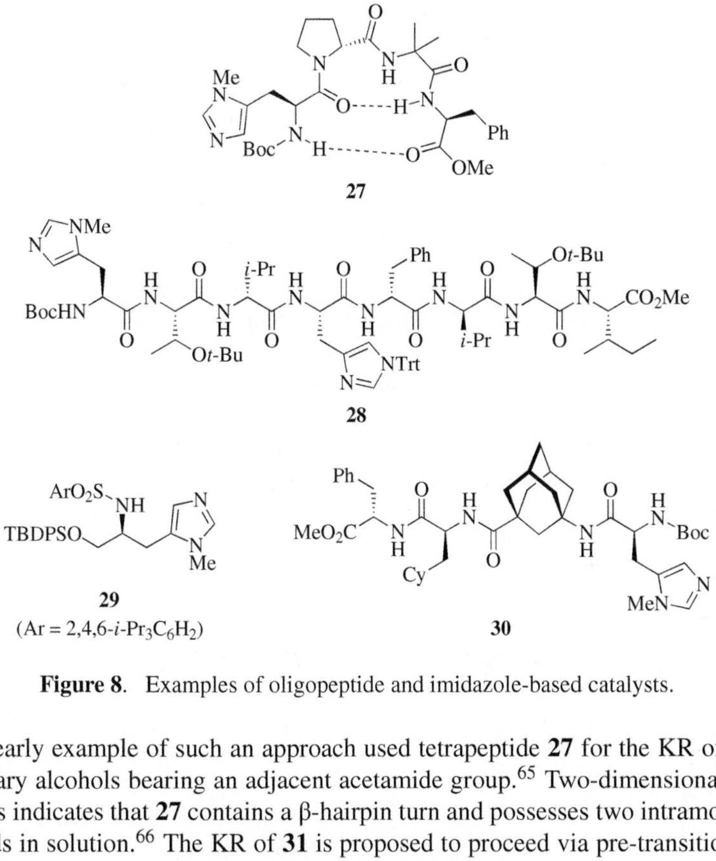

Figure 8. Examples of oligopeptide and imidazole-based catalysts.

An early example of such an approach used tetrapeptide **27** for the KR of cyclic secondary alcohols bearing an adjacent acetamide group.[65] Two-dimensional NMR analysis indicates that **27** contains a β-hairpin turn and possesses two intramolecular H-bonds in solution.[66] The KR of **31** is proposed to proceed via pre-transition-state assembly **32** in which a H-bond is formed between the acetamide of the substrate and the D-proline amide moiety for the fast-reacting enantiomer (Scheme 13).[66] Evidence for this interaction comes from isosteric replacement of the D-proline amide functional group with an alkene, which results in a catalyst that adopts a similar conformation in solution, but is nonselective in the KR of alcohol **31**. A more constrained β-hairpin conformation can be obtained by substituting the D-proline amide within **27** with a thioamide, which results in higher selectivity for the KR.[67] Octapeptide catalyst **28**, containing two histidine residues, is selective for the KR of secondary alcohols that do not contain an adjacent acetamide.[68] Structure-selectivity relationships that entail the replacement of each amino acid residue within **28** with an alanine suggests a bifunctional catalysis mechanism, in which one histidine acts as the acyl transfer agent and the other as the base, in conjunction with a possible conformational change during the reaction coordinate.[69]

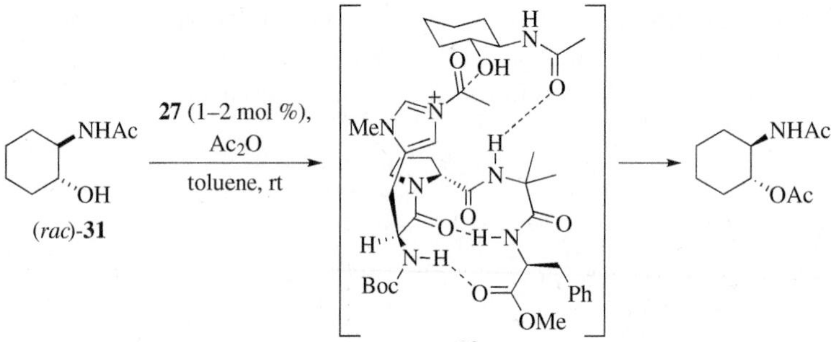

Scheme 13

Catalyst design based upon a single histidine such as **29** can be used for the acylative KR of monoprotected diols (Scheme 14).[70] Computational analysis suggests that the favored conformation of the *N*-acyl catalyst **29** has a stabilizing electrostatic interaction between the acyl oxygen atoms and the imidazoyl 2-proton.[71] The fast-reacting enantiomer of **33** is proposed to have a H-bonding interaction between the carbamoyl oxygen of the substrate and the sulfonylamino proton of the *N*-acyl catalyst in the pre-transition-state assembly **34**. A similar model is used to rationalize the observed selectivity in the acylative desymmetrization of carbamoyl-protected, *meso*-glycerol derivatives using catalyst **29**.[72]

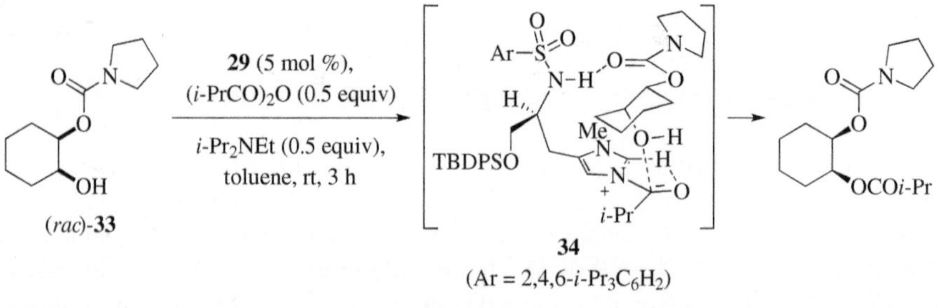

$(Ar = 2,4,6-i\text{-}Pr_3C_6H_2)$

Scheme 14

An alternative catalyst design uses more rigid and lipophilic oligopeptides such as **30** for the KR of 1,2-diols (Scheme 15).[73] Computational studies suggest that the intermediate *N*-acyl imidazolium ion discriminates between the enantiomers of diol **35** through a dynamic binding event that involves a combination of H-bonding and dispersion interactions in the favored pre-transition-state assembly **36**.[73−75]

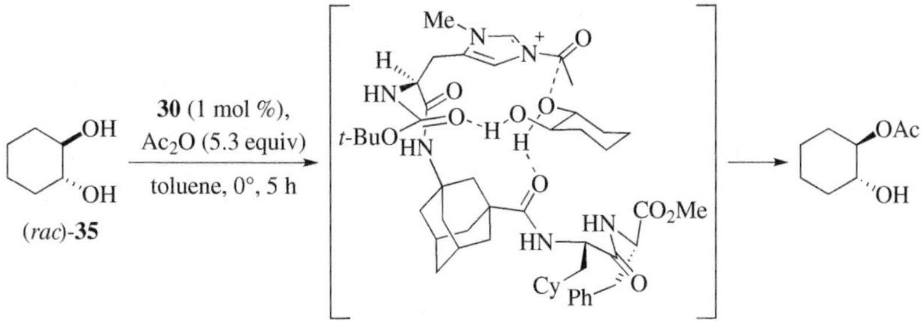

Scheme 15

N-Heterocyclic Carbenes. The use of NHCs as acylation catalysts is less well explored compared with other common, organocatalytic Lewis bases. Although chiral NHCs react with acylating agents at the carboxylic acid oxidation level to generate *N*-acyl azolium species that are capable of resolving secondary alcohols,[76,77] it is their unique reactivity with aldehydes that provides an alternative method of acylation. For example, NHCs promote redox reactions of reducible, α-substituted aldehydes such as α,β-unsaturated aldehydes, and α-bromo aldehydes to generate acyl azolium species that participate in the acylative KR of alcohols.[78,79] Since the racemic substrate cannot react directly with the reducible, α-substituted aldehyde, there is no possibility for undesirable, noncatalyzed acylation reactions with other acylating agents. Mechanistically, the NHC reacts with the α,β-unsaturated aldehyde to form adduct **37**, which can tautomerize to form Breslow intermediate **38** (Scheme 16).[78] Protonation in the β-position of **38** generates enol **39**, which can tautomerize into the required acyl azolium ion **40**. Alternatively, a strong oxidant can be used to oxidize the adduct formed between an NHC and an aldehyde to generate acyl azolium species **40** without the necessity for an α-substituent.[80−84] In both cases, the addition of a base is required to facilitate proton transfer and, in many cases, to generate the active NHC catalysts from its corresponding salt precatalyst.

N-Heterocyclic carbene catalysis can be used for the acylative KR of secondary amines, minimizing the problems associated with noncatalyzed acylation of the racemic substrate. Since the acyl azolium species reacts slowly with amines, a second Lewis base is required to promote a tandem catalytic KR process (Scheme 17).[34,85] First, the α′-hydroxyenone **6** acts as a masked acylating agent, reacting with achiral NHC **41** to promote a retro-benzoin reaction prior to an NHC-catalyzed redox process to form acyl azolium ion **42**. This species reacts with hydroxylamine **43** to form the key chiral acylating agent **44** for the KR. ^{1}H NMR spectroscopic studies

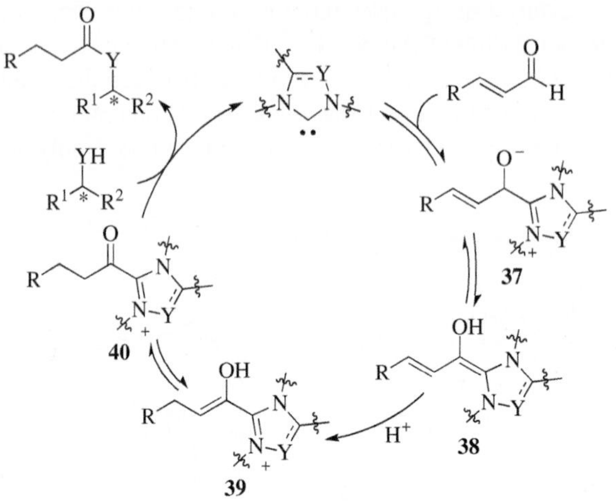

Scheme 16

show that *O*-acyl hydroxylamine **44** forms rapidly under the reaction conditions, and its independent preparation demonstrates that it is a competent stoichiometric reagent for the KR of secondary, cyclic amines such as **45**.[34]

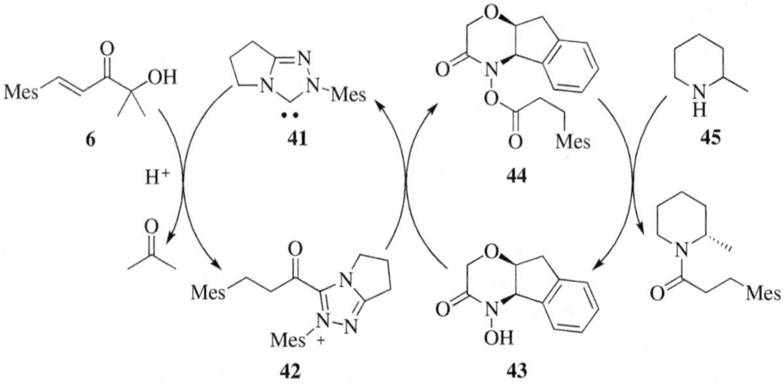

Scheme 17

Miscellaneous. The majority of Lewis base catalysts for acylative KR and desymmetrization employ nucleophilic sp^2-hybridized nitrogen atoms to react with the achiral, acylating reactant to form the active, chiral, acylating agent. Nevertheless, tertiary amines such as compound **46**, containing an sp^3-hybridized nitrogen atom as the reactive center, can also be used for KR reactions and the desymmetrization of 1,2-diols (Fig. 9).[86] Nucleophilic, phosphine-based catalysts such as **47** and **48** can also be used for the selective KR of various alcohols using anhydrides as the acylating agent.[87-89]

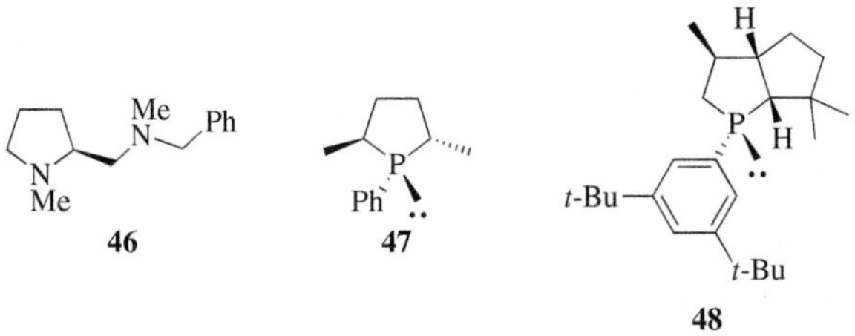

Figure 9. Examples of tertiary amine- and phosphine-based catalysts.

Lewis Acid Catalysis. Chiral copper and zinc complexes are the most commonly reported Lewis acid catalysts for the acylative KR and desymmetrization of alcohols. A general catalytic cycle for such Lewis acid catalyzed processes is outlined in Scheme 18. The catalyst typically coordinates with the alcohol substrate to form an intermediate alkoxide that is more susceptible to O-acylation. The reaction selectivity is controlled by the presence of chiral ligands, in which the enantiodiscrimination depends on either selective binding of the substrate [Scheme 18a(a)] or the different rates of acylation for the two diastereomeric alkoxide species [Scheme 18a(b)]. Substrates used for these Lewis acid catalyzed processes typically

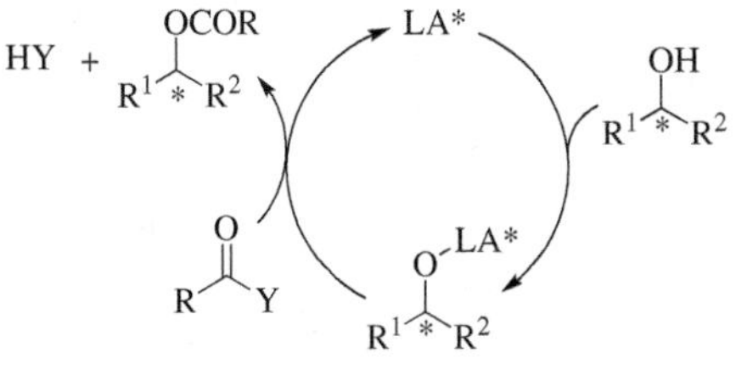

Scheme 18

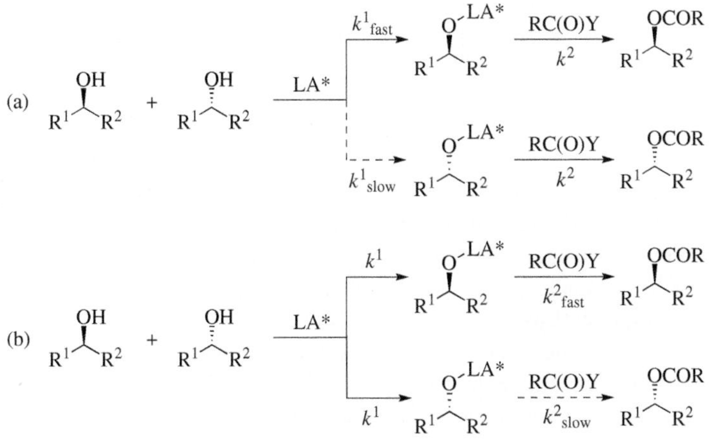

Scheme 18a

have two heteroatoms in close proximity to facilitate bidentate coordination to the metal center. As such, Lewis acid catalysts are particularly effective for the desymmetrization and/or KR of 1,3-diol substrates.

For example, the dinuclear zinc complex **49** has been successfully used for the desymmetrization of a range of 1,3-diols using vinyl benzoate as the acylating agent (Scheme 19).[90] The proposed mechanism involves the coordination of the 1,3-diol substrate with **49** to form a zinc alkoxide, in which the metal center also binds to the vinyl benzoate. The catalyst, therefore, provides an amphoteric environment in which one zinc atom forms an alkoxide, while the other acts as a Lewis acid to activate the acylating agent. Subsequent benzoylation within the coordination sphere of intermediate **50**, followed by decomplexation, completes the catalytic cycle. The diarylcarbinol moieties within catalyst **49** are critical for enforcing the chiral environment that accounts for the high levels of selectivity observed in the desymmetrization processes.[90,91]

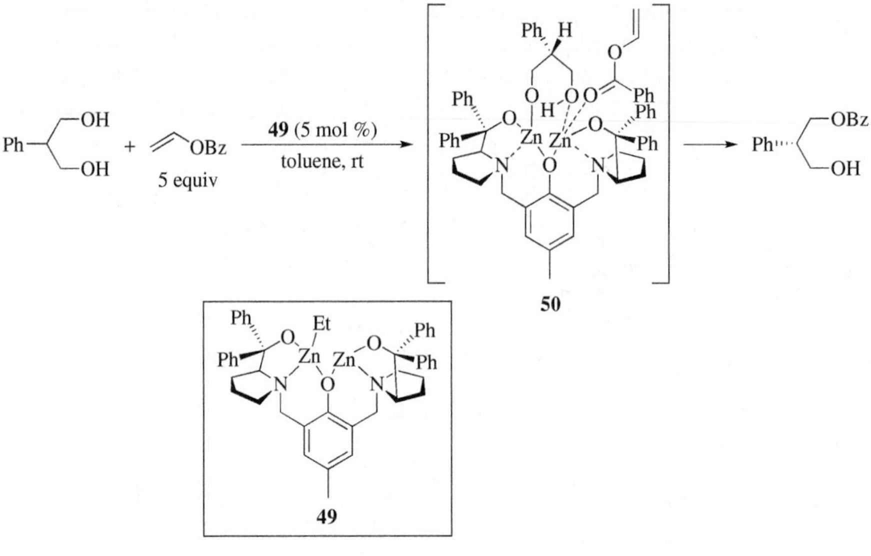

Scheme 19

Copper catalysts bearing C_2-symmetric bisoxazoline (Box) and pyridinebisoxazoline (Pybox) ligands have also been used for the desymmetrization of 1,3-diols. For example, Pybox ligand **51** reacts with $CuCl_2$ to form an active Lewis acid catalyst for the desymmetrization of 1,3-diols bearing prochiral quaternary centers.[92] The pre-transition-state assembly is thought to involve an octahedral copper complex **52**, where the tridentate Pybox coordinates in the equatorial plane and the benzoyl chloride acylating agent is axial (Scheme 20).[92] The 1,3-diol substrate then coordinates an axial and equatorial site in the metal center such that the substituents on the prochiral carbon minimize their steric interactions with the ligand backbone. Selective acylation of the proximal equatorial oxygen atom results in desymmetrization.

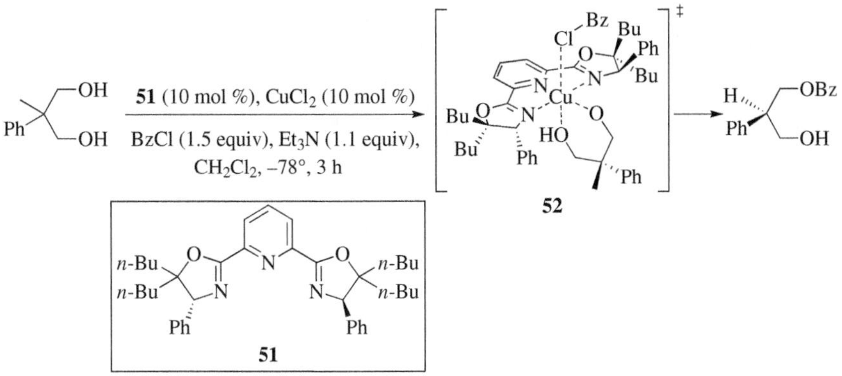

Scheme 20

Similar pre-transition-state assemblies are proposed to rationalize the stereo-selectivity observed in desymmetrizations using Cu(II)-Box complexes.[93,94] In these cases, the 1,3-diol substrates have an additional heteroatom-containing functional group on the prochiral carbon center (e.g., amide or carbamate) such that the substrate can coordinate in a tridentate manner. Lewis acid catalyzed KR processes using Cu(II)-Box complexes are also thought to proceed through similar mechanisms, with suitable substrates able to bind in a bidentate fashion to the copper center.[95−99] Computational studies on the KR of α-hydroxy lactones using a Cu-Box Lewis acid catalyst and *n*-propyl isocyanate as the *O*-acylating agent suggest that the difference in energy between the two diastereomeric intermediates is minimal and that the reaction selectivity is kinetically determined upon irreversible carbamoylation.[97]

Lewis acid catalysts based on chiral tin, yttrium, and scandium complexes have also been reported for the KR of alcohols.[100−103] In each case, the alcohol sub-strate is proposed to bind to the metal center prior to *O*-acylation, and the selectivity is determined by minimizing the steric interaction of the substrate with the ligand backbone.

Brønsted Acid Catalysis. The use of Brønsted acid catalysis in acylative KR is less well explored compared with Lewis acid and Lewis base catalysis. Brønsted acid catalyzed KR typically involves the H-bonding of the catalyst with the acylating agent to provide a chiral environment that can discriminate between the two enantiomers of the incoming nucleophile. For example, DFT calculations suggest that the resolu-tion of **53** using phosphoric acid **54** proceeds through the pre-transition-state assem-bly **55**, with H-bonding interactions with both the anhydride and the fast-reacting enantiomer of **53** (Scheme 21).[104] The selectivity and reaction conversion in this process also correlate with the basicity of the reaction solvent, which provides exper-imental evidence for the importance of H-bonding interactions in the reaction.

A dual-catalytic system using chiral H-bonding Brønsted acid catalyst **56** in combination with DMAP has been successfully used for the KR[31] and

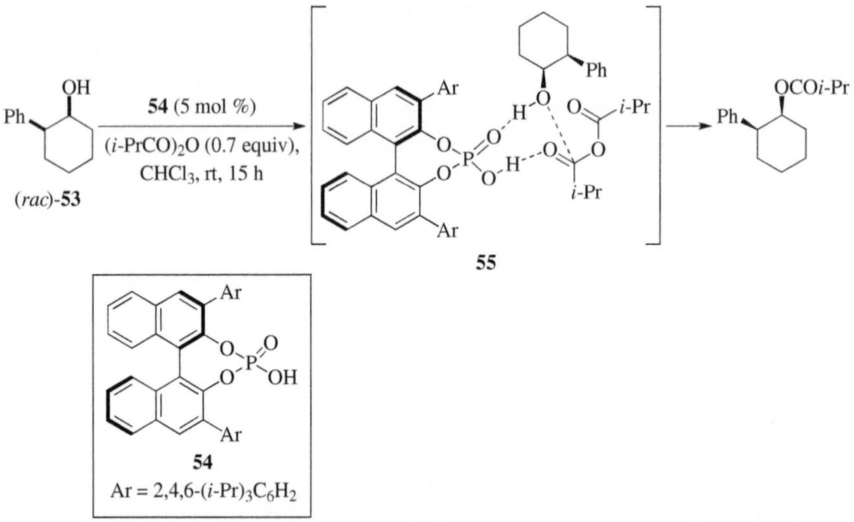

Scheme 21

desymmetrization[32] of various amines. Experimental and computational evidence
suggests that the reaction proceeds through the formation of ion-pair **57**, consisting
of an *N*-benzoylpyridinium and its benzoate counterion complexed with the chiral
thiourea catalyst (Scheme 22).[105] DFT studies suggest that, in the presence of the
favored (*R*)-configured amine, a quaternary complex is formed prior to acylation
that includes a three-fold π-π stacking interaction between one of the thiourea aryl
groups, the pyridinium core, and phenethylamine. The corresponding quaternary
complex with the (*S*) configured amine is higher in energy. The ion pair is thought to
not only provide the asymmetric environment required for enantiodiscrimination of

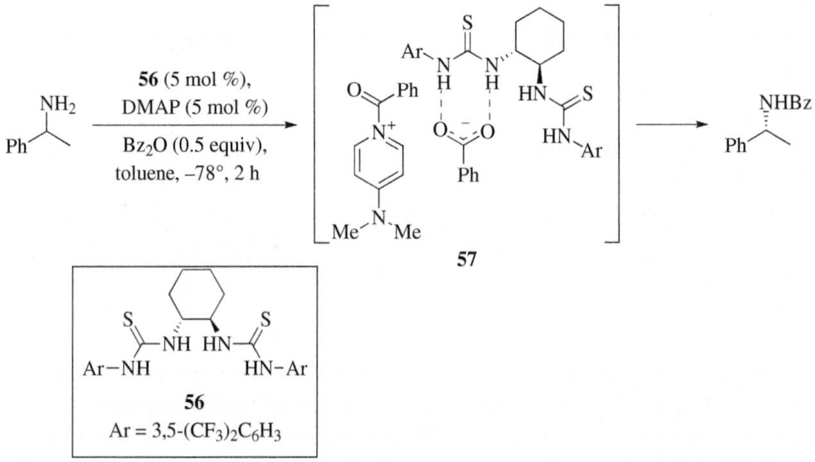

Scheme 22

the racemic amines, but it may also alter the equilibrium concentrations of DMAP and its N-benzoylpyridinium salt.[31] The reaction selectivity is very much dependent upon the nature of the counterion bound to the thiourea[105] and is also affected by the nature of the achiral Lewis base employed.[106]

SCOPE AND LIMITATIONS

Desymmetrization of Prochiral Diols

Nonenzymatic desymmetrization of *meso*- and/or prochiral diols through enantio-selective O-acylation is an attractive method for the synthesis of enantiomerically enriched, small-molecule building blocks. Since the desymmetrization of prochiral systems involves breaking the inherent symmetry, it is theoretically possible to achieve quantitative conversion into a single enantiomer of the product.[6,13,107−111] A range of O-acylative approaches have been developed for the desymmetrization of diol substrates, with many methods allowing the synthesis of synthetically useful diol derivatives with high levels of enantioselectivity.

1,2-Diols. The desymmetrization of *meso*-1,2-diols has been widely investigated via several different methods, including those that use stoichiometric, chiral O-acylating agents, which includes using either Lewis base or Lewis acid catalyzed stereoselective O-acylations.

Stoichiometric Methods. One of the first examples of the desymmetrization of *meso*-1,2-diols involved the monoacylation of the corresponding cyclic tin alkoxides using the enantiomerically pure (1*S*)-(+)-ketopinic acid chloride (**59**).[112] For example, reacting *meso*-dimethyl tartrate (**58**) with Bu$_2$SnO followed by **59** (1.34 equiv) affords the corresponding monoester in 80% yield and with 95.0:5.0 dr (Scheme 23).[112] This method supported the concept of stereoselective desymmetrization and laid the foundation for the development of catalytic procedures. An alternative stoichiometric O-acylating agent based upon a thiazolidine-2-thione auxiliary core has also been evaluated, but only low levels of stereoselectivity were

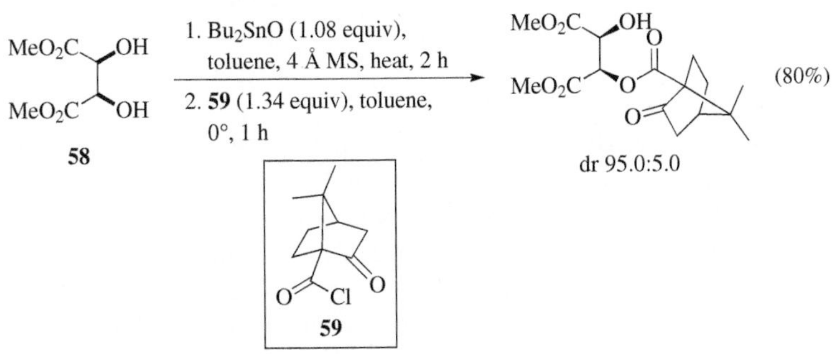

Scheme 23

observed for 1,2-diol substrates.[25,26] In practice, stoichiometric methods for desymmetrization are less commonly used because of the advances in the development of catalytic methods, which offer a broader substrate scope and higher levels of selectivity.

Lewis Base Catalysis.   Chiral phosphine- and phosphinite-based catalysts such as **47** and **60–62** have been employed in 1,2-diol desymmetrization reactions (Fig. 10). An early example of catalytic desymmetrization of *meso*-1,2-diols used chiral phosphine **47** and benzoic anhydride as the acyl source.[87] *meso*-Hydrobenzoin (**63**) was successfully monobenzoylated to form the ester **64** in 84% yield and with modest 84.0:16.0 er (Scheme 24).[87] The selectivity was subsequently improved by using phosphine **60** as a catalyst, allowing ester **64** to be isolated in 97.0:3.0 er.[113] The scope of this process was extended to two cyclic diols, with the product enantiomeric ratio being dependent on the ring-size.[87] For example, the reaction of *meso*-cyclohexane-1,2-diol with acetic anhydride in the presence of **47** (16 mol %) gave the corresponding ester in 83.5:16.5 er, whereas the reaction using *meso*-cyclopentane-1,2-diol formed the monoester with a reduced 76.0:24.0 er.

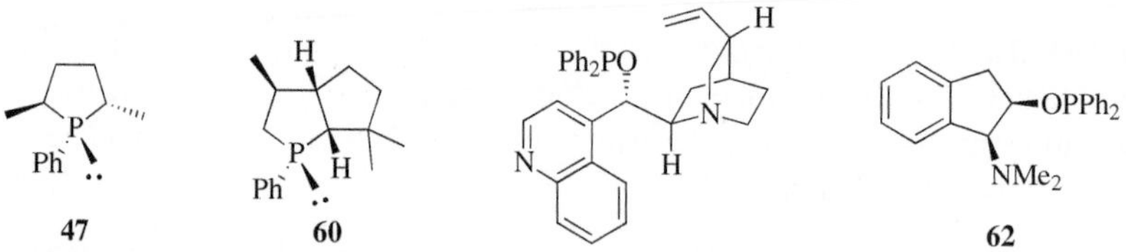

Figure 10.   Phosphine- and phosphinite-based catalysts.

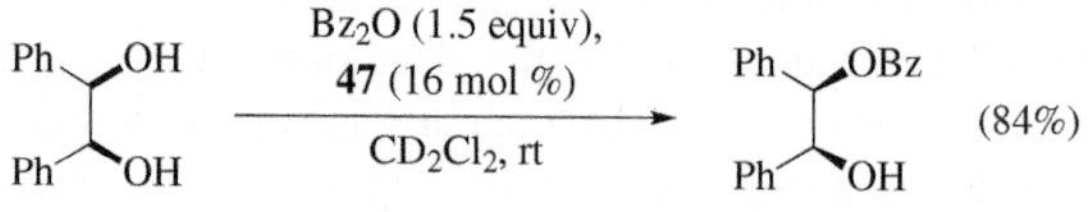

Scheme 24

Bifunctional phosphinite catalyst **61** derived from cinchonine allows a range of *meso*-1,2-diols to be desymmetrized through enantioselective acylation with benzoyl chloride to form the monobenzoylated esters in high yield and with excellent levels of enantioselectivity (Scheme 25).[114] In this case, the benzoylation of *meso*-cyclopentane-1,2-diol and *meso*-cyclohexane-1,2-diol proceeds with both similar yield and enantiomeric excess. Bifunctional amino-phosphinite **62** can also be used as a catalyst at much lower loading (5 mol % for **62** compared with 30 mol % for **61**) for the desymmetrization of the same range of diols with comparable levels of stereoselectivity using 4-*tert*-butylbenzoyl chloride as the optimal *O*-acylating agent.[115]

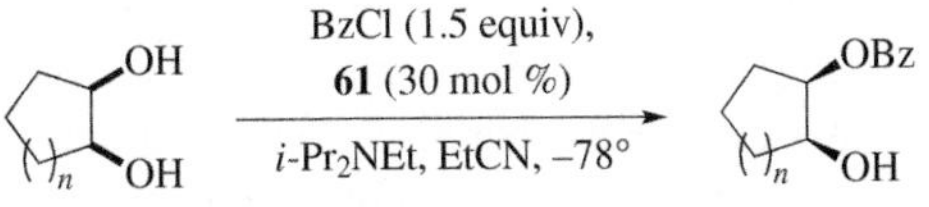

n	Yield (%)	er
1	80	96.5:3.5
2	85	97.0:3.0

Scheme 25

Enantiomerically enriched proline derivatives have been employed as efficient catalysts for the desymmetrization of a range of *meso*-1,2-diols. Although an initial report used stoichiometric amounts of the chiral diamine with benzoyl chloride as the *O*-acylating agent,[116] it was subsequently found that adding an equivalent of triethylamine as base permits the diamine to be employed as a catalyst.[86,117] For example, *meso*-cyclohexane-1,2-diol (**65**) undergoes desymmetrization with benzoyl chloride and triethylamine at low temperature using only 0.5 mol % of chiral diamine **46** to afford the benzoyl ester **66** in 87% yield and with 98.5:1.5 er (Scheme 26).[86]

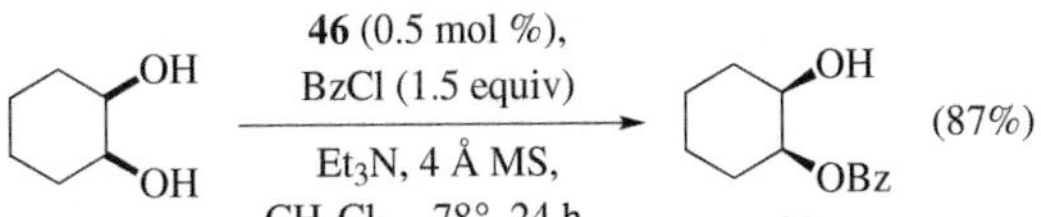

Scheme 26

The truncated cinchona alkaloid diamine **68** (2 mol %) catalyzes the asymmetric monobenzoylation of a range of cyclic and acyclic *meso*-1,2-diols.[118] These reactions use benzoyl chloride as the acylating agent in combination with triethylamine as a stoichiometric base in either THF or EtOAc at –60°, which affords the benzoyl ester products in generally good yields and with moderate to excellent levels of stereoselectivity. For example, the functionalized acyclic *meso*-diol **67** is benzoylated to furnish ester **69** in 51% yield and with 96.5:3.5 er (Scheme 27).[118]

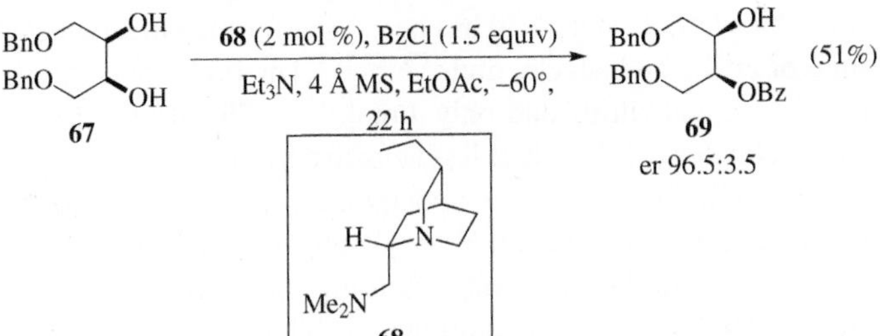

Scheme 27

Various enantiomerically enriched DMAP derivatives have been evaluated as catalysts for desymmetrization reactions of common *meso*-diols such as *meso*-cyclohexane-1,2-diol and/or *meso*-hydrobenzoin, as exemplified by **9**, **12**, **70**, and **71** (Fig. 11).[41,44,119,120] The reactions typically use the sterically demanding isobutyric anhydride as an *O*-acylating agent and a tertiary amine as a stoichiometric base. However, these catalysts have only been examined on a limited number of *meso*-1,2-diols and often afford inferior stereoselectivities compared with tertiary amine catalysts such as **46** and **68**. Moreover, these DMAP derivatives require multistep syntheses for their preparation and are therefore of limited preparative value.

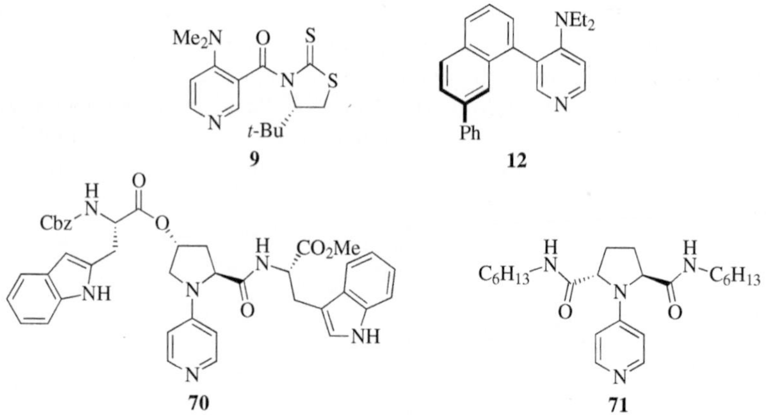

Figure 11.   DMAP derivatives for the desymmetrization of 1,2-diols.

Recently, the acylative desymmetrization of *meso*-1,2-diols using binaphthyl-based DMAP derivative **72** has been reported for a range of acyclic *meso*-1,2-diols and *meso*-cyclohexane 1,2-diol (**65**), amongst others (Scheme 28).[121] Low catalyst loadings (0.1 mol %) can be employed to facilitate the desymmetrization in a relatively short reaction time (3 h) to afford the enantioenriched monoacylated products in moderate to good yields (up to 85%) with high enantioselectivities (up to 97.0:3.0 er).

Peptide-based acylation catalyst **30** has been used for the efficient desymmetrization of a range of cyclic and acyclic diols.[73,122,123] The protocol employs an excess of inexpensive acetic anhydride and only 1 mol % of **30**. In this case, the desymmetrization of cyclic *meso*-1,2-diols is less sensitive to ring size, with five-, six-, and eight-membered cyclic *meso*-1,2-diols undergoing selective acetylation with good levels of stereoselectivity (Scheme 29).[123] Aryl-substituted acyclic diols are less well tolerated, with *meso*-hydrobenzoin (**63**) affording the corresponding monoacetate in 45% yield and 69.0:31.0 er. An additional problem with this particular process is the propensity of the monoacetate products to undergo partial racemization upon workup. Therefore, a one-pot desymmetrization-oxidation procedure has been

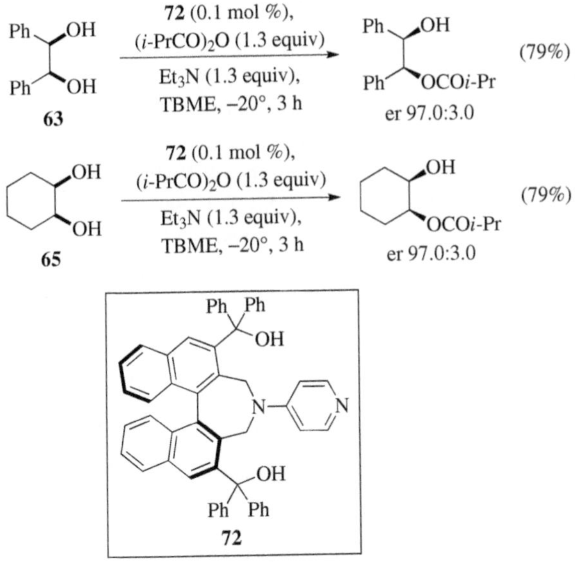

Scheme 28

developed to avoid the need to isolate the sensitive monoacetate intermediates. For instance, treating the crude reaction mixture of the desymmetrization of **65** with TEMPO (60 mol %) and m-CPBA (8 equiv) gives the α-acetyl ketone **73** in 70% yield and with 94.0:6.0 er (Scheme 30).[123]

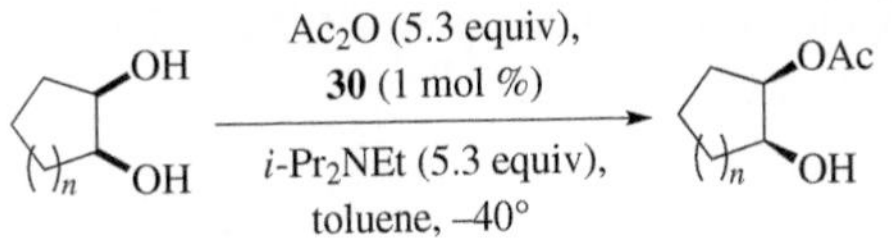

n	Time (h)	Yield (%)	er
1	7	89	90.0:10.0
2	48	81	94.0:6.0
4	22	97	92.0:8.0

Scheme 29

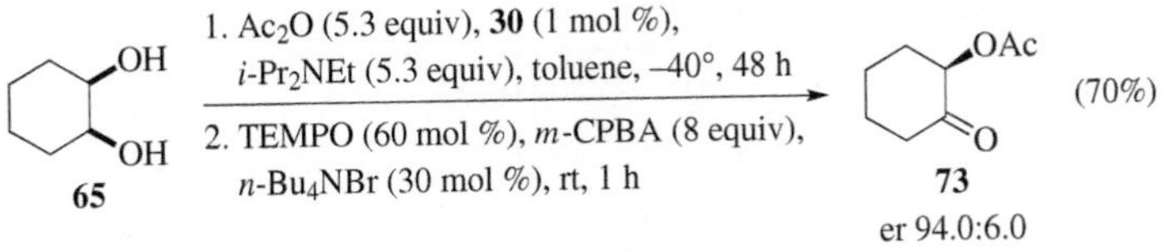

Scheme 30

N-Heterocyclic carbene catalysts have also been evaluated for the desymmetrization of *meso*-cyclohexane-1,2-diol **65**, using aldehydes as O-acylating agents under oxidative conditions.[82,80,124] To date, these procedures have been examined with one substrate and afford significantly lower selectivity compared with the other Lewis base catalyzed processes.

Lewis Acid Catalysis.   Several metal-catalyzed, stereoselective monoacylations of *meso*-1,2-diols have been developed. In most cases, copper salts, in conjunction with various chiral ligands, provide excellent catalytic systems for desymmetrization reactions using aroyl chlorides as acylating agents.

For example, *meso*-hydrobenzoin (**63**) undergoes stereoselective benzoylation using CuCl$_2$ (5 mol %) in combination with the (*R,R*)-Ph-Box ligand **74** to give ester *ent*-**64** in 79% yield and with excellent 97.0:3.0 er (Scheme 31).[125] The reaction can also be performed in water, albeit higher catalyst loadings of CuCl$_2$ (20 mol %) and chiral ligand **74** (30 mol %) are required along with a stoichiometric amount of the peptide coupling reagent DMT-MM and DMAP (10 mol %) as a cocatalyst.[126]

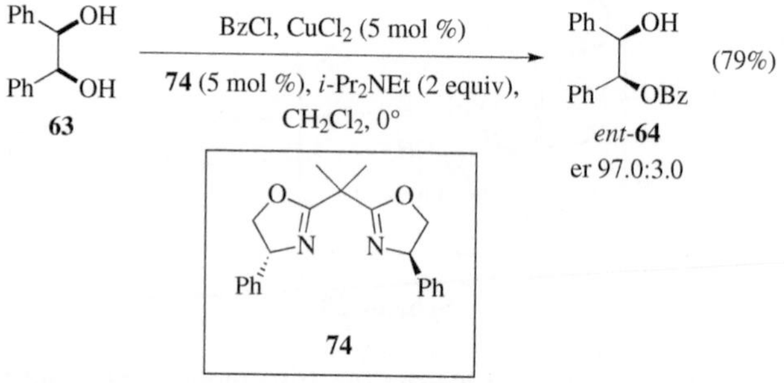

Scheme 31

The related desymmetrization of *meso*-dibenzyl tartrate (**75**) has been reported with CuCl$_2$/AgOPiv and the (*S,S*)-Ph-Box ligand *ent*-**74** in combination with 2-pyridyl ester **76** as the acylating agent. The acylated product is obtained in high yield (94%) and with excellent 98.0:2.0 er (Scheme 32).[127]

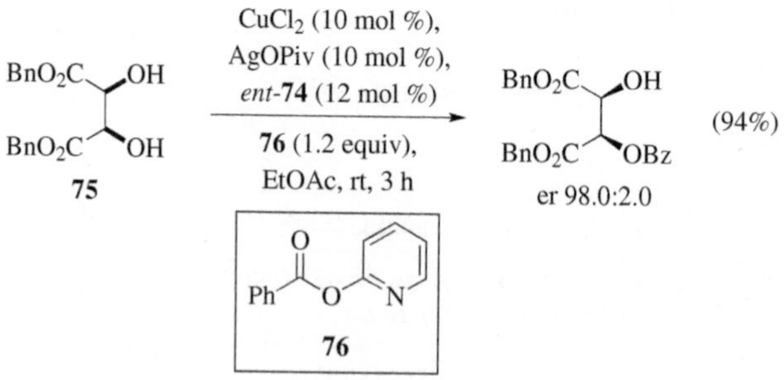

Scheme 32

The same catalytic system can also be used for the asymmetric monocarbamoylation of a wide range of *meso*-1,2-diols. For example, treatment of various cyclic *meso*-1,2-diols with phenyl isocyanate in the presence of Cu(OTf)$_2$ (10 mol %) and *ent*-**74** (10 mol %) leads to efficient desymmetrization in good yield and with high levels of selectivity for each ring size examined (Scheme 33).[128] This catalytic system is effective for several cyclic and acyclic *meso*-1,2-diols, using benzoyl chloride, phenyl isocyanate, and 4-toluenesulfonyl chloride for the benzoylation, carbamoylation, and sulfonylation, respectively.[129] The level of stereoselectivity observed is dependent on the reagent used, with monotosylation generally the most efficient protocol across the range of *meso*-1,2-diols. For example, the benzoylation of cyclic *meso*-1,2-diol **77** under the aforementioned catalytic conditions leads to a racemic benzoyl ester product. In contrast, the analogous process with phenyl isocyanate affords the corresponding carbamate in an improved 86.0:14.0 er, and 4-tosyl chloride further improves the stereoselectivity to furnish the corresponding 4-tosyl ester in 97.0:3.0 er (Scheme 34).[129]

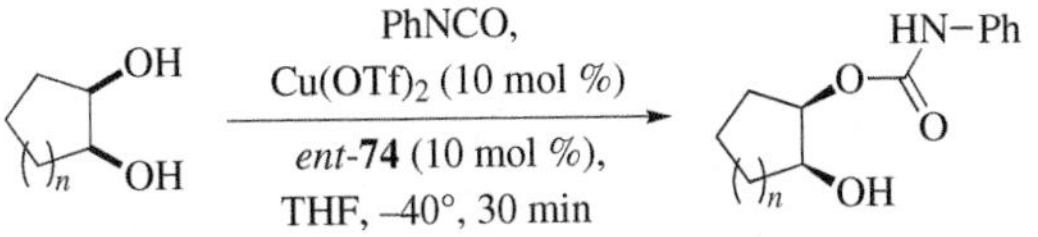
PhNCO,
$Cu(OTf)_2$ (10 mol %)
ent-**74** (10 mol %),
THF, –40°, 30 min
OH
OH
HN–Ph
O
O
OH

n	Yield (%)	er
1	82	93.0:7.0
2	69	93.0:7.0
3	83	95.5:4.5
4	72	96.5:3.5

Scheme 33

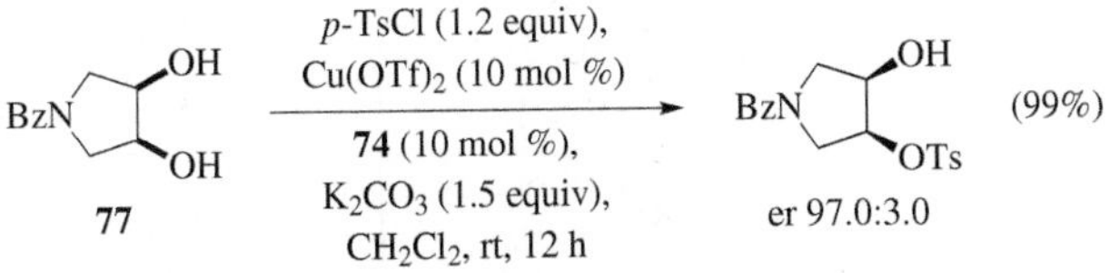
p-TsCl (1.2 equiv),
$Cu(OTf)_2$ (10 mol %)
74 (10 mol %),
K_2CO_3 (1.5 equiv),
CH_2Cl_2, rt, 12 h
BzN
OH
OH
77
BzN
OH
OTs
er 97.0:3.0
(99%)

Scheme 34

Anionic bisoxazoline derivatives containing a tetrasubstituted boron atom bridge have been used as ligands in desymmetrization reactions. The neutral zwitterionic complex formed from the combination of $CuCl_2$ with the chiral ligand **78** permits the selective monobenzoylation of a limited range of *meso*-1,2-diols at low catalyst loading (1 mol %), to afford the benzoyl esters in good yields and with modest to high levels of stereoselectivity (Scheme 35).[130]

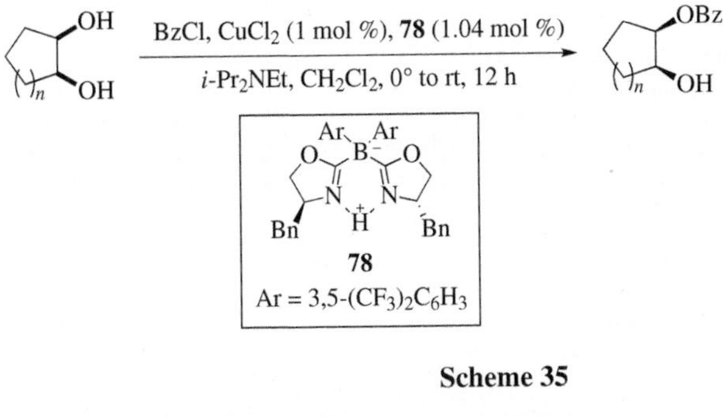
BzCl, $CuCl_2$ (1 mol %), **78** (1.04 mol %)
i-Pr_2NEt, CH_2Cl_2, 0° to rt, 12 h
OH
OH
OBz
OH
Ar Ar
B
O O
N N
H
Bn Bn
78
Ar = 3,5-$(CF_3)_2C_6H_3$

n	Yield (%)	er
1	73	88.0:12.0
2	83	95.0:5.0

Scheme 35

Solid-supported, *N*-tethered bis(imidazoline) ligand **80** has been used to optimize the reaction conditions for the desymmetrization of *meso*-butane-2,3-diol (**79**).[131] High-throughput screening using circular dichroism to monitor the level of asymmetric induction directly led to the development of highly efficient catalytic conditions, which could be successfully transferred to solution phase. For example, using the chiral ligand **81** with CuCl (5 mol %) permits the diol **79** to be selectively *O*-acylated with benzoyl chloride to provide ester **82** in an excellent 97% yield and 97.5:2.5 er (Scheme 36).[131] Although the reaction is highly stereoselective for a range of aroyl chlorides as *O*-acylating agents, cyclic *meso*-1,2-diols lead to lower selectivities under the optimized conditions.

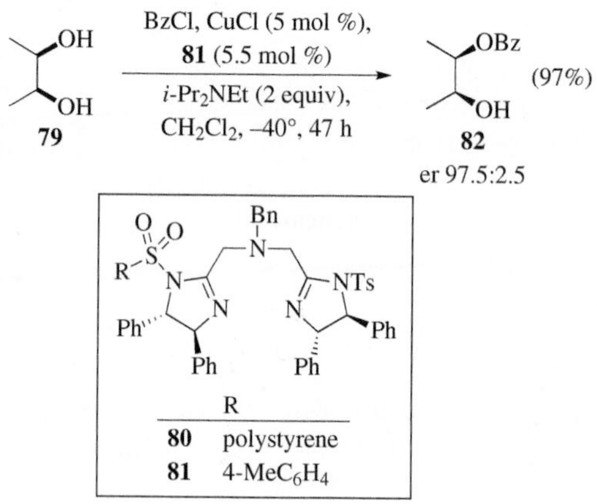

Scheme 36

C_2-Symmetric diamine ligands can also be used for the asymmetric, Cu-catalyzed monoacylation of *meso*-1,2-diols.[132,133] For example, the diamine **83**, which is prepared in four steps from L-proline, can be used in conjunction with CuCl$_2$ (3 mol %) and benzoyl chloride to effect the efficient desymmetrization of a range of cyclic and acyclic *meso*-1,2-diols (Scheme 37).[133] The direct comparison of the results with the bisoxazoline ligand **74** under the same reaction conditions illustrates that the diamine **83** provides higher levels of stereoselectivity in all cases.

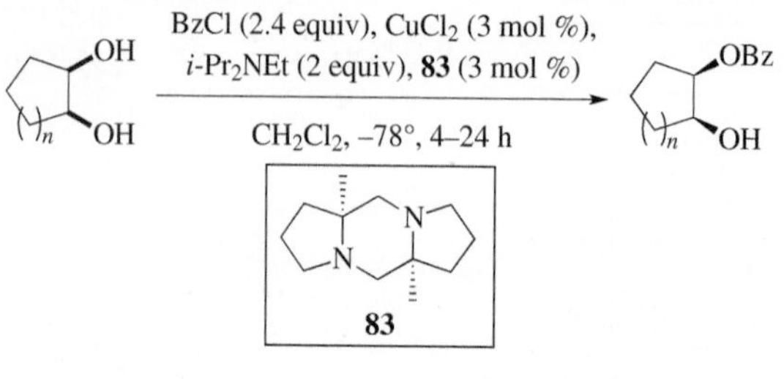

n	Yield (%)	er
1	73	89.5:10.5
2	100	96.5:3.5
4	82	85.0:15.0

Scheme 37

1,3-Diols. The desymmetrization of prochiral and/or *meso*-1,3-diols has been widely explored using a range of different methods. In particular, the desymmetrization of prochiral 2-substituted 1,3-propanediol and glycerol derivatives has received significant attention, with the development of many efficient methods.

Stoichiometric Methods. An early example of the desymmetrization of prochiral 1,3-diols was performed on protected glycerol derivative **84**. Treatment of the 1,3-diol **84** with (methylcyclopentadienyl)tin(II) chloride to facilitate the in situ formation of the corresponding tin alkoxide, followed by addition of chiral diamine **85** (1 equiv) and benzoyl chloride (2 equiv), affords the monobenzoylation adduct **86** with 92.0:8.0 er but only 20% yield (Scheme 38).[134]

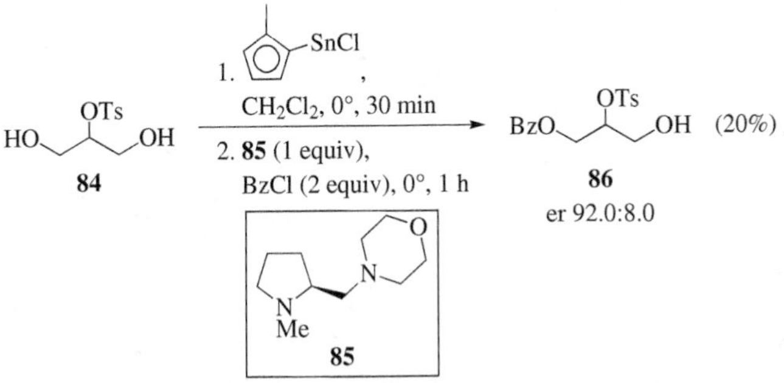

Scheme 38

The desymmetrization of *meso*-1,3-diols was first demonstrated using stoichiometric amounts of *O*-benzoyl quinidine **88** (2 equiv) and benzoyl chloride (2 equiv) for the acylation of *meso*-cyclopent-4-ene-1,3-diol (**87**), to form the ester **89** in a promising 70:30 er, albeit very low yield (Scheme 39).[135] Stoichiometric *O*-acylating agents based upon a thiazolidine-2-thione auxiliary core have also been investigated for the desymmetrization of *meso*-1,3-diols, but low levels of stereoselectivity were observed.[25,26]

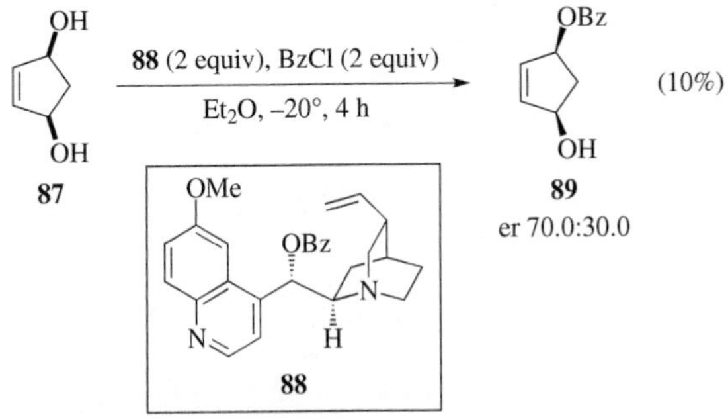

Scheme 39

Lewis Base Catalysis.   The desymmetrization of cyclic *meso*-1,3-diols using small-molecule Lewis base catalysis is challenging, which is exemplified by the relatively limited number of successful catalysts that have been reported. For instance, the chiral proline derivative **90** facilitates the desymmetrization of two cyclic *meso*-1,3-diols using benzoyl chloride as the *O*-acylating agent.[117,136] The reaction proceeds with a very low catalyst loading (0.5 mol %), forming the corresponding monobenzoyl esters in good yields and with excellent enantioselectivities (Scheme 40).[136]

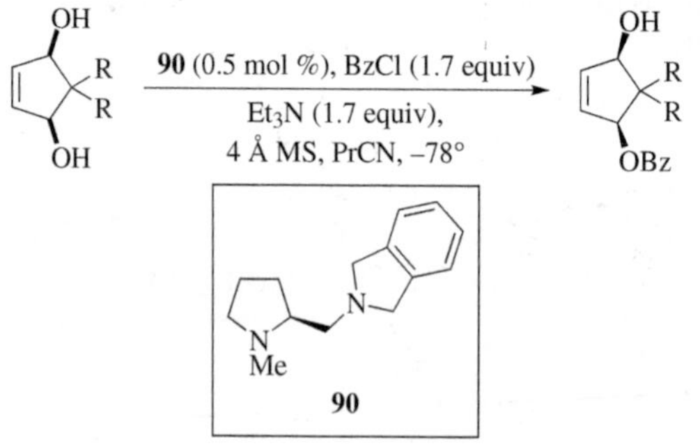

R	Time (h)	Yield (%)	er
H	3	37	99.0:1.0
Me	8	87	>99.5:0.5

Scheme 40

Bifunctional phosphinite catalyst **91**, derived from commercially available quinidine, can also be used for the selective benzoylation of a limited number of cyclic *meso*-1,3-diols. In each case, the ester products are formed in high yields with excellent levels of stereoselectivity (Scheme 41).[137] However, the same catalytic system with prochiral 1,3-diols affords significantly lower yields, and the monoester products are essentially racemic in all cases.

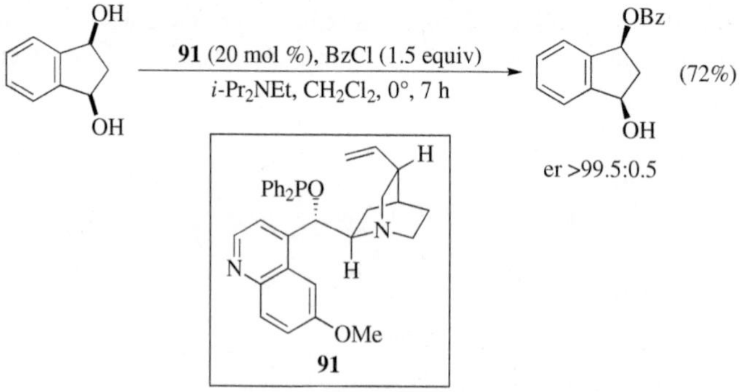

Scheme 41

Simple proline derivatives such as **46** and **90** have been employed at very low catalyst loadings (0.5 mol %) for the desymmetrization of prochiral 1,3-diols.[138] Prochiral diols bearing various substituents that permit further functionalization are

O-acylated using aroyl chlorides, to afford the corresponding monoesters in typically low yields (<33%), but with excellent enantioselectivities. For example, epoxide **92** is acylated using benzoyl chloride and **46** (0.5 mol %) to prepare ester **93** in 31% yield and 98.0:2.0 er (Scheme 42).[138]

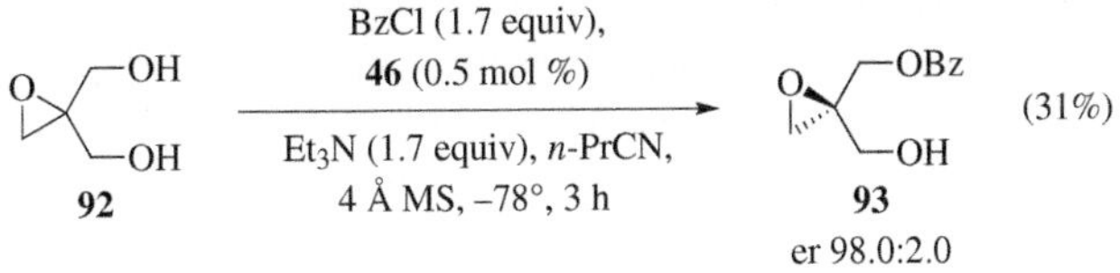

Scheme 42

Bifunctional amino phosphinite catalyst **62** has also been used for the O-aroylation of prochiral 1,3-diols. A limited number of diols, such as **94**, react with 4-*tert*-butylbenzoyl chloride in the presence of **62** (5 mol %) to form esters, such as **95** in good yields, but with only modest levels of stereoselectivity (Scheme 43).[115]

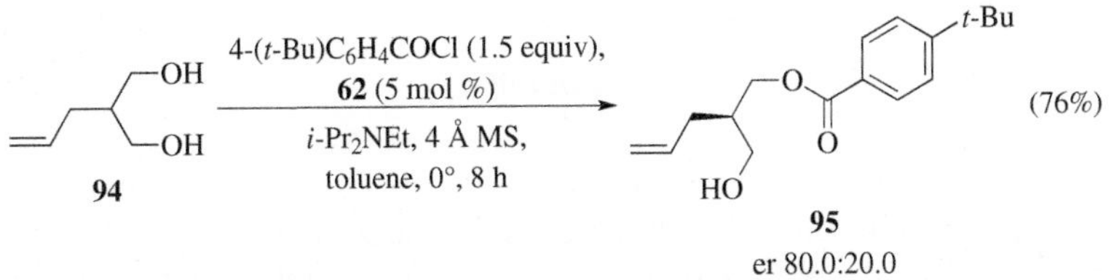

Scheme 43

Pentapeptide **96** is an effective catalyst for the selective acylation of O-benzyl-protected glycerol derivatives (Scheme 44).[139] The reactions use acetic anhydride as the O-acylating agent and are performed at low temperature to form the monoacety-lated products in typically low yields but with high levels of stereoselectivity that

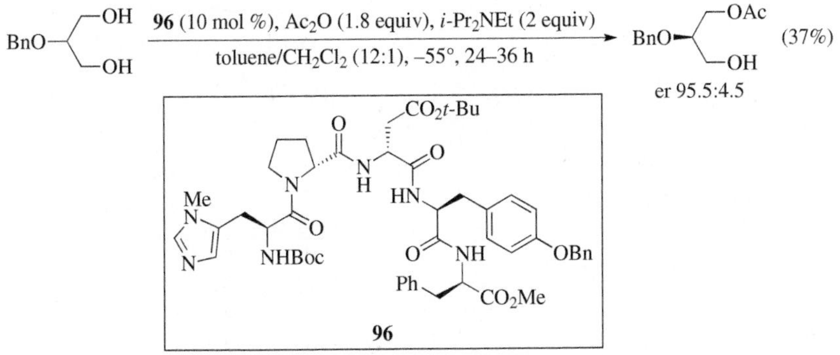

Scheme 44

are comparable with those observed from enzymatic desymmetrization of the same substrates. The system has also been evaluated for the desymmetrization of acyclic *meso*-1,3-diols, which afford significantly lower levels of enantioselectivity.

The bifunctional catalyst **98**, which contains a Brønsted acidic sulfonamide and a Lewis basic imidazole, has been used to facilitate the desymmetrization of a small range of glycerol derivatives. For example, the protected glycerol **97** could be selectively acylated with cyclohexanoyl anhydride in the presence of **98** (5 mol %) at room temperature to form ester **99** in 74% yield and with 96.5:3.5 er (Scheme 45).[72]

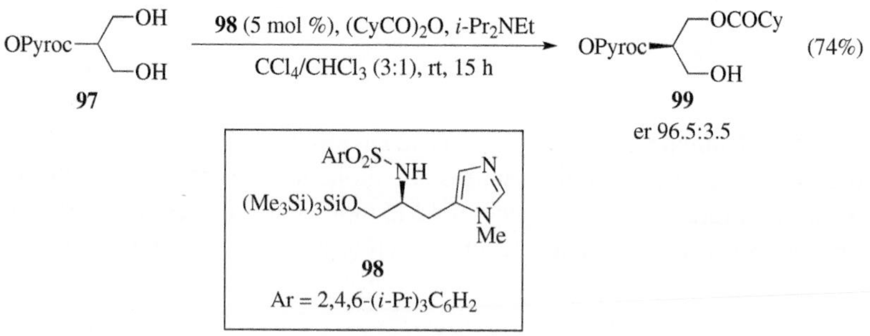

Scheme 45

The desymmetrization of 1,3-diols through oxidative lactonization has been achieved using a combination of NHC precursor **100** and quinone **101** (Scheme 46).[140] This protocol proved general for the enantioselective preparation of a range of

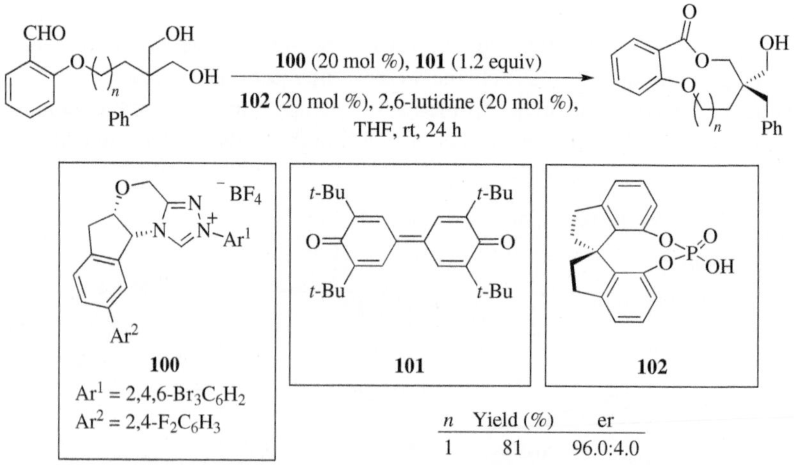

n	Yield (%)	er
1	81	96.0:4.0
2	68	94.0:6.0
3	51	92.0:8.0
4	55	93.0:7.0

Scheme 46

nine- to twelve-membered lactones with up to 96.0:4.0 er, requiring the addition of phosphoric acid **102** as an additive for optimal enantioselectivity.

Lewis Acid Catalysis.   Lewis acid catalyzed desymmetrization reactions have primarily focused on the development of methods for the stereoselective mono-*O*-acylation of prochiral 1,3-diols. In particular, using $CuCl_2$ as the Lewis acid catalyst in combination with a range of enantiomerically pure ligands (e.g., **51**, **103–107**) has been widely explored (Fig. 12). Dinuclear zinc complexes obtained from the reaction of Et_2Zn with chiral ligands have also been investigated as desymmetrization catalysts.

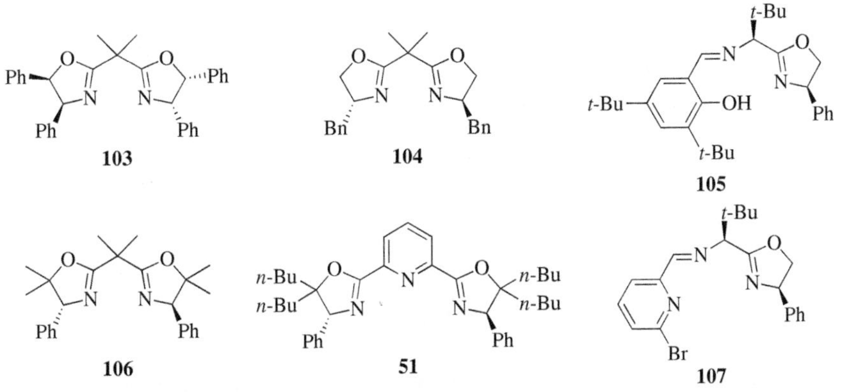

Figure 12. Ligands for copper-catalyzed desymmetrization of 1,3-diols.

Substituted glycerol derivatives undergo efficient desymmetrization using benzoyl chloride as the *O*-acylating agent. The reaction is both practical and straightforward, employing $CuCl_2$ and the requisite chiral ligand as the catalyst in combination with a stoichiometric amount of triethylamine in THF at room temperature. A range of different ligands has been surveyed in this process, with Box ligand **104** and imine **107** proving to be optimal (Scheme 47).[141,142] Importantly, unlike with the Lewis base catalyzed protocols, the tertiary alcohol within the glycerol derivatives does not require a protecting group. The reaction has been examined on a wide

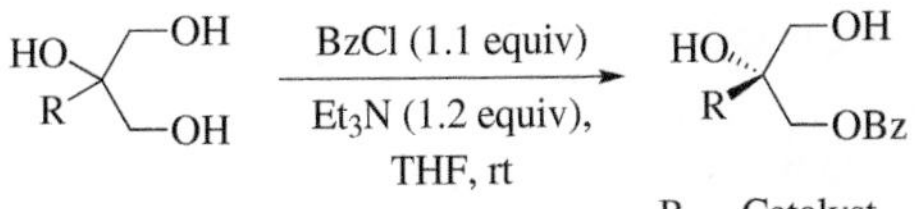

R	Catalyst	Time (h)	Yield (%)	er
Me	**104•CuCl₂** (5%)	1	96	97.0:3.0
Me	**107•CuCl₂** (20%)	0.5	99	97.5:2.5
Ph	**104•CuCl₂** (5%)	1	67	65.0:35.0
Ph	**107•CuCl₂** (20%)	0.5	99	96.0:4.0

Scheme 47

range of 2-substituted glycerol derivatives bearing a range of different functional groups that are amenable to further functionalization. In general, catalytic systems employing either **104** or **107** as chiral ligands afford the chiral tertiary alcohol products in high yields and with excellent levels of stereoselectivity. However, for more challenging substrates, such as 2-vinyl-, phenyl-, and benzyl-substituted glycerols, complex **107•CuCl₂** (20 mol %) provides excellent yields and appreciably higher enantioselectivities in all cases.[141,142]

This method has also been extended to the desymmetrization of several 2-substituted 2-amino 1,3-propanediols. In this case, Box ligand **103** is optimal and thereby permits a series of highly functionalized prochiral 1,3-diols to undergo desymmetrization with benzoyl chloride to form the tertiary amine products in excellent yields and with high enantioselectivities (Scheme 48).[93] Nevertheless, the choice of the *N*-protecting group is essential for obtaining high levels of stereoselectivity, with *N*-benzoyl the most effective among those that were examined.

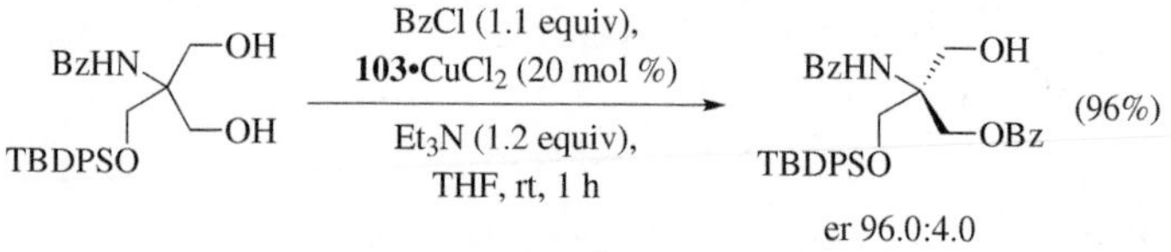

Scheme 48

Prochiral 1,3-propanediols containing a carbamate undergo stereoselective intramolecular acylation in the presence of the catalyst derived from CuCl₂ with Box ligand **106** (5 mol %) to form oxazolidin-2-ones. The process has been extensively optimized in terms of ligand, stoichiometric base, and solvent, providing access to oxazolidin-2-one products in excellent yields and with high levels of enantioselectivity (Scheme 49).[94] The enantiomeric purity of the oxazolidin-2-one products can be further enhanced by performing an in situ KR of the enantiomerically enriched alcohol using benzoyl chloride. Importantly, the same catalyst can be used for both the desymmetrization and KR processes, leading to an efficient one-pot procedure for the synthesis of a wide range of 4,4-disubstituted oxazolidin-2-one products with high levels of enantioselectivity (Scheme 50).[94]

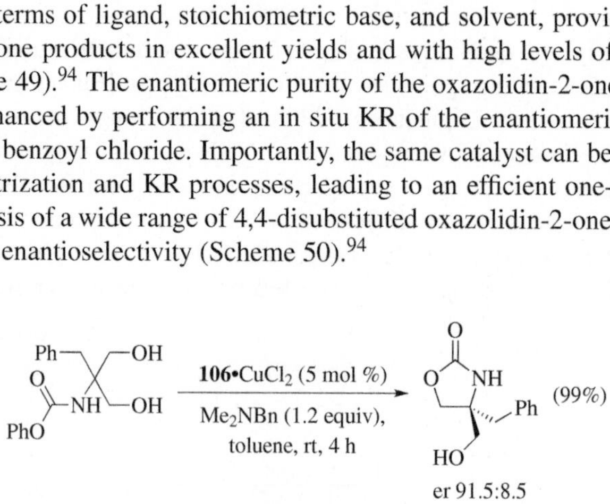

Scheme 49

This simple catalytic system can be further extended to allow an impressive range of 2,2-disubstituted 1,3-propanediols to undergo desymmetrization using the

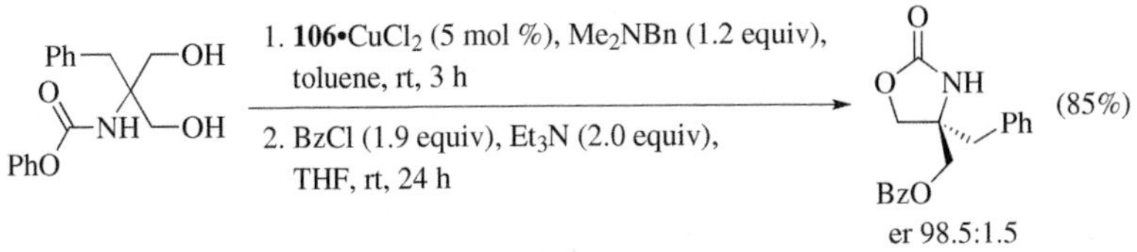

Scheme 50

complex derived from $CuCl_2$ (10 mol %) and PyBox ligand **51** to form challenging quaternary stereocenters in good yields and with generally high levels of enantioselectivity (Scheme 51).[92] Several different carbon substituents are tolerated, including alkyl, acyl, substituted alkenyl, and cyano that all deliver good levels of stereoselectivity, whereas terminal alkenyl and aryl substituents generally produce slightly reduced enantioselectivity. The process is particularly impressive given the simplicity of the catalytic system used and highlights the potential synthetic utility of desymmetrization processes to overcome the challenges associated with the generation of quaternary stereocenters.

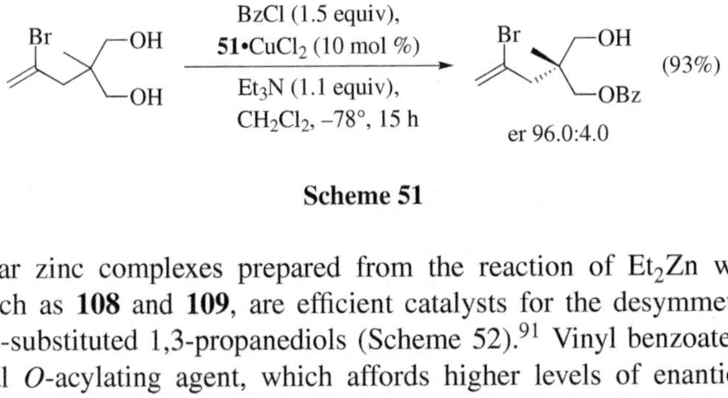

Scheme 51

Dinuclear zinc complexes prepared from the reaction of Et_2Zn with phenol ligands, such as **108** and **109**, are efficient catalysts for the desymmetrization of prochiral 2-substituted 1,3-propanediols (Scheme 52).[91] Vinyl benzoate is used as the optimal O-acylating agent, which affords higher levels of enantioselectivity compared with vinyl acetate. Ligand **108** is the most effective for 2-heteroaryl- and

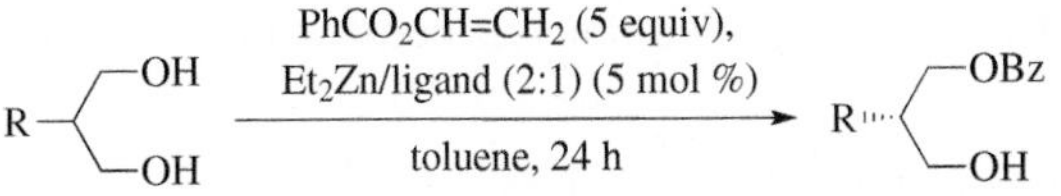

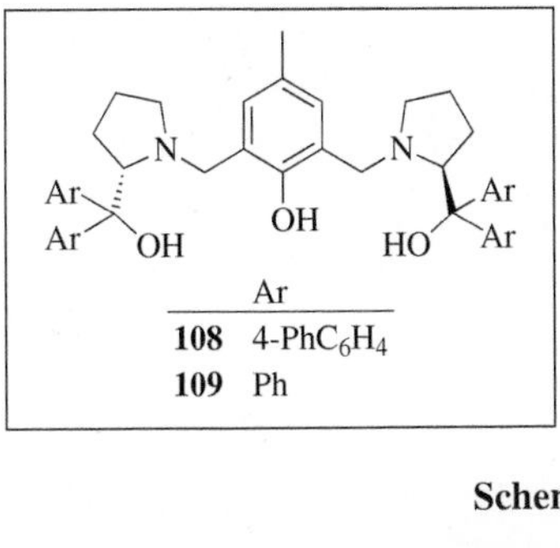

R	Ligand	Temp (°)	Yield (%)	er
Ph	108	−15	94	95.5:4.5
$PhCH_2$	109	−20	85	95.0:5.0

	Ar
108	4-PhC_6H_4
109	Ph

Scheme 52

2-aryl-substituted 1,3-propanediols.[91] Arenes bearing electron-donating substituents are suitable substrates, giving the corresponding monoesters in good yields and excellent levels of stereoselectivity. However, electron-withdrawing substituents require increased catalysts loadings (10 mol %) to obtain high yields and enantio-selectivities. 2-Alkyl- and alkenyl-substituted 1,3-propanediols most effectively undergo desymmetrization using ligand **109**, again allowing the ester products to be formed in good yield with high levels of stereoselectivity.[90]

Quinidine-derived sulfonamide **110** has been used as a ligand for Et$_2$Zn in the catalytic desymmetrization of prochiral 2-substituted glycerol derivatives using acetic anhydride as the O-acylating agent (Scheme 53).[143] A wide range of O-protecting groups within the prochiral triol are suitable, as are the different alkyl, alkenyl, and aryl 2-substituents, which furnish the monoacetyl products in high yield and with moderate to good levels of enantioselectivity. Notably, the opposite enantiomers of the ester products are accessible with comparable levels of selectivity using a pseudo-enantiomeric sulfonamide ligand derived from quinine.

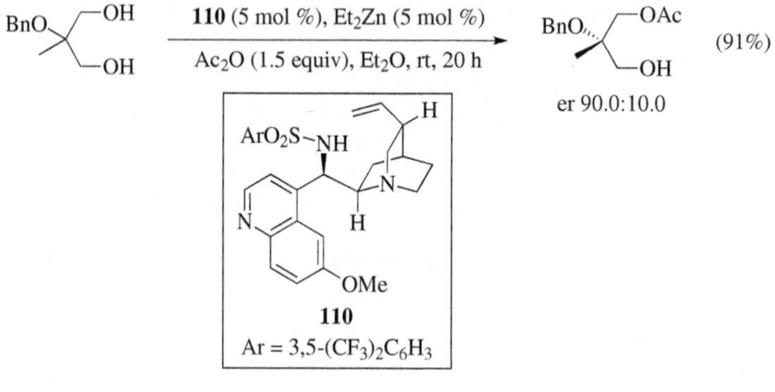

Scheme 53

The dimeric sulfonamide complex **111** provides an excellent catalyst for the desymmetrization of N-protected, 2-substituted 2-amino 1,3-propanediols (Scheme 54).[144] Optimization of the reaction conditions demonstrated that N-Boc

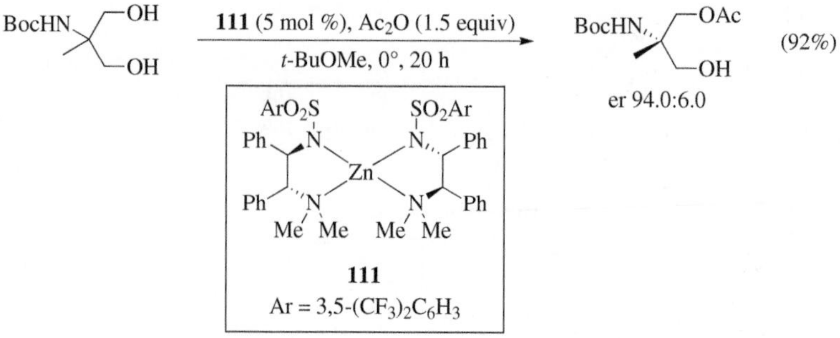

Scheme 54

derivatives afford higher levels of enantioselectivity compared with N-Cbz, and that t-BuOMe is the optimal solvent. Consequently, a limited range of 2-alkyl-substituted diols affords the tertiary amine products in good yields and with good levels of enantioselectivity using this approach.

Miscellaneous Diols. Methods for the desymmetrization of *meso*-diols with a greater than 1,3-relationship between the two alcohols have been less well investigated. Nevertheless, a few reports of such desymmetrization reactions make use of stoichiometric chiral acylating agents as well as Lewis base and Lewis acid catalyzed promoted reactions.

(1S,4R)-Camphanoyl iodide (**112**) is an effective stoichiometric O-acylating agent for the desymmetrization of a range of bicyclic *meso*-1,4-diols (Scheme 55).[27] In this case, the use of 2,6-dimethoxypyridine (**113**) in THF at low temperature is essential for the formation of the desired monoester products with good levels of diastereoselectivity.

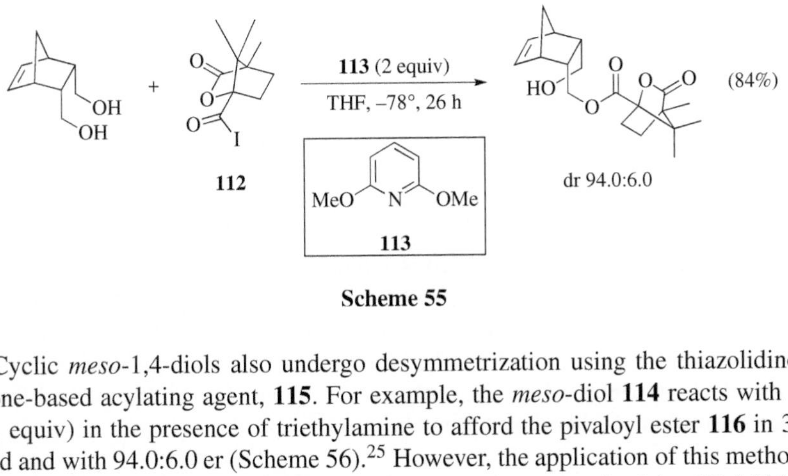

Scheme 55

Cyclic *meso*-1,4-diols also undergo desymmetrization using the thiazolidine-2-thione-based acylating agent, **115**. For example, the *meso*-diol **114** reacts with **115** (1.1 equiv) in the presence of triethylamine to afford the pivaloyl ester **116** in 34% yield and with 94.0:6.0 er (Scheme 56).[25] However, the application of this method to monocyclic *meso*-1,4-diols results in lower levels of enantioselectivity. Diol **114** can also undergo catalytic desymmetrization using bifunctional phosphinite catalyst **91** and benzoyl chloride as the O-acylating agent to form the benzoyl ester **117** in 55% yield and with 91.0:9.0 er (Scheme 57).[137]

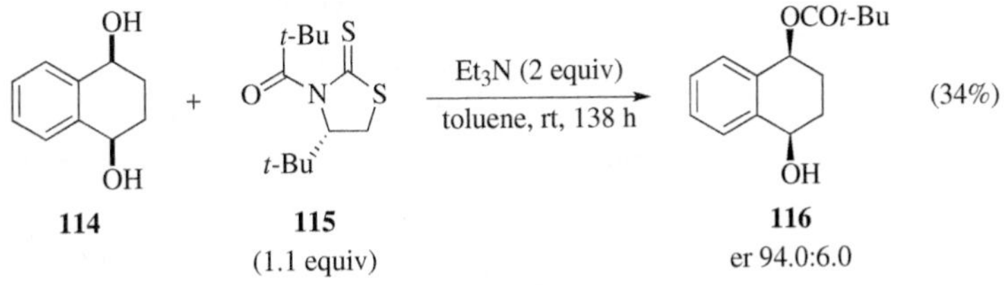

Scheme 56

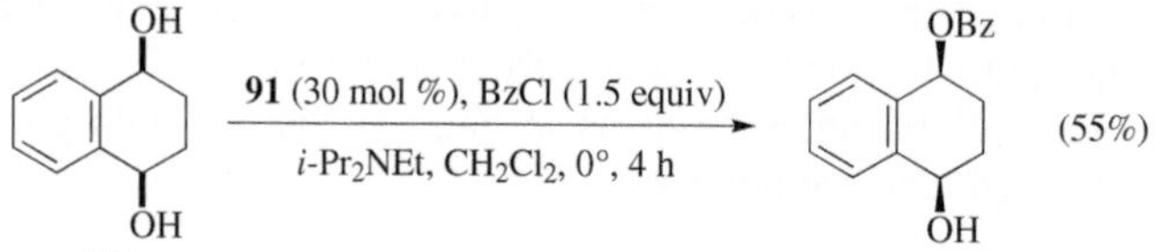

114 **117**

er 91.0:9.0

Scheme 57

The quincorine-derived amine **68** is an efficient catalyst for the desymmetrization of *meso*-chromium complex **118** with benzoyl chloride. The reaction proceeds with low catalysts loadings (2 mol %), allowing the formation of benzoyl ester **119** in 83% yield and with excellent 99.5:0.5 er (Scheme 58).[145] Importantly, the pseudo-enantiomeric form of the catalyst derived from quincoridine permits access to the enantiomer of **119** with the same level of stereoselectivity.

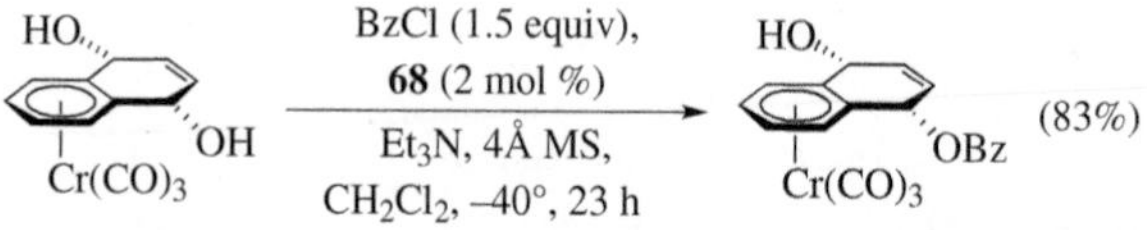

118 **119**

er 99.5:0.5

Scheme 58

Planar-chiral, DMAP derivative **10** is a highly efficient Lewis base catalyst for the enantioselective acylation of *meso*-1,5-diol **120** (Scheme 59).[146] This particular protocol uses acetic anhydride as the *O*-acylating agent and requires very low catalysts loading (1 mol %) to form the monoester **121** in 91% yield with excellent enantioselectivity (99.5:0.5 er). Nevertheless, the generality of this method for the desymmetrization of other diols has not been investigated.

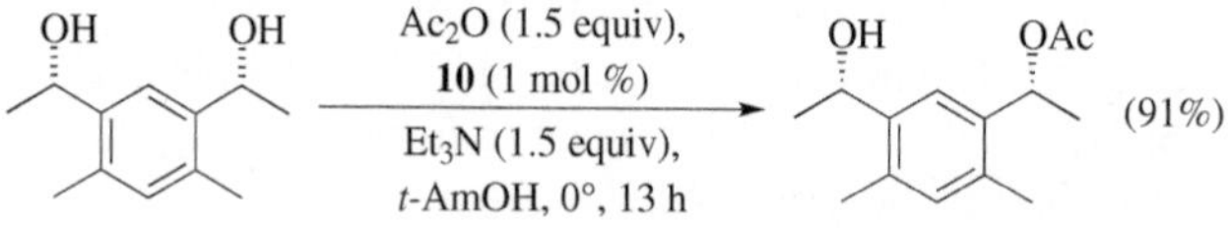

120 **121**

er >99.5:0.5

Scheme 59

4-(Dimethylamino)pyridine derivative **9** catalyzes the desymmetrization of a limited number of *meso*-1,4-, 1,5-, and 1,6-diols using isobutyric anhydride as the *O*-acylating agent to form the corresponding esters with generally high levels of enantioselectivity. For example, diol **122** undergoes stereoselective acylation using **9** (5 mol %) to furnish ester **123** in 69% yield and with 98.0:2.0 er (Scheme 60).[41]

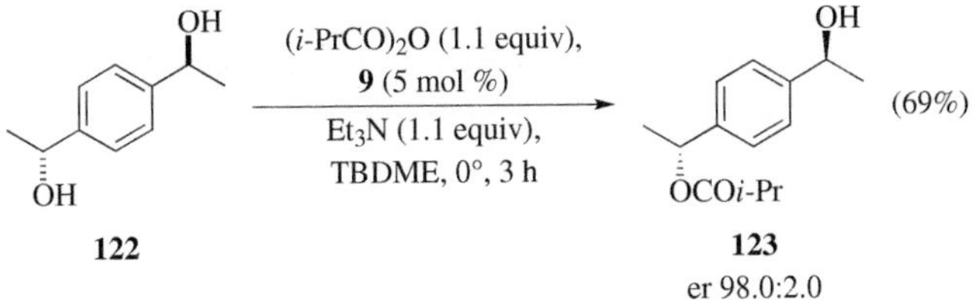

Scheme 60

A dinuclear, zinc complex obtained from the reaction of Et$_2$Zn with the chiral phenol ligand **108** is an effective Lewis acid catalyst for the desymmetrization of *meso*-1,4-diol **124** (Scheme 61).[91] The reaction employs vinyl benzoate as the *O*-acylating agent, to generate the monobenzoyl ester **125** in an excellent 93% yield and with high enantioselectivity (95.5:4.5 er).

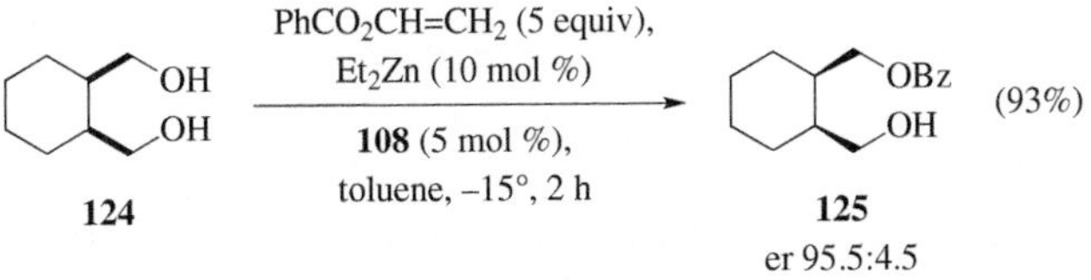

Scheme 61

An oxidative NHC-catalyzed acylative desymmetrization of prochiral bisphenol containing phosphinates has been used to prepare enantiomerically enriched *P*-stereogenic phosphinates. When the quinone **101** is the oxidant, the catalytic desymmetrization can be performed on gram-scale using 5 mol % of the NHC precatalyst **126** (Scheme 62).[147]

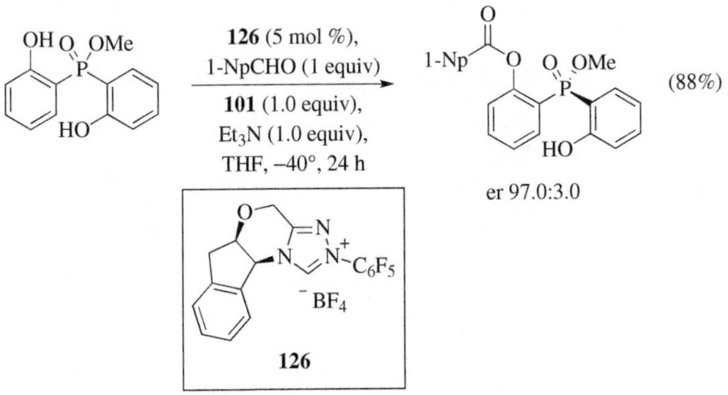

Scheme 62

Using NHC precatalyst **127** (10 mol %) and α-aroyloxyaldehyde **128** (1 equiv), the desymmetrization of bisphenolic triarylmethanes can be achieved through enantioselective acylation to afford the enantioenriched products in high yields and enantioselectivities (Scheme 63). The extension of this concept to the desymmetrization of 1,2-diarylalkanes under similar conditions also proceeded with excellent enantioselectivity.[148]

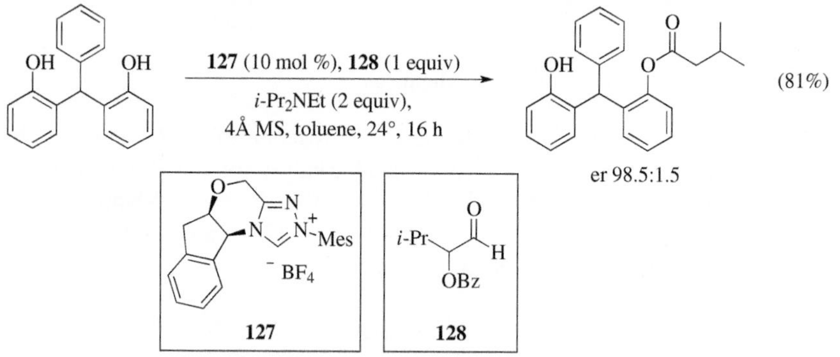

Scheme 63

Oxidative NHC catalysis has been used for the atroposelective desymmetrization of a series of biaryl amino alcohols through enantioselective acylation. For example, with the NHC precatalyst **130** (10 mol %) and isobutyraldehyde (1.5 equiv) in the presence of oxidant **101** (1.2 equiv) and K_2CO_3, the desymmetrization of **129** generated **131** in 92% yield and with >99.0:1.0 er (Scheme 64).[149] This process proceeds through a cooperative enantioselective acylation event, followed by a kinetic resolution process, resulting in enhancement of product enantiomeric ratio as the reaction proceeds. A related enantioselective NHC-catalyzed redox process has also been reported.[150]

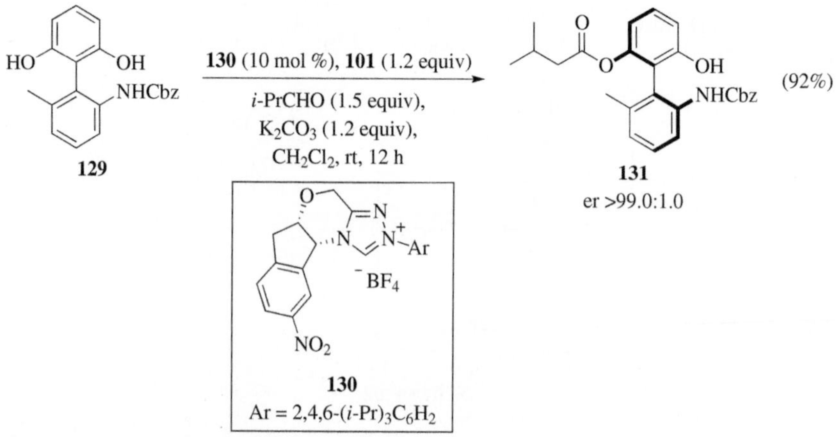

Scheme 64

Conclusions and Outlook. Several excellent methods are now available for the acylative desymmetrization of a variety of diols. For example, a range of highly stereoselective methods for the catalytic desymmetrization of prochiral 1,3-diols is available for the preparation of challenging ternary and quaternary stereocenters, which provide significant synthetic utility. Despite the advances made in this area, the development of improved acylative desymmetrization processes is still required. In many cases, the observed stereoselectivity of catalytic desymmetrization methods is lower than that of comparable KRs of structurally similar substrates. Furthermore, the development of new methods that are compatible with synthetically useful, nonparticipating functional groups that will permit their incorporation into larger synthetic efforts would also be valuable. In general, the desymmetrization of *meso*-diols with a greater than 1,3-relationship between the two alcohols is not as well established, possibly because of the increased complexity in the synthesis of the required starting materials. Nevertheless, the limited number of *meso*-1,4-, 1,5-, and 1,6-diols that have been investigated have provided promising results with generally high levels of stereoselectivity. As such, the continued development of the acylative desymmetrization of synthetically complex diols is likely to be an important area of research.

Kinetic Resolution of Alcohols

Benzylic Alcohols. The catalytic acylative KR of benzylic alcohols is often used for benchmarking the efficiency of new KR methods, as such a wide variety of methods has been reported. The majority of examples use Lewis base catalysis; however, efficient methods using Lewis acid and Brønsted acid catalysts have also been developed.

Stoichiometric Methods. A limited number of alkyl-substituted benzylic alcohols can be efficiently resolved using *N*-benzoyl oxazolidin-2-one (**132**) as a stoichiometric acylating agent (Scheme 65).[151] The optimal conditions require an excess of the racemic alcohol (10 equiv) with an equivalent of methylmagnesium bromide to facilitate the in situ formation of the corresponding magnesium alkoxide. Selective acylation of the (*R*)-alcohol permits the corresponding benzoyl ester to be isolated in good yield and with high enantioselectivity.

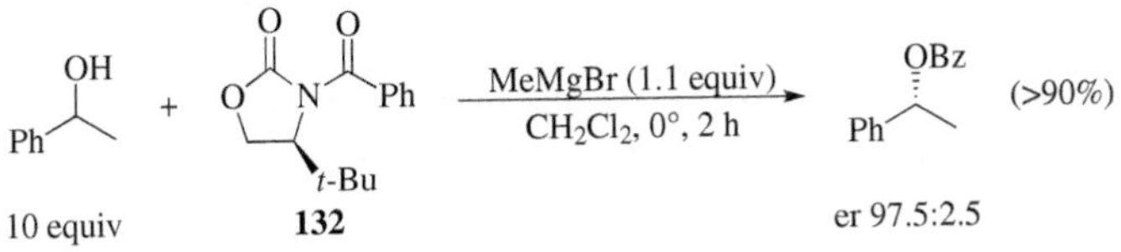

Scheme 65

The chiral *N*-acyl DMAP derivative **133** can also be used as a stoichiometric acyl source for the KR of benzylic alcohols (Scheme 66).[152] The process requires the addition of zinc chloride and an organic base to promote acylation, albeit the reaction rate is slow, and extended times are required to achieve moderate conversion into the ester products. The reaction was examined on a small number of alcohols, with 1-naphthyl-substituted alcohols providing optimal selectivity.[152,153]

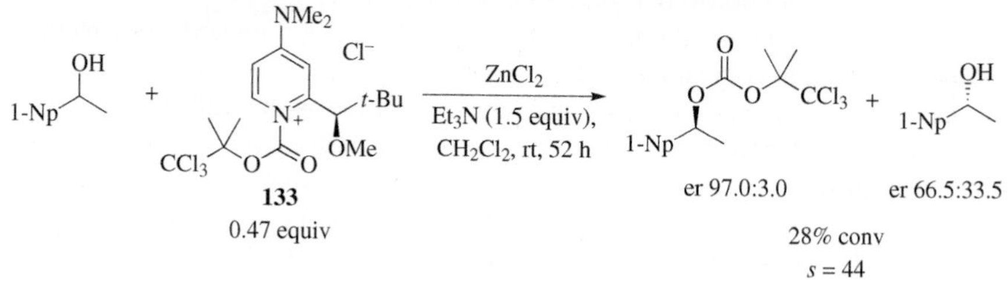

Scheme 66

Isolated *N*-acyl *N′*-methyl imidazolium salts derived from the reaction of (*S*)-2-phenylpropanoyl chloride with *N*-methylimidazole can also be employed in stoichiometric KRs. Alternatively, the parent acid chloride can be used directly in combination with a catalytic amount of *N*-methylimidazole (Scheme 67).[28] However, the method is highly substrate-specific, with only naphthyl-substituted benzylic alcohols providing useful selectivity.

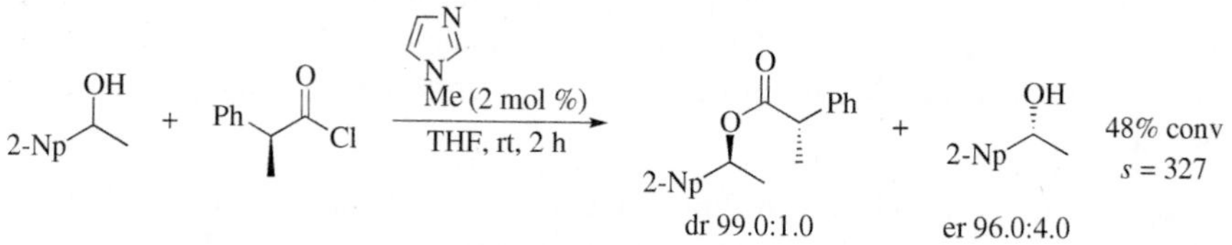

Scheme 67

Lewis Base Catalysis. One of the first reported examples of Lewis base catalyzed, acylative KR of benzylic alcohols used monocyclic tertiary phosphine catalyst **47**.[87] Subsequently, the design optimization of the catalyst led to the 5,5-fused bicyclic phosphine **48**, which gave improved acylation rates and selectivity factors.[88,154,155] The high activity of **48** allowed the KR of benzylic alcohols to be conducted at low temperature with sterically hindered anhydrides, such as isobutyric anhydride. Good selectivity factors were obtained for a range of benzylic alcohols, with exceptional levels of selectivity obtained for aryl derivatives that contain *ortho* substituents (Scheme 68).[155] The inherent air sensitivity of phosphine catalysts can be circumvented using the corresponding air-stable HBF$_4$ salt analogues, which are neutralized in situ with triethylamine to release the free phosphine catalyst.[156] Although operationally simple, the selectivity factors obtained were generally slightly reduced compared to the method using the free phosphine.

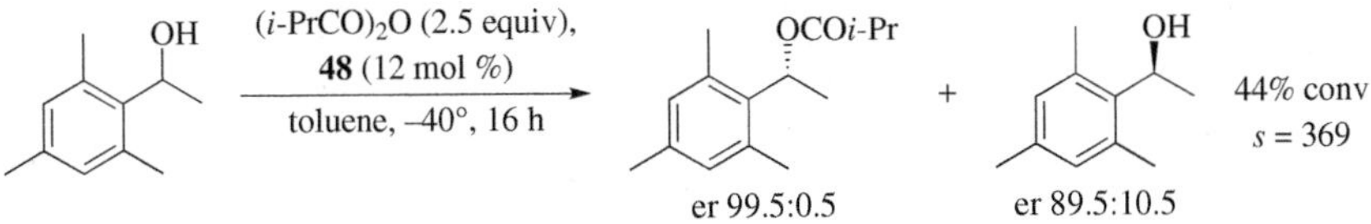

Scheme 68

Another early report employed the proline-derived tertiary amine catalyst **90** (Scheme 69).[157] To this end, using benzoyl chloride as the acylating agent in the presence of triethylamine and only 0.3 mol % of catalyst **90**, the KR of two benzylic alcohols with acceptable levels of selectivity was reported. Notably, the presence of an *ortho* substituent on the aromatic ring of the benzylic substrate was beneficial for selectivity.[157,158]

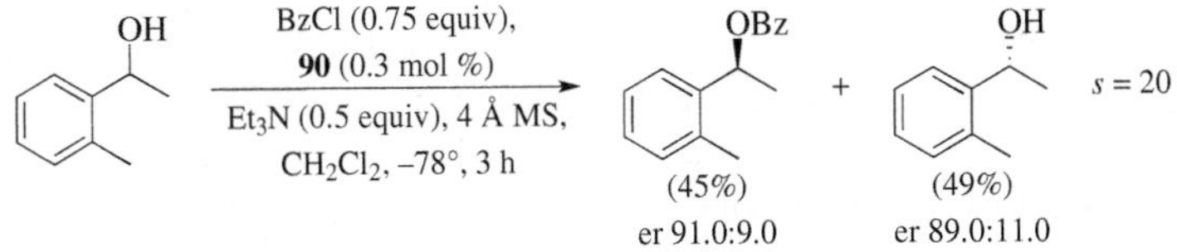

Scheme 69

Probably the most widely explored class of Lewis base catalysts for the KR of benzylic alcohols is based on the 4-(dialkylamino)pyridine scaffold. Although 4-(dialkylamino)pyridine derivatives are excellent acyl transfer catalysts, the development of chiral analogues is not straightforward because of the inherent planarity of the catalyst. The first effective catalyst design was based upon π-complexation of a DMAP derivative to a cyclopentadienyl iron fragment to give planar-chiral ferrocene derivative **10**.[146,159] Optimal conditions for the KR of benzylic alcohols used acetic anhydride as the acyl donor in *tert*-amyl alcohol at 0° (Scheme 70).[146] The choice of solvent had a pronounced effect on both conversion and the selectivity factor. Sterically hindered substrates bearing bulky alkyl groups or *ortho*-substituted aryl groups afford the highest selectivity factors. Interestingly, replacing the η^5-C_5Ph_5 component of the ligand with η^5-C_5Me_5 gave a catalyst that was effectively unselective for the acylation of 1-phenylethanol (**13**).

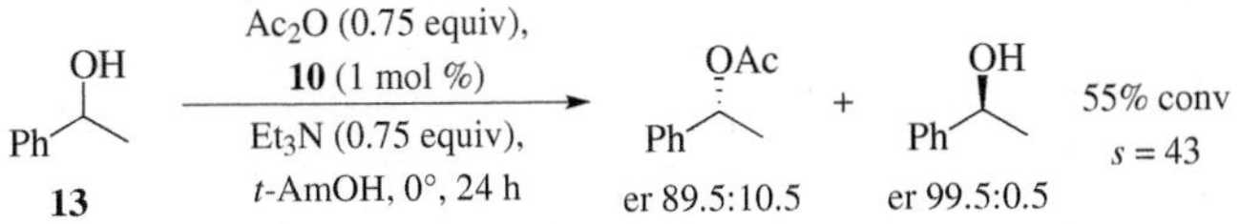

Scheme 70

The planar-chiral DMAP catalyst **10** has also been used for the efficient KR of β-functionalized benzylic alcohol derivatives, such as 2-hydroxy-2-arylphosphonates[160] and 1,2-azido alcohols (Scheme 71),[161] to provide useful synthetic building blocks in enantioenriched form.

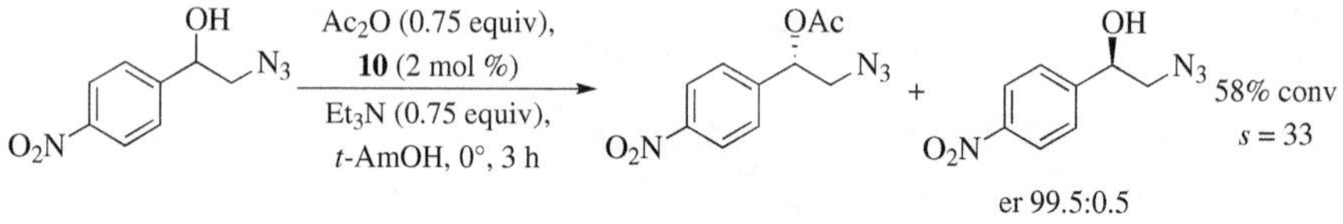

Scheme 71

The DMAP analogue **11**, which contains helical chirality, has also been developed for the KR of benzylic alcohols.[42,162] The helicene **11** is prepared in racemic form in 41% yield over 6 steps, and the racemate is resolved by preparative, chiral stationary phase HPLC. In keeping with methods using planar-chiral DMAP catalyst **10**, the optimal selectivity factors are obtained in *tert*-amyl alcohol at 0°. The resolution of benzylic alcohols is effected with good to excellent selectivity factors, with the highest selectivities observed using 2- and 2,6-disubstituted benzylic alcohols. Notably, the catalyst loading can be reduced to 0.05 mol % for the KR of 1-(naphthalen-1-yl)ethanol (Scheme 72).[42]

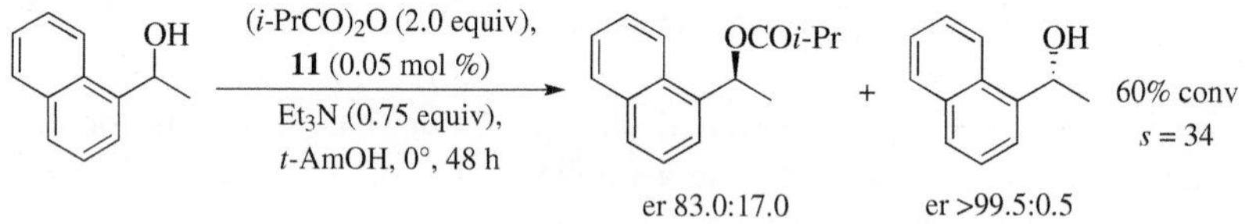

Scheme 72

The preparation of chiral 2- and 3-substituted DMAP derivatives has also been investigated. Although 2-substitution may appear to be an attractive approach because of the introduction of a stereodirecting group proximal to the reactive pyridine nitrogen, these analogues invariably result in a poor catalytic activity because of increased steric hindrance. The design of 3-substituted DMAP derivatives, in which a fixed or induced (fluxional) chiral axis is present, has proven much more successful (Fig. 13).

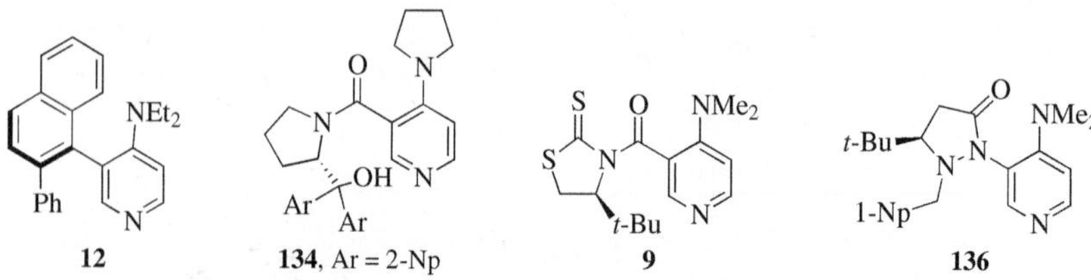

Figure 13. DMAP derivatives possessing a fluxional chiral axis.

The 3-aryl-substituted DMAP derivative **12**, which possesses a chiral axis because of restricted rotation about an aryl-aryl bond, has been successfully employed for the KR of a limited number of benzylic alcohols using isobutyric anhydride in toluene at −78° (Scheme 73).[163] Good selectivity factors are obtained, with sterically hindered substrates (bulky alkyl and *ortho*-substituted aryl) giving the highest selectivities. Notably, variations in the *N*-alkyl substituents and 2-arylnaphthalenyl unit afforded no significant improvements.[163−165]

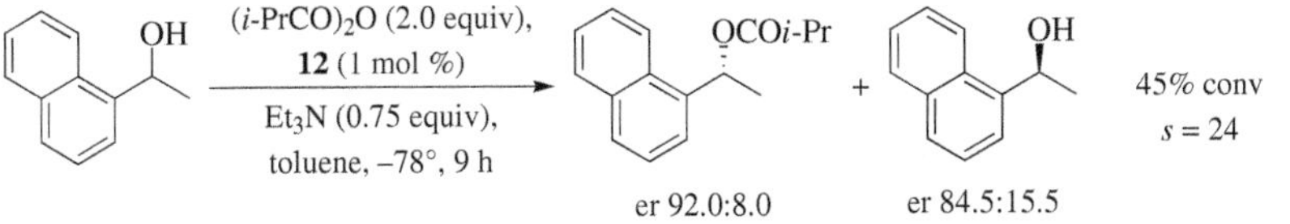

Scheme 73

Proline-derived PPY derivatives **134** and **135** have been used for the KR of both simple benzylic alcohols and Morita−Baylis−Hillman adducts.[39,166,167] Although only modest selectivity factors are obtained for the KR of simple benzylic alcohols, the resolution of Morita−Baylis−Hillman adducts can be achieved, in some cases, with synthetically useful selectivity factors (Scheme 74).[167]

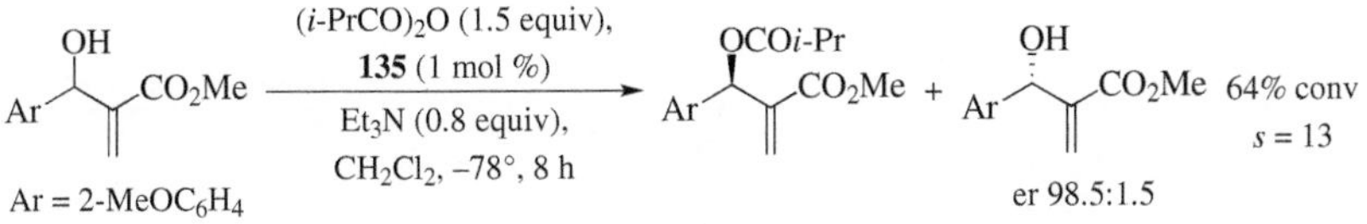

Scheme 74

Chiral DMAP derivatives **9** and **136** have also been successfully applied for the KR of benzylic alcohols and give selectivity factors of up to 13 and 37, respectively.[168,169] Notably, only chiral DMAP derivatives that contain a fixed or (proposed) induced chiral axis have been successful, in which the derivatives that contain a simple point chiral group (which does not efficiently induce a chiral axis) afford low selectivity factors in general.[170−174]

Enantiomerically pure PPY derivatives containing stereogenic centers within their pyrrolidine moieties have also been investigated for KR. In general, very low selectivity factors have been reported,[175−179] which is presumably because of the remote position of the stereocontrolling elements relative to the pyridine nitrogen. The single example that affords an acceptable level of selectivity is the KR of the cyclic amino alcohol **137** with the chiral PPY derivative **138** (Scheme 75).[180] The reaction uses isobutyric anhydride as the acylating agent in combination with 2,4,6-collidine as an auxiliary base in chloroform to give a selectivity factor of 21. Nevertheless, the resolution of simple benzylic alcohols has not reported.

The octapeptide **28** was developed following a fluorescence-based assay procedure, which was used for the efficient acylative KR of a limited range of benzylic

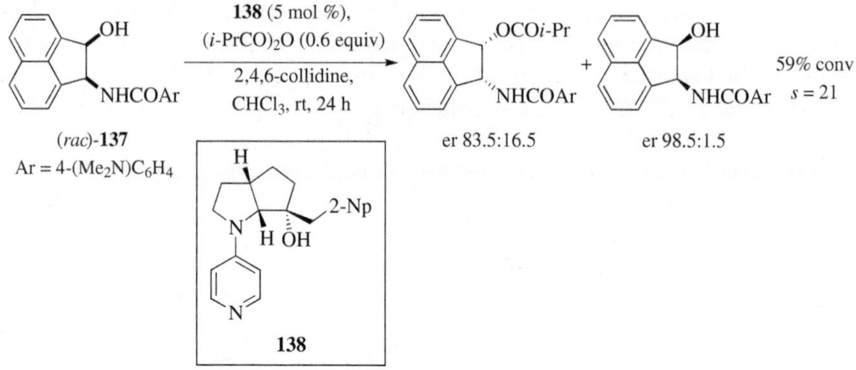

Scheme 75

alcohols.[68] The two histidine residues at the i and $i+3$ positions are essential for catalytic activity, while either shorter peptides or the inversion of the configuration of the $i+3$ histidine residue results in a reduction in both reactivity and selectivity.[69] Following an alanine scan, a second-generation octapeptide **139** was identified in which the replacement of the $i+2$ D-valine residue with D-alanine provides a more active and selective catalyst (Scheme 76).[69]

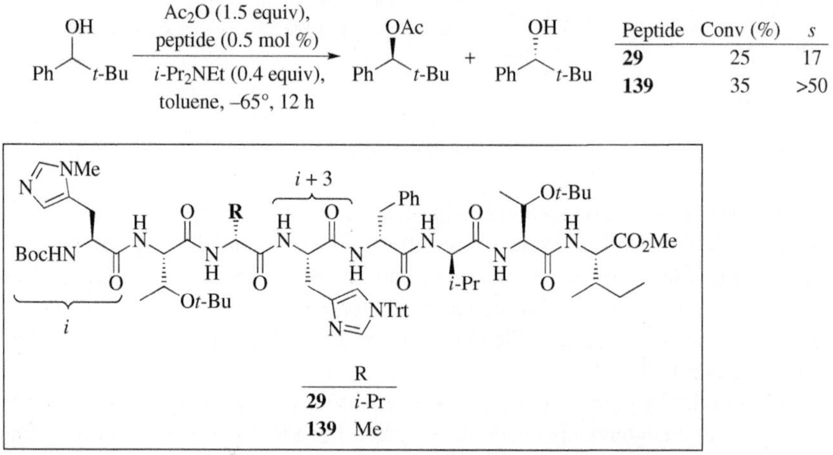

Scheme 76

Inspired by this work, bifunctional L-histidine-derived catalyst **98** was developed and applied to the acylative KR of alcohols.[70,71] Substrates are limited to those bearing an H-bond-accepting amide in the β-position relative to the alcohol (Scheme 77).[71] The potential utility of the products is expanded by demonstrating that 3-pyrroline amides can be applied, with subsequent hydrolysis providing access to enantioenriched α-hydroxy carboxylic acid derivatives.[181]

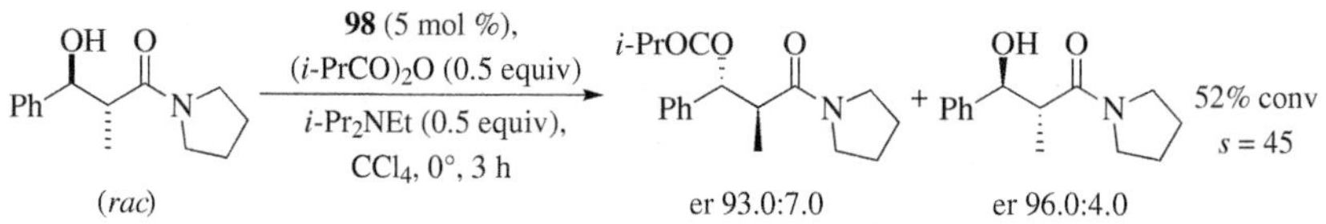

Scheme 77

In the course of further developing *N*-alkylated imidazoles as acyl transfer catalysts, the pyrrolidine-substituted chiral imidazole **140** effects the KR of a range of simple benzylic alcohols, including heteroaromatic carbinols (Scheme 78).[182] Optimal selectivity factors are obtained using 10 mol % of **140** in toluene at −40° and isobutyric anhydride as the acyl donor. The highest selectivity factors are obtained for benzylic alcohols bearing *ortho* substituents or bulky alkyl groups.

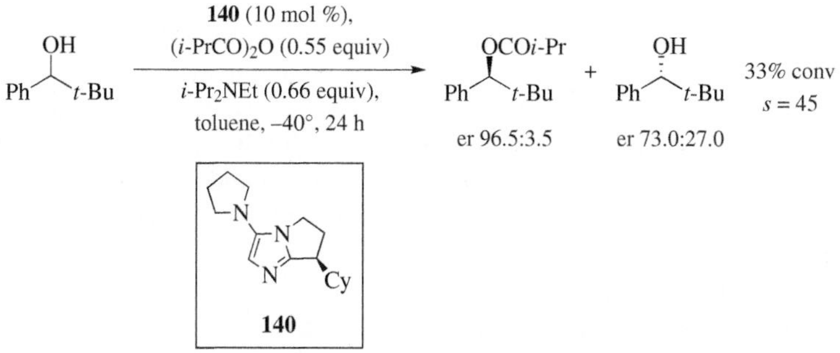

Scheme 78

The design of catalysts based upon the amidine (and related bases) scaffold benefits from the inherent advantage that amidines bear an sp^3-hybridized carbon atom adjacent to the nucleophilic nitrogen atom, allowing the introduction of a stereogenic center (Fig. 14). Initial studies identified (*R*)-phenylglycinol-derived amidine **15** as an efficient acyl transfer catalyst for the KR of benzylic alcohols.[182] The presence of an electron-withdrawing group on the catalyst core is essential for promoting both efficient acyl transfer and inducing high enantiodiscrimination. A second-generation amidine catalyst **16** bearing an extended aromatic core displays higher reactivity and increased selectivity factors for all benzylic alcohols reported (Scheme 79).[51]

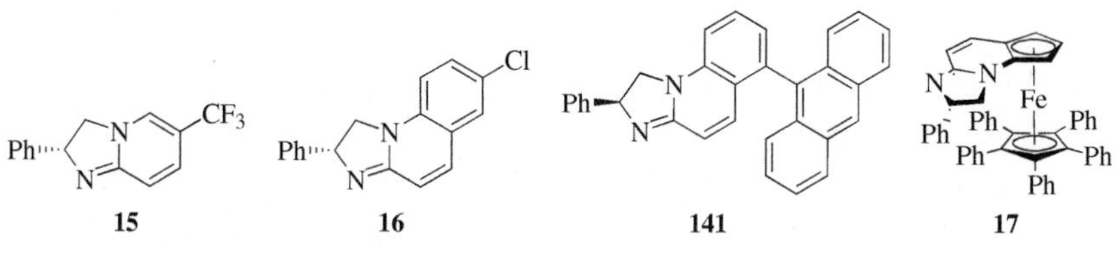

Figure 14. Amidine-based catalysts for the KR of benzylic alcohols.

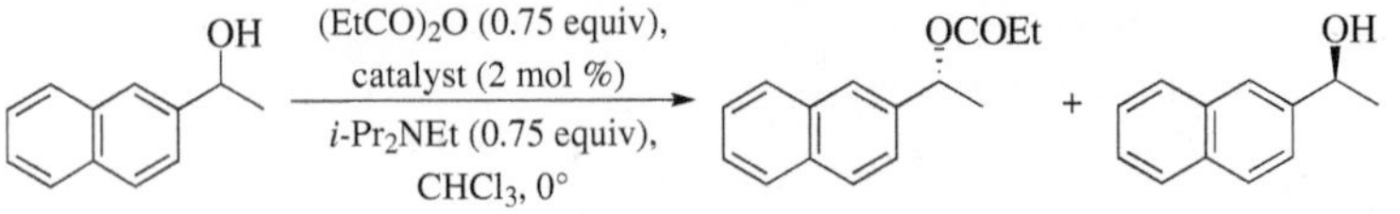

Catalyst	Time (h)	Conv (%)	er_{ester}	$er_{alcohol}$	s
15	8	50	93.5:6.5	94.5:5.5	43
16	2	51	95.5:4.5	97.0:3.0	76

Scheme 79

The amidine catalyst **141** bearing a 9-anthracenyl unit at the 6-position is particularly effective for the KR of benzylic alcohols and provides high selectivity factors for sterically hindered substrates (Scheme 80).[49,183] Although slightly higher catalyst loadings and longer reaction times are required relative to catalyst **16**, excellent levels of selectivity are reported for a range of benzylic alcohols bearing both electron-donating and electron-withdrawing groups (18 examples, $s = 43–450$). Planar-chiral amidine **17**, containing both planar and point chirality, is employed in the KR of a wide range of benzylic alcohols, which include a variety of hetero-aromatic carbinols (24 examples, $s = 31–1892$).[49,184] The alternative diastereoisomer of the catalyst, in which either the point or planar chirality is inverted, provides no catalytic activity. The highest selectivity factors are obtained for sterically hindered substrates; however, all the examples demonstrate synthetically useful resolutions.

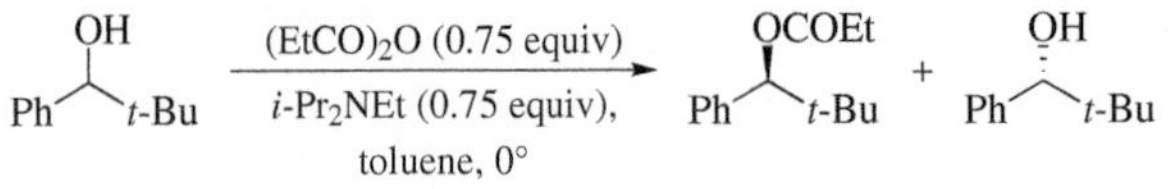

Catalyst	Time (h)	Conv (%)	er_{ester}	$er_{alcohol}$	s
141 (5%)	46	48	99.5:0.5	95.5:4.5	530
17 (2%)	10	45	99.5:0.5	90.5:9.5	534

Scheme 80

(*R*)-Phenylglycinol-derived guanidine **18** has also been reported for the acyla-tive KR of benzylic alcohols (Scheme 81).[50] By combining diphenylacetic acid and pivalic anhydride, a diphenylacetic acid/pivalic acid mixed anhydride is formed in situ, which permits the transfer of the diphenylacetic acid group as the acyl donor in the KR process. In the absence of an exogenous base, higher selectivity factors are obtained, with a relatively small negative impact on conversion. For sterically hin-dered alcohols, the use of a less sterically demanding acyl donor, 3-phenylpropanoic acid, is required to achieve high conversions and selectivity factors.

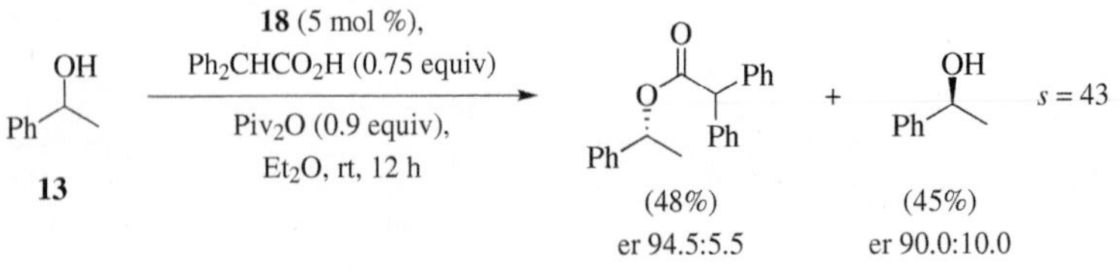

Scheme 81

The acylative KR of racemic β-hydroxy esters containing aromatic β-substituents can be achieved using the guanidine catalyst **142** and cyclohexanecarboxylic anhydride. The introduction of a branched N-substituent containing a stereogenic center is crucial for high selectivity (Scheme 82).[185]

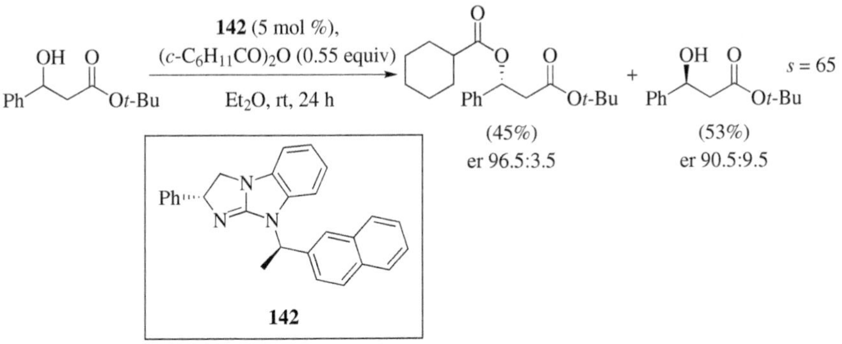

Scheme 82

Structurally related chiral isothiourea catalysts have been widely deployed in KR processes (Fig. 15). The initial report focused on the use of commercially available tetramisole (**22**) and its benzannulated analogue **23**.[186] Although the KR of benzylic alcohols using **22** results in good levels of selectivity, the use of benzannulated analogue **23** gives exceptionally high selectivity factors.

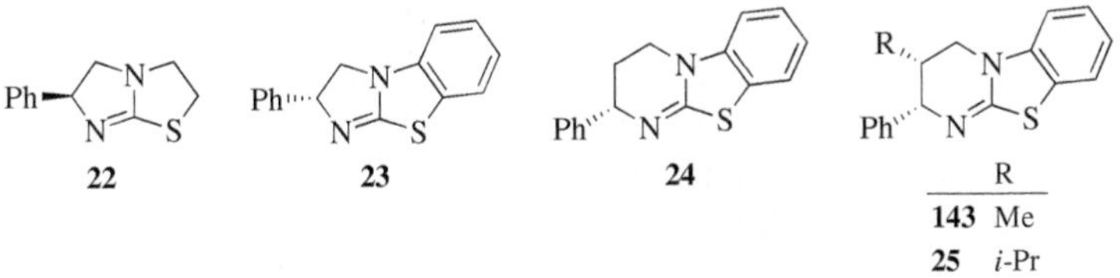

Figure 15. Isothiourea-based catalysts for the KR of benzylic alcohols.

Isothiourea catalyst **23** has since been used for the KR of a wide range of benzylic alcohols. In general, benzylic alcohol substrates with increased steric hindrance (large alkyl groups or *ortho* substitution within the aryl ring) lead to high selectivity factors. However, the use of benzylic alcohols bearing 2,6-disubstituted aryl rings results in much lower conversions and selectivity factors (Scheme 83).[186] The optimal conditions for KRs using isothiourea catalysts often require either propionic or isobutyric anhydride as the acylating agent in chloroform at 0°.

Isothiourea catalysis is also compatible with the use of in situ formed mixed anhydrides.[55,187,188] For example, a range of heteroaromatic benzylic alcohols undergo KR using **23** (5 mol %) with the mixed anhydride formed in situ from 2,2-diphenylacetic acid and pivalic anhydride (Scheme 84).[59] Heteroaromatic

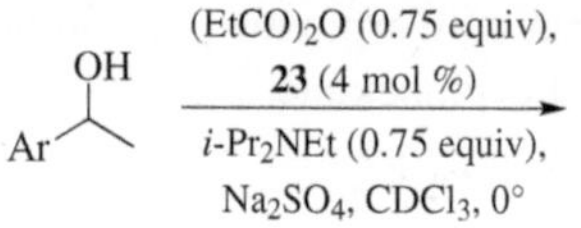

Ar	Time (h)	Conv (%)	er_{ester}	$er_{alcohol}$	s
Mes	24	17	74.5:25.5	55.0:45.0	3
Ph	24	49	96.5:3.5	94.5:5.5	81
2-MeC$_6$H$_4$	33	50	98.0:2.0	98.5:1.5	223

Scheme 83

benzylic alcohols incorporating furyl, thienyl, (benzo)thiazole, and benzoxazole groups are resolved with good to excellent selectivity factors in all cases ($s = 28$–142). The benzoxazole-derived carbinol products can be further transformed using a two-step reduction/oxidation process to enantiomerically enriched 1,2-amino alcohols.[189]

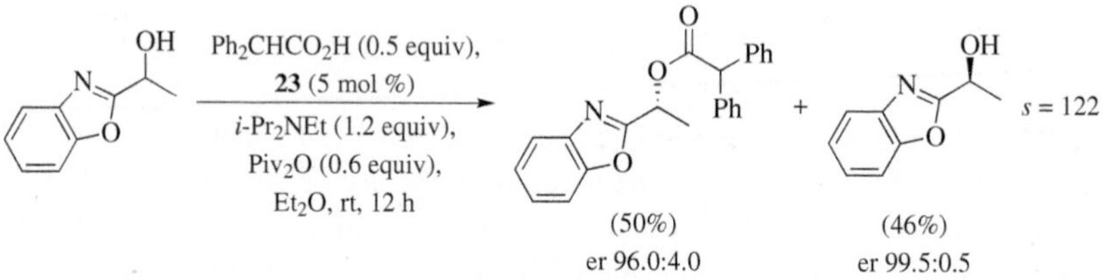

Scheme 84

Isothiourea **23** has also been used for the KR of a range of 2,2-difluoro-3-hydroxy-3-arylpropionate esters and 2,2,2-trifluoro-1-arylethanol derivatives.[190,191] For the resolution of 2,2,2-trifluoro-1-arylethanol derivatives, the choice of solvent is significant, in which the reactions in i-Pr$_2$O provide significantly improved selectivity factors compared to CHCl$_3$ (Scheme 85).[191] Benzylic alcohols bearing both electron-donating and electron-withdrawing groups are generally resolved with good to high selectivity factors.

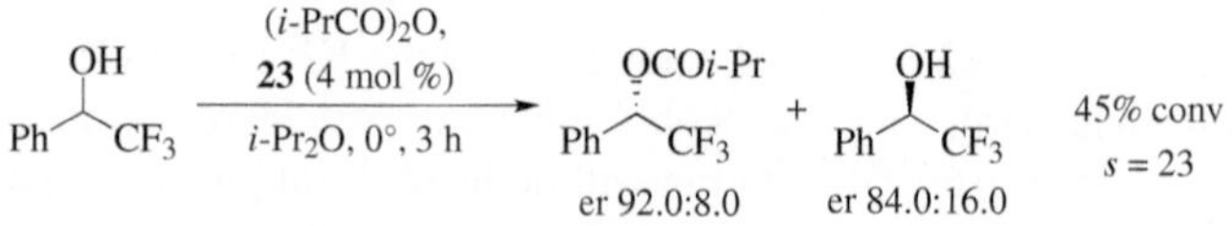

Scheme 85

Based on preliminary investigations into the rates of acylation using achiral isothiourea catalysts with varying ring sizes,[192,193] the catalysts **24**, **25**, and **143** were designed and investigated in KR processes.[60,194,195] In line with the achiral series, fast rates of acylation are observed, allowing reactions to be conducted in shorter reaction times, at lower catalyst loadings, and lower temperatures. The KR of a limited number of benzylic alcohols has been reported using **24**, **25**, and **143**; however, the selectivity factors obtained are generally lower than those obtained using **23** for this substrate class (Scheme 86).[60,186]

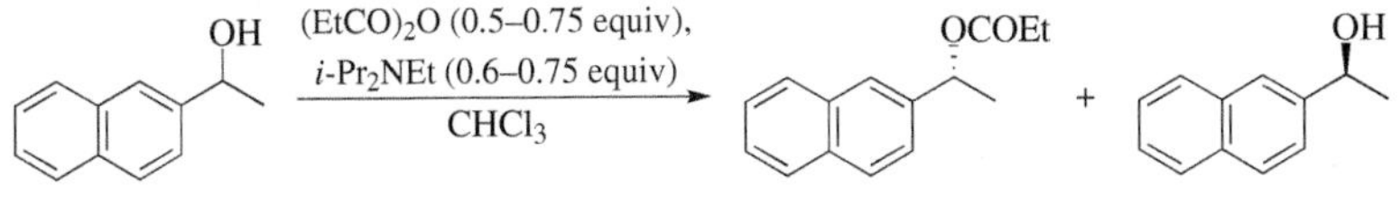

Catalyst	Temp (°)	Time (h)	Conv (%)	er_{ester}	$er_{alcohol}$	s
23 (4%)	0	11	49	97.0:3.0	95.0:5.0	107
25 (0.75%)	−20	1	43	97.0:3.0	85.5:14.5	65

Scheme 86

A polystyrene-supported isothiourea catalyst **144**, based on the homogeneous catalyst HyperBTM **25**, was prepared and used for the acylative KR of secondary alcohols.[196] A range of alcohols, including benzylic, allylic, propargylic, and cycloalkyl derivatives, including a 1,2-diol, were resolved with good to excellent selectivity factors obtained (*s* values up to 600). The immobilized catalyst was recovered and recycled 15 times and was also applied to the KR of secondary alcohols in a continuous-flow process (Scheme 87).

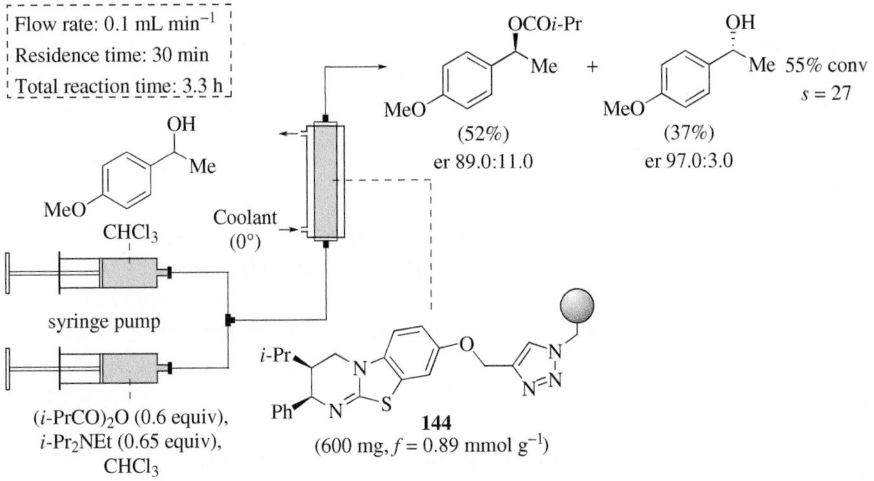

Scheme 87

The KR of benzylic alcohols has also been reported using NHCs as Lewis base catalysts. These methods typically use either a vinyl acetate derivative as the acylating agent,[76,77] or an aldehyde as a latent acyl donor that is formed by NHC-redox catalysis.[78,81,82] The NHC-catalyzed KR of benzylic alcohols has been reported with synthetically useful selectivity factors using **145** (5 mol %) and related derivatives, in combination with vinyl diphenylacetate (Scheme 88).[76] The use of an acylating agent with a sterically hindered acyl group is imperative to obtain high selectivity factors because the use of vinyl acetate under analogous conditions leads to low selectivity ($s = 2$).

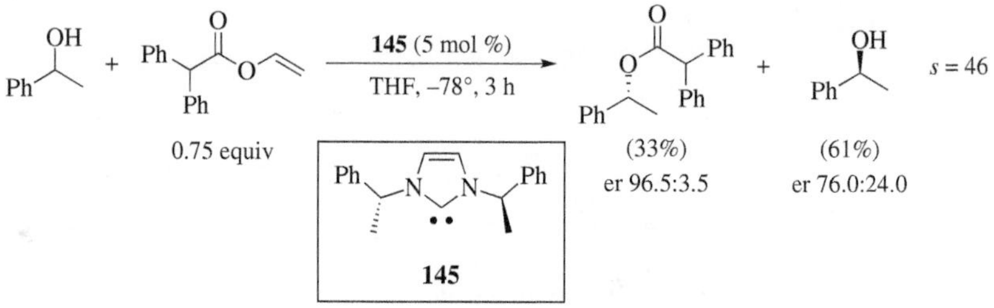

Scheme 88

The KR of secondary, aryl, cycloalkyl carbinols by their benzoylation using phosphinite **62** as the catalyst proceeds with s values of up to 44.[197] Interestingly, **62** is more effective for the kinetic resolution of aryl, cycloalkyl carbinols than aryl, acyclic-alkyl carbinols, such as 1-phenylethanol (**13**) (Scheme 89).

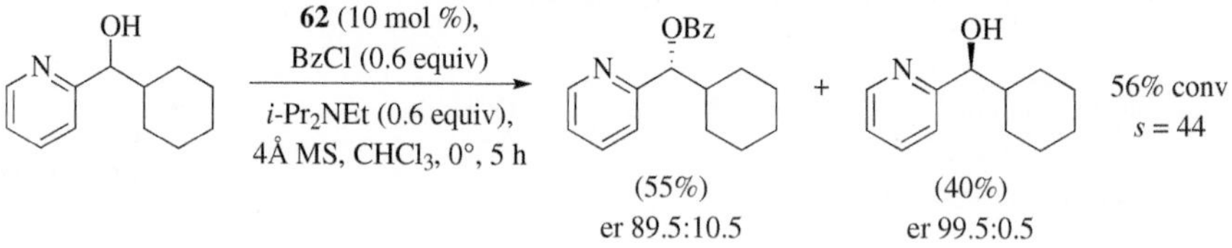

Scheme 89

Lewis Acid Catalysis. Early examples of Lewis acid catalyzed KR of simple benzylic alcohols were generally unsuccessful, in which only low selectivity factors were reported.[101,198] However, one example of note is the use of yttrium salen complex **146** for the KR of indanol using isopropenyl acetate (Scheme 90).[101] Although only a modest selectivity factor was obtained ($s = 5$), indanol is typically difficult to resolve by acylative methods, and as such, this remains one of the better methods available for this substrate.

The most efficient Lewis acid catalyzed KRs have been reported for benzylic alcohol substrates bearing an additional Lewis basic functional group in either the α- or β-position relative to the alcohol, which presumably provides additional coordination to the Lewis acidic metal catalysts. An early method to exploit this effect was the benzoylation of benzoin and a mandelic acid derivative using a copper catalyst with the chiral aza(bisoxazoline) ligand **147** (Scheme 91).[199] Although only two examples were reported, they proceed with synthetically useful selectivity factors. A chiral copper coordination polymer has also been employed for the KR of benzoin with excellent selectivity factors (up to 50).[96] The polymer could be recovered and recycled at least three times without significant loss in activity or selectivity.

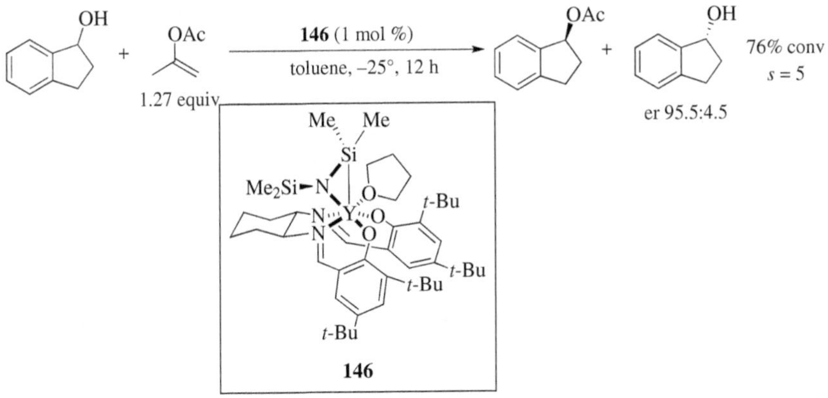

Scheme 90

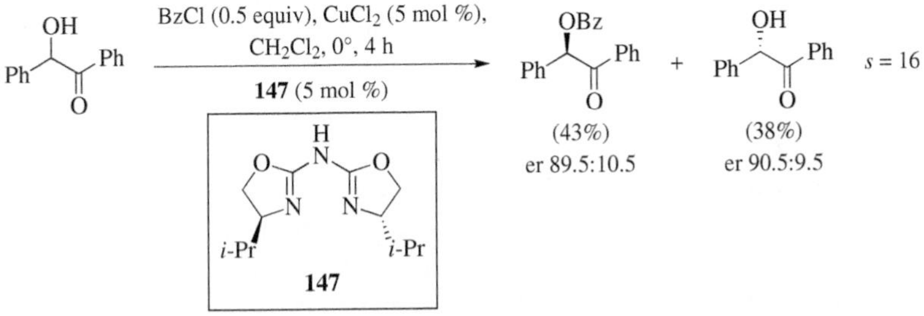

Scheme 91

Substrates in which the second point of ligation to the Lewis acidic metal center is four bonds from the alcohol can also undergo KR using copper catalysis. For example, the copper(II) complex bearing the chiral bisoxazoline ligand **74** promotes the KR of a range of β-aryl-β-hydroxyalkanephosphonates, such as **148**, which are resolved with acceptable levels of selectivity in general (Scheme 92).[98] Among various acylating agents investigated, 2-fluorobenzoyl chloride provides optimal selectivity factors in comparison with the others examined. The resolution of substrates bearing a β-aryl group containing an electron-donating substituent also results in good selectivity factors, whereas the presence of an electron-withdrawing substituent generally results in lower selectivity. The same system has also been used

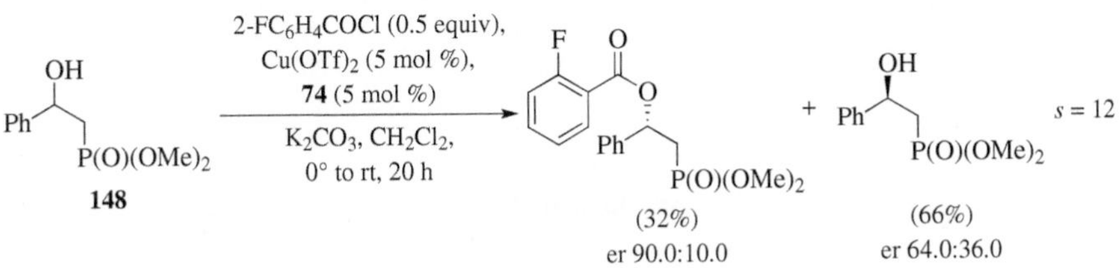

Scheme 92

for the KR of 3-hydroxyalanamides. In this case, the only example using a benzylic alcohol affords modest selectivity ($s = 8$).[200]

Brønsted Acid Catalysis.   Brønsted acids have only been recently reported as catalysts for the KR of benzylic alcohols. Using a combination of phosphoric acid **54**, 1,4-diazabicyclo[2.2.2]octane (DABCO), and acetyl chloride, five benzylic alcohols, including one Morita–Baylis–Hillman adduct, can be resolved with good selectivity factors (Scheme 93).[201] A structurally similar, chiral, phosphoric acid catalyst can be used for the KR of alcohols in the absence of DABCO using isobutyric anhydride as the acyl donor.[104] The phosphoric acid is proposed to act as a bifunctional Lewis base/Brønsted acid catalyst, albeit only a single example of a benzylic alcohol was reported.

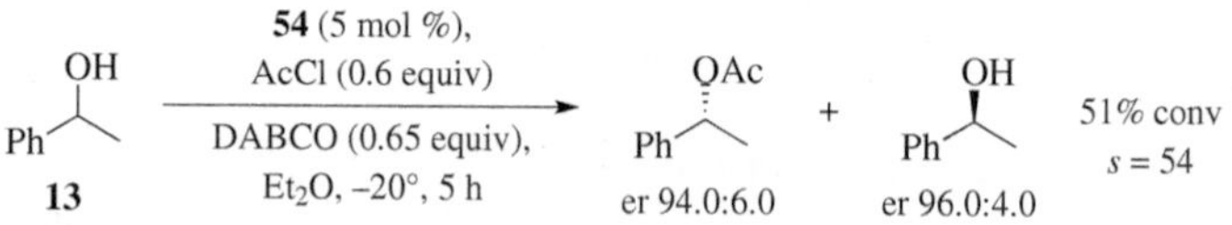

Scheme 93

Allylic Alcohols. The KR of allylic alcohols is a synthetically valuable process given the array of transformation that is associated with the recovered enantiomerically enriched substrates. Typically, secondary allylic alcohols bearing an alkyl substituent at the stereogenic carbon center are examined in catalytic KR protocols. The π-bond of the alkene acts as the recognition motif for the catalyst, similarly to the aforementioned benzylic derivatives.

Lewis Base Catalysis.   The planar-chiral DMAP analogue **10** (1 mol %) is an effective catalyst for the KR of allylic alcohols using acetic anhydride as the acylating agent.[159,202] The reaction has been examined with a range of alkene substitution patterns, in which 1,1,2-trisubstituted allylic alcohols afford the highest selectivity (Scheme 94).[202] The process is compatible with various alkenyl and alkyl substituents around the stereogenic alcohol center, giving synthetically useful selectivity in most cases. The preparative utility of this method has also been demonstrated through the KR of allylic alcohols that can be used as intermediates in the synthesis of natural products and important biologically active compounds.

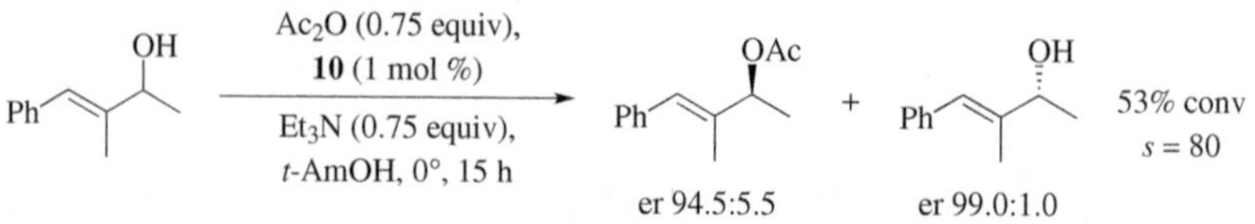

Scheme 94

The amidine-based catalyst **16** (2 mol %) can also be used for the KR of various secondary cinnamyl alcohol derivatives with propionic anhydride. The highest selectivities are obtained with more highly substituted alkenes (Scheme 95),[51] while allylic alcohols without either a conjugated aryl or alkenyl substituent are less selective.[57] Related isothiourea catalysts **23** and **25** can be used for the KR of cinnamyl alcohol derivatives with comparable levels of selectivity.[57,60] Secondary cinnamyl alcohol derivatives also undergo KR with good selectivities using a C_2-symmetric, chiral, imidazolium NHC catalyst in combination with vinyl diphenylacetate as the acylating agent.[76]

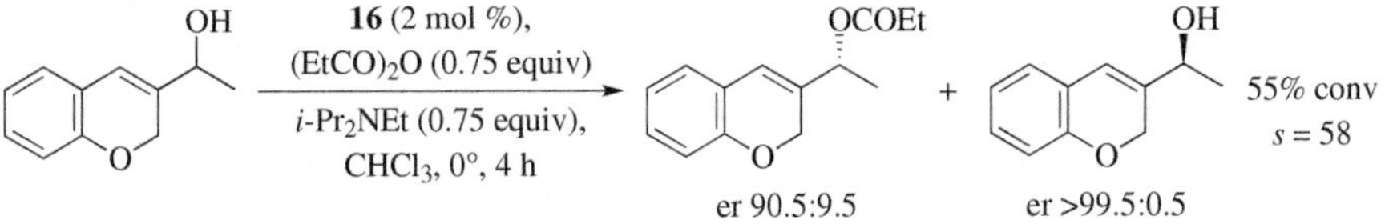

OH

16 (2 mol %),
(EtCO)$_2$O (0.75 equiv)

i-Pr$_2$NEt (0.75 equiv),
CHCl$_3$, 0°, 4 h

OCOEt

er 90.5:9.5

OH

er >99.5:0.5

55% conv
s = 58

Scheme 95

A series of allylic alcohols bearing 1,1-disubstituted alkenes can undergo efficient KR using amidine catalyst **149** (5 mol %) and propionic anhydride in a mixed solvent system (Scheme 96).[203] The reaction scope has been demonstrated for a range of aryl substituents on the alkene, with synthetically useful selectivity factors obtained in all cases.

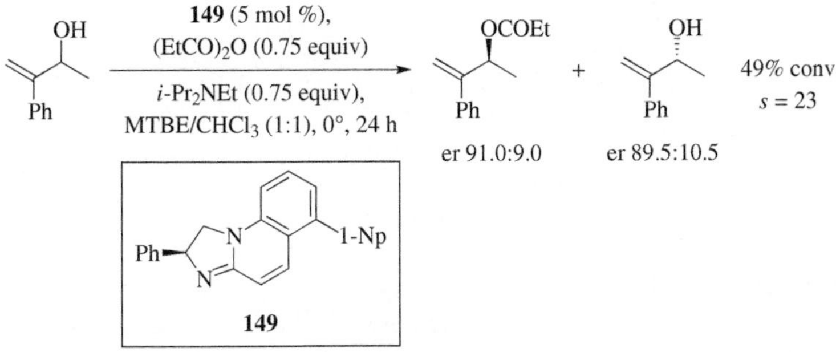

OH

Ph

149 (5 mol %),
(EtCO)$_2$O (0.75 equiv)

i-Pr$_2$NEt (0.75 equiv),
MTBE/CHCl$_3$ (1:1), 0°, 24 h

OCOEt

Ph

er 91.0:9.0

OH

Ph

er 89.5:10.5

49% conv
s = 23

Ph

1-Np

149

Scheme 96

The related amidine catalyst **141** (5 mol %) can be used for the KR of Morita–Baylis–Hillman (MBH) adducts (Scheme 97).[53] The presence of a sterically demanding benzhydryl ester substituent results in the highest selectivity, and the reaction can also be performed in alternative solvents using different organic bases without overly compromising the stereoselectivity. The process can accommodate a range of alkyl substituents, including those containing protected alcohol moieties on the stereogenic carbon center, which provide excellent selectivity factors in most cases. The KR of MBH adducts bearing aryl substituents on the stereogenic carbon center is more challenging, in which only moderate selectivity is obtained using a chiral DMAP analogue as the catalyst.[167]

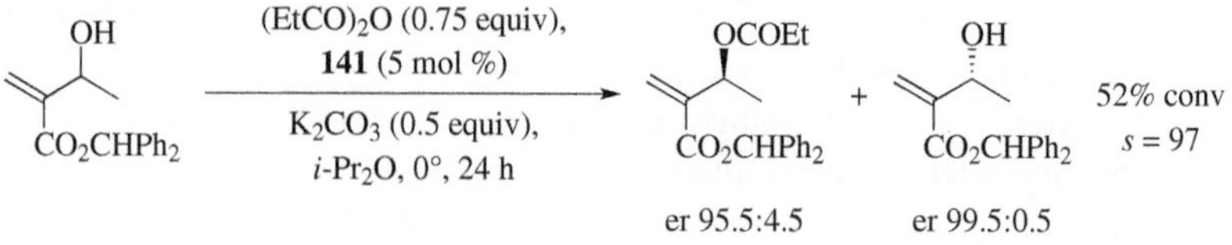

OH   (EtCO)$_2$O (0.75 equiv), **141** (5 mol %)   OCOEt   OH   52% conv

CO$_2$CHPh$_2$ K$_2$CO$_3$ (0.5 equiv), i-Pr$_2$O, 0°, 24 h   CO$_2$CHPh$_2$ CO$_2$CHPh$_2$ s = 97

er 95.5:4.5   er 99.5:0.5

Scheme 97

Secondary alcohols that contain both alkenyl and aryl substituents on the stereogenic carbon atom are particularly challenging substrates for KR as the catalyst must differentiate between two sp^2-hybridized centers to achieve selectivity. Such substrates can be resolved using planar-chiral amidine **17** as the catalyst and propionic anhydride as the acyl source to provide good selectivity across a range of aryl substituents.[49,204] Isothiourea catalyst **25** can also be used for the efficient KR of secondary alcohols bearing both an aryl and a vinyl substituent at low catalyst loadings (typically 1 mol %) using isobutyric anhydride as the acylating agent.[205] Substrates containing aryl rings that have electron-donating substituents lead to high selectivities. In contrast, electron-withdrawing aryl substituents lead to lower selectivity factors. The KR of 1-(naphthalen-2-yl)prop-2-en-1-ol (**150**) proceeds with exceptionally high selectivity ($s = 1980$), providing both the ester product and recovered alcohol in high enantioselectivity at 49% conversion (Scheme 98).[205] Notably, this protocol features in the total synthesis of the natural product descurainolide A.[206]

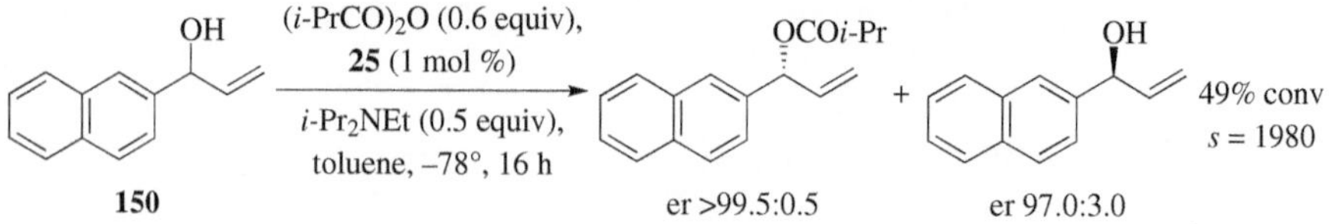

OH   (i-PrCO)$_2$O (0.6 equiv), **25** (1 mol %)   OCOi-Pr   OH   49% conv

150   i-Pr$_2$NEt (0.5 equiv), toluene, −78°, 16 h   er >99.5:0.5   er 97.0:3.0   s = 1980

Scheme 98

The bicyclic phosphine **48** (5 mol %) catalyzes the KR of a range of cyclic alcohols using isobutyric anhydride at low temperature.[154,156,207] The reaction provides high selectivities in the KR of a range of substituted cyclic allylic alcohols (Scheme 99),[207] albeit acyclic secondary allylic alcohols afford significantly lower selectivities under similar conditions.

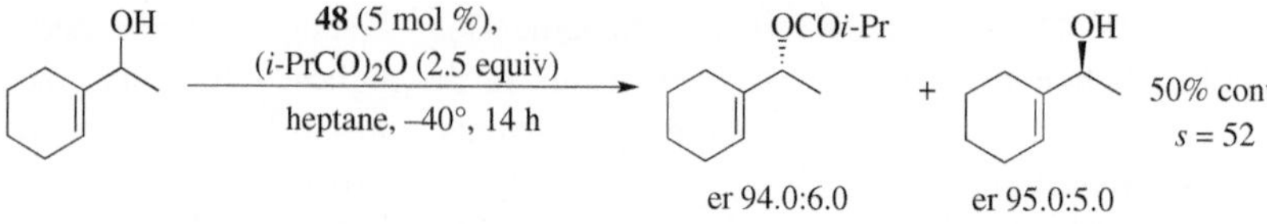

OH   **48** (5 mol %), (i-PrCO)$_2$O (2.5 equiv)   OCOi-Pr   OH   50% conv

heptane, −40°, 14 h   er 94.0:6.0   er 95.0:5.0   s = 52

Scheme 99

Lewis Acid Catalysis. The only reported example of a Lewis acid catalyzed KR of allylic alcohols uses the chiral complex derived from copper(II) triflate

(5 mol %) with the Box ligand, **74** (5 mol %) to resolve α-hydroxyallylphosphonate **151** (Scheme 100).[208] Although the overall reactivity is low (only 18% conversion), the selectivity is excellent, affording the corresponding ester product as a single enantiomer.

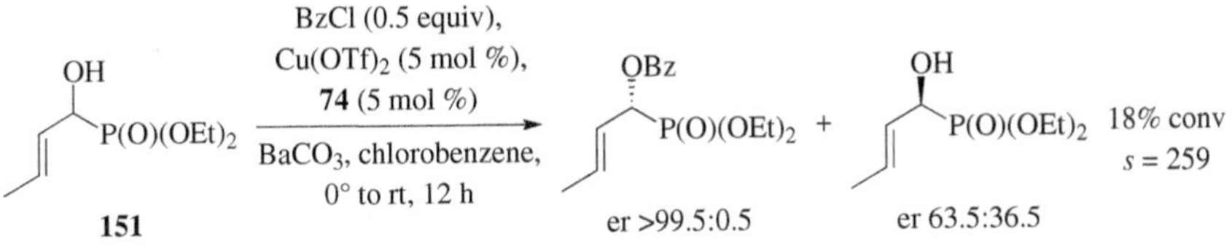

OH
‖
$P(O)(OEt)_2$

151

BzCl (0.5 equiv),
Cu(OTf)$_2$ (5 mol %),
74 (5 mol %)
───────────────→
BaCO$_3$, chlorobenzene,
0° to rt, 12 h

OBz
‖
$P(O)(OEt)_2$ +
er >99.5:0.5

OH
‖
$P(O)(OEt)_2$ 18% conv
er 63.5:36.5   s = 259

Scheme 100

Brønsted Acid Catalysis.   The chiral phosphoric acid *ent*-**54** catalyzes the KR of a simple MBH secondary allylic alcohol **152** using acetyl chloride and DABCO (Scheme 101).[201] The reactions proceed with promising selectivities; however, only two examples within this class of substrate have been reported providing significant potential for further studies.

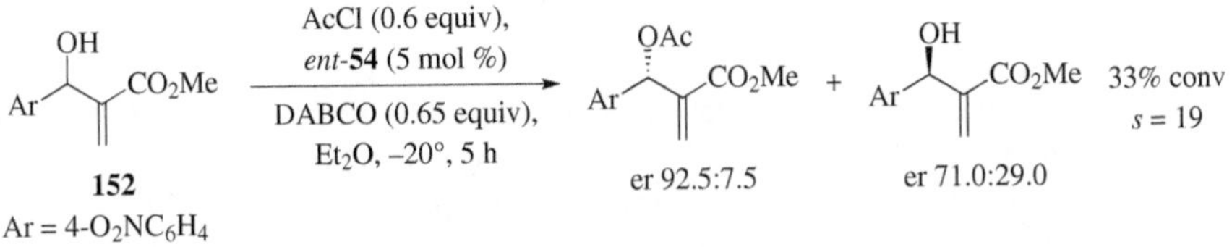

OH
‖
Ar CO_2Me

152
Ar = 4-O$_2$NC$_6$H$_4$

AcCl (0.6 equiv),
ent-**54** (5 mol %)
───────────────→
DABCO (0.65 equiv),
Et$_2$O, –20°, 5 h

OAc
‖
Ar CO_2Me +
er 92.5:7.5

OH
‖
Ar CO_2Me 33% conv
er 71.0:29.0   s = 19

Scheme 101

Propargylic Alcohols.

Lewis Base Catalysis.   The first example of the catalytic acylative KR of propargylic alcohols used the planar-chiral DMAP catalyst **10**,[209] in which the highest selectivity was obtained using less sterically hindered anhydrides, such as acetic anhydride. In contrast to many Lewis base catalyzed KR processes, the addition of triethylamine induces a base-promoted acylation and decreases the selectivity. Interestingly, the reaction selectivity also decreases with increased steric demand of the alkyl substituent on the stereogenic carbon atom, which is again in contrast to the trend typically observed for benzylic alcohols. For example, a simple methyl substituent in propargylic alcohol **153** gives s = 20 (Scheme 102).[209]

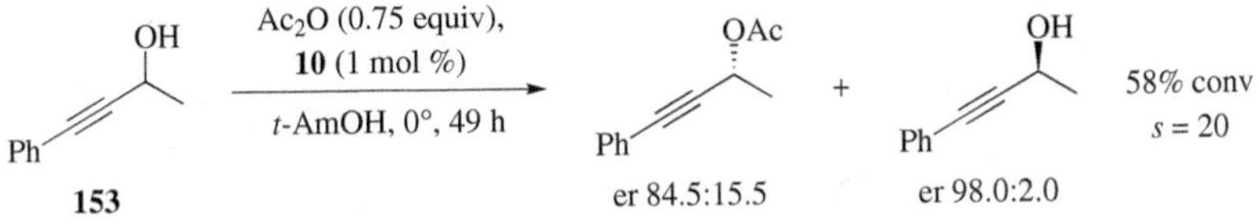

OH
‖
Ph

153

Ac$_2$O (0.75 equiv),
10 (1 mol %)
───────────────→
t-AmOH, 0°, 49 h

OAc
‖
Ph
er 84.5:15.5

+

OH
‖
Ph   58% conv
er 98.0:2.0   s = 20

Scheme 102

whereas the substrate with a *t*-Bu substituent requires increased reaction times and leads to reduced selectivity (*s* = 4). The efficiency of the KR increases when the alkyne is further conjugated with an aryl or alkynyl substituent. Although alternative DMAP-derived catalysts have also been investigated for the KR of propargylic alcohols,[168,169,173] they universally provide low selectivity.

Isothiourea catalyst **23** can also be used for the KR of a range of propargylic alcohols, using propionic anhydride as the acyl source in chloroform at 0° (Scheme 103).[210] The same trends in selectivity as with the use of the planar-chiral DMAP catalyst **10** are observed, in which the presence of more sterically demanding alkyl substituents lead to decreased selectivity. In accordance with the aforementioned process, the presence of conjugated alkynyl substituents also leads to shorter reaction times and increased selectivity.

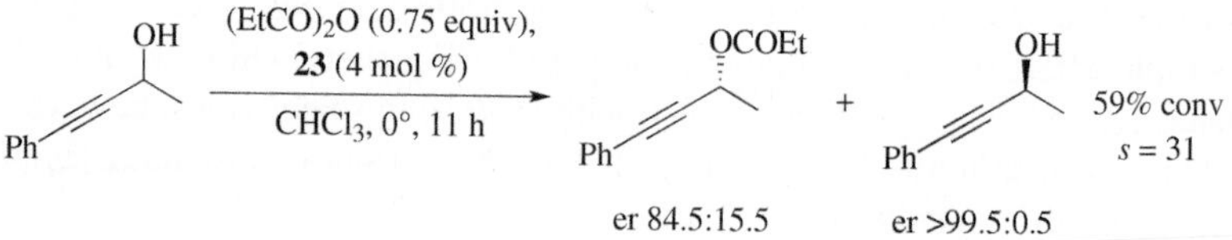

Scheme 103

Brønsted Acid Catalysis.   To date, there is only a single report of a Brønsted acid catalyzed KR of propargylic alcohols that provides synthetically useful selectivity. The chiral phosphoric acid catalyst *ent*-**54** (5 mol %) in combination with acetyl chloride as the acylating agent delivers good selectivity for the KR of **153** at –20° in diethyl ether (Scheme 104).[201] Interestingly, the KR of propargylic alcohol **154** bearing alkynyl and phenyl substituents on the stereogenic carbon atom results in a switch in the sense of enantioselectivity under otherwise identical conditions (Scheme 105).[201] Nonetheless, despite these intriguing results, the substrate scope has not been explored beyond these two examples.

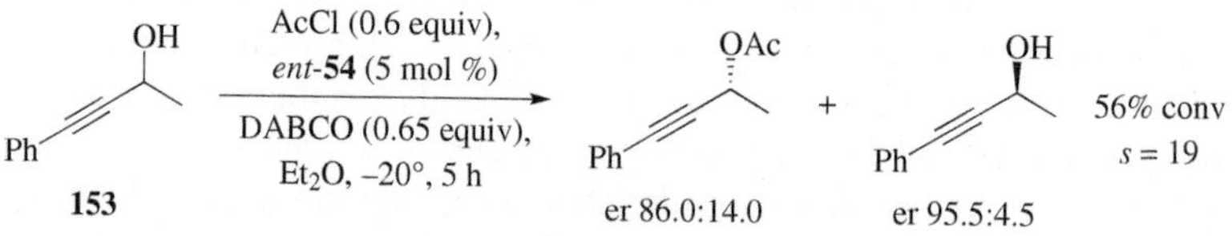

Scheme 104

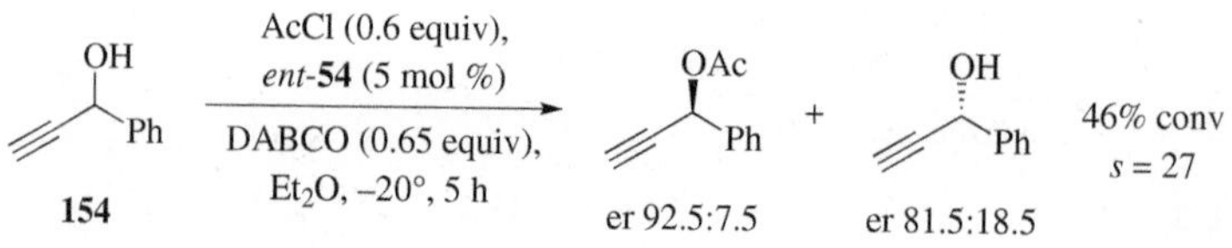

Scheme 105

Cycloalkanols. The acylative KR of various substituted cycloalkanols has been investigated using both Lewis base and Lewis acid catalysts. Individual catalytic systems are generally dependent on the cycloalkanol ring size and often require either a sterically demanding or heteroatom substituent adjacent to the carbinol center to obtain high selectivity. In such cases, the relative configuration of the two adjacent substituents is also important for the efficiency of the KR.

Lewis Base Catalysis. The DMAP derivative **138** (5 mol %) catalyzes the KR of protected cyclic *cis*-1,2-amino alcohols, such as **155**, using isobutyric anhydride as the acylating agent (Scheme 106).[180] The reaction gives moderate selectivity for a range of substrates, including those containing five-, six-, and seven-membered carbocyclic cores, as well as benzannulated, protected amino indanol derivatives. In each case, the reactions require high conversion to permit the corresponding alcohols to be isolated as essentially single enantiomers.

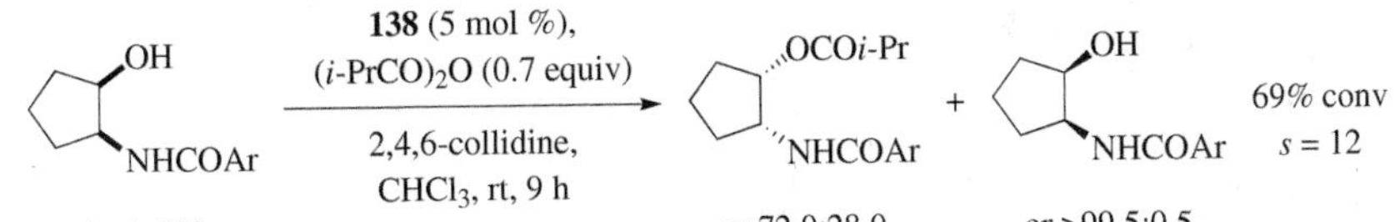

Scheme 106

The related cyclic *cis*-1,2-amino alcohols can be resolved using catalyst **156** (5 mol %) bearing a substituted proline-derived backbone (Scheme 107).[178] The protected amino substituent is essential for obtaining reasonable selectivity factors, in which the KR of *cis*-2-phenylcyclohexan-1-ol under the same conditions essentially affords no stereoselectivity. A polymer-supported variant of **156** has also been investigated but provides lower selectivity compared with the solution phase catalyst.[177]

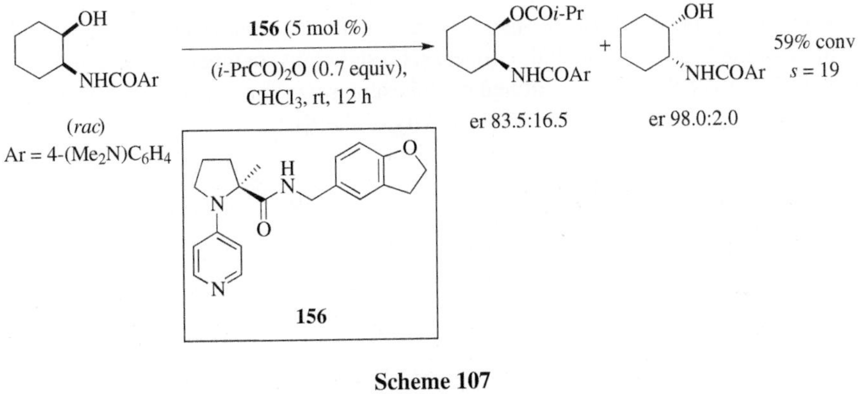

Scheme 107

Oligopeptide catalysts have also been investigated for the KR of a small number of cyclic *trans*-1,2-amino alcohols using acetic anhydride as the acylating agent

(Scheme 108).[211] Octapeptide **157** (1 mol %) is optimal, providing good selectivities for the KR of both five- and six-membered ring derivatives ($s = 27$ and 51). Interestingly, the seven-membered ring variant affords reduced selectivity ($s = 15$), and the related *cis*-1,3-amino alcohol gives essentially no selectivity under the same conditions.

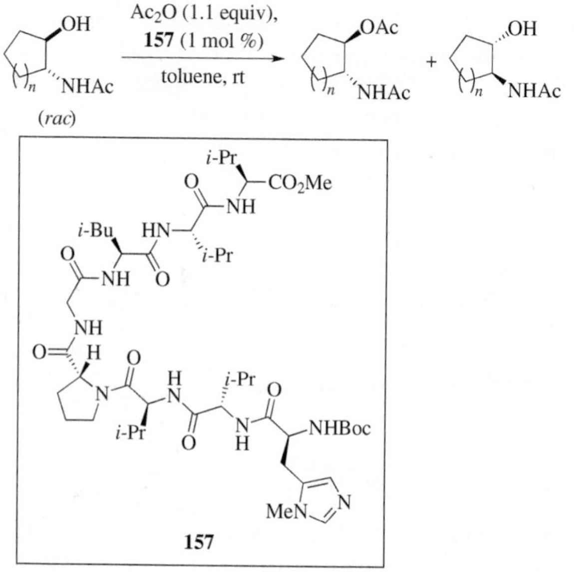

n	Conv (%)	s
1	49	27
2	50	51
3	45	15

Scheme 108

The diamine catalyst **90** (0.3 mol %) and benzoyl chloride can be used to facilitate the KR of a range of *trans*-1,2-disubstituted cycloalkanols. For example, the KR of *trans*-2-bromocyclohexan-1-ol (**158**) proceeds with excellent selectivity to give both the ester and recovered alcohol in high enantiomeric excesses (Scheme 109).[157] The reaction also proceeds for *trans*-cycloalkanols with either an ester or a phenyl substituent in the 2-position, including five-, six-, and eight-membered rings, with high selectivities in each case. Substituted cyclohexene rings can also be utilized without diminishing either the reactivity or selectivity of the KR.

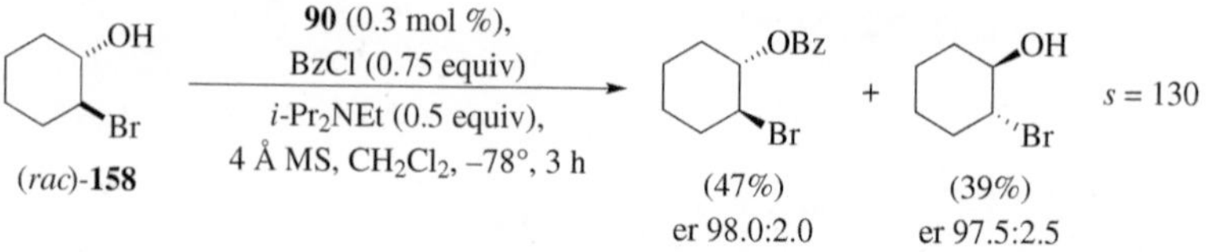

Scheme 109

The same reaction conditions can also be used for the KR of cyclic *trans*-β-hydroxy sulfides (Scheme 110).[212] Excellent selectivity is obtained for a range of sulfides that contain sterically demanding aryl or alkyl substituents. *trans*-β-Hydroxy sulfides on six-, seven-, and eight-membered cycloalkane rings are also viable substrates, albeit the KR of substituted cyclopentane derivatives results in significantly lower selectivities.

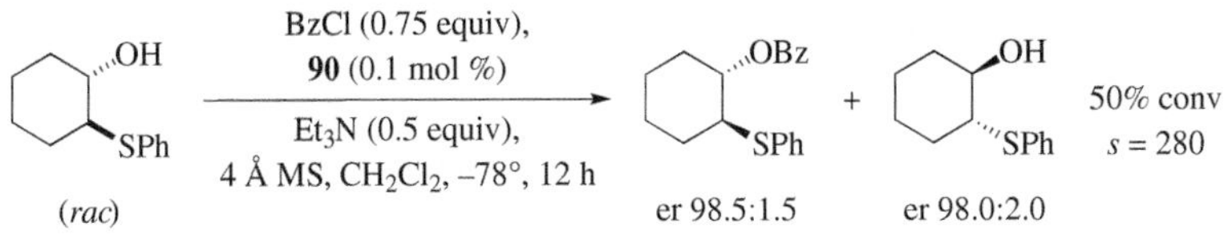

Scheme 110

A JandaJel™-supported diamine derivative **160** catalyzes the KR of various cycloalkanols with similar trends in selectivity compared with reactions using an analogous solution-phase catalyst. For example, *trans*-2-phenylcyclohexan-1-ol (**159**) undergoes an efficient KR in which the ester product and recovered alcohol are obtained with excellent enantiomeric ratios (Scheme 111).[213] In this case, the catalyst is effectively recycled up to six times without reduction in either reactivity or selectivity. In contrast, the KR of substituted cyclopentane and cyclooctane derivatives, in addition to benzannulated 1-indanol and 1,2,3,4-tetrahydro-1-naphthol, proceed with lower selectivity (*s* = 3 and 6, respectively).

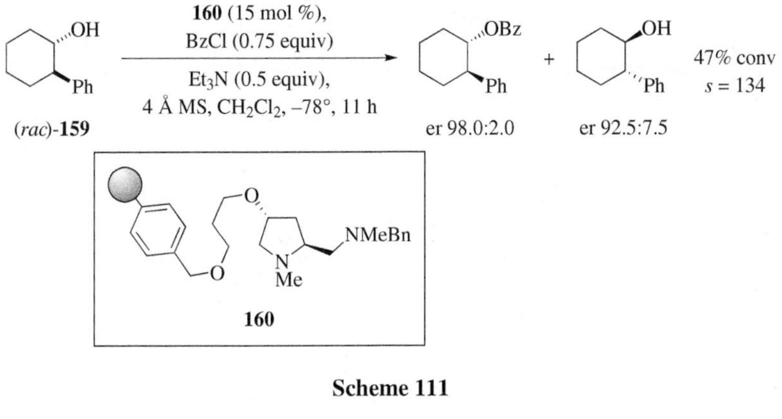

Scheme 111

Isothiourea catalyst **24** (4 mol %), in combination with propionic anhydride, also facilitates the KR of 1,2-substituted *trans*-cycloalkanols with high selectivity (Scheme 112).[194] Optimal selectivity factors are obtained by performing the reactions at low temperature, but synthetically useful KR is also possible at room temperature. The process is selective for 2-aryl- and 2-heteroaryl-substituted *trans*-cycloalkanols, whereas the presence of either a 2-alkyl or 2-azido substituent results in lower selectivity. Interestingly, the corresponding *cis*-2-phenylcyclohexan-1-ol also undergoes KR under the optimized reaction conditions (*s* = 28).

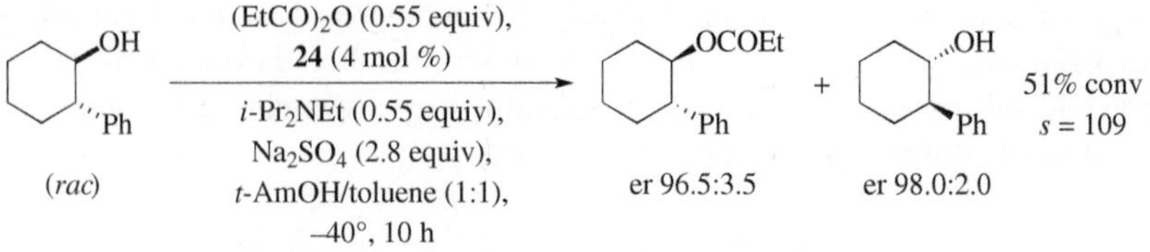

$(EtCO)_2O$ (0.55 equiv),
24 (4 mol %)
i-Pr$_2$NEt (0.55 equiv),
Na$_2$SO$_4$ (2.8 equiv),
t-AmOH/toluene (1:1),
–40°, 10 h

er 96.5:3.5   er 98.0:2.0

51% conv
s = 109

Scheme 112

A range of substituted α-hydroxy-γ-lactones undergo KR with the isothiourea catalyst, **23** (5 mol %), and the mixed anhydride formed in situ from the reaction of diphenylacetic acid with pivaloyl anhydride (Scheme 113).[63] Excellent selectivity factors are obtained for α-hydroxy-γ-lactones that contain either different alkyl substituents or fused carbocyclic rings in the β-position.

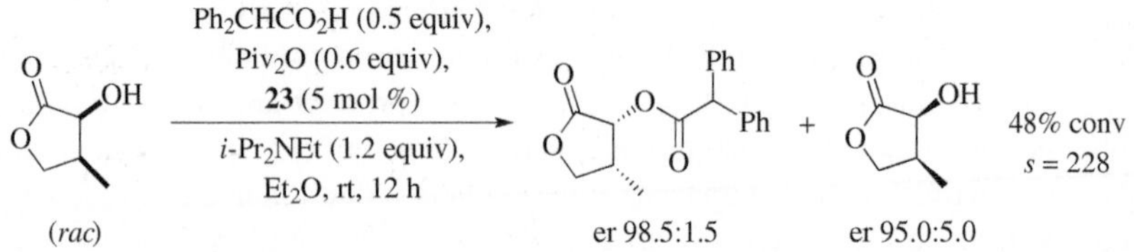

Ph$_2$CHCO$_2$H (0.5 equiv),
Piv$_2$O (0.6 equiv),
23 (5 mol %)
i-Pr$_2$NEt (1.2 equiv),
Et$_2$O, rt, 12 h

er 98.5:1.5   er 95.0:5.0

48% conv
s = 228

Scheme 113

Lewis Acid Catalysis. The few examples of acylative KR of substituted cycloalkanols using Lewis acid catalysis that employ either a chiral palladium-[214] or yttrium-based[101] catalyst result in low stereoselectivities. In contrast, the chiral complex derived from the combination of CuCl$_2$ (1 mol %) with the boraBox ligand **78** (1 mol %) permits the KR of a small number of pyridyl alcohols using benzoyl chloride as the acylating agent (Scheme 114).[215] The reaction selectivity is highly dependent on the cycloalkanol ring size, in which the fused cyclohexanols give higher selectivity factors than the corresponding fused cyclopentanol derivatives. The KR is also sensitive to substitution at the 2-position in the pyridine. For instance, the presence of a phenyl substituent results in higher selectivity (*s* = 125) compared with the unsubstituted substrate (*s* = 19).

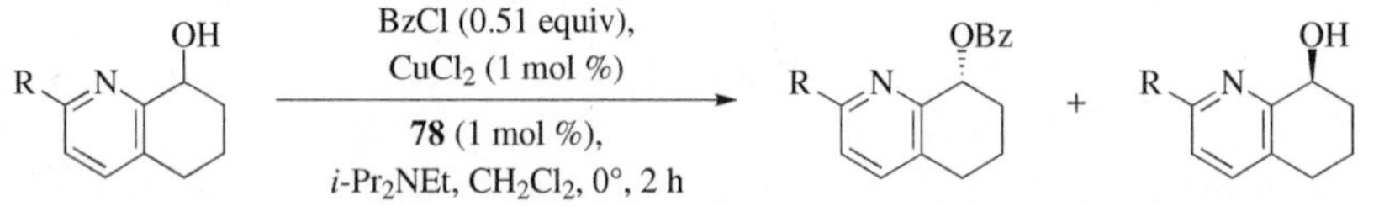

BzCl (0.51 equiv),
CuCl$_2$ (1 mol %)
78 (1 mol %),
i-Pr$_2$NEt, CH$_2$Cl$_2$, 0°, 2 h

R	Conv (%)	er$_{ester}$	er$_{alcohol}$	s
H	52	88.0:12.0	91.5:8.5	19
Ph	42	98.5:1.5	85.0:15.0	125

Scheme 114

The pre-prepared chiral Cu(II)-Box complex **162** can be used for the KR of various α-hydroxy-γ-lactones at low catalyst loading (0.05 mol %) using *n*-propyl isocyanate as the acylating agent.[97] The process affords high selectivity factors for a range of α-hydroxy lactones that contain β,β-disubstitution. The KR of *cis*-α-hydroxy lactone **161** also proceeds with excellent selectivity (Scheme 115);[97] however, the KR of the corresponding *trans*-diastereoisomer gives essentially no stereoselectivity ($s = 2$) under the same reaction conditions.

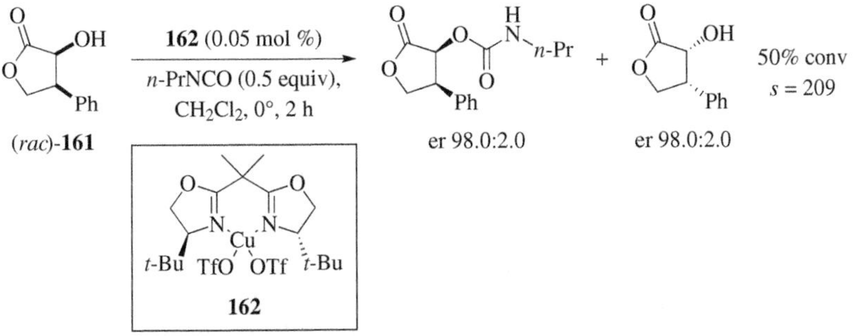

Scheme 115

***rac*-1,2-Diols.** A limited number of 1,2-diol substrates have been evaluated in acylative KR in which most studies focus on the use of either hydrobenzoin (**163**) or simple cyclic *anti*-1,2-diols. An early example used copper-Box complex **164** (5 mol %) and benzoyl chloride as the acylating agent to provide excellent selectivity for the KR of hydrobenzoin (**163**) (Scheme 116).[125] The enantioselectivity is highly dependent on the nature of the acylating agent, with simple, alkyl acid chlorides affording markedly lower selectivity factors. This method is also applicable to a limited range of cyclic and acyclic diols, with synthetically useful selectivity obtained in all cases.

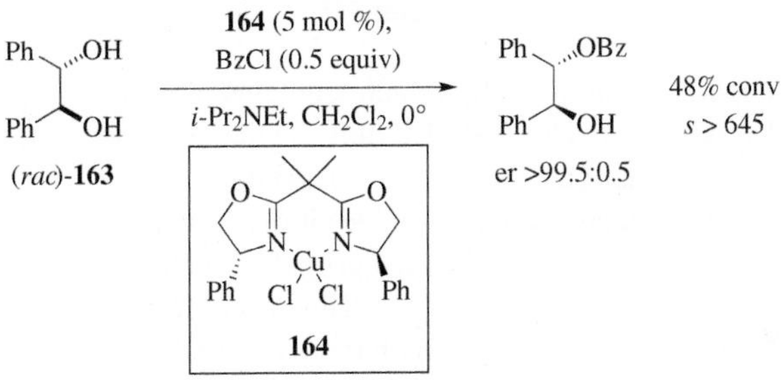

Scheme 116

Various ligands and/or copper complexes have been evaluated for the KR of simple diols. For example, boraBox ligand **78** (1 mol %), in combination with $CuCl_2$ (1 mol %), provides a particularly efficient catalyst for the KR of hydrobenzoin (**163**), providing both the monoester product and recovered diol in high enantioselectivity at 51% conversion (Scheme 117).[215] Various polymer-supported copper complexes have also been employed in the acylative KR of hydrobenzoin (**163**), and many provide excellent reactivities and selectivities.[96,199,216,217] Copper complexes immobilized on magnetic nanoparticles have also been evaluated for the KR of hydrobenzoin (**163**) under both batch and continuous-flow conditions, again with exceptional levels of selectivity.[218,219]

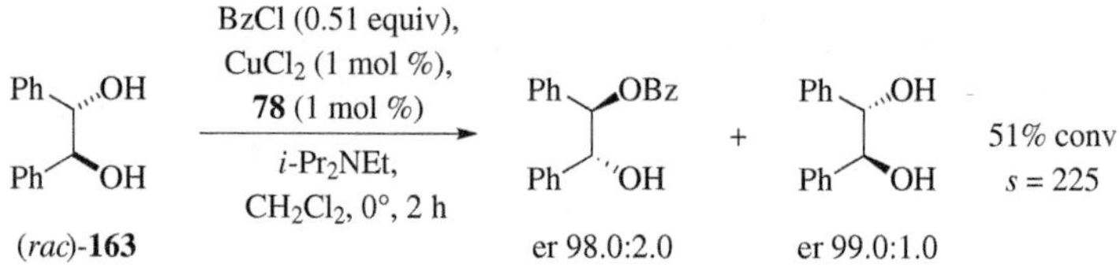

Ph⟍⟋OH
Ph⟋⟍OH
(*rac*)-**163**

BzCl (0.51 equiv),
CuCl$_2$ (1 mol %),
78 (1 mol %)
⟶
i-Pr$_2$NEt,
CH$_2$Cl$_2$, 0°, 2 h

Ph⟍OBz
Ph⟋OH
er 98.0:2.0

+

Ph⟍OH
Ph⟋OH
er 99.0:1.0

51% conv
s = 225

Scheme 117

The Lewis basic phosphinite derivative of quinidine **91** provides exceptional selectivity for the KR of hydrobenzoin (**163**) using an aryl acid chloride as the acylating agent to give both the ester product and recovered 1,2-diol as single enantiomers at 50% conversion (Scheme 118).[220] This method has also been applied to a small range of cyclic and acyclic 1,2-diols, in which all the other substrates examined afford selectivity factors an order of magnitude lower than that obtained for hydrobenzoin (**163**).

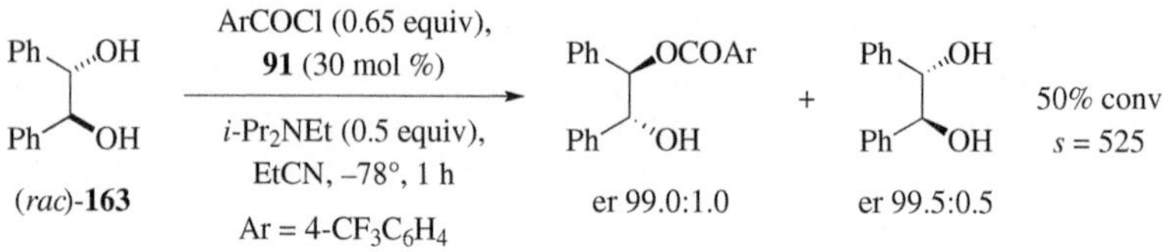

Ph⟍OH
Ph⟋OH
(*rac*)-**163**

ArCOCl (0.65 equiv),
91 (30 mol %)
⟶
i-Pr$_2$NEt (0.5 equiv),
EtCN, −78°, 1 h
Ar = 4-CF$_3$C$_6$H$_4$

Ph⟍OCOAr
Ph⟋OH
er 99.0:1.0

+

Ph⟍OH
Ph⟋OH
er 99.5:0.5

50% conv
s = 525

Scheme 118

Lipophilic peptide analogue **30**, which contains a rigid adamantyl core, is effective for the KR of cyclic *anti*-1,2-diols using excess acetic anhydride as the acyl source. The reaction selectivity is dependent on the ring size, in which five-membered cyclopentane-1,2-diol provides low selectivity (*s* = 8), whereas six-, seven-, and eight-membered analogues all provide excellent selectivities (*s* > 50) under similar conditions (Scheme 119).[74] The reaction efficiency is also highly dependent on the nature of the solvent, with more polar solvents, such as acetonitrile or dichloromethane, requiring longer reaction times and affording lower selectivities compared with toluene. The KR process can also be performed in a one-pot process, starting from the parent cyclic alkene.[221] Oxidation to the corresponding epoxide is followed by ring-opening to provide the racemic *anti*-diol in situ before the addition of the peptide catalyst. Carboxylic acids, in combination with diisopropylcarbodiimide, can also be used as the acyl sources for the KR of cyclic 1,2-diols using **30**,[222] in which both the reactivities and selectivities are affected by the steric demand of the carboxylic acid employed. Alternatively, the acylative KR of diols using structurally related lipophilic peptide catalysts can be performed under oxidative conditions allowing either aldehydes[223] or alcohols[224] to be used as the acyl source.

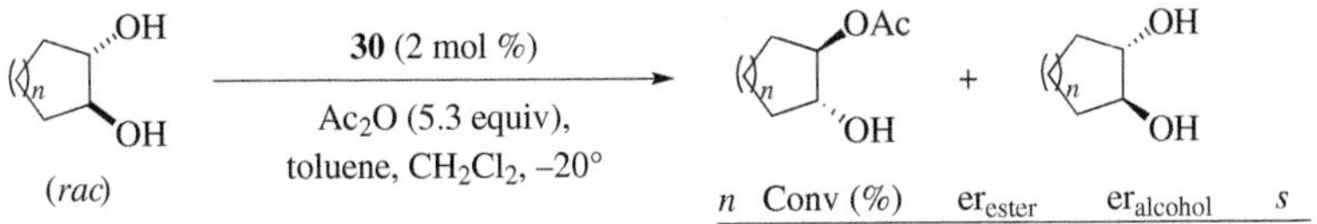

n	Conv (%)	er$_{ester}$	er$_{alcohol}$	s
1	63	74.5:25.5	92.5:7.5	8
2	57	87.5:12.5	>99.5:0.5	>50

Scheme 119

N-Heterocyclic carbene redox catalysis using α-bromo aldehyde **165** as the acyl source can be used for the KR of a range of cyclic *anti*-1,2-diols. The optimal conditions employ NHC **166** at low catalyst loading (0.5 mol %) in conjunction with a substituted benzoic acid co-catalyst and Proton Sponge, which react in situ to form the corresponding carboxylate (Scheme 120).[79] The nature of the co-catalytic acid and base used has a dramatic effect on both reactivity and selectivity. Excellent enantioselectivities are obtained for the KR of six- to eight-membered cyclic *anti*-1,2-diols, which again contrasts the KR of the *anti*-cyclopentane-1,2-diol that affords lower, but still synthetically useful, selectivity (s = 18) under the same conditions.

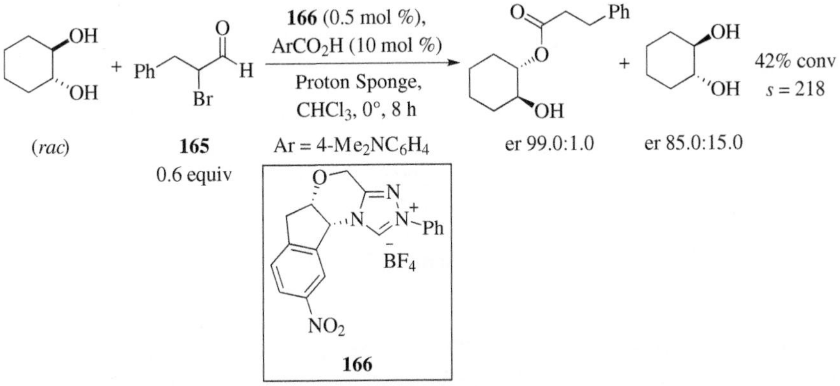

Scheme 120

The acylative KR of 1,2-diols using the axially chiral DMAP derivative **72** has been developed with s values of up to 180.[225] Key mechanistic experiments revealed that hydrogen bonding between the tertiary alcohols in the catalyst and the 1,2-diol unit of the substrate is critical for accelerating the rate of monoacylation and achieving high levels of enantioselectivity (Scheme 121). The catalytic system applies to a wide range of substrates, including racemic acyclic and cyclic 1,2-diols, which provide high selectivity for monoacylation. Notably, the process was demonstrated on a 10 g scale to illustrate its utility for preparative-scale applications.

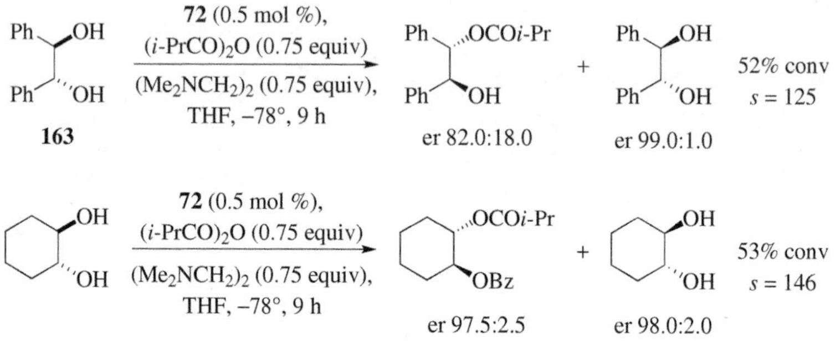

Scheme 121

The isothiourea-catalyzed acylative KR of acyclic 1,2-diols using 1 mol % of HyperBTM **25** has also been reported.[226] In contrast to the above example, the bifunctional nature of the 1,2-diol was exploited in a sequential double kinetic resolution, in which both kinetic resolutions operate synergistically to provide access to highly enantioenriched products. For example, the KR of **163** in CHCl$_3$ at 0° affords the diester (1*S*,2*S*)-**167** in 50% yield with 97.0:3.0 er and the monoester (1*R*,2*R*)-**163** and the enantiomerically enriched (1*R*,2*R*)-diol **168** in a 39% combined yield with >99.0:1.0 er at 51% conversion (Scheme 122).

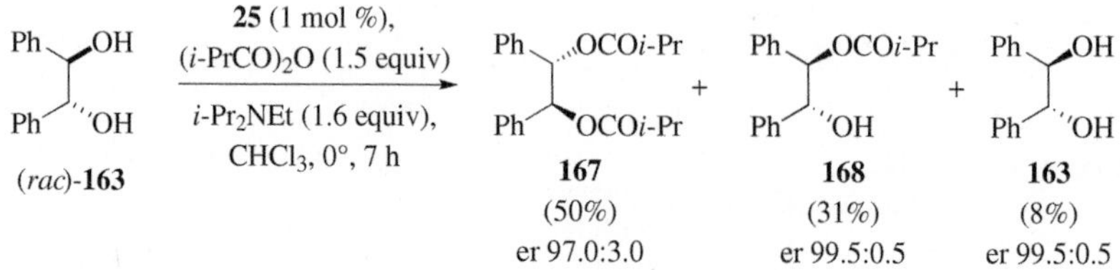

Scheme 122

Monoprotected *meso*-1,2-Diols. The current methods available for the acylative KR of monoprotected *meso*-diols typically rely on Lewis base catalysis. The most commonly employed catalysts for this type of investigation include chiral DMAP derivatives, imidazoles, and isothioureas, including some solid-supported variants. Additionally, a protocol that employs an achiral Lewis base in conjunction with a chiral Brønsted acid catalyst has also been reported.

Lewis Base Catalysis. An early example of the catalytic KR of monoprotected *meso*-diols used the chiral DMAP analogue, *ent*-**138** (Scheme 123).[227] The optimal conditions involve isobutyric anhydride as the acylating agent in conjunction with 2,4,6-collidine as the base in toluene at room temperature. A range of cyclic substrates was evaluated in the context of the alcohol protecting group and ring size. Consequently, the efficiency of the KR increases with the steric bulk of the protecting group, in which substrates that contain a 4-N,N-dimethylaminophenyl ester afford the highest levels of selectivity. Additional studies with this protecting group examined the effect of the ring size, in which six-membered rings provide the highest selectivities, whereas the five-, seven-, and eight-membered rings afford lower selectivities ($s =$ 8, 7, and 6, respectively).

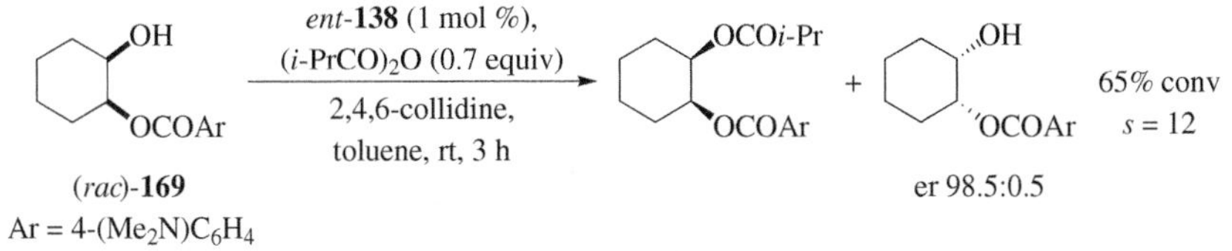

Scheme 123

4-(Dimethylamino)pyridine analogues that have appended α-methyl proline derivatives have also been investigated as catalysts for the KR of the protected diol **169** (Scheme 124).[178] The α-methyl substituent is essential to prevent catalyst epimerization under the reaction conditions, in which the secondary amide **156** affords the highest selectivity. Under the optimal KR reaction conditions using isobutyric anhydride, the alcohol **169** is recovered in 97.5:2.5 er at 62% conversion. Notably, the Merrifield resin-supported version of **156** also permits the KR of protected diol **169**, with comparable levels of selectivity to those obtained in solution.[177] In this case, the reaction requires an increase in the amount of anhydride and extended reaction times to compensate for the decreased reactivity of the solid-phase system. However, the catalyst is recovered and recycled four times while maintaining reactivity and selectivity.

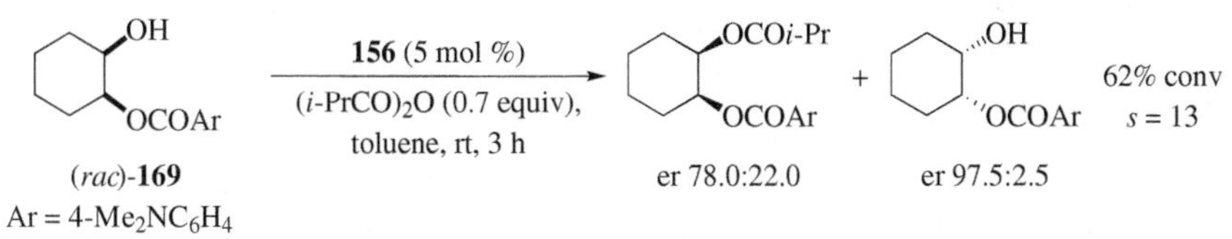

Scheme 124

The resolution of protected diols can also be achieved using PPY derivatives substituted on the 4-amino pyridine ring. For example, the PPY derivative **170** catalyzes the KR of a limited range of cyclic, monoprotected diols using acetic anhydride as the acyl donor (Scheme 125).[228] The pendant N,N-dimethylamino functional group in **170** is essential since alternative hydrogen-bond acceptors or donors provide less effective selectivity.[39,166] The KR of cyclic, monoprotected diols containing five- and six-membered rings proceeds with comparable levels of selectivity ($s =$ 10 and 8, respectively), whereas larger rings tend to be less selective.

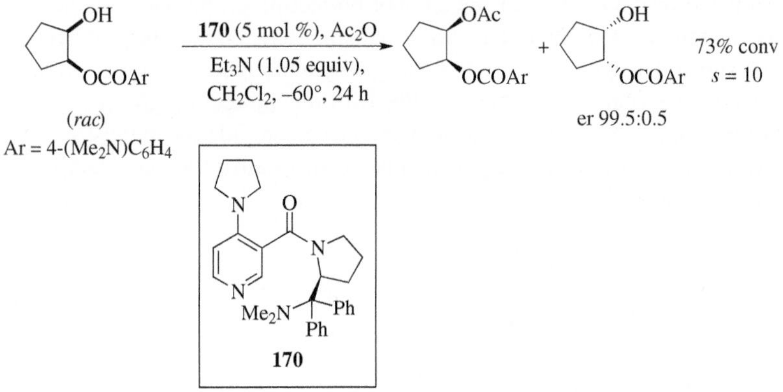

Scheme 125

The magnetic-nanoparticle-immobilized DMAP derivative **171** facilitates the KR of a limited number of monoprotected diols (Scheme 126).[229] Nevertheless, the catalyst is recycled through twenty consecutive KRs using the protected diol **169** and affords comparable reactivity and selectivity in all cases. The same catalyst is then employed for the KR of other substrates without loss in activity, demonstrating that the system is at least capable of a minimum of 32 consecutive cycles.

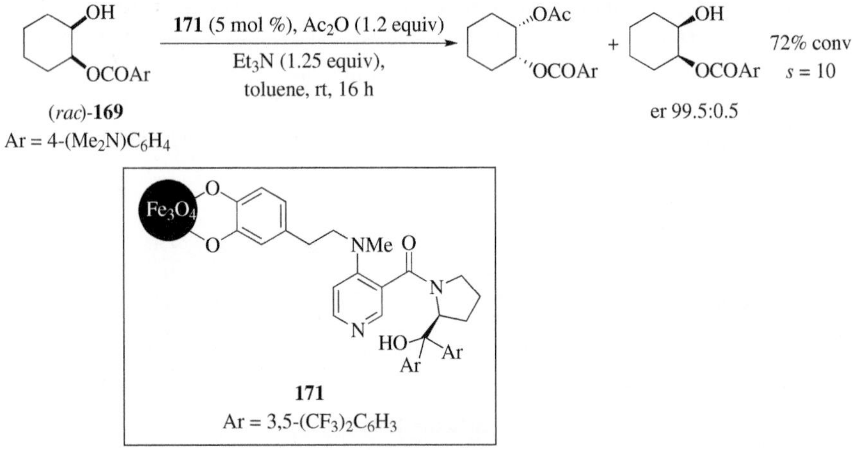

Scheme 126

Histidine-derived catalysts currently represent the state-of-the-art for the KR of protected diols. For example, the histidine derivative **29** is an efficient catalyst for the KR of a range of substrates with excellent selectivity using isobutyric anhydride as the acylating agent.[70,71,181] The substrates bearing a pyrrolidine-based carbamate protecting group affords the highest selectivity, in which both cyclic and acyclic protected diols undergo KR with synthetically usefully selectivity factors (s up to 93).

For instance, the protected diol **172** undergoes efficient resolution under the optimal reaction conditions to furnish the ester product in 96.5:3.5 er, whereas the alcohol **172** is recovered in 91.0:9.0 er (Scheme 127).[70] A polymer-supported analogue of **29**, which can be efficiently recycled, gives levels of selectivity comparable to the solution variant.

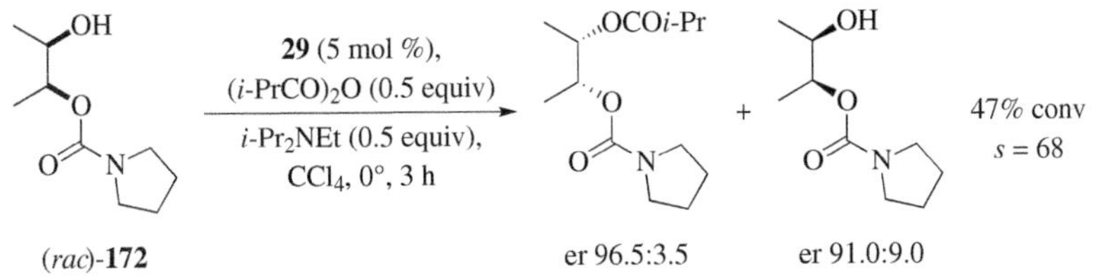

Scheme 127

Histidine analogue **173** is also an effective catalyst for the KR of pyrrolidine carbamate-protected diols under the same conditions as those reported for **29** (Scheme 128).[230] The thioamide in **173** is important since the corresponding amide affords much lower selectivity. The optimal reaction conditions were tested on a range of monoprotected cyclic and acyclic diols, and synthetically useful levels of selectivity were obtained in all cases. Notably, cyclic substrates containing seven- and eight-membered rings are efficiently resolved, even though the selectivity factors are lower than those for the corresponding six-membered analogue.

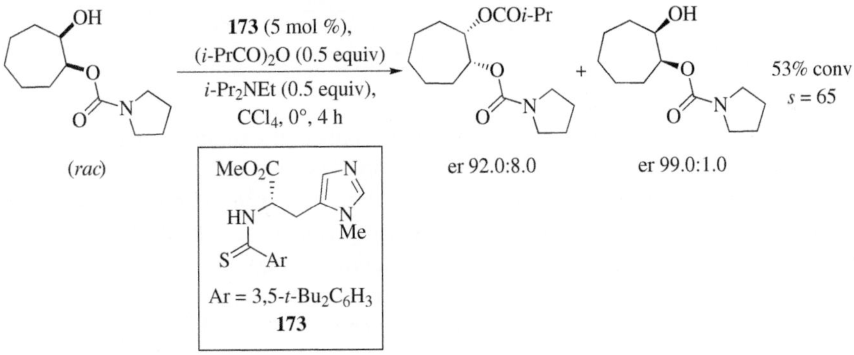

Scheme 128

Brønsted Acid Catalysis.   The benzoyl-protected diol **174** can be effectively resolved using the chiral phosphoric acid *ent-***54** (5 mol %) using acetyl chloride as the acyl donor (Scheme 129).[201] The reaction proceeds with excellent selectivity, allowing the ester product to be obtained in 95.5:4.5 er and the alcohol **174** to be recovered in >99.5:0.5 er. However, this catalyst system has not been further tested on other substrates within this class to delineate its generality.

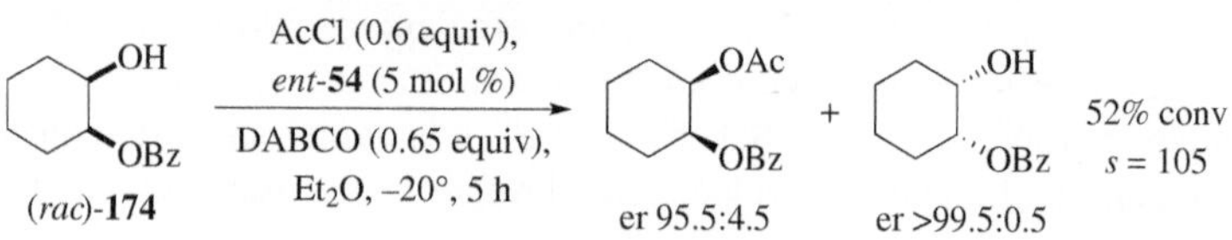

Scheme 129

1,3-Diols. A synergistic double catalytic kinetic resolution (DoCKR) of racemic acyclic *anti*-1,3-diols using isothiourea **25** leads to diesters and recovered diols with high enantiopurities (Scheme 130).[231] This protocol exploits an additive Horeau amplification involving two successive enantioselective acylation reactions and has been applied to both C_2-symmetric and non-C_2-symmetric *anti*-1,3-diols.

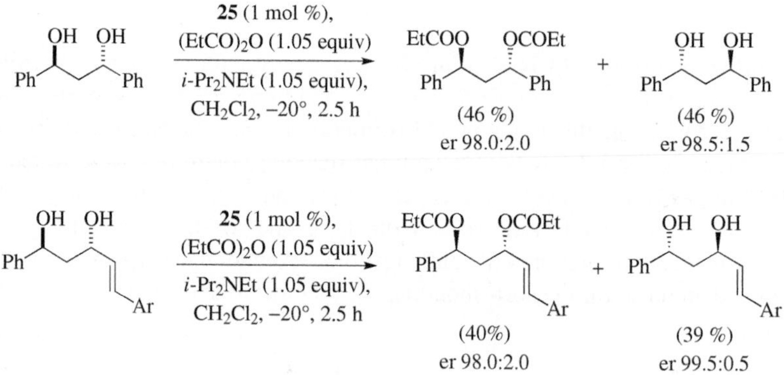

Scheme 130

Atropisomeric Alcohols. Only a limited number of examples of the acylative KR of atropisomeric alcohols have been reported, which all involve using Lewis base catalysis. The fluxionally chiral DMAP derivative **136** (15 mol %) is used to resolve monoprotected BINOL derivatives using isobutyric anhydride as the acyl source (Scheme 131).[169] Various phenol protecting groups are compatible, which leads to efficient resolution with good selectivities. Nevertheless, more extensive structural variations of the steric and electronic properties of the naphthyl rings have not been explored.

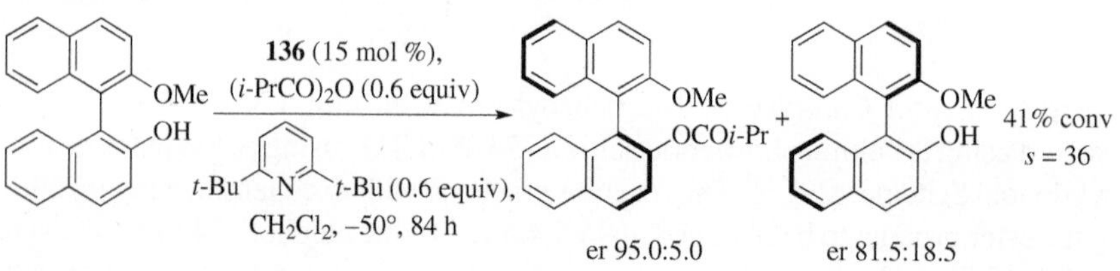

Scheme 131

N-Heterocyclic-carbene-catalyzed redox catalysis can also be used for the KR of various atropisomeric diols. The optimal conditions employ the NHC **127** (10 mol %) in combination with α-*O*-benzoyl aldehyde **128** as the redox-active acylating agent (Scheme 132).[84] This method applies to a wide range of 2,2'-biaryl-1,1'-diols containing various substitution patterns, which exhibit good reactivity and high selectivity in all cases. Notably, the KR of highly sterically hindered substrates such as **175** provide higher selectivity factors without compromising the overall reactivity.

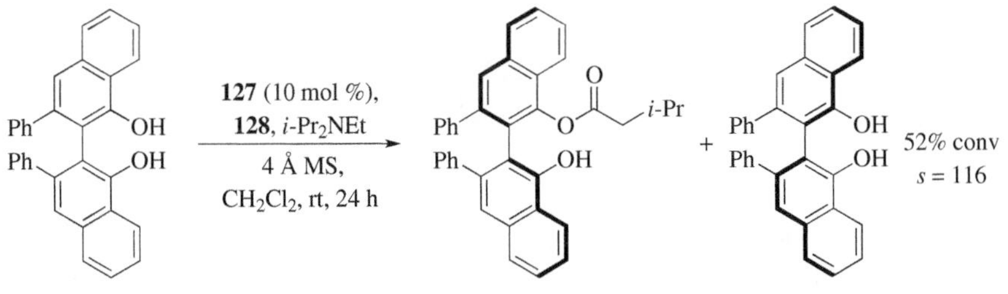

127 (10 mol %),
128, *i*-Pr₂NEt

4 Å MS,
CH₂Cl₂, rt, 24 h

175

er 96.0:4.0

er 99.5:0.5

52% conv
s = 116

Scheme 132

An isothiourea-catalyzed acylative KR of unprotected BINOL derivatives has been developed using **23** as the Lewis base catalyst (*s* values up to 190).[232] The investigation of the reaction scope and limitations results in three key observations: i) the diol present in the substrate is essential for good conversion and high *s* values; ii) the use of (2,2-diphenyl)acetic pivalic anhydride **176** is critical to minimize diacylation and afford high selectivity; and iii) the presence of substituents at the 3,3'-positions of the diol hinders effective acylation. This final observation has been exploited for the regioselective acylative KR of unsymmetrical biaryl diol substrates that contain a single 3-substituent, such as **177** (Scheme 133).

The NHC-catalyzed acylative KR of a series of axially chiral anilides has been reported, where the starting racemic anilide is present as a single diastereoisomer.[233] The generation of an α,β-unsaturated acyl azolium species in situ from the addition of the NHC catalyst **178** to the ynal **179**, followed by enantioselective acylation, provides axially chiral isoindolinones in high yields and with excellent enantioselectivities (Scheme 134).

α-Hydroxy Esters and α-Hydroxy Amides. The KR of α-hydroxy esters is a potentially valuable process given the synthetic utility of this class of substrate. However, the KR is challenging because of the presence of the carbonyl substituent adjacent to the stereogenic carbinol center, which can lead to epimerization. As such, only a few examples of effective acylative KR protocols for the KR of α-hydroxy esters have been reported.

Lewis Base Catalysis. The acylative KR of mandelic acid methyl ester (**180**) was first investigated using bicyclic phosphine **60** (4.8 mol %) as the catalyst and

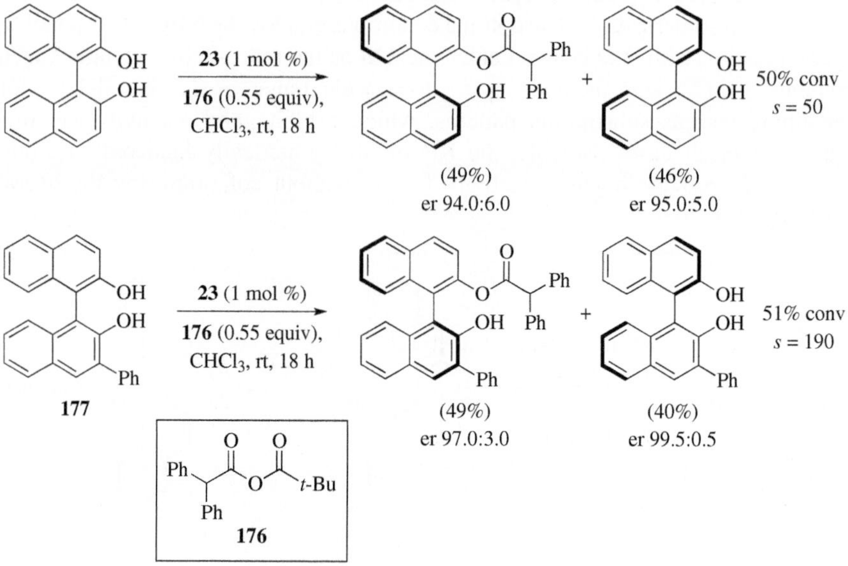

(49%)
er 94.0:6.0

(46%)
er 95.0:5.0

50% conv
s = 50

23 (1 mol %)
176 (0.55 equiv),
CHCl₃, rt, 18 h

(49%)
er 97.0:3.0

(40%)
er 99.5:0.5

51% conv
s = 190

177

176

Scheme 133

178 (10 mol %),
179 (1.25 equiv)
NaOAc (1 equiv),
4 Å MS, CHCl₃, rt, 84 h

*single
diastereoisomer*

(49%)
er 97.0:3.0

(48%)
er 98.5:1.5

51% conv
s = 136

178
Ar = 2,6-(*i*-Pr)₂C₆H₃

179

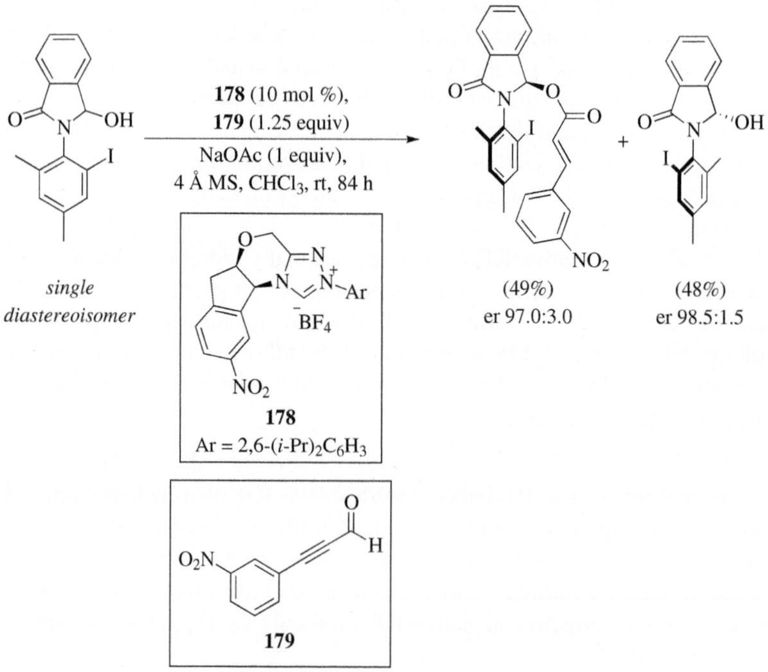

Scheme 134

isobutyric anhydride as the acylating agent (Scheme 135).[155] Nevertheless, only low selectivity was obtained ($s = 3$) for this challenging reaction, since the catalyst needs to differentiate between two sp^2-hybridized substituents on the stereogenic carbinol center.

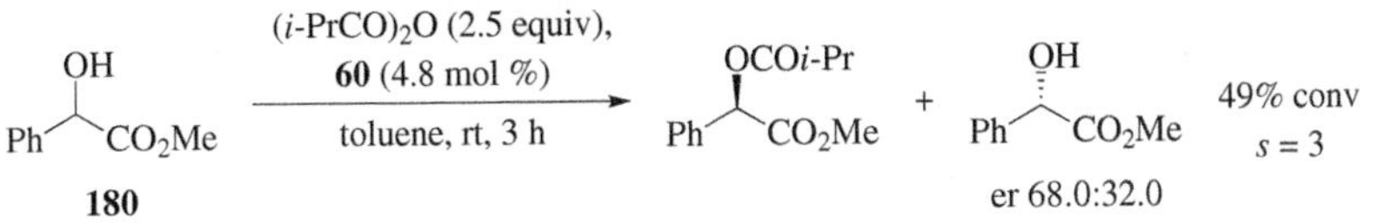

Scheme 135

The KR of α-hydroxy esters bearing sp^3-hybridized alkyl substituents on the stereogenic carbinol center proceeds with significantly higher stereoselectivities, using isothiourea **23** (5 mol %) as the catalyst (Scheme 136).[62] The optimal conditions employ a mixed anhydride generated in situ from the reaction of diphenylacetic acid and pivaloyl anhydride. The method is effective for the KR of a range of alkyl substituents, including those containing pendant heteroatom functional groups, to provide synthetically useful levels of selectivity in all cases. As an alternative to the mixed anhydride system, diphenylacetic anhydride can be used directly as the acylating agent for the efficient KR of a wide range of α-hydroxy esters under otherwise identical reaction conditions.[61]

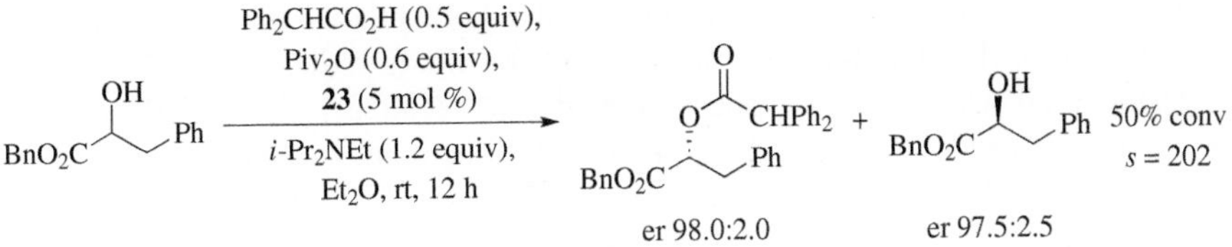

Scheme 136

Through the same approach, the isothiourea **23** catalyzes the KR of 2-hydroxyamides using the mixed anhydride generated in situ from diphenylacetic acid and pivaloyl anhydride as the acylating agent (Scheme 137).[234] Optimal selectivity is observed with either dimethyl amide or the Weinreb amide derivatives, which exhibit very good tolerance to variations in the β-alkyl substituent.

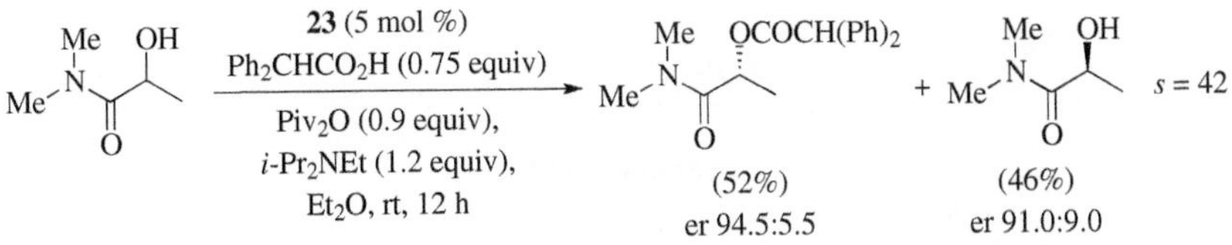

Scheme 137

Lewis Acid Catalysis.   The chiral Cu(II)-Box complex **162** (0.2 mol %) facilitates the KR of α-hydroxy esters using *n*-propyl isocyanate as the acylating agent
(Scheme 138).[99] The process is effective for a range of substrates bearing highly
sterically demanding β-substituents to provide the carbamate products and recovered alcohols with high enantioselectivities. This method is potentially synthetically
valuable, given the difficulty of preparing such substrates in an enantiomerically
enriched form using conventional methods.

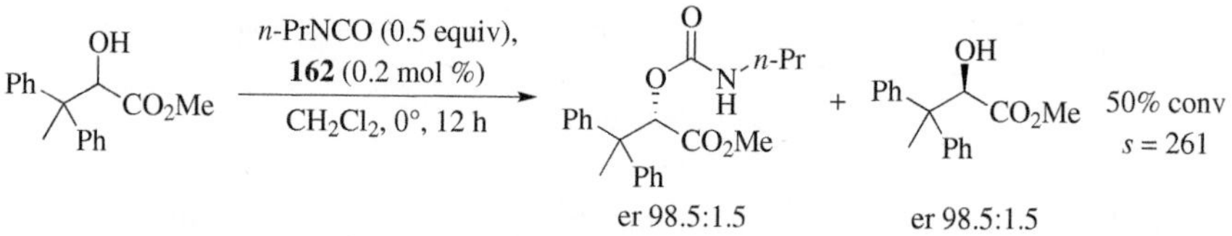

Scheme 138

The acylative KR of mandelic acid ester substrates has been achieved using
Sc(OTf)$_3$ (5 mol %) in combination with chiral *N,N'*-dioxide ligand **181** (5 mol %)
and acetic anhydride (Scheme 139).[103] An initial study into the resolution of
mandelic acid esters bearing different ester groups demonstrated that while the *t*-Bu
ester is optimal, all the examples are resolved with good to excellent selectivity
factors (11 examples, *s* = 33–91). Further substrate variation demonstrated that
ortho-, *meta-*, and *para*-substituted aromatic groups bearing both electron-donating
and electron-withdrawing groups are viable substrates, giving generally high
selectivity factors (*s* = 24–247). In contrast with other methods, the presence of alkyl
substitution on the stereogenic carbinol center results in reduced stereoselectivity.
The attempted resolution of 1-phenylethanol (**13**) resulted in low conversion and
essentially no selectivity, indicating that the ester functional group is needed for both
reactivity and selectivity.

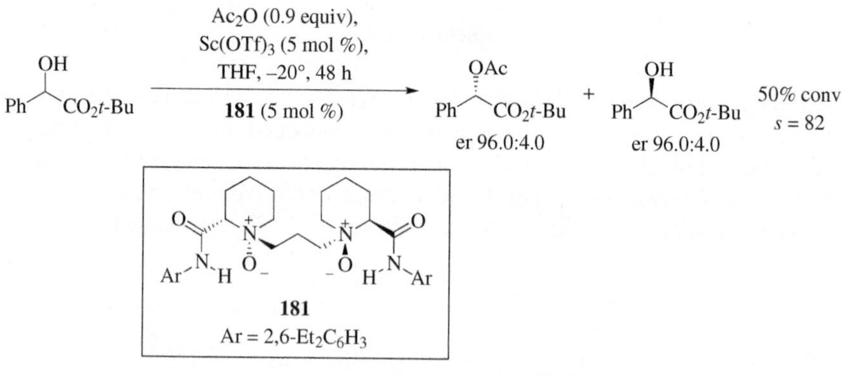

Scheme 139

Primary Alcohols. The KR of substrates that involve the acylation of primary alcohols is inherently challenging, as the stereocenter of interest is remote

from the site of reactivity, and thus only a limited number of examples have been reported.

The first nonenzymatic KR of unprotected 1,2-diols through regioselective acylation of a primary alcohol employed the chiral tin catalyst **182** (0.25 mol %) using benzoyl chloride (Scheme 140).[100] The reaction selectivity is highly dependent on the nature of the base, in which sodium carbonate is optimal. Furthermore, the addition of a small amount of water is essential for obtaining reasonable selectivity. Notably, the amount of water is critical since either the absence or excess of water (e.g., 50 equiv) results in a significant decrease in selectivity. The reaction scope comprises a limited number of 1,2-diols bearing either an alkyl or aryl 2-substituent, in which only moderate levels of selectivity are achieved.

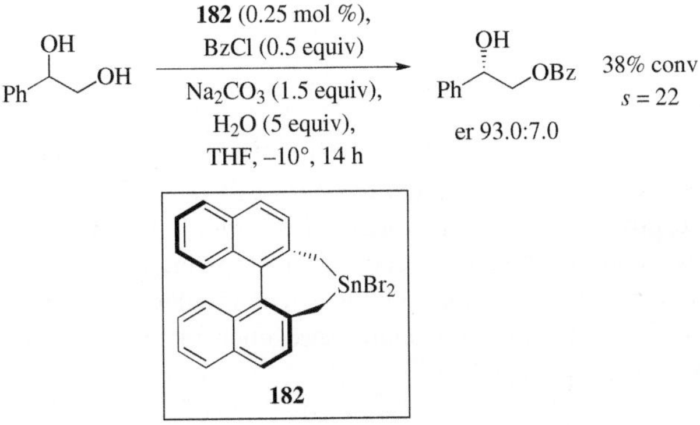

Scheme 140

The KR of a substrate that has both a benzylic secondary and a primary alcohol motif has also been accomplished using oxidative NHC catalysis.[235] Consequently, the combination of the NHC precatalyst **183**, benzaldehyde, and quinone **101** as the oxidant, permits the KR of a selection of diols with s values of up to 22 (Scheme 141).

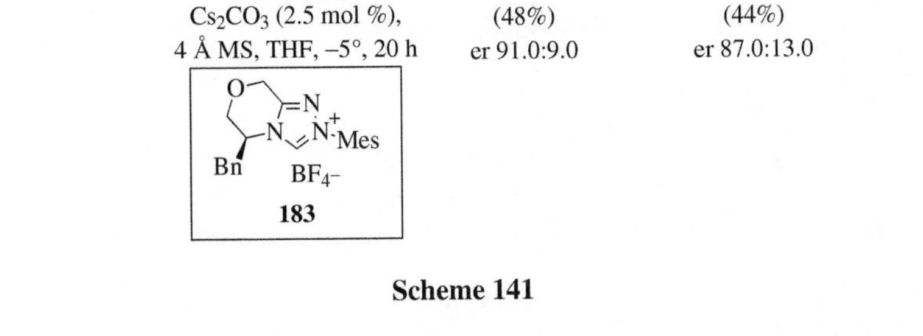

Scheme 141

Monoprotected 1,2-diols such as **184** can be effectively resolved using the histidine-based catalyst **173** (5 mol %) and isobutyric anhydride (Scheme 142).[230]

The reaction is suitable for a limited number of aryl and alkyl 2-substituents, which afford good reactivities and good to excellent selectivities in all cases. Primary alcohols containing a protected 2-amino substituent can also be efficiently resolved using the related histidine-based catalyst **29** under similar conditions.[70,71] The KR of *N*-protected serine derivatives through acylation of the free hydroxy group has also been attempted using a DMAP-derived catalyst, albeit only moderate selectivity factors have been obtained to date.[177,178,236]

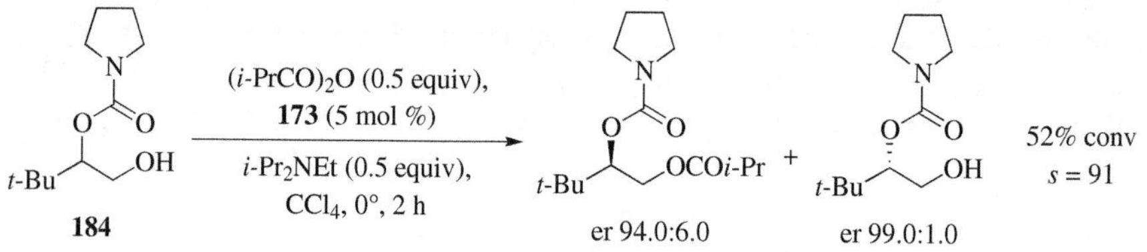

Scheme 142

The KR of a few primary alcohols has been investigated using the isothiourea catalyst **24**, with low selectivity reported ($s = 4$) (Scheme 143).[237] This method was extrapolated to provide an empirical method of predicting the absolute configuration of primary alcohols with a β-stereocenter. Comparing the rates of reaction of an enantiomerically enriched alcohol with either enantiomer of catalyst **24** allows the absolution configuration to be predicted based upon the stereochemical model for acylation with this class of catalyst.

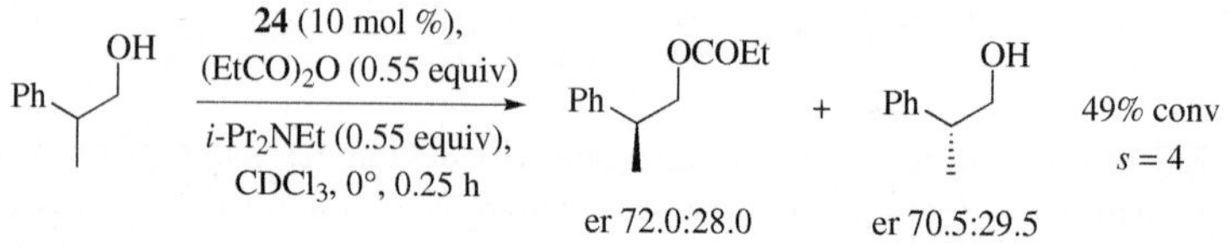

Scheme 143

The KR of *N*-protected amino alcohols can also be achieved using a copper-Box Lewis acid catalyst with benzoyl chloride as the acyl source. Good enantio-selectivities are observed for the KR of various acyclic, 2-substituted amino alcohols, whereas much higher selectivity factors are obtained using cyclic amino alcohols such as *N*-benzoyl prolinol **185** (Scheme 144).[238]

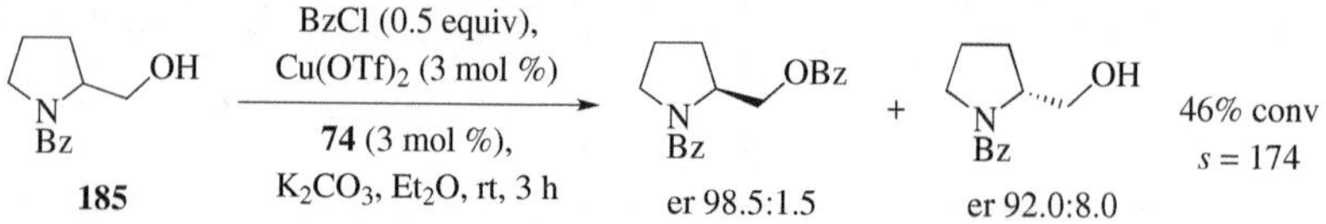

Scheme 144

A conceptually distinct approach to KR involves chiral Brønsted acid catalyzed, intramolecular cyclization of primary alcohols onto pendant ester substituents bearing α-stereocenters to form enantioenriched γ-lactones. Although the process affords only modest selectivities for the KR of pendant tertiary stereocenters, the quaternary-substituted substrate **186** affords synthetically useful $s = 17$ using phosphoric acid *ent*-**54** as the catalyst (Scheme 145).[239] The γ-lactone products and unreacted starting materials are readily separated, and the enantiomerically enriched alcohols are easily derivatized using the primary hydroxyl group.

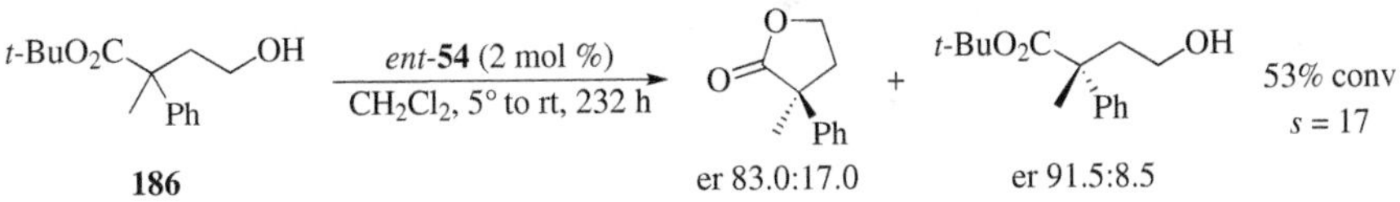

Scheme 145

Tertiary Alcohols. The acylative KR of tertiary alcohols is challenging as the system must differentiate between the three non-hydrogen substituents on the stereogenic alcohol. Furthermore, because of increased steric hindrance around the alcohol, the overall rate of acylation of tertiary alcohols is much slower than with secondary alcohols. As such, only a few reports of tertiary alcohol KR exist, and currently, no general solution to this challenging problem has been reported.

High-throughput screening using a fluorescence-based assay aided the discovery of peptide analogue **187**, which is employed for the KR of *N*-acetyl 1,2-amino alcohol substrates containing a tertiary alcohol stereocenter using excess acetic anhydride as the acyl source (Scheme 146).[240] Good reactivities and selectivities are observed for a narrow range of aryl substituents. However, the attempted KR of a substrate containing an α-hydroxy ester motif gave essentially no selectivity under these conditions.

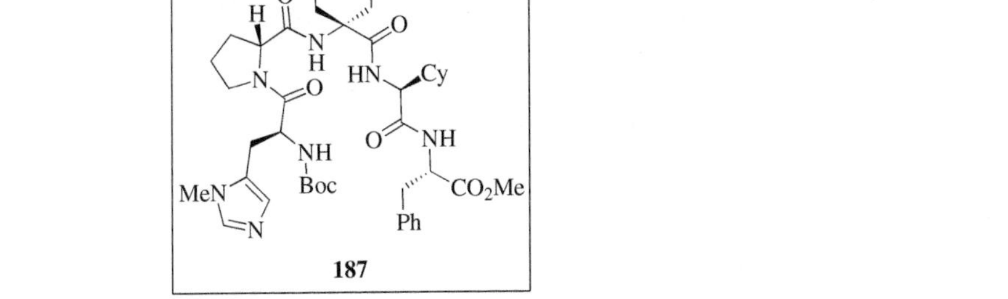

Scheme 146

Oxidative NHC-catalysis can be used to resolve 3-hydroxy-3-substituted oxindole derivatives with good selectivities. The optimized conditions use NHC **126**

(10 mol %) as the catalyst and cinnamaldehyde as the acyl source in combination with excess manganese dioxide as an oxidant and DBU as the base (Scheme 147).[83] The method requires magnesium triflate, which serves as a Lewis acidic co-catalyst to improve both reactivity and selectivity. The addition of sodium tetrafluoroborate was also reported to further improve selectivity. The reaction scope has been investigated for a range of oxindole derivatives, including those containing both alkyl and aryl 3-substituents, which furnish good to excellent reactivities and selectivities in all cases. Substitution around the benzenoid ring of the oxindole core and different N-protecting groups are also accommodated. However, the broader applicability of these conditions for the KR of acyclic derivatives has not been reported.

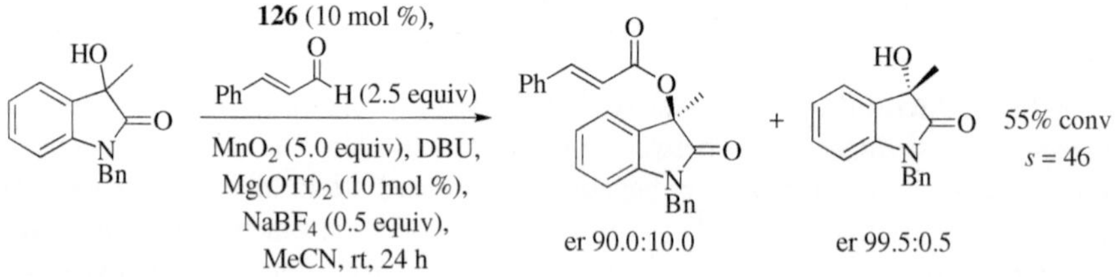

Scheme 147

The isothiourea HyperBTM **25** catalyzes the acylative KR of two classes of tertiary heterocyclic alcohols, namely, 3-hydroxy-3-substituted oxindoles and α-hydroxy-γ-lactams (Scheme 148).[241] The reaction proceeds at low catalyst

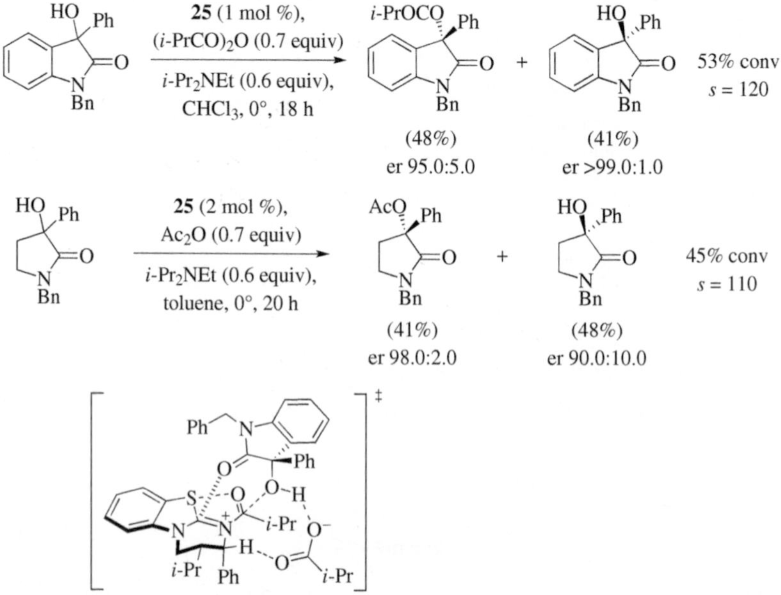

Scheme 148

loadings (generally 1–2 mol %) with either isobutyric or acetic anhydride used as the acylating agent. A variety of substitutions around the oxindole and lactam cores are tolerated, in addition to different *N*-protecting groups. A combination of experimental and computational studies identified a critical C=O•••isothiouronium interaction for the efficient enantiodiscrimination in the KR.

Immobilized isothiourea catalysts have also been used for the KR of these types of substrates, in which the Merrifield-supported HyperBTM **144** showing optimal selectivity in batch using industrially preferred solvents, such as EtOAc and DMC.[242] This process was subsequently conducted in a packed-bed reactor to demonstrate the KR of tertiary heterocyclic alcohols in a continuous-flow process. High selectivities were obtained for the resolution of 3-hydroxyoxindole derivatives in EtOAc (*s* up to 70) and 3-hydroxypyrrolidinone derivatives in toluene (*s* up to 42) (Scheme 149).

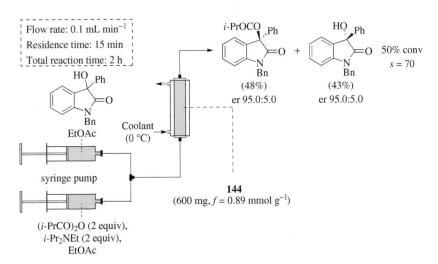

Scheme 149

The KR of tertiary 2-alkoxycarboxamido allylic alcohols can be achieved through a phosphoric acid catalyzed, intramolecular transesterification reaction (Scheme 150).[243] Alkyl-, aryl-, and dialkyl-substituted tertiary allylic alcohols are resolved with excellent selectivity, affording both the recovered tertiary alcohols and the carbamate products with high enantioselectivities (*s* values up to 165).

Miscellaneous Alcohols. The resolution of acyclic secondary alcohols bearing two sp³-hybridized substituents on the stereogenic carbon center is particularly challenging. To allow effective discrimination between the two enantiomers in these cases, one of the substituents typically contains a pendant functional group proximal to the reacting alcohol center.

For example, the histidine-derived catalyst **29** (5 mol %) is used to resolve the protected threonine derivative **189** using isobutyric anhydride with good selectivity

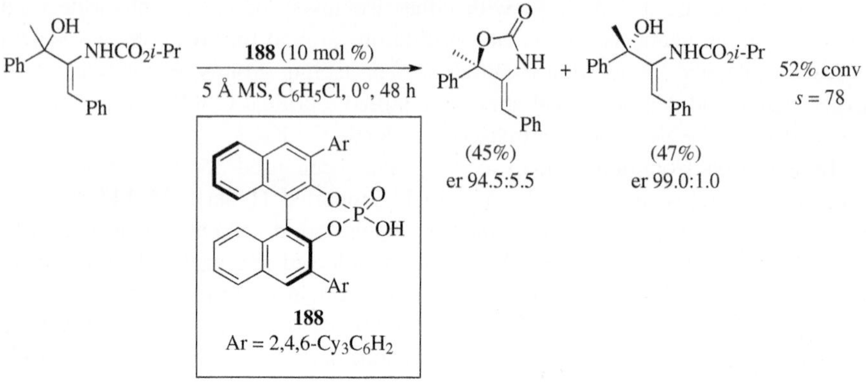

Scheme 150

(Scheme 151).[70] The reaction proceeds with synthetically useful selectivity and demonstrates good functional group tolerance, although this is the only such example of this substrate class reported.

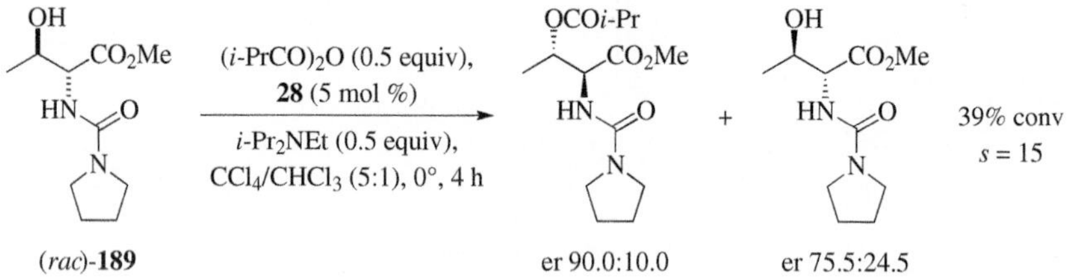

Scheme 151

Simple β-hydroxy ester **190** is effectively resolved using amidine catalyst **141** and propionic anhydride as the acylating agent (Scheme 152).[53] Although potentially useful, the reaction selectivity is considerably lower than that obtained for the KR of Morita-Baylis-Hillman adducts bearing an sp^2-hybridized α-methylene substituent (Scheme 97) under otherwise identical conditions.

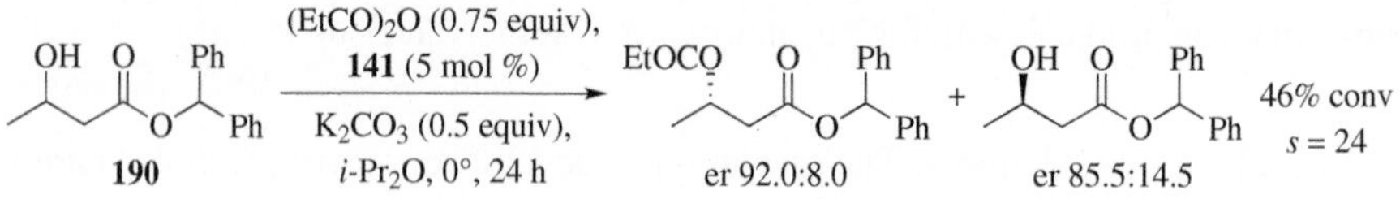

Scheme 152

A more general approach to the KR of β-hydroxy amides has been devised using a copper-based Lewis acid catalyst (10 mol %) in combination with Box ligand **74** (10 mol %). The reaction is suitable for a range of secondary and tertiary amides that afford good reactivities and high selectivities (Scheme 153).[200] However, the presence of extended alkyl substituents on the stereogenic carbon atom (other than methyl) results in lower selectivities.

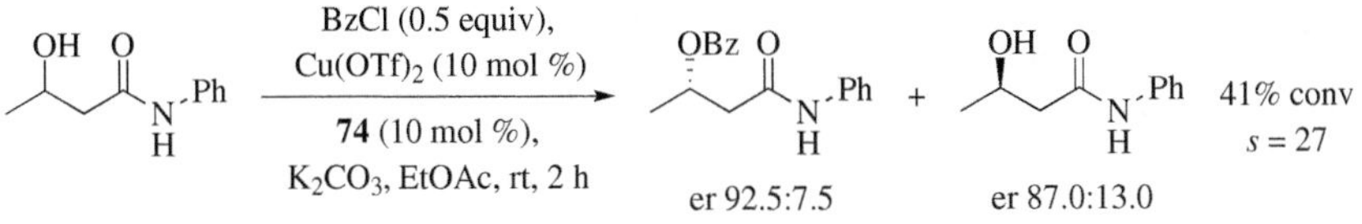

Scheme 153

The same copper-Box Lewis acid catalyst system applies to the resolution of a range of α-hydroxyphosphonates with high selectivities (Scheme 154).[208] The optimal conditions use benzoyl chloride as the acylating agent in conjunction with barium carbonate as the base in chlorobenzene. The process accommodates a variety of alkyl substituents, including those bearing pendant functional groups, such as benzyl ethers and protected amines. Although the reaction selectivity is generally high, the overall reactivity toward acylation is limited, and low conversions are observed in some cases. In contrast, the KR of β-hydroxyphosphonates proceeds with lower selectivities using the same Lewis acid catalyst system under similar reaction conditions.[98]

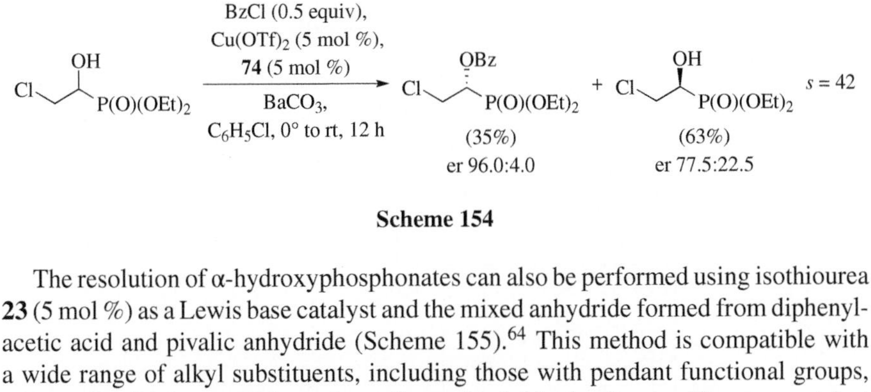

Scheme 154

The resolution of α-hydroxyphosphonates can also be performed using isothiourea **23** (5 mol %) as a Lewis base catalyst and the mixed anhydride formed from diphenylacetic acid and pivalic anhydride (Scheme 155).[64] This method is compatible with a wide range of alkyl substituents, including those with pendant functional groups, which afford high reactivities and excellent selectivities in all cases. The same catalyst can also be employed for the KR of an alkyl cyanohydrin using propionic anhydride as the acyl source, with good selectivity.[244]

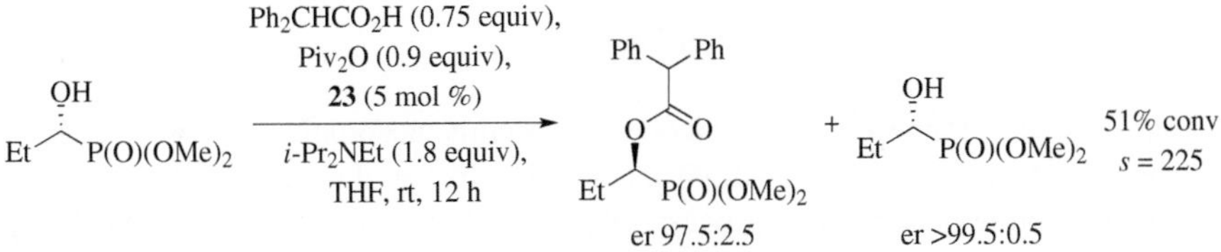

Scheme 155

The acylative KR of secondary alcohols bearing two alkyl substituents on the stereogenic carbon atom is particularly challenging. Although various Lewis base catalysts have been evaluated in such reactions, they generally give low selectivities, and thus no general solution currently exists for this process.[68,157,158,194,245,246] The most promising results to date have been reported using isothiourea ent-**23** (8 mol %) and propionic anhydride as the acyl source, which permits 1-phenylpropan-2-ol (**191**) to be resolved with moderate selectivity, albeit with low reactivity (Scheme 156).[57]

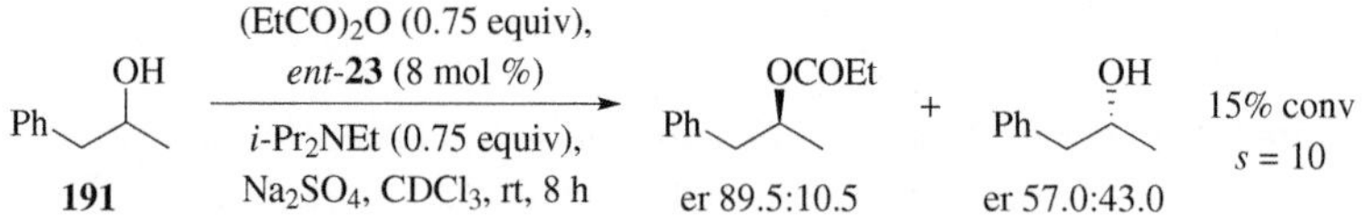

Scheme 156

Brønsted acid catalysis features in the KR of planar-chiral PHANOL derivatives, which uses a chiral phosphoric acid to catalyze the enantioselective acylation of one of the phenol groups (Scheme 157).[247] The inclusion of bulky substituents at the 3,3'-positions within the phosphoric acid **192** is critical for optimal enantioselectivity, as is the presence of two hydroxy groups within the PHANOL substrate that facilitate H-bonding within the favored transition state. The KR of planar-chiral cyclophanes bearing only one free hydroxy group gave lower reactivities and enantioselectivities (Scheme 157).

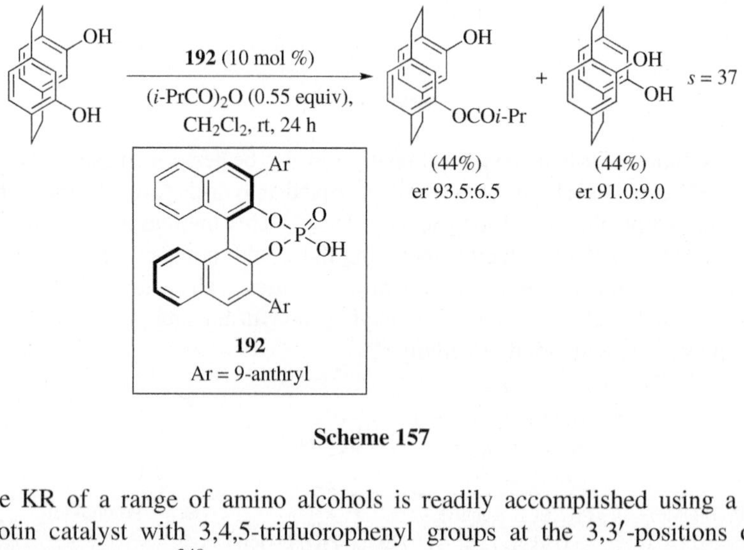

Scheme 157

The KR of a range of amino alcohols is readily accomplished using a chiral organotin catalyst with 3,4,5-trifluorophenyl groups at the 3,3'-positions of the binaphthyl framework.[248] The process tolerates alkyl- and aryl-substituted amino alcohols, including cyclic substrates, to afford the corresponding products in high enantioselectivities and with s values of up to >500 (Scheme 158).

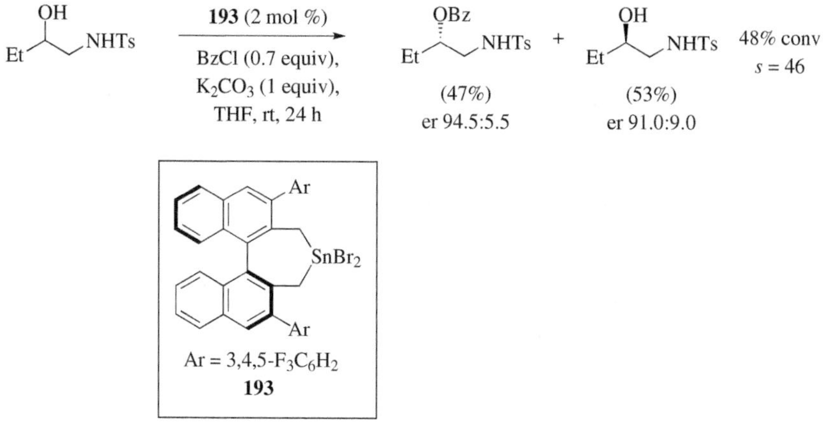

Scheme 158

Conclusions and Outlook. The nonenzymatic, acylative KR of alcohols has witnessed great advances over the last twenty years. An array of highly efficient catalytic methods is available that permits the KR of a wide range of alcohols with synthetically useful selectivity factors, and many of these processes have established synthetic value. However, despite many advances, there is still an opportunity for further developments in the field of acylative KR of alcohols.

Although there are some reports of acylative KR proceeding at low catalyst loadings (<1 mol %), the majority of methods rely on significantly higher catalyst loadings (ca. 5–10 mol % is common). Hence, the development of more efficient catalytic systems that can proceed at low loadings, under mild conditions and using readily available acylating agents is desirable.

Many methods have been investigated for the KR of secondary alcohol substrates bearing two electronically and/or sterically distinct substituents on the stereogenic carbinol center, for example, an sp^2-hybridized aryl substituent with an sp^3-hybridized alkyl substituent. However, the development of protocols that permit the KR of more challenging alcohols, in which the catalyst must differentiate between two similar substituents on the stereogenic carbinol center, is an ongoing challenge within the area. Other classes of substrate, including tertiary and atropisomeric alcohols, are also underexplored in the acylative KR. In all cases, increased functional-group tolerance and the complexity of substrates will enable the incorporation of the acylative KR into target-directed synthesis.

Overall, these remaining challenges will ensure that nonenzymatic acylative KR of alcohols will remain an active area of interest and research in the coming years.

Desymmetrization of Diamines

1,2-Diamines. The only example to date of a desymmetrization reaction of *meso*-diamines through an enantioselective *N*-acylation uses dual catalysis. For example, *meso*-diamine **194** undergoes desymmetrization to afford the *N*-benzoyl derivative **196** in 82% yield and with 97.5:2.5 er using the H-bonding catalyst **195** (10 mol %) in combination with DMAP (10 mol %) (Scheme 159).[32] Importantly, an equivalent of triethylamine is required to achieve complete conversion, and performing the reaction in the absence of 4 Å MS leads to a dramatic and unexpected reduction in the enantiomeric ratio (75% yield of **196** in 75.5:24.5 er).

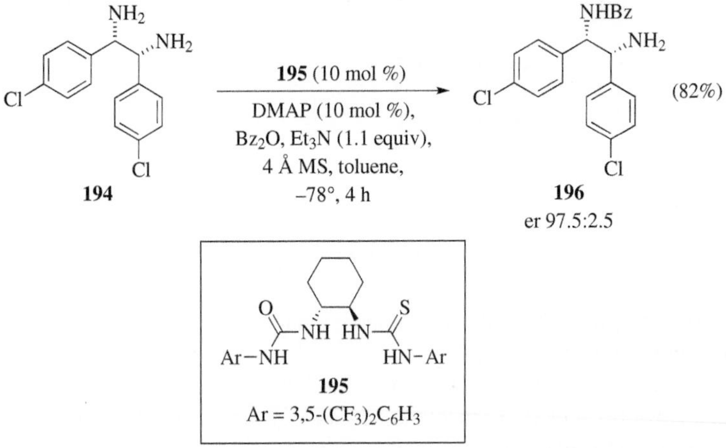

Scheme 159

Kinetic Resolution of Amines and Lactams

The nonenzymatic, catalytic KR of amines through *N*-acylation is a recognized synthetic challenge. The direct, rapid, uncatalyzed reaction of many amines with common *N*-acylating agents, such as acid chlorides and acid anhydrides, results in a competitive, racemic background process in many cases. Because of this inherent reactivity, the use of enantiomerically enriched acylating agents for the stoichiometric KR of amines is still the most common approach. Nevertheless, several creative and effective strategies for the catalytic acylative KR of amines have been developed that overcome the challenge posed by the rapid, uncatalyzed *N*-acylation reaction. However, the scope of these processes remains limited compared with the KR of alcohols.

Benzylic Amines. As with the acylative KR of alcohols, the KR of amines is most commonly examined on racemic, benzylic amines.

Stoichiometric Methods. The KR of amines was demonstrated using the enantiomerically pure *N,N*-diacetyl 1,1′-binaphthyl-2,2′-diamine (**197**) as a stoichiometric *N*-acylating agent (Scheme 160).[249] Racemic benzylic and alkylamines react with 0.25 equivalents of **197** in DMSO to afford enantiomerically enriched amides with poor enantioselectivities. For example, the resolution of 1-phenylethylamine (**2**) with **197** gives (*S*)-*N*-(1-phenylethyl)acetamide (**3**) in 24% yield and with 65.0:35.0 er.

Although this method has little practical value, it paved the way for the further development of enantioselective N-acylating agents.

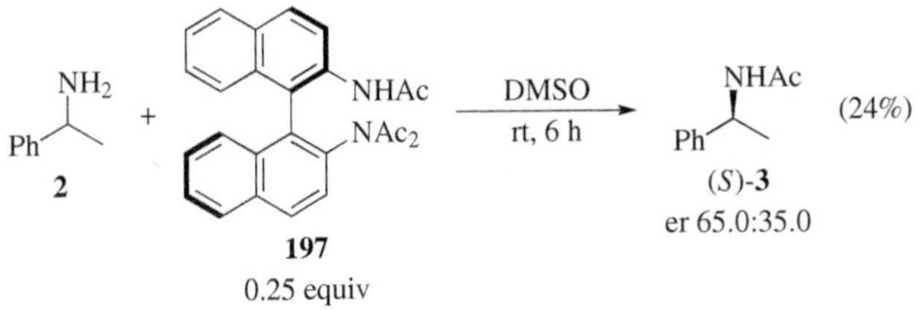

Scheme 160

Notably, N,N-diacylaminoquinazolinone derivatives, such as **198**,[250–252] and the N-acetyl planar-chiral PPY derivative **199**[253] afford improved selectivities for the KR of benzylamines (Figure 16). However, the necessity to employ super-stoichiometric amounts of **199** (8 equiv) to achieve high levels of enantioselectivity and the synthetic complexity of these agents make this approach unlikely to be of practical value for large-scale applications.

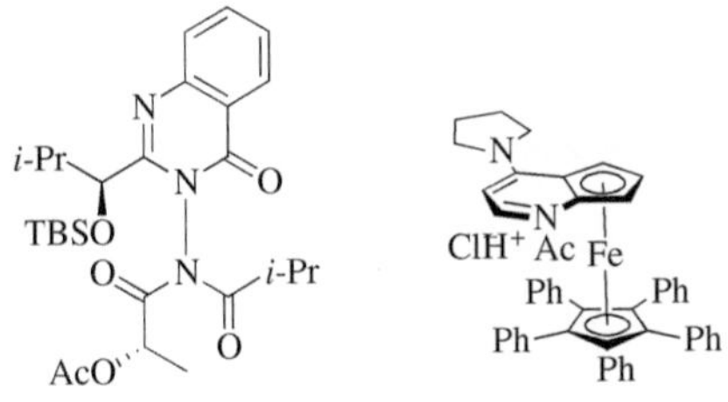

Figure 16. Reagents for the stoichiometric KR of benzylic amines.

The most significant advance in terms of substrate scope and enantioselectivity in the stoichiometric KR of amines is the development of enantiomerically pure N-acetyl-1,2-bis(trifluoromethanesulfonamide) (**1**) as an N-acylating agent.[30] This bench-stable reagent is prepared in two steps from commercially available (1S,2S)-diaminocyclohexane and is highly chemoselective for N-acylation over O-acylation. For example, **1** is stable in ethanol and acetylates 5-amino-1-pentanol exclusively on nitrogen. Moreover, a range of benzylic amines can be resolved with good levels of enantioselectivity using 0.5 equivalents of **1**. The stereoselectivity of this resolution process is solvent dependent such that the resolution of 1-phenylethylamine (**2**) in toluene affords (R)-N-(1-phenylethyl)acetamide, whereas reaction in DMPU preferentially forms the (S)-enantiomer (Scheme 161).[30] The ability to access either enantiomer of product selectively through a simple solvent switch is another practical advantage of this method.

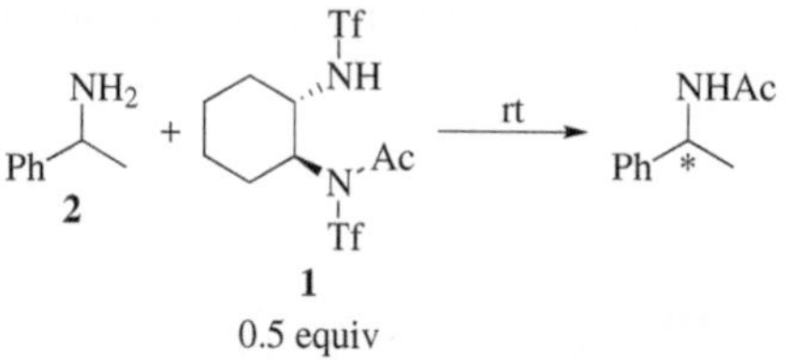

Solvent	Time (h)	Configuration	er	*s*
toluene	2	(*R*)	79.0:21.0	7
DMPU	3	(*S*)	92.0:8.0	30

Scheme 161

The stereoselectivity observed for the resolution of benzylic amines using **1** as the *N*-acylating agent can be further improved through the addition of the quaternary ammonium salt *n*-Oct_3NMeCl (1 M solution in THF). The overall concentration of the salt significantly impacts enantioselectivity, in which lower concentrations lead to lower selectivity. Under optimized conditions, combining benzylic amine **2** with **1** (0.5 equiv) in the presence of *n*-Oct_3NMeCl (1 M solution in THF) affords the corresponding acetamide in 97.0:3.0 er at 50% conversion (Scheme 162).[254] This selectivity factor is one of the highest reported for this class of substrate and demonstrates the potential synthetic value of this method. Stoichiometric acylating agent **1** has also been successfully used in ionic liquids,[29] and it has also been supported on Merrifield resin to form a recyclable *N*-acylating agent for the resolution of benzylamines.[255]

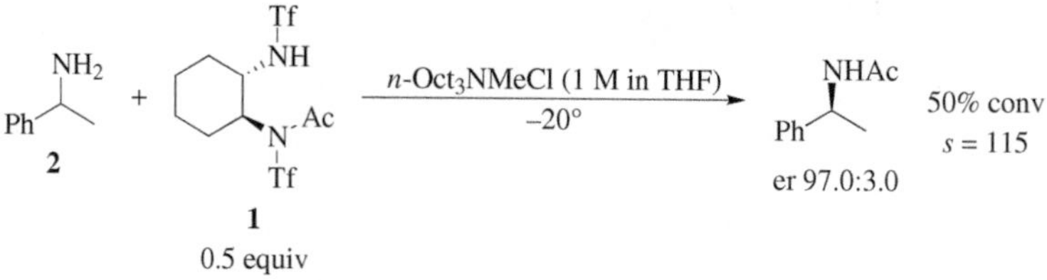

Scheme 162

Enantiomerically pure *N*-benzoyl benzimidazole derivative **200** (0.5 equiv) is effective for the KR of **2** in THF at −10°, which affords the isolated benzamide product in 95.0:5.0 er, and the unreacted amine in 94.5:5.5 er at 50% conversion (Scheme 163).[256] Moreover, **200** can be used to resolve aliphatic amines with synthetically useful levels of enantioselectivity (Scheme 164).[256] This feature is particularly important given that the resolution of racemic amines that contain two similar aliphatic substituents is highly challenging, and few methods can accomplish this type of KR with high levels of enantioselectivity.

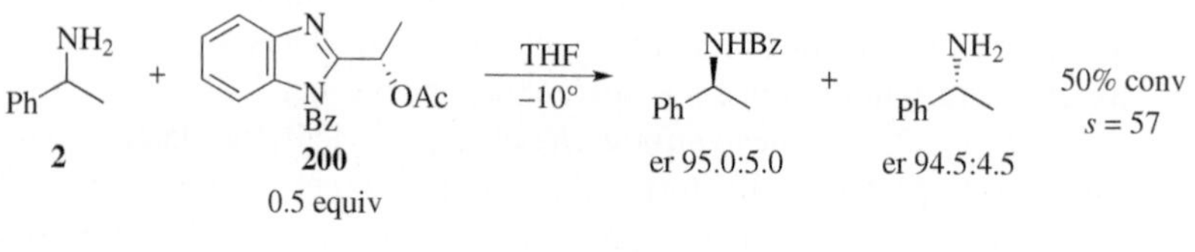

Scheme 163

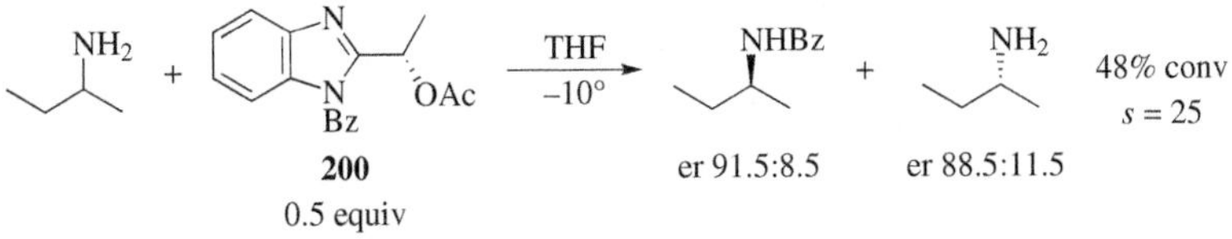

Scheme 164

Catalytic Methods. The first effective KR of primary benzylic amines used planar-chiral PPY derivative **201** as the catalyst. The critical breakthrough came from using the *O*-acyl azlactone **5** as the acylating agent because it preferentially reacts with the catalyst **201** instead of the racemic amine (Scheme 165).[33] Further improvements in selectivity were achieved by performing the reaction at −50° and adding *O*-acyl azlactone **5** in two batches. For example, 1-phenylethylamine reacts with *O*-acyl azlactone **5** (0.6 equiv) and 10 mol % of planar-chiral PPY **201** to afford the corresponding carbamate with reasonable selectivity. A range of aryl substituents with a variety of substitution patterns are accommodated in this process, including both electron-withdrawing and electron-donating substituents, to give the corresponding carbamates with useful levels of enantioselectivity.

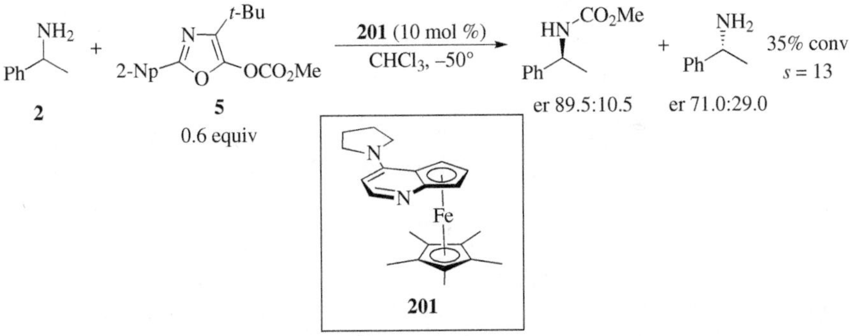

Scheme 165

A range of secondary benzylamine derivatives can be effectively resolved using a dual catalytic approach. Good levels of stereoselectivity are obtained from the reaction of benzoic anhydride with various benzylamine derivatives that have different aryl substituents, in the presence of the enantiomerically pure thiourea catalyst **56** and DMAP (20 mol %). For example, 1,2-diphenylethan-1-amine (**202**), which is a particularly challenging substrate because of the similarity between the two *C*-substituents on the stereogenic center, undergoes KR with *s* = 15 at 47% conversion (Scheme 166).[31] This dual-catalytic process has been further improved through the modification of the achiral DMAP derivative.[106] For instance, 4-di-*n*-propylaminopyridine (**204**) (5 mol %) in combination with the chiral thiourea-urea catalyst **195** improves the selectivity for the resolution of a range of secondary benzylamine derivatives.

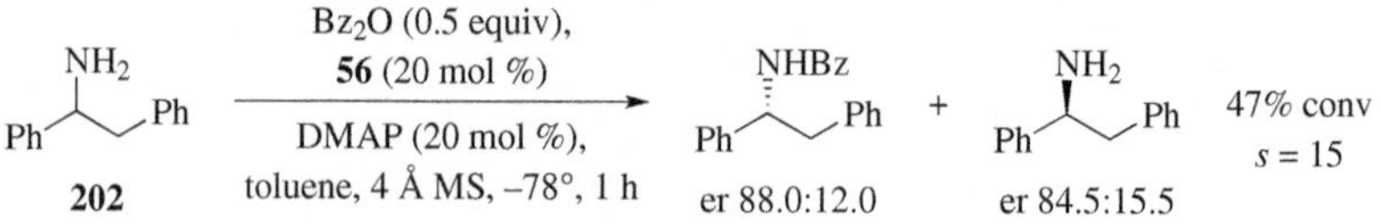

Scheme 166

This dual catalytic method has been further extended to the KR of C_2-symmetric 1,2-diamines. Levels of selectivity comparable with the α-alkylbenzylamine derivatives described above are observed. For example, (1R,2R)-1,2-diphenylethane-1,2-diamine (**203**) is effectively resolved using the dual catalytic system of **195** (10 mol %) and **204** (10 mol %) to afford the products with good enantiomeric excess ratios at 47% conversion (Scheme 167).[257] In this case, the addition of excess TrocCl upon completion of the KR is used to facilitate purification and HPLC analysis of the resolved diamines.

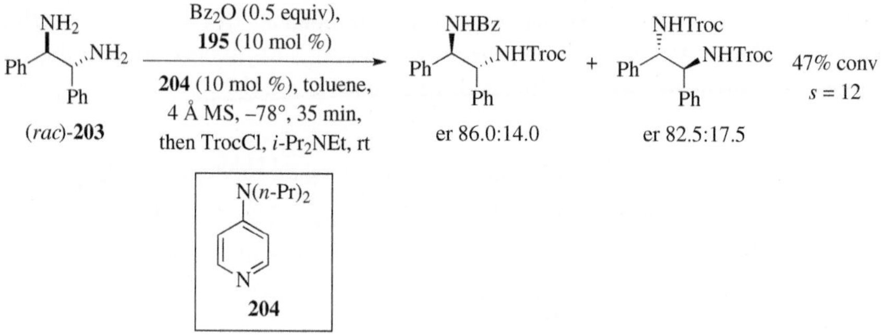

Scheme 167

A conceptually different approach to the KR of α-methylbenzylamine uses a bifunctional planar-chiral boronic acid catalyst **205** and benzoic acid as the N-acylating agent. The reaction gives only modest enantioselectivity, forming the benzamide product in 70.5:29.5 er at 21% conversion ($s = 3$) after heating in fluorobenzene for 48 h (Scheme 168).[258] All attempts to improve the reactivity and

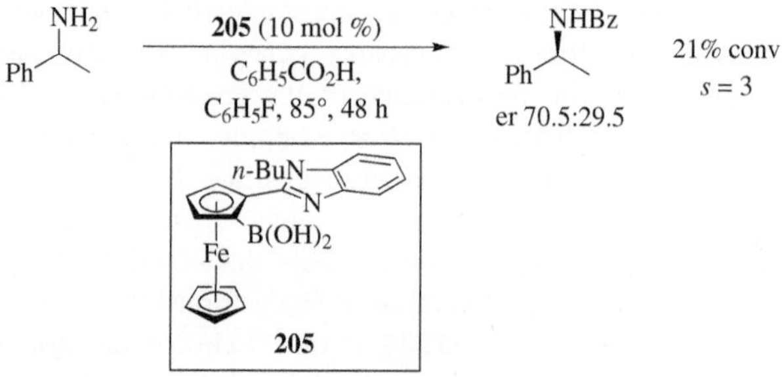

Scheme 168

stereoselectivity through variations in the substrate and reaction conditions were unsuccessful. Although this current procedure is not synthetically useful, the concept of using carboxylic acids as N-acylating agents in KRs to furnish water as the only byproduct, is an attractive area in need of further development.

Propargylic Amines.

Stoichiometric Methods. Several substituted propargylic amines undergo KR with excellent levels of enantioselectivity using N-acetyl-1,2-bis(trifluoromethane-sulfonamide) (**1**) as the stoichiometric acylating agent in the presence of the quaternary ammonium salt, n-Oct$_3$NMeCl (Scheme 169).[259] A range of alkyl substituents are accommodated, which afford the (S)-acetamides with excellent levels of enantioselectivity at 50% conversion. Hence this approach compares favorably with those reported for similar substrates under catalytic KR conditions. However, compound **1** cannot generally differentiate efficiently between enantiomeric amines bearing both alkynyl and aromatic substituents on the stereogenic carbon center, which results in virtually no selectivity.

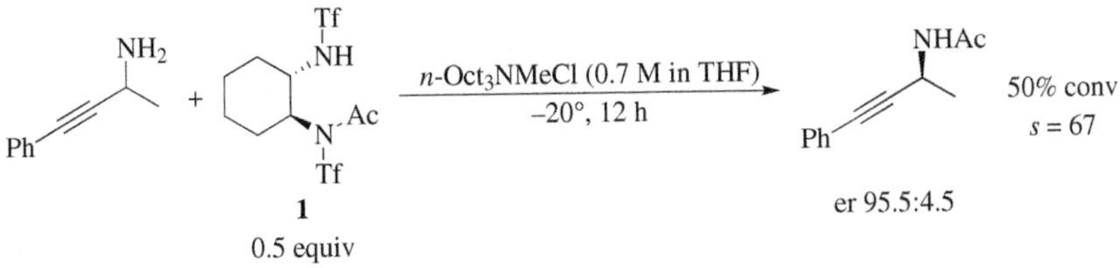

Scheme 169

Catalytic Methods. To date, there is only a single report of a catalytic KR of propargylic amines.[260] The process employs benzoic anhydride as the N-acylating agent in the presence of DMAP (5 mol %) and the chiral thiourea **195** (5 mol %) in toluene at $-78°$, which permits a range of propargylic amines to be resolved with synthetically useful levels of selectivity. Good selectivities are observed for substrates bearing simple alkyl and various alkynyl substituents on the stereogenic carbon atom. Impressively, the catalytic system can differentiate between enantiomeric amines containing both an alkynyl and a phenyl substituent on the stereogenic center. For example, 1,3-diphenylprop-2-yn-1-amine (**206**) affords the corresponding benzoyl amide in 89.5:10.5 er at 34% conversion using this protocol (Scheme 170).[260]

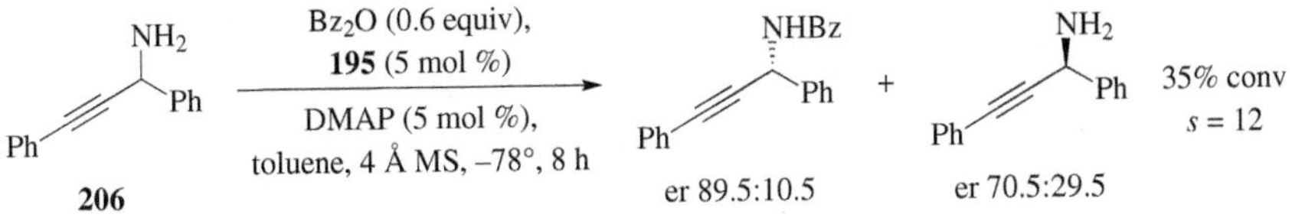

Scheme 170

Allylic Amines.

Stoichiometric Methods. N-Acetyl-1,2-bis(trifluoromethanesulfonamide) (**1**) is also the only stoichiometric acylating agent reported for the KR of allylic amines (Scheme 171).[261] The stereoselectivity is optimal using a polymer-supported, quaternary ammonium salt as an additive at –20° in THF. The reaction selectivity is highly dependent upon the substitution of the allylic amine such that the highest *s* values are obtained with 1,1-disubstituted allylic amines where R^1 is an aryl substituent. Notably, the process is significantly less selective for the corresponding 1,2-disubstituted and 1,1,2-trisubstituted allylic amines, in which the corresponding amides are isolated with low enantioselectivities. Interestingly, the selectivity for the KR of allylic amines with **1** is generally lower than the analogous resolutions of benzylic and propargylic amines with the same reagent.

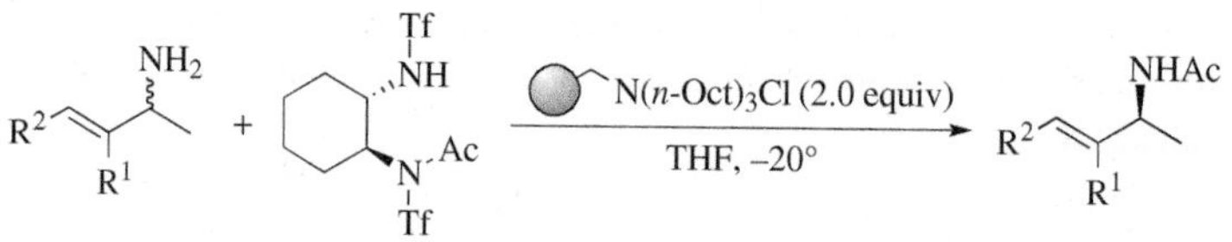

R^1	R^2	Conv (%)	er	s
Ph	H	50	91.5:8.5	28
H	Ph	50	71.0:29.0	4
Me	Ph	50	81.0:19.0	8

Scheme 171

Catalytic Methods. A dual-catalytic system employing thiourea **195** and PPY as a co-catalyst facilitates the KR of allylic amines using benzoic anhydride as the acyl source (Scheme 172).[262] In general, the selectivity of the process is lower than that observed for propargylic amines. As allylic amines are more nucleophilic than the corresponding propargylic amines, the former are prone to a more prominent background reaction. This problem is minimized by shortening the reaction times, to afford the optimal reaction selectivity. A range of aryl-substituted allylic amines are viable substrates that furnish the corresponding benzoyl protected amines with reasonable selectivity factors. The process also works for 1,1,2-trisubstituted allylic amines, although lower selectivities are observed with the corresponding 1,2-substituted allylic amines. As with the resolution of propargylic amines, the catalytic system can distinguish between alkenyl and aromatic π-system substituents on the stereogenic carbon, albeit with reduced selectivity.

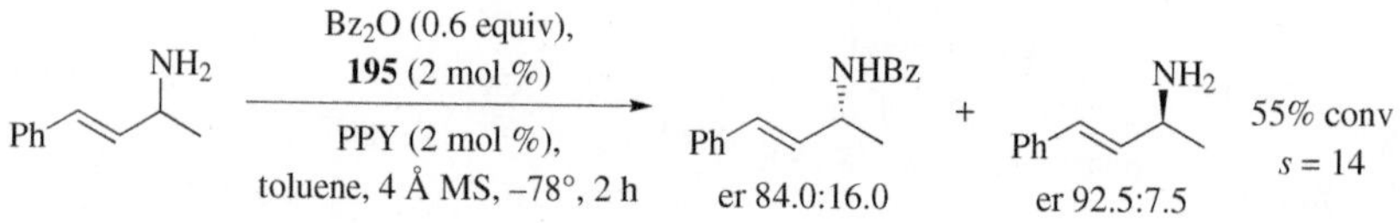

Scheme 172

Cyclic Secondary Amines.

Stoichiometric Methods. Various enantiomerically enriched acid chlorides containing α-stereocenters are employed for the diastereoselective resolution of a limited number of cyclic, secondary amines with modest levels of stereoselectivity.[263–267] For example, N-phthaloyl amino acid chlorides have been investigated for the resolution of the secondary amine **207** to form the corresponding N-acylated products with moderate levels of diastereoselectivity (Scheme 173).[266] This method has been demonstrated for only a limited range of secondary amines related to **207**, which affords the products with lower levels of stereoselectivity than alternative methods. The potential difficulties in handling amino acid chlorides in conjunction with their propensity to undergo epimerization at the α-stereocenter suggest that this approach is unlikely to be widely applicable and of practical value. Nevertheless, this approach has been extended to the KR of 3-*tert*-butyl-3,4-dihydro-2*H*-[1,4]-benzoxazine through acylation with N-phthaloyl-(*S*)-phenylalanyl or (*R*)-2-phenoxypropionyl chlorides, followed by acidic hydrolysis of the corresponding diastereomerically pure amides.[268]

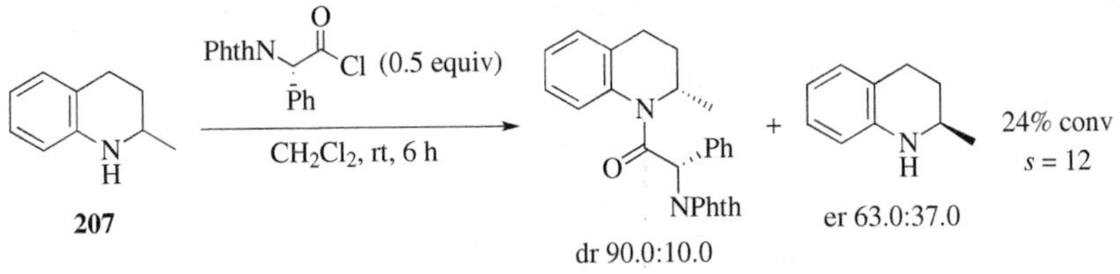

Scheme 173

N,N-Diacylaminoquinazolinone derivative **208** resolves 2-methylpiperidine (**45**) to furnish the corresponding benzoyl amide product and recovered amine with excellent enantioselectivity (>95.5:4.5 er) in good yield (Scheme 174).[269] Interestingly, despite these encouraging results, the scope of the reaction has not been further explored.

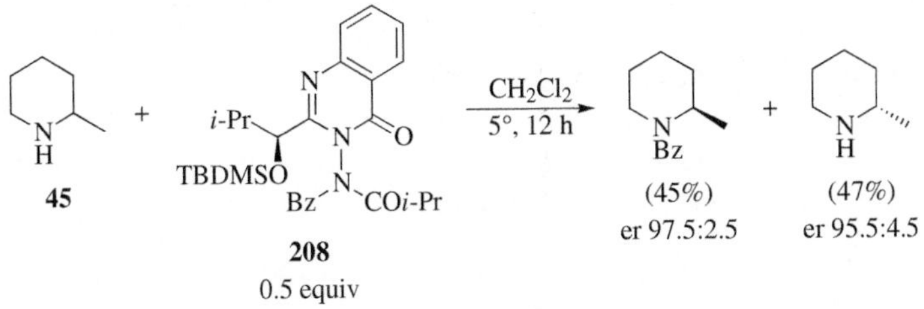

Scheme 174

Enantiomerically pure polystyrene-supported, *O*-acyl hydroxamic acid derivatives have recently been reported for the efficient KR of an impressive range of cyclic secondary amines.[270] The protocol is operationally simple, albeit with relatively

long reaction times (up to 48 h) required at room temperature. A wide range of
racemic 2-substituted amines, including piperidines, piperazines, morpholines,
tetrahydroisoquinolines, and diazepanones, are effectively resolved with high levels
of selectivity. The polystyrene-supported hydroxamic acid is readily recovered
by filtration and can be recycled without loss in efficiency through O-acylation
using the required acid anhydride. Importantly, unlike the related catalytic procedure, this method permits the enantioselective transfer of N-acyl groups that
can be cleaved under mild conditions. For example, 209 can be resolved using
polystyrene-supported N-pent-4-enoyl hydroxamic acid (210) with a selectivity
factor of 20 at 49% conversion (Scheme 175).[270] The resulting enantiomerically
enriched amide is readily deprotected using iodine to afford (S)-209 without any
detectable racemization. A related, ROMP-gel supported, chiral hydroxamic acid
derivative also facilitates the efficient KR of a range of complex, substituted, cyclic,
secondary amines on a multi-gram scale with comparable levels of selectivity,
thereby making this an attractive approach to this problem.[271]

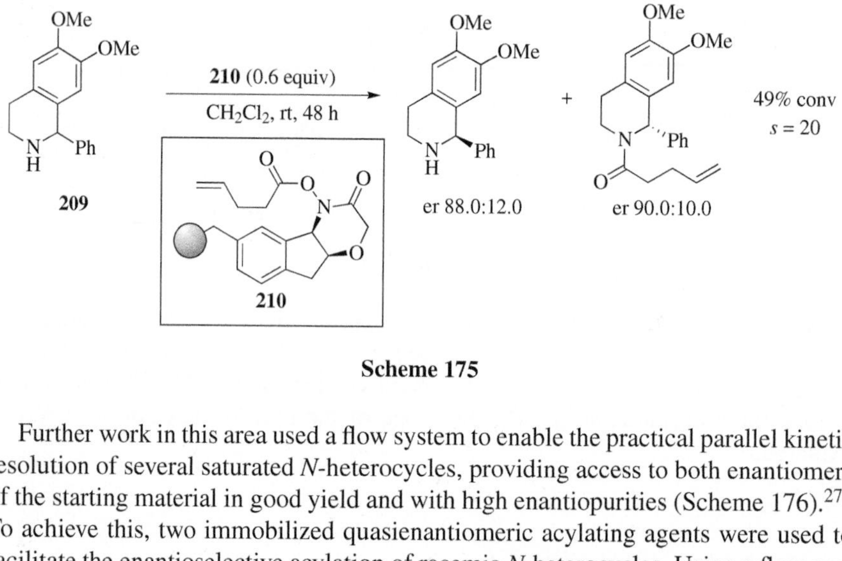

Scheme 175

Further work in this area used a flow system to enable the practical parallel kinetic
resolution of several saturated N-heterocycles, providing access to both enantiomers
of the starting material in good yield and with high enantiopurities (Scheme 176).[272]
To achieve this, two immobilized quasienantiomeric acylating agents were used to
facilitate the enantioselective acylation of racemic N-heterocycles. Using a flow system permits the effective separation, recovery, and reuse of the polymer-supported
reagents. The amide products can then be readily separated and hydrolyzed to the
corresponding amines without epimerization.

   Catalytic Methods.   Planar-chiral PPY derivative 211 (5 mol %) is an effective
catalyst for the KR of 2-substituted indolines using O-acyl azlactone 212 as the
acylating agent (Scheme 177).[273] The additives LiBr (1.5 equiv) and 18-crown-6
(0.75 equiv) are essential for obtaining high selectivity factors, albeit their precise
role is unknown at present. A range of functionalized 2-substituted indolines, which
includes 2,3-dialkyl-substituted derivatives, are viable, and the resulting N-acyl
indolines are formed in high selectivities. Notably, the nature of the 2-substituent is

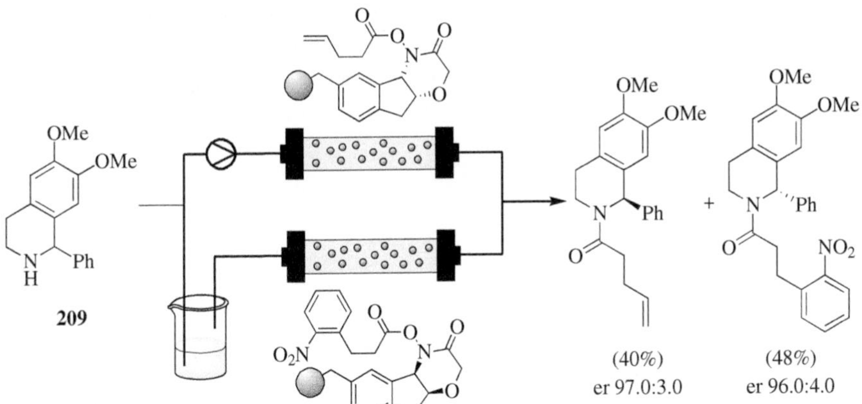

Scheme 176

important, in which sp^2-hybridized groups lead to lower selectivities. The highest selectivities are obtained when reactions are performed between −10° and 0° with reaction times of up to five days. The reaction time can be reduced (two days) if the reaction is performed at room temperature, but with lower stereoselectivity. The specialist nature of PPY derivative **211** and the extended times required are likely to make this protocol suitable for only small-scale applications.

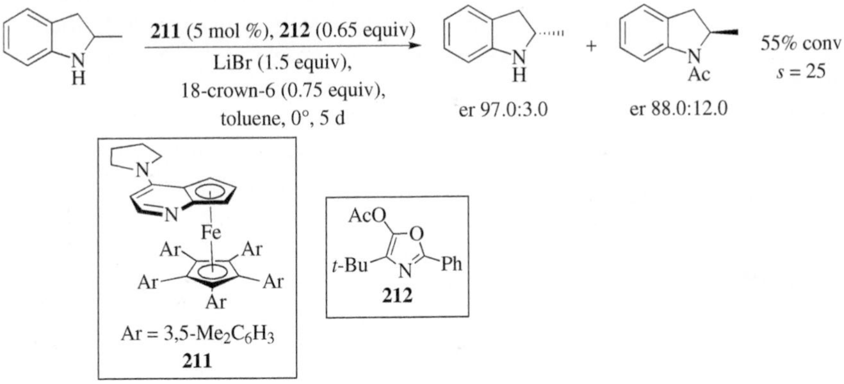

Scheme 177

An alternative method for the KR of cyclic secondary amines using α′-hydroxyenone **6** as a masked acylating agent represents the current state of the art in this area.[34,85] The optimized protocol uses the achiral NHC **214** (10 mol %) in conjunction with enantiomerically pure hydroxylamine **213** (10 mol %) and K$_2$CO$_3$ as the base in isopropyl acetate at room temperature (Scheme 178).[85] The KR is effective for an extensive range of 2-substituted cyclic secondary amines, including piperidines, piperazines, morpholines, tetrahydroisoquinolines, and azepanes, forming the enantiomerically enriched products with synthetically useful selectivity factors. Moreover, the process is compatible with pendant functional groups, namely, esters, ethers, lactams, and carbamates. The mild conditions and the impressive reaction scope make this resolution protocol particularly attractive for this substrate class.

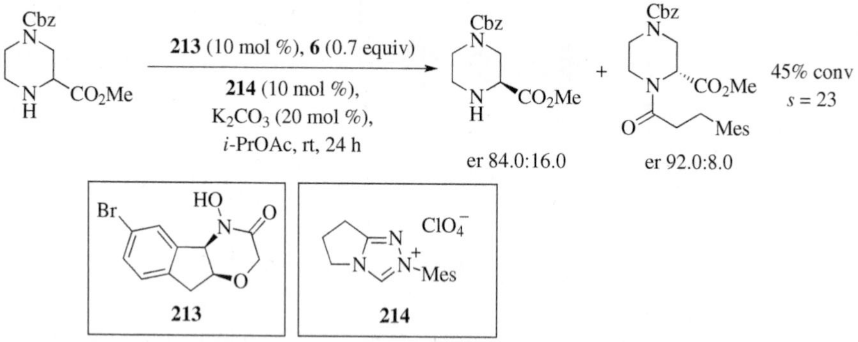

Scheme 178

Amino Esters. A few stoichiometric N-acylating reagents have been employed in the KR of α-amino esters;[30,249,252] however, the scope is typically limited to proteinogenic amino acid derivatives. Hence, given the commercial availability of both enantiomers of these substrates, the need for efficient resolution procedures is limited. Nevertheless, the N-benzoyl benzimidazole derivative **200** efficiently resolves α-amino esters with good levels of stereoselectivity (Scheme 179).[274] The process has been applied to amino esters of alanine, valine, and leucine containing alkyl side chains as well as esters of phenylalanine and phenylglycine, which afford the corresponding N-benzoyl derivatives and unreacted amines in high enantioselectivities with synthetically useful s values (typically >30) at 50% conversion.

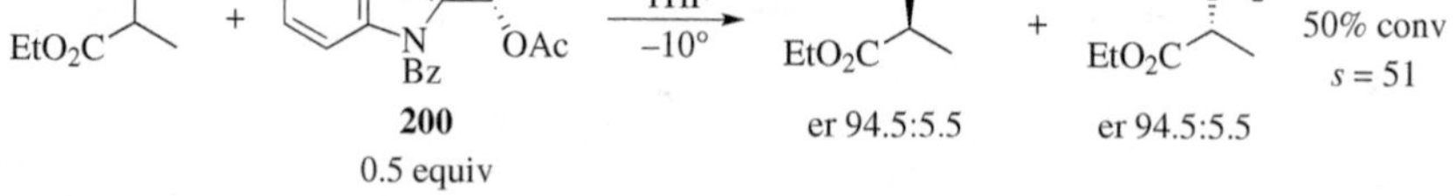

Scheme 179

Atropisomeric Amines. A single example of the catalytic KR of an atropiso-meric aniline **215** has been reported using N-methyl-5-azaindoline **216** (1 mol %) and excess isobutyric anhydride in CH_2Cl_2 at $-50°$, which affords only modest selec-tivity at 31% conversion (Scheme 180).[275] Interestingly, the attempted resolution of the 1-phenylnaphthyl congener proceeds without enantioselectivity, emphasizing the highly specific nature of this process.

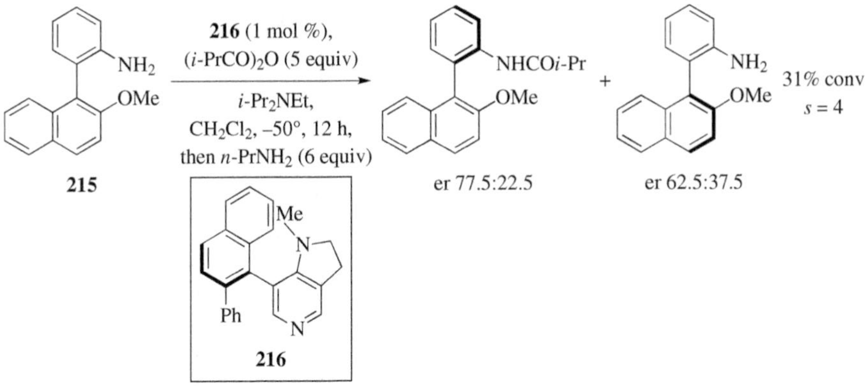

Scheme 180

Oxazolidinones, Lactams, and Formamides. One strategy to minimize the challenges associated with the competitive background reaction in amine KRs is to reduce the nucleophilicity of an amine by conjugation with a carbonyl or thiocarbonyl unit. For example, isothioureas have been employed for the KR of oxazolidinones, with isothiourea **23** being optimal for this process (s up to 520) using isobutyric anhydride as the acylating reagent.[52,276] Importantly, a C(4)-aryl stereodirecting unit within the oxazolidinone is necessary for high selectivity, and 5,5-*gem*-dimethyl substitution also promotes high enantioselectivity (Scheme 181).[276] The utility of this protocol is also evident from the ability to conduct it on a preparative scale.

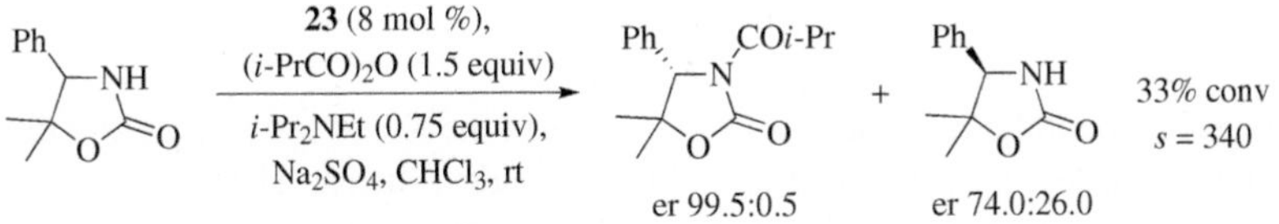

Scheme 181

The related KR of γ-lactams and γ-thiolactams has also been reported using isothiourea **23**. Interestingly, although pyrrolidinone and imidazolidinone substrates are unreactive under standard reaction conditions, substrates that contain an additional heteroatom in the ring (and thus increased N−H acidity) such as oxazolidin-4-one **217**, thiazolidin-4-one, and imidazolidin-4-one undergo rapid and selective acylation in the presence of **23** (4 mol %) with excellent selectivity factors (Scheme 182).[52] Structurally related γ-thiolactams are also readily resolved with high selectivity under the same reaction conditions.

Building on this work, the enantioselective synthesis of *erythro*-sphingosine was achieved through the acylative KR of functionalized oxazolidinone **218** using isothiourea **23** (4 mol %) with excellent selectivity ($s = 117$) (Scheme 183).[277]

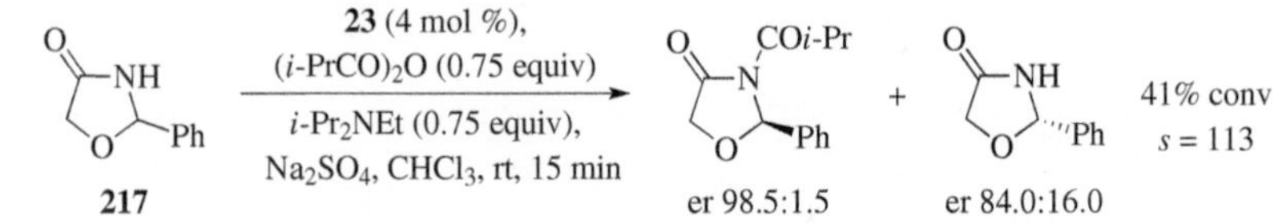

Scheme 182

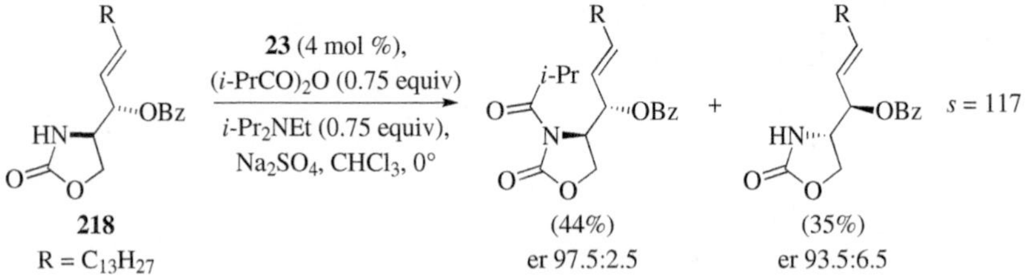

Scheme 183

The KR of a series of β-lactams bearing a C(4)-aryl or heteroaryl subunit has been accomplished through *N*-acylation using the amidine catalyst **16** in conjunction with isobutyric anhydride (Scheme 184).[278] A range of aryl substituents is accommodated such that electron-rich and naphthyl substituents afford the highest selectivities.

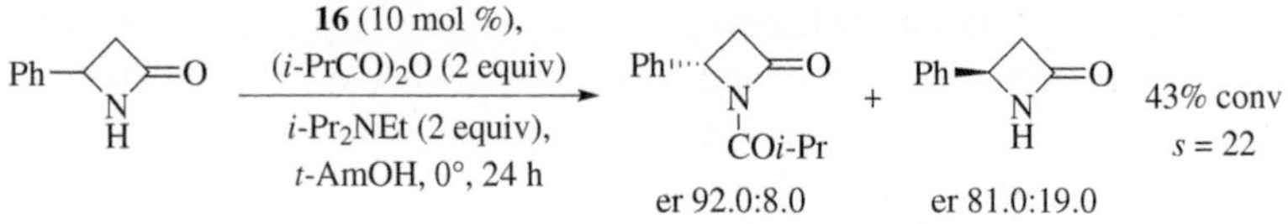

Scheme 184

The use of peptidic catalysts has been applied to the KR of formamide and thioformamide substrates using histidine-containing peptides, such as **219** (Scheme 185).[279] The method applies to a wide range of thioformyl benzylic amines bearing various

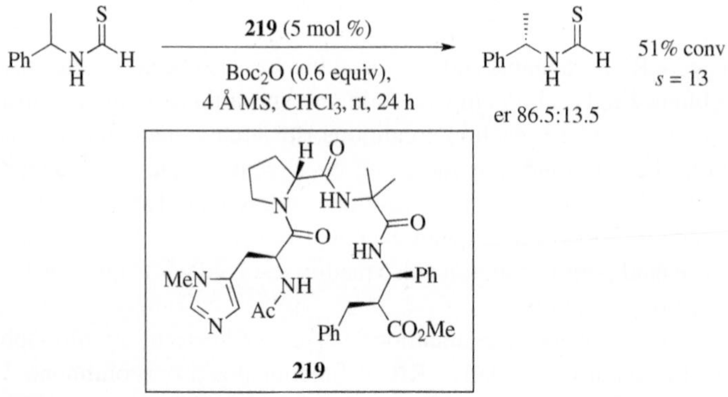

Scheme 185

aryl substituents, in which high selectivities are obtained using Boc_2O as the N-acylating reagent.

Conclusions and Outlook. Over the last fifteen years, significant progress has been accomplished on the KR of amines through N-acylation; however, in comparison with the KRs of alcohols, there is a paucity of efficient, catalytic methods for the resolution of amines. In many cases, it remains practical to use an enantiomerically pure, stoichiometric N-acylating reagent for optimal reaction selectivity. The reported scope of both the catalytic and stoichiometric methods for the resolution of amines is generally limited to a few substrate classes. In particular, there is a lack of methods that have been tested on molecules containing pendant functional groups that may be of further use in natural product synthesis or for industrial applications. The further development of small-molecule catalysts that can effectively resolve amines using simple and widely available N-acylating agents under mild reaction conditions is significant and a worthwhile challenge.

APPLICATIONS TO SYNTHESIS

Synthesis of Pharmaceuticals

Azithromycin. Azithromycin (**228**) is a potent, semisynthetic, macrolide antibiotic derived from erythromycin A. The first asymmetric total synthesis of azithromycin (**228**) employed a desymmetrization reaction of a prochiral 1,3-diol as the key step in the synthesis of both fragments. The prochiral diol **220** undergoes efficient desymmetrization using the chiral complex derived from $CuCl_2$ (20 mol %) and imine **107** in the presence of benzoyl chloride as the O-acylating agent. Monobenzoyl ester **221** is obtained in an excellent 95.5:4.5 er in almost quantitative yield in 30 minutes at room temperature (Scheme 186).[280] Regioselective mesylation of the primary alcohol followed by DBU-promoted cyclization affords an epoxide. Hydrolysis of the benzoyl ester provides the primary alcohol **222** in good yield over the two steps—further elaboration of **222** permits access to key synthetic intermediate **223**.

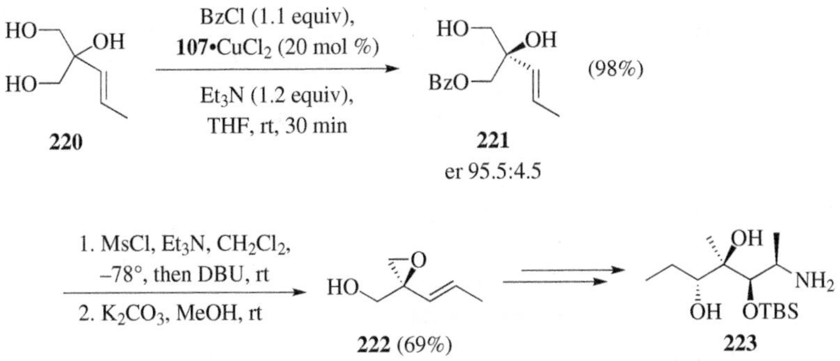

Scheme 186

A similar desymmetrization reaction is employed for the synthesis of fragment **226**. In this case, *ent*-**107**•CuCl$_2$ catalyzes the stereoselective monobenzoylation of prochiral 1,3-diol **224** to afford ester **225** in 94% yield and 96.0:4.0 er (Scheme 187).[280] A ten-step linear sequence is then used to convert ester **225** into **226**. Chemoselective oxidation of the residual primary alcohol in **226** followed by a subsequent reductive amination with amine **223** under hydrogenation conditions and methylation of the nitrogen atom provides **227** in 70% yield over the three steps (Scheme 188).[280] Macrolactonization of **227** followed by glycosylation and deprotection completes the synthesis of azithromycin **228**.

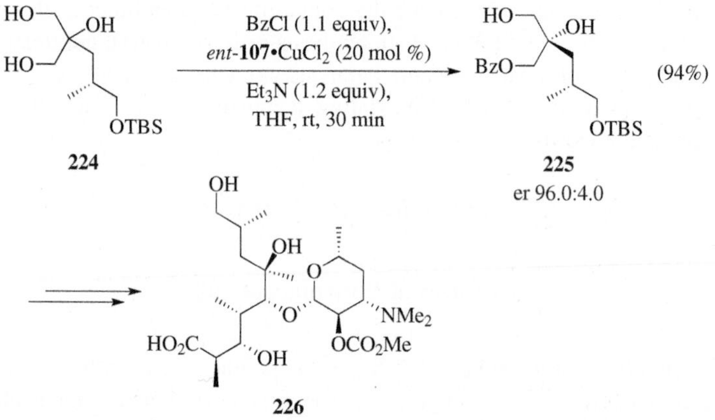

Scheme 187

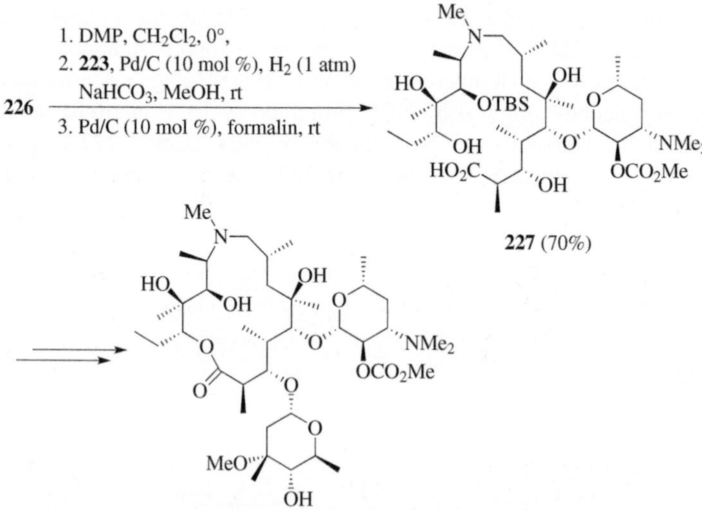

Scheme 188

(R)-(–)-Baclofen. Lewis base catalyzed KR of secondary allylic alcohols has been used to prepare intermediates for the synthesis of pharmaceutical targets. For example, the KR of allylic alcohol **229** is performed on a gram-scale using planar-chiral DMAP derivative **10** (1 mol %) and acetic anhydride as the acylating agent to provide the enantiomerically pure alcohol (S)-**229** in 40% yield (Scheme 189).[202] Allylic alcohol **229** has previously been used as an intermediate in the synthesis of the muscle relaxant (R)-(–)-baclofen **230**.[281]

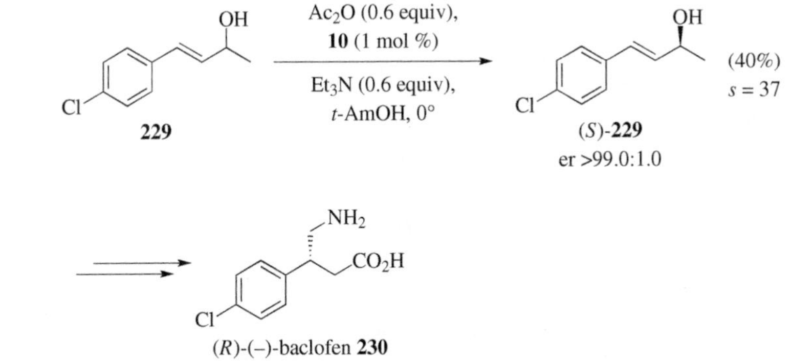

Scheme 189

Epothilone A. The same resolution method has also been employed to construct the enantiomerically pure allylic alcohol **231**,[202] which serves as a key intermediate in the total synthesis of the potent anticancer agent, epothilone A **232**.[282] Allylic alcohol **231** undergoes efficient KR using *ent-***10** (1 mol %) and acetic anhydride to provide alcohol (4R,5R)-**231** in 99.0:1.0 er with s = 107 (Scheme 190).[202]

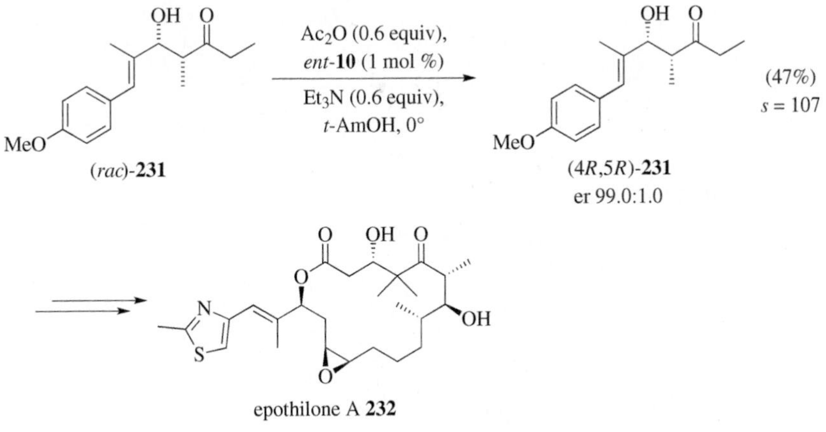

Scheme 190

(R,S)-Mefloquine. Quinine derivative mefloquine (**233**) is an effective malaria treatment currently sold as a racemate, albeit one of the enantiomers is suspected of causing undesirable neurological side effects. Polymer-supported hydroxamic acid derivative **234** has been used for the stoichiometric KR of **233** on a multigram scale to access enantiomerically pure (*R,S*)-**233** (Scheme 191).[271] The supported reagent can be readily recovered and regenerated, and thereby permit multiple cycles of KR, each on approximately 20 g of **233**, without significant loss in efficiency and selectivity. Overall, a single batch of the parent-supported hydroxamic acid is reported to resolve 150 g of **233** over seven cycles, allowing approximately 50 g of enantiomerically pure **233** to be isolated.

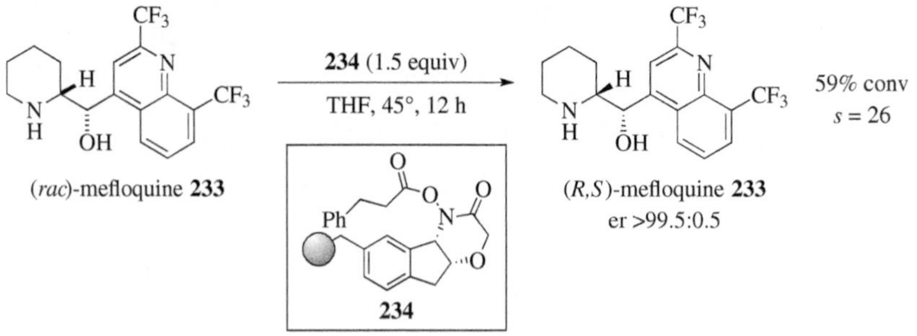

Scheme 191

Synthesis of Natural Products

(–)-Lobeline. The alkaloid (–)-lobeline (**238**) is the principal active component of Indian tobacco (*Lobelia inflata*) and has been prescribed as an antiasthmatic, an expectorant, a respiratory stimulant, and a smoking-cessation aid. Lewis base catalyzed desymmetrization has been used as a key step in the asymmetric synthesis of (–)-lobeline (Scheme 192).[283] Lobelanine (**235**) is prepared in one step using

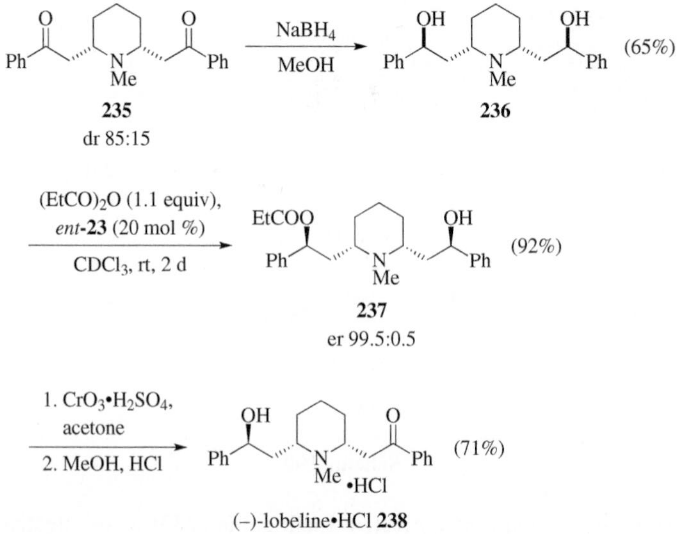

Scheme 192

a reported procedure with modest diastereocontrol (85.0:15.0 dr). Diastereo-selective reduction of **235** using NaBH$_4$ in MeOH affords lobelanidine **236** as a single diastereoisomer in 65% yield after purification by column chromatography. Desymmetrization of *meso*-lobelanidine (**236**) is achieved using isothiourea catalyst *ent*-**23** (20 mol %) and propionic anhydride to generate ester **237** in 92% yield with excellent enantioselectivity (>99.5:0.5 er). Oxidation of the remaining alcohol followed by hydrolysis of the ester provides (−)-lobeline **238** in 71% yield. The unnatural enantiomer (+)-lobeline is also prepared in an analogous manner using **23** as the desymmetrization catalyst.

Iso- and Bongkrekic Acids. Isobongkrekic acid (**242**) and its isomer bongkrekic acid (**243**) are complex fatty triacids produced by *Pseudomonas cocovenenans*. These compounds are toxic antibiotics that have potent antiapoptotic activity and have been widely employed to study apoptosis mechanisms.[284] Kinetic resolution has been used to improve the enantiomeric excess of a key fragment in a highly convergent total synthesis of both natural products (Scheme 193).[284,285] The synthesis of fragment **241** involves an indium-mediated addition of propargyl bromide to aldehyde **239** using (1*R*,2*S*)-2-amino-1,2-diphenylethan-1-ol (**240**) as a chiral auxiliary to form homopropargylic alcohol **241** in 93% yield and with modest

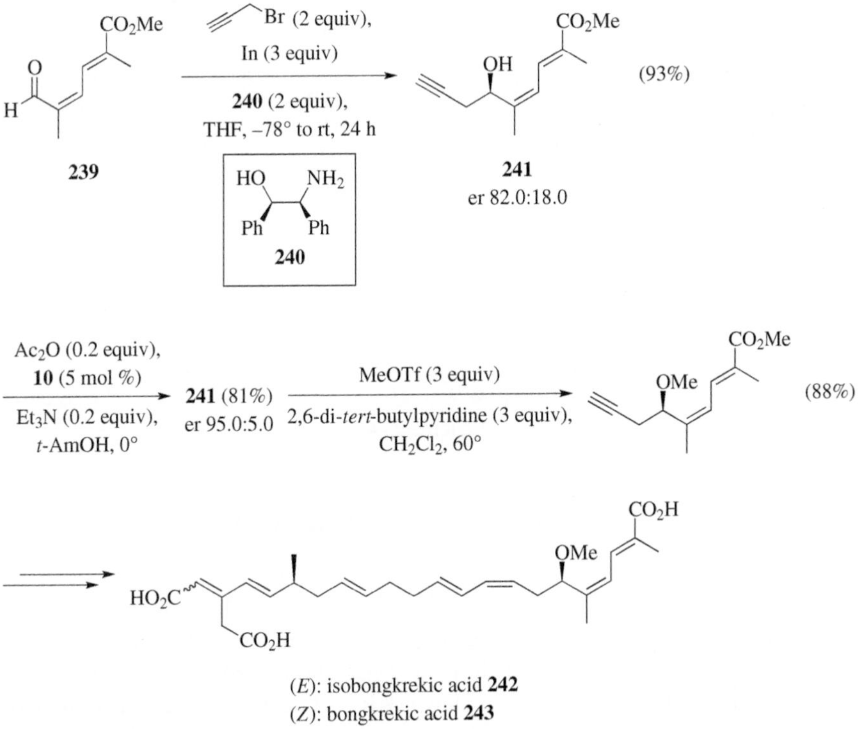

Scheme 193

enantioselectivity (82.0:18.0 er). Subsequently, the secondary alcohol **241** was subjected to a KR using planar-chiral DMAP derivative **10** (5 mol %) as the catalyst to afford improved selectivity 95.0:5.0 er by the selective acylation of the undesired enantiomer.

Mitosane Core. The mitomycin family of natural products are important antitumor agents, which have attracted significant interest from the context of chemical synthesis and biological studies. The mitosane core **245** has been prepared using the KR of the cyclic allylic alcohol **244** (Scheme 194).[286,287] A series of oligopeptide catalysts were prepared and surveyed in the KR of **244**, which identified **27** as the optimum catalyst.[286] Treatment of **244** with acetic anhydride in the presence of the peptide catalyst **27** (2 mol %) permits the allylic alcohol (*S*)-**244** to be recovered in 40% yield and 95.0:5.0 er (*s* = 27 at 53% conversion). The enantiomeric enrichment could be further enhanced to >99.5:0.5 er after a single recrystallization. In addition, the undesired enantiomeric ester from the KR was readily recycled by hydrolysis into (*R*)-**244**, followed by its oxidation and reduction to afford **244**. The enantiomerically enriched alcohol (*S*)-**244** is readily converted into the mitosane core **245**, which permits further elaboration into a series of synthetically useful derivatives.[287]

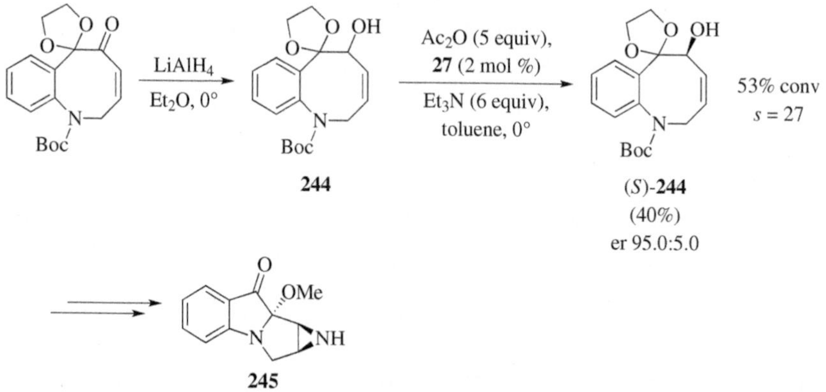

Scheme 194

COMPARISON WITH OTHER METHODS

Overview

Resolution strategies have played an important role in the preparation of enantiomerically enriched compounds. This review has showcased the strategies within this area that are available for the acylative KR and desymmetrization of alcohols and amines using synthetic small-molecule catalysts. In many cases, the challenges associated with the preparation of the products in enantiomerically pure form through an acylation strategy have stimulated the generation of new catalyst structures and allowed the elucidation of new principles and strategies in enantioselective catalysis that are arguably more valuable than the products of these single transformations. In this light, the comparison of the methods available for the preparation of a given compound in enantiomerically pure form should, therefore, be treated with caution and on an individual basis. It is therefore advised that a range of issues should be considered when considering the so-called "best" or "most practical" route to a given enantiomerically enriched product of interest. Direct comparisons of alternative, catalytic, enantioselective methods available to make any given enantiomerically enriched compound, let alone further comparison with stoichiometric methods, are rare and further complicated by the plethora of routes available to these compounds of interest.

As such, a variety of factors will often dictate the best synthetic route to an enantiomerically enriched compound of interest. The following simple guidelines should be considered when evaluating a KR or desymmetrization approach if an acylative strategy using a small-molecule catalyst is to be competitive with alternative routes.

1. The compound of interest should be easy to prepare in racemic or prochiral form from readily available and inexpensive reagents.
2. The catalyst is either readily prepared or cheap and commercially available in both enantiomeric forms. The catalyst should be able to be used on the desired scale at low catalyst loadings and ideally conveniently recycled.
3. The acylating reagent should be simple and commercially available on a large scale at low cost.
4. For a KR process, the resolved starting material and product must be easily separated, and in the ideal case should both be isolated in high enantiomeric enrichment.

Comparison with Acylative and Hydrolytic Enzymatic Kinetic Resolution

By any measure, enzymes are highly efficient and important catalysts for KR processes, with numerous acylative enzymes widely used in a synthetic context.[15-17] Lipases and esterases typically display broad substrate tolerance, with complementary enzymes often available that give opposite enantioselectivities. The acyl donor typically used with a lipase is vinyl acetate, while with an esterase, water is used to effect selective hydrolysis of a racemic ester substrate, both of which are inexpensive, and the reactions are easy to perform by nonspecialists.

One field in which the ability of enzymes currently outrivals that of small-molecule acylative catalysis is their participation in dynamic kinetic resolution (DKR) processes.[288-296] Within the last decade, a range of racemization catalysts based on ruthenium and iridium have been developed that are compatible with numerous classes of alcohols under mild reaction conditions. To date, only a single example showcasing the DKR of a benzylic alcohol in which this racemization process has been coupled with a nonenzymatic small-molecule acylative catalyst has been reported.[297] In contrast, a range of enzymes are compatible with these racemization catalysts, resulting in a broad scope of substrates capable of undergoing DKR, including secondary alcohols, chlorohydrins, homoallylic alcohols, allylic alcohols, diols, and heterocyclic amino alcohols. Furthermore, the enzymes associated with these resolutions are generally commercially available and have high thermostability. Although still relatively rare, DKR systems involving amines have also been achieved through coupling enzymatic catalysis with racemization of the slow-reacting enantiomer.[298–306] Despite the challenges associated with the racemization of these substrates and the requirement for high reaction temperatures to function efficiently, the enzymatic reaction still allows the effective DKR of substrates such as benzylic amines.

EXPERIMENTAL CONDITIONS

Racemic alcohols and amine substrates suitable for acylative KR or desymmetrization reactions are often readily prepared using standard synthetic techniques. Acylative KR and desymmetrization reactions can be performed under a wide range of conditions, with given methods often optimized for either the specific class of substrate, the catalyst, or the acylating agent of interest to maximize reaction selectivity and/or product enantioselectivity. Kinetic resolution reactions necessarily lead to a mixture of acylated product and unreacted starting material, which are most often separated using silica gel chromatography.

Acylating Agent

The choice of acylating agent is often highly important and can impact both the reactivity and selectivity of a process. For catalytic KR and desymmetrization reactions, acid chlorides and symmetric anhydrides are the most commonly employed acylating agents. Such reagents can sometimes be prone to hydrolysis, and therefore the use of anhydrous solvents can be required in some cases. A key consideration is the rate of noncatalyzed acylation of the substrate of interest, which should be minimized to achieve the highest possible selectivity. Although a KR theoretically requires the use of 0.5 equivalent of the acylating agent relative to the substrate, in practice more is required to achieve the required reaction conversion. As such, it is often important to quench any remaining acylating agent upon completion of the reaction to avoid unwanted acylation of the unreacted starting material during workup.

Catalyst

Many classes of catalyst can be employed, and the choice is often dependent on the degree of selectivity achievable for the substrate of interest. The complexity of catalysts available is highly variable; some require multistep syntheses to prepare, whereas others are commercially available. Some catalysts are sensitive to oxidation and therefore require an inert atmosphere.

Base

Many optimized procedures require the addition of base to sequester the acid released during the acylation step, although this is not always essential. In general, the base can have a significant effect on reactivity, but it often only has minimal effect on the reaction selectivity. Simple, commercially available organic bases such as triethylamine and diisopropylethylamine are the most commonly employed across the range of procedures available; the number of equivalents used often matches the amount of acylating agent present in the reaction.

Solvent

The reaction solvent is often important for both reactivity and selectivity and is often optimized for a given substrate/catalyst system. In many cases, reactions are performed at low temperature to maximize selectivity, which in turn places some restrictions on possible solvents.

EXPERIMENTAL PROCEDURES

(R)-3-Hydroxy-2-phenylpropyl Benzoate [Lewis Acid Catalyzed Desymmetrization of 2-Phenylpropane-1,3-diol].[91] Diethylzinc (1.1 M in toluene, 22.5 µL, 0.025 mmol) was added dropwise to a stirred solution of ligand **108** (11.8 mg, 0.0125 mmol) in anhydrous toluene (1 mL) under argon, and the resulting solution was stirred for 30 min. The catalyst solution was then added to a solution of 2-phenylpropane-1,3-diol (38.1 mg, 0.25 mmol) and vinyl benzoate (0.173 mL, 1.25 mmol) in anhydrous toluene (2.5 mL) at −15° under argon. After 24 h the reaction was quenched with aqueous KH_2PO_4 (5%) and diluted with Et_2O before the phases were separated, and the aqueous layer was extracted with Et_2O (2 × 20 mL). The combined organics were washed with H_2O, aqueous $NaHCO_3$, and brine before being dried (anhydrous $MgSO_4$) and concentrated under reduced pressure. The residue was purified by silica gel chromatography (petrol ether/EtOAc) to give the ester (60.1 mg, 94%, 95.5:4.5 er) as a colorless oil: R_f 0.34 (petrol ether/EtOAc, 70:30); HPLC t_R (R) 15.3 min (95.5%), t_R (S) 23.5 min (4.5%) [Chiralcel OJ, heptane/i-PrOH (90:10), 1.0 mL/min, 254 nm]; $[\alpha]_D^{26}$ − 8.1 (c 0.2, $CHCl_3$); FTIR (neat) 3420, 2360, 2341, 1716, 1273, 760 cm^{-1}; ^{1}H NMR (300 MHz, $CDCl_3$) δ 8.02 −7.98 (m, 2H), 7.59–7.53 (m, 1H), 7.46–7.27 (m, 7H), 4.64 (dq, J = 4.2, 6.8 Hz, 2H), 3.95 (dq, J = 1.4, 6.3 Hz, 2H), 3.31 (dt, J = 6.3, 12.9 Hz, 1H), 2.15 (s, 1H),

1.79 (s, 1H); ^{13}C NMR (75 MHz, CDCl$_3$) δ 166.8, 139.0, 133.1, 129.9, 129.6, 128.8, 128.4, 128.2, 127.4, 65.4, 63.8, 47.5.

(36%)   (58%)

er 97.0:3.0

(R)-2-(Benzyloxy)-3-hydroxypropyl Acetate [Peptide-Catalyzed Desymmetrization of 2-(Benzyloxy)propane-1,3-diol].[139] 2-(Benzyloxy)propane-1,3-diol (20.2 mg, 0.111 mmol) was dissolved in toluene (4.0 mL) before i-Pr$_2$NEt (38.6 μL, 0.222 mmol) was added. The peptide catalyst **96** (0.0111 mmol) was added from a stock solution in CH$_2$Cl$_2$ (0.228 mL, 48 mM), and the reaction was cooled to −55°. Freshly distilled Ac$_2$O (19.9 μL, 0.211 mmol) was added, and the reaction mixture was stirred at −55° for 36 h. The reaction was quenched through addition of MeOH, the mixture was warmed to rt, and concentrated under reduced pressure. The residue was purified by silica gel chromatography (hexane/EtOAc, 1:1) to give the monoester (9.4 mg, 36%, 97.0:3.0 er) as a clear oil and the bisester (17.1 mg, 58%) as a clear oil. Monoester: R_f 0.33 (hexane/EtOAc, 1:1); HPLC t_R (R) 59.6 min (97%), t_R (S) 68.9 min (3%) [Chiralpak AD, hexane/i-PrOH (98:2), 0.5 mL/min]; $[\alpha]_D^{20}$ + 18.4 (c 2.0, CHCl$_3$); FTIR (film) 3452, 2930, 1740, 1457, 1369, 1237, 1104, 1041 cm^{-1}; ^{1}H NMR (400 MHz, CDCl$_3$) δ 7.35–7.31 (m, 5H), 4.72 (d, J = 12.0 Hz, 1H), 4.61 (d, J = 11.6 Hz, 1H), 4.23 (d, J = 4.8 Hz, 2H), 3.72 (m, 2H), 3.66 (m, 1H), 2.07 (s, 3H); ^{13}C NMR (100 MHz, CDCl$_3$) δ 170.9, 128.6, 128.0, 127.9, 77.3, 72.4, 63.1, 61.2, 21.1; LRMS-ESI (m/z): [M + Na]$^+$ 247.03; HRMS-EI (m/z): [M]$^+$ calcd for C$_{12}$H$_{16}$O$_4$, 224.1049; found, 224.1051. Bisester: R_f 0.68 (hexane/EtOAc, 1:1); FTIR (film) 3031, 2955, 2880, 1740, 1501, 1457, 1369, 1237, 1117, 1054, 916, 746, 708 cm^{-1}; ^{1}H NMR (400 MHz, CDCl$_3$) δ 7.36–7.31 (m, 5H), 4.66 (s, 2H), 4.24 (dd, J = 11.6, 4.8 Hz, 2H), 4.16 (dd, J = 11.2, 5.2 Hz, 2H), 3.81 (m, 2H), 2.06 (s, 6H); ^{13}C NMR (100 MHz, CDCl$_3$) δ 170.6, 137.7, 128.4, 127.9, 127.8, 74.3, 72.2, 63.2, 21.0; HRMS-EI (m/z): [M]$^+$ calcd for C$_{14}$H$_{18}$O$_5$, 266.1154; found, 266.1160.

(*R*)-1-(4-Methoxyphenyl)allyl Isobutyrate [Lewis Base Catalyzed Kinetic Resolution of 1-(4-Methoxyphenyl)prop-2-en-1-ol].[205] 1-(4-Methoxyphenyl) prop-2-en-1-ol (53 mg, 0.32 mmol) was dissolved in toluene (0.8 mL), and the solution was cooled to −78°. Isothiourea **25** (1 mol %, 100 μL of a 0.032 M solution in toluene), *i*-Pr$_2$NEt (28 μL, 0.16 mmol), and isobutyric anhydride (26 μL, 0.16 mmol) were added, and the solution was stirred at −78° for 16 h. The reaction was quenched with 1 M HCl, the mixture was diluted with EtOAc, and washed successively with 1 M HCl (×2), aq NaHCO$_3$ (×2), and brine. The organic layer was dried (anhydrous Na$_2$SO$_4$), filtered, and concentrated under reduced pressure. The residue was purified by silica gel chromatography (petrol ether/EtOAc, 80:20) to give the ester (26 mg, 34%, 92.0:8.0 er) as a colorless oil and the alcohol (21 mg, 40%, 92.0:8.0 er) as a colorless oil. Ester: R_f 0.64 (petrol ether/EtOAc, 80:20); HPLC t_R (*R*) 8.7 min (92%), t_R (*S*) 11.2 min (8%) [Chiralcel OJ-H, hexane/*i*-PrOH (99:1), 1.0 mL/min, 211 nm, 30°]; $[\alpha]_D^{20}$ + 27.5 (*c* 1.0, CHCl$_3$); ^{1}H NMR (400 MHz, CDCl$_3$) δ 7.30–7.26 (m, 2H), 6.92–6.83 (m, 2H), 6.21 (d, *J* = 5.6 Hz, 1H), 6.00 (ddd, *J* = 17.1, 10.5, 5.6 Hz, 1H), 5.27 (dt, *J* = 17.2, 1.4 Hz, 1H), 5.22 (dt, *J* = 10.5, 1.3 Hz, 1H), 3.80 (s, 3H), 2.59 (hept, *J* = 7.0 Hz, 1H), 1.19 (d, *J* = 7.0 Hz, 3H), 1.16 (d, *J* = 7.0 Hz, 3H). Alcohol: R_f 0.28 (petrol ether/EtOAc, 80:20); $[\alpha]_D^{20}$ − 6.5 (*c* 0.4, CHCl$_3$); HPLC t_R (*R*) 11.0 min (8%), t_R (*S*) 13.8 min (92%) [Chiralcel OD-H, hexane/*i*-PrOH (95:5), 1.0 mL/min, 211 nm, 30°]; ^{1}H NMR (500 MHz, CDCl$_3$) δ 7.30 (d, *J* = 8.7 Hz, 2H), 6.89 (d, *J* = 8.7 Hz, 2H), 6.05 (ddd, *J* = 16.8, 10.3, 5.9 Hz, 1H), 5.34 (d, *J* = 17.1 Hz, 1H), 5.19 (d, *J* = 10.3 Hz, 1H), 5.17 (s, 1H), 3.81 (s, 3H), 1.85 (d, *J* = 3.6 Hz, 1H).

(1*S*,2*S*)-2-Bromocyclohexyl Benzoate [Lewis Base Catalyzed Kinetic Resolution of *trans*-2-Bromocyclohexan-1-ol].[117] A 100-mL, two-necked flask containing 4 Å MS (1.0 g) was flame-dried under reduced pressure before being placed under an atmosphere of argon. (*S*)-1-Methyl-2-[(dihydroisoindol-2-yl)methyl]pyrrolidine (**90**, 65 mg, 0.28 mmol) in CH$_2$Cl$_2$ (5 mL), triethylamine (7.6 mL, 55 mmol) in

CH_2Cl_2 (15 mL), and (rac)-trans-2-bromocyclohexanol (17.91 g, 100 mmol) in CH_2Cl_2 (40 mL) were added, and the solution was cooled to −78°. Benzoyl chloride (7.5 mL, 65 mmol) in CH_2Cl_2 (20 mL) was added slowly over 30 min, and the resulting mixture was stirred for 3 h at −78°. The reaction was quenched with a phosphate buffer (pH 7), the layers were separated, and the aqueous layer was extracted with Et_2O (3 × 50 mL). The combined organics were washed with H_2O, dried (anhydrous Na_2SO_4), filtered, and concentrated under reduced pressure. The residue was purified by silica gel chromatography (hexanes/EtOAc, 50:1) to give the ester (14.25 g, 50%, 97.5:2.5 er) and the alcohol (6.60 g, 37%, >99.5:0.5 er). Ester: HPLC t_R (1S,2S) 14.6 min (97.5%), t_R (1R,2R) 16.7 min (2.5%) [Chiralcel OD, hexanes/i-PrOH (1000:1), 1.0 mL/min, 254 nm); $[\alpha]_D^{20}$ + 104.6 (c 1.0, $CHCl_3$); FTIR (neat) 2941, 1720, 1450, 1274, 1104, 1027, 945, 711 cm^{-1}; ^{1}H NMR (500 MHz, $CDCl_3$) δ 8.07 (m, 2H), 7.57 (m, 1H), 7.46 (t, J = 7.5 Hz, 2H), 5.14 (dt, J = 9.0, 4.5 Hz, 1H), 4.16 (m, 1H), 2.45–2.39 (m, 1H), 2.33–2.25 (m, 1H), 1.99–1.91 (m, 1H), 1.85–1.75 (m, 2 H), 1.57–1.48 (m, 2H), 1.44–1.35 (m, 1H); ^{13}C NMR (125 MHz, $CDCl_3$) δ 165.8, 133.2, 130.4, 129.9, 128.5, 76.5, 52.8, 35.7, 31.3, 25.6, 23.5; HRMS-FAB (m/z): [M + H]$^+$ calcd for $C_{13}H_{15}BrO_2$, 283.0334; found, 283.0308. Anal. Calcd for $C_{13}H_{15}BrO_2$: C, 55.14; H, 5.34. Found: C, 55.05; H, 5.42. Alcohol: HPLC (after conversion to the corresponding benzoate) t_R (1S,2S) 14.3 min (<0.5%), t_R (1R,2R) 16.8 min (>99.5%) [Chiralcel OD, hexanes/i-PrOH (1000:1), 1.0 mL/min, 254 nm]; $[\alpha]_D^{24}$ − 33.0 (c 1.0, $CHCl_3$); FTIR (neat) 3350, 2835, 1450, 1070, 955 cm^{-1}; ^{1}H NMR (500 MHz, $CDCl_3$) δ 3.89 (ddd, J = 12.1, 9.5, 4.5 Hz, 1H), 3.60 (dt, J = 9.9, 4.5 Hz, 1H), 2.62 (s, 1H), 2.36–2.30 (m, 1H), 2.16–2.10 (m, 1H), 1.86–1.77 (m, 2H), 1.70–1.65 (m, 1H), 1.40–1.22 (m, 3H); ^{13}C NMR (125 MHz, $CDCl_3$) δ 75.4, 62.0, 36.4, 33.7, 26.8, 24.3.

(1S,2S)-2-Phenylcyclohexyl Isobutyrate [Brønsted Acid Catalyzed Kinetic Resolution of cis-2-Phenylcyclohexan-1-ol].[104] Isobutyric anhydride (0.12 mL, 0.7 mmol) was added to a solution of cis-2-phenylcyclohexan-1-ol (176 mg, 1.0 mmol) and **54** (42 mg, 0.05 mmol) in $CHCl_3$ (0.1 mL) at rt. After 15 h the reaction was quenched with the addition of aqueous $NaHCO_3$ (4 mL), MeOH (0.2 mL), and aqueous NH_4Cl (0.2 mL), and the resulting solution was stirred for 15 min before being diluted with H_2O and EtOAc. The layers were separated, and the aqueous phase was extracted with EtOAc. The combined organics were dried (anhydrous Na_2SO_4) and concentrated under reduced pressure. The residue was purified by silica gel chromatography (hexane/EtOAc, 1:0 to 1:4) to give the ester (130 mg, 52%, 95.0:5.0 er) as a colorless oil and the alcohol (84 mg, 48%, >99.0:1.0 er) as a colorless oil. Ester: HPLC t_R (1S,2S) 8.8 min (95%), t_R (1R,2R) 9.1 min (5%) [Chiralcel OD-3, hexane/i-PrOH (300:1), 1.5 mL/min, 254 nm); $[\alpha]_D^{23}$ + 71.9 (c 1.02, CHCl$_3$); FTIR (neat) 2936, 1724, 1195, 1157 cm^{-1}; ^{1}H NMR (500 MHz, CDCl$_3$) δ 7.27–7.15 (m, 5H), 5.15 (m, 1H), 2.78 (ddd, J = 12.9, 3.2, 3.0 Hz, 1H), 2.42 (qq, J = 7.2, 6.9 Hz, 1H), 2.12–2.02 (m, 2H), 1.92 (m, 1H), 1.78 (m, 1H), 1.66–1.39 (m, 4H), 1.05 (d, J = 7.2 Hz, 3H), 0.99 (d, J = 6.9 Hz, 3H); ^{13}C NMR (125 MHz, CDCl$_3$) δ 176.1, 143.1, 128.0, 127.8, 126.2, 72.7, 46.3, 34.3, 30.7, 25.9, 25.8, 20.2, 19.0, 18.9; LRMS-EI (m/z): [M]$^+$ 246, 203, 175, 159. Anal. Calcd for C$_{16}$H$_{22}$O$_2$: C, 78.01; H, 9.00. Found: C, 78.20; H, 9.06. Alcohol: $[\alpha]_D^{23}$ − 109.1 (c 1.06, MeOH); HPLC t_R (1R,2R) 23.3 min (>99%), t_R (1R,2R) 28.6 min (<1%) [Chiralcel OD-3, hexane/i-PrOH (300:1), 1.5 mL/min, 254 nm].

BzCl (0.5 equiv), 164 (5 mol %)

i-Pr$_2$NEt, CH$_2$Cl$_2$, 0°

(rac)

(48%)

er >99.5:0.5

(1S,2S)-2-Hydroxy-1,2-diphenylethyl Benzoate [Lewis Acid Catalyzed Kinetic Resolution of *trans*-2-Hydroxy-1,2-diphenylethyl Benzoate].[125] Benzoyl chloride (58.0 μL, 0.500 mmol) was added to a solution of *trans*-2-hydroxy-1,2-diphenylethyl benzoate (214 mg, 1.00 mmol), i-Pr$_2$NEt (174 μL, 1.00 mmol) and **164** (23.4 mg, 0.050 mmol) in CH$_2$Cl$_2$ (5 mL) at 0°, and the reaction mixture was stirred until complete by TLC analysis. The mixture was poured onto H$_2$O and extracted with CH$_2$Cl$_2$ (×3). The combined organics were dried (anhydrous MgSO$_4$) and concentrated under reduced pressure. The residue was purified by silica gel chromatography (hexane/EtOAc, 3:1) to give the ester (153 mg, 48%, >99.5:0.5 er) as a white solid. mp 146–148°; HPLC t_R (1S,2S) 13.2 min (>99.5%), t_R (1R,2R) 27.7 min (<0.5%) [Chiralpak OJ, hexane/i-PrOH (5:1), 1 mL/min, 254 nm]; $[\alpha]_D^{23}$− 66.3 (c 1.00, MeOH); FTIR (neat) 3472, 3034, 1721, 1453, 1273, 1113, 704 cm^{-1}; ^{1}H NMR (500 MHz, CDCl$_3$) δ 8.13–8.10 (m, 2H), 7.62–7.57 (m, 1H), 7.51–7.45 (m, 2H), 7.30–7.18 (m, 10H), 6.11 (d, J = 7.3 Hz, 1H), 5.10 (d, J = 7.3 Hz, 1H), 2.63–2.56 (s, 1H).

Bz$_2$O (0.5 equiv),
56 (20 mol %)

Ph⟋ NH$_2$ → Ph⟋ NHBz + Ph⟋ NH$_2$

DMAP (20 mol %),
4 Å MS, toluene,
–78°, 1 h

(42%) (—)
er 85.0:15.0   er 79.0:21.0

(R)-N-(1-Phenylethyl)benzamide [H-Bonding Catalyzed Kinetic Resolution of 1-Phenylethylamine].[31] 4-(Dimethylamino)pyridine (6.1 mg, 0.05 mmol), benzoic anhydride (23.0 mg, 0.125 mmol) and 4 Å MS (100 mg) were added to a flame-dried flask. Freshly distilled toluene (23.0 mL) was added, and the reaction mixture was stirred at rt for 15 min. The solution was cooled to –78° over 15 min before a solution of **56** (33.0 mg, 0.05 mmol) in toluene (2 mL) was added. After stirring for 15 min 1-phenylethylamine (30.3 mg, 0.25 mmol) was added, and the reaction mixture was stirred at –78° for 1 h. The reaction was quenched by adding MeMgCl (3.0 M in THF, 0.167 mL, 0.500 mmol) at –78° and stirring for 10 min. Excess Grignard reagent was quenched with 1 M HCl (5 mL). The reaction mixture was warmed to rt and extracted with Et$_2$O (3 × 20 mL). The combined organics were washed successively with 1 M HCl (5 mL) and brine (5 mL) before being dried (anhydrous Na$_2$SO$_4$) and concentrated under reduced pressure. The residue was purified by silica gel chromatography (hexanes/Et$_2$O) to give the amide (24.0 mg, 42%, 85.0:15.0 er) as a white solid. The unreacted amine was isolated by basifying the aqueous layer to pH 10 with 15% NaOH and subsequent extraction with Et$_2$O (3 × 20 mL). The combined organics were washed with brine before being dried (anhydrous Na$_2$SO$_4$) and concentrated under reduced pressure. The crude material was benzoylated following a standard procedure and analyzed by HPLC. Amide: mp 104–105 °C; R$_f$ 0.17 (hexanes/Et$_2$O, 7:3); HPLC t_R (R) 12.2 min (85%), t_R (S) 17.8 min (15%) [Chiralpak OD-H, hexane/i-PrOH (9:1), 1 mL/min, 230 nm]; [α]$_D^{20}$ + 14.0 (c 1.0, CHCl$_3$); FTIR (KBr) 3356, 1633, 1518, 1488, 717 cm^{-1}; ^{1}H NMR (500 MHz, CDCl$_3$) δ 7.81–7.73 (m, 2H), 7.53–7.46 (m, 1H), 7.45–7.33 (m, 6H), 7.32–7.25 (m, 1H), 6.35 (d, J = 5.5 Hz, 1H), 5.41–5.29 (m, 1H), 1.62 (d, J = 6.9 Hz, 3H); ^{13}C NMR (125 MHz, CDCl$_3$) δ 166.5, 143.1, 134.6, 131.4, 128.7, 128.5, 127.4, 126.9, 126.2, 49.2, 21.7.; LRMS-ESI (m/z): [M + H]$^+$ 226.1. Amine: HPLC analysis (after benzoylation) t_R (R) 12.2 min (21%), t_R (S) 17.8 min (79%) [Chiralpak OD-H, hexane/i-PrOH (9:1), 1 mL/min, 230 nm].

(S)-N-(4-Phenylbut-3-yn-2-yl)acetamide [Stoichiometric Kinetic Resolution of 4-Phenylbut-3-yn-2-amine].[259] Aliquat™ 336 (0.8 mL) was added to a solution of 4-phenylbut-3-yn-2-amine (20.3 mg, 0.14 mmol) in THF (1.2 mL) at −20°. A solution of (1S,2S)-**1** (29.4 mg, 0.07 mmol) in THF (0.6 mL) was added over 1 h using a syringe pump, and the resulting mixture was stirred overnight. The solution was concentrated under reduced pressure, and the residue was purified by silica gel chromatography to give the amide (13 mg, 50%, 95.5:4.5 er). SFC t_R (R) 2.4 min (4.5%), t_R (S) 2.8 min (95.5%) [Chiralpak OD-H, CO_2/i-PrOH (9:1), 150 bar, 5 mL/min, 254 nm]; $[\alpha]_D^{20}$ − 163 (c 0.6, $CHCl_3$); FTIR (neat) 3290, 3057, 2981, 2932, 1643, 1542, 1372, 1134, 757, 691 cm^{-1}; ^{1}H NMR (400 MHz, $CDCl_3$) δ 7.46–7.38 (m, 2H), 7.35–7.28 (m, 3H), 5.84 (s, 1H), 5.05 (dq, J = 13.8, 6.9 Hz, 1H), 2.02 (s, 3H), 1.49 (d, J = 6.8 Hz, 3H); ^{13}C NMR (100 MHz, $CDCl_3$) δ 169.2, 131.8, 128.4, 128.3, 122.7, 89.5, 82.2, 37.7, 23.3, 22.6. HRMS-ESI (m/z): [M + Na]$^+$ calcd for $C_{12}H_{13}NaNO$, 210.0889; found, 210.0887.

(S)-3-Mesityl-1-(2-methylpiperidin-1-yl)propan-1-one [Lewis Base Catalyzed Kinetic Resolution of 2-Methylpiperidine].[34] Hydroxamic acid **213** (5.1 mg, 0.025 mmol), NHC **214** (8.2 mg, 0.025 mmol), α′-hydroxyenone **6** (40.7 mg, 0.175 mmol), and 2-methylpiperidine (24.8 mg, 0.250 mmol) were dissolved in CH_2Cl_2 (2.5 mL) before DBU (7.5 μL, 0.050 mmol) was added, and the reaction mixture was stirred for 18 h at rt. The reaction was quenched by addition of DBU (37.4 μL, 0.250 mmol) and CbzCl (37.6 μL, 0.250 mmol), and the mixture was stirred for 2 h at rt before aqueous $NaHCO_3$ (10 mL) was added. The aqueous phase was extracted with CH_2Cl_2 (3 × 10 mL), and the combined organics were dried (anhydrous Na_2SO_4) and concentrated under reduced pressure. The residue was purified by silica gel chromatography to give the amide (38 mg, 55%, 86.0:14.0 er) and the corresponding Cbz-protected amine (23 mg, 39%, 94.0:6.0 er). Amide: SFC

t_R (*R*) 3.55 min (14%), t_R (*S*) 3.95 min (86%) [Chiralpak AS-H, CO_2/*i*-PrOH (95:5), 3 mL/min, 254 nm]; $[\alpha]_D^{29}$ + 18.6 (*c* 6.3, $CHCl_3$); FTIR (neat) 2936, 2862, 1725, 1638, 1424, 1270, 1180 cm^{-1}; ^{1}H NMR (400 MHz, $CDCl_3$) δ 6.84 (s, 2H), 4.96 (s, 0.5H), 4.55 (d, *J* = 12.4 Hz, 0.5H), 4.54 (d, *J* = 12.0 Hz, 0.5H), 4.06 (s, 0.5H), 3.05 (t, *J* = 13.3 Hz, 0.5H), 3.00 (m, 2H), 2.67 (t, *J* = 13.1 Hz, 0.5H), 2.55–2.35 (m, 2H), 2.30 (s, 6H), 2.25 (s, 3H), 1.73–1.25 (m, 6H), 1.21–1.13 (m, 3H); ^{13}C NMR (100 MHz, $CDCl_3$) δ 170.7, 136.1, 135.3, 135.0, 128.9, 48.1, 43.6, 40.5, 36.2, 32.8, 32.3, 30.5, 29.8, 26.2, 25.4, 25.1, 20.7, 19.7, 18.7, 16.5, 15.5; HRMS-ESI (*m/z*): [M + H]$^+$ calcd for $C_{18}H_{28}NO$, 274.2165; found, 274.2166. Cbz-protected amine: HPLC t_R (*S*) 15.7 min (6%), t_R (*R*) 16.5 min (94%) [Chiralcel AD-H, hexane/*i*-PrOH (99:1), 1 mL/min, 254 nm); $[\alpha]_D^{25}$ − 42.6 (*c* 0.8, $CHCl_3$); ^{1}H NMR (300 MHz, $CDCl_3$) δ 7.42–7.26 (m, 5H), 5.15 (d, *J* = 12.9 Hz, 1H), 5.10 (d, *J* = 12.9 Hz, 1H), 4.55–4.40 (m, 1H), 4.01 (dd, *J* = 13.3, 3.0 Hz, 1H), 2.89 (td, *J* = 13.3, 2.0 Hz, 1H), 1.75–1.30 (m, 6H), 1.16 (d, *J* = 7.0 Hz, 3H).

TABULAR SURVEY

The tables are organized by reaction classification in the order: desymmetrization of diols (Table 1), KR of alcohols (Table 2), desymmetrization of diamines (Table 3), KR of amines (Table 4), and KR of lactams and thiolactams (Table 5). Further subdivisions group similar structures together, following the same order and classifications as in the Scope and Limitations section. However, unlike in the Scope and Limitations, further divisions according to the class of catalyst used are not made to allow direct comparison of all available methods for individual substrates. Entries within each table are organized according to increasing size of the carbon skeleton of the substrate undergoing desymmetrization or KR. Unless otherwise stated, the stoichiometry of the reactants and/or reagents is one equivalent. Where possible, the reaction conversion (%), product enantiomeric ratios (er), and reaction selectivity factor (s) are given as reported in the primary literature.

A series of charts depicting catalysts, ligands, and common additives precedes the tables. The charts are arranged by common structural classes, with bold numbers used to identify each structure within the tables.

The tables cover the literature through 2019.

The following abbreviations (in addition to those included in "*The Journal of Organic Chemistry* Standard Abbreviations and Acronyms") are used in the text and Tabular Survey:

t-AmOH	*tert*-amyl alcohol
β-CD	β-cyclodextrin
Cy	cyclohexyl
DIC	*N,N'*-diisopropylcarbodiimide
DMC	dimethyl carbonate
DMT-MM	4-(4,6-dimethoxy-1,3,5-triazin-2-yl)-4-methylmorpholinium chloride
Mes	mesityl
MS	molecular sieves
MTBE	*tert*-butyl methyl ether
Np	naphthyl
Oct	octyl
Phth	phthaloyl
PPY	4-pyrrolidinopyridine
Py	pyridyl
Pyroc	3-pyrrolidine-1-carbonyl
s	selectivity factor
TrocCl	2,2,2-trichloroethyl chloroformate
TIPS-β-CD	6-*O*-triisopropylsilyl-β-cyclodextrin

CHART 1. STOICHIOMETRIC ACYLATING AGENTS

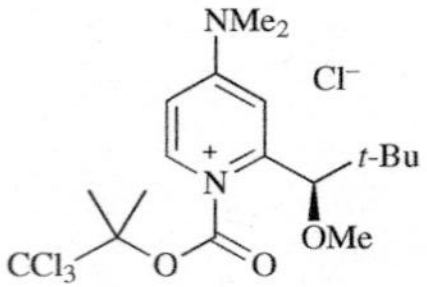

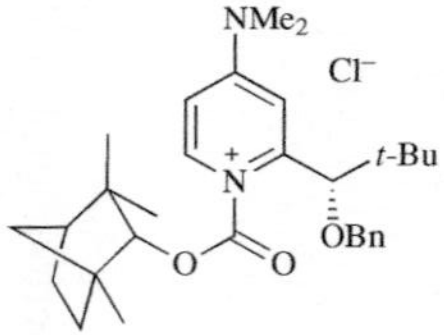

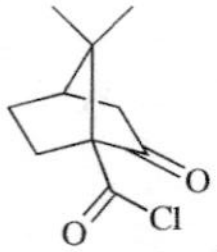

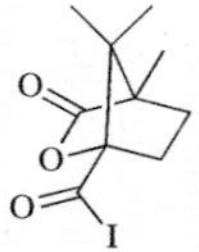

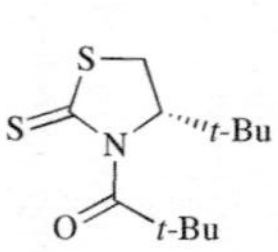

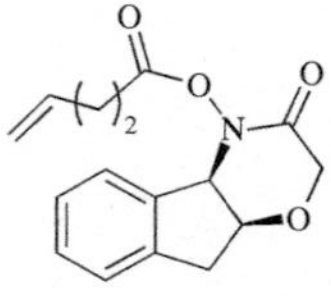

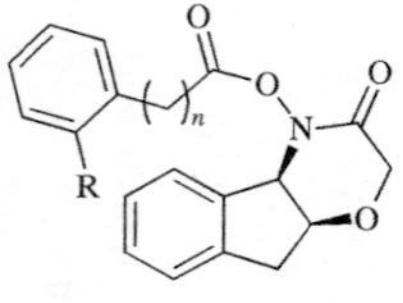

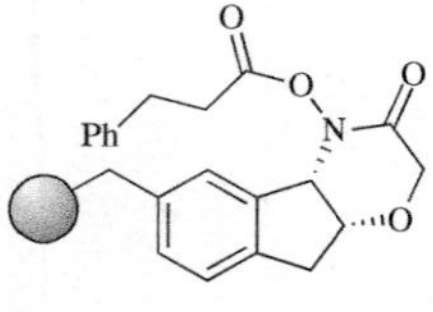

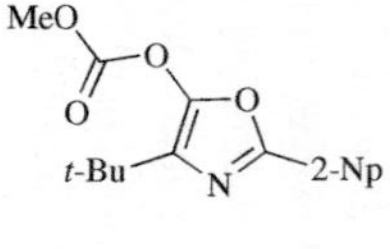

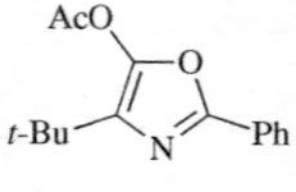

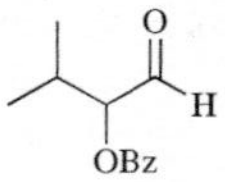

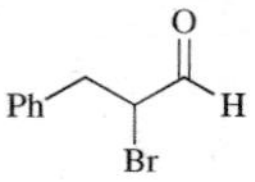

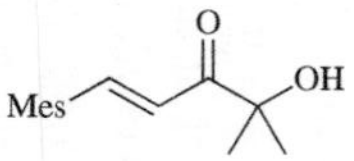

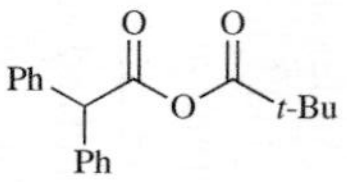

CHART 2. LEWIS ACID CATALYSTS AND LIGANDS

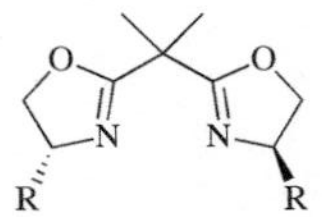

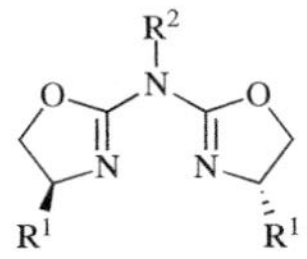

	R¹	R²
19	i-Pr	H
20	i-Pr	MeOPEG$_{5000}$CH$_2$
21	Ph	H
22	PhCH$_2$	Me
23	PhCH$_2$	MeOPEG$_{5000}$CH$_2$

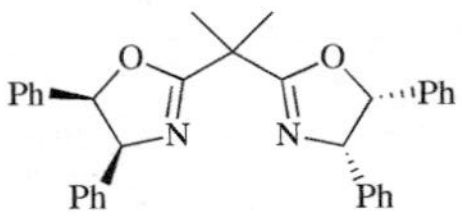

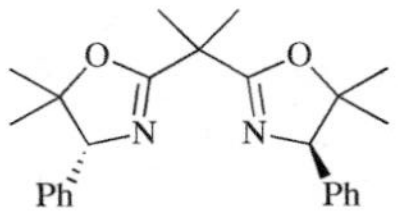

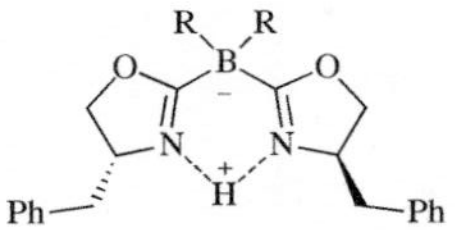

15 R = Ph
16 R = PhCH$_2$

17

18

24 R = Et
25 R = 3,5-(CF$_3$)$_2$C$_6$H$_3$

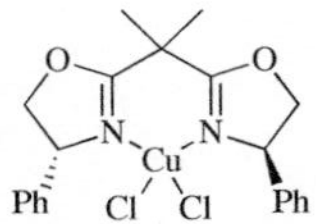

26

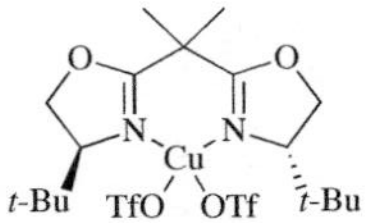

27

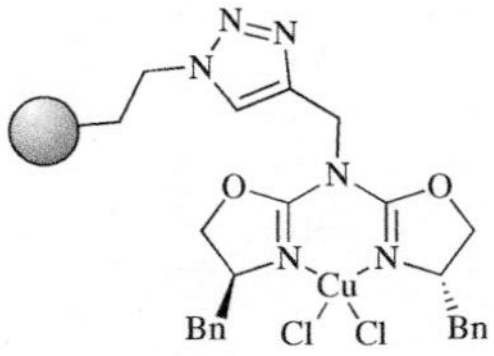

28 magnetite nanoparticle

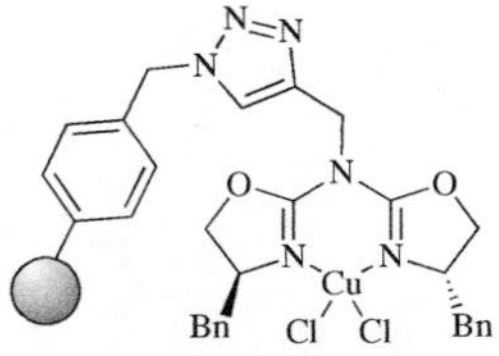

29 cobalt nanoparticle

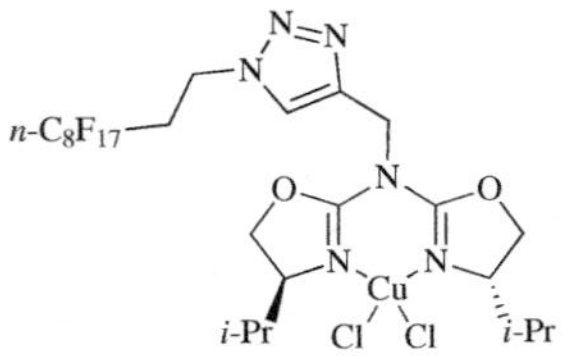

30

CHART 2. LEWIS ACID CATALYSTS AND LIGANDS (*Continued*)

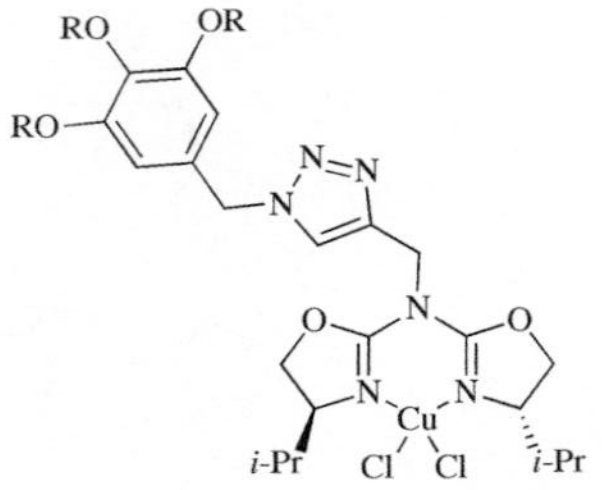

31

R = 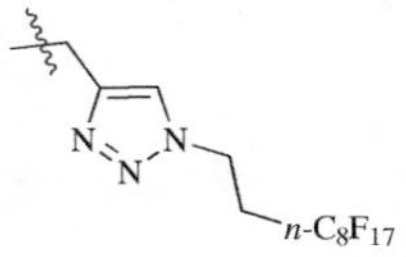

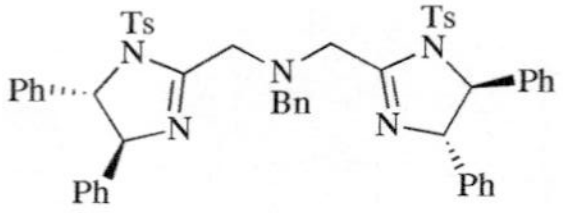

32

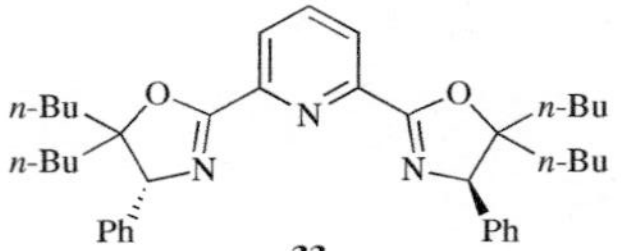

33

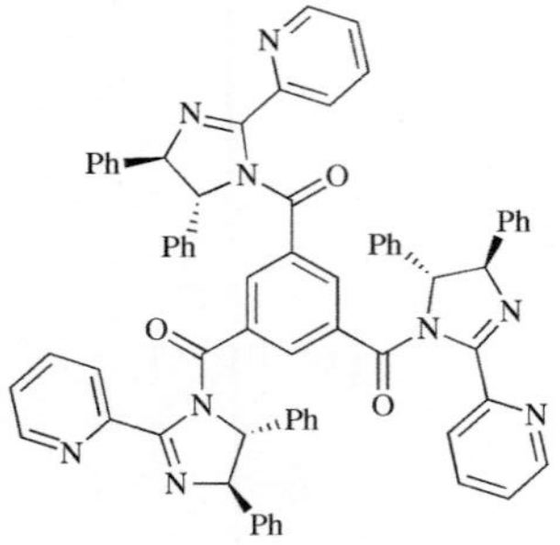

34

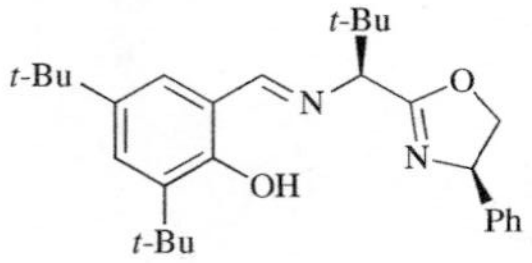

35

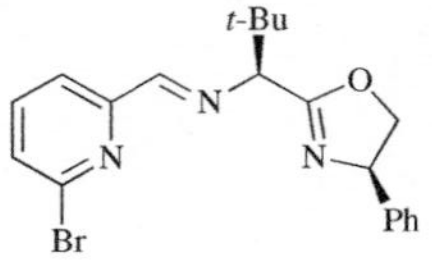

36

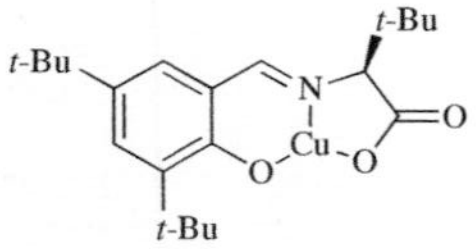

37

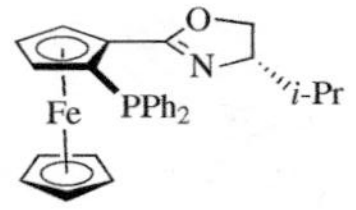

38

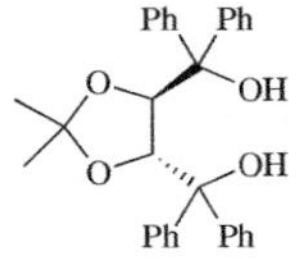

R

O

HN

R

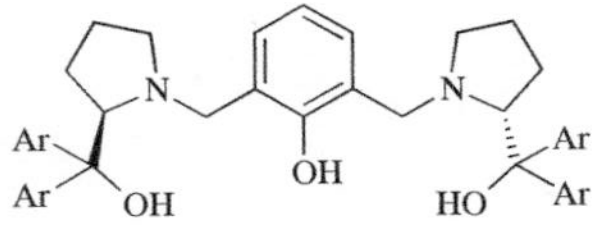

N+—O⁻

N+—O⁻

NH

O

R

R

	R
42a	Et
42b	i-Pr

45

SnCl

48

Ar

Ar

HO

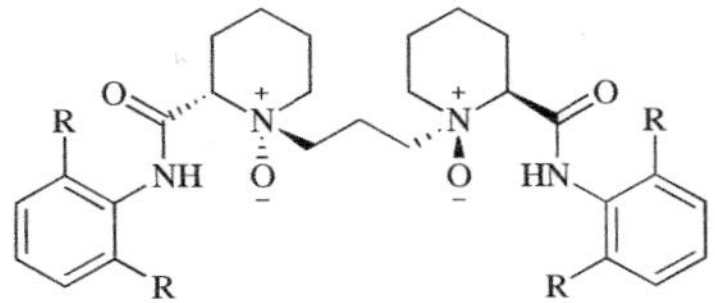

N

OH

N

Ar Ar

OH

	Ar
40	Ph
41	4-PhC₆H₄

SO₂Ar

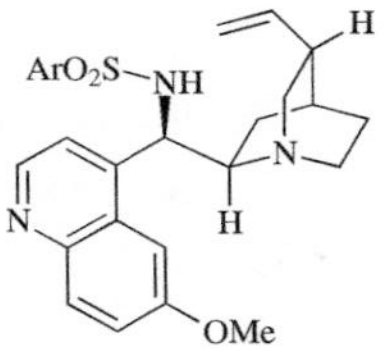

N

H

N
H

OMe

N

44

Ar = 3,5-(CF₃)₂C₆H₃

R

SnBr₂

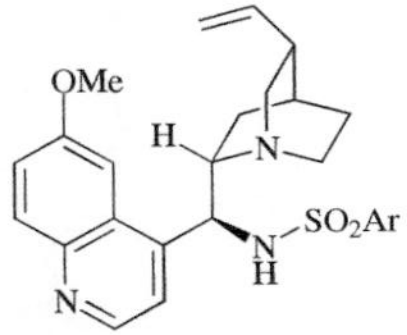

R

	R
47a	H
47b	3,4,5-F₃C₆H₂

Ph Ph

OH

OH

Ph Ph

O   O

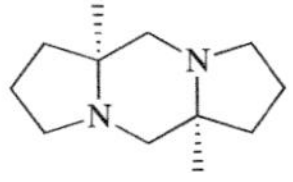

39

H

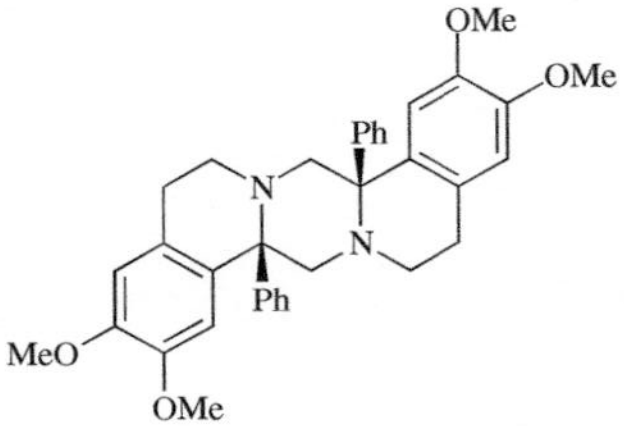

N

H

OMe

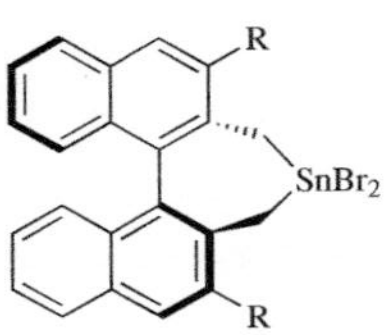

ArO₂S—NH

N

43

Ar = 3,5-(CF₃)₂C₆H₃

OMe

OMe

Ph

N

Ph

N

MeO

OMe

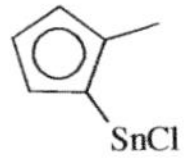

46

CHART 2. LEWIS ACID CATALYSTS AND LIGANDS (*Continued*)

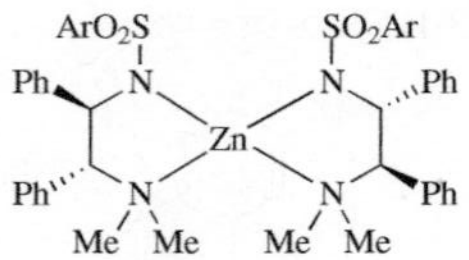

51

50

Ar = 3,5-(CF₃)₂C₆H₃

49

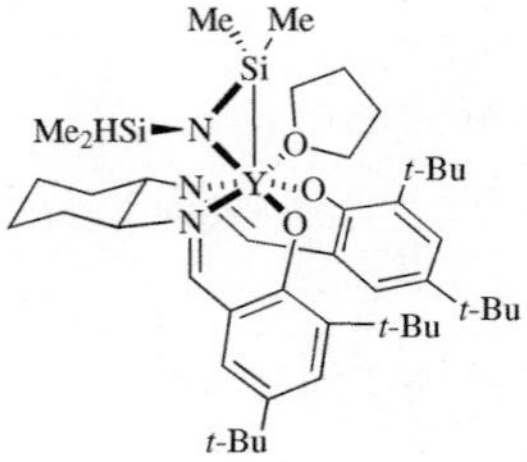

122

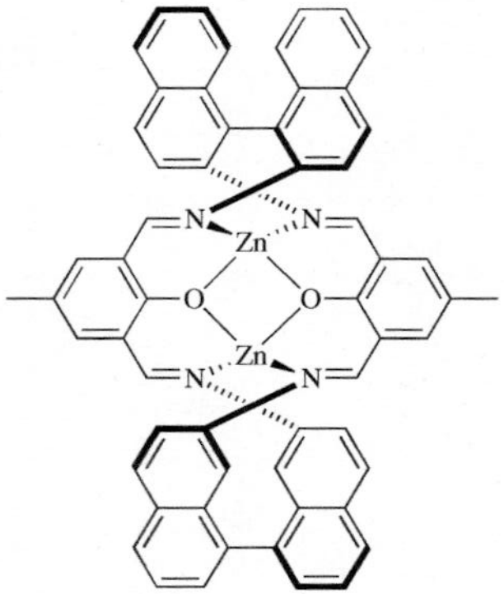

CHART 3. LEWIS BASE CATALYSTS

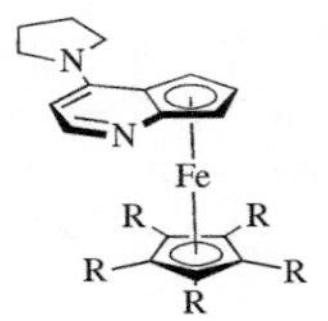

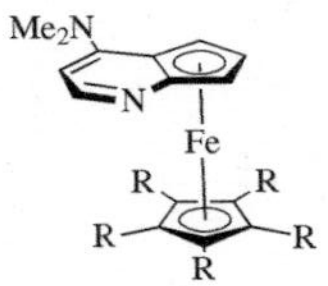

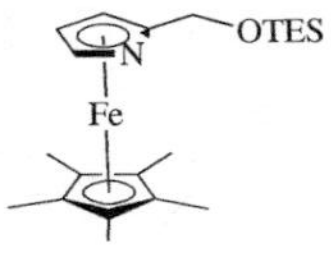

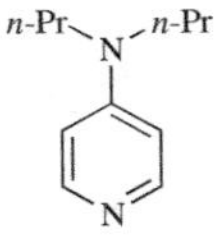

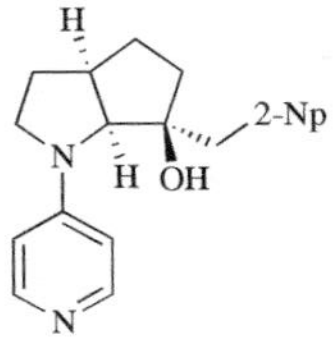

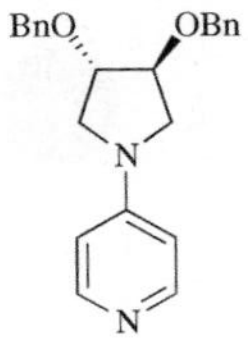

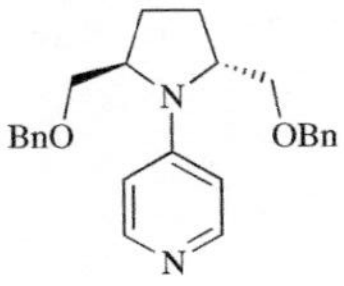

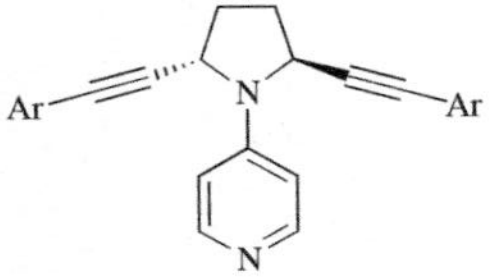

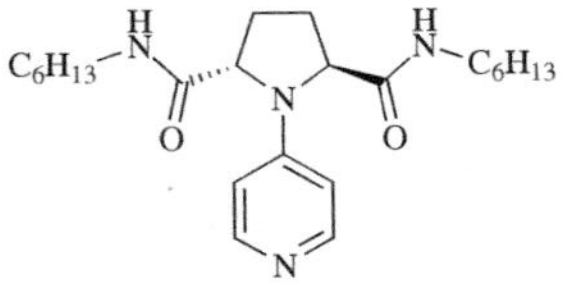

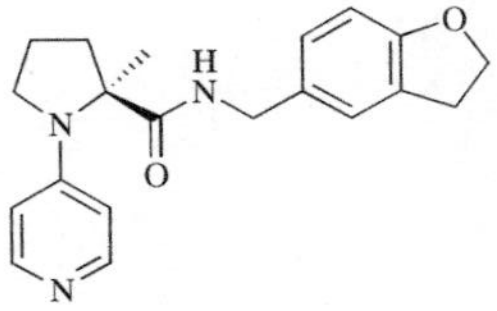

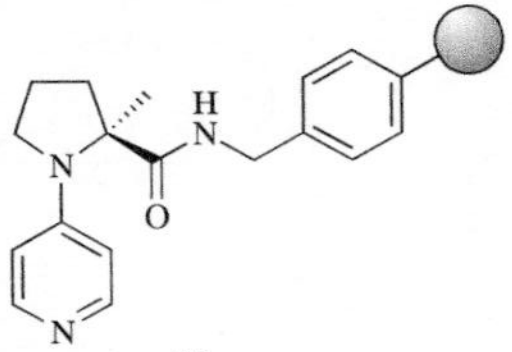

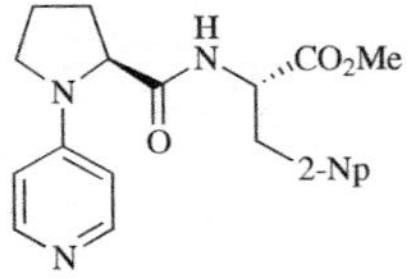

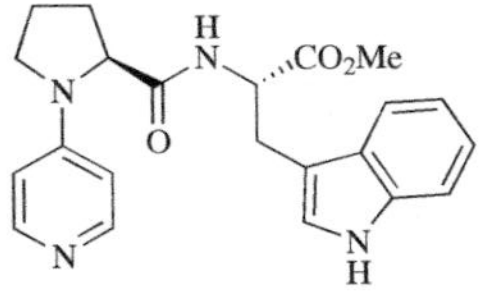

CHART 3. LEWIS BASE CATALYSTS (*Continued*)

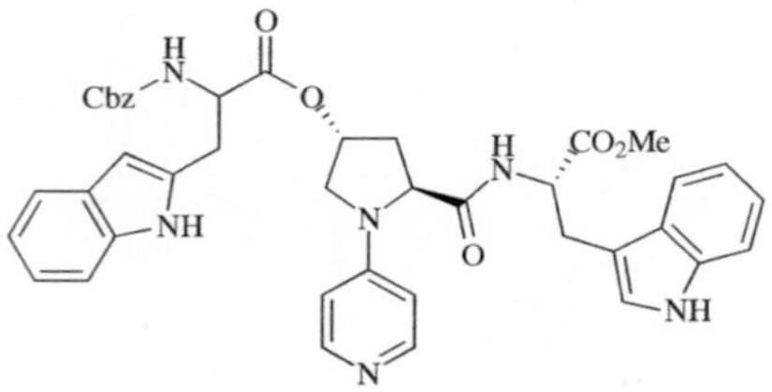

68

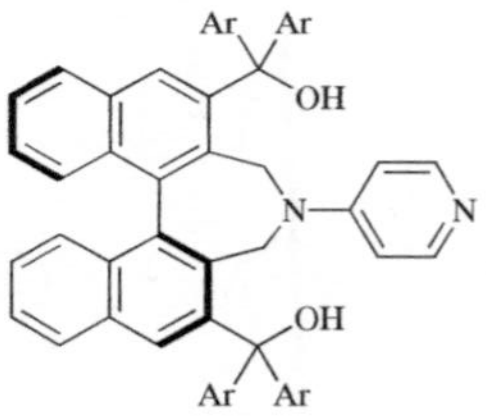

69a Ar = 4-*t*-BuC$_6$H$_4$
69b Ar = Ph

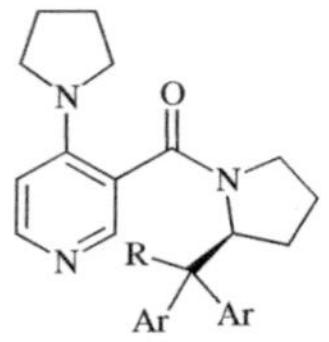

	Ar	R
70	Ph	HO
71	2-Np	HO
72	3,5-(CF$_3$)$_2$C$_6$H$_3$	HO
73	3,5-Ph$_2$C$_6$H$_3$	HO
74	Ph	Me$_2$N

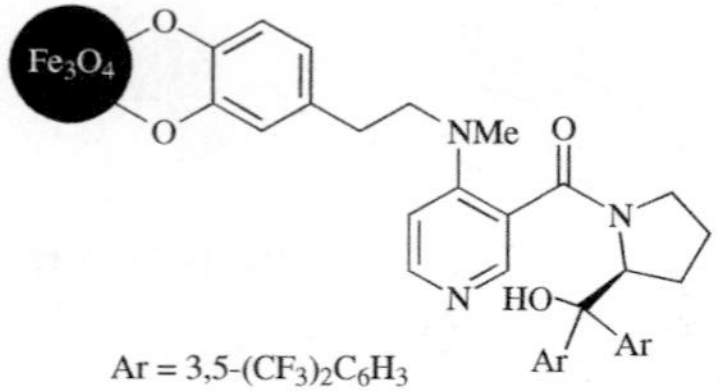

75 Ar = 3,5-(CF$_3$)$_2$C$_6$H$_3$

Fe$_3$O$_4$

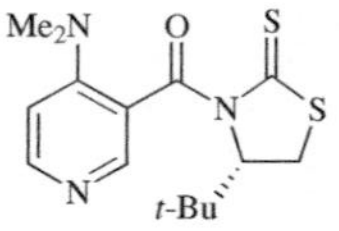

76

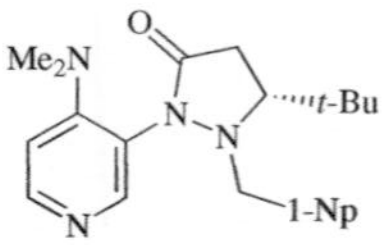

77

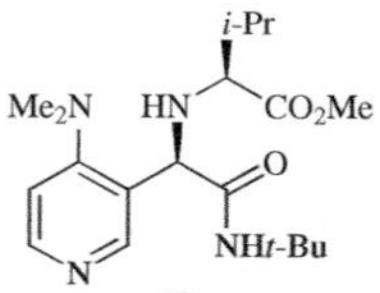

78

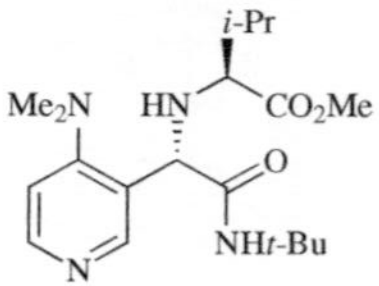

79

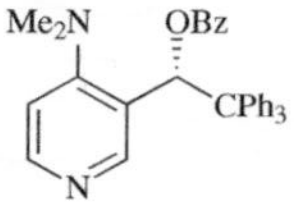

80

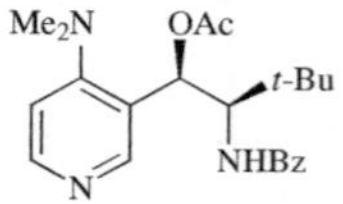

81

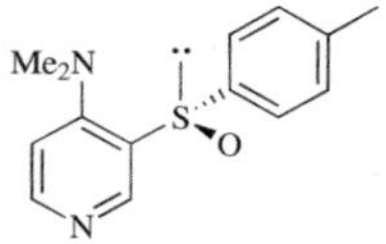

82

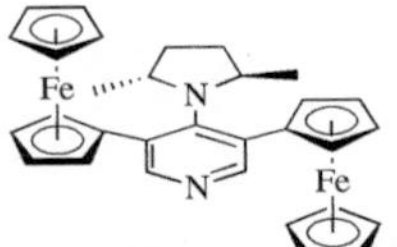

83

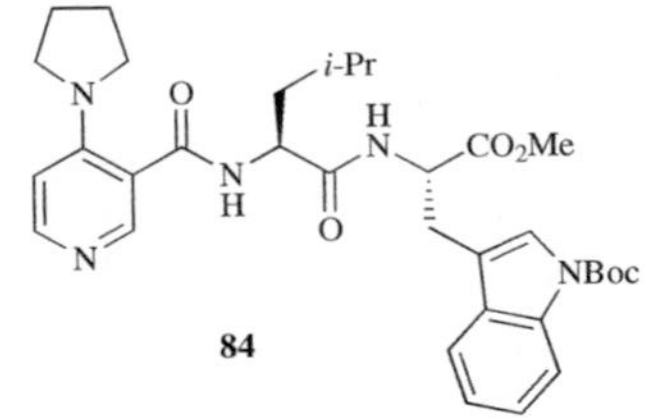

84

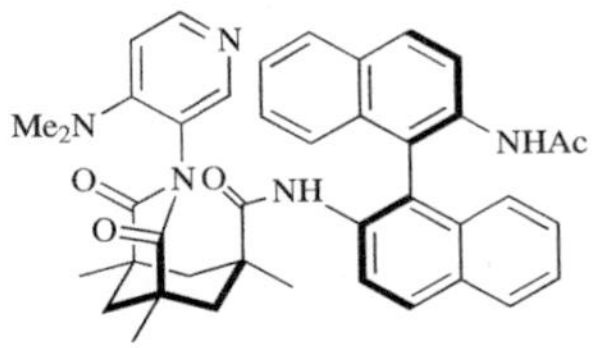

85

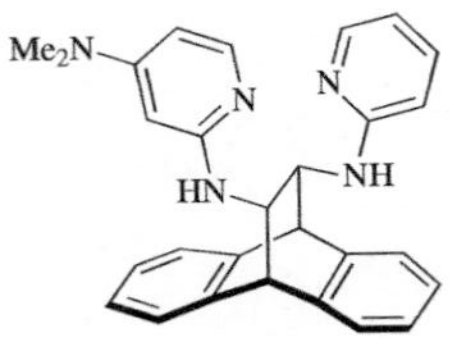

86

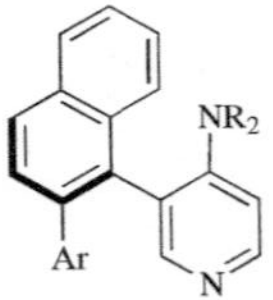

	Ar	R
87	Ph	Et
88	Ph	*n*-Bu
89	3,5-$Ph_2C_6H_3$	Et

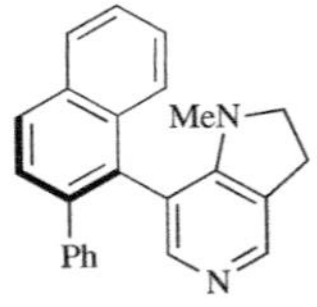

90

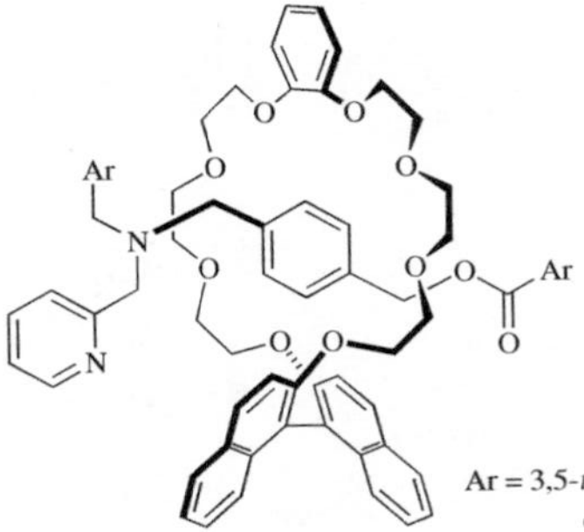

Ar = 3,5-*t*-$Bu_2C_6H_3$
91

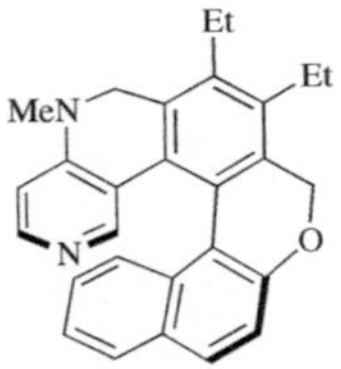

92

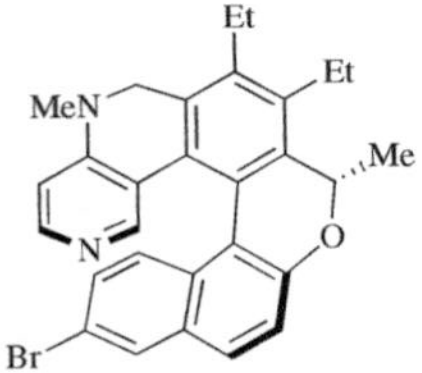

93

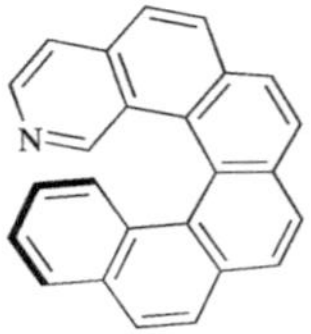

94

CHART 3. LEWIS BASE CATALYSTS (*Continued*)

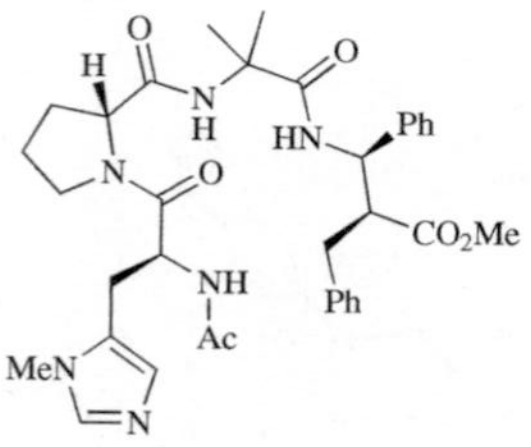

95

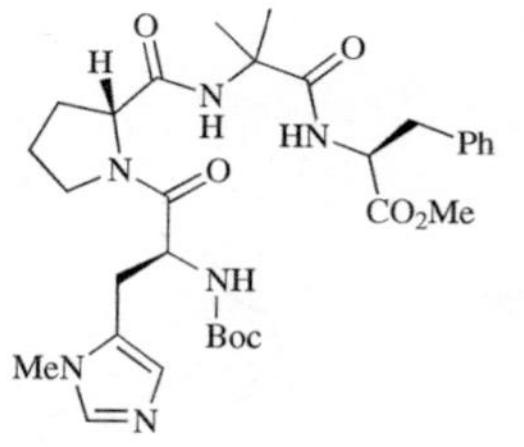

96

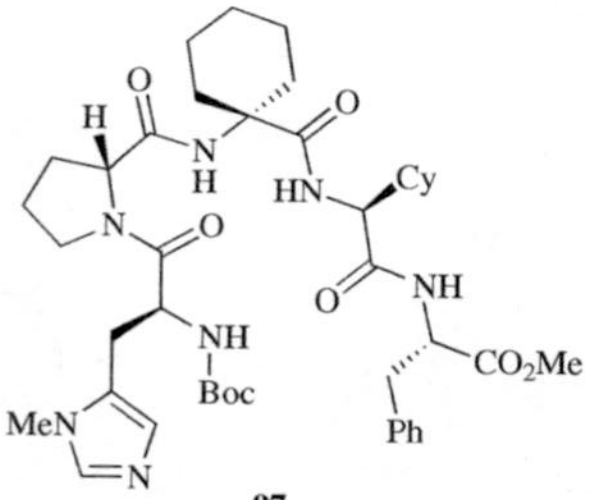

97

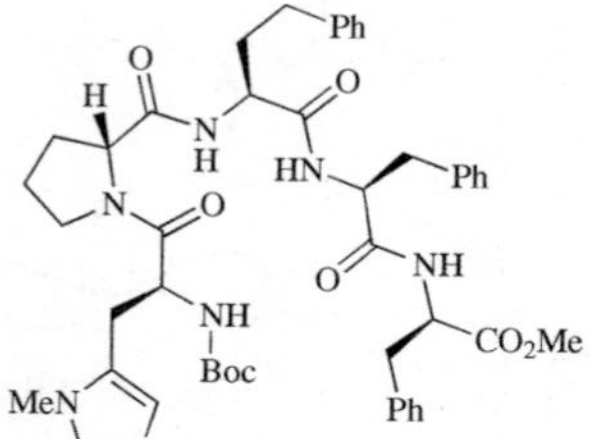

98

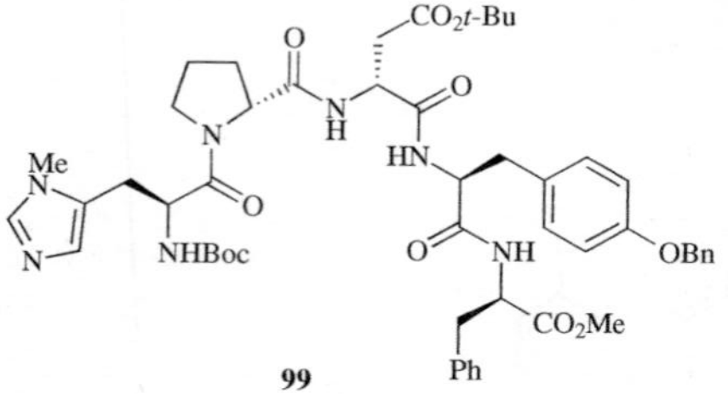

99

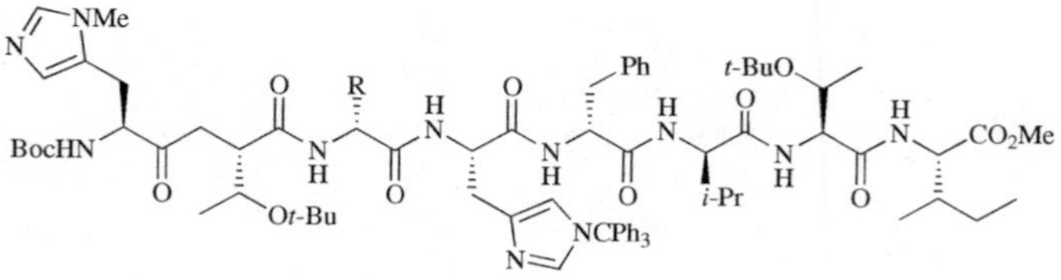

	R
100	Me
101	*i*-Pr

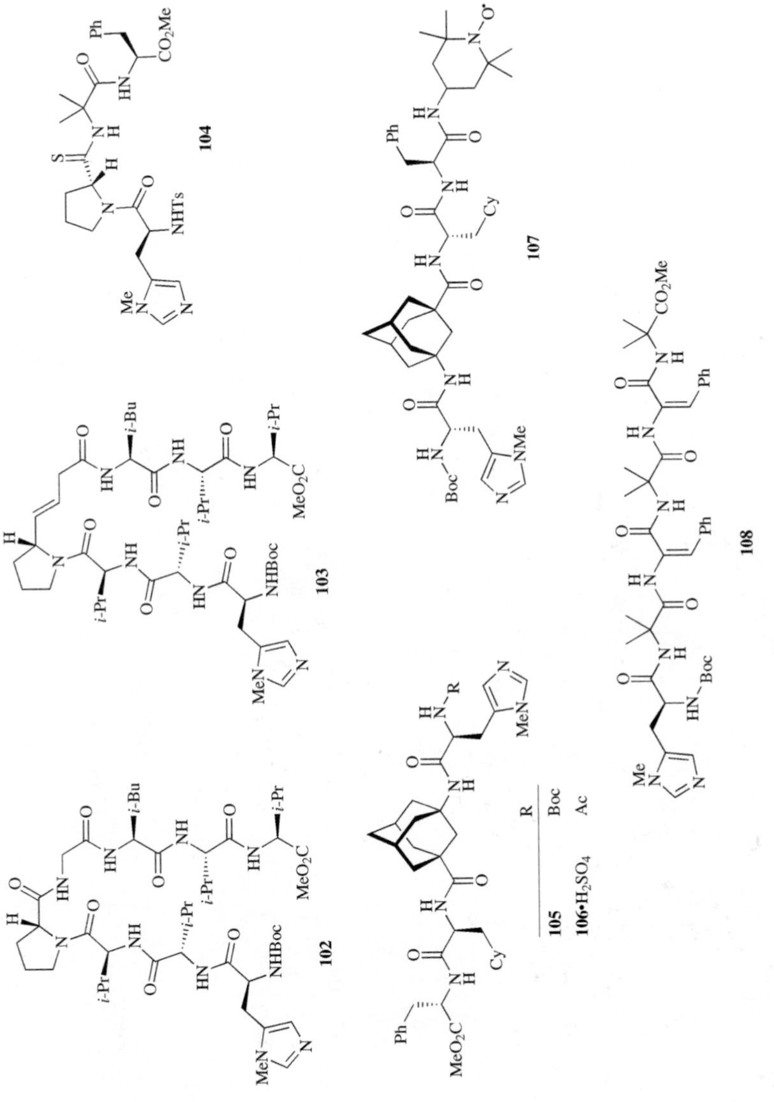

104

107

103

102

105
106·H₂SO₄

108

CHART 3. LEWIS BASE CATALYSTS (*Continued*)

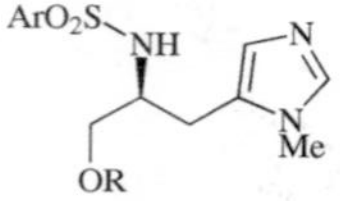

Ar = 2,4,6-i-$Pr_3C_6H_2$

	R
109	DPS
110	$(TMS)_3Si$

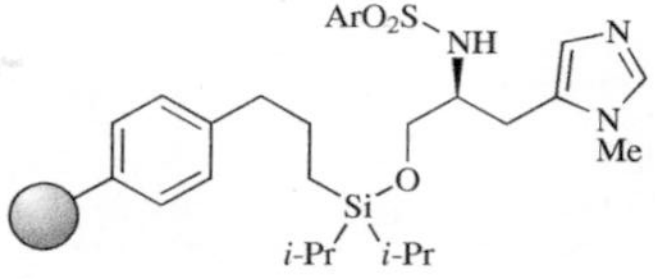

Ar = 2,4,6-i-$Pr_3C_6H_2$

111

 = polystyrene

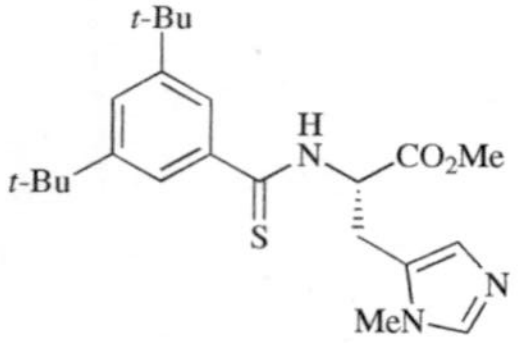

112

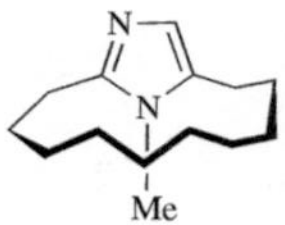

113

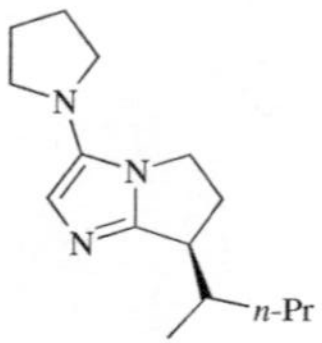

114

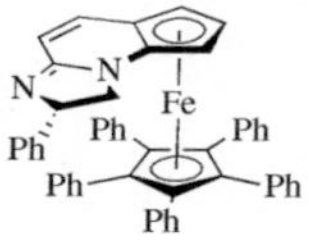

115

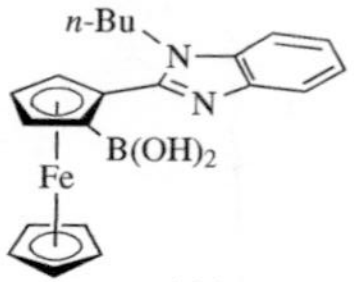

116

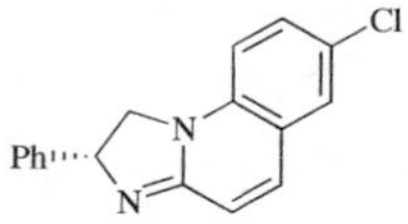

117

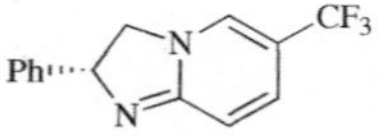

118

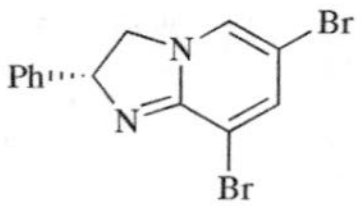

119

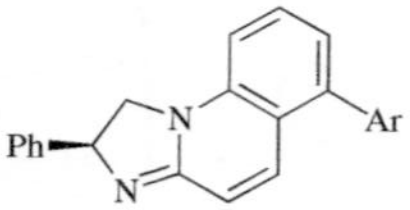

	Ar
120	1-Np
121	9-anthracenyl

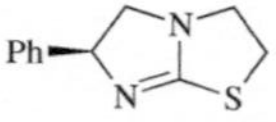

122

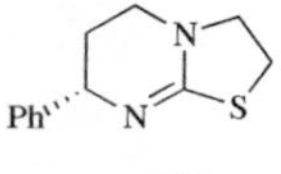

123

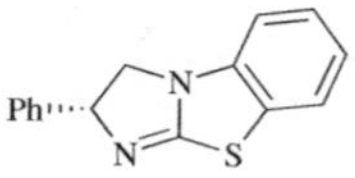

124

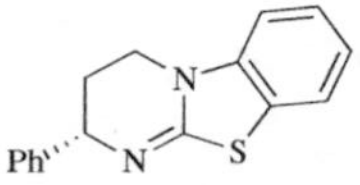

125

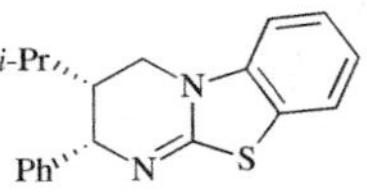

126

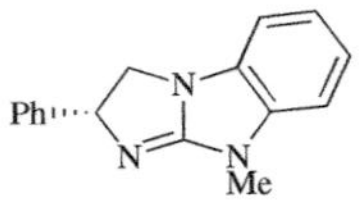

i-Pr

Ph

127

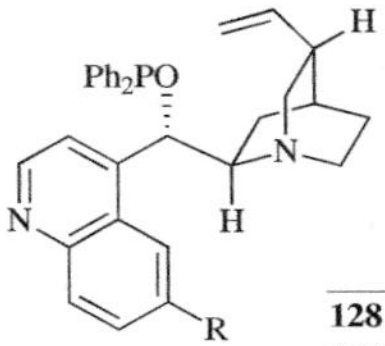

Me

Ph

128 | R: H
129 | MeO

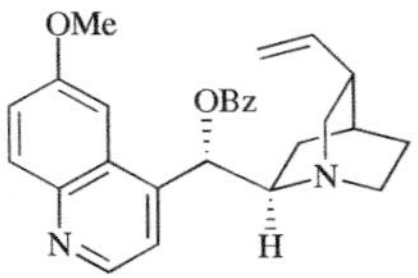

Ph₂PO

130

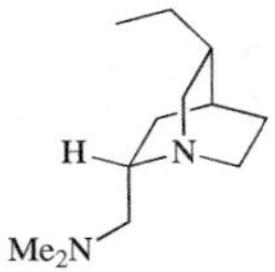

OMe

OBz

H

OMe

131

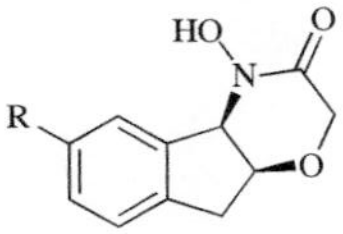

Me₂N

H

132 | R: H
133 | Br

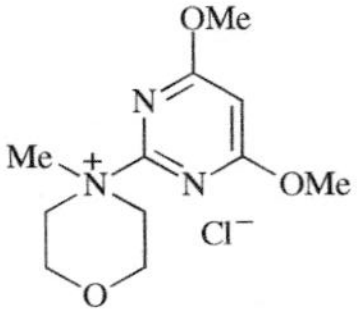

HO

R

O

134

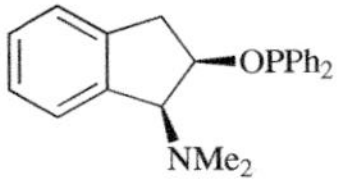

OMe

OMe

Me

N⁺

Cl⁻

135

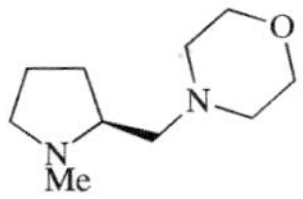

OPPh₂

NMe₂

136

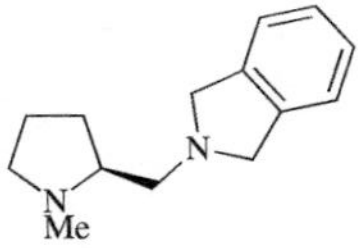

Me

137

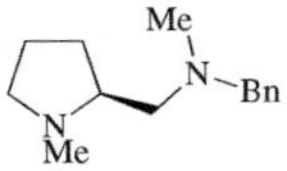

N

Me

138

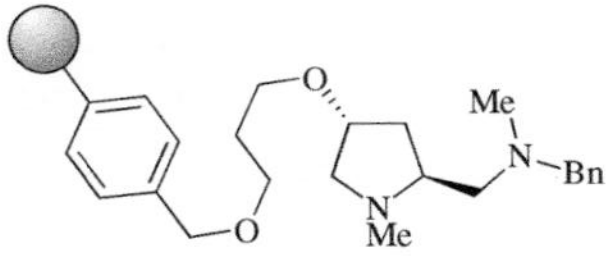

Me

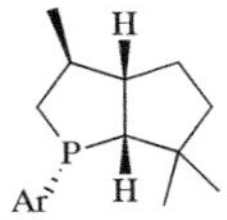

N–Bn

Me

139

Me

N–Bn

Me

= J_{anda}-el

140 | Ph
141 | 3,5-t-Bu₂C₆H₃
142 | 3,5-t-Bu₂-4-MeOC₆H₂

Ar

H

H

P

Ar

129

CHART 3. LEWIS BASE CATALYSTS (*Continued*)

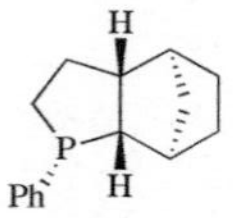

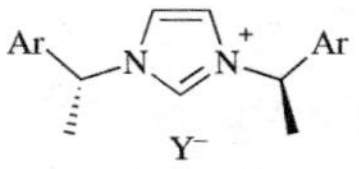

	Ar	Y
146	Ph	BF$_4$
147	1-Np	BF$_4$
148	1-Np	Cl
149	1-pyrenyl	BF$_4$

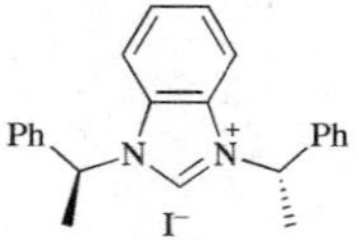

153

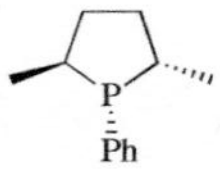

144

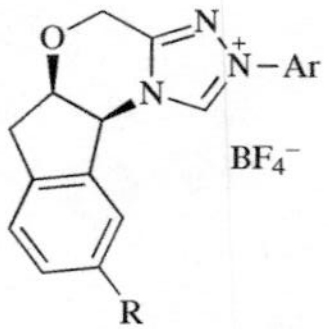

	Ar	R
145a	C$_6$F$_5$	H
145b	Ph	O$_2$N
145c	2,4,6-Cl$_3$C$_6$H$_2$	H
145d	Mes	H
145e	Mes	Br
145f	2,6-i-Pr$_2$C$_6$H$_3$	O$_2$N
145g	2,4,6-i-Pr$_3$C$_6$H$_2$	O$_2$N

152

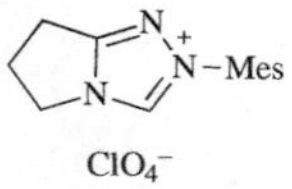

151

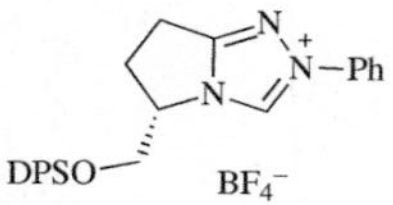

143

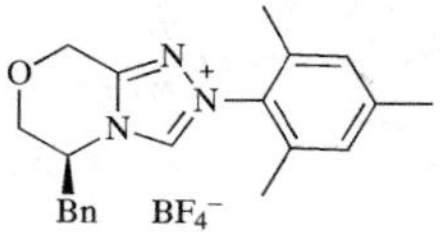

150

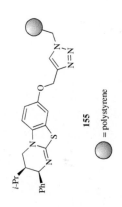

154

155

= polystyrene

156

CHART 4. BRØNSTED ACID CATALYSTS

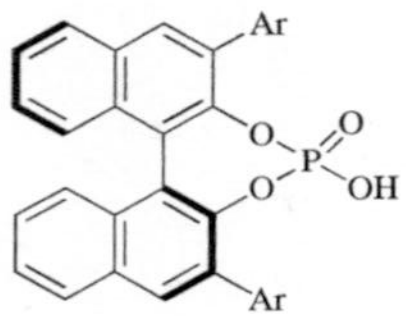

Y	
158a	S
158b	O

	Ar
157a	2,4,6-*i*-Pr₃C₆H₂
157b	2,4,6-Cy₃C₆H₂

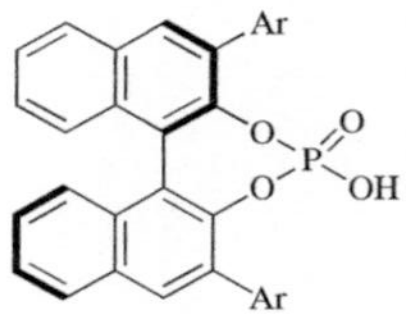

Ar = 9-anthracenyl **157c**

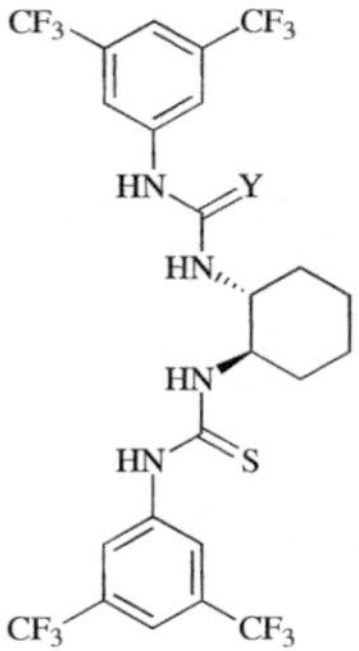

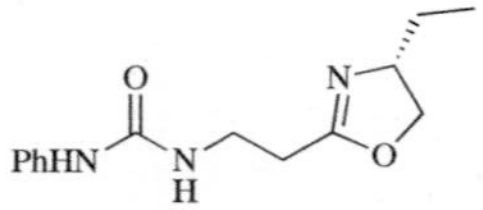

159

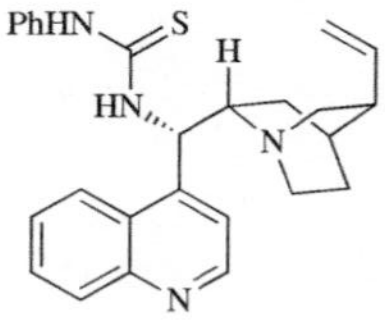

160

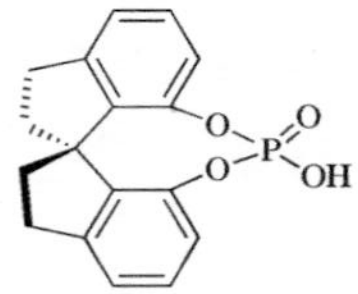

161

CHART 5. MISCELLANEOUS OXIDANTS AND ADDITIVES

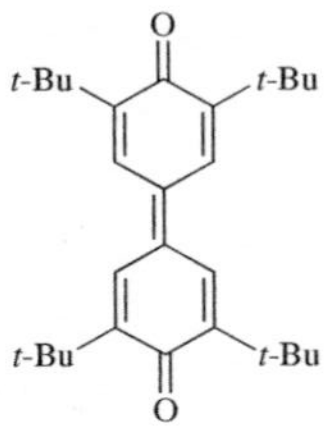

162

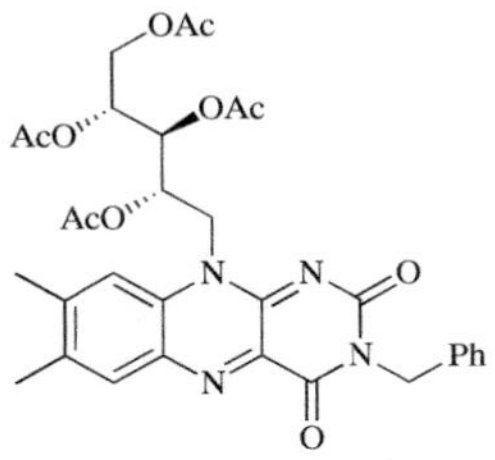

163

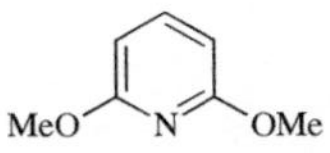

164

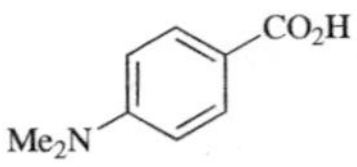

165

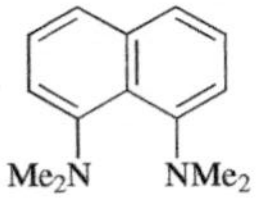

166

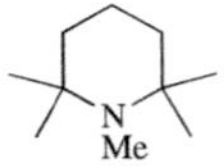

167

TABLE 1A. DESYMMETRIZATION OF *MESO*-1,2-DIOLS

*Please refer to the charts preceding the tables for the structures indicated by the **bold** numbers.*

Diol	Conditions	Product(s) and Yield(s) (%)	Refs.
C4			
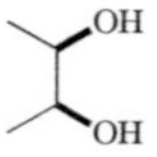	Ac$_2$O (5.3 eq), **105** (1 mol %), *i*-Pr$_2$NEt (5.3 eq), toluene, −40°, 24 h	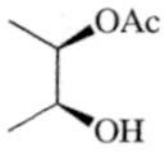 (85) er 92.0:8.0	123
	(*i*-PrCO)$_2$O (1.3 eq), **63** (5 mol %), collidine (1.4 eq), CHCl$_3$, −60°, 24 h	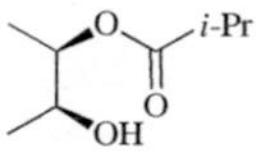 (72) er 93.5:6.5	120
C4–16	(*i*-PrCO)$_2$O (1.3 eq), **69b** (0.1 mol %), Et$_3$N (1.3 eq), MTBE, 3 h		121

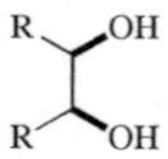

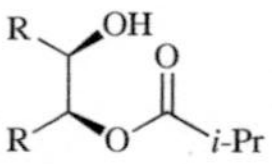

R	Temp (°)		er
Me	−40	(85)	97.0:3.0
ClCH$_2$	−20	(46)	86.0:14.0
N$_3$CH$_2$	−20	(50)	86.0:14.0
PivOCH$_2$	−40	(51)	91.0:9.0
BnOCH$_2$	−20	(63)	94.5:5.5
CH$_2$=CHCH$_2$	−20	(81)	91.0:9.0
Ph	−20	(79)	97.0:3.0
4-FC$_6$H$_4$	−20	(78)	95.0:5.0
4-BrC$_6$H$_4$	−40	(64)	95.0:5.0
4-MeC$_6$H$_4$	−20	(73)	97.0:3.0

Diol	Conditions	Product(s) and Yield(s) (%)	Refs.
C4			
	BzCl (2.4 eq), **46**•CuCl$_2$ (3 mol %), *i*-Pr$_2$NEt (2 eq), CH$_2$Cl$_2$, −78°	(78) er 86.0:14.0	132

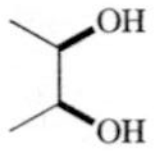

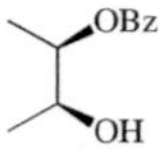

134

Conditions	Product	Yield (er)	Refs.
BzCl (1.5 eq), **131** (2 mol %), Et$_3$N, 4 Å MS, EtOAc, –60°, 22 h	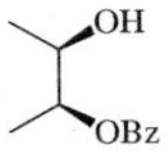	(82) er 95.0:5.0	118
BzCl (2.4 eq), CuCl$_2$ (3 mol %), **45** (3 mol %), i-Pr$_2$NEt (2 eq), CH$_2$Cl$_2$, –78°	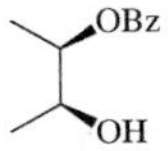	(77) er 96.0:4.0	133
BzCl (1.5 eq), **138** (0.5 mol %), Et$_3$N, 4 Å MS, CH$_2$Cl$_2$, –78°, 3 h	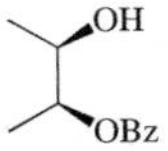	(85) er 97.0:3.0	86
BzCl (1.5 eq), **137**, 4 Å MS, CH$_2$Cl$_2$, –78°, 24 h	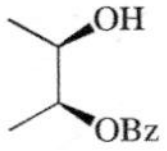	(80) er 95.5:4.5	116
BzCl (1.5 eq), **128** (30 mol %), i-Pr$_2$NEt, EtCN, –78°, 4 h	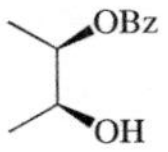	(99) er 93.0:7.0	114
BzCl, CuCl$_2$ (1 mol %), **25** (1 mol %), i-Pr$_2$NEt, CH$_2$Cl$_2$, 0°	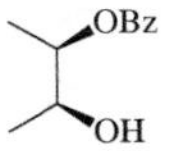	(65) er 97.0:3.0	130
BzCl, Cu(OTf)$_2$ (10 mol %), **15** (10 mol %), K$_2$CO$_3$ (1.5 eq), CH$_2$Cl$_2$, rt, 3 h	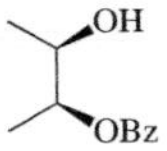	(78) er 98.5:1.5	129

TABLE 1A. DESYMMETRIZATION OF *MESO*-1,2-DIOLS (*Continued*)

*Please refer to the charts preceding the tables for the structures indicated by the **bold** numbers.*

Diol	Conditions	Product(s) and Yield(s) (%)	Refs.
C₄			
(structure: 2,3-butanediol, OH/OH)	BzCl, **51** (2 mol %), *i*-Pr₂NEt (2 eq), CH₂Cl₂, 0°, 8 h	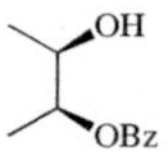 (structure OH/OBz) (96) er 70.0:30.0	307
	4-*t*-BuC₆H₄COCl (1.5 eq), **135** (5 mol %), *i*-Pr₂NEt, 4 Å MS, toluene, 0°, 24 h	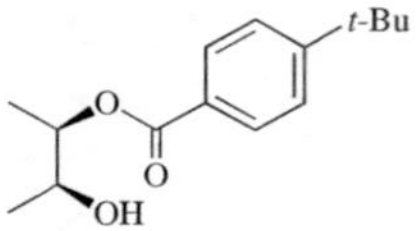 (structure ester with *t*-Bu aryl, OH) (82) er 94.5:5.5	115
	ArCOCl, CuCl (5 mol %), **32** (5.5 mol %), *i*-Pr₂NEt (2 eq), CH₂Cl₂, –40°	(structure: O–Ar ester, OH) see table below	131
	PhNCO, Cu(OTf)₂ (10 mol %), **15** (10 mol %), THF, rt, 0.5 h	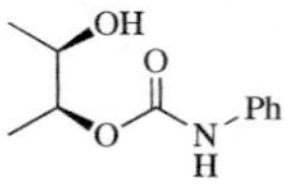 (structure: carbamate N–Ph, OH) (94) er 85.0:15.0	129
	PhNCO, Cu(OTf)₂ (10 mol %), *ent*-**15** (10 mol %), THF, rt, 0.5 h	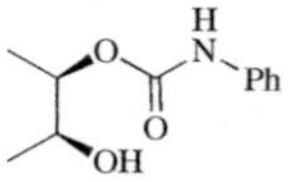 (structure: carbamate N–Ph, OH) (94) er 85.0:15.0	128

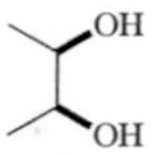

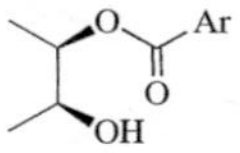

Ar	Time (h)		er
Ph	47	(97)	97.5:2.5
4-MeOC₆H₄	26	(87)	96.5:3.5
4-BrC₆H₄	50	(74)	96.0:4.0
2-MeC₆H₄	50	(98)	97.0:3.0
3-MeC₆H₄	51	(91)	94.5:5.5
4-MeC₆H₄	50	(89)	96.5:3.5

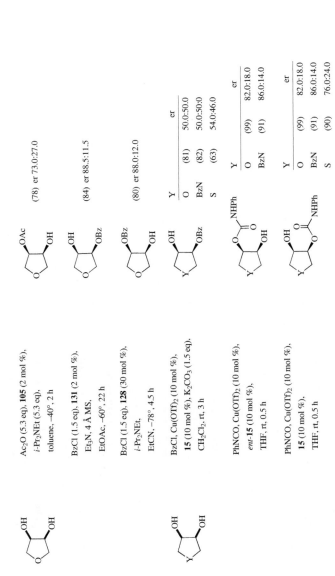

Substrate	Conditions	Product	Yield (er)	Ref.
(tetrahydrofuran-3,4-diol)	Ac$_2$O (5.3 eq), **105** (2 mol %), i-Pr$_2$NEt (5.3 eq), toluene, −40°, 2 h	OAc / OH	(78) er 73.0:27.0	123
	BzCl (1.5 eq), **131** (2 mol %), Et$_3$N, 4 Å MS. EtOAc, −60°, 22 h	OH / OBz	(84) er 88.5:11.5	118
	BzCl (1.5 eq), **128** (30 mol %), i-Pr$_2$NEt, EtCN, −78°, 4.5 h	OBz / OH	(80) er 88.0:12.0	114
(Y-heterocycle diol)	BzCl, Cu(OTf)$_2$ (10 mol %), **15** (10 mol %), K$_2$CO$_3$ (1.5 eq). CH$_2$Cl$_2$, rt, 3 h	OH / OBz	Y — (yield) — er: O (81) 50.0:50.0; BzN (82) 50.0:50.0; S (63) 54.0:46.0	129
	PhNCO, Cu(OTf)$_2$ (10 mol %), *ent*-**15** (10 mol %), THF, rt, 0.5 h	OH / O–C(=O)–NHPh	Y — (yield) — er: O (99) 82.0:18.0; BzN (91) 86.0:14.0	128
	PhNCO, Cu(OTf)$_2$ (10 mol %), **15** (10 mol %), THF, rt, 0.5 h	OH / O–C(=O)–NHPh	Y — (yield) — er: O (99) 82.0:18.0; BzN (91) 86.0:14.0; S (90) 76.0:24.0	129

TABLE 1A. DESYMMETRIZATION OF *MESO*-1,2-DIOLS (*Continued*)

*Please refer to the charts preceding the tables for the structures indicated by the **bold** numbers.*

C_4

Diol	Conditions	Product(s) and Yield(s) (%)	Refs.
 BnO⟍⟍OH BnO⟍⟍OH	BzCl, Cu(OTf)$_2$ (10 mol %), **15** (10 mol %), K$_2$CO$_3$ (1.5 eq), CH$_2$Cl$_2$, rt, 3 h	BnO⟍⟍OH / BnO⟍⟍OBz (36) er 98.0:2.0	129
	BzCl (1.5 eq), **131** (2 mol %), Et$_3$N, 4 Å MS, EtOAc, –60°, 22 h	BnO⟍⟍OH / BnO⟍⟍OBz (51) er 96.5:3.5	118
	BzCl (1.5 eq), **138** (0.5 mol %), Et$_3$N, 4 Å MS, CH$_2$Cl$_2$, –78°, 24 h	BnO⟍⟍OH / BnO⟍⟍OBz (73) er 91.0:9.0	86
	PhNCO, Cu(OTf)$_2$ (10 mol %), **15** (10 mol %), THF, rt, 0.5 h	BnO⟍⟍OH / BnO⟍⟍O—C(=O)—N(H)Ph (91) er 91.0:9.0	129
 RO$_2$C⟍⟍OH RO$_2$C⟍⟍OH	BzCl, CuCl$_2$ (10 mol %), *ent*-**15** (12 mol %), AgOPiv (10 mol %), EtOAc, rt	RO$_2$C⟍⟍OH / RO$_2$C⟍⟍OBz | R	Time (h)		er
Me	18	(80)	92.0:8.0
t-Bu	24	(80)	61.5:38.5
Bn	3	(94)	98.0:2.0		127
 MeO$_2$C⟍⟍OH MeO$_2$C⟍⟍OH	1. Bu$_2$SnO (1.1 eq), 4 Å MS, toluene, 115°, 2 h 2. **3** (1.3 eq), toluene, 0°, 1 h	MeO$_2$C⟍⟍OH / MeO$_2$C⟍⟍O—(camphor ester) (80) dr 95.0:5.0	112

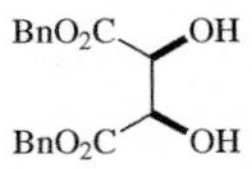

C₅

RCO₂-2-Py, CuCl₂ (10 mol %),
ent-**15** (12 mol %),
AgOPiv (10 mol %), EtOAc, rt, 6 h

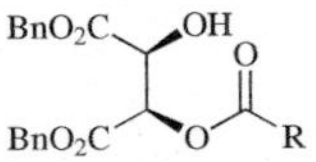

R		er
(*E*)-MeCH=CH	(81)	65.5:34.5
Ph	(94)	98.0:2.0
(*E*)-PhCH=CH	(89)	91.0:9.0
(*E*)-3,4-(BnO)₂C₆H₃CH=CH	(89)	94.5:5.5
n-C₁₁H₂₃	(70)	56.5:43.5
3,5-*t*-Bu₂C₆H₃	(80)	96.5:3.5

127

C₅₋₈

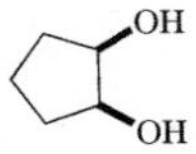

Ac₂O (1.5 eq), **144** (16 mol %),
CD₂Cl₂, rt, 5 h

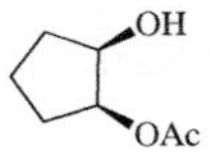

(10) er 76.0:24.0

87

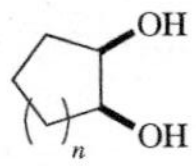

Ac₂O (5.3 eq), **105** (1 mol %),
i-Pr₂NEt (5.3 eq),
toluene, –40°

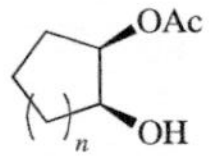

n	Time (h)		er
1	7	(89)	90.0:10.0
2	48	(81)	94.0:6.0
4	22	(97)	92.0:8.0

123

C₅

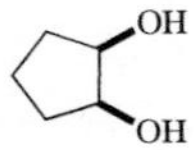

BzCl, CuCl₂ (1 mol %),
25 (1 mol %),
i-Pr₂NEt, CH₂Cl₂, 0°

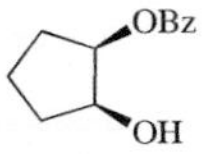

(73) er 88.0:12.0

130

139

TABLE 1A. DESYMMETRIZATION OF *MESO*-1,2-DIOLS (*Continued*)

Diol	Conditions	Product(s) and Yield(s) (%)	Refs.

*Please refer to the charts preceding the tables for the structures indicated by the **bold** numbers.*

C$_5$

| | BzCl (1.5 eq), **128** (30 mol %), | (80) er 96.5:3.5 | 114 |
| | *i*-Pr$_2$NEt, EtCN, –78°, 4 h | | |

C$_{5-8}$

	BzCl (2.4 eq), CuCl$_2$ (3 mol %),		133
	45 (3 mol %), *i*-Pr$_2$NEt (2 eq),		
	CH$_2$Cl$_2$, –78°		

n		er
1	(73)	89.5:10.5
2	(100)	96.5:3.5
4	(82)	85.0:15.0

	PhNCO, Cu(OTf)$_2$ (10 mol %),		128
	ent-**15** (10 mol %),		
	THF, –40°, 0.5 h		

n		er
1	(82)	93.0:7.0
2	(69)	93.0:7.0
3	(83)	95.5:4.5
4	(72)	96.5:3.5

	BzCl (1.5 eq), **131** (2 mol %),		118
	Et$_3$N, 4 Å MS,		
	EtOAc, –60°, 22 h		

n		er
1	(87)	89.0:11.0
2	(92)	98.5:1.5
4	(86)	88.5:11.5

	BzCl, Cu(OTf)$_2$ (10 mol %),		129
	15 (10 mol %), K$_2$CO$_3$ (1.5 eq),		
	CH$_2$Cl$_2$, rt, 3 h		

n		er
1	(47)	51.5:48.5
3	(88)	79.0:21.0
4	(85)	82.5:17.5

	n		er
	1	(97)	72.0:28.0
	2	(95)	85.0:15.0
	4	(93)	65.0:35.0

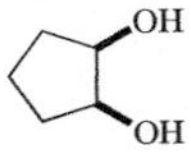

BzCl, **51** (2 mol %),

i-Pr₂NEt (2 eq), CH₂Cl₂, 0°, 8 h

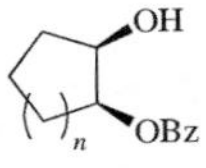

307

4-*t*-BuC₆H₄COCl (1.5 eq),
135 (5 mol %), *i*-Pr₂NEt,
4 Å MS, toluene, 0°, 12 h

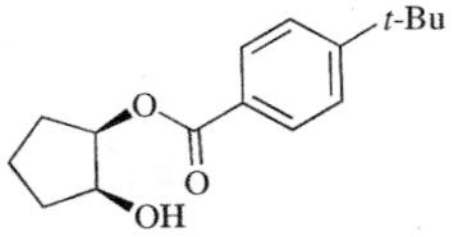

(96) er 79.5:20.5

115

4-MeOC₆H₄COCl,
CuCl (5 mol %), **32** (5.5 mol %),
i-Pr₂NEt (2 eq), CH₂Cl₂, –40°, 55 h

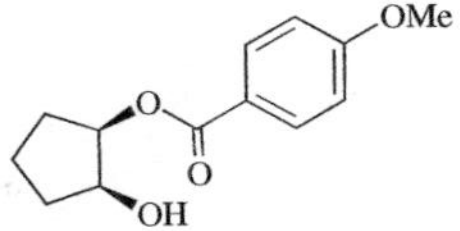

(69) er 80.5:19.5

131

1. Bu₂SnO (1.1 eq).
4 Å MS, toluene, 115°, 2 h
2. **3** (1.3 eq), toluene, 0°, 1 h

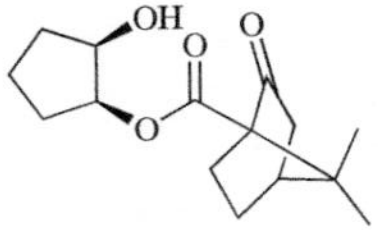

(79) dr 53.5:46.5

112

	n		er
	1	(91)	86.0:14.0
	3	(83)	91.5:8.5
	4	(96)	93.0:7.0

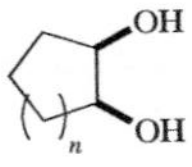

PhNCO, Cu(OTf)₂ (10 mol %),
15 (10 mol %),
THF, rt, 0.5 h

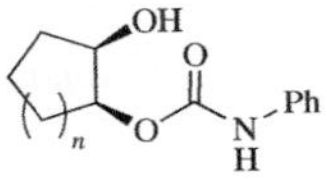

129

C₅

C₅₋₈

TABLE 1A. DESYMMETRIZATION OF *MESO*-1,2-DIOLS (*Continued*)

*Please refer to the charts preceding the tables for the structures indicated by the **bold** numbers.*

Diol	Conditions	Product(s) and Yield(s) (%)	Refs.
C₆			
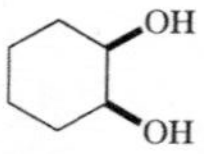 (cyclohexane-1,2-diol, OH/OH)	Ac_2O (5.3 eq), **107** (5 mol %), toluene, 0°, 7 h	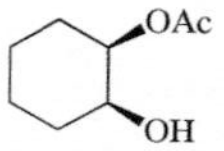 (77) er 88.0:12.0	122
	144 (16 mol %), Ac_2O (1.5 eq), CD_2Cl_2, rt, 5 h	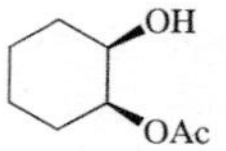 (66) er 83.5:16.5	87
	Ac_2O (5.3 eq), **105** (2 mol %), toluene, −40°, 48 h	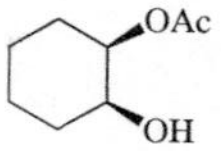 (99) er 94.0:6.0	73
	$(i\text{-PrCO})_2O$ (1.3 eq), **68** (5 mol %), collidine (1.4 eq), toluene, rt, 4 h	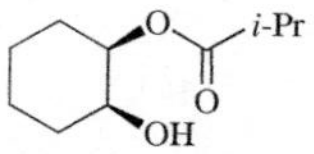 (72) er 77.0:23.0	119
	$(i\text{-PrCO})_2O$ (1.3 eq), **63** (5 mol %), collidine (1.4 eq), $CHCl_3$, rt, 4 h	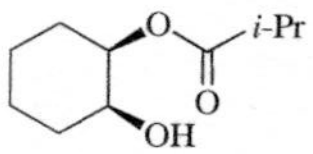 (75) er 93.5:6.5	120
	$(i\text{-PrCO})_2O$ (2 eq), **87** (1 mol %), Et_3N (0.75 eq), toluene, −78°, 9 h	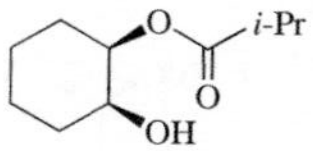 (20) er 89.0:11.0	44
	$(i\text{-PrCO})_2O$ (1.3 eq), **69b** (0.1 mol %), Et_3N (1.3 eq), MTBE, −20°, 3 h	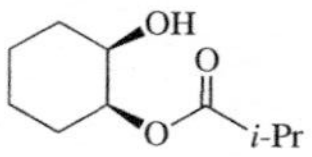 (79) er 97.0:3.0	121

5 (1.1 eq), toluene, 80°, 15 h

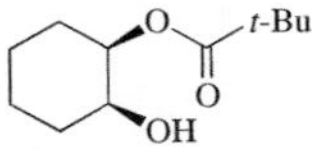

(55) er 71.0:29.0   25, 26

PhCHO, *ent*-**145a** (10 mol %), **163** (10 mol %), Et$_3$N, 4 Å MS, CHCl$_3$, rt, 4 h

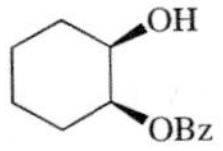

(30) er 82.0:18.0   82

BzCl (2.4 eq), **46**•CuCl$_2$ (3 mol %), *i*-Pr$_2$NEt (2 eq), CH$_2$Cl$_2$, –78°

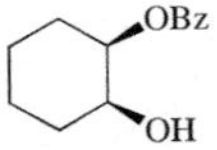

(63) er 80.0:20.0   132

BzCl (1.5 eq), **128** (30 mol %), *i*-Pr$_2$NEt, EtCN, –78°, 6 h

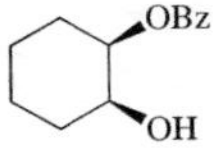

(85) er 97.0:3.0   114

BzCl (1.5 eq), **138** (0.5 mol %), Et$_3$N, 4 Å MS, CH$_2$Cl$_2$, –78°, 24 h

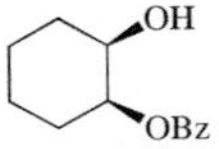

(87) er 98.5:1.5   86

BzCl (1.5 eq), **137** (1 eq), 4 Å MS, EtCN, –78°, 24 h

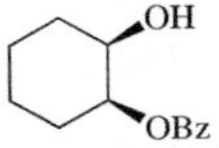

(89) er 96.5:3.5   116

BzCl, CuCl$_2$ (1 mol %), **25** (1 mol %), *i*-Pr$_2$NEt, CH$_2$Cl$_2$, 0°

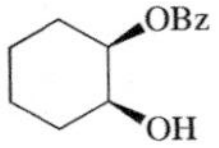

(83) er 95.0:5.0   130

TABLE 1A. DESYMMETRIZATION OF MESO-1,2-DIOLS (*Continued*)

*Please refer to the charts preceding the tables for the structures indicated by the **bold** numbers.*

Diol	Conditions	Product(s) and Yield(s) (%)	Refs.
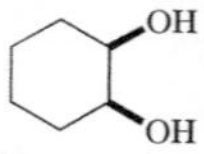 C$_6$ (cyclohexane-1,2-diol, OH/OH)	BzCl, CuCl$_2$ (10 mol %), *ent*-**15** (12 mol %), AgOPiv (10 mol %), EtOAc, rt, 28 h	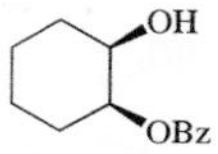 (80) er 64.0:36.0	127
	4-MeC$_6$H$_4$COCl (1.5 eq), **137** (0.5 mol %), Et$_3$N, 4 Å MS, CH$_2$Cl$_2$, –78°, 3 h	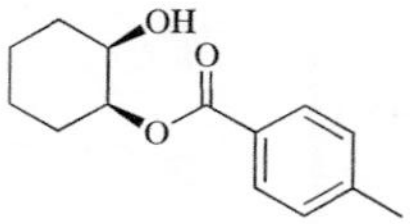 (92) er 99.0:1.0	117
	4-MeC$_6$H$_4$COCl, CuCl (5 mol %), **32** (5.5 mol %), *i*-Pr$_2$NEt (2 eq), CH$_2$Cl$_2$, 0°, 19 h	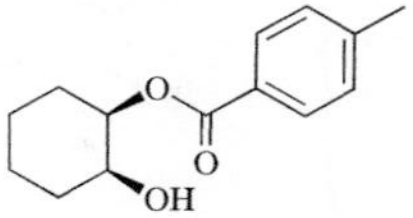 (72) er 90.0:10.0	131
	4-*t*-BuC$_6$H$_4$COCl (1.5 eq), **135** (5 mol %), *i*-Pr$_2$NEt, 4 Å MS, toluene, 0°, 12 h	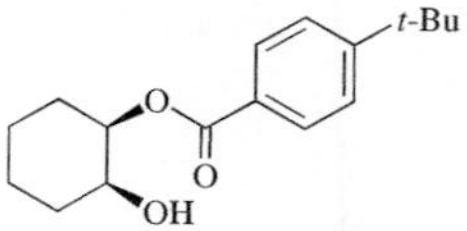 (83) er 96.5:3.5	115
	(*E*)-PhCH=CHCHO, **145d** (30 mol %), K$_2$CO$_3$ (30 mol %), 18-crown-6 (15 mol %), Proton Sponge, MnO$_2$ (15 eq), CH$_2$Cl$_2$, –30°	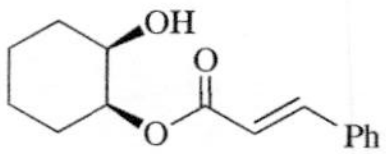 (58) er 90.0:10.0	80, 124

144

Starting material (left): cyclohexene-diol, OH / OH

Conditions	Product	R		er	Ref
RNCO, Cu(OTf)$_2$ (10 mol %), ent-15 (10 mol %), THF, rt, 0.5 h	(O–C(=O)–NHR, OH)	t-Bu	(50)	87.5:12.5	128
		Ph	(92)	88.0:12.0	
		4-ClC$_6$H$_4$	(93)	88.0:12.0	
		4-MeOC$_6$H$_4$	(85)	84.0:16.0	
		1-adamantyl	(68)	81.0:19.0	
		1-Np	(77)	88.5:11.5	
(i-PrCO)$_2$O (1.3 eq), 69b (0.1 mol %), Et$_3$N (1.3 eq), MTBE, −40°, 3 h	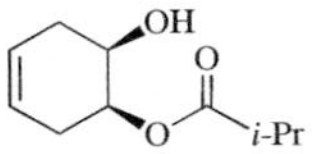 (OH, O–C(=O)–i-Pr)	(86) er 94.0:6.0			121
BzCl (1.5 eq), 137 (1 eq), 4 Å MS, CH$_2$Cl$_2$, −78°, 24 h	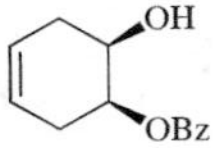 (OH, OBz)	(78) er 98.0:2.0			116
BzCl (1.5 eq), 131 (2 mol %), Et$_3$N, 4 Å MS, EtOAc, −60°, 22 h	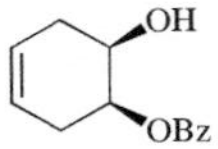 (OH, OBz)	(79) er 92.0:8.0			118
BzCl (1.5 eq), 138 (0.5 mol %), Et$_3$N, 4 Å MS, CH$_2$Cl$_2$, −78°, 3 h	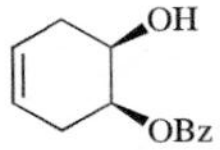 (OH, OBz)	(81) er 95.0:5.0			86
BzCl, Cu(OTf)$_2$ (10 mol %), 15 (10 mol %), K$_2$CO$_3$ (1.5 eq), CH$_2$Cl$_2$, rt, 3 h	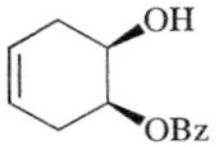 (OH, OBz)	(68) er 96.5:3.5			129

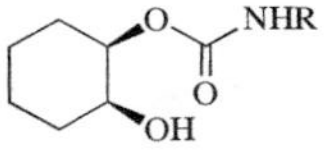

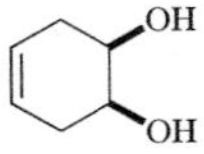

145

TABLE 1A. DESYMMETRIZATION OF *MESO*-1,2-DIOLS (*Continued*)

Diol	Conditions	Product(s) and Yield(s) (%)	Refs.

*Please refer to the charts preceding the tables for the structures indicated by the **bold** numbers.*

C$_6$

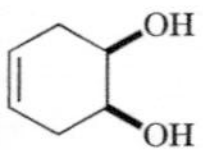

PhNCO, Cu(OTf)$_2$ (10 mol %),
15 (10 mol %), THF, rt, 0.5 h

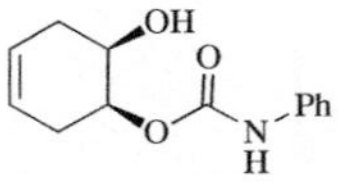

(96) er 79.5:20.5

129

C$_8$

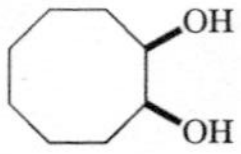

BzCl (2.4 eq), **46**·CuCl$_2$ (3 mol %),
i-Pr$_2$NEt (2 eq), CH$_2$Cl$_2$, –78°

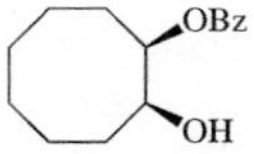

(78) er 74.0:26.0

132

BzCl (1.2 eq), **134** (1.2 eq),
CuCl$_2$ (20 mol %), **15** (30 mol %),
DMAP (10 mol %), *i*-Pr$_2$NEt,
H$_2$O, rt, 5 d

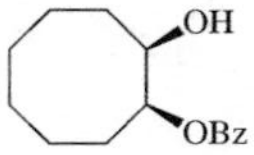

(59) er 80.0:20.0

126

BzCl, CuCl$_2$ (10 mol %),
ent-**15** (12 mol %),
AgOPiv (10 mol %), EtOAc, rt, 28 h

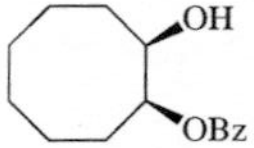

(79) er 51.5:48.5

127

4-MeC$_6$H$_4$COCl (1.5 eq),
137 (0.5 mol %), Et$_3$N, 4 Å MS,
CH$_2$Cl$_2$, –78°, 3 h

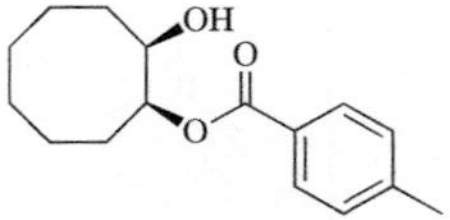

(92) er 99.5:0.5

117

4-MeOC$_6$H$_4$COCl, CuCl (5 mol %),
32 (5.5 mol %), *i*-Pr$_2$NEt (2 eq),
CH$_2$Cl$_2$, –78°, 70 h

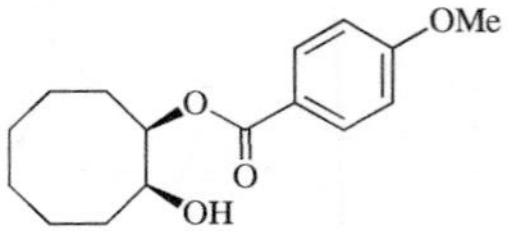

(76) er 93.0:7.0

131

146

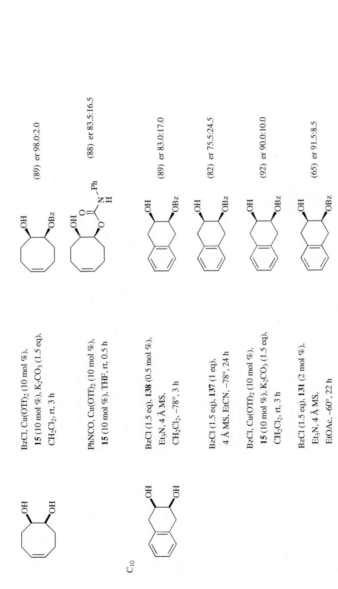

C$_{10}$

Substrate	Conditions	Product	Ref.
	BzCl, Cu(OTf)$_2$ (10 mol %), **15** (10 mol %), K$_2$CO$_3$ (1.5 eq), CH$_2$Cl$_2$, rt, 3 h	(89) er 98.0:2.0	129
	PhNCO, Cu(OTf)$_2$ (10 mol %), **15** (10 mol %), THF, rt, 0.5 h	(88) er 83.5:16.5	129
	BzCl (1.5 eq), **138** (0.5 mol %), Et$_3$N, 4 Å MS, CH$_2$Cl$_2$, –78°, 3 h	(89) er 83.0:17.0	86
	BzCl (1.5 eq), **137** (1 eq), 4 Å MS, EtCN, –78°, 24 h	(82) er 75.5:24.5	116
	BzCl, Cu(OTf)$_2$ (10 mol %), **15** (10 mol %), K$_2$CO$_3$ (1.5 eq), CH$_2$Cl$_2$, rt, 3 h	(92) er 90.0:10.0	129
	BzCl (1.5 eq), **131** (2 mol %), Et$_3$N, 4 Å MS, EtOAc, –60°, 22 h	(65) er 91.5:8.5	118

TABLE 1A. DESYMMETRIZATION OF MESO-1,2-DIOLS (*Continued*)

*Please refer to the charts preceding the tables for the structures indicated by the **bold** numbers.*

Diol	Conditions	Product(s) and Yield(s) (%)	Refs.
C$_{10}$	PhNCO, Cu(OTf)$_2$ (10 mol %), **15** (10 mol %), THF, rt, 0.5 h	(86) er 75.0:25.0	129
C$_{12}$	1. Bu$_2$SnO (1.1 eq), 4 Å MS, toluene, 115°, 2 h 2. **3** (1.3 eq), toluene, 0°, 1 h	(90) dr 60.0:40.0	112
C$_{14}$	Ac$_2$O (5.3 eq), **105** (2 mol %), *i*-Pr$_2$NEt (5.3 eq), toluene, –20°, 24 h	(45) er 69.0:31.0	123
	RCOCl, **51** (2 mol %), *i*-Pr$_2$NEt (2 eq), CH$_2$Cl$_2$, 0°, 8 h	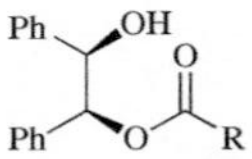	307

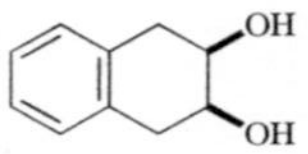

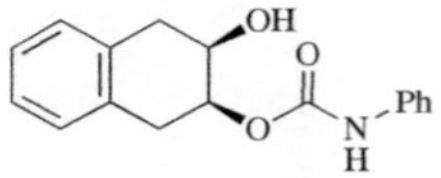

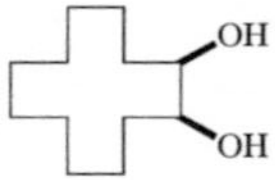

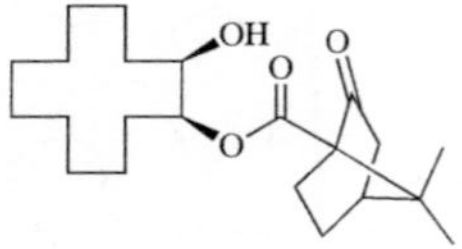

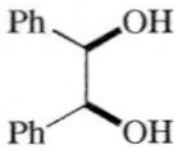

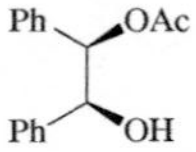

	er	
Ar		
Me	(96)	59.0:41.0
Cy	(94)	67.0:33.0
Ph	(98)	96.0:4.0
2-FC$_6$H$_4$	(93)	89.0:11.0
4-FC$_6$H$_4$	(96)	91.0:9.0
4-O$_2$NC$_6$H$_4$	(95)	55.0:45.0
4-MeC$_6$H$_4$	(95)	94.0:6.0

148

Conditions	Product	Yield (er)	Ref.
$(i\text{-PrCO})_2O$ (1.1 eq), **76** (5 mol %), Et$_3$N (1.1 eq), THF, 0°, 3 h	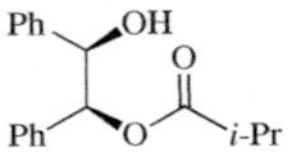 Ph–OH / ester i-Pr (Ph)	(31) er 86.0:14.0	41
$(i\text{-PrCO})_2O$ (2 eq), **87** (1 mol %), Et$_3$N (0.75 eq), toluene, –78°, 9 h	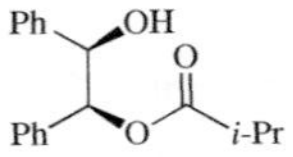 Ph–OH / ester i-Pr (Ph)	(26) er 72.5:27.5	44
BzCl (2.4 eq), **46**•CuCl$_2$ (3 mol %), i-Pr$_2$NEt (2 eq), CH$_2$Cl$_2$, –78°	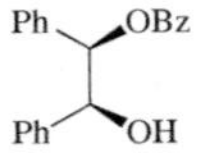 Ph–OBz / Ph–OH	(56) er 87.0:13.0	132
BzCl (1.5 eq), **131** (2 mol %), Et$_3$N, 4 Å MS, THF, –60°, 22 h	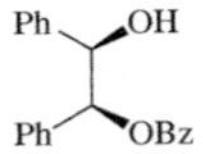 Ph–OH / Ph–OBz	(68) er 67.0:33.0	118
BzCl, Cu(OTf)$_2$ (5 mol %), **15** (5 mol %), i-Pr$_2$NEt (2 eq), CH$_2$Cl$_2$, 0°	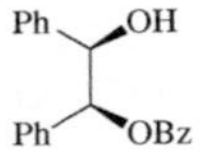 Ph–OH / Ph–OBz	(79) er 97.0:3.0	125
BzCl (1.5 eq), **128** (30 mol %), i-Pr$_2$NEt, EtCN, –78°, 1.5 h	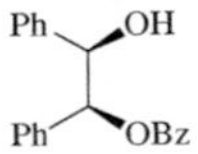 Ph–OH / Ph–OBz	(98) er 95.5:4.5	114
BzCl (1.2 eq), **134** (1.2 eq), CuCl$_2$ (20 mol %), **15** (30 mol %), DMAP (10 mol %), i-Pr$_2$NEt, H$_2$O, rt, 5 d	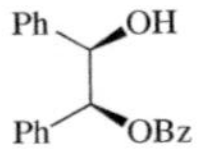 Ph–OH / Ph–OBz	(44) er 99.5:0.5	126

TABLE 1A. DESYMMETRIZATION OF *MESO*-1,2-DIOLS (*Continued*)

Diol	Conditions	Product(s) and Yield(s) (%)	Refs.

*Please refer to the charts preceding the tables for the structures indicated by the **bold** numbers.*

C$_{14}$

Diol	Conditions	Product(s) and Yield(s) (%)	Refs.
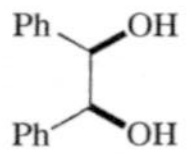 Ph—OH Ph—OH	BzCl (2.4 eq), CuCl$_2$ (3 mol %), **45** (3 mol %), *i*-Pr$_2$NEt (2 eq), CH$_2$Cl$_2$, –78°	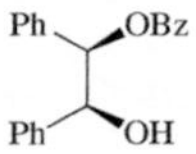 Ph—OBz Ph—OH (73) er 93.0:7.0	133
	BzCl (1.5 eq), **138** (0.5 mol %), Et$_3$N, 4 Å MS, CH$_2$Cl$_2$, –78°, 24 h	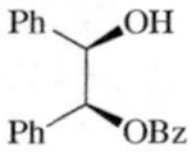 Ph—OH Ph—OBz (80) er 80.0:20.0	86
	BzCl (1.5 eq), **137**, 4 Å MS, CH$_2$Cl$_2$, –78°, 24 h	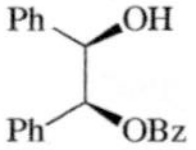 Ph—OH Ph—OBz (68) er 82.0:18.0	116
	BzCl (2 eq), **91** (10 mol %), *i*-Pr$_2$NEt (2 eq), CH$_2$Cl$_2$, –80°, 24 h	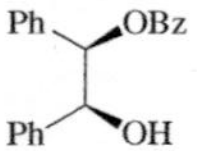 Ph—OBz Ph—OH (90) er 99.0:1.0	308
	Bz$_2$O (1.5 eq), **144** (16 mol %), CD$_2$Cl$_2$, rt, 5 h	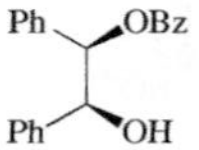 Ph—OBz Ph—OH (84) er 84.0:16.0	87
	BzCl, CuCl$_2$ (10 mol %), *ent*-**15** (12 mol %), AgOPiv (10 mol %), EtOAc, rt, 27 h	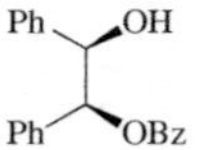 Ph—OH Ph—OBz (71) er 76.5:23.5	127
	Bz$_2$O (1.5 eq), **140** (10 mol %), CH$_2$Cl$_2$, –30°, 22 h	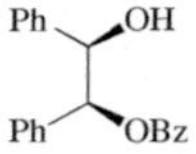 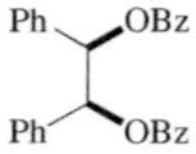 Ph—OH Ph—OBz **I** + Ph—OBz Ph—OBz **II** **I + II** (97) er **I** 97.0:3.0	113

150

ArCOCl, **34**·Cu(BF₄)₂ (5 mol %),
i-Pr₂NEt (2 eq),
CH₂Cl₂, –40°

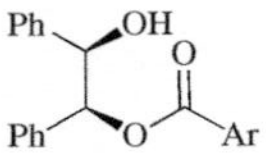

Ar	Time (h)	er	
			309
Ph	32	(92)	91.0:19.0
2-BrC₆H₄	25	(90)	81.0:19.0
4-BrC₆H₄	21	(94)	80.5:19.5
2-IC₆H₄	26	(92)	79.0:21.0
2-O₂NC₆H₄	21	(>99)	83.5:16.5
2-MeC₆H₄	43	(79)	89.5:10.5
3-MeC₆H₄	32	(90)	80.0:20.0
4-MeC₆H₄	21	(>99)	83.5:16.5
4-t-BuC₆H₄	37	(82)	92.5:7.5

4-t-BuC₆H₄COCl (1.5 eq),
135 (5 mol %), i-Pr₂NEt.
4 Å MS, toluene, 0°, 12 h

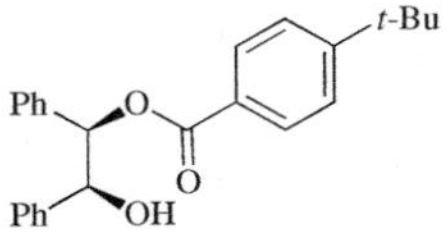

(97) er 97.5:2.5   115

4-BrC₆H₄COCl, CuCl (5 mol %),
32 (5.5 mol %), i-Pr₂NEt (2 eq).
CH₂Cl₂, –78°, 91 h

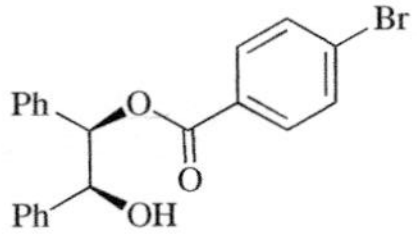

(59) er 90.0:10.0   131

1. Bu₂SnO (1.1 eq), 4 Å MS,
toluene, 115°, 2 h
2. **3** (1.3 eq).
CH₂Cl₂, 0° to rt, 24 h

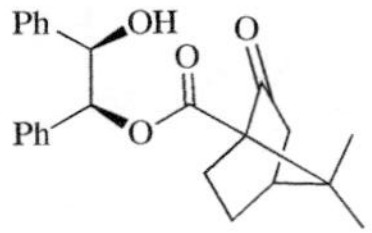

(89) dr 73.0:27.0   112

TABLE 1B. DESYMMETRIZATION OF *MESO*-1,3-DIOLS

*Please refer to the charts preceding the tables for the structures indicated by the **bold** numbers.*

Diol	Conditions	Product(s) and Yield(s) (%)	Refs.
C_3			
[HO—/—OTs diol]	1. **48**, CH$_2$Cl$_2$, 0°, 0.5 h 2. BzCl (2 eq), **136**, 0°, 1 h	[HO—/—OTs, BzO] (20) er 92.0:8.0	134
[HO—/—O—Ar diol]	Ac$_2$O (1.8 eq), **99** (10 mol %), *i*-Pr$_2$NEt (2 eq). toluene/CH$_2$Cl$_2$ (12:1), –55°	[HO—/—O—Ar, AcO]	139
		Ar	er
		Ph (37)	95.5:4.5
		4-FC$_6$H$_4$ (40)	84.0:16.0
		4-MeOC$_6$H$_4$ (48)	94.0:6.0
		3,5-(MeO)$_2$C$_6$H$_3$ (34)	97.5:2.5
		4-CF$_3$C$_6$H$_4$ (37)	86.5:13.5
[HO—/—OPyroc diol]	(CyCO)$_2$O, **110** (5 mol %), *i*-Pr$_2$NEt, CCl$_4$/CHCl$_3$ (3:1), rt, 15 h	[Cy—C(O)O—/—OPyroc, HO] (74) er 96.5:3.5	72
[HO—/—N(Pyroc)R diol]	(CyCO)$_2$O (1.2 eq), **110** (5 mol %), *i*-Pr$_2$NEt (3 eq). CCl$_4$/CHCl$_3$ (3:1), 10°, 24 h	[Cy—C(O)O—/—N(Pyroc)R, HO]	72
		R	er
		H (64)	79.5:20.5
		Me (60)	77.5:22.5
		Bn (15)	55.5:44.5
C_{3-10}			
[HO—/—R diol]	PhCO$_2$CH=CH$_2$ (5 eq). Et$_2$Zn/**40** (2:1) (5 mol %). toluene, –20°, 24 h	[BzO—/—R, HO]	90
		R	er
		TIPSO (22)	80.0:20.0
		BnO (36)	70.0:30.0
		Me (88)	95.0:5.0
		Et (86)	93.0:7.0
		CH$_2$=CHCH$_2$ (84)	92.0:8.0
		=CH$_2$ (90)	90.0:10.0
		PhCH$_2$ (85)	95.0:5.0

C$_4$

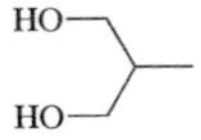

PhCO$_2$CH=CH$_2$ (5 eq),
Et$_2$Zn/**41** (2:1) (5 mol %),
toluene, −20°, 20 h

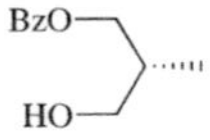

(89) er 91.0:9.0

91

4-t-BuC$_6$H$_4$COCl (1.5 eq),
137 (0.5 mol %),
i-Pr$_2$NEt (1.5 eq), 4 Å MS,
n-PrCN, −78°, 3 h

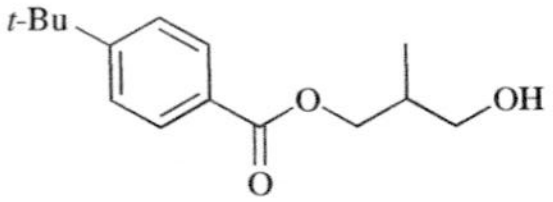

(33) er 98.0:2.0

117

C$_{4-10}$

4-t-BuC$_6$H$_4$COCl (1.5 eq),
135 (5 mol %), i-Pr$_2$NEt,
4 Å MS, toluene, 0°, 9 h

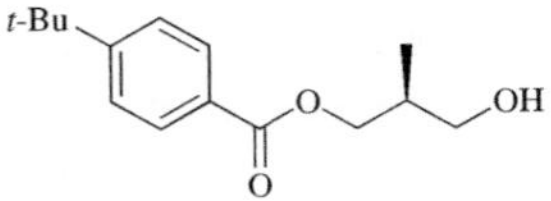

(75) er 73.5:26.5

115

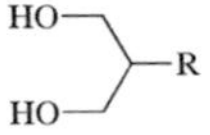

4-t-BuC$_6$H$_4$COCl (1.5 eq),
137 (0.5 mol %),
i-Pr$_2$NEt (1.5 eq),
solvent, 4 Å MS, −78°, 3 h

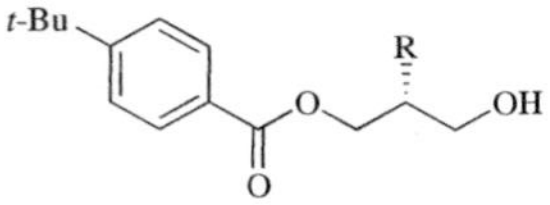

R	Solvent		er
Me	n-PrCN	(33)	98.0:2.0
TBSOCH$_2$	CH$_2$Cl$_2$	(24)	94.5:5.5
CH$_2$=CHCH$_2$	n-PrCN	(30)	99.0:1.0
PhCH$_2$	n-PrCN	(22)	96.5:3.5

138

C$_4$

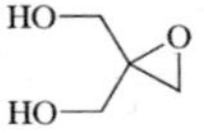

BzCl (1.7 eq), **138** (0.5 mol %),
Et$_3$N (1.7 eq),
n-PrCN, 4 Å MS, −78°, 3 h

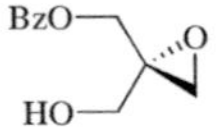

(31) er 98.0:2.0

138

TABLE 1B. DESYMMETRIZATION OF *MESO*-1,3-DIOLS (*Continued*)

Diol	Conditions	Product(s) and Yield(s) (%)		Refs.

*Please refer to the charts preceding the tables for the structures indicated by the **bold** numbers.*

C$_{4-11}$

Diol structure: HO, HO, OH, R

Conditions:
BzCl (1.1 eq),
12•CuCl$_2$ (5 mol %),
Et$_3$N (1.2 eq), THF, rt, 1 h

Product: HO, BzO, OH, ''R

Refs. 141

R		er
Me	(96)	97.0:3.0
TBSOCH$_2$	(94)	94.5:5.5
CH$_2$=CH	(97)	63.5:36.5
TBSO(CH$_2$)$_2$	(94)	95.5:4.5
n-Pr	(97)	96.0:4.0
i-Pr	(94)	90.0:10.0
CH$_2$=CHCH$_2$	(98)	96.0:4.0
Me$_2$CH(CH$_2$)$_2$	(98)	95.5:4.5
Ph	(67)	65.0:35.0
PhCH$_2$	(68)	83.5:16.5
Ph(CH$_2$)$_2$	(97)	96.0:4.0

C$_{4-10}$

Diol structure: HO, HO, OH, R

Conditions:
BzCl (1.1 eq),
36•CuCl$_2$ (20 mol %),
Et$_3$N (1.2 eq), THF, rt, 0.5 h

Product: HO, BzO, OH, ''R

Refs. 142

R		er
Me	(99)	97.5:2.5
BnOCH$_2$	(99)	96.5:3.5
BnO(CH$_2$)$_2$	(99)	96.5:3.5
CH$_2$=CH	(99)	95.0:5.0
TMS—≡	(81)	93.5:6.5
i-Pr	(99)	95.0:5.0
CH$_2$=C(Me)	(99)	91.5:8.5
CH$_2$=CHCH$_2$	(99)	96.5:3.5
n-Bu	(99)	97.5:2.5
i-Bu	(99)	96.0:4.0
Ph	(99)	96.0:4.0
4-MeOC$_6$H$_4$	(99)	95.5:4.5
4-ClC$_6$H$_4$	(99)	97.0:3.0
2,4-F$_2$C$_6$H$_3$	(99)	98.0:2.0
PhCH$_2$	(99)	97.0:3.0

C$_{4-11}$

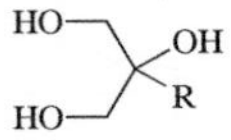

BzCl (1.1 eq),
35-CuCl$_2$ (30 mol %),
Et$_3$N (1.2 eq), THF, rt, 1 h

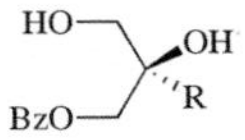

141

R		er
TBSOCH$_2$	(90)	96.5:3.5
CH$_2$=CH	(94)	90.5:9.5
i-Pr	(85)	91.5:8.5
Ph	(94)	90.0:10.0
PhCH$_2$	(91)	93.0:7.0
Ph(CH$_2$)$_2$	(91)	95.0:5.0

C$_4$

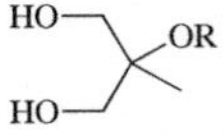

Ac$_2$O (1.5 eq), Et$_2$Zn (5 mol %),
43 (5 mol %),
Et$_2$O, rt, 20 h

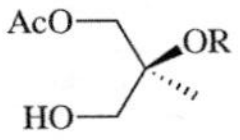

143

R		er
TBS	(55)	76.0:24.0
PhCH$_2$	(91)	90.0:10.0
PMB	(86)	86.5:13.5
3,4-(MeO)$_2$C$_6$H$_3$CH$_2$	(94)	87.0:13.0
3,5-(MeO)$_2$C$_6$H$_3$CH$_2$	(94)	90.0:10.0
4-CF$_3$C$_6$H$_4$CH$_2$	(88)	83.0:17.0
2,4,6-Me$_3$C$_6$H$_2$CH$_2$	(76)	85.0:15.0
1-NpCH$_2$	(96)	89.0:11.0
2-NpCH$_2$	(93)	88.5:11.5
4-PhC$_6$H$_4$CH$_2$	(74)	87.0:13.0

TABLE 1B. DESYMMETRIZATION OF *MESO*-1,3-DIOLS (*Continued*)

Diol	Conditions	Product(s) and Yield(s) (%)	Refs.

*Please refer to the charts preceding the tables for the structures indicated by the **bold** numbers.*

C$_{4-9}$

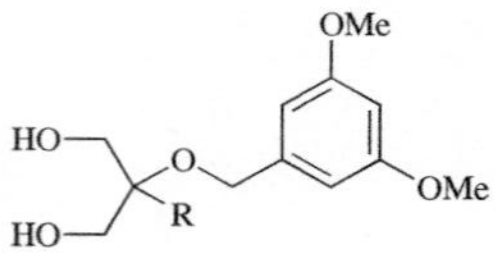

Ac$_2$O (1.5 eq), Et$_2$Zn (5 mol %), **43** (5 mol %), Et$_2$O, 0°, 20 h

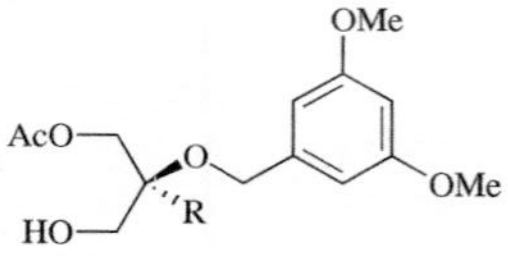

R		er
Me	(78)	93.0:7.0
Et	(68)	88.5:11.5
i-Pr	(69)	90.0:10.0
CH$_2$=CHCH$_2$	(69)	87.0:13.0
Ph	(72)	92.0:8.0

143

Ac$_2$O (1.5 eq), Et$_2$Zn (5 mol %), **44** (5 mol %), Et$_2$O, 0°, 20 h

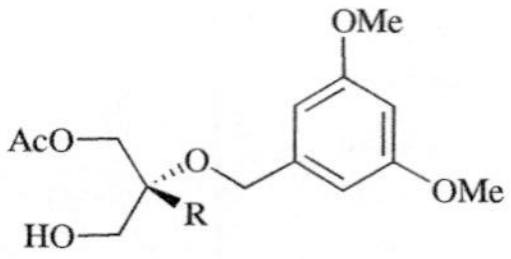

R		er
Me	(65)	91.5:8.5
Et	(56)	89.5:10.5
i-Pr	(66)	90.0:10.0
CH$_2$=CHCH$_2$	(66)	89.5:10.5
Ph	(68)	92.0:8.0

143

C$_{4-10}$

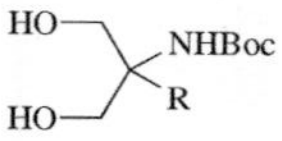

Ac$_2$O (1.5 eq), **49** (5 mol %), MTBE, 0°, 20 h

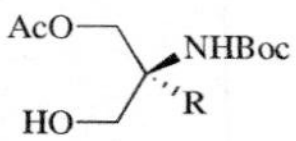

R		er
Me	(92)	94.0:6.0
Et	(87)	93.0:7.0
CH$_2$=CHCH$_2$	(70)	91.0:9.0
n-C$_6$H$_{13}$	(84)	91.5:8.5
PhCH$_2$	(70)	85.0:15.0

144

Top scheme (93)

C_{4-11} HO⟍⟋NHBz R / HO

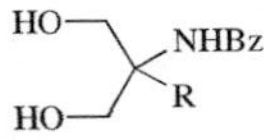

1. $CuCl_2$ (20 mol %), **17** (20 mol %), THF, rt
2. BzCl (1.1 eq), Et_3N (1.2 eq), rt, 1 h

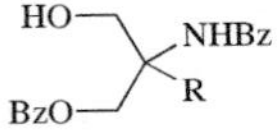

HO⟍⟋NHBz R / BzO

R		er	Config.	
Me	(96)	95.5:4.5	(−)	93
$TBSOCH_2$	(96)	96.0:4.0	(R)	
Et	(95)	97.5:2.5	(S)	
$BnO(CH_2)_3$	(92)	97.0:3.0	(−)	
i-Bu	(98)	95.0:5.0	(−)	
$CyCH_2$	(95)	96.0:4.0	(−)	
$PhCH_2$	(94)	95.0:5.0	(S)	
$Ph(CH_2)_2$	(94)	96.0:4.0	(S)	

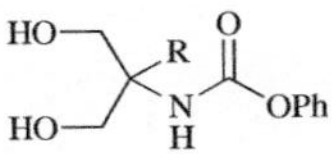

Bottom scheme (94)

HO⟍⟋ R—N(H)—OPh (O) / HO

1. **18**•$CuCl_2$ (5 mol %), Me_2NBn (1.2 eq), toluene, rt, time
2. BzCl (1.9 eq), Et_3N (2.0 eq), THF, rt, 24 h

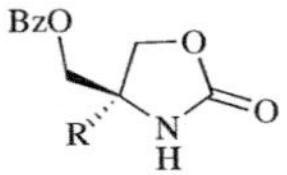

$BzOCH_2$ oxazolidinone   94

R	Time (h)		er
Me	11	(85)	98.5:1.5
Et	3	(74)	99.0:1.0
CH_2=CH	4	(82)	98.5:1.5
TMS—≡	23	(70)	97.0:3.0
$BnO(CH_2)_3$	4	(73)	99.0:1.0
CH_2=CHCH_2	5	(84)	99.0:1.0
(E)-EtO_2CCH=CH	12	(75)	98.5:1.5
$CyCH_2$	4	(73)	99.0:1.0
Ph	6	(87)	99.5:0.5
4-ClC_6H_4	19	(93)	97.5:2.5
4-MeC_6H_4	23	(80)	98.5:1.5
$PhCH_2$	3	(85)	98.5:1.5
$Ph(CH_2)_2$	5	(71)	99.0:0.5

TABLE 1B. DESYMMETRIZATION OF *MESO*-1,3-DIOLS (*Continued*)

Diol	Conditions	Product(s) and Yield(s) (%)	Refs.

*Please refer to the charts preceding the tables for the structures indicated by the **bold** numbers.*

C₅

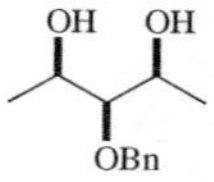

	Ac₂O (1.8 eq), **99** (10 mol %), *i*-Pr₂NEt (2 eq). toluene/CH₂Cl₂ (12:1), rt	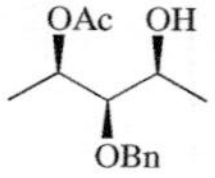 (49) er 71.5:28.5	139
	Ac₂O (1.8 eq), **99** (10 mol %), *i*-Pr₂NEt (2 eq). toluene/CH₂Cl₂ (12:1), rt	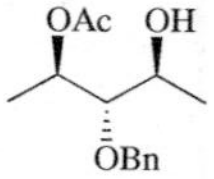 (38) er 58.5:41.5	139
	(CyCO)₂O, **110** (5 mol %), *i*-Pr₂NEt (3 eq). CCl₄/CHCl₃ (9:1), rt, 4 h	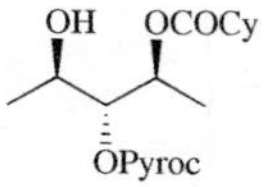 (71) er 97.0:3.0	72
	PhCO₂CH=CH₂, (5 eq), Et₂Zn/**40** (2:1, 5 mol %), toluene, –20°, 36 h	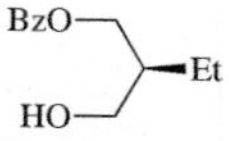 (99) er 93.0:7.0	310

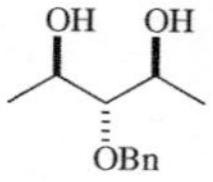

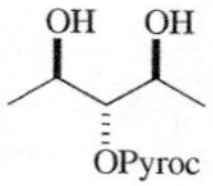

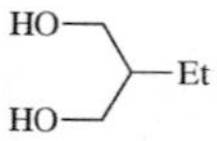

C₅₋₁₂

| 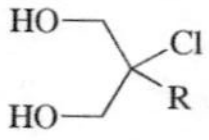 | 1-NpCHO (2 eq), **145c** (5 mol %),
162 (1 eq), Cs₂CO₃ (10 mol %).
THF, –45°, 16 h | 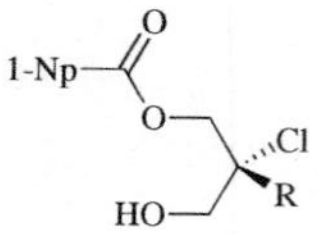 | 311 |

R		er
Et	(86)	85.0:15.0
i-Pr	(78)	65.0:35.0
CH₂=CHCH₂	(85)	90.0:10.0
Ph(CH₂)₂	(83)	81.0:19.0
Ph——CH₂	(82)	95.0:5.0

R	Time (h)		er	
Me	5	(73)	99.5:0.5	92
Et	12	(67)	99.0:1.0	
CH₂=CHCH₂	18	(72)	99.0:1.0	

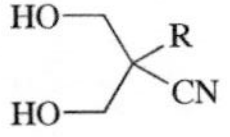

BzCl (1.5 eq), **33**•CuCl₂ (10 mol %),
i-Pr₂NEt (1.2 eq),
toluene/CH₂Cl₂ (20:1), –78°

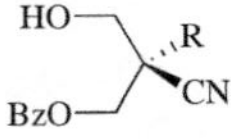

R		er	
H	(93)	96.0:4.0	93
EtCO₂	(95)	94.0:6.0	

1. CuCl₂ (20 mol %),
17 (20 mol %), THF, rt
2. BzCl (1.1 eq),
Et₃N (1.2 eq), rt, 1 h

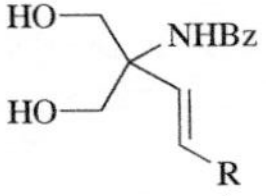

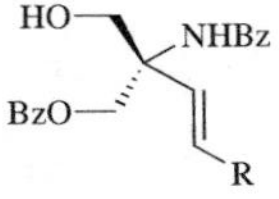

(96) er 93.5:6.5   93

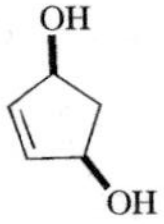

1. CuCl₂ (20 mol %),
17 (20 mol %), THF, rt
2. BzCl (1.1 eq),
Et₃N (1.2 eq), rt, 1 h

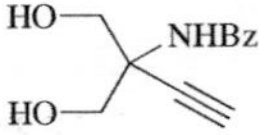

(63) er 52.0:48.0   73

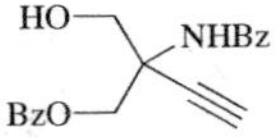

Ac₂O (5.3 eq), **105** (2 mol %),
toluene, –20°, 24 h

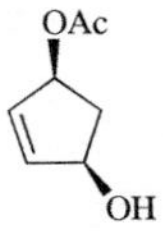

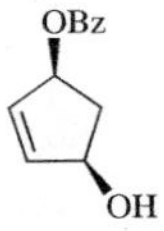

(10) er 70.0:30.0   135

BzCl (2 eq), **130** (2 eq),
Et₂O, –20°, 4 h

C₅₋₇

C₅₋₆

C₅

Diol	Conditions	Product(s) and Yield(s) (%)	Refs.

*Please refer to the charts preceding the tables for the structures indicated by the **bold** numbers.*

C$_5$

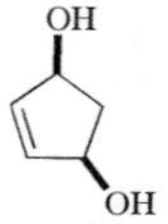

	BzCl (1.7 eq), **137** (0.5 mol %), Et$_3$N (1.7 eq), 4 Å MS, *n*-PrCN, −78°, 3 h	(38) er 99.0:1.0	117
	BzCl (1.6 eq), **137** (0.5 mol %), Et$_3$N (1.6 eq), 4 Å MS, *n*-PrCN, −78°, 3 h	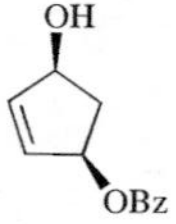 (45) er 98.5:1.5	136
	BzCl (1.5 eq), **129** (30 mol %), *i*-Pr$_2$NEt, CH$_2$Cl$_2$, 0°, 4 h	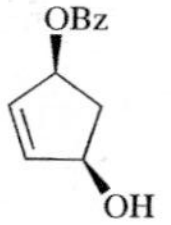 (82) er 90.5:9.5	137

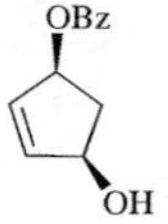

C$_6$

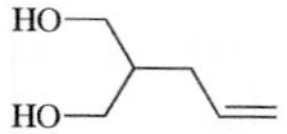

| | 4-*t*-BuC$_6$H$_4$COCl (1.5 eq), **135** (5 mol %), *i*-Pr$_2$NEt, 4 Å MS, toluene, 0°, 8 h | (76) er 80.0:20.0 | 115 |

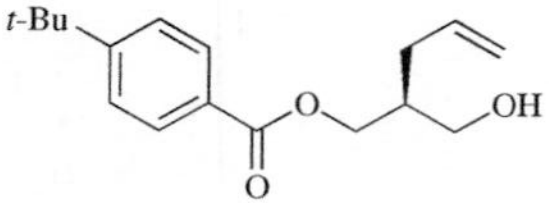

C$_{6-11}$

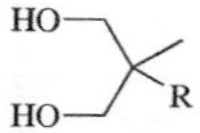

| | BzCl (1.5 eq), **33**·CuCl$_2$ (10 mol %), Et$_3$N (1.1 eq), CH$_2$Cl$_2$, −78° | | 92 |

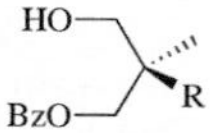

R	Time (h)		er
Et	12	(95)	92.0:8.0
Ac	12	(85)	97.5:2.5
CH$_2$=CH	20	(91)	77.0:23.0
Br$_2$C=CH	3	(99)	98.5:1.5
i-Pr	5	(96)	97.0:3.0
CH$_2$=C(Me)	12	(97)	98.0:2.0
Ph	3	(98)	91.5:8.5
PhCH$_2$	3	(99)	97.5:2.5

Substrate	Conditions	Product	Refs.
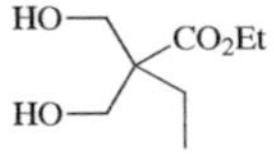 C₆ (HO, HO, CO₂Et, Et)	PhNCO, Cu(OTf)₂ (10 mol %), *ent*-**15** (10 mol %), THF, 0°, 1 h	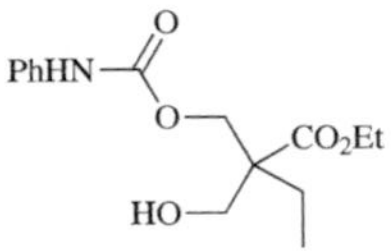 PhHN–C(=O)–O–CH₂C(CO₂Et)(Et)CH₂OH (88) er 77.5:22.5	128
(HO, HO, OH, propenyl)	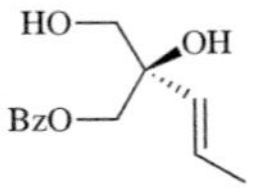 BzCl (1.1 eq), **36**•CuCl₂ (20 mol %), Et₃N (1.2 eq), THF, rt, 0.5 h	(HO, BzO, OH, propenyl) (98) er 95.5:4.5	312, 280
(HO, HO, OH, propenyl)	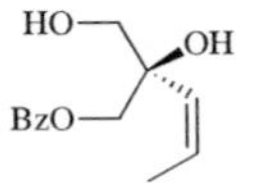 BzCl (1.1 eq), **36**•CuCl₂ (20 mol %), Et₃N (1.2 eq), THF, rt, 0.5 h	(HO, BzO, OH, propenyl) (90) er 91.0:9.0	312
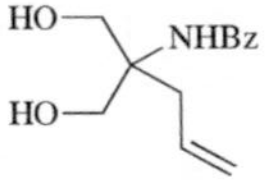 (HO, HO, NHBz, allyl)	1. CuCl₂ (20 mol %), **17** (20 mol %), THF, rt 2. BzCl (1.1 eq), Et₃N (1.2 eq), rt, 1 h	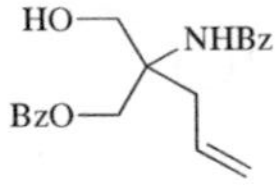 (HO, BzO, NHBz, allyl) (96) er 96.0:4.0	93
cyclohexane-1,3-diol (OH, OH)	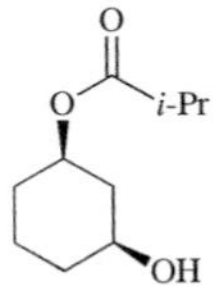 (*i*-PrCO)₂O (1.3 eq), **68** (5 mol %), collidine (1.4 eq), toluene, rt, 4 h	isobutyrate ester of cyclohexanol (OH) (48) er 76.0:24.0	119

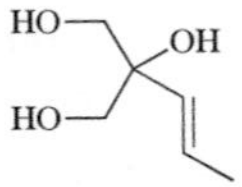

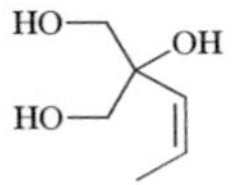

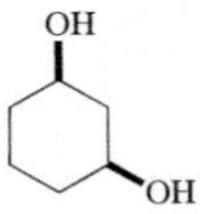

161

TABLE 1B. DESYMMETRIZATION OF *MESO*-1,3-DIOLS (*Continued*)

*Please refer to the charts preceding the tables for the structures indicated by the **bold** numbers.*

Diol	Conditions	Product(s) and Yield(s) (%)	Refs.
C$_6$			
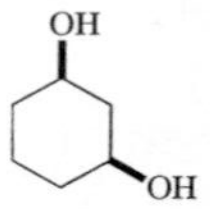	(i-PrCO)$_2$O (1.3 eq), **63** (5 mol %), collidine (1.4 eq), CHCl$_3$, rt, 4 h	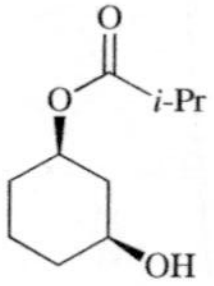 (69) er 65.5:34.5	120
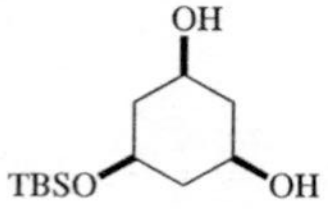	BzCl (1.5 eq), **129** (30 mol %), i-Pr$_2$NEt, CH$_2$Cl$_2$, 0°, 6 h	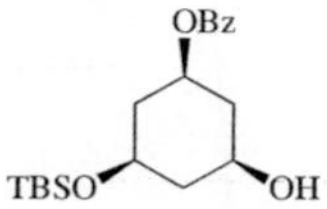 (73) er 85.0:15.0	137
C$_7$			
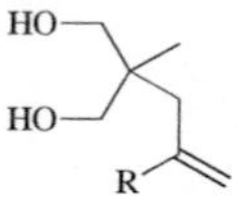	BzCl (1.5 eq), **33·**CuCl$_2$ (10 mol %), Et$_3$N (1.1 eq), CH$_2$Cl$_2$, −78°	 | 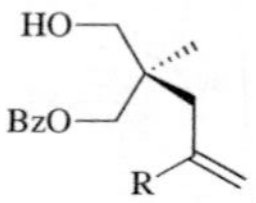 R	Time (h)		er
H	12	(98)	94.5:5.5
Br	15	93	96.0:4.0		92
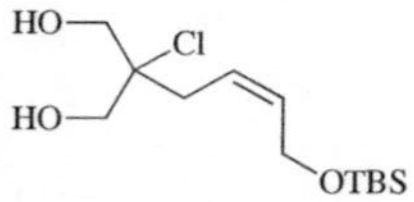	1-NpCHO (2 eq), **145c** (5 mol %), **162** (1 eq), Cs$_2$CO$_3$ (10 mol %), THF, −45°, 16 h	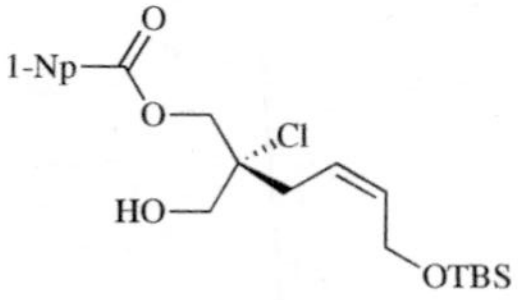 (80) er 93.0:7.0	311
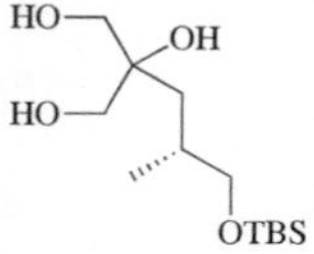	BzCl (1.1 eq), *ent*-**36·**CuCl$_2$ (20 mol %), Et$_3$N (1.2 eq), THF, rt, 0.5 h	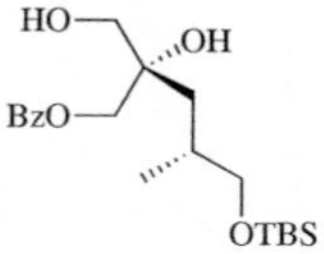 (94) dr 96.0:4.0	280

162

C_{7-8}		

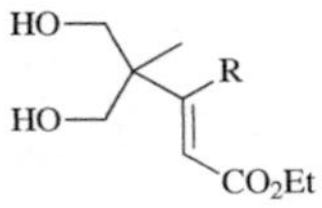

BzCl (1.5 eq),
33·CuCl₂ (10 mol %),
Et₃N (1.1 eq),
CH₂Cl₂, −78°, 12 h

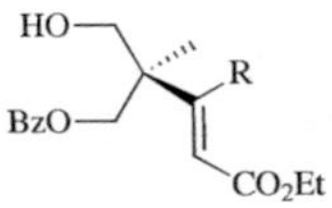

R		er
H	(95)	92.5:7.5
Me	(95)	99.0:1.0

92

C_{7-15}	

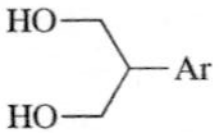

PhCO₂CH=CH₂ (5 eq),
Et₂Zn/41 (2:1, x mol %),
toluene, −15°, 24 h

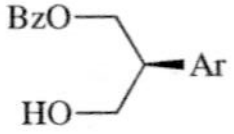

Ar	x		er
2-thienyl	5	(78)	85.0:15.0
3-thienyl	10	(88)	87.0:13.0
Ph	5	(94)	95.5:4.5
PMB	5	(99)	96.5:3.5
4-ClC₆H₄	10	(89)	95.0:5.0
4-MeC₆H₄	5	(94)	95.5:4.5
2-Np	10	(97)	96.5:3.5
4-PhC₆H₄	10	(83)	93.0:7.0

91, 90

C_7	

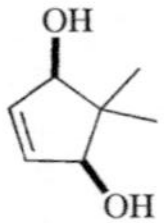

BzCl (1.7 eq), 137 (0.5 mol %),
Et₃N (1.7 eq), 4 Å MS,
n-PrCN, −78°, 8 h

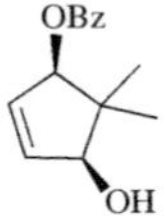

(87) er 99.5:0.5

117

BzCl (1.6 eq), 137 (0.5 mol %),
Et₃N (1.6 eq), 4 Å MS,
n-PrCN, −78°, 8 h

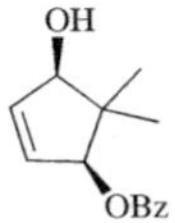

(87) er >99.5:0.5

136

TABLE 1B. DESYMMETRIZATION OF *MESO*-1,3-DIOLS (*Continued*)

Diol	Conditions	Product(s) and Yield(s) (%)	Refs.

*Please refer to the charts preceding the tables for the structures indicated by the **bold** numbers.*

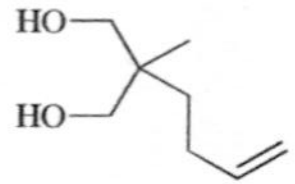

C$_8$

BzCl (1.5 eq), **33**·CuCl$_2$ (10 mol %),
Et$_3$N (1.1 eq),
CH$_2$Cl$_2$, –78°, 12 h

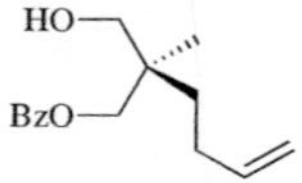

(95) er 97.0:3.0

92

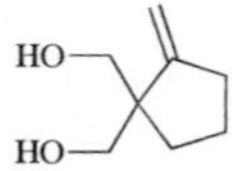

BzCl (1.5 eq), **33**·CuCl$_2$ (10 mol %),
Et$_3$N (1.1 eq),
CH$_2$Cl$_2$, –78°, 20 h

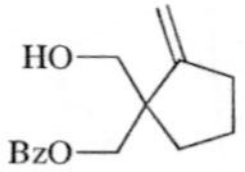

(92) er 93.5:6.5

92

C$_{8–14}$

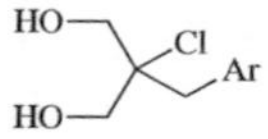

1-NpCHO (2 eq), **145c** (5 mol %),
162 (1 eq), Cs$_2$CO$_3$ (10 mol %),
THF, –45°, 16 h

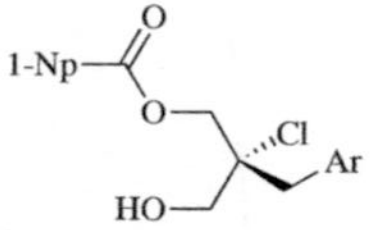

Ar		er
2-thienyl	(90)	94.0:6.0
Ph	(95)	95.0:5.0
PMP	(89)	95.0:5.0
4-TBSOC$_6$H$_4$	(80)	90.0:10.0
4-MeC$_6$H$_4$	(92)	95.0:5.0
4-CF$_3$C$_6$H$_4$	(91)	94.0:6.0
1-Np	(90)	93.0:7.0

311

C$_9$

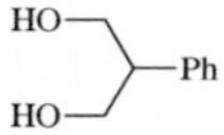

4-*t*-BuC$_6$H$_4$COCl (1.5 eq),
135 (5 mol %), *i*-Pr$_2$NEt,
4 Å MS, toluene, 0°, 7 h

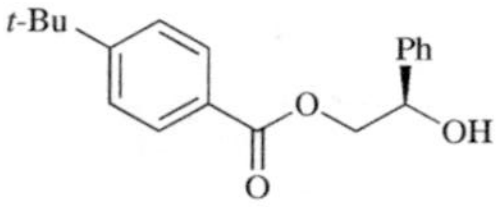

(78) er 82.0:18.0

115

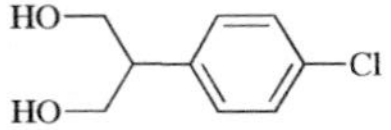

C$_{9-10}$

4-t-BuC$_6$H$_4$COCl (1.5 eq),
135 (5 mol %), i-Pr$_2$NEt,
4 Å MS, toluene, 0°, 6 h

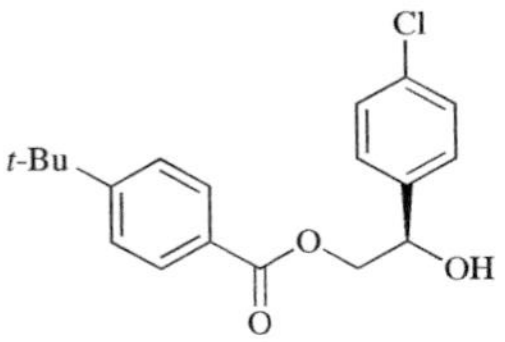

(72) er 83.5:16.5

115

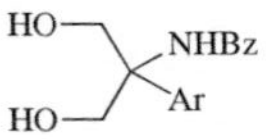

1. CuCl$_2$ (20 mol %),
17 (20 mol %), THF, rt
2. BzCl (4 eq),
Et$_3$N (3 eq), rt, 1 h

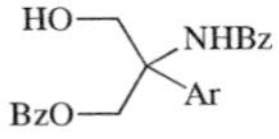

Ar		er	Config.
Ph	(87)	91.5:8.5	(S)
4-ClC$_6$H$_4$	(97)	92.5:7.5	(—)
4-MeC$_6$H$_4$	(97)	94.0:6.0	(—)

93

C$_{9-19}$

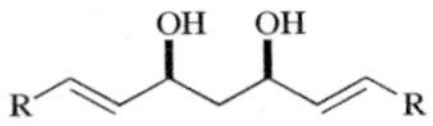

(EtCO)$_2$O (1.2 eq), **126** (2 mol %),
i-Pr$_2$NEt (1.2 eq),
CH$_2$Cl$_2$, –20°, 3 h

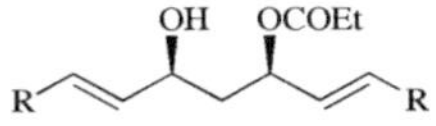

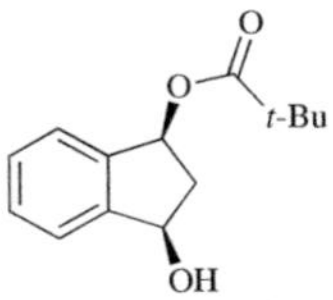

R		er
Me	(46)	92.0:8.0
2-furyl	(81)	98.5:1.5
Ph	(78)	99.0:1.0
2-MeOC$_6$H$_4$	(81)	97.0:3.0
4-MeOC$_6$H$_4$	(76)	97.5:2.5
4-ClC$_6$H$_4$	(77)	95.5:4.5

313

C$_3$

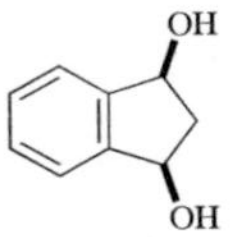

5 (1.1 eq), toluene, 80°, 14 h

(78) er 66.5:33.5

25, 26

TABLE 1B. DESYMMETRIZATION OF *MESO*-1,3-DIOLS (*Continued*)

Diol	Conditions	Product(s) and Yield(s) (%)	Refs.
	*Please refer to the charts preceding the tables for the structures indicated by the **bold** numbers.*		
C$_9$	BzCl (1.5 eq), **129** (20 mol %), *i*-Pr$_2$NEt, CH$_2$Cl$_2$, 0°, 7 h	(72) er 99.5:0.5	137
C$_{10}$	BzCl (1.5 eq), **33**·CuCl$_2$ (10 mol %), Et$_3$N (1.1 eq), toluene, −78°, 12 h	(91) er 75.5:24.5	92
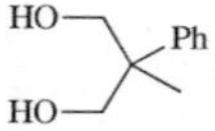	4-*t*-BuC$_6$H$_4$COCl (1.5 eq), **135** (5 mol %), *i*-Pr$_2$NEt, 4 Å MS, toluene, 0°, 12 h	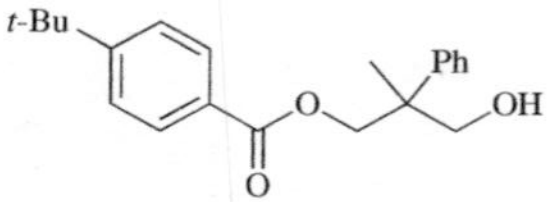 (55) er 71.0:29.0	115
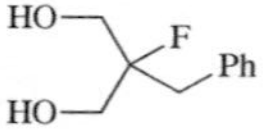	1-NpCHO (2 eq), **145c** (5 mol %), **162** (1 eq), Cs$_2$CO$_3$ (10 mol %), THF, −45°, 16 h	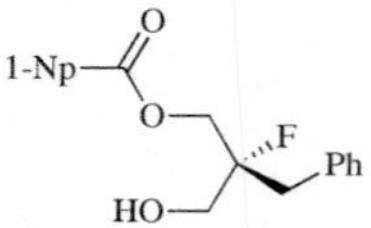 (95) er 85.0:15.0	311
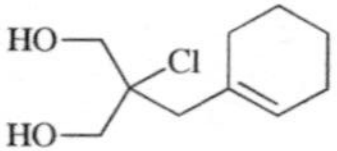	1-NpCHO (2 eq), **145c** (5 mol %), **162** (1 eq), Cs$_2$CO$_3$ (10 mol %), THF, −45°, 16 h	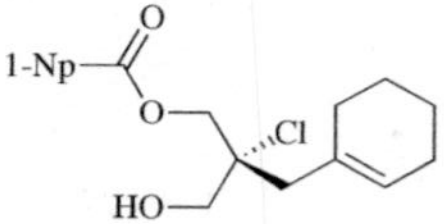 (80) er 95.0:5.0	311

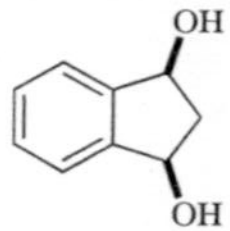

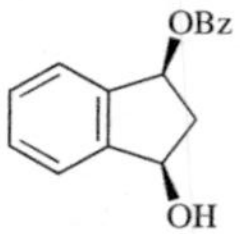

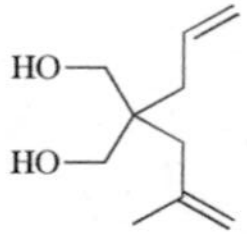

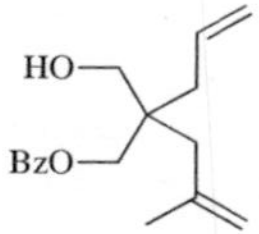

C$_{11}$

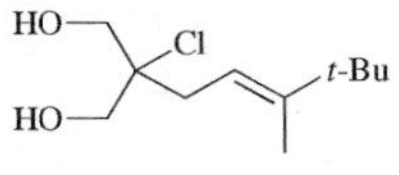

1-NpCHO (2 eq), **145c** (5 mol %), **162** (1 eq), Cs$_2$CO$_3$ (10 mol %), THF, –45°, 16 h

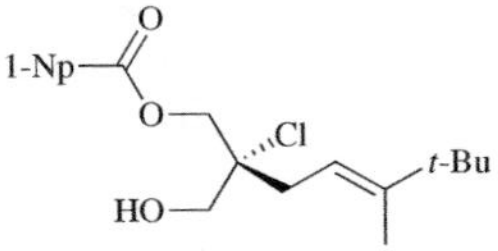

(85) er 95.0:5.0   311

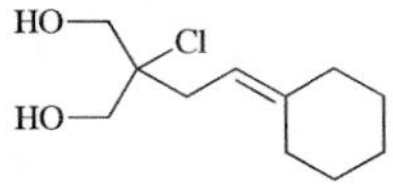

1-NpCHO (2 eq), **145c** (5 mol %), **162** (1 eq), Cs$_2$CO$_3$ (10 mol %), THF, –45°, 16 h

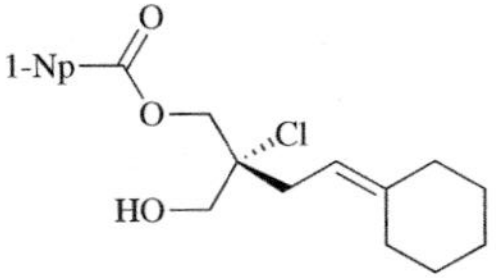

(88) er 93.0:7.0   311

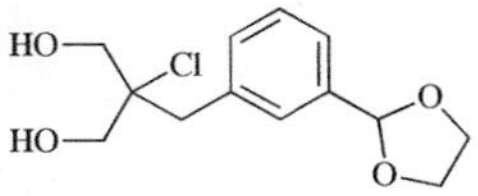

1-NpCHO (2 eq), **145c** (5 mol %), **162** (1 eq), Cs$_2$CO$_3$ (10 mol %), THF, –45°, 16 h

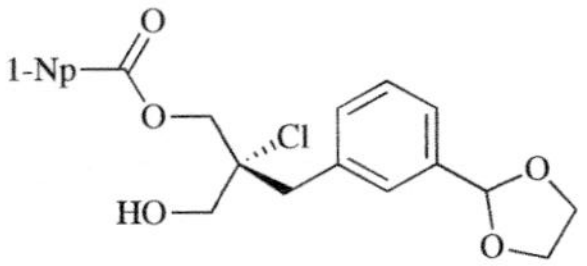

(85) er 95.5:4.5   311

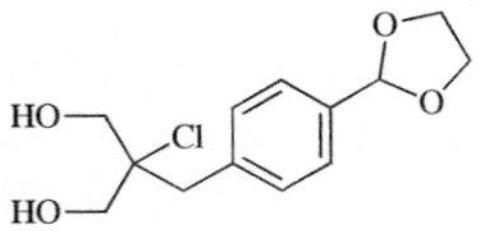

1-NpCHO (2 eq), **145c** (5 mol %), **162** (1 eq), Cs$_2$CO$_3$ (10 mol %), THF, –45°, 16 h

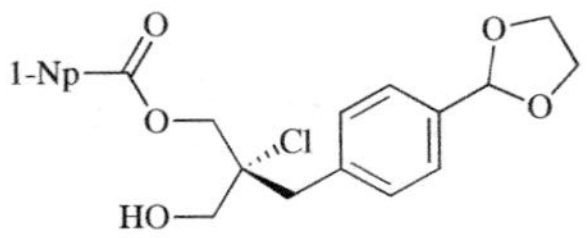

(85) er 94.0:6.0   311

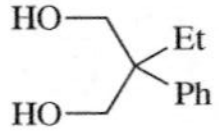

BzCl (1.5 eq), **33**•CuCl$_2$ (10 mol %), Et$_3$N (1.1 eq), toluene, –78°, 12 h

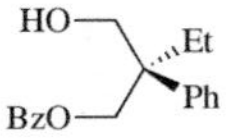

(98) er 74.5:25.5   92

TABLE 1B. DESYMMETRIZATION OF *MESO*-1,3-DIOLS (*Continued*)

*Please refer to the charts preceding the tables for the structures indicated by the **bold** numbers.*

Diol	Conditions	Product(s) and Yield(s) (%)	Refs.
C$_{11-17}$	(EtCO)$_2$O (1.2 eq), **126** (2 mol %), *i*-Pr$_2$NEt (1.2 eq), CH$_2$Cl$_2$, –20°, 3 h	 Ar er 2-thienyl (83) 98.0:2.0 Ph (85) 99.0:1.0 2-MeOC$_6$H$_4$ (86) 97.5:2.5 4-FC$_6$H$_4$ (59) 98.0:2.0 4-MeC$_6$H$_4$ (61) 95.5:4.5	313
C$_{12}$	BzCl (1.5 eq), **33**-CuCl$_2$ (10 mol %), Et$_3$N (1.1 eq), toluene, –78°, 12 h	 (88) er 92.0:8.0	92
C$_{12-13}$	1-NpCHO (2 eq), **145c** (5 mol %), **162** (1 eq), Cs$_2$CO$_3$ (10 mol %), THF, –45°, 16 h	 R er H (86) 95.0:5.0 Me (80) 94.0:6.0	311
	1-NpCHO (2 eq), **145c** (5 mol %), **162** (1 eq), Cs$_2$CO$_3$ (10 mol %), THF, –45°, 16 h	 R er H (86) 93.0:7.0 Me (86) 95.5:4.5	311
C$_{12-25}$	**156** (20 mol %), **161** (20 mol %), 2,6-lutidine (20 mol %), **162** (1.2 eq), 4 Å MS, THF, 24 h	 R er H (80) 71.0:29.0 Cl (78) 89.0:11.0 Me (72) 80.0:20.0 Ph (58) 90.0:10.0 2-ClC$_6$H$_4$CH$_2$ (83) 90.0:10.0 2-MeC$_6$H$_4$CH$_2$ (80) 91.0:9.0 (E)-PhCH$_2$=CHCH$_2$ (67) 78.0:22.0 2-NpCH$_2$ (82) 91.0:9.0	140

C₁₂

Substrate	Conditions	Product	Yield (er)	Ref
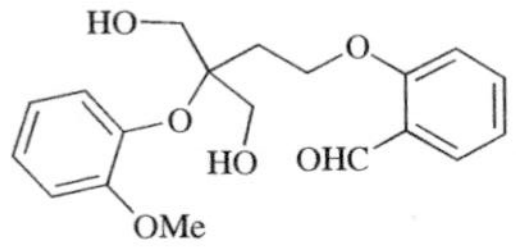 (structure: HO, O–OHC, HO, OMe)	156 (20 mol %), 161 (20 mol %), 2,6-lutidine (20 mol %), 162 (1.2 eq), 4 Å MS, THF, 24 h	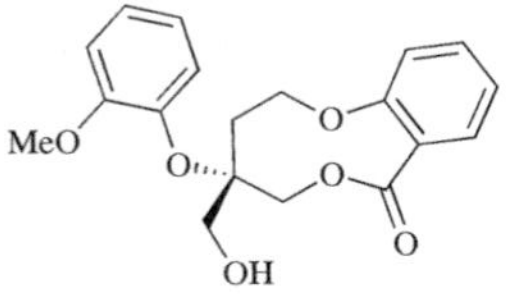 (benzo-fused lactone, MeO, OH)	(81) er 96.0:4.0	140
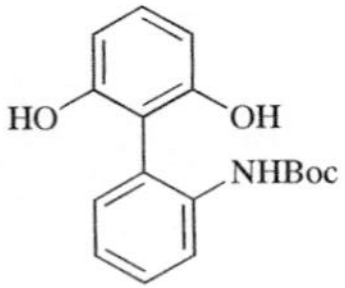 (OH, NHBoc, HO biaryl)	11 (2.5 eq), 145e (20 mol %), i-Pr₂NEt (2.0 eq), THF, 0°, 24 h	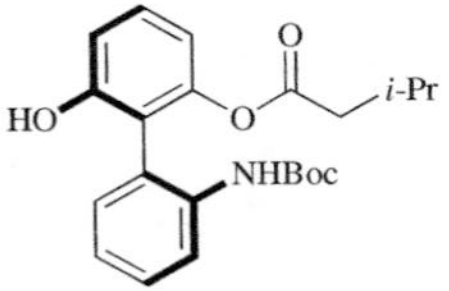 (i-Pr, O, NHBoc, HO)	(72) er 97.5:2.5	150
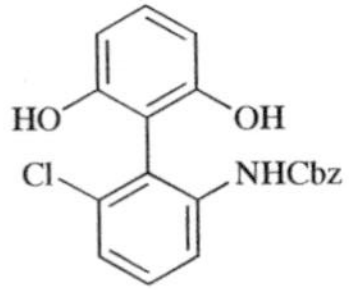 (OH, NHCbz, HO, Cl biaryl)	i-PrCH₂CHO (1.5 eq), 145g (10 mol %), 162 (1.2 eq), K₂CO₃ (1.2 eq), CH₂Cl₂, rt	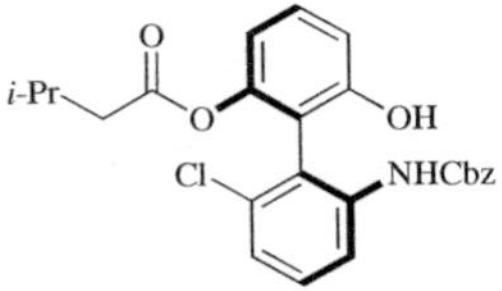 (OH, NHCbz, Cl, i-Pr, O)	(90) er 99.0:1.0	149
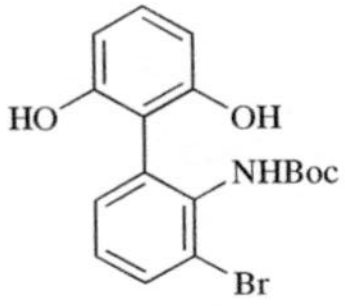 (OH, NHBoc, HO, Br biaryl)	11 (2.5 eq), 145e (20 mol %), i-Pr₂NEt (2.0 eq), THF, 0°, 24 h	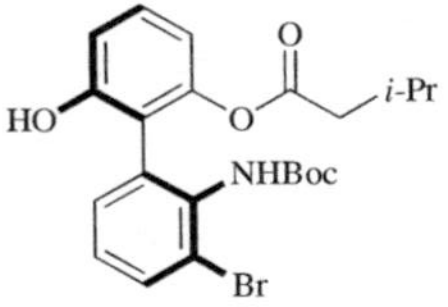 (i-Pr, O, NHBoc, Br, HO)	(81) er 97.0:3.0	150

169

TABLE 1B. DESYMMETRIZATION OF *MESO*-1,3-DIOLS (*Continued*)

Diol	Conditions	Product(s) and Yield(s) (%)	Refs.

*Please refer to the charts preceding the tables for the structures indicated by the **bold** numbers.*

C_{12}

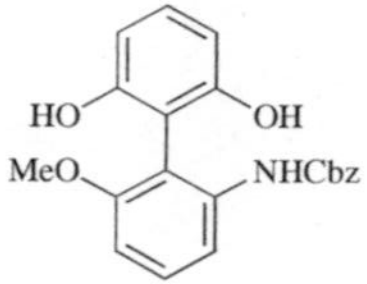

11 (2.5 eq), **145e** (20 mol %),
i-Pr$_2$NEt (2.0 eq), THF, 0°, 24 h

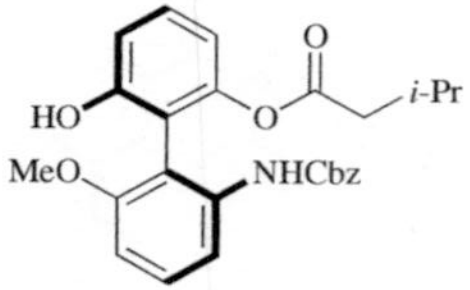

(50) er 99.5:0.5

150

i-PrCH$_2$CHO (1.5 eq),
145g (10 mol %), **162** (1.2 eq),
K$_2$CO$_3$ (1.2 eq), CH$_2$Cl$_2$, rt

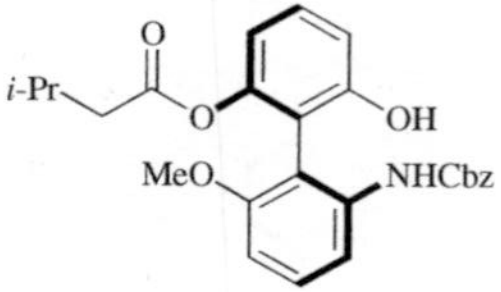

(90) er 99.0:1.0

149

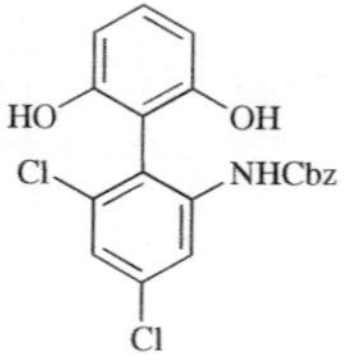

i-PrCH$_2$CHO (1.5 eq),
145g (10 mol %), **162** (1.2 eq),
K$_2$CO$_3$ (1.2 eq), CH$_2$Cl$_2$, rt

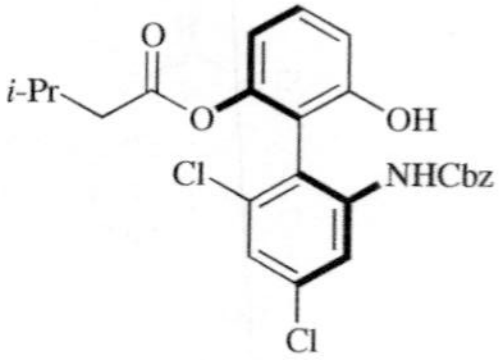

(93) er 99.0:1.0

149

C_{13}

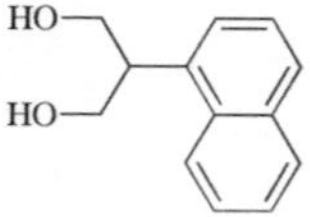

PhCO$_2$CH=CH$_2$ (5 eq),
Et$_2$Zn/*ent*-**41** (2:1, 5 mol %),
toluene, –15°, 24 h

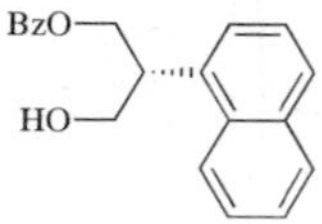

(80), er 83.5:16.5

90

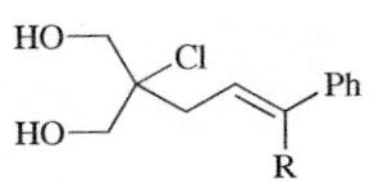

1-NpCHO (2 eq), **145c** (5 mol %),
162 (1 eq), Cs$_2$CO$_3$ (10 mol %),
THF, –45°, 16 h

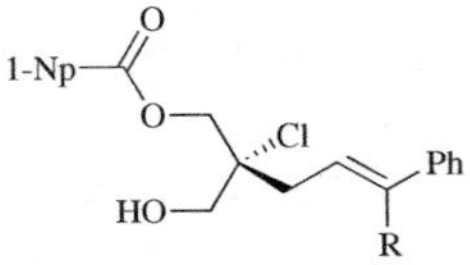

R		er
Me	(88)	96.0:4.0
Et	(86)	95.0:5.0
Ph	(81)	95.0:5.0

311

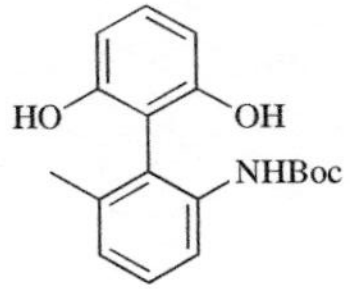

11 (2.5 eq), **145e** (20 mol %),
i-Pr$_2$NEt (2.0 eq), THF, 0°, 24 h

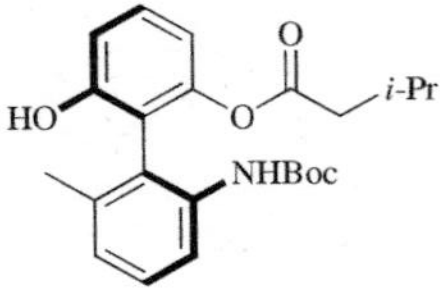

(81) er 99.0:1.0

150

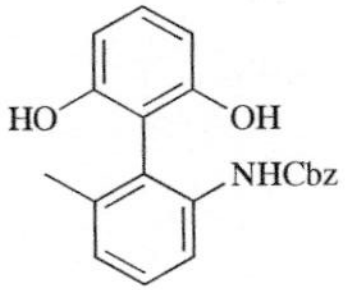

RCHO (1.5 eq),
145g (10 mol %), **162** (1.2 eq),
K$_2$CO$_3$ (1.2 eq), CH$_2$Cl$_2$, rt

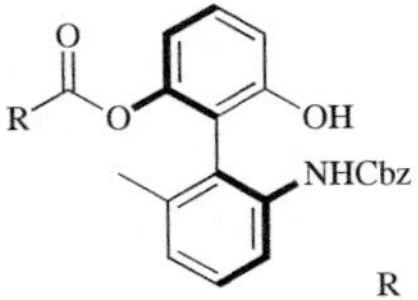

R		er
n-Pr	(83)	93.0:7.0
c-C$_3$H$_5$	(84)	95.0:5.0
BnO(CH$_2$)$_2$CH$_2$	(89)	90.0:10.0
i-PrCH$_2$	(92)	>99.5:0.5
t-BuCH$_2$	(86)	>99.5:0.5
c-C$_5$H$_9$	(91)	97.5:2.5
c-C$_6$H$_{11}$	(82)	99.0:1.0
CH$_3$(CH$_2$)$_5$CH$_2$	(90)	94.0:6.0
CH$_2$=CH(CH$_2$)$_7$-CH$_2$	(90)	91.5:8.5

149

C$_{13-18}$

C$_{13}$

171

TABLE 1B. DESYMMETRIZATION OF *MESO*-1,3-DIOLS (*Continued*)

*Please refer to the charts preceding the tables for the structures indicated by the **bold** numbers.*

C₁₃

Diol	Conditions	Product(s) and Yield(s) (%)	Refs.
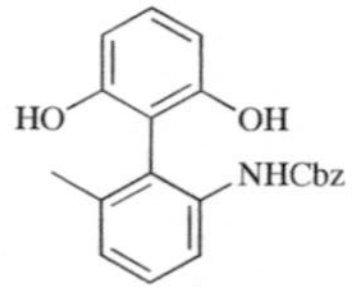	**11** (2.5 eq), **145e** (20 mol %), *i*-Pr₂NEt (2.0 eq), THF, 0°, 24 h	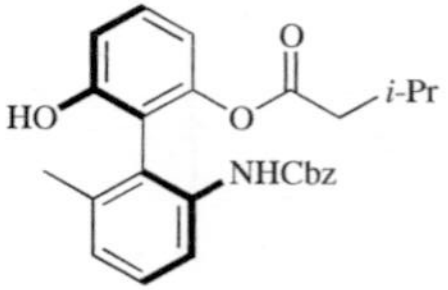 (70) er 99.5:0.5	150
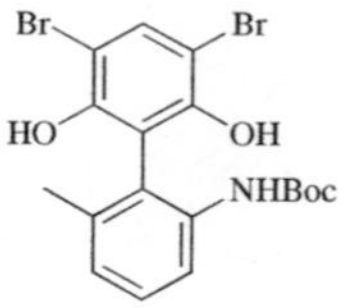	**11** (2.5 eq), **145e** (20 mol %), *i*-Pr₂NEt (2.0 eq), THF, 0°, 24 h	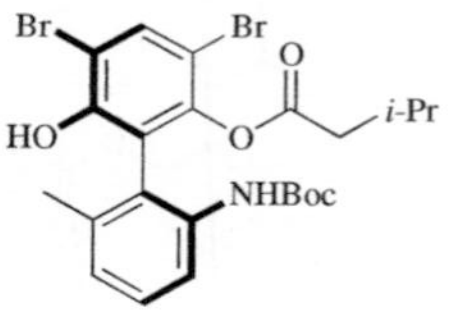 (83) er 99.5:0.5	150
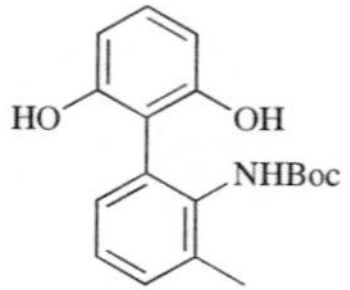	**11** (2.5 eq), **145e** (20 mol %), *i*-Pr₂NEt (2.0 eq), THF, 0°, 24 h	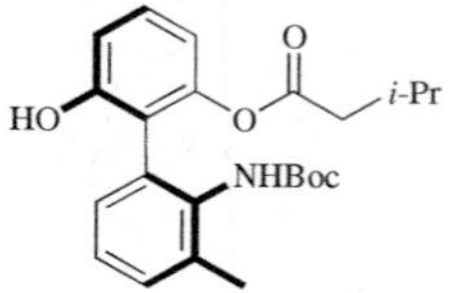 (80) er 99.5:0.5	150
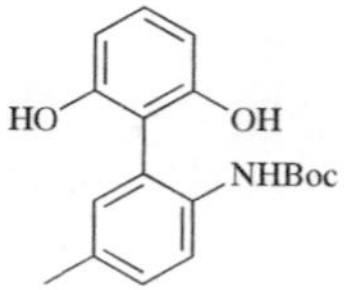	**11** (2.5 eq), **145e** (20 mol %), *i*-Pr₂NEt (2.0 eq), THF, 0°, 24 h	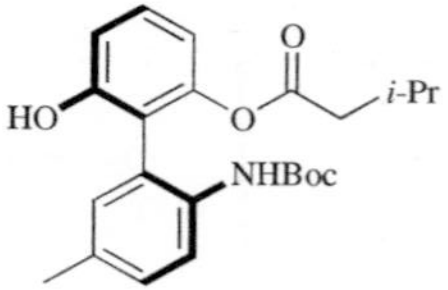 (64) er 97.5:2.5	150

Substrate	Conditions	Product	Yield (er)	Ref
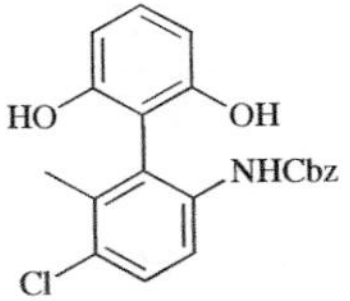 (structure: C$_{13-19}$, biaryl with OH, NHCbz, HO, Cl, CH$_3$)	i-PrCH$_2$CHO (1.5 eq), **145g** (10 mol %), **162** (1.2 eq), K$_2$CO$_3$ (1.2 eq), CH$_2$Cl$_2$, rt	(structure)	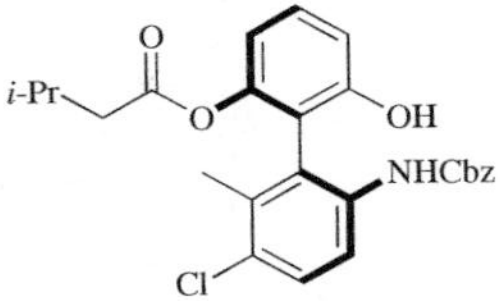 (93) er 99.0:1.0	149
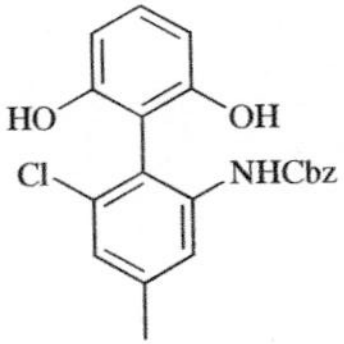 (structure: biaryl with OH, NHCbz, HO, Cl, CH$_3$)	i-PrCH$_2$CHO (1.5 eq), **145g** (10 mol %), **162** (1.2 eq), K$_2$CO$_3$ (1.2 eq), CH$_2$Cl$_2$, rt	(structure)	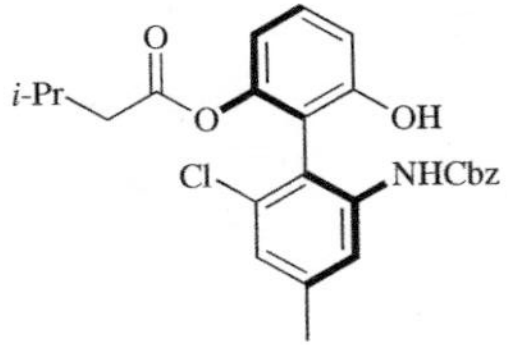 (94) er 99.0:1.0	149
(structure C$_{13-19}$: 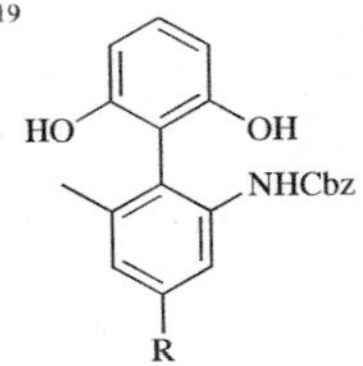 biaryl with OH, NHCbz, HO, R, CH$_3$)	i-PrCH$_2$CHO (1.5 eq), **145g** (10 mol %), **162** (1.2 eq), K$_2$CO$_3$ (1.2 eq), CH$_2$Cl$_2$, rt	(structure)	see table below	149
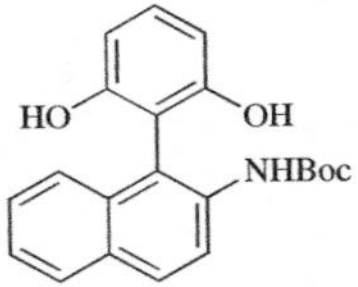 (structure C$_{13}$: naphthalene biaryl with OH, NHBoc, HO)	**11** (2.5 eq), **145e** (20 mol %), i-Pr$_2$NEt (2.0 eq), THF, 0°, 24 h	(structure, i-Pr, NHBoc, HO)	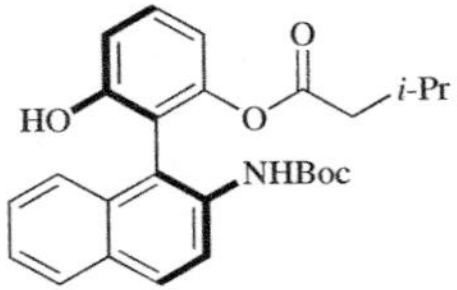 (85) er 98.5:1.5	150

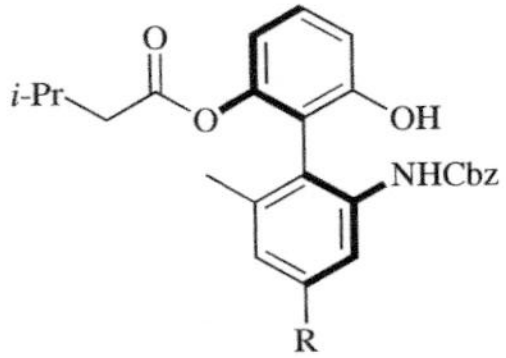

R		er
Cl	(92)	99.0:1.0
Me	(89)	>99.5:0.5
NC–	(85)	99.0:1.0
Ph	(89)	>99.5:0.5

173

TABLE 1B. DESYMMETRIZATION OF *MESO*-1,3-DIOLS (*Continued*)

Diol	Conditions	Product(s) and Yield(s) (%)		Refs.
Please refer to the charts preceding the tables for the structures indicated by the ***bold*** *numbers.*				
C_{14}				
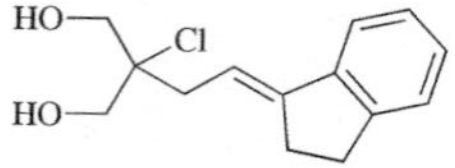	1-NpCHO (2 eq), **145c** (5 mol %), **162** (1 eq), Cs_2CO_3 (10 mol %), THF, –45°, 16 h	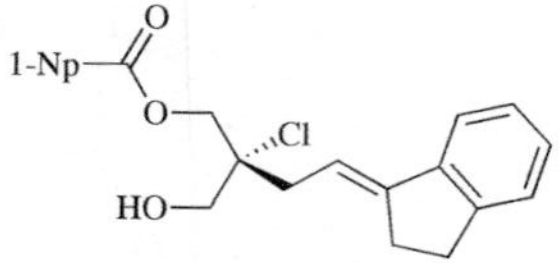	(80) er 96.0:4.0	311
C_{14-15}				
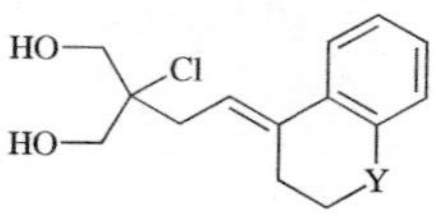	1-NpCHO (2 eq), **145c** (5 mol %), **162** (1 eq), Cs_2CO_3 (10 mol %), THF, –45°, 16 h	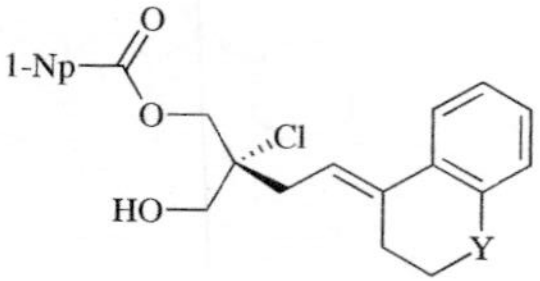	Y / er: O (79) 95.0:5.0 CH_2 (85) 95.0:5.0	311
C_{14}				
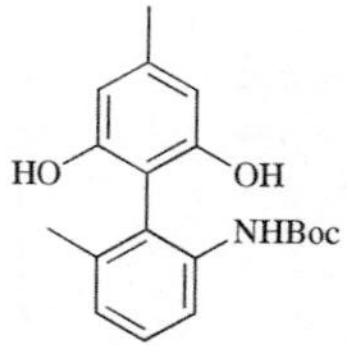	**11** (2.5 eq), **145e** (20 mol %), *i*-Pr_2NEt (2.0 eq), THF, 0°, 24 h	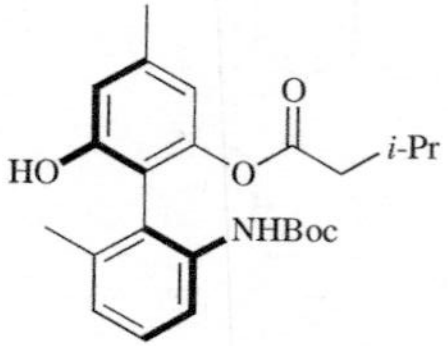	(85) er 97.5:2.5	150
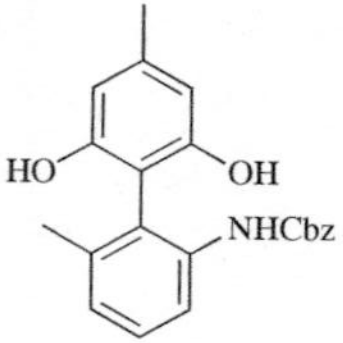	*i*-$PrCH_2CHO$ (1.5 eq), **145g** (10 mol %), **162** (1.2 eq), K_2CO_3 (1.2 eq), CH_2Cl_2, rt	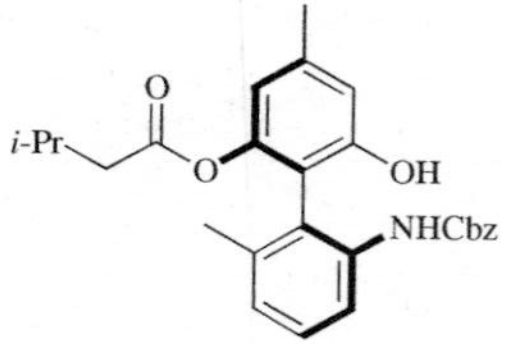	(93) er 98.0:2.0	149

Substrate	Conditions	Product	(Yield) er	Ref.
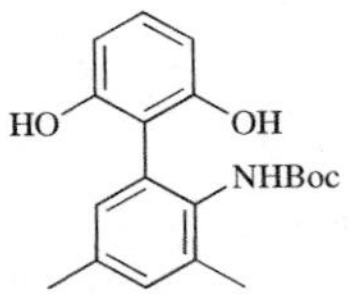	11 (2.5 eq), **145e** (20 mol %), $i\text{-Pr}_2\text{NEt}$ (2.0 eq), THF, 0°, 24 h		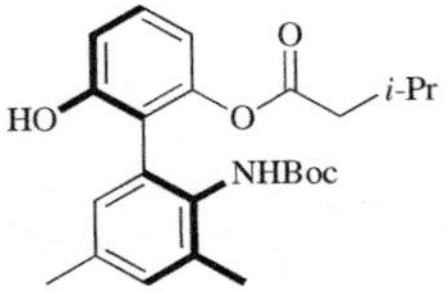 (86) er 99.0:1.0	150
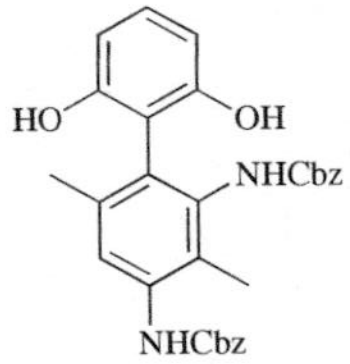	$i\text{-PrCH}_2\text{CHO}$ (1.5 eq), **145g** (10 mol %), **162** (1.2 eq), K_2CO_3 (1.2 eq), CH_2Cl_2, rt		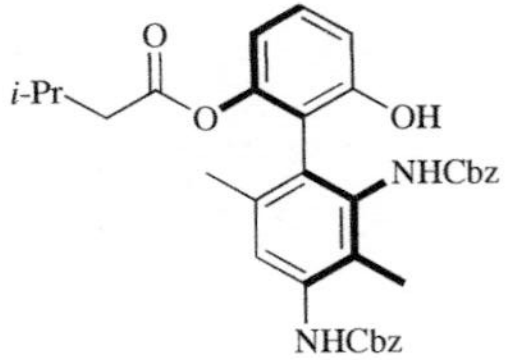 (87) er 99.5:0.5	149
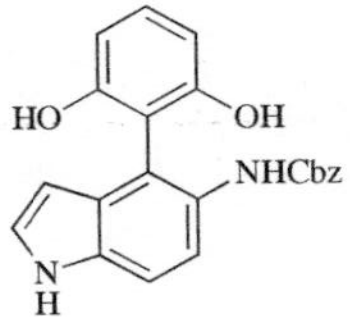	$i\text{-PrCH}_2\text{CHO}$ (1.5 eq), **145g** (10 mol %), **162** (1.2 eq), K_2CO_3 (1.2 eq), CH_2Cl_2, rt		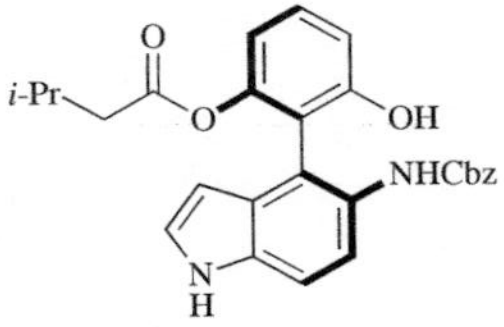 (90) er 99.0:1.0	149
C_{15}	$(i\text{-PrCO})_2\text{O}$ (1.1 eq), **76** (5 mol %), Et_3N (1.1 eq), MTBE, 0°, 3 h		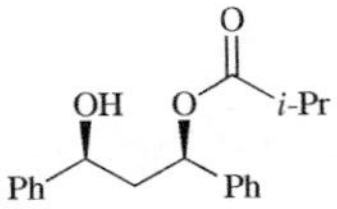 (61) er 79.5:20.5	41
C_{16}	**156** (20 mol %), **161** (20 mol %), 2,6-lutidine (20 mol %), **162** (1.2 eq), 4 Å MS, THF, 24 h		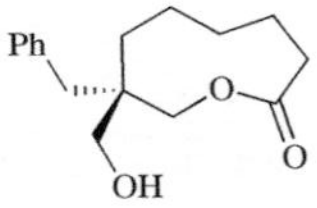 (53) er 90.0:10.0	140

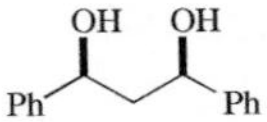

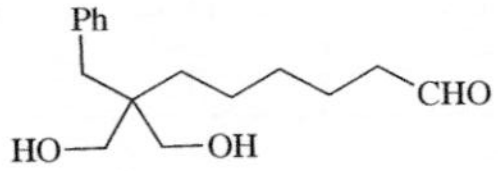

Please refer to the charts preceding the tables for the structures indicated by the **bold** numbers.

Diol	Conditions	Product(s) and Yield(s) (%)	Refs.
C₁₆			

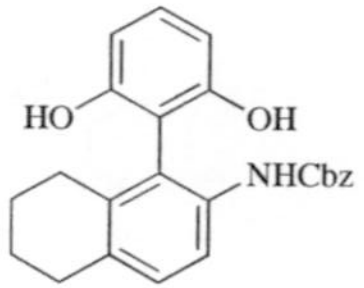

	i-PrCH₂CHO (1.5 eq), **145g** (10 mol %), **162** (1.2 eq), K₂CO₃ (1.2 eq), CH₂Cl₂, rt	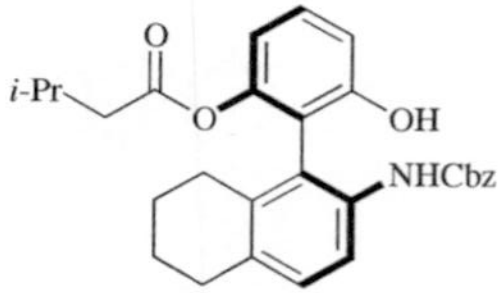 (91) er 98.0:2.0	149
	i-PrCH₂CHO (1.5 eq), **145g** (10 mol %), **162** (1.2 eq), K₂CO₃ (1.2 eq), CH₂Cl₂, rt	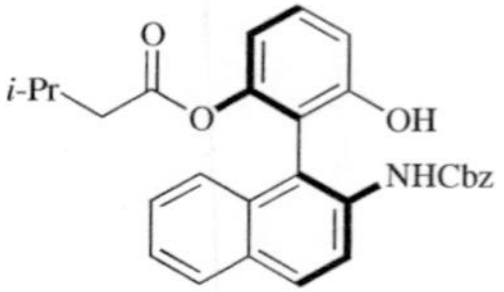 (84) er 99.5:0.5	149

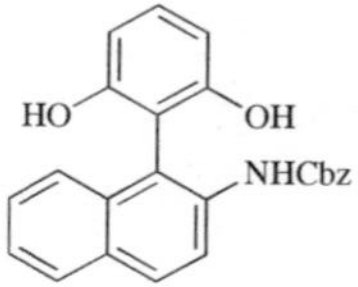

C₁₉			
	(EtCO)₂O (1.2 eq), **126** (2 mol %), *i*-Pr₂NEt (1.2 eq), CH₂Cl₂, −20°, 3 h	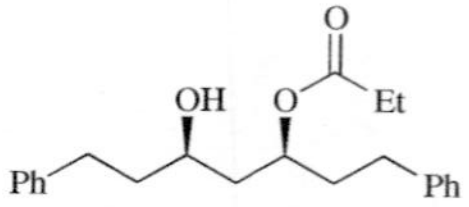 (38) er 60.5:39.5	313

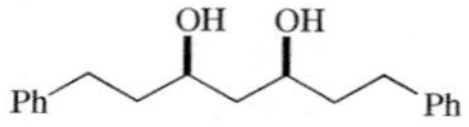

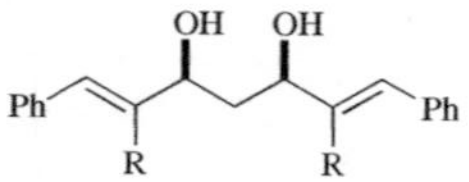

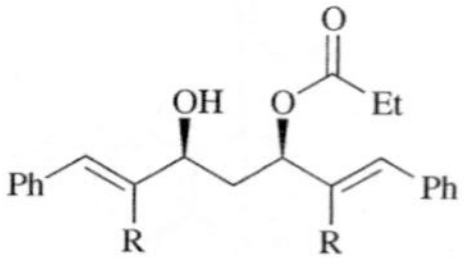

R		er
H	(83)	98.5:1.5
Br	(91)	99.0:1.0

| | (EtCO)₂O (1.2 eq), **126** (2 mol %), *i*-Pr₂NEt (1.2 eq), CH₂Cl₂, −20°, 3 h | | 313 |
| 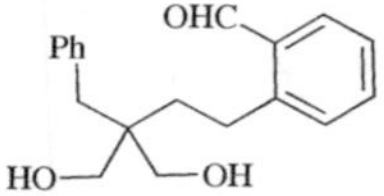 | **156** (20 mol %), **161** (20 mol %), 2,6-lutidine (20 mol %), **162** (1.2 eq), 4 Å MS, THF, 24 h | 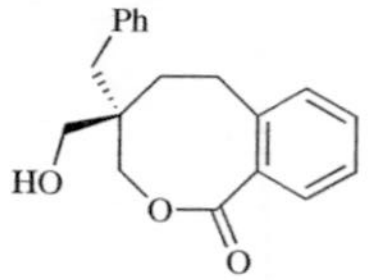 (64) er 90.0:10.0 | 140 |

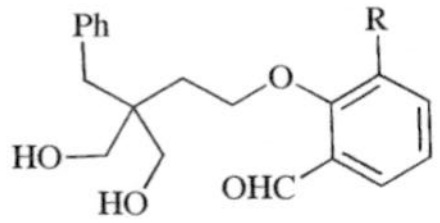

Row 1

156 (20 mol %), 161 (20 mol %),
2,6-lutidine (20 mol %),
162 (1.2 eq), 4 Å MS, THF, 24 h

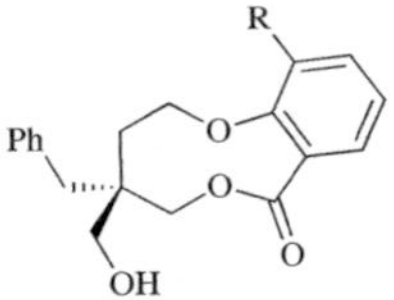

R		er
H	(81)	96.0:4.0
F	(72)	96.0:4.0
EtO	(75)	92.0:8.0

140

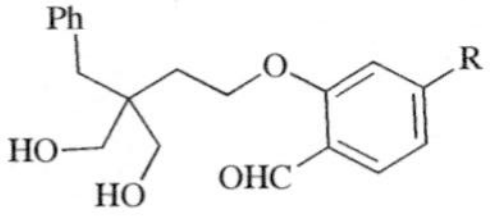

Row 2

156 (20 mol %), 161 (20 mol %),
2,6-lutidine (20 mol %),
162 (1.2 eq), 4 Å MS, THF, 24 h

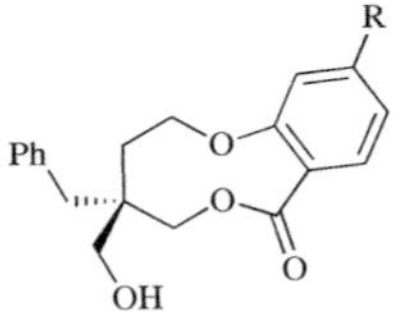

R		er
F	(80)	95.0:5.0
MeO	(82)	96.0:4.0
Me$_2$N	(69)	94.0:6.0

140

Row 3

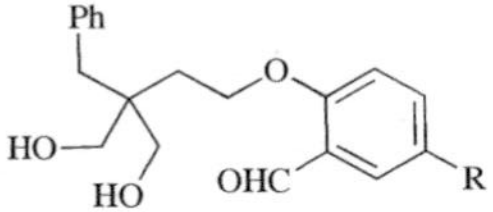

156 (20 mol %), 161 (20 mol %),
2,6-lutidine (20 mol %),
162 (1.2 eq), 4 Å MS, THF, 24 h

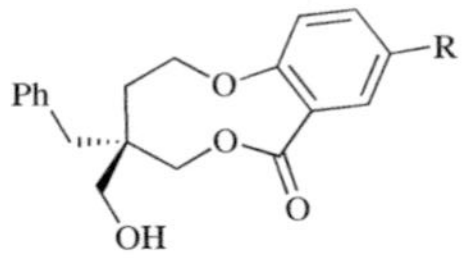

R		er
F	(65)	94.0:6.0
Cl	(83)	94.0:6.0
Br	(85)	96.0:4.0
MeO	(67)	95.0:5.0
Me	(80)	96.0:4.0

140

C$_{19-20}$

Row 4

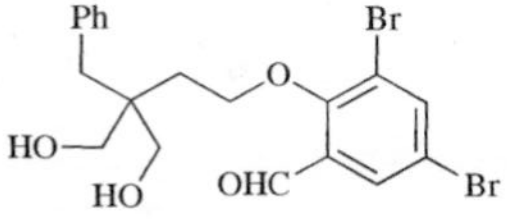

156 (20 mol %), 161 (20 mol %),
2,6-lutidine (20 mol %),
162 (1.2 eq), 4 Å MS, THF, 24 h

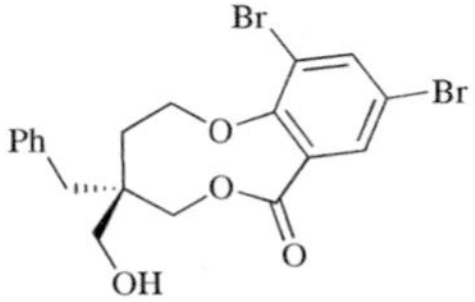

(63) er 90.0:10.0

140

C$_{19}$

TABLE 1B. DESYMMETRIZATION OF *MESO*-1,3-DIOLS (*Continued*)

Diol	Conditions	Product(s) and Yield(s) (%)	Refs.

*Please refer to the charts preceding the tables for the structures indicated by the **bold** numbers.*

C₁₉

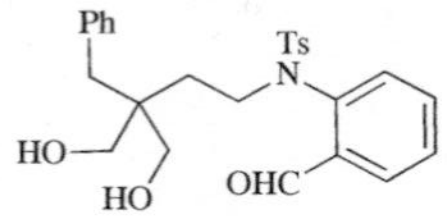

156 (20 mol %), 161 (20 mol %),
2,6-lutidine (20 mol %),
162 (1.2 eq), 4 Å MS, THF, 24 h

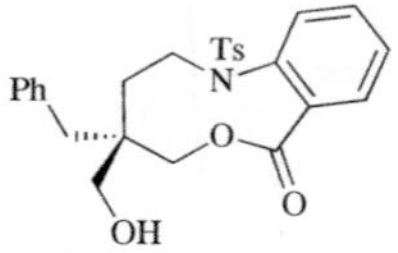

(75) er 98.0:2.0

140

C₂₀₋₂₂

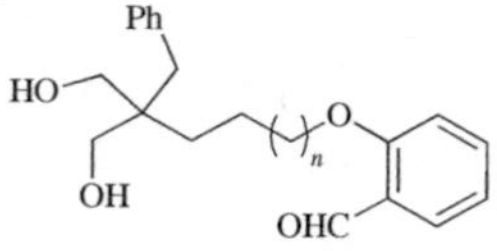

156 (20 mol %), 161 (20 mol %),
2,6-lutidine (20 mol %),
162 (1.2 eq), 4 Å MS, THF, 24 h

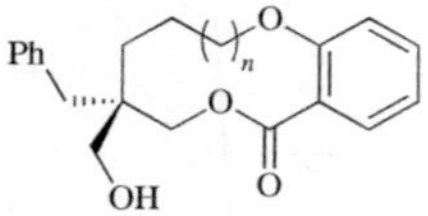

n		er
1	(68)	94.0:6.0
2	(51)	92.0:8.0
3	(55)	93.0:7.0

140

C₂₁

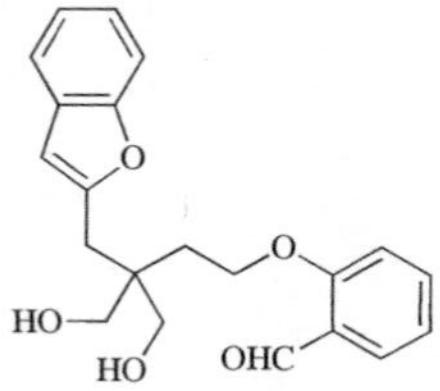

156 (20 mol %), 161 (20 mol %),
2,6-lutidine (20 mol %),
162 (1.2 eq), 4 Å MS, THF, 24 h

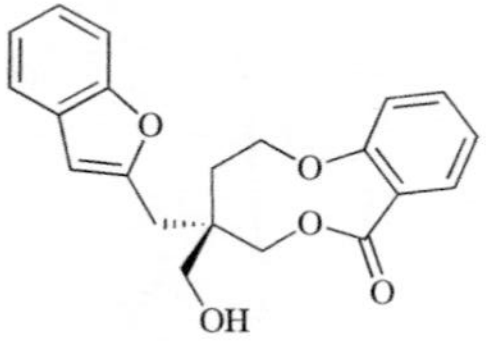

(74) er 92.0:8.0

140

C₂₃

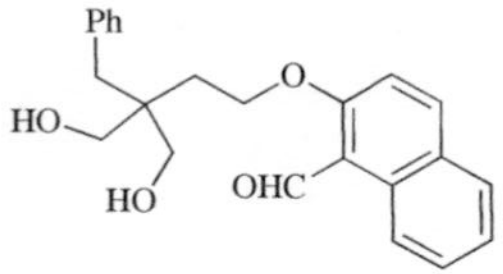

156 (20 mol %), 161 (20 mol %),
2,6-lutidine (20 mol %),
162 (1.2 eq), 4 Å MS, THF, 24 h

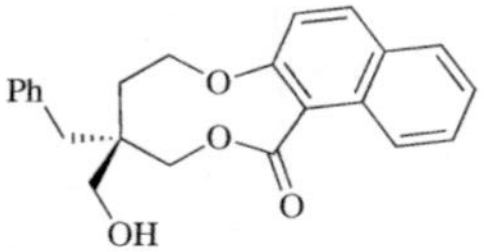

(68) er 94.0:6.0

140

C_{25}	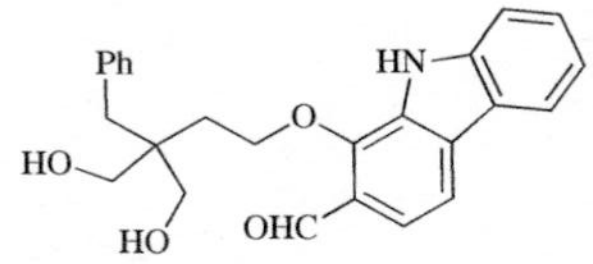 **156** (20 mol %), **161** (20 mol %), 2,6-lutidine (20 mol %), **162** (1.2 eq), 4 Å MS, THF, 24 h	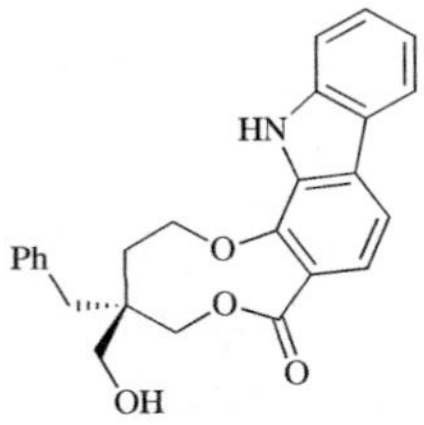 (58) er 98.0:2.0	140
C_{27}	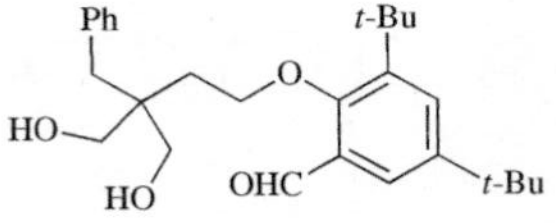 **156** (20 mol %), **161** (20 mol %), 2,6-lutidine (20 mol %), **162** (1.2 eq), 4 Å MS, THF, 24 h	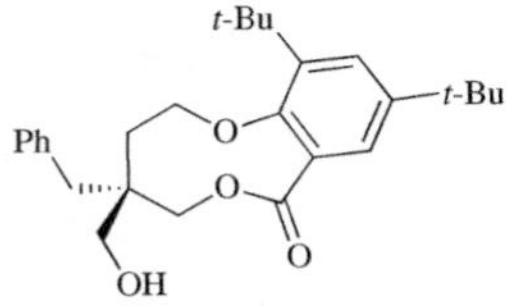 (62) er 90.0:10.0	140
C_{28}	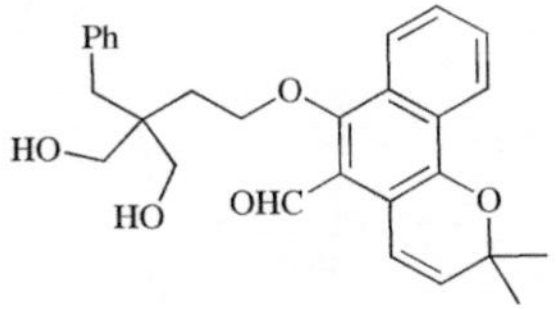 **156** (20 mol %), **161** (20 mol %), 2,6-lutidine (20 mol %), **162** (1.2 eq), 4 Å MS, THF, 24 h	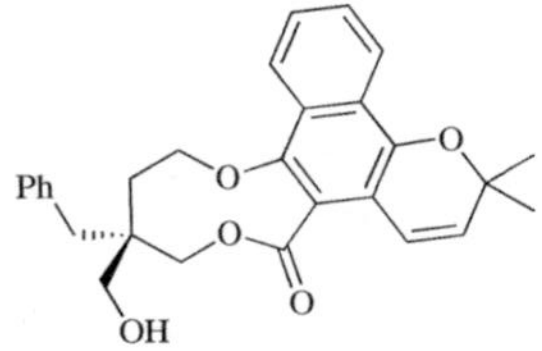 (75) er 99.0:1.0	140

TABLE 1C. DESYMMETRIZATION OF MISCELLANEOUS DIOLS

Diol	Conditions	Product(s) and Yield(s) (%)	Refs.

*Please refer to the charts preceding the tables for the structures indicated by the **bold** numbers.*

C4

4-t-BuC$_6$H$_4$COCl (1.5 eq),
138 (0.5 mol %),
i-Pr$_2$NEt (1.5 eq), 4 Å MS,
CH$_2$Cl$_2$/DMF (9:1), –78°, 3 h

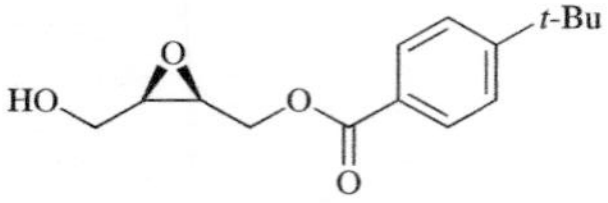

(57) er 99.5:0.5

117

C5

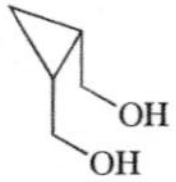

4 (1 eq), **164** (2 eq), THF, –78°, 23 h

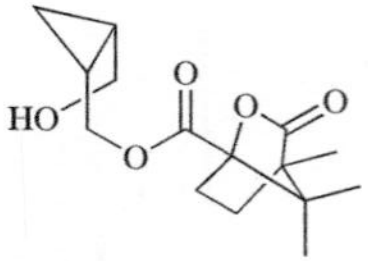

(54) dr 89.5:10.5

27

C6

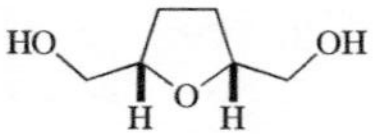

Ac$_2$O (5.3 eq), **105** (2 mol %),
toluene, 0°, 5 h

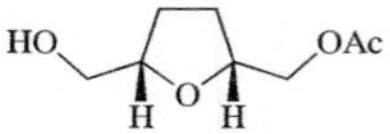

(99) er 69.5:30.5

73

C8

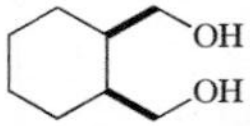

PhCO$_2$CH=CH$_2$ (5 eq).
Et$_2$Zn/**41** (2:1) (5 mol %),
toluene, –15°, 2 h

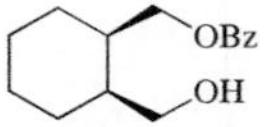

(93) er 95.5:4.5

91

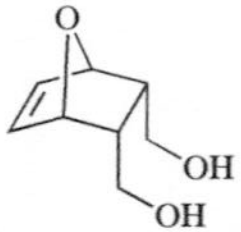

4, **164** (2 eq), THF, –78°, 23 h;
then –40°, 4 h

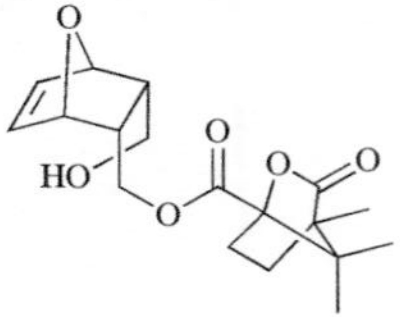

(51) dr 90.0:10.0

27

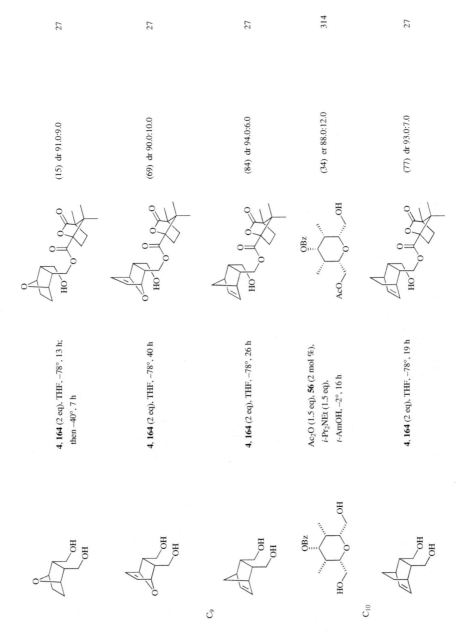

				27
				27
				27
				314
				27

(15) dr 91.0:9.0

4, **164** (2 eq), THF, –78°, 13 h; then –40°, 7 h

(69) dr 90.0:10.0

4, **164** (2 eq), THF, –78°, 40 h

(84) dr 94.0:6.0

4, **164** (2 eq), THF, –78°, 26 h

(34) er 88.0:12.0

Ac₂O (1.5 eq), **56** (2 mol %), *i*-Pr₂NEt (1.5 eq), *t*-AmOH, –2°, 16 h

(77) dr 93.0:7.0

4, **164** (2 eq), THF, –78°, 19 h

C₉

C₁₀

181

TABLE 1C. DESYMMETRIZATION OF MISCELLANEOUS DIOLS (*Continued*)

*Please refer to the charts preceding the tables for the structures indicated by the **bold** numbers.*

Diol	Conditions	Product(s) and Yield(s) (%)	Refs.
C_{10}			
	BzCl (1.5 eq), **129** (30 mol %), *i*-Pr$_2$NEt, CH$_2$Cl$_2$, 0°, 4 h	(55) er 91.0:9.0	137
	5 (1.1 eq), Et$_3$N (2 eq), toluene, rt, 138 h	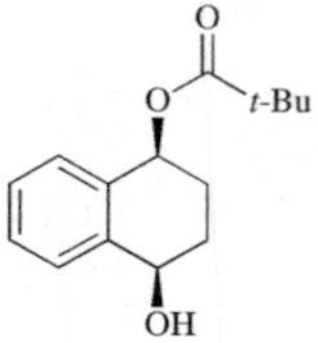 (34) er 94.0:6.0	25
	5 (1.1 eq), toluene, 80°, 20 h	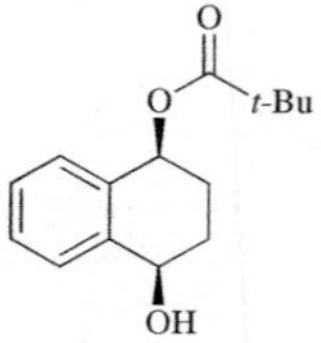 (79) er 89.5:10.5	26
	(*i*-PrCO)$_2$O (1.1 eq), **76** (5 mol %), Et$_3$N (1.1 eq), MTBE, 0°, 5 h	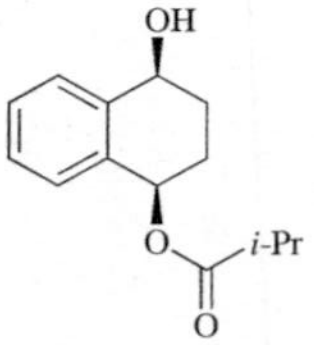 (44) er 59.0:41.0	41

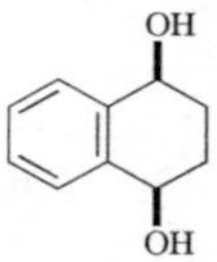

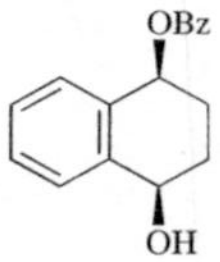

182

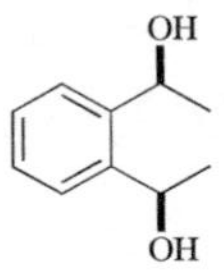

(*i*-PrCO)$_2$O (1.1 eq), **76** (5 mol %), Et$_3$N (1.1 eq), MTBE, 0°, 2 h

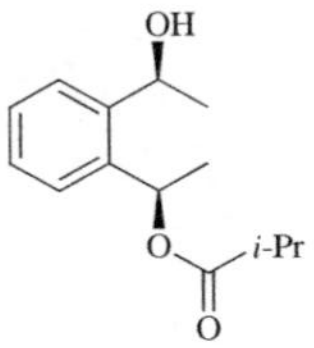

(87) er 98.5:1.5 — 41

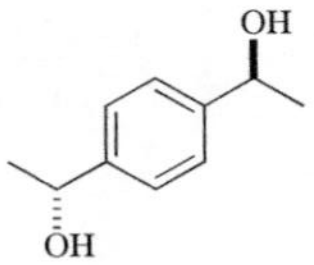

(*i*-PrCO)$_2$O (1.1 eq), **76** (5 mol %), Et$_3$N (1.1 eq), MTBE, 0°, 3 h

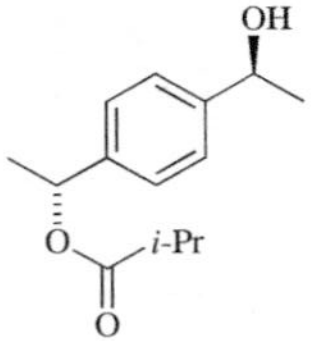

(69) er 98.0:2.0 — 41

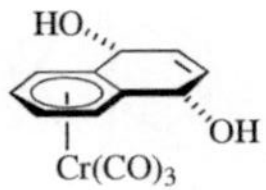

BzCl (1.5 eq), **131** (*x* mol %), Et$_3$N, 4 Å MS, CH_2Cl_2

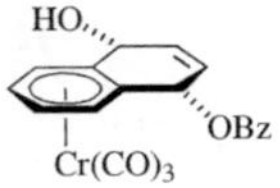

x	Temp (°)	Time (h)		er	
2	–40	23	(83)	99.5:0.5	145
10	–60	22	(76)	99.5:0.5	315

C_{11}

Ac_2O (5.3 eq), **105** (2 mol %), toluene, 0°, 6 h

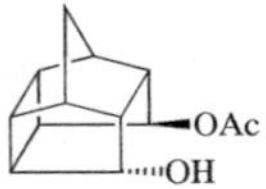

(90) er 52.5:47.5 — 73

TABLE 1C. DESYMMETRIZATION OF MISCELLANEOUS DIOLS (*Continued*)

Please refer to the charts preceding the tables for the structures indicated by the **bold** numbers.

Diol	Conditions	Product(s) and Yield(s) (%)	Refs.
C$_{11}$			
	Ac$_2$O (5.3 eq), **105** (2 mol %), toluene, 0°, 5 h	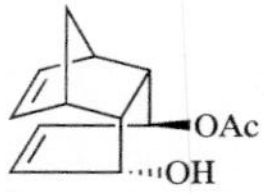 (2) er 55.5:45.0	73
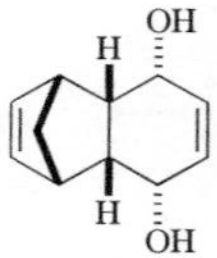	**5** (1.1 eq), toluene, 80°, 14 h	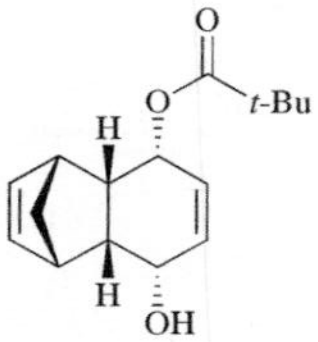 (68) er 72.0:28.0	25, 26
C$_{11-13}$			
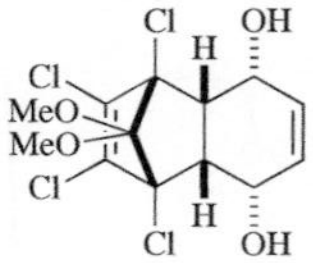	**5** (1.1 eq), heptane, 80°, 96 h	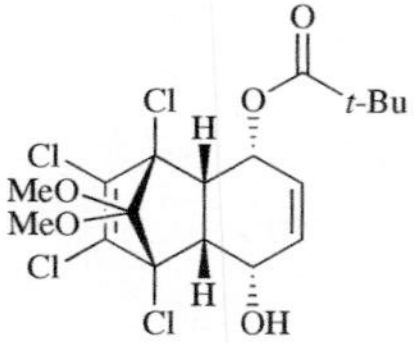 (45) er 80.5:19.5	25, 26
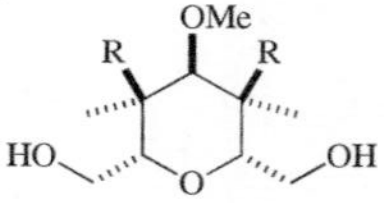	Ac$_2$O (1.5 eq), **56** (2 mol %), *i*-Pr$_2$NEt (1.5 eq), *t*-AmOH, −2°, 16 h	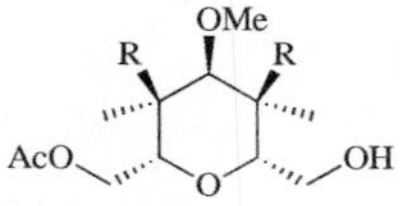 R Me (65) er 88.0:12.0 Et (68) er 94.0:6.0	314
C$_{12}$			
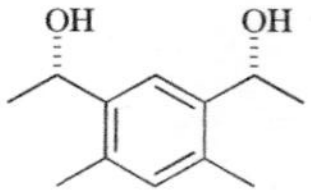	Ac$_2$O (1.5 eq), **56** (1 mol %), Et$_3$N (1.5 eq), *t*-AmOH, 0°, 13 h	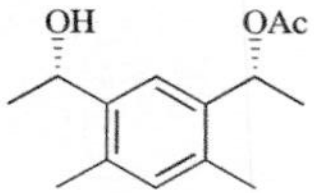 (91) er 99.5:0.5	146

(i-PrCO)$_2$O (1.1 eq), **76** (5 mol %), Et$_3$N (1.1 eq), MTBE, 0°, 3 h

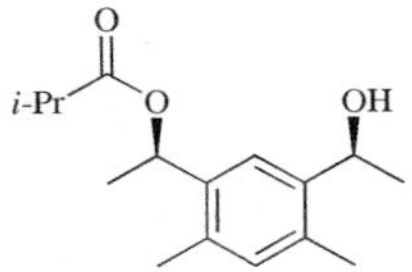

(87) er 94.0:6.0

41

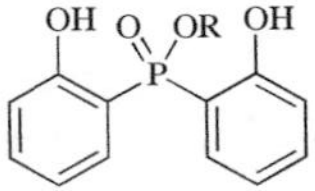

1-NpCHO, **145a** (5 mol %), **162** (1.0 eq), Et$_3$N (1.0 eq), THF, −40°, 24 h

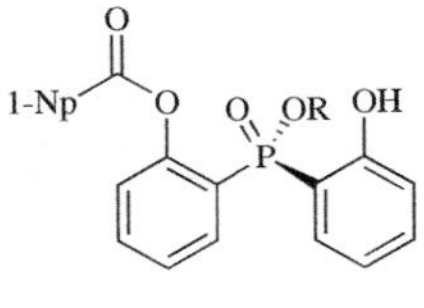

R		er
Me	(88)	97.0:3.0
Et	(75)	97.0:3.0
i-Pr	(87)	95.0:5.0

147

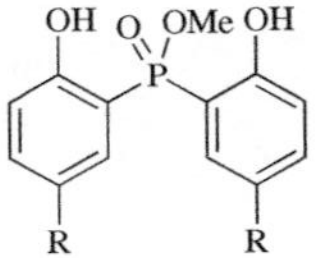

1-NpCHO, **145a** (5 mol %), **162** (1.0 eq), Et$_3$N (1.0 eq), THF, −40°, 24 h

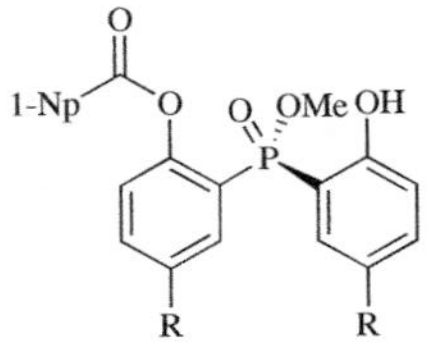

R		er
F	(62)	99.0:1.0
Cl	(57)	98.0:2.0
Br	(73)	97.0:3.0
MeO	(89)	92.0:8.0
Me	(90)	95.0:5.0

147

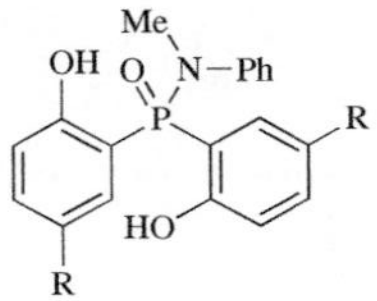

MesCHO, **145c** (5 mol %), **162** (1.0 eq), Et$_3$N (1.0 eq), THF, rt, 24 h

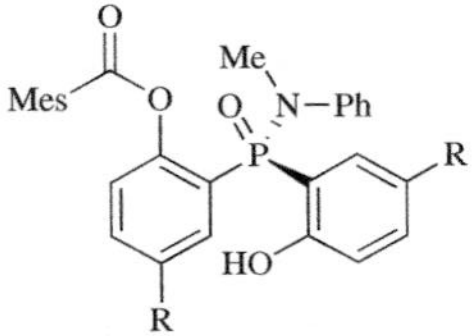

R		er
H	(79)	94.0:6.0
Br	(71)	94.0:6.0
Me	(90)	95.0:5.0

147

C$_{12-14}$

TABLE 1C. DESYMMETRIZATION OF MISCELLANEOUS DIOLS (*Continued*)

Diol	Conditions	Product(s) and Yield(s) (%)	Refs.

*Please refer to the charts preceding the tables for the structures indicated by the **bold** numbers.*

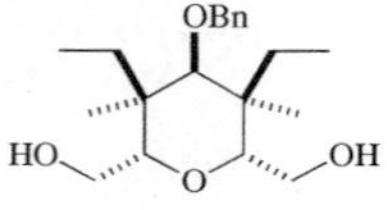

C13

Ac2O (1.5 eq), **56** (2 mol %), i-Pr2NEt (1.5 eq), t-AmOH, –2°, 16 h

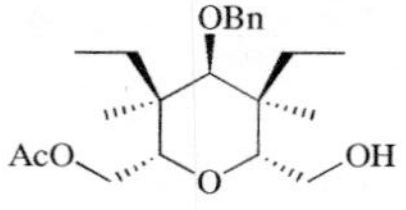

(51) er 89.0:11.0

314

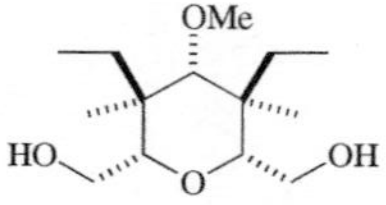

Ac2O (1.5 eq), **56** (2 mol %), i-Pr2NEt (1.5 eq), t-AmOH, –2°, 16 h

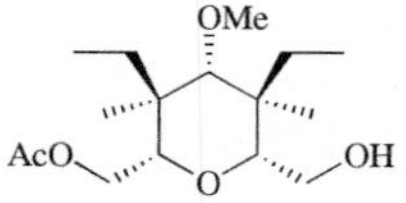

(53) er 69.0:31.0

314

C13–25

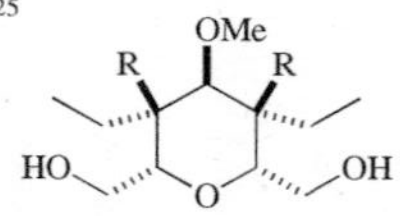

Ac2O (1.5 eq), **56** (2 mol %), i-Pr2NEt (1.5 eq), t-AmOH, –2°, 16 h

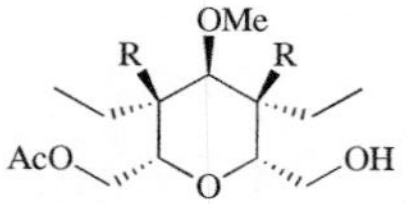

R		er
Me	(69)	97.0:3.0
PhCH2	(47)	94.0:6.0
PMB	(71)	86.0:14.0

314

C14–25

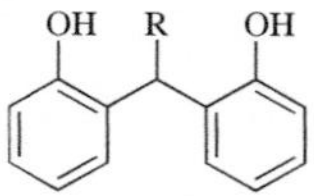

11 (1.0 eq), **145d** (10 mol %), i-Pr2NEt (2.0 eq), 4 Å MS, toluene, rt, 16 h

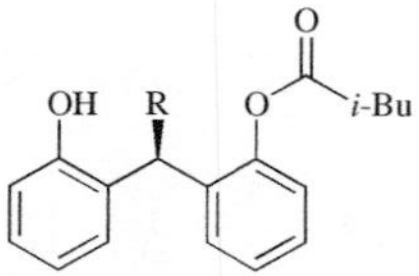

R		er
Me	(81)	96.5:3.5
Ph	(81)	98.5:1.5
4-FC6H4	(85)	98.5:1.5
4-ClC6H4	(85)	98.5:1.5
2-MeC6H4	(62)	99.5:0.5
4-MeC6H4	(87)	97.5:2.5
3,5-Me2C6H3	(82)	98.0:2.0
4-t-BuC6H4	(93)	97.0:3.0
2-Np	(72)	98.0:2.0
4-PhC6H4	(93)	98.5:1.5

148

C_{14-22}

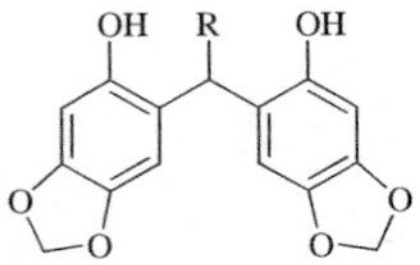

11 (1.0 eq), 145d (10 mol %),
i-Pr$_2$NEt (2.0 eq),
4 Å MS, toluene, rt, 16 h

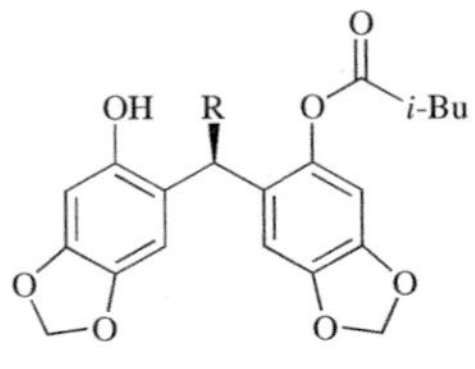

R		er	148
Me	(80)	93.5:6.5	
2-furyl	(85)	93.5:6.5	
4-MeOC$_6$H$_4$	(90)	98.5:1.5	
4-i-PrC$_6$H$_4$	(90)	98.0:2.0	

C_{16-21}

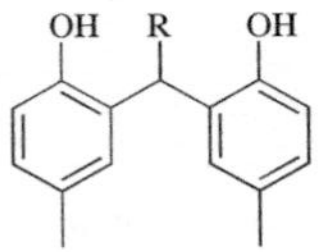

11 (1.0 eq), 145d (10 mol %),
i-Pr$_2$NEt (2.0 eq),
4 Å MS, toluene, rt, 16 h

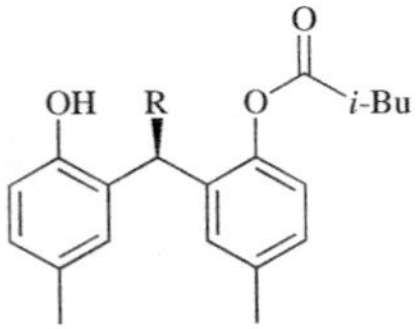

R		er	148
Me	(98)	95.5:4.5	
t-Bu	(85)	98.5:1.5	
n-C$_6$H$_{13}$	(88)	93.0:7.0	

C_{18}

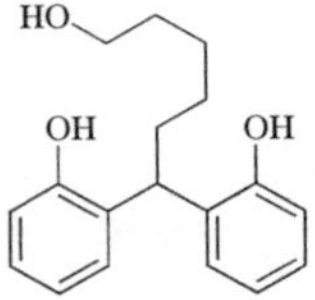

11 (1.0 eq), 145d (10 mol %),
i-Pr$_2$NEt (2.0 eq),
4 Å MS, THF, rt, 16 h

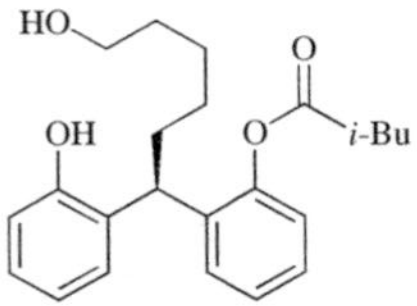

(82) er 96.0:4.0   148

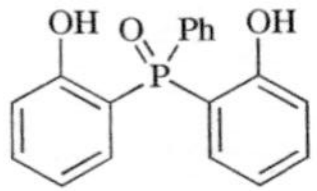

MesCHO, 145c (5 mol %),
162 (1.0 eq), Et$_3$N (1.0 eq),
THF, rt, 24 h

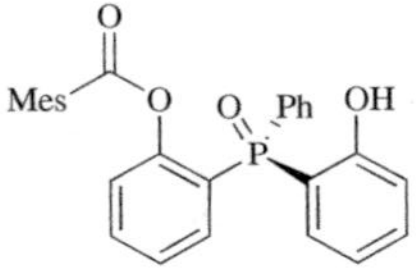

(68) er 71.0:29.0   147

TABLE 1C. DESYMMETRIZATION OF MISCELLANEOUS DIOLS (*Continued*)

Diol	Conditions	Product(s) and Yield(s) (%)	Refs.

*Please refer to the charts preceding the tables for the structures indicated by the **bold** numbers.*

C₁₈

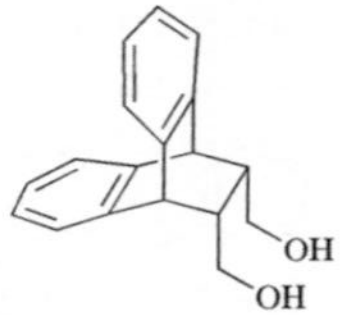

4, **164** (2 eq), THF, –78°, 22 h

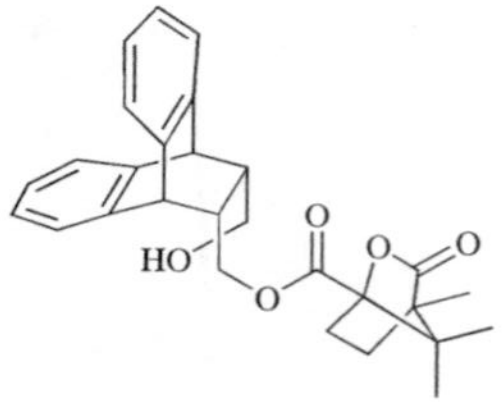

(79) dr 87.0:13.0

27

C₁₉

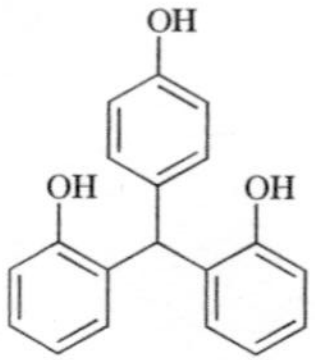

11 (1.0 eq), **145d** (10 mol %),
i-Pr₂NEt (2.0 eq),
4 Å MS, THF, rt, 16 h

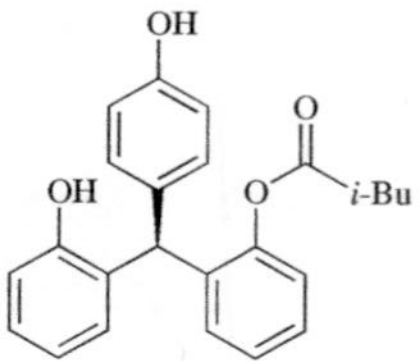

(72) er 97.5:2.5

148

C₁₉–₂₇

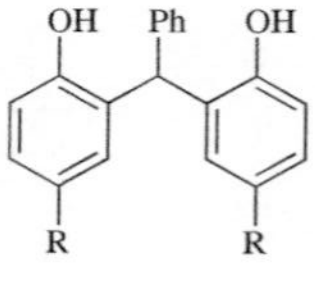

11 (1.0 eq), **145d** (10 mol %),
i-Pr₂NEt (2.0 eq),
4 Å MS, toluene, rt, 16 h

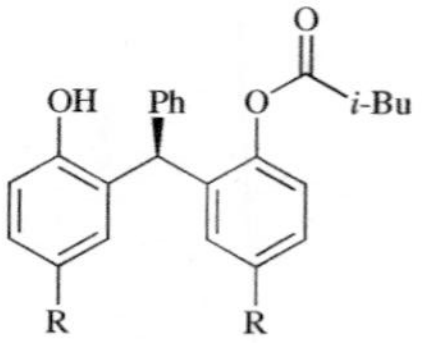

R		er
Cl	(94)	94.5:5.5
MeO	(95)	98.0:2.0
Me	(96)	98.5:1.5
t-Bu	(80)	98.0:2.0

148

11 (1.0 eq), **145d** (10 mol %),
i-Pr₂NEt (2.0 eq),
4 Å MS, toluene, rt, 16 h

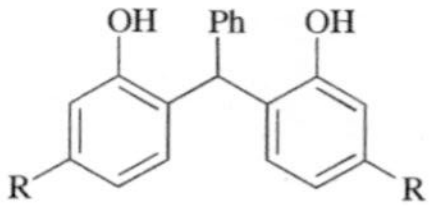

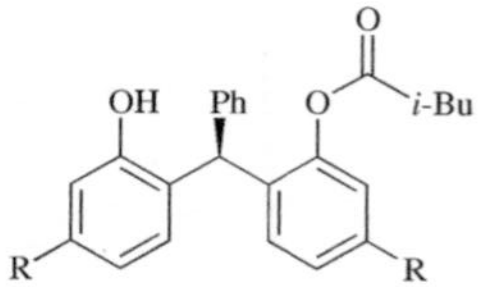

R		er
F	(74)	96.5:3.5
Me	(93)	98.0:2.0
t-Bu	(74)	98.0:2.0

148

188

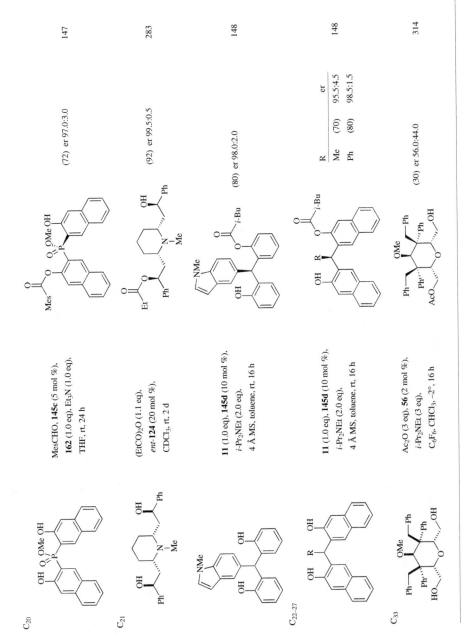

Substrate	Conditions	Product	Yield (er)	Ref
C$_{20}$	MesCHO, **145c** (5 mol %), **162** (1.0 eq), Et$_3$N (1.0 eq), THF, rt, 24 h		(72) er 97.0:3.0	147
C$_{21}$	(EtCO)$_2$O (1.1 eq), *ent*-**124** (20 mol %), CDCl$_3$, rt, 2 d		(92) er 99.5:0.5	283
C$_{22–27}$	**11** (1.0 eq), **145d** (10 mol %), *i*-Pr$_2$NEt (2.0 eq), 4 Å MS, toluene, rt, 16 h		(80) er 98.0:2.0	148
	11 (1.0 eq), **145d** (10 mol %), *i*-Pr$_2$NEt (2.0 eq), 4 Å MS, toluene, rt, 16 h		R er Me (70) 95.5:4.5 Ph (80) 98.5:1.5	148
C$_{33}$	Ac$_2$O (3 eq), **56** (2 mol %), *i*-Pr$_2$NEt (3 eq), C$_6$F$_6$, CHCl$_3$, –2°, 16 h		(30) er 56.0:44.0	314

189

TABLE 2A. STOICHIOMETRIC KINETIC RESOLUTION OF SECONDARY ALCOHOLS

*Please refer to the charts preceding the tables for the structures indicated by the **bold** numbers.*

Alcohol	Acylating Agent	Conditions	Product(s) and Conversion(s) (%)	Refs.
C₈ (1-cyclohexylethanol) 10 eq	oxazolidinone (t-Bu, N–Bz)	MeMgBr (1.1 eq), CH₂Cl₂, 0°, 20 h	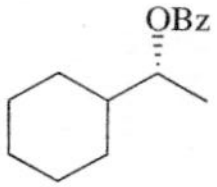 (>90) er 52.5:47.5	151
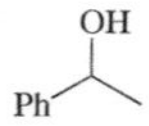 **C₈₋₉** (Ph, OH) 2.4 eq	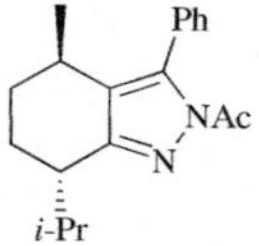 pyrazole (Ph, NAc, i-Pr)	AlCl₃, i-Pr₂NEt, toluene, hexane, –5°, 17 h	 OAc (89) er 82.0:18.0 + OH (60) er 70.5:29.5	316
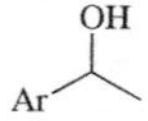 (Ar, OH) 2.4 eq	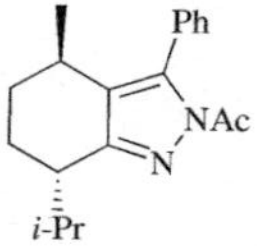 pyrazole (Ph, NAc, i-Pr)	AlCl₃, toluene, hexane, –5°, 17 h	see table below	316
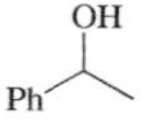 **C₈** (Ph, OH) 2.4 eq	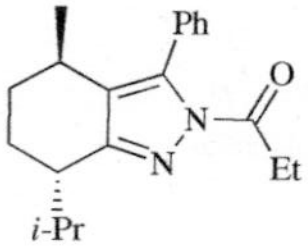 pyrazole (Ph, C(=O)Et, i-Pr)	AlCl₃, toluene, hexane, –5°, 17 h	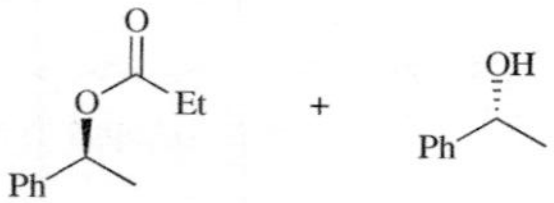 propionate (54) er 83.0:17.0 + OH (40) er 78.5:21.5	316

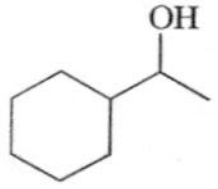

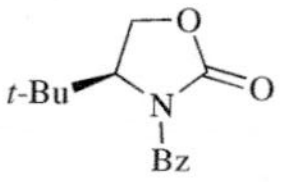

Ar	I	er I	II	er II
Ph	(99)	83.0:17.0	(64)	73.0:27.0
4-ClC₆H₄	(85)	80.0:20.0	(81)	65.0:35.0
2-MeC₆H₄	(29)	84.5:15.5	(58)	58.0:42.0
4-MeC₆H₄	(79)	76.5:23.5	(24)	56.5:43.5

ZnCl$_2$ (1 eq),
Et$_3$N (1.5 eq),
CH$_2$Cl$_2$, rt

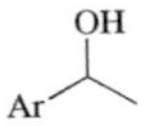

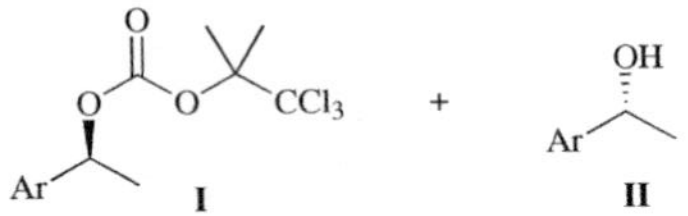

152

Ar	Time (h)		er I	er II	s
Ph	40	(25)	96.5:3.5	(—)	38
4-ClC$_6$H$_4$	40	(20)	95.5:5.0	60.5:39.5	24
1-Np	52	(28)	97.0:3.0	66.5:33.5	44
2-Np	52	(24)	97.0:3.0	65.0:35.0	45

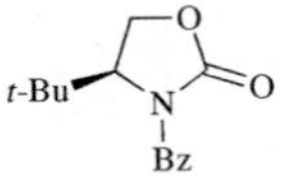

MeMgBr (1.1 eq),
CH$_2$Cl$_2$, 0°

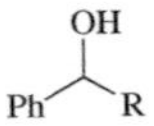

151

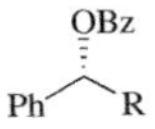

(>90)

R	Time (h)	er
Me	2	97.5:2.5
Et	2	95.0:5.0
n-Pr	18	96.5:3.5
i-Pr	18	82.5:17.5
n-Bu	18	96.5:3.5

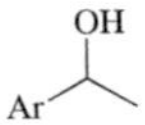

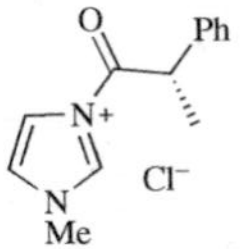

THF, rt

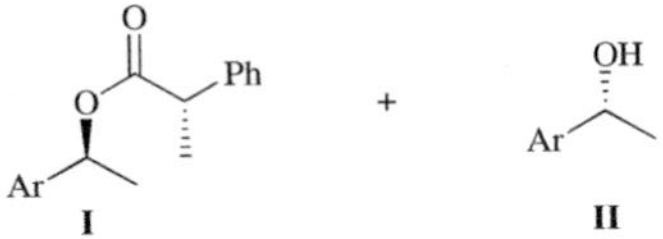

28

Ar		dr I	er II	s
Ph	(33)	60.5:39.5	55.5:44.5	1
1-Np	(48)	98.0:2.0	95.0:5.0	151
2-Np	(47)	99.0:1.0	95.5:4.5	316

C$_{8-12}$

0.47 eq

1

C$_{8-11}$

10 eq

C$_{8-12}$

191

TABLE 2A. STOICHIOMETRIC KINETIC RESOLUTION OF SECONDARY ALCOHOLS (*Continued*)

*Please refer to the charts preceding the tables for the structures indicated by the **bold** numbers.*

Alcohol	Acylating Agent	Conditions	Product(s) and Conversion(s) (%)	Refs.
C_{8-9}				
(OH, Ar) 0.47 eq	**1**	ZnCl$_2$ (1 eq), **167** (1.5 eq), CH$_2$Cl$_2$, rt	**I** (carbonate) + **II** (OH, Ar)	152

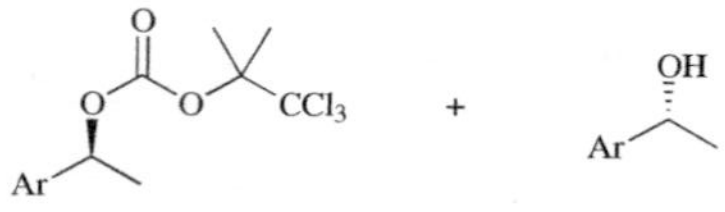

Ar	Time (h)		er I	er II	s
2-ClC$_6$H$_4$	38	(43)	95.0:5.0	83.0:17.0	4
2-MeC$_6$H$_4$	43	(39)	96.5:3.5	79.5:20.5	53

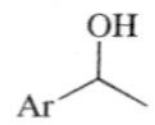

Alcohol	Acylating Agent	Conditions	Product(s) and Conversion(s) (%)	Refs.
C_9				
(OH, Ph) 2.4 eq	(Ph, NAc, *i*-Pr)	AlCl$_3$, toluene, hexane, –5°, 17 h	(OAc, Ph) (89) er 83.0:17.0 + (OH, Ph) (65) er 71.0:29.0	316
(OH, Ph)	**1** 0.47 eq	ZnCl$_2$ (1 eq), Et$_3$N (1.5 eq), CH$_2$Cl$_2$, rt, 62 h	**I** (CCl$_3$ carbonate, Ph) er 94.5:5.5 + (OH, Ph) er 61.0:39.0 (20), s = 22	152
(OH, Ph) 10 eq	(oxazolidinone, Ph, *t*-Bu)	MeMgBr (1.1 eq), CH$_2$Cl$_2$, 0°, 8 h	(OBz, Ph) (>90) er 57.5:42.5	151

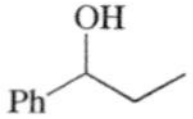

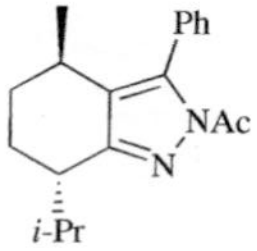

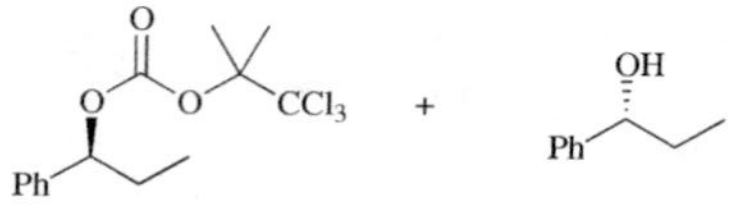

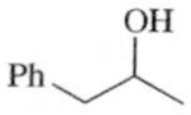

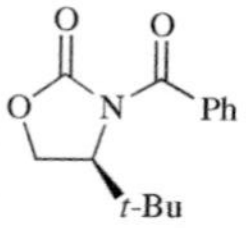

C$_{9-12}$

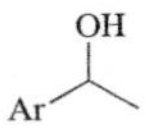

OH / Ar

1, 2
0.56 eq each

MgBr$_2$ (2 eq),
Et$_3$N (3.4 eq),
CH$_2$Cl$_2$, rt, 36 h

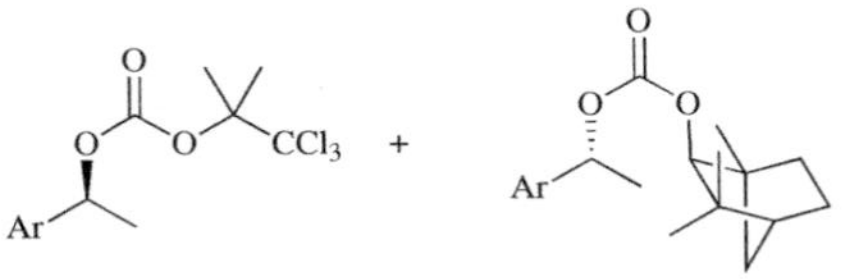

I

II

153

Ar	**I**	er **I**	**II**	dr **II**
2-MeC$_6$H$_4$	(46)	94.0:6.0	(49)	97.5:2.5
1-Np	(49)	93.0:7.0	(43)	96.5:3.5
2-Np	(46)	91.5:8.5	(46)	97.0:3.0

C$_{10}$

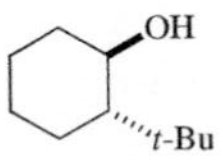

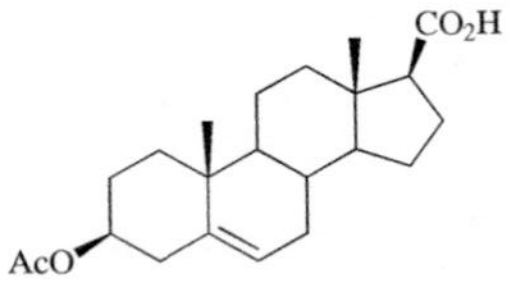

OH / t-Bu

1.5 eq

CO$_2$H

DCC (2 eq),
DMAP (2 eq),
CH$_2$Cl$_2$, rt, 24 h

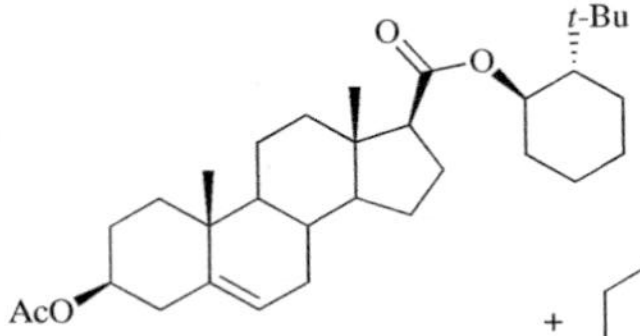

I dr 85.5:14.5

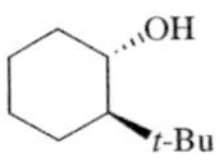

I + II (—) 317

II er 50.5:49.5

TABLE 2A. STOICHIOMETRIC KINETIC RESOLUTION OF SECONDARY ALCOHOLS (*Continued*)

Alcohol	Acylating Agent	Conditions	Product(s) and Conversion(s) (%)	Refs.

*Please refer to the charts preceding the tables for the structures indicated by the **bold** numbers.*

C_{10}

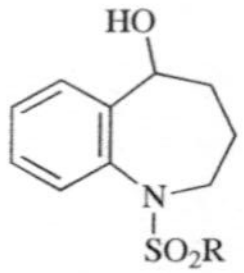

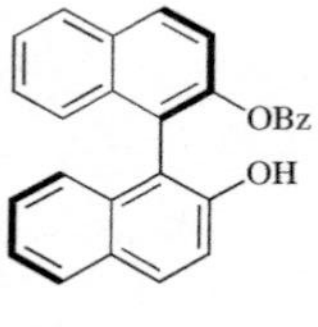

0.8 eq

MeMgBr (1.8 eq), CH$_2$Cl$_2$, rt, 48 h

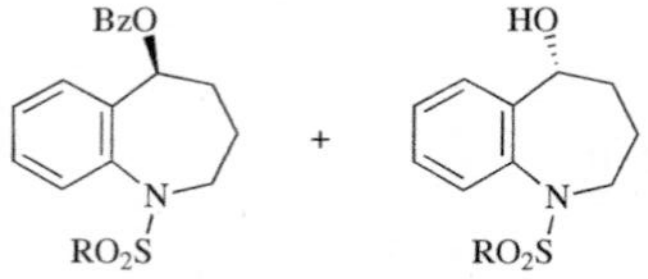

BzO — I   +   HO — II

R		er **I**	s
Me	(24)	77.0:23.0	4
Ph	(24)	87.5:12.5	9
4-MeOC$_6$H$_4$	(22)	83.0:17.0	6
4-BrC$_6$H$_4$	(28)	83.0:17.0	6
4-MeC$_6$H$_4$	(22)	88.0:12.0	9

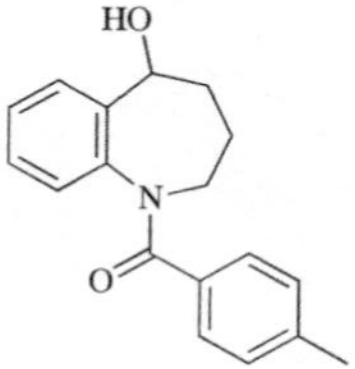

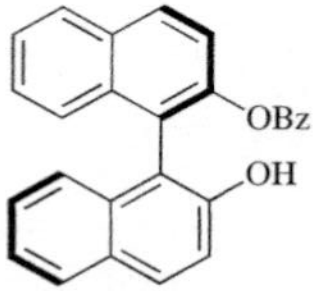

318

0.8 eq

MeMgBr (1.8 eq), CH$_2$Cl$_2$, rt, 96 h

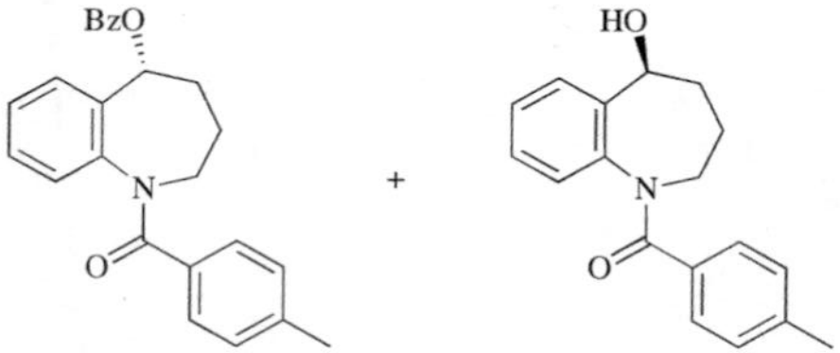

I er 64.0:36.0   +   **II**

(7), s = 2

318

C_{11-16}

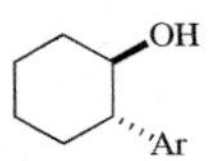

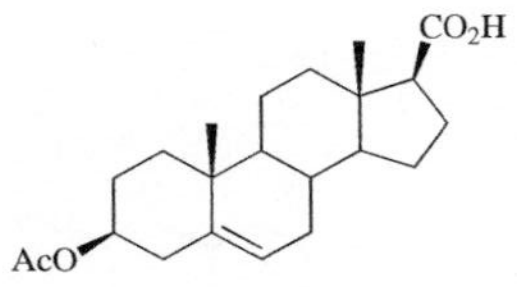

1.5 eq

DCC (2 eq), DMAP (2 eq), CH_2Cl_2, rt, 24 h

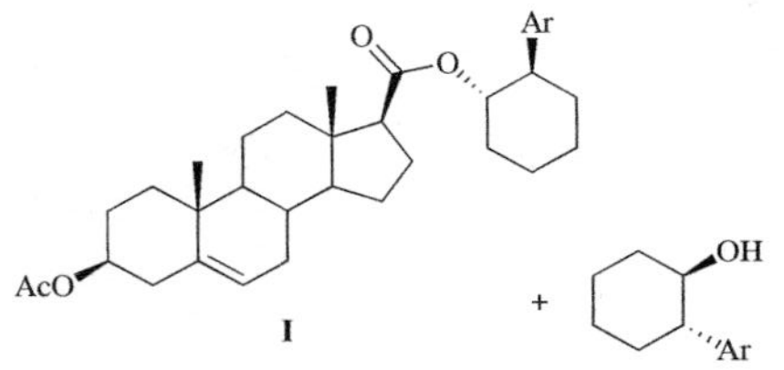

I **II**

317

Ar		dr **I**	er **II**	*s*
2-pyridyl	(—)	75.5:24.5	75.0:25.0	5
Ph	(—)	74.5:25.5	87.0:13.0	7
4-$MeOC_6H_4$	(—)	84.0:16.0	89.0:11.0	28
4-ClC_6H_4	(—)	67.0:33.0	>95.0:5.0	9
4-MeC_6H_4	(—)	79.0:21.0	80.5:19.5	7
2-quinolyl	(—)	88.5:11.5	96.0:4.0	40
1-Np	(—)	78.0:22.0	98.5:1.5	14
2-Np	(—)	76.0:24.0	>95.0:5.0	31

C_{11}

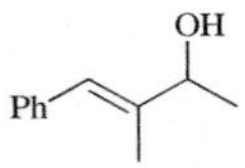

1

0.47 eq

$MgBr_2$ (1 eq), Et_3N (1.5 eq), CH_2Cl_2, rt, 60 h

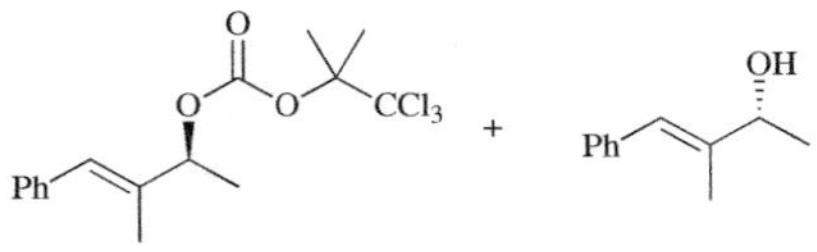

er 89.0:11.0 er 68.5:31.5

(40), *s* = 14

152

TABLE 2A. STOICHIOMETRIC KINETIC RESOLUTION OF SECONDARY ALCOHOLS (*Continued*)

Alcohol	Acylating Agent	Conditions	Product(s) and Conversion(s) (%)	Refs.

*Please refer to the charts preceding the tables for the structures indicated by the **bold** numbers.*

C_{11}

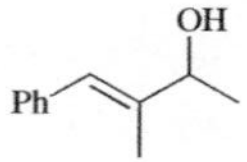

10 eq

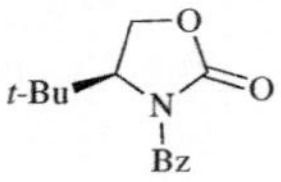

MeMgBr (1.1 eq),
CH_2Cl_2, 0°, 20 h

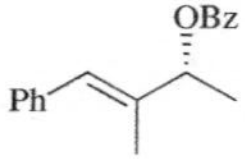

(>90) er 92.5:7.5

151

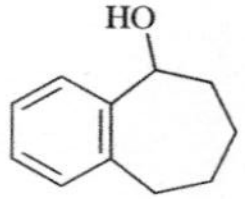

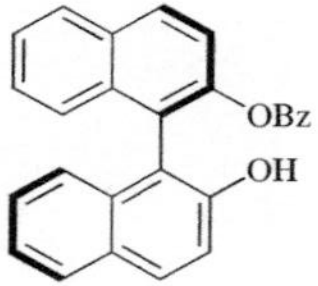

0.8 eq

MeMgBr (1.8 eq),
CH_2Cl_2, rt, 48 h

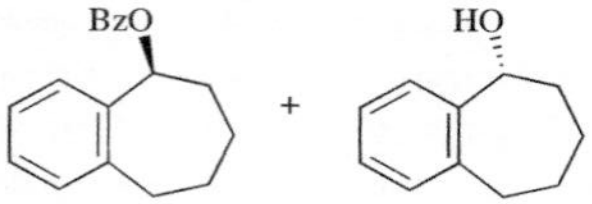

er 58.0:42.0

(26), s = 2

318

C_{12}

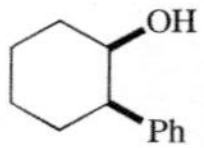

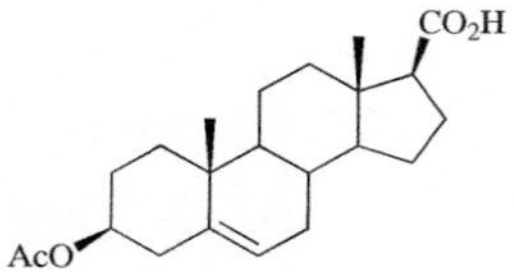

1.5 eq

DCC (2 eq),
DMAP (2 eq),
CH_2Cl_2, rt, 24 h

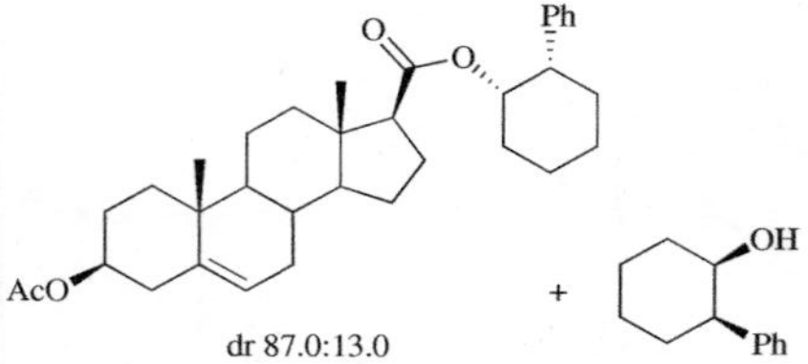

dr 87.0:13.0

er 96.0:4.0

(—), s = 10

317

C$_{12-16}$

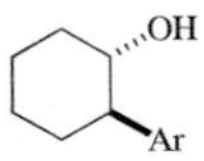

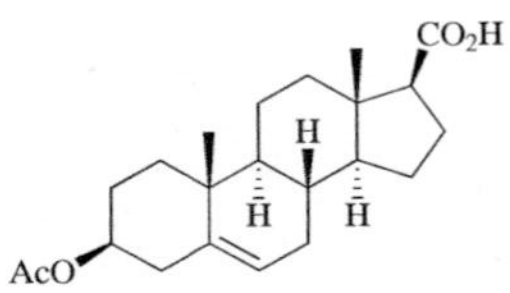

PS-DCC (2 eq),
DMAP (2 eq)
CH$_2$Cl$_2$, rt, 24 h

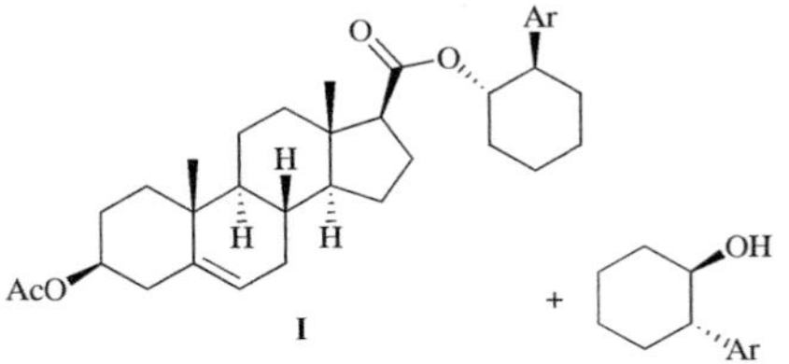

319

Ar	I	dr I	II	er II	s
Ph	(49)	83.0:17.0	(50)	78.0:22.0	6
4-ClC$_6$H$_4$	(62)	78.0:22.0	(31)	83.5:16.5	4
4-MeOC$_6$H$_4$	(48)	86.5:13.5	(40)	81.5:18.5	10
4-MeC$_6$H$_4$	(61)	84.5:15.5	(38)	79.0:21.0	4
1-Np	(61)	84.5:15.5	(39)	99.5:0.5	22
2-Np	(57)	94.0:6.0	(40)	99.5:0.5	35

C$_{12}$

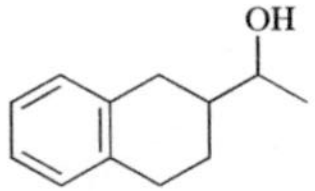

10 eq

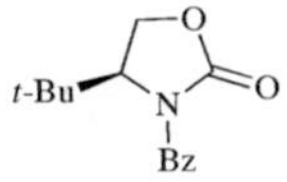

MeMgBr (1.1 eq),
CH$_2$Cl$_2$, 0°, 8 h

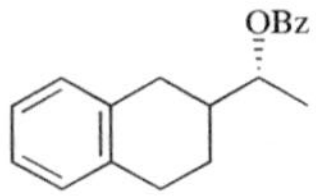

(>90) er 76.5:23.5

151

TABLE 2A. STOICHIOMETRIC KINETIC RESOLUTION OF SECONDARY ALCOHOLS (*Continued*)

Alcohol	Acylating Agent	Conditions	Product(s) and Conversion(s) (%)	Refs.

*Please refer to the charts preceding the tables for the structures indicated by the **bold** numbers.*

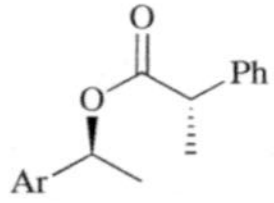

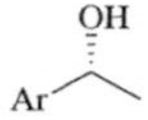

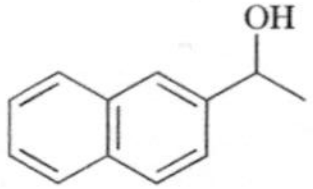

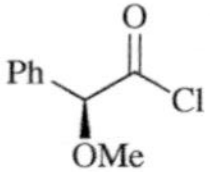

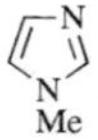

C$_{12}$

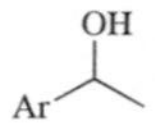

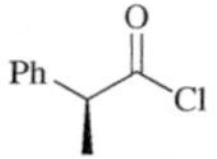

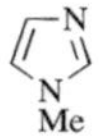

Ar		dr I	er II	s
1-Np	(48)	97.5:2.5	95.5:4.5	124
2-Np	(48)	99.0:1.0	96.0:4.0	327

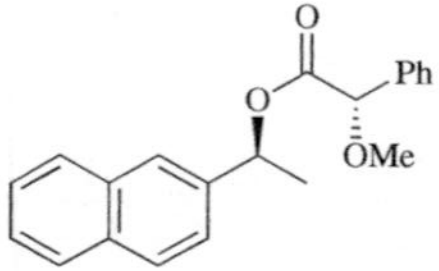

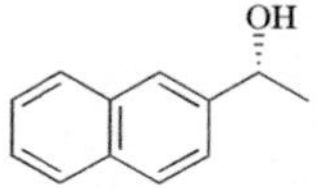

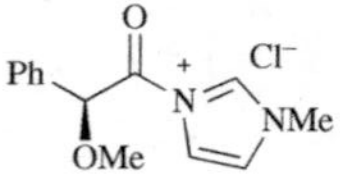

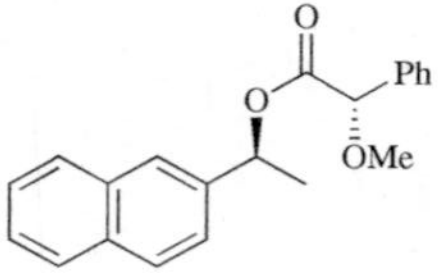

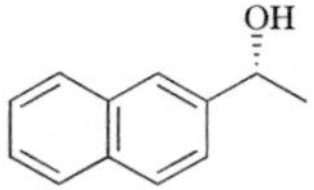

dr 97.0:3.0   er 97.5:2.5

(47), s = 120

dr 96.5:3.5   er 97.0:3.0

(47), s = 98

TABLE 2B. CATALYTIC KINETIC RESOLUTION OF BENZYLIC ALCOHOLS

Alcohol	Conditions	Product(s) and Conversion(s) (%)	Refs.

Please refer to the charts preceding the tables for the structures indicated by the **bold** numbers.

C₃

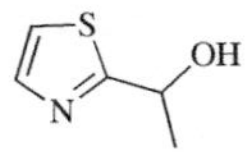

Ph₂CHCO₂H (0.5 eq),
Piv₂O (0.6 eq), **124** (5 mol %),
i-Pr₂NEt (1.2 eq), Et₂O, rt, 12 h

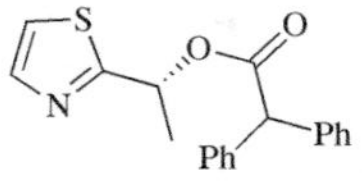

er 93.5:6.5   +   er 90.5:9.5   (51), s = 37

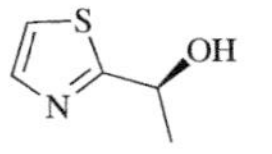

59

C₃₋₅

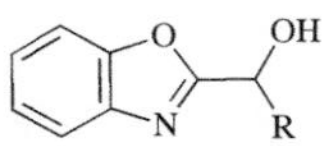

Ph₂CHCO₂H (0.5 eq),
Piv₂O (0.6 eq), **124** (5 mol %),
i-Pr₂NEt (1.2 eq), Et₂O, rt

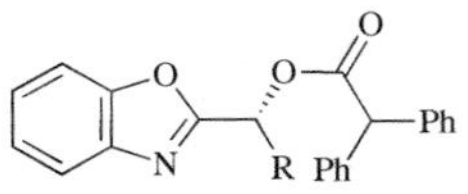

I   +   **II**

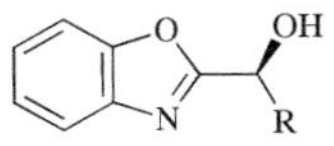

R	Time (h)		er **I**	er **II**	s
Me	12	(52)	96.0:4.0	99.5:0.5	122
Et	12	(49)	97.0:3.0	95.5:4.5	100
i-Pr	18	(50)	96.5:3.5	97.0:3.0	96

59

C₃

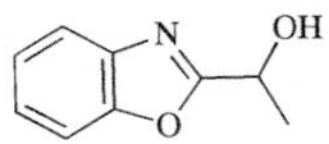

Ph(CH₂)₂CO₂H (0.5 eq),
Piv₂O (0.6 eq), **124** (5 mol %),
i-Pr₂NEt (1.2 eq), CH₂Cl₂, rt, 12 h

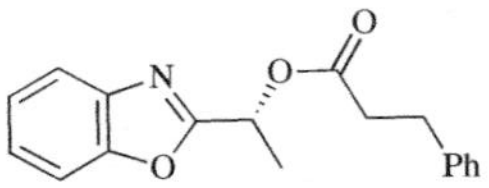

er 94.0:6.0   +   (—), s = 22   er 69.0:31.0

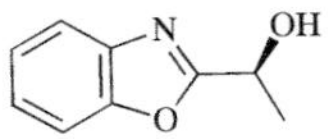

189

TABLE 2B. CATALYTIC KINETIC RESOLUTION OF BENZYLIC ALCOHOLS (*Continued*)

Alcohol	Conditions	Product(s) and Conversion(s) (%)	Refs.

*Please refer to the charts preceding the tables for the structures indicated by the **bold** numbers.*

C$_{3-5}$

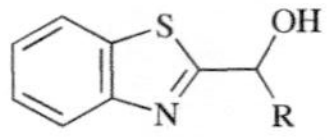

Ph$_2$CHCO$_2$H (0.5 eq),
Piv$_2$O (0.6 eq), **124** (5 mol %),
i-Pr$_2$NEt (1.2 eq), Et$_2$O, rt, 12 h

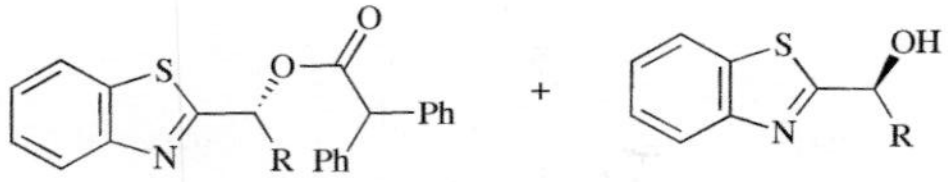

I   +   **II**

R		er I	er II	s
Me	(49)	95.5:4.5	93.0:7.0	57
Et	(46)	98.0:2.0	85.5:14.5	93
i-Pr	(47)	98.0:2.0	93.0:7.0	142

59

C$_3$

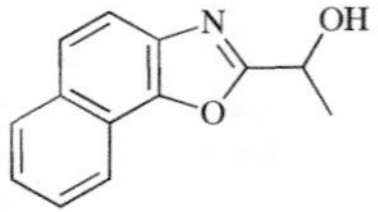

Ph(CH$_2$)$_2$CO$_2$H (0.5 eq),
Piv$_2$O (0.6 eq), **124** (5 mol %),
i-Pr$_2$NEt (1.2 eq), CH$_2$Cl$_2$, rt, 12 h

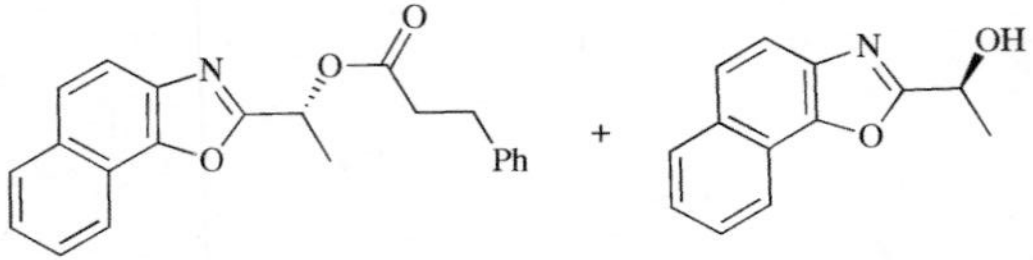

er 93.0:7.0   (—), s = 19   er 79.0:21.0

189

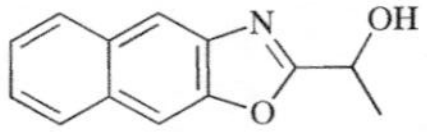

Ph(CH$_2$)$_2$CO$_2$H (0.5 eq),
Piv$_2$O (0.6 eq), **124** (5 mol %),
i-Pr$_2$NEt (1.2 eq), CH$_2$Cl$_2$, rt, 12 h

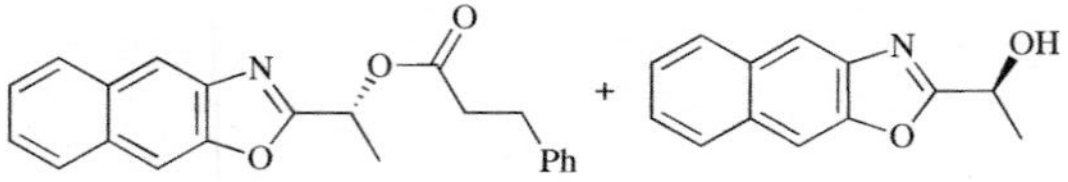

er 91.5:8.5   (—), s = 20   er 73.5:26.5

189

C_{3-10}

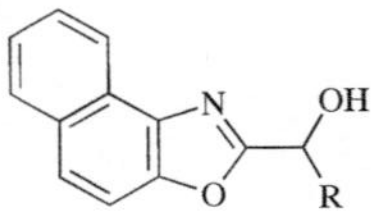

$Ph(CH_2)_2CO_2H$ (0.5 eq),
Piv_2O (0.6 eq), **124** (5 mol %),
i-Pr_2NEt (1.2 eq), CH_2Cl_2, rt, 12 h

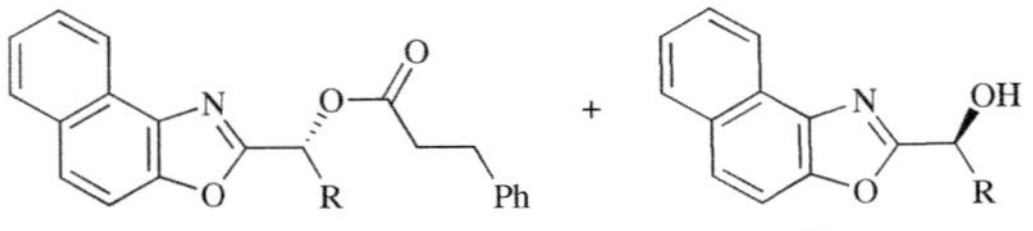

I + **II**

189

R		er **I**	er **II**	*s*
Me	(—)	93.5:6.5	86.5:13.5	31
$TBSOCH_2$	(—)	94.5:5.5	89.5:10.5	41
i-Pr	(—)	91.5:8.5	85.5:14.5	22
CH_2=$CHCH_2$	(—)	90.5:9.5	86.5:13.5	20
t-Bu	(—)	94.0:6.0	86.5:13.5	34
$Ph(CH_2)_2$	(—)	93.0:7.0	87.5:12.5	30

C_{6-9}

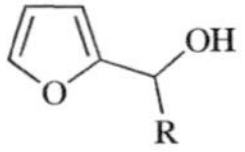

Ph_2CHCO_2H (0.5 eq),
Piv_2O (0.6 eq), **124** (5 mol %),
i-Pr_2NEt (1.2 eq), Et_2O, rt, 12 h

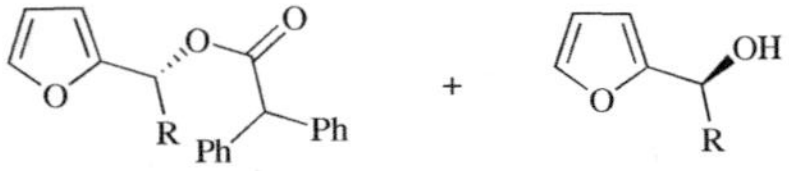

I + **II**

59

R		er **I**	er **II**	*s*
Me	(39)	95.0:5.0	78.5:21.5	34
Et	(41)	94.5:5.5	81.0:19.0	31
i-Pr	(42)	97.0:3.0	83.5:16.5	69
t-Bu	(19)	95.5:4.5	61.0:39.0	28

TABLE 2B. CATALYTIC KINETIC RESOLUTION OF BENZYLIC ALCOHOLS (*Continued*)

Alcohol	Conditions	Product(s) and Conversion(s) (%)	Refs.

*Please refer to the charts preceding the tables for the structures indicated by the **bold** numbers.*

C$_{6-9}$

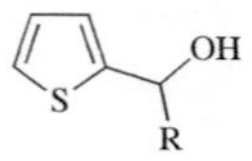

Ph$_2$CHCO$_2$H (0.5 eq),
Piv$_2$O (0.6 eq), **124** (5 mol %),
i-Pr$_2$NEt (1.2 eq), Et$_2$O, rt, 12 h

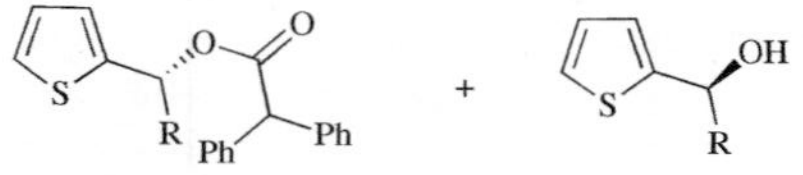

I   **II**

59

R		er **I**	er **II**	s
Me	(44)	94.5:5.5	84.5:15.5	35
Et	(46)	96.5:3.5	90.5:9.5	71
i-Pr	(46)	98.0:2.0	91.0:9.0	118
t-Bu	(19)	98.0:2.0	61.5:38.5	69

C$_6$

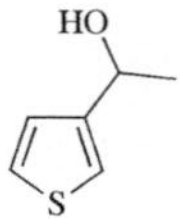

Ph$_2$CHCO$_2$H (0.5 eq),
Piv$_2$O (0.6 eq), **124** (5 mol %),
i-Pr$_2$NEt (1.2 eq), Et$_2$O, rt, 12 h

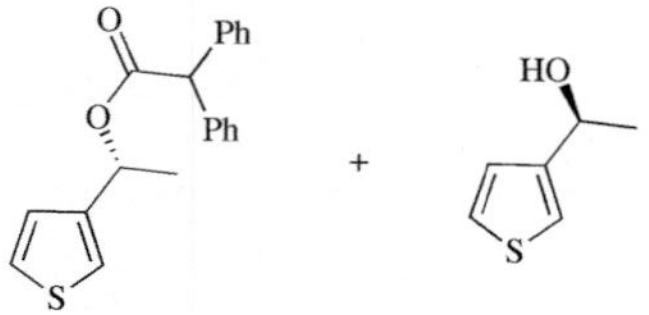

er 92.5:7.5   er 94.0:6.0

(51), s = 37

59

C$_{6-7}$

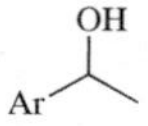

(EtCO)$_2$O (0.75 eq), **124** (10 mol %),
i-Pr$_2$NEt (0.75 eq), CDCl$_3$, 0°, 30 h

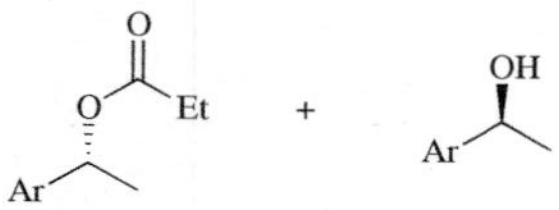

I   **II**

57

Ar		er **I**	er **II**	s
2-thienyl	(38)	92.0:8.0	76.0:24.0	19
2-pyridyl	(42)	92.0:8.0	80.5:19.5	21

C_{6-16}

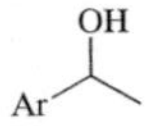

OH / Ar

(EtCO)$_2$O (0.75 eq), **117** (x mol %),
i-Pr$_2$NEt (0.75 eq), CDCl$_3$, 0°

I + II

57

Ar	x	Time (h)		er **I**	er **II**	s
2-thienyl	2	6	(45)	89.0:11.0	81.5:18.5	16
2-pyridyl	5	35	(47)	87.0:13.0	83.5:16.5	14

(i-PrCO)$_2$O (0.55 eq),
155 (1 mol %),
i-Pr$_2$NEt (0.6 eq), CHCl$_3$, 0°, 7 h

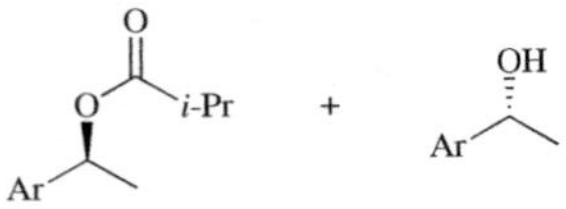

I + II

196

Ar		er **I**	er **II**	s
2-thienyl	(49)	91.0:9.0	90.0:10.0	25
2-pyridyl	(30)	63.0:37.0	56.0:44.0	2
4-FC$_6$H$_4$	(45)	92.0:8.0	83.0:17.0	22
4-MeOC$_6$H$_4$	(39)	95.0:5.0	79.0:21.0	31
2,5-(MeO)$_2$C$_6$H$_3$	(42)	89.0:11.0	78.0:22.0	14
4-CF$_3$C$_6$H$_4$	(53)	86.0:14.0	91.0:9.0	15
1-Np	(47)	94.0:6.0	90.0:10.0	41
2-Np	(47)	98.0:2.0	92.0:8.0	100
9-anthracenyl	(44)	97.0:3.0	88.0:12.0	80

TABLE 2B. CATALYTIC KINETIC RESOLUTION OF BENZYLIC ALCOHOLS (*Continued*)

Alcohol	Conditions	Product(s) and Conversion(s) (%)	Refs.

*Please refer to the charts preceding the tables for the structures indicated by the **bold** numbers.*

C$_{6-12}$

Alcohol:

OH / Ar–CF$_3$

Conditions:

(i-PrCO)$_2$O (1 eq),
124 (4 mol %),
i-Pr$_2$O, 0°

Products: **I** (Ar–CF$_3$ ester of i-Pr) + **II** (OH / Ar–CF$_3$)

Refs.: 191

Ar	Time (h)		er I	er II	s
2-thienyl	1	(50)	84.5:15.5	84.5:15.5	12
Ph	3	(45)	91.5:8.5	84.0:16.0	23
2-FC$_6$H$_4$	2	(58)	80.5:19.5	93.0:7.0	11
3-FC$_6$H$_4$	1	(51)	72.5:27.5	73.5:26.5	4
4-FC$_6$H$_4$	6	(51)	85.5:14.5	86.5:13.5	13
4-ClC$_6$H$_4$	5	(53)	88.0:12.0	93.0:7.0	20
4-MeOC$_6$H$_4$	3	(43)	94.5:5.5	84.5:15.5	35
3-O$_2$NC$_6$H$_4$	1	(57)	84.5:15.5	95.5:4.5	17
4-O$_2$NC$_6$H$_4$	1	(66)	75.0:25.0	99.0:1.0	13
4-MeSC$_6$H$_4$	1	(50)	91.0:9.0	91.0:9.0	25
2-MeC$_6$H$_4$	2	(43)	93.5:6.5	82.5:17.5	28
4-MeC$_6$H$_4$	1	(45)	93.5:6.5	85.0:15.0	31
4-NCC$_6$H$_4$	1	(53)	86.5:13.5	91.5:8.5	16
1-Np	2	(52)	93.0:7.0	96.0:4.0	44
2-Np	1	(52)	95.0:5.0	98.0:2.0	71

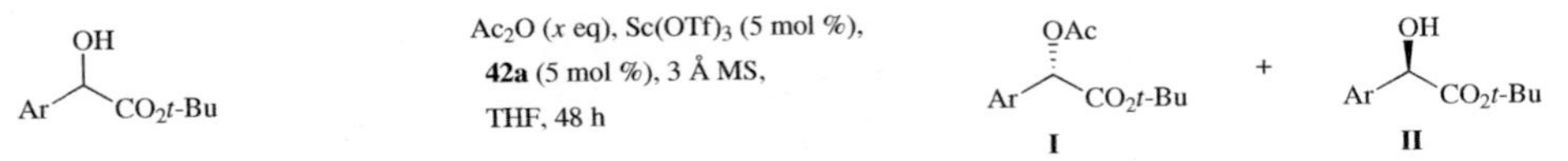

Ac$_2$O (x eq), Sc(OTf)$_3$ (5 mol %),
42a (5 mol %), 3 Å MS,
THF, 48 h

OH / Ar—CO$_2$t-Bu → OAc / Ar—CO$_2$t-Bu (**I**) + OH / Ar—CO$_2$t-Bu (**II**)

Ar	x	Temp (°)		er **I**	er **II**	s
2-furyl	0.8	−20	(50)	98.5:1.5	97.5:2.5	247
4-FC$_6$H$_4$	0.9	−20	(50)	96.0:4.0	95.0:5.0	71
2-ClC$_6$H$_4$	1.0	−20	(50)	95.0:5.0	95.0:5.0	60
3-ClC$_6$H$_4$	0.9	−20	(48)	96.5:3.5	92.0:8.0	79
4-ClC$_6$H$_4$	0.9	−20	(48)	97.0:3.0	93.0:7.0	86
2-BrC$_6$H$_4$	0.9	−20	(50)	95.5:4.5	94.5:5.5	60
3-MeOC$_6$H$_4$	0.9	−20	(49)	97.5:2.5	95.0:5.0	112
4-MeOC$_6$H$_4$	0.9	−20	(52)	93.5:6.5	96.5:3.5	51
4-O$_2$NC$_6$H$_4$	1.1	−20	(47)	96.0:4.0	90.5:9.5	58
2-MeC$_6$H$_4$	0.7	0	(52)	90.0:10.0	92.5:7.5	24
3-MeC$_6$H$_4$	0.9	−20	(49)	95.5:4.5	93.0:7.0	58
4-MeC$_6$H$_4$	0.9	−20	(50)	94.5:5.5	94.0:6.0	48
1-Np	1.3	−20	(48)	97.5:2.5	95.0:5.0	130

OH / Ar—CO$_2$t-Bu

TABLE 2B. CATALYTIC KINETIC RESOLUTION OF BENZYLIC ALCOHOLS (*Continued*)

Alcohol	Conditions	Product(s) and Conversion(s) (%)	Refs.

*Please refer to the charts preceding the tables for the structures indicated by the **bold** numbers.*

C₇

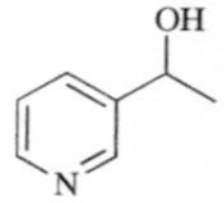

(i-PrCO)₂O (0.8 eq), **76** (0.5 mol %),
Et₃N (0.9 eq), CHCl₃, –30°, 48 h

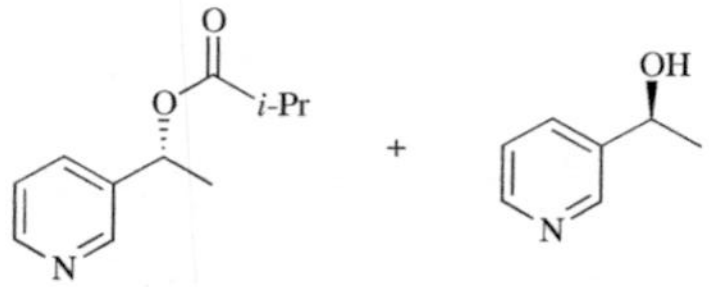

er 73.5:26.5   er 96.0:4.0

(66), *s* = 8   168

C₇₋₁₂

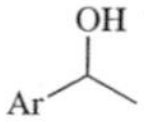

(i-PrCO)₂O (0.55 eq), **114** (10 mol %),
i-Pr₂NEt (0.66 eq), toluene, –40°, 48 h

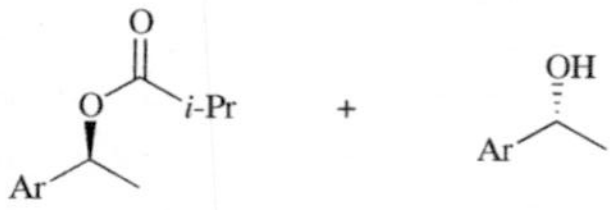

I   **II**

182

Ar		er I	er II	s
3-pyridyl	(45)	90.0:10.0	83.0:17.0	18
Ph	(45)	90.0:10.0	82.5:17.5	17
4-FC₆H₄	(49)	87.5:12.5	86.5:13.5	15
2-ClC₆H₄	(42)	90.0:10.0	78.5:21.5	16
4-ClC₆H₄	(43)	90.0:10.0	80.5:19.5	16
4-MeOC₆H₄	(45)	90.5:9.5	82.5:17.5	19
2-MeC₆H₄	(49)	93.5:6.5	91.5:8.5	38
4-MeC₆H₄	(41)	89.5:10.5	77.5:22.5	15
2,4-Me₂C₆H₃	(40)	95.0:5.0	80.5:19.5	34
2,4,6-Me₃C₆H₂	(34)	97.5:2.5	75.0:25.0	67
1-Np	(44)	93.0:7.0	84.5:15.5	28
2-Np	(46)	90.0:10.0	84.0:16.0	19

190

C$_{7-13}$

Ar—CH(OH)—CF$_2$—CO$_2$Et

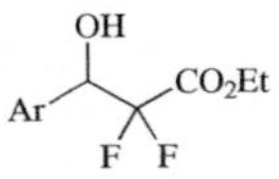

(EtCO)$_2$O (0.66 eq),
124 (4 mol %),
CHCl$_3$, rt,

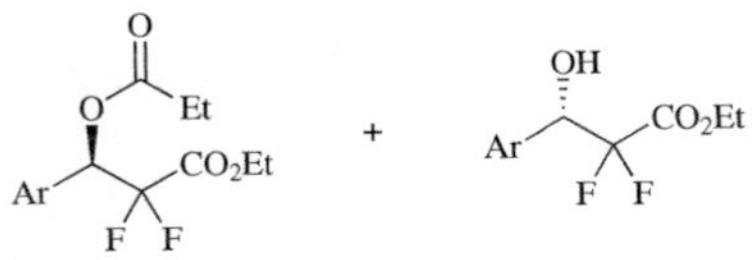

Ar—CH(OCOEt)—CF$_2$—CO$_2$Et
I

+

Ar—CH(OH)—CF$_2$—CO$_2$Et
II

Ar	Time (h)	er **I**	er **II**	s
2-furyl	1 (54)	83.5:16.5	88.0:12.0	11
2-thienyl	1 (47)	82.0:18.0	78.0:22.0	8
Ph	5 (62)	92.5:7.5	99.0:1.0	20
2-FC$_6$H$_4$	1 (48)	84.5:15.5	81.5:18.5	10
3-FC$_6$H$_4$	6 (14)	89.5:10.5	56.5:43.5	10
4-FC$_6$H$_4$	1 (42)	86.0:14.0	76.0:24.0	10
C$_6$F$_5$	3 (79)	53.0:47.0	61.5:38.5	1
4-MeOC$_6$H$_4$	3 (48)	90.5:9.5	87.5:12.5	21
4-MeSC$_6$H$_4$	3 (40)	92.5:7.5	79.0:21.0	22
4-MeC$_6$H$_4$	3 (47)	91.0:9.0	85.5:14.5	20
1-Np	3 (45)	94.0:6.0	86.0:14.0	34
2-Np	3 (49)	94.5:5.5	92.5:7.5	47

C$_8$

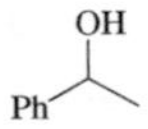

Ph—CH(OH)—CH$_3$

Ac$_2$O, TaCl$_2$ (15 mol %),
39 (15 mol %), CH$_2$Cl$_2$, rt, 40 h

Ph—CH(OAc)—CH$_3$

+

Ph—CH(OH)—CH$_3$ (60)

er 62.5:37.5

198

207

TABLE 2B. CATALYTIC KINETIC RESOLUTION OF BENZYLIC ALCOHOLS (*Continued*)

*Please refer to the charts preceding the tables for the structures indicated by the **bold** numbers.*

Alcohol	Conditions	Product(s) and Conversion(s) (%)	Refs.
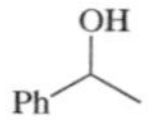 C$_8$ OH Ph	Ac$_2$O (0.75 eq), **61** (2 mol %), Et$_3$N (0.75 eq), THF, −78°, 2 h	 OAc Ph er 57.5:42.5 + OH Ph (46), s = 2 er 74.0:26.0	176
	Vinyl acetate, **148** (3 mol %), t-BuOK (2.5 eq), Et$_2$O, 0°, 12 h	 OAc Ph er 65.5:34.5 + OH Ph (39), s = 2 er 60.0:40.0	77
	Ac$_2$O (0.75 eq), **60** (2 mol %), Et$_3$N (0.75 eq), THF, −78°, 9 h	 OAc Ph er 61.0:39.0 + OH Ph (62), s = 2 er 51.0:49.0	179
	Ac$_2$O (0.75 eq), **78** (5 mol %), Et$_3$N (0.75 eq), toluene, −60°, 15 h	 OAc Ph er 64.0:36.0 + OH Ph (43), s = 2 er 61.5:38.5	172
	Ac$_2$O (0.7 eq), **86** (2 mol %), Cs$_2$CO$_3$ (0.7 eq), THF, rt, 24 h	 OAc Ph er 70.0:30.0 + OH Ph (40), s = 3 er 63.5:36.5	174

C_{8-11}

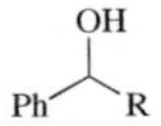

Ac$_2$O (1.2 eq), **56** (2 mol %),
Et$_3$N, Et$_2$O, rt

R	Time (h)		er **I**	er **II**	s
Me	25	(62)	79.5:20.5	97.5:2.5	14
ClCH$_2$	27	(69)	72.5:27.5	99.5:0.5	12
Et	29	(62)	81.0:19.0	99.5:0.5	20
i-Pr	26	(55)	89.0:11.0	99.0:1.0	36
t-Bu	49	(51)	94.0:6.0	96.0:4.0	52

159

Ac$_2$O (0.75 eq), **56** (1 mol %),
Et$_3$N (0.75 eq), t-AmOH, rt

R	Time (h)		er **I**	er **II**	s
Me	24	(56)	89.5:10.5	99.5:0.5	43
ClCH$_2$	6	(56)	88.5:11.5	98.5:1.5	32
t-Bu	111	(51)	96.0:4.0	98.0:2.0	98

146

Ac$_2$O (0.7 eq), **83** (0.5 mol %),
i-Pr$_2$NEt (0.7 eq),
toluene, –40°, 24 h

R		er **II**	s
Me	(23)	60.0:40.0	6
t-Bu	(13)	53.0:47.0	3

320

TABLE 2B. CATALYTIC KINETIC RESOLUTION OF BENZYLIC ALCOHOLS (*Continued*)

Alcohol	Conditions	Product(s) and Conversion(s) (%)	Refs.

*Please refer to the charts preceding the tables for the structures indicated by the **bold** numbers.*

C$_{8-11}$

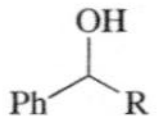

OH / Ph / R

Ac$_2$O (1 eq), **85** (1 mol %), t-AmOH, 0°

OAc / Ph / R (**I**) + OH / Ph / R (**II**)

321

R	Time (h)		er **I**	er **II**	s
Me	11	(70)	67.0:33.0	89.5:10.5	4
Et	16	(67)	71.5:28.5	92.5:7.5	6
i-Pr	17	(77)	65.5:34.5	99.5:0.5	8
t-Bu	12	(59)	82.0:18.0	95.0:5.0	13

C$_8$

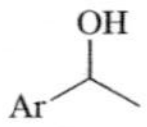

OH / Ar

Ac$_2$O (2.0 eq), **82** (5 mol %), Et$_3$N (0.6 eq), acetone, −78°, 18 h

OAc / Ar (**I**) + OH / Ar (**II**)

170

Ar		er **I**	er **II**	s
Ph	(16)	75.5:24.5	54.5:45.5	3
4-ClC$_6$H$_4$	(22)	72.0:28.0	56.5:43.5	3
2-MeOC$_6$H$_4$	(17)	80.0:20.0	56.0:44.0	5

C$_{8-12}$

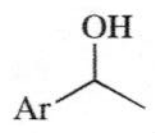

OH / Ar

Ac$_2$O (0.75 eq), **89** (1 mol %),
Et$_3$N (0.75 eq),
toluene, −78°   165

OAc / Ar (**I**) + OH / Ar (**II**)

Ar	Time (h)		er I	er II	s
Ph	6	(6)	95.0:5.0	53.0:47.0	22
2-MeOC$_6$H$_4$	7	(6)	95.5:4.5	53.0:47.0	22
2-MeC$_6$H$_4$	7	(19)	96.0:4.0	61.0:39.0	31
2,6-Me$_2$C$_6$H$_3$	7	(13)	95.0:5.0	57.0:43.0	22
1-Np	7	(15)	97.0:3.0	58.5:41.5	39

AcCl (0.6 eq), **157a** (5 mol %),
DABCO (0.65 eq),
Et$_2$O, −20°, 5 h   201

OAc / Ar (**I**) + OH / Ar (**II**)

Ar		er I	er II	s
Ph	(51)	94.0:6.0	96.0:4.0	54
1-Np	(47)	93.0:7.0	88.5:11.5	30
2-Np	(55)	90.0:10.0	98.0:2.0	34

Ac$_2$O (1.5 eq), **101** (2.5 mol %),
toluene, −65°, 24 h   68

OAc / Ar (**I**) + OH / Ar (**II**)

Ar		s
Ph	(—)	20
4-FC$_6$H$_4$	(—)	11
4-MeOC$_6$H$_4$	(—)	16
1-Np	(—)	>50

TABLE 2B. CATALYTIC KINETIC RESOLUTION OF BENZYLIC ALCOHOLS (*Continued*)

Alcohol	Conditions	Product(s) and Conversion(s) (%)	Refs.

*Please refer to the charts preceding the tables for the structures indicated by the **bold** numbers.*

C8–12

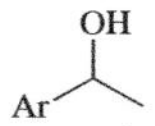

Ac₂O (0.75 eq), **79** (5 mol %),
Et₃N (0.75·eq),
toluene, −78°, 15 h

Ar		er **I**	er **II**	s
Ph	(17)	87.0:13.0	58.0:42.0	8
1-Np	(14)	89.0:11.0	(—)	9
2-Np	(17)	89.0:11.0	(—)	9

173

AcOC(Me)=CH₂ (1.27 eq),
50 (1 mol %),
toluene, −3°

Ar	Time (h)		er **I**	er **II**	s
Ph	6	(65)	92.5:7.5	61.5:38.5	2
1-Np	8	(61)	94.5:5.5	68.0:32.0	2

101

C8

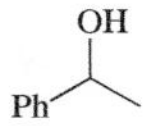

RCO₂H (0.5 eq), **124** (5 mol %),
Piv₂O (0.6 eq), *i*-Pr₂NEt (1.2 eq),
CH₂Cl₂, rt, 12 h

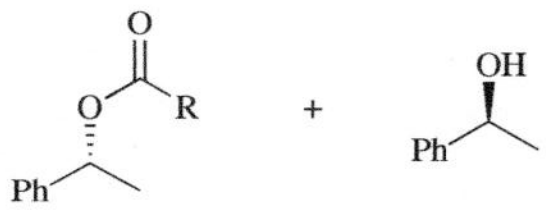

R		er **I**	er **II**	s
Et	(45)	92.5:7.5	85.0:15.0	26
Ph(CH₂)₂	(42)	94.5:5.5	82.5:17.5	33

55

212

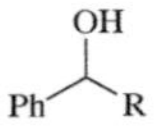

RCO$_2$H (0.55 eq), **124** (5 mol %),
(3-PyCO)$_2$O (0.55 eq),
CH$_2$Cl$_2$, 0°, 12 h

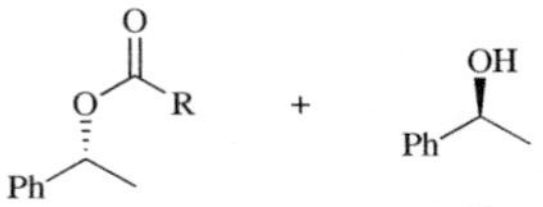

R		er I	er II	s
Et	(41)	95.5:4.5	81.5:18.5	42
Ph(CH$_2$)$_2$	(44)	95.5:4.5	85.5:14.5	45

(EtCO)$_2$O (0.75 eq),
118 (2 mol %),
i-PrNEt$_2$ (0.75 eq), CHCl$_3$, 0°

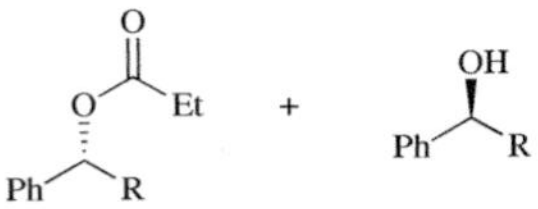

R	Time (h)		er I	er II	s
Me	8	(32)	94.5:5.5	75.0:25.0	26
Et	8	(39)	95.5:4.5	71.5:28.5	36
i-Pr	30	(55)	90.5:9.5	65.0:35.0	41
t-Bu	52	(48)	97.0:3.0	92.0:8.0	85

C$_{8-11}$

TABLE 2B. CATALYTIC KINETIC RESOLUTION OF BENZYLIC ALCOHOLS (*Continued*)

Alcohol	Conditions	Product(s) and Conversion(s) (%)	Refs.

*Please refer to the charts preceding the tables for the structures indicated by the **bold** numbers.*

C$_{8-11}$

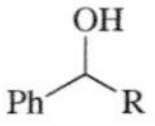

Alcohol: OH, Ph–CH(R)

(EtCO)$_2$O (0.75 eq), 117 (2 mol %), i-Pr$_2$NEt (0.75 eq), CHCl$_3$, 0° — Refs. 51

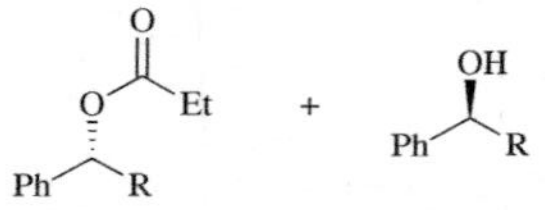

Products: **I** (ester) + **II** (OH)

R	Time (h)	(%)	er I	er II	s
Me	8	(55)	89.0:11.0	98.5:1.5	33
Et	8	(53)	92.0:8.0	97.5:2.5	42
i-Pr	4	(50)	95.0:5.0	95.5:4.5	61
t-Bu	8	(42)	98.5:1.5	85.0:15.0	124

(EtCO)$_2$O (0.75 eq), 120 (5 mol %), i-Pr$_2$NEt (0.75 eq), toluene, 0° — Refs. 183

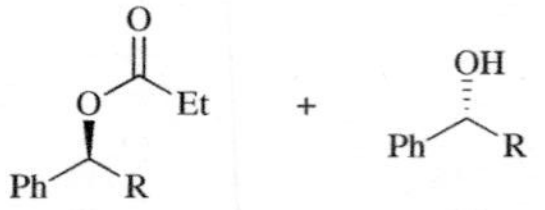

Products: **I** (ester) + **II** (OH)

R	Time (h)	(%)	er I	er II	s
Me	10	(49)	96.5:3.5	94.5:5.5	89
Et	10	(47)	94.5:5.5	90.5:9.5	43
i-Pr	15	(49)	97.5:2.5	96.0:4.0	124
t-Bu	35	(41)	99.5:0.5	84.0:16.0	450

C$_8$

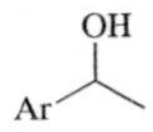

(EtCO)$_2$O (0.5 eq), **126** (0.75 mol %),
i-Pr$_2$NEt (0.6 eq), CHCl$_3$, 0°, 1 h

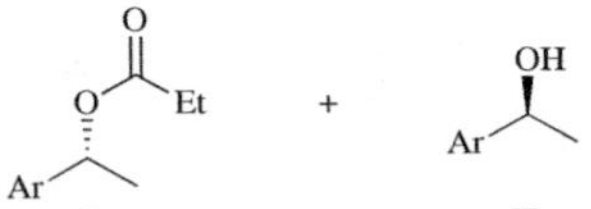

Ar		er I	er II	s
Ph	(46)	93.5:6.5	87.5:12.5	35
4-FC$_6$H$_4$	(45)	88.0:12.0	87.0:13.0	17
4-MeOC$_6$H$_4$	(34)	94.5:5.5	73.5:26.5	35

60

C$_{8-12}$

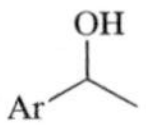

(EtCO)$_2$O (0.75 eq), **115** (2 mol %),
i-Pr$_2$NEt (0.75 eq), toluene, 0°

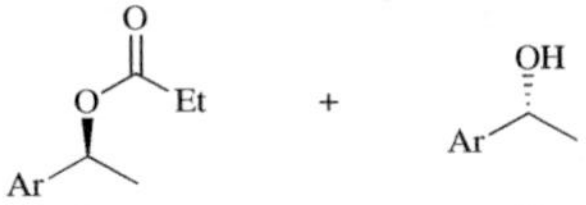

Ar	Time (h)		er I	er II	s
Ph	8	(51)	91.5:8.5	94.0:6.0	31
4-ClC$_6$H$_4$	8	(52)	92.5:7.5	96.5:3.5	41
4-BrC$_6$H$_4$	8	(53)	91.5:8.5	97.5:2.5	37
2-MeOC$_6$H$_4$	8	(37)	97.0:3.0	77.5:22.5	57
3-MeOC$_6$H$_4$	8	(51)	93.5:6.5	95.0:5.0	44
4-MeOC$_6$H$_4$	8	(49)	93.0:7.0	92.0:8.0	35
1-Np	7	(52)	93.0:7.0	97.0:3.0	47

49

TABLE 2B. CATALYTIC KINETIC RESOLUTION OF BENZYLIC ALCOHOLS (*Continued*)

Alcohol	Conditions	Product(s) and Conversion(s) (%)	Refs.

*Please refer to the charts preceding the tables for the structures indicated by the **bold** numbers.*

C_8

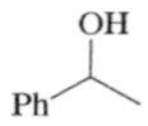

OH / Ph (structure)

(*i*-PrCO)$_2$O (0.7 eq),
83 (0.5 mol %), *i*-Pr$_2$NEt (0.7 eq),
toluene, –40°, 24 h

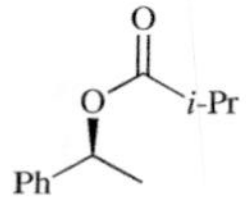

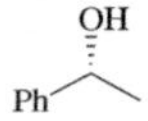

Product I: ester, (29) + OH/Ph (32), *s* = 4
er 62.5:37.5

320

(*i*-PrCO)$_2$O (1.0 eq), **113** (5 mol %),
i-Pr$_2$NEt (1.0 eq), CH$_2$Cl$_2$, rt, 24 h

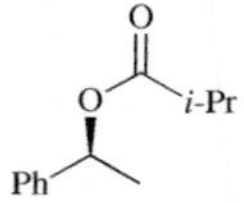

(47) + (61), *s* = 3
er 66.0:34.0   er 75.0:25.0

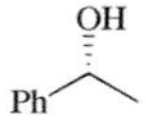

322

(*i*-PrCO)$_2$O (0.75 eq), **94** (20 mol %),
i-Pr$_2$NEt (0.75 eq), CHCl$_3$, rt, 68 h

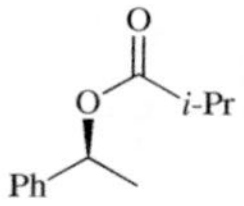

(51) + (65), *s* = 9
er 75.0:25.0   er 96.0:4.0

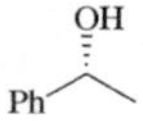

323

C_{8-11}

OH / Ph, R

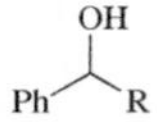

(*i*-PrCO)$_2$O (2.5 eq),
141 (x mol %), heptane, –40°

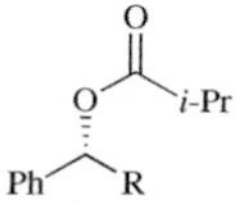

154

Product I and II table:

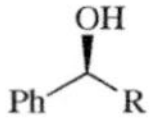

R	x	Time (h)	I	er I	II	er II	s
Me	2.5	4	(29)	96.5:3.5		69.0:31.0	42
i-Pr	2.8	42	(47)	97.5:2.5		92.0:8.0	100
n-Bu	3.9	8	(51)	94.5:5.5		96.5:3.5	57
i-Bu	3.5	7	(42)	94.0:6.0		82.0:18.0	31

216

(i-PrCO)$_2$O (2.0 eq), **87** (1 mol %),
Et$_3$N (0.75 eq), toluene, –78°

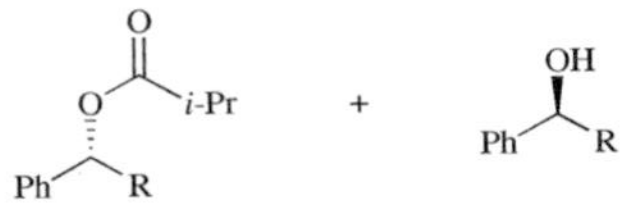

I + **II**

R	Time (h)		er I	er II	s
Me	8	(39)	89.0:11.0	75.0:25.0	13
Et	10	(35)	89.5:10.5	71.5:28.5	13
i-Pr	10	(29)	86.5:13.5	65.0:35.0	8
t-Bu	11	(18)	94.5:5.5	59.5:40.5	20

(i-PrCO)$_2$O (0.75 eq),
124 (4 mol %), i-Pr$_2$NEt (0.75 eq),
CHCl$_3$, Na$_2$SO$_4$, 0°

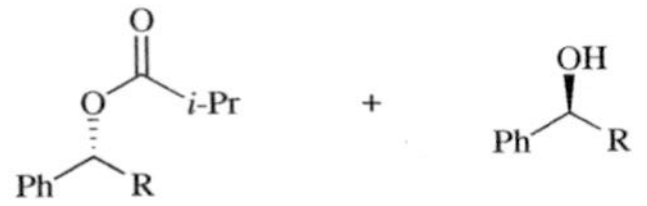

I + **II**

R	Time (h)		er I	er II	s
Me	33	(46)	97.5:2.5	77.5:22.5	104
Et	36	(45)	98.5:1.5	89.5:10.5	146
t-Bu	48	(31)	99.0:1.0	72.0:28.0	196

TABLE 2B. CATALYTIC KINETIC RESOLUTION OF BENZYLIC ALCOHOLS (*Continued*)

Alcohol	Conditions	Product(s) and Conversion(s) (%)	Refs.

*Please refer to the charts preceding the tables for the structures indicated by the **bold** numbers.*

C_{8-11}

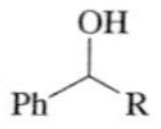

Alcohol: OH, Ph, R

Entry 1 — Conditions: (i-PrCO)$_2$O (0.6 eq), **77** (10 mol %), Et$_2$O, –50° Refs. 169

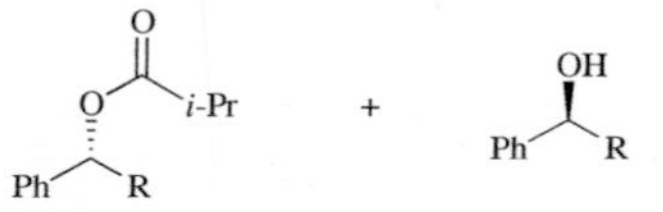

Products: **I** (O=C, i-Pr, O, Ph, R) + **II** (OH, Ph, R)

R	Time (h)	(%)	er I	er II	s
Me	120	(53)	77.5:22.5	80.5:19.5	6
Et	168	(49)	85.5:14.5	83.5:16.5	12
t-Bu	168	(45)	85.0:15.0	78.5:21.5	10

Entry 2 — Conditions: (i-PrCO)$_2$O (0.75 eq), **69a** (5 mol %), Cs$_2$CO$_3$ (0.75 eq), toluene, –60°, 15 h Refs. 324

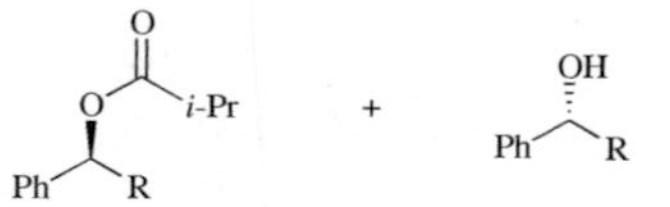

Products: **I** (O=C, i-Pr, O, Ph, R) + **II** (OH, Ph, R)

R	(%)	er I	er II	s
Me	(53)	84.0:16.0	88.0:12.0	11
ClCH$_2$	(73)	68.0:32.0	99.5:0.5	15
Et	(35)	93.0:7.0	72.5:27.5	20
t-Bu	(14)	93.0:7.0	57.0:43.0	15

C$_{8-10}$

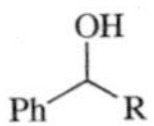

OH / Ph, R

(i-PrCO)$_2$O (0.55 eq),
155 (1 mol %),
i-Pr$_2$NEt (0.6 eq), CHCl$_3$, 0°, 7 h

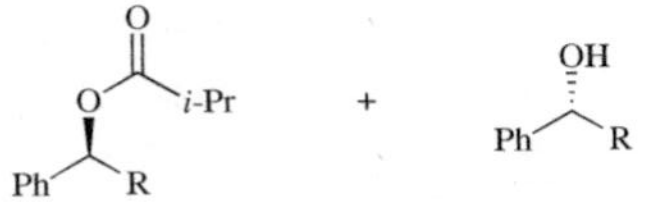

O, i-Pr, R, Ph — **I**   +   OH, Ph, R — **II**

196

R		er **I**	er **II**	s
Me	(43)	96.0:4.0	85.0:15.0	46
ClCH$_2$	(51)	92.0:8.0	93.0:7.0	31
CF$_3$	(48)	73.0:27.0	72.0:28.0	4
Et	(47)	96.0:4.0	92.0:8.0	70
i-Pr	(40)	98.0:2.0	82.0:18.0	80

C$_{8-12}$

OH, Ar

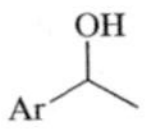

(i-PrCO)$_2$O (0.75 eq),
70 (1 mol %), Et$_3$N (0.75 eq),
CH$_2$Cl$_2$, −78°, 6 h

O, i-Pr, Ar — **I**   +   OH, Ar — **II**

39

Ar		er **II**	s
Ph	(28)	62.5:37.5	6
2-MeOC$_6$H$_4$	(18)	59.0:41.0	11
4-MeOC$_6$H$_4$	(20)	59.5:40.5	9
2,4-(MeO)$_2$C$_6$H$_3$	(40)	70.0:30.0	6
2-O$_2$NC$_6$H$_4$	(28)	53.0:47.0	2
4-O$_2$NC$_6$H$_4$	(27)	53.0:47.0	2
2-MeC$_6$H$_4$	(17)	57.0:43.0	6
2,6-Me$_2$C$_6$H$_3$	(15)	50.5:49.5	1
1-Np	(23)	60.0:40.0	7
2-Np	(37)	70.0:30.0	8

TABLE 2B. CATALYTIC KINETIC RESOLUTION OF BENZYLIC ALCOHOLS (*Continued*)

Alcohol	Conditions	Product(s) and Conversion(s) (%)	Refs.

*Please refer to the charts preceding the tables for the structures indicated by the **bold** numbers.*

C$_{8-12}$

OH / Ar

(*i*-PrCO)$_2$O (0.7 eq).
84 (5 mol %), Et$_3$N (0.9 eq).
CH$_2$Cl$_2$, –78°, 3 h

I + II

246

Ar		er I	er II	s
Ph	(42)	84.5:15.5	75.0:25.0	9
4-BrC$_6$H$_4$	(45)	71.5:28.5	68.0:32.0	4
4-MeOC$_6$H$_4$	(30)	84.0:16.0	64.5:35.5	7
4-O$_2$NC$_6$H$_4$	(36)	62.5:37.5	57.0:43.0	2
2-MeC$_6$H$_4$	(44)	85.0:15.0	77.0:23.0	10
Mes	(44)	86.0:14.0	78.5:21.5	11
1-Np	(45)	85.5:14.5	79.5:20.5	11
2-Np	(50)	81.0:19.0	81.5:18.5	8

(*i*-PrCO)$_2$O (0.8 eq).
76 (0.5 mol %), Et$_3$N (0.9 eq).
CHCl$_3$, rt, 12 h

I + II

168

Ar		er I	er II	s
Ph	(65)	74.0:26.0	94.5:5.5	8
4-MeOC$_6$H$_4$	(68)	73.0:27.0	98.5:1.5	10
4-O$_2$NC$_6$H$_4$	(72)	69.0:31.0	99.0:1.0	9
1-Np	(65)	76.0:24.0	98.5:1.5	13

C$_{8-14}$

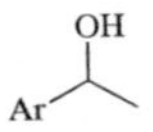

OH / Ar

(i-PrCO)$_2$O (0.75 eq),
92 (0.5 mol %), Et$_3$N (0.75 eq),
t-AmOH

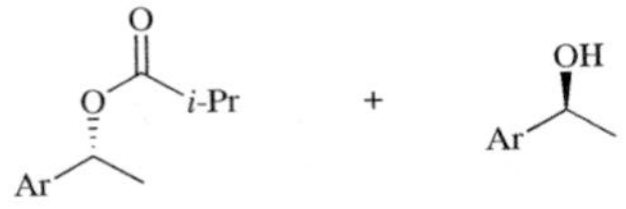

O
i-Pr
O / Ar
I

+

OH / Ar
II

42

Ar	Time (h)		er I	er II	s
Ph	5	(46)	89.5:10.5	82.5:17.5	17
4-FC$_6$H$_4$	6	(47)	89.5:10.5	85.5:14.5	18
2-BrC$_6$H$_4$	20	(42)	93.0:7.0	81.0:19.0	23
2-MeOC$_6$H$_4$	24	(45)	95.0:5.0	87.5:12.5	43
2-MeC$_6$H$_4$	6	(41)	94.5:5.5	81.5:18.5	32
Mes	12	(45)	96.0:4.0	87.5:12.5	52
1-Np	4	(48)	93.0:7.0	90.0:10.0	33
2-Np	7	(51)	91.5:8.5	95.0:5.0	33
2-PhC$_6$H$_4$	48	(56)	88.0:12.0	98.0:2.0	27

C$_{8-12}$

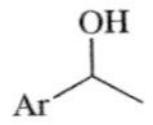

OH / Ar

Diketene (1.2 eq),
57 (10 mol %),
benzene, rt

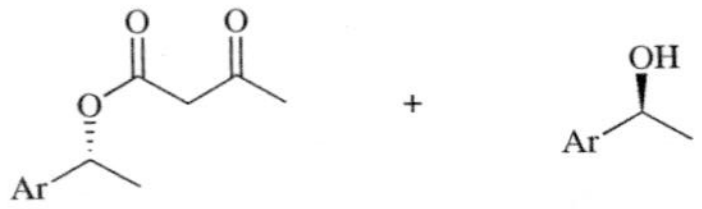

O O
O / Ar
I

+

OH / Ar
II

325

Ar	Time (h)		er I	er II	s
Ph	4	(58)	69.0:31.0	77.0:23.0	4
1-Np	5	(67)	71.5:28.5	93.5:6.5	7

TABLE 2B. CATALYTIC KINETIC RESOLUTION OF BENZYLIC ALCOHOLS (*Continued*)

Alcohol	Conditions	Product(s) and Conversion(s) (%)	Refs.

*Please refer to the charts preceding the tables for the structures indicated by the **bold** numbers.*

C$_{8-12}$

Alcohol: OH, Ar

Conditions: Diketene (1.2 eq), *ent*-**57** (10 mol %), benzene, rt

Products: **I** (Ar ester) + **II** (OH, Ar)

Ar	Time (h)		er I	er II	s
Ph	4	(58)	69.5:30.5	76.5:23.5	4
1-Np	5	(67)	71.5:28.5	93.5:6.5	7

Refs.: 325

C$_{8-10}$

Conditions: BzCl (0.6 eq), **135** (10 mol %), *i*-Pr$_2$NEt (0.6 eq), 4 Å MS, CHCl$_3$, 0°, 5 h

Products: **I** (OBz, Ph, R) + **II** (OH, Ph, R)

R		er I	er II	s
Me	(61)	63.0:37.0	70.0:30.0	2
Et	(—)	51.5:48.5	51.0:49.0	1
c-C$_3$H$_5$	(52)	66.0:34.0	67.5:32.5	3

Refs.: 197

C$_{8-9}$

Conditions: BzBr (1.0 eq), **137** (30 mol %), SnBr$_2$ (0.3 eq), 4 Å MS, CH$_2$Cl$_2$, −78°

Products: **I** (OBz, Ar) + **II** (OH, Ar)

Ar	Time (h)		er I	er II	s
Ph	3	(51)	79.5:20.5	80.5:19.5	6
2-MeC$_6$H$_4$	4	(51)	88.5:11.5	89.5:10.5	12

Refs.: 158

C$_{8-12}$

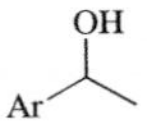

Ar—CH(OH)CH$_3$ (OH structure)

BzCl (0.75 eq), **137** (0.3 mol %),
Et$_3$N (0.5 eq), 4 Å MS,
CH$_2$Cl$_2$, –78°, 3 h

OBz **I** + OH **II**

157

Ar		er **I**	er **II**	s
Ph	(49)	84.5:15.5	83.5:16.5	9
2-MeC$_6$H$_4$	(49)	91.0:9.0	89.0:11.0	20

38•PdCl$_2$ (10 mol %),
Ph$_3$Bi(OAc)$_2$ (0.6 eq),
AgOAc (3 eq), CO (5 atm),
THF, 30°, 48 h

OBz **I** + OH **II**

214

Ar		er **I**	er **II**	s
Ph	(—)	59.5:40.5	54.0:46.0	—
2-MeC$_6$H$_4$	(—)	62.5:37.5	56.5:43.5	—
1-Np	(—)	59.5:40.5	56.5:43.5	—
2-Np	(—)	59.0:41.0	55.5:44.5	—

TABLE 2B. CATALYTIC KINETIC RESOLUTION OF BENZYLIC ALCOHOLS (Continued)

Alcohol	Conditions	Product(s) and Conversion(s) (%)	Refs.

*Please refer to the charts preceding the tables for the structures indicated by the **bold** numbers.*

C$_{8-12}$

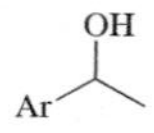

PhCHO (10 eq),
ent-**145a** (10 mol %),
163 (10 mol %),
Et$_3$N (0.5 eq), 4 Å MS,
toluene, rt, 4 h

I (OBz, Ar) + II (OH, Ar)

Ar		er II	s
Ph	(65)	71.5:28.5	2
1-Np	(55)	72.0:28.0	3
2-Np	(72)	83.0:17.0	3

82

C$_{8-11}$

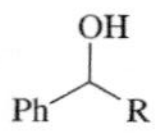

(3-ClC$_6$H$_4$CO)$_2$O (2.5 eq),
144 (16 mol %), CD$_2$Cl$_2$, rt

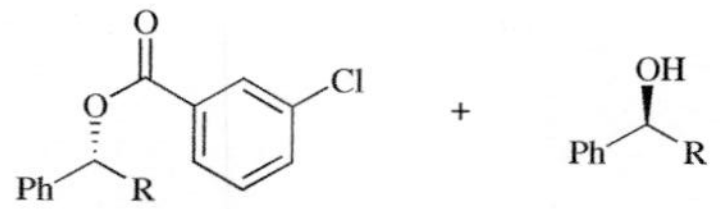

I + II

R	Time (d)		er I	er II	s
Me	—	(51)	67.0:33.0	—	3
t-Bu	18	(25)	91.0:10.0	64.5:35.5	13

87

C$_8$

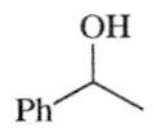

PhCH=CHCHO (3.0 eq),
150 (5 mol %), DBU (5 mol %),
PhOH (1.0 eq), toluene, 80°, 144 h

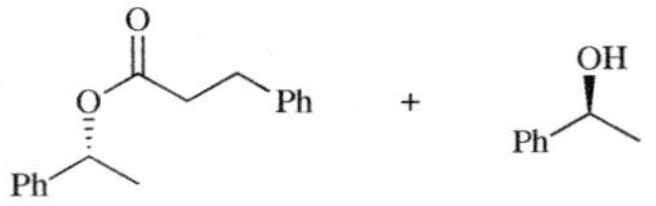

+

(40), s = 5

78

C$_{8-11}$

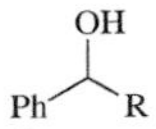

Ph(CH$_2$)$_2$CO$_2$H (0.55 eq),
124 (5 mol %),
(3-PyCO)$_2$O (0.55 eq),
CHCl$_3$, 0°, 12 h

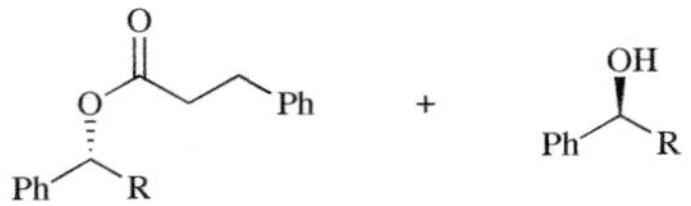

I + **II**

R	er **I**	er **II**	s
Me	95.5:4.5 (44)	85.5:14.5	45
Et	96.5:3.5 (46)	89.5:10.5	63
t-Bu	98.0:2.0 (36)	77.5:22.5	94

58

C$_{8-12}$

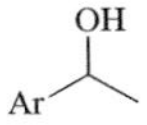

PhCH=CHCHO (0.33 eq),
152 (1.7 mol %), **162** (0.33 eq),
DBU (0.37 eq), toluene

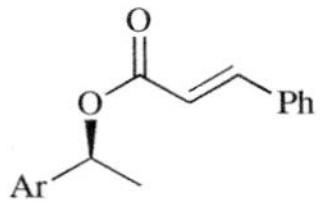

Ar	Temp (°)	Time (h)	Yield (%)	er
Ph	rt	6	(19)	55.0:45.0
2-MeOC$_6$H$_4$	rt	12	(20)	55.0:45.0
2-Np	–20	12	(24)	71.5:28.5

81

TABLE 2B. CATALYTIC KINETIC RESOLUTION OF BENZYLIC ALCOHOLS (Continued)

Alcohol	Conditions	Product(s) and Conversion(s) (%)	Refs.

Please refer to the charts preceding the tables for the structures indicated by the bold numbers.

C$_{8-12}$

Alcohol:

OH / Ar (with isopropyl)

Ph$_2$CHCO$_2$CH=CH$_2$ (1.5 eq), **147** (5 mol %), THF, −78°

76

Products: I (ester, Ar–O–C(=O)–CHPh$_2$ ... Ph) + II (OH / Ar)

Ar	Time (h)		er I	er II	s
Ph	3	(—)	98.0:2.0	75.5:24.5	80
4-FC$_6$H$_4$	4	(—)	95.5:4.5	81.0:19.0	42
4-MeOC$_6$H$_4$	2	(—)	97.0:3.0	68.0:32.0	48
1-Np	4	(—)	97.0:3.0	69.5:30.5	47
2-Np	6	(—)	97.5:2.5	68.0:32.0	56

Ph$_2$CHCO$_2$H (0.75 eq), Piv$_2$O (0.9 eq), **127** (5 mol %), Et$_2$O, rt, 12 h

50

Products: I (ester) + II (OH / Ar)

Ar		er I	er II	s
Ph	(47)	94.5:5.5	90.0:10.0	43
4-FC$_6$H$_4$	(48)	90.5:9.5	87.5:12.5	21
4-MeOC$_6$H$_4$	(50)	93.0:7.0	92.5:7.5	37
2-MeC$_6$H$_4$	(49)	95.0:5.0	93.5:6.5	54
4-MeC$_6$H$_4$	(47)	94.5:5.5	89.0:11.0	42
1-Np	(50)	94.0:6.0	93.5:6.5	44
2-Np	(49)	92.5:7.5	93.5:6.5	32

C_8

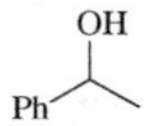

ArCHO (10 eq), *ent*-**145a** (10 mol %), **163** (10 mol %), NEt_3 (0.5 eq), 4 Å MS, toluene, rt, 19 h

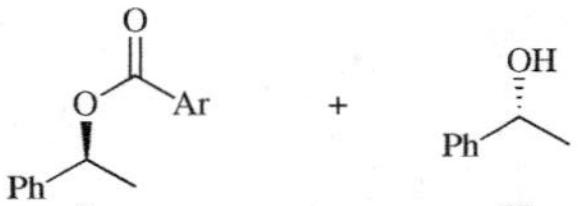

I + II

Ar		er **II**	*s*
1-Np	(50)	66.0:34.0	3
2-Np	(47)	69.5:30.5	4

82

C_{8-12}

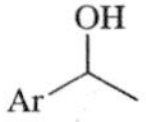

Ac_2O (1.2 eq), **56** (2 mol %), Et_3N, Et_2O, rt

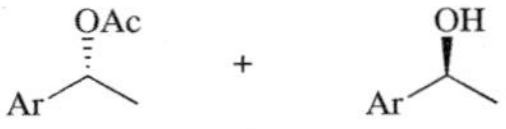

I + II

Ar	Time (h)		er **I**	er **II**	*s*
4-FC_6H_4	29	(64)	78.0:22.0	99.5:0.5	18
4-$MeOC_6H_4$	25	(60)	81.5:18.5	97.5:2.5	15
1-Np	27	(63)	79.0:21.0	99.5:0.5	22

159

Ac_2O (0.75 eq), **56** (1 mol %), Et_3N (0.75 eq), *t*-AmOH, rt

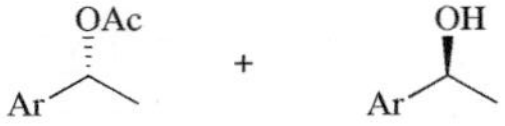

I + II

Ar	Time (h)		er **I**	er **II**	*s*
4-FC_6H_4	22	(54)	93.0:7.0	99.5:0.5	71
2-MeC_6H_4	6	(53)	94.0:6.0	99.0:1.0	74
1-Np	25	(52)	94.5:5.5	97.5:2.5	66

146

TABLE 2B. CATALYTIC KINETIC RESOLUTION OF BENZYLIC ALCOHOLS (*Continued*)

Alcohol	Conditions	Product(s) and Conversion(s) (%)	Refs.

Please refer to the charts preceding the tables for the structures indicated by the bold numbers.

C$_8$

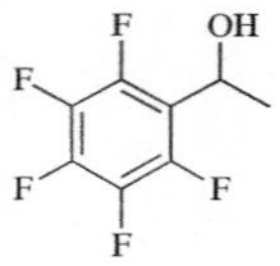

(*i*-PrCO)$_2$O (2.5 eq),
140 (3.9 mol %),
toluene, rt, 3 h

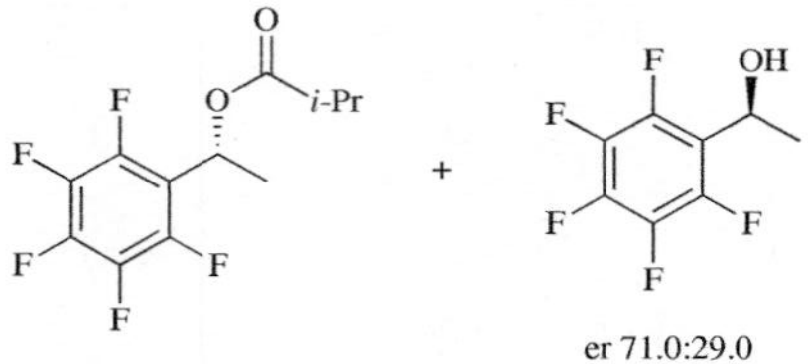

er 71.0:29.0

(48), *s* = 4   155

C$_{8-12}$

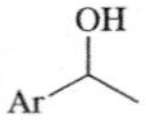

(EtCO)$_2$O (0.75 eq), **120** (5 mol %),
i-Pr$_2$NEt (0.75 eq),
toluene, 0°, 10 h

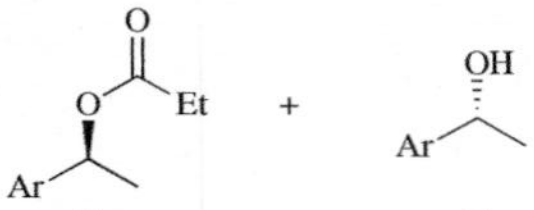

Ar		er **I**	er **II**	*s*
4-ClC$_6$H$_4$	(50)	95.5:4.5	96.0:4.0	73
4-BrC$_6$H$_4$	(51)	95.5:4.5	97.5:2.5	85
2-MeOC$_6$H$_4$	(50)	95.5:4.5	95.5:4.5	70
4-MeOC$_6$H$_4$	(42)	97.0:3.0	84.5:15.5	68
4-O$_2$NC$_6$H$_4$	(54)	93.0:7.0	>99.5:0.5	95
2-MeC$_6$H$_4$	(45)	97.5:2.5	89.0:11.0	95
4-MeC$_6$H$_4$	(50)	96.5:3.5	95.5:4.5	91
2,4-Me$_2$C$_6$H$_3$	(49)	97.0:3.0	95.0:5.0	95
1-Np	(50)	96.5:3.5	96.0:4.0	88
2-Np	(50)	97.0:3.0	96.5:3.5	113

183

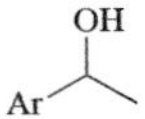

BzCl (0.6 eq), **135** (10 mol %),
i-Pr₂NEt (0.6 eq),
4 Å MS, CHCl₃, 0°, 5 h

Ar/OBz **I** + Ar/OH **II**

197

Ar		er **I**	er **II**	s
4-ClC₆H₄	(60)	68.0:32.0	77.0:23.0	4
4-MeOC₆H₄	(57)	61.5:38.5	65.0:35.0	2
2-MeC₆H₄	(51)	85.5:14.5	87.0:13.0	13
4-MeC₆H₄	(53)	66.0:34.0	68.0:32.0	3
4-CF₃C₆H₄	(57)	73.0:27.0	80.5:19.5	5
2,6-Me₂C₆H₃	(58)	84.5:15.5	97.5:2.5	20
2,4,6-Me₃C₆H₂	(50)	92.5:7.5	92.5:7.5	33
2-Np	(51)	60.5:39.5	66.5:33.5	2

(EtCO)₂O (0.75 eq), **118** (2 mol %),
i-Pr₂NEt (0.75 eq), CDCl₃, 0°

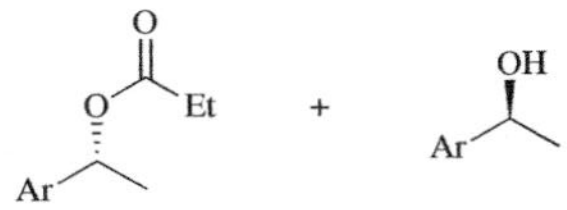

ester **I** + Ar/OH **II**

57

Ar	Time (h)		er **I**	er **II**	s
3-BrC₆H₄	8	(48)	90.0:10.0	86.0:14.0	19
3-MeOC₆H₄	8	(46)	94.0:6.0	87.0:13.0	37
3-O₂NC₆H₄	8	(33)	89.0:11.0	69.5:30.5	12
3-Me₂NC₆H₄	8	(43)	95.0:5.0	85.0:15.0	39
2-MeC₆H₄	8	(35)	97.0:3.0	75.0:25.0	57
3-MeC₆H₄	8	(20)	95.0:5.0	61.0:39.0	24
2,4,6-Me₃C₆H₂	25	(50)	82.5:17.5	82.0:18.0	9

C₈₋₁₁

Ar/OH

TABLE 2B. CATALYTIC KINETIC RESOLUTION OF BENZYLIC ALCOHOLS (*Continued*)

Alcohol	Conditions	Product(s) and Conversion(s) (%)	Refs.

*Please refer to the charts preceding the tables for the structures indicated by the **bold** numbers.*

C$_{8-12}$

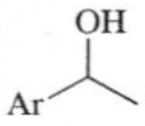

(EtCO)$_2$O (0.75 eq), **118** (2 mol %), i-PrNEt$_2$ (0.75 eq), CHCl$_3$, 0°

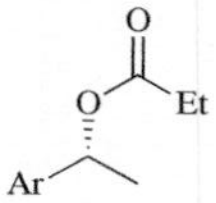

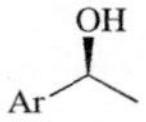

Ar	Time (h)		er I	er II	s
3-BrC$_6$H$_4$	8	(44)	94.0:6.0	84.5:15.5	32
3-MeOC$_6$H$_4$	8	(40)	94.5:5.5	81.5:18.5	34
2-MeC$_6$H$_4$	8	(44)	93.0:7.0	83.0:17.0	26
3-MeC$_6$H$_4$	8	(36)	94.5:5.5	74.5:25.5	27
2,4,6-Me$_3$C$_6$H$_2$	30	(53)	88.0:12.0	94.0:6.0	20
1-Np	8	(51)	95.5:4.5	94.5:5.5	56

47

C$_8$

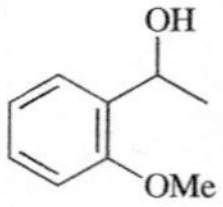

(EtCO)$_2$O (0.75 eq), **124** (5 mol %), i-Pr$_2$NEt (0.75 eq), CDCl$_3$, 0°, 20 h

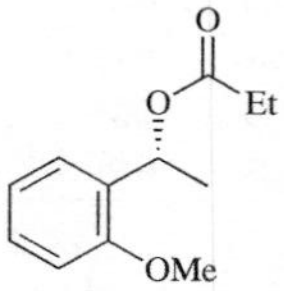

er 95.5:4.5   +   er 99.5:0.5

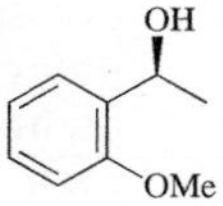

(52), s = 105   57

C8-11

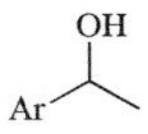

(EtCO)$_2$O (0.75 eq), **117** (x mol %),
i-Pr$_2$NEt (0.75 eq), CDCl$_3$, 0°

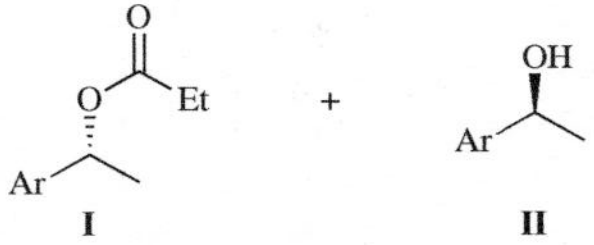

57

Ar	x	Time (h)	er I	er II	s
2-MeOC$_6$H$_4$	5	5 (49)	97.0:3.0	96.0:4.0	115
2-MeC$_6$H$_4$	2	6 (48)	96.0:4.0	92.5:7.5	61
2,4,6-Me$_3$C$_6$H$_2$	2	7 (48)	87.0:13.0	83.5:16.5	13

C8-12

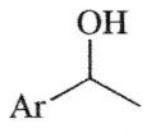

(i-PrCO)$_2$O (2.5 eq),
141 (x mol %),
heptane, –40°

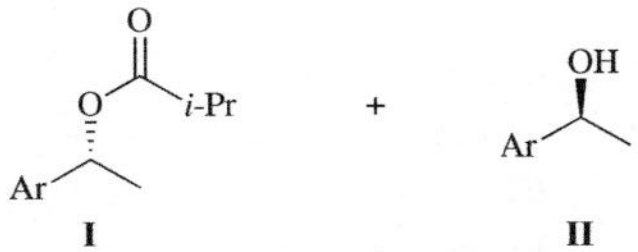

154

Ar	x	Time (h)	er I	er II	s
2-MeOC$_6$H$_4$	6.8	10 (29)	98.5:1.5	69.5:30.5	81
2-MeC$_6$H$_4$	3.5	4 (50)	97.5:2.5	97.5:2.5	145
2,4,6-Me$_3$C$_6$H$_2$	12.1	16 (44)	99.5:0.5	89.5:10.5	369
1-Np	3.9	7 (30)	98.5:1.5	70.5:29.5	99

231

TABLE 2B. CATALYTIC KINETIC RESOLUTION OF BENZYLIC ALCOHOLS (*Continued*)

Alcohol	Conditions	Product(s) and Conversion(s) (%)	Refs.

*Please refer to the charts preceding the tables for the structures indicated by the **bold** numbers.*

C_{8–12}

OH / Ar (structure)

(*i*-PrCO)₂O (2.0 eq), **87** (1 mol %),
Et₃N (0.75 eq), toluene, –78°

163

Ar	Time (h)		er I	er II	s
2-MeOC₆H₄	12	(33)	90.8:9.2	70.1:29.9	15
2-MeC₆H₄	10	(41)	93.0:7.0	80.4:19.6	25
2,6-Me₂C₆H₃	8	(19)	95.5:4.5	60.5:39.5	25
1-Np	9	(45)	92.0:8.0	84.5:15.5	24

(*i*-PrCO)₂O (0.75 eq).
69a (5 mol %), Cs₂CO₃ (0.75 eq).
toluene, –60°, 15 h

324

Ar		er I	er II	s
2-MeOC₆H₄	(18)	86.0:14.0	58.0:42.0	7
3-MeOC₆H₄	(37)	76.0:24.0	65.0:35.0	15
4-MeOC₆H₄	(61)	82.0:18.0	>99.5:0.5	30
4-CF₃C₆H₄	(44)	92.0:8.0	83.0:17.0	23
4-OHCC₆H₄	(47)	81.0:19.0	77.0:23.0	7
Mes	(18)	90.0:10.0	59.0:41.0	11
1-Np	(35)	92.0:8.0	73.0:27.0	19
2-Np	(40)	98.0:2.0	84.0:16.0	80

2-PyCHO (0.33 eq), **145a** (1.7 mol %),
162 (0.33 eq), DBU (0.37 eq),
toluene, rt

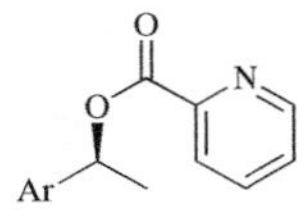

81

Ar	Time (h)	Yield (%)	er
2-MeOC$_6$H$_4$	4	26	57.5:42.5
1-Np	3	27	81.5:18.5

Ac$_2$O (0.7 eq), **83** (0.5 mol %),
i-Pr$_2$NEt (0.7 eq),
toluene, –40°, 24 h

320

I + **II**

Ar		er **II**	s
4-MeOC$_6$H$_4$	(69)	91.0:9.0	5
4-O$_2$NC$_6$H$_4$	(35)	59.0:41.0	2
1-Np	(79)	85.5:14.5	3

Ac$_2$O (1 eq), **85** (1 mol %),
t-AmOH, 0°, 15 h

321

I + **II**

Ar		er **I**	er **II**	s
4-MeOC$_6$H$_4$	(72)	66.0:34.0	92.0:8.0	5
2-Np	(72)	69.0:31.0	99.0:1.0	8

TABLE 2B. CATALYTIC KINETIC RESOLUTION OF BENZYLIC ALCOHOLS (*Continued*)

Alcohol	Conditions	Product(s) and Conversion(s) (%)	Refs.

*Please refer to the charts preceding the tables for the structures indicated by the **bold** numbers.*

C_8

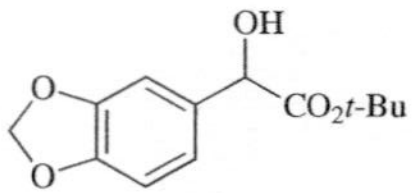

Ac$_2$O (0.7 eq), Sc(OTf)$_3$ (5 mol %), **42a** (5 mol %), 3 Å MS, THF, 0°, 48 h

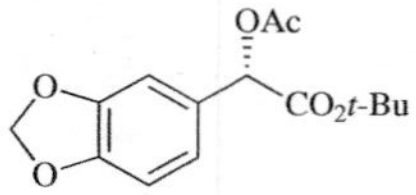

+

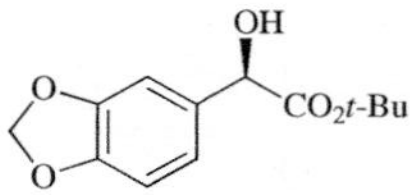

er 91.5:8.5   (53), s = 37   er 97.0:3.0   103

C_{8-14}

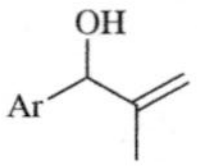

(EtCO)$_2$O (0.75 eq), **115** (2 mol %), *i*-Pr$_2$NEt (0.75 eq), toluene, 0°

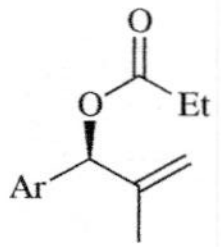

I

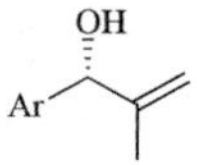

+

II

204

Ar	Time (h)		er I	er II	s
2-thienyl	6	(50)	65.0:35.0	65.5:34.5	3
Ph	7	(51)	76.5:23.5	78.0:22.0	6
4-ClC$_6$H$_4$	12	(69)	71.5:28.5	98.0:2.0	9
3-BrC$_6$H$_4$	10	(66)	71.5:28.5	90.5:9.5	6
2-MeC$_6$H$_4$	13	(62)	80.5:19.5	>99.5:0.5	24
4-MeC$_6$H$_4$	12	(59)	80.0:20.0	93.0:7.0	11
1-Np	8	(62)	80.5:19.5	99.5:0.5	21
2-Np	10	(62)	78.0:22.0	96.0:4.0	11

C_8

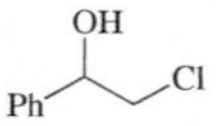

(*i*-PrCO)$_2$O (2.5 eq), **141** (4.2 mol %), heptane, −20°, 1 h

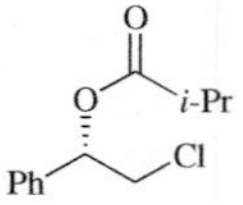

+

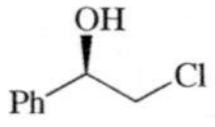

er 90.0:10.0   er 85.0:15.0   (47), s = 19   155

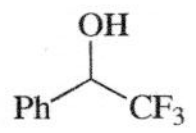

OH
Ph—CF₃

(EtCO)₂O (2.5 eq), **124** (4 mol %),
toluene, rt, 5 h

$\longrightarrow$

O, Et, CF₃, Ph
er 89.0:11.0

+

OH, CF₃, Ph
er 56.5:43.5

(14), s = 9

244

(i-PrCO)₂O (2.5 eq), **141** (0.8 mol %),
heptane, −20°, 1 h

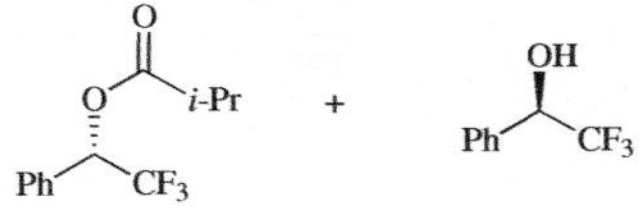

i-Pr, CF₃, Ph
er 89.0:11.0

+

OH, CF₃, Ph
er 69.0:31.0

(33), s = 12

155

OH
Ph—CO₂R

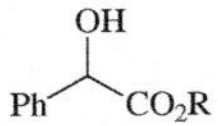

Ac₂O (x eq), Sc(OTf)₃ (5 mol %),
42 (5 mol %), 3 Å MS,
THF, 48 h

OAc, Ph, CO₂R
I

+

OH, Ph, CO₂R
II

103

R	x	Temp (°)		er **I**	er **II**	s
Me	0.8	0	(49)	95.5:4.5	93.5:6.5	57
Et	0.7	0	(49)	95.5:4.5	94.0:6.0	63
n-Pr	0.75	0	(50)	95.0:5.0	94.0:6.0	54
i-Pr	0.8	0	(52)	93.0:7.0	96.0:4.0	42
allyl	0.9	0	(51)	94.5:5.5	96.5:3.5	57
n-Bu	0.8	0	(49)	94.0:6.0	92.5:7.5	45
i-Bu	0.8	0	(50)	92.5:7.5	92.0:8.0	34
t-Bu	0.9	−20	(50)	96.0:4.0	96.0:4.0	82
Cy	0.7	0	(53)	91.0:9.0	96.0:4.0	33
PhCH₂	0.95	0	(51)	96.0:4.0	98.0:2.0	91
1-adamantyl	0.95	−20	(52)	92.0:8.0	96.0:4.0	36

TABLE 2B. CATALYTIC KINETIC RESOLUTION OF BENZYLIC ALCOHOLS (*Continued*)

Alcohol	Conditions	Product(s) and Conversion(s) (%)	Refs.
*Please refer to the charts preceding the tables for the structures indicated by the **bold** numbers.*			
C_8			
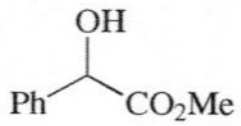	$(i\text{-}PrCO)_2O$ (2.5 eq), **140** (4.8 mol %), toluene, rt, 3 h	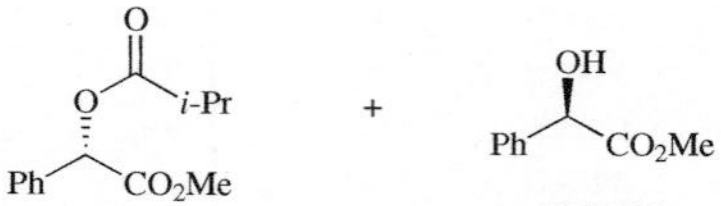 er 68.0:32.0 (49), $s = 3$	155
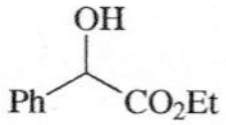	$(EtCO)_2O$ (2.5 eq), **124** (4 mol %), toluene, rt, 24 h	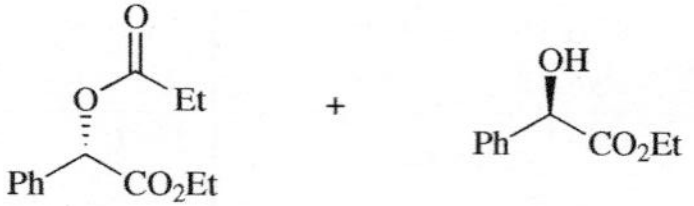 er 61.5:38.5 + er 55.0:45.0 (31), $s = 2$	244
C_{8-14}			
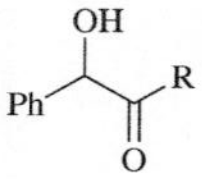	BzCl (0.5 eq), $CuCl_2$ (5 mol %), **19** (5 mol %), CH_2Cl_2, 0°, 4 h	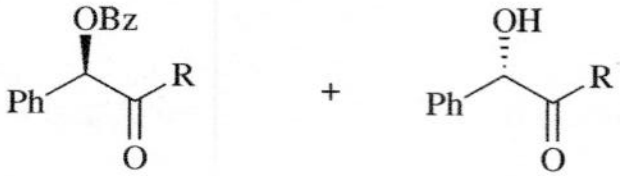 **I** + **II**	199

R		er **I**	er **II**	*s*
EtO	(—)	87.5:12.5	—	13
Ph	(—)	89.5:10.5	—	16

C_{8-12}

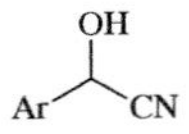

(EtCO)$_2$O (2.5 eq), **124** (4 mol %), toluene, rt

+

I   **II**

Ar	Time (h)		er **I**	er **II**	s
Ph	5	(51)	80.5:19.5	82.0:18.0	8
4-FC$_6$H$_4$	48	(31)	61.5:38.5	55.0:45.0	2
4-MeSC$_6$H$_4$	3	(70)	65.0:35.0	85.5:14.5	4
1-Np	1	(82)	61.0:39.0	99.0:1.0	6

C_8

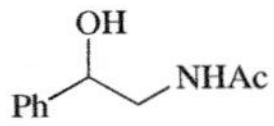

Ac$_2$O (0.84 eq), **104** (2.5 mol %), i-Pr$_2$NEt, toluene, rt, 4 h

+

er 56.5:43.5   er 56.5:43.5

(50), $s = 2$

67

TABLE 2B. CATALYTIC KINETIC RESOLUTION OF BENZYLIC ALCOHOLS (*Continued*)

Alcohol	Conditions	Product(s) and Conversion(s) (%)	Refs.

*Please refer to the charts preceding the tables for the structures indicated by the **bold** numbers.*

C_{8-12}

OH / Ar—NHTs

BzCl (0.7 eq), **47b** (2 mol %), K$_2$CO$_3$ (1 eq), THF, rt, 16 h

OBz / Ar—NHTs (**I**) + OH / Ar—NHTs (**II**)

248

Ar	er I	er II	s
Ph	(51) 95.5:4.5	96.5:3.5	71
4-ClC$_6$H$_4$	(45) 94.0:6.0	86.5:13.5	34
4-MeOC$_6$H$_4$	(49) 96.5:3.5	95.0:5.0	84
3,4-(MeO)$_2$C$_6$H$_3$	(47) 96.5:3.5	91.5:8.5	73
2-MeC$_6$H$_4$	(53) 91.5:8.5	96.5:3.5	36
4-MeC$_6$H$_4$	(50) 95.5:4.5	96.0:4.0	68
2-Np	(49) 96.0:4.0	93.0:7.0	65

OH / Ar—N$_3$

Ac$_2$O (0.75 eq), **56** (2 mol %), Et$_3$N (0.75 eq), *t*-AmOH, 0°

OAc / Ar—N$_3$ (**I**) + OH / Ar—N$_3$ (**II**)

161

Ar	Time (h)	er II	s
Ph	24	(73) 99.5:0.5	17
3-BrC$_6$H$_4$	9	(61) 99.5:0.5	20
4-BrC$_6$H$_4$	4	(63) 98.5:1.5	13
4-MeOC$_6$H$_4$	13	(65) 99.5:0.5	14
2,5-(MeO)$_2$C$_6$H$_3$	13	(60) 99.5:0.5	24
4-O$_2$NC$_6$H$_4$	3	(58) 99.5:0.5	33
4-BzOC$_6$H$_4$	20	(58) 96.5:3.5	16
4-MeC$_6$H$_4$	5	(70) 90.5:9.5	4
4-NCC$_6$H$_4$	10	(62) >99.5:0.5	35
2-Np	3	(60) >99.5:0.5	45

C$_8$

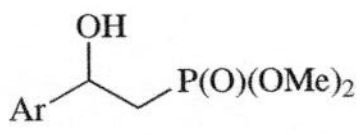

Ac$_2$O (0.75 eq), **56** (1 mol %),
t-AmOH, 0°, 4 h

Ar		er II	s
Ph	(31)	69.0:31.0	24
3-MeOC$_6$H$_4$	(31)	68.0:32.0	16
4-O$_2$NC$_6$H$_4$	(16)	61.5:38.5	5

160

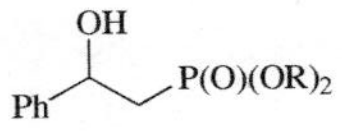

BzCl (0.5 eq), **15** (5 mol %),
Cu(OTf)$_2$ (5 mol %), K$_2$CO$_3$ (1 eq),
CH$_2$Cl$_2$, 0° to rt, 20 h

R		er I	er II	s
Me	(—)	81.0:19.0	62.0:38.0	5
Et	(—)	83.0:17.0	70.5:29.5	7
i-Pr	(—)	83.0:17.0	68.0:32.0	7
n-Bu	(—)	78.0:22.0	60.0:40.0	4

98

Alcohol	Conditions	Product(s) and Conversion(s) (%)	Refs.

*Please refer to the charts preceding the tables for the structures indicated by the **bold** numbers.*

C$_8$

Conditions: 2-FC$_6$H$_4$COCl (0.5 eq), **15** (5 mol %), Cu(OTf)$_2$ (5 mol %), K$_2$CO$_3$ (1 eq), CH$_2$Cl$_2$, 0° to rt, 20 h

98

R		er I	er II	s
Me	(—)	90.0:10.0	64.0:36.0	12
Et	(—)	91.5:8.5	68.0:32.0	15
i-Pr	(—)	88.5:11.5	64.5:35.5	10
n-Bu	(—)	91.0:9.0	64.0:36.0	13

C$_{8-12}$

Conditions: 2-FC$_6$H$_4$COCl (0.5 eq), **15** (5 mol %), Cu(OTf)$_2$ (5 mol %), K$_2$CO$_3$ (1 eq), CH$_2$Cl$_2$, 0° to rt, 20 h

98

Ar		er I	er II	s
4-FC$_6$H$_4$	(—)	86.0:14.0	70.5:29.5	9
4-ClC$_6$H$_4$	(—)	83.0:17.0	68.5:31.5	7
4-BrC$_6$H$_4$	(—)	80.0:20.0	74.5:25.5	6
2-MeOC$_6$H$_4$	(—)	87.0:13.0	69.0:31.0	10
4-MeOC$_6$H$_4$	(—)	92.0:8.0	64.5:35.5	15
4-O$_2$NC$_6$H$_4$	(—)	78.5:21.5	64.5:35.5	5
2-MeC$_6$H$_4$	(—)	93.5:6.5	69.0:31.0	21
3-MeC$_6$H$_4$	(—)	91.5:8.5	65.5:34.5	15
4-MeC$_6$H$_4$	(—)	87.5:12.5	69.0:31.0	10
1-Np	(—)	89.5:10.5	71.5:28.5	13
2-Np	(—)	86.0:14.0	73.5:26.5	10

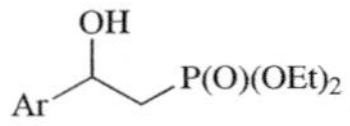

OH, Ar, P(O)(OEt)$_2$

Ac$_2$O (0.75 eq), **56** (1 mol %), t-AmOH, 0°, 4 h

I (OAc, Ar, P(O)(OEt)$_2$) + II (OH, Ar, P(O)(OEt)$_2$)

160

Ar		er II	s
Ph	(25)	66.0:34.0	43
4-FC$_6$H$_4$	(38)	69.0:31.0	11
2,4-Cl$_2$C$_6$H$_3$	(28)	67.5:32.5	35
3-MeOC$_6$H$_4$	(20)	61.0:39.0	19
4-O$_2$NC$_6$H$_4$	(25)	65.5:34.5	62
4-MeC$_6$H$_4$	(14)	57.5:42.5	25
2-Np	(37)	77.5:22.5	68

2-FC$_6$H$_4$COCl (0.5 eq), **15** (5 mol %), Cu(OTf)$_2$ (5 mol%), K$_2$CO$_3$ (1 eq), CH$_2$Cl$_2$, 0° to rt, 20 h

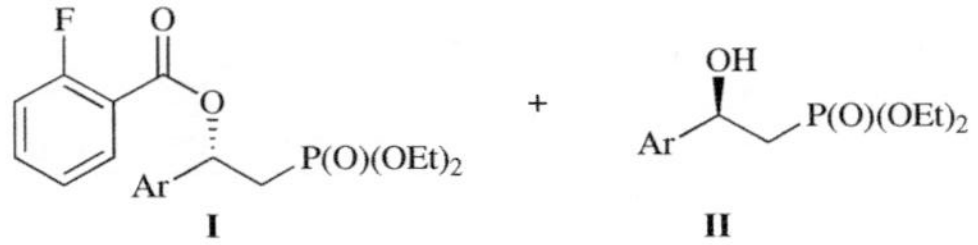

I (2-F-C$_6$H$_4$CO$_2$, Ar, P(O)(OEt)$_2$) + II (OH, Ar, P(O)(OEt)$_2$)

98

Ar		er I	er II	s
4-MeOC$_6$H$_4$	(—)	88.5:11.5	79.5:20.5	14
2-MeC$_6$H$_4$	(—)	89.0:11.0	80.0:20.0	15
1-Np	(—)	80.5:19.5	73.5:26.5	6

TABLE 2B. CATALYTIC KINETIC RESOLUTION OF BENZYLIC ALCOHOLS (*Continued*)

Alcohol	Conditions	Product(s) and Conversion(s) (%)	Refs.

*Please refer to the charts preceding the tables for the structures indicated by the **bold** numbers.*

C_{8-14}

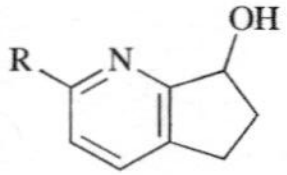

BzCl (0.51 eq), **25** (1 mol %),
CuCl$_2$ (1 mol %), i-Pr$_2$NEt (1.0 eq),
CH$_2$Cl$_2$, 0°, 2 h

I　　　　　**II**

R		er **I**	er **II**	s
H	(46)	52.5:47.5	52.5:47.5	1
Ph	(45)	95.5:4.5	88.0:12.0	51

215

C_9

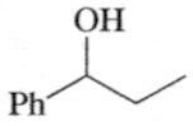

Ac$_2$O (2.0 eq), **82** (5 mol %),
Et$_3$N (0.6 eq),
acetone, −78°, 18 h

er 73.0:27.0　　　er 55.0:45.0　　(18), s = 3

170

Ac$_2$O (0.75 eq), **89** (1 mol %),
Et$_3$N (0.75 eq),
toluene, −78°, 9 h

er 91.5:8.5　　　er 58.5:41.5　　(17), s = 13

165

C_{9-11}

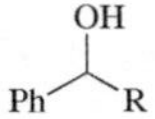

Ac$_2$O (1.5 eq), **101** (2.5 mol %),
toluene, −65°, 24 h

R		s
Et	(—)	8
t-Bu	(—)	30

68

C₉

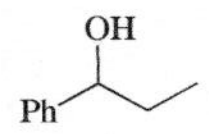

RCO₂H (0.75 eq), **124** (5 mol %),
Bz₂O (0.90 eq), i-Pr₂NEt (1.8 eq),
CH₂Cl₂, rt, 12 h

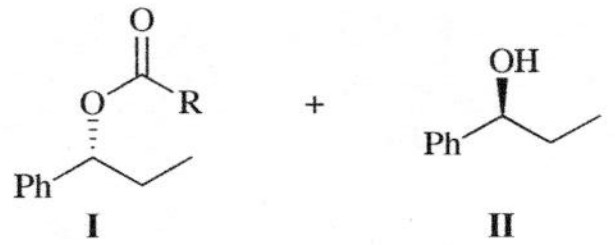

R		er **I**	er **II**	s
MeOCH₂	(64)	77.5:22.5	99.0:1.0	14
Et	(50)	93.0:7.0	93.5:6.5	38
CH₂=CH(CH₂)₂	(55)	88.5:11.5	97.5:2.5	28
i-Pr(CH₂)₂	(47)	91.0:9.0	86.0:14.0	22
Cy	(50)	81.5:18.5	82.0:18.0	8
Ph(CH₂)₂	(49)	94.5:5.5	92.5:7.5	46
Ph(CH₂)₃	(52)	91.0:9.0	95.0:5.0	31

RCO₂H (0.75 eq), **124** (5 mol %),
(4-MeOC₆H₄CO)₂O (0.90 eq),
i-Pr₂NEt (1.8 eq), CH₂Cl₂, rt, 12 h

R		er **I**	er **II**	s
MeOCH₂	(32)	91.0:9.0	69.0:31.0	15
Et	(46)	94.5:5.5	88.0:12.0	39
CH₂=CH(CH₂)₂	(51)	93.0:7.0	95.5:4.5	42
i-Pr(CH₂)₂	(46)	91.5:8.5	85.5:14.5	23
Cy	(40)	88.0:12.0	75.5:24.5	12
Ph(CH₂)₂	(45)	95.0:5.0	87.5:12.5	43
Ph(CH₂)₃	(43)	95.0:5.0	84.5:15.5	39

TABLE 2B. CATALYTIC KINETIC RESOLUTION OF BENZYLIC ALCOHOLS (*Continued*)

Alcohol	Conditions	Product(s) and Conversion(s) (%)	Refs.

Please refer to the charts preceding the tables for the structures indicated by the ***bold*** *numbers.*

C_9

RCO_2H (0.55 eq), **124** (5 mol %), $(3\text{-PyCO})_2O$ (0.55 eq), CH_2Cl_2, 0°, 12 h

I + II

Refs. 188

R		er **I**	er **II**	*s*
$MeOCH_2$	(46)	91.0:9.0	85.0:15.0	22
Et	(41)	96.5:3.5	82.0:18.0	56
$CH_2{=}CH(CH_2)_2$	(38)	95.5:4.5	77.5:22.5	38
$i\text{-}Pr(CH_2)_2$	(43)	96.0:4.0	85.0:15.0	50
Cy	(39)	96.5:3.5	79.5:20.5	50
$Ph(CH_2)_2$	(46)	96.5:3.5	89.5:10.5	63
$Ph(CH_2)_3$	(38)	96.5:3.5	78.5:21.5	49

RCO_2H (0.5 eq), **124** (5 mol %), Piv_2O (0.6 eq), $i\text{-}Pr_2NEt$ (1.2 eq), CH_2Cl_2, rt, 12 h

I + II

Refs. 55

R		er **I**	er **II**	*s*
$MeOCH_2$	(41)	93.5:6.5	80.0:20.0	26
Et	(47)	94.5:5.5	89.5:10.5	43
$CH_2{=}CH(CH_2)_2$	(45)	94.5:5.5	87.0:13.0	39
$i\text{-}Pr(CH_2)_2$	(44)	95.0:5.0	85.5:14.5	40
Cy	(39)	97.5:2.5	80.0:20.0	73
$Ph(CH_2)_2$	(46)	95.5:4.5	88.5:11.5	47
$Ph(CH_2)_3$	(44)	95.0:5.0	86.0:14.0	41

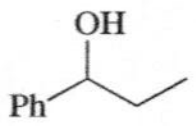

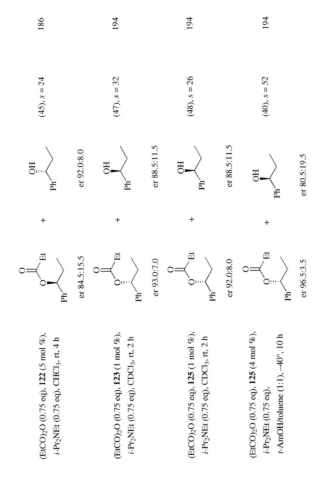

(EtCO)₂O (0.75 eq), **122** (5 mol %), *i*-Pr₂NEt (0.75 eq), CHCl₃, rt, 4 h

er 84.5:15.5   er 92.0:8.0   (45), *s* = 24   186

(EtCO)₂O (0.75 eq), **123** (1 mol %), *i*-Pr₂NEt (0.75 eq), CDCl₃, rt, 2 h

er 93.0:7.0   er 88.5:11.5   (47), *s* = 32   194

(EtCO)₂O (0.75 eq), **125** (1 mol %), *i*-Pr₂NEt (0.75 eq), CDCl₃, rt, 2 h

er 92.0:8.0   er 88.5:11.5   (48), *s* = 26   194

(EtCO)₂O (0.75 eq), **125** (4 mol %), *i*-Pr₂NEt (0.75 eq), *t*-AmOH/toluene (1:1), –40°, 10 h

er 96.5:3.5   er 80.5:19.5   (40), *s* = 52   194

TABLE 2B. CATALYTIC KINETIC RESOLUTION OF BENZYLIC ALCOHOLS (*Continued*)

Alcohol	Conditions	Product(s) and Conversion(s) (%)	Refs.

*Please refer to the charts preceding the tables for the structures indicated by the **bold** numbers.*

C$_{9-11}$

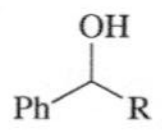

OH / Ph / R

(EtCO)$_2$O (0.75 eq), **115** (2 mol %), i-Pr$_2$NEt (0.75 eq), toluene

I + II

49

R	Temp (°)	Time (h)		er I	er II	s
Et	0	7	(53)	93.0:7.0	99.0:1.0	58
i-Pr	0	8	(50)	96.0:4.0	96.0:4.0	84
t-Bu	–40	24	(38)	>99.5:0.5	81.0:19.0	1892

C$_9$

OH / Ph

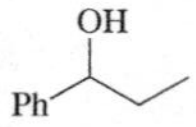

(i-PrCO)$_2$O (0.75 eq), **92** (0.5 mol %), Et$_3$N (0.75 eq), t-AmOH, 7 h

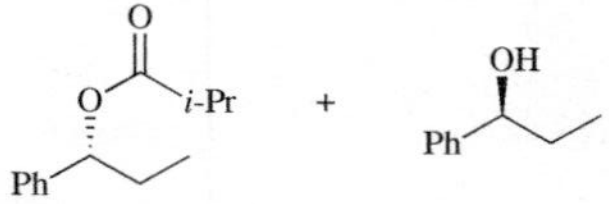

er 92.5:7.5 + er 83.0:17.0

(44), s = 23

42

Ph$_2$CHCO$_2$CH=CH$_2$ (1.5 eq), **147** (5 mol %), THF, –78°, 0.5 h

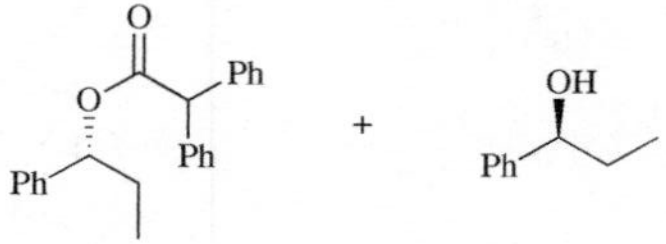

er 96.0:4.0 + er 71.0:29.0

(—), s = 38

76

C$_{9-13}$

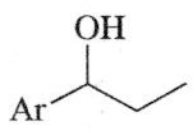

Ph$_2$CHCO$_2$H (0.75 eq),
Piv$_2$O (0.9 eq), **127** (5 mol %),
Et$_2$O, rt, 12 h

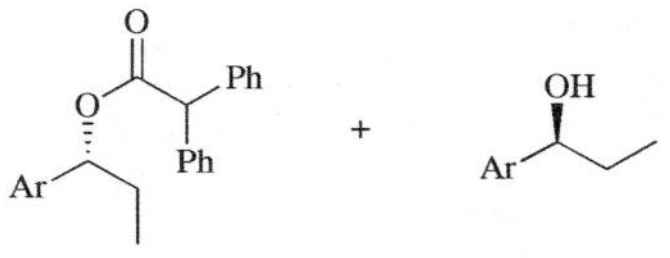

I + II

50

Ar		er I	er II	s
Ph	(49)	96.0:4.0	93.5:6.5	68
4-FC$_6$H$_4$	(48)	92.0:8.0	89.0:11.0	27
4-MeOC$_6$H$_4$	(49)	94.5:5.5	93.0:7.0	50
2-MeC$_6$H$_4$	(47)	96.0:4.0	91.5:8.5	66
4-MeC$_6$H$_4$	(46)	96.5:3.5	90.0:10.0	64
1-Np	(46)	97.5:2.5	91.0:9.0	98
2-Np	(50)	95.0:5.0	95.5:4.5	60

C$_{9-11}$

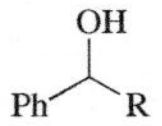

2-PyCHO (0.33 eq), **145a** (1.7 mol %),
162 (0.33 eq), DBU (0.37 eq),
toluene, rt, 4 h

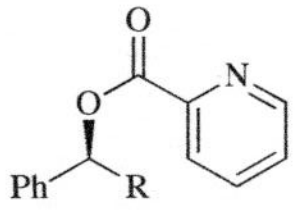

+

er 81.0:19.0

81

R	Yield (%)	er
Et	(28)	60.0:40.0
t-Bu	(13)	65.0:35.0

C$_9$

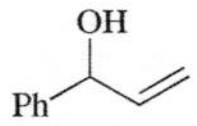

(EtCO)$_2$O (0.75 eq), **115** (2 mol %),
i-Pr$_2$NEt (0.75 eq), toluene, 0°, 8 h

+

er 77.5:22.5

(55), s = 10

204

TABLE 2B. CATALYTIC KINETIC RESOLUTION OF BENZYLIC ALCOHOLS (*Continued*)

Alcohol	Conditions	Product(s) and Conversion(s) (%)	Refs.

*Please refer to the charts preceding the tables for the structures indicated by the **bold** numbers.*

C_9

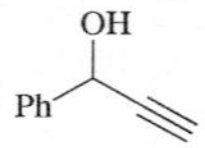

AcCl (0.6 eq), **157a** (5 mol %),
DABCO (0.65 eq),
Et$_2$O, –20°, 5 h

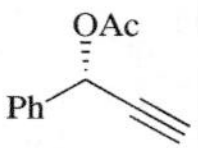

OAc / Ph er 92.5:7.5   +   OH / Ph (46), s = 27   201
er 81.5:18.5

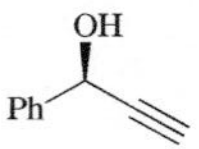

Ac$_2$O (0.75 eq), **79** (5 mol %),
Et$_3$N (0.75 eq), toluene, –78°, 15 h

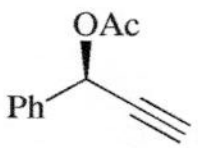

OAc / Ph er 72.0:28.0   +   OH / Ph (18), s = 3   173

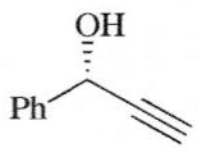

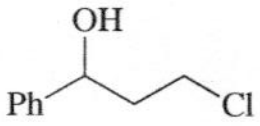

(*i*-PrCO)$_2$O (0.75 eq), **124** (4 mol %),
i-Pr$_2$NEt (0.75 eq), CDCl$_3$, 0°, 24 h

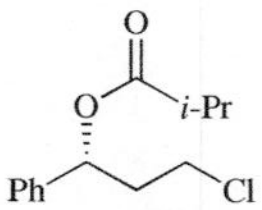

I er 96.0:4.0   +   II er 93.0:7.0   (48), s = 66   57

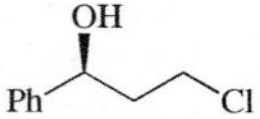

C_{9-11}

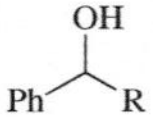

(*i*-PrCO)$_2$O (2.0 eq), **88** (1 mol %),
Et$_3$N (0.75 eq), toluene, –95°, 15 h

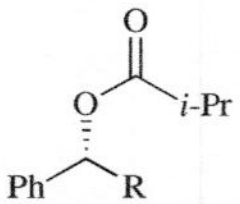

I + II   164

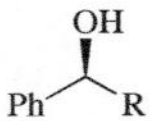

R		er I	er II	s
ClCH$_2$CH$_2$	(53)	79.5:20.5	83.5:16.5	8
t-Bu	(16)	95.5:4.5	59.0:41.0	26

C$_9$

Substrate	Conditions	Products		Ref.
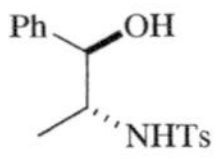 Ph–OH, –NHTs	BzCl (0.7 eq), **47b** (2 mol %), K$_2$CO$_3$ (1 eq), THF, rt, 24 h	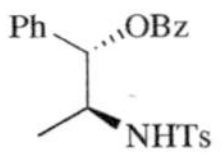 Ph–OBz, –NHTs er 99.5:0.5	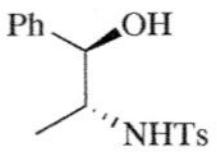 Ph–OH, –NHTs er 91.0:9.0 (45), s > 500	248
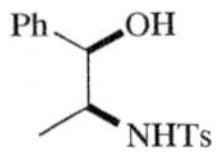 Ph–OH, –NHTs	BzCl (0.7 eq), **47b** (2 mol %), K$_2$CO$_3$ (1 eq), THF, rt, 24 h	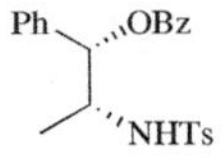 Ph–OBz, –NHTs er 83.0:17.0	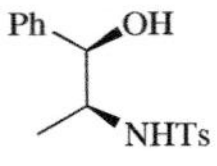 Ph–OH, –NHTs er 76.5:23.5 (44), s = 8	248
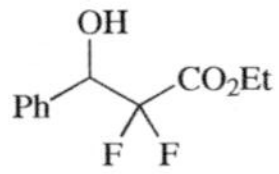 OH, CO$_2$Et, F, F, Ph	(EtCO)$_2$O (2.5 eq), **124** (4 mol %), toluene, rt, 2 h	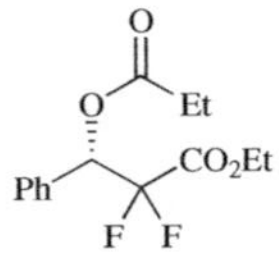 ester, CO$_2$Et, F, F, Ph er 85.5:14.5	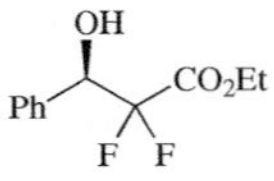 OH, CO$_2$Et, F, F, Ph er 84.0:16.0 (49), s = 12	244
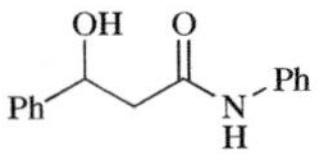 OH, O, N–H, Ph	BzCl (0.5 eq), **15** (10 mol %), Cu(OTf)$_2$ (10 mol %), K$_2$CO$_3$, CH$_2$Cl$_2$, 0° to rt, 24 h	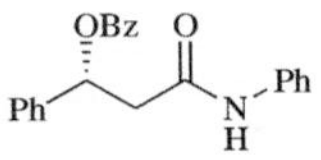 OBz, O, N–H, Ph er 87.0:13.0	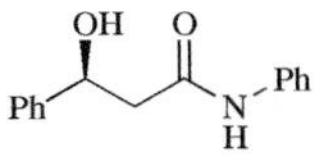 OH, O, N–H, Ph er 60.0:40.0 (18), s = 8	200

249

TABLE 2B. CATALYTIC KINETIC RESOLUTION OF BENZYLIC ALCOHOLS (Continued)

*Please refer to the charts preceding the tables for the structures indicated by the **bold** numbers.*

C_9

Alcohol	Conditions	Product(s) and Conversion(s) (%)	Refs.
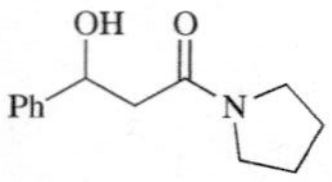	$(i\text{-PrCO})_2\text{O}$ (0.5 eq), **109** (5 mol %), $i\text{-Pr}_2\text{NEt}$ (0.5 eq), CCl_4, 0°, 3 h	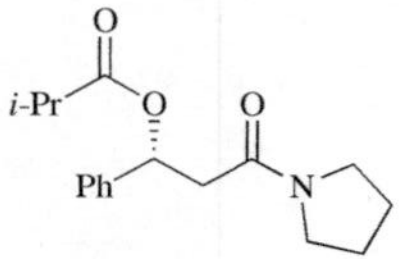 er 91.0:9.0 + (44), $s = 19$ er 82.0:18.0	70
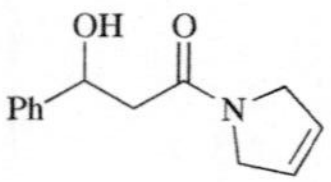	$(i\text{-PrCO})_2\text{O}$ (0.5 eq), **109** (5 mol %), $i\text{-Pr}_2\text{NEt}$ (0.5 eq), CCl_4, 0°, 3 h	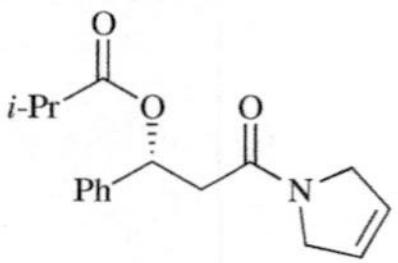 er 88.5:11.5 + (54), $s = 25$ er 96.0:4.0	181
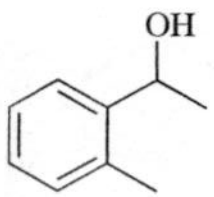	$(i\text{-PrCO})_2\text{O}$ (2.5 eq), **142** (1 mol %), Et_3N (2 mol %), heptane, −40°, 16 h	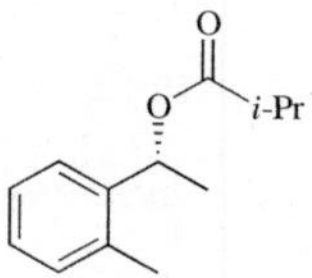 er 96.0:4.0 + (48), $s = 65$ er 92.5:7.5	156

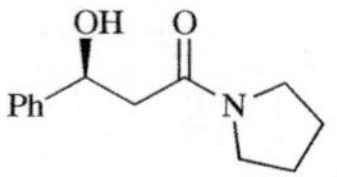

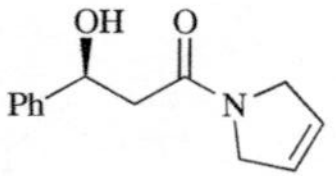

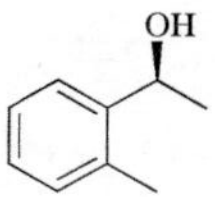

250

C$_{9-12}$

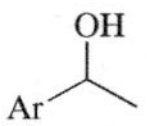

(i-PrCO)$_2$O (2.0 eq), **88** (1 mol %),
Et$_3$N (0.75 eq), toluene, −95°, 15 h

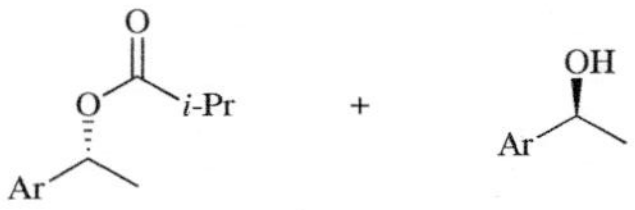

I + II

164

Ar		er I	er II	s
2-MeC$_6$H$_4$	(55)	88.0:12.0	97.0:3.0	26
2,6-Me$_2$C$_6$H$_3$	(58)	86.0:14.0	99.5:0.5	29
1-Np	(37)	95.5:4.5	77.0:23.0	36

(i-PrCO)$_2$O (0.75 eq), **124** (4 mol %),
i-Pr$_2$NEt (0.75 eq),
Na$_2$SO$_4$, CHCl$_3$, 0°

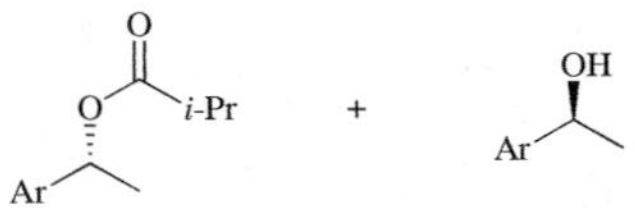

I + II

186

Ar	Time (h)		er I	er II	s
2-MeC$_6$H$_4$	33	(47)	99.5:0.5	94.0:6.0	378
1-Np	9	(49)	98.5:1.5	96.0:4.0	184
2-Np	11	(47)	98.5:1.5	95.5:4.5	226

(i-PrCO)$_2$O (1 eq), **ent-81** (1 mol %),
Et$_3$N (1.5 eq), toluene, −40°, 4 h

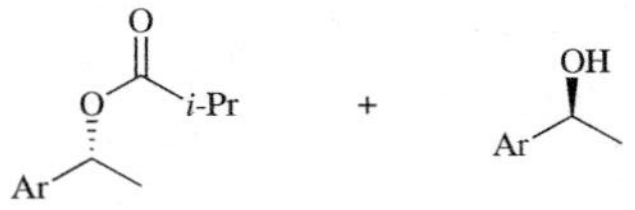

I + II

171

Ar		er I	er II	s
2-MeC$_6$H$_4$	(33)	87.0:13.0	68.0:32.0	10
1-Np	(36)	93.0:7.0	74.0:26.0	21

TABLE 2B. CATALYTIC KINETIC RESOLUTION OF BENZYLIC ALCOHOLS (*Continued*)

Alcohol	Conditions	Product(s) and Conversion(s) (%)	Refs.

*Please refer to the charts preceding the tables for the structures indicated by the **bold** numbers.*

C$_{9-12}$

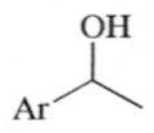

OH
|
Ar — CH(CH$_3$)

(3-PyCO)$_2$O (1 eq),
140 (2 mol %), Et$_3$N (1.5 eq),
t-AmOH, CH$_2$Cl$_2$, –25°

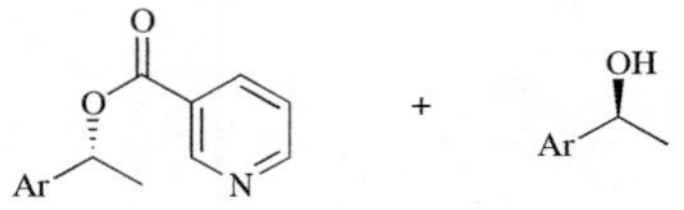

I + **II**

171

Ar	Time (h)		er I	er II	s
2-MeC$_6$H$_4$	20	(37)	91.5:8.5	74.5:25.5	17
1-Np	3	(68)	73.0:27.0	99.0:1.0	11

(3-PyCO)$_2$O (1 eq),
80 (2 mol %), Et$_3$N (1.5 eq),
t-AmOH, CH$_2$Cl$_2$, –25°

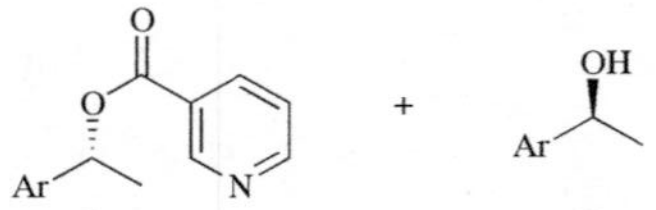

I + **II**

171

Ar	Time (h)		er I	er II	s
2-MeC$_6$H$_4$	64	(50)	84.5:15.5	85.0:15.0	14
1-Np	37	(27)	83.0:17.0	62.0:38.0	6

252

C9-15

(EtCO)2O (0.75 eq), **115** (1 mol %),
i-Pr2NEt (0.75 eq), toluene, 0°

I + **II**

Ar	Time (h)		er **I**	er **II**	s
2-furyl	93	(55)	90.5:9.5	>99.5:0.5	48
2-thienyl	48	(54)	93.0:7.0	>99.5:0.5	73
Ph	11	(49)	>99.5:0.5	96.5:3.5	690
4-ClC6H4	12	(52)	96.0:4.0	>99.5:0.5	120
4-MeC6H4	11	(52)	94.0:6.0	98.0:2.0	61
1-Np	48	(54)	92.5:7.5	>99.5:0.5	71
2-Np	19	(52)	96.5:3.5	>99.5:0.5	142

184

C9-10

(c-C6H11CO)2O (0.75 eq),
127 (5 mol %),
Et2O, rt, 24 h

I + **II**

Ar		er **I**	er **II**	s
Ph	(51)	94.5:5.5	95.5:4.5	54
2-ClC6H4	(47)	91.5:8.5	87.5:12.5	24
2-MeOC6H4	(14)	88.5:11.5	56.5:43.5	9
3-MeOC6H4	(51)	93.5:6.5	94.5:5.5	44
4-MeOC6H4	(49)	93.5:6.5	91.5:8.5	39
2-MeC6H4	(48)	95.5:4.5	92.0:8.0	55
3-MeC6H4	(46)	94.5:5.5	88.5:11.5	41
4-MeC6H4	(49)	95.5:4.5	94.0:6.0	59

326

TABLE 2B. CATALYTIC KINETIC RESOLUTION OF BENZYLIC ALCOHOLS (Continued)

*Please refer to the charts preceding the tables for the structures indicated by the **bold** numbers.*

C$_{9-13}$

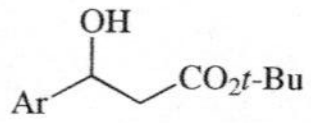

Alcohol	Conditions	Product(s) and Conversion(s) (%)	Refs.
	(c-C$_6$H$_{11}$CO)$_2$O (0.65 eq). **154** (5 mol %), Et$_2$O, rt, 24 h		185

Ar		er I	er II	s
Ph	(—)	96.5:3.5	90.5:9.5	65
2-ClC$_6$H$_4$	(—)	90.0:10.0	96.0:4.0	29
3-ClC$_6$H$_4$	(—)	91.5:8.5	93.5:6.5	31
4-ClC$_6$H$_4$	(—)	92.0:8.0	95.5:4.5	37
2-MeOC$_6$H$_4$	(—)	95.0:5.0	98.0:2.0	71
3-MeOC$_6$H$_4$	(—)	94.0:6.0	97.0:3.0	52
4-MeOC$_6$H$_4$	(—)	95.0:5.0	95.0:5.0	56
2-MeC$_6$H$_4$	(—)	91.5:8.5	97.0:3.0	38
3-MeC$_6$H$_4$	(—)	95.5:4.5	95.0:5.0	62
4-MeC$_6$H$_4$	(—)	94.5:5.5	95.0:5.0	55
1-Np	(—)	95.0:5.0	95.5:4.5	62
2-Np	(—)	94.5:5.5	97.5:2.5	63

C9

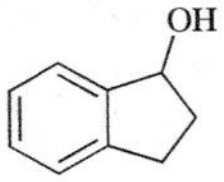

(c-C6H11CO)2O (0.55 eq),
127 (5 mol %),
Et2O, rt, 24 h

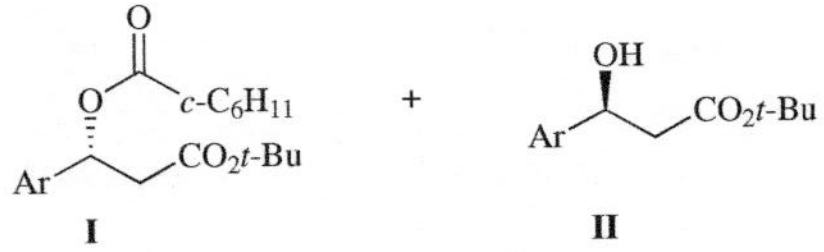

I + **II**

326

Ar		er I	er II	s
3-ClC6H4	(51)	94.0:6.0	95.5:4.5	48
4-ClC6H4	(48)	94.0:6.0	96.0:4.0	53
1-Np	(52)	95.0:5.0	99.5:0.5	93
2-Np	(51)	94.5:5.5	96.5:3.5	60

AcOC(Me)=CH2 (1.27 eq),
50 (1 mol %), toluene, −25°, 12 h

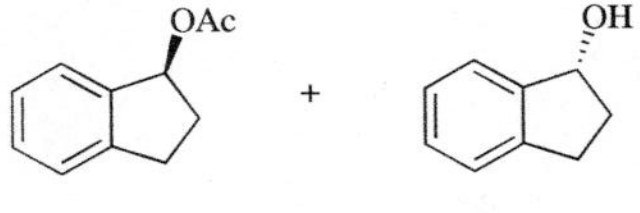

(76), s = 5   101

er 95.5:4.5

(EtCO)2O (0.75 eq), **124** (8 mol %),
i-Pr2NEt (0.75 eq), CDCl3, rt, 22 h

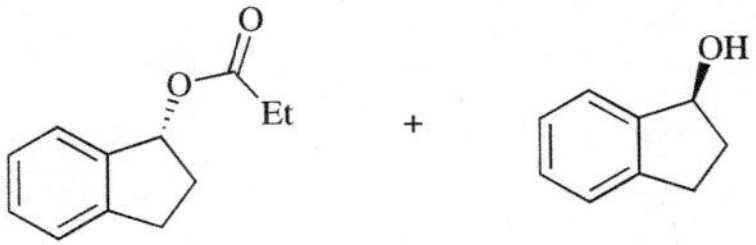

(15), s = 1   57

er 55.0:45.0   er 51.0:49.0

(i-PrCO)2O (2.0 eq),
88 (1 mol %), Et3N (0.75 eq),
toluene, −98°, 15 h

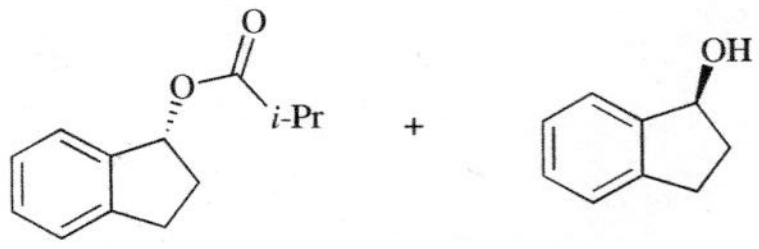

(32), s = 3   164

er 68.5:31.5   er 58.5:41.5

TABLE 2B. CATALYTIC KINETIC RESOLUTION OF BENZYLIC ALCOHOLS (*Continued*)

*Please refer to the charts preceding the tables for the structures indicated by the **bold** numbers.*

Alcohol	Conditions	Product(s) and Conversion(s) (%)	Refs.
C_{9-10}	(i-PrCO)$_2$O (2.5 eq), **141** (5.3 mol %), heptane, rt		155

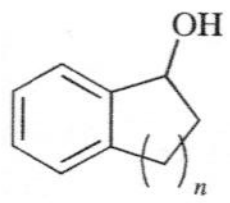

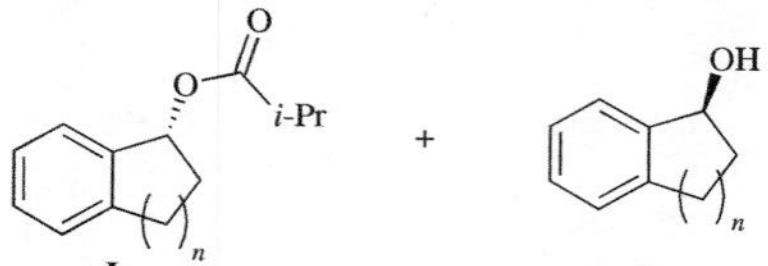

n	Time (h)		er **II**	s
1	3	(54)	55.5:44.5	1
2	24	(42)	51.5:48.5	1

Alcohol	Conditions	Product(s) and Conversion(s) (%)	Refs.
C_9	BzCl (0.75 eq), **139** (15 mol %), Et$_3$N (0.5 eq), 4 Å MS, CH$_2$Cl$_2$, −78°, 11 h	er 69.0:31.0 (44), $s = 3$ er 65.0:35.0	213
	BzCl (0.75 eq), **138** (15 mol %), Et$_3$N (0.5 eq), 4 Å MS, CH$_2$Cl$_2$, −78°, 11 h	er 68.5:31.5 (59), $s = 4$ er 76.5:23.5	213

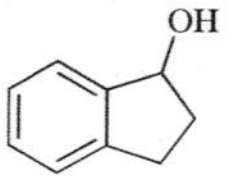

C_{9-15}

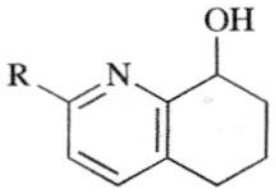

BzCl (0.51 eq), **25** (1 mol %),
CuCl$_2$ (1 mol %), i-Pr$_2$NEt (1.0 eq),
CH$_2$Cl$_2$, 0°, 2 h

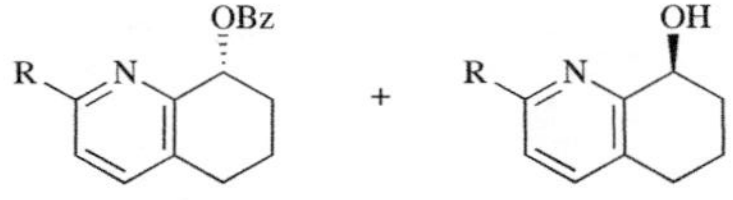

I + **II**

R		er **I**	er **II**	s
H	(52)	88.0:12.0	91.5:8.5	19
Cl	(44)	82.5:17.5	79.0:21.0	8
Ph	(42)	98.5:1.5	85.0:15.0	125

215

C_{10}

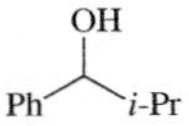

EtCO$_2$H (0.55 eq), **124** (5 mol %),
(3-PyCO)$_2$O (0.55 eq),
CHCl$_3$, 0°, 12 h

I
er 97.5:2.5

+

II
er 84.5:15.5

(42), s = 84

58

RCO$_2$H (0.55 eq), **124** (5 mol %),
(3-PyCO)$_2$O (0.55 eq),
CH$_2$Cl$_2$, 0°, 12 h

I + **II**

R		er **I**	er **II**	s
Et	(42)	97.5:2.5	84.5:15.5	84
Ph(CH$_2$)$_2$	(45)	95.5:4.5	86.5:13.5	49

188

TABLE 2B. CATALYTIC KINETIC RESOLUTION OF BENZYLIC ALCOHOLS (Continued)

Alcohol	Conditions	Product(s) and Conversion(s) (%)	Refs.

*Please refer to the charts preceding the tables for the structures indicated by the **bold** numbers.*

C_{10}

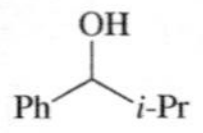

RCO_2H (0.5 eq), **124** (5 mol %),
Piv_2O (0.6 eq), $i\text{-}Pr_2NEt$ (1.2 eq),
CH_2Cl_2, rt, 12 h

R		er I	er II	s
Et	(47)	95.5:4.5	91.0:9.0	53
Cy	(35)	94.5:5.5	73.5:26.5	28
Ph(CH₂)₂	(35)	>99.5:0.5	88.5:11.5	1580

55

C_{10-11}

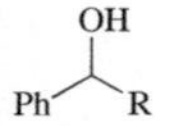

$EtCO_2H$ (0.75 eq), **124** (5 mol %),
(4-MeOC₆H₄CO)₂O (0.90 eq),
$i\text{-}Pr_2NEt$ (1.8 eq), CH_2Cl_2, rt, 12 h

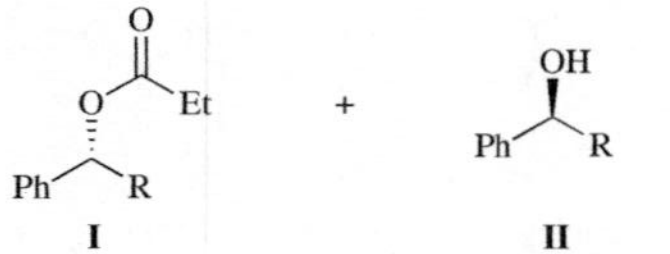

R		er I	er II	s
i-Pr	(47)	95.0:5.0	90.5:9.5	47
t-Bu	(32)	96.5:3.5	72.0:28.0	42

187

$EtCO_2H$ (0.75 eq), **124** (5 mol %),
Bz_2O (0.90 eq), $i\text{-}Pr_2NEt$ (1.8 eq),
CH_2Cl_2, rt, 12 h

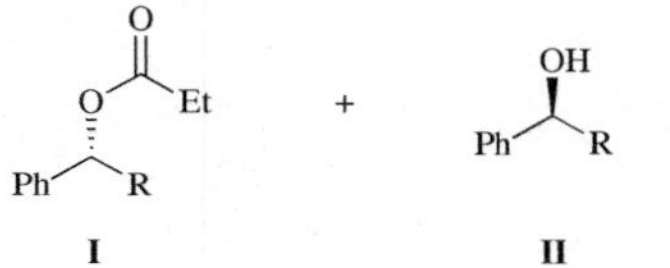

R		er I	er II	s
i-Pr	(52)	91.5:8.5	95.5:4.5	34
t-Bu	(33)	93.0:7.0	71.5:28.5	20

187

C$_{10}$

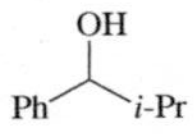

(i-PrCO)$_2$O (0.7 eq), **73** (1 mol %),
Et$_3$N (0.8 eq), CH$_2$Cl$_2$, 0°, 6 h

er 72.0:28.0

+

er 87.0:13.0

(63), s = 5   167

(i-PrCO)$_2$O (0.7 eq), **84** (5 mol %),
Et$_3$N (0.9 eq), CH$_2$Cl$_2$, −78°, 3 h

er 84.5:15.5

+

er 84.5:15.5

(50), s = 11   246

C$_{10-11}$

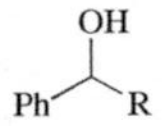

(i-PrCO)$_2$O (2.5 eq), **142** (2 mol %),
Et$_3$N (4 mol %), heptane, −40°

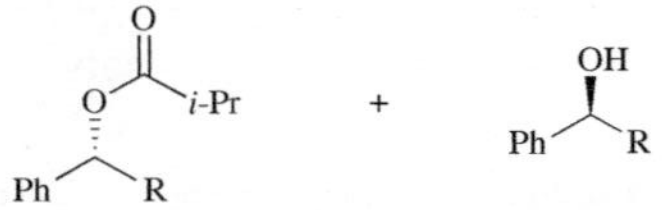

I

+

II

156

R	Time (h)		er **I**	er **II**	s
i-Pr	46	(46)	97.0:3.0	90.0:10.0	82
n-Bu	10	(22)	97.0:3.0	63.0:37.0	40

TABLE 2B. CATALYTIC KINETIC RESOLUTION OF BENZYLIC ALCOHOLS (*Continued*)

Alcohol	Conditions	Product(s) and Conversion(s) (%)	Refs.

*Please refer to the charts preceding the tables for the structures indicated by the **bold** numbers.*

C_{10-11}

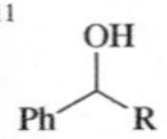

OH / Ph R

Bz$_2$O (2.5 eq), **143** (6 mol %), toluene, rt

OBz / Ph⋯R   **I**   +   OH / Ph R   **II**

R	Time (h)		er I	er II	s
i-Pr	12	(37)	87.0:13.0	72.0:28.0	10
n-Bu	11	(46)	85.0:15.0	80.0:20.0	10

89

C_{10-14}

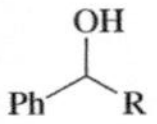

OH / Ph R

BzCl (0.6 eq), **135** (10 mol %), *i*-Pr$_2$NEt (0.6 eq), 4 Å MS, CHCl$_3$, 0°, 5 h

OBz / Ph⋯R   **I**   +   OH / Ph R   **II**

R		er I	er II	s
i-Pr	(57)	86.0:14.0	98.5:1.5	25
c-C$_5$H$_9$	(45)	88.5:11.5	81.5:18.5	15
c-C$_6$H$_{11}$	(61)	81.5:18.5	99.5:0.5	22
c-C$_7$H$_{13}$	(54)	89.0:11.0	96.0:4.0	26

197

C$_{10-11}$

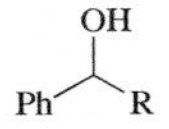

OH
Ph R

Ph(CH$_2$)$_2$CO$_2$H (0.75 eq),
124 (5 mol %),
(4-MeOC$_6$H$_4$CO)$_2$O (0.90 eq),
i-Pr$_2$NEt (1.8 eq), CH$_2$Cl$_2$, rt, 12 h

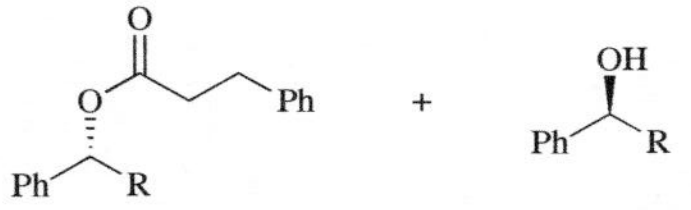

O
‖
Ph
R
Ph   **I**

+

OH
R
Ph   **II**

R		er **I**	er **II**	s
i-Pr	(41)	96.0:4.0	82.0:18.0	46
t-Bu	(38)	98.0:2.0	79.0:21.0	88

187

Ph(CH$_2$)$_2$CO$_2$H (0.75 eq),
124 (5 mol %),
Bz$_2$O (0.90 eq), i-Pr$_2$NEt (1.8 eq),
CH$_2$Cl$_2$, rt, 12 h

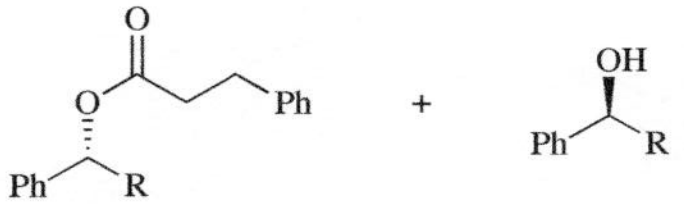

O
‖
Ph
R
Ph   **I**

+

OH
R
Ph   **II**

R		er **I**	er **II**	s
i-Pr	(52)	91.5:8.5	95.0:5.0	33
t-Bu	(37)	93.0:7.0	75.0:25.0	22

187

Ph(CH$_2$)$_2$CO$_2$H (0.75 eq),
Piv$_2$O (0.9 eq), **127** (5 mol %),
Et$_2$O, rt, 24 h

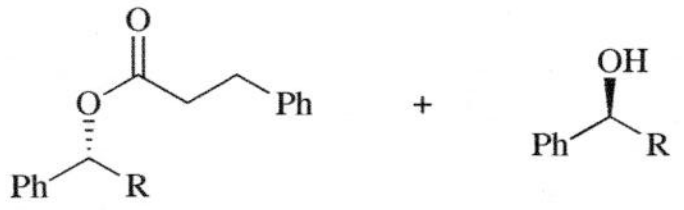

O
‖
Ph
R
Ph   **I**

+

OH
R
Ph   **II**

R		er **I**	er **II**	s
i-Pr	(52)	95.0:5.0	98.5:1.5	85
t-Bu	(43)	97.0:3.0	86.0:14.0	67

50

TABLE 2B. CATALYTIC KINETIC RESOLUTION OF BENZYLIC ALCOHOLS (*Continued*)

Alcohol	Conditions	Product(s) and Conversion(s) (%)	Refs.

*Please refer to the charts preceding the tables for the structures indicated by the **bold** numbers.*

C_{10}

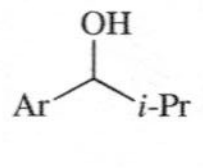

(*i*-PrCO)$_2$O (0.75 eq), **70** (1 mol %),
Et$_3$N (0.75 eq), CH$_2$Cl$_2$, –78°, 6 h

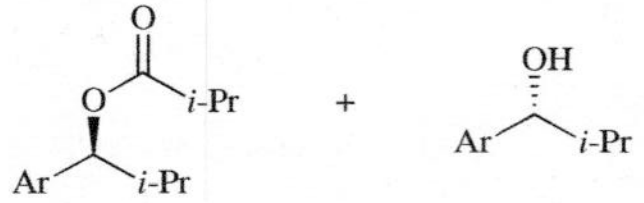

Ar		er **II**		s
Ph	(19)	59.5:40.5		14
4-MeOC$_6$H$_4$	(15)	57.0:43.0		10

39

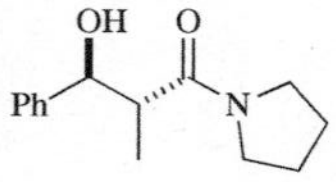

(*i*-PrCO)$_2$O (0.5 eq), **109** (5 mol %),
i-Pr$_2$NEt (0.5 eq), CCl$_4$, 0°, 3 h

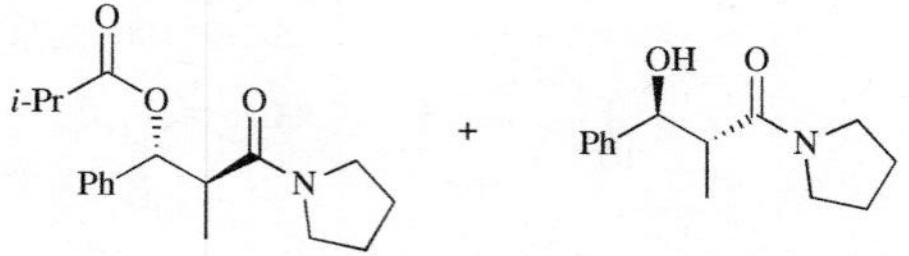

er 91.0:9.0   (49), *s* = 25   er 90.0:10.0

70, 71

(*i*-PrCO)$_2$O (0.5 eq), **110** (5 mol %),
i-Pr$_2$NEt (0.5 eq), CCl$_4$, 0°, 3 h

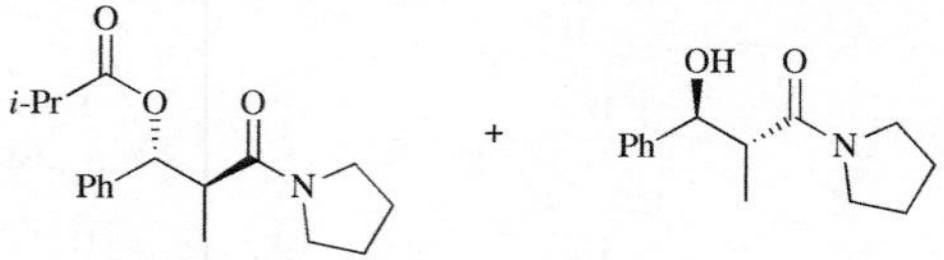

er 93.0:7.0   (52), *s* = 45   er 96.0:4.0

71

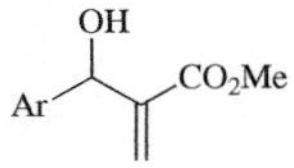

(i-PrCO)$_2$O (1.5 eq), **72** (1 mol %),
Et$_3$N (0.8 eq), CH$_2$Cl$_2$, −78°, 8 h

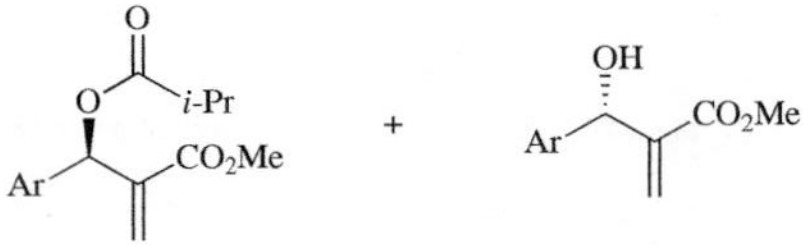

167

Ar		er **II**	s
Ph	(83)	97.5:2.5	4
2-MeOC$_6$H$_4$	(64)	98.5:1.5	13

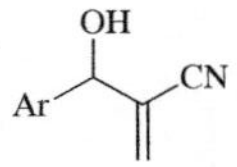

(i-PrCO)$_2$O (1.5 eq), **72** (1 mol %),
Et$_3$N (0.8 eq), CH$_2$Cl$_2$, −78°, 8 h

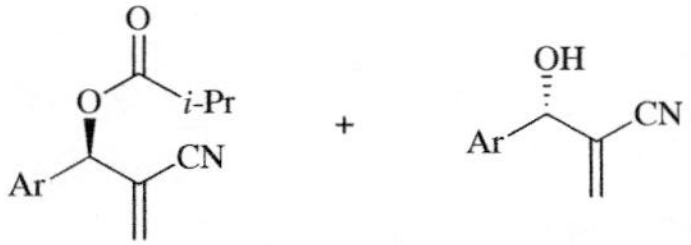

167

Ar		er **II**	s
Ph	(85)	96.5:3.5	4
2-MeOC$_6$H$_4$	(78)	91.0:9.0	4

Alcohol	Conditions	Product(s) and Conversion(s) (%)	Refs.

*Please refer to the charts preceding the tables for the structures indicated by the **bold** numbers.*

C_{10}

Alcohol	Conditions	Product(s) and Conversion(s) (%)	Refs.
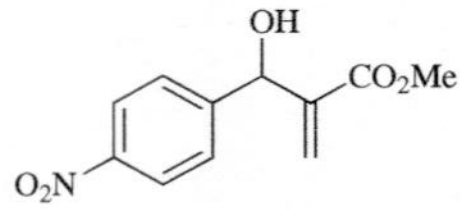 (4-O_2N-phenyl, CO_2Me allylic alcohol)	AcCl (0.6 eq), **157a** (5 mol %), DABCO (0.65 eq), Et_2O, –20°, 5 h	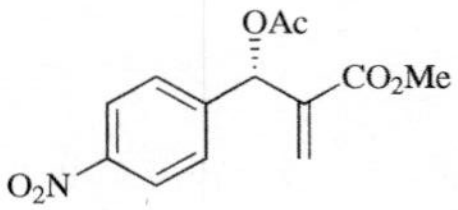 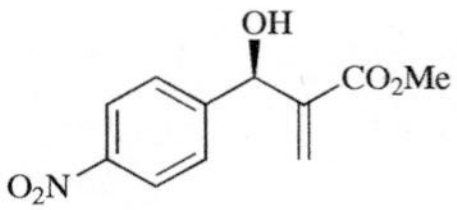 OAc product er 92.5:7.5 + OH product (33), s = 19, er 71.0:29.0	201
1-(3,5-bis-CF_3-phenyl)ethanol	$(i\text{-PrCO})_2O$ (0.75 eq), **124** (4 mol %), $i\text{-Pr}_2NEt$ (0.75 eq), $CDCl_3$, 0°, 24 h	i-Pr ester er 95.5:4.5 + OH (45), s = 48, er 87.5:12.5	57
(2-Cl-pyridin-4-yl)(t-Bu)methanol	$(EtCO)_2O$ (0.75 eq), **115** (1 mol %), $i\text{-Pr}_2NEt$ (0.75 eq), toluene, 0°, 16 h	Et ester er 93.5:6.5 + OH (54), s = 74, er >99.5:0.5	184

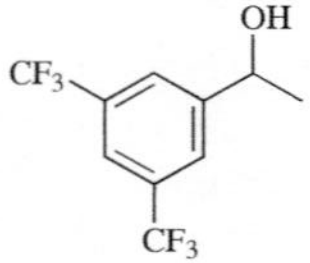

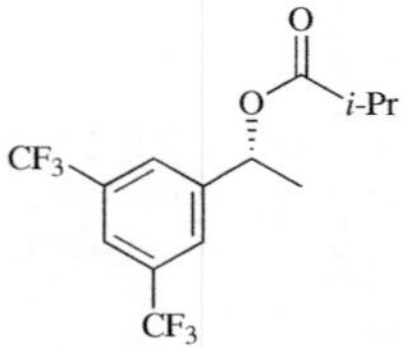

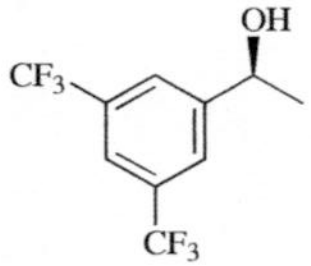

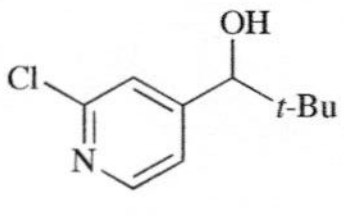

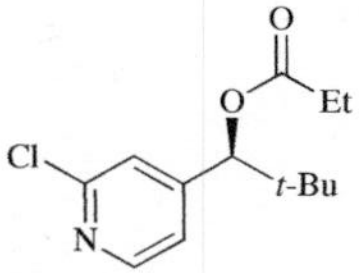

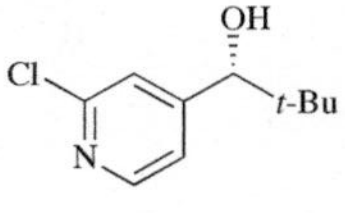

(EtCO)$_2$O (0.75 eq), **120** (5 mol %),
i-Pr$_2$NEt (0.75 eq), toluene, 0°, 51 h

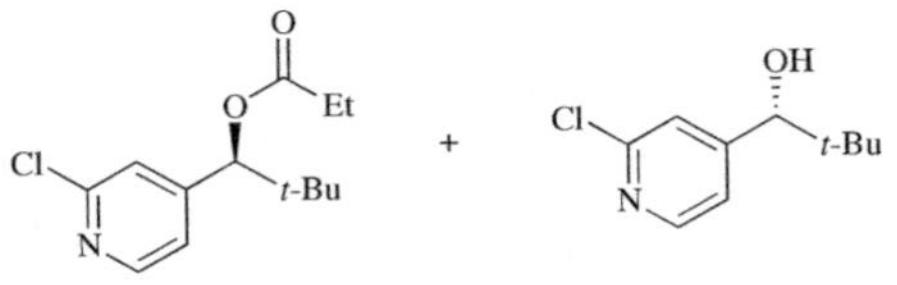

er 95.0:5.0   (53), s = 118   er 99.5:0.5

183

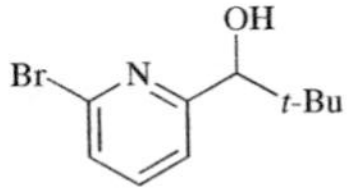

(EtCO)$_2$O (0.75 eq), **115** (15 mol %),
i-Pr$_2$NEt (0.75 eq), toluene, 0°, 78 h

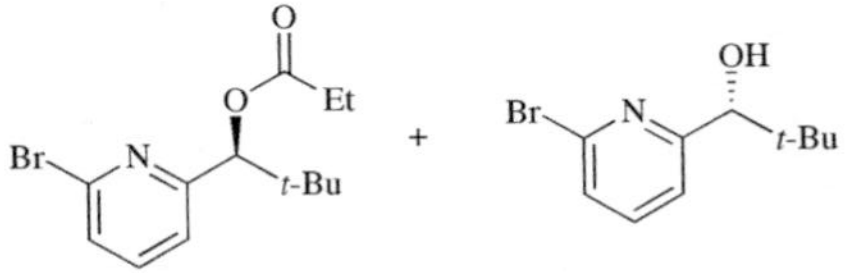

er 88.5:11.5   (56), s = 39   er >99.5:0.5

184

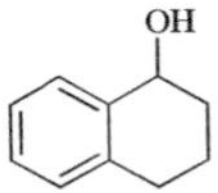

AcOC(Me)=CH$_2$ (1.27 eq),
50 (2 mol %), toluene, –10°, 8 h

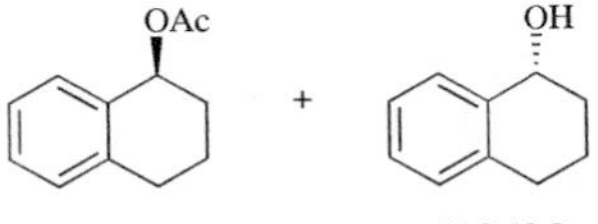

er 79.0:21.0   (39), s = 2   er 57.0:43.0

101

BzCl (0.75 eq), **139** (15 mol %),
Et$_3$N (0.5 eq), 4 Å MS,
CH$_2$Cl$_2$, –78°, 11 h

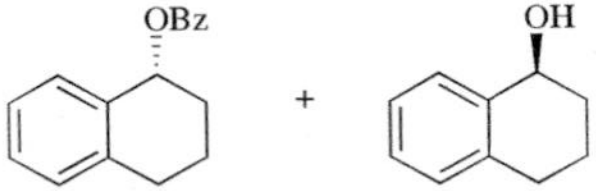

er 79.0:21.0   (46), s = 6   er 75.0:25.0

213

265

TABLE 2B. CATALYTIC KINETIC RESOLUTION OF BENZYLIC ALCOHOLS (*Continued*)

*Please refer to the charts preceding the tables for the structures indicated by the **bold** numbers.*

Alcohol	Conditions	Product(s) and Conversion(s) (%)	Refs.

C$_{11}$

OH
Ph — t-Bu

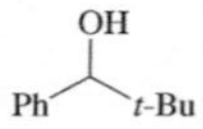

Ac$_2$O (1.5 eq), **100** (0.5 mol %),
i-Pr$_2$NEt (0.4 eq),
toluene, –65°, 12 h

OAc
Ph — t-Bu
+
OH
Ph — t-Bu
(35), s >50

69

Ac$_2$O (0.75 eq), **79** (5 mol %),
Et$_3$N (0.75 eq),
toluene, –78°, 15 h

OAc
Ph — t-Bu
er 92.0:8.0
+
OH
Ph — t-Bu
(8), s = 12

173

(EtCO$_2$)$_2$O (0.5 eq), **126** (0.75 mol %),
i-Pr$_2$NEt (0.6 eq), CHCl$_3$, 0°, 1 h

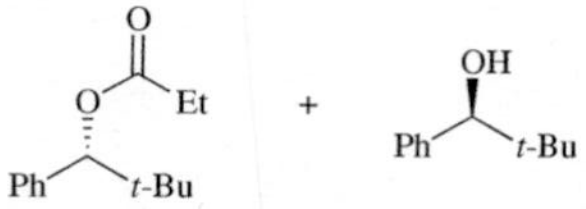

O
‖
Et — C — O
Ph — t-Bu
er 99.5:0.5
+
OH
Ph — t-Bu
er 99.0:1.0
(50), s >100

60

RCO$_2$H (0.55 eq), **124** (5 mol %),
(3-PyCO)$_2$O (0.55 eq),
CH$_2$Cl$_2$, 0°, 12 h

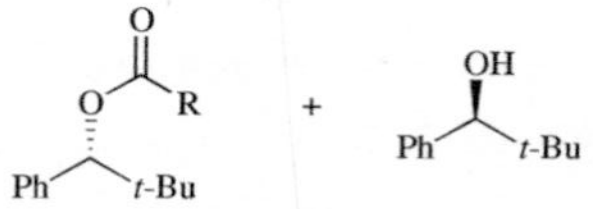

O
‖
R — C — O
Ph — t-Bu
I
+
OH
Ph — t-Bu
II

188

R		er I	er II	s
Et	(21)	96.5:3.5	62.5:37.5	36
Ph(CH$_2$)$_2$	(36)	98.0:2.0	77.5:22.5	94

RCO$_2$H (0.5 eq), **124** (5 mol %),
Piv$_2$O (0.6 eq), i-Pr$_2$NEt (1.2 eq),
CH$_2$Cl$_2$, rt, 12 h

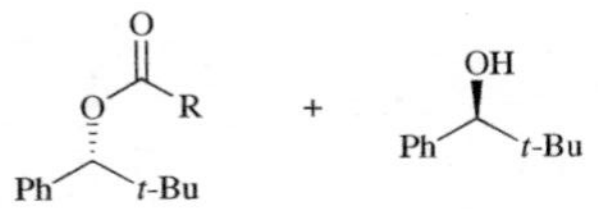

R	er I	er II	s
Et	(33) 98.5:1.5	74.0:26.0	101
Ph(CH$_2$)$_2$	(42) 97.5:2.5	85.0:15.0	81

55

(i-PrCO)$_2$O (2.5 eq), **142** (15 mol %),
Et$_3$N (30 mol %), toluene, rt, 139 h

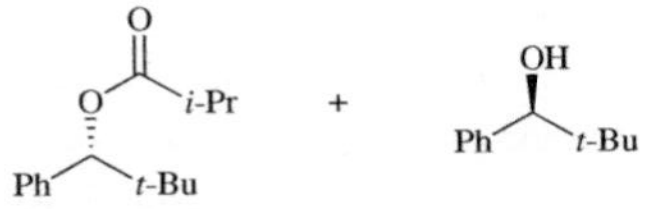

er 91.5:8.5   er 58.5:41.5   (19), s = 13

156

(i-PrCO)$_2$O (0.8 eq), **76** (0.5 mol %),
Et$_3$N (0.9 eq), CHCl$_3$, rt, 72 h

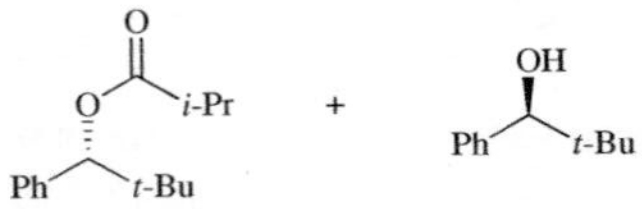

er 77.0:23.0   er 94.0:6.0   (62), s = 10

168

TABLE 2B. CATALYTIC KINETIC RESOLUTION OF BENZYLIC ALCOHOLS (*Continued*)

Alcohol	Conditions	Product(s) and Conversion(s) (%)	Refs.
*Please refer to the charts preceding the tables for the structures indicated by the **bold** numbers.*			

C₁₁ — C_{11}

Alcohol:

OH / Ph — *t*-Bu

Conditions	Product(s) and Conversion(s) (%)	Refs.
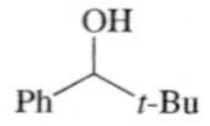 (*i*-PrCO)₂O (1 eq), *ent*-**81** (1 mol %), Et₃N (1.5 eq), toluene, −40°, 4 h	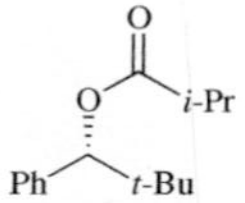 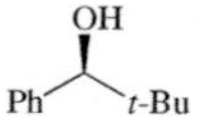 er 62.5:37.5 + er 55.0:45.0 (29), *s* = 2	171
(*i*-PrCO)₂O (0.75 eq), **92** (0.5 mol %), Et₃N (0.75 eq), *t*-AmOH, 14 h	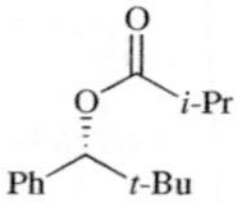 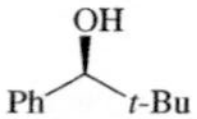 er 93.0:7.0 + er 83.5:16.5 (44), *s* = 26	42
(*i*-PrCO)₂O (0.7 eq), **157a** (5 mol %), CHCl₃, rt, 24 h	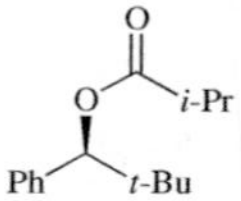 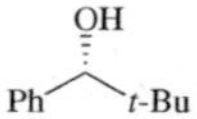 er 90.0:10.0 + er 92.0:8.0 (51), *s* = 23	104
(*i*-PrCO)₂O (0.55 eq), **114** (10 mol %), *i*-Pr₂NEt (0.66 eq), toluene, −40°, 48 h	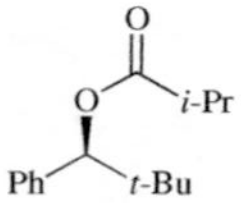 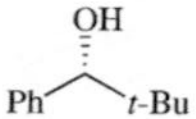 er 96.5:3.5 + er 73.0:27.0 (33), *s* = 45	182
(*i*-PrCO)₂O (0.55 eq), **155** (5 mol %), *i*-Pr₂NEt (0.6 eq), CHCl₃, rt, 30 h	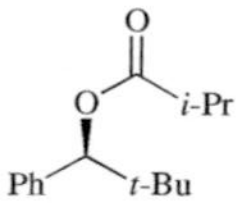 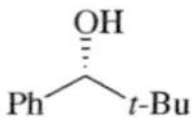 er 98.0:2.0 + er 81.0:19.0 (39), *s* = 110	196

Products in each row: *i*-Pr–C(=O)–O–CH(Ph)(*t*-Bu) (ester) and OH–CH(Ph)(*t*-Bu) (alcohol).

154

171

197

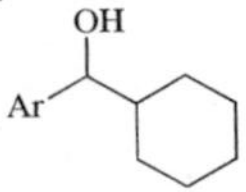

Bz₂O (2.5 eq), **140** (4.9 mol %), toluene, −40°, 65 h

er 96.5:3.5 + er 89.5:10.5 (46), s = 67

(3-PyCO)₂O (1 eq), catalyst (2 mol %), Et₃N (1.5 eq), t-AmOH, CH₂Cl₂, −25°

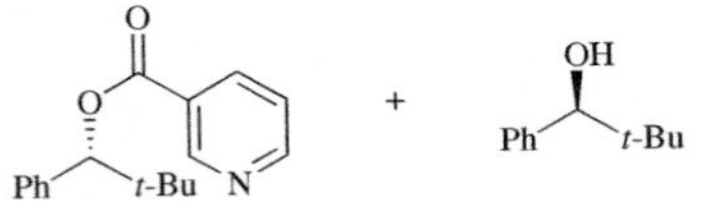

I + **II**

Catalyst	Time (h)		er I	er II	s
80	64	(43)	83.5:16.5	75.5:24.5	10
140	20	(34)	97.5:2.5	74.5:25.5	57

C₁₁₋₁₂

BzCl (0.6 eq), **135** (10 mol %), i-Pr₂NEt (0.6 eq), 4 Å MS, CHCl₃, 0°, 5 h

I + **II**

Ar		er I	er II	s
2-furyl	(61)	81.5:18.5	99.5:0.5	22
2-thienyl	(64)	78.5:21.5	99.5:0.5	18
2-pyridyl	(56)	89.5:10.5	99.5:0.5	44
3-pyridyl	(—)	52.5:47.5	51.0:49.0	1

TABLE 2B. CATALYTIC KINETIC RESOLUTION OF BENZYLIC ALCOHOLS (Continued)

Alcohol	Conditions	Product(s) and Conversion(s) (%)	Refs.

*Please refer to the charts preceding the tables for the structures indicated by the **bold** numbers.*

C_{11–15}

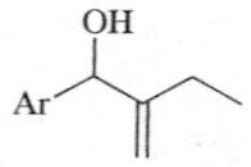

(EtCO)$_2$O (0.75 eq), **115** (2 mol %), i-Pr$_2$NEt (0.75 eq), toluene, 0°

204

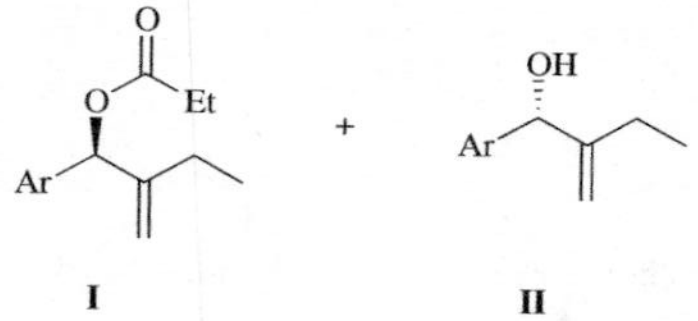

Ar	Time (h)		er I	er II	s
Ph	9	(60)	73.5:26.5	85.0:15.0	6
2-MeC$_6$H$_4$	3	(61)	81.5:18.5	99.5:0.5	23
1-Np	7	(64)	78.0:22.0	99.5:0.5	18
2-Np	8	(65)	75.5:24.5	97.5:2.5	11

C$_{11–17}$

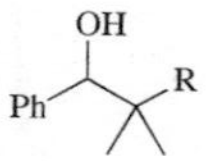

(EtCO)$_2$O (0.75 eq), **115** (1 mol %), i-Pr$_2$NEt (0.75 eq), toluene, 0°

184

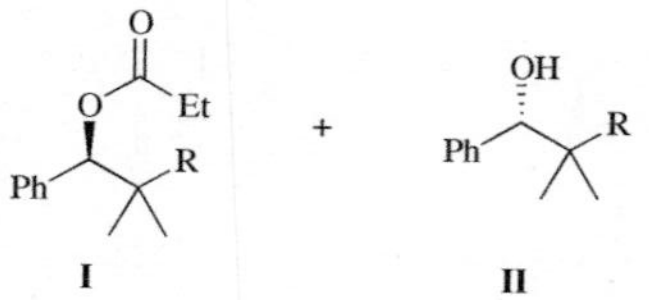

R	Time (h)		er I	er II	s
TBSOCH$_2$	93	(53)	92.5:7.5	>99.5:0.5	86
MeO$_2$C	36	(54)	92.5:7.5	>99.5:0.5	85
Bz	93	(54)	94.5:5.5	>99.5:0.5	68

270

C$_{11-15}$

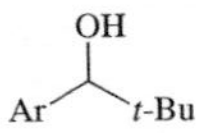

(EtCO)$_2$O (0.75 eq), **120** (5 mol %),
i-Pr$_2$NEt (0.75 eq),
toluene, 0°, 48 h

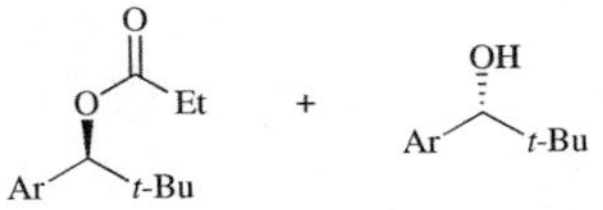

I   +   **II**

Ar	er **I**	er **II**	s
4-ClC$_6$H$_4$	(52) 96.5:3.5	99.5:0.5	152
4-MeC$_6$H$_4$	(43) 99.5:0.5	88.0:12.0	456
2-Np	(50) 98.5:1.5	98.0:2.0	258

C$_{11}$

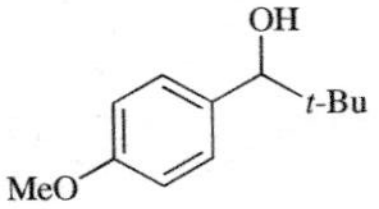

Bz$_2$O (2.5 eq), **140** (5.1 mol %),
toluene, rt, 10 h

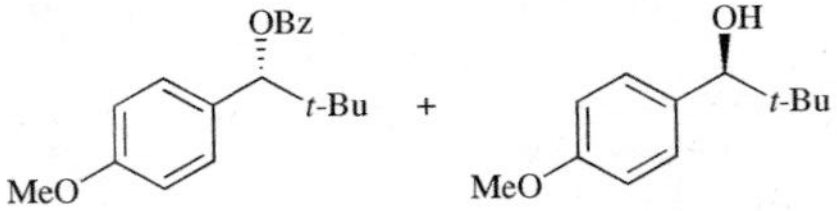

er 91.5:8.5   +   er 83.0:17.0   (44), s = 22   155

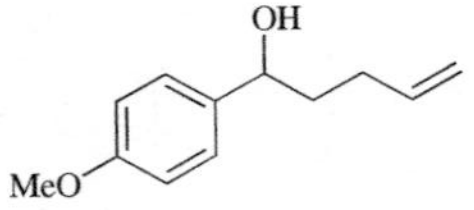

4-BnOC$_6$H$_4$(CH$_2$)$_2$CO$_2$H (0.9 eq),
127 (6 mol %), Piv$_2$O (1.1 eq),
Et$_2$O, rt, 24 h

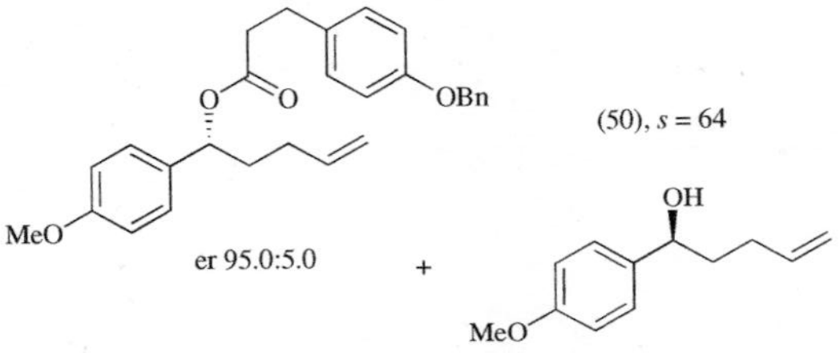

er 95.0:5.0   +   (50), s = 64   327

er 95.5:4.5

TABLE 2B. CATALYTIC KINETIC RESOLUTION OF BENZYLIC ALCOHOLS (*Continued*)

Alcohol	Conditions	Product(s) and Conversion(s) (%)	Refs.
*Please refer to the charts preceding the tables for the structures indicated by the **bold** numbers.*			
C_{12}			
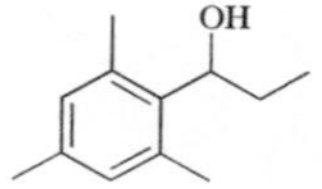	BzCl (0.6 eq), **135** (10 mol %), *i*-Pr_2NEt (0.6 eq), 4 Å MS, $CHCl_3$, 0°, 5 h	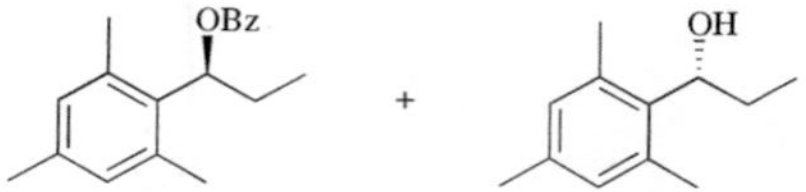 er 97.5:2.5 + er 92.5:7.5 (47), *s* = 106	197
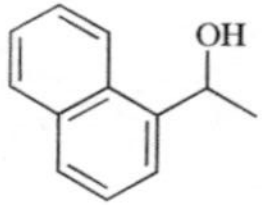	Vinyl acetate, **149** (3 mol %), *t*-BuOK (2.5 eq), Et_2O, rt, 72 h	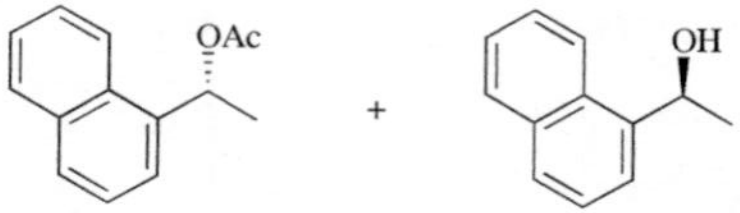 er 74.5:25.5 + er 60.0:40.0 (29), *s* = 4	77
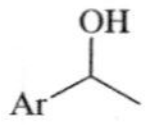	$(EtCO)_2O$ (0.75 eq), **117** (2 mol %), *i*-Pr_2NEt (0.75 eq), $CHCl_3$, 0°, 8 h	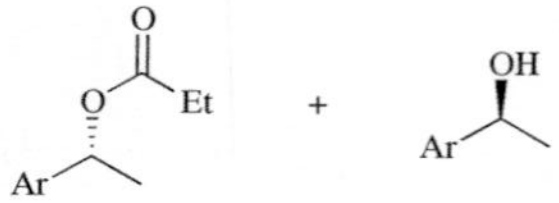 **I** + **II**	51

Ar		er **I**	er **II**	*s*
1-Np	(55)	90.0:10.0	99.5:0.5	56
2-Np	(51)	95.5:4.5	97.0:3.0	76

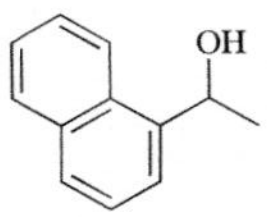

(EtCO)$_2$O (0.5 eq), **126** (0.75 mol %),
i-Pr$_2$NEt (0.6 eq), CHCl$_3$, 1 h

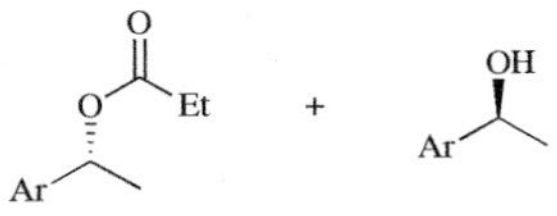

I + II

Ar	Temp (°)		er **I**	er **II**	*s*
1-Np	−40	(39)	95.5:4.5	78.5:21.5	38
2-Np	−20	(43)	97.0:3.0	85.5:14.5	65

(*i*-PrCO)$_2$O (0.75 eq), **93** (5 mol %),
Et$_3$N (0.75 eq), *t*-AmOH, 0°, 3 h

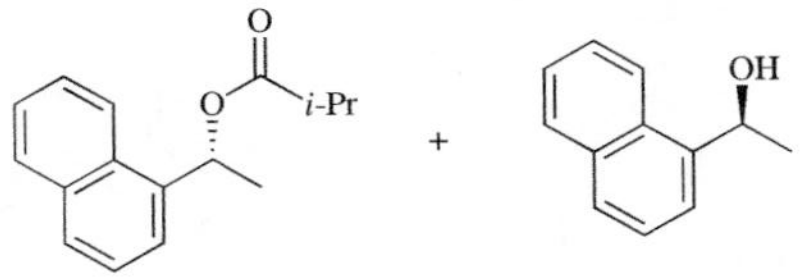

er 81.5:18.5 + er 78.5:21.5

(—), *s* = 33   162

(*i*-PrCO)$_2$O (1.5 eq), **71** (1 mol %),
Et$_3$N (0.75 eq), CH$_2$Cl$_2$, −78°, 8 h

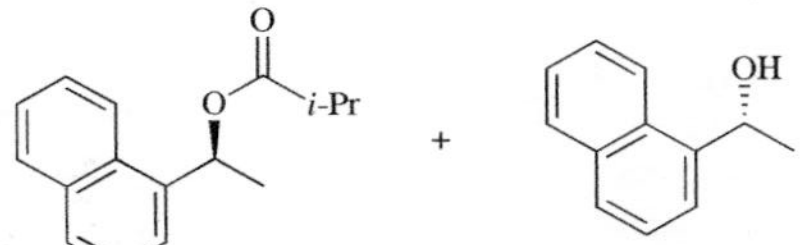

er 75.5:24.5

(43), *s* = 9   166

TABLE 2B. CATALYTIC KINETIC RESOLUTION OF BENZYLIC ALCOHOLS (*Continued*)

Alcohol	Conditions	Product(s) and Conversion(s) (%)	Refs.

*Please refer to the charts preceding the tables for the structures indicated by the **bold** numbers.*

C$_{12}$

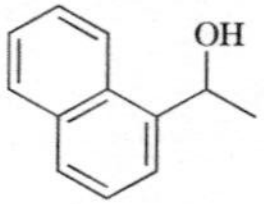

(*i*-PrCO)$_2$O (2.5 eq), **142** (6 mol %),
Et$_3$N (12 mol %), toluene, rt, 2 h

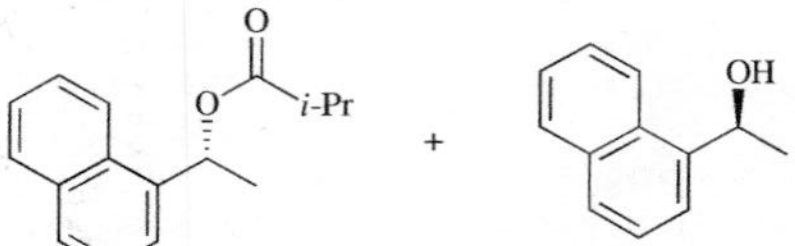

er 95.0:5.0   +   er 74.5:25.5

(35), *s* = 30   156

(*i*-PrCO)$_2$O (2.5 eq),
143 (2.5 mol %),
toluene, –25°, 93 h

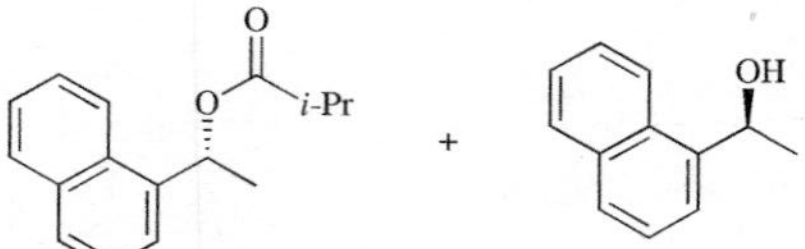

er 73.5:26.5   +   er 87.0:13.0

(63), *s* = 6   89

(*i*-PrCO)$_2$O (1.5 eq),
65 (5 mol %),
CHCl$_3$, rt, 24 h

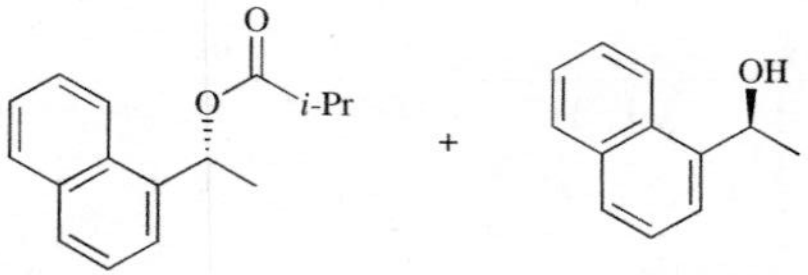

+   er 55.5:44.5

(59), *s* = 1   177

(*i*-PrCO)$_2$O (1.5 eq),
64 (5 mol %),
CHCl$_3$, rt, 24 h

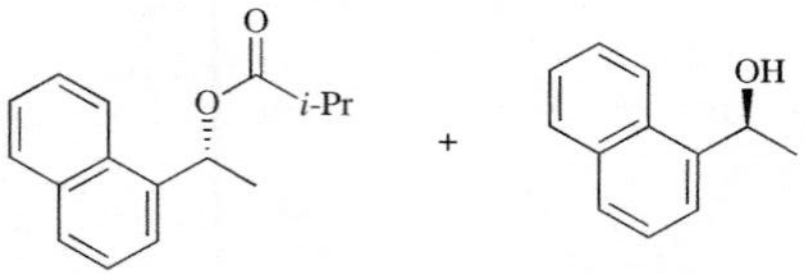

+   er 61.0:39.0

(74), *s* = 1   177

274

(i-PrCO)$_2$O (0.7 eq), **64** (5 mol %), toluene, rt, 3 h

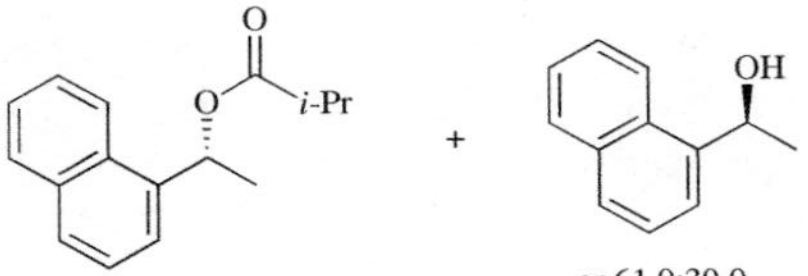

er 92.0:8.0 + er 61.0:39.0 (74), s = 1   178

(i-PrCO)$_2$O (0.75 eq), **87** (1 mol %), Et$_3$N (0.75 eq), toluene, –78°, 9 h

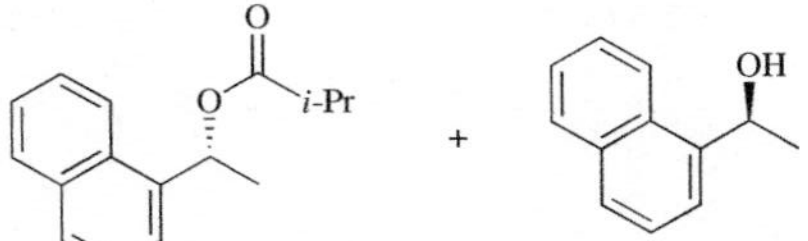

+ er 84.5:15.5 (45), s = 24   328

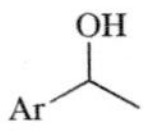

(i-PrCO)$_2$O (0.6 eq), **77** (10 mol %), Et$_2$O, –50°, 48 h

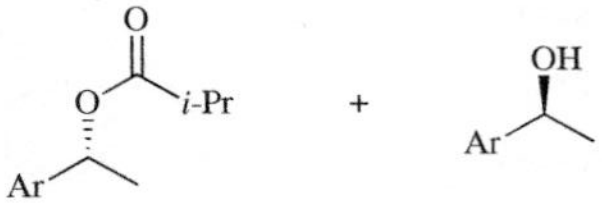

I + **II**

Ar		er **I**	er **II**	s
1-Np	(42)	80.5:19.5	88.5:11.5	6
2-Np	(47)	94.0:6.0	72.0:28.0	37

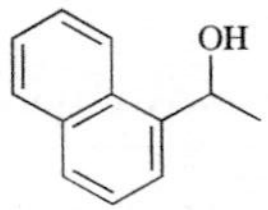

169

Bz$_2$O (2.5 eq), **143** (6 mol %), toluene, –25°, 15 h

er 92.5:7.5 + er 72.5:27.5 (35), s = 19   89

TABLE 2B. CATALYTIC KINETIC RESOLUTION OF BENZYLIC ALCOHOLS (*Continued*)

Alcohol	Conditions	Product(s) and Conversion(s) (%)	Refs.

*Please refer to the charts preceding the tables for the structures indicated by the **bold** numbers.*

C$_{12}$

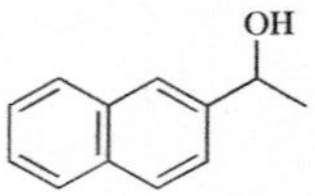

	Conditions	Product(s) and Conversion(s) (%)	Refs.
	Vinyl acetate, **147** (3 mol %), t-BuOK (2.5 eq), Et$_2$O, 0°, 18 h	OAc structure (er 68.5:31.5) + OH structure (er 72.0:28.0) (29), s = 4	77
	Ac$_2$O (0.7 eq), **62** (10 mol %), Et$_3$N (0.7 eq), toluene, rt	 OAc structure + OH structure (er 68.5:31.5) (56), s = 2	175
	(EtCO)$_2$O (0.75 eq), **125** (4 mol %), i-Pr$_2$NEt (0.75 eq), t-AmOH/toluene (1:1), –40°, 5 h	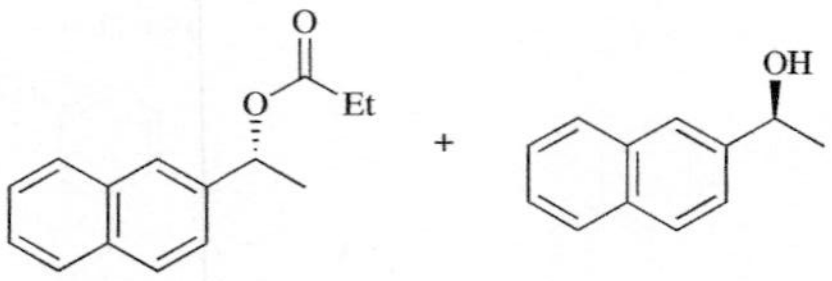 OC(=O)Et ester (er 96.5:3.5) + OH structure (er 77.5:22.5) (49), s = 49	194
	(i-PrCO)$_2$O (2.5 eq), **141** (6.6 mol %), toluene, rt, 1 h	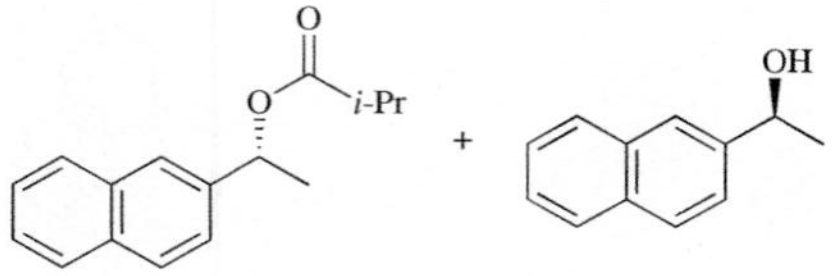 OC(=O)i-Pr ester (er 90.0:10.0) + OH structure (er 83.0:17.0) (45), s = 17	155

C_{12-14}

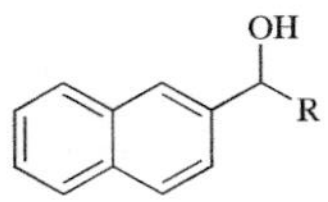

(i-PrCO)$_2$O (0.55 eq),
155 (5 mol %),
i-Pr$_2$NEt (0.6 eq),
(MeO$_2$CO, rt, 24 h)

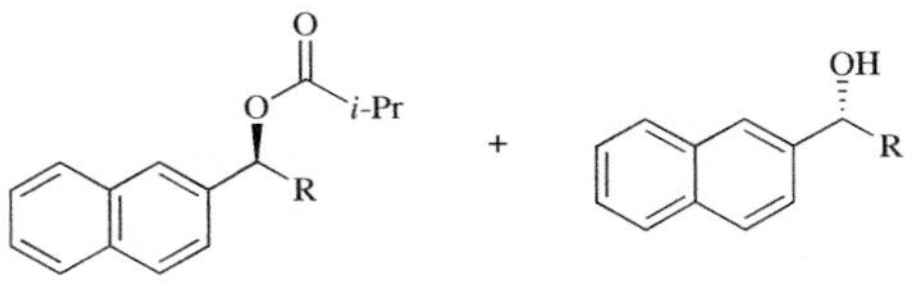

242

R		er I	er II	s
	I		**II**	
Me	(50)	94.0:6.0	94.0:6.0	45
N$_3$CH$_2$	(52)	90.0:10.0	94.0:6.0	25
i-Pr	(52)	96.0:4.0	99.5:0.5	140

C_{12}

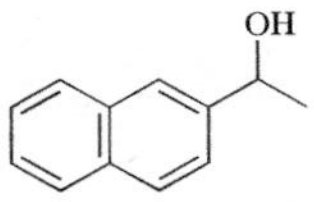

BzCl (0.75 eq), **139** (15 mol %),
Et$_3$N (0.5 eq), 4 Å MS,
CH$_2$Cl$_2$, –78°, 11 h

+

er 58.0:42.0   er 58.0:42.0

(50), s = 2   213

BzCl (0.75 eq), **138** (15 mol %),
Et$_3$N (0.5 eq), 4 Å MS,
CH$_2$Cl$_2$, –78°, 11 h

+

er 61.5:38.5   er 66.5:33.5

(59), s = 2   213

Bz$_2$O (2.5 eq), **140** (2.6 mol %),
toluene, rt, 1 h

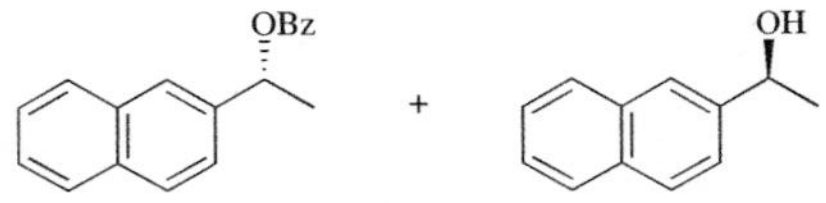

+

er 81.0:19.0   er 76.0:24.0

(46), s = 7   155

TABLE 2B. CATALYTIC KINETIC RESOLUTION OF BENZYLIC ALCOHOLS (Continued)

Alcohol	Conditions	Product(s) and Conversion(s) (%)	Refs.

*Please refer to the charts preceding the tables for the structures indicated by the **bold** numbers.*

C$_{12-14}$

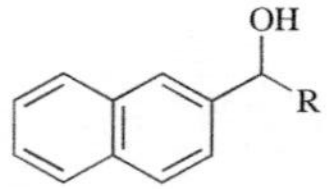

(i-PrCO)$_2$O (0.55 eq),
155 (1 mol %),
i-Pr$_2$NEt (0.6 eq),
CHCl$_3$, 0°, 7 h

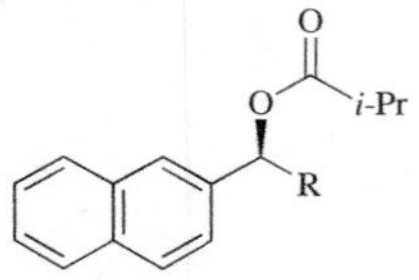

I + II

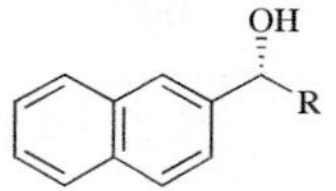

196

C$_{12}$

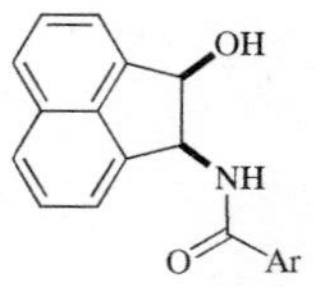

Ar = 4-Me$_2$NC$_6$H$_4$

(i-PrCO)$_2$O (0.6 eq),
ent-**59** (5 mol %),
2,4,6-collidine, CHCl$_3$, 20°, 24 h

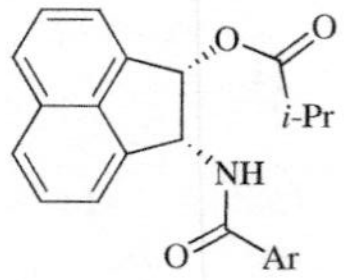

R	er I		er II		s
N$_3$CH$_2$	(57)	87.5:12.5	(59)	99.5:0.5	60
i-Pr	(45)	99.5:0.5		91.0:9.0	600

er 83.5:16.5 + er 98.5:1.5 (59), s = 21

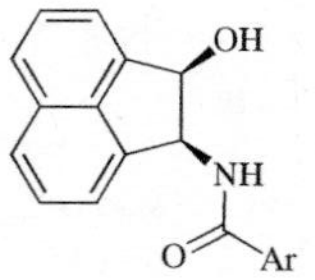

180

C$_{13}$

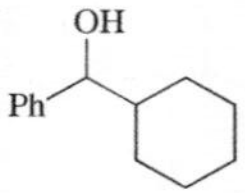

Cinnamaldehyde (0.33 eq),
152 (1.7 mol %), **162** (0.33 eq),
DBU (0.37 eq), toluene, –20°, 12 h

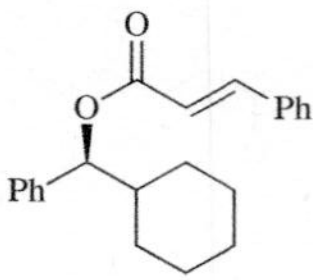

(17) er 59.5:40.5

81

278

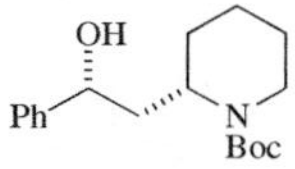

(EtCO)$_2$O (0.75 eq), **124** (16 mol %), *i*-Pr$_2$NEt (0.75 eq), CDCl$_3$, rt, 24 h

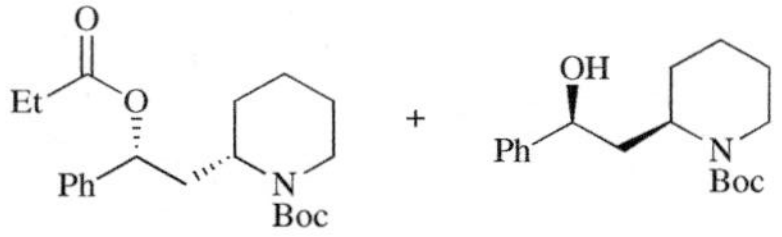

er 94.5:5.5   +   er 90.5:9.5   (47), *s* = 44   57

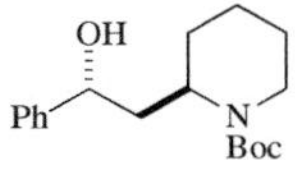

(*i*-PrCO)$_2$O (0.75 eq), **124** (8 mol %), *i*-Pr$_2$NEt (0.75 eq), CDCl$_3$, rt, 18 h

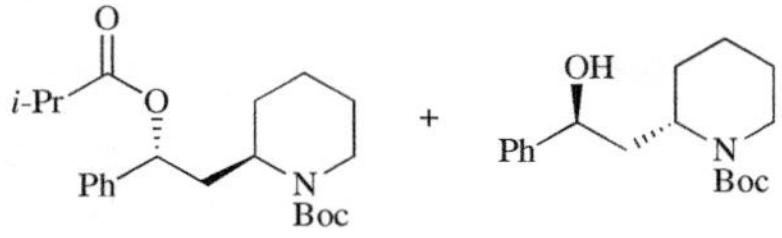

er 97.0:3.0   +   er 91.0:9.0   (47), *s* = 79   57

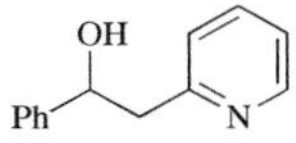

(*i*-PrCO)$_2$O (0.75 eq), **124** (16 mol %), *i*-Pr$_2$NEt (0.75 eq), CDCl$_3$, 0°, 22 h

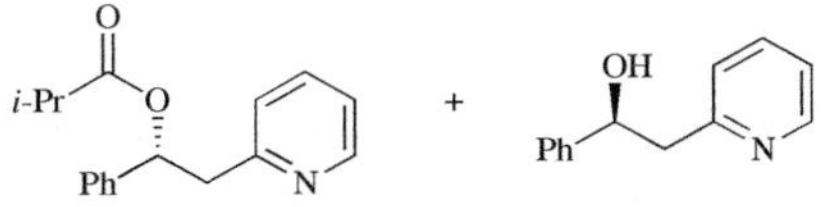

er 93.0:7.0   +   er 85.0:15.0   (45), *s* = 27   57

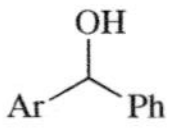

2-PyCHO (0.33 eq), **145a** (1.7 mol %), **162** (0.33 eq), DBU (0.37 eq), toluene, rt

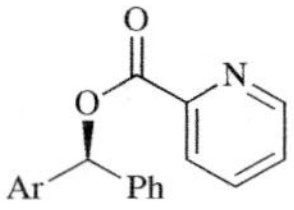

81

C$_{13-17}$

Ar	Time (h)	Yield (%)	er
2-BrC$_6$H$_4$	2	(29)	67.5:32.5
2-MeC$_6$H$_4$	3	(26)	55.5:44.5
1-Np	6	(17)	53.5:46.5

TABLE 2B. CATALYTIC KINETIC RESOLUTION OF BENZYLIC ALCOHOLS (Continued)

Alcohol	Conditions	Product(s) and Conversion(s) (%)	Refs.

*Please refer to the charts preceding the tables for the structures indicated by the **bold** numbers.*

C_{13-17}

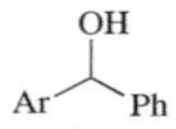

$(i\text{-PrCO})_2O$ (0.75 eq),
69a (5 mol %), Cs_2CO_3 (0.75 eq),
toluene, $-60°$, 15 h

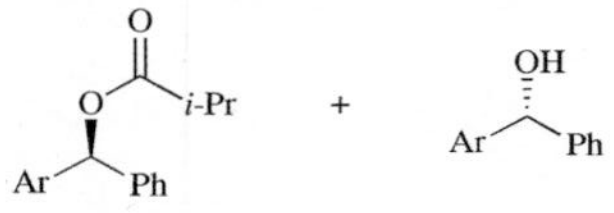

Ar		er **I**	er **II**	s
4-MeOC$_6$H$_4$	(45)	74.0:26.0	70.0:30.0	4
4-CF$_3$C$_6$H$_4$	(45)	64.0:36.0	61.0:39.0	2
2-Np	(18)	83.0:17.0	57.0:43.0	6

324

C_{13}

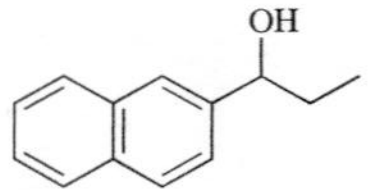

$(i\text{-PrCO})_2O$ (0.6 eq),
126 (1 mol %), $i\text{-Pr}_2$NEt (0.5 eq),
toluene, $-78°$, 16 h

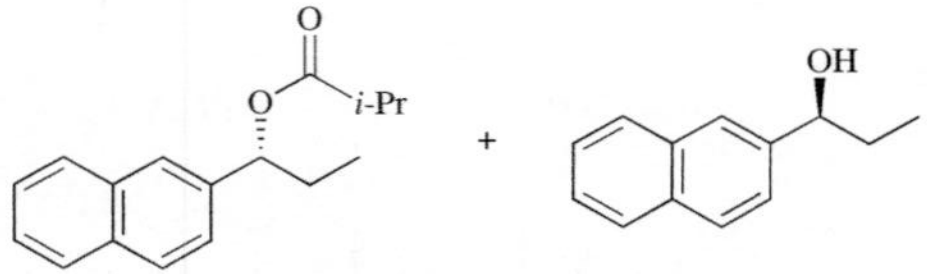

(52), s = 152   er 99.5:0.5

er 96.0:4.0

205

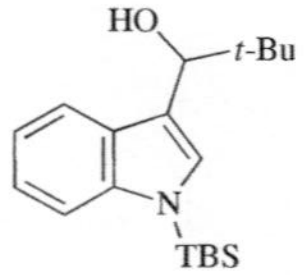

$(EtCO)_2O$ (0.75 eq), **115** (1 mol %),
$i\text{-Pr}_2$NEt (0.75 eq), toluene, rt, 72 h

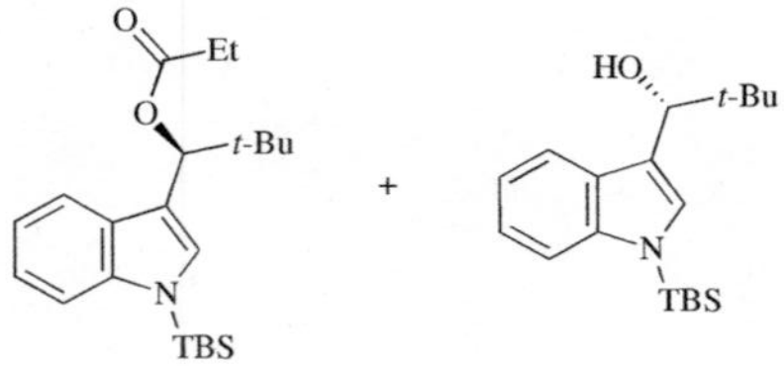

(57), s = 37   er >99.5:0.5

er 88.0:12.0

184

C_{14} 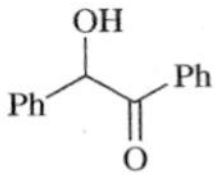	Ac_2O (0.5 eq), **37** (5 mol %), Et_3N (1.0 eq), Et_2O, –25°, 15 h	 er 95.0:5.0 + er 91.0:9.0 (48), $s = 50$	96
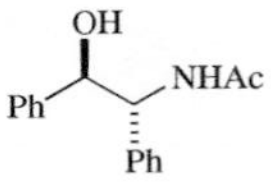	Ac_2O (0.84 eq), **104** (2.5 mol %), *i*-Pr_2NEt, toluene, rt, 3 h	 er 62.5:37.5 + er 71.5:28.5 (63), $s = 2$	67
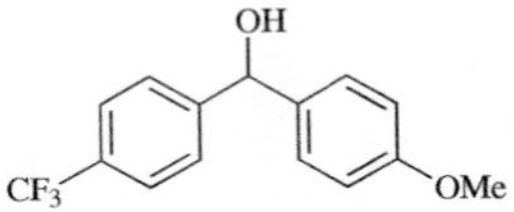	(*i*-$PrCO)_2O$ (0.75 eq), **69a** (5 mol %), Cs_2CO_3 (0.75 eq), toluene, –60°, 15 h	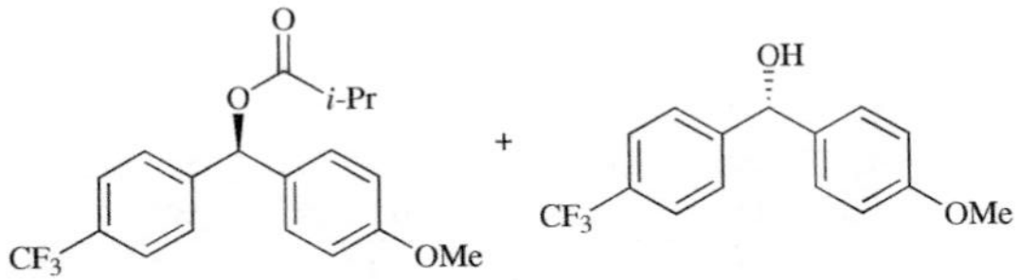 er 60.0:40.0 (52), $s = 2$ er 61.0:39.0	324
C_{15} 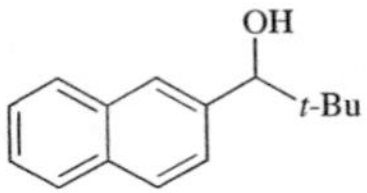	Ac_2O (1 eq), **85** (1 mol %), *t*-AmOH, 0°, 3 h	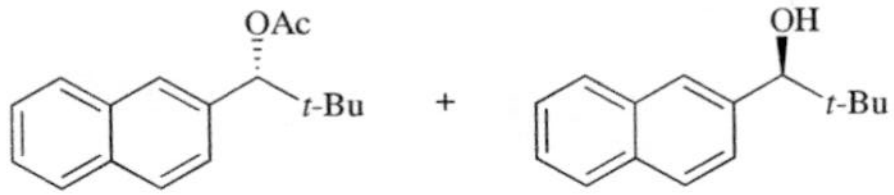 er 78.5:21.5 (63), $s = 12$ er 97.5:2.5	321

TABLE 2B. CATALYTIC KINETIC RESOLUTION OF BENZYLIC ALCOHOLS (Continued)

Alcohol	Conditions	Product(s) and Conversion(s) (%)	Refs.

Please refer to the charts preceding the tables for the structures indicated by the bold numbers.

C₁₅

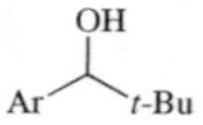

Conditions: (i-PrCO)₂O (0.75 eq), **124** (4 mol %), i-Pr₂NEt (0.75 eq), CHCl₃, Na₂SO₄, 0°

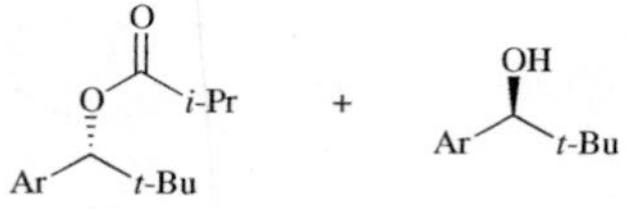

I + **II**

Ar	Time (h)		er I	er II	s
1-Np	48	(34)	99.0:1.0	75.5:24.5	214
2-Np	32	(38)	99.5:0.5	79.5:20.5	309

186

C₁₆

Conditions: (EtCO)₂O (0.75 eq), **117** (2 mol %), i-Pr₂NEt (0.75 eq), CDCl₃, 0°, 5 h

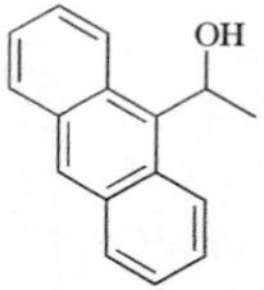

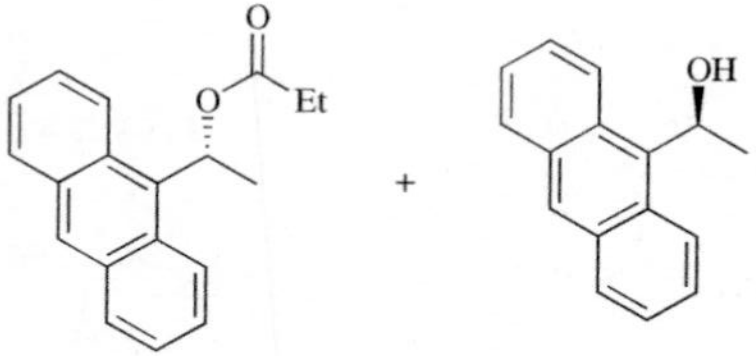

er 91.5:8.5   +   er 98.5:1.5

(54), s = 47

57

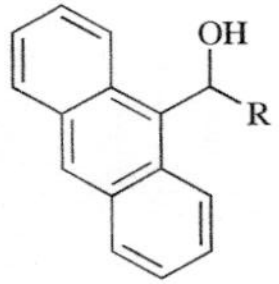

57

R	Time (h)		er I	er II	s
Me	24	(46)	99.0:1.0	91.5:8.5	207
CF$_3$	1	(49)	98.5:1.5	97.0:3.0	243

(EtCO)$_2$O (0.75 eq),
ent-**124** (4 mol %),
i-Pr$_2$NEt (0.75 eq), CDCl$_3$, 0°

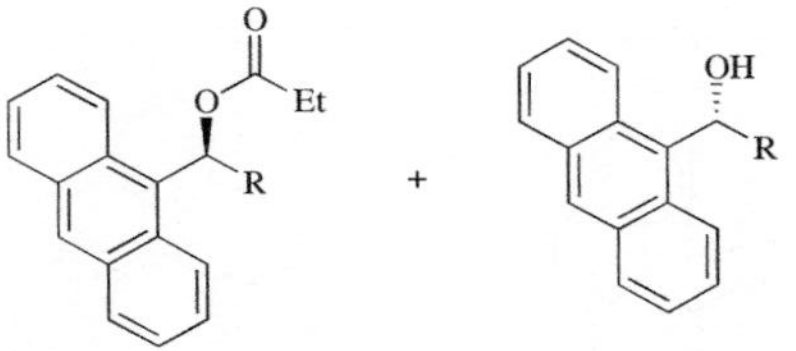

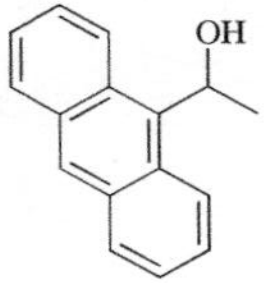

(*i*-PrCO)$_2$O (0.75 eq),
92 (0.5 mol %),
Et$_3$N (0.75 eq), *t*-AmOH, 6 h

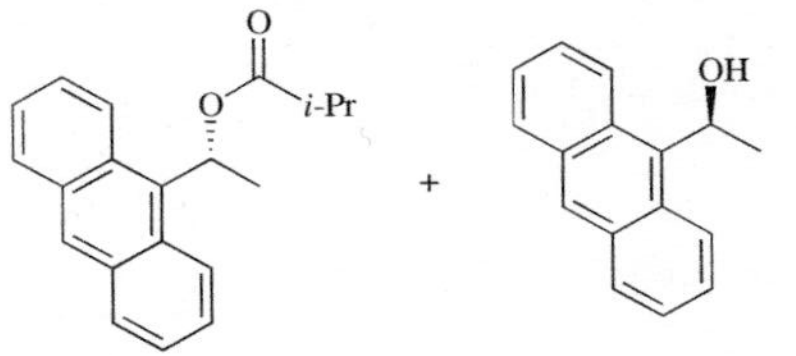

er 98.0:2.0   er 92.0:8.0   (47), *s* = 116   42

(*i*-PrCO)$_2$O (0.75 eq),
69a (5 mol %), Cs$_2$CO$_3$ (0.75 eq),
toluene, –60°, 15 h

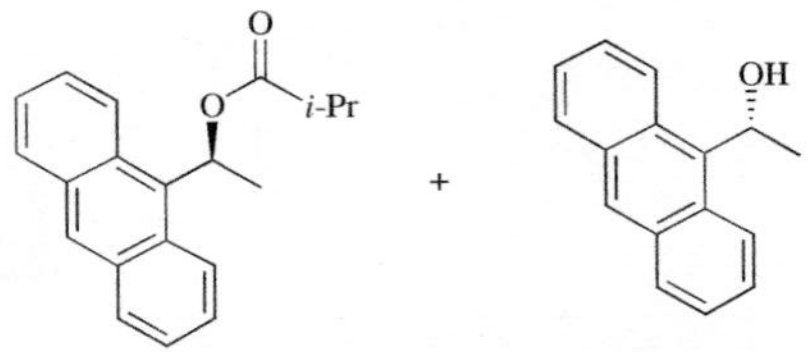

er 71.0:29.0   er 52.5:47.5   (11), *s* = 3   324

TABLE 2B. CATALYTIC KINETIC RESOLUTION OF BENZYLIC ALCOHOLS (*Continued*)

*Please refer to the charts preceding the tables for the structures indicated by the **bold** numbers.*

Alcohol	Conditions	Product(s) and Conversion(s) (%)	Refs.
C₁₆	(*i*-PrCO)₂O (0.7 eq), **84** (5 mol %), Et₃N (0.9 eq), CH₂Cl₂, –78°, 3 h	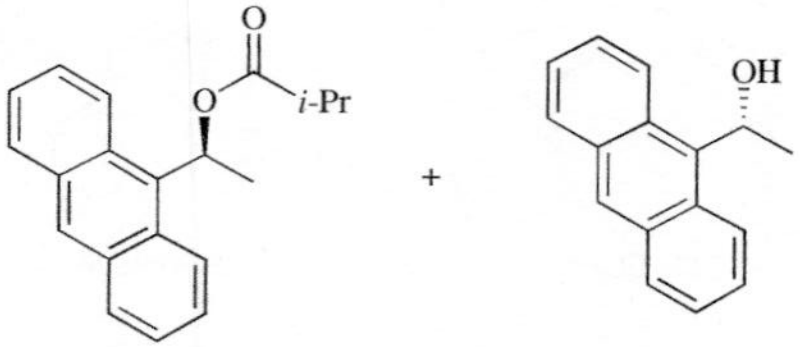 er 70.0:30.0 er 80.0:20.0 (56), *s* = 4	246
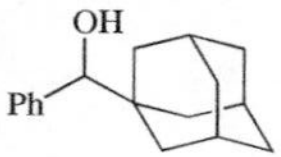 C₁₇	(PhCO)₂O (2.5 eq), **140** (7.1 mol %), toluene, rt, 12 h	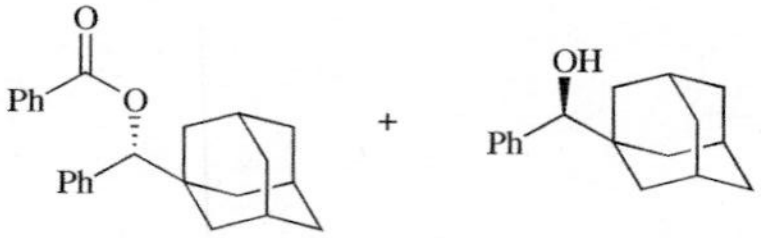 er 89.0:11.0 er 93.5:6.5 (53), *s* = 23	155
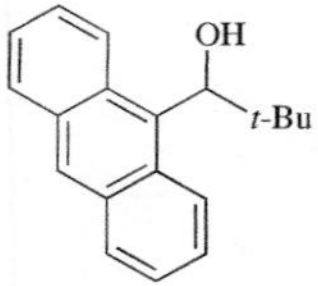 C₁₉	(EtCO)₂O (0.75 eq), **115** (1 mol %), *i*-Pr₂NEt (0.75 eq), toluene, 0°, 96 h	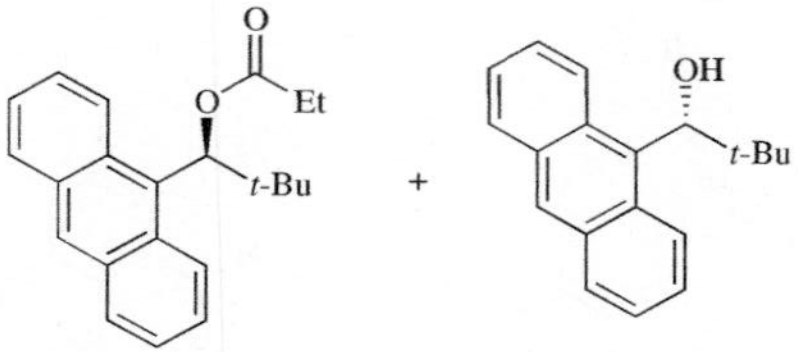 er 88.0:12.0 er >99.5:0.5 (57), *s* = 37	184

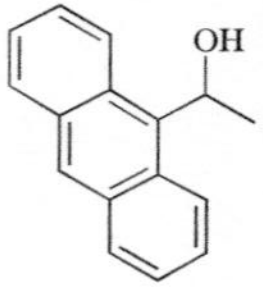

TABLE 2C. CATALYTIC KINETIC RESOLUTION OF ALLYLIC ALCOHOLS

Please refer to the charts preceding the tables for the structures indicated by the **bold** numbers.

Alcohol	Conditions	Product(s) and Conversion(s) (%)	Refs.
C₄			
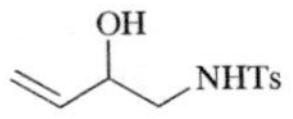	BzCl (0.7 eq), **47b** (2 mol %), K₂CO₃ (1 eq), THF, rt, 16 h	 er 93.0:7.0 + er 90.5:9.5 (49), $s = 32$	248
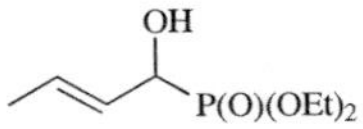	BzCl (0.5 eq), **15** (5 mol %), Cu(OTf)₂ (5 mol%), BaCO₃ (1 eq), chlorobenzene, 0° to rt, 12 h	 er >99.5:0.5 + er 63.5:36.5 (18), $s = 259$	208
C₅			
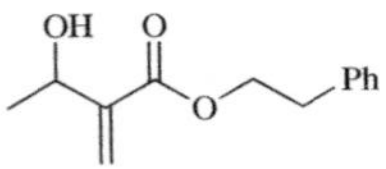	(EtCO)₂O (0.75 eq), **121** (5 mol %), K₂CO₃ (0.5 eq), i-Pr₂O, 0°, 24 h	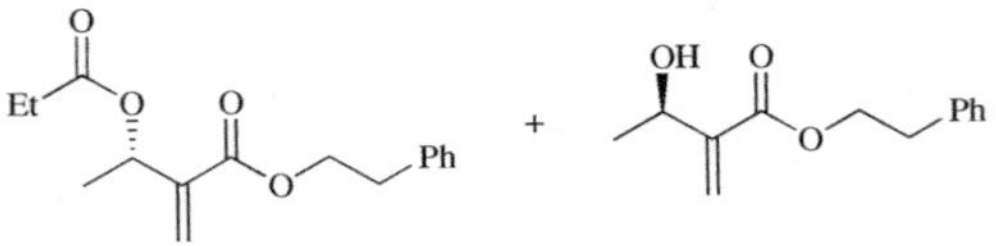 er 95.0:5.0 + er 90.5:9.5 (47), $s = 50$	53
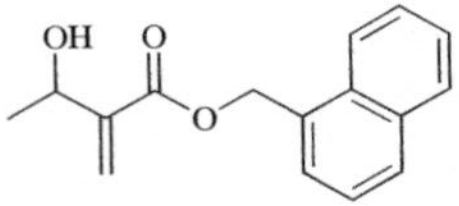	(EtCO)₂O (0.75 eq), **121** (5 mol %), K₂CO₃ (0.5 eq), i-Pr₂O, 0°, 24 h	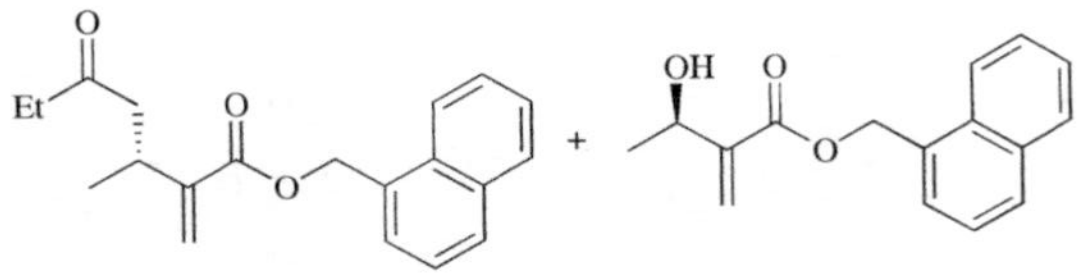 er 92.5:7.5 + er 80.5:19.5 (42), $s = 22$	53

TABLE 2C. CATALYTIC KINETIC RESOLUTION OF ALLYLIC ALCOHOLS (*Continued*)

Alcohol	Conditions	Product(s) and Conversion(s) (%)	Refs.

*Please refer to the charts preceding the tables for the structures indicated by the **bold** numbers.*

C$_{5-12}$

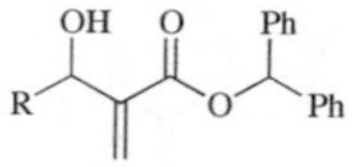

(EtCO)$_2$O (0.75 eq),
121 (5 mol %), K$_2$CO$_3$ (0.5 eq),
i-Pr$_2$O, 0°, 24 h

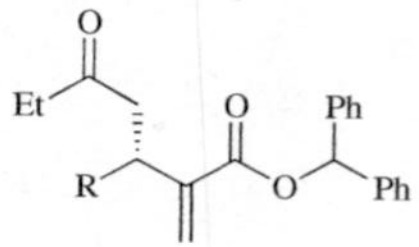

R	(I)	er I	er II	s
Me	(52)	95.5:4.5	99.5:0.5	97
BnOCH$_2$	(61)	67.5:32.5	77.0:23.0	3
TBSOCH$_2$	(50)	93.0:7.0	92.5:7.5	37
Et	(52)	95.0:5.0	98.5:1.5	76
BnO(CH$_2$)$_2$	(60)	82.5:17.5	99.5:0.5	24
TBSO(CH$_2$)$_2$	(52)	95.0:5.0	98.0:2.0	75
n-Pr	(54)	93.0:7.0	99.5:0.5	76
i-Pr	(41)	95.5:4.5	81.5:18.5	40
n-Bu	(53)	94.5:5.5	99.5:0.5	82
i-Bu	(53)	93.5:6.5	>99.5:0.5	108
c-C$_6$H$_{11}$	(42)	94.0:6.0	82.0:18.0	29
BnCH$_2$	(55)	91.5:8.5	99.5:0.5	59

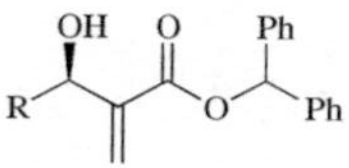

53

C$_6$

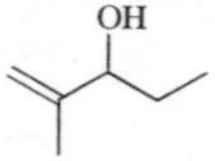

Ac$_2$O (0.75 eq), *ent*-**56** (2 mol %),
Et$_3$N (0.75 eq),
t-AmOH, 0°, 13.5 days

(63), s = 11

er 96.5:3.5

202

Substrate	Conditions	Product (ester)	Product (alcohol)	ref
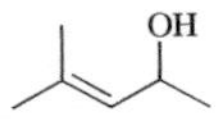 4-methylpent-3-en-2-ol (OH)	(EtCO$_2$O (0.75 eq), **124** (10 mol %), i-Pr$_2$NEt (0.75 eq), CDCl$_3$, 0°, 35 h	ethyl ester, er 67.0:33.0	OH, (40), s = 3, er 61.0:39.0	57
4-methylpent-3-en-2-ol (OH)	(EtCO$_2$O (0.75 eq), **117** (10 mol %), i-Pr$_2$NEt (0.75 eq), CDCl$_3$, 0°, 24 h	ethyl ester, er 84.5:15.5	OH, (46), s = 10, er 79.5:20.5	57
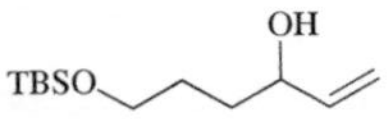 TBSO-substituted homoallylic alcohol (OH)	AcCl (0.6 eq), **157a** (5 mol %), DABCO (0.65 eq), Et$_2$O, –20°, 5 h	TBSO OAc, er 86.0:14.0	TBSO OH, (46), s = 11, er 84.5:15.5	201
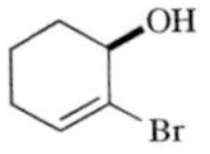 2-bromo-cyclohex-2-enol (OH, Br)	BzCl (0.75 eq), **137** (0.3 mol %), i-Pr$_2$NEt (0.5 eq), 4 Å MS, CH$_2$Cl$_2$, –78°, 3 h	 OBz Br, er 93.5:6.5	OH Br, (—), s = 35, er 98.0:2.0	245
C$_7$ 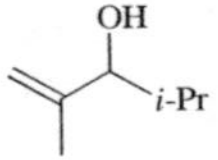 — i-Pr methallyl alcohol (OH, i-Pr)	Ac$_2$O (0.75 eq), *ent*-**56** (2 mol %), Et$_3$N (0.75 eq), t-AmOH, 0°, 7 d	 OAc i-Pr, er 96.5:3.5	OH i-Pr, (58), s = 17, er 96.5:3.5	202

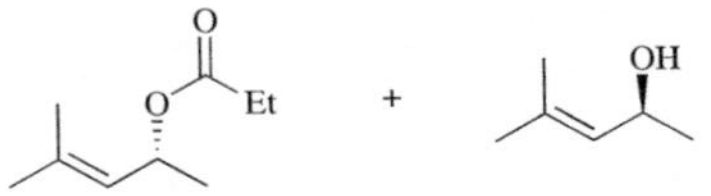

Alcohol	Conditions	Product(s) and Conversion(s) (%)	Refs.

*Please refer to the charts preceding the tables for the structures indicated by the **bold** numbers.*

C_7

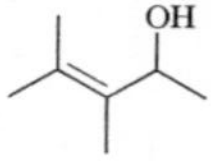

Ac$_2$O (0.75 eq), *ent*-**56** (1 mol %),
Et$_3$N (0.75 eq),
t-AmOH, 0°, 7 days

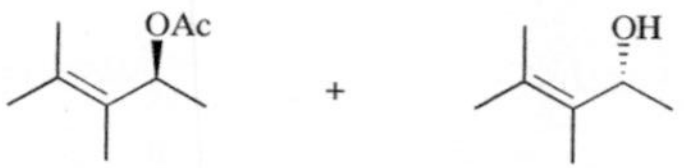

+

(59), s = 29

er 99.5:0.5

202

C_{7-12}

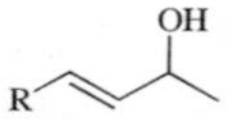

(EtCO)$_2$O (0.75 eq),
124 (10 mol %),
i-Pr$_2$NEt (0.75 eq), CDCl$_3$, 0°

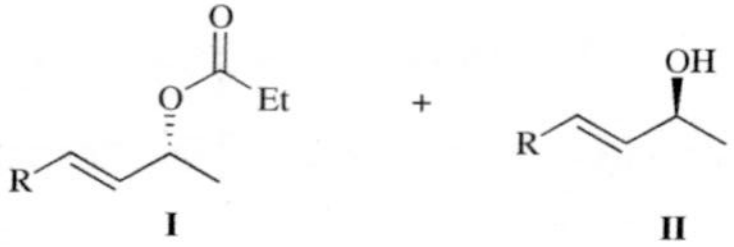

I + **II**

R	Time (h)		er I	er II	s
MeCH=CH	120	(35)	88.5:11.5	71.0:29.0	11
Ph(CH$_2$)$_2$	30	(32)	81.5:18.5	65.0:35.0	6

57

(EtCO)$_2$O (0.75 eq),
117 (*x* mol %),
i-Pr$_2$NEt (0.75 eq), CDCl$_3$, 0°

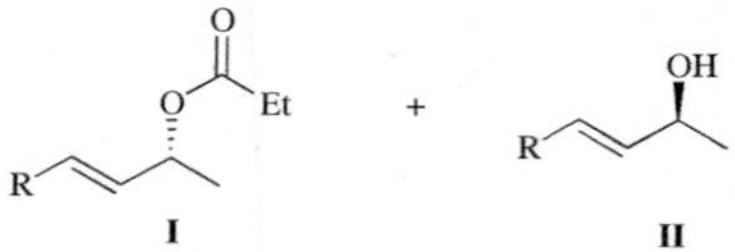

I + **II**

R	*x*	Time (h)		er I	er II	s
MeCH=CH	5	7	(52)	89.0:11.0	92.0:8.0	21
Ph(CH$_2$)$_2$	10	12	(43)	84.0:16.0	76.0:24.0	9

57

C_{7-8}

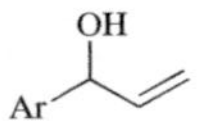

$(i\text{-PrCO})_2O$ (0.6 eq), **126** (1 mol %), $i\text{-Pr}_2NEt$ (0.5 eq), toluene, –78°, 16 h

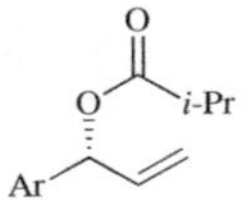

I

\+

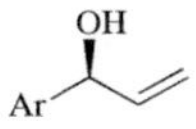

II

205

Ar		er **I**	er **II**	*s*
2-thienyl	(44)	84.0:16.0	76.0:24.0	9
2-pyridyl	(49)	68.0:32.0	67.0:33.0	3
3-pyridyl	(46)	73.0:27.0	69.0:31.0	4

C_8

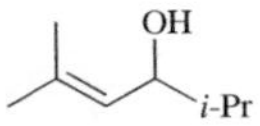

Ac_2O (0.75 eq), *ent*-**56** (2 mol %), Et_3N (0.75 eq), *t*-AmOH, 0°, 7 d

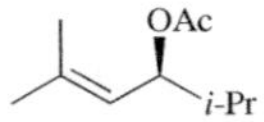

\+

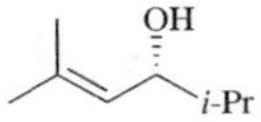

er 98.5:1.5

(60), *s* = 18

202

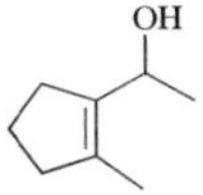

$(i\text{-PrCO})_2O$ (2.5 eq), **141** (5 mol %), heptane, –40°, 46 h

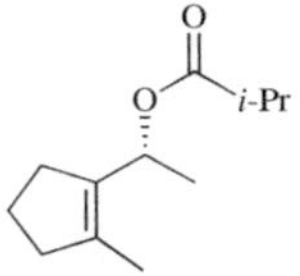

er 74.5:25.5

\+

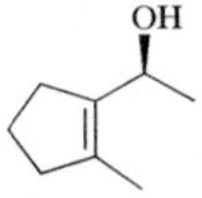

er 99.5:0.5

(67), *s* = 25

207

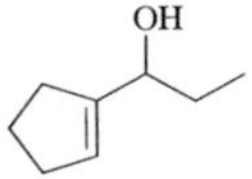

$(i\text{-PrCO})_2O$ (2.5 eq), **141** (5 mol %), toluene, –40°, 46 h

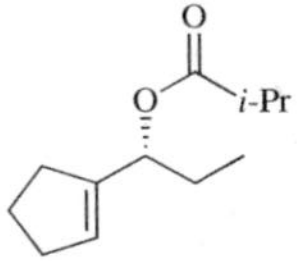

er 97.0:3.0

\+

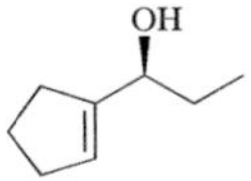

er 78.0:22.0

(38), *s* = 52

207

TABLE 2C. CATALYTIC KINETIC RESOLUTION OF ALLYLIC ALCOHOLS (*Continued*)

Alcohol	Conditions	Product(s) and Conversion(s) (%)	Refs.

*Please refer to the charts preceding the tables for the structures indicated by the **bold** numbers.*

C₈

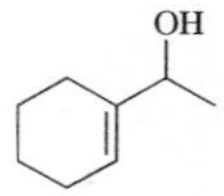

(EtCO)₂O (0.75 eq),
117 (10 mol %),
i-Pr₂NEt (0.75 eq),
CDCl₃, 0°, 11 h

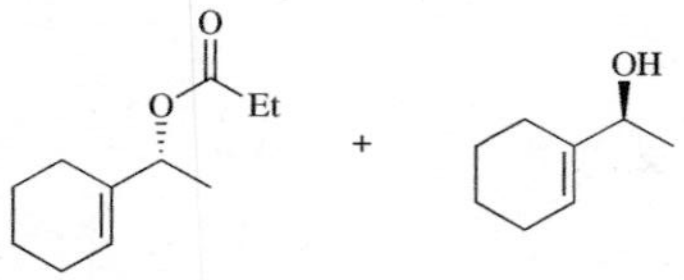

er 89.5:10.5 + er 85.5:14.5

(47), *s* = 17

51

C₈₋₉

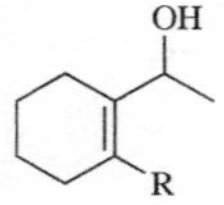

(EtCO)₂O (0.75 eq),
124 (10 mol %),
i-Pr₂NEt (0.75 eq), CDCl₃, 0°

I + **II**

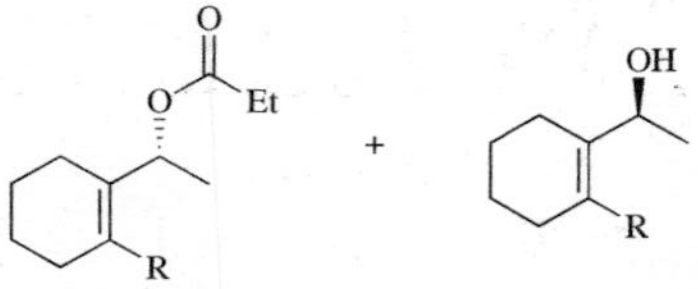

57

R	Time (h)		er I	er II	s
H	47	(25)	89.0:11.0	62.5:37.5	10
Me	35	(23)	62.5:37.5	54.0:46.0	2

C₈

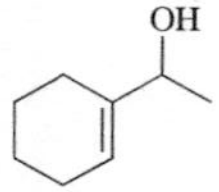

(*i*-PrCO)₂O (2.5 eq),
141 (6.6 mol %),
heptane, –40°, 14 h

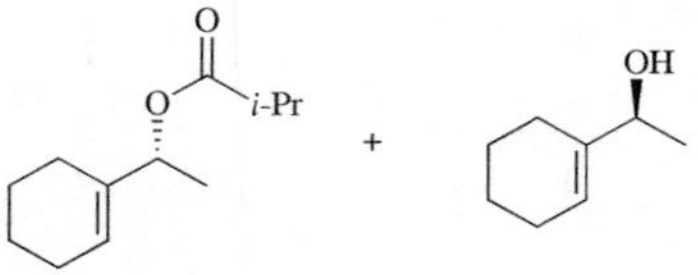

er 94.0:6.0 + er 95.0:5.0

(50), *s* = 49

154

290

(*i*-PrCO)$_2$O (2.5 eq),
142 (5 mol %), Et$_3$N (10 mol %),
heptane, −40°, 13 h

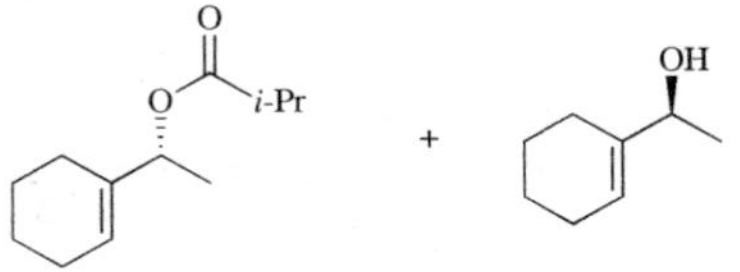

er 91.5:8.5   +   er 95.5:4.5   (52), *s* = 34   156

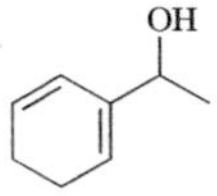

(*i*-PrCO)$_2$O (2.5 eq),
141 (5 mol %),
heptane, −40°, 7 h

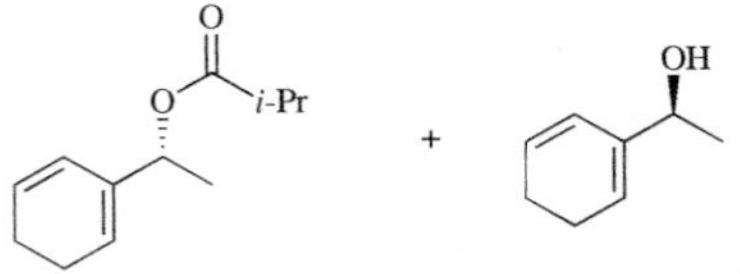

er 90.5:9.5   +   er 95.0:5.0   (53), *s* = 34   207

160 (20 mol %), CHCl$_3$, −20°

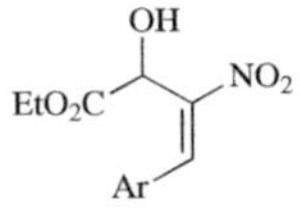

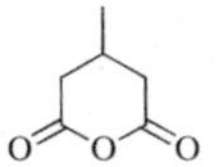

(0.6 eq),

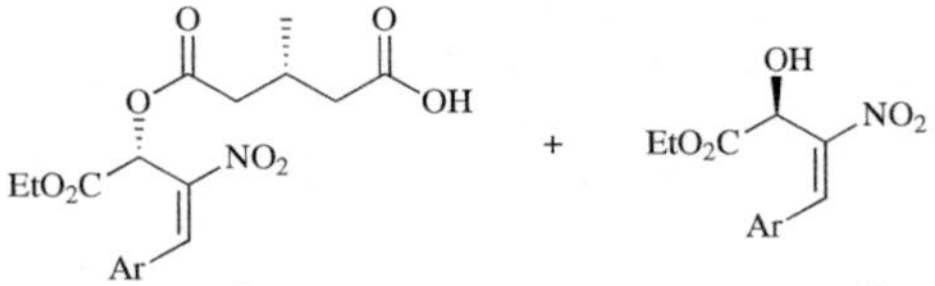

I   +   II   329

Ar	Time (h)	I	dr I	er I	II	er II
2-thienyl	72	(36)	>95.0:5.0	91.0:9.0	(55)	91.5:8.5
Ph	6	(39)	90.0:10.0	96.0:4.0	(49)	97.0:3.0
4-MeC$_6$H$_4$	96	(47)	92.0:8.0	91.5:8.5	(45)	94.0:6.0

C$_{8-11}$

Alcohol	Conditions	Product(s) and Conversion(s) (%)	Refs.

*Please refer to the charts preceding the tables for the structures indicated by the **bold** numbers.*

C_{8–11}

160 (20 mol %), CHCl₃, –20°

Ph lactone (0.6 eq).

I + II

329

Ar	Time (h)	I		dr I	er I	II		er II
2-thienyl	4	(39)		91.0:9.0	96.0:4.0	(37)		87.0:13.0
4-ClC₆H₄	144	(43)		93.0:7.0	>99.5:0.5	(49)		91.0:9.0
2-BrC₆H₄	120	(34)		88.0:12.0	90.0:10.0	(52)		80.5:19.5
3-BrC₆H₄	4	(43)		92.0:8.0	92.0:8.0	(53)		97.5:2.5
4-BrC₆H₄	120	(51)		89.0:11.0	98.0:2.0	(42)		89.5:10.5
4-MeC₆H₄	120	(42)		>95.0:5.0	99.5:0.5	(47)		93.0:7.0

C_{8–14}

(EtCO)₂O (0.75 eq), **115** (2 mol %), i-Pr₂NEt (0.75 eq), toluene, 0°

I + II

204

Ar	Time (h)	I	er I	II	er II	s
2-thienyl	6	(50)	65.0:35.0		65.5:34.5	3
Ph	7	(51)	76.5:23.5		78.0:22.0	6
4-ClC₆H₄	12	(69)	71.5:28.5		98.0:2.0	9
3-BrC₆H₄	10	(66)	71.5:28.5		90.5:9.5	6
2-MeC₆H₄	13	(62)	80.5:19.5		>99.5:0.5	24
4-MeC₆H₄	12	(59)	80.0:20.0		93.0:7.0	11
1-Np	8	(62)	80.5:19.5		99.5:0.5	21
2-Np	10	(62)	78.0:22.0		96.0:4.0	11

C9

Substrate	Conditions	Product(s)	Refs.
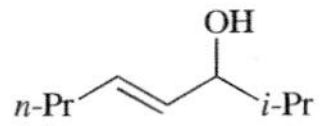	Ac$_2$O (0.9 eq), *ent*-**56** (2 mol %), Et$_3$N (0.9 eq), *t*-AmOH, 0°, 28 d	 (75), *s* = 5 er 96.0:4.0	202
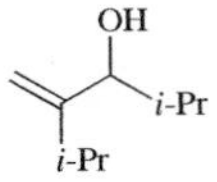	Ac$_2$O (1.5 eq), *ent*-**56** (2 mol %), Et$_3$N (0.8 eq), *t*-AmOH, 0°, 14.5 d	(63), *s* = 10 er 96.0:4.0	202
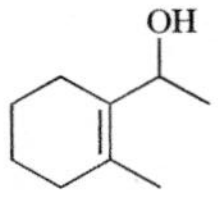	(EtCO)$_2$O (0.75 eq), **117** (10 mol %), *i*-Pr$_2$NEt (0.75 eq), CDCl$_3$, 0°, 45 h	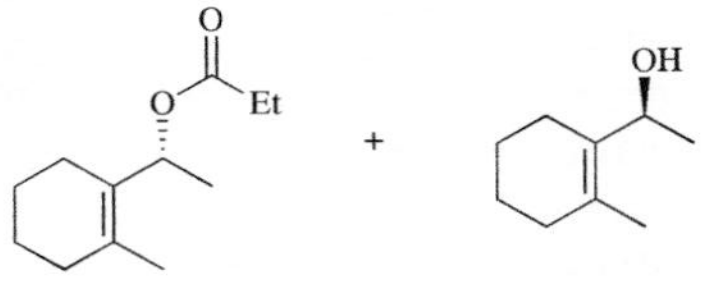 (35), *s* = 14 er 90.0:10.0 er 71.5:28.5	57
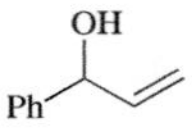	(EtCO)$_2$O (0.75 eq), **115** (2 mol %), *i*-Pr$_2$NEt (0.75 eq), toluene, 0°, 8 h	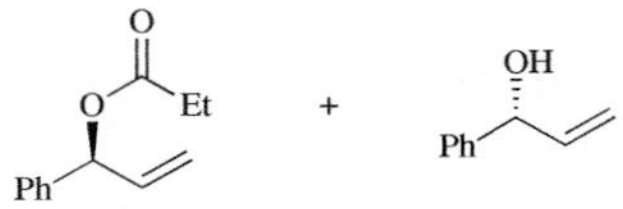 (55), *s* = 10 er 81.0:19.0 er 77.5:22.5	204

TABLE 2C. CATALYTIC KINETIC RESOLUTION OF ALLYLIC ALCOHOLS (*Continued*)

Alcohol	Conditions	Product(s) and Conversion(s) (%)	Refs.

*Please refer to the charts preceding the tables for the structures indicated by the **bold** numbers.*

C_{9-15}

Alcohol: OH, Ar

Conditions:
(i-PrCO)$_2$O (0.6 eq),
126 (1 mol %), i-Pr$_2$NEt (0.5 eq),
toluene, −78°, 16 h

Products: **I** (ester, i-Pr-C(O)O, Ar) + **II** (OH, Ar)

Refs.: 205

Ar		er I	er II	s
Ph	(41)	95.0:5.0	82.0:18.0	35
2-MeOC$_6$H$_4$	(52)	—	95.0:5.0	36
3-MeOC$_6$H$_4$	(43)	96.0:4.0	86.0:14.0	59
4-MeOC$_6$H$_4$	(50)	92.0:8.0	92.0:8.0	29
2,6-(MeO)$_2$C$_6$H$_3$	(37)	99.0:1.0	78.0:22.0	110
3,4-(MeO)$_2$C$_6$H$_3$	(60)	80.0:20.0	99.5:0.5	44
3,4,5-(MeO)$_3$C$_6$H$_2$	(51)	92.0:8.0	94.0:6.0	33
3-FC$_6$H$_4$	(54)	84.0:16.0	89.0:11.0	12
4-ClC$_6$H$_4$	(48)	91.0:9.0	87.0:13.0	17
4-BrC$_6$H$_4$	(35)	84.0:16.0	68.0:32.0	8
2-CF$_3$C$_6$H$_4$	(37)	82.0:18.0	68.0:32.0	7
3-CF$_3$C$_6$H$_4$	(50)	88.0:12.0	86.0:14.0	15
4-CF$_3$C$_6$H$_4$	(52)	82.0:18.0	83.0:17.0	8
3-CH$_2$=CHC$_6$H$_4$	(48)	92.0:8.0	89.0:11.0	26
Mes	(22)	90.0:10.0	61.0:39.0	11
1-Np	(46)	98.0:2.0	92.0:8.0	108
2-Np	(49)	99.5:0.5	97.0:3.0	1980
4-PhC$_6$H$_4$	(42)	88.0:12.0	78.0:22.0	13

C₉

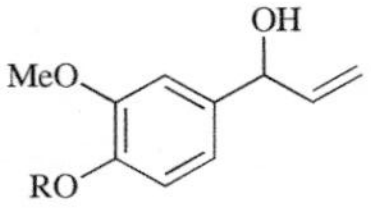

(i-PrCO)₂O (0.55–0.7 eq),
126 (1 mol %), i-Pr₂NEt (0.6 eq),
toluene, −78°, 16 h

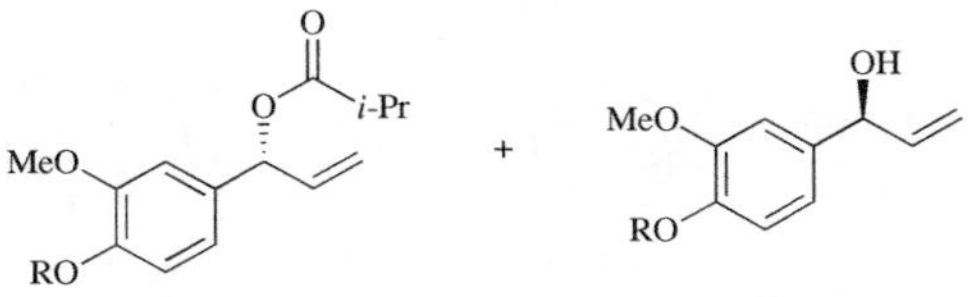

206

R		er **I**	er **II**	s
TBS	(52)	90.5:9.5	94.5:5.5	29
Piv	(53)	92.5:7.5	98.5:1.5	49

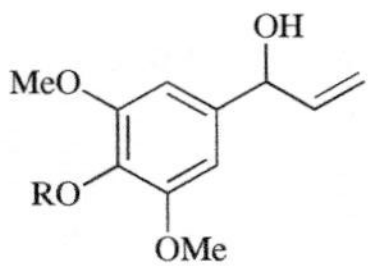

(i-PrCO)₂O (0.55–0.7 eq),
126 (1 mol %), i-Pr₂NEt (0.6 eq),
solvent, −78°, 16 h

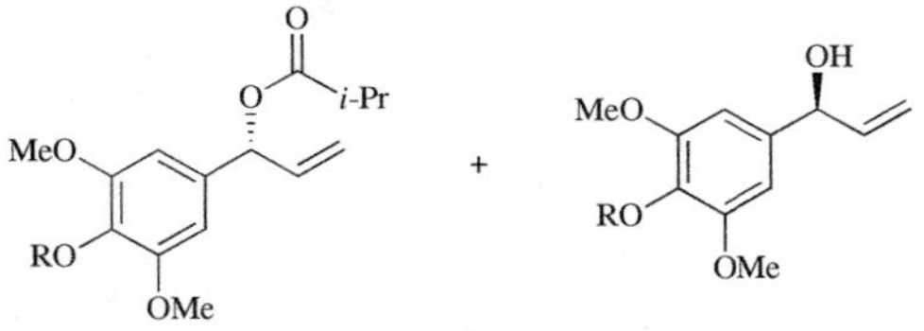

206

R	Solvent		er **I**	er **II**	s
TBS	toluene	(56)	89.0:11.0	99.5:0.5	43
Piv	toluene	(57)	87.5:12.5	99.5:0.5	64
Tf	toluene	(44)	85.5:14.5	77.5:22.5	10
Ts	THF	(56)	86.0:14.0	96.5:3.5	20

TABLE 2C. CATALYTIC KINETIC RESOLUTION OF ALLYLIC ALCOHOLS (*Continued*)

*Please refer to the charts preceding the tables for the structures indicated by the **bold** numbers.*

Alcohol	Conditions	Product(s) and Conversion(s) (%)	Refs.
C$_{10}$			
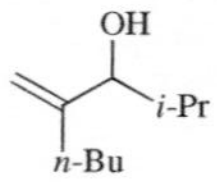 (OH, *i*-Pr, *n*-Bu)	Ac$_2$O (0.75 eq), *ent*-**56** (1 mol %), Et$_3$N (0.75 eq), *t*-AmOH, 0°, 12 d	 (OAc, *i*-Pr, *n*-Bu) + (OH, *i*-Pr, *n*-Bu) (55), *s* = 25 — er 97.0:3.0	202
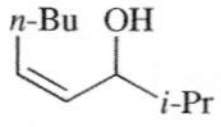 (*n*-Bu, OH, *i*-Pr)	Ac$_2$O (0.9 eq), *ent*-**56** (2 mol %), Et$_3$N (0.9 eq), *t*-AmOH, 0°, 21 d	 (*n*-Bu, OAc, *i*-Pr) + (*n*-Bu, OH, *i*-Pr) (73), *s* = 5 — er 95.0:5.0	202
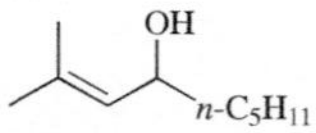 (OH, *n*-C$_5$H$_{11}$)	Ac$_2$O (0.74 eq), *ent*-**56** (2.4 mol %), Et$_3$N (0.74 eq), *t*-AmOH, 0°, 7 d	 (OAc, *n*-C$_5$H$_{11}$) + (OH, *n*-C$_5$H$_{11}$) (66), *s* = 12 — er 98.5:1.5	202
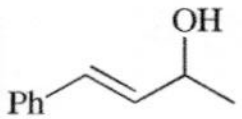 (OH, Ph)	Ac$_2$O (1.2 eq), **56** (2 mol %), Et$_3$N, Et$_2$O, rt, 31 h	 (OAc, Ph) er 74.5:25.5 + (OH, Ph) er 99.5:0.5 (67), *s* = 14	159
(OH, Ph)	Ac$_2$O (0.75 eq), *ent*-**56** (1 mol %), Et$_3$N (0.75 eq), *t*-AmOH, 0°, 15 h	 (OAc, Ph) + (OH, Ph) (54), *s* = 64 — er 99.5:0.5	202

C_{10-12}

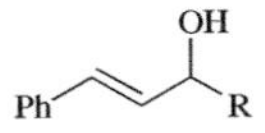

AcOC(Me)=CH$_2$ (1.27 eq),
50 (2 mol %),
toluene, –10°, 9 h

er 56.5:43.5

(42), s = 2

101

(EtCO)$_2$O (0.75 eq), **115** (2 mol %),
i-Pr$_2$NEt (0.75 eq),
toluene, 0°, 8 h

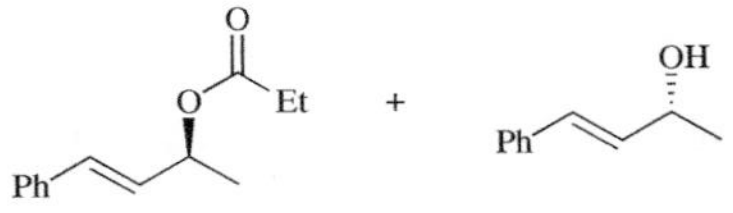

er 92.5:7.5   er 68.0:32.0

(30), s = 18

49

(EtCO)$_2$O (0.5 eq), **126** (0.75 mol %),
i-Pr$_2$NEt (0.6 eq),
CHCl$_3$, 0°, 1 h

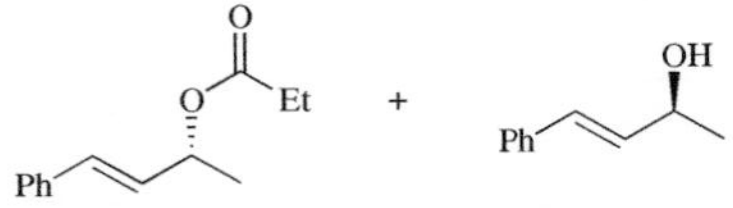

er 91.5:8.5   er 86.5:13.5

(47), s = 22

60

(EtCO)$_2$O (0.75 eq), **117** (2 mol %),
i-Pr$_2$NEt (0.75 eq),
CHCl$_3$, 0°, 8 h

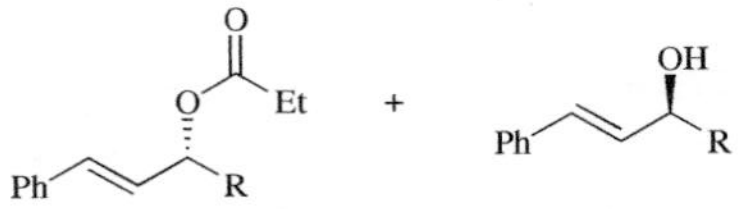

I   **II**

R		er **I**	er **II**	s
Me	(44)	93.0:7.0	83.0:17.0	27
i-Pr	(51)	90.5:9.5	92.5:7.5	26

51

TABLE 2C. CATALYTIC KINETIC RESOLUTION OF ALLYLIC ALCOHOLS (Continued)

Alcohol	Conditions	Product(s) and Conversion(s) (%)	Refs.

*Please refer to the charts preceding the tables for the structures indicated by the **bold** numbers.*

C_{10}

Alcohol	Conditions	Product(s) and Conversion(s) (%)	Refs.
(OH allylic alcohol, Ph)	$(i\text{-PrCO})_2O$ (0.8 eq), **76** (0.5 mol %), Et_3N (0.9 eq), $CHCl_3$, –30°, 48 h	(ester, i-Pr) er 75.5:24.5 (15) + (OH) er 97.0:3.0 (65), $s = 10$	168
	$(i\text{-PrCO})_2O$ (0.55 eq), **155** (1 mol %), $i\text{-Pr}_2NEt$ (0.6 eq), $CHCl_3$, 0°, 7 h	(ester, i-Pr) er 89.0:11.0 (42) + (OH) er 85.0:15.0 (47), $s = 17$	196
	$Ph_2CHCO_2CH=CH_2$ (0.75 eq), **146** (5 mol %), THF, –78°, 3.5 h	(ester, Ph Ph) er 92.0:8.0 + (OH) er 65.5:34.5 (—), $s = 16$	76

C_{10-12}

Alcohol	Conditions	Product(s) and Conversion(s) (%)	Refs.
(OH allylic alcohol, R)	$(EtCO)_2O$ (0.75 eq), **124** (10 mol %), $i\text{-Pr}_2NEt$ (0.75 eq), $CDCl_3$, rt	**I** (Et ester) + **II** (OH)	57

R	Time (h)		er I	er II	s
Ph	35	(15)	79.0:21.0	55.0:45.0	4
$n\text{-C}_8H_{17}$	24	(42)	78.5:21.5	69.5:30.5	5

57

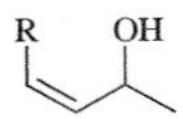

(EtCO)₂O (0.75 eq), **117** (x mol %),
i-Pr₂NEt (0.75 eq), CDCl₃, 0°

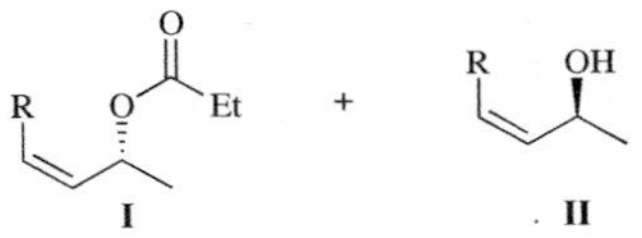

I   +   II

R	x	Time (h)	er I	er II	s
Ph	5	23	(43) 91.5:8.5	(56) 83.0:17.0	22
n-C₈H₁₇	10	6	(47) 84.0:16.0	80.5:19.5	10

$n\text{-}C_8H_{17}$

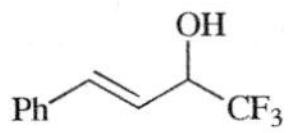

(i-PrCO)₂O (1 eq), **124** (4 mol %),
i-Pr₂O, 0°, 1.5 h

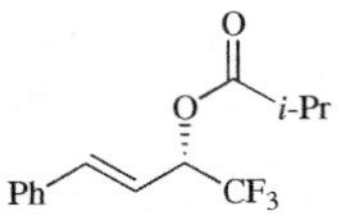

er 72.5:27.5 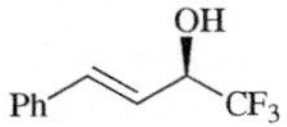   +   er 78.5:21.5   (56), s = 5   191

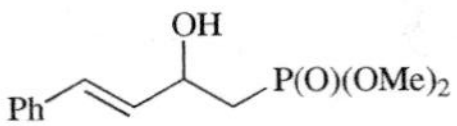

2-FC₆H₄COCl (0.5 eq), **15** (5 mol %),
Cu(OTf)₂ (5 mol%), K₂CO₃ (1 eq),
CH₂Cl₂, 0° to rt, 20 h

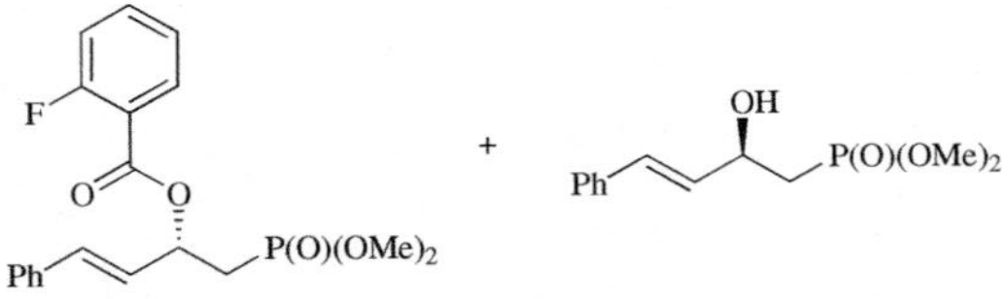

er 79.0:21.0   +   er 69.0:31.0   98

(—), s = 5

C₁₀

TABLE 2C. CATALYTIC KINETIC RESOLUTION OF ALLYLIC ALCOHOLS (*Continued*)

Alcohol	Conditions	Product(s) and Conversion(s) (%)	Refs.

*Please refer to the charts preceding the tables for the structures indicated by the **bold** numbers.*

C$_{10-14}$

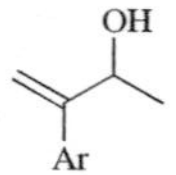

Conditions: (EtCO)$_2$O (0.75 eq), **120** (5 mol %), *i*-Pr$_2$NEt (0.75 eq), MTBE/CHCl$_3$ (1:1), 0°, 24 h

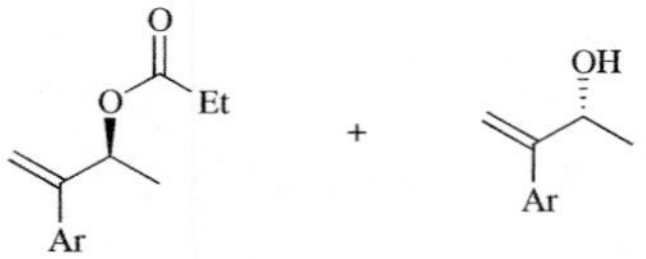

Refs. 203

Ar		er I	er II	s
Ph	(49)	91.0:9.0	89.5:10.5	23
2-MeOC$_6$H$_4$	(22)	92.0:8.0	62.0:38.0	15
3-MeOC$_6$H$_4$	(37)	93.0:7.0	75.0:25.0	22
4-MeOC$_6$H$_4$	(47)	94.0:6.0	89.0:11.0	37
3,4-(MeO)$_2$C$_6$H$_3$	(42)	94.0:6.0	82.0:18.0	30
3,4,5-(MeO)$_3$C$_6$H$_2$	(43)	91.5:8.5	81.5:18.5	21
4-ClC$_6$H$_4$	(48)	91.0:9.0	87.0:13.0	22
4-O$_2$NC$_6$H$_4$	(53)	86.0:14.0	91.5:8.5	16
4-MeC$_6$H$_4$	(37)	93.0:7.0	75.0:25.0	22
1-Np	(39)	88.0:12.0	74.0:26.0	12
2-Np	(45)	93.0:7.0	85.5:14.5	28

C$_{10}$

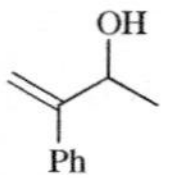

Conditions: (*i*-PrCO)$_2$O (2.5 eq), **141** (5 mol %), toluene, −40°, 19 h

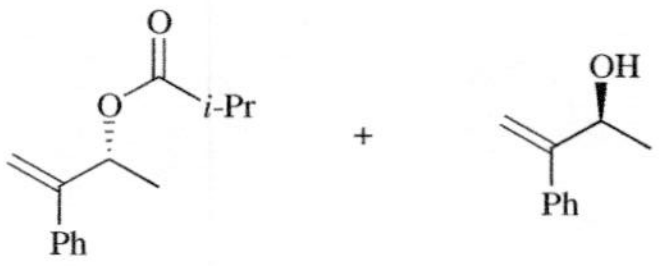

er 87.0:13.0   +   er 73.0:27.0   (38), *s* = 11

Refs. 207

300

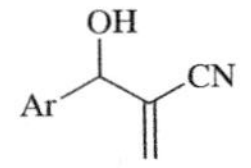

(i-PrCO)₂O (1.5 eq),
72 (1 mol %), Et₃N (0.8 eq),
CH₂Cl₂, −78°, 8 h

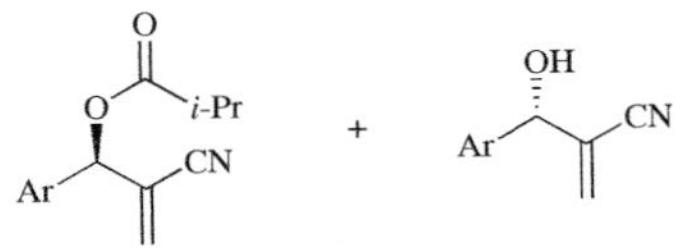

167

Ar		er **II**	s
Ph	(85)	96.5:3.5	4
2-MeOC₆H₄	(78)	91.0:9.0	4

(i-PrCO)₂O (1.5 eq), **72** (1 mol %),
Et₃N (0.8 eq), CH₂Cl₂, −78°, 8 h

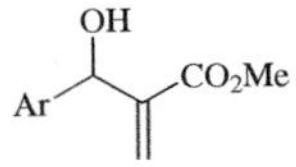

167

Ar		er **II**	s
Ph	(83)	97.5:2.5	4
2-MeOC₆H₄	(64)	98.5:1.5	13

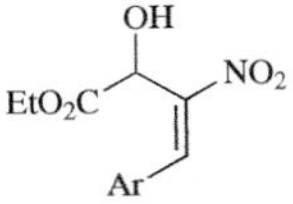

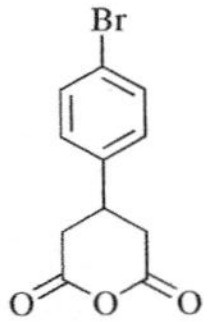

(0.6 eq),

160 (20 mol %), CHCl₃, −20°

329

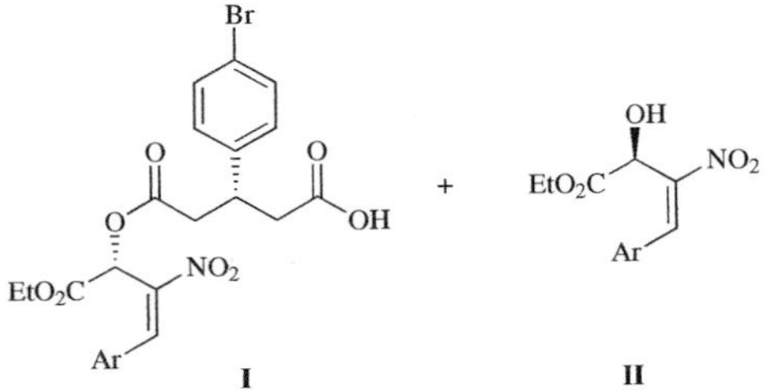

Ar	Time (h)	**I**	dr **I**	er **I**	**II**	er **II**
Ph	96	(53)	94.0:6.0	96.5:3.5	(40)	93.5:6.5
4-BrC₆H₄	120	(54)	90.0:10.0	96.0:4.0	(36)	97.0:3.0

301

TABLE 2C. CATALYTIC KINETIC RESOLUTION OF ALLYLIC ALCOHOLS (*Continued*)

Alcohol	Conditions	Product(s) and Conversion(s) (%)	Refs.

*Please refer to the charts preceding the tables for the structures indicated by the **bold** numbers.*

C$_{10}$

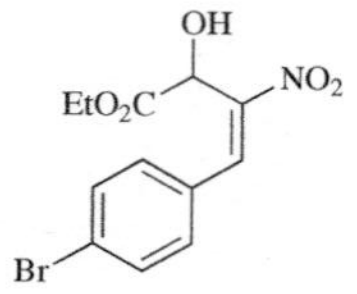

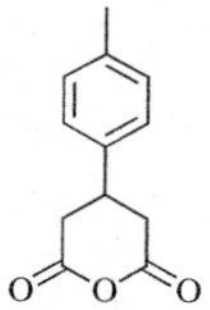

160 (20 mol %),
CHCl$_3$, −20°, 120 h

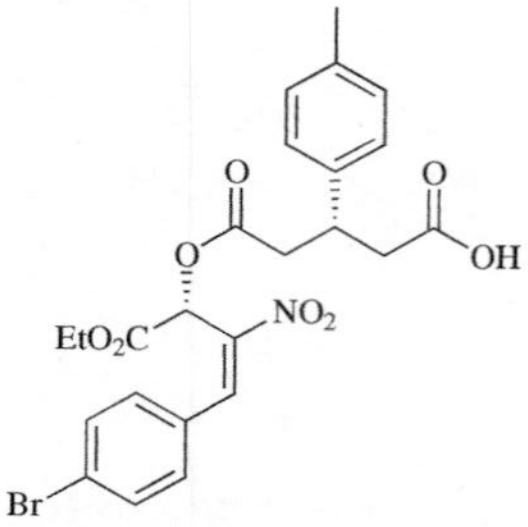

(50), dr 91.0:9.0, er 97.5:2.5

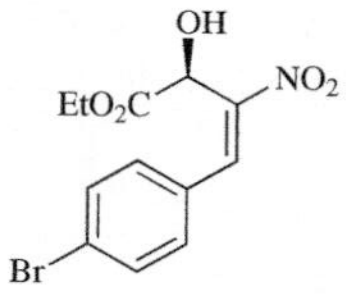

(41), er 92.0:8.0

329

C$_{10–11}$

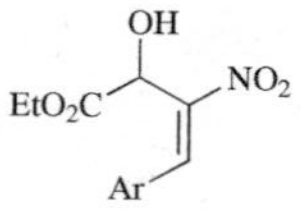

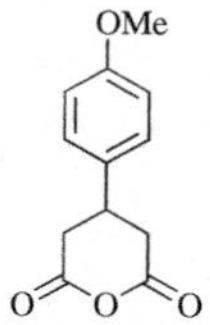

160 (20 mol %),
CHCl$_3$, −20°, 120 h

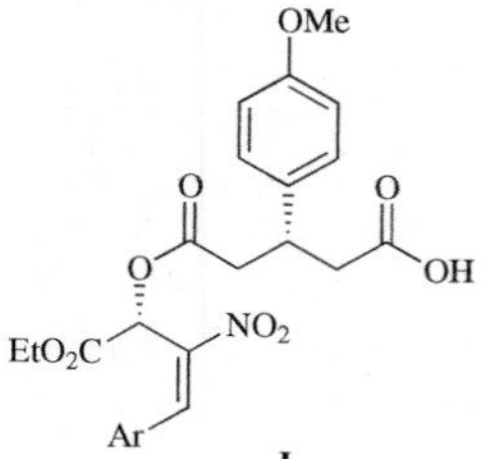

Ar	I	dr I	er I	II	er II
3-BrC$_6$H$_4$	(42)	91.0:9.0	96.0:4.0	(45)	96.0:4.0
4-MeC$_6$H$_4$	(40)	>95.0:5.0	98.0:2.0	(49)	91.0:9.0

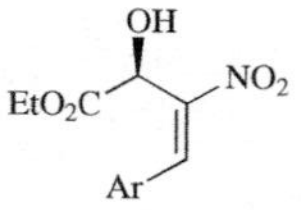

329

C$_{10}$

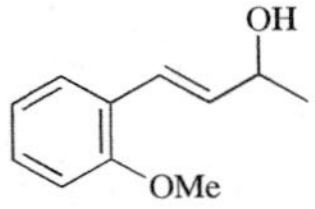

(EtCO)$_2$O (0.75 eq), **117** (2 mol %),
i-Pr$_2$NEt (0.75 eq), CHCl$_3$, 0°, 8 h

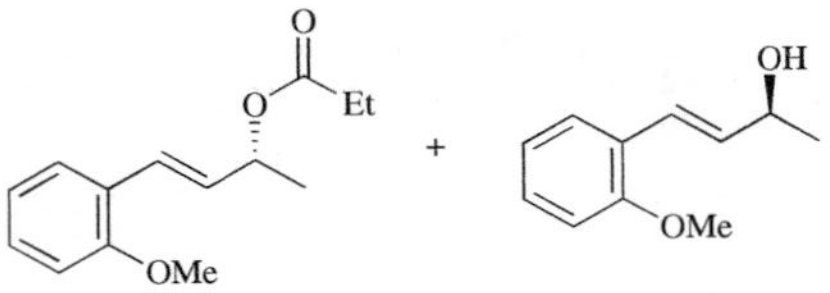

er 89.0:11.0   er 98.5:1.5

(56), *s* = 33   51

C$_{10–14}$

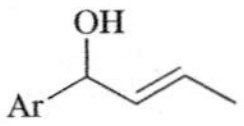

(*i*-PrCO)$_2$O (0.6 eq),
126 (1 mol %), *i*-Pr$_2$NEt (0.5 eq),
toluene, −78°, 16 h

I   **II**

Ar	er **I**	er **II**	*s*
PMP	(57) 65.0:35.0	73.0:27.0	—
2-Np	(47) 85.0:15.0	81.0:19.0	11

205

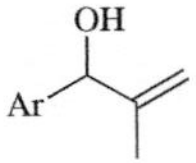

C$_{10–14}$ *(structure)*

(*i*-PrCO)$_2$O (0.6 eq),
126 (1 mol %), *i*-Pr$_2$NEt (0.5 eq),
toluene, −78°, 16 h

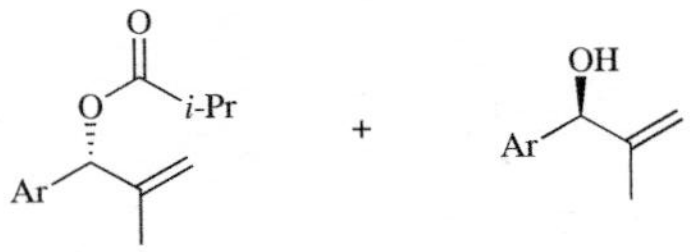

I   **II**

Ar	er **I**	er **II**	*s*
PMP	(45) 84.0:16.0	79.0:21.0	10
2-Np	(47) 92.0:8.0	86.0:14.0	24

205

TABLE 2C. CATALYTIC KINETIC RESOLUTION OF ALLYLIC ALCOHOLS (Continued)

Please refer to the charts preceding the tables for the structures indicated by the bold numbers.

Alcohol	Conditions	Product(s) and Conversion(s) (%)	Refs.
C10			
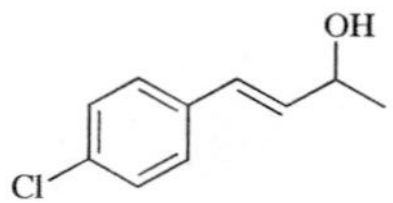	(i-PrCO)₂O (2.5 eq), **141** (5 mol %), toluene, –40°, 12 h	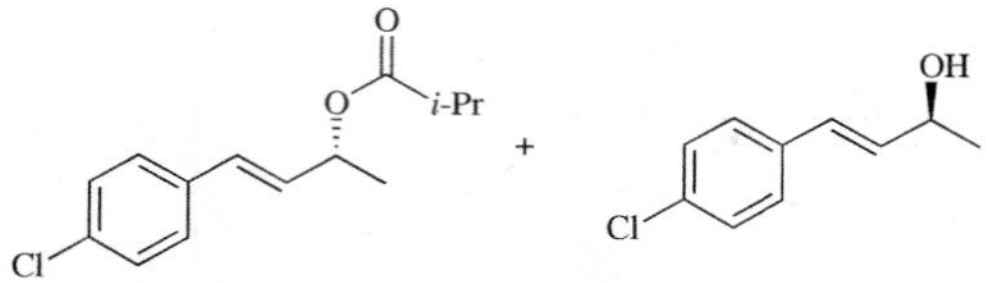 er 81.5:18.5 (55), s = 10 / er 89.0:11.0	207
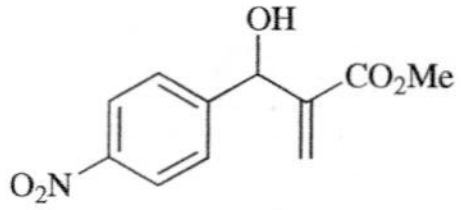	AcCl (0.6 eq), **157a** (5 mol %), DABCO (0.65 eq), Et₂O, –20°, 5 h	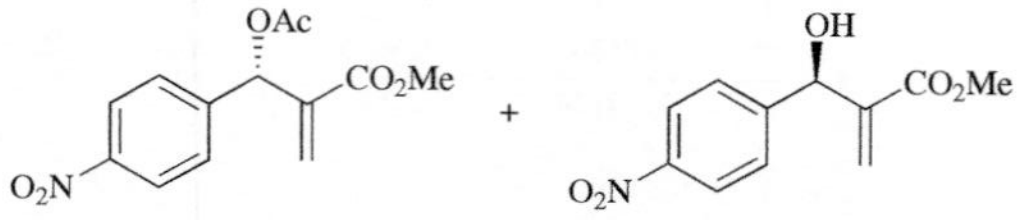 er 92.5:7.5 (33), s = 19 / er 71.0:29.0	201
C11			
	Ac₂O (0.9 eq), **ent-56** (2 mol %), Et₃N (0.9 eq), t-AmOH, 0°, 28 d	 er 95.0:5.0 (7), s = 5	202
	Ac₂O (1.2 eq), **56** (2 mol %), Et₃N, Et₂O, rt, 25 h	 er 81.5:18.5 / er 99.5:0.5 (61), s = 22	159

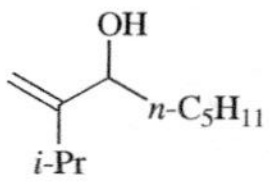

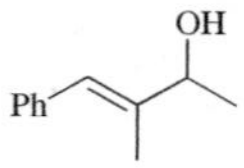

C$_{11-13}$

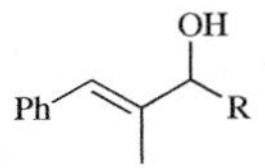

Ac$_2$O (0.75 eq), *ent*-**56** (1 mol %), Et$_3$N (0.75 eq), *t*-AmOH, 0°, 15 h

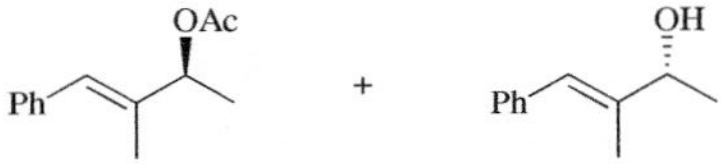

er 99.0:1.0

(53), $s = 80$ — 202

(EtCO)$_2$O (0.75 eq), **117** (2 mol %), *i*-Pr$_2$NEt (0.75 eq), CHCl$_3$, 0°, 8 h

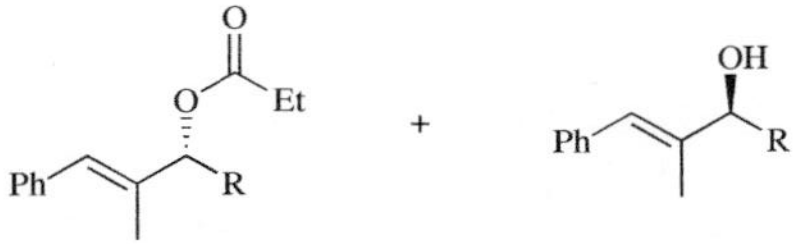

I + **II** — 51

R		er **I**	er **II**	*s*
Me	(39)	91.0:9.0	76.5:23.5	17
i-Pr	(31)	94.0:6.0	70.0:30.0	24

C$_{11}$

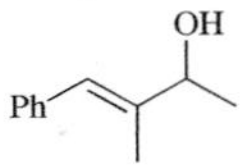

(*i*-PrCO)$_2$O (2.5 eq), **141** (5 mol %), toluene, –40°, 27 h

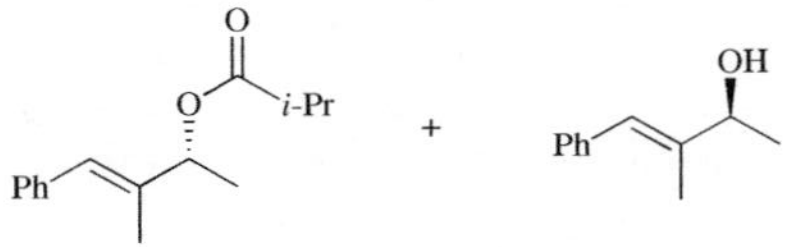

er 91.0:9.0 + er 83.5:16.5

(45), $s = 21$ — 207

Ph$_2$CHCO$_2$CH=CH$_2$ (0.75 eq), **146** (5 mol %), THF, –78°, 1 h

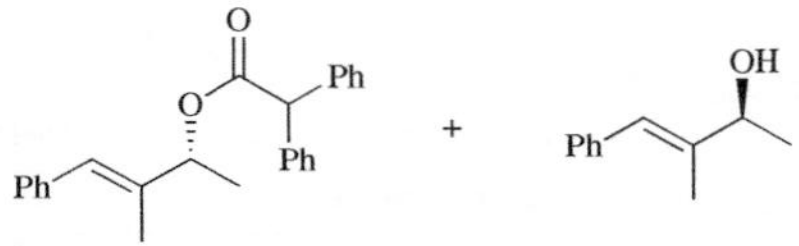

er 93.5:6.5 + er 71.0:29.0

(—), $s = 22$ — 76

Alcohol	Conditions	Product(s) and Conversion(s) (%)	Refs.

*Please refer to the charts preceding the tables for the structures indicated by the **bold** numbers.*

C₁₁

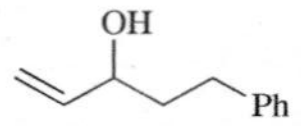

(*i*-PrCO)₂O (0.6 eq),
126 (1 mol %), *i*-Pr₂NEt (0.5 eq),
toluene, −78°, 16 h

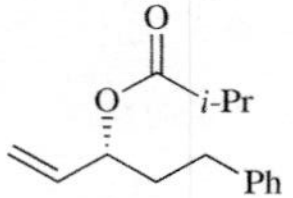

er 68.0:32.0   er 61.0:39.0

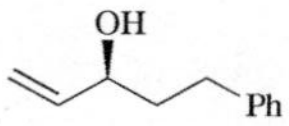

(40), *s* = 3   205

C₁₁₋₁₂

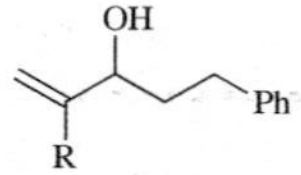

(*i*-PrCO)₂O (2.5 eq),
141 (5 mol %),
toluene, −40°

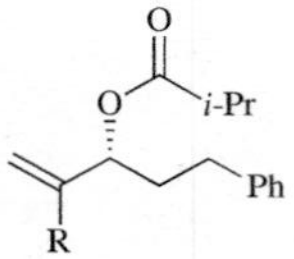

R	Time (h)		er **I**	er **II**	*s*
H	41	(48)	72.5:27.5	71.0:29.0	4
Me	19	(48)	86.0:14.0	83.0:17.0	12

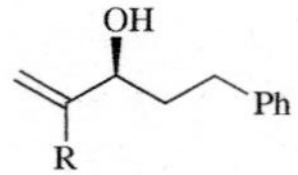

207

C₁₁

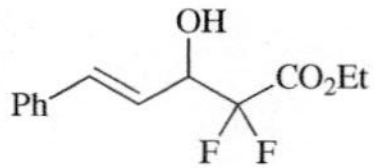

(EtCO)₂O (0.66 eq),
124 (4 mol %),
CHCl₃, rt, 3 h

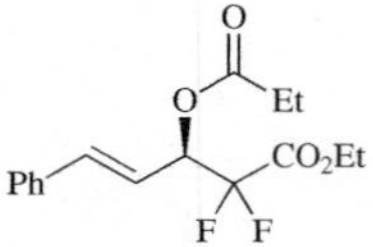

er 68.0:32.0   er 69.5:30.5

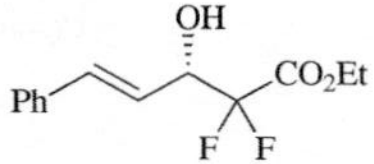

(52), *s* = 3   190

C₁₁₋₁₅

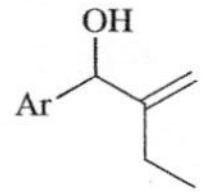

(EtCO)₂O (0.75 eq), **115** (2 mol %),
i-Pr₂NEt (0.75 eq),
toluene, 0°

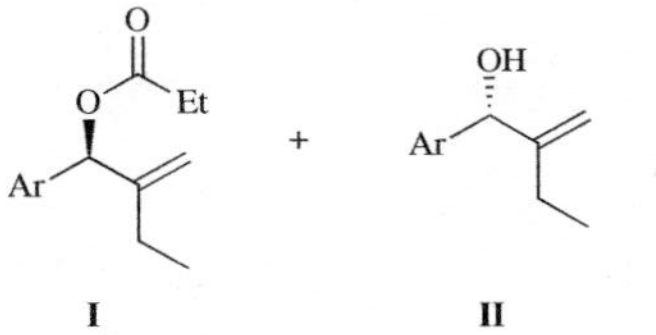

I + **II**

Ar	Time (h)		er **I**	er **II**	s
Ph	9	(60)	73.5:26.5	85.0:15.0	6
2-MeC₆H₄	3	(61)	81.5:18.5	99.5:0.5	23
1-Np	7	(64)	78.0:22.0	99.5:0.5	18
2-Np	8	(65)	75.5:24.5	97.5:2.5	11

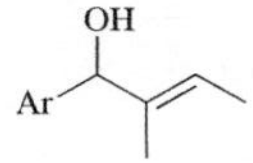

(i-PrCO)₂O (0.6 eq),
126 (1 mol %), i-Pr₂NEt (0.5 eq),
toluene, −78°, 16 h

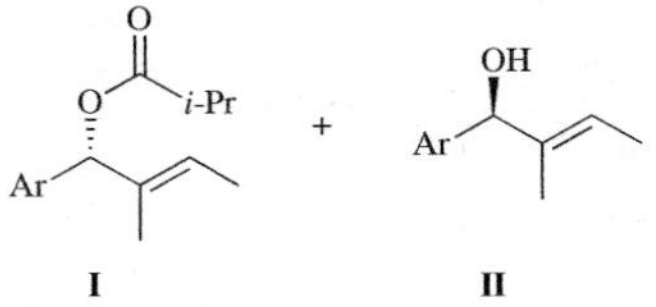

I + **II**

Ar		er **I**	er **II**	s
PMP	(38)	70.0:30.0	62.0:38.0	3
2-Np	(53)	80.0:20.0	84.0:16.0	8

TABLE 2C. CATALYTIC KINETIC RESOLUTION OF ALLYLIC ALCOHOLS (*Continued*)

Alcohol	Conditions	Product(s) and Conversion(s) (%)	Refs.

*Please refer to the charts preceding the tables for the structures indicated by the **bold** numbers.*

C₁₁

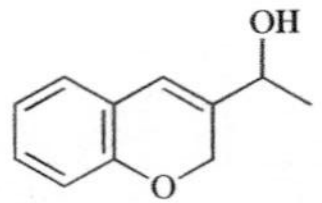

(EtCO)₂O (0.75 eq), **117** (2 mol %),
i-Pr₂NEt (0.75 eq), CHCl₃, 0°, 4 h

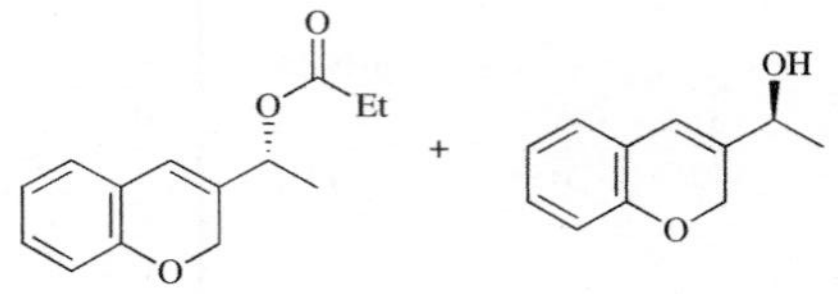

er 71.0:29.0

er 90.5:9.5 + er 99.5:0.5

(55), s = 58   51

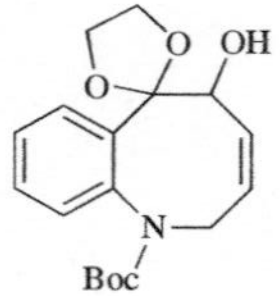

Ac₂O (5 eq), **98** (2 mol %),
Et₃N (6 eq),
toluene, 0°, 72 h

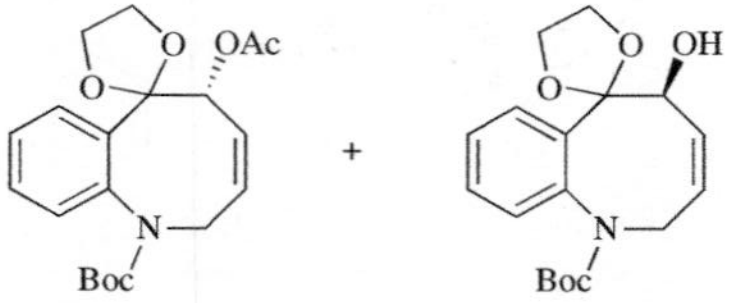

er 95.0:5.0 + er 99.5:0.5

(53), s = 27   287, 286

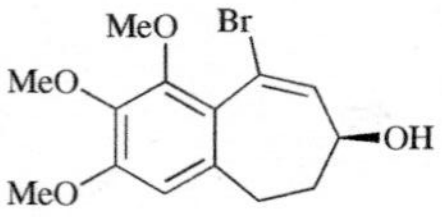

Ac₂O (1.5 eq), **56** (2 mol %),
Et₃N (1.8 eq),
t-AmOH, 0°, 15 h

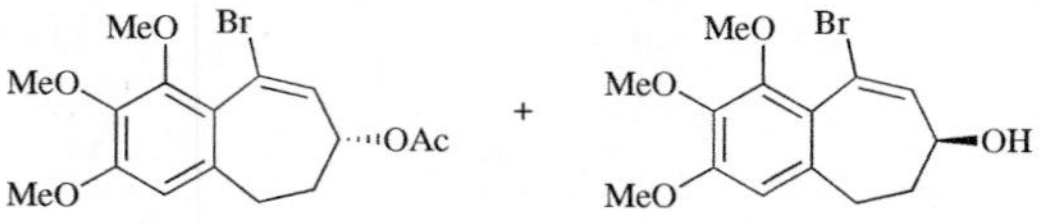

(49), s = 13 + er 99.5:0.5

330

C₁₂

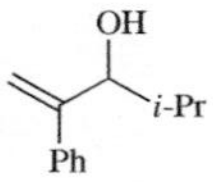

Ac₂O (0.75 eq), *ent*-**56** (1 mol %),
Et₃N (0.75 eq),
t-AmOH, 0°, 7 d

er 96.5:3.5 + (59), s = 14

202

203

(EtCO)$_2$O (0.75 eq), **120** (5 mol %),
i-Pr$_2$NEt (0.75 eq),
MTBE/CHCl$_3$ (1:1), 0°, 24 h

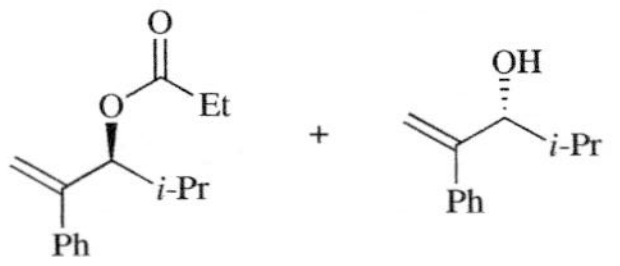

er 94.5:5.5   +   (23), *s* = 21
er 63.0:37.0

53

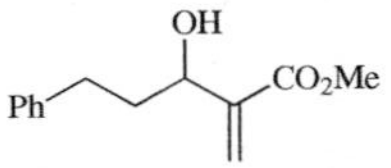

(EtCO)$_2$O (0.75 eq),
121 (5 mol %), K$_2$CO$_3$ (0.5 eq),
i-Pr$_2$O, 0°, 24 h

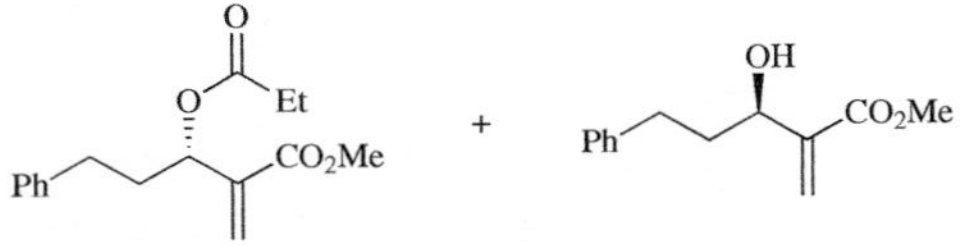

er 90.0:10.0   +   (47), *s* = 20
er 86.0:14.0

207

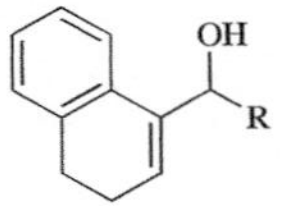

(*i*-PrCO)$_2$O (2.5 eq),
141 (5 mol %),
toluene, −40°

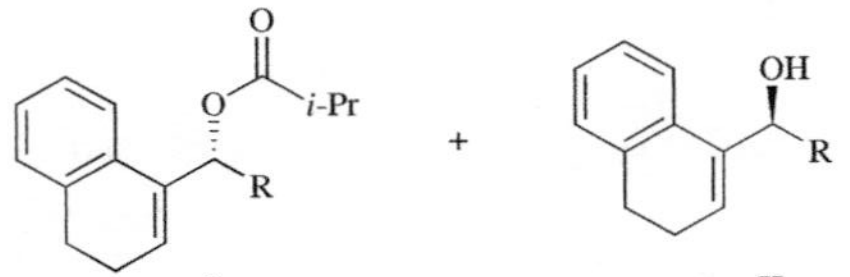

I   **II**

R	Time (h)		er **I**	er **II**	*s*
Me	72	(53)	93.5:6.5	98.0:2.0	55
Et	128	(34)	97.5:2.5	74.5:25.5	61

C$_{12-13}$

TABLE 2C. CATALYTIC KINETIC RESOLUTION OF ALLYLIC ALCOHOLS (*Continued*)

Alcohol	Conditions	Product(s) and Conversion(s) (%)	Refs.

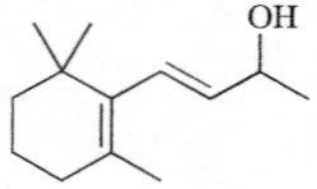

Please refer to the charts preceding the tables for the structures indicated by the bold numbers.

C₁₃

(EtCO)₂O (0.75 eq), **124** (10 mol %),
i-Pr₂NEt (0.75 eq), CDCl₃, 0°, 47 h

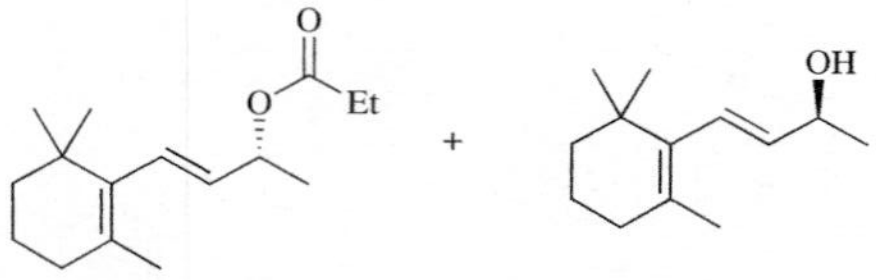

er 67.0:33.0   er 57.5:42.5   (30), *s* = 2   57

(EtCO)₂O (0.75 eq), **117** (5 mol %),
i-Pr₂NEt (0.75 eq), CDCl₃, 0°, 23 h

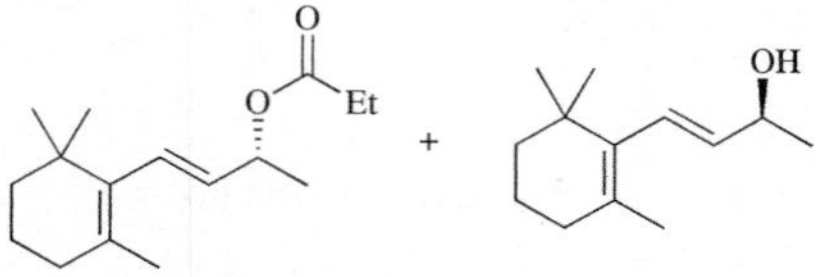

er 66.0:34.0   er 64.0:36.0   (47), *s* = 3   57

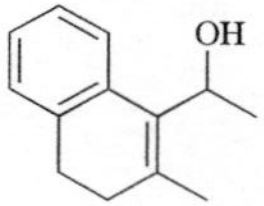

(*i*-PrCO)₂O (2.5 eq),
141 (5 mol %),
heptane, –40°, 25 h

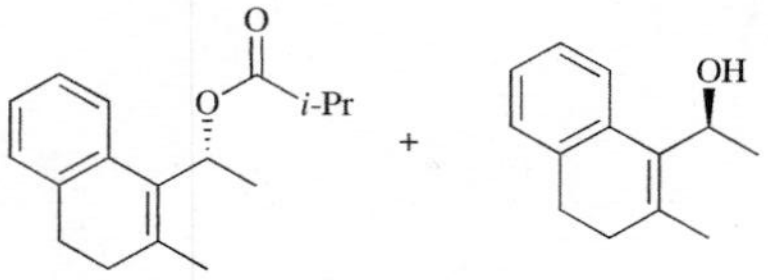

er 97.5:2.5   er 82.0:18.0   (40), *s* = 82   207

(*i*-PrCO)₂O (2.5 eq),
142 (7 mol %), Et₃N (14 mol %),
heptane, –40°, 64 h

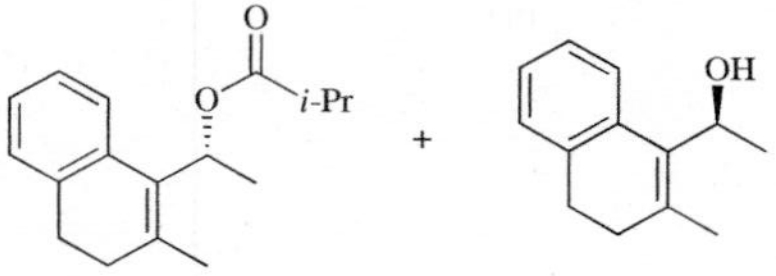

er 90.5:9.5   er 99.5:0.5   (55), *s* = 49   156

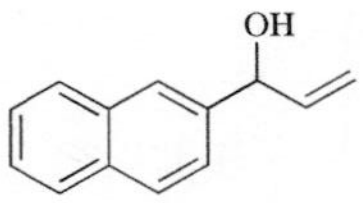

242

er 85.0:15.0

er 87.0:13.0   (48), $s = 14$

(i-PrCO)$_2$O (0.55 eq),
155 (5 mol %), i-Pr$_2$NEt (0.6 eq),
(MeO)$_2$CO, rt, 24 h

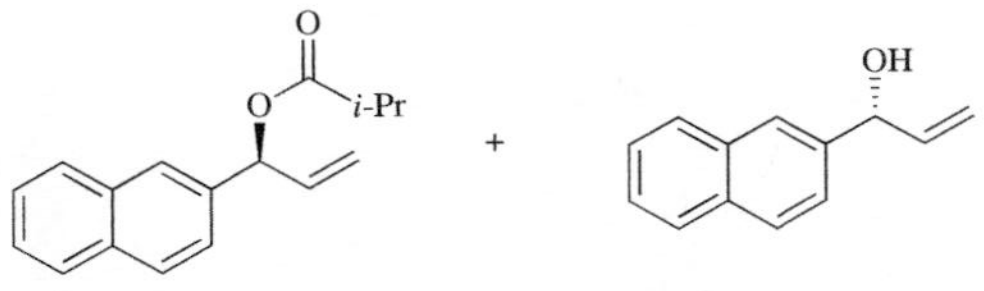

196

er 97.0:3.0

er 88.0:12.0   (55), $s = 25$

(i-PrCO)$_2$O (0.55 eq),
155 (1 mol %), i-Pr$_2$NEt (0.6 eq),
CHCl$_3$, 0°, 7 h

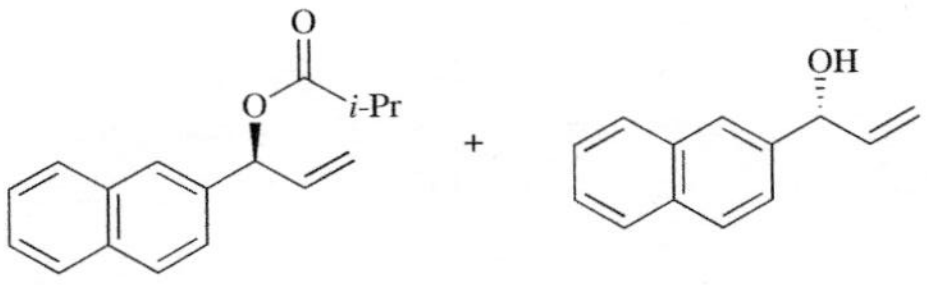

205

er 72.0:28.0

er 75.0:25.0   (47), $s = 5$

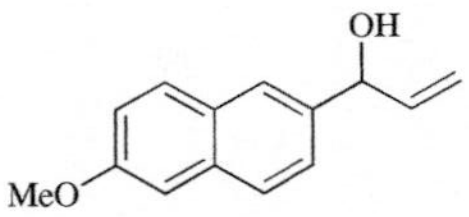

(i-PrCO)$_2$O (0.6 eq),
126 (1 mol %), i-Pr$_2$NEt (0.5 eq),
toluene, −78°, 16 h

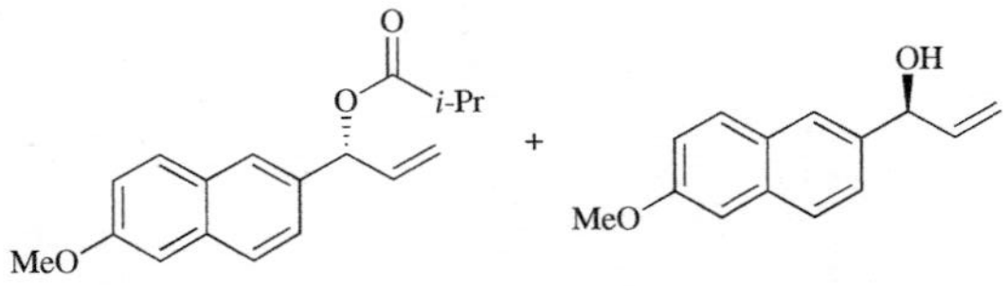

TABLE 2D. CATALYTIC KINETIC RESOLUTION OF PROPARGYLIC ALCOHOLS

Alcohol	Conditions	Product(s) and Conversion(s) (%)	Refs.

*Please refer to the charts preceding the tables for the structures indicated by the **bold** numbers.*

C_6

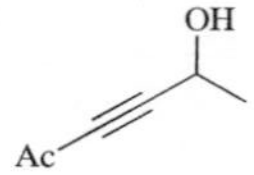

Ac₂O (0.75 eq),
55 (1 mol %),
t-AmOH, 0°, 17 h

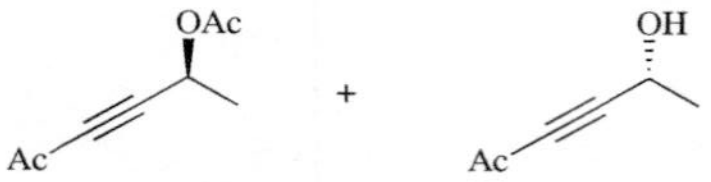

er 77.5:22.5   +   er 97.5:2.5

(64), *s* = 12

209

C_{6-10}

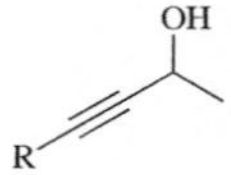

(EtCO)₂O (0.75 eq),
124 (4 mol %),
CHCl₃, 0°, 2 h

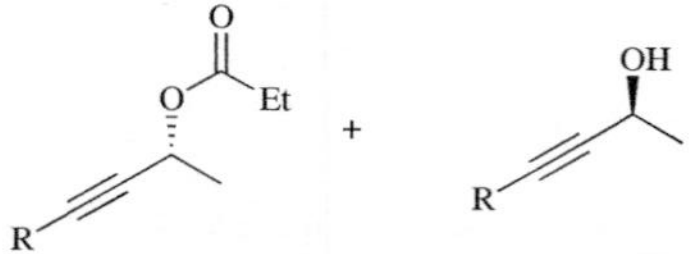

I   +   **II**

R		er **I**	er **II**	*s*
Ac	(55)	89.0:11.0	96.5:3.5	26
1-hexynyl	(52)	91.5:8.5	94.5:5.5	32

210

C_{8-10}

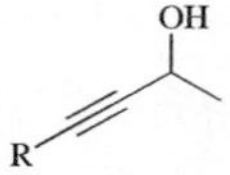

(EtCO)₂O (0.75 eq),
124 (10 mol %),
CHCl₃, 0°

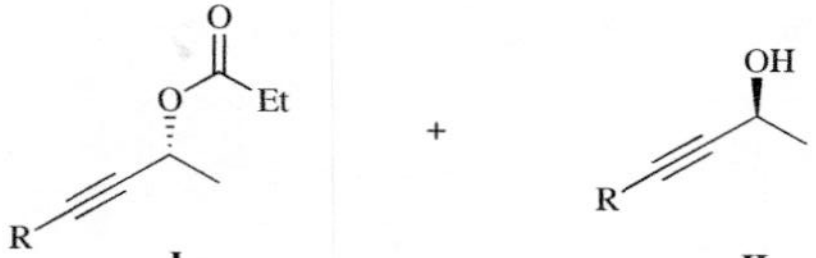

I   +   **II**

R	Time (h)		er **I**	er **II**	*s*
n-Bu	25	(42)	88.5:11.5	77.5:22.5	13
t-Bu	19	(48)	80.0:20.0	78.0:22.0	7
c-C₆H₁₁	23	(60)	80.0:20.0	95.0:5.0	11

210

312

C$_8$

(EtCO)$_2$O (0.75 eq),
124 (10 mol %),
CHCl$_3$, 0°

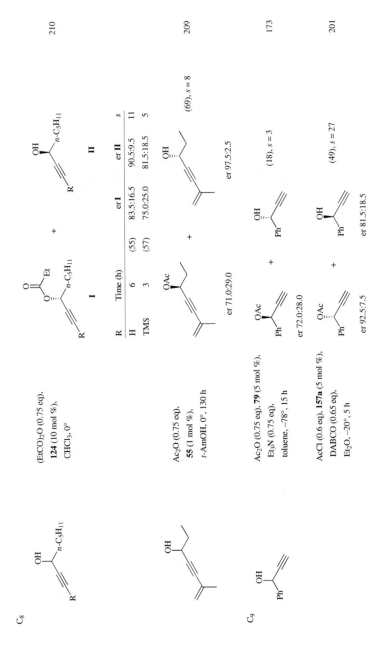

R	Time (h)		erI	erII	s
H	6	(55)	83.5:16.5	90.5:9.5	11
TMS	3	(57)	75.0:25.0	81.5:18.5	5

210

Ac$_2$O (0.75 eq),
55 (1 mol %),
t-AmOH, 0°, 130 h

er 71.0:29.0   er 97.5:2.5   (69), *s* = 8

209

Ac$_2$O (0.75 eq), **79** (5 mol %),
Et$_3$N (0.75 eq),
toluene, −78°, 15 h

er 72.0:28.0   (18), *s* = 3

173

C$_9$

AcCl (0.6 eq), **157a** (5 mol %),
DABCO (0.65 eq),
Et$_2$O, −20°, 5 h

er 92.5:7.5   er 81.5:18.5   (49), *s* = 27

201

313

TABLE 2D. CATALYTIC KINETIC RESOLUTION OF PROPARGYLIC ALCOHOLS (Continued)

Alcohol	Conditions	Product(s) and Conversion(s) (%)	Refs.

Please refer to the charts preceding the tables for the structures indicated by the **bold** numbers.

C₉

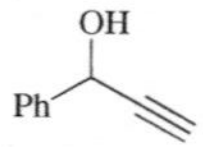	(i-PrCO)₂O (0.55 eq), **155** (1 mol %), i-Pr₂NEt (0.6 eq), CHCl₃, 0°, 7 h	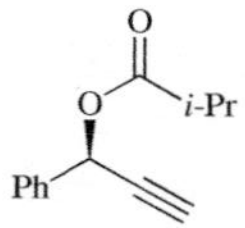 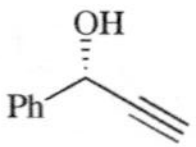 (52), s = 3 er 67.0:33.0	196
	(3-ClC₆H₄CO₂O (2.5 eq), **141** (9 mol %), toluene, rt, 15 h er 53.5:46.5	+ (41), s = 1 er 55.0:45.0	155
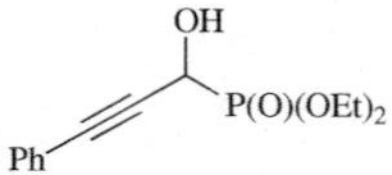	BzCl (0.5 eq), **15** (5 mol %), Cu(OTf)₂ (5 mol%), BaCO₃ (1 eq), chlorobenzene, 0° to rt, 12 h er 71.0:29.0	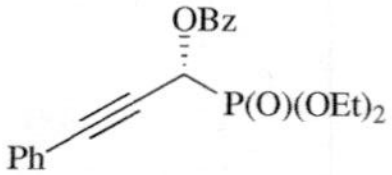 + (45), s = 4 er 70.5:29.5	208

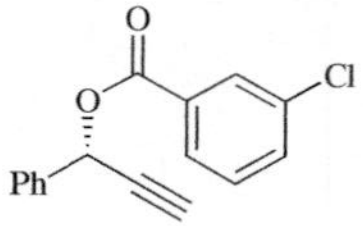

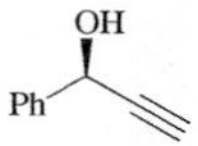

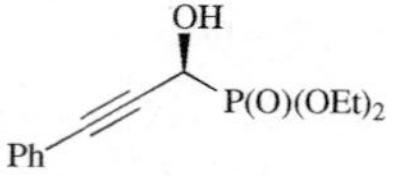

C₁₀

| 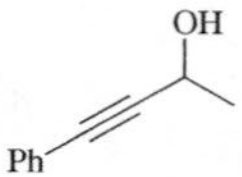 | AcCl (0.6 eq), **157a** (5 mol %), DABCO (0.65 eq), Et₂O. −20°, 5 h er 86.0:14.0 | 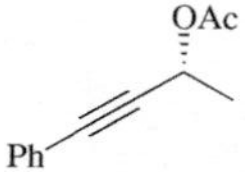 + (56), s = 19 er 95.5:4.5 | 201 |
| | 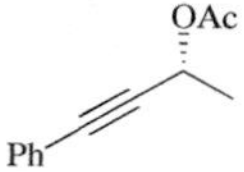 Ac₂O (0.75 eq), **79** (5 mol %), Et₃N (0.75 eq), toluene, −78°, 15 h er 77.0:23.0 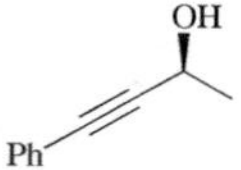 | + (19), s = 4 | 173 |

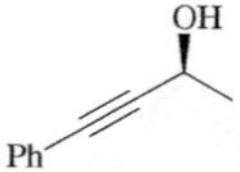

314

C_{10-13}

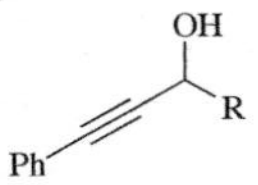

Ac_2O (0.75 eq), **55** (1 mol %), *t*-AmOH, 0°

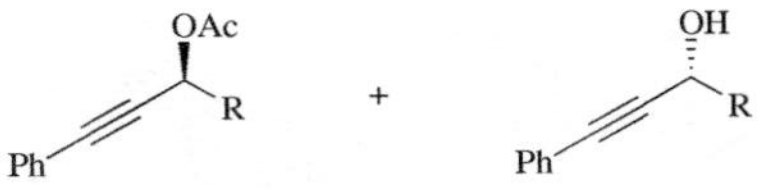

I + II

R	Time (h)		er **I**	er **II**	*s*
Me	49	(58)	84.5:15.5	98.0:2.0	20
Et	39	(58)	83.5:16.5	97.5:2.5	18
i-Pr	86	(63)	77.5:22.5	96.5:3.5	11
t-Bu	479	(86)	81.5:18.5	96.0:4.0	4

209

C_{10}

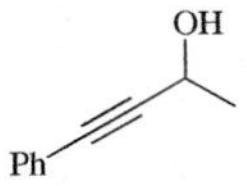

$(EtCO)_2O$ (0.5 eq), **126** (0.75 mol %), *i*-Pr_2NEt (0.6 eq), $CHCl_3$, 0°, 1 h

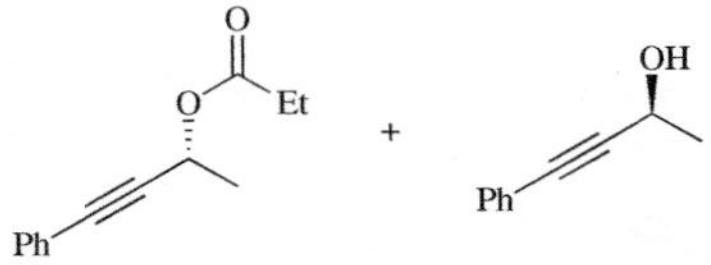

er 91.5:8.5 + er 77.5:22.5

(40), *s* = 18

60

C_{10-13}

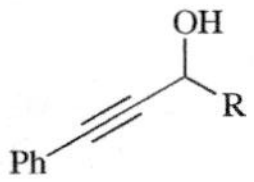

$(EtCO)_2O$ (0.75 eq), **124** (4 mol %), $CHCl_3$, 0°, 11 h

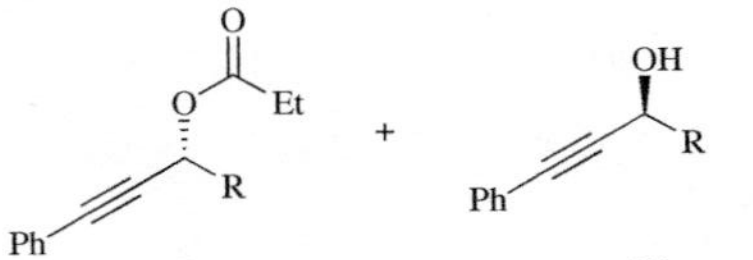

I + II

R		er **I**	er **II**	*s*
Me	(59)	84.5:15.5	>99.5:0.5	31
Et	(56)	87.5:12.5	98.0:2.0	27
i-Pr	(56)	85.5:14.5	95.5:4.5	18
t-Bu	(43)	85.0:15.0	77.0:23.0	10

210

TABLE 2D. CATALYTIC KINETIC RESOLUTION OF PROPARGYLIC ALCOHOLS (*Continued*)

Alcohol	Conditions	Product(s) and Conversion(s) (%)	Refs.

*Please refer to the charts preceding the tables for the structures indicated by the **bold** numbers.*

C$_{10}$

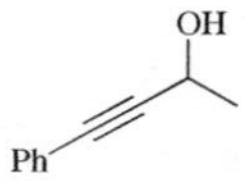

(*i*-PrCO)$_2$O (0.6 eq),
77 (10 mol %),
Et$_2$O, –50°, 48 h

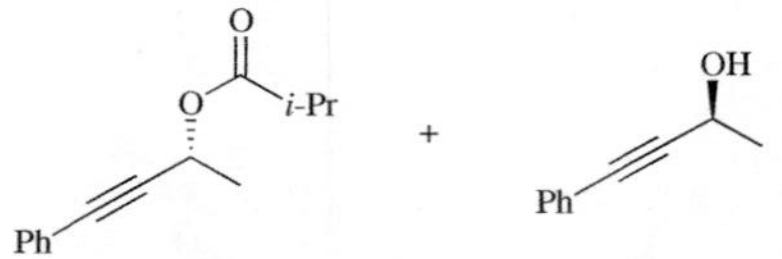

er 89.0:11.0

+

er 86.5:13.5

(48), *s* = 17

169

(*i*-PrCO)$_2$O (0.8 eq),
76 (0.5 mol %), Et$_3$N (0.9 eq),
CHCl$_3$, –30°, 48 h

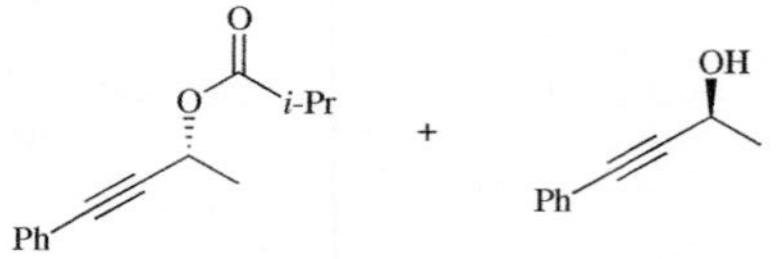

er 75.0:25.0

+

er 89.0:11.0

(61), *s* = 7

168

(*i*-PrCO)$_2$O (0.7 eq),
84 (5 mol %), Et$_3$N (0.9 eq),
CH$_2$Cl$_2$, –78°, 3 h

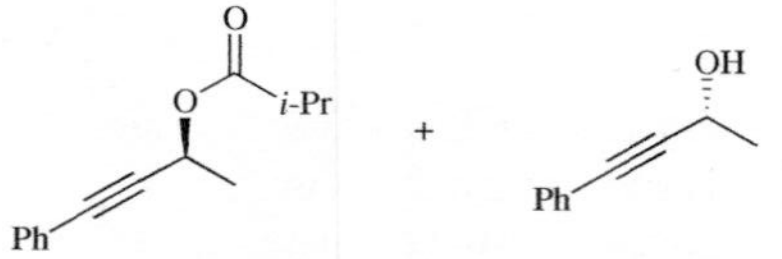

er 65.0:35.0

+

er 52.0:48.0

(11), *s* = 2

246

(*i*-PrCO)$_2$O (0.55 eq),
155 (5 mol %),
i-Pr$_2$NEt (0.6 eq),
(MeO)$_2$CO, rt, 24 h

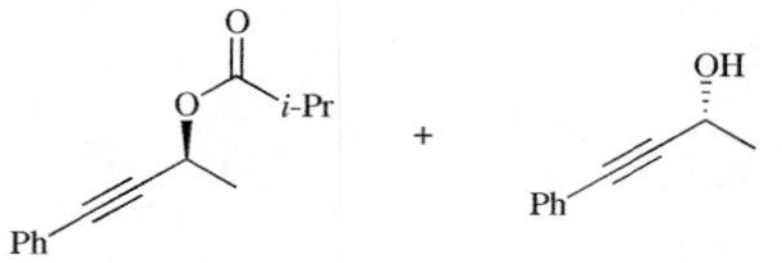

er 81.0:19.0

+

er 89.0:11.0

(55), *s* = 10

242

C_{10-12}

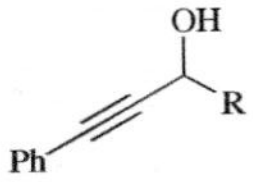

$(i\text{-PrCO})_2O$ (0.55 eq),
155 (1 mol %),
$i\text{-Pr}_2NEt$ (0.6 eq),
$CHCl_3$, 0°, 7 h

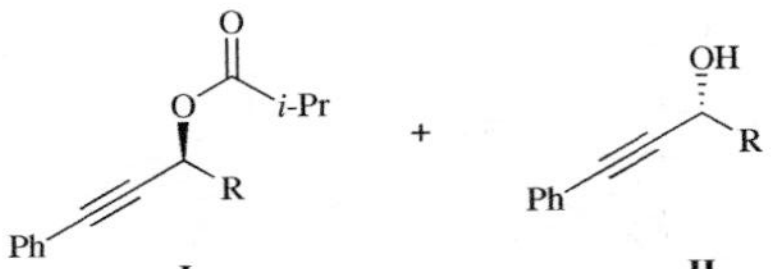

I + II

196

R		er **I**	er **II**	*s*
Me	(54)	89.0:11.0	95.0:5.0	23
i-Pr	(54)	89.0:11.0	95.0:5.0	26

C_{10-11}

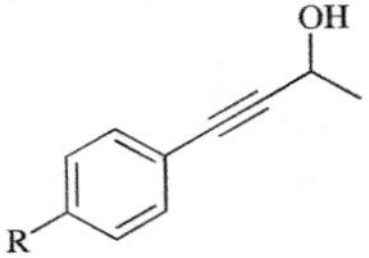

Ac_2O (0.75 eq),
55 (1 mol %),
t-AmOH, 0°

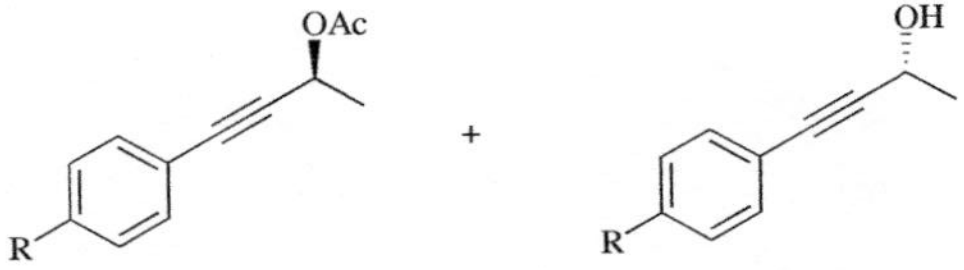

I + II

209

R	Time (h)		er **I**	er **II**	*s*
F	84	(65)	76.5:23.5	98.5:1.5	13
MeO	92	(60)	81.5:18.5	97.0:3.0	15
CF_3	47	(71)	70.0:30.0	99.5:0.5	11

TABLE 2D. CATALYTIC KINETIC RESOLUTION OF PROPARGYLIC ALCOHOLS (*Continued*)

*Please refer to the charts preceding the tables for the structures indicated by the **bold** numbers.*

Alcohol	Conditions	Product(s) and Conversion(s) (%)	Refs.
C$_{10}$			
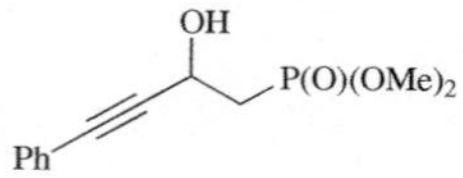	2-FC$_6$H$_4$COCl (0.5 eq), **15** (5 mol %), Cu(OTf)$_2$ (5 mol%), K$_2$CO$_3$ (1 eq), CH$_2$Cl$_2$, 0° to rt, 20 h	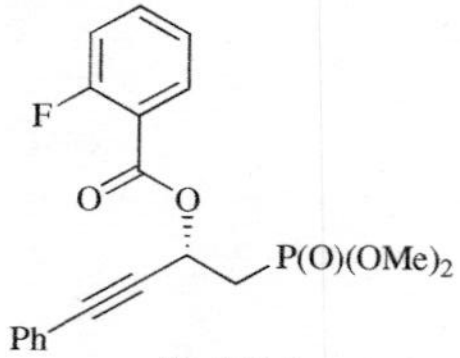 er 73.5:25.5 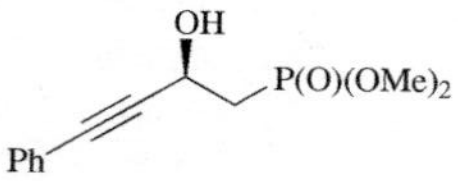 (—), s = 4 + er 75.0:25.0	98
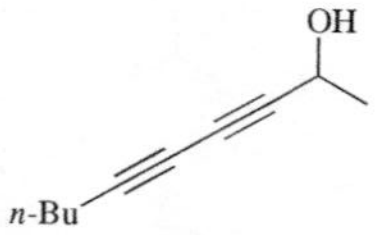	Ac$_2$O (0.75 eq), **55** (1 mol %), t-AmOH, 0°, 41 h	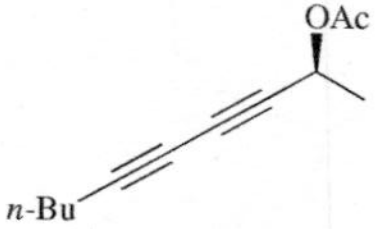 er 71.0:29.0 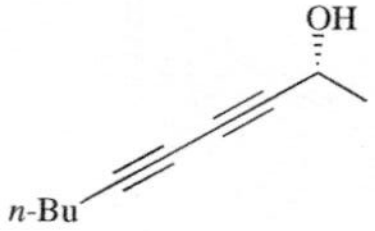 + er 99.5:0.5 (70), s = 11	209
C$_{13}$			
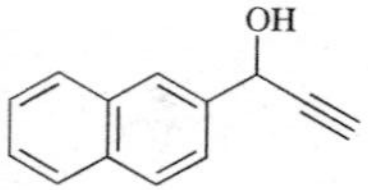	(i-PrCO)$_2$O (0.6 eq), **126** (1 mol %), i-Pr$_2$NEt (0.5 eq), toluene, –78°, 16 h	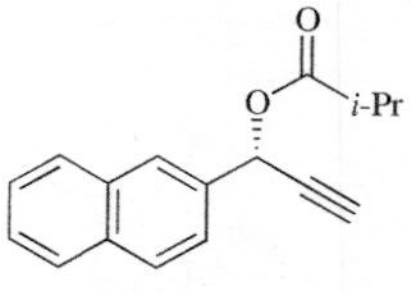 er 64.0:36.0 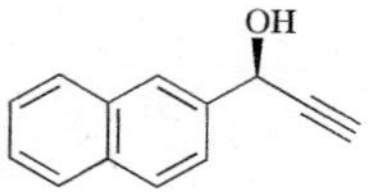 + er 66.0:34.0 (53), s = 3	205

318

TABLE 2E. CATALYTIC KINETIC RESOLUTION OF CYCLOALKANOLS

*Please refer to the charts preceding the tables for the structures indicated by the **bold** numbers.*

Alcohol	Conditions	Product(s) and Conversion(s) (%)	Refs.
C$_4$ 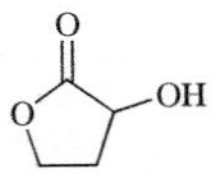 	(Ph$_2$CHCO)$_2$O (0.5 eq), **124** (5 mol %), *i*-Pr$_2$NEt (0.5 eq), Et$_2$O, rt, 12 h	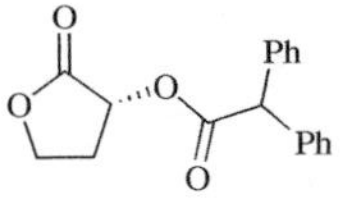 + er 93.0:7.0 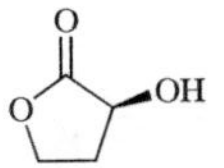 er 94.5:5.5 (51), *s* = 40	63
C$_{4-8}$ 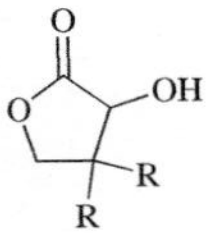 	*n*-PrNCO (0.5 eq), **27** (0.05 mol %), CH$_2$Cl$_2$, 0°, 4 h	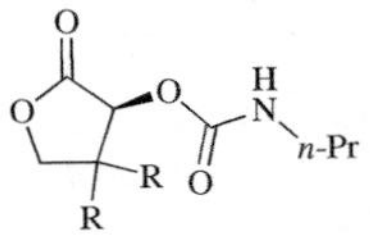 **I** 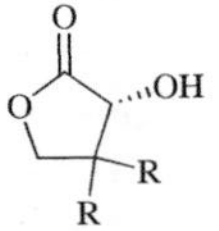 + **II** 	97

R	er **I**	er **II**	*s*
H	(47) 70.5:29.5	68.0:32.0	3
Me	(50) 96.5:3.5	96.5:3.5	95
Et	(46) 97.5:2.5	92.5:7.5	114

Alcohol	Conditions	Product(s) and Conversion(s) (%)	Refs.
C$_5$ 	BzCl (0.75 eq), **137** (0.3 mol %), *i*-Pr$_2$NEt (0.5 eq), 4 Å MS, CH$_2$Cl$_2$, –78°, 3 h	+ er 78.0:28.0 er 96.5:3.5 (—), *s* = —	245
C$_{5-8}$ 	Ac$_2$O (1.1 eq), **102** (0.4 mol %), toluene, rt	+	211

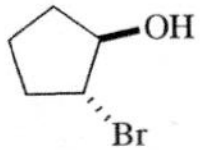

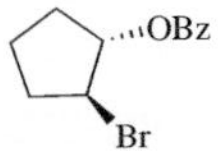

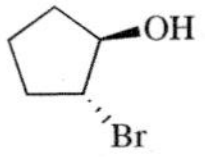

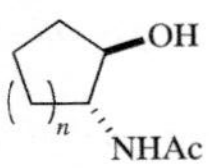

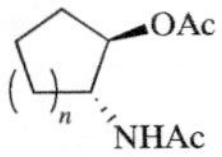

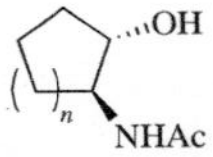

n		*s*
1	(49)	27
2	(50)	51
3	(45)	15

TABLE 2E. CATALYTIC KINETIC RESOLUTION OF CYCLOALKANOLS (*Continued*)

Alcohol	Conditions	Product(s) and Conversion(s) (%)	Refs.

*Please refer to the charts preceding the tables for the structures indicated by the **bold** numbers.*

C5-8

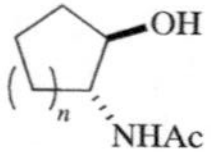

Ac$_2$O (1.1 eq), **103** (2 mol %), toluene, rt

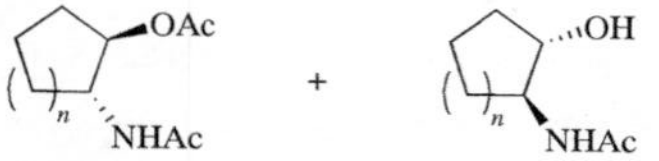

I + **II**

n		s
1	(53)	26
2	(55)	50
3	(47)	31

66

Ac$_2$O (0.84 eq), **104** (2.5 mol %), i-Pr$_2$NEt, toluene, rt

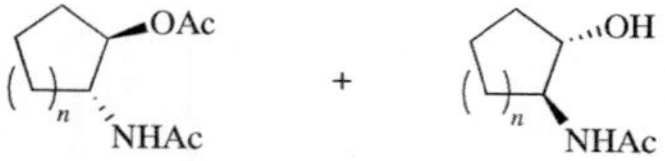

I + **II**

n	Time (min)		er I	er II	s
1	120	(44)	84.0:16.0	76.5:23.5	9
2	40	(50)	97.0:3.0	96.5:3.5	109
3	70	(48)	96.5:3.5	93.5:6.5	75
4	40	(43)	94.5:5.5	84.0:16.0	34

67

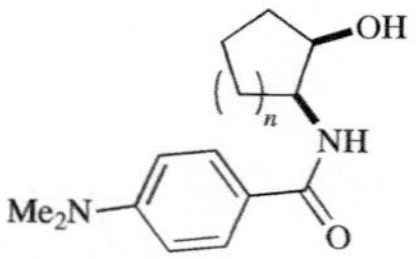

C5-7

(i-PrCO)$_2$O (0.7 eq), ent-**59** (5 mol %), 2,4,6-collidine, CHCl$_3$, 20°, 9 h

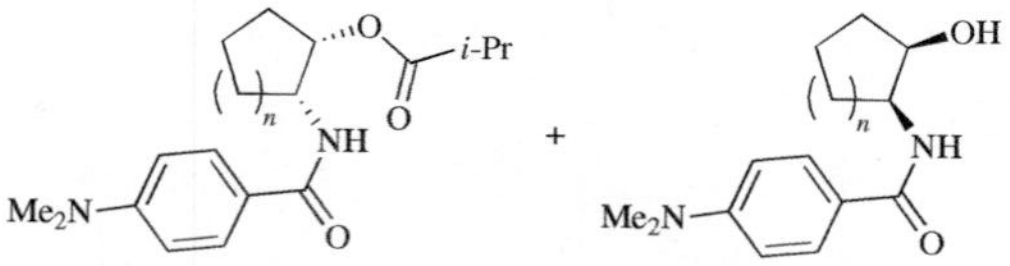

I + **II**

n		er I	er II	s
1	(69)	72.0:28.0	99.5:0.5	12
2	(68)	74.0:26.0	99.5:0.5	14
3	(69)	73.0:27.0	98.5:1.5	10

180

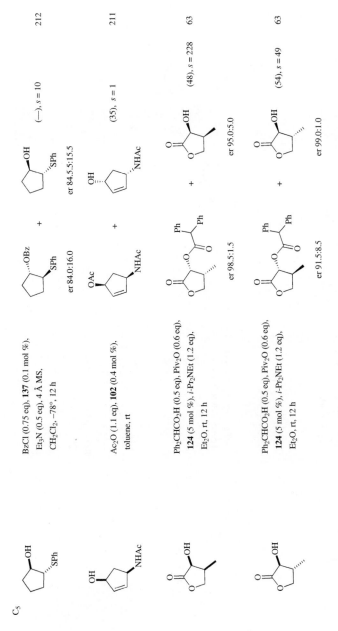

C$_5$

BzCl (0.75 eq), **137** (0.1 mol %),
Et$_3$N (0.5 eq), 4 Å MS,
CH$_2$Cl$_2$, −78°, 12 h

er 84.0:16.0   er 84.5.5:15.5

(—), s = 10   212

Ac$_2$O (1.1 eq), **102** (0.4 mol %),
toluene, rt

(35), s = 1   211

Ph$_2$CHCO$_2$H (0.5 eq), Piv$_2$O (0.6 eq),
124 (5 mol %), i-Pr$_2$NEt (1.2 eq),
Et$_2$O, rt, 12 h

er 98.5:1.5   er 95.0:5.0

(48), s = 228   63

Ph$_2$CHCO$_2$H (0.5 eq), Piv$_2$O (0.6 eq),
124 (5 mol %), i-Pr$_2$NEt (1.2 eq),
Et$_2$O, rt, 12 h

er 91.5:8.5   er 99.0:1.0

(54), s = 49   63

TABLE 2E. CATALYTIC KINETIC RESOLUTION OF CYCLOALKANOLS (*Continued*)

Alcohol	Conditions	Product(s) and Conversion(s) (%)	Refs.

*Please refer to the charts preceding the tables for the structures indicated by the **bold** numbers.*

C₆

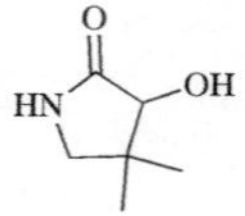

n-PrNCO (0.5 eq), **27** (0.05 mol %), CH₂Cl₂, rt, 6 h

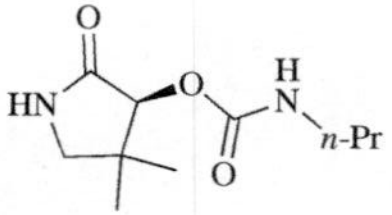

er 70.5:29.5 + er 62.0:38.0

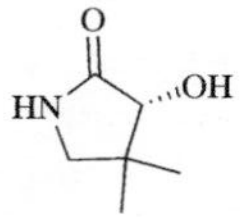

(37), *s* = 3   97

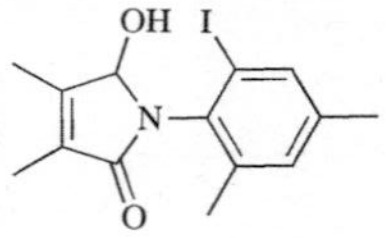

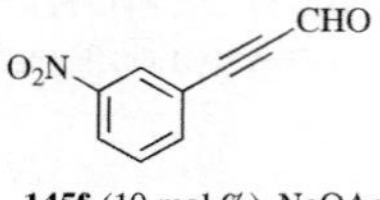

145f (10 mol %), NaOAc (1 eq), 4 Å MS, CHCl₃, rt, 120 h

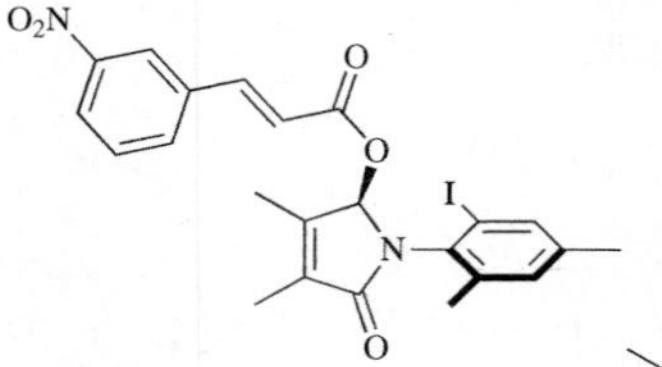

(1.25 eq).

er 90.0:10.0

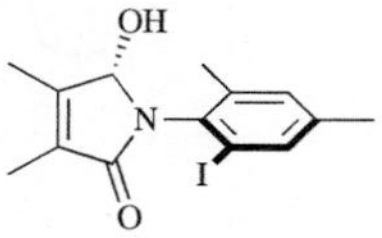

(53), *s* = 28   233

RNCO (0.5 eq), **27** (0.05 mol %), CH₂Cl₂, rt, 1 h

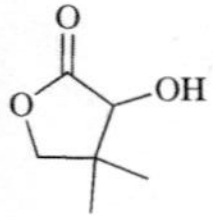

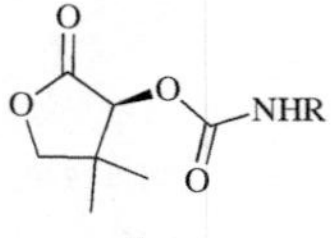

er 95.5:4.5

I + **II**

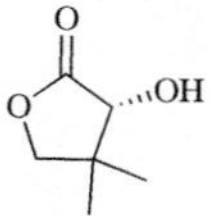

97

R		er **I**	er **II**	*s*
n-Pr	(50)	95.0:5.0	95.5:4.5	62
t-Bu	(30)	88.0:12.0	66.0:34.0	10
Cy	(45)	93.5:6.5	86.0:14.0	32
Ph	(50)	94.0:6.0	94.0:6.0	43
PhCH₂	(42)	93.0:7.0	81.5:18.5	25

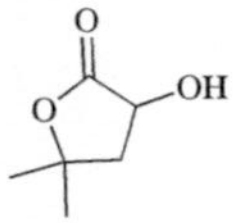

Ph$_2$CHCO$_2$H (0.5 eq),
Piv$_2$O (0.6 eq), **124** (5 mol %),
i-Pr$_2$NEt (1.2 eq), Et$_2$O, rt, 12 h

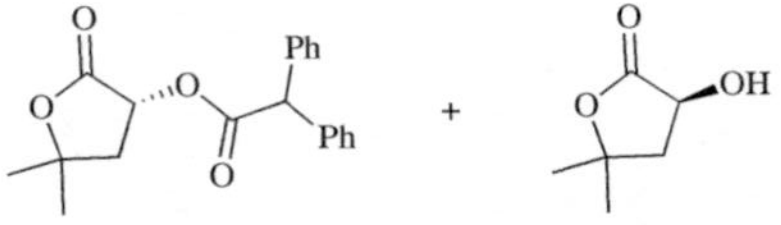

er 93.5:6.5   +   er 96.5:3.5   (52), *s* = 49   63

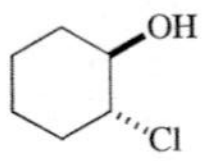

BzCl (0.75 eq), **137** (0.3 mol %),
i-Pr$_2$NEt (0.5 eq), 4 Å MS,
CH$_2$Cl$_2$, –78°, 3 h

er 97.5:2.5   +   er 91.0:9.0   (—), *s* = 92   245

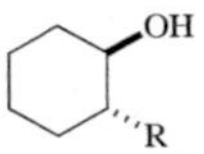

BzCl (0.75 eq), **137** (0.3 mol %),
Et$_3$N (0.5 eq), 4 Å MS,
CH$_2$Cl$_2$, –78°, 3 h

I   +   **II**   157

R	er I		er II	*s*
Br	(50)	98.0:2.0	97.5:2.5	130
EtCO$_2$	(49)	92.5:7.5	90.5:9.5	27
i-PrCO$_2$	(52)	92.0:8.0	95.0:5.0	27

C$_{6-7}$

TABLE 2E. CATALYTIC KINETIC RESOLUTION OF CYCLOALKANOLS (*Continued*)

*Please refer to the charts preceding the tables for the structures indicated by the **bold** numbers.*

Alcohol	Conditions	Product(s) and Conversion(s) (%)	Refs.
C₆			
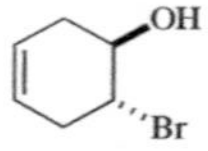	BzCl (0.75 eq), **137** (0.3 mol %), Et₃N (0.5 eq), 4 Å MS, CH₂Cl₂, −78°, 3 h	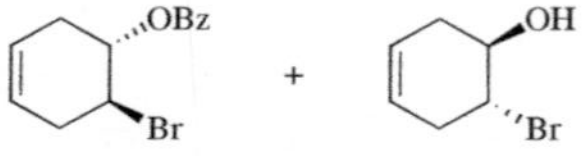 (48), *s* = 170	157
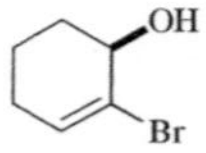	BzCl (0.75 eq), **137** (0.3 mol %), *i*-Pr₂NEt (0.5 eq), 4 Å MS, CH₂Cl₂, −78°, 3 h	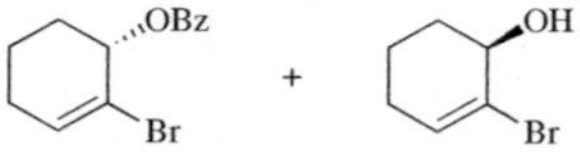 (—), *s* = 35	245
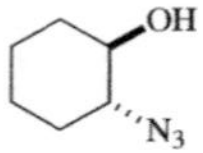	(EtCO)₂O (0.75 eq), **125** (4 mol %), *i*-Pr₂NEt (0.75 eq), *t*-AmOH/toluene (1:1), −40°, 10 h	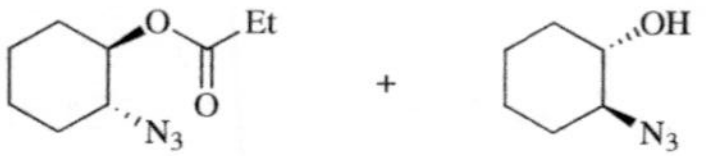 (26), *s* = 10	194
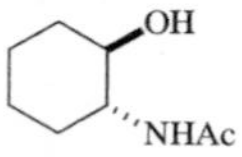	Ac₂O (2.9 eq), **96** (5 mol %), toluene, rt, 12 h	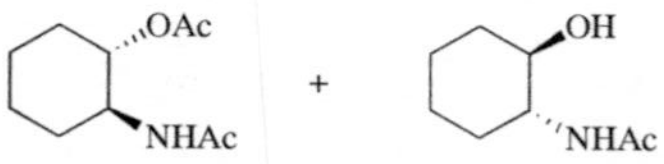 (58), *s* = 28	65

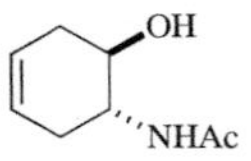

OH, NHAc (cyclohexenyl substrate)

Ac$_2$O (0.84 eq).
104 (2.5 mol %), i-Pr$_2$NEt,
toluene, rt, 50 min

OAc, NHAc + OH, NHAc   (51), $s = 77$ 67

er 95.5:4.5

OH, NH–CO–CF$_3$ (cyclohexyl substrate)

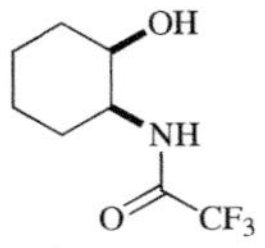

(i-PrCO)$_2$O (0.7 eq).
64 (5 mol %),
CHCl$_3$, rt, 16 h

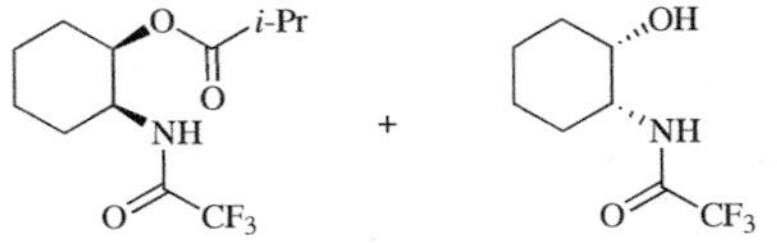

i-Pr–CO–O, NH–CO–CF$_3$ + OH, NH–CO–CF$_3$ (68), $s = 10$ 236

er 98.5:1.5

OH, NH–CO–C$_6$H$_4$–NMe$_2$ (cyclohexyl substrate)

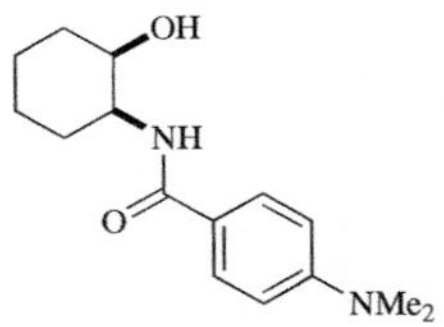

(i-PrCO)$_2$O (0.7 eq).
catalyst (5 mol %), 2,4,6-collidine,
CHCl$_3$, rt, 12 h

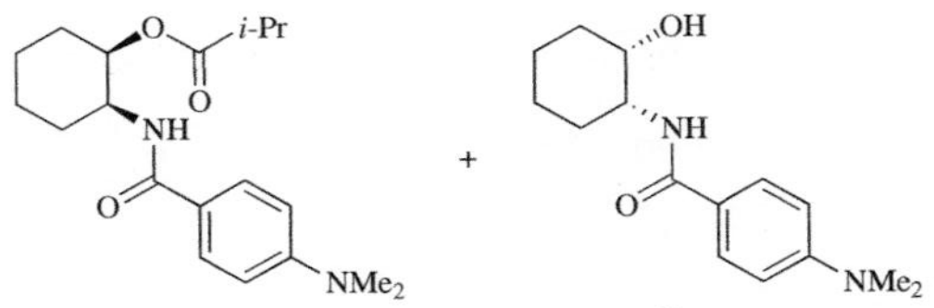

I   +   **II**   331

Catalyst	er **I**	er **II**	s
67	(67) 73.0:27.0	96.0:4.0	8
66	(70) 71.0:29.0	99.5:0.5	11

TABLE 2E. CATALYTIC KINETIC RESOLUTION OF CYCLOALKANOLS (*Continued*)

Alcohol	Conditions	Product(s) and Conversion(s) (%)	Refs.

*Please refer to the charts preceding the tables for the structures indicated by the **bold** numbers.*

C₆

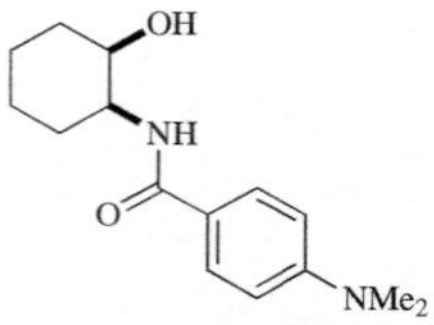	(*i*-PrCO)₂O (1.5 eq), **65** (5 mol %), CH₂Cl₂, rt, 24 h	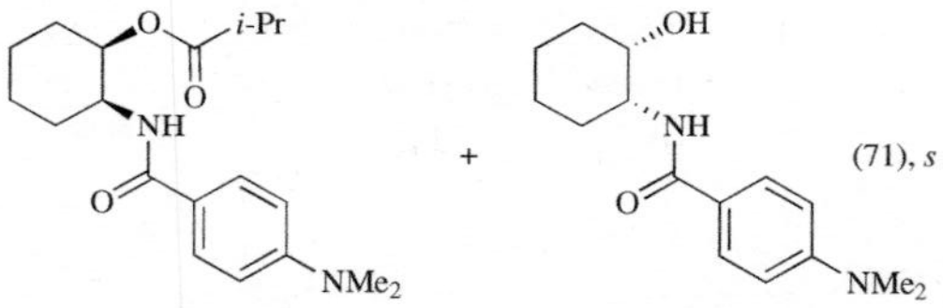 (71), *s* = 6 er 95.5:4.5	177
	(*i*-PrCO)₂O (1.5 eq), **64** (5 mol %), CHCl₃, rt, 24 h	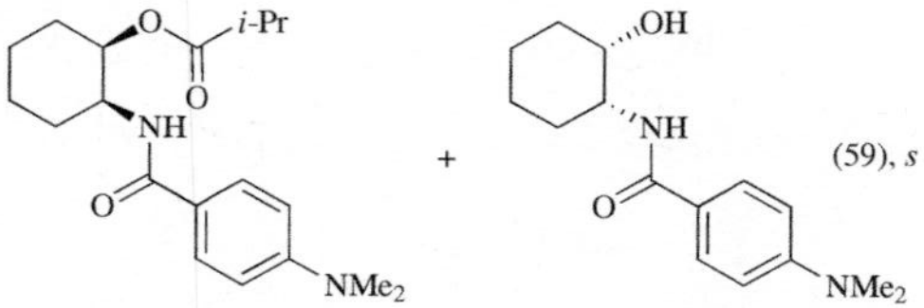 (59), *s* = 19 er 98.0:2.0	177
	(*i*-PrCO)₂O (0.7 eq), **64** (5 mol %), toluene, rt, 3 h	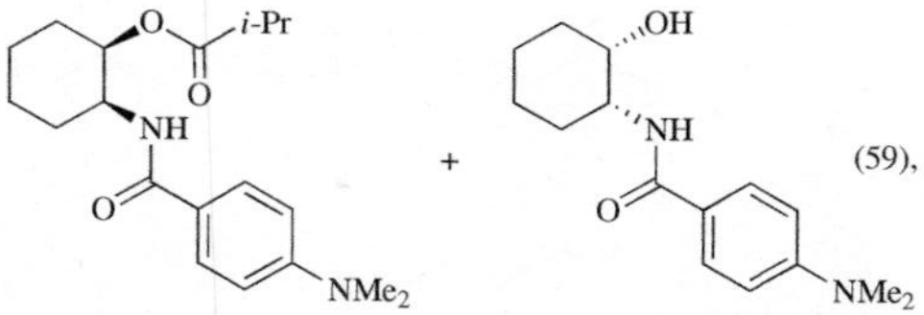 (59), *s* = 19 er 98.0:2.0	178

326

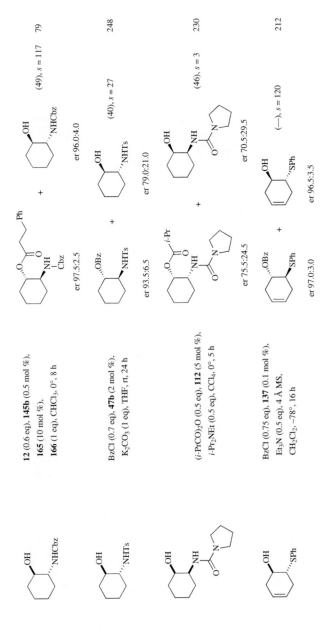

12 (0.6 eq), **145b** (0.5 mol %), **165** (10 mol %), **166** (1 eq). CHCl₃, 0°, 8 h

er 97.5:2.5

er 96.0:4.0

(49), s = 117   79

BzCl (0.7 eq), **47b** (2 mol %), K₂CO₃ (1 eq), THF, rt, 24 h

er 93.5:6.5

er 79.0:21.0

(40), s = 27   248

(*i*-PrCO)₂O (0.5 eq), **112** (5 mol %), *i*-Pr₂NEt (0.5 eq), CCl₄, 0°, 5 h

er 75.5:24.5

er 70.5:29.5

(46), s = 3   230

BzCl (0.75 eq), **137** (0.1 mol %), Et₃N (0.5 eq), 4 Å MS, CH₂Cl₂, −78°, 16 h

er 97.0:3.0

er 96.5:3.5

(—), s = 120   212

TABLE 2E. CATALYTIC KINETIC RESOLUTION OF CYCLOALKANOLS (*Continued*)

Alcohol	Conditions	Product(s) and Conversion(s) (%)	Refs.

*Please refer to the charts preceding the tables for the structures indicated by the **bold** numbers.*

C$_6$

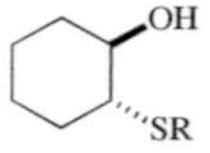

| BzCl (0.75 eq), **137** (0.3 mol %), Et$_3$N (0.5 eq), 4 Å MS, CH$_2$Cl$_2$, −78° | **I** + **II** | 212 |

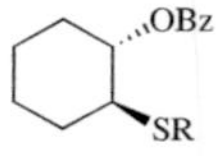

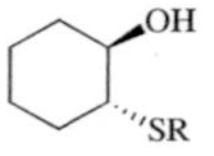

R	Time (h)		er I	er II	s
n-Bu	12	(—)	96.0:4.0	96.5:3.5	57
t-Bu	42	(—)	99.0:1.0	86.5:13.5	210
Ph	12	(—)	98.5:1.5	98.0:2.0	280
4-ClC$_6$H$_4$	13	(—)	99.0:1.0	97.0:3.0	360
4-MeC$_6$H$_4$	13	(—)	98.0:2.0	98.5:1.5	160
PhCH$_2$	12	(—)	98.0:2.0	99.5:0.5	160

C$_7$

| Ph$_2$CHCO$_2$H (0.5 eq), Piv$_2$O (0.6 eq), **124** (5 mol %), *i*-Pr$_2$NEt (1.2 eq), Et$_2$O, rt, 12 h | er 98.5:1.5 + er 98.0:2.0 (50), *s* = 313 | 63 |

| 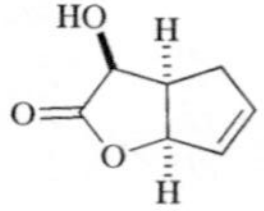 Ph$_2$CHCO$_2$H (0.5 eq), Piv$_2$O (0.6 eq), **124** (5 mol %), *i*-Pr$_2$NEt (1.2 eq), Et$_2$O, rt, 12 h | 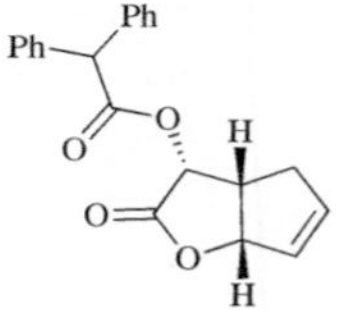 er 94.5:5.5 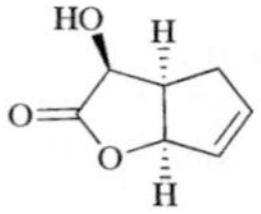 + er 99.5:0.5 (53), *s* = 85 | 63 |

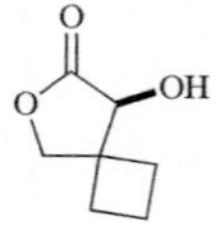

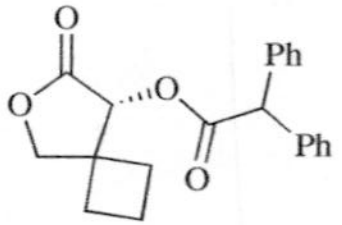

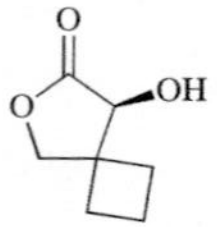

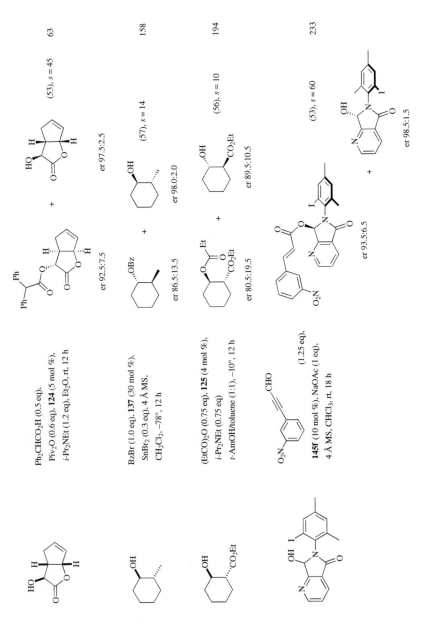

Ph₂CHCO₂H (0.5 eq), Piv₂O (0.6 eq), **124** (5 mol %), *i*-Pr₂NEt (1.2 eq), Et₂O, rt, 12 h

(53), *s* = 45

63

er 97.5:2.5

er 92.5:7.5

BzBr (1.0 eq), **137** (30 mol %), SnBr₂ (0.3 eq.), 4 Å MS, CH₂Cl₂, –78°, 12 h

(57), *s* = 14

158

er 98.0:2.0

er 86.5:13.5

(EtCO)₂O (0.75 eq), **125** (4 mol %), *i*-Pr₂NEt (0.75 eq) *t*-AmOH/toluene (1:1), –10°, 12 h

(56), *s* = 10

194

er 89.5:10.5

er 80.5:19.5

(1.25 eq),

145f (10 mol %), NaOAc (1 eq), 4 Å MS, CHCl₃, rt, 18 h

(53), *s* = 60

233

er 98.5:1.5

er 93.5:6.5

TABLE 2E. CATALYTIC KINETIC RESOLUTION OF CYCLOALKANOLS (*Continued*)

Alcohol	Conditions	Product(s) and Conversion(s) (%)	Refs.

*Please refer to the charts preceding the tables for the structures indicated by the **bold** numbers.*

C₇

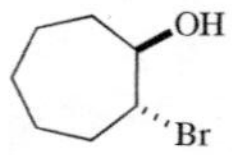

BzCl (0.75 eq), **137** (0.3 mol %),
i-Pr₂NEt (0.5 eq), 4 Å MS,
CH₂Cl₂, –78°, 3 h

er 96.5:3.5   +   er 93.5:6.5

(—), *s* = 58   245

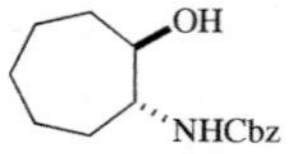

12 (0.6 eq), **145b** (0.5 mol %),
165 (10 mol %), **166** (1 eq),
CHCl₃, 0°, 18 h

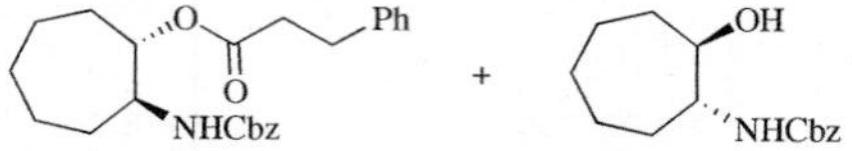

er 97.0:3.0   +   er 90.0:10.0

(46), *s* = 82   79

C₇₋₈

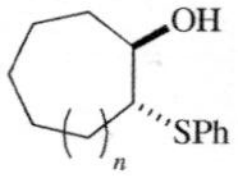

BzCl (0.75 eq), **137** (0.5 mol %),
Et₃N (0.5 eq), 4 Å MS,
CH₂Cl₂, –78°, 5 h

I   +   **II**

n	er I	er II	s
1	(—) 98.5:1.5	98.5:1.5	200
2	(—) 97.5:2.5	99.5:0.5	160

212

C₈

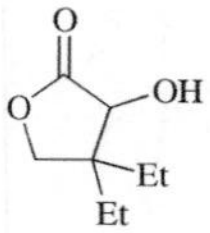

Ph₂CHCO₂H (0.5 eq),
Piv₂O (0.6 eq), **124** (5 mol %),
i-Pr₂NEt (1.2 eq), Et₂O, rt, 12 h

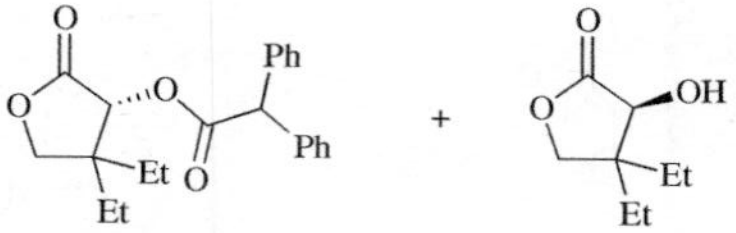

er 99.0:1.0   +   er 99.5:0.5

(50), *s* = 552   63

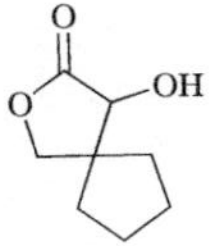

n-PrNCO (0.5 eq), **27** (0.05 mol %),
CH$_2$Cl$_2$, 0°, 4 h

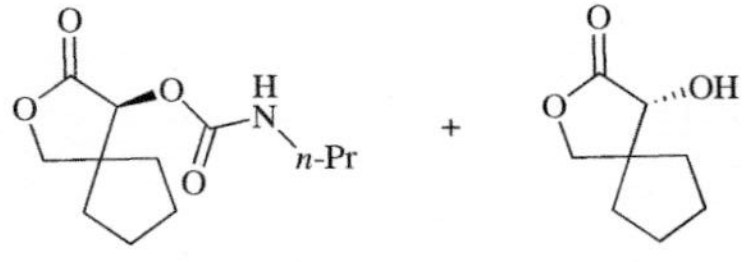

(51), s = 119

97

er 98.5:1.5

er 96.5:3.5

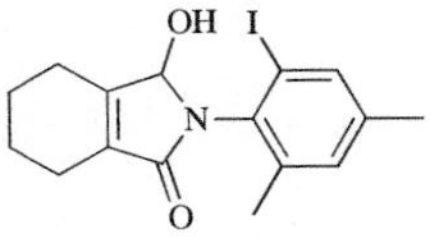

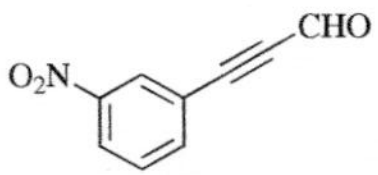

145f (10 mol %), NaOAc (1 eq),
4 Å MS, CHCl$_3$, rt, 96 h

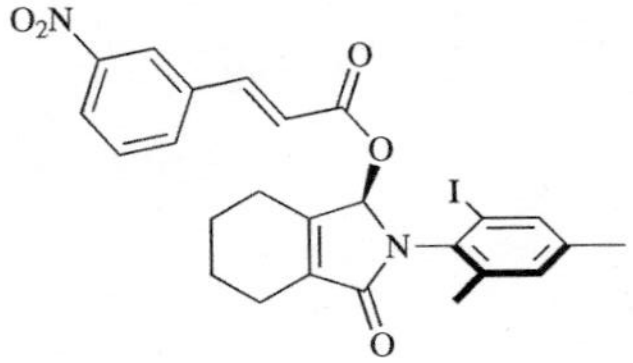

(1.25 eq),

(49), s = 26

233

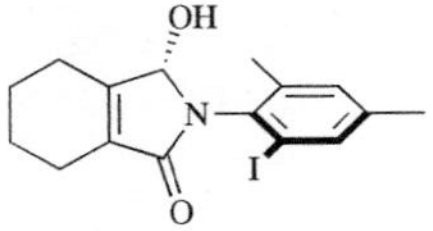

er 95.5:4.5

er 96.5:3.5

TABLE 2E. CATALYTIC KINETIC RESOLUTION OF CYCLOALKANOLS (Continued)

Alcohol	Conditions	Product(s) and Conversion(s) (%)	Refs.

Please refer to the charts preceding the tables for the structures indicated by the bold numbers.

C₈

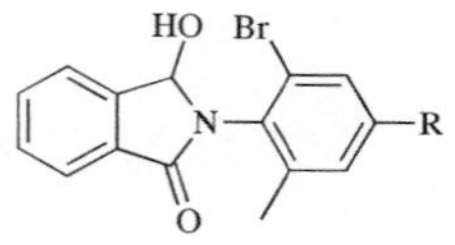

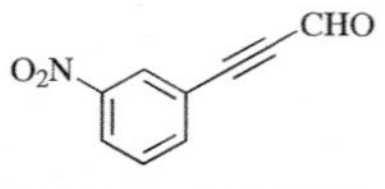

145f (10 mol %), NaOAc (1 eq), 4 Å MS, CHCl₃, rt

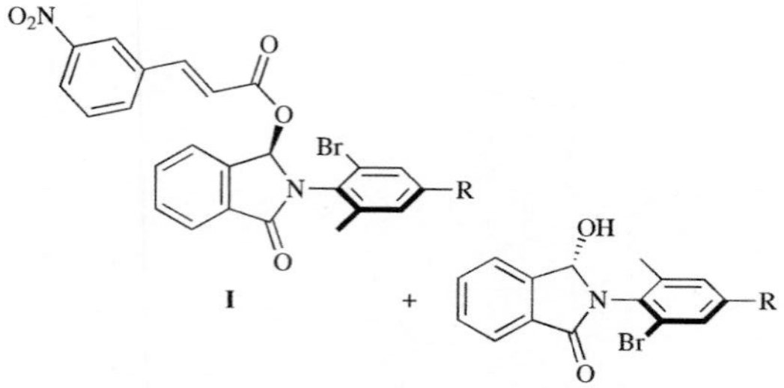

R	Time (h)		er I	er II	s
H	36	(51)	95.5:4.5	97.0:3.0	75
Br	48	(48)	97.0:3.0	94.0:6.0	92
Me	36	(49)	97.0:3.0	95.5:4.5	103

233

332

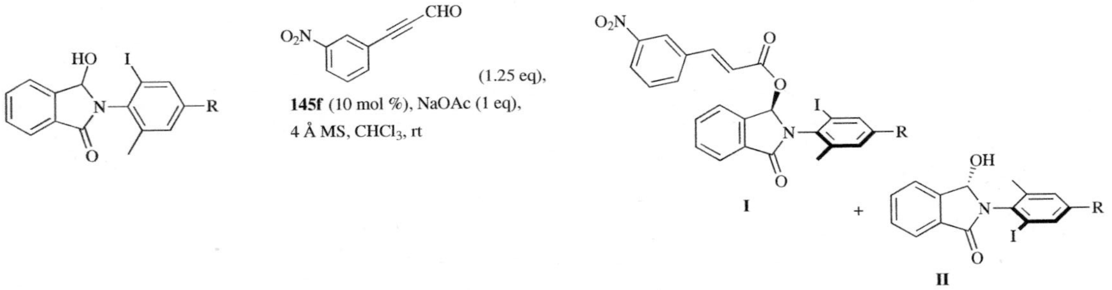

145f (10 mol %), NaOAc (1 eq),
4 Å MS, CHCl$_3$, rt

(1.25 eq.)

I + **II**

R	Time (h)		er **I**	er **II**	s
H	72	(47)	97.5:2.5	91.5:8.5	102
Cl	72	(47)	96.0:4.0	90.5:9.5	60
Br	72	(51)	90.5:9.5	91.5:8.5	25
I	36	(49)	97.0:3.0	95.5:4.5	103

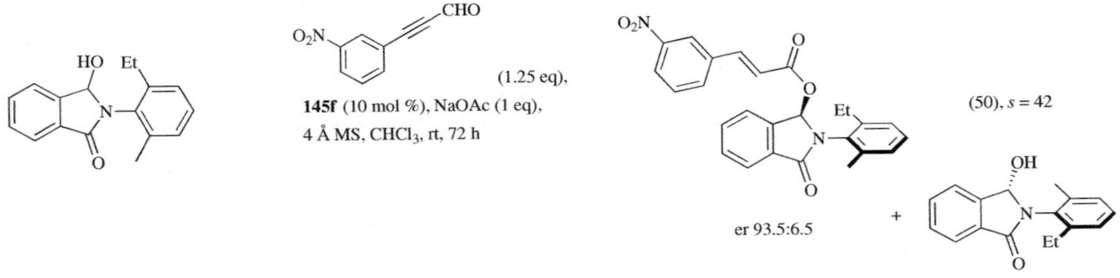

145f (10 mol %), NaOAc (1 eq),
4 Å MS, CHCl$_3$, rt, 72 h

(1.25 eq.)

er 93.5:6.5   +   (50), s = 42

er 94.0:6.0

333

TABLE 2E. CATALYTIC KINETIC RESOLUTION OF CYCLOALKANOLS (*Continued*)

Alcohol	Conditions	Product(s) and Conversion(s) (%)	Refs.

*Please refer to the charts preceding the tables for the structures indicated by the **bold** numbers.*

C$_8$

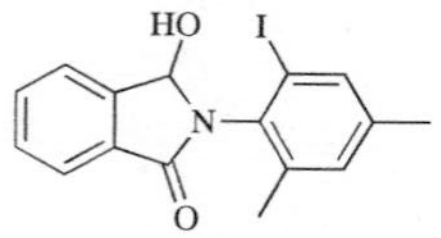

Ar$\equiv$CHO (1.25 eq),
145f (10 mol %), NaOAc (1 eq),
4 Å MS, CHCl$_3$, rt

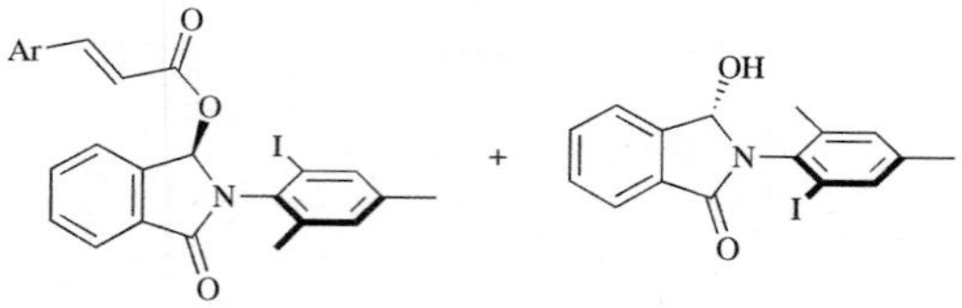

I + **II**

233

Ar	Time (h)		er I	er II	s
3-thienyl	84	(54)	91.0:9.0	98.0:2.0	39
Ph	84	(52)	92.5:7.5	96.5:3.5	42
2-FC$_6$H$_4$	72	(51)	90.5:9.5	93.0:7.0	26
3-FC$_6$H$_4$	72	(49)	95.0:5.0	93.0:7.0	53
4-FC$_6$H$_4$	72	(48)	94.5:5.5	91.0:9.0	44
3-ClC$_6$H$_4$	72	(51)	96.0:4.0	97.5:2.5	89
4-ClC$_6$H$_4$	72	(47)	96.0:4.0	90.5:9.5	60
4-MeOC$_6$H$_4$	72	(46)	93.5:6.5	86.5:13.5	31
3-O$_2$NC$_6$H$_4$	72	(51)	97.0:3.0	98.5:1.5	136
4-O$_2$NC$_6$H$_4$	72	(48)	95.5:4.5	92.5:7.5	58
3-CF$_3$C$_6$H$_4$	72	(49)	97.0:3.0	95.0:5.0	100
1-Np	72	(49)	95.0:5.0	93.5:6.5	54
2-Np	96	(52)	90.0:10.0	93.0:7.0	25

334

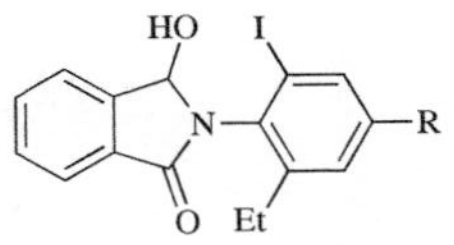

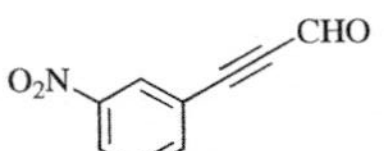

233

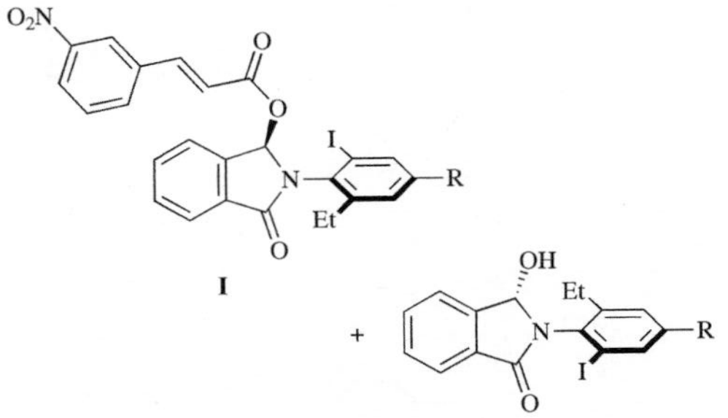

I + **II**

R	Time (h)		er **I**	er **II**	s
I	36	(49)	92.5:7.5	91.0:9.0	31
Et	96	(49)	95.5:4.5	93.0:7.0	59

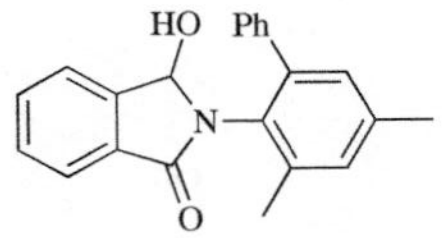

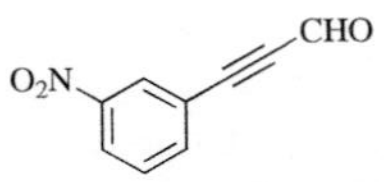

(1.25 eq),

145f (10 mol %), NaOAc (1 eq),

4 Å MS, CHCl$_3$, rt

233

(47), $s = 42$

er 95.0:5.0 + er 90.0:10.0

(1.25 eq),

145f (10 mol %), NaOAc (1 eq),

4 Å MS, CHCl$_3$, rt, 72 h

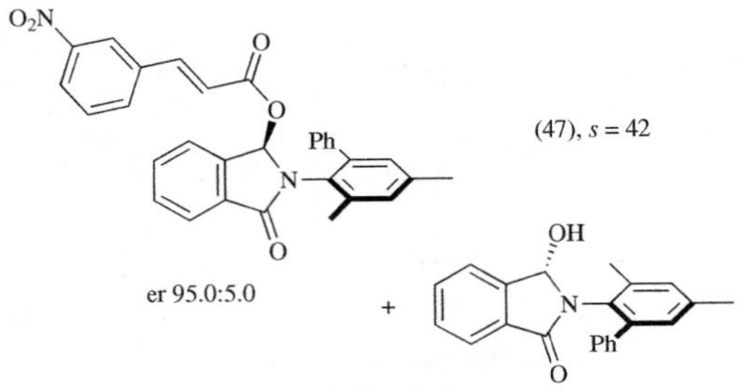

TABLE 2E. CATALYTIC KINETIC RESOLUTION OF CYCLOALKANOLS (*Continued*)

Alcohol	Conditions	Product(s) and Conversion(s) (%)	Refs.

*Please refer to the charts preceding the tables for the structures indicated by the **bold** numbers.*

C_8

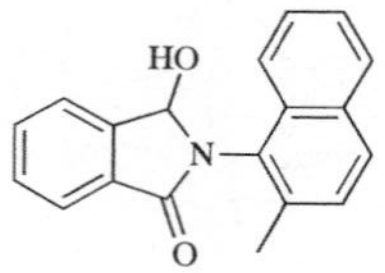

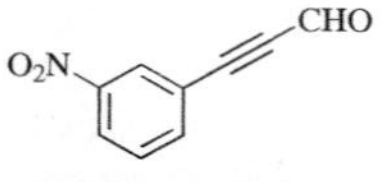

145f (10 mol %), NaOAc (1 eq),
(1.25 eq),
4 Å MS, CHCl$_3$, rt, 72 h

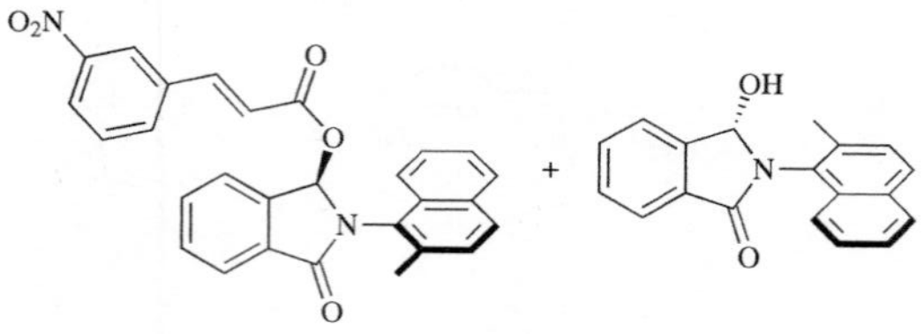

er 90.5:9.5 (52)   +   (53), s = 32   er 96.5:3.5

145f (10 mol %), NaOAc (1 eq),
(1.25 eq),
4 Å MS, CHCl$_3$, rt

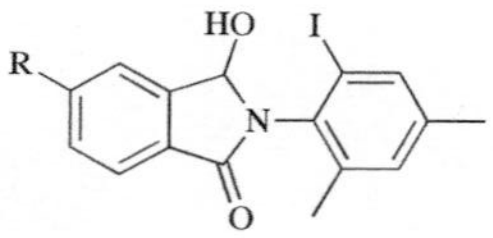

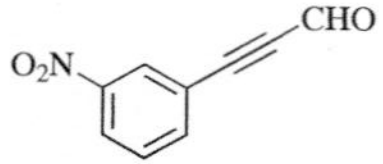

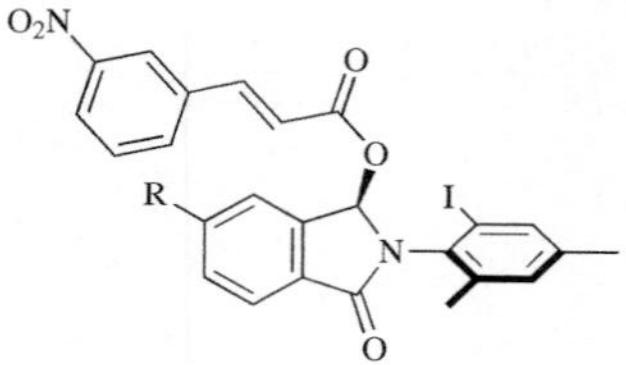

I 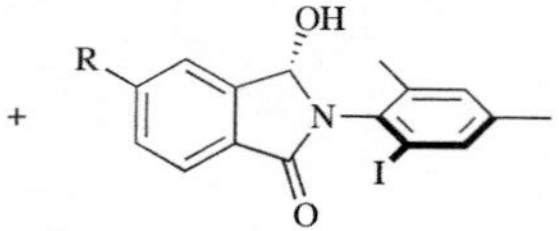   +   **II**

233

233

R	Time (h)		er I	er II	s
Cl	24	(52)	95.0:5.0	98.5:1.5	80
MeO	72	(51)	95.5:4.5	97.5:2.5	79
O$_2$N	48	(50)	95.5:4.5	95.0:5.0	65

336

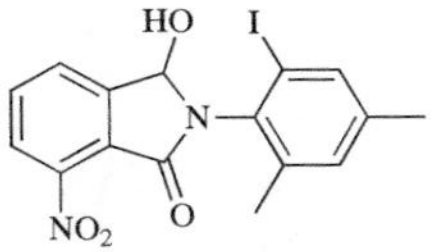

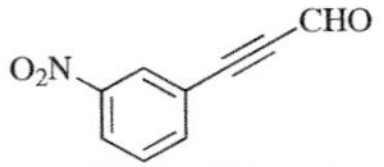

233

(54), s = 56

er 92.5:7.5

er 99.0:1.0

NO₂

(1.25 eq),

145f (10 mol %), NaOAc (1 eq),

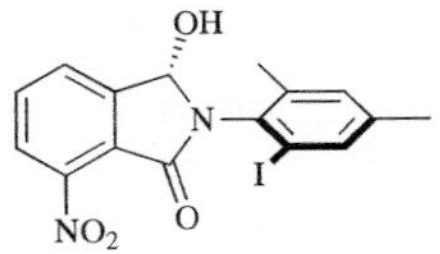

4 Å MS, CHCl₃, rt, 72 h

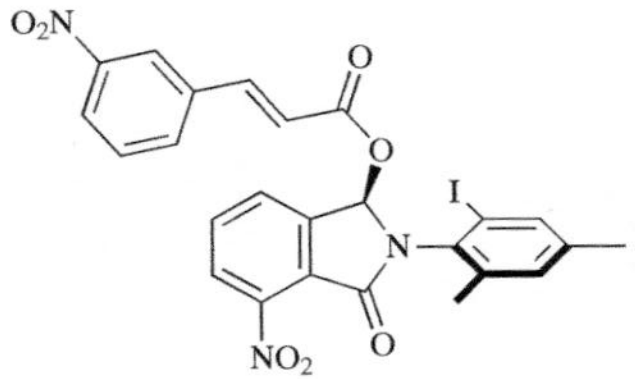

233

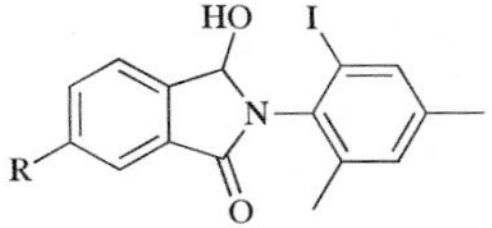

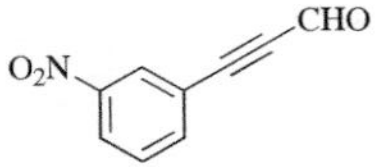

II

I

(1.25 eq),

145f (10 mol %), NaOAc (1 eq),

4 Å MS, CHCl₃, rt, 72 h

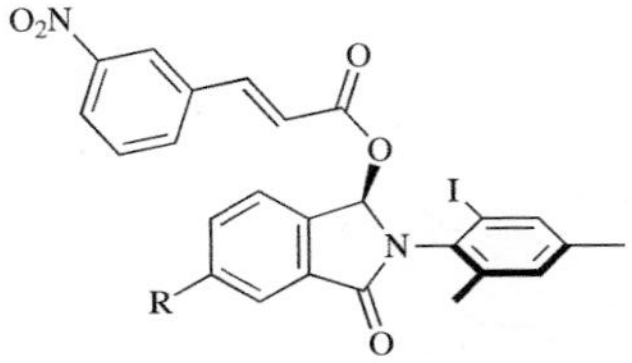

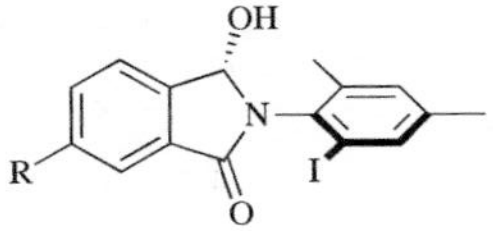

R	Time (h)		er **I**	er **II**	s
O₂N	12	(52)	95.0:5.0	98.0:2.0	74
Me	72	(51)	96.5:3.5	98.0:2.0	108

TABLE 2E. CATALYTIC KINETIC RESOLUTION OF CYCLOALKANOLS (*Continued*)

Alcohol	Conditions	Product(s) and Conversion(s) (%)	Refs.

*Please refer to the charts preceding the tables for the structures indicated by the **bold** numbers.*

C$_8$

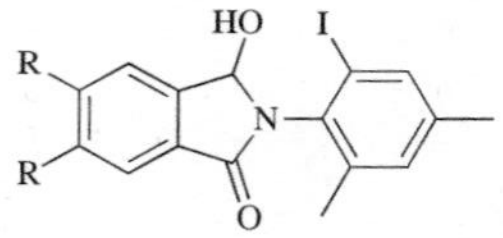

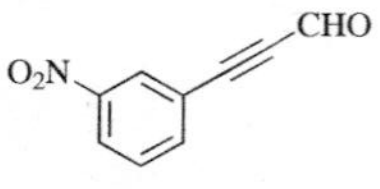

(1.25 eq),

145f (10 mol %), NaOAc (1 eq),

4 Å MS, CHCl$_3$, rt

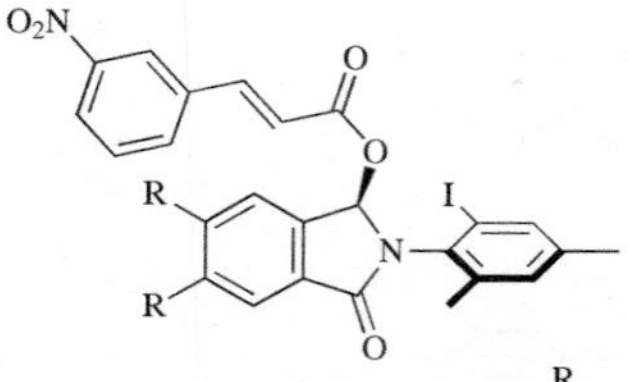

I 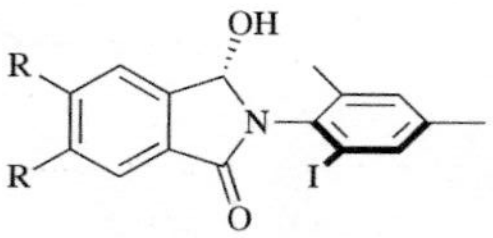 + **II**

233

R	Time (h)		er **I**	er **II**	s
Cl	72	(49)	95.0:5.0	93.0:7.0	54
Br	36	(51)	95.5:4.5	96.5:3.5	65
MeO	72	(50)	96.0:4.0	96.0:4.0	79
Me	72	(51)	95.5:4.5	96.5:3.5	72

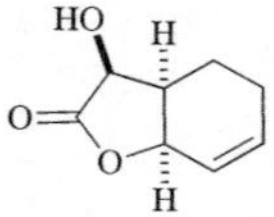

Ph$_2$CHCO$_2$H (0.5 eq),

Piv$_2$O (0.6 eq), **124** (5 mol %),

i-Pr$_2$NEt (1.2 eq), Et$_2$O, rt, 12 h

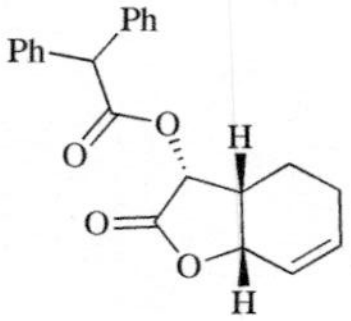

er 95.0:5.0 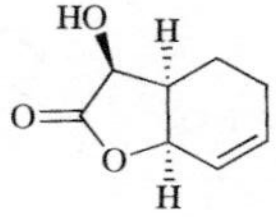   +   er 94.5:5.5

(50), s = 54

63

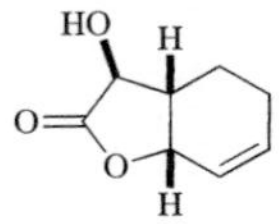

C$_{8-14}$

Ph$_2$CHCO$_2$H (0.5 eq),
Piv$_2$O (0.6 eq), **124** (5 mol %),
i-Pr$_2$NEt (1.2 eq), Et$_2$O, rt, 12 h

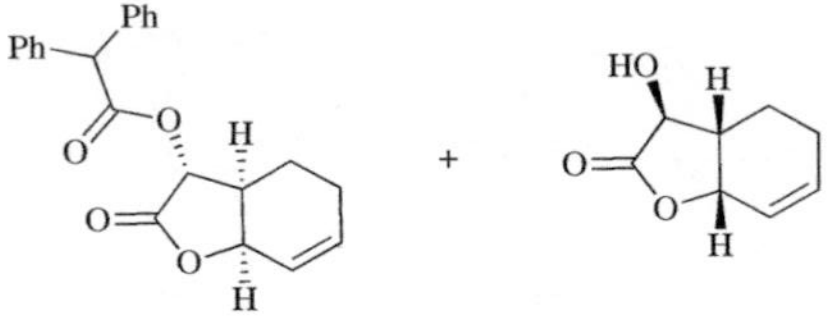

+

er 90.5:9.5   er 92.0:8.0

(51), s = 24   63

C$_{8-14}$

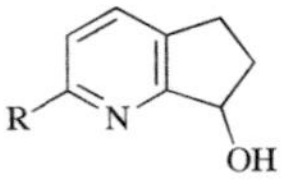

BzCl (0.51 eq), **25** (1 mol %),
CuCl$_2$ (1 mol %), i-Pr$_2$NEt (1.0 eq),
CH$_2$Cl$_2$, 0°, 2 h

+

I   **II**

215

R		er I	er II	s
H	(46)	52.5:47.5	52.5:47.5	1
Ph	(45)	95.5:4.5	88.0:12.0	51

C$_8$

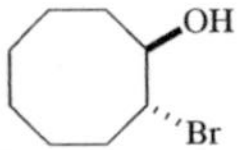

BzCl (0.75 eq), **137** (0.3 mol %),
i-Pr$_2$NEt (0.5 eq), 4 Å MS,
CH$_2$Cl$_2$, −78°, 3 h

+

er 96.5:3.5   er 97.0:3.0

(—), s = 71   245

339

TABLE 2E. CATALYTIC KINETIC RESOLUTION OF CYCLOALKANOLS (*Continued*)

Alcohol	Conditions	Product(s) and Conversion(s) (%)	Refs.
*Please refer to the charts preceding the tables for the structures indicated by the **bold** numbers.*			
C_8			
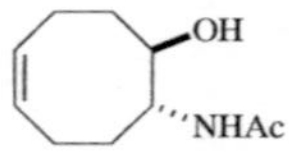	Ac_2O (0.84 eq), **104** (2.5 mol %), *i*-Pr_2NEt, toluene, rt, 110 min	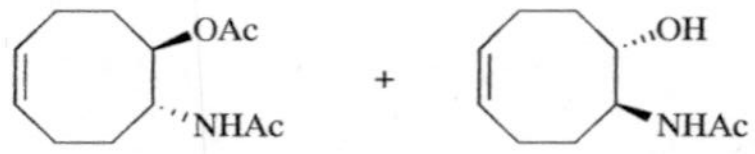 er 73.5:26.5 + er 67.5:32.5 (43), *s* = 4	67
C_9			
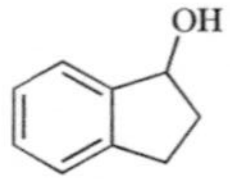	AcOC(Me)=CH_2 (1.27 eq), **50** (1 mol %), toluene, –25°, 12 h	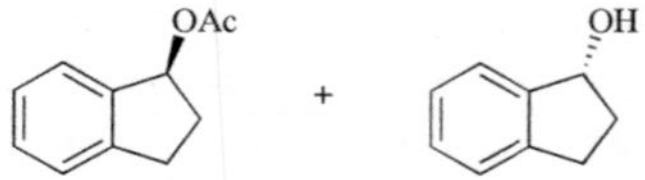 + er 95.5:4.5 (76), *s* = 5	101
	$(EtCO)_2O$ (0.75 eq), **124** (8 mol %), *i*-Pr_2NEt (0.75 eq), $CDCl_3$, rt, 22 h	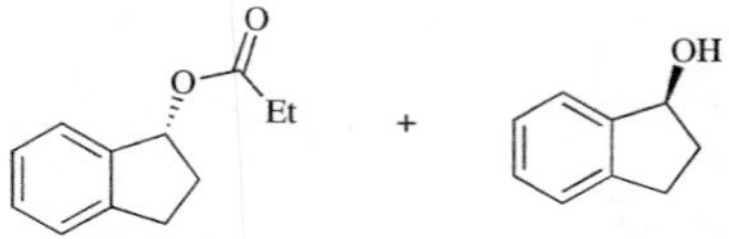 er 55.0:45.0 + er 51.0:49.0 (15), *s* = 1	57
	(*i*-$PrCO)_2O$ (2.0 eq), **88** (1 mol %), Et_3N (0.75 eq), toluene, –98°, 15 h	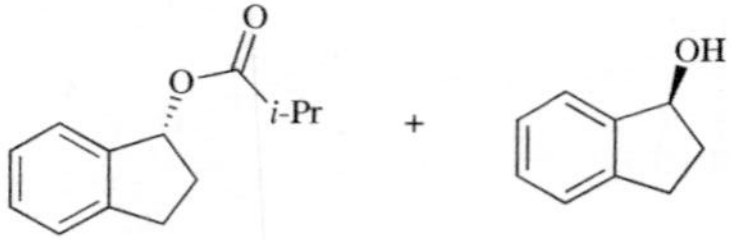 er 68.5:31.5 + er 58.5:41.5 (32), *s* = 3	164

C_{9-10}

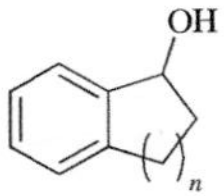

$(i\text{-PrCO})_2O$ (2.5 eq),
141 (5.3 mol %),
heptane, rt

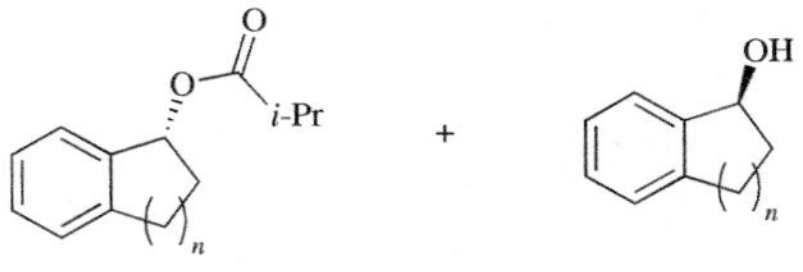

155

I   +   **II**

n	Time (h)		er **II**	s
1	3	(54)	55.5:44.5	1
2	24	(42)	51.5:48.5	1

C_9

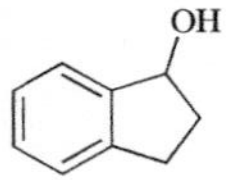

BzCl (0.75 eq), **139** (15 mol %),
Et_3N (0.5 eq), 4 Å MS,
CH_2Cl_2, −78°, 11 h

er 69.0:31.0   +   er 65.0:35.0

(44), $s = 3$ 213

BzCl (0.75 eq), **138** (15 mol %),
Et_3N (0.5 eq), 4 Å MS,
CH_2Cl_2, −78°, 11 h

er 68.5:31.5   +   er 76.5:23.5

(59), $s = 4$ 213

TABLE 2E. CATALYTIC KINETIC RESOLUTION OF CYCLOALKANOLS (*Continued*)

Alcohol	Conditions	Product(s) and Conversion(s) (%)	Refs.

*Please refer to the charts preceding the tables for the structures indicated by the **bold** numbers.*

C$_9$

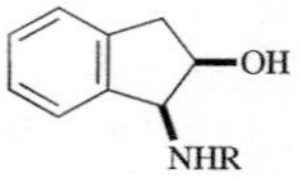

(*i*-PrCO)$_2$O (0.7 eq),
64 (5 mol %),
CHCl$_3$, rt, 16 h

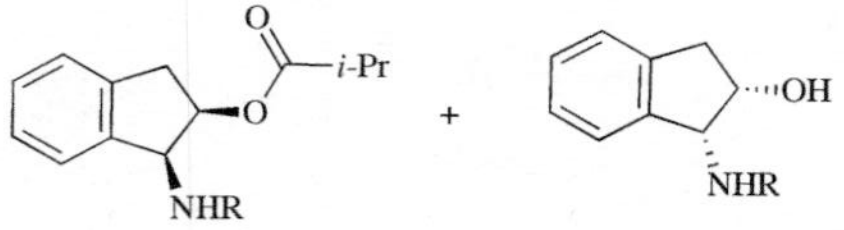

R	(I)	er II	s
CF$_3$CO	(69)	92.0:8.0	5
Boc	(77)	91.5:8.5	4
C$_6$F$_5$CO	(51)	67.5:32.5	3
4-Me$_2$NC$_6$H$_4$CO	(74)	99.5:0.5	9
Cbz	(49)	55.0:45.0	1
Phth	(61)	89.5:10.5	7

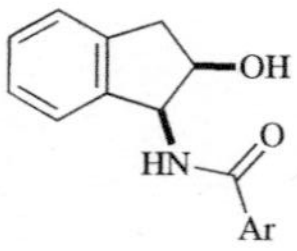

236

(*i*-PrCO)$_2$O (0.7 eq),
ent-59 (5 mol %),
2,4,6-collidine, CH$_2$Cl$_2$, 20°, 3 h

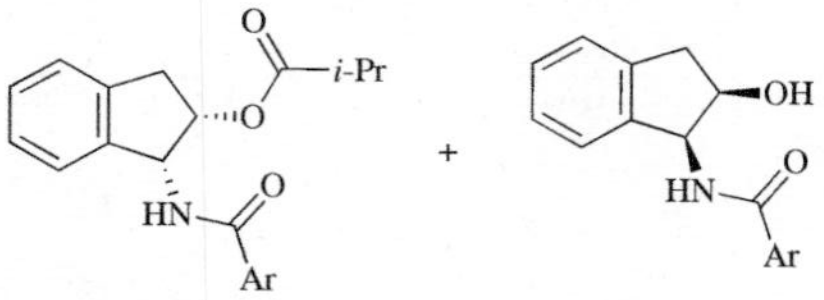

er 78.0:22.0   er 99.5:0.5   (64), *s* = 17   180

Ar = 4-Me$_2$NC$_6$H$_4$

(*i*-PrCO)$_2$O (1.5 eq),
65 (5 mol %),
CHCl$_3$, rt, 24 h

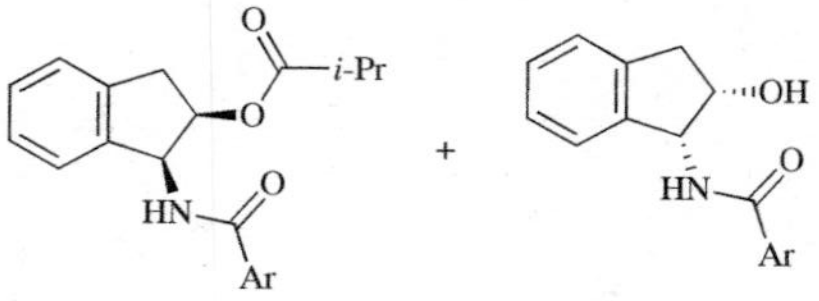

er 98.5:1.5   (72), *s* = 8   177

(i-PrCO)$_2$O (1.5 eq),
64 (5 mol %),
CHCl$_3$, rt, 24 h

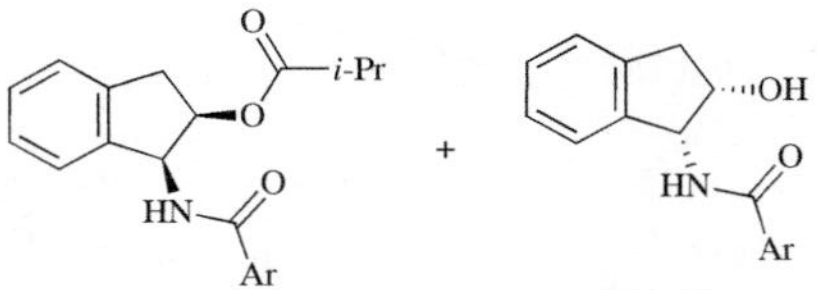

er 99.0:1.0

(74), s = 8   177

(i-PrCO)$_2$O (0.7 eq),
64 (5 mol %),
toluene, rt, 3 h

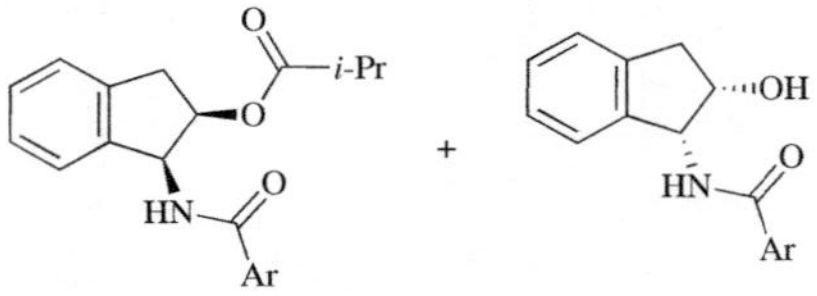

er 99.5:0.5

(74), s = 9   178

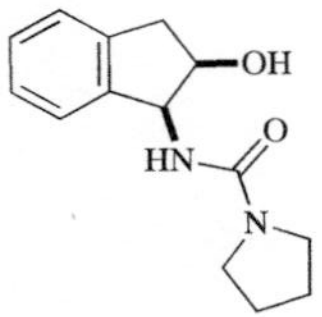

(i-PrCO)$_2$O (0.5 eq),
109 (5 mol %), i-Pr$_2$NEt (0.5 eq),
CHCl$_3$/CCl$_4$ (2:3), 0°, 3 h

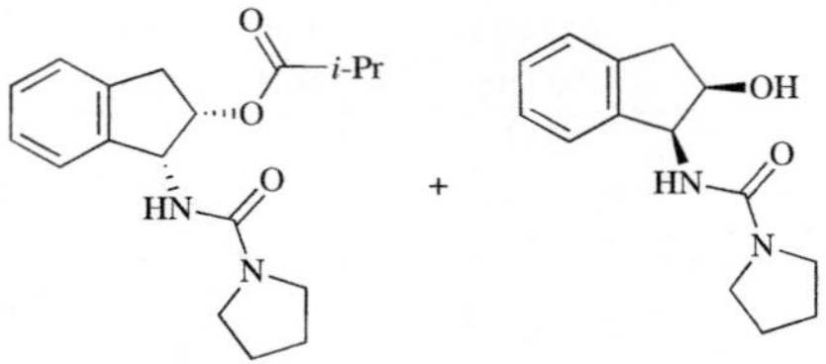

er 96.5:3.5   er 83.5:16.5

(42), s = 51   70

TABLE 2E. CATALYTIC KINETIC RESOLUTION OF CYCLOALKANOLS (*Continued*)

Alcohol	Conditions	Product(s) and Conversion(s) (%)	Refs.

*Please refer to the charts preceding the tables for the structures indicated by the **bold** numbers.*

C_{9-15}

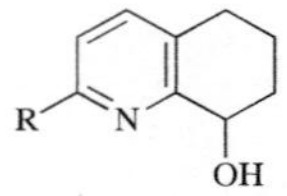

BzCl (0.51 eq), **25** (1 mol %),
CuCl$_2$ (1 mol %),
i-Pr$_2$NEt (1.0 eq), CH$_2$Cl$_2$, 0°, 2 h

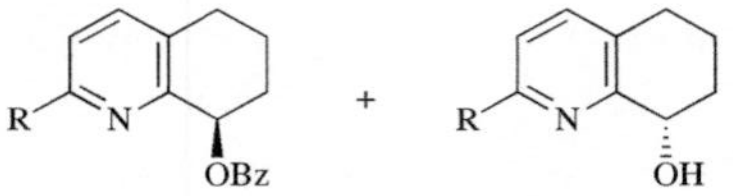

+

I + II

R		er I	er II	s
H	(52)	88.0:12.0	(50) 91.5:8.5	19
Cl	(44)	82.5:17.5	79.0:21.0	8
Ph	(42)	98.5:1.5	85.0:15.0	125

215

C_{10}

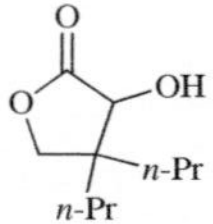

Ph$_2$CHCO$_2$H (0.5 eq),
Piv$_2$O (0.6 eq), **124** (5 mol %),
i-Pr$_2$NEt (1.2 eq), Et$_2$O, rt, 12 h

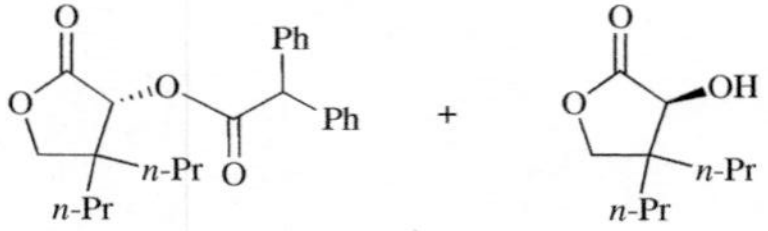

+

er >99.5:0.5   er >99.5:0.5

(50), *s* >1000   63

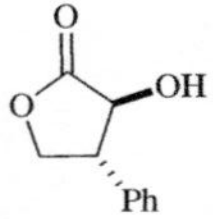

n-PrNCO (0.5 eq),
27 (0.05 mol %),
CH$_2$Cl$_2$, rt, 7 h

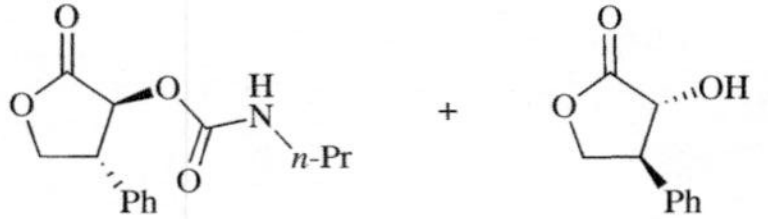

+

er 68.5:31.5   er 66.0:34.0

(46), *s* = 3   97

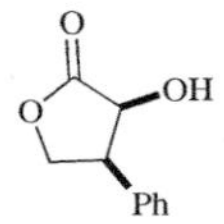

n-PrNCO (0.5 eq),
27 (0.05 mol %),
CH₂Cl₂, 0°, 2 h

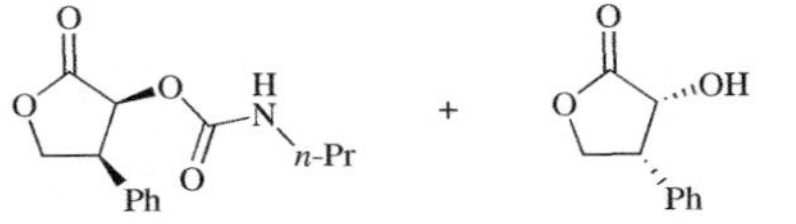

er 98.0:2.0

(50), *s* = 209

er 98.0:2.0

97

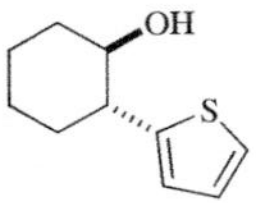

(EtCO)₂O (0.75 eq), **125** (4 mol %),
i-Pr₂NEt (0.75 eq),
t-AmOH/toluene (1:1), –40°, 10 h

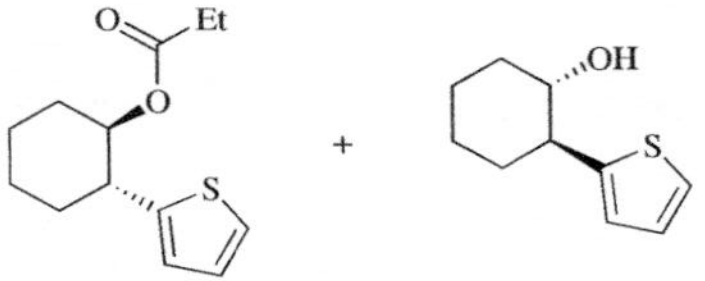

er 95.0:5.0

(45), *s* = 44

er 86.5:13.5

194

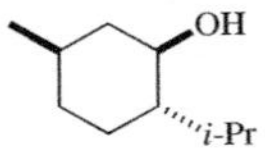

(EtCO)₂O (0.75 eq), **125** (4 mol %),
i-Pr₂NEt (0.75 eq),
t-AmOH/toluene (1:1), –10°, 12 h

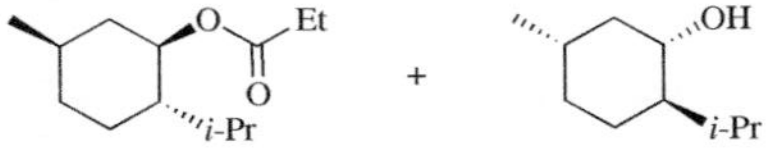

er 72.5:27.5

(32), *s* = 3

er 60.5:39.5

194

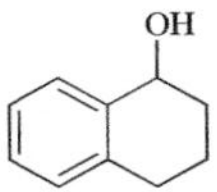

AcOC(Me)=CH₂ (1.27 eq),
50 (2 mol %),
toluene, –10°, 8 h

(39), *s* = 2

er 57.0:43.0

101

Ac₂O, TaCl₂ (5 mol %),
39 (10 mol %),
CH₂Cl₂, rt, 40 h

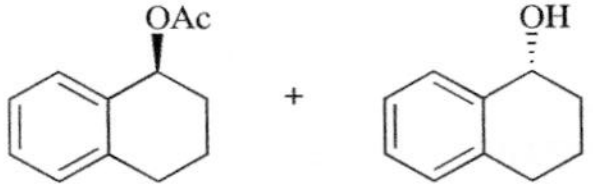

(55)

er 70.0:30.0

198

TABLE 2E. CATALYTIC KINETIC RESOLUTION OF CYCLOALKANOLS (*Continued*)

Alcohol	Conditions	Product(s) and Conversion(s) (%)	Refs.
Please refer to the charts preceding the tables for the structures indicated by the **bold** numbers.			
C$_{10}$			
	BzCl (0.75 eq), **139** (15 mol %), Et$_3$N (0.5 eq), 4 Å MS, CH$_2$Cl$_2$, –78°, 11 h	er 79.0:21.0 + (46), *s* = 6 er 75.0:25.0	213
	BzCl (0.75 eq), **137** (0.5 mol %), Et$_3$N (0.5 eq), 4 Å MS, CH$_2$Cl$_2$, –78°, 3 h	er 93.0:7.0 + (—), *s* = 34 er 90.5:9.5	212
C$_{11}$			
	(EtCO)$_2$O (0.75 eq), **125** (4 mol %), *i*-Pr$_2$NEt (0.75 eq), *t*-AmOH/toluene (1:1), –40°, 7 h	er 95.0:5.0 + (51), *s* = 65 er 97.0:3.0	194
	(*i*-PrCO)$_2$O (0.55 eq), **155** (5 mol %), *i*-Pr$_2$NEt (0.6 eq), (MeO)$_2$CO, rt, 24 h	er 90.0:10.0 + (53), *s* = 27 er 95.0:5.0	242
	(*i*-PrCO)$_2$O (0.55 eq), **155** (1 mol %), *i*-Pr$_2$NEt (0.6 eq), CHCl$_3$, rt, 7 h	er 95.0:5.0 + (51), *s* = 70 er 98.0:2.0	196

346

C$_{11-13}$

OH Ph (structure)

OH Ph (structure)

104

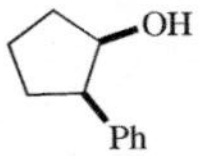

(i-PrCO)$_2$O (0.7 eq),
157a (5 mol %),
CHCl$_3$, rt, 24 h

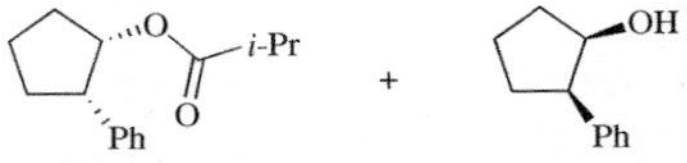

(structure with O, i-Pr ester, Ph) er 94.0:6.0 + (structure OH Ph) er 96.0:4.0

(51), s = 55

213

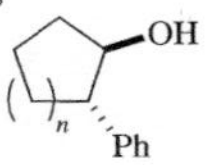

BzCl (0.75 eq), **139** (15 mol %),
Et$_3$N (0.5 eq), 4 Å MS,
CH$_2$Cl$_2$, –78°, 11 h

(structure OBz Ph) **I** + (structure OH Ph) **II**

n	er I	er II	s
1	(46)	92.5:7.5	86.0:14.0	27
2	(47)	98.0:2.0	92.5:7.5	134
3	(15)	89.0:11.0	57.0:43.0	9

158

BzBr (1.0 eq), **137** (30 mol %),
SnBr$_2$ (0.3 eq), 4 Å MS,
CH$_2$Cl$_2$, –78°, 24 h

(structure OBz Ph) **I** + (structure OH Ph) **II**

n	er I	er II	s
1	(50)	93.0:7.0	93.0:7.0	27
2	(46)	98.5:1.5	92.0:8.0	>100
3	(52)	94.5:5.5	97.5:2.5	41

TABLE 2E. CATALYTIC KINETIC RESOLUTION OF CYCLOALKANOLS (*Continued*)

Alcohol	Conditions	Product(s) and Conversion(s) (%)	Refs.

Please refer to the charts preceding the tables for the structures indicated by the bold numbers.

C₁₁₋₁₃

Alcohol	Conditions	Product(s) and Conversion(s) (%)	Refs.
[cyclopentanol structure with OH, Ph]	BzCl (0.75 eq), **137** (0.3 mol %), Et₃N (0.5 eq), 4 Å MS, CH₂Cl₂, −78°, 3 h	[structure I, OBz, Ph] + [structure II, OH, Ph]	157

n		er **I**	er **II**	s
1	(50)	94.5:5.5	94.0:6.0	37
2	(50)	98.0:2.0	97.5:2.5	160
3	(45)	97.5:2.5	89.5:10.5	88

C₁₁

Alcohol	Conditions	Product(s) and Conversion(s) (%)	Refs.
[piperidine structure, OH, Ph, N–Boc]	(*i*-PrCO)₂O (0.7 eq), **157a** (5 mol %), CHCl₃, rt, 24 h	[ester structure, O–*i*-Pr, Ph, N–Boc] er 94.0:6.0 + [structure, OH, Ph, N–Boc] er >99.0:1.0 (53), $s = 128$	104
[lactam structure, OH, Ph, N–H, C=O]	(*i*-PrCO)₂O (1.5 eq), **157a** (5 mol %), MeNO₂, rt, 24 h	[structure, OAc, Ph, N–H, C=O] er 95.0:5.0 + [structure, OH, Ph, N–H, C=O] er 85.0:15.0 (44), $s = 39$	104
[bicyclic structure, MeO, MeO, Br, OH] er 71.0:29.0	Ac₂O (1.5 eq), **56** (2 mol %), Et₃N (1.8 eq), *t*-AmOH, 0°, 15 h	[structure, MeO, MeO, Br, OAc] (49), $s = 13$ + [structure, MeO, MeO, Br, OH] er 99.5:0.5	330

348

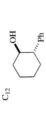

C$_{12}$

Ac$_2$O (0.7 eq), **62** (10 mol %), Et$_3$N (0.7 eq), toluene, rt	OAc / Ph	+	OH / Ph er 62.0:38.0	(51), $s = 2$ 175
38•PdCl$_2$ (10 mol %), Ph$_3$Bi(OAc)$_2$ (0.6 eq), AgOAc (3 eq), CO (5 atm), THF, 30°, 48 h	OAc / Ph er 69.5:30.5	+	OH / Ph er 61.5:38.5	(—) 214
Ac$_2$O (0.75 eq), **79** (5 mol %), Et$_3$N (0.75 eq), toluene, –78°, 15 h	OAc / Ph er 91.0:9.0	+	OH / Ph	(14), $s = 10$ 173
Ac$_2$O (1.8 eq), **75** (5 mol %), Et$_3$N (1.85 eq), toluene, rt, 16 h	OAc / Ph er 99.5:0.5	+	OH / Ph	(89), $s = 5$ 229
Ac$_2$O (1.5 eq), **101** (2.5 mol %), toluene, –65°, 24 h	OAc / Ph	+	OH / Ph	(—), $s = >50$ 68

TABLE 2E. CATALYTIC KINETIC RESOLUTION OF CYCLOALKANOLS (*Continued*)

Alcohol	Conditions	Product(s) and Conversion(s) (%)	Refs.

*Please refer to the charts preceding the tables for the structures indicated by the **bold** numbers.*

C$_{12}$

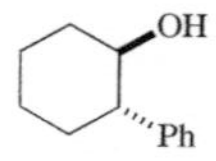

Ac$_2$O (1.5 eq), **100** (0.5 mol %),
i-Pr$_2$NEt (0.4 eq),
toluene, –65°, 12 h

(37), *s* > 50

69

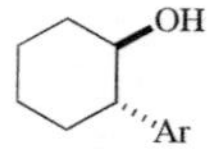

Ac$_2$O (1 eq), **85** (1 mol %),
t-AmOH, 0°

I   **II**

Ar	Time (h)		er I	er II	*s*
Ph	2	(62)	81.0:19.0	99.5:0.5	21
PMP	4	(68)	68.0:32.0	87.5:12.5	5

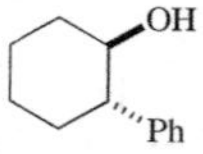

321

(EtCO)$_2$O (0.75 eq),
catalyst (*x* mol %),
i-Pr$_2$NEt (0.75 eq), CDCl$_3$, rt

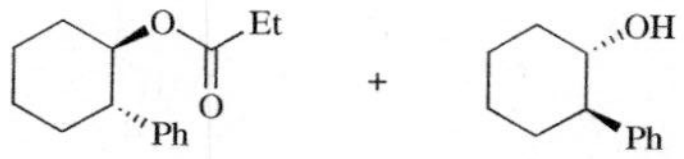

I   +   **II**

Catalyst	*x*	Time (h)		er I	er II	*s*
117	8	12	(44)	87.5:12.5	79.0:21.0	12
123	2	2	(44)	88.0:12.0	80.0:20.0	14
124	8	6	(50)	91.0:9.0	90.5:9.5	25
125	2	1	(50)	92.0:8.0	80.0:20.0	29

194

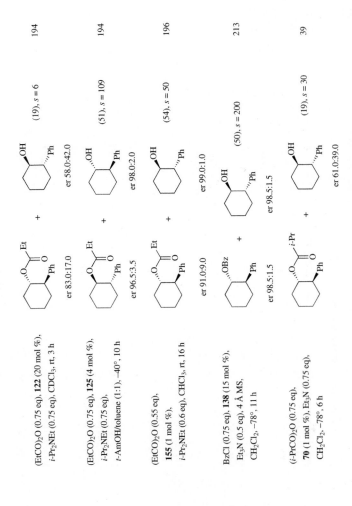

(EtCO)₂O (0.75 eq), **122** (20 mol %),
i-Pr₂NEt (0.75 eq), CDCl₃, rt, 3 h

er 83.0:17.0 + er 58.0:42.0   (19), s = 6   194

(EtCO)₂O (0.75 eq), **125** (4 mol %),
i-Pr₂NEt (0.75 eq),
t-AmOH/toluene (1:1), –40°, 10 h

er 96.5:3.5 + er 98.0:2.0   (51), s = 109   194

(EtCO)₂O (0.55 eq),
155 (1 mol %),
i-Pr₂NEt (0.6 eq), CHCl₃, rt, 16 h

er 91.0:9.0 + er 99.0:1.0   (54), s = 50   196

BzCl (0.75 eq), **138** (15 mol %),
Et₃N (0.5 eq), 4 Å MS,
CH₂Cl₂, –78°, 11 h

er 98.5:1.5 + er 98.5:1.5   (50), s = 200   213

(i-PrCO)₂O (0.75 eq),
70 (1 mol %), Et₃N (0.75 eq),
CH₂Cl₂, –78°, 6 h

er 61.0:39.0 +   (19), s = 30   39

Alcohol	Conditions	Product(s) and Conversion(s) (%)	Refs.

*Please refer to the charts preceding the tables for the structures indicated by the **bold** numbers.*

C_{12}

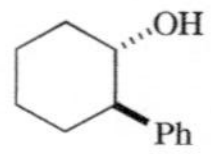

Alcohol	Conditions	Product(s) and Conversion(s) (%)	Refs.
(trans-2-phenylcyclohexanol)	(*i*-PrCO)$_2$O (0.7 eq), **64** (5 mol %), toluene, rt, 3 h	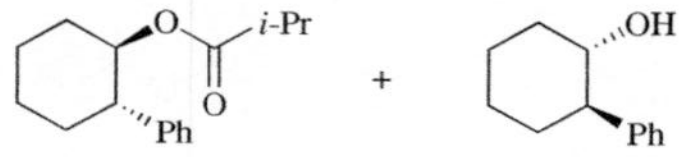 ester (*i*-Pr) er 55.5:44.5 + OH/Ph (65), *s* = 1	178
	(*i*-PrCO)$_2$O (1.5 eq), **65** (5 mol %), CH$_2$Cl$_2$, rt, 24 h	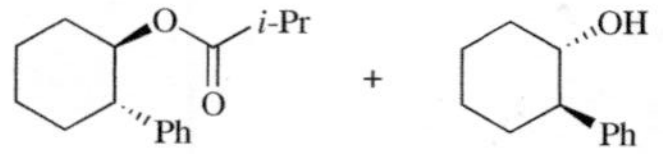 ester (*i*-Pr) er 53.0:47.0 + OH/Ph (42), *s* = 1	177
	(*i*-PrCO)$_2$O (1.5 eq), **64** (5 mol %), toluene, rt, 24 h	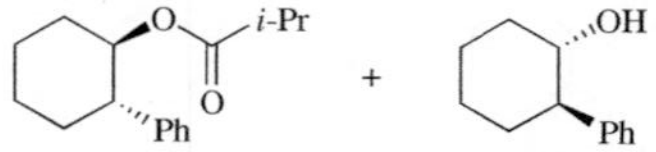 ester (*i*-Pr) er 55.5:44.5 + OH/Ph (65), *s* = 1	177
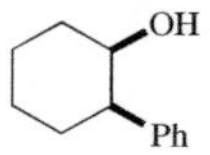 (cis-2-phenylcyclohexanol)	(EtCO)$_2$O (0.75 eq), **125** (4 mol %), *i*-Pr$_2$NEt (0.75 eq), *t*-AmOH/toluene (1:1), –10°, 12 h	 ester (Et) er 93.0:7.0 + OH/Ph er 86.0:14.0 (46), *s* = 28	194
	(EtCO)$_2$O (0.55 eq), **155** (1 mol %), *i*-Pr$_2$NEt (0.6 eq), CHCl$_3$, rt, 16 h	 ester (Et) er 84.0:16.0 + OH/Ph er 84.0:16.0 (50), *s* = 11	196

C$_{12-14}$

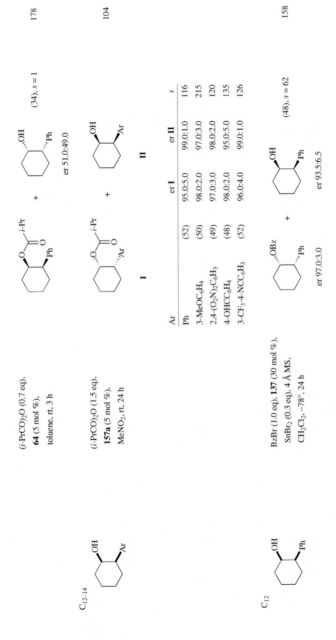

(i-PrCO)$_2$O (0.7 eq),
64 (5 mol %),
toluene, rt, 3 h

(34), s = 1

er 51.0:49.0

178

(i-PrCO)$_2$O (1.5 eq),
157a (5 mol %),
MeNO$_2$, rt, 24 h

104

Ar	er I		er II	s
Ph	(52)	95.0:5.0	99.0:1.0	116
3-MeOC$_6$H$_4$	(50)	98.0:2.0	97.0:3.0	215
2,4-(O$_2$N)$_2$C$_6$H$_3$	(49)	97.0:3.0	98.0:2.0	120
4-OHCC$_6$H$_4$	(48)	98.0:2.0	95.0:5.0	135
3-CF$_3$-4-NCC$_6$H$_3$	(52)	96.0:4.0	99.0:1.0	126

C$_{12}$

BzBr (1.0 eq), **137** (30 mol %),
SnBr$_2$ (0.3 eq), 4 Å MS,
CH$_2$Cl$_2$, –78°, 24 h

er 97.0:3.0

(48), s = 62

er 93.5:6.5

158

TABLE 2E. CATALYTIC KINETIC RESOLUTION OF CYCLOALKANOLS (*Continued*)

Alcohol	Conditions	Product(s) and Conversion(s) (%)	Refs.

*Please refer to the charts preceding the tables for the structures indicated by the **bold** numbers.*

C$_{12}$

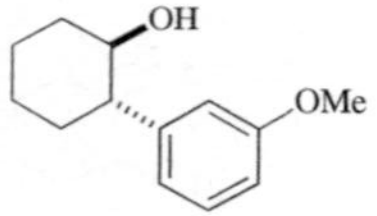

Alcohol	Conditions	Product(s) and Conversion(s) (%)	Refs.
[cyclohexanol with 3-OMe-phenyl substituent, OH]	(*i*-PrCO)$_2$O (1.5 eq), **157a** (5 mol %), MeNO$_2$, rt, 24 h	[*i*-PrCO ester of cyclohexanol with 3-OMe-phenyl] er 90.0:10.0 + [cyclohexanol, OH, OMe] (55), *s* = 39 er 99.0:1.0	104
[acenaphthylene, OH, NH–C(O)Ar] Ar = 4-Me$_2$NC$_6$H$_4$	(*i*-PrCO)$_2$O (0.6 eq), **ent-59** (5 mol %), 2,4,6-collidine, CHCl$_3$, 20°, 24 h	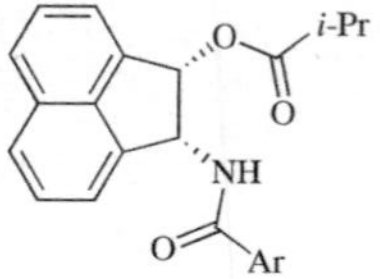 [acenaphthylene *i*-Pr ester, NH–C(O)Ar] er 83.5:16.5 + [acenaphthylene OH, NH–C(O)Ar] (59), *s* = 21 er 98.5:1.5	180
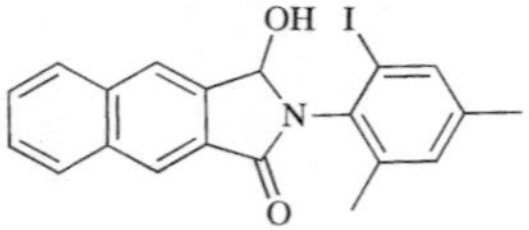 [isoindolinone with OH and 2-iodo-4,6-dimethylphenyl]	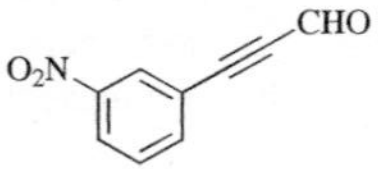 [3-O$_2$N-phenyl propynal] CHO (1.25 eq), **145f** (10 mol %), NaOAc (1 eq), 4 Å MS, CHCl$_3$, rt, 72 h	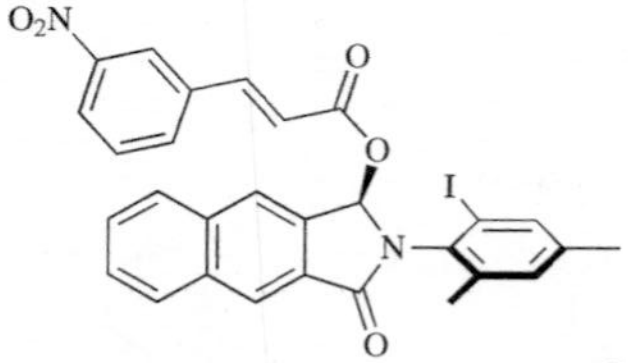 [isoindolinone ester, 3-nitrocinnamate] er 93.5:6.5 + [isoindolinone OH] (49), *s* = 39 er 92.5:7.5	233

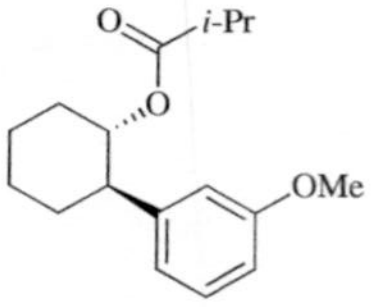

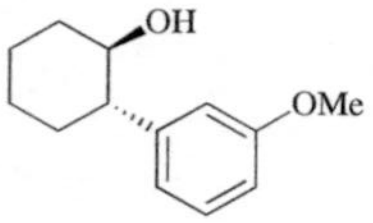

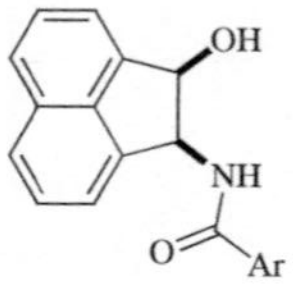

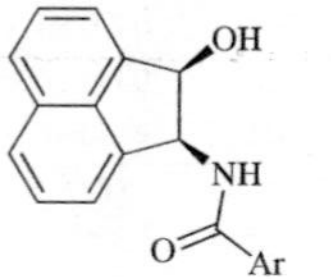

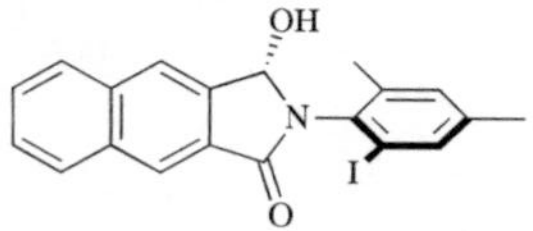

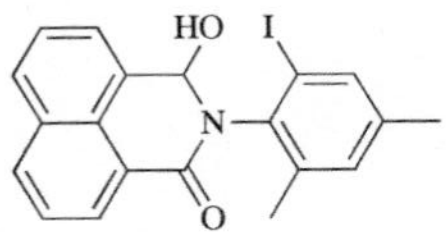

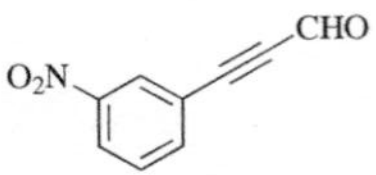

233

(48), $s = 117$

er 94.5:5.5

er 95.0:5.0

(1.25 eq),

145f (10 mol %), NaOAc (1 eq),
4 Å MS, CHCl$_3$, rt, 36 h

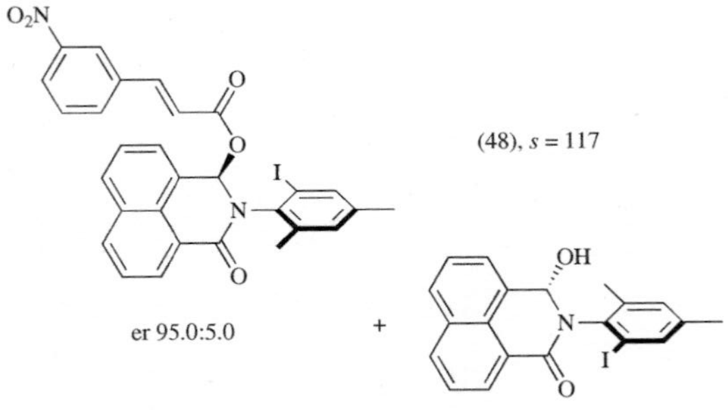

C$_{13}$

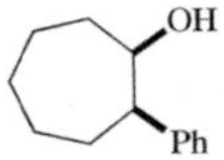

(*i*-PrCO)$_2$O (0.7 eq),
157a (5 mol %),
CHCl$_3$, rt, 24 h

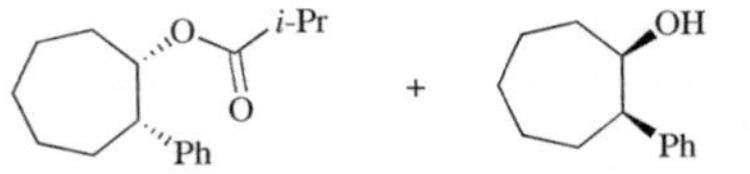

(50), $s = 151$

er >99.0:1.0

er 96.0:4.0

104

C$_{14}$

(EtCO$_2$)$_2$O (0.75 eq), **125** (2 mol %),
i-Pr$_2$NEt (0.75 eq), CDCl$_3$, rt, 26 h

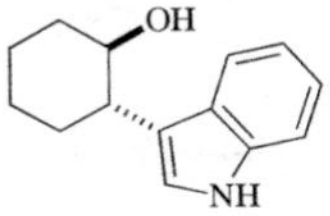

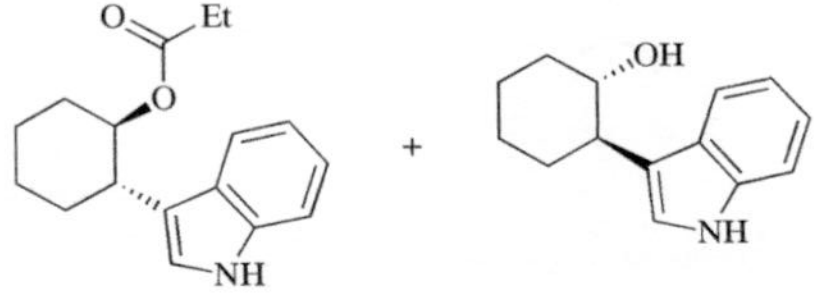

(41), $s = 49$

er 82.5:17.5

er 96.0:4.0

194

TABLE 2E. CATALYTIC KINETIC RESOLUTION OF CYCLOALKANOLS (*Continued*)

Alcohol	Conditions	Product(s) and Conversion(s) (%)	Refs.

*Please refer to the charts preceding the tables for the structures indicated by the **bold** numbers.*

C$_{14}$

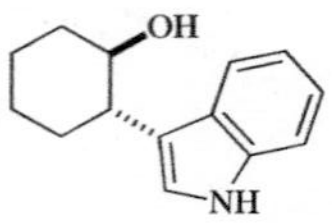

(EtCO)$_2$O (0.55 eq),
155 (1 mol %),
i-Pr$_2$NEt (0.6 eq), CHCl$_3$, rt, 16 h

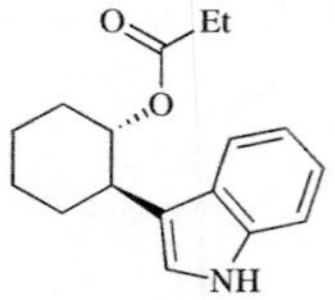

er 96.0:4.0   +   er 89.0:11.0   (46), *s* = 50   196

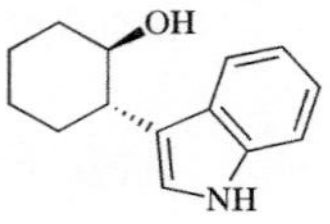

C$_{16}$

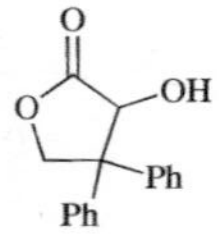

n-PrNCO (0.5 eq),
27 (0.05 mol %),
CH$_2$Cl$_2$, rt, 9 h

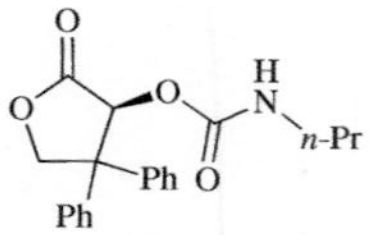

er 97.0:3.0   +   er 96.5:3.5   (50), *s* = 113   97

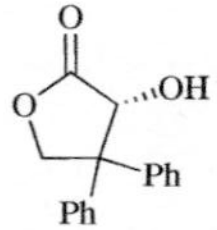

356

TABLE 2F. CATALYTIC KINETIC RESOLUTION OF rac-DIOLS

Alcohol	Conditions	Product(s) and Conversion(s) (%)	Refs.

*Please refer to the charts preceding the tables for the structures indicated by the **bold** numbers.*

C$_4$

| 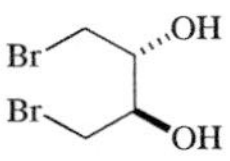 | 4-CF$_3$C$_6$H$_4$COCl (0.65 eq). **129** (30 mol %), i-Pr$_2$NEt (0.5 eq), EtCN, −78°, 1 h | 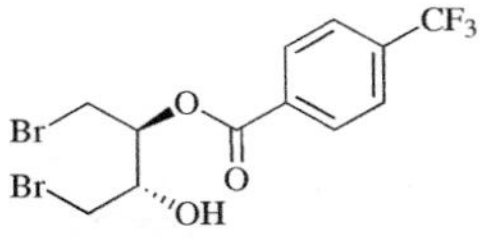 er 84.0:16.0 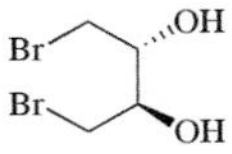   +   er 84.5:15.5   (50), s = 11 | 220 |

C$_{4-8}$

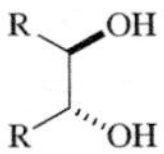

| | (i-PrCO)$_2$O (0.75 eq), **69b** (0.5 mol %), (Me$_2$NCH$_2$)$_2$ (0.75 eq), THF, −78°, 9 h | **I** + **II** | 225 |

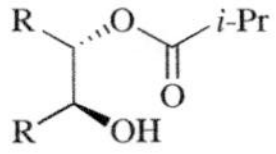

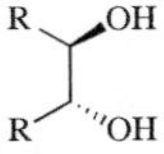

R	er **I**		er **II**	s
BnOCH$_2$	(39)	77.5:22.5	68.5:31.5	2
BnO$_2$C	(40)	62.5:37.5	60.0:40.0	2
Et	(46)	92.0:8.0	84.5:15.5	23
CH$_2$=CH	(40)	85.5:14.5	74.0:26.0	10
CH$_2$=CHCH$_2$	(52)	92.0:8.0	96.0:4.0	36

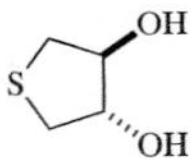

C$_4$

| | (i-PrCO)$_2$O (0.75 eq), **69b** (0.5 mol %), (Me$_2$NCH$_2$)$_2$ (0.75 eq), THF, −78°, 9 h | er 81.5:18.5   +   er 69.5:30.5   (43), s = 6 | 225 |

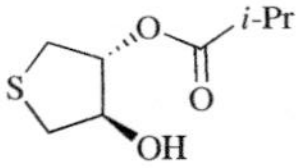

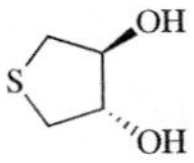

TABLE 2F. CATALYTIC KINETIC RESOLUTION OF rac-DIOLS (*Continued*)

Alcohol	Conditions	Product(s) and Conversion(s) (%)	Refs.

*Please refer to the charts preceding the tables for the structures indicated by the **bold** numbers.*

C₄

| 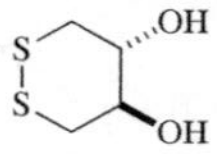 | (i-PrCO)₂O (0.75 eq), **69b** (0.5 mol %), (Me₂NCH₂)₂ (0.75 eq), THF, −78°, 9 h | 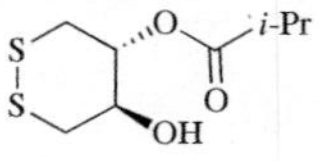 er 86.0:14.0 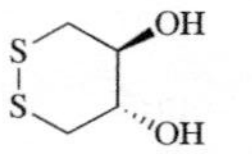   +   er 96.5:3.5   (56), s = 20 | 225 |
| | 4-CF₃C₆H₄COCl (0.65 eq), **129** (30 mol %), i-Pr₂NEt (0.5 eq), EtCN, −78°, 1 h | 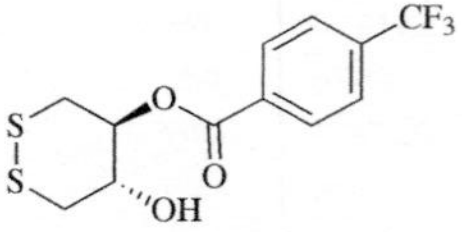 er 87.0:13.0 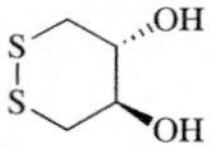   +   er 76.0:24.0   (41), s = 11 | 220 |

C₅

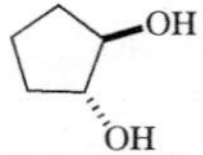

| | Ac₂O (5.3 eq), **105** (2 mol %), toluene, CH₂Cl₂, −20°, 9 h | 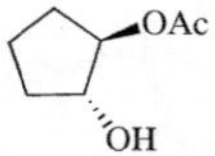 er 74.5:25.5 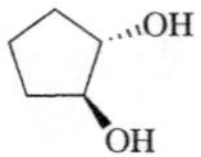   +   er 92.5:7.5   (63), s = 8 | 74 |

C₅₋₈

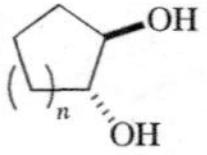

| | AcOH (2 eq), DIC (1.2 eq), **105** (2 mol %), toluene, 0° | I   +   II | 222 |

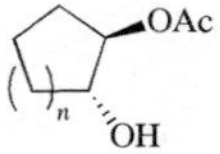

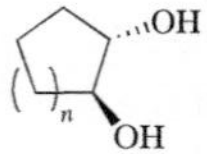

n	Time (h)	er I	er II	s
1	10 (76)	66.0:34.0	99.5:0.5	11
3	16 (55)	91.0:9.0	99.5:0.5	>50
4	18 (55)	90.5:9.5	99.5:0.5	>50

221

Ac₂O (5.3 eq),
i-Pr₂NEt (5.3 eq),
106 (5 mol %),
toluene, 0°

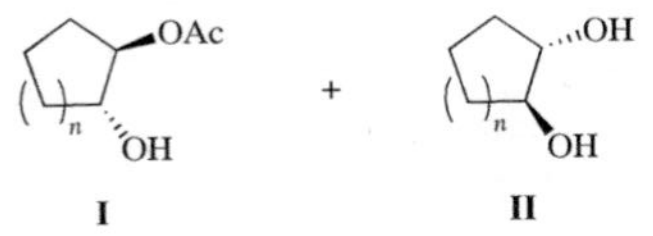

I + **II**

n	Time (h)		er **I**	er **II**	s
1	48	(71)	66.0:34.0	88.0:12.0	4
2	2	(57)	87.0:13.0	99.5:0.5	48
3	2	(58)	85.0:15.0	99.5:0.5	40
4	3	(62)	80.5:19.5	99.5:0.5	29

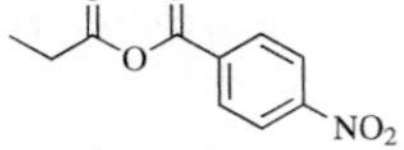

223

105 (5 mol %), i-Pr₂NEt
toluene, 0°, 24 h

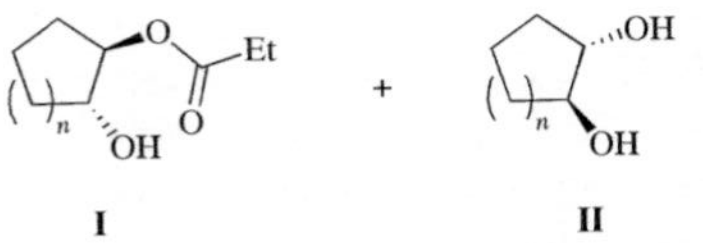

I + **II**

n		er **I**	er **II**	s
1	(70)	65.5:34.5	88.0:12.0	4
2	(47)	94.0:6.0	90.5:9.5	39
3	(49)	94.0:6.0	93.0:7.0	43
4	(50)	96.5:3.5	97.0:3.0	>50

TABLE 2F. CATALYTIC KINETIC RESOLUTION OF rac-DIOLS (*Continued*)

Alcohol	Conditions	Product(s) and Conversion(s) (%)	Refs.

*Please refer to the charts preceding the tables for the structures indicated by the **bold** numbers.*

C_{5-8}

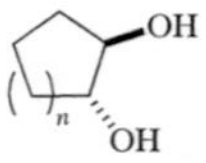

(i-PrCO)$_2$O (0.75 eq), **69b** (0.5 mol %), (Me_2NCH_2)$_2$ (0.75 eq), THF, –78°, 9 h

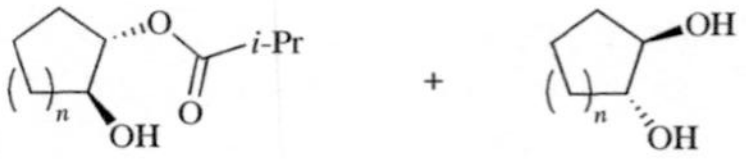

I + II

225

n		er **I**	er **II**	s
1	(33)	69.5:30.5	64.0:36.0	3
2	(53)	97.5:2.5	97.5:2.5	146
3	(52)	91.0:9.0	92.5:7.5	28
4	(58)	88.5:11.5	99.5:0.5	75

12 (0.6 eq), **145b** (0.5 mol %), **165** (10 mol %), **166** (1 eq), $CHCl_3$, 0°

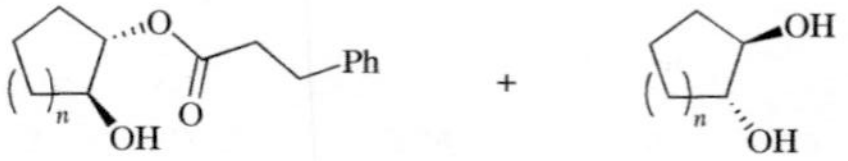

I + II

79

n	Time (h)	**I** + **II**	er **I**	er **II**	s
1	14	(40)	91.0:9.0	77.0:23.0	18
2	8	(42)	99.0:1.0	85.0:15.0	218
3	8	(44)	98.5:1.5	88.5:11.5	149
4	8	(41)	99.0:1.0	84.5:15.5	196

C$_5$

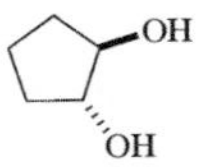

ArCOCl (0.65 eq),
129 (30 mol %),
i-Pr$_2$NEt (0.5 eq),
EtCN, –78°, 1 h

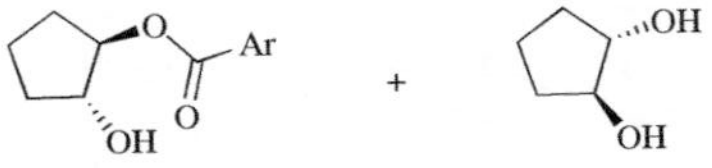

I + **II**

Ar	**I + II**	er **I**	er **II**	*s*
4-ClC$_6$H$_4$	(48)	90.5:9.5	87.0:13.0	21
4-CF$_3$C$_6$H$_4$	(44)	90.5:9.5	81.5:18.5	18

220

C$_{6–8}$

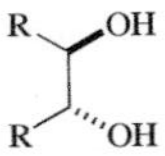

Ac$_2$O (5.3 eq),
i-Pr$_2$NEt (5.3 eq),
106 (5 mol %),
toluene, 0°

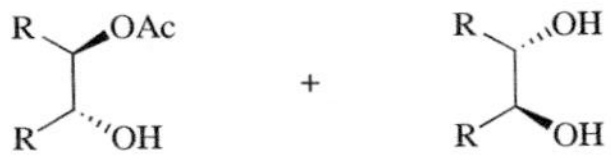

I + **II**

R	Time (h)	**I + II**	er **I**	er **II**	*s*
Et	4	(68)	74.0:26.0	99.5:0.5	19
n-Pr	5	(67)	74.0:26.0	99.5:0.5	19

221

C$_{6–14}$

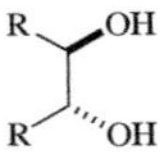

1. EtOH, **107** (2.5 mol %),
m-CPBA (2 eq),
toluene, rt, 15 h
2. DIC (2 eq), rt, 2 h
3. (±)-alcohol (0.5 eq), 0°, time

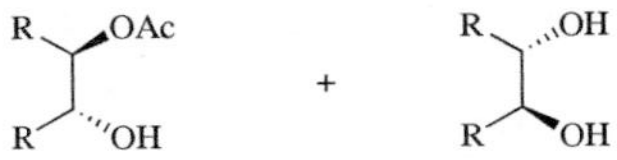

I + **II**

R	Time (h)		er **I**	er **II**
Et	12	(—)	86.5:13.5	86.5:13.5
n-Pr	18	(—)	88.0:12.0	82.0:18.0
Ph	4	(—)	69.0:31.0	87.5:12.5

224

TABLE 2F. CATALYTIC KINETIC RESOLUTION OF rac-DIOLS (*Continued*)

Alcohol	Conditions	Product(s) and Conversion(s) (%)	Refs.

*Please refer to the charts preceding the tables for the structures indicated by the **bold** numbers.*

C_{6-16}

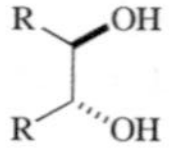

Conditions: BzCl (0.5 eq), **26** (5 mol %), i-Pr$_2$NEt, CH$_2$Cl$_2$, 0°

Refs. 125

I + II

R	I + II	er I	s
Et	(44)	88.5:11.5	14
Ph	(48)	>99.5:0.5	>645
4-MeOC$_6$H$_4$	(49)	99.0:1.0	356
4-ClC$_6$H$_4$	(48)	>99.5:0.5	>645
4-MeC$_6$H$_4$	(47)	98.5:1.5	183

C$_6$

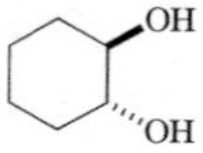

Conditions: RCO$_2$H (2 eq), DIC (1.2 eq), **105** (2 mol %), toluene, 0°

Refs. 222

I + II

R	Time (h)	I + II	er I	er II	s
H	7	(67)	70.5:29.5	91.5:8.5	6
Me	15	(55)	91.5:8.5	99.5:0.5	>50
Et	15	(55)	91.5:8.5	99.5:0.5	>50
i-Pr	24	(54)	92.5:7.5	99.5:0.5	>50
t-Bu	48	(4)	93.0:7.0	52.0:48.0	14
Ph	48	(38)	80.0:20.0	68.0:32.0	6
4-ClC$_6$H$_4$	48	(58)	80.0:20.0	92.0:8.0	10
PhCH$_2$	2	(57)	87.5:12.5	99.5:0.5	>50

Ac$_2$O (5.3 eq),
105 (2 mol %),
toluene, −20°, 4 h

er 87.5:12.5 + er 99.5:0.5

(57), s >50

74

Ac$_2$O (0.5 eq),
37 (5 mol %), Et$_3$N,
Et$_2$O, −25°, 15 h

er 86.0:14.0 + er 84.0:16.0

(48), s = 12

96

1. RCH$_2$OH, **107** (2.5 mol %),
m-CPBA (2 eq),
toluene, rt, 15 h
2. DIC (2 eq), rt, 2 h
3. (±)-Alcohol (0.5 eq), 0°, 6 h

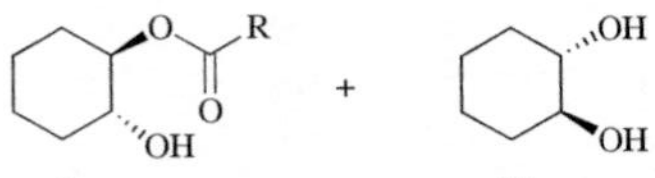

I + II

224

R	er I	er II
Me	92.0:8.0	98.5:1.5
Et	91.0:9.0	95.5:4.5
i-Pr	87.0:13.0	93.5:6.5
i-Bu	93.5:6.5	89.5:10.5
t-Bu	75.0:25.0	65.5:34.5
Cy	76.0:24.0	90.5:9.5
Ph	74.0:26.0	65.0:35.0
PhCH$_2$	83.5:16.5	92.0:8.0
Ph(CH$_2$)$_2$	80.0:20.0	95.5:4.5
n-C$_9$H$_{19}$	90.0:10.0	94.5:5.5

TABLE 2F. CATALYTIC KINETIC RESOLUTION OF rac-DIOLS (*Continued*)

Alcohol	Conditions	Product(s) and Conversion(s) (%)	Refs.

*Please refer to the charts preceding the tables for the structures indicated by the **bold** numbers.*

C₆

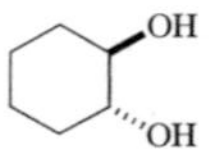

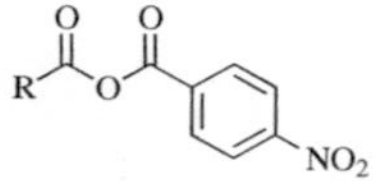

105 (5 mol %),

i-Pr₂NEt, toluene, 0°

223

R	Time (h)		er I	er II	s
i-Pr	6	(44)	95.0:5.0	86.0:14.0	40
i-Bu	24	(48)	91.0:9.0	88.0:12.0	24
t-Bu	48	(4)	96.0:4.0	52.0:48.0	25
Cy	18	(46)	81.0:19.0	76.0:24.0	7
Ph	48	(7)	89.0:11.0	53.0:47.0	9
PhCH₂	6	(35)	96.5:3.5	75.0:25.0	47
Ph(CH₂)₂	6	(50)	91.0:9.0	92.5:7.5	27
n-C₉H₁₉	24	(47)	94.0:6.0	89.5:10.5	38

PhCHO (10 eq).

ent-**145a** (10 mol %),

163 (10 mol %), NEt₃ (0.5 eq),

4 Å MS, CHCl₃, rt, 1 h

(62), s = 6

er 87.5:12.5

82

BzCl (0.51 eq), **25** (1 mol %), CuCl$_2$ (1 mol %), *i*-Pr$_2$NEt, CH$_2$Cl$_2$, 0°, 2 h

OBz / ''OH
er 91.5:8.5

+

OH / ''OH
(44), *s* = 21
er 82.0:18.0

215

BzCl (0.5 eq), **26** (5 mol %), *i*-Pr$_2$NEt, CH$_2$Cl$_2$, 0°

OBz / OH
er 90.0:10.0

+

OH / ''OH
(37), *s* = 14

125

ArCOCl (0.65 eq), **129** (30 mol %), *i*-Pr$_2$NEt (0.5 eq), EtCN, –78°, 1 h

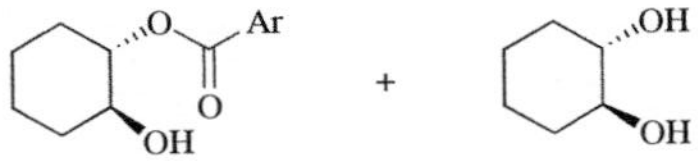

I

+

II

Ar		er **I**	er **II**	*s*
4-ClC$_6$H$_4$	(10)	96.0:4.0	55.0:45.0	26
4-CF$_3$C$_6$H$_4$	(14)	94.5:5.5	57.5:42.5	20

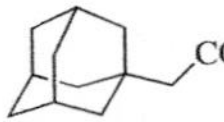

220

CO$_2$H (2 eq), DIC (1.2 eq), **105** (2 mol %), toluene, 0°, 48 h

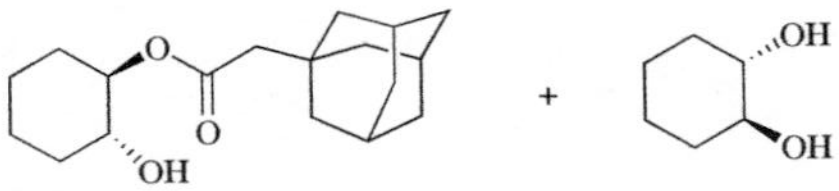

er 78.5:21.5

+

''OH / ''OH
(60), *s* = 9
er 92.5:7.5

222

365

TABLE 2F. CATALYTIC KINETIC RESOLUTION OF rac-DIOLS (*Continued*)

Alcohol	Conditions	Product(s) and Conversion(s) (%)	Refs.

*Please refer to the charts preceding the tables for the structures indicated by the **bold** numbers.*

C$_6$

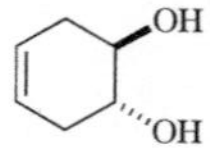

(*i*-PrCO)$_2$O (0.75 eq), **69b** (0.5 mol %),
(Me$_2$NCH$_2$)$_2$ (0.75 eq),
THF, −78°, 9 h

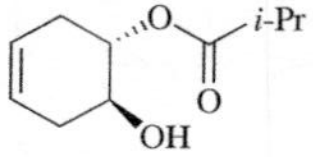

er 90.5:9.5   er 99.5:0.5

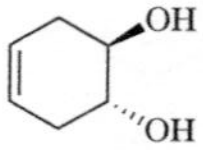

(55), s = 49

225

12 (0.6 eq), **145b** (0.5 mol %),
165 (10 mol %), **166**,
CHCl$_3$, 0°, 8 h

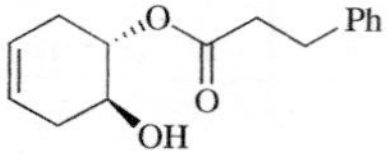

er 98.5:1.5   er 84.5:15.5

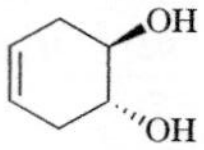

(41), s = 136

79

C$_7$

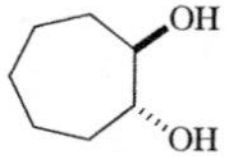

1. EtOH, **107** (2.5 mol %),
   m-CPBA (2 eq),
   toluene, rt, 15 h
2. DIC (2 eq), rt, 2 h
3. (±)-Alcohol (0.5 eq), 0°, 7 h

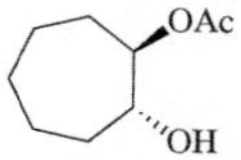

er 92.0:8.0   er 96.0:4.0

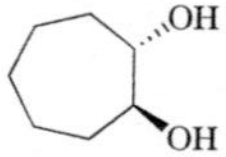

(—)

224

Ac$_2$O (5.3 eq),
105 (1 mol %),
toluene, 0°, 5 h

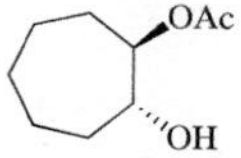

er 89.5:10.5   er 99.5:0.5

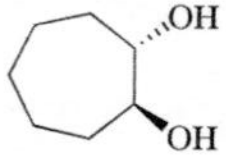

(59), s >50

74

BzCl (0.5 eq), **26** (5 mol %),
i-Pr$_2$NEt, CH$_2$Cl$_2$, 0°

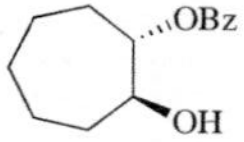

er 92.0:8.0

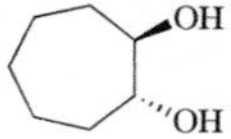

(49), s = 28

125

C$_8$

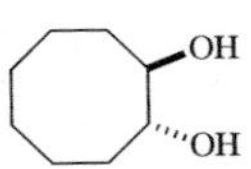

1. EtOH, **107** (2.5 mol %), *m*-CPBA (2 eq), toluene, rt, 15 h
2. DIC (2 eq), rt, 2 h
3. (±)-Alcohol (0.5 eq), 0°, 8 h

er 96.0:4.0   er 90.5:9.5   (—)   224

Ac$_2$O (5.3 eq), **105** (1 mol %), toluene, 0°, 5 h

er 99.5:0.5   er 92.5:7.5   (56), *s* >50   74

C$_{13}$

(EtCO)$_2$O (1.05 eq), *ent*-**126** (1 mol %), *i*-Pr$_2$NEt (1.05 eq), CH$_2$Cl$_2$, −20°, 2.5 h

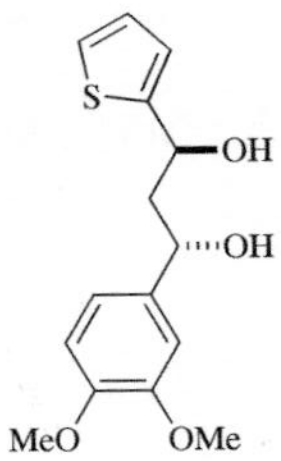

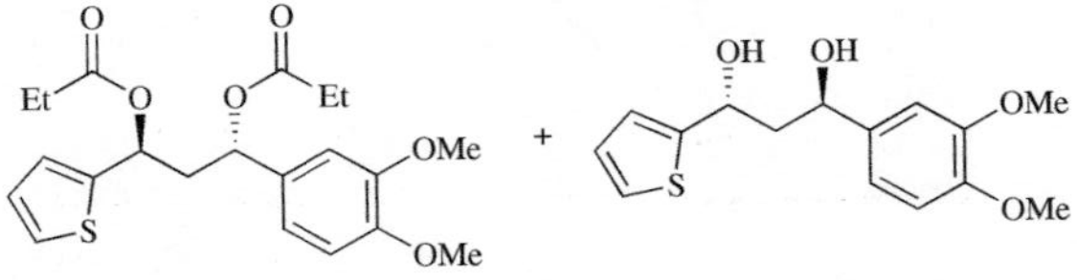

er 98.5:1.5   (—), *s* = 30   er 99.0:1.0   231

C$_{13–15}$

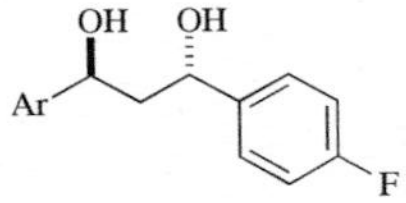

(EtCO)$_2$O (1.05 eq), *ent*-**126** (1 mol %), *i*-Pr$_2$NEt (1.05 eq), CH$_2$Cl$_2$, −20°, 2.5 h

231

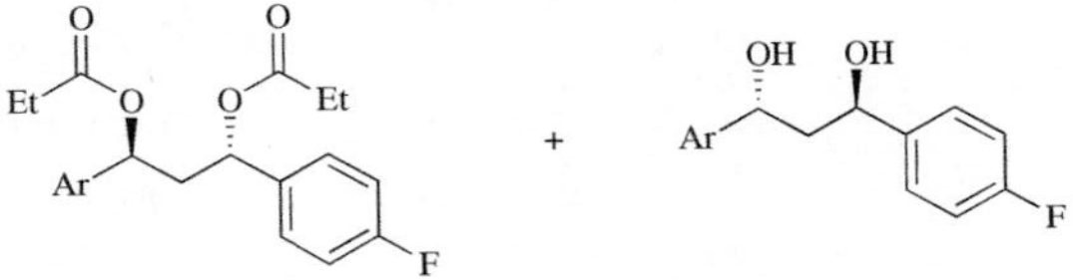

I   **II**

Ar	I	II	er **I**	er **II**	*s*
2-furyl	(—)	(—)	98.0:2.0	97.5:2.5	79
4-ClC$_6$H$_4$	(—)	(—)	96.5:3.5	97.5:2.5	18

TABLE 2F. CATALYTIC KINETIC RESOLUTION OF *rac*-DIOLS (*Continued*)

Alcohol	Conditions	Product(s) and Conversion(s) (%)	Refs.

*Please refer to the charts preceding the tables for the structures indicated by the **bold** numbers.*

C₁₄

Ph–⟨OH⟩ / Ph–⟨OH⟩   Ac₂O (0.5 eq), **37** (5 mol %), Et₃N, Et₂O, –25°, 15 h   (Ph–OAc / Ph–OH, er 82.0:18.0) + (Ph–OH / Ph–OH, er 77.5:22.5)   (46), *s* = 8   96

(*i*-PrCO)₂O (1.5 eq), **155** (1 mol %), *i*-Pr₂NEt (0.6 eq), CHCl₃, 0°, 7 h   (diester I, er 99.5:0.5) + (diester, er 82.0:18.0) + (Ph–OH / Ph–OH, er 99.5:0.5)   (80)   196

C₁₄₋₂₂

Ar–⟨OH⟩ / Ar–⟨OH⟩   (*i*-PrCO)₂O (0.75 eq), **69b** (0.5 mol %), (Me₂NCH₂)₂ (0.75 eq), THF, –78°, 9 h   I + II   225

Ar		er **I**	er **II**	*s*
Ph	(52)	96.5:3.5	99.0:1.0	125
2-BrC₆H₄	(42)	98.5:1.5	85.0:15.0	146
3-BrC₆H₄	(52)	94.5:5.5	96.0:4.0	54
4-BrC₆H₄	(51)	95.5:4.5	97.0:3.0	76
2-MeC₆H₄	(51)	96.5:3.5	98.0:2.0	114
4-MeC₆H₄	(47)	98.0:2.0	91.0:9.0	139
4-CF₃C₆H₄	(55)	96.0:4.0	99.0:1.0	106
4-MeO₂CC₆H₄	(32)	98.0:2.0	70.0:30.0	75
1-Np	(44)	98.5:1.5	88.0:12.0	180
2-Np	(52)	84.0:16.0	62.5:37.5	7

368

(i-PrCO)$_2$O (1.5 eq),
126 (1 mol %),
i-Pr$_2$NEt (1.6 eq),
CHCl$_3$, 0°, 7 h

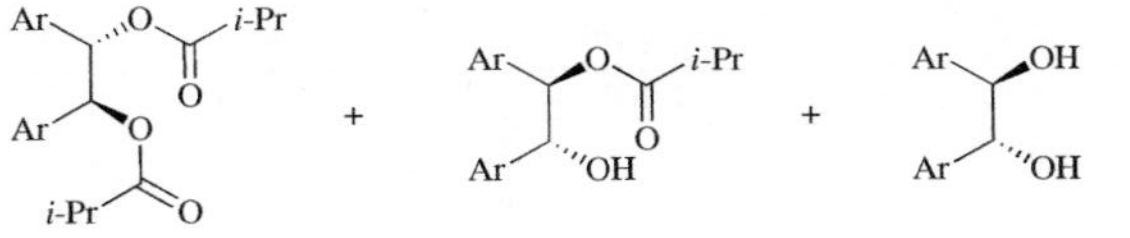

226

I + **II** + **III**

Ar		er I	er II	er III
Ph	(87)	97.0:3.0	99.5:0.5	99.5:0.5
4-ClC$_6$H$_4$	(92)	95.0:5.0	99.0:1.0	99.5:0.5
4-MeOC$_6$H$_4$	(84)	98.0:2.0	99.0:1.0	99.5:0.5
4-MeC$_6$H$_4$	(91)	96.0:4.0	99.5:0.5	99.5:0.5
4-CF$_3$C$_6$H$_4$	(93)	96.0:4.0	99.0:1.0	99.5:0.5
4-MeO$_2$CC$_6$H$_4$	(88)	90.0:10.0	94.0:6.0	87.0:13.0
1-Np	(87)	95.0:5.0	99.5:0.5	99.5:0.5

C$_{14}$

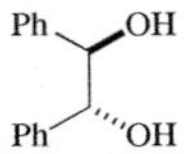

BzCl (0.5 eq), CuCl$_2$ (0.5 mol %),
22 (0.5 mol %), CH$_2$Cl$_2$, 0°, 2 h

199

er 99.5:0.5   (—), s = 751

TABLE 2F. CATALYTIC KINETIC RESOLUTION OF rac-DIOLS (*Continued*)

Alcohol	Conditions	Product(s) and Conversion(s) (%)	Refs.

*Please refer to the charts preceding the tables for the structures indicated by the **bold** numbers.*

C$_{14}$

Ph⌁OH
Ph⌁OH

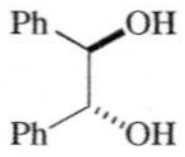 BzCl (0.5 eq), CuCl$_2$ (0.7 mol %), **23** (0.7 mol %), CH$_2$Cl$_2$, 0°, 2 h	 Ph⌁OBz / Ph⌁OH (er >99.5:0.5) + Ph⌁OH / Ph⌁OH	(—), s > 411	199
BzCl (0.5 eq), CuCl$_2$ (5 mol %), **159** (10 mol %), *i*-Pr$_2$NEt, CH$_2$Cl$_2$, 0°, 3 h	 Ph⌁OBz / Ph⌁OH (er 93:0:7.0) + Ph⌁OH / Ph⌁OH	(—), s = 28	332
BzCl (0.51 eq), **25** (1 mol %), CuCl$_2$ (1 mol %), *i*-Pr$_2$NEt, CH$_2$Cl$_2$, 0°, 2 h	Ph⌁OBz / Ph⌁OH (er 98.0:2.0) + Ph⌁OH / Ph⌁OH (er 99.0:1.0)	(51), s = 225	215
BzCl (0.5 eq), catalyst (x mol %) *i*-Pr$_2$NEt, CH$_2$Cl$_2$, 0°	Ph⌁OBz / Ph⌁OH + Ph⌁OH / Ph⌁OH		216

I   **II**

Catalyst	x	Time (h)		er **I**
20•CuCl$_2$	2	6	(28)	81.5:18.5
30	1	2	(46)	99.5:0.5
31	3	6	(42)	97.0:3.0

370

BzCl (0.5 eq), **29** (1 mol %), i-Pr$_2$NEt, CH$_2$Cl$_2$, 0°, 3 h

Ph OBz / Ph OH + Ph OH / Ph OH

er 99.5:0.5 (—), s = 536 219

BzCl (0.5 eq), **22·**CuCl$_2$ (0.5 mol %), i-Pr$_2$NEt, CH$_2$Cl$_2$, 0°, 3 h

er 99.5:0.5 (—), s = 751 218

BzCl (0.5 eq), Cu(OTf)$_2$ (1 mol %), **15** (1 mol %), i-Pr$_2$NEt, CH$_2$Cl$_2$, 0°, 24 h

(—), s = 581 217

BzCl (0.5 eq), **28** (1 mol%), i-Pr$_2$NEt, CH$_2$Cl$_2$, 0°, 2 h

er 99.5:0.5 (—), s = 311 218

er 98.0:2.0

4-CF$_3$C$_6$H$_4$COCl (0.65 eq), **129** (30 mol %), i-Pr$_2$NEt (0.5 eq), EtCN, −78°, 1 h 220

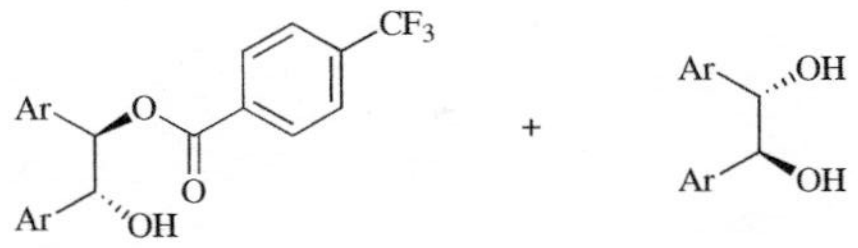

I + II

Ar		er I	er II	s
Ph	(50)	99.0:1.0	99.5:0.5	525
4-ClC$_6$H$_4$	(51)	92.5:7.5	95.0:5.0	38

TABLE 2F. CATALYTIC KINETIC RESOLUTION OF rac-DIOLS (*Continued*)

Alcohol	Conditions	Product(s) and Conversion(s) (%)	Refs.

*Please refer to the charts preceding the tables for the structures indicated by the **bold** numbers.*

C_{14-19}

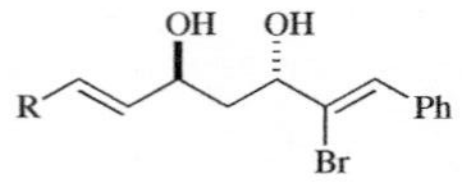

$(EtCO)_2O$ (1.05 eq), *ent*-**126** (1 mol %), *i*-Pr_2NEt (1.05 eq), CH_2Cl_2, –20°, 2.5 h

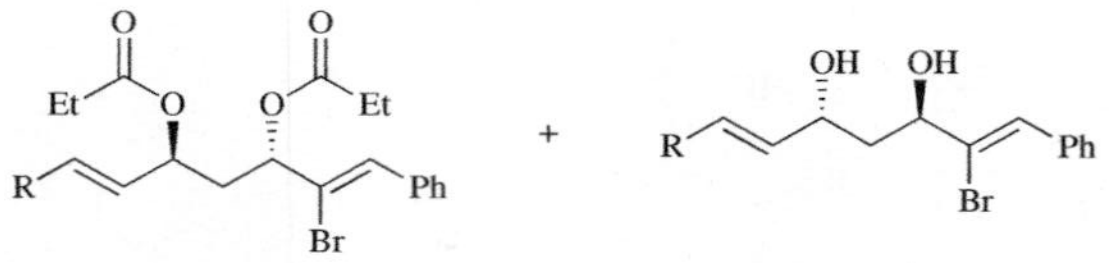

I + **II**

R		er **I**	er **II**	*s*
Me	(—)	88.5:11.5	99.5:0.5	21
Ph	(—)	98.0:2.0	99.5:0.5	22

231

C_{15}

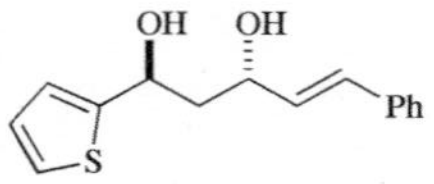

$(EtCO)_2O$ (1.05 eq), *ent*-**126** (1 mol %), *i*-Pr_2NEt (1.05 eq), CH_2Cl_2, –20°, 2.5 h

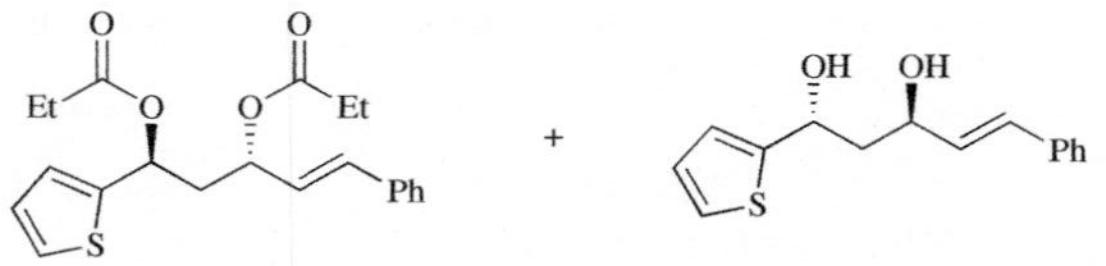

er 90.5:9.5 (—), $s = 24$ er 99.5:0.5

231

C_{15-17}

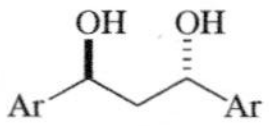

$(EtCO)_2O$ (1.05 eq), *ent*-**126** (1 mol %), *i*-Pr_2NEt (1.05 eq), CH_2Cl_2, –20°, 2.5 h

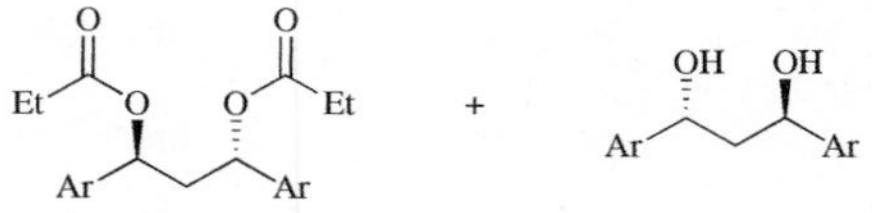

I + **II**

Ar		er **I**	er **II**	*s*
Ph	(—)	98.0:2.0	98.5:1.5	46
4-FC_6H_4	(—)	98.5:1.5	99.5:0.5	73
2-$MeOC_6H_4$	(—)	99.5:0.5	99.5:0.5	60
4-MeC_6H_4	(—)	97.5:2.5	99.5:0.5	51

231

C_{17}

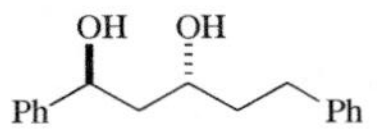

(EtCO)$_2$O (1.05 eq),
ent-**126** (1 mol %),
i-Pr$_2$NEt (1.05 eq),
CH$_2$Cl$_2$, −20°, 2.5 h

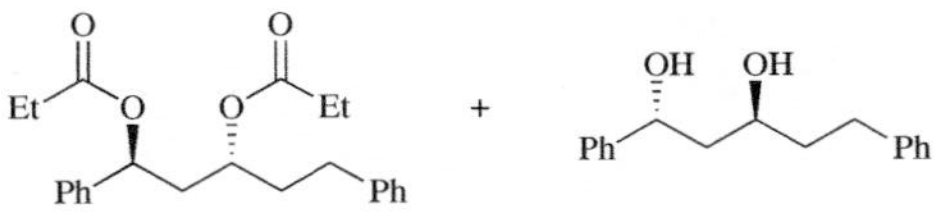

er 87.0:13.0   +   er 99.5:0.5

(—), s = 27   231

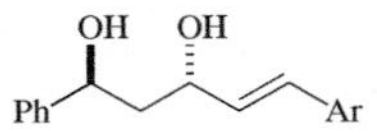

(EtCO)$_2$O (1.05 eq),
ent-**126** (1 mol %),
i-Pr$_2$NEt (1.05 eq),
CH$_2$Cl$_2$, −20°, 2.5 h

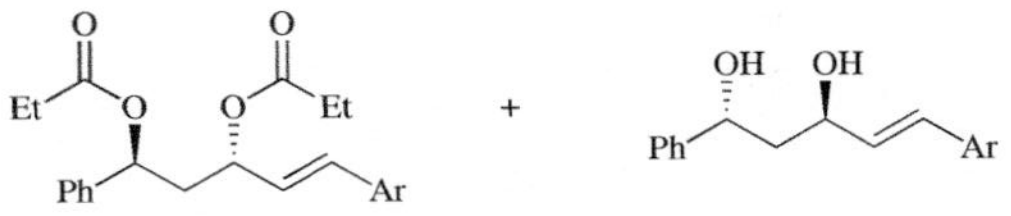

I   +   II

Ar		er I	er II	s
Ph	(—)	98.0:2.0	99.5:0.5	29
2-MeOC$_6$H$_4$	(—)	96.0:4.0	99.5:0.5	24

231

C_{17-18}

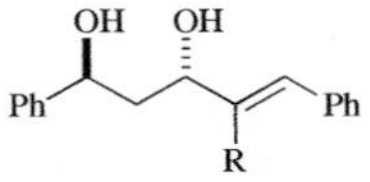

(EtCO)$_2$O (1.05 eq),
ent-**126** (1 mol %),
i-Pr$_2$NEt (1.05 eq),
CH$_2$Cl$_2$, −20°, 2.5 h

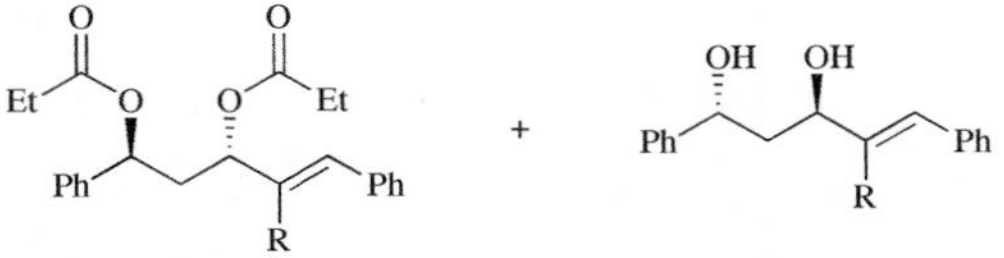

I   +   II

R		er I	er II	s
Br	(—)	98.5:1.5	99.5:0.5	39
Me	(—)	98.0:2.0	99.5:0.5	25

231

373

TABLE 2F. CATALYTIC KINETIC RESOLUTION OF rac-DIOLS (Continued)

Alcohol	Conditions	Product(s) and Conversion(s) (%)	Refs.

*Please refer to the charts preceding the tables for the structures indicated by the **bold** numbers.*

C$_{18}$

| 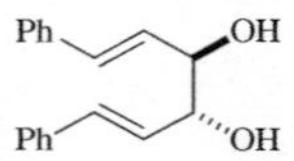 | (i-PrCO)$_2$O (1.5 eq), **126** (1 mol %), i-Pr$_2$NEt (1.6 eq), CHCl$_3$, 0°, 7 h | 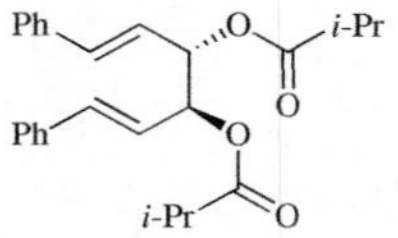 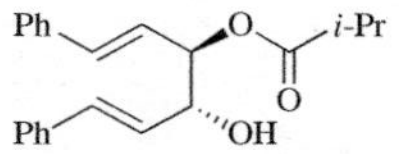 (100) er 77.0:23.0 + er 79.0:21.0 | 226 |
| 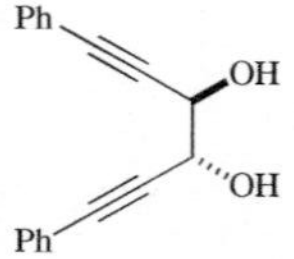 | (i-PrCO)$_2$O (1.5 eq), **126** (1 mol %), i-Pr$_2$NEt (1.6 eq), CHCl$_3$, 0°, 7 h | 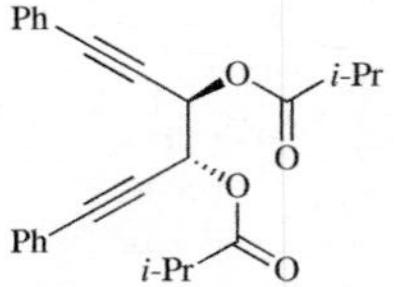 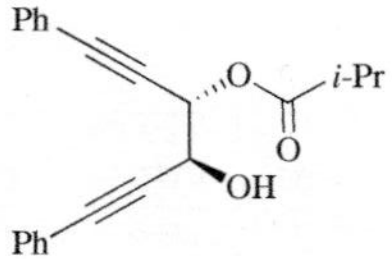 (92) er 85.0:15.0 + er 83.0:17.0 + er 99.5:0.5 | 226 |

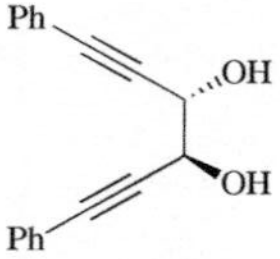

C$_{19}$

| 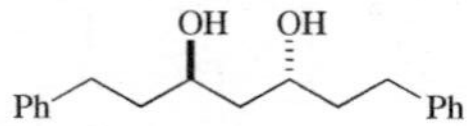 | (EtCO)$_2$O (1.05 eq), *ent*-**126** (1 mol %), i-Pr$_2$NEt (1.05 eq), CH$_2$Cl$_2$, −20°, 2.5 h | 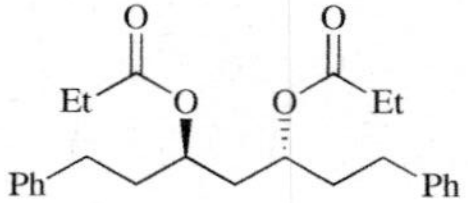 er 99.5:0.5 + er 61.0:39.0 (—), s = 7 | 231 |

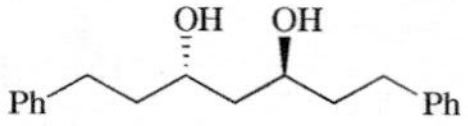

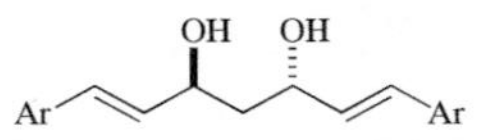

(EtCO)$_2$O (1.05 eq),
ent-**126** (1 mol %),
i-Pr$_2$NEt (1.05 eq),
CH$_2$Cl$_2$, –20°, 2.5 h

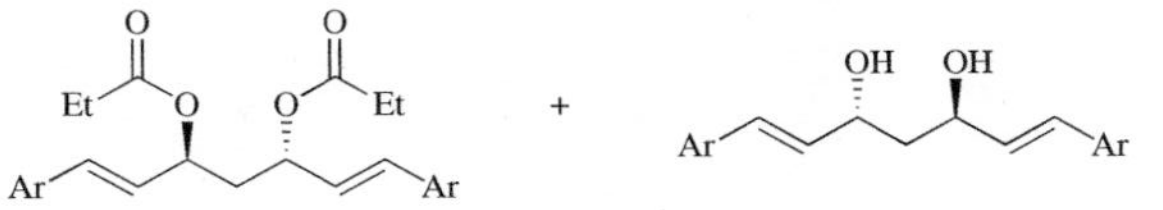

I + **II**   231

Ar		er **I**	er **II**	s
Ph	(—)	99.5:0.5	98.5:1.5	28
4-ClC$_6$H$_4$	(—)	99.0:1.0	97.5:2.5	16

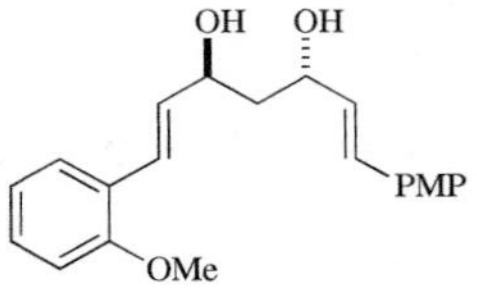

(EtCO)$_2$O (1.05 eq),
ent-**126** (1 mol %),
i-Pr$_2$NEt (1.05 eq),
CH$_2$Cl$_2$, –20°, 2.5 h

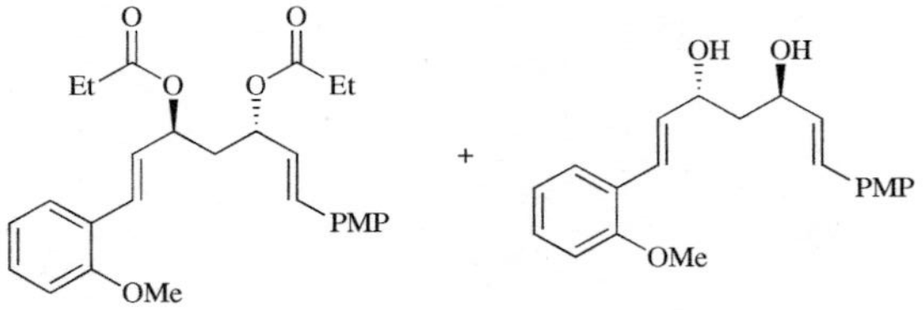

+   231

er 96.0:4.0   (—), s = 22   er 99.5:0.5

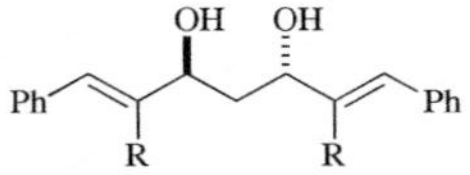

(EtCO)$_2$O (1.05 eq),
ent-**126** (1 mol %),
i-Pr$_2$NEt (1.05 eq),
CH$_2$Cl$_2$, –20°, 2.5 h

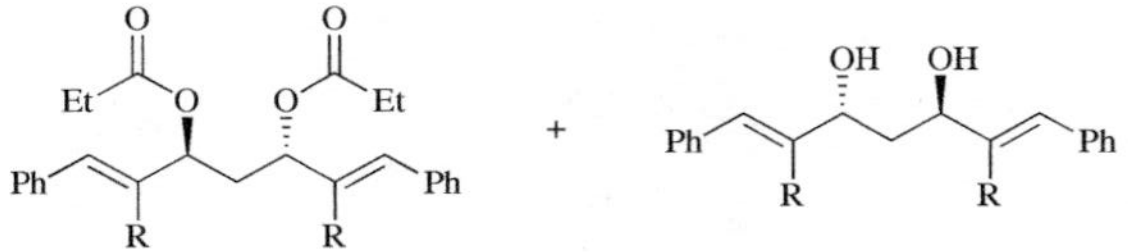

I + **II**   231

R		er **I**	er **II**	s
Br	(—)	98.5:1.5	97.0:3.0	15
Me	(—)	99.5:0.5	99.5:0.5	51

C$_{19-21}$

TABLE 2G. CATALYTIC KINETIC RESOLUTION OF MONOPROTECTED *MESO*-DIOLS

*Please refer to the charts preceding the tables for the structures indicated by the **bold** numbers.*

Alcohol	Conditions	Product(s) and Conversion(s) (%)	Refs.
C₄			
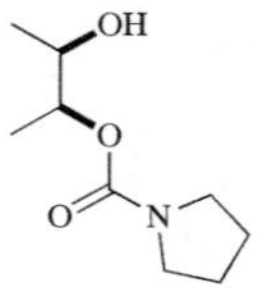	(*i*-PrCO)₂O (0.5 eq), **112** (5 mol %), *i*-Pr₂NEt (0.5 eq), CCl₄, 0°, 2 h	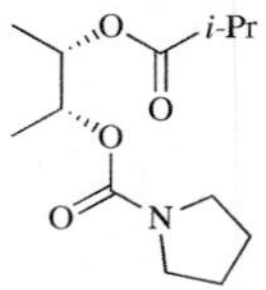 er 93.0:7.0 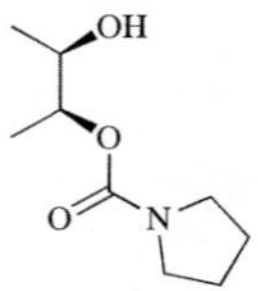 + er 84.0:16.0 (44), *s* = 28	230
	(*i*-PrCO)₂O (0.5 eq), **109** (5 mol %), *i*-Pr₂NEt (0.5 eq), CCl₄, 0°, 3 h	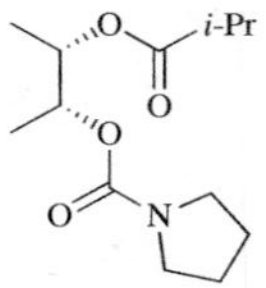 er 96.5:3.5 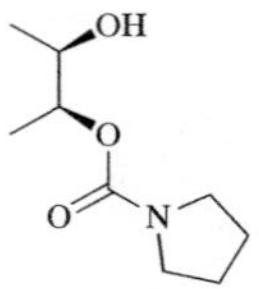 + er 91.0:9.0 (47), *s* = 68	70
	Ac₂O (0.7 eq), **74** (5 mol %), Et₃N (0.75 eq), CH₂Cl₂, –60°, 24 h	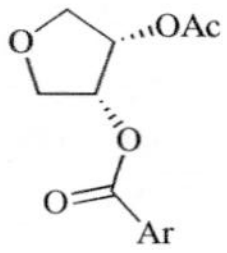 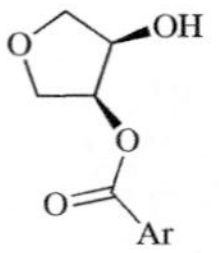 er 68.5:31.5 + (52), *s* = 3	228

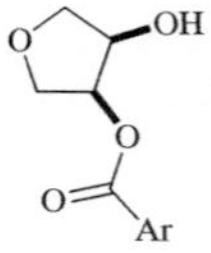

Ar = 4-Me₂NC₆H₄

376

227

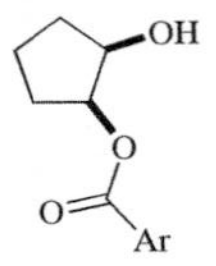

227

229

(71), s = 8

(73), s = 10

(i-PrCO)$_2$O (0.7 eq), **59** (1 mol %), 2,4,6-collidine, toluene, rt, 4 h

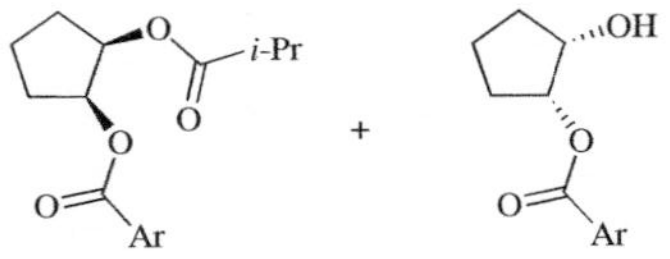

Ar = 4-Me$_2$NC$_6$H$_4$

C$_5$

er 98.5:1.5

Ac$_2$O, **74** (5 mol %), Et$_3$N (1.05 eq), CH$_2$Cl$_2$, –60°, 24 h

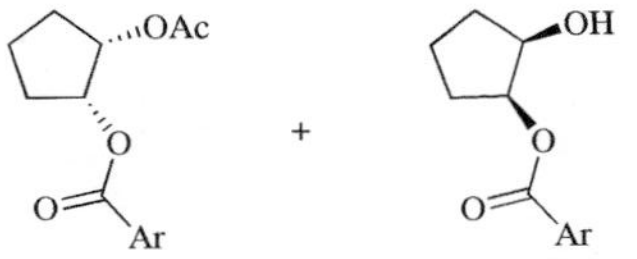

er 99.5:0.5

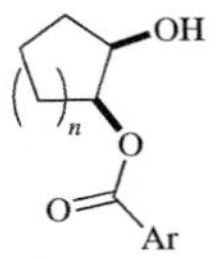

Ac$_2$O (x eq), **75** (1 mol %), Et$_3$N (y eq), toluene, rt, 16 h

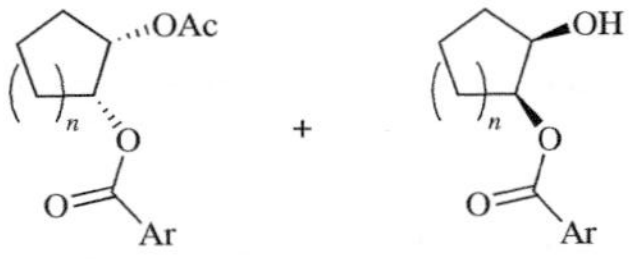

Ar = 4-Me$_2$NC$_6$H$_4$

C$_{5-8}$

n	I			II		
	x	y		er II		s
1	1.2	1.85	(87)	96.5:3.5		3
2	1.8	1.25	(72)	99.5:0.5		10
3	2.0	2.10	(63)	85.0:15.0		5
4	1.8	1.85	(69)	94.5:5.5		6

TABLE 2G. CATALYTIC KINETIC RESOLUTION OF MONOPROTECTED *MESO*-DIOLS (*Continued*)

Alcohol	Conditions	Product(s) and Conversion(s) (%)	Refs.

*Please refer to the charts preceding the tables for the structures indicated by the **bold** numbers.*

C₅

(*i*-PrCO)₂O (0.5 eq),
109 (5 mol %),
i-Pr₂NEt (0.5 eq), CCl₄, 3 h

er 97.0:3.0   +   er 95.0:5.0

(49), *s* = 93   70

C_{5–8}

(*i*-PrCO)₂O (0.5 eq),
112 (5 mol %),
i-Pr₂NEt (0.5 eq), CCl₄, 0°

I   +   **II**

n	Time (h)		er I	er II	*s*
1	4	(54)	90.0:10.0	98.0:2.0	40
3	4	(53)	92.0:8.0	99.0:1.0	65
4	5	(62)	80.0:20.0	99.0:1.0	25

230

C_6

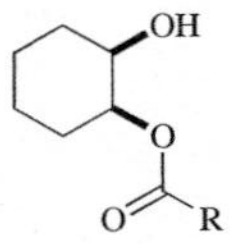

(i-PrCO)$_2$O (0.5 eq),
112 (5 mol %),
i-Pr$_2$NEt (0.5 eq), CCl$_4$, 0°

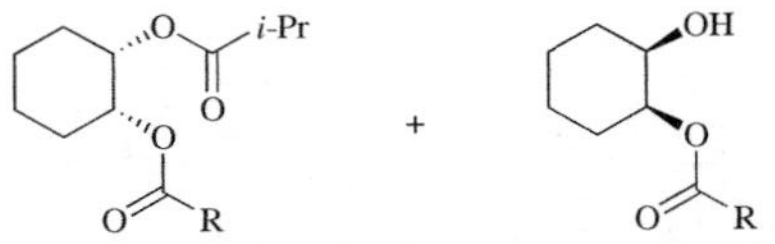

230

R	Time (h)		er I	er II	s
Me	6	(58)	78.0:22.0	89.0:11.0	8
Me$_2$N	4	(47)	90.5:9.5	86.0:14.0	21
1-piperidyl	4	(52)	82.5:17.5	86.5:13.5	11
4-Me$_2$NC$_6$H$_4$	5	(50)	80.5:19.5	81.0:19.0	8

(i-PrCO)$_2$O (0.7 eq),
59 (1 mol %),
2,4,6-collidine, toluene, rt

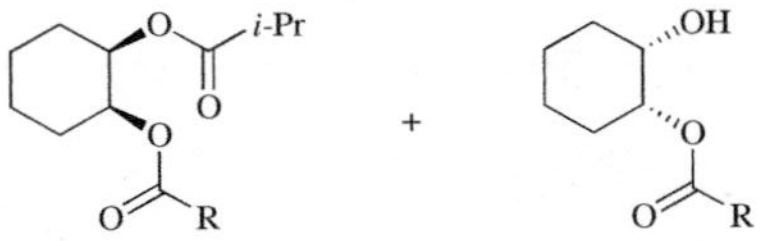

227

R	Time (h)		er II	s
i-Pr	5	(69)	88.0:12.0	4
t-Bu	4	(68)	97.0:3.0	8
Ph	5	(71)	90.5:9.5	5
4-MeOC$_6$H$_4$	2	(70)	92.5:7.5	5
4-O$_2$NC$_6$H$_4$	5	(73)	77.0:23.0	2
4-Me$_2$NC$_6$H$_4$	3	(65)	98.5:1.5	12

TABLE 2G. CATALYTIC KINETIC RESOLUTION OF MONOPROTECTED *MESO*-DIOLS (*Continued*)

Alcohol	Conditions	Product(s) and Conversion(s) (%)	Refs.

*Please refer to the charts preceding the tables for the structures indicated by the **bold** numbers.*

C_6

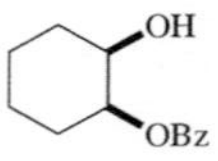	Ac$_2$O (0.75 eq), **79** (5 mol %), Et$_3$N (0.75 eq), toluene, −78°, 15 h	 OAc/OBz + OH/OBz er 68.0:32.0 (34), *s* = 3	173
	AcCl (0.6 eq), **157a** (5 mol %), DABCO (0.65 eq), Et$_2$O, −20°, 5 h	 OAc/OBz er 95.5:4.5 + OH/OBz er 99.5:0.5 (52), *s* = 105	201
	Ac$_2$O (0.7 eq), **62** (10 mol %), Et$_3$N (0.7 eq), toluene, rt	 OAc/OBz er 72.5:27.5 + OH/OBz (64), *s* = 3	175
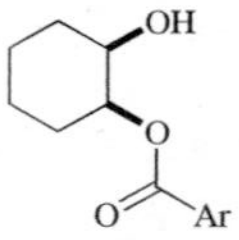	(*i*-PrCO)$_2$O (0.8 eq), **71** (1 mol %), Et$_3$N (0.8 eq), CH$_2$Cl$_2$, rt, 16 h	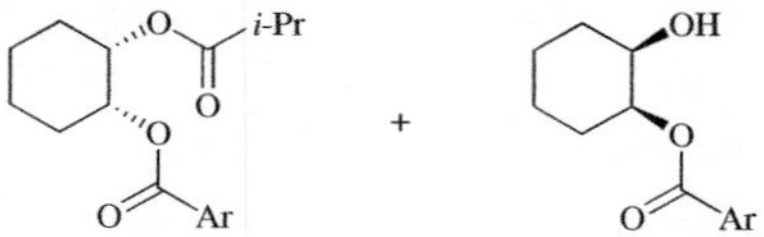 **I** + **II**	166

Ar		er **II**	*s*
Ph	(88)	97.5:2.5	4
4-MeOC$_6$H$_4$	(71)	90.0:10.0	4
4-Me$_2$NC$_6$H$_4$	(74)	95.0:5.0	5

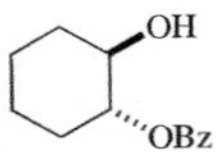

(EtCO)$_2$O (0.75 eq), **125** (4 mol %),
i-Pr$_2$NEt (0.75 eq),
t-AmOH/toluene (1:1), –40°, 10 h

(31), s = 6   194

er 81.5:18.5   +   er 64.0:36.0

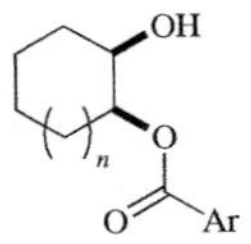

Ac$_2$O (2.5 eq), **74** (5 mol %),
Et$_3$N (2.55 eq),
CH$_2$Cl$_2$, –60°, 24 h

228

I   +   **II**

n		er **II**	s
1	(65)	95.5:4.5	8
2	(70)	90.5:9.5	5
3	(74)	95.0:5.0	5

(i-PrCO)$_2$O (1.5 eq),
70 (1 mol %),
Et$_3$N (0.8 eq),
CH$_2$Cl$_2$, –78°, 8 h

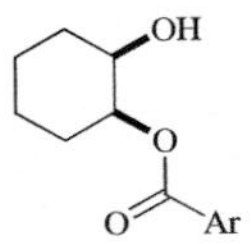

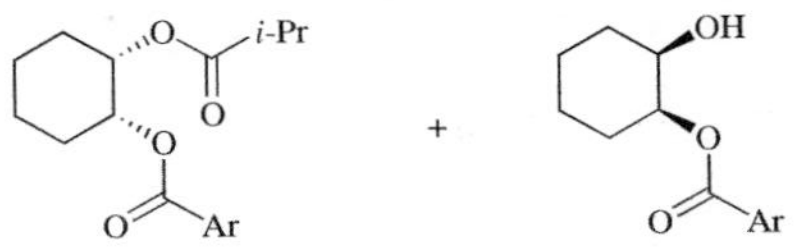

(69), s = 9   166

+

er 72.0:28.0   er 96.5:3.5

Ar = 4-Me$_2$NC$_6$H$_4$

Ar = 4-Me$_2$NC$_6$H$_4$

TABLE 2G. CATALYTIC KINETIC RESOLUTION OF MONOPROTECTED *MESO*-DIOLS (*Continued*)

Alcohol	Conditions	Product(s) and Conversion(s) (%)	Refs.

*Please refer to the charts preceding the tables for the structures indicated by the **bold** numbers.*

C_6

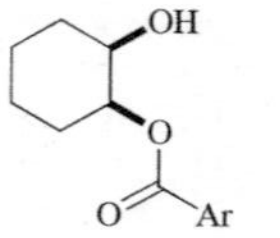

Ar = 4-Me$_2$NC$_6$H$_4$

	(*i*-PrCO)$_2$O (0.5 eq), **111** (5 mol %), *i*-Pr$_2$NEt (0.5 eq), toluene, rt, 6 h	er 81.0:19.0 + er 93.0:7.0	(42), *s* = 26	71
	(*i*-PrCO)$_2$O (0.7 eq), **72** (1 mol %), Et$_3$N (0.8 eq), CH$_2$Cl$_2$, −78°, 24 h	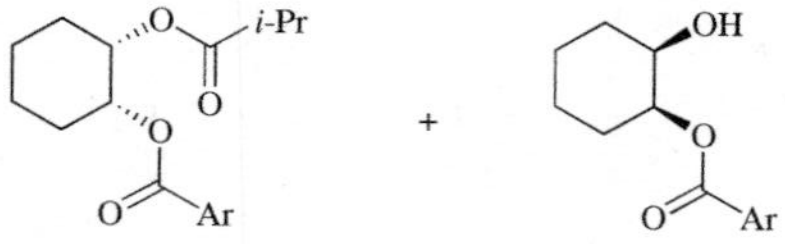 er 75.0:25.0 + er 98.0:2.0	(66), *s* = 11	167
	(*i*-PrCO)$_2$O (1.5 eq), **65** (5 mol %), Et$_3$N (0.8 eq), CH$_2$Cl$_2$, rt, 24 h	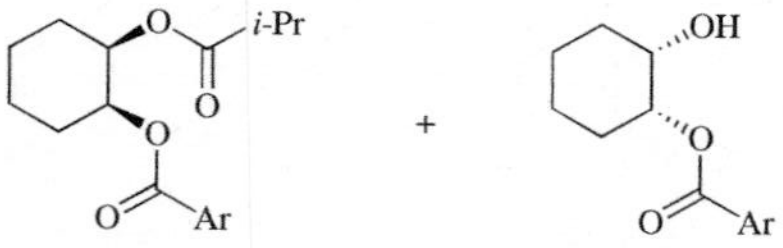 + er 92.0:8.0	(55), *s* = 13	177
	(*i*-PrCO)$_2$O (0.7 eq), **64** (5 mol %), toluene, rt, 3 h	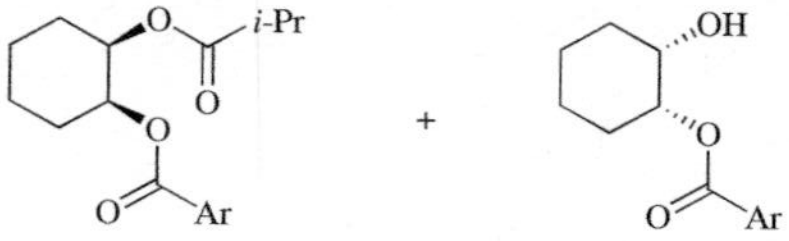 er 78.0:22.0 + er 97.5:2.5	(62), *s* = 13	178

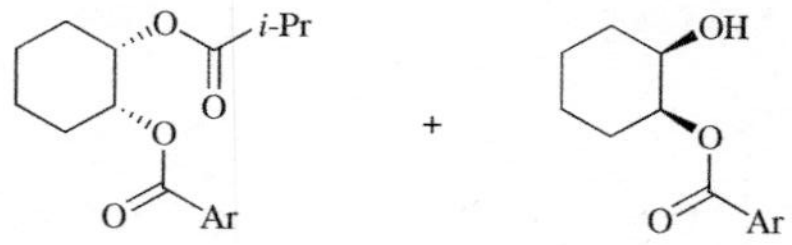

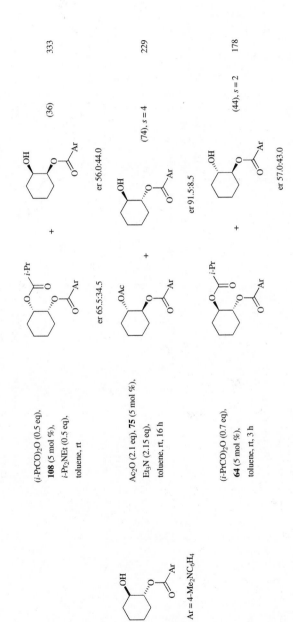

(i-PrCO)₂O (0.5 eq),
108 (5 mol %),
i-Pr₂NEt (0.5 eq),
toluene, rt

er 65.5:34.5   +   er 56.0:44.0   (36)   333

Ac₂O (2.1 eq), **75** (5 mol %),
Et₃N (2.15 eq),
toluene, rt, 16 h

er 91.5:8.5   (74), s = 4   229

(i-PrCO)₂O (0.7 eq),
64 (5 mol %),
toluene, rt, 3 h

+   er 57.0:43.0   (44), s = 2   178

Ar = 4-Me₂NC₆H₄

383

TABLE 2G. CATALYTIC KINETIC RESOLUTION OF MONOPROTECTED *MESO*-DIOLS (Continued)

*Please refer to the charts preceding the tables for the structures indicated by the **bold** numbers.*

Alcohol	Conditions	Product(s) and Conversion(s) (%)	Refs.

C_6

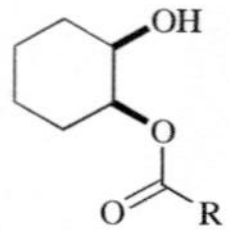

(*i*-PrCO)$_2$O (0.5 eq),
109 (5 mol %),
i-Pr$_2$NEt (0.5 eq), CCl$_4$, 3 h

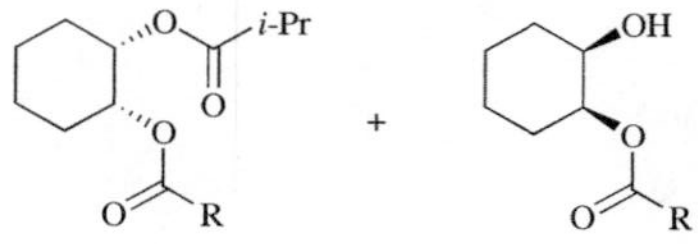

70

R	Temp (°)		er I	er II	s
Me$_2$N	0	(54)	91.5:8.5	99.5:0.5	64
C$_4$H$_8$N	0	(52)	95.0:5.0	98.5:1.5	87
4-Me$_2$NC$_6$H$_4$	rt	(49)	91.5:8.5	90.5:9.5	27

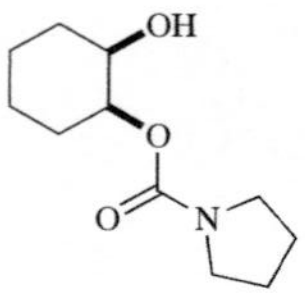

(*i*-PrCO)$_2$O (0.5 eq),
111 (5 mol %),
i-Pr$_2$NEt (0.5 eq), CCl$_4$, 0°, 7 h

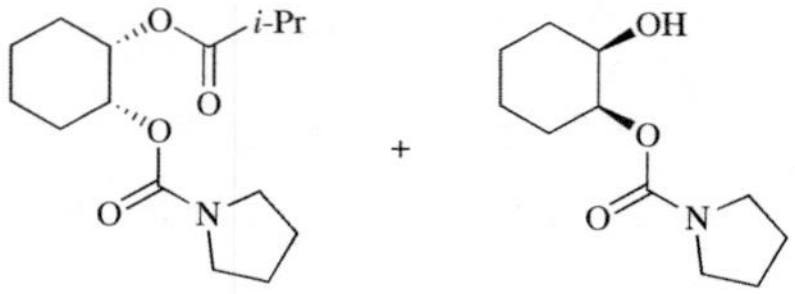

er 95.0:5.0

(49), s = 37

70

(*i*-PrCO)$_2$O (0.75 eq), **70** (1 mol %),
Et$_3$N (0.75 eq), CH$_2$Cl$_2$, −78°, 16 h

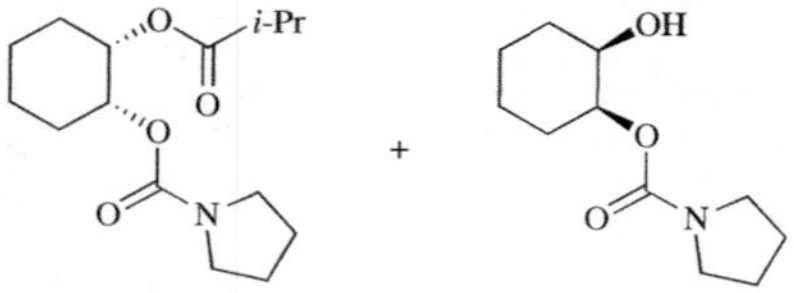

er 86.5:13.5

(27), s = 9

39

384

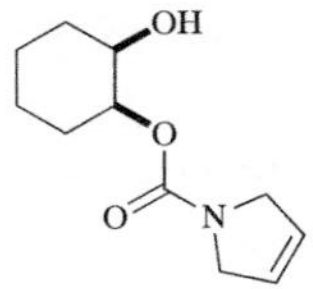

(i-PrCO)$_2$O (0.5 eq), **109** (5 mol %),
i-Pr$_2$NEt (0.5 eq), CCl$_4$, 0°, 3 h

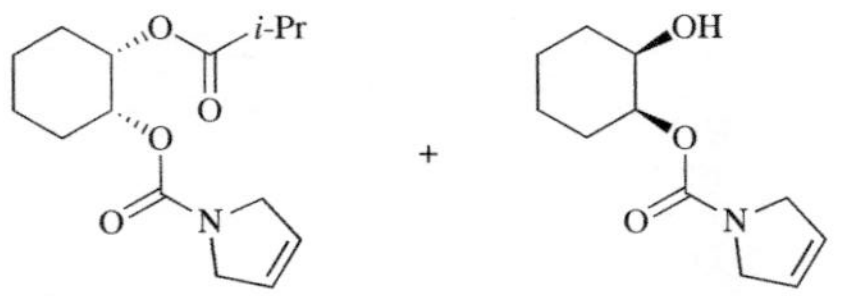

er 93.5:6.5   +   er 99.0:1.0

(53), s = 65   181

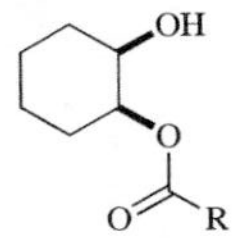

(i-PrCO)$_2$O (0.5 eq),
112 (5 mol %),
i-Pr$_2$NEt (0.5 eq), CCl$_4$, 0°

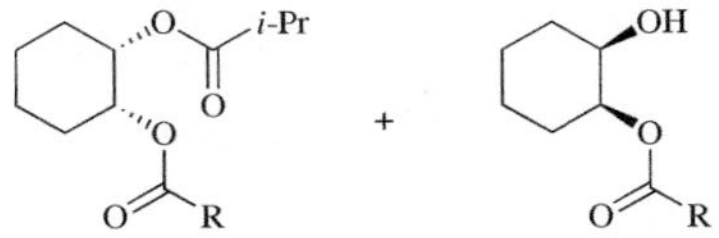

R	Time (h)		er I	er II	s
1-pyrrolidyl	3	(52)	95.0:5.0	98.5:1.5	80
c-C$_5$H$_9$	10	(48)	82.0:18.0	80.0:20.0	9

230

(i-PrCO)$_2$O (0.7 eq),
62 (10 mol %),
Et$_3$N (0.7 eq), toluene, rt

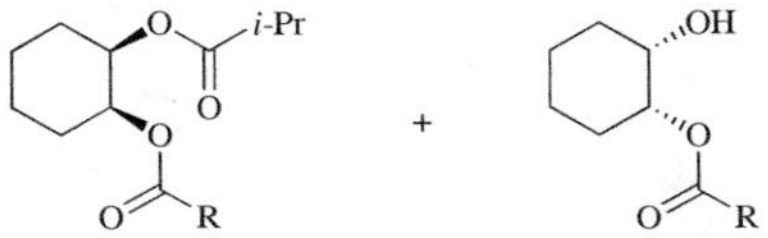

R		er II	s
Ph$_2$N	(65)	99.0:1.0	7
4-Me$_2$NC$_6$H$_4$	(66)	94.5:5.5	14

175

TABLE 2G. CATALYTIC KINETIC RESOLUTION OF MONOPROTECTED *MESO*-DIOLS (*Continued*)

Alcohol	Conditions	Product(s) and Conversion(s) (%)	Refs.

*Please refer to the charts preceding the tables for the structures indicated by the **bold** numbers.*

C₇

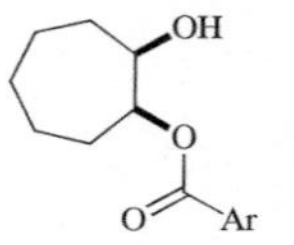

Ar = 4-Me₂NC₆H₄

(i-PrCO)₂O (0.7 eq), **59** (1 mol %),
2,4,6-collidine, toluene, rt, 4 h

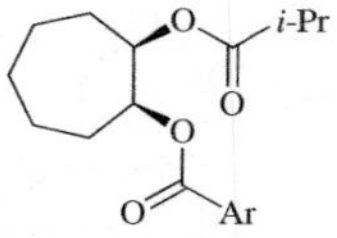

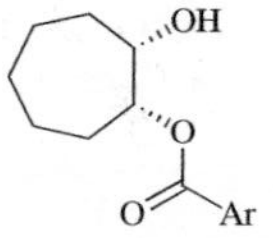

+

er 96.0:4.0

(70), *s* = 7

227

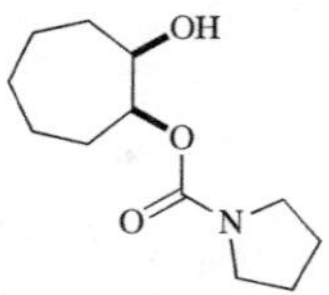

(i-PrCO)₂O (0.5 eq),
109 (5 mol %),
i-Pr₂NEt (0.5 eq), CCl₄, 0°, 3 h

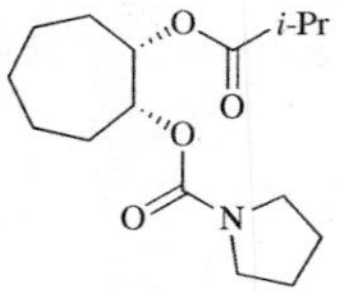

+

er 96.5:3.5

er 96.0:4.0

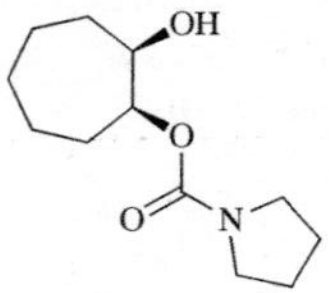

(50), *s* = 83

227

C₈

Ar = 4-Me₂NC₆H₄

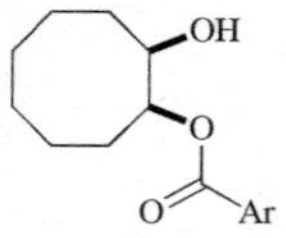

(i-PrCO)₂O (0.7 eq), **59** (1 mol %),
2,4,6-collidine, toluene, rt, 5 h

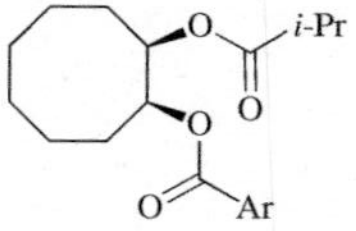

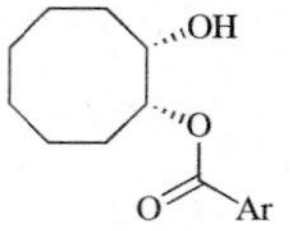

+

er 96.0:4.0

(73), *s* = 6

227

386

C$_{14}$

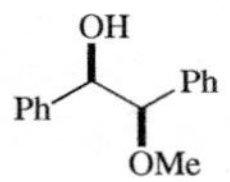	(i-PrCO)$_2$O (0.55 eq), **126** (1 mol %), i-Pr$_2$NEt (0.6 eq), CHCl$_3$, 0°, 7 h	 er 82.0:18.0 + er 78.0:22.0	(45), s = 10	226
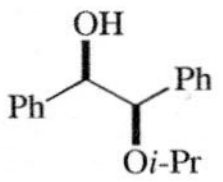	(i-PrCO)$_2$O (0.55 eq), **126** (1 mol %), i-Pr$_2$NEt (0.6 eq), CHCl$_3$, 0°, 7 h	 er 96.0:4.0 + er 61.0:39.0	(20), s = 34	226

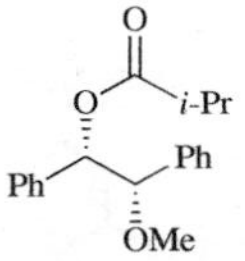

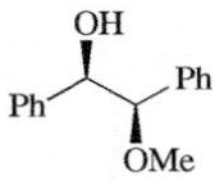

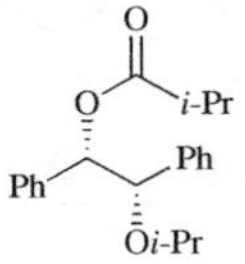

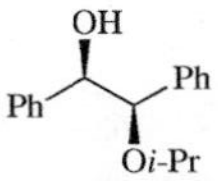

TABLE 2H. CATALYTIC KINETIC RESOLUTION OF ATROPISOMERIC ALCOHOLS

Alcohol	Conditions	Product(s) and Conversion(s) (%)	Refs.

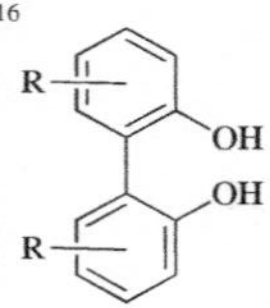

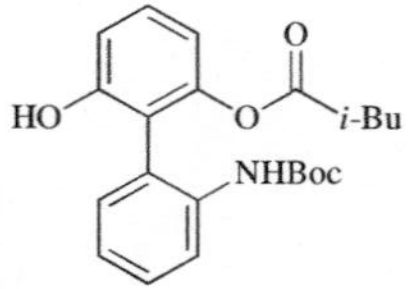

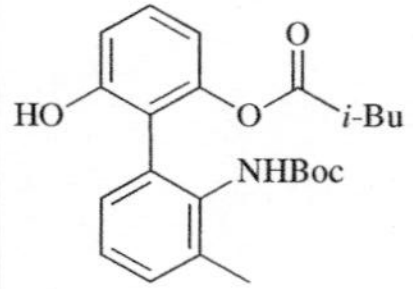

Please refer to the charts preceding the tables for the structures indicated by the bold numbers.

C$_{12-16}$

11 (0.7 eq), 145d (10 mol %),
i-Pr$_2$NEt (1.0 eq), 4 Å MS,
CH$_2$Cl$_2$, rt, 24 h

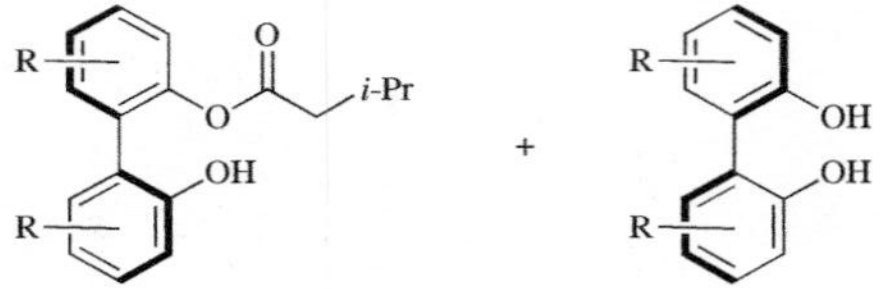

R	I	er I	er II	s
6-MeO	(62)	80.0:20.0	99.5:0.5	22
5,6-Me$_2$	(59)	84.0:16.0	99.5:0.5	30
5-Cl-4,6-Me$_2$	(60)	83.0:17.0	99.5:0.5	26

84

C$_{12}$

11 (0.7 eq), 145e (10 mol %),
i-Pr$_2$NEt (1.0 eq),
THF, 0°, 24 h

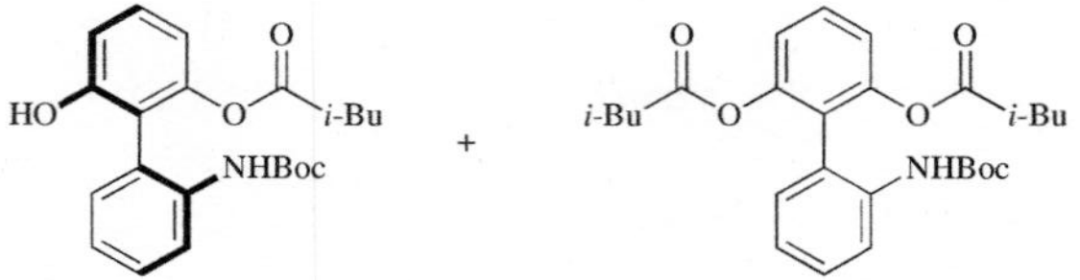

er 90.0:10.0   (38), s = 8

150

C$_{13}$

11 (0.7 eq), 145e (10 mol %),
i-Pr$_2$NEt (1.0 eq),
THF, 0°, 24 h

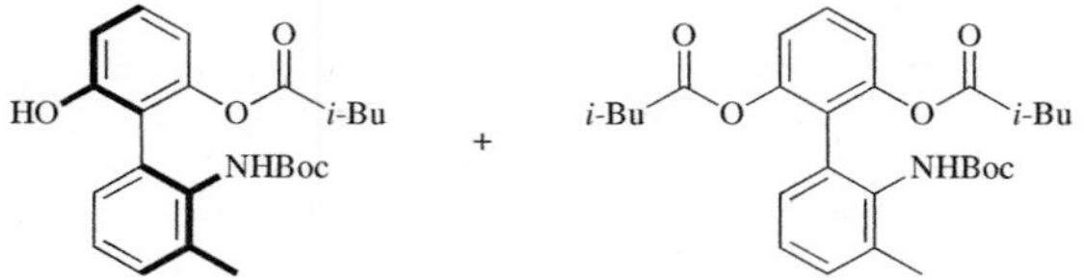

er 99.5:0.5   (33), s = 17

150

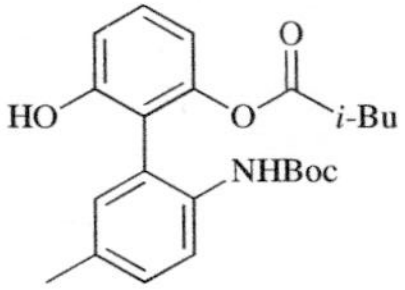

C$_{14}$

11 (0.7 eq), **145e** (10 mol %),
i-Pr$_2$NEt (1.0 eq),
THF, 0°, 24 h

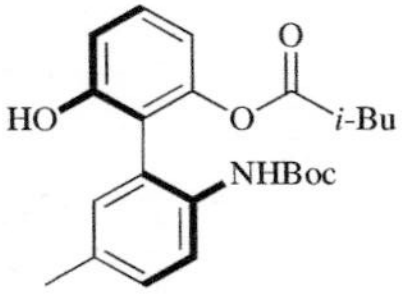

er 88.0:12.0

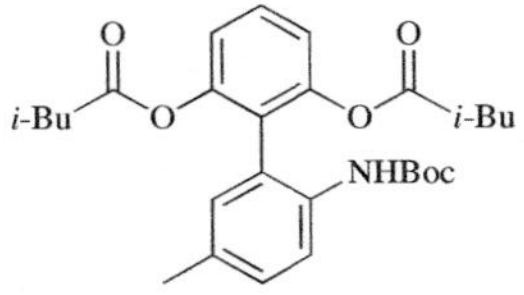

(38), $s = 7$

150

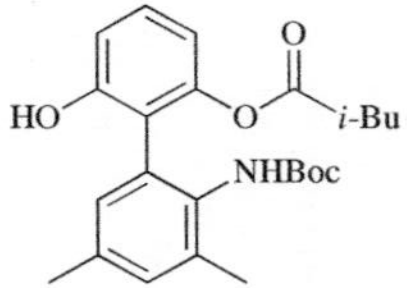

11 (0.7 eq), **145e** (10 mol %),
i-Pr$_2$NEt (1.0 eq),
THF, 0°, 24 h

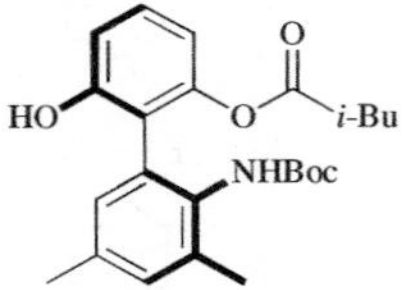

er 99.5:0.5

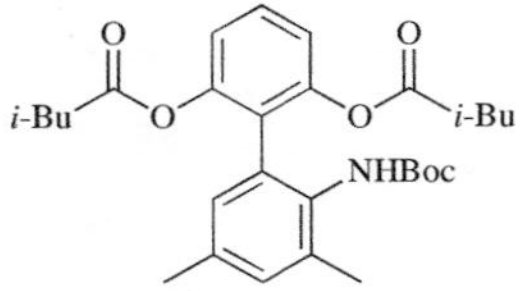

(35), $s = 18$

150

C$_{16}$

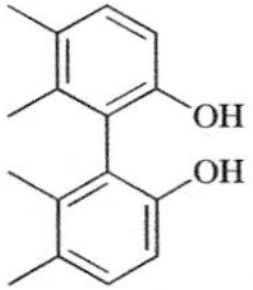

14 (0.55 eq), **124** (1 mol %),
CHCl$_3$, rt, 18 h

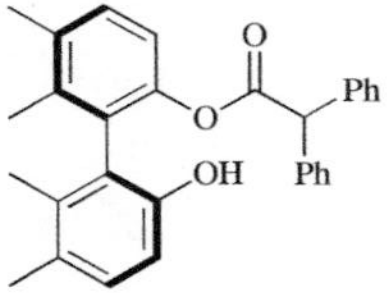

er 94.0:6.0

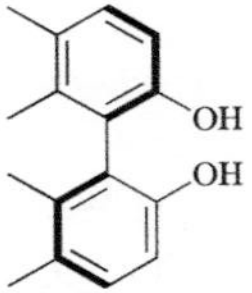

er 91.0:9.0

(48), $s = 37$

232

TABLE 2H. CATALYTIC KINETIC RESOLUTION OF ATROPISOMERIC ALCOHOLS (*Continued*)

Alcohol	Conditions	Product(s) and Conversion(s) (%)	Refs.

*Please refer to the charts preceding the tables for the structures indicated by the **bold** numbers.*

C$_{16}$

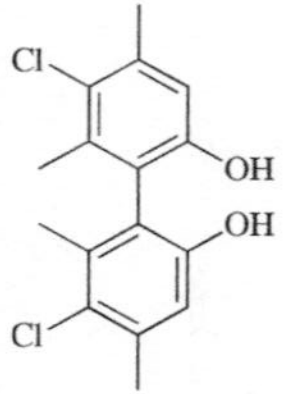

14 (0.55 eq), 124 (1 mol %),
CHCl$_3$, rt, 18 h

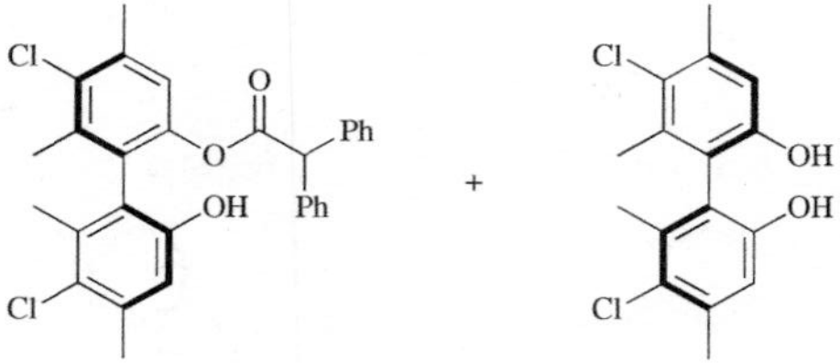

er 87.0:13.0

er 85.0:15.0

(48), *s* = 13   232

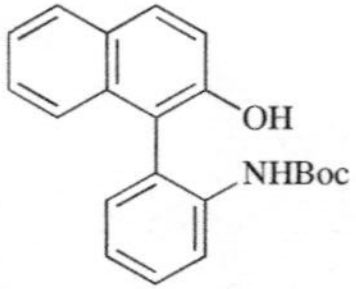

11 (0.7 eq), 145e (10 mol %),
i-Pr$_2$NEt, 4 Å MS,
THF, 0°, 24 h

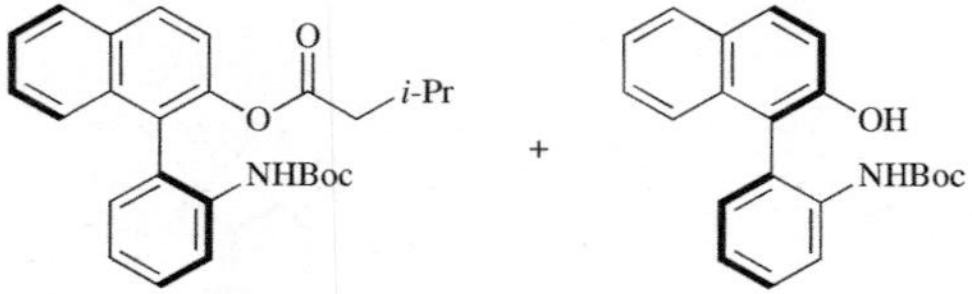

er 79.0:21.0

er 99.5:0.5

(63), *s* = 18   84

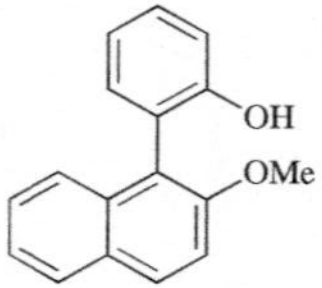

(*i*-PrCO)$_2$O (0.6 eq),
77 (15 mol %),
2,6-di-*tert*-butylpyridine (0.6 eq),
CH$_2$Cl$_2$, –50°, 84 h

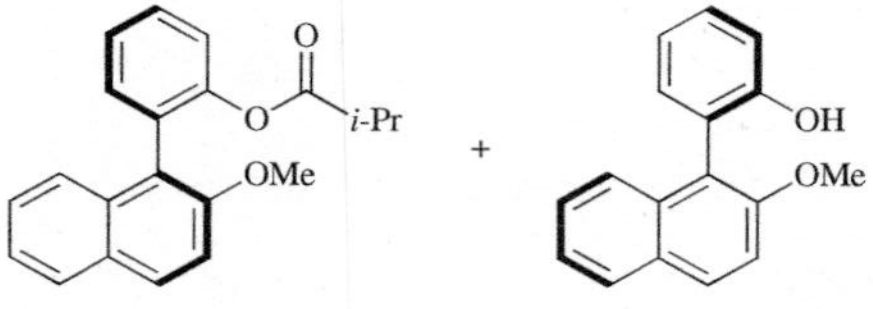

er 94.5:5.5

er 95.0:5.0

(50), *s* = 51   169

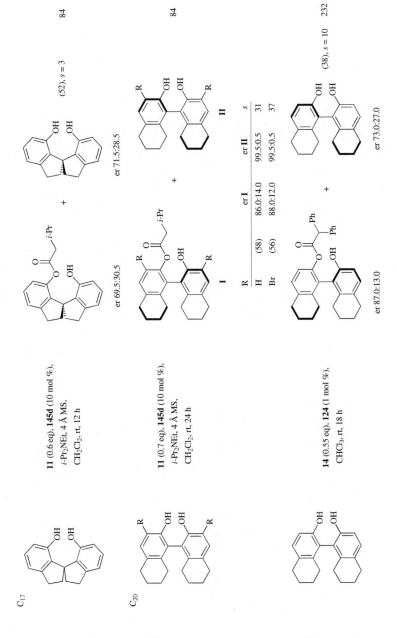

R		er **I**	er **II**	s
H	(58)	86.0:14.0	99.5:0.5	31
Br	(56)	88.0:12.0	99.5:0.5	37

C$_{17}$

11 (0.6 eq), **145d** (10 mol %),
i-Pr$_2$NEt, 4 Å MS,
CH$_2$Cl$_2$, rt, 12 h

er 69.5:30.5   er 71.5:28.5   (52), s = 3   84

C$_{20}$

11 (0.7 eq), **145d** (10 mol %),
i-Pr$_2$NEt, 4 Å MS,
CH$_2$Cl$_2$, rt, 24 h

I   +   **II**   84

14 (0.55 eq), **124** (1 mol %),
CHCl$_3$, rt, 18 h

er 87.0:13.0   +   er 73.0:27.0   (38), s = 10   232

391

TABLE 2H. CATALYTIC KINETIC RESOLUTION OF ATROPISOMERIC ALCOHOLS (*Continued*)

Alcohol	Conditions	Product(s) and Conversion(s) (%)	Refs.

*Please refer to the charts preceding the tables for the structures indicated by the **bold** numbers.*

C$_{20}$

First entry

Conditions: (*i*-PrCO)$_2$O (0.6 eq), **77** (15 mol %), 2,6-di-*tert*-butylpyridine (0.6 eq), CH$_2$Cl$_2$, –50°, 96 h.

Products: er 81.5:18.5 (**I**) + er 86.5:13.5, (54), *s* = 10. Refs. 169

Second entry

Conditions: (RCO)$_2$O (0.55 eq), **124** (1 mol %), CHCl$_3$, rt, 18 h.

Products: er 81.5:18.5 (**I**) + er 86.5:13.5 (**II**). Refs. 232

R	er **I**		er **II**	*s*
Et	80.5:19.5	(59)	95.5:4.5	14
i-Pr	93.5:6.5	(49)	91.5:8.5	36
i-Bu	82.0:18.0	(48)	79.0:21.0	8
i-PrCH$_2$CH$_2$	90.0:10.0	(30)	67.0:33.0	11
Et$_2$CH	96.5:3.5	(15)	58.0:42.0	34
c-C$_5$H$_9$	89.0:11.0	(53)	94.0:6.0	23
3-pyridyl	89.5:10.5	(54)	94.5:5.5	21
c-C$_6$H$_{11}$	94.5:5.5	(45)	86.0:14.0	34
Ph	90.5:9.5	(36)	72.5:27.5	15
4-BrC$_6$H$_4$	86.0:14.0	(54)	92.0:8.0	16
4-O$_2$NC$_6$H$_4$	87.5:12.5	(51)	88.5:11.5	16
(*n*-Pr)$_2$CH	97.0:3.0	(12)	56.5:43.5	37
PhCH$_2$	92.0:8.0	(48)	87.5:12.5	21
PhCH=CH	72.5:27.5	(33)	60.5:39.5	3
PhCH$_2$CH$_2$	87.5:12.5	(42)	76.5:23.5	11
Ph$_2$CH	92.5:7.5	(52)	96.5:3.5	43

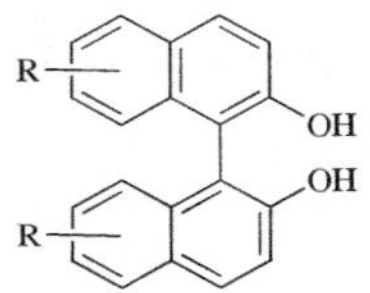

11 (0.7 eq), 145d (10 mol %),
i-Pr$_2$NEt, 4 Å MS,
CH$_2$Cl$_2$, rt, 24 h

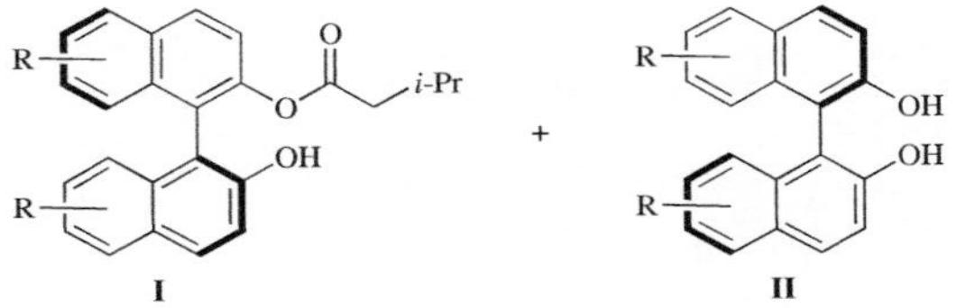

84

R		er I	er II	s
H	(55)	91.0:9.0	99.5:0.5	52
3-Br	(56)	88.5:11.5	99.5:0.5	38
6-Br	(55)	90.5:9.5	99.5:0.5	49
7-MeO	(56)	88.5:11.5	99.5:0.5	40
7-i-PrO	(57)	88.0:12.0	99.5:0.5	41

14 (0.55 eq), 124 (1 mol %),
CHCl$_3$, rt, 18 h

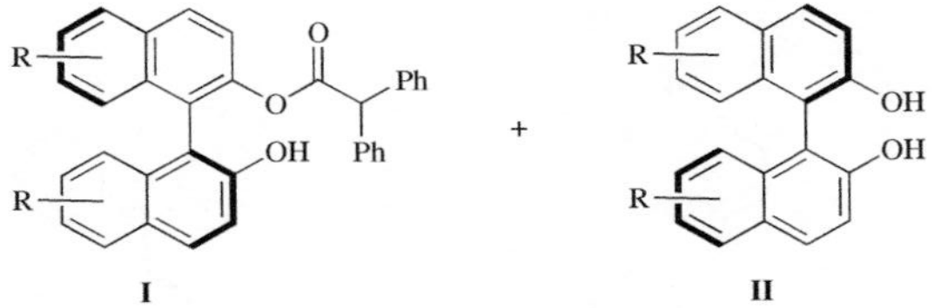

232

R		er I	er II	s
H	(50)	94.0:6.0	95.0:5.0	50
6-Br	(55)	88.0:12.0	96.0:4.0	25
7-Br	(54)	82.0:18.0	88.0:12.0	10
7-MeO	(52)	94.0:6.0	98.0:2.0	60

Alcohol	Conditions	Product(s) and Conversion(s) (%)	Refs.

*Please refer to the charts preceding the tables for the structures indicated by the **bold** numbers.*

C$_{20}$

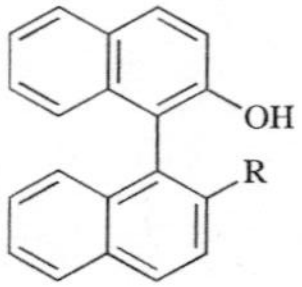

11 (0.6 eq), **145d** (10 mol %), i-Pr$_2$NEt, 4 Å MS, CH$_2$Cl$_2$, rt, 24 h

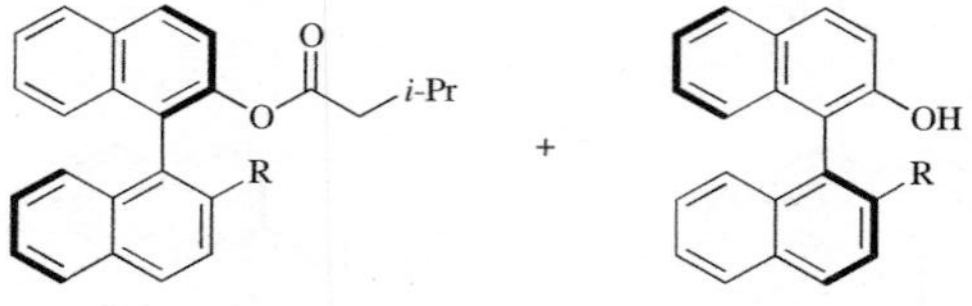

I + **II**

R		er I	er II	s
MeO	(31)	59.0:41.0	54.0:46.0	2
Me$_2$N	(14)	65.0:35.0	52.5:47.5	2

84

14 (0.55 eq), **124** (1 mol %), CHCl$_3$, rt, 18 h

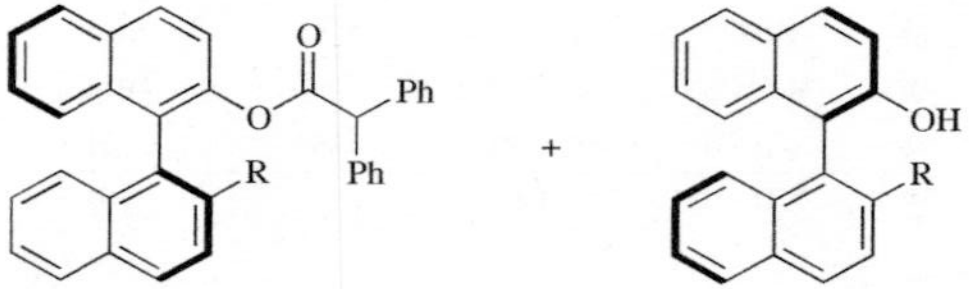

I + **II**

R		er I	er II	s
MeO	(39)	72.0:28.0	64.0:36.0	3
BocHN	(49)	89.0:11.0	86.0:14.0	17

232

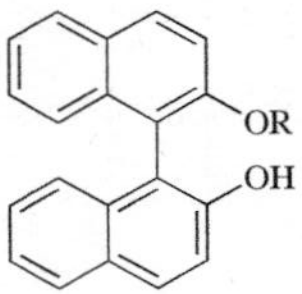

$(i\text{-PrCO})_2\text{O}$ (0.6 eq),
77 (15 mol %),
2,6-di-*tert*-butylpyridine (0.6 eq).
CH_2Cl_2, –50°

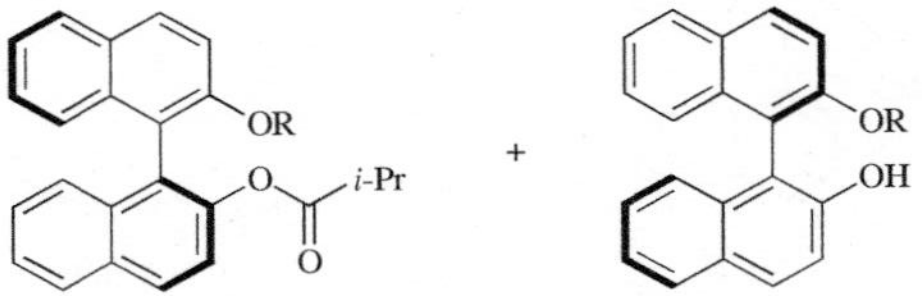

R	Time (h)		er I	er II	s
Ms	96	(57)	80.0:20.0	92.5:7.5	11
Me	84	(41)	95.0:5.0	81.5:18.5	36
MOM	120	(42)	92.5:7.5	81.0:19.0	23
allyl	120	(44)	92.0:8.0	84.0:16.0	25
Bn	120	(47)	91.5:8.5	86.0:14.0	23

$(i\text{-PrCO})_2\text{O}$ (0.6 eq),
77 (15 mol %), pyridine (0.6 eq),
CH_2Cl_2, –50°, 144 h

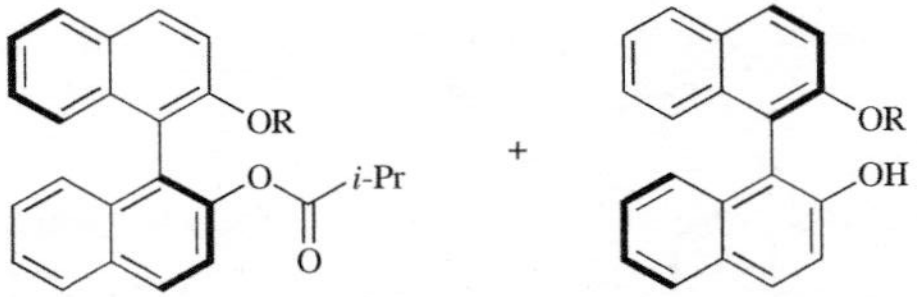

R		er I	er II	s
i-Pr	(45)	90.0:10.0	83.5:16.5	18
Boc	(47)	85.5:14.5	81.5:18.5	11

TABLE 2H. CATALYTIC KINETIC RESOLUTION OF ATROPISOMERIC ALCOHOLS (Continued)

*Please refer to the charts preceding the tables for the structures indicated by the **bold** numbers.*

Alcohol	Conditions	Product(s) and Conversion(s) (%)	Refs.
C$_{20-26}$ 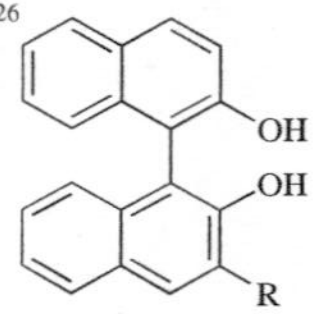	**14** (0.55 eq), **124** (1 mol %), CHCl$_3$, rt, 18 h	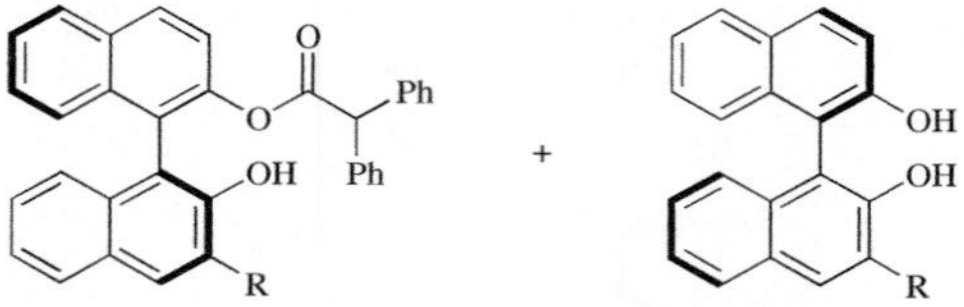 **I** + **II**	232

R		er I	er II	s
Br	(52)	93.0:7.0	97.0:3.0	49
MeO$_2$C	(41)	74.0:26.0	67.0:33.0	4
Ph	(51)	97.0:3.0	99.5:0.5	190

Alcohol	Conditions	Product(s) and Conversion(s) (%)	Refs.
C$_{20}$ 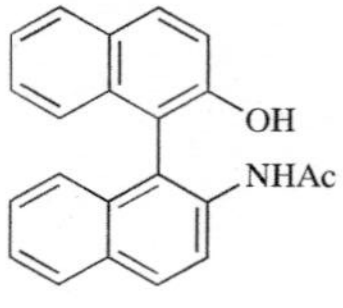	**11** (0.7 eq), **145d** (10 mol %), *i*-Pr$_2$NEt, 4 Å MS, THF, rt, 24 h	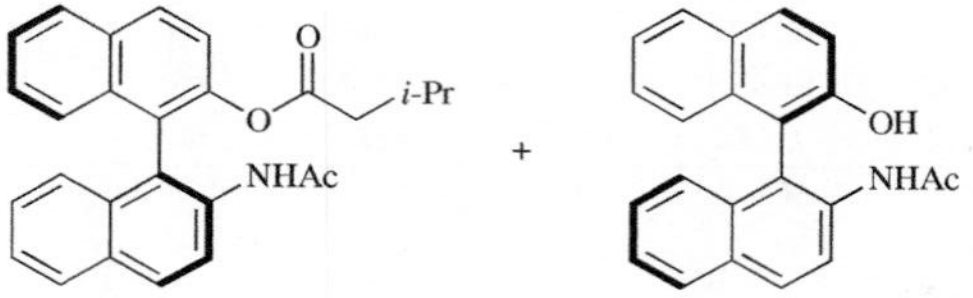 er 81.5:18.5 + er 97.5:2.5 (60), *s* = 15	84
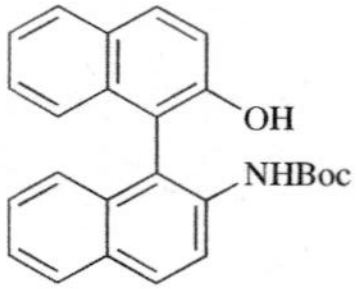	**11** (0.7 eq), **145e** (10 mol %), *i*-Pr$_2$NEt, 4 Å MS, THF, 0°, 24 h	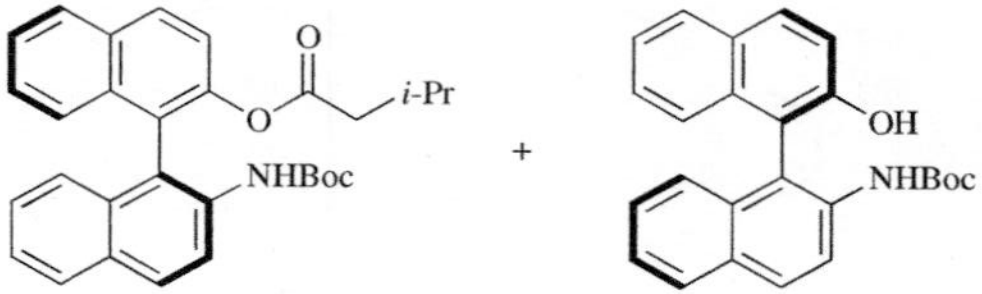 er 87.0:13.0 + er 99.5:0.5 (57), *s* = 34	84

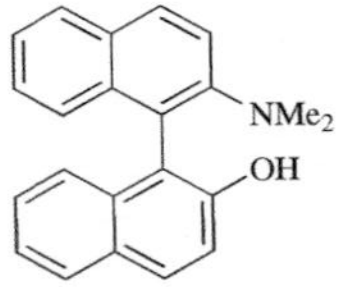

C_{28}

(i-PrCO)$_2$O (0.6 eq),
77 (15 mol %),
2,6-di-$tert$-butylpyridine (0.6 eq),
CH$_2$Cl$_2$, –50°, 144 h

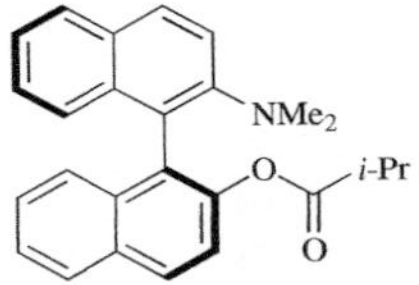

er 93.0:7.0   +   er 73.0:27.0   (35), $s = 21$ 169

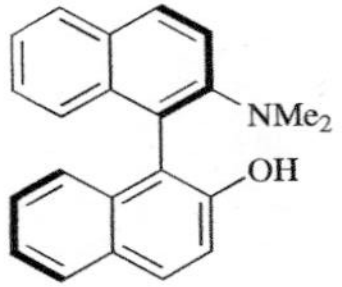

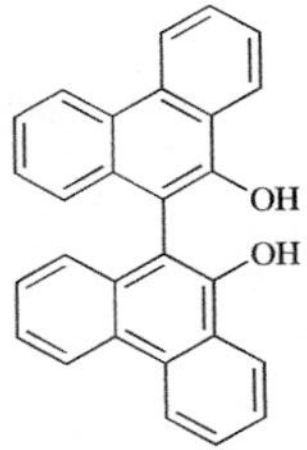

14 (0.55 eq), **124** (1 mol %),
CHCl$_3$, rt, 18 h

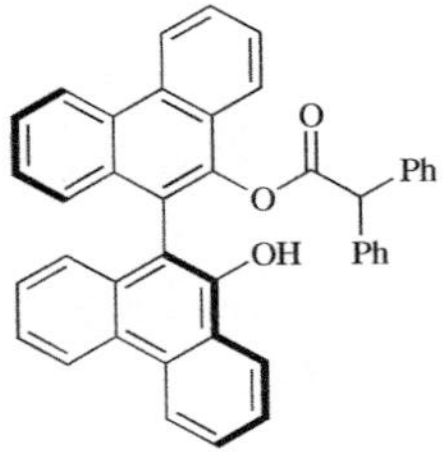

er 90.0:10.0   +   er 62.0:38.0   (23), $s = 11$ 232

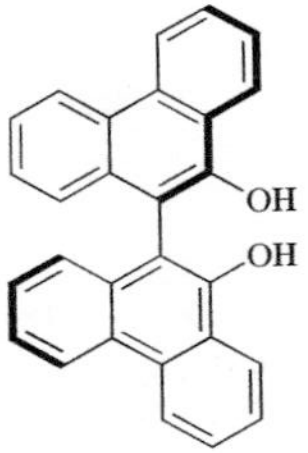

C_{32}

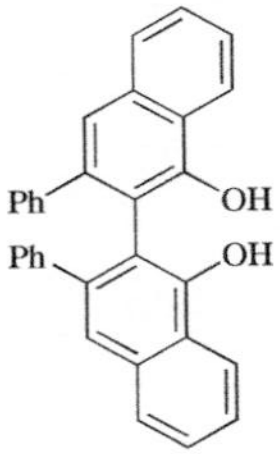

11 (0.7 eq), **145d** (10 mol %),
i-Pr$_2$NEt, 4 Å MS,
CH$_2$Cl$_2$, rt, 24 h

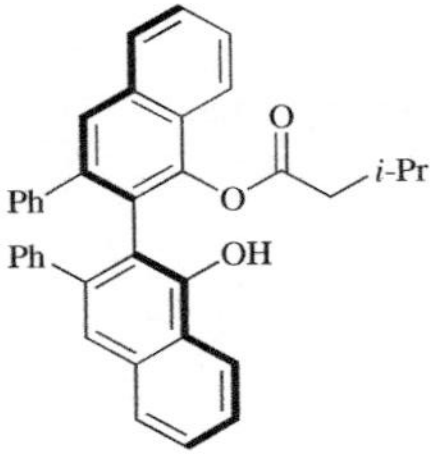

er 96.0:4.0   +   er 99.5:0.5   (52), $s = 116$ 84

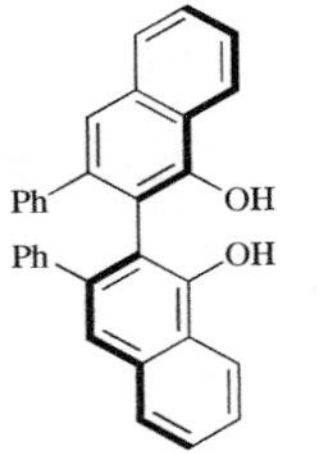

TABLE 21. CATALYTIC KINETIC RESOLUTION OF α-HYDROXY ESTERS

Ester	Conditions	Product(s) and Conversion(s) (%)	Refs.

*Please refer to the charts preceding the tables for the structures indicated by the **bold** numbers.*

C3–8

Ester: structure with OH, R, CO2Bn

Conditions: Ac2O (x eq), Sc(OTf)3 (5 mol %), **42b** (5 mol %), 3 Å MS. THF, 30°, 48 h

Products: I (OAc, R, CO2Bn) + II (OH, R, CO2Bn)

Refs.: 103

R	x		er I	er II	s
Me	0.7	(48)	87.5:12.5	84.0:16.0	14
Cy	0.8	(46)	86.0:14.0	80.5:19.5	11

C3

Ester: structure with OH, CO2Bn

Conditions: (Ph2CHCO)2O (0.275 eq), Piv2O (0.275 eq), **124** (5 mol %), i-Pr2NEt (0.55 eq), Et2O, rt, 12 h

Products: ester er 98.0:2.0 + alcohol er 98.0:2.0

(50), s = 182

Refs.: 61

C3–10

Ester: structure with OH, R, CO2Bn

Conditions: Ph2CHCO2H (0.5 eq), Piv2O (0.6 eq), **124** (5 mol %), i-Pr2NEt (1.2 eq), Et2O, rt, 12 h

Products: I + II

Refs.: 62

R		er I	er II	s
Me	(46)	98.5:1.5	91.0:9.0	146
TBSOCH2	(48)	96.5:3.5	93.5:6.5	80
Et	(50)	97.5:2.5	97.0:3.0	127
TBSO(CH2)2	(47)	98.0:2.0	93.5:6.5	146
n-Pr	(50)	97.5:2.5	98.5:1.5	171
i-Pr	(44)	96.0:4.0	86.5:13.5	53
n-Bu	(43)	98.0:2.0	94.0:6.0	128
i-Bu	(51)	97.0:3.0	98.5:1.5	140
Cy	(45)	95.5:4.5	87.5:12.5	47
Ph(CH2)2	(50)	98.0:2.0	97.5:2.5	202

61

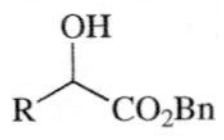

(Ph$_2$CHCO)$_2$O (0.3 eq),
Piv$_2$O (0.3 eq), **124** (5 mol %),
i-Pr$_2$NEt (0.6 eq), Et$_2$O, rt, 12 h

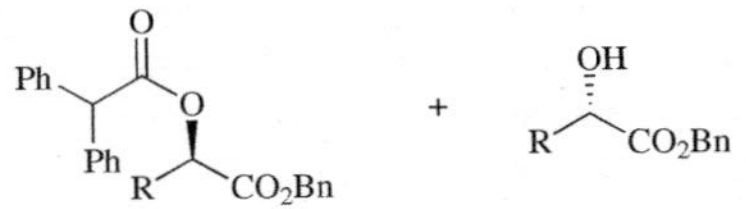

I + II

R		er I	er II	s
TBSOCH$_2$	(43)	95.5:4.5	99.5:0.5	102
TBSO(CH$_2$)$_2$	(49)	97.5:2.5	95.5:4.5	129
n-Pr	(51)	97.0:3.0	99.5:0.5	162
TBSO(CH$_2$)$_3$	(49)	98.5:1.5	97.0:3.0	208
n-Bu	(50)	97.0:3.0	98.0:2.0	126
Ph(CH$_2$)$_2$	(49)	98.0:2.0	94.0:6.0	144

(Ph$_2$CHCO)$_2$O (0.5 eq),
124 (5 mol %),
i-Pr$_2$NEt (0.5 eq), Et$_2$O, rt, 12 h

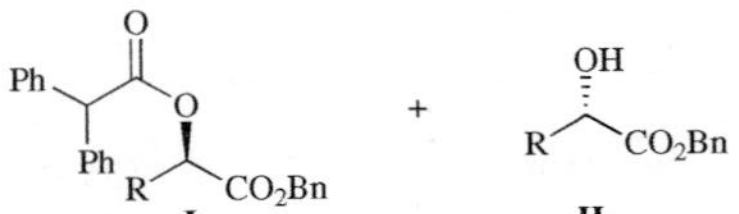

I + II

R		er I	er II	s
Et	(47)	98.0:2.0	93.0:7.0	152
i-Pr	(42)	96.0:4.0	83.0:17.0	46
i-Bu	(47)	98.5:1.5	93.5:6.5	167
Cy	(41)	95.5:4.5	81.5:18.5	42
TBSO(CH$_2$)$_{12}$	(51)	97.5:2.5	95.5:4.5	119

C$_{4-14}$

TABLE 2I. CATALYTIC KINETIC RESOLUTION OF α-HYDROXY ESTERS (Continued)

Ester	Conditions	Product(s) and Conversion(s) (%)	Refs.

Please refer to the charts preceding the tables for the structures indicated by the **bold** numbers.

C₅

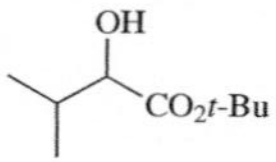

n-PrNCO (0.5 eq),
27 (0.2 mol %),
CH₂Cl₂, 0°, 1 h

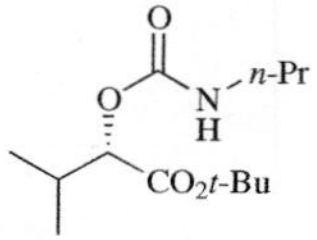

er 84.5:15.5

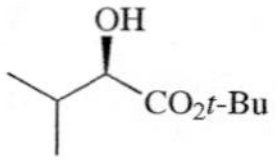

+

er 86.0:14.0

(51), s = 12

99

C₆

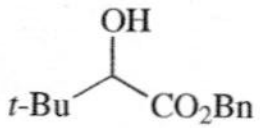

Ac₂O (x eq), Sc(OTf)₃ (5 mol %),
42a (5 mol %), 3 Å MS,
THF, 30°, 48 h

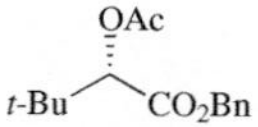

er 85.0:15.0

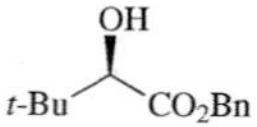

+

er 84.5:15.5

(50), s = 11

103

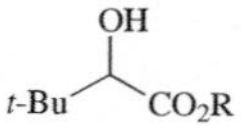

n-PrNCO (0.5 eq),
27 (0.2 mol %),
CH₂Cl₂, 0°, 1 h

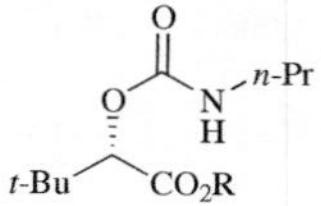

I

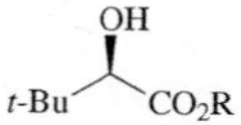

+

II

99

R		er I	er II	s
Me	(50)	98.0:2.0	97.5:2.5	170
Et	(51)	97.5:2.5	98.5:1.5	153
i-Pr	(50)	97.5:2.5	98.0:2.0	164
t-Bu	(50)	98.0:2.0	98.5:1.5	195

C_{6-12}

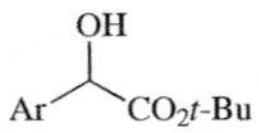

(structure: Ar–CH(OH)–CO_2t-Bu)

Ac$_2$O (x eq),
Sc(OTf)$_3$ (5 mol %),
42a (5 mol %),
3 Å MS, THF, 48 h

(structure **I**: Ar–CH(OAc)–CO$_2t$-Bu) + (structure **II**: Ar–CH(OH)–CO$_2t$-Bu)

103

Ar	x	Temp (°)		er **I**	er **II**	s
2-furyl	0.8	−20	(50)	98.5:1.5	97.5:2.5	247
4-FC$_6$H$_4$	0.9	−20	(50)	96.0:4.0	95.0:5.0	71
2-ClC$_6$H$_4$	1.0	−20	(50)	95.5:5.0	95.0:5.0	60
3-ClC$_6$H$_4$	0.9	−20	(48)	96.5:3.5	92.0:8.0	79
4-ClC$_6$H$_4$	0.9	−20	(48)	97.0:3.0	93.0:7.0	86
2-BrC$_6$H$_4$	0.9	−20	(50)	95.5:4.5	94.5:5.5	60
3-MeOC$_6$H$_4$	0.9	−20	(49)	97.5:2.5	95.0:5.0	112
4-MeOC$_6$H$_4$	0.9	−20	(52)	93.5:6.5	96.5:3.5	51
4-O$_2$NC$_6$H$_4$	1.1	−20	(47)	96.0:4.0	90.5:9.5	58
2-MeC$_6$H$_4$	0.7	0	(52)	90.0:10.0	92.5:7.5	24
3-MeC$_6$H$_4$	0.9	−20	(49)	95.5:4.5	93.0:7.0	58
4-MeC$_6$H$_4$	0.9	−20	(50)	94.5:5.5	94.0:6.0	48
1-Np	1.3	−20	(48)	97.5:2.5	95.0:5.0	130

TABLE 21. CATALYTIC KINETIC RESOLUTION OF α-HYDROXY ESTERS (*Continued*)

Ester	Conditions	Product(s) and Conversion(s) (%)	Refs.

*Please refer to the charts preceding the tables for the structures indicated by the **bold** numbers.*

C$_8$

Ester: OH / Ph / CO$_2$R

Conditions: Ac$_2$O (x eq), Sc(OTf)$_3$ (5 mol %), **42a** (5 mol %), 3 Å MS, THF, 48 h

Products: I (OAc / Ph / CO$_2$R) + II (OH / Ph / CO$_2$R)

Refs.: 103

R	x	Temp (°)		er I	er II	s
Me	0.8	0	(49)	95.5:4.5	93.5:6.5	57
Et	0.7	0	(49)	95.5:4.5	94.0:6.0	63
n-Pr	0.75	0	(50)	95.0:5.0	94.0:6.0	54
i-Pr	0.8	0	(52)	93.0:7.0	96.0:4.0	42
allyl	0.9	0	(51)	94.5:5.5	96.5:3.5	57
n-Bu	0.8	0	(49)	94.0:6.0	92.5:7.5	45
i-Bu	0.8	0	(50)	92.5:7.5	92.0:8.0	34
t-Bu	0.9	−20	(50)	96.0:4.0	96.0:4.0	82
Cy	0.7	0	(53)	91.0:9.0	96.0:4.0	33
PhCH$_2$	0.95	0	(51)	96.0:4.0	98.0:2.0	91
1-adamantyl	0.95	−20	(52)	92.0:8.0	96.0:4.0	36

Ester: OH / Ph / CO$_2$Me

Conditions: (i-PrCO)$_2$O (2.5 eq), **140** (4.8 mol %), toluene, rt, 3 h

Products: (i-Pr–C(=O)–O / Ph / CO$_2$Me) + (OH / Ph / CO$_2$Me), er 68.0:32.0, (49), $s = 3$

Refs.: 155

C₃

Substrate: OH, Ph, CO₂Et

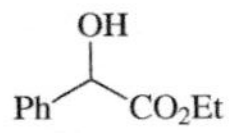

(EtCO)₂O (2.5 eq), **124** (4 mol %), toluene, rt, 24 h

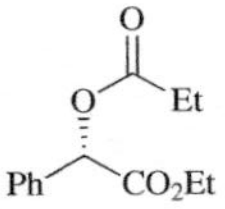

$$\text{O=, Et, O, Ph, CO}_2\text{Et} \quad \text{er } 61.5{:}38.5 \quad + \quad \text{OH, Ph, CO}_2\text{Et} \quad (31),\ s = 2 \quad \text{er } 55.0{:}45.0$$

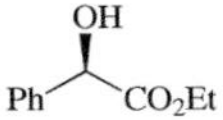

244

Substrate: OH, CO₂t-Bu (methylenedioxyphenyl)

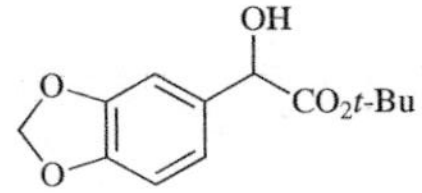

Ac₂O (0.7 eq), Sc(OTf)₃ (5 mol %), **42a** (5 mol %), 3 Å MS, THF, 0°, 48 h

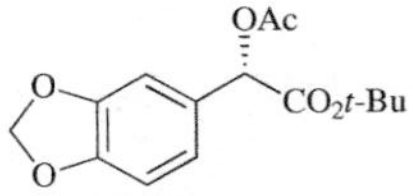

$$\text{OAc, CO}_2\text{t-Bu} \quad \text{er } 91.5{:}8.5 \quad + \quad \text{OH, CO}_2\text{t-Bu} \quad (53),\ s = 37 \quad \text{er } 97.0{:}3.0$$

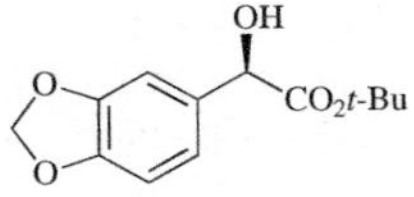

103

Substrate: OH, Ph, CO₂t-Bu

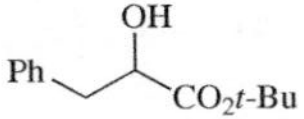

Ac₂O (0.9 eq), Sc(OTf)₃ (5 mol %), **42b** (5 mol %), 3 Å MS, THF, 30°, 48 h

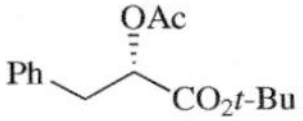

$$\text{OAc, Ph, CO}_2\text{t-Bu} \quad \text{er } 86.0{:}14.0 \quad + \quad \text{OH, Ph, CO}_2\text{t-Bu} \quad (50),\ s = 13 \quad \text{er } 86.5{:}13.5$$

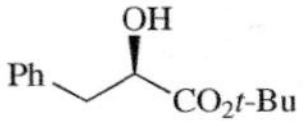

103

Substrate: OH, CO₂Me (methylcyclohexyl)

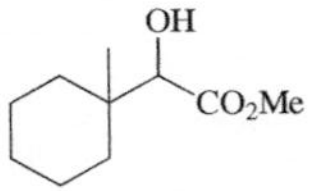

n-PrNCO (0.5 eq), **27** (0.2 mol %), CH₂Cl₂, 0°, 3 h

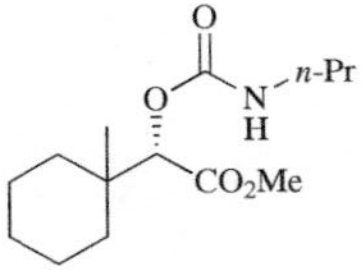

$$\text{O, N-}n\text{-Pr, H, CO}_2\text{Me} \quad \text{er } 97.5{:}2.5 \quad + \quad \text{OH, CO}_2\text{Me} \quad (50),\ s = 146 \quad \text{er } 97.5{:}2.5$$

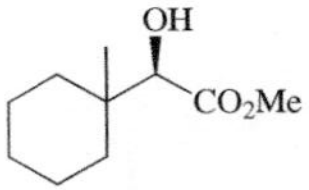

99

TABLE 2I. CATALYTIC KINETIC RESOLUTION OF α-HYDROXY ESTERS (*Continued*)

Ester	Conditions	Product(s) and Conversion(s) (%)	Refs.

*Please refer to the charts preceding the tables for the structures indicated by the **bold** numbers.*

C_{11}

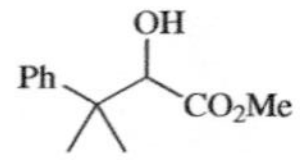

n-PrNCO (0.5 eq),
27 (0.2 mol %),
CH_2Cl_2, 0°, 3 h

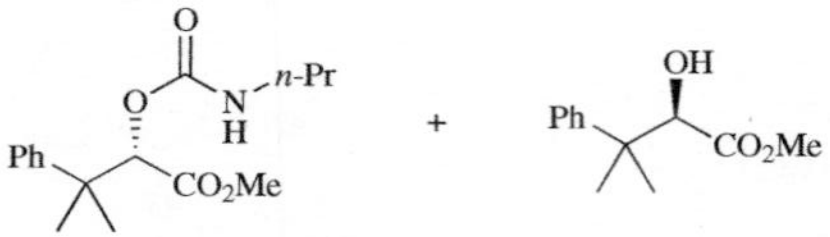

er 97.5:2.5   +   er 98.5:2.5   (50), *s* = 182   99

C_{15-16}

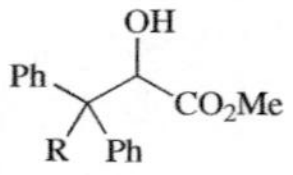

n-PrNCO (0.5 eq),
27 (0.2 mol %),
CH_2Cl_2, 0°

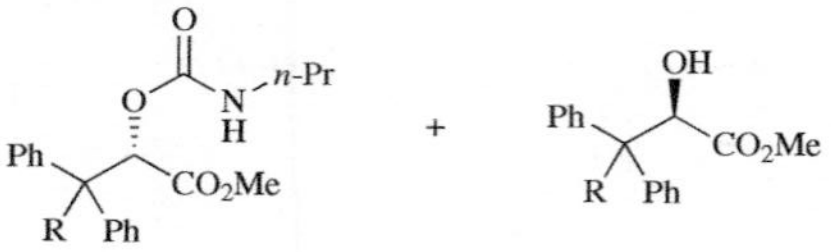

I   +   **II**   99

R	Time (h)		er **I**	er **II**	*s*
H	2	(50)	92.0:8.0	91.5:8.5	29
MeO	3	(38)	95.0:5.0	75.5:24.5	21
Me	12	(50)	98.5:1.5	98.5:1.5	261

TABLE 2J. CATALYTIC KINETIC RESOLUTION OF PRIMARY ALCOHOLS

Alcohol	Conditions	Product(s) and Conversion(s) (%)	Refs.

Please refer to the charts preceding the tables for the structures indicated by the **bold** numbers.

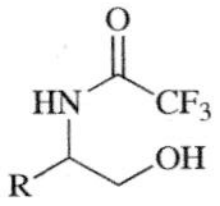

C₃₋₈

| | | 236 |

(i-PrCO)₂O (0.7 eq),
64 (5 mol %),
CHCl₃, rt, 16 h

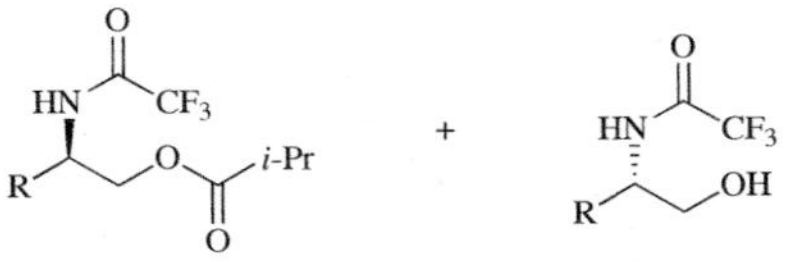

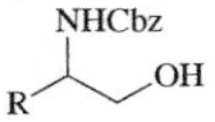

R		er **II**	s
MeO₂C	(71)	91.0:9.0	5
4-ClC₆H₄	(69)	98.5:1.5	10

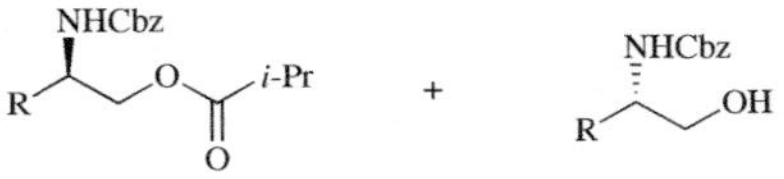

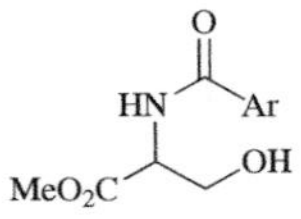

(i-PrCO)₂O (0.7 eq),
64 (5 mol %),
CHCl₃, rt, 16 h

236

R		er **II**	s
MeO₂C	(56)	56.5:43.5	1
4-ClC₆H₄	(62)	77.5:22.5	3

(i-PrCO)₂O (1.5 eq),
64 (5 mol %),
CHCl₃, rt, 24 h

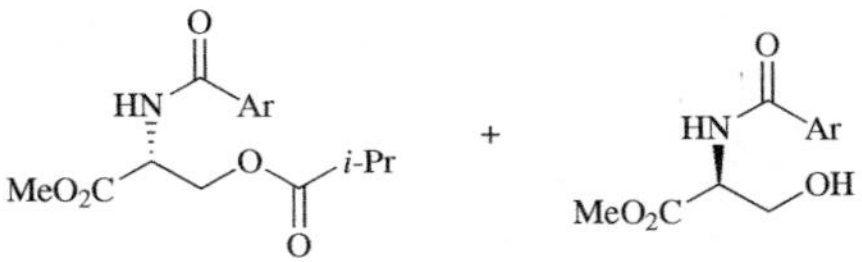

(69), s = 12   177, 178

er 99.5:0.5

Ar = 4-Me₂NC₆H₄

Alcohol	Conditions	Product(s) and Conversion(s) (%)	Refs.

*Please refer to the charts preceding the tables for the structures indicated by the **bold** numbers.*

C₃

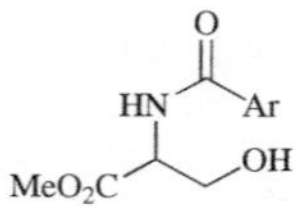

$(i\text{-PrCO})_2O$ (1.5 eq),
65 (5 mol %),
CH_2Cl_2, rt, 24 h

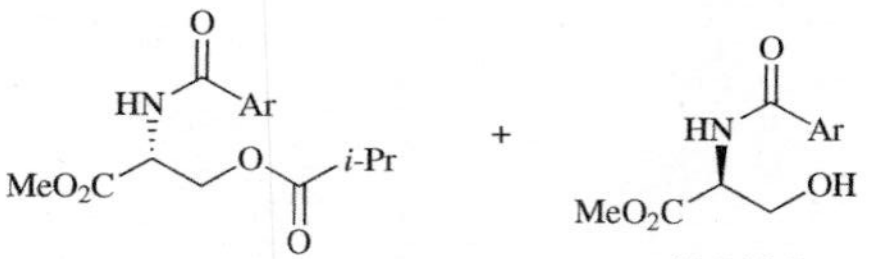

(71), s = 3

177

Ar = 4-Me₂NC₆H₄

C₃₋₉

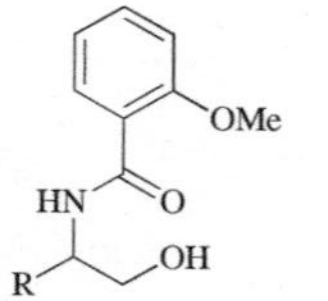

BzCl (0.5 eq),
Cu(OTf)₂ (3 mol %),
15 (3 mol %), K₂CO₃,
THF, rt, 3 h

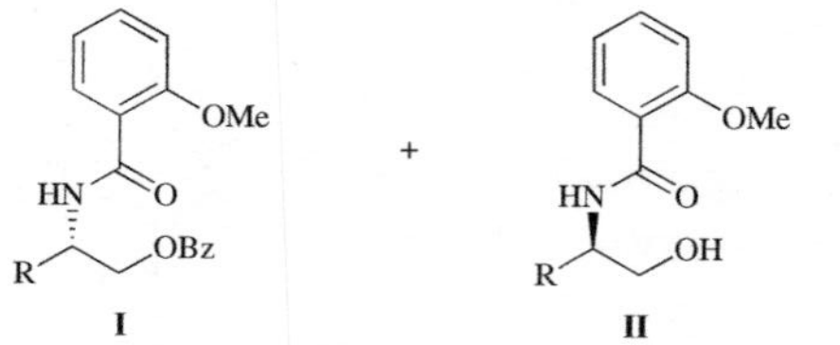

238

R		er I	er II	s
Me	(—)	94.5:5.5	77.0:23.0	29
Et	(—)	93.0:7.0	84.0:16.0	28
Ph	(—)	92.5:7.5	68.0:32.0	18
4-ClC₆H₄CH₂	(—)	93.5:6.5	90.0:10.0	35
PMB	(—)	92.5:7.5	65.0:35.0	16

C₃

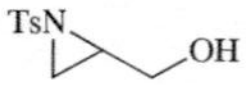

BzCl (0.6 eq), **138** (0.3 mol %),
$i\text{-Pr}_2$NEt (0.5 eq),
4 Å MS, CH₂Cl₂, −78°, 3 h

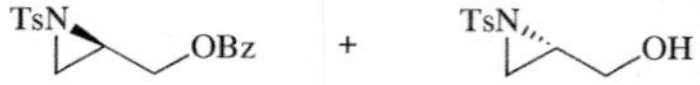

er 66.0:34.0

(—), s = 2

334

334

335

235

C_{3-8}

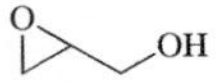

1-NpCOCl (0.75 eq),
138 (0.3 mol %),
i-Pr$_2$NEt (0.5 eq),
4 Å MS, CH$_2$Cl$_2$, −78°, 3 h

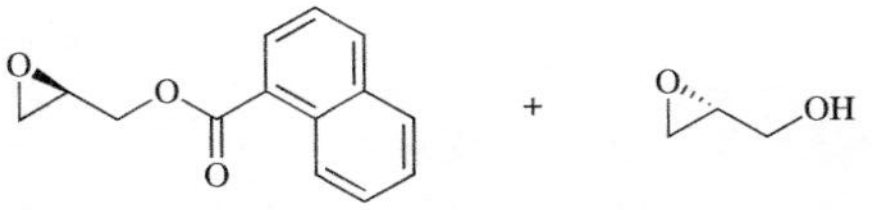

er 82.0:18.0 + er 73.0:27.0

(—), s = 6

C_{3-8}

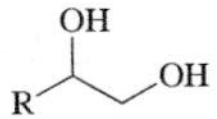

BzCl (0.5 eq), **32** (5.5 mol %),
CuCl (5 mol %), *i*-Pr$_2$NEt,
CH$_2$Cl$_2$, −40°

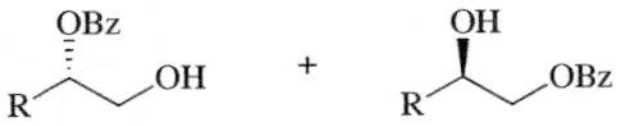

I + **II**

R	Time (h)	**I**	er **I**	**II**	er **II**
Me	33	(14)	86.5:13.5	(76)	73.5:26.5
Et	22	(8)	86.5:13.5	(80)	72.0:28.0
n-Bu	20	(12)	96.0:4.0	(78)	71.5:28.5
Ph	22	(12)	—	(80)	71.0:29.0

C_{3-5}

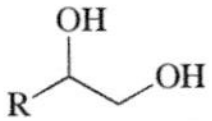

PhCHO (0.5 eq), **153** (2.5 mol %),
162 (0.5 eq), Cs$_2$CO$_3$ (0.025 eq),
4 Å MS, THF, −5°, 20 h

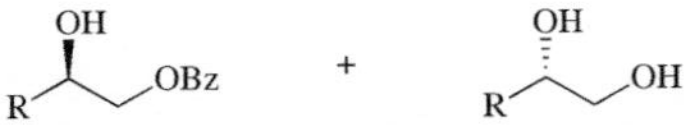

I + **II**

R		er **I**	er **II**	*s*
Me	(—)	87.0:13.0	75.0:25.0	11
Et	(—)	88.0:12.0	81.0:19.0	14
n-Pr	(—)	90.0:10.0	70.0:30.0	13

Alcohol	Conditions	Product(s) and Conversion(s) (%)	Refs.

*Please refer to the charts preceding the tables for the structures indicated by the **bold** numbers.*

C$_{3-8}$

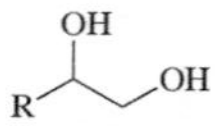

2-MeC$_6$H$_4$COCl (0.5 eq), **32** (5.5 mol%), CuCl (5 mol %), i-Pr$_2$NEt, CH$_2$Cl$_2$, –40°, 20 h

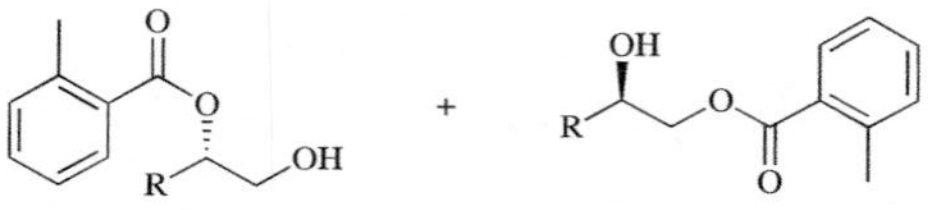

R	I	er I	II	er II
Me	(14)	94.5:5.5	(62)	89.0:11.0
Et	(20)	94.5:5.5	(78)	89.5:10.5
n-Bu	(22)	98.5:1.5	(68)	87.5:12.5
t-Bu	(—)	—	(86)	86.5:13.5
Ph	(14)	99.0:1.0	(72)	80.5:19.5

335

C$_3$

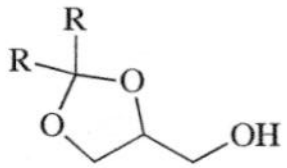

BzCl (0.75 eq), **138** (0.3 mol %), i-Pr$_2$NEt (0.5 eq), 4 Å MS, CH$_2$Cl$_2$, –78°, 3 h

R	I	er I	er II	s
Me	(—)	92.5:7.5	86.0:14.0	5
n-Pr	(—)	98.0:2.0	92.5:7.5	7
i-Pr	(—)	79.5:20.5	87.5:12.5	7
n-Bu	(—)	89.0:11.0	57.0:43.0	6
-(CH$_2$)$_5$-	(—)	76.5:23.5	82.5:17.5	5
Cy	(—)	71.5:28.5	91.0:9.0	6
Ph	(—)	77.0:23.0	81.0:19.0	6
PhCH$_2$	(—)	61.5:38.5	64.5:35.5	2

334

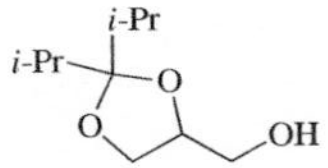

ArCOCl (0.75 eq),
138 (0.3 mol %),
i-Pr_2NEt (0.5 eq),
4 Å MS, CH_2Cl_2, –78°, 3 h

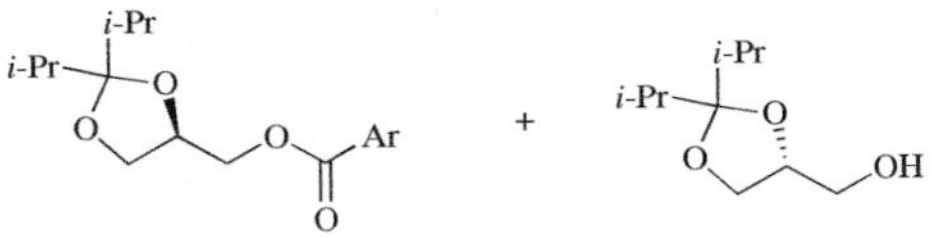

I + II

Ar		er **I**	er **II**	s
Ph	(—)	79.5:20.5	87.5:12.5	7
4-ClC_6H_4	(—)	75.5:24.5	88.0:12.0	5
4-$MeOC_6H_4$	(—)	85.0:15.0	72.0:28.0	9
2-MeC_6H_4	(—)	80.5:19.5	57.5:42.5	5
3-MeC_6H_4	(—)	75.5:24.5	89.5:10.5	6
4-MeC_6H_4	(—)	84.0:16.0	90.5:9.5	10
2-t-BuC_6H_4	(—)	78.5:21.5	92.0:8.0	7
1-Np	(—)	88.5:11.5	84.0:16.0	16
2-Np	(—)	89.0:11.0	55.0:45.0	9

334

C_{4-12}

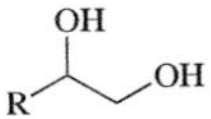

BzCl (0.5 eq), **15** (0.25 mol %),
Na_2CO_3 (1.5 eq), H_2O (5 eq),
THF, –10°, 14 h

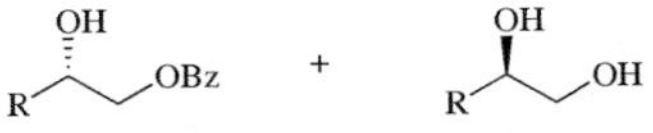

I + II

R		er **I**	s
Et	(35)	72.0:28.0	3
Ph	(38)	93.0:7.0	22
n-$C_{10}H_{21}$	(35)	79.5:20.5	5
1-Np	(41)	86.0:14.0	10
2-Np	(25)	82.0:18.0	6

100

TABLE 2J. CATALYTIC KINETIC RESOLUTION OF PRIMARY ALCOHOLS (*Continued*)

Alcohol	Conditions	Product(s) and Conversion(s) (%)	Refs.

*Please refer to the charts preceding the tables for the structures indicated by the **bold** numbers.*

C₄

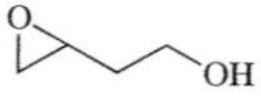

4-MeC₆H₄COCl (0.75 eq), **138** (0.3 mol %), *i*-Pr₂NEt (0.5 eq), 4 Å MS, CH₂Cl₂, −78°, 3 h

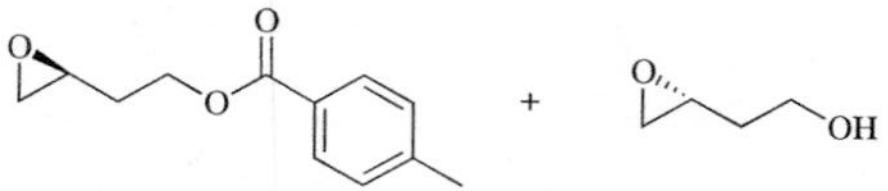

er 55.5:45.5 + er 56.0:44.0   (−), s = 1

334

C₅

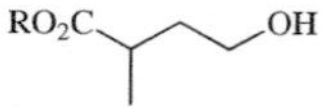

157a (2 mol %), CH₂Cl₂

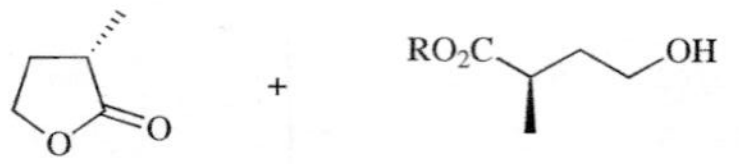

I + II

R	Temp (°)	Time (h)		er I	er II	s
i-Pr	5	68	(68)	62.0:38.0	60.5:39.5	1
t-Bu	5	24	(40)	65.5:34.5	79.0:21.0	25
(*i*-Pr)₂CH	35	144	(50)	63.5:36.5	57.0:43.0	2

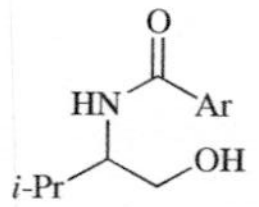

239

BzCl (0.5 eq), Cu(OTf)₂ (3 mol %), **15** (3 mol %), K₂CO₃, THF, rt, 3 h

I + II

R		er I	er II	s
Ph	(—)	89.0:11.0	66.0:34.0	11
2-MeOC₆H₄	(—)	96.5:3.5	77.5:22.5	50
4-MeOC₆H₄	(—)	91.5:8.5	81.0:19.0	20
2-ClC₆H₄	(—)	88.5:11.5	65.0:35.0	10
4-ClC₆H₄	(—)	73.0:27.0	58.0:42.0	3

238

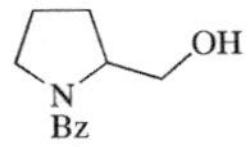

OH
│
N
Bz

BzCl (0.5 eq), Cu(OTf)₂ (3 mol %),
15 (3 mol %), K₂CO₃,
THF, rt, 3 h

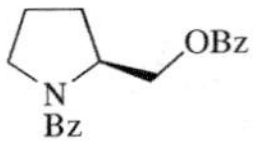

OBz
│
N
Bz
er 98.5:1.5

+

OH
│
N
Bz
(43), s = 174
er 92.0:8.0

238

OH
(THF)

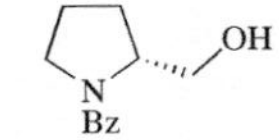

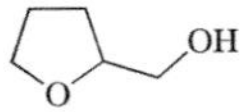

4-MeC₆H₄COCl (0.75 eq),
138 (0.3 mol %),
i-Pr₂NEt (0.5 eq),
4 Å MS, CH₂Cl₂, −78°, 3 h

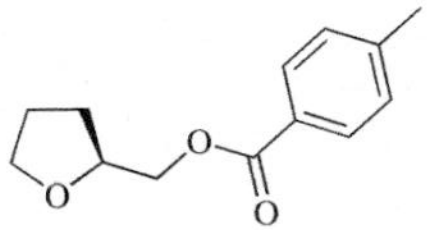

(4-tolyl ester of THF-CH₂)
er 75.0:25.0

+

OH
(THF)
er 73.0:27.0
(—), s = 5

334

pyrrolidine–C(O)–NH–(i-Pr)–CH₂OH

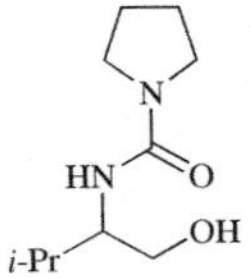

(i-PrCO)₂O (0.5 eq),
109 (5 mol %),
i-Pr₂NEt (0.5 eq), CCl₄, 0°, 3 h

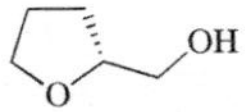

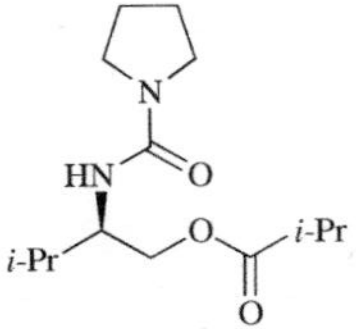

pyrrolidine–C(O)–NH–(i-Pr)–CH₂–O–C(O)–i-Pr
er 93.0:7.0

+

pyrrolidine–C(O)–NH–(i-Pr)–CH₂OH
er 94.0:6.0
(50), s = 39

70

NHCbz
(4-Cl-C₆H₄)–CH–CH₂OH

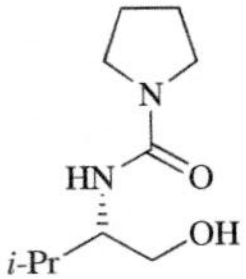

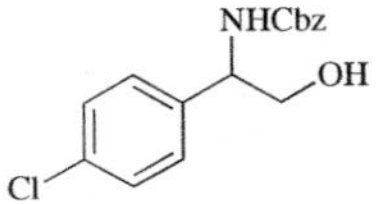

(i-PrCO)₂O (0.7 eq),
64 (5 mol %),
CHCl₃, rt, 16 h

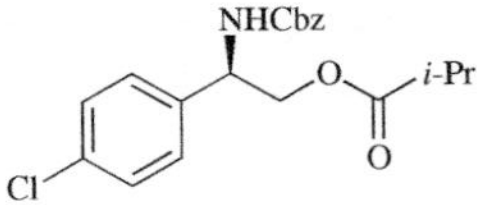

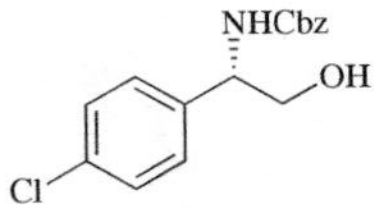

NHCbz
(4-Cl-C₆H₄)–CH–CH₂–O–C(O)–i-Pr
er 61.5:38.5

+

NHCbz
(4-Cl-C₆H₄)–CH–CH₂OH
(63), s = 2

236

C₆

TABLE 2J. CATALYTIC KINETIC RESOLUTION OF PRIMARY ALCOHOLS (*Continued*)

Alcohol	Conditions	Product(s) and Conversion(s) (%)	Refs.

*Please refer to the charts preceding the tables for the structures indicated by the **bold** numbers.*

C6

4-BrC6H4COCl (0.5 eq), **32** (5.5 mol%), CuCl (5 mol %), i-Pr2NEt, CH2Cl2, –40°

(14) er 90.0:10.0   (58) er 75.5:24.5

335

C6–12

PhCHO (0.5 eq), **153** (2.5 mol %), **162** (0.5 eq), Cs2CO3 (0.025 eq), 4 Å MS, THF, –5°, 20 h

I   II

235

Ar		er I	er II	s
2-furyl	(—)	74.0:26.0	84.0:16.0	6
2-thienyl	(—)	84.0:16.0	68.0:32.0	7
Ph	(—)	91.0:9.0	87.0:13.0	22
4-FC6H4	(—)	87.0:13.0	89.0:11.0	16
2-ClC6H4	(—)	86.0:14.0	79.0:21.0	11
4-ClC6H4	(—)	86.0:14.0	89.0:11.0	14
3-BrC6H4	(—)	89.0:11.0	88.0:12.0	18
4-BrC6H4	(—)	85.0:15.0	87.0:13.0	12
2-MeOC6H4	(—)	88.0:12.0	92.0:8.0	19
3-MeOC6H4	(—)	85.0:15.0	90.0:10.0	14
4-MeOC6H4	(—)	85.0:15.0	89.0:11.0	13
2-MeC6H4	(—)	90.0:10.0	85.0:15.0	19
3-MeC6H4	(—)	91.0:9.0	77.0:23.0	17
4-MeC6H4	(—)	89.0:11.0	87.0:13.0	18
1-Np	(—)	89.0:11.0	89.0:11.0	19
2-Np	(—)	87.0:13.0	92.0:8.0	17

239

C$_{6-10}$

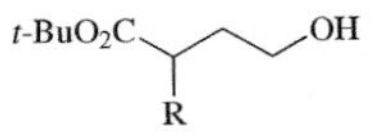

t-BuO$_2$C —— R, OH

157a (2 mol %), solvent

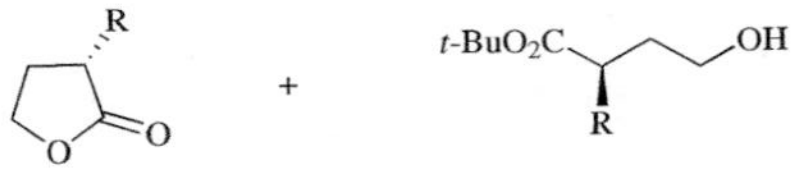

I (lactone, R) + II (t-BuO$_2$C, R, OH)

R	Solvent	Temp (°)	Time (h)		er I	er II	s
Et	CH$_2$Cl$_2$	rt	24	(67)	—	92.5:7.5	6
i-Pr	hexane	5	72	(56)	68.5:31.5	76.0:24.0	4
CH$_2$=CHCH$_2$	CH$_2$Cl$_2$	5	72	(63)	75.0:25.0	93.0:7.0	8
Ph	toluene	5	40	(71)	63.0:37.0	75.0:25.0	2

230

C$_{6-12}$

pyrrolidine-carbamate O, R, OH

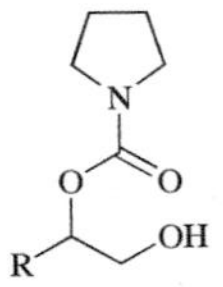

(i-PrCO)$_2$O (0.5 eq),
112 (5 mol %),
i-Pr$_2$NEt (0.5 eq), CCl$_4$, 0°

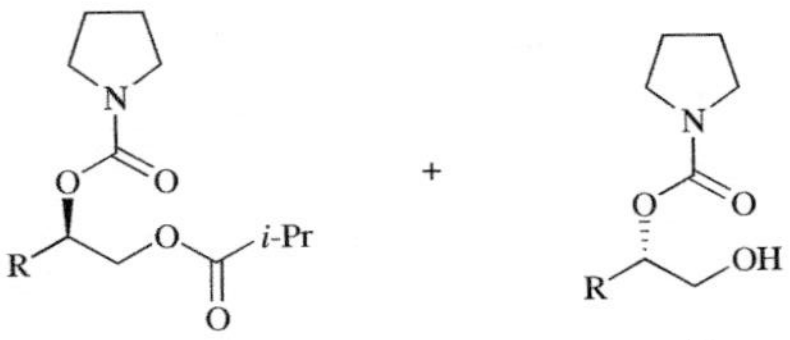

I (pyrrolidine-carbamate, O, i-Pr, R) + II (pyrrolidine-carbamate, O, R, OH)

R	Time (h)		er I	er II	s
t-Bu	2	(52)	94.0:6.0	99.0:1.0	91
Cy	1	(53)	89.0:11.0	96.0:4.0	32
Ph	4	(53)	88.0:12.0	94.0:6.0	23
4-MeC$_6$H$_4$	3	(43)	91.5:8.5	81.5:18.5	21
4-t-BuC$_6$H$_4$	3	(53)	88.5:11.5	93.0:7.0	22

TABLE 2J. CATALYTIC KINETIC RESOLUTION OF PRIMARY ALCOHOLS (*Continued*)

Alcohol	Conditions	Product(s) and Conversion(s) (%)	Refs.

*Please refer to the charts preceding the tables for the structures indicated by the **bold** numbers.*

C$_6$

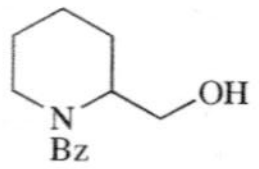

BzCl (0.5 eq),
Cu(OTf)$_2$ (3 mol %), **15** (3 mol %),
K$_2$CO$_3$, THF, rt, 3 h

(49), s = 177

er 97.5:2.5   +   er 99.0:1.0

238

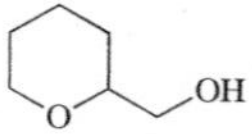

4-MeC$_6$H$_4$COCl (0.75 eq),
138 (0.3 mol %), *i*-Pr$_2$NEt (0.5 eq),
4 Å MS, CH$_2$Cl$_2$, –78°, 3 h

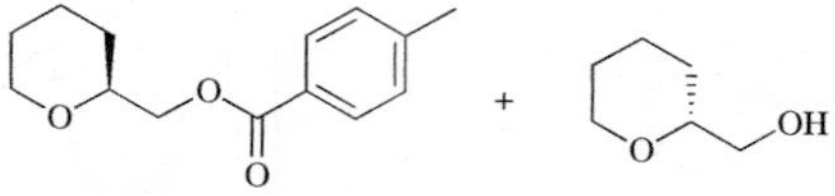

(—), s = 4

er 70.5:29.5   +   er 70.5:29.5

334

C$_8$

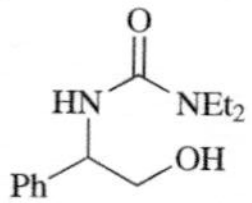

(*i*-PrCO)$_2$O (0.5 eq),
109 (5 mol %), *i*-Pr$_2$NEt (0.5 eq),
CHCl$_3$/CCl$_4$ (2:5), 0°, 3 h

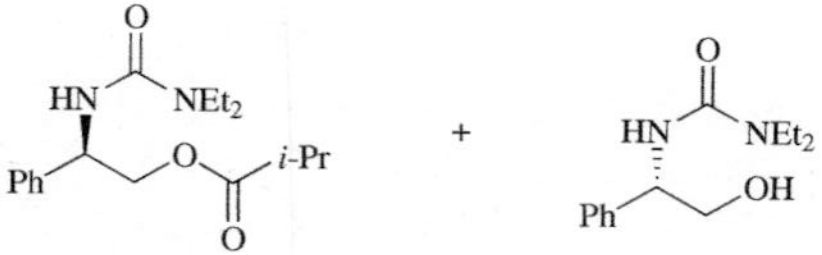

(53), s = 17

er 87.0:13.0   +   er 91.5:8.5

71

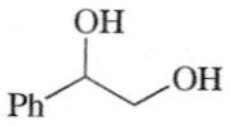

Ac$_2$O (0.5 eq), **37** (5 mol %),
Et$_3$N, Et$_2$O, –25°, 15 h

(45), s = 2

er 61.0:39.0   +   er 59.0:41.0

96

C_{9-13}

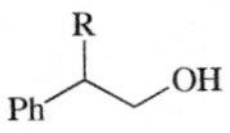

(EtCO)₂O (0.55 eq), **125** (10 mol %),
i-Pr₂NEt (0.55 eq),
CDCl₃, 0°, 0.25 h

I + **II**

R		er I	er II	s
Me	(49)	72.0:28.0	70.5:29.5	4
C₅H₁₁	(35)	75.5:24.5	63.5:36.5	4

237

C_9

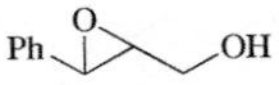

4-*t*-BuC₆H₄COCl (0.6 eq),
138 (0.3 mol %),
i-Pr₂NEt (0.5 eq),
4 Å MS, CH₂Cl₂, –78°, 3 h

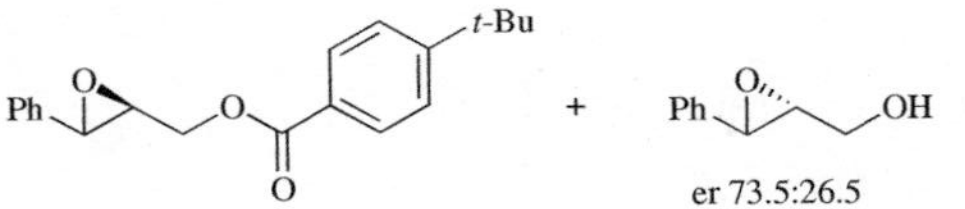

er 72.0:28.0   er 73.5:26.5

(—), *s* = 4   334

C_{10}

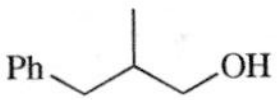

(EtCO)₂O (0.55 eq),
125 (10 mol %),
i-Pr₂NEt (0.55 eq),
CDCl₃, 0°, 0.25 h

er 53.0:47.0   er 51.5:48.5

(30), *s* = 1   237

415

Alcohol	Conditions	Product(s) and Conversion(s) (%)	Refs.

*Please refer to the charts preceding the tables for the structures indicated by the **bold** numbers.*

C₁₀

BzCl (0.6 eq), **138** (0.3 mol %), *i*-Pr₂NEt (0.5 eq), 4 Å MS, CH₂Cl₂, –78°, 3 h

er 51.0:49.0   +   er 51.5:48.5   (—), s = 1   334

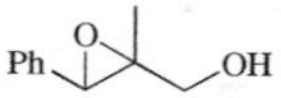

BzCl (0.6 eq), **138** (0.3 mol %), *i*-Pr₂NEt (0.5 eq), 4 Å MS, CH₂Cl₂, –78°, 3 h

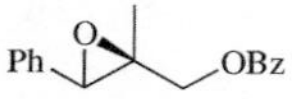

er 71.5:28.5   +   er 74.0:26.0   (—), s = 4   334

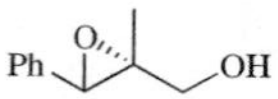

C₁₁

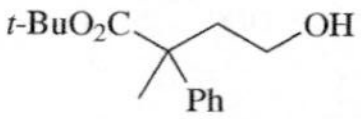

157a (2 mol %), CH₂Cl₂, 5° to rt, 232 h

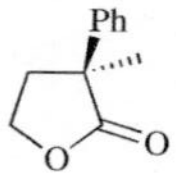

er 83.0:17.0   +   (53), s = 17   239

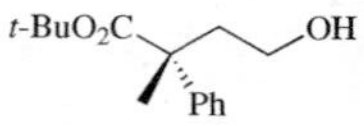

er 91.5:8.5

C₁₃

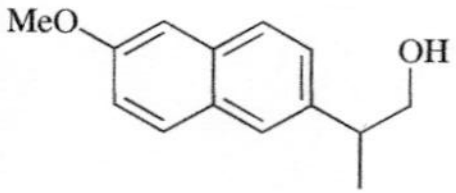

(EtCO)₂O (0.55 eq), **125** (10 mol %), *i*-Pr₂NEt (0.55 eq), CDCl₃, 0°, 0.25 h

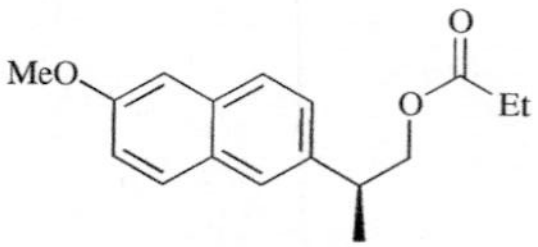

er 70.0:30.0   +   er 76.0:24.0   (52), s = 4   237

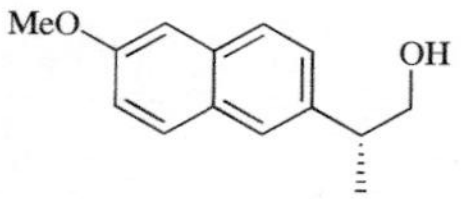

TABLE 2K. CATALYTIC KINETIC RESOLUTION OF TERTIARY ALCOHOLS

*Please refer to the charts preceding the tables for the structures indicated by the **bold** numbers.*

Alcohol	Conditions	Product(s) and Conversion(s) (%)	Refs.
C_8	Ac_2O (0.7 eq), **126** (2 mol %), i-Pr_2NEt (0.6 eq), toluene, 0°, 20 h	 er 97.0:3.0 + er 99.5:0.5 (52), $s = 200$	241
C_{8-10}	Ac_2O (1 eq), **155** (600 mg), i-Pr_2NEt (1 eq), toluene, 0°, 2 h		242

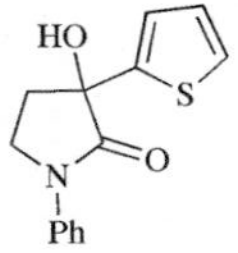

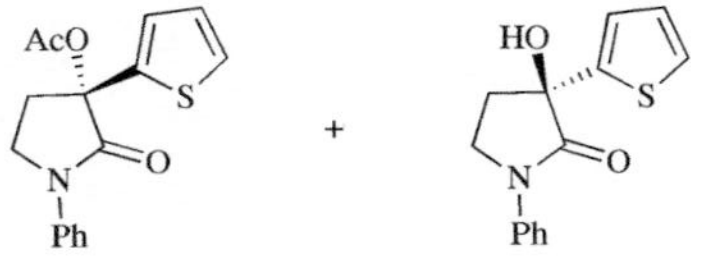

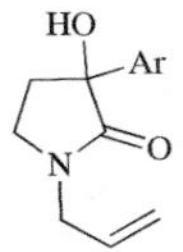

Ar		er I	er II	s
2-thienyl	(52)	90.0:10.0	94.0:6.0	25
3-thienyl	(45)	93.0:7.0	85.0:15.0	29
4-ClC_6H_4	(40)	96.0:4.0	80.0:20.0	42
4-$MeOC_6H_4$	(53)	87.0:13.0	92.0:8.0	18

C_9	Ac_2O (5.0 eq), **97** (10 mol %), Et_3N (2 eq), CH_2Cl_2, toluene, –20°, 164 h		240

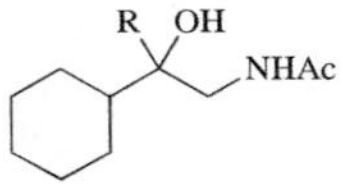

R			er I	er II	s
Me	(35)	(51)	94.0:6.0	74.0:26.0	19
MeO_2C			(—)	(—)	1

417

TABLE 2K. CATALYTIC KINETIC RESOLUTION OF TERTIARY ALCOHOLS (*Continued*)

Alcohol	Conditions	Product(s) and Conversion(s) (%)	Refs.

*Please refer to the charts preceding the tables for the structures indicated by the **bold** numbers.*

C$_{9-13}$

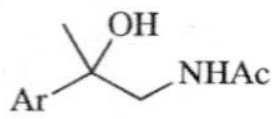

Ac$_2$O (5.0 eq), **97** (10 mol %), Et$_3$N (2 eq), CH$_2$Cl$_2$, toluene, –20°, 82 h

I + II

Ar		er I	er II	s
Ph	(37)	98.0:2.0	74.5:25.5	40
4-O$_2$NC$_6$H$_4$	(40)	98.0:2.0	73.5:26.5	32
4-MeC$_6$H$_4$	(48)	95.5:4.5	94.0:6.0	>50
2-Np	(35)	93.0:7.0	67.5:32.5	40

240

C$_9$

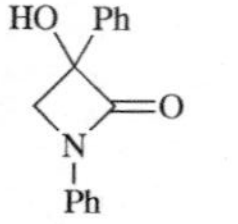

Ac$_2$O (0.7 eq), **126** (2 mol %), i-Pr$_2$NEt (0.6 eq), toluene, 0°, 20 h

I (er 78.0:22.0) + II (er 95.0:5.0)

(61), s = 10

241

C$_{9-14}$

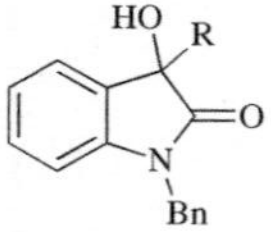

(i-PrCO)$_2$O (0.7 eq), **126** (1 mol %), i-Pr$_2$NEt (0.6 eq), CHCl$_3$, 0°, 18 h

I + II

R		er I	er II	s
Me	(53)	92.0:8.0	99.0:1.0	50
CH$_2$=CH	(53)	91.0:9.0	99.5:0.5	50
CH$_2$=CHCH$_2$	(47)	97.5:2.5	93.0:7.0	106
CH$_2$=C(CH$_3$)	(40)	98.0:2.0	83.0:17.0	80
2-furyl	(46)	97.0:3.0	90.0:10.0	70
Ph	(53)	95.0:5.0	99.5:0.5	120

241

PhCH=CHCHO (2.5 eq).
DBU (1.0 eq), **145a** (10 mol %),
MnO$_2$ (5.0 eq), NaBF$_4$ (0.5 eq),
Mg(OTf)$_2$ (10 mol %),
4 Å MS, THF, rt

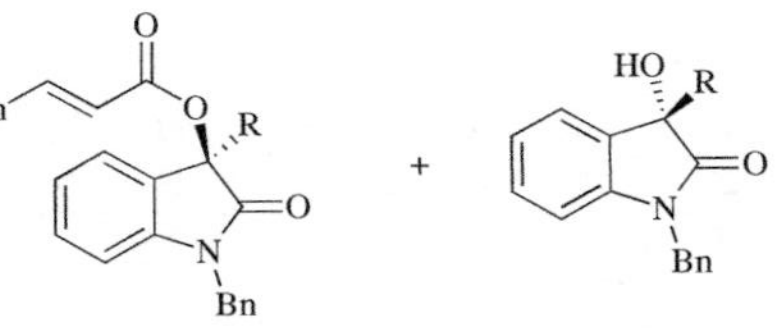

I + **II**

R	Time (h)		er I	er II	s
Me	24	(55)	90.0:10.0	99.5:0.5	46
Et	24	(56)	89.0:11.0	99.5:0.5	41
CH$_2$=CH	36	(57)	86.0:14.0	99.0:1.0	27
HC≡C	24	(63)	79.0:21.0	99.5:0.5	18
i-Pr	24	(27)	97.5:2.5	67.5:32.5	56
CH$_2$=CHCH$_2$	24	(53)	93.5:6.5	99.0:1.0	70
2-pyridyl	48	(34)	91.5:8.5	71.5:28.5	16
n-C$_6$H$_{13}$	24	(46)	96.0:4.0	90.0:10.0	59
Ph	48	(60)	82.0:18.0	99.0:1.0	20
4-FC$_6$H$_4$	36	(39)	77.5:22.5	77.5:22.5	23
4-MeOC$_6$H$_4$	36	(53)	96.0:4.0	96.0:4.0	29

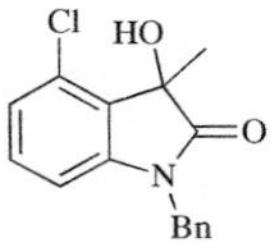

(i-PrCO)$_2$O (0.7 eq).
126 (10 mol %),
i-Pr$_2$NEt (0.6 eq),
CHCl$_3$, 0°, 118 h

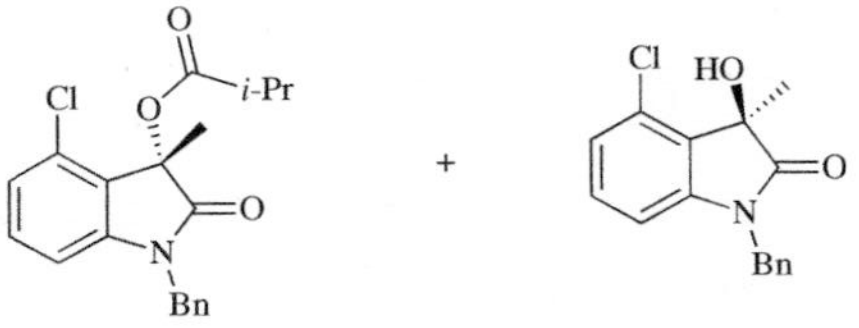

er 85.0:15.0 + er 99.5:0.5

(59), s = 32   241

C$_9$

TABLE 2K. CATALYTIC KINETIC RESOLUTION OF TERTIARY ALCOHOLS (*Continued*)

Alcohol	Conditions	Product(s) and Conversion(s) (%)	Refs.

*Please refer to the charts preceding the tables for the structures indicated by the **bold** numbers.*

C_9

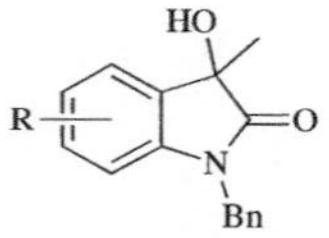

$(i\text{-PrCO})_2O$ (0.7 eq),
155 (5 mol %),
$i\text{-Pr}_2NEt$ (0.6 eq),
$(MeO)_2CO$, rt, 24 h

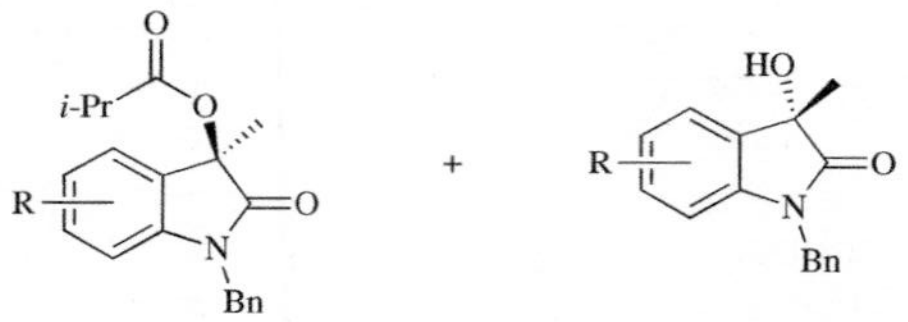

I + II

R		er **I**	er **II**	*s*
4-Cl	(41)	90.0:10.0	78.0:22.0	16
7-Cl	(62)	76.0:24.0	91.0:9.0	7

242

$(i\text{-PrCO})_2O$ (0.7 eq),
126 (1 mol %),
$i\text{-Pr}_2NEt$ (0.6 eq), $CHCl_3$, 0°

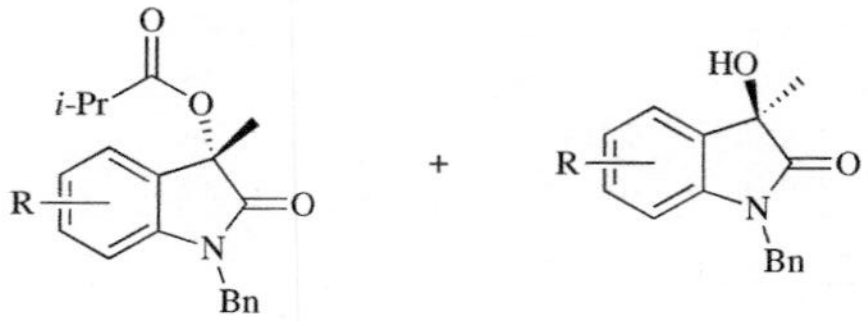

I + II

R	Time (h)		er **I**	er **II**	*s*
7-Cl	87	(54)	89.0:11.0	99.5:0.5	44
5-MeO	18	(54)	90.0:10.0	98.0:2.0	34

241

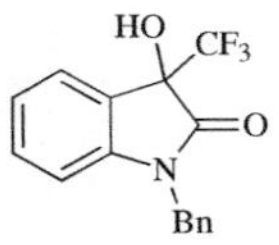

241

242

(59), $s = 32$

er 99.5:0.5

er 85.0:15.0

(*i*-PrCO)$_2$O (0.7 eq),
126 (1 mol %),
i-Pr$_2$NEt (0.6 eq),
CHCl$_3$, –40°, 18 h

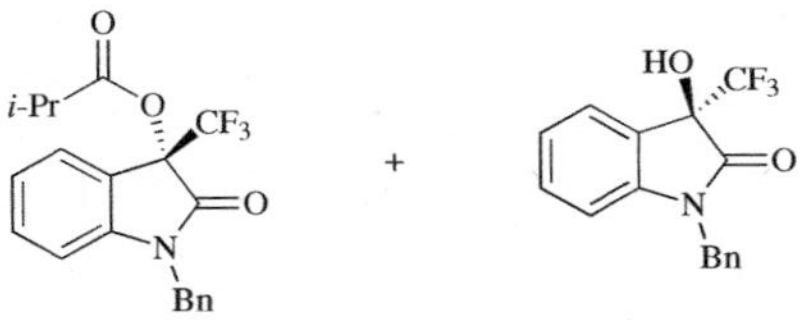

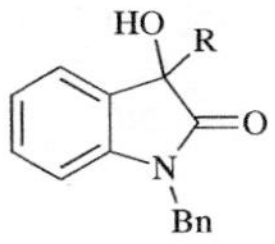

(*i*-PrCO)$_2$O (0.7 eq),
155 (5 mol %),
i-Pr$_2$NEt (0.6 eq),
(MeO)$_2$CO, rt, 24 h

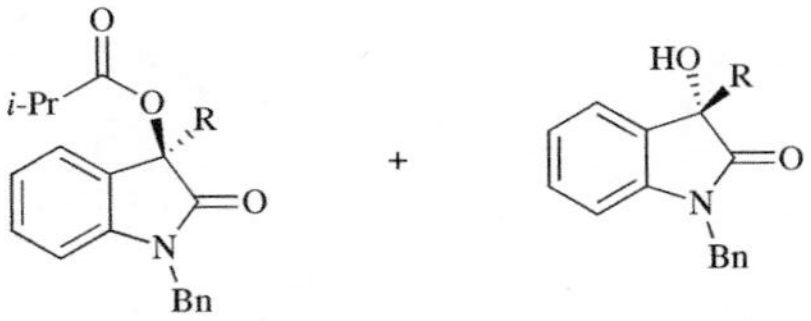

C$_{9-14}$

R		er I	er II		s
CF$_3$	(62)	80.0:20.0	99.5:0.5		21
i-Pr	(45)	87.0:13.0	80.0:20.0		13
CH$_2$=CHCH$_2$	(53)	88.0:12.0	93.0:7.0		20
(CH$_3$)$_2$C=CH	(44)	92.0:8.0	83.0:17.0		24
Ph	(51)	90.0:10.0	96.0:4.0		60

I

II

TABLE 2K. CATALYTIC KINETIC RESOLUTION OF TERTIARY ALCOHOLS *(Continued)*

Alcohol	Conditions	Product(s) and Conversion(s) (%)	Refs.

Please refer to the charts preceding the tables for the structures indicated by the **bold** numbers.

C$_{9-16}$

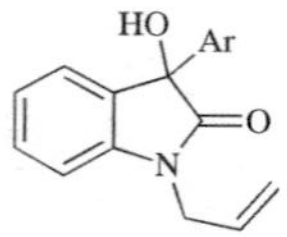

$(i\text{-PrCO})_2O$ (0.7 eq),
126 (1 mol %),
$i\text{-Pr}_2NEt$ (0.6 eq),
$CHCl_3$, 0°, 18 h

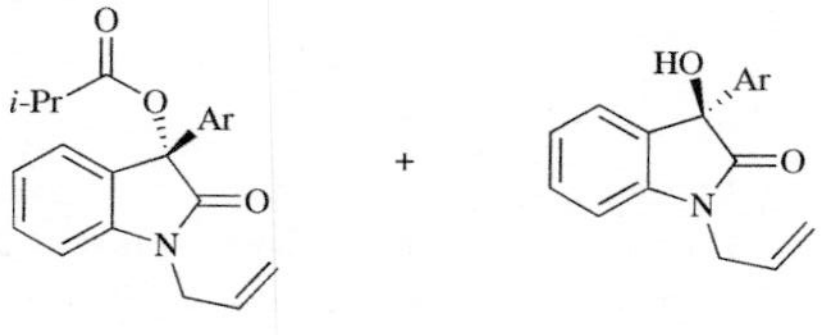

Ar	er I		er II	s
benzoxazol-2-yl	(51)	87.0:13.0	89.0:11.0	15
3-pyridyl	(52)	94.0:6.0	98.0:2.0	70
3,5-(CF$_3$)$_2$C$_6$H$_3$	(51)	95.0:5.0	96.0:4.0	60

241

C$_9$

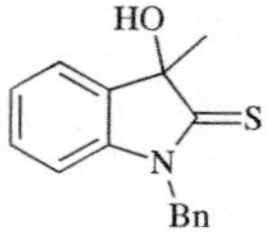

$(i\text{-PrCO})_2O$ (0.7 eq),
126 (5 mol %),
$i\text{-Pr}_2NEt$ (0.6 eq),
$CHCl_3$, 0°, 112 h

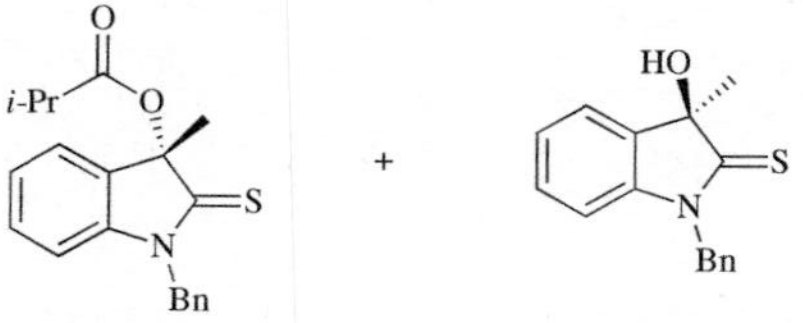

er 95.0:5.0   +   er 81.0:19.0   (41), s = 39

241

C$_{10}$

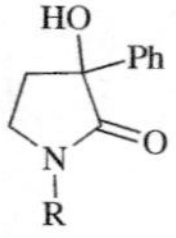

Ac_2O (1 eq), **155** (600 mg),
$i\text{-Pr}_2NEt$ (1 eq),
toluene, 0°, 2 h

R	er I		er II	s
Ph	(52)	91.0:9.0	94.0:6.0	29
Bn	(52)	91.0:9.0	93.0:7.0	27

242

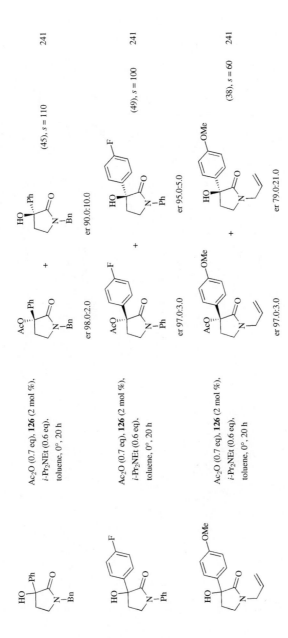

	Ac_2O (0.7 eq), **126** (2 mol %), i-Pr_2NEt (0.6 eq), toluene, 0°, 20 h	er 98.0:2.0	+ er 90.0:10.0	(45), $s = 110$	241
	Ac_2O (0.7 eq), **126** (2 mol %), i-Pr_2NEt (0.6 eq), toluene, 0°, 20 h	er 97.0:3.0	+ er 95.0:5.0	(49), $s = 100$	241
	Ac_2O (0.7 eq), **126** (2 mol %), i-Pr_2NEt (0.6 eq), toluene, 0°, 20 h	er 97.0:3.0	+ er 79.0:21.0	(38), $s = 60$	241

TABLE 2K. CATALYTIC KINETIC RESOLUTION OF TERTIARY ALCOHOLS (Continued)

Alcohol	Conditions	Product(s) and Conversion(s) (%)	Refs.

*Please refer to the charts preceding the tables for the structures indicated by the **bold** numbers.*

C_{10}

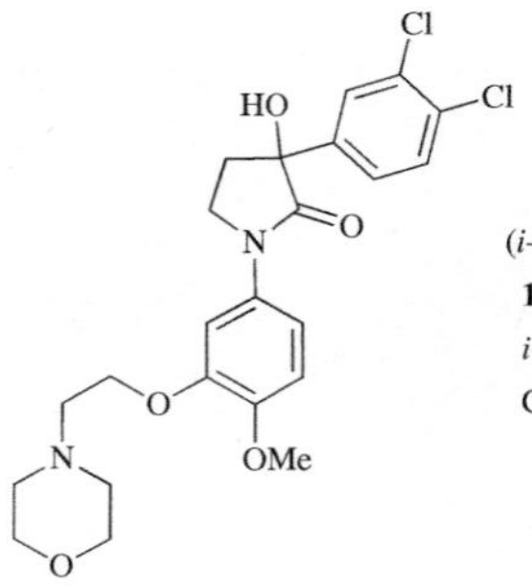

(i-PrCO)$_2$O (1.2 eq),
126 (5 mol %),
i-Pr$_2$NEt (0.6 eq),
CHCl$_3$, 0°, 35 h

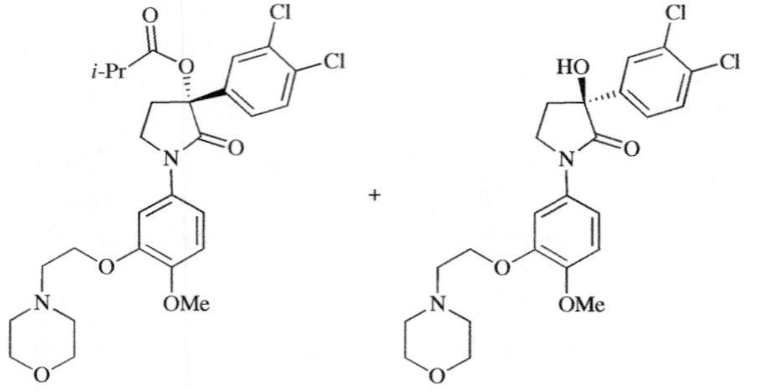

er 90.0:10.0   (54), s = 33   er 97.0:3.0

241

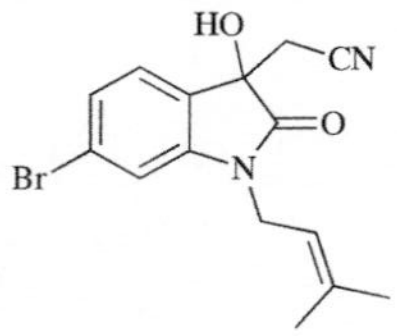

(i-PrCO)$_2$O (0.6 eq),
126 (1 mol %),
i-PrCO$_2$H (0.5 eq),
CHCl$_3$, 0°, 18 h

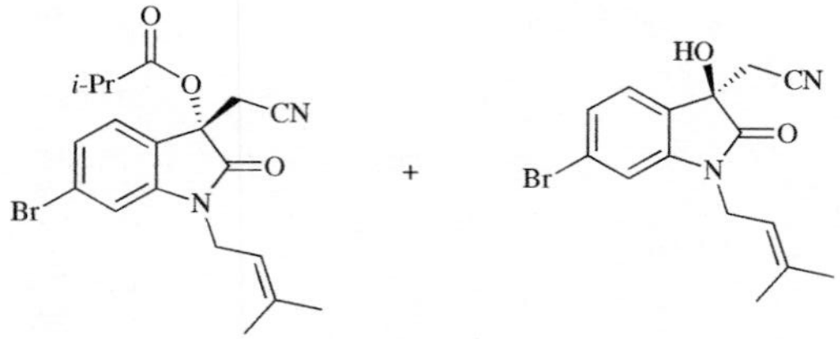

er 86.0:14.0   (58), s = 30   er 99.5:0.5

241

C$_{10-12}$

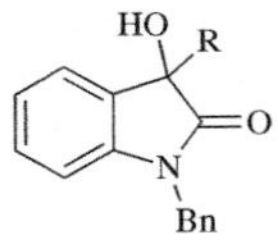

(i-PrCO)$_2$O (0.7 eq),
126 (5 mol %),
i-Pr$_2$NEt (0.6 eq), CHCl$_3$, 0°

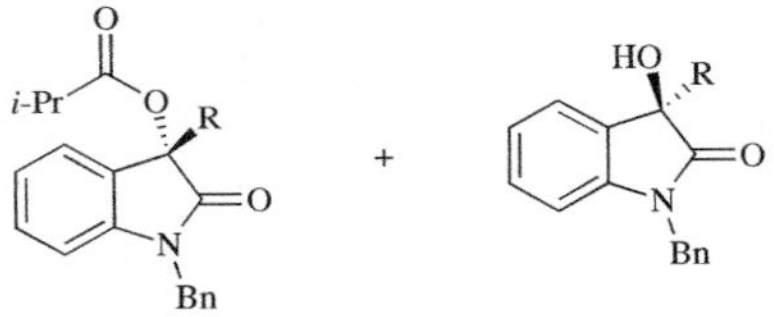

241

R	Time (h)		er **I**	er **II**	s
(CH$_3$)$_2$C=CH	72	(44)	97.0:3.0	87.0:13.0	60
Et$_2$NCOCH$_2$	26	(45)	93.0:7.0	84.0:16.0	26

C$_{11}$

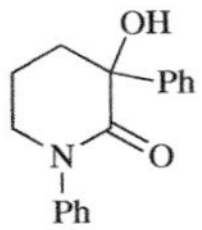

Ac$_2$O (1.5 eq), **126** (20 mol %),
i-Pr$_2$NEt (1.5 eq),
toluene, 90°, 24 h

er 92.0:8.0

+

(48), s = 25

er 89.0:11.0

241

C$_{11}$

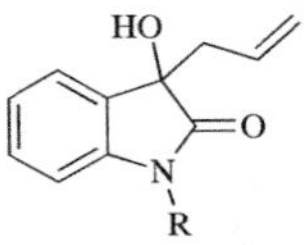

PhCH=CHCHO (2.5 eq),
DBU (1.0 eq), **145a** (10 mol %),
MnO$_2$ (5.0 eq), NaBF$_4$ (0.5 eq),
Mg(OTf)$_2$ (10 mol %),
4 Å MS, THF, rt, 55 h

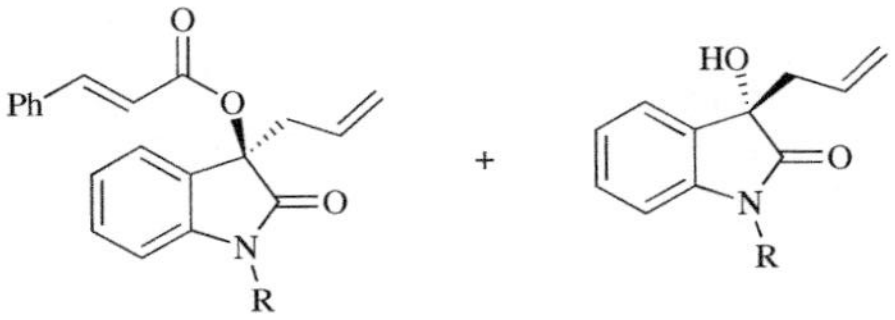

83

R		er **I**		er **II**	s
H	(29)	92.5:7.5		67.5:32.5	17
Me	(28)	90.5:9.5		65.5:34.5	12
PMB	(52)	94.5:5.5		98.5:1.5	68

TABLE 2K. CATALYTIC KINETIC RESOLUTION OF TERTIARY ALCOHOLS (*Continued*)

Alcohol	Conditions	Product(s) and Conversion(s) (%)	Refs.

*Please refer to the charts preceding the tables for the structures indicated by the **bold** numbers.*

C₁₁

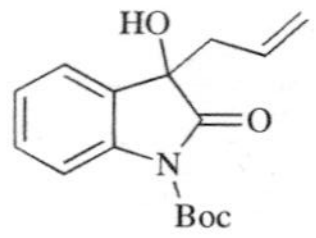

(*i*-PrCO₂O (0.7 eq),
126 (1 mol %),
i-Pr₂NEt (0.6 eq),
CHCl₃, 0°, 18 h

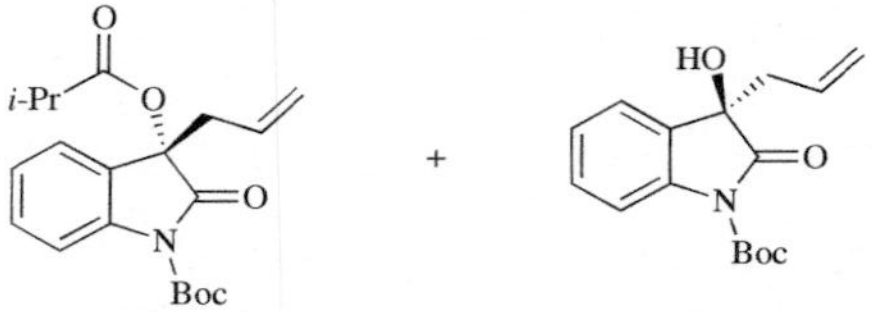

+

er 99.0:1.0   er 96.0:4.0   (52), *s* = 110   241

(*i*-PrCO₂O (0.7 eq),
155 (5 mol %),
i-Pr₂NEt (0.6 eq),
(MeO)₂CO, rt, 24 h

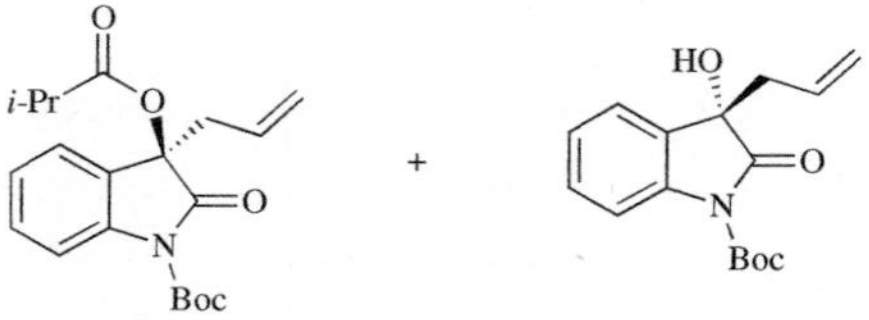

+

er 77.0:23.0   (64), *s* = 14   242

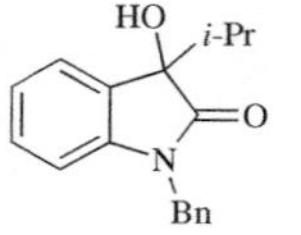

(*i*-PrCO₂O (0.7 eq),
126 (2 mol %),
i-Pr₂NEt (0.6 eq),
CHCl₃, 0°, 115 h

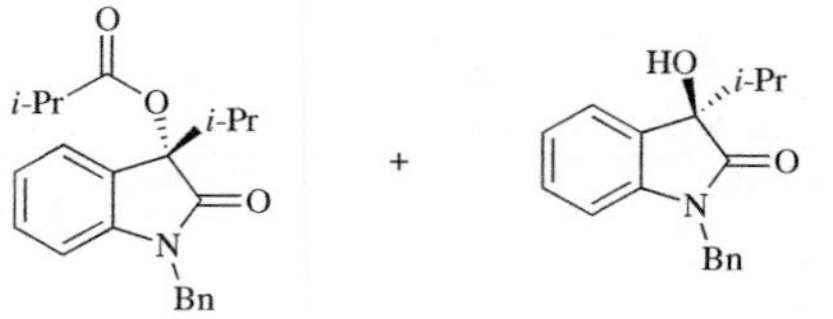

+

er 97.0:3.0   er 99.0:1.0   (42), *s* = 60   241

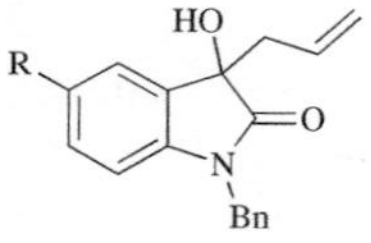

PhCH=CHCHO (2.5 eq),
DBU (1.0 eq), **145a** (10 mol %),
MnO₂ (5.0 eq), NaBF₄ (0.5 eq),
Mg(OTf)₂ (10 mol %),
4 Å MS, THF, rt, 72 h

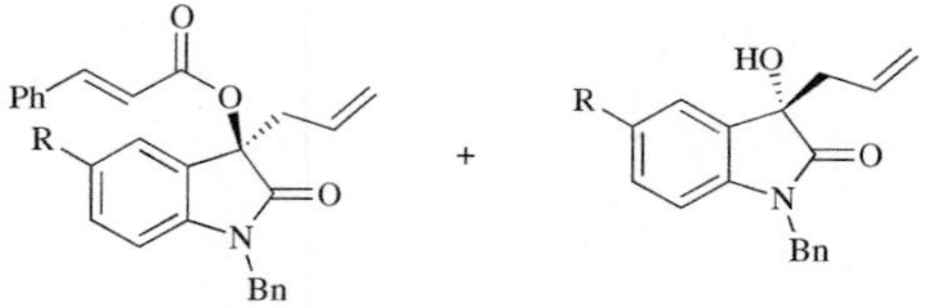

+

I   **II**   er 85.0:15.0   83

R		er I	er II	s
Cl	(52)	91.5:8.5	96.0:4.0	34
Br	(54)	90.5:9.5	97.5:2.5	35
MeO	(52)	95.5:4.5	99.0:1.0	78

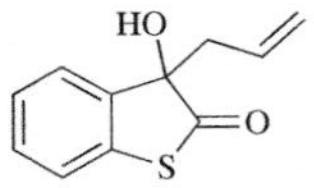

(i-PrCO)$_2$O (0.7 eq), **126** (1 mol %), i-Pr$_2$NEt (0.6 eq), CHCl$_3$, 0°, 18 h

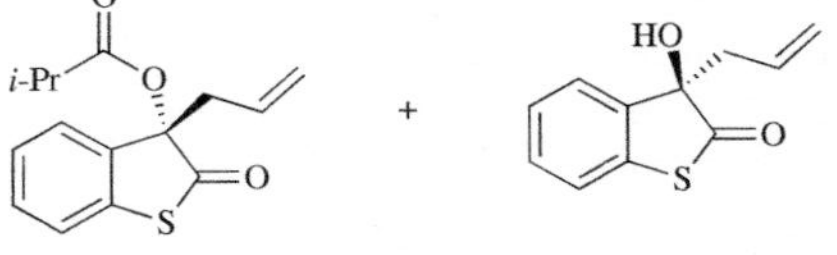

er 92.0:8.0   +   er 97.0:3.0   (54), s = 41   241

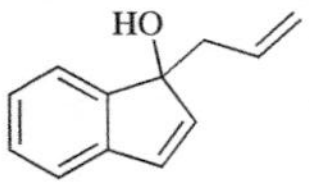

Ac$_2$O (0.7 eq), **126** (5 mol %), i-Pr$_2$NEt (0.6 eq), CHCl$_3$, 0°, 24 h

er 70.0:30.0   +   er 86.0:14.0   (64), s = 5   241

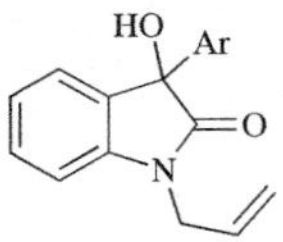

(i-PrCO)$_2$O (0.7 eq), **155** (5 mol %), i-Pr$_2$NEt (0.6 eq), (MeO)$_2$CO, rt, 24 h

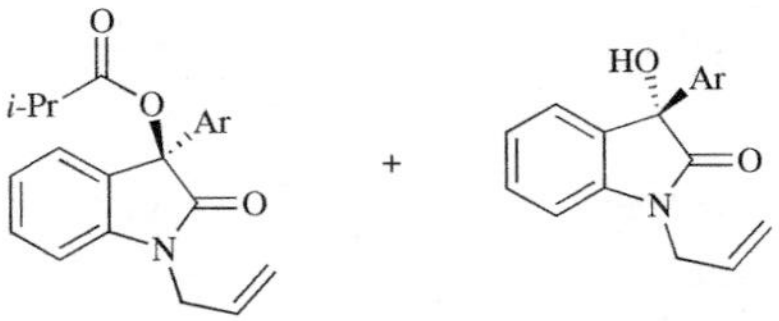

I   +   **II**

Ar		er I	er II	s
2-thienyl	(57)	88.0:12.0	99.0:1.0	34
3-pyridyl	(59)	84.0:16.0	99.0:1.0	26
Ph	(54)	92.0:8.0	99.5:0.5	70
3,5-(CF$_3$)$_2$C$_6$H$_3$	(59)	82.0:18.0	95.0:5.0	13
6-MeO-naphth-2-yl	(57)	87.0:13.0	99.5:0.5	36

242

C$_{12}$

C$_{12-18}$

TABLE 2K. CATALYTIC KINETIC RESOLUTION OF TERTIARY ALCOHOLS (*Continued*)

Alcohol	Conditions	Product(s) and Conversion(s) (%)	Refs.

*Please refer to the charts preceding the tables for the structures indicated by the **bold** numbers.*

C$_{13}$

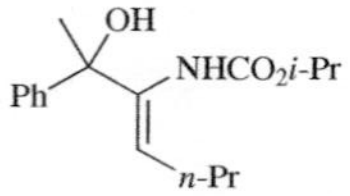

157b (10 mol %), 5 Å MS, chlorobenzene, rt

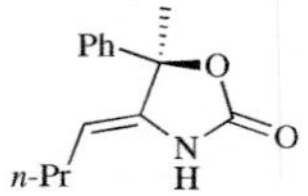

er 88.5:11.5 + er 90.0:10.0

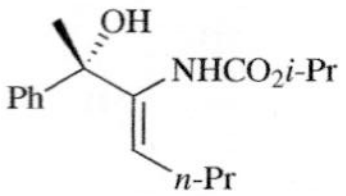

(—), s = 19   243

C$_{13-17}$

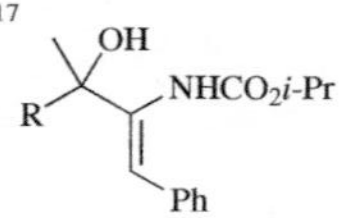

157b (10 mol %), 5 Å MS, chlorobenzene, 0°

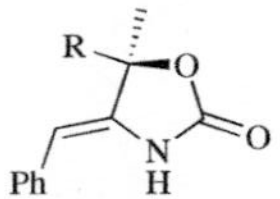

I + **II**

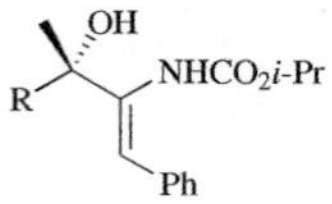

R		er **I**	er **II**	s
i-Pr	(—)	92.5:7.5	95.0:5.0	38
i-Bu	(—)	92.5:7.5	95.5:4.5	39
t-Bu	(—)	97.5:2.5	98.5:1.5	165
2-thienyl	(—)	89.5:10.5	93.0:7.0	23
n-C$_5$H$_{11}$	(—)	92.0:8.0	99.0:1.0	52
c-C$_6$H$_{11}$	(—)	94.0:6.0	99.0:1.0	82
4-FC$_6$H$_5$	(—)	91.5:8.5	94.0:6.0	31
4-MeC$_6$H$_5$	(—)	96.5:3.5	96.5:3.5	94

243

428

C$_{13-14}$

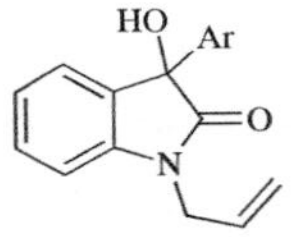

(i-PrCO)$_2$O (0.7 eq),
126 (2 mol %),
i-Pr$_2$NEt (0.6 eq),
CHCl$_3$, 0°, 64 h

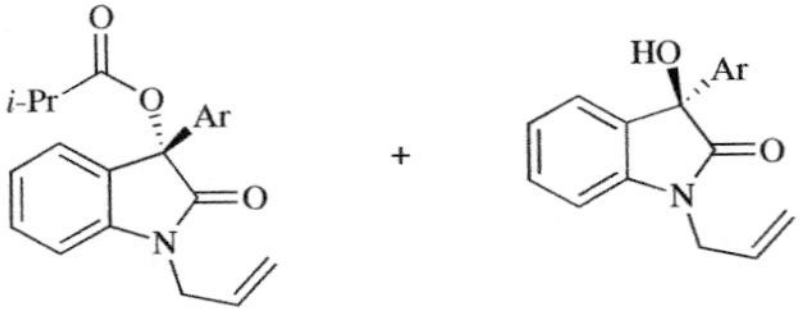

241

I + II

Ar		er I	er II	s
2-pyridyl	(42)	96.0:4.0	86.0:14.0	60
4-MeOC$_6$H$_4$	(49)	98.0:2.0	97.0:3.0	200

C$_{13}$

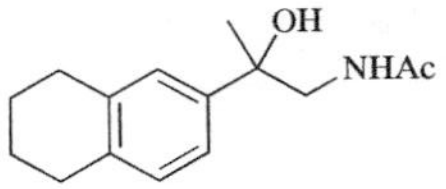

Ac$_2$O (5.0 eq), **97** (10 mol %),
Et$_3$N (2 eq), CH$_2$Cl$_2$,
toluene, –20°, 82 h

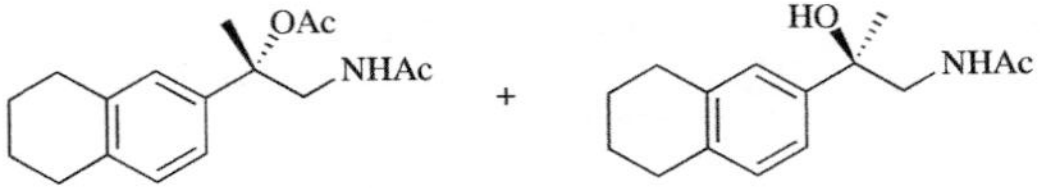

(38), s = 39

er 94.5:5.5   +   er 70.0:30.0

240

C$_{14}$

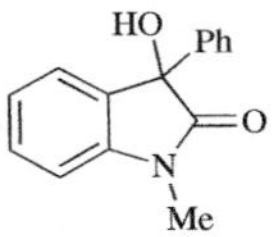

(i-PrCO)$_2$O (0.7 eq),
155 (5 mol %),
i-Pr$_2$NEt (0.6 eq),
(MeO)$_2$CO, rt, 24 h

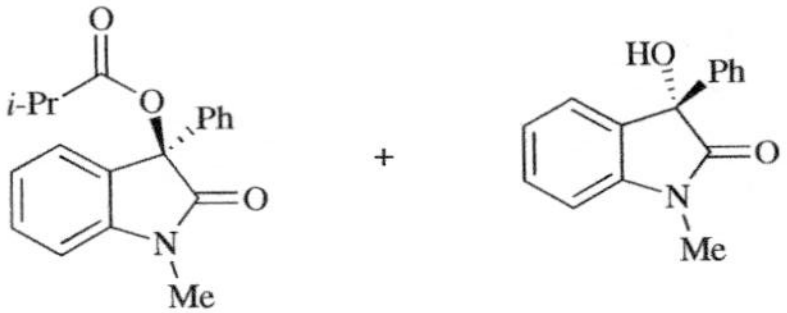

(53), s = 70

er 94.0:6.0   +   er 99.0:1.0

242

TABLE 2K. CATALYTIC KINETIC RESOLUTION OF TERTIARY ALCOHOLS (*Continued*)

Alcohol	Conditions	Product(s) and Conversion(s) (%)	Refs.

*Please refer to the charts preceding the tables for the structures indicated by the **bold** numbers.*

C_{14}

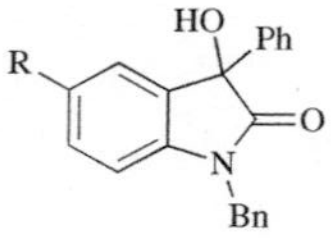

$(i\text{-PrCO})_2O$ (0.7 eq), **126** (2 mol %), $i\text{-Pr}_2NEt$ (0.6 eq), $CHCl_3$, 0°, 24 h

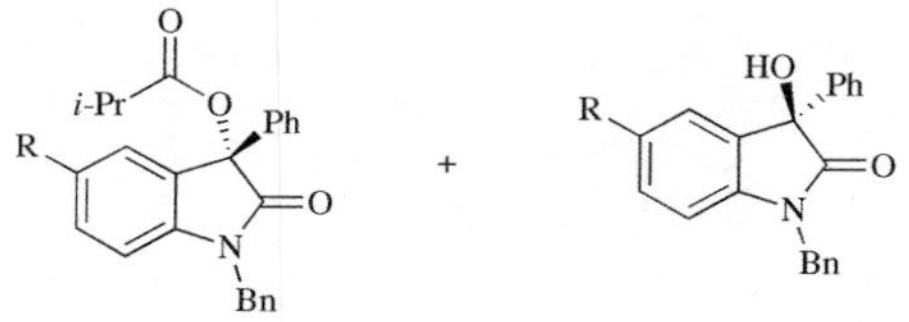

I **II**

R		er **I**	er **II**	*s*
Cl	(49)	94.0:6.0	93.0:7.0	44
Br	(49)	93.0:7.0	91.0:9.0	31

241

C_{14-15}

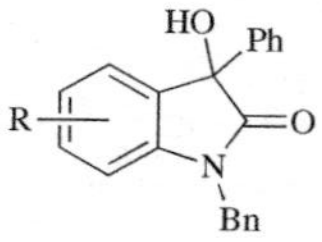

$(i\text{-PrCO})_2O$ (0.7 eq), **155** (5 mol %), $i\text{-Pr}_2NEt$ (0.6 eq), $(MeO)_2CO$, rt, 24 h

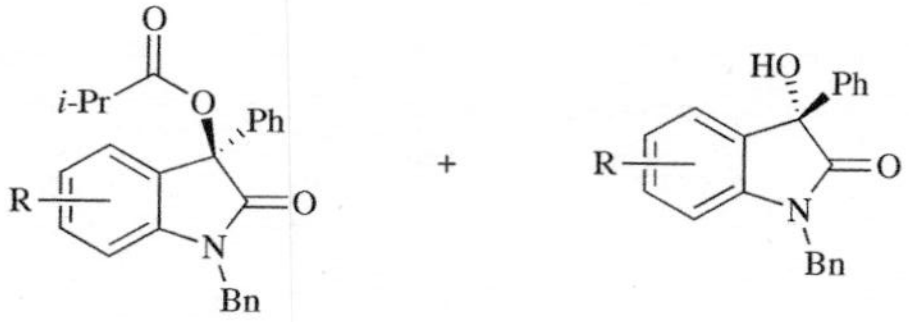

I **II**

R		er **I**	er **II**	*s*
6-Cl	(52)	79.0:21.0	81.0:19.0	9
5-Me	(53)	90.0:10.0	95.0:5.0	27

242

C14

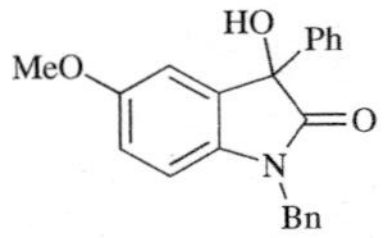

(i-PrCO)$_2$O (0.7 eq), 126 (5 mol %), i-Pr$_2$NEt (0.6 eq), CHCl$_3$, 0°, 40 h

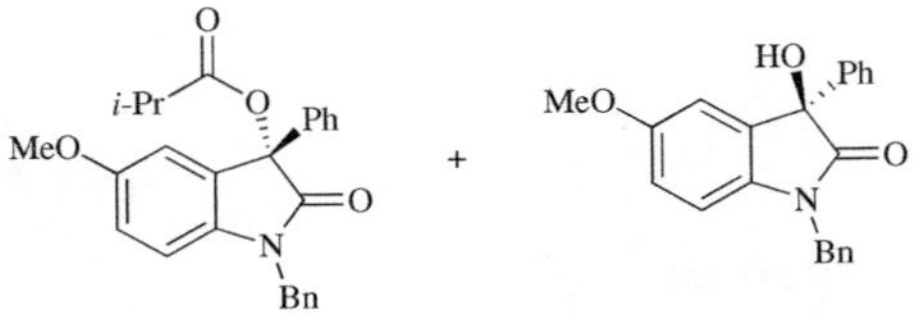

er 98.0:2.0   er 89.0:11.0   (45), s = 100   241

C14-15

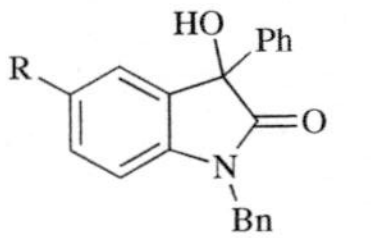

(i-PrCO)$_2$O (0.7 eq), 126 (1 mol %), i-Pr$_2$NEt (0.6 eq), CHCl$_3$, 0°

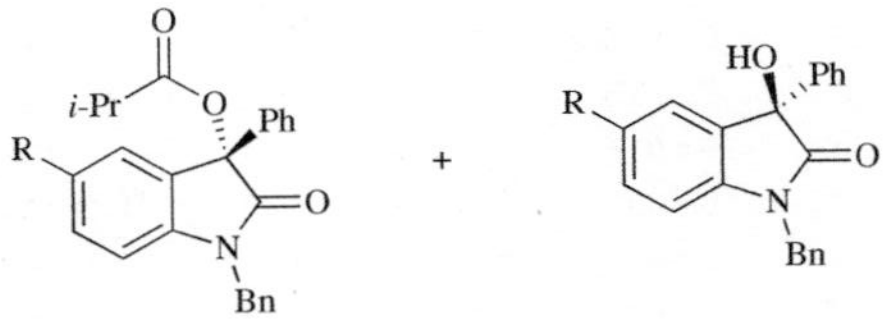

241

R	Time (h)		er **I**	er **II**	s
O$_2$N	18	(62)	78.0:22.0	95.0:5.0	11
Me$_2$N	42	(47)	98.0:2.0	92.0:8.0	140
Me	42	(47)	97.0:3.0	91.0:9.0	90

I   **II**

C14

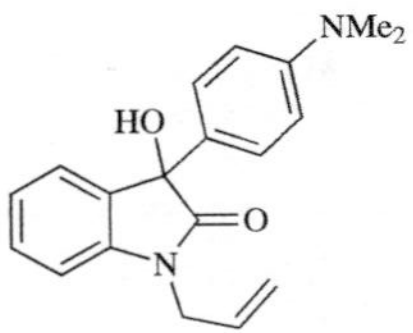

(i-PrCO)$_2$O (0.7 eq), 126 (10 mol %), i-Pr$_2$NEt (0.6 eq), CHCl$_3$, 0°, 18 h

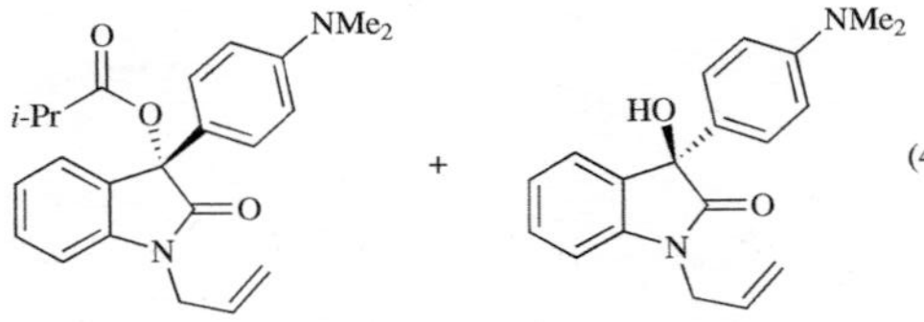

er 52.0:48.0   er 97.0:3.0   (49), s > 200   241

431

TABLE 2K. CATALYTIC KINETIC RESOLUTION OF TERTIARY ALCOHOLS (*Continued*)

Alcohol	Conditions	Product(s) and Conversion(s) (%)	Refs.

*Please refer to the charts preceding the tables for the structures indicated by the **bold** numbers.*

C_{16-20}

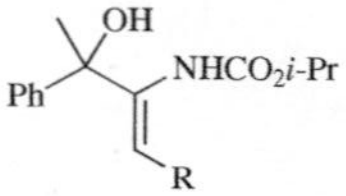

157b (10 mol %), 5 Å MS, chlorobenzene, 0°

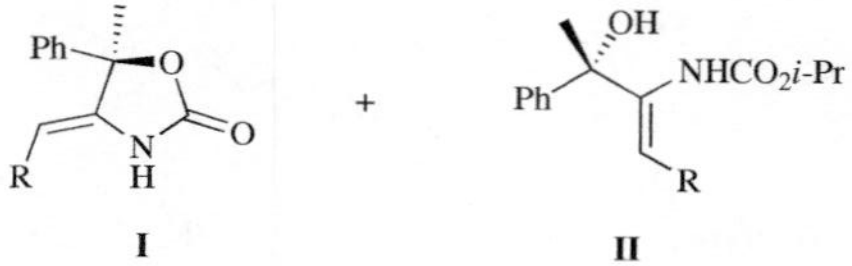

243

R		er I	er II	s
Ph	(—)	94.5:5.5	99.0:1.0	78
3-FC$_6$H$_4$	(—)	93.5:6.5	96.0:4.0	47
4-ClC$_6$H$_4$	(—)	95.5:4.5	94.5:5.5	64
3-BrC$_6$H$_4$	(—)	92.5:7.5	96.0:4.0	40
3-MeOC$_6$H$_4$	(—)	95.5:4.5	94.0:6.0	62
4-MeC$_6$H$_4$	(—)	96.5:3.5	96.0:4.0	94
4-CF$_3$C$_6$H$_4$	(—)	92.5:7.5	92.5:7.5	33
2-Np	(—)	97.5:2.5	96.5:3.5	133

C_{16}

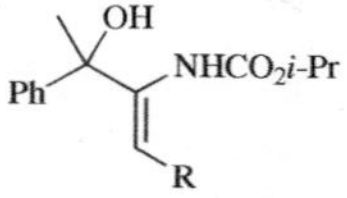

157b (10 mol %), 5 Å MS, chlorobenzene, 10°

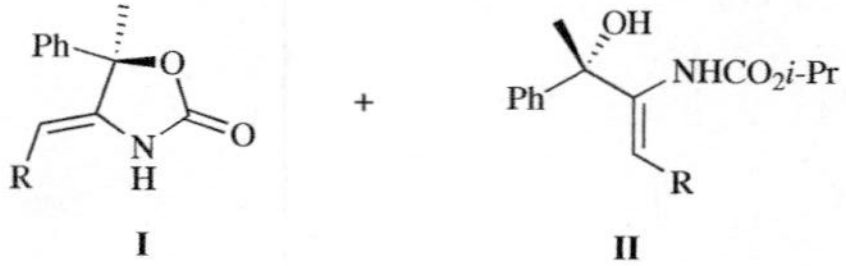

243

R		er I	er II	s
2-ClC$_6$H$_5$	(—)	89.0:11.0	93.5:6.5	23
4-MeOC$_6$H$_5$	(—)	97.0:3.0	97.5:2.5	121

C_{16-20}

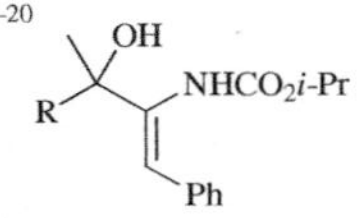

157b (10 mol %), 5 Å MS, chlorobenzene, 10°

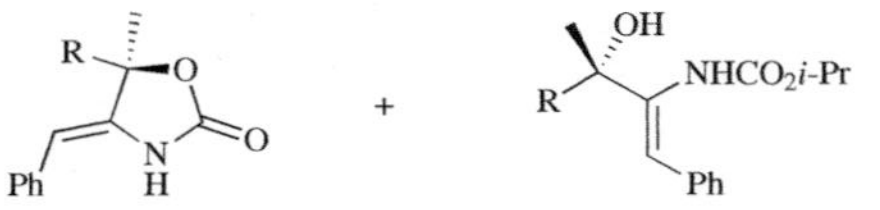

243

I + II

R		er I	er II	s
3-ClC$_6$H$_4$	(—)	87.0:13.0	94.5:5.5	20
2-MeOC$_6$H$_4$	(—)	97.5:2.5	97.0:3.0	139
3-MeOC$_6$H$_4$	(—)	96.5:3.5	97.0:3.0	98
4-MeOC$_6$H$_4$	(—)	96.5:3.5	99.0:1.0	127
4-CF$_3$C$_6$H$_4$	(—)	88.0:12.0	95.5:4.5	23
2-Np	(—)	88.5:11.5	97.0:3.0	27

C_{17-18}

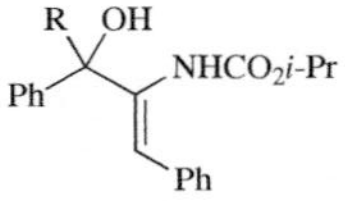

157b (10 mol %), 5 Å MS, chlorobenzene

243

I + II

R	Temp(°)		er I	er II	s
Et	10	(—)	95.5:4.5	97.0:3.0	75
i-Pr	rt	(—)	84.0:16.0	94.5:5.5	15

C_{18}

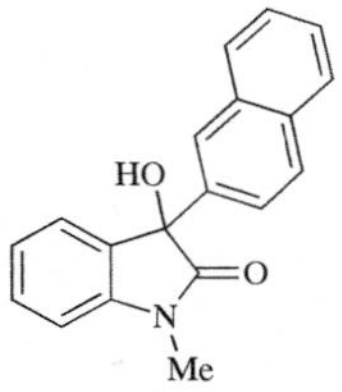

(i-PrCO)$_2$O (0.7 eq). 126 (1 mol %), i-Pr$_2$NEt (0.6 eq). CHCl$_3$, 0°, 18 h

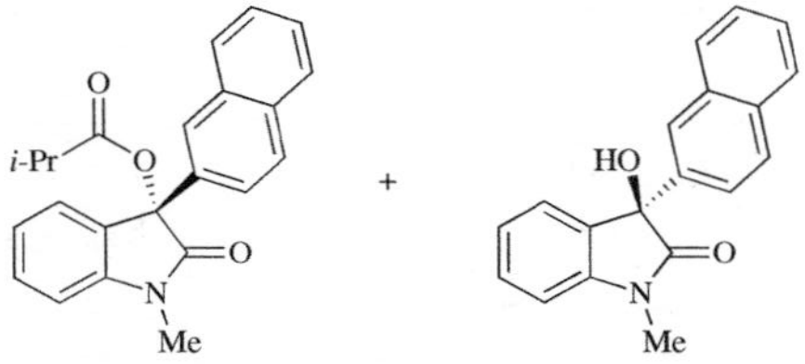

241

er 98.0:2.0   +   er 99.0:1.0

(51), s = 180

TABLE 2L. CATALYTIC KINETIC RESOLUTION OF MISCELLANEOUS ALCOHOLS

Alcohol	Conditions	Product(s) and Conversion(s) (%)				Refs.

*Please refer to the charts preceding the tables for the structures indicated by the **bold** numbers.*

C₂₋₉

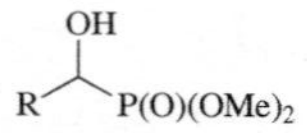

Ph₂CHCO₂H (0.75 eq),
Piv₂O (0.9 eq),
124 (5 mol %),
i-Pr₂NEt (1.8 eq),
THF, rt, 12 h

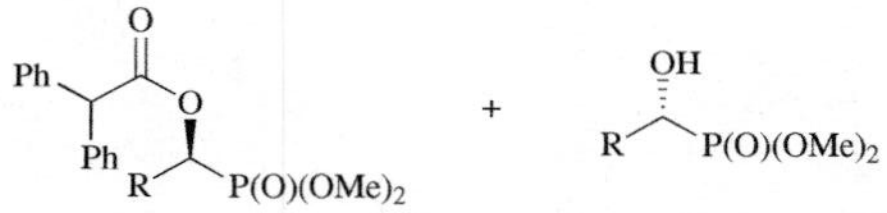

64

R		er I	er II	s
TBDPSOCH₂	(52)	95.5:4.5	99.5:0.5	112
BocHNCH₂	(57)	86.5:13.5	99.5:0.5	30
BnO(CH₂)₂	(51)	97.0:3.0	>99.5:0.5	203
CbzHN(CH₂)₂	(53)	93.5:6.5	>99.5:0.5	84
Et	(51)	97.5:2.5	>99.5:0.5	225
n-Pr	(51)	97.0:3.0	>99.5:0.5	204
i-Pr	(23)	97.5:2.5	64.0:36.0	51
t-Bu	(3)	98.0:2.0	51.5:48.5	53
Cy	(44)	99.0:1.0	88.0:12.0	191
Bn	(51)	97.5:2.5	>99.5:0.5	243
Ph(CH₂)₂	(50)	99.0:1.0	>99.5:0.5	518

434

C$_{2-8}$

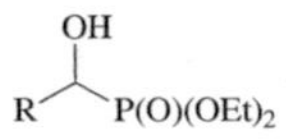

R$\diagdown$ OH / P(O)(OEt)$_2$

BzCl (0.5 eq), **15** (5 mol %),
Cu(OTf)$_2$ (5 mol%), BaCO$_3$,
chlorobenzene, 0° to rt, 12 h

R$\diagdown$ OBz / P(O)(OEt)$_2$ **I**   +   R$\diagdown$ OH / P(O)(OEt)$_2$ **II**

R		er I	er II	s
Me	(37)	90.0:10.0	82.5:17.5	18
ClCH$_2$	(35)	96.0:4.0	77.5:22.5	42
Et	(26)	94.0:6.0	73.5:26.5	25
BnO(CH$_2$)$_2$	(30)	97.5:2.5	69.5:30.5	57
CbzHN(CH$_2$)$_2$	(13)	90.5:9.5	53.5:46.5	10
BocHN(CH$_2$)$_2$	(29)	97.0:3.0	70.0:30.0	48
n-Pr	(28)	>99.5:0.5	68.5:31.5	286
i-Pr	(52)	87.0:13.0	93.5:6.5	32
BnO(CH$_2$)$_3$	(27)	94.0:6.0	73.0:27.0	25
i-Bu	(20)	97.0:3.0	66.0:34.0	44
2-furyl	(38)	83.0:17.0	62.0:38.0	6
Cy	(32)	94.0:6.0	71.0:29.0	24
Bn	(38)	95.0:5.0	89.5:10.5	46

TABLE 2L. CATALYTIC KINETIC RESOLUTION OF MISCELLANEOUS ALCOHOLS (*Continued*)

Alcohol	Conditions	Product(s) and Conversion(s) (%)	Refs.

*Please refer to the charts preceding the tables for the structures indicated by the **bold** numbers.*

C$_{3-10}$ — Alcohol: OH, R, NHTs

Conditions: BzCl (0.7 eq), **47b** (2 mol %), K$_2$CO$_3$ (1 eq), THF, rt, 24 h

Products: I (OBz, R, NHTs) + II (OH, R, NHTs)

Refs: 248

R		er I	er II	s
Me	(53)	90.0:10.0	95.0:5.0	28
PhOCH$_2$	(38)	94.5:5.5	77.0:23.0	30
Et	(48)	94.5:5.5	91.0:9.0	46
n-Pr	(51)	94.0:6.0	95.5:4.5	46
c-C$_3$H$_5$	(54)	88.5:11.5	95.0:5.0	23
i-Bu	(54)	90.0:10.0	96.5:3.5	30
t-Bu	(48)	99.0:1.0	95.5:4.5	317
c-C$_6$H$_{11}$	(52)	94.5:5.5	98.0:2.0	70
Ph(CH$_2$)$_2$	(52)	92.0:8.0	96.0:4.0	39

C$_{3-8}$ — Alcohol: OH, R, C(=O)NMe$_2$

Conditions: (Ph$_2$CHCO)$_2$O (0.6 eq), **124** (5 mol %), i-Pr$_2$NEt (0.6 eq), Et$_2$O, rt, 12 h

Products: I (O-C(=O)CHPh$_2$, R, C(=O)NMe$_2$) + II (OH, R, C(=O)NMe$_2$)

Refs: 234

R		er I	er II	s
TBSOCH$_2$	(—)	90.0:10.0	84.5:15.5	18
TBSO(CH$_2$)$_3$	(—)	98.0:2.0	96.0:4.0	176
i-Bu	(—)	98.5:1.5	97.0:3.0	208
c-C$_6$H$_{11}$	(—)	87.0:13.0	84.0:16.0	13

234

C_{3-10}

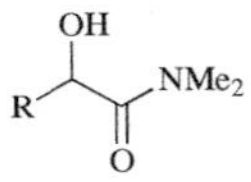

Ph$_2$CHCO$_2$H (0.75 eq),
Piv$_2$O (0.9 eq), **124** (5 mol %),
i-Pr$_2$NEt (1.2 eq),
Et$_2$O, rt, 12 h

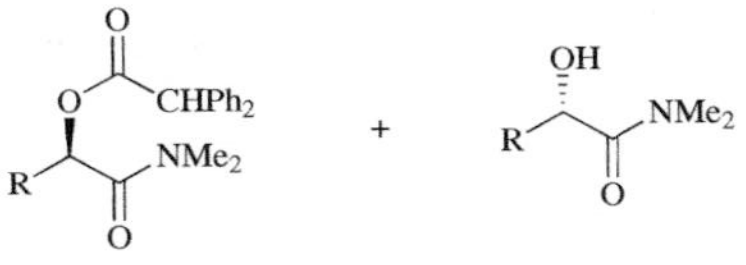

I   +   **II**

R		er I	er II	s
Me	(±)	94.5:5.5	91.0:9.0	42
Et	(±)	96.5:3.5	77.0:23.0	45
TBSO(CH$_2$)$_2$	(±)	96.5:3.5	97.5:2.5	103
n-Pr	(±)	97.0:3.0	82.5:17.5	65
i-Pr	(±)	91.0:9.0	57.0:43.0	12
n-Bu	(±)	97.5:2.5	84.5:15.5	75
Ph(CH$_2$)$_2$	(±)	96.0:4.0	99.5:0.5	254

234

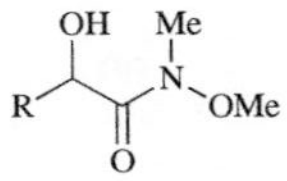

Ph$_2$CHCO$_2$H (0.5 eq),
Piv$_2$O (0.6 eq), **124** (5 mol %),
i-Pr$_2$NEt (1.2 eq),
Et$_2$O, rt, 12 h

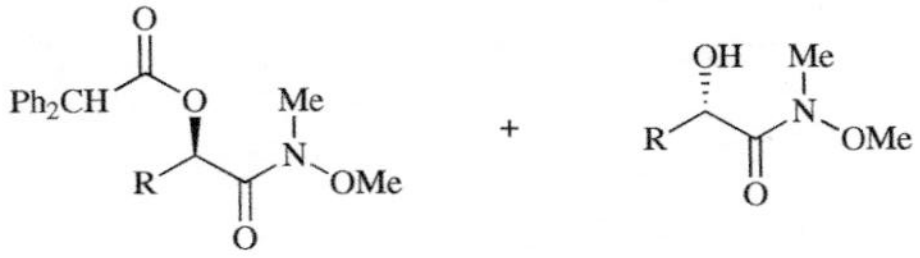

I   +   **II**

R		er I	er II	s
Me	(±)	96.5:3.5	99.5:0.5	130
TBSOCH$_2$	(±)	93.0:7.0	75.5:24.5	22
Et	(±)	98.0:2.0	92.5:7.5	118
TBSO(CH$_2$)$_2$	(±)	97.5:2.5	95.5:4.5	115
n-Pr	(±)	98.0:2.0	97.0:3.0	176
TBSO(CH$_2$)$_3$	(±)	97.0:3.0	99.5:0.5	118
i-Pr	(±)	83.0:17.0	53.5:46.5	5
n-Bu	(±)	98.0:2.0	84.5:15.5	113
i-Bu	(±)	98.5:1.5	94.5:5.5	168
c-C$_6$H$_{11}$	(±)	79.5:20.5	52.5:47.5	4
Ph(CH$_2$)$_2$	(±)	97.0:3.0	99.5:0.5	156

TABLE 2L. CATALYTIC KINETIC RESOLUTION OF MISCELLANEOUS ALCOHOLS (*Continued*)

Alcohol	Conditions	Product(s) and Conversion(s) (%)	Refs.

*Please refer to the charts preceding the tables for the structures indicated by the **bold** numbers.*

C_3

Alcohol: OH ... $P(O)(OR)_2$

Conditions: 2-FC_6H_4COCl (0.5 eq), **15** (5 mol %), $Cu(OTf)_2$ (5 mol %), K_2CO_3, CH_2Cl_2, 0° to rt, 12 h

Product: **I** (...$P(O)(OR)_2$) + **II** (OH ...$P(O)(OR)_2$)

R		er **I**	er **II**	s
Me	(—)	78.5:21.5	55.0:45.0	4
Et	(—)	78.0:22.0	74.5:25.5	6

Refs: 98

C_4

Alcohol: OH

Conditions: Ac_2O (1.5 eq), **101** (2.5 mol %), toluene, −65°, 24 h

Product: OAc + OH, (—), s = 4

Refs: 68

Alcohol: OH O NHR

Conditions: BzCl (0.5 eq), **15** (10 mol %), $Cu(OTf)_2$ (10 mol %), K_2CO_3, EtOAc, rt

Product: **I** (OBz ...O NHR) + **II** (OH ...O NHR)

R	Time (h)		er **I**	er **II**	s
Ph	2	(41)	92.5:7.5	87.0:13.0	27
4-ClC_6H_4	2	(46)	89.5:10.5	83.0:17.0	17
2-MeC_6H_4	2	(37)	94.5:5.5	78.0:22.0	30
4-MeC_6H_4	3	(44)	94.0:6.0	86.5:13.5	34
Bn	3	(48)	89.0:11.0	82.5:17.5	16
3,5-$Me_2C_6H_3$	24	(47)	89.0:11.0	82.5:17.5	16
3,5-$(CF_3)_2C_6H_3$	24	(30)	78.5:21.5	63.5:36.5	5

Refs: 200

438

200

R	Time (h)	I	er I	II	er II	s
Me	3	(46)	88.0:12.0	(39)	65.5:34.5	10
Ph	12	(37)	92.0:8.0		80.0:20.0	21

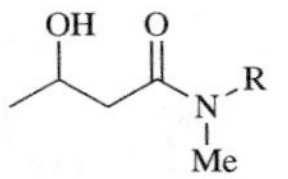

I: OBz / O / N–R / Me (+) II: OH / O / N–R / Me

BzCl (0.5 eq), **15** (10 mol %), Cu(OTf)$_2$ (10 mol %), K$_2$CO$_3$, EtOAc, rt

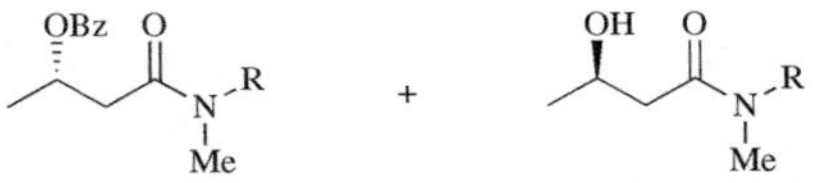

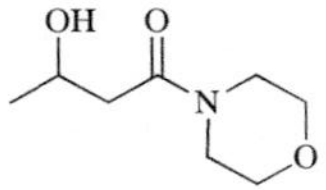

OH / O / N–R / Me

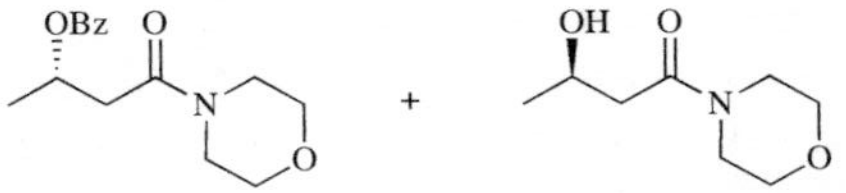

er 91.0:9.0 + er 78.5:21.5 (39), s = 18 200

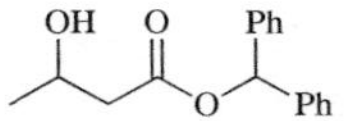

BzCl (0.5 eq), **15** (10 mol %), Cu(OTf)$_2$ (10 mol%), K$_2$CO$_3$, CH$_2$Cl$_2$, 0° to rt, 24 h

---

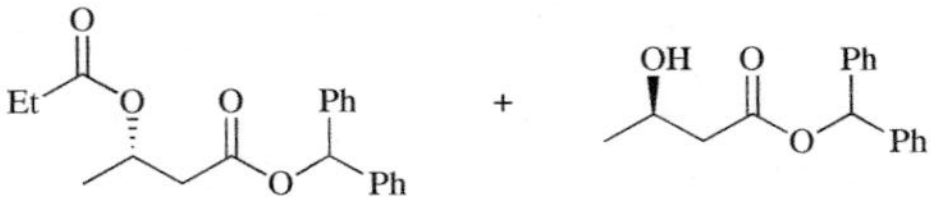

er 92.0:8.0 + er 85.5:14.5 (46), s = 24 53

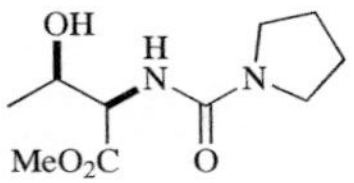

(EtCO)$_2$O (0.75 eq), **121** (5 mol %), K$_2$CO$_3$ (0.5 eq), i-Pr$_2$O, 0°, 24 h

---

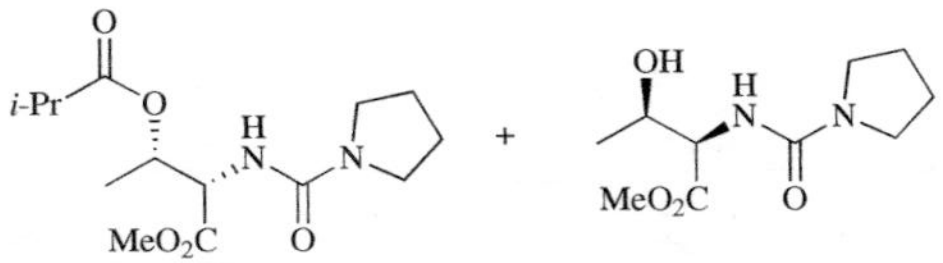

er 90.0:10.0 + er 75.5:24.5 (39), s = 15 70

(i-PrCO)$_2$O (0.5 eq), **109** (5 mol %), i-Pr$_2$NEt (0.5 eq), CCl$_4$, 0°, 4 h

439

TABLE 2L. CATALYTIC KINETIC RESOLUTION OF MISCELLANEOUS ALCOHOLS (Continued)

Please refer to the charts preceding the tables for the structures indicated by the bold numbers.

Alcohol	Conditions	Product(s) and Conversion(s) (%)	Refs.

C₄

Alcohol: BnO, OH, Br substituted chain

Conditions: BzCl (0.75 eq), **137** (0.3 mol %), i-Pr₂NEt (0.5 eq), 4 Å MS, CH₂Cl₂, −78°, 3 h

Products: (BnO, OBz, Br) er 91.5:8.5 + (BnO, OH, Br) er 52.5:47.5   (—), s = 12

Refs. 245

C₅₋₇

Alcohol: OH, O, NHPh, R chain

Conditions: BzCl (0.5 eq), **15** (10 mol %), Cu(OTf)₂ (10 mol %), K₂CO₃, CH₂Cl₂, 0° to rt

Products: I (OBz, O, NHPh, R) + II (OH, O, NHPh, R)

R	Time (h)		er I	er II	s
Et	12	(38)	83.5:16.5	70.5:29.5	8
n-Pr	24	(34)	84.0:16.0	72.5:27.5	8
i-Pr	24	(20)	82.0:18.0	62.0:38.0	6
BocHN(CH₂)₃	12	(40)	91.0:9.0	77.5:22.5	18
i-Bu	24	(23)	79.0:21.0	68.5:31.5	5
Br(CH₂)₄	2	(40)	90.0:10.0	79.0:21.0	16

Refs. 200

C₆

Alcohol: cyclohexane (inositol) with OH, OBn, OH substituents and R‑C(O)O group

Conditions: BzCl (0.5 eq), **26** (5 mol %), i-Pr₂NEt, 0°, 3 h

Products: I (cyclohexane with OBz, OBn, OH groups) + II (cyclohexane with OH, OBn groups)

R		er I	er II
Me	(—)	80.0:20.0	—
i-Pr	(—)	83.5:16.5	—
i-Bu	(—)	87.0:13.0	—
Ph	(—)	—	85.0:15.0
4-PhC₆H₄	(—)	95.5:4.5	—

Refs. 95

C$_8$

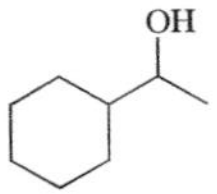

Structure: 1-cyclohexylethanol (OH)

Structure: Ph–CH(OH)–CH$_2$–P(O)(OR)$_2$

Ac$_2$O (1.5 eq), **101** (2.5 mol %), toluene, −65°, 24 h

Product: 1-cyclohexylethyl acetate (OAc), er 60.5:39.5 + 1-cyclohexylethanol (OH), (—), s = 9 — 68

(EtCO)$_2$O (0.75 eq), **117** (20 mol %), i-Pr$_2$NEt (0.75 eq), CDCl$_3$, 0°, 53 h

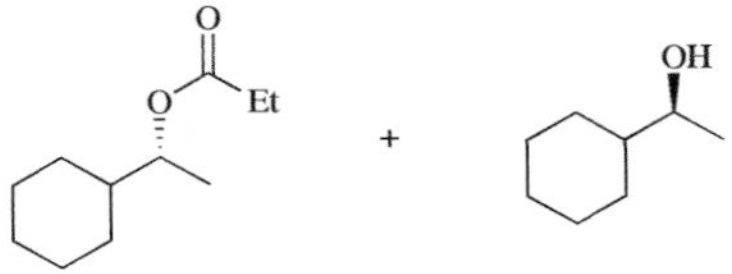

Product: propionate ester, er 60.5:39.5 + 1-cyclohexylethanol (OH), er 57.0:43.0, (39), s = 2 — 57

(EtCO)$_2$O (0.75 eq), **124** (20 mol %), i-Pr$_2$NEt (0.75 eq), CDCl$_3$, rt, 52 h

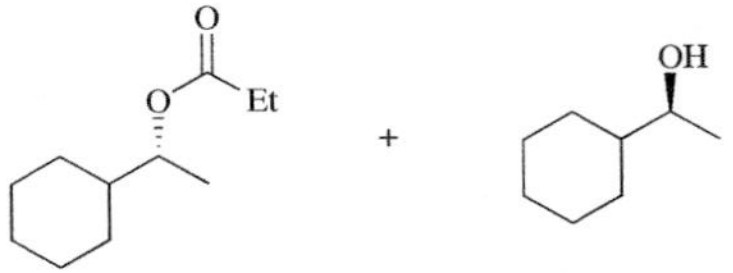

Product: propionate ester, er 61.5:38.5 + 1-cyclohexylethanol (OH), er 67.0:33.0, (59), s = 2 — 57

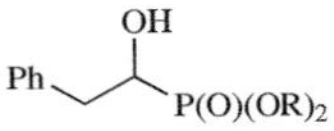

BzCl (0.5 eq), **15** (5 mol %), Cu(OTf)$_2$ (5 mol%), K$_2$CO$_3$, CH$_2$Cl$_2$, 0° to rt, 12h

I: Ph–CH(OBz)–CH$_2$–P(O)(OR)$_2$ + **II**: Ph–CH(OH)–CH$_2$–P(O)(OR)$_2$ — 208

R	I	er I	er II	s
Me	(45)	82.5:17.5	82.5:17.5	9
Et	(39)	91.5:8.5	76.0:24.0	18
i-Pr	(32)	84.0:16.0	69.0:31.0	8
Bn	(38)	75.0:25.0	67.5:32.5	4

TABLE 2L. CATALYTIC KINETIC RESOLUTION OF MISCELLANEOUS ALCOHOLS (*Continued*)

*Please refer to the charts preceding the tables for the structures indicated by the **bold** numbers.*

Alcohol	Conditions	Product(s) and Conversion(s) (%)	Refs.
C₈ Ph — CH(OH) — P(O)(OEt)₂	BzCl (0.5 eq), **15** (5 mol %), Cu(OTf)₂ (5 mol%), K₂CO₃, CH₂Cl₂, 0° to rt, 12 h	Ph — CH(OBz) — P(O)(OEt)₂ er 91.5:8.5 + Ph — CH(OH) — P(O)(OEt)₂ er 76.0:24.0 (—), s = 18	98
C₉ Ph — CH₂CH(OH)CH₃	(EtCO)₂O (0.75 eq), **125** (4 mol %), i-Pr₂NEt (0.75 eq), t-AmOH/toluene (1:1), −40°, 10 h	Et — C(O)O — CH(CH₃)CH₂Ph er 85.5:14.5 + Ph — CH₂CH(OH)CH₃ er 67.0:33.0 (32), s = 8	194
	(EtCO)₂O (0.75 eq), **124** (8 mol %), i-Pr₂NEt (0.75 eq), CDCl₃, 0°, 8 h	Et — C(O)O — CH(CH₃)CH₂Ph er 89.5:10.5 + Ph — CH₂CH(OH)CH₃ er 57.0:43.0 (15), s = 10	57
	(EtCO)₂O (1.5 eq), **155** (1 mol %), i-Pr₂NEt (0.6 eq), CHCl₃, rt, 7 h	Et — C(O)O — CH(CH₃)CH₂Ph er 80.0:20.0 + Ph — CH₂CH(OH)CH₃ er 77.0:23.0 (47), s = 7	196
	(i-PrCO)₂O (0.7 eq), **84** (5 mol %), Et₃N (0.9 eq), CH₂Cl₂, −78°, 3 h	i-Pr — C(O)O — CH(CH₃)CH₂Ph er 82.5:17.5 + Ph — CH₂CH(OH)CH₃ er 58.0:42.0 (20), s = 6	246

NHTs
![structure]
Ph ⟍⟍ OH

BzBr, **137** (30 mol %),
SnBr₂ (0.3 eq), 4 Å MS,
CH₂Cl₂, −78°, 12 h

OBz
Ph ⟍⟍
er 77.5:22.5

$+$

OH
Ph ⟍⟍
er 75.5:24.5

(48), $s = 5$ 158

BzCl (0.75 eq),
137 (0.3 mol %),
Et₃N (0.5 eq), 4 Å MS,
CH₂Cl₂, −78°, 3 h

OBz
Ph ⟍⟍
er 73.0:27.0

$+$

OH
Ph ⟍⟍
er 75.5:24.5

(53), $s = 4$ 157

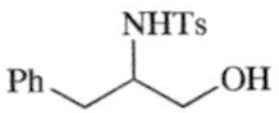

4-MeOC₆H₄COCl (0.7 eq),
47b (2 mol %), K₂CO₃ (1 eq),
THF, rt, 24 h

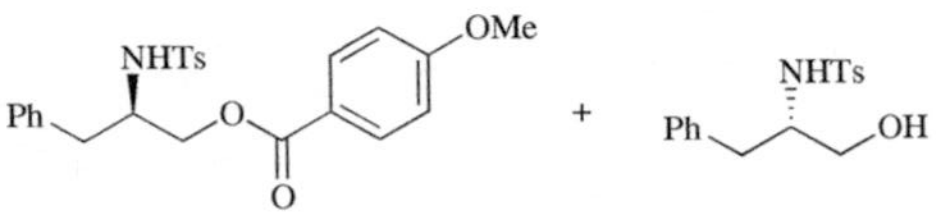

NHTs
Ph ⟍⟍
O
OMe benzoate
er 88.5:11.5

$+$

NHTs
Ph ⟍⟍ OH

(52), $s = 20$ 248

OH
Ph ⟍⟍ P(O)(OMe)₂

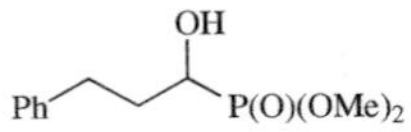

RCO₂H (0.5 eq), Piv₂O (0.6 eq),
124 (5 mol %), i-Pr₂NEt (1.2 eq),
THF, rt, 12 h

O
R ⟍ O
Ph ⟍⟍ P(O)(OMe)₂
I

$+$

OH
Ph ⟍⟍ P(O)(OMe)₂
II

64

R		er **I**	er **II**	s
Me	(45)	97.5:2.5	89.0:11.0	88
Et	(49)	97.5:2.5	94.5:5.5	110
i-Pr	(48)	99.0:1.0	93.5:6.5	286
Cy	(46)	97.5:2.5	92.0:8.0	96
Ph(CH₂)₂	(48)	97.0:3.0	95.0:5.0	96

443

TABLE 2L. CATALYTIC KINETIC RESOLUTION OF MISCELLANEOUS ALCOHOLS (Continued)

Alcohol	Conditions	Product(s) and Conversion(s) (%)	Refs.

*Please refer to the charts preceding the tables for the structures indicated by the **bold** numbers.*

C₉

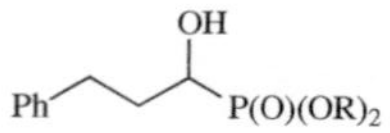

Ph₂CHCO₂H (0.5 eq),
Piv₂O (0.6 eq), **124** (5 mol %),
i-Pr₂NEt (1.2 eq),
THF, rt, 12 h

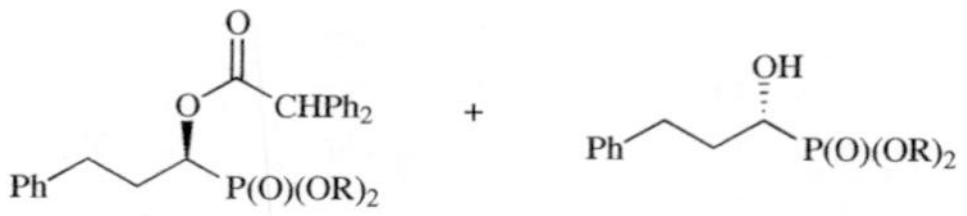

R	I	er I	er II	s
Me	(46)	99.5:0.5	94.5:5.5	419
Et	(46)	99.0:1.0	93.5:6.5	337
Ph	(47)	97.0:3.0	96.5:3.5	106

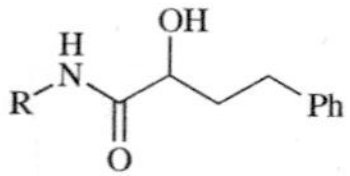

64

C₁₀

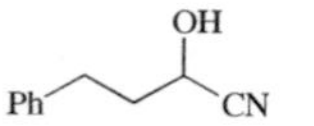

(EtCO)₂O (2.5 eq),
124 (4 mol %),
toluene, rt, 23 h

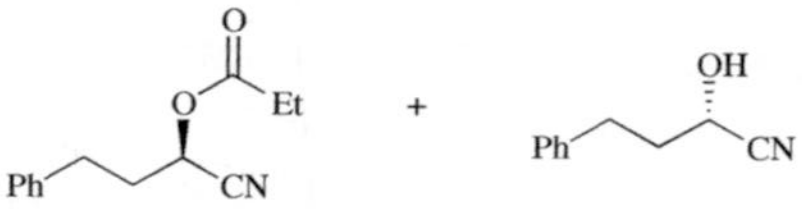

(63), s = 13

er 78.5:21.5   er 98.0:2.0

244

Ph₂CHCO₂H (0.5 eq),
Piv₂O (0.6 eq), **124** (5 mol %),
i-Pr₂NEt (1.2 eq),
Et₂O, rt, 12 h

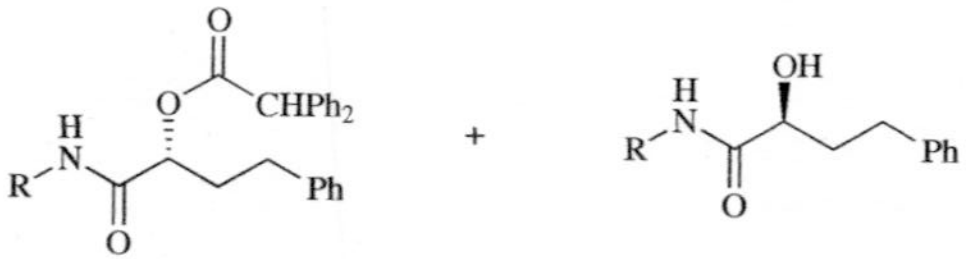

R	I	er I	er II	s
Me	(—)	56.0:44.0	57.0:43.0	1
Ph	(—)	89.5:10.5	86.5:13.5	18
Bn	(—)	58.5:41.5	57.5:42.5	2

234

C₁₆

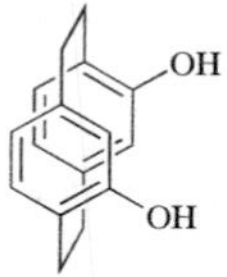

(*i*-PrCO)₂O (0.55 eq),
157c (10 mol %),
CH₂Cl₂, rt, 24 h

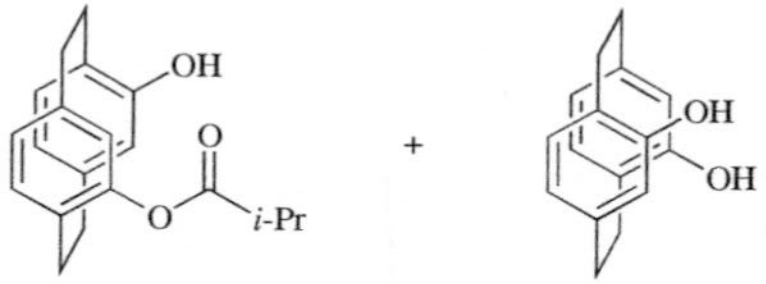

(—), s = 37

er 93.5:6.5   er 91.0:9.0

247

TABLE 3. DESYMMETRIZATION OF MESO-1,2-DIAMINES

Diamine	Conditions	Product(s) and Yield(s) (%)				Refs.

Please refer to the charts preceding the tables for the structures indicated by the bold numbers.

C_{14-16}

Conditions:
Bz$_2$O, **158b** (10 mol %), DMAP (10 mol %), Et$_3$N (1.1 eq), toluene, 4 Å MS, −78°

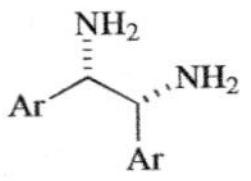

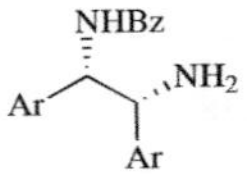

Ar	Time (h)		er	Refs.
4-FC$_6$H$_4$	4	(71)	96.0:4.0	32
2-ClC$_6$H$_4$	11	(65)	85.5:14.5	
3-ClC$_6$H$_4$	8	(61)	90.5:9.5	
3,4-Cl$_2$C$_6$H$_3$	4	(68)	93.5:6.5	
4-ClC$_6$H$_4$	4	(82)	97.5:2.5	
4-BrC$_6$H$_4$	5	(81)	90.5:9.5	
2-MeC$_6$H$_4$	4	(77)	93.0:7.0	
4-MeC$_6$H$_4$	8	(68)	86.0:14.0	
3-Br-4-MeC$_6$H$_3$	6	(60)	89.5:10.5	

445

TABLE 4A. STOICHIOMETRIC KINETIC RESOLUTION OF BENZYLIC AMINES

Amine	Acylating Agent	Conditions	Product(s) and Conversion(s) (%)	Refs.

*Please refer to the charts preceding the tables for the structures indicated by the **bold** numbers.*

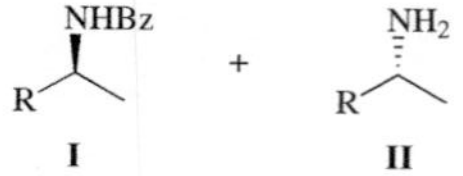

C$_{4-8}$

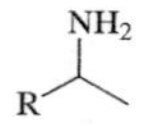

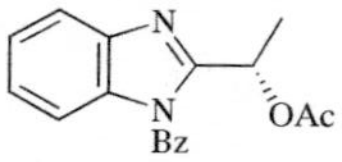

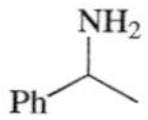

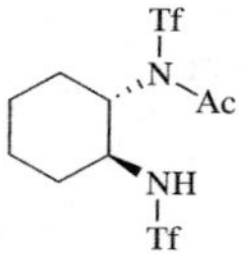

THF, –10°, 12 h

0.5 eq

I + **II**

R	er **I**		er **II**	s
Et	91.5:8.5	(48)	88.5:11.5	25
i-Pr	91.5:8.5	(49)	89.5:10.5	25
n-C$_5$H$_{11}$	91.5:8.5	(50)	91.5:8.5	28
Ph	95.0:5.0	(50)	94.5:5.5	57

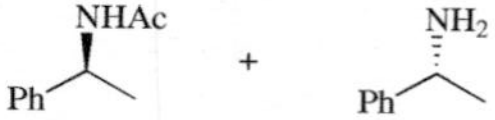

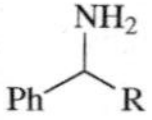

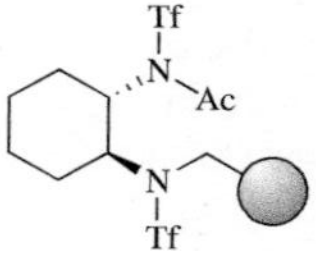

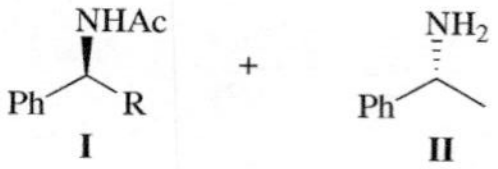

256

C$_8$

MeN(n-Oct)$_3$Cl (1 M), THF, –20°

0.5 eq

I + **II**

er 97.0:3.0   (50), s = 115

254

C$_{8-9}$

Benzene, rt, 3 h

0.2 M

I + **II**

R	er **I**	s
Me	91.0:9.0	12
Et	84.5:15.5	6

255

446

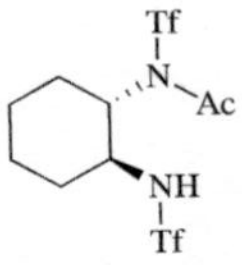

Scheme 1 (C₈₋₁₀)

Catalyst (cyclohexane-1,2-diyl with Tf–N–Ac / NH–Tf), 0.5 eq

MeN(n-Oct)₃Cl, 30°

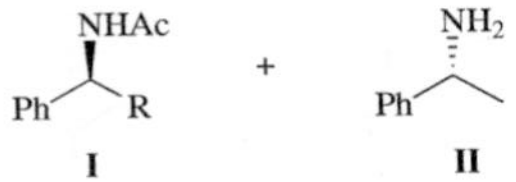

NHAc, Ph, R (**I**) + NH₂, Ph, R (**II**)

R	er **I**		s
Me	(50)	92.0:8.0	30
Et	(50)	90.0:10.0	22

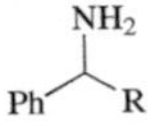

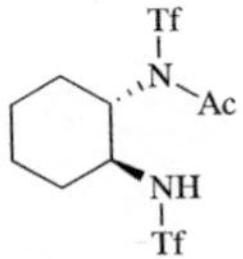

29

Scheme 2 (C₈₋₁₀)

Catalyst (cyclohexane-1,2-diyl with Tf–N–Ac / NH–Tf), 0.33 eq

DMPU, –20°, 24 h

NHAc, Ph, R (**I**) + NH₂, Ph, R (**II**)

R		er **I**
Me	(33)	95.0:5.0
Et	(33)	93.0:7.0
i-Pr	(33)	86.0:14.0

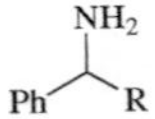

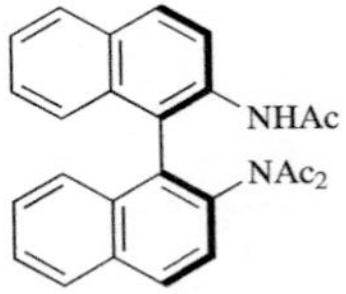

30

Scheme 3 (C₈₋₁₄)

Binaphthyl (NHAc, NAc₂), 0.25 eq

DMSO, rt, 6 h

NH₂, Ph, R (**II**)

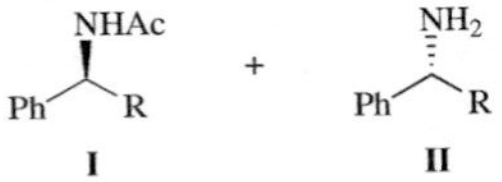

R		er **I**
Me	(24)	65.0:35.0
Bn	(23)	55.5:44.5

249

TABLE 4A. STOICHIOMETRIC KINETIC RESOLUTION OF BENZYLIC AMINES (*Continued*)

Amine	Acylating Agent	Conditions	Product(s) and Conversion(s) (%)	Refs.

*Please refer to the charts preceding the tables for the structures indicated by the **bold** numbers.*

C8–12

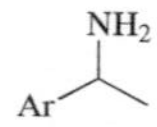

8 eq

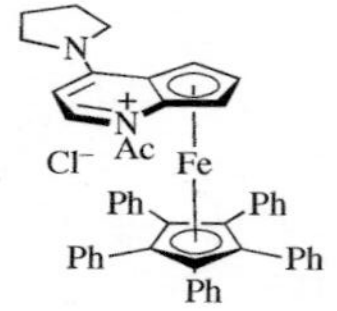

CH₂Cl₂, –78°

I   II

Ar	er I	
Ph	(—)	93.5:6.5
4-MeOC₆H₄	(—)	90.5:9.5
2-MeC₆H₄	(—)	95.5:4.5
4-CF₃C₆H₄	(—)	92.5:7.5
1-Np	(—)	95.0:5.0

253

C9

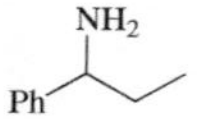

8 eq

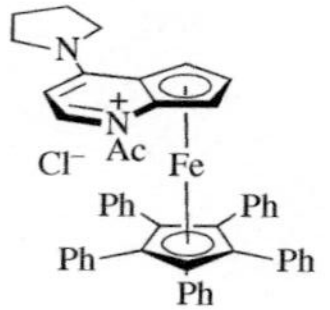

CH₂Cl₂, –78°

I   II

er 83.0:17.0

(—)

253

C10–12

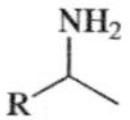

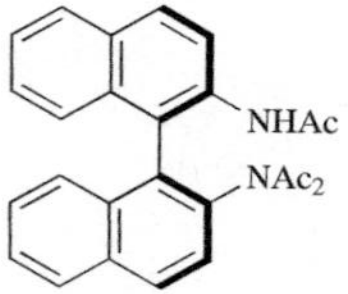

0.25 eq

DMSO, rt, 15 h

I   II

R		er I	
Ph(CH₂)₂	(25)	54.0:46.0	
1-Np	(25)	72.0:28.0	

249

C10

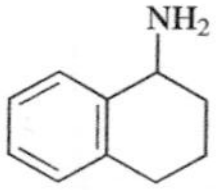

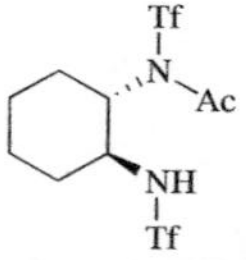

0.33 eq

DMPU, –20°, 24 h

(33)

er 84.5:15.5

30

448

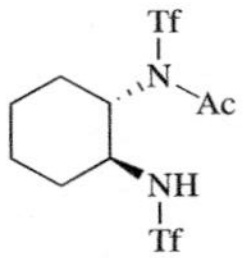

C_{12}

MeN(n-Oct)$_3$Cl, 30°

0.5 eq

er 70.0:30.0

(50), $s = 3$ 29

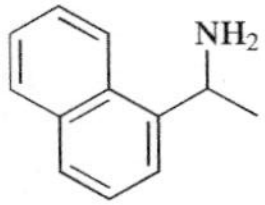

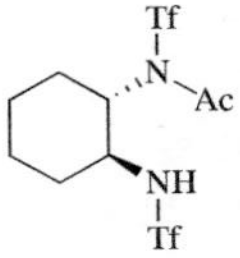

MeN(n-Oct)$_3$Cl, 30°

0.5 eq

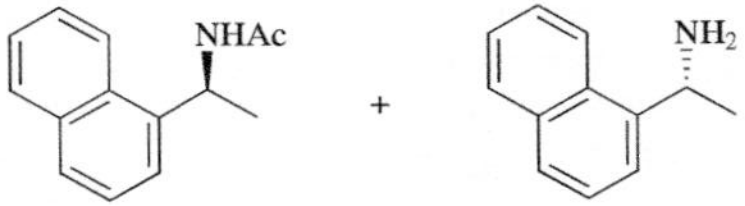

er 89.0:11.0

(50), $s = 19$ 29

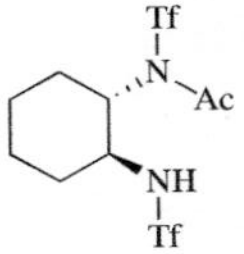

DMPU, −20°, 24 h

0.33 eq

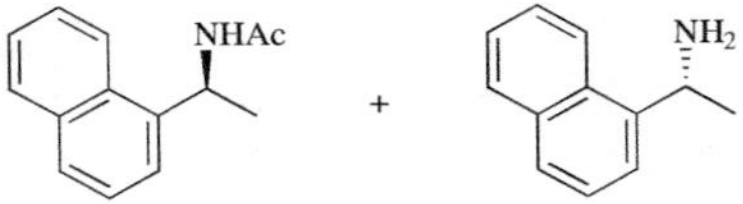

er 86.5:13.5

(33)   30

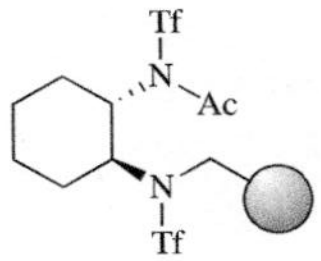

Benzene, rt, 3 h

0.2 M

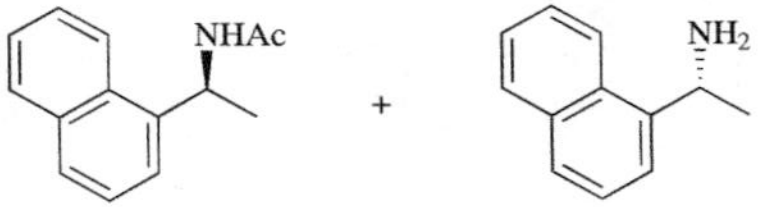

er 89.0:11.0

(—), $s = 10$ 255

Amine	Conditions	Product(s) and Conversion(s) (%)			Refs.

*Please refer to the charts preceding the tables for the structures indicated by the **bold** numbers.*

C8–12

Amine: NH$_2$, Ar

Conditions: **9** (0.6 eq in 2 batches), **52** (10 mol %), CHCl$_3$, –50°, 24 h

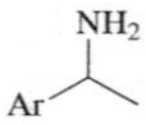

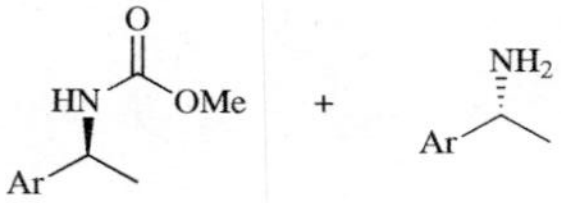

Products: HN–C(=O)OMe, Ar + NH$_2$, Ar

Ar		s
Ph	(—)	12
3-MeOC$_6$H$_4$	(—)	22
4-MeOC$_6$H$_4$	(—)	11
2-MeC$_6$H$_4$	(—)	16
4-CF$_3$C$_6$H$_4$	(—)	13
1-Np	(—)	27

Refs: 33

C8

Amine: NH$_2$, Ph

Conditions: BzOH, **116** (10 mol %), C$_6$H$_5$F, 85°, 48 h

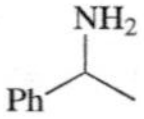

Products: NHBz, Ph + NH$_2$, Ph (21)

er 70.5:29.5

Refs: 258

C8–12

Amine: NH$_2$, Ar

Conditions: Bz$_2$O (0.5 eq), **158b** (5 mol %), **58** (5 mol %), 4 Å MS, toluene, –78°, 2 h

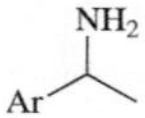

Products: NHBz, Ar + NH$_2$, Ar

Ar		s
Ph	(49)	27
2-ClC$_6$H$_4$	(49)	32
4-BrC$_6$H$_4$	(50)	37
4-MeC$_6$H$_4$	(47)	15
1-Np	(49)	64

Refs: 106

Bz$_2$O (0.5 eq), **158a** (20 mol %),
DMAP (20 mol %), 4 Å MS,
toluene, −78°, 1 h

I: Ar with NHBz   +   II: Ar with NH$_2$

Ar		er I	er II	s
Ph	(45)	85.0:15.0	79.0:21.0	10
4-FC$_6$H$_4$	(43)	90.0:10.0	80.0:20.0	16
2-ClC$_6$H$_4$	(48)	87.0:13.0	84.5:15.5	13
3-ClC$_6$H$_4$	(49)	89.0:11.0	87.0:13.0	18
4-ClC$_6$H$_4$	(47)	91.5:8.5	85.5:14.5	20
4-BrC$_6$H$_4$	(42)	91.5:8.5	80.5:19.5	20
2-MeC$_6$H$_4$	(49)	88.5:11.5	86.5:13.5	17
3-MeC$_6$H$_4$	(46)	81.0:19.0	76.5:23.5	7
4-MeC$_6$H$_4$	(45)	81.5:18.5	75.5:24.5	7
1-Np	(48)	91.0:9.0	88.5:11.5	24
2-Np	(46)	87.5:12.5	82.5:17.5	13

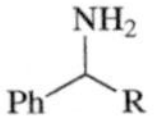

Bz$_2$O (0.5 eq), **158b** (5 mol %),
58 (5 mol %), 4 Å MS,
toluene, −78°, 2 h

C$_{9-14}$ NH$_2$ R Ph

NHBz R Ph   +   NH$_2$ R Ph

R		s
Et	(49)	30
i-Pr	(49)	67
t-Bu	(34)	21
PhCH$_2$	(49)	24

TABLE 4B. CATALYTIC KINETIC RESOLUTION OF BENZYLIC AMINES (Continued)

Amine	Conditions	Product(s) and Conversion(s) (%)	Refs.

*Please refer to the charts preceding the tables for the structures indicated by the **bold** numbers.*

C₉

Amine	Conditions	Product(s) and Conversion(s) (%)	Refs.
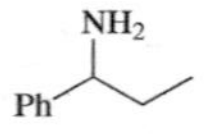 (structure: NH₂, Ph)	**9** (0.6 eq in 2 batches), **52** (10 mol %), CHCl₃, –50°, 24 h	 (structure) + (structure) (—), s = 16	33

C₉₋₁₄

| 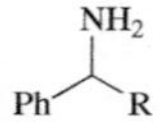 | B₂O (0.5 eq), **158a** (20 mol %), DMAP (20 mol %), 4 Å MS, toluene, –78°, 1 h |  (structures **I** and **II**) | 31 |

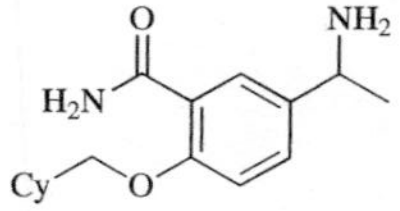

R		er **I**	er **II**	s
Et	(45)	88.5:11.5	81.0:19.0	15
n-Bu	(45)	82.0:18.0	76.5:23.5	8
PhCH₂	(47)	88.0:12.0	84.5:15.5	15

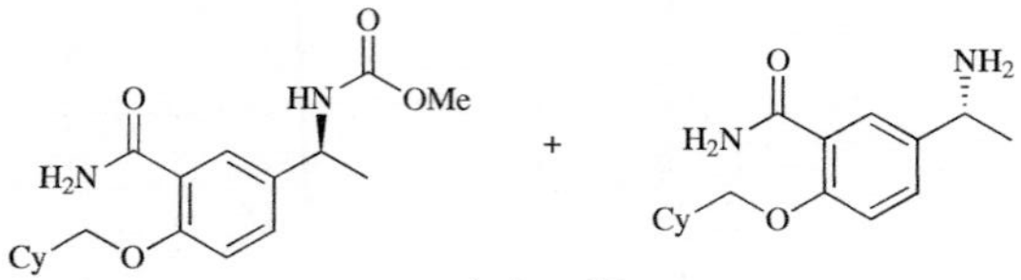

C₉

| | **9** (0.6 eq in 2 batches), **52** (10 mol %), CHCl₃, –50°, 24 h | (structures) + (structure)   (—), s = 11 | 33 |

C₁₂

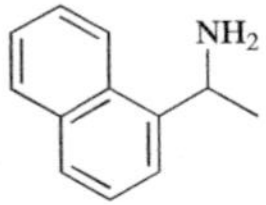

(RCO)$_2$O (0.5 eq),
TIPS-CD (5 eq),
Et$_3$N, cyclohexane, 10°

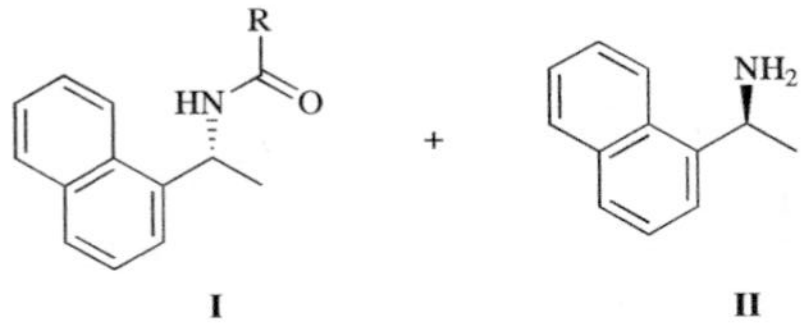

I + II

R	Time (h)		er I	s
Me	1	(42)	79.5:20.5	6
Ph	40	(49)	87.5:12.5	15

336

C_{14–16}

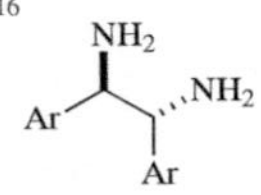

Bz$_2$O (0.5 eq), **158b** (10 mol %),
58 (10 mol %), 4 Å MS,
toluene, −78°;
then TrocCl, *i*-Pr$_2$NEt, rt

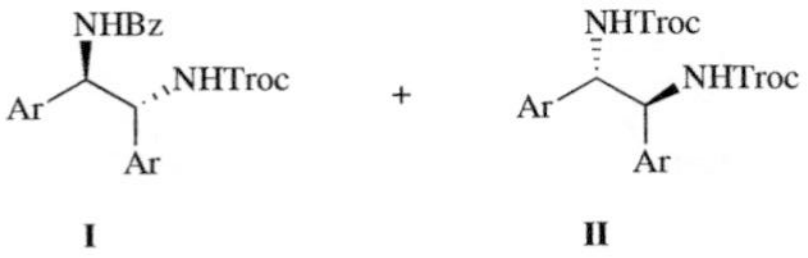

I + II

Ar		er I	er II	s
Ph	(47)	86.0:14.0	82.5:17.5	12
4-MeOC$_6$H$_4$	(53)	85.0:15.0	90.0:10.0	14
4-FC$_6$H$_4$	(48)	91.5:8.5	89.0:11.0	25
3-ClC$_6$H$_4$	(46)	91.0:9.0	85.0:15.0	21
4-ClC$_6$H$_4$	(51)	91.5:8.5	93.0:7.0	30
4-BrC$_6$H$_4$	(46)	91.5:8.5	85.0:15.0	22
2-MeC$_6$H$_4$	(50)	88.0:12.0	87.5:12.5	16
3-MeC$_6$H$_4$	(49)	72.5:27.5	72.0:28.0	4
4-MeC$_6$H$_4$	(49)	84.5:15.5	83.5:16.5	11

257

TABLE 4C. STOICHIOMETRIC KINETIC RESOLUTION OF PROPARGYLIC AMINES

Amine	Acylating Agent	Conditions	Product(s) and Conversion(s) (%)	Refs.
C$_8$				
	0.5 eq	MeN(Oct)$_3$Cl, THF, −20°, 12 h	er 95.5:4.5 (51), s = 78 **I** + **II**	259
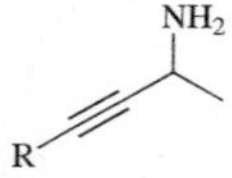	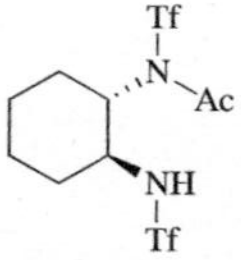 0.5 eq	MeN(Oct)$_3$Cl, THF, −20°, 12 h	 **I** + **II**	259

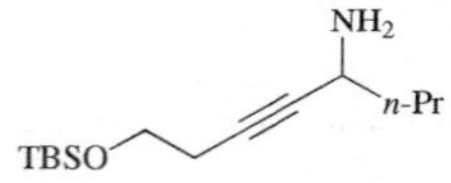

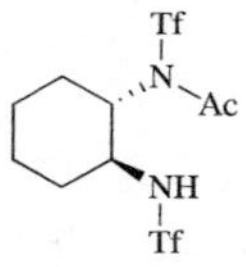

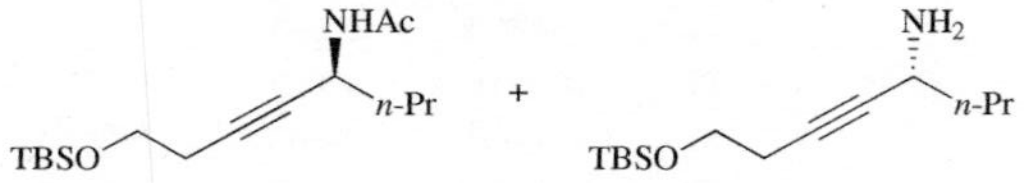

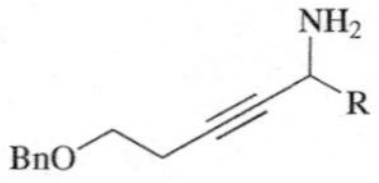

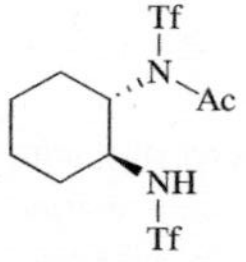

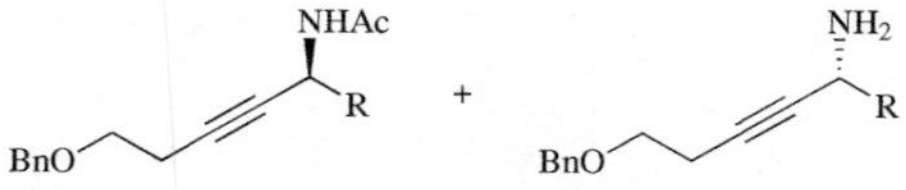

R		er I	s
n-Pr	(50)	97.0:3.0	—
BnO(CH$_2$)$_3$	(48)	95.0:5.0	49

C$_{8-14}$	0.5 eq	MeN(Oct)$_3$Cl, THF, −20°, 12 h	**I** + **II**	259

C_{10-16}

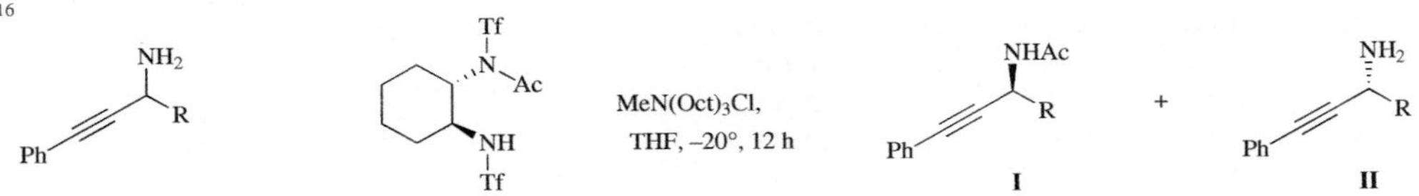

R		er **I**	s
$CH_3CH_2CH=CH$	(50)	94.0:6.0	45
2-py	(50)	95.5:4.5	67
2-MeOC_6H_4	(50)	94.5:5.5	67
3-MeOC_6H_4	(50)	94.0:6.0	45
4-MeOC_6H_4	(50)	96.5:3.5	94
$2,4\text{-F}_2C_6H_3$	(50)	95.5:4.5	67
4-MeC_6H_4	(50)	95.5:4.5	67
$3\text{-CF}_3\text{-C}_6H_4$	(50)	94.0:6.0	45
1-Np	(50)	86.0:14.0	14
2-Np	(50)	98.0:2.0	193

R		er **I**	s
Me	(50)	95.5:4.5	67
Et	(50)	96.0:4.0	79
i-Pr	(50)	95.0:5.0	58
i-Bu	(50)	93.5:6.5	51
4-MeOC_6H_4	(50)	53.5:46.5	1
$PhCH_2$	(50)	93.0:7.0	33

MeN(Oct)₃Cl,
THF, −20°, 12 h

0.5 eq

TABLE 4D. CATALYTIC KINETIC RESOLUTION OF PROPARGYLIC AMINES

Amine	Conditions	Product(s) and Conversion(s) (%)	Refs.

*Please refer to the charts preceding the tables for the structures indicated by the **bold** numbers.*

C$_{10-14}$

Amine structure: NH$_2$, Ar (propargylic amine)

Conditions: Bz$_2$O (0.6 eq), **158b** (5 mol %), DMAP (5 mol %), 4 Å MS, toluene, –78°, 3 h

Products: **I** (NHBz, Ar) + **II** (NH$_2$, Ar)

Ar		er I	er II	s
Ph	(48)	94.0:6.0	90.5:9.5	39
2-ClC$_6$H$_4$	(45)	91.0:9.0	83.5:16.5	21
3-ClC$_6$H$_4$	(47)	96.0:4.0	90.0:10.0	56
4-ClC$_6$H$_4$	(46)	94.0:6.0	87.5:12.5	38
2-MeC$_6$H$_4$	(43)	91.0:9.0	83.0:17.0	19
3-MeC$_6$H$_4$	(48)	94.5:5.5	92.0:8.0	44
4-MeC$_6$H$_4$	(43)	94.4:5.5	85.0:15.0	35
4-CF$_3$C$_6$H$_4$	(45)	93.5:6.5	83.5:16.5	29
1-Np	(48)	90.0:10.0	86.0:14.0	19

Refs. 260

C$_{11-15}$

Amine structure: NH$_2$, R, Ph

Conditions: Bz$_2$O (0.6 eq), **158b** (5 mol %), DMAP (5 mol %), 4 Å MS, toluene, –78°

Products: **I** (NHBz, R, Ph) + **II** (NH$_2$, R, Ph)

Ar	Time (h)		er I	er II	s
Et	3	(48)	95.0:5.0	90.5:9.5	46
i-Pr	3	(48)	95.5:4.5	92.0:8.0	56
Ph	8	(35)	89.5:10.5	70.5:29.5	12

Refs. 260

C$_{12}$

Amine structure: NH$_2$, n-C$_8$H$_{17}$

Conditions: Bz$_2$O (0.6 eq), **158b** (5 mol %), DMAP (5 mol %), 4 Å MS, toluene, –78°, 3 h

Products: NHBz, n-C$_8$H$_{17}$, er 84.5:15.5 + NH$_2$, n-C$_8$H$_{17}$, er 86.0:14.0

(52), s = 12

Refs. 260

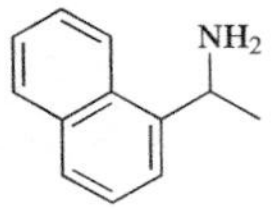

RCOCl (0.5 eq), Et$_3$N,
TIPS-CD (5 eq),
cyclohexane, 10°

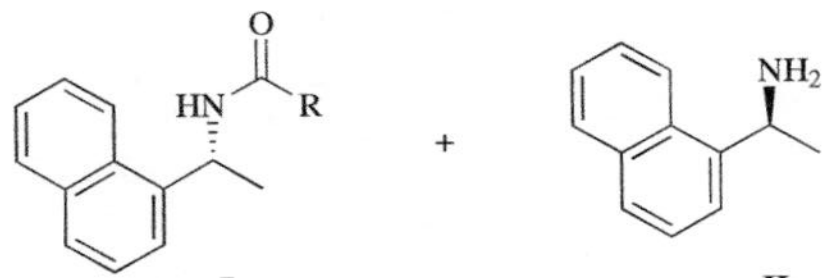

I + II

R		er I	s
Me	(45)	79.0:21.0	6
CH$_2$=CH	(42)	82.0:18.0	7
t-Bu	(46)	82.5:17.5	8
Cy	(45)	87.5:12.5	13
Ph	(44)	88.5:11.5	14
2,4-Cl$_2$C$_6$H$_4$	(50)	85.0:15.0	12
n-C$_7$H$_{15}$	(43)	87.0:13.0	12
1-Np	(15)	82.5:17.5	5
1-adamantyl	(41)	87.5:12.5	12

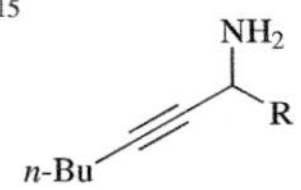

260

Bz$_2$O (0.6 eq), **158b** (5 mol %),
DMAP (5 mol %), 4 Å MS,
toluene, –78°, 3 h

C$_{14-15}$

I + II

R		er I	er **II**	s
n-C$_7$H$_{15}$	(50)	86.0:14.0	86.5:13.5	14
Ph(CH$_2$)$_2$	(46)	88.0:12.0	83.0:17.0	15

TABLE 4E. STOICHIOMETRIC KINETIC RESOLUTION OF ALLYLIC AMINES

Amine	Acylating Agent	Conditions	Product(s) and Conversion(s) (%)	Refs.

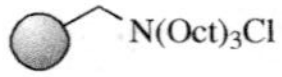

C$_{5-11}$

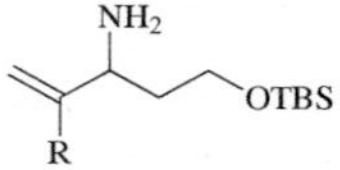

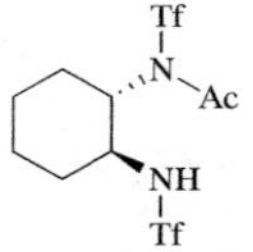

N(Oct)$_3$Cl (2.0 eq).

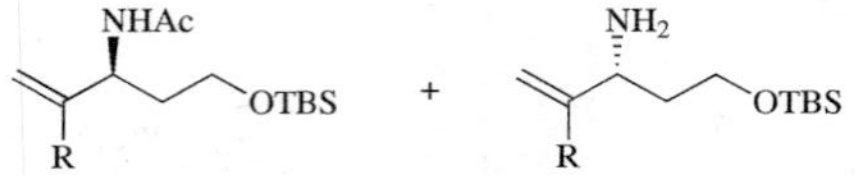

THF, –20°, 24 h

0.5 eq

I + **II**

R		er **I**	s
H	(48)	80.0:20.0	7
TMS	(49)	92.0:8.0	28
Me	(43)	85.5:14.5	10
Ph	(48)	91.5:8.5	25

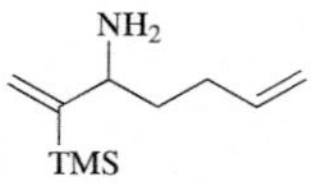

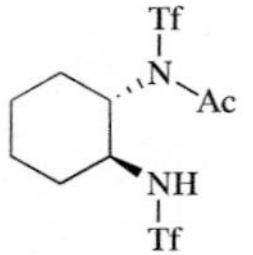

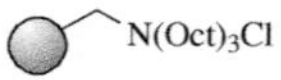

261

C$_7$

N(Oct)$_3$Cl (2.0 eq),

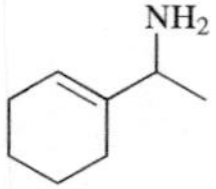

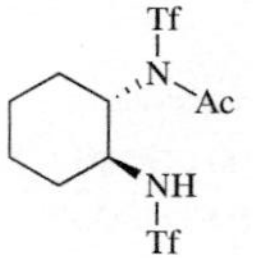

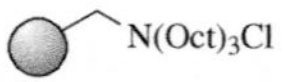

THF, –20°, 24 h

0.5 eq

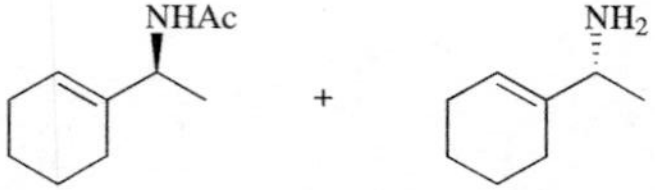

er 92.0:8.0   (49), s = 28

261

C$_8$

N(Oct)$_3$Cl (2.0 eq),

THF, –20°, 24 h

0.5 eq

er 82.0:18.0   (50), s = 9

261

C$_{10}$

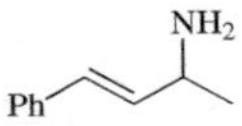

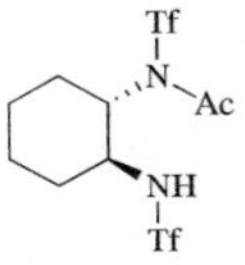

0.5 eq

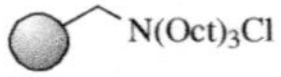

Tf–N(Ac)⋯NH–Tf

N(Oct)$_3$Cl (2.0 eq),

THF, –20°, 24 h

er 71.0:29.0

(50), s = 4   261

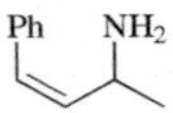

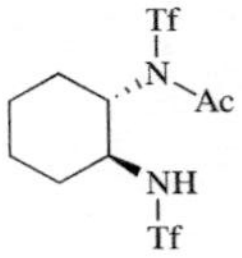

0.5 eq

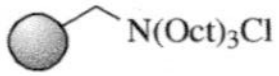

Tf–N(Ac)⋯NH–Tf

N(Oct)$_3$Cl (2.0 eq),

THF, –20°, 24 h

er 82.5:17.5

(44), s = 8   261

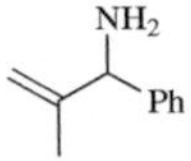

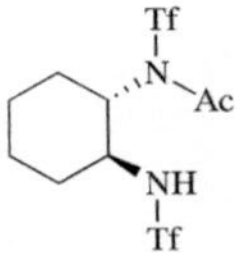

0.5 eq

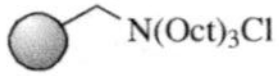

Tf–N(Ac)⋯NH–Tf

N(Oct)$_3$Cl (2.0 eq),

THF, –20°, 24 h

er 62.5:37.5

(50), s = 2   261

TABLE 4E. STOICHIOMETRIC KINETIC RESOLUTION OF ALLYLIC AMINES (*Continued*)

Amine	Acylating Agent	Conditions	Product(s) and Conversion(s) (%)	Refs.
C_{10–16}	0.5 eq	N(Oct)₃Cl THF, –20°, 24 h	**I** + **II**	261

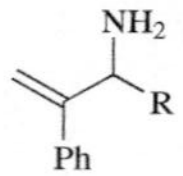

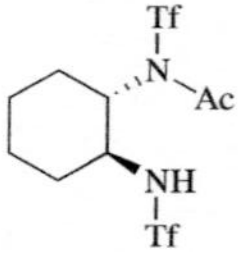

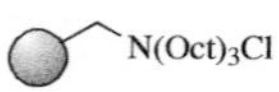

R		er I	s
Me	(50)	91.5:8.5	28
Et	(50)	92.0:8.0	30
i-Pr	(46)	87.5:12.5	14
i-Bu	(42)	92.0:8.0	21
CH₂=CH(CH₂)₂	(49)	92.5:7.5	31
PhCH₂	(49)	81.5:18.5	8

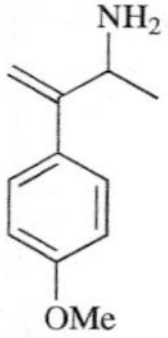

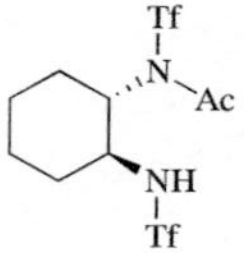

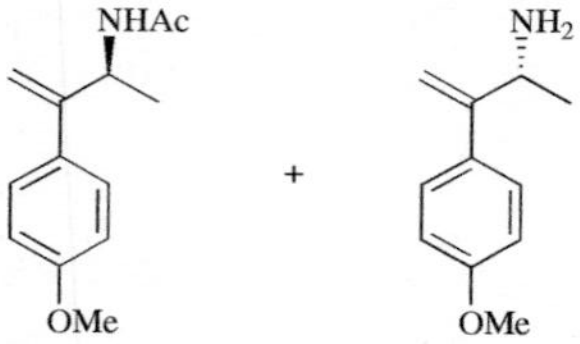

Amine	Acylating Agent	Conditions	Product(s) and Conversion(s) (%)	Refs.
C₁₀	0.5 eq	N(Oct)₃Cl (2.0 eq), THF, –20°, 24 h	er 94.0:6.0 + (45), s = 34	261

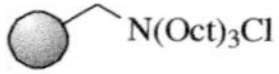

C$_{11}$

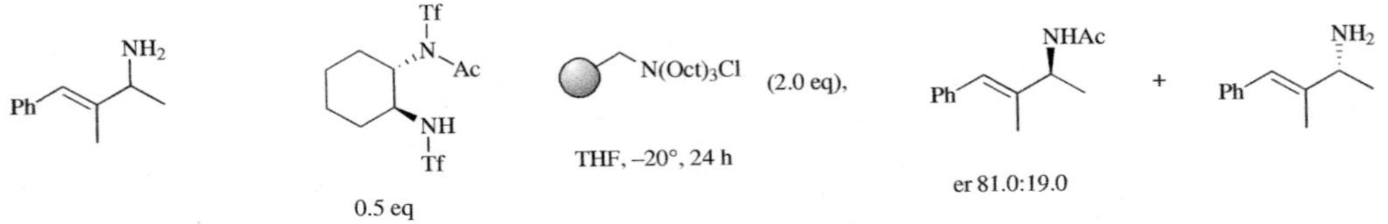

(structure: NH$_2$, Ph, methyl) + Tf–N(Ac)–/NH–Tf cyclohexane 0.5 eq + bead–N(Oct)$_3$Cl (2.0 eq), THF, –20°, 24 h → NHAc (Ph, methyl) er 81.0:19.0 + NH$_2$ (Ph, methyl) (50), s = 8   261

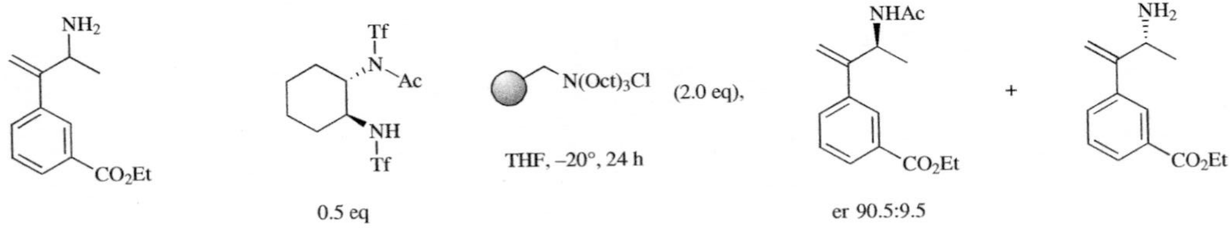

(structure: NH$_2$, CO$_2$Et) + Tf–N(Ac)–/NH–Tf cyclohexane 0.5 eq + bead–N(Oct)$_3$Cl (2.0 eq), THF, –20°, 24 h → NHAc CO$_2$Et er 90.5:9.5 + NH$_2$ CO$_2$Et (46), s = 20   261

C$_{12}$

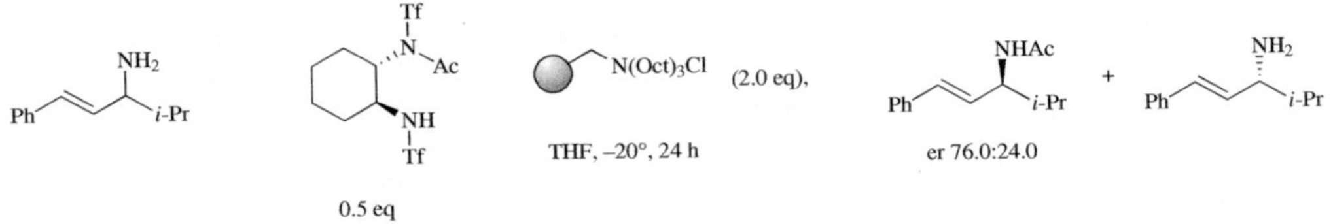

(structure: NH$_2$, i-Pr, Ph) + Tf–N(Ac)–/NH–Tf cyclohexane 0.5 eq + bead–N(Oct)$_3$Cl (2.0 eq), THF, –20°, 24 h → NHAc i-Pr Ph er 76.0:24.0 + NH$_2$ i-Pr Ph (50), s = 5   261

TABLE 4F. CATALYTIC KINETIC RESOLUTION OF ALLYLIC AMINES

Please refer to the charts preceding the tables for the structures indicated by the bold numbers.

Amine	Conditions	Product(s) and Conversion(s) (%)	Refs.
C₈	Bz₂O (0.6 eq), **158b** (2 mol %), PPY (2 mol %), 4 Å MS, toluene, −78°, 2 h	(49), s = 4	262
C₁₀₋₁₄	Bz₂O (0.6 eq), **158b** (2 mol %), PPY (2 mol %), 4 Å MS, toluene, −78°, 2 h		262

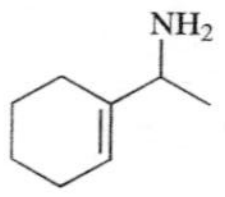

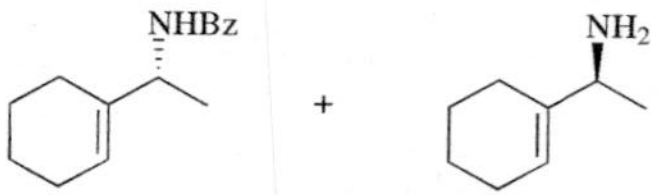

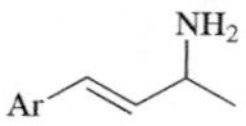

Ar		er I	er II	s
Ph	(55)	84.0:16.0	92.5:7.5	14
2-ClC₆H₄	(46)	79.0:21.0	82.5:17.5	7
3-ClC₆H₄	(54)	82.0:18.0	87.5:12.5	10
4-ClC₆H₄	(51)	87.5:12.5	88.0:12.0	17
2-MeC₆H₄	(47)	79.5:20.5	74.5:25.5	7
3-MeC₆H₄	(46)	81.5:18.5	78.5:21.5	8
4-MeC₆H₄	(50)	81.5:18.5	82.0:18.0	9
2-Np	(55)	81.0:19.0	88.0:12.0	9

C11   Bz₂O (0.6 eq), **158b** (2 mol %),   NHBz   +   NH₂   (48), s = 6   262
       PPY (2 mol %), 4 Å MS,
       toluene, −78°, 2 h   er 77.5:22.5   er 75.5:24.5

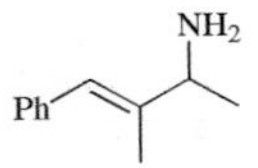

C11–13   Bz₂O (0.6 eq), **158b** (2 mol %),   NHBz   +   NH₂   262
          PPY (2 mol %), 4 Å MS,   R   R
          toluene, −78°, 2 h   **I**   **II**

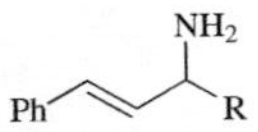

R		er **I**	er **II**	s
Et	(53)	84.5:15.5	88.0:12.0	12
i-Pr	(50)	89.5:10.5	88.5:11.5	20
t-Bu	(53)	88.0:12.0	89.5:10.5	18

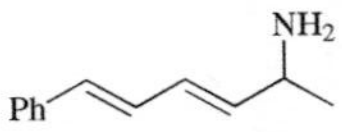

C12   Bz₂O (0.6 eq), **158b** (2 mol %),   NHBz   +   NH₂   (51), s = 9   262
       PPY (2 mol %), 4 Å MS,
       toluene, −78°, 2 h   er 83.0:17.0   er 83.5:16.5

TABLE 4G. STOICHIOMETRIC KINETIC RESOLUTION OF CYCLIC SECONDARY AMINES

Please refer to the charts preceding the tables for the structures indicated by the **bold numbers**.

Amine	Acylating Agent	Conditions	Product(s) and Conversion(s) (%)	Refs.

C_5

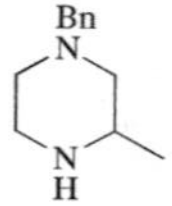

7a (0.6–0.7 eq)   CH$_2$Cl$_2$, rt, 48 h

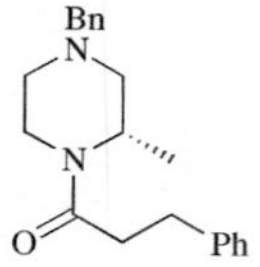

er 75.0:25.0   +   er 99.5:0.5   (66), s = 14   270

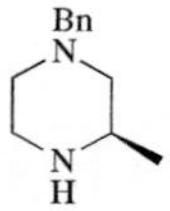

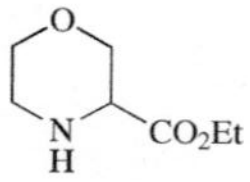

7b (0.6–0.7 eq)   CH$_2$Cl$_2$, rt, 48 h

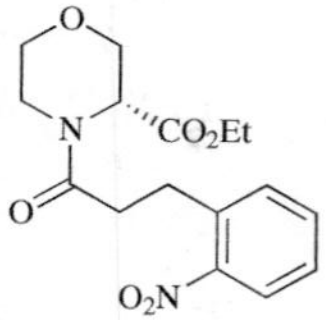

er 90.0:10.0   +   er 80.0:20.0   (43), s = 17   270

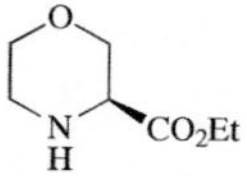

$C_{6–8}$

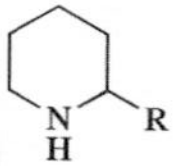

7a (0.6–0.7 eq)   CH$_2$Cl$_2$, rt, 48 h

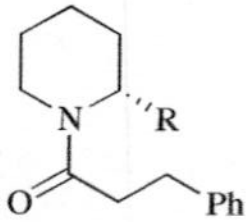

I   +   II   270

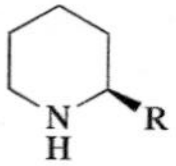

R		er I	er II	s
EtO$_2$C	(58)	84.0:16.0	96.0:4.0	17
Et	(62)	80.0:20.0	98.0:2.0	15
n-Pr	(54)	86.0:14.0	93.0:7.0	17

464

C7

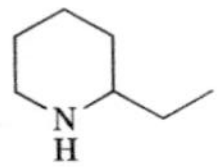

8 (0.7 eq)

THF, 45°, 16 h

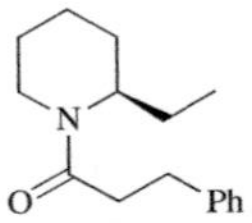

er 87.0:13.0

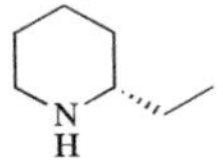

+ er 97.0:3.0 (57), s = 13 271

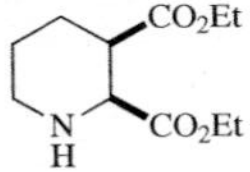

8 (0.7 eq)

THF, 45°, 16 h

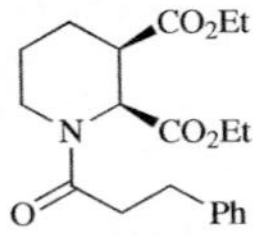

er 86.0:14.0

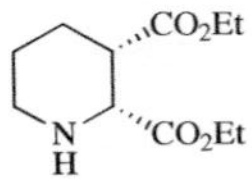

+ er 93.0:7.0 (54), s = 18 271

C10

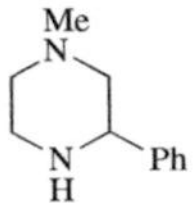

7a (0.6–0.7 eq)

CH₂Cl₂, rt, 48 h

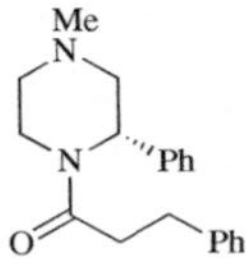

er 81.0:19.0

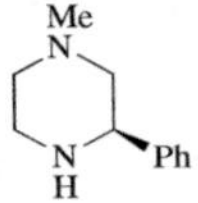

+ er 96.0:4.0 (60), s = 13 270

C11

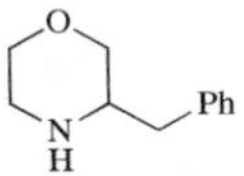

6 (0.6–0.7 eq)

CH₂Cl₂, rt, 15 h

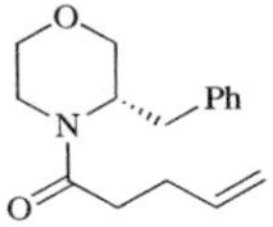

er 93.0:7.0

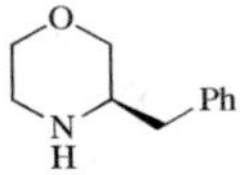

+ er 65.0:35.0 (26), s = 18 270

TABLE 4G. STOICHIOMETRIC KINETIC RESOLUTION OF CYCLIC SECONDARY AMINES (Continued)

Amine	Acylating Agent	Conditions	Product(s) and Conversion(s) (%)	Refs.

*Please refer to the charts preceding the tables for the structures indicated by the **bold** numbers.*

C₁₁

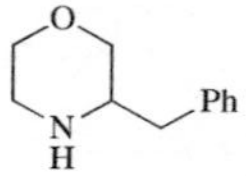

7a (0.6–0.7 eq) CH₂Cl₂, rt, 48 h

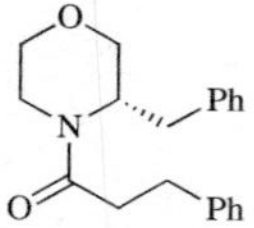

er 88.0:12.0   +   (53), s = 20   270

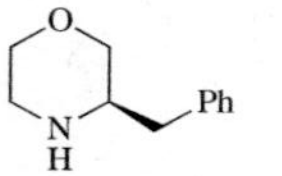

er 93.0:7.0

7b (0.6–0.7 eq) CH₂Cl₂, rt, 48 h

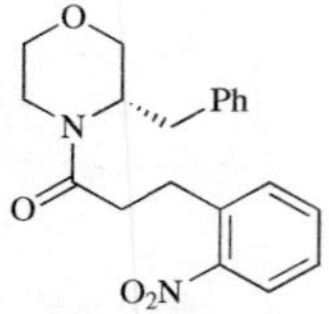

er 92.0:8.0   +   (47), s = 25   270

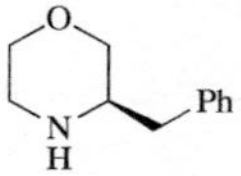

er 87.0:13.0

C₁₂

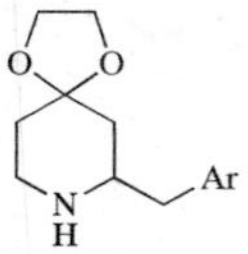

8 (0.7 eq) THF, 45°, 16 h

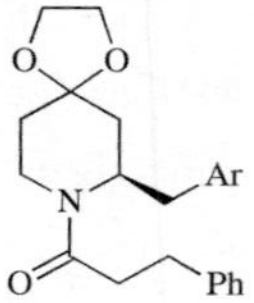

I   +   **II**   271

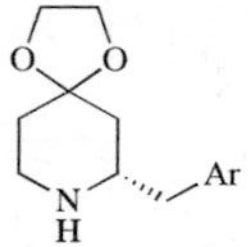

Ar	er I		er II	s
Ph	(50)	86.0:14.0	87.0:13.0	13
PMP	(49)	88.0:12.0	86.0:14.0	16

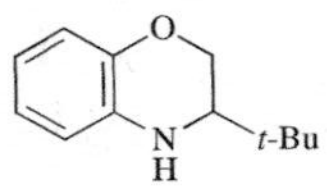

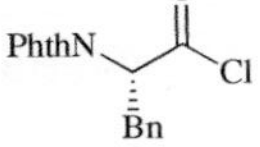

(40), *s* = 5   268

er 69.0:31.0

(39), *s* = 8   268

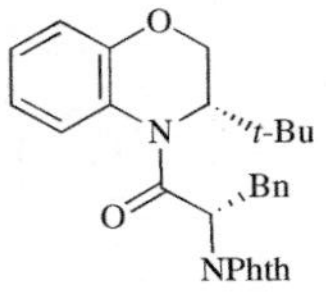

er 72.5:27.5

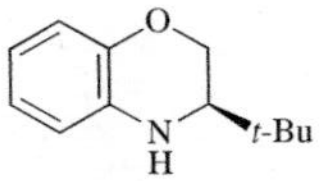

(43), *s* > 200   268

er 87.5:12.5

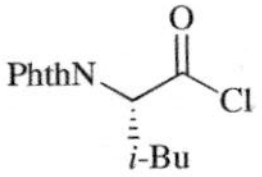

dr 78.5:21.5

dr 82.5:17.5

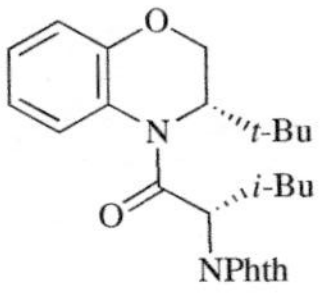

dr 99.0:1.0

CH_2Cl_2, rt, 6 h

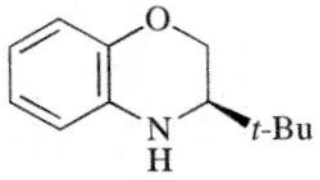

CH_2Cl_2, rt, 6 h

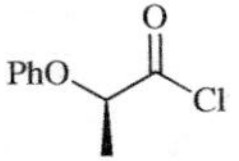

Toluene, rt, 6 h

0.5 eq

0.5 eq

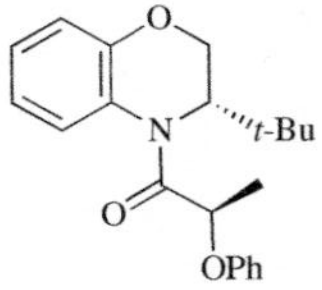

0.5 eq

t-Bu

Bn

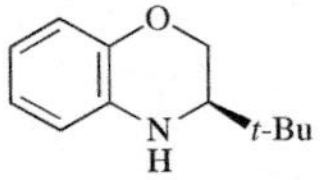

NPhth

i-Bu

OPh

PhthN

PhO

TABLE 4G. STOICHIOMETRIC KINETIC RESOLUTION OF CYCLIC SECONDARY AMINES (*Continued*)

Amine	Acylating Agent	Conditions	Product(s) and Conversion(s) (%)	Refs.

*Please refer to the charts preceding the tables for the structures indicated by the **bold** numbers.*

C₁₂

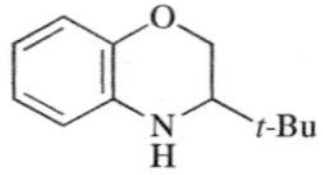

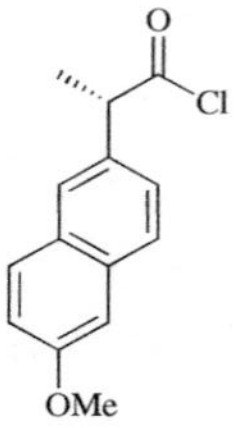

CH₂Cl₂, rt, 6 h

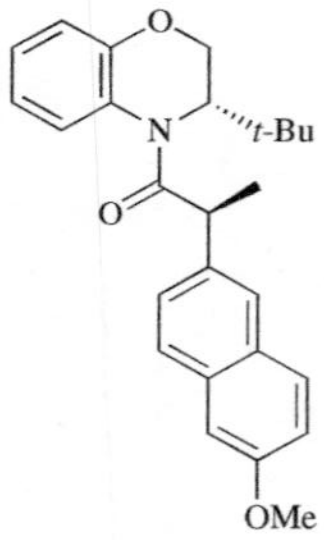

er 90.5:9.5   +   er 87.5:12.5   (48), *s* = 21   268

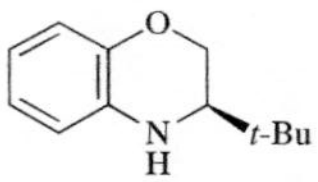

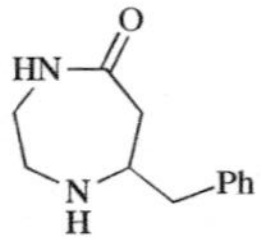

0.5 eq

7a (0.6–0.7 eq)

CH₂Cl₂, rt, 48 h

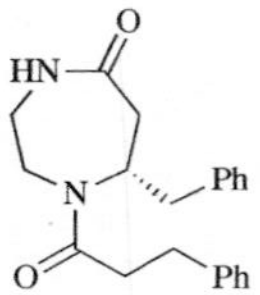

er 94.0:6.0   +   er 62.0:38.0   (20), *s* = 19   270

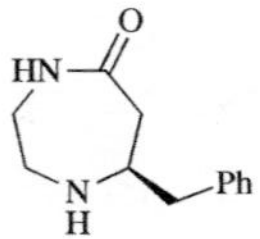

C₁₃

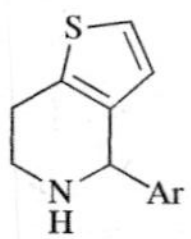

8 (0.7 eq)

THF, 45°, 16 h

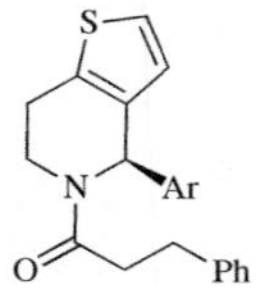

I   +   II   271

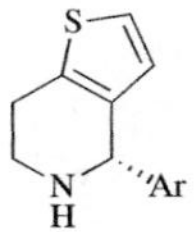

Ar	er I		er II		*s*
Ph	(55)	87.0:13.0	96.0:4.0		21
4-FC₆H₄	(60)	83.0:17.0	99.5:0.5		34

468

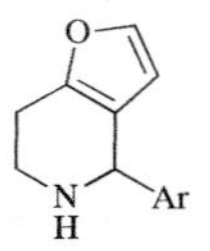

C$_{13-14}$

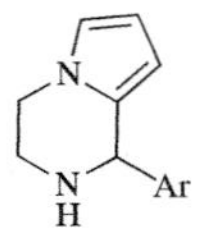

8 (0.7 eq) THF, 45°, 16 h (→) 271

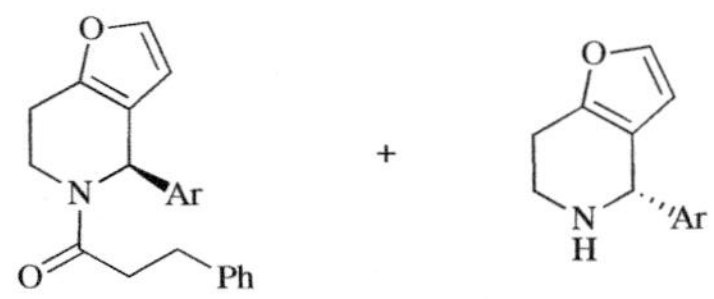

I + II er 99.0:1.0

8 (0.7 eq) THF, 45°, 16 h 271

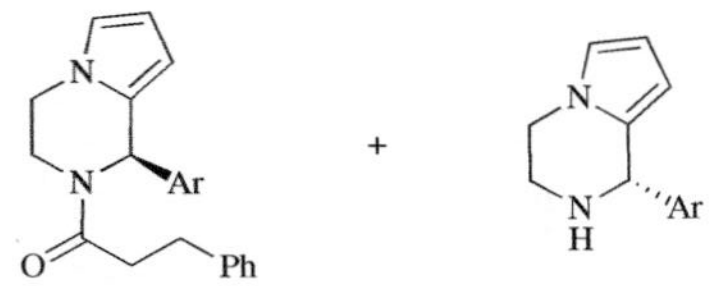

Ar		er I	er II	s
Ph	(54)	84.0:16.0	93.0:7.0	9
PhCH$_2$	(46)	84.0:16.0	79.0:21.0	17

I + II

C$_{13-17}$

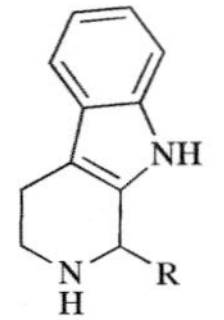

8 (0.7 eq) THF, 45°, 16 h 271

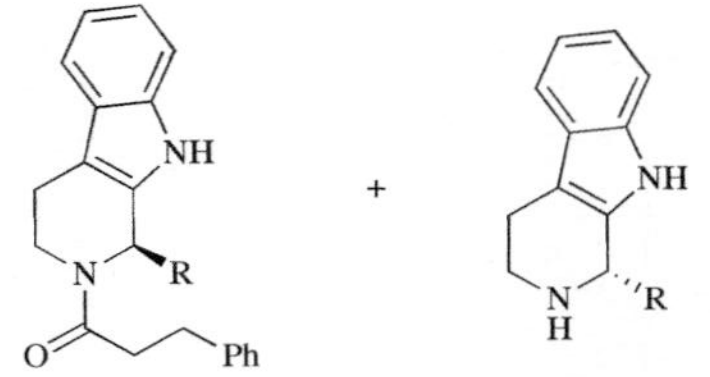

R		er I	er II	s
Et	(70)	70.0:30.0	96.0:4.0	7
Ph	(61)	81.0:19.0	99.0:1.0	19
4-FC$_6$H$_4$	(64)	78.0:22.0	99.5:0.5	19

I + II

TABLE 4G. STOICHIOMETRIC KINETIC RESOLUTION OF CYCLIC SECONDARY AMINES (*Continued*)

Please refer to the charts preceding the tables for the structures indicated by the ***bold*** *numbers.*

Amine	Acylating Agent	Conditions	Product(s) and Conversion(s) (%)	Refs.
C_{15}				
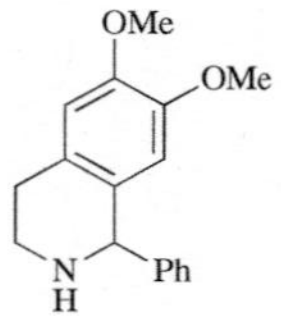	**6** (0.6–0.7 eq)	CH_2Cl_2, rt, 48 h	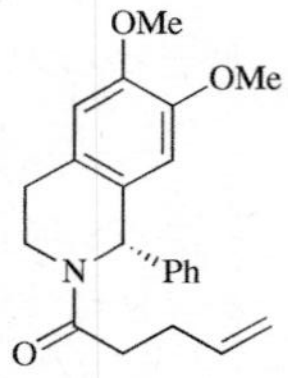 er 90.0:10.0 + 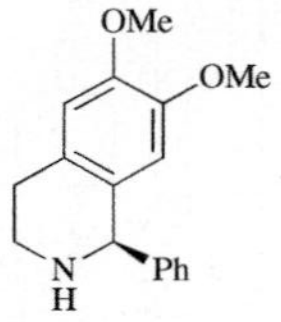 er 88.0:12.0 (49), *s* = 20	270
	8 (0.7 eq)	THF, 45°, 16 h	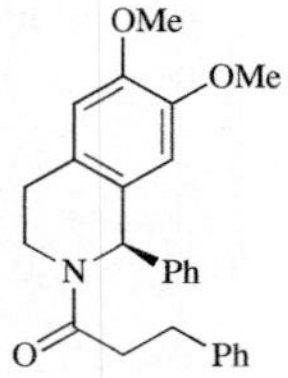 er 70.0:30.0 + 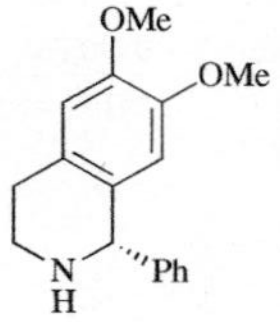 er 99.5:0.5 (70), *s* = 23	271

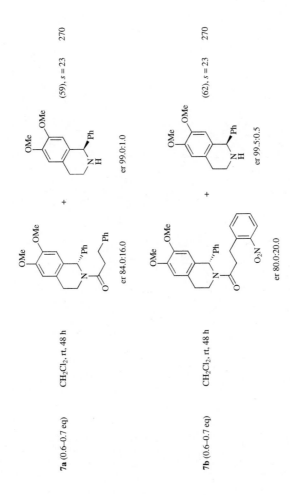

7a (0.6–0.7 eq)　　CH$_2$Cl$_2$, rt, 48 h

er 84.0:16.0

er 99.0:1.0

(59), $s = 23$　　270

7b (0.6–0.7 eq)　　CH$_2$Cl$_2$, rt, 48 h

er 80.0:20.0

er 99.5:0.5

(62), $s = 23$　　270

TABLE 4G. STOICHIOMETRIC KINETIC RESOLUTION OF CYCLIC SECONDARY AMINES (*Continued*)

Amine	Acylating Agent	Conditions	Product(s) and Conversion(s) (%)	Refs.

*Please refer to the charts preceding the tables for the structures indicated by the **bold** numbers.*

C₁₇

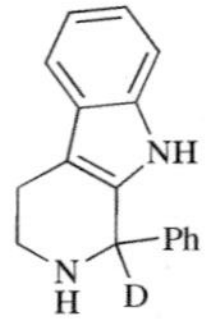

8 (0.7 eq)

THF, 45°, 16 h

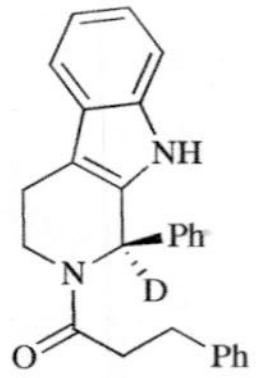

er 71.0:29.0   er 99.5:0.5

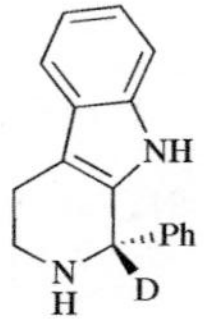

(63), s = 20   271

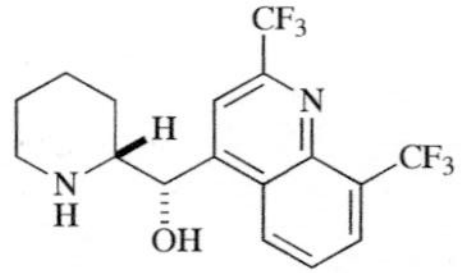

8 (1.5 eq)

THF, 45°, 12 h

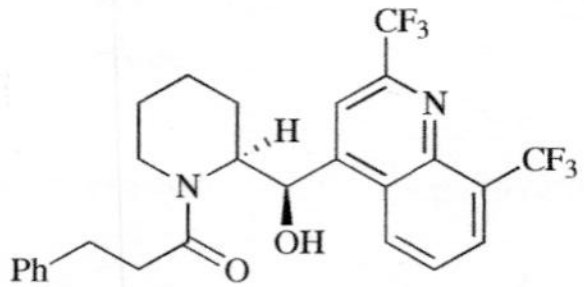

er 84.0:16.0

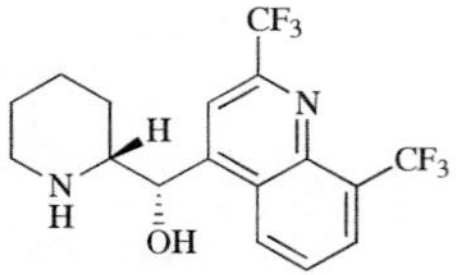

(59), s = 26   271

er 99.5:0.5

Amine	Conditions	Product(s) and Conversion(s) (%)	Refs.

*Please refer to the charts preceding the tables for the structures indicated by the **bold** numbers.*

C₅

13 (0.7 eq), 133 (5 mol %),
151 (10 mol %),
K₂CO₃ (20 mol %),
i-PrOAc, rt, 48 h

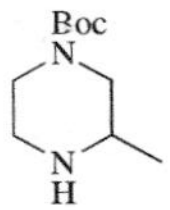

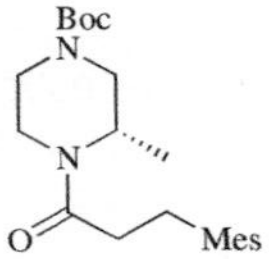

er 86.0:14.0   +   er 97.0:3.0

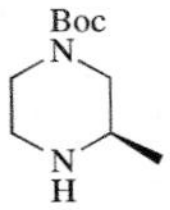

(57), *s* = 21

85

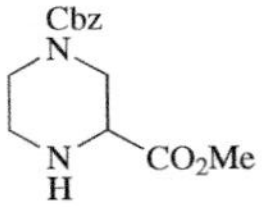

13 (0.7 eq), 133 (5 mol %),
151 (10 mol %),
K₂CO₃ (20 mol %),
i-PrOAc, rt, 48 h

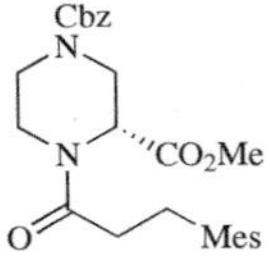

er 92.0:8.0   +   er 84.0:16.0

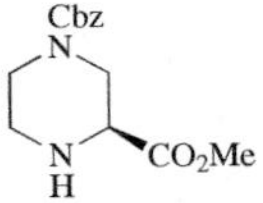

(45), *s* = 23

85

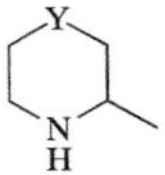

13 (0.7 eq), 132 (10 mol %),
151 (10 mol %),
DBU (20 mol %),
CH₂Cl₂, rt, 18 h

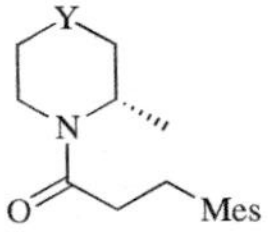

I   +   II

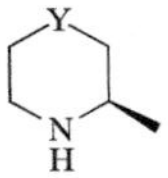

Y		er **I**	er **II**	*s*
BnN	(45)	89.0:11.0	82.0:18.0	16
O	(41)	87.0:13.0	76.0:24.0	11

34

473

TABLE 4H. CATALYTIC KINETIC RESOLUTION OF CYCLIC SECONDARY AMINES (*Continued*)

Amine	Conditions	Product(s) and Conversion(s) (%)	Refs.

*Please refer to the charts preceding the tables for the structures indicated by the **bold** numbers.*

C₅

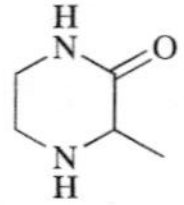

Amine	Conditions	Product(s) and Conversion(s) (%)	Refs.
	13 (0.7 eq), **133** (5 mol %), **151** (10 mol %), K$_2$CO$_3$ (20 mol %), *i*-PrOAc, rt, 48 h	er 79.0:21.0 + (58), *s* = 9 er 90.0:10.0	85

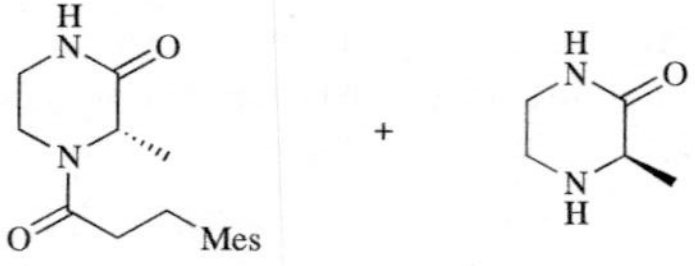

C₆₋₁₃

Amine	Conditions	Product(s) and Conversion(s) (%)	Refs.
	13 (0.7 eq), **132** (10 mol %), **151** (10 mol %), DBU (20 mol %), CH$_2$Cl$_2$, rt, 18 h	**I** + **II**	34

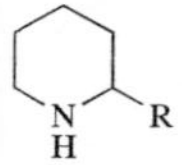

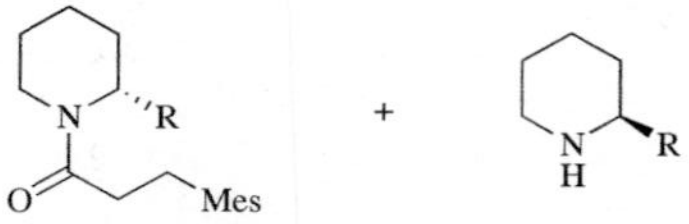

R		er I	er II	*s*
Me	(55)	86.0:14.0	94.0:6.0	17
EtO$_2$C	(46)	91.0:9.0	83.0:17.0	20
TBSOCH$_2$	(29)	91.0:9.0	67.0:33.0	14
Et	(53)	84.0:16.0	88.0:12.0	12
n-Pr	(48)	91.0:9.0	88.0:12.0	23
PhCH$_2$	(34)	94.0:6.0	70.0:30.0	23
3-FC$_6$H$_4$CH$_2$	(30)	93.0:7.0	68.0:32.0	19
3-CF$_3$C$_6$H$_4$CH$_2$	(34)	93.0:7.0	72.0:28.0	21
Ph(CH$_2$)$_2$	(41)	92.0:8.0	79.0:21.0	21

C$_{6-12}$

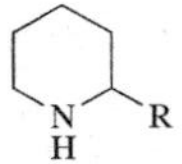

13 (0.7 eq), **133** (5 mol %),
151 (10 mol %),
K$_2$CO$_3$ (20 mol %),
i-PrOAc, rt, 48 h   85

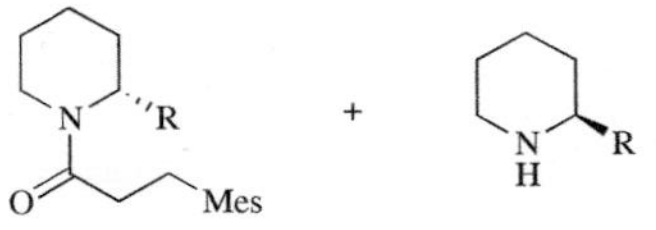

R		er **I**	er **II**	s
MeO$_2$C	(33)	93.0:7.0	71.0:29.0	20
Et	(57)	86.0:14.0	97.0:3.0	21
MeO(CH$_2$)$_2$	(52)	90.0:10.0	94.0:6.0	26
n-Pr	(61)	82.0:18.0	99.5:0.5	22
PhCH$_2$	(52)	91.0:9.0	94.0:6.0	29

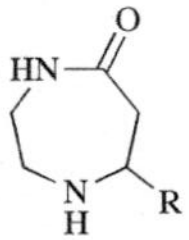

13 (0.7 eq), **133** (5 mol %),
151 (10 mol %),
K$_2$CO$_3$ (20 mol %),
i-PrOAc, rt, 48 h   85

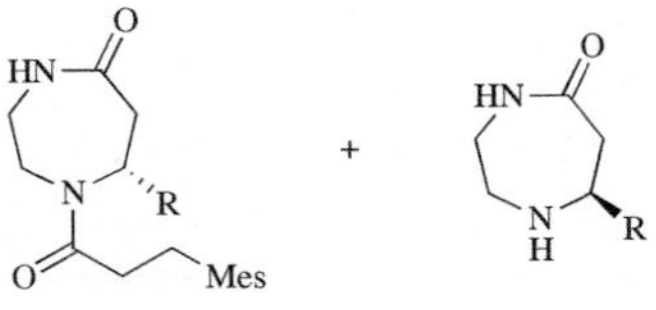

R		er **I**	er **II**	s
Me	(54)	88.0:12.0	95.0:5.0	22
Et	(64)	78.0:22.0	99.5:0.5	20
PhCH$_2$	(41)	94.0:6.0	81.0:19.0	30

TABLE 4H. CATALYTIC KINETIC RESOLUTION OF CYCLIC SECONDARY AMINES (Continued)

Amine	Conditions	Product(s) and Conversion(s) (%)	Refs.

*Please refer to the charts preceding the tables for the structures indicated by the **bold** numbers.*

C7

Amine	Conditions	Product(s) and Conversion(s) (%)	Refs.
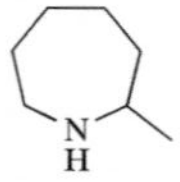 (2-methylazepane)	**13** (0.7 eq), **133** (5 mol %), **151** (10 mol %), K_2CO_3 (20 mol %), i-PrOAc, rt, 48 h	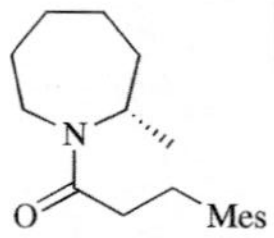 er 84.0:16.0 + (46), s = 9 er 79.0:21.0	85
	13 (0.7 eq), **132** (10 mol %), **151** (10 mol %), DBU (20 mol %), CH_2Cl_2, rt, 18 h	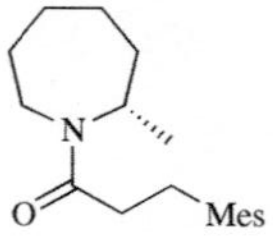 er 82.0:18.0 + (47), s = 8 er 78.0:22.0	34

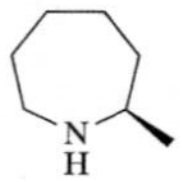

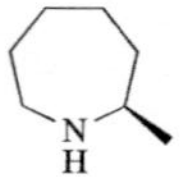

C9

Amine	Conditions	Product(s) and Conversion(s) (%)	Refs.
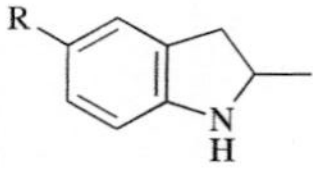 (R-substituted 2-methylindoline)	**10** (0.65 eq), **54** (5 mol %), LiBr (1.5 eq), 18-crown-6 (0.75 eq), toluene	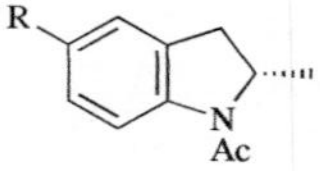 **I** 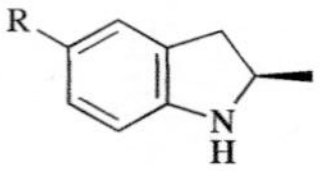 + **II** (see table below)	273

R	Temp (°)	Time (d)		er I	er II	s
MeO	–10	2	(60)	80.0:20.0	96.0:4.0	12
Br	rt	23	(57)	82.0:18.0	91.5:8.5	11

C$_{9-16}$

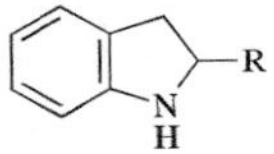

10 (0.65 eq), **54** (5 mol %),
LiBr (1.5 eq),
18-crown-6 (0.75 eq),
toluene

273

R	Temp (°)	Time (d)		er **I**	er **II**	s
Me	0	5	(55)	88.0:12.0	97.0:3.0	25
TBSOCH$_2$	10	19	(57)	81.0:19.0	92.0:8.0	11
n-Pr	0	10	(50)	91.0:9.0	91.0:9.0	25
Ph(CH$_2$)$_2$	rt	14	(60)	82.0:18.0	99.0:1.0	20

I   +   **II**

C$_{10}$

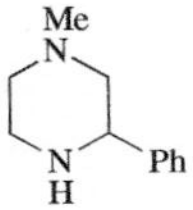

13 (0.7 eq), **133** (5 mol %),
151 (10 mol %),
K$_2$CO$_3$ (20 mol %),
i-PrOAc, rt, 48 h

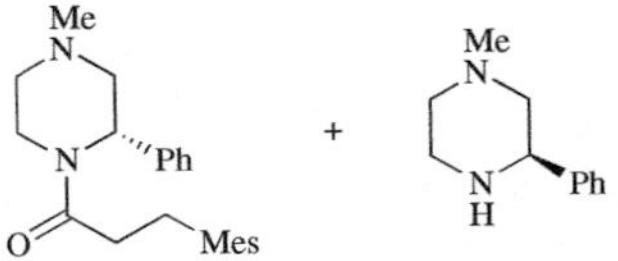

er 90.0:10.0   +   er 99.0:1.0   (46), s = 46

85

13 (0.7 eq), **132** (10 mol %),
151 (10 mol %),
DBU (20 mol %),
CH$_2$Cl$_2$, rt, 18 h

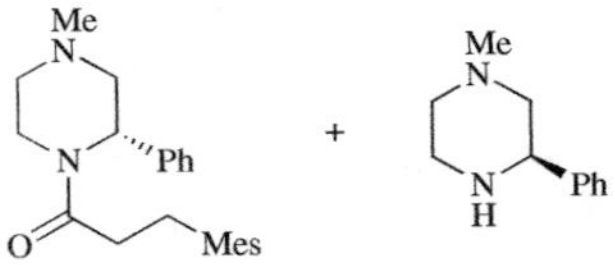

er 93.0:7.0   +   er 79.0:21.0   (40), s = 24

34

TABLE 4H. CATALYTIC KINETIC RESOLUTION OF CYCLIC SECONDARY AMINES (*Continued*)

Amine	Conditions	Product(s) and Conversion(s) (%)	Refs.

*Please refer to the charts preceding the tables for the structures indicated by the **bold** numbers.*

C₁₀

| 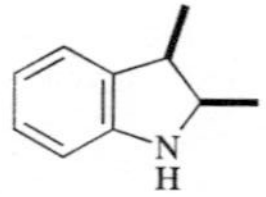 | **10** (0.65 eq), **54** (5 mol %), LiBr (1.5 eq), 18-crown-6 (0.75 eq), toluene, 0°, 7 d | 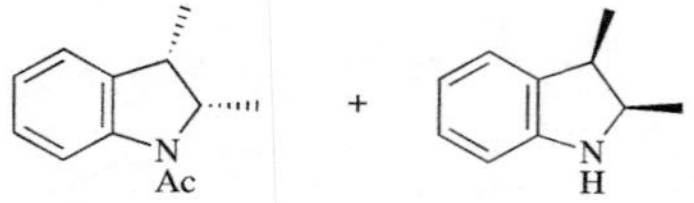 er 87.0:13.0   +   er 95.5:4.5   (55), *s* = 21 | 273 |

| 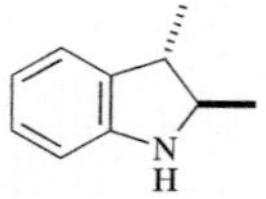 | **10** (0.65 eq), **54** (5 mol %), LiBr (1.5 eq), 18-crown-6 (0.75 eq), toluene, rt, 20 d | 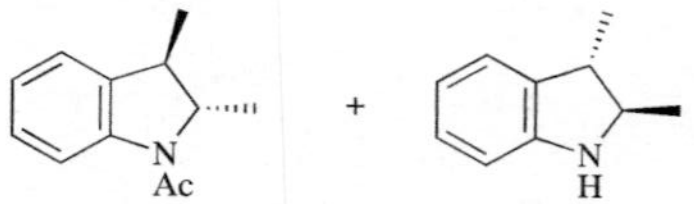 er 76.5:23.5   +   er 97.0:3.0   (64), *s* = 11 | 273 |

C₁₀₋₁₅

| 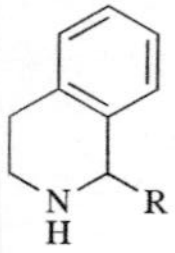 | **13** (0.7 eq), **132** (10 mol %), **151** (10 mol %), DBU (20 mol %), CH₂Cl₂, rt, 18 h | 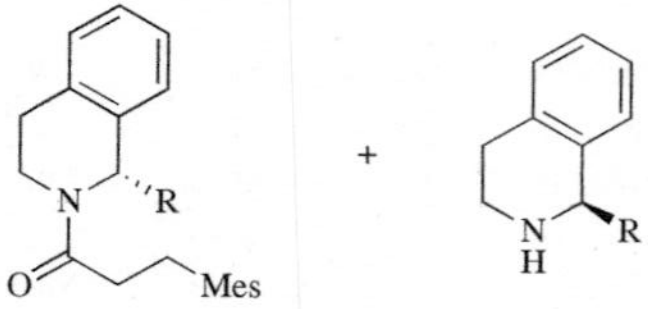 | 34 |

R		er I	er II	*s*
Me	(56)	83.0:17.0	93.0:7.0	13
Ph	(49)	95.0:5.0	95.0:5.0	53

478

C_{11}

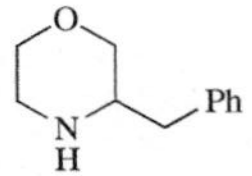

13 (0.7 eq), **133** (5 mol %), **151** (10 mol %), K_2CO_3 (20 mol %), i-PrOAc, rt, 48 h

er 93.0:7.0   +   (29), $s = 29$
er 87.0:13.0   85

13 (0.7 eq), **132** (10 mol %), **151** (10 mol %), DBU (20 mol %), CH_2Cl_2, rt, 18 h

er 90.0:10.0   +   (51), $s = 23$
er 91.0:9.0   34

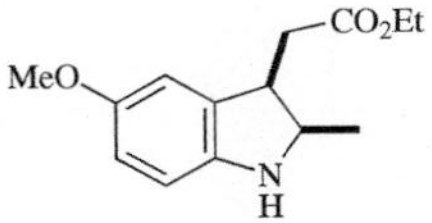

10 (0.65 eq), **54** (5 mol %), LiBr (1.5 eq), 18-crown-6 (0.75 eq), toluene, −10°, 4 d

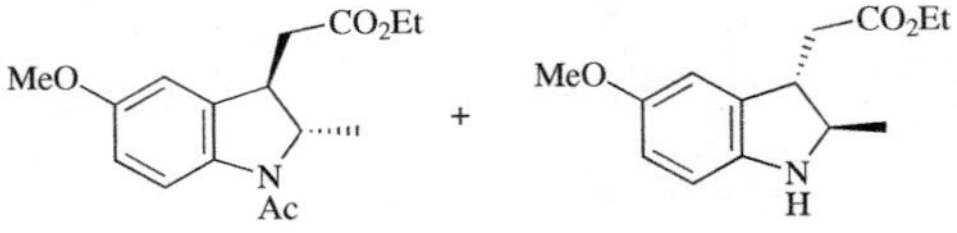

er 87.5:12.5   **I**   +   er 92.5:7.5   (53), $s = 18$ 273

C_{12-13}

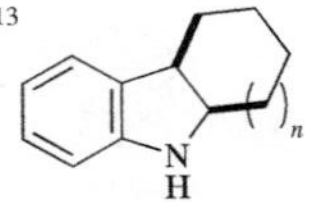

10 (0.65 eq), **54** (5 mol %), LiBr (1.5 eq), 18-crown-6 (0.75 eq), toluene, 0°

I   +   **II**

n	Time (d)		er **I**	er **II**	s
1	6	(58)	76.5:23.5	92.0:8.0	10
2	10	(52)	90.0:10.0	93.0:7.0	24

273

TABLE 4H. CATALYTIC KINETIC RESOLUTION OF CYCLIC SECONDARY AMINES (Continued)

Amine	Conditions	Product(s) and Conversion(s) (%)	Refs.

*Please refer to the charts preceding the tables for the structures indicated by the **bold** numbers.*

C₁₅

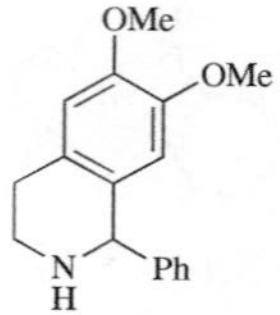

13 (0.7 eq), **133** (5 mol %),
151 (10 mol %),
K₂CO₃ (20 mol %),
i-PrOAc, rt, 48 h

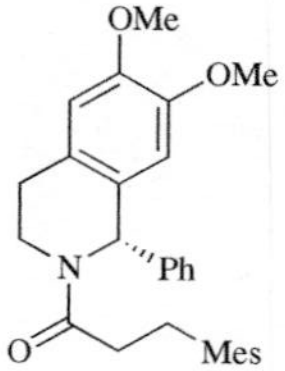

er 97.0:3.0

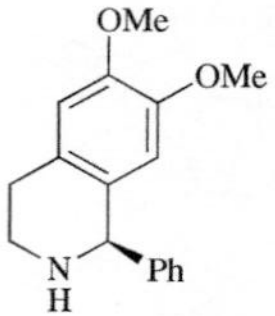

+

er 98.0:2.0

(51), *s* = 127

85

13 (0.7 eq), **132** (10 mol %),
151 (10 mol %),
DBU (20 mol %),
CH₂Cl₂, rt, 18 h

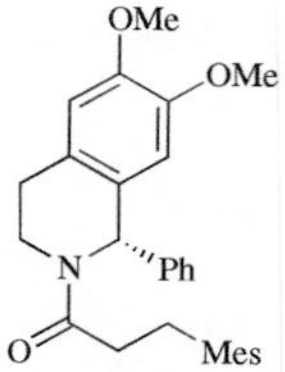

er 96.0:4.0

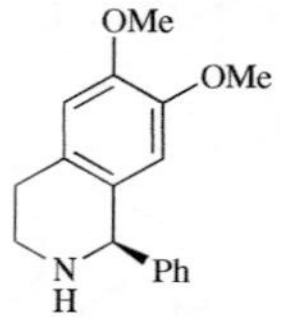

+

er 95.0:5.0

(50), *s* = 74

34

C₁₆

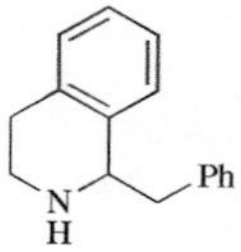

13 (0.7 eq), **133** (5 mol %),
151 (10 mol %),
K₂CO₃ (20 mol %),
i-PrOAc, rt, 48 h

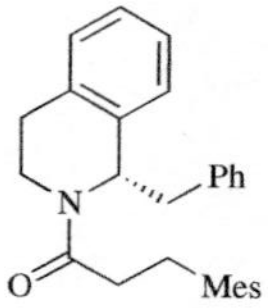

er 89.0:11.0

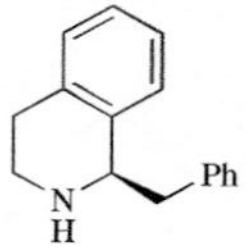

+

er 83.0:17.0

(46), *s* = 16

85

480

TABLE 4I. STOICHIOMETRIC KINETIC RESOLUTION OF AMINO ESTERS

Ester	Acylating Agent	Conditions	Product(s) and Conversion(s) (%)	Refs.

C$_{3-9}$

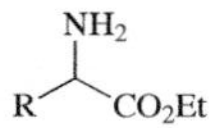

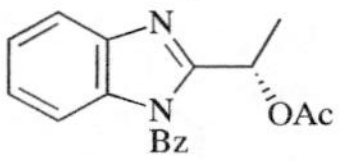

Ester: R–CH(NH$_2$)–CO$_2$Et

Acylating Agent: (benzimidazole)–CH(CH$_3$)–OAc, N–Bz; 0.5 eq

Conditions: THF, –10°, 12 h

Products: I [R–CH(NHBz)–CO$_2$Et] + II [R–CH(NH$_2$)–CO$_2$Et]

R		er I	er II	s
Me	(50)	94.5:5.5	94.5:5.5	51
PhCH$_2$	(50)	96.0:4.0	95.5:4.5	73

Refs. 274

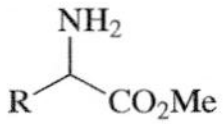

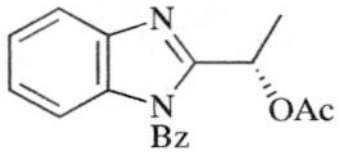

C$_{5-8}$

Ester: R–CH(NH$_2$)–CO$_2$Me

Acylating Agent: (benzimidazole)–CH(CH$_3$)–OAc, N–Bz; 0.5 eq

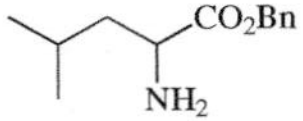

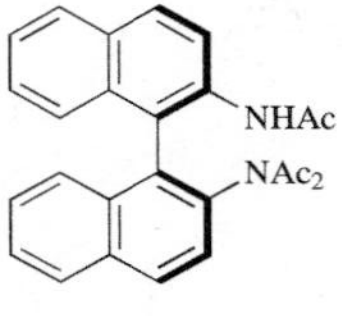

Conditions: THF, –10°, 12 h

Products: I [R–CH(NHBz)–CO$_2$Me] + II [R–CH(NH$_2$)–CO$_2$Me]

R		er I	er II	s
i-Pr	(50)	95.0:5.0	95.0:5.0	58
i-Bu	(50)	95.0:5.0	95.0:5.0	61
Ph	(50)	96.0:4.0	95.5:4.5	74

Refs. 274

C$_6$

Ester: (i-Bu)CH(NH$_2$)–CO$_2$Bn

Acylating Agent: binaphthyl NHAc / NAc$_2$; 0.25 eq

Conditions: DMSO, rt, 72 h

Products: (i-Bu)CH(NHAc)–CO$_2$Bn + (i-Bu)CH(NH$_2$)–CO$_2$Bn, er 55.5:44.5 (21)

Refs. 249

481

TABLE 4I. STOICHIOMETRIC KINETIC RESOLUTION OF AMINO ESTERS (*Continued*)

Ester	Acylating Agent	Conditions	Product(s) and Conversion(s) (%)	Refs.
C₉				

C₉

Ester: (structure: Ph–CH₂–CH(NH₂)–CO₂Me)

Acylating Agent	Conditions	Product(s) and Conversion(s) (%)	Refs.
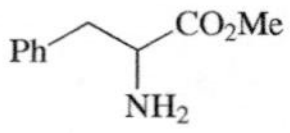 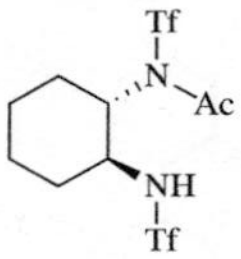 (cyclohexane bearing N(Tf)Ac and NH–Tf groups) 0.5 eq	MeN(n-Oct)₃Cl, 30°	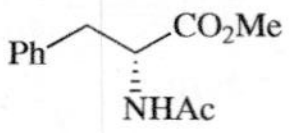 (structure: Ph–CH₂–CH(NHAc)–CO₂Me) er 91.0:9.0 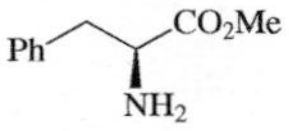 + (structure: Ph–CH₂–CH(NH₂)–CO₂Me) (50), s = 25	29
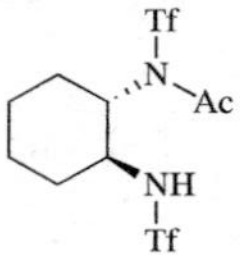 (cyclohexane bearing N(Tf)Ac and NH–Tf groups) 0.33 eq	DMPU, –20°, 24 h	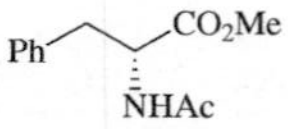 (structure: Ph–CH₂–CH(NHAc)–CO₂Me) er 90.0:10.0 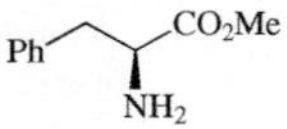 + (structure: Ph–CH₂–CH(NH₂)–CO₂Me) (33)	30
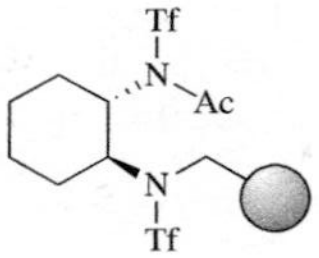 (cyclohexane bearing N(Tf)Ac and N–Tf with polymer bead) 0.2 M	Benzene, rt, 3 h	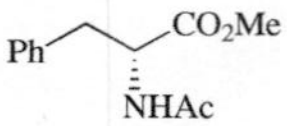 (structure: Ph–CH₂–CH(NHAc)–CO₂Me) er 85.0:15.0 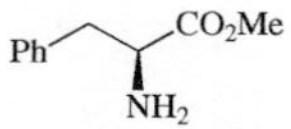 + (structure: Ph–CH₂–CH(NH₂)–CO₂Me) (—), s = 7	255

Ester: (structure: Ph–CH₂–CH(NH₂)–CO₂Bn)

Acylating Agent	Conditions	Product(s) and Conversion(s) (%)	Refs.
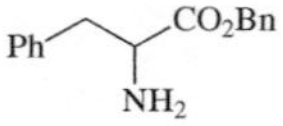 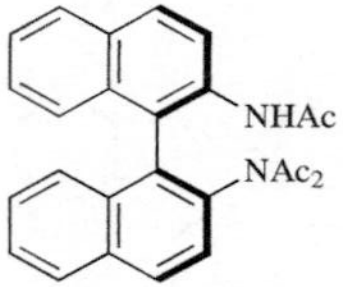 (binaphthyl bearing NHAc and NAc₂ groups) 0.25 eq	DMSO, rt, 72 h	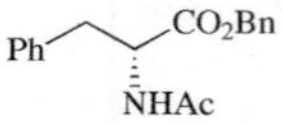 (structure: Ph–CH₂–CH(NHAc)–CO₂Bn) er 74.0:26.0 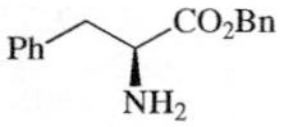 + (structure: Ph–CH₂–CH(NH₂)–CO₂Bn) (24)	249

TABLE 4J. CATALYTIC KINETIC RESOLUTION OF ATROPISOMERIC AMINES

Amine	Conditions	Product(s) and Conversion(s) (%)	Refs.

*Please refer to the charts preceding the tables for the structures indicated by the **bold** numbers.*

C$_{16}$

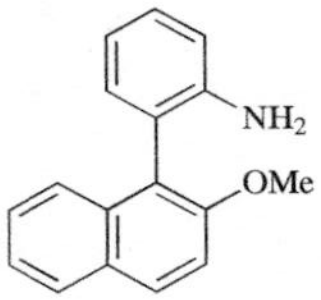

(*i*-PrCO)$_2$O (5 eq), **90** (1 mol %),
i-Pr$_2$NEt, CH$_2$Cl$_2$, –50°, 12 h;
then *n*-PrNH$_2$ (6 eq)

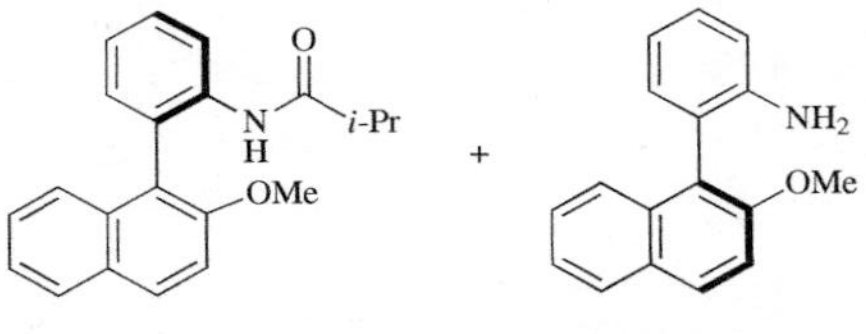

er 77.5:22.5

+

(31), *s* = 4

er 62.5:37.5

275

| Lactam or Thiolactam | Conditions | Product(s) and Conversion(s) (%) | | | | | Refs. |

*Please refer to the charts preceding the tables for the structures indicated by the **bold** numbers.*

C$_{4-13}$

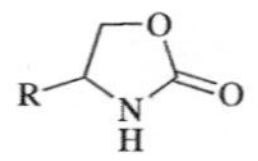

$(i\text{-PrCO})_2$O (0.75 eq), **124** (4 mol %),
$i\text{-Pr}_2$NEt (0.75 eq), Na$_2$SO$_4$, CHCl$_3$, rt

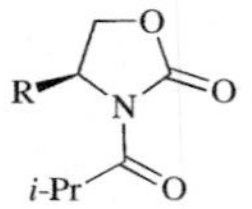

I 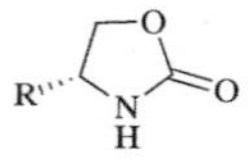   **II**

R	Time (h)	er I	er II	s	
BnO$_2$C	2	(52)	95.5:4.5	99.0:1.0	95	52
2-furyl	9	(42)	98.0:2.0	85.5:14.5	96	276
2-thienyl	6	(49)	99.0:1.0	98.0:2.0	430	276
Ph	9	(36)	99.0:1.0	77.0:23.0	200	276
PhC≡C	2	(45)	96.0:4.0	88.0:12.0	58	52
1-Np	21	(42)	99.5:0.5	86.0:14.0	450	276
2-Np	14	(47)	99.0:1.0	94.0:6.0	260	276

C$_5$

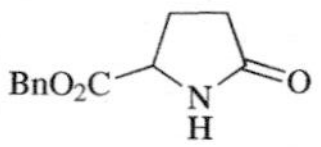

$(i\text{-PrCO})_2$O (0.75 eq), **117** (10 mol %),
$i\text{-Pr}_2$NEt (0.75 eq), t-AmOH, rt, 72 h

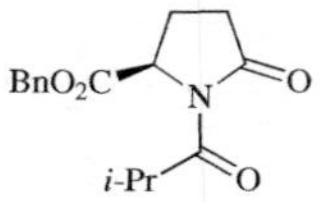

er 98.0:2.0 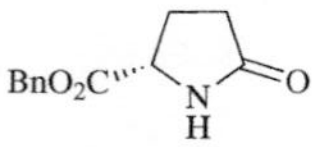   er 90.5:9.5

(46), s = 131   52

C$_7$

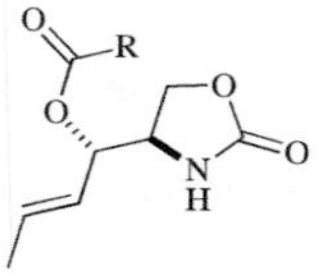

$(i\text{-PrCO})_2$O (0.75 eq),
124 (4 mol %), $i\text{-Pr}_2$NEt (0.75 eq),
Na$_2$SO$_4$, CHCl$_3$, rt

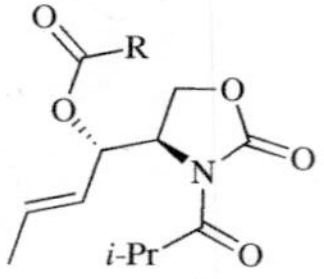

I 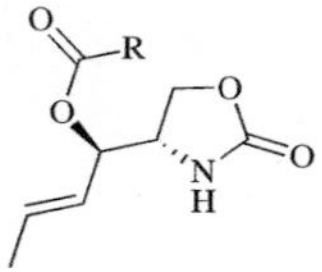   **II**

277

R	Time (h)		er I	er II	s
Me	5	(46)	96.0:4.0	89.5:10.5	60
t-Bu	6	(42)	94.5:5.5	82.5:17.5	35
Ph	4	(43)	98.0:2.0	86.5:13.5	108
4-MeOC6H4	4	(46)	95.5:4.5	89.5:10.5	50

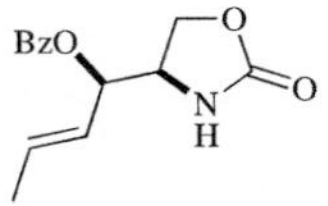

er 92.5:7.5   +   er 80.5:19.5

(42), s = 26   277

(i-PrCO)₂O (0.75 eq), 124 (4 mol %), i-Pr₂NEt (0.75 eq), Na₂SO₄, CHCl₃, 0°

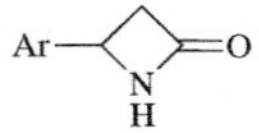

(i-PrCO)₂O (2 eq), ent-117 (10 mol %), i-Pr₂NEt (2 eq), t-AmOH, 0°

Ar	Time (h)		er I	er II	s
2-thienyl	30	(59)	82.5:17.5	97.5:2.5	17
Ph	24	(43)	92.0:8.0	81.5:18.5	22
4-MeOC6H4	30	(42)	95.0:5.0	83.0:17.0	38
2-ClC6H4	30	(14)	74.0:26.0	54.0:46.0	3
4-ClC6H4	30	(54)	90.0:10.0	97.0:3.0	30
1-Np	72	(41)	91.5:8.5	78.5:21.5	19
2-Np	30	(53)	92.5:7.5	98.5:1.5	54

278

C₇₋₁₃

TABLE 5A. KINETIC RESOLUTION OF LACTAMS AND THIOLACTAMS (*Continued*)

Lactam or Thiolactam	Conditions	Product(s) and Conversion(s) (%)	Refs.

*Please refer to the charts preceding the tables for the structures indicated by the **bold** numbers.*

C_{9-15}

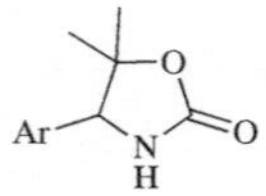

(i-PrCO)$_2$O (1.5 eq), **124** (8 mol %), i-Pr$_2$NEt (0.75 eq), Na_2SO_4, $CHCl_3$, rt

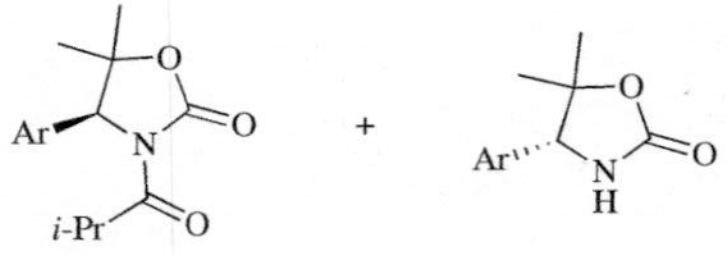

I + **II**

276

Ar	Time (h)		er **I**	er **II**	*s*
2-furyl	7	(43)	97.5:2.5	85.5:14.5	88
2-thienyl	7	(50)	98.5:1.5	99.5:0.5	390
Ph	12	(33)	99.5:0.5	74.0:26.0	340
1-Np	7	(45)	99.5:0.5	90.0:10.0	520
2-Np	9	(37)	99.0:1.0	78.5:21.5	200

C_9

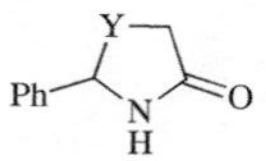

(i-PrCO)$_2$O (0.75 eq), **124** (4 mol %), i-Pr$_2$NEt (0.75 eq), Na_2SO_4, $CHCl_3$, rt

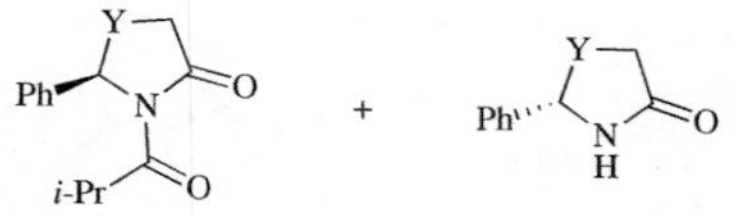

I + **II**

52

Y	Time (min)		er **I**	er **II**	*s*
O	15	(41)	98.5:1.5	83.0:17.0	131
BnO_2CN	15	(42)	98.5:1.5	85.5:14.5	151
S	45	(47)	99.0:1.0	93.5:6.5	294

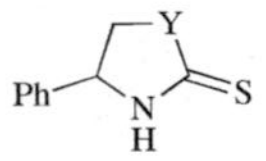

(i-PrCO)₂O (0.75 eq), **124** (4 mol %), i-Pr₂NEt (0.75 eq), Na₂SO₄, CHCl₃, rt

52

I + **II**

Y	Time (h)		er **I**	er **II**	s
O	1	(49)	96.5:3.5	95.0:5.0	82
S	4	(47)	93.0:7.0	87.5:12.5	30

(i-PrCO)₂O (0.75 eq), **124** (4 mol %), i-Pr₂NEt (0.75 eq), Na₂SO₄, CHCl₃, rt, 6 h

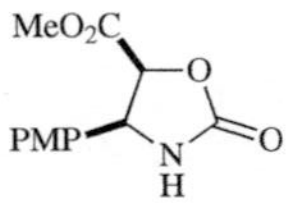

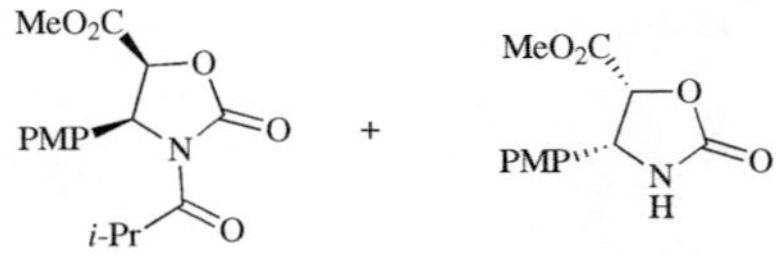

er 96.5:3.5 + er 98.0:2.0

(51), s = 110   276

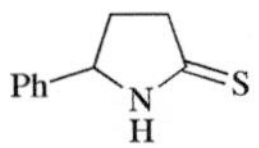

(i-PrCO)₂O (0.75 eq), **124** (8 mol %), i-Pr₂NEt (0.75 eq), Na₂SO₄, CHCl₃, rt, 4 h

er 98.5:1.5 + er 88.0:12.0

(44), s = 141   52

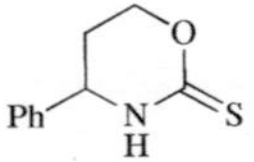

(i-PrCO)₂O (0.75 eq), **124** (10 mol %), i-Pr₂NEt (0.75 eq), Na₂SO₄, CHCl₃, rt, 24 h

er 54.5:45.5 +

(14), s = 1   52

C₁₀

487

TABLE 5A. KINETIC RESOLUTION OF LACTAMS AND THIOLACTAMS (*Continued*)

Lactam or Thiolactam	Conditions	Product(s) and Conversion(s) (%)	Refs.

*Please refer to the charts preceding the tables for the structures indicated by the **bold** numbers.*

C$_{10}$

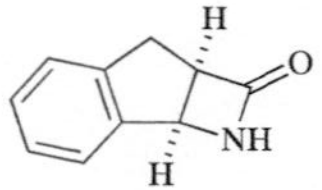

(*i*-PrCO)$_2$O (2 eq), *ent*-**117** (10 mol %),

i-Pr$_2$NEt (2 eq), *t*-AmOH, 0°, 30 h

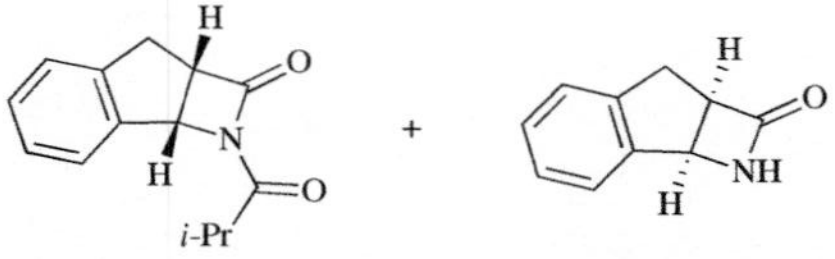

er 89.0:11.0   +   er 78.0:22.0   (42), *s* = 14   278

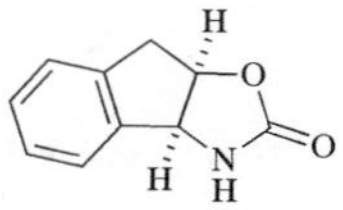

(*i*-PrCO)$_2$O (0.75 eq), **117** (10 mol %),

i-Pr$_2$NEt (0.75 eq), *t*-AmOH, 0°, 30 h

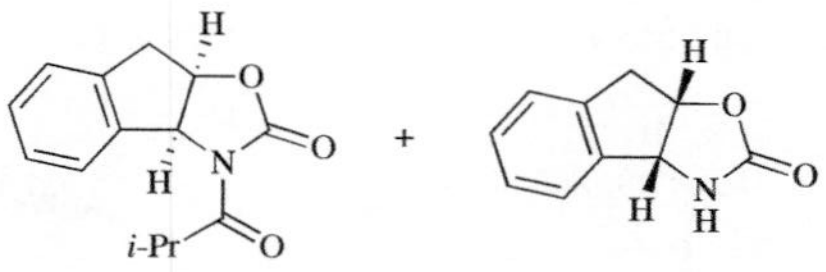

er 95.5:4.5   +   er 75.0:25.0   (36), *s* = 36   52

C$_{11}$

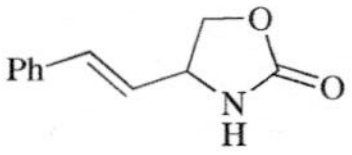

(*i*-PrCO)$_2$O (0.75 eq), *ent*-**117** (4 mol %),

i-Pr$_2$NEt (0.75 eq), *t*-AmOH, 0°, 24 h

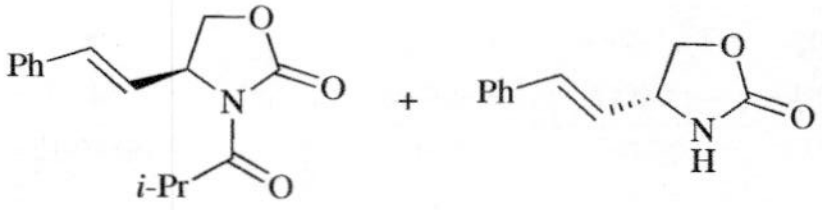

er 97.5:2.5   +   er 86.5:13.5   (43), *s* = 92   52

C$_{13}$

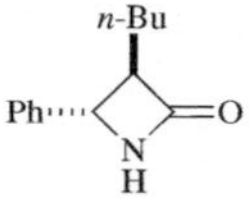

(*i*-PrCO)$_2$O (4 eq), **117** (10 mol %),

i-Pr$_2$NEt (4 eq), *t*-AmOH, 0°, 72 h

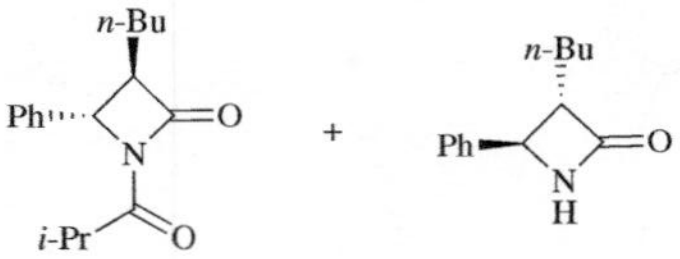

er 91.0:9.0   +   er 74.5:25.5   (38), *s* = 16   278

C15

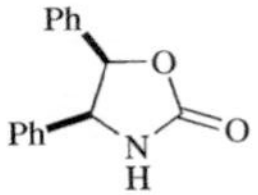

(i-PrCO)2O (0.75 eq), **124** (4 mol %), i-Pr2NEt (0.75 eq), CHCl3, rt, 6 h

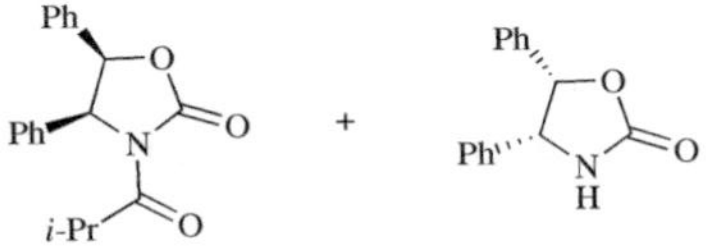

er 99.5:0.5

+

er 86.0:14.0

(43), s = 300

276

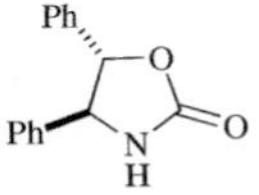

C15

(i-PrCO)2O (0.75 eq), **124** (4 mol %), i-Pr2NEt (0.75 eq), Na2SO4, CHCl3, rt, 6 h

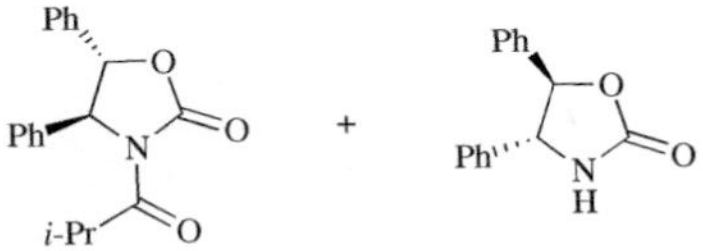

er 96.0:4.0

+

er 86.0:14.0

(44), s = 50

276

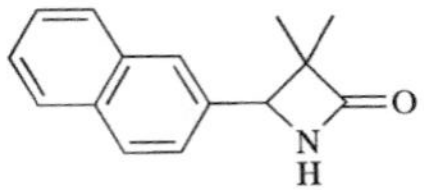

(i-PrCO)2O (4 eq), *ent*-**117** (10 mol %), i-Pr2NEt (4 eq), t-AmOH, 0°, 72 h

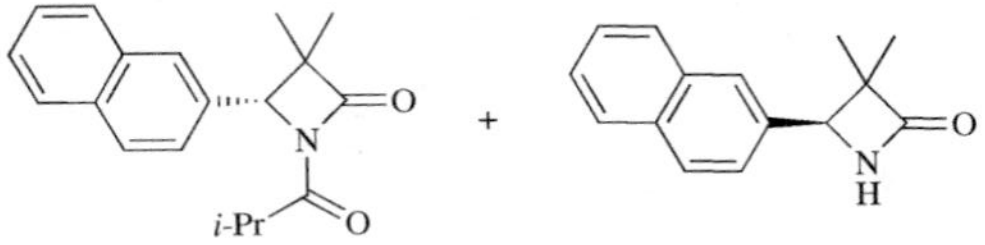

er 90.5:9.5

+

er 64.5:35.5

(26), s = 13

278

C18

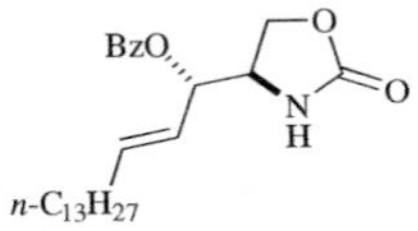

(i-PrCO)2O (0.75 eq), **124** (4 mol %), i-Pr2NEt (0.75 eq), Na2SO4, CHCl3, 0°

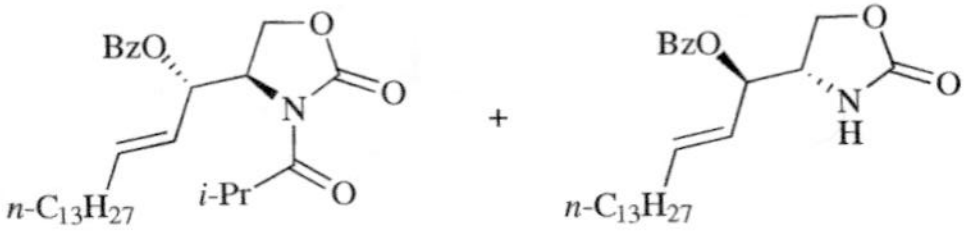

er 97.5:2.5

+

er 93.5:6.5

(46), s = 117

277

TABLE 5B. KINETIC RESOLUTION OF THIOAMIDES

*Please refer to the charts preceding the tables for the structures indicated by the **bold** numbers.*

	Thioamide	Conditions	Product(s) and Conversion(s) (%)	Refs.
C$_8$		(i-PrCO)$_2$O (0.75 eq), **124** (10 mol %), i-Pr$_2$NEt (0.75 eq), Na$_2$SO$_4$, CHCl$_3$, rt, 48 h	+ er 54.5:45.5 (24), s = 1	52
C$_{8-14}$		Boc$_2$O (0.6 eq), **95** (5 mol %), CHCl$_3$, rt, 24 h	**I** + **II**	279

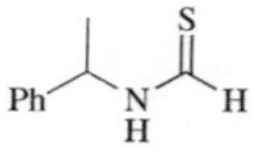

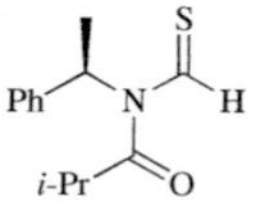

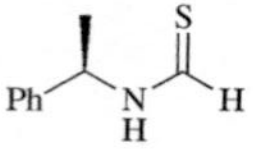

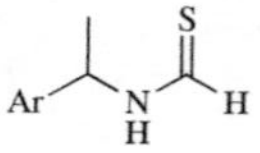

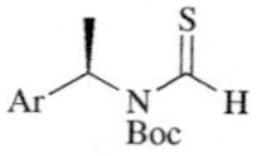

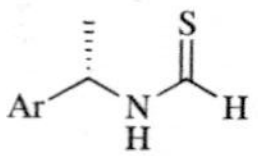

Ar		er **I**	s
Ph	(51)	86.5:13.5	13
3-MeOC$_6$H$_4$	(52)	95.5:4.5	33
4-MeOC$_6$H$_4$	(52)	88.0:12.0	12
3,5-(MeO)$_2$C$_6$H$_3$	(52)	96.5:3.5	44
3-PhOC$_6$H$_4$	(58)	97.5:2.5	25
3-BrC$_6$H$_4$	(52)	88.0:12.0	13
2-Np	(54)	91.0:9.0	14
4-PhC$_6$H$_4$	(53)	86.5:13.5	10

C$_9$

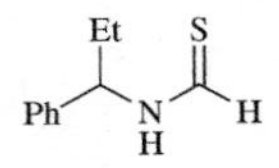

Boc$_2$O (0.6 eq),
95 (5 mol %),
CHCl$_3$, rt, 24 h

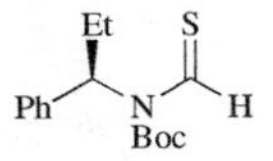

er 85.0:15.0

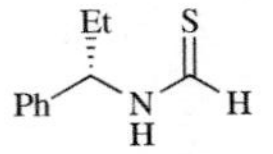

+

(52), s = 9

279

C$_{12}$

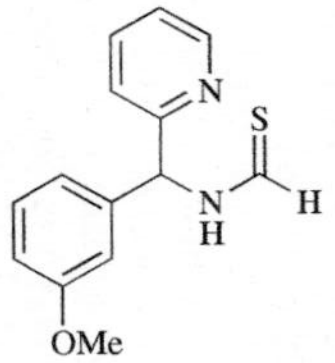

Boc$_2$O (0.65 eq),
95 (5 mol %),
CHCl$_3$, rt, 24 h

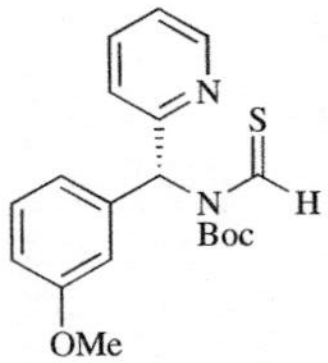

er 79.0:21.0

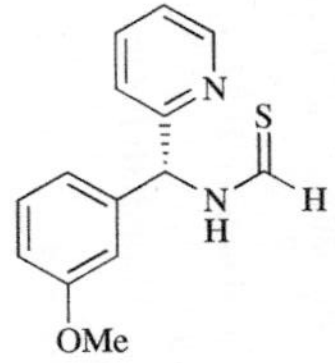

+

(53), s = 6

279

REFERENCES

[1] Kagan, H. B.; Fiaud, J.-C. *Kinetic Resolution*; Wiley: New York, 1988; Vol. 18.
[2] Spivey, A. C.; Maddaford, A.; Redgrave, A. J. *Org. Prep. Proced. Int.* **2000**, *32*, 331.
[3] Robinson, D. E. J. E.; Bull, S. D. *Tetrahedron: Asymmetry* **2003**, *14*, 1407.
[4] Spivey, A. C.; McDaid, P. In *Enantioselective Organocatalysis*; Dalco, P. I., Ed.; Wiley-VCH: Weinheim, 2007; pp 287–329.
[5] Spivey, A. C.; Arseniyadis, S. *Top. Curr. Chem.* **2006**, *291*, 233.
[6] Müller, C. E.; Schreiner, P. R. *Angew. Chem., Int. Ed.* **2011**, *50*, 6012.
[7] Mandai, H.; Fujii, K.; Suga, S. *Tetrahedron Lett.* **2018**, *59*, 1787.
[8] De Risi, C.; Bortolini, O.; Di Carmine, G.; Ragno, D.; Massi, A. *Synthesis* **2019**, *51*, 1871.
[9] Kagan, H. B. *Tetrahedron* **2001**, *57*, 2449.
[10] Greenhalgh, M. D.; Taylor, J. E.; Smith, A. D. *Tetrahedron* **2018**, *74*, 5554.
[11] Ismagilov, R. F. *J. Org. Chem.* **1998**, *63*, 3772.
[12] Blackmond, D. G. *J. Am. Chem. Soc.* **2001**, *123*, 545.
[13] Enriquez-Garcia, A.; Kündig, E. P. *Chem. Soc. Rev.* **2012**, *41*, 7803.
[14] Suzuki, T. *Tetrahedron Lett.* **2017**, *58*, 4731.
[15] Ghanem, A.; Aboul-Enein, H. Y. *Tetrahedron: Asymmetry* **2004**, *15*, 3331.
[16] Sih, C. J.; Wu, S.-H. *Top. Stereochem.* **2007**, *19*, 63.
[17] Williams, J. M. J.; Parker, R. J.; Neri, C. In *Enzyme Catalysis in Organic Synthesis*, 2nd Ed.; Drauz, K., Waldmann, H.; Wiley-VCH: 2008; pp 287–312.
[18] Dehli, J. R.; Gotor, V. *Chem. Soc. Rev.* **2002**, *31*, 365.
[19] Pellissier, H. *Tetrahedron* **2003**, *59*, 8291.
[20] Pellissier, H. *Tetrahedron* **2016**, *72*, 3133.
[21] Pellissier, H. *Tetrahedron* **2011**, *67*, 3769.
[22] Pellissier, H. *Tetrahedron* **2008**, *64*, 1563.
[23] Zhang, Z. J.; Mao, J. C.; Wan, B. S.; Chen, H. L. *Progress in Chemistry* **2004**, *16*, 574.
[24] Huerta, F. F.; Minidis, A. B. E.; Bäckvall, J. E. *Chem. Soc. Rev.* **2001**, *30*, 321.
[25] Yamada, S.; Katsumata, H. *Chem. Lett.* **1998**, 995.
[26] Yamada, S.; Katsumata, H. *J. Org. Chem.* **1999**, *64*, 9365.
[27] Ishihara, K.; Kubota, M.; Yamamoto, H. *Synlett* **1994**, 611.
[28] Leclercq, L.; Suisse, I.; Agbossou-Niedercorn, F. *Eur. J. Org. Chem.* **2010**, 2696.
[29] Sabot, C.; Subhash, P. V.; Valleix, A.; Arseniyadis, S.; Mioskowski, C. *Synlett* **2008**, 268.
[30] Arseniyadis, S.; Valleix, A.; Wagner, A.; Mioskowski, C. *Angew. Chem., Int. Ed.* **2004**, *43*, 3314.
[31] De, C. K.; Klauber, E. G.; Seidel, D. *J. Am. Chem. Soc.* **2009**, *131*, 17060.
[32] De, C. K.; Seidel, D. *J. Am. Chem. Soc.* **2011**, *133*, 14538.
[33] Arai, S.; Bellemin-Laponnaz, S.; Fu, G. C. *Angew. Chem., Int. Ed.* **2001**, *40*, 234.
[34] Binanzer, M.; Hsieh, S.-Y.; Bode, J. W. *J. Am. Chem. Soc.* **2011**, *133*, 19698.
[35] Höfle, G.; Steglich, W.; Vorbrüggen, H. *Angew. Chem., Int. Ed.* **1978**, *17*, 569.
[36] Spivey, A. C.; Arseniyadis, S. *Angew. Chem., Int. Ed.* **2004**, *43*, 5436.
[37] Lutz, V.; Glatthaar, J.; Würtele, C.; Serafin, M.; Hausmann, H.; Schreiner, P. R. *Chem.—Eur. J.* **2009**, *15*, 8548.
[38] Xu, S. J.; Held, I.; Kempf, B.; Mayr, H.; Steglich, W.; Zipse, H. *Chem.—Eur. J.* **2005**, *11*, 4751.
[39] Dalaigh, C. O.; Hynes, S. J.; O'Brien, J. E.; McCabe, T.; Maher, D. J.; Watson, G. W.; Connon, S. J. *Org. Biomol. Chem.* **2006**, *4*, 2785.
[40] Wei, Y.; Held, I.; Zipse, H. *Org. Biomol. Chem.* **2006**, *4*, 4223.
[41] Yamada, S.; Misono, T.; Iwai, Y.; Masumizu, A.; Akiyama, Y. *J. Org. Chem.* **2006**, *71*, 6872.
[42] Crittall, M.; Rzepa, H. S.; Carbery, D. R. *Org. Lett.* **2011**, *13*, 1250.
[43] Mesas-Sanchez, L.; Diner, P. *Chem.—Eur. J.* **2015**, *21*, 5623.
[44] Spivey, A. C.; Zhu, F. J.; Mitchell, M. B.; Davey, S. G.; Jarvest, R. L. *J. Org. Chem.* **2003**, *68*, 7379.
[45] Larionov, E.; Mahesh, M.; Spivey, A. C.; Wei, Y.; Zipse, H. *J. Am. Chem. Soc.* **2012**, *134*, 9390.
[46] Taylor, J. E.; Bull, S. D.; Williams, J. M. J. *Chem. Soc. Rev.* **2012**, *41*, 2109.
[47] Birman, V. B.; Uffman, E. W.; Hui, J.; Li, X. M.; Kilbane, C. J. *J. Am. Chem. Soc.* **2004**, *126*, 12226.
[48] Li, X.; Liu, P.; Houk, K. N.; Birman, V. B. *J. Am. Chem. Soc.* **2008**, *130*, 13836.
[49] Hu, B.; Meng, M.; Wang, Z.; Du, W.-T.; Fossey, J. S.; Hu, X.-Q.; Deng, W.-P. *J. Am. Chem. Soc.* **2010**, *132*, 17041.
[50] Nakata, K.; Shiina, I. *Org. Biomol. Chem.* **2011**, *9*, 7092.
[51] Birman, V. B.; Jiang, H. *Org. Lett.* **2005**, *7*, 3445.
[52] Yang, X.; Bumbu, V. D.; Liu, P.; Li, X.; Jiang, H.; Uffman, E. W.; Guo, L.; Zhang, W.; Jiang, X.; Houk, K. N.; Birman, V. B. *J. Am. Chem. Soc.* **2012**, *134*, 17605.

53 Jiang, S.-S.; Xu, Q.-C.; Zhu, M.-Y.; Yu, X.; Deng, W.-P. *J. Org. Chem.* **2015**, *80*, 3159.
54 Wagner, A. J.; Rychnovsky, S. D. *Org. Lett.* **2013**, *15*, 5504.
55 Shiina, I.; Nakata, K.; Sugimoto, M.; Onda, Y.-s.; Iizumi, T.; Ono, K. *Heterocycles* **2009**, *77*, 801.
56 Liu, P.; Yang, X.; Birman, V. B.; Houk, K. N. *Org. Lett.* **2012**, *14*, 3288.
57 Li, X.; Jiang, H.; Uffman, E. W.; Guo, L.; Zhang, Y.; Yang, X.; Birman, V. B. *J. Org. Chem.* **2012**, *77*, 1722.
58 Shiina, I.; Nakata, K.; Ono, K.; Mukaiyama, T. *Helv. Chim. Acta* **2012**, *95*, 1891.
59 Shiina, I.; Ono, K.; Nakata, K. *Chem. Lett.* **2011**, *40*, 147.
60 Belmessieri, D.; Joannesse, C.; Woods, P. A.; MacGregor, C.; Jones, C.; Campbell, C. D.; Johnston, C. P.; Duguet, N.; Concellon, C.; Bragg, R. A.; Smith, A. D. *Org. Biomol. Chem.* **2011**, *9*, 559.
61 Nakata, K.; Sekiguchi, A.; Shiina, I. *Tetrahedron: Asymmetry* **2011**, *22*, 1610.
62 Shiina, I.; Nakata, K.; Ono, K.; Sugimoto, M.; Sekiguchi, A. *Chem.—Eur. J.* **2010**, *16*, 167.
63 Nakata, K.; Gotoh, K.; Ono, K.; Futami, K.; Shiina, I. *Org. Lett.* **2013**, *15*, 1170.
64 Shiina, I.; Ono, K.; Nakahara, T. *Chem. Commun.* **2013**, *49*, 10700.
65 Copeland, G. T.; Jarvo, E. R.; Miller, S. J. *J. Org. Chem.* **1998**, *63*, 6784.
66 Vasbinder, M. M.; Jarvo, E. R.; Miller, S. J. *Angew. Chem., Int. Ed.* **2001**, *40*, 2824.
67 Chen, P.; Qu, J. *J. Org. Chem.* **2011**, *76*, 2994.
68 Copeland, G. T.; Miller, S. J. *J. Am. Chem. Soc.* **2001**, *123*, 6496.
69 Fierman, M. B.; O'Leary, D. J.; Steinmetz, W. E.; Miller, S. J. *J. Am. Chem. Soc.* **2004**, *126*, 6967.
70 Ishihara, K.; Kosugi, Y.; Akakura, M. *J. Am. Chem. Soc.* **2004**, *126*, 12212.
71 Kosugi, Y.; Akakura, M.; Ishihara, K. *Tetrahedron* **2007**, *63*, 6191.
72 Sakakura, A.; Umemura, S.; Ishihara, K. *Adv. Synth. Catal.* **2011**, *353*, 1938.
73 Müller, C. E.; Zell, D.; Hrdina, R.; Wende, R. C.; Wanka, L.; Schuler, S. M. M.; Schreiner, P. R. *J. Org. Chem.* **2013**, *78*, 8465.
74 Mueller, C. E.; Wanka, L.; Jewell, K.; Schreiner, P. R. *Angew. Chem., Int. Ed.* **2008**, *47*, 6180.
75 Shinisha, C. B.; Sunoj, R. B. *Org. Lett.* **2009**, *11*, 3242.
76 Kano, T.; Sasaki, K.; Maruoka, K. *Org. Lett.* **2005**, *7*, 1347.
77 Suzuki, Y.; Yamauchi, K.; Muramatsu, K.; Sato, M. *Chem. Commun.* **2004**, 2770.
78 Chan, A.; Scheidt, K. A. *Org. Lett.* **2005**, *7*, 905.
79 Kuwano, S.; Harada, S.; Kang, B.; Oriez, R.; Yamaoka, Y.; Takasu, K.; Yamada, K.-i. *J. Am. Chem. Soc.* **2013**, *135*, 11485.
80 Maki, B. E.; Chan, A.; Phillips, E. M.; Scheidt, K. A. *Org. Lett.* **2006**, *9*, 371.
81 De, S. S.; Biswas, A.; Song, C. H.; Studer, A. *Synthesis* **2011**, 1974.
82 Iwahana, S.; Iida, H.; Yashima, E. *Chem.—Eur. J.* **2011**, *17*, 8009.
83 Lu, S.; Poh, S. B.; Siau, W.-Y.; Zhao, Y. *Angew. Chem., Int. Ed.* **2013**, *52*, 1731.
84 Lu, S.; Poh, S. B.; Zhao, Y. *Angew. Chem., Int. Ed.* **2014**, *53*, 11041.
85 Hsieh, S.-Y.; Binanzer, M.; Kreituss, I.; Bode, J. W. *Chem. Commun.* **2012**, *48*, 8892.
86 Oriyama, T.; Imai, K.; Sano, T.; Hosoya, T. *Tetrahedron Lett.* **1998**, *39*, 3529.
87 Vedejs, E.; Daugulis, O.; Diver, S. T. *J. Org. Chem.* **1996**, *61*, 430.
88 Vedejs, E.; Daugulis, O.; Harper, L. A.; MacKay, J. A.; Powell, D. R. *J. Org. Chem.* **2003**, *68*, 5020.
89 MacKay, J. A.; Vedejs, E. *J. Org. Chem.* **2006**, *71*, 498.
90 Trost, B. M.; Malhotra, S.; Mino, T.; Rajapaksa, N. S. *Chem.—Eur. J.* **2008**, *14*, 7648.
91 Trost, B. M.; Mino, T. *J. Am. Chem. Soc.* **2003**, *125*, 2410.
92 Lee, J. Y.; You, Y. S.; Kang, S. H. *J. Am. Chem. Soc.* **2011**, *133*, 1772.
93 Hong, M. S.; Kim, T. W.; Jung, B.; Kang, S. H. *Chem.—Eur. J.* **2008**, *14*, 3290.
94 You, Y. S.; Kim, T. W.; Kang, S. H. *Chem. Commun.* **2013**, *49*, 9669.
95 Matsumura, Y.; Maki, T.; Tsurumaki, K.; Onomura, O. *Tetrahedron Lett.* **2004**, *45*, 9131.
96 Jammi, S.; Rout, L.; Saha, P.; Akkilagunta, V. K.; Sanyasi, S.; Punniyamurthy, T. *Inorg. Chem.* **2008**, *47*, 5093.
97 Kurono, N.; Kondo, T.; Wakabayashi, M.; Ooka, H.; Inoue, T.; Tachikawa, H.; Ohkuma, T. *Chem.—Asian J.* **2008**, *3*, 1289.
98 Moriyama, A.; Matsumura, S.; Kuriyama, M.; Onomura, O. *Tetrahedron: Asymmetry* **2010**, *21*, 810.
99 Kurono, N.; Ohtsuga, K.; Wakabayashi, M.; Kondo, T.; Ooka, H.; Ohkuma, T. *J. Org. Chem.* **2011**, *76*, 10312.
100 Iwasaki, F.; Maki, T.; Nakashima, W.; Onomura, O.; Matsumura, Y. *Org. Lett.* **1999**, *1*, 969.
101 Lin, M.-H.; RajanBabu, T. V. *Org. Lett.* **2002**, *4*, 1607.
102 Sanan, T. T.; RajanBabu, T. V.; Hadad, C. M. *J. Org. Chem.* **2010**, *75*, 2369.
103 Zhang, Y.; Liu, X.; Zhou, L.; Wu, W.; Huang, T.; Liao, Y.; Lin, L.; Feng, X. *Chem.—Eur. J.* **2014**, *20*, 15884.

[104] Harada, S.; Kuwano, S.; Yamaoka, Y.; Yamada, K.-i.; Takasu, K. *Angew. Chem., Int. Ed.* **2013**, *52*, 10227.

[105] Mittal, N.; Lippert, K. M.; De, C. K.; Klauber, E. G.; Emge, T. J.; Schreiner, P. R.; Seidel, D. *J. Am. Chem. Soc.* **2015**, *137*, 5748.

[106] Mittal, N.; Sun, D. X.; Seidel, D. *Org. Lett.* **2012**, *14*, 3084.

[107] Bode, S. E.; Wolberg, M.; Müller, M. *Synthesis* **2006**, 557.

[108] Diaz-de-Villegas, M. D.; Galvez, J. A.; Badorrey, R.; Lopez-Ram-de-Viu, M. P. *Chem.—Eur. J.* **2012**, *18*, 13920.

[109] Garcia-Urdiales, E.; Alfonso, I.; Gotor, V. *Chem. Rev.* **2005**, *105*, 313.

[110] Hudlicky, T.; Reed, J. W. *Chem. Soc. Rev.* **2009**, *38*, 3117.

[111] Willis, M. C. *J. Chem. Soc., Perkin Trans. 1* **1999**, 1765.

[112] Mukaiyama, T.; Tomioka, I.; Shimizu, M. *Chem. Lett.* **1984**, 49.

[113] Vedejs, E.; Daugulis, O.; Tuttle, N. *J. Org. Chem.* **2004**, *69*, 1389.

[114] Mizuta, S.; Sadamori, M.; Fujimoto, T.; Yamamoto, I. *Angew. Chem., Int. Ed.* **2003**, *42*, 3383.

[115] Aida, H.; Mori, K.; Yamaguchi, Y.; Mizuta, S.; Moriyama, T.; Yamamoto, I.; Fujimoto, T. *Org. Lett.* **2012**, *14*, 812.

[116] Oriyama, T.; Imai, K.; Hosoya, T.; Sano, T. *Tetrahedron Lett.* **1998**, *39*, 397.

[117] Terakado, D.; Oriyama, T. *Org. Synth.* **2006**, *83*, 70.

[118] Kundig, E. P.; Enriquez Garcia, A.; Lomberget, T.; Perez Garcia, P.; Romanens, P. *Chem. Commun.* **2008**, 3519.

[119] Kawabata, T.; Stragies, R.; Fukaya, T.; Nagaoka, Y.; Schedel, H.; Fuji, K. *Tetrahedron Lett.* **2003**, *44*, 1545.

[120] Schedel, H.; Kan, K.; Ueda, Y.; Mishiro, K.; Yoshida, K.; Furuta, T.; Kawabata, T. *Beilstein J. Org. Chem.* **2012**, *8*, 1778.

[121] Mandai, H.; Yasuhara, H.; Fujii, K.; Shimomura, Y.; Mitsudo, K.; Suga, S. *J. Org. Chem.* **2017**, *82*, 6846.

[122] Müller, C. E.; Hrdina, R.; Wende, R. C.; Schreiner, P. R. *Chem.—Eur. J.* **2011**, *17*, 6309.

[123] Müller, C. E.; Zell, D.; Schreiner, P. R. *Chem.—Eur. J.* **2009**, *15*, 9647.

[124] Maki, B. E.; Chan, A.; Phillips, E. M.; Scheidt, K. A. *Tetrahedron* **2009**, *65*, 3102.

[125] Matsumura, Y.; Maki, T.; Murakami, S.; Onomura, O. *J. Am. Chem. Soc.* **2003**, *125*, 2052.

[126] Muramatsu, W.; William, J. M.; Onomura, O. *J. Org. Chem.* **2012**, *77*, 754.

[127] Hashimoto, Y.; Michimuko, C.; Yamaguchi, K.; Nakajima, M.; Sugiura, M. *J. Org. Chem.* **2019**, *84*, 9313.

[128] Matsumoto, K.; Mitsuda, M.; Ushijima, N.; Demizu, Y.; Onomura, O.; Matsumura, Y. *Tetrahedron Lett.* **2006**, *47*, 8453.

[129] Demizu, Y.; Matsumoto, K.; Onomura, O.; Matsumura, Y. *Tetrahedron Lett.* **2007**, *48*, 7605.

[130] Mazet, C.; Kohler, V.; Pfaltz, A. *Angew. Chem., Int. Ed.* **2005**, *44*, 4888.

[131] Arai, T.; Mizukami, T.; Yanagisawa, A. *Org. Lett.* **2007**, *9*, 1145.

[132] Kałuża, Z.; Bielawski, K.; Ćwiek, R.; Niedziejko, P.; Kaliski, P. *Tetrahedron: Asymmetry* **2013**, *24*, 1435.

[133] Nakamura, D.; Kakiuchi, K.; Koga, K.; Shirai, R. *Org. Lett.* **2006**, *8*, 6139.

[134] Ichikawa, J.; Asami, M.; Mukaiyama, T. *Chem. Lett.* **1984**, 949.

[135] Duhamel, L.; Herman, T. *Tetrahedron Lett.* **1985**, *26*, 3099.

[136] Oriyama, T.; Hosoya, T.; Sano, T. *Heterocycles* **2000**, *52*, 1065.

[137] Mizuta, S.; Tsuzuki, T.; Fujimoto, T.; Yamamoto, I. *Org. Lett.* **2005**, *7*, 3633.

[138] Oriyama, T.; Taguchi, H.; Terakado, D.; Sano, T. *Chem. Lett.* **2002**, 26.

[139] Lewis, C. A.; Sculimbrene, B. R.; Xu, Y. J.; Miller, S. J. *Org. Lett.* **2005**, *7*, 3021.

[140] Wu, Z.; Wang, J. *ACS Catal.* **2017**, *7*, 7647.

[141] Jung, B.; Hong, M. S.; Kang, S. H. *Angew. Chem., Int. Ed.* **2007**, *46*, 2616.

[142] Jung, B.; Kang, S. H. *Proc. Natl. Acad. Sci. U. S. A.* **2007**, *104*, 1471.

[143] Sano, S.; Tsumura, T.; Tanimoto, N.; Honjo, T.; Nakao, M.; Nagao, Y. *Synlett* **2010**, 256.

[144] Honjo, T.; Nakao, M.; Sano, S.; Shiro, M.; Yamaguchi, K.; Sei, Y.; Nagao, Y. *Org. Lett.* **2007**, *9*, 509.

[145] Kundig, E. P.; Lomberget, T.; Bragg, R.; Poulard, C.; Bernardinelli, G. *Chem. Commun.* **2004**, 1548.

[146] Ruble, J. C.; Tweddell, J.; Fu, G. C. *J. Org. Chem.* **1998**, *63*, 2794.

[147] Huang, Z.; Huang, X.; Li, B.; Mou, C.; Yang, S.; Song, B.-A.; Chi, Y. R. *J. Am. Chem. Soc.* **2016**, *138*, 7524.

[148] Lu, S.; Song, X.; Poh, S. B.; Yang, H.; Wong, M. W.; Zhao, Y. *Chem.—Eur. J.* **2017**, *23*, 2275.

[149] Yang, G.; Guo, D.; Meng, D.; Wang, J. *Nature Commun.* **2019**, *10*, 3062.

[150] Lu, S.; Poh, S. B.; Rong, Z.-Q.; Zhao, Y. *Org. Lett.* **2019**, *21*, 6169.

[151] Evans, D. A.; Anderson, J. C.; Taylor, M. K. *Tetrahedron Lett.* **1993**, *34*, 5563.

152 Vedejs, E.; Chen, X. H. *J. Am. Chem. Soc.* **1996**, *118*, 1809.
153 Vedejs, E.; Chen, X. H. *J. Am. Chem. Soc.* **1997**, *119*, 2584.
154 Vedejs, E.; Daugulis, O. *J. Am. Chem. Soc.* **1999**, *121*, 5813.
155 Vedejs, E.; Daugulis, A. *J. Am. Chem. Soc.* **2003**, *125*, 4166.
156 MacKay, J. A.; Vedejs, E. *J. Org. Chem.* **2004**, *69*, 6934.
157 Sano, T.; Imai, K.; Ohashi, K.; Oriyama, T. *Chem. Lett.* **1999**, 265.
158 Oriyama, T.; Hori, Y.; Imai, K.; Sasaki, R. *Tetrahedron Lett.* **1996**, *37*, 8543.
159 Ruble, J. C.; Latham, H. A.; Fu, G. C. *J. Am. Chem. Soc.* **1997**, *119*, 1492.
160 Mesas-Sanchez, L.; Diaz-Alvarez, A. E.; Koukal, P.; Diner, P. *Tetrahedron* **2014**, *70*, 3807.
161 Mesas-Sanchez, L.; Diaz-Alvarez, A. E.; Diner, P. *Tetrahedron* **2013**, *69*, 753.
162 Crittall, M. R.; Fairhurst, N. W. G.; Carbery, D. R. *Chem. Commun.* **2012**, *48*, 11181.
163 Spivey, A. C.; Fekner, T.; Spey, S. E. *J. Org. Chem.* **2000**, *65*, 3154.
164 Spivey, A. C.; Leese, D. P.; Zhu, F. J.; Davey, S. G.; Jarvest, R. L. *Tetrahedron* **2004**, *60*, 4513.
165 Spivey, A. C.; Arseniyadis, S.; Fekner, T.; Maddaford, A.; Leese, D. P. *Tetrahedron* **2006**, *62*, 295.
166 Dalaigh, C. O.; Hynes, S. J.; Maher, D. J.; Connon, S. J. *Org. Biomol. Chem.* **2005**, *3*, 981.
167 O Dalaigh, C.; Connon, S. J. *J. Org. Chem.* **2007**, *72*, 7066.
168 Yamada, S.; Misono, T.; Iwai, Y. *Tetrahedron Lett.* **2005**, *46*, 2239.
169 Ma, G.; Deng, J.; Sibi, M. P. *Angew. Chem., Int. Ed.* **2014**, *53*, 11818.
170 Poisson, T.; Penhoat, M.; Papamicael, C.; Dupas, G.; Dalla, V.; Marsais, F.; Levacher, V. *Synlett* **2005**, 2285.
171 Duffey, T. A.; MacKay, J. A.; Vedejs, E. *J. Org. Chem.* **2010**, *75*, 4674.
172 Mandai, H.; Irie, S.; Mitsudo, K.; Suga, S. *Molecules* **2011**, *16*, 8815.
173 Mandai, H.; Irie, S.; Akehi, M.; Yuri, K.; Yoden, M.; Mitsudo, K.; Suga, S. *Heterocycles* **2013**, *87*, 329.
174 Yazicioglu, E. Y.; Tanyeli, C. *Tetrahedron: Asymmetry* **2012**, *23*, 1694.
175 Naraku, G.; Shimomoto, N.; Hanamoto, T.; Inanaga, J. *Enantiomer* **2000**, *5*, 135.
176 Spivey, A. C.; Maddaford, A.; Fekner, T.; Redgrave, A. J.; Frampton, C. S. *J. Chem. Soc., Perkin Trans. 1* **2000**, 3460.
177 Pelotier, B.; Priem, G.; Campbell, I. B.; Macdonald, S. J. F.; Anson, M. S. *Synlett* **2003**, 679.
178 Priem, G.; Pelotier, B.; Macdonald, S. J. F.; Anson, M. S.; Campbell, I. B. *J. Org. Chem.* **2003**, *68*, 3844.
179 Diez, D.; Gil, M. J.; Moro, R. F.; Garrido, N. M.; Marcos, I. S.; Basabe, P.; Sanz, F.; Broughton, H. B.; Urones, J. G. *Tetrahedron: Asymmetry* **2005**, *16*, 2980.
180 Kawabata, T.; Yamamoto, K.; Momose, Y.; Yoshida, H.; Nagaoka, Y.; Fuji, K. *Chem. Commun.* **2001**, 2700.
181 Sakakura, A.; Umemura, S.; Ishihara, K. *Synlett* **2009**, 1647.
182 Zhang, Z.; Wang, M.; Xie, F.; Sun, H.; Zhang, W. *Adv. Synth. Catal.* **2014**, *356*, 3164.
183 Wang, Z.; Ye, J.; Wu, R.; Liu, Y.-Z.; Fossey, J. S.; Cheng, J.; Deng, W.-P. *Catal. Sci. Technol.* **2014**, *4*, 1909.
184 Hu, X.; Meng, M.; Fossey, J. S.; Mo, W.; Hu, X.; Deng, W.-P. *Chem. Commun.* **2011**, *47*, 10632.
185 Yamada, A.; Nakata, K.; Shiina, I. *Tetrahedron: Asymmetry* **2017**, *28*, 516.
186 Birman, V. B.; Li, X. M. *Org. Lett.* **2006**, *8*, 1351.
187 Shiina, I.; Nakata, K. *Tetrahedron Lett.* **2007**, *48*, 8314.
188 Nakata, K.; Shiina, I. *Heterocycles* **2010**, *80*, 169.
189 Nakata, K.; Ono, K.; Shiina, I. *Heterocycles* **2011**, *82*, 1171.
190 Zhou, H.; Xu, Q.; Chen, P. *Tetrahedron* **2008**, *64*, 6494.
191 Xu, Q.; Zhou, H.; Geng, X.; Chen, P. *Tetrahedron* **2009**, *65*, 2232.
192 Kobayashi, M.; Okamoto, S. *Tetrahedron Lett.* **2006**, *47*, 4347.
193 Birman, V. B.; Li, X. M.; Han, Z. F. *Org. Lett.* **2007**, *9*, 37.
194 Birman, V. B.; Li, X. *Org. Lett.* **2008**, *10*, 1115.
195 Zhang, Y.; Birman, V. B. *Adv. Synth. Catal.* **2009**, *351*, 2525.
196 Neyyappadath, R. M.; Chisholm, R.; Greenhalgh, M. D.; Rodríguez-Escrich, C.; Pericàs, M. A.; Hähner, G.; Smith, A. D. *ACS Catal.* **2018**, *8*, 1067.
197 Hara, N.; Fujisawa, S.; Fujita, M.; Miyazawa, M.; Ochiai, K.; Katsuda, S.; Fujimoto, T. *Tetrahedron* **2018**, *74*, 296.
198 Chandrasekhar, S.; Ramachander, T.; Takhi, M. *Tetrahedron Lett.* **1998**, *39*, 3263.
199 Gissibl, A.; Finn, M. G.; Reiser, O. *Org. Lett.* **2005**, *7*, 2325.
200 Demizu, Y.; Kubo, Y.; Matsumura, Y.; Onomura, O. *Synlett* **2008**, 433.
201 Mandai, H.; Murota, K.; Mitsudo, K.; Suga, S. *Org. Lett.* **2012**, *14*, 3486.

202   Bellemin-Laponnaz, S.; Tweddell, J.; Ruble, J. C.; Breitling, F. M.; Fu, G. C. *Chem. Commun.* **2000**, 1009.

203   Jiang, S.-S.; Gu, B.-Q.; Zhu, M.-Y.; Yu, X.; Deng, W.-P. *Tetrahedron* **2015**, *71*, 1187.

204   Hu, B.; Meng, M.; Jiang, S.; Deng, W. *Chin. J. Chem.* **2012**, *30*, 1289.

205   Musolino, S. F.; Ojo, O. S.; Westwood, N. J.; Taylor, J. E.; Smith, A. D. *Chem.—Eur. J.* **2016**, *22*, 18916.

206   Ojo, O. S.; Nardone, B.; Musolino, S. F.; Neal, A. R.; Wilson, L.; Lebl, T.; Slawin, A. M. Z.; Cordes, D. B.; Taylor, J. E.; Naismith, J. H.; Smith, A. D.; Westwood, N. J. *Org. Biomol. Chem.* **2018**, *16*, 266.

207   Vedejs, E.; MacKay, J. A. *Org. Lett.* **2001**, *3*, 535.

208   Demizu, Y.; Moriyama, A.; Onomura, O. *Tetrahedron Lett.* **2009**, *50*, 5241.

209   Tao, B.; Ruble, J. C.; Hoic, D. A.; Fu, G. C. *J. Am. Chem. Soc.* **1999**, *121*, 5091.

210   Birman, V. B.; Guo, L. *Org. Lett.* **2006**, *8*, 4859.

211   Jarvo, E. R.; Copeland, G. T.; Papaioannou, N.; Bonitatebus, P. J.; Miller, S. J. *J. Am. Chem. Soc.* **1999**, *121*, 11638.

212   Kawamata, Y.; Oriyama, T. *Chem. Lett.* **2010**, *39*, 382.

213   Clapham, B.; Cho, C. W.; Janda, K. D. *J. Org. Chem.* **2001**, *66*, 868.

214   Miyake, Y.; Iwata, T.; Chung, K. G.; Nishibayashi, Y.; Uemura, S. *Chem. Commun.* **2001**, 2584.

215   Mazet, C.; Roseblade, S.; Köhler, V.; Pfaltz, A. *Org. Lett.* **2006**, *8*, 1879.

216   Rasappan, R.; Olbrich, T.; Reiser, O. *Adv. Synth. Catal.* **2009**, *351*, 1961.

217   Carneiro, L.; Silva, A. R.; Lourenço, M. A. O.; Mayoral, A.; Diaz, I.; Ferreira, P. *Eur. J. Inorg. Chem.* **2016**, *2016*, 413.

218   Schätz, A.; Hager, M.; Reiser, O. *Adv. Funct. Mater.* **2009**, *19*, 2109.

219   Schatz, A.; Grass, R. N.; Kainz, Q.; Stark, W. J.; Reiser, O. *Chem. Mat.* **2010**, *22*, 305.

220   Mizuta, S.; Ohtsubo, Y.; Tsuzuki, T.; Fujimoto, T.; Yamamoto, I. *Tetrahedron Lett.* **2006**, *47*, 8227.

221   Hrdina, R.; Mueller, C. E.; Wende, R. C.; Wanka, L.; Schreiner, P. R. *Chem. Commun.* **2012**, *48*, 2498.

222   Hrdina, R.; Mueller, C. E.; Schreiner, P. R. *Chem. Commun.* **2010**, *46*, 2689.

223   Hofmann, C.; Schuler, S. M. M.; Wende, R. C.; Schreiner, P. R. *Chem. Commun.* **2014**, *50*, 1221.

224   Hofmann, C.; Schuemann, J. M.; Schreiner, P. R. *J. Org. Chem.* **2015**, *80*, 1972.

225   Fujii, K.; Mitsudo, K.; Mandai, H.; Suga, S. *Adv. Synth. Catal.* **2017**, *359*, 2778.

226   Harrer, S.; Greenhalgh, M. D.; Neyyappadath, R. M.; Smith, A. D. *Synlett* **2019**, *30*, 1555.

227   Kawabata, T.; Nagato, M.; Takasu, K.; Fuji, K. *J. Am. Chem. Soc.* **1997**, *119*, 3169.

228   Gleeson, O.; Gun'ko, Y. K.; Connon, S. J. *Synlett* **2013**, *24*, 1728.

229   Gleeson, O.; Tekoriute, R.; Gun'ko, Y. K.; Connon, S. J. *Chem.—Eur. J.* **2009**, *15*, 5669.

230   Geng, X.-L.; Wang, J.; Li, G.-X.; Chen, P.; Tian, S.-F.; Qu, J. *J. Org. Chem.* **2008**, *73*, 8558.

231   Merad, J.; Borkar, P.; Caijo, F.; Pons, J.-M.; Parrain, J.-L.; Chuzel, O.; Bressy, C. *Angew. Chem., Int. Ed.* **2017**, *56*, 16052.

232   Qu, S.; Greenhalgh, M. D.; Smith, A. D. *Chem.—Eur. J.* **2019**, *25*, 2816.

233   Bie, J.; Lang, M.; Wang, J. *Org. Lett.* **2018**, *20*, 5866.

234   Murata, T.; Kawanishi, T.; Sekiguchi, A.; Ishikawa, R.; Ono, K.; Nakata, K.; Shiina, I. *Molecules* **2018**, *23*, 2003.

235   Liu, B.; Yan, J.; Huang, R.; Wang, W.; Jin, Z.; Zanoni, G.; Zheng, P.; Yang, S.; Chi, Y. R. *Org. Lett.* **2018**, *20*, 3447.

236   Pelotier, B.; Priem, G.; Macdonald, S. J. F.; Anson, M. S.; Upton, R. J.; Campbell, I. B. *Tetrahedron Lett.* **2005**, *46*, 9005.

237   Burns, A. S.; Wagner, A. J.; Fulton, J. L.; Young, K.; Zakarian, A.; Rychnovsky, S. D. *Org. Lett.* **2017**, *19*, 2953.

238   Mitsuda, M.; Tanaka, T.; Demizu, Y.; Onomura, O.; Matsumura, Y. *Tetrahedron Lett.* **2006**, *47*, 8073.

239   Qabaja, G.; Wilent, J. E.; Benavides, A. R.; Bullard, G. E.; Petersen, K. S. *Org. Lett.* **2013**, *15*, 1266.

240   Jarvo, E. R.; Evans, C. A.; Copeland, G. T.; Miller, S. J. *J. Org. Chem.* **2001**, *66*, 5522.

241   Greenhalgh, M. D.; Smith, S. M.; Walden, D. M.; Taylor, J. E.; Brice, Z.; Robinson, E. R. T.; Fallan, C.; Cordes, D. B.; Slawin, A. M. Z.; Richardson, H. C.; Grove, M. A.; Cheong, P. H.-Y.; Smith, A. D. *Angew. Chem., Int. Ed.* **2018**, *57*, 3200.

242   Guha, N. R.; Neyyappadath, R. M.; Greenhalgh, M. D.; Chisholm, R.; Smith, S. M.; McEvoy, M. L.; Young, C. M.; Rodríguez-Escrich, C.; Pericàs, M. A.; Hähner, G.; Smith, A. D. *Green Chem.* **2018**, *20*, 4537.

243   Rajkumar, S.; He, S.; Yang, X. *Angew. Chem., Int. Ed.* **2019**, *58*, 10315.

244   Chen, P. R.; Zhang, Y.; Zhou, H.; Xu, Q. *Acta Chim. Sin.* (Engl. Ed.) **2010**, *68*, 1431.

245   Sano, T.; Miyata, H.; Oriyama, T. *Enantiomer* **2000**, *5*, 119.

[246] Cozett, R. E.; Venter, G. A.; Gokada, M. R.; Hunter, R. *Org. Biomol. Chem.* **2016**, *14*, 10914.

[247] Mori, K.; Kishi, H.; Akiyama, T. *Synthesis* **2017**, *49*, 365.

[248] Yang, H.; Zheng, W.-H. *Angew. Chem., Int. Ed.* **2019**, *58*, 16177.

[249] Kondo, K.; Kurosaki, T.; Murakami, Y. *Synlett* **1998**, *1998*, 725.

[250] Al-Sehemi, A. G.; Atkinson, R. S.; Fawcett, J.; Russell, D. R. *Tetrahedron Lett.* **2000**, *41*, 2243.

[251] Al-Sehemi, A. G.; Atkinson, R. S.; Fawcett, J.; Russell, D. R. *Chem. Commun.* **2000**, 43.

[252] Al-Sehemi, A. G.; Atkinson, R. S.; Fawcett, J. *J. Chem. Soc., Perkin Trans. 1* **2002**, 257.

[253] Ie, Y.; Fu, G. C. *Chem. Commun.* **2000**, 119.

[254] Arseniyadis, S.; Subhash, P. V.; Valleix, A.; Mathew, S. P.; Blackmond, D. G.; Wagner, A.; Mioskowski, C. *J. Am. Chem. Soc.* **2005**, *127*, 6138.

[255] Arseniyadis, S.; Subhash, P. V.; Valleix, A.; Wagner, A.; Mioskowski, C. *Chem. Commun.* **2005**, 3310.

[256] Karnik, A. V.; Kamath, S. S. *Tetrahedron: Asymmetry* **2008**, *19*, 45.

[257] Min, C.; Mittal, N.; De, C. K.; Seidel, D. *Chem. Commun.* **2012**, *48*, 10853.

[258] Arnold, K.; Davies, B.; Hérault, D.; Whiting, A. *Angew. Chem., Int. Ed.* **2008**, *47*, 2673.

[259] Kolleth, A.; Christoph, S.; Arseniyadis, S.; Cossy, J. *Chem. Commun.* **2012**, *48*, 10511.

[260] Klauber, E. G.; De, C. K.; Shah, T. K.; Seidel, D. *J. Am. Chem. Soc.* **2010**, *132*, 13624.

[261] Kolleth, A.; Cattoen, M.; Arseniyadis, S.; Cossy, J. *Chem. Commun.* **2013**, *49*, 9338.

[262] Klauber, E. G.; Mittal, N.; Shah, T. K.; Seidel, D. *Org. Lett.* **2011**, *13*, 2464.

[263] Charushin, V. N.; Krasnov, V. P.; Levit, G. L.; Korolyova, M. A.; Kodess, M. I.; Chupakhin, O. N.; Kim, M. H.; Lee, H. S.; Park, Y. J.; Kim, K. C. *Tetrahedron: Asymmetry* **1999**, *10*, 2691.

[264] Krasnov, V. P.; Levit, G. L.; Bukrina, I. M.; Andreeva, I. N.; Sadretdinova, L. S.; Korolyova, M. A.; Kodess, M. I.; Charushin, V. N.; Chupakhin, O. N. *Tetrahedron: Asymmetry* **2003**, *14*, 1985.

[265] Krasnov, V. P.; Levit, G. L.; Kodess, M. I.; Charushin, V. N.; Chupakhin, O. N. *Tetrahedron: Asymmetry* **2004**, *15*, 859.

[266] Gruzdev, D. A.; Levit, G. L.; Krasnov, V. P.; Chulakov, E. N.; Sadretdinova, L. S.; Grishakov, A. N.; Ezhikova, M. A.; Kodess, M. I.; Charushin, V. N. *Tetrahedron: Asymmetry* **2010**, *21*, 936.

[267] Gruzdev, D. A.; Levit, G. L.; Krasnov, V. P. *Tetrahedron: Asymmetry* **2012**, *23*, 1640.

[268] Vakarov, S. A.; Gruzdev, D. A.; Chulakov, E. N.; Levit, G. L.; Krasnov, V. P. *Russ. Chem. Bull.* **2019**, *68*, 841.

[269] Al-Sehemi, A. G.; Atkinson, R. S.; Fawcett, J.; Russell, D. R. *Tetrahedron Lett.* **2000**, *41*, 2239.

[270] Kreituss, I.; Murakami, Y.; Binanzer, M.; Bode, J. W. *Angew. Chem., Int. Ed.* **2012**, *51*, 10660.

[271] Kreituss, I.; Chen, K.-Y.; Eitel, S. H.; Adam, J.-M.; Wuitschik, G.; Fettes, A.; Bode, J. W. *Angew. Chem., Int. Ed.* **2016**, *55*, 1553.

[272] Kreituss, I.; Bode, J. W. *Nat. Chem.* **2017**, *9*, 446.

[273] Arp, F. O.; Fu, G. C. *J. Am. Chem. Soc.* **2006**, *128*, 14264.

[274] Karnik, A. V.; Kamath, S. S. *J. Org. Chem.* **2007**, *72*, 7435.

[275] Arseniyadis, S.; Mahesh, M.; McDaid, P.; Hampel, T.; Davey, S. G.; Spivey, A. C. *Collect. Czech. Chem. Commun.* **2011**, *76*, 1239.

[276] Birman, V. B.; Jiang, H.; Li, X.; Guo, L.; Uffman, E. W. *J. Am. Chem. Soc.* **2006**, *128*, 6536.

[277] Guasch, J.; Giménez-Nueno, I.; Funes-Ardoiz, I.; Bernús, M.; Matheu, M. I.; Maseras, F.; Castillón, S.; Díaz, Y. *Chem.—Eur. J.* **2018**, *24*, 4635.

[278] Yang, X.; Bumbu, V. D.; Birman, V. B. *Org. Lett.* **2011**, *13*, 4755.

[279] Fowler, B. S.; Mikochik, P. J.; Miller, S. J. *J. Am. Chem. Soc.* **2010**, *132*, 2870.

[280] Kim, H. C.; Kang, S. H. *Angew. Chem., Int. Ed.* **2009**, *48*, 1827.

[281] Brenna, E.; Caraccia, N.; Fuganti, C.; Fuganti, D.; Grasselli, P. *Tetrahedron: Asymmetry* **1997**, *8*, 3801.

[282] Sinha, S. C.; Barbas, C. F.; Lerner, R. A. *Proc. Natl. Acad. Sci. U. S. A.* **1998**, *95*, 14603.

[283] Birman, V. B.; Jiang, H.; Li, X. *Org. Lett.* **2007**, *9*, 3237.

[284] Francais, A.; Leyva, A.; Etxebarria-Jardi, G.; Ley, S. V. *Org. Lett.* **2010**, *12*, 340.

[285] Français, A.; Leyva-Pérez, A.; Etxebarria-Jardi, G.; Peña, J.; Ley, S. V. *Chem.—Eur. J.* **2011**, *17*, 329.

[286] Papaioannou, N.; Evans, C. A.; Blank, J. T.; Miller, S. J. *Org. Lett.* **2001**, *3*, 2879.

[287] Papaioannou, N.; Blank, J. T.; Miller, S. J. *J. Org. Chem.* **2003**, *68*, 2728.

[288] El Gihani, M. T.; Williams, J. M. J. *Curr. Opin. Chem. Biol.* **1999**, *3*, 11.

[289] Kim, M.-J.; Ahn, Y.; Park, J. *Curr. Opin. Biotechnol.* **2002**, *13*, 578.

[290] Pàmies, O.; Bäckvall, J.-E. *Trends Biotechnol.* **2004**, *22*, 130.

[291] Turner, N. J. *Curr. Opin. Chem. Biol.* **2004**, *8*, 114.

[292] Ahmed, M.; Kelly, T.; Ghanem, A. *Tetrahedron* **2012**, *68*, 6781.

[293] Kim, C.; Park, J.; Kim, M. J. In *Comprehensive Chirality*; Yamamoto, H., Carreira, E., Eds.; Elsevier: Amsterdam, 2012; pp 156–180.

294 Rachwalski, M.; Vermue, N.; Rutjes, F. P. J. T. *Chem. Soc. Rev.* **2013**, *42*, 9268.
295 Verho, O.; Bäckvall, J.-E. *J. Am. Chem. Soc.* **2015**, *137*, 3996.
296 Lee, J. H.; Han, K.; Kim, M.-J.; Park, J. *Eur. J. Org. Chem.* **2010**, *2010*, 999.
297 Lee, S. Y.; Murphy, J. M.; Ukai, A.; Fu, G. C. *J. Am. Chem. Soc.* **2012**, *134*, 15149.
298 Paetzold, J.; Bäckvall, J. E. *J. Am. Chem. Soc.* **2005**, *127*, 17620.
299 Blacker, A. J.; Stirling, M. J.; Page, M. I. *Org. Process Res. Dev.* **2007**, *11*, 642.
300 Gastaldi, S.; Escoubet, S.; Vanthuyne, N.; Gil, G.; Bertrand, M. P. *Org. Lett.* **2007**, *9*, 837.
301 Kim, M.-J.; Kim, W.-H.; Han, K.; Choi, Y. K.; Park, J. *Org. Lett.* **2007**, *9*, 1157.
302 Nechab, M.; Azzi, N.; Vanthuyne, N.; Bertrand, M.; Gastaldi, S.; Gil, G. *J. Org. Chem.* **2007**, *72*, 6918.
303 Choi, Y. K.; Kim, Y.; Han, K.; Park, J.; Kim, M.-J. *J. Org. Chem.* **2009**, *74*, 9543.
304 Truppo, M. D.; Rozzell, J. D.; Turner, N. J. *Org. Process Res. Dev.* **2010**, *14*, 234.
305 Poulhès, F.; Vanthuyne, N.; Bertrand, M. P.; Gastaldi, S.; Gil, G. *J. Org. Chem.* **2011**, *76*, 7281.
306 El Blidi, L.; Nechab, M.; Vanthuyne, N.; Gastaldi, S.; Bertrand, M. P.; Gil, G. *J. Org. Chem.* **2009**, *74*, 2901.
307 Chinnaraja, E.; Arunachalam, R.; Suresh, E.; Sen, S. K.; Natarajan, R.; Subramanian, P. S. *Inorg. Chem.* **2019**, *58*, 4465.
308 Xu, K.; Nakazono, K.; Takata, T. *Chem. Lett.* **2016**, *45*, 1274.
309 Arai, T.; Sakagami, K. *Eur. J. Org. Chem.* **2012**, 1097.
310 Trost, B. M.; Michaelis, D. J.; Malhotra, S. *Org. Lett.* **2013**, *15*, 5274.
311 Li, B.-S.; Wang, Y.; Proctor, R. S. J.; Jin, Z.; Chi, Y. R. *Chem. Commun.* **2016**, *52*, 8313.
312 Kim, H. C.; Youn, J.-H.; Kang, S. H. *Synlett* **2008**, 2526.
313 Merad, J.; Borkar, P.; Bouyon Yenda, T.; Roux, C.; Pons, J.-M.; Parrain, J.-L.; Chuzel, O.; Bressy, C. *Org. Lett.* **2015**, *17*, 2118.
314 Roux, C.; Candy, M.; Pons, J.-M.; Chuzel, O.; Bressy, C. *Angew. Chem., Int. Ed.* **2014**, *53*, 766.
315 Kündig, E. P.; Garcia, A. E.; Lomberget, T.; Bernardinelli, G. *Angew. Chem., Int. Ed.* **2006**, *45*, 98.
316 Kashima, C.; Mizuhara, S.; Miwa, Y.; Yokoyama, Y. *Tetrahedron: Asymmetry* **2002**, *13*, 1713.
317 Matsugi, M.; Hagimoto, Y.; Nojima, M.; Kita, Y. *Org. Process Res. Dev.* **2003**, *7*, 583.
318 Matsubara, J.; Otsubo, K.; Kawano, Y.; Kitano, K.; Ohtani, T.; Yamashita, H.; Morita, S.; Uchida, M. *Heterocycles* **2000**, *52*, 81.
319 Shamoto, K.; Miyazaki, A.; Matsukura, M.; Kobayashi, Y.; Shioiri, T.; Matsugi, M. *Synth. Commun.* **2013**, *43*, 1425.
320 Nguyen, H. V.; Motevalli, M.; Richards, C. J. *Synlett* **2007**, 725.
321 Jeong, K. S.; Kim, S. H.; Park, H. J.; Chang, K. J.; Kim, K. S. *Chem. Lett.* **2002**, 1114.
322 Ishida, Y.; Iwasa, E.; Matsuoka, Y.; Miyauchi, H.; Saigo, K. *Chem. Commun.* **2009**, 3401.
323 Samal, M.; Misek, J.; Stara, I. G.; Stary, I. *Collect. Czech. Chem. Commun.* **2009**, *74*, 1151.
324 Fujii, K.; Mitsudo, K.; Mandai, H.; Suga, S. *Bull. Chem. Soc. Jpn.* **2016**, *89*, 1081.
325 Ruble, J. C.; Fu, G. C. *J. Org. Chem.* **1996**, *61*, 7230.
326 Yamada, A.; Nakata, K. *Tetrahedron Lett.* **2016**, *57*, 4697.
327 Nakata, K.; Tokumaru, T.; Iwamoto, H.; Nishigaichi, Y.; Shiina, I. *Asian J. Org. Chem.* **2013**, *2*, 920.
328 Spivey, A. C.; Maddaford, A.; Leese, D. P.; Redgrave, A. J. *J. Chem. Soc., Perkin Trans. 1* **2001**, 1785.
329 Roy, S.; Chen, K.-F.; Chen, K. *J. Org. Chem.* **2014**, *79*, 8955.
330 Vorogushin, A. V.; Wulff, W. D.; Hansen, H.-J. *Tetrahedron* **2008**, *64*, 949.
331 Kawabata, T.; Stragies, R.; Fukaya, T.; Fuji, K. *Chirality* **2003**, *15*, 71.
332 Durini, M.; Russotto, E.; Pignataro, L.; Reiser, O.; Piarulli, U. *Eur. J. Org. Chem.* **2012**, *2012*, 5451.
333 Kamikawa, K.; Shiromoto, T.; Yamaguchi, J.; Takemoto, S.; Matsuzaka, H. *Res. Chem. Intermed.* **2009**, *35*, 931.
334 Terakado, D.; Koutaka, H.; Oriyama, T. *Tetrahedron: Asymmetry* **2005**, *16*, 1157.
335 Arai, T.; Mizukami, T.; Mishiro, A.; Yanagisawa, A. *Heterocycles* **2008**, *76*, 995.
336 Kida, T.; Iwamoto, T.; Asahara, H.; Hinoue, T.; Akashi, M. *J. Am. Chem. Soc.* **2013**, *135*, 3371.
337 Asahara, H.; Kida, T.; Iwamoto, T.; Hinoue, T.; Akashi, M. *Tetrahedron* **2014**, *70*, 197.

CHAPTER 2

THE PIANCATELLI REACTION

Lucile Marin, Emmanuelle Schulz, David Lebœuf, and Vincent Gandon

Institute for Molecular and Materials Chemistry of Orsay, CNRS UMR 8182, Université Paris-Sud, Université Paris-Saclay, Bâtiment 420, 91405 Orsay cedex, France

Edited by Steven M. Weinreb

CONTENTS

vincent.gandon@u-psud.fr

How to cite: Marin L.; Schulz E.; Lebœuf D.; and Gandon V. The Piancatelli Reaction. *Org. React.* **2020**, *104*, 499–612.

ACKNOWLEDGMENTS

We thank CNRS, Ministère de l'Enseignement Supérieur et de la Recherche, Université Paris-Sud, and Institut Universitaire de France (IUF) for the support of our work in this area.

INTRODUCTION

Cyclopentenones are important intermediates for the synthesis of natural products and unnatural bioactive compounds. Classical methods for their synthesis involve the Pauson–Khand reaction,[1–6] the Nazarov reaction,[7–16] and the Rautenstrauch rearrangement.[17–20] In both the Nazarov and the Rautenstrauch reactions, the formation of the five-membered ring is achieved by a conrotatory 4π-electrocyclization of a transient pentadienyl carbocation. In 1976, Piancatelli and coworkers reported a new method to access 4-hydroxycyclopent-2-enones that also relies on this type of stereoselective mechanistic step (see "Mechanism and Stereochemistry"). Hence, the treatment of 2-furylcarbinols with aqueous acids, such as formic acid, stereoselectively leads to the *trans*-cyclopentenone product (Scheme 1).[21] This process is a rare type of reaction that directly transforms a heterocycle into a carbocycle. The same group later reported that substoichiometric amounts of zinc(II) chloride also promote the same rearrangement under milder conditions, which permits more reactive substrates that are prone to decomposition with Brønsted acids to be employed.[22] Although the original Brønsted or Lewis acid promoted Piancatelli reaction has led to several applications in the synthesis of compounds of biological interest (see "Applications to Synthesis"), recent years have witnessed the development of new variations in which nucleophiles other than water are employed to trap one of the carbocationic intermediates. This approach permits the preparation of a variety of *trans*-4-substituted cyclopent-2-enones and related compounds. Consequently, the combination of anilines as nucleophiles with dysprosium(III) trifluoromethanesulfonate as the catalyst provided the first catalytic aza-Piancatelli reactions.[23] Subsequently, an intramolecular oxa-Piancatelli reaction was also described in which the nucleophile is a hydroxyalkyl substituent on the furan ring. The Piancatelli rearrangement was also extended to an intramolecular carba-Piancatelli reaction using aromatic tertiary amides as nucleophiles.[24] Sulfur nucleophiles have not been reported in the Piancatelli reaction, but an alternative method for preparing mercaptocyclopentenones from furaldehydes is available (see "Comparison with Other Methods").[25]

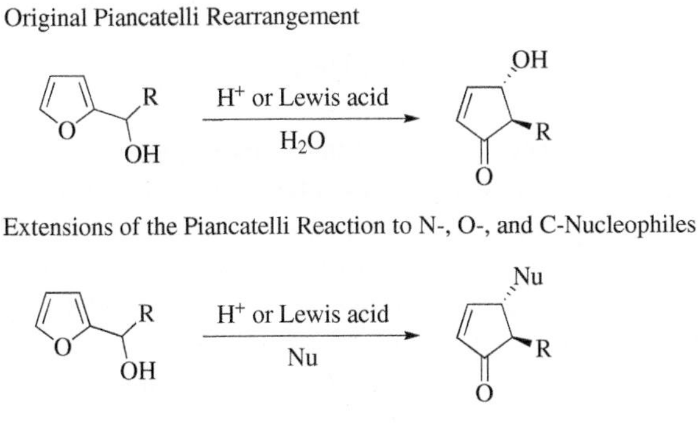

Original Piancatelli Rearrangement

Extensions of the Piancatelli Reaction to N-, O-, and C-Nucleophiles

Scheme 1

These aspects and other applications, such as the incorporation of Piancatelli reactions into cascade processes, are described in this chapter. This topic has also

been the subject of a number of previous reviews.[15,26–34] This chapter aims to criti-
cally examine the use of this chemistry for the rapid construction of complex—and
often enantioenriched—molecules from simple, readily available furylcarbinols. An
exhaustive compilation of Piancatelli reactions covering the period from 1976 to
early 2020 is presented in the Tabular Survey.

MECHANISM AND STEREOCHEMISTRY

Electrocyclic Ring Closure

Original Mechanistic Proposal. The proposed mechanism summarized in
Scheme 2 accounts for the *trans*-stereoselectivity in the rearrangement to form the
4-hydroxycyclopent-2-enone.[21] Protonation of the starting furylcarbinol promotes
the formation of carbocation **1**. Addition of water to resonance structure **2** affords
the protonated cyclic hemiacetal **3**, which undergoes ring opening to compound **4**.
4π-Conrotatory electrocyclization of resonance structure **5** forms the five-membered
ring in compound **6** with exclusive *trans* stereoselectivity, and loss of a proton
delivers 4-hydroxycyclopent-2-enone **7** and regenerates the acid catalyst.

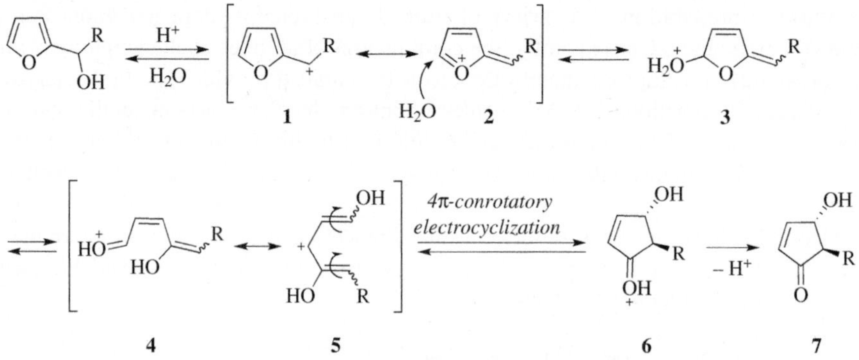

Scheme 2

DFT Computations.   The critical step in the above mechanism, the electrocyclic
ring closure, has been studied by B3LYP/6-311G* calculations using simple five- and
six-carbon model compounds.[35] A comparison of the Natural Bond Orbital (NBO)
analysis with computations on the aromaticity of the transition structures results in
the conclusion that the Piancatelli rearrangements of these model systems are indeed
pericyclic reactions. Similar to the Nazarov reaction,[7–15] cyclizations of *out,out* isomers (Scheme 3) have low energy barriers (ΔG = 6.0 kcal/mol for the specific case
shown in Scheme 3). The *in,in* isomers also lead to *trans*-products, but the corresponding transition state is much higher in energy (24.8 kcal/mol). The unobserved
cis-products that arise from two possible *in,out* isomers are disfavored with transition
states that are at 12.1 and 17.3 kcal/mol on the energy surface.

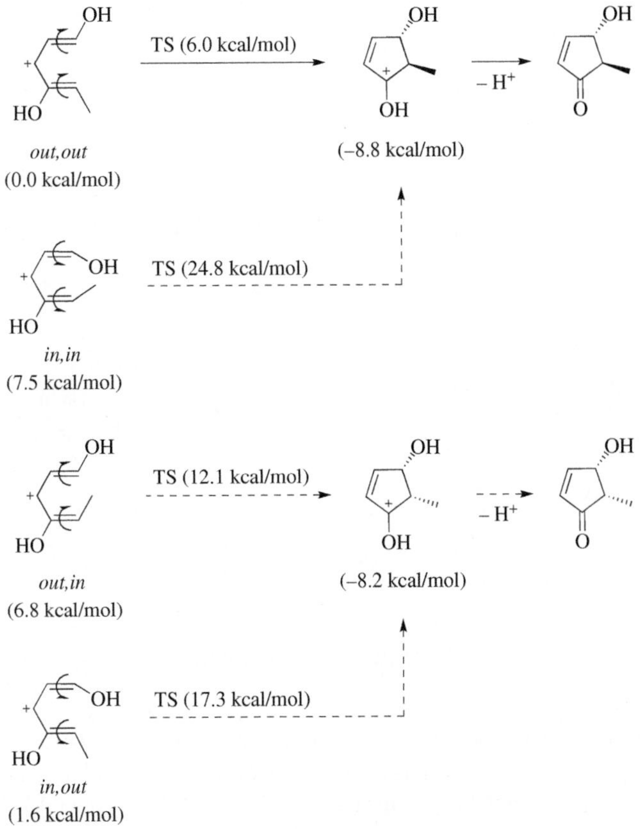

Scheme 3

Mechanism of the Aza-Piancatelli Reaction. The mechanism of the Piancatelli rearrangement shown in Scheme 2 remains mostly valid with nucleophiles other than water. For example, the mechanism of $Dy(OTf)_3$- and TFA-catalyzed aza-Piancatelli reactions with anilines has been studied (Scheme 4).[36,37] In the case involving a Lewis acid (LA), the abstraction of hydroxide affords carbocation **1/2**, which reacts with an aniline to give protonated hemiaminal **8**. Proton shift and ring-opening provide iminium ion **9**, which can be represented as the pentadienyl carbocation **10**. Conrotatory 4π-electrocyclization furnishes *trans*-intermediate **11**, which loses the proton to furnish the final product. This proton transfer results in cleavage of the LA–O bond, regenerating the catalyst and forming water as the only byproduct of the reaction. The rate of the reaction is controlled by the competitive binding of the catalyst with the furylcarbinol or with the aniline, as outlined in Scheme 4. The formation of off-cycle species **12** impacts the rate-determining step; this finding explains why only less basic amines like anilines can be used as nucleophilic reagents in aza-Piancatelli reactions. More basic amines tend to shift the complexation equilibrium toward amine–catalyst adduct **12**.

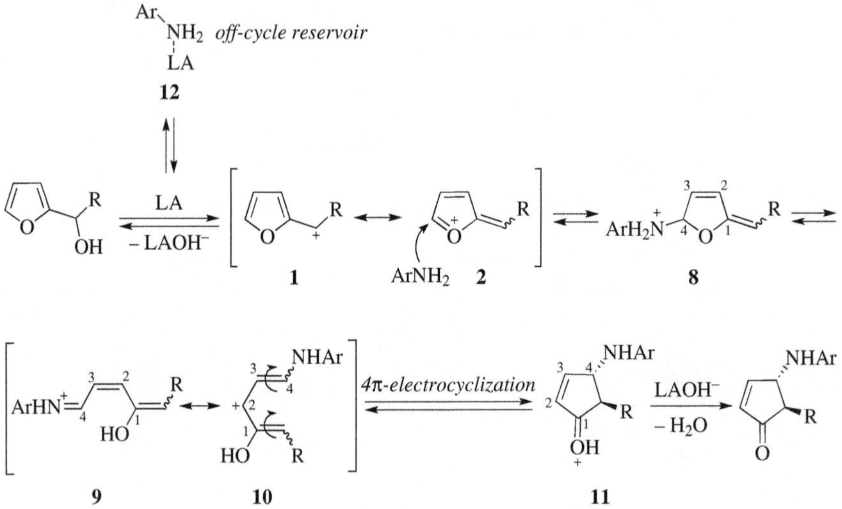

Scheme 4

Alternative Mechanisms

Ionic Stepwise Mechanism. The original mechanistic proposal involving an electrocyclization step is widely accepted; however, an alternative mechanism has been proposed to account for the formation of Piancatelli product **13** (32% yield) when an alkyl-substituted 2-furylcarbinol is heated in refluxing water in the absence of an acid catalyst (Scheme 5).[38] The major component of the mixture is the thermodynamically more stable isomer **14**, which is obtained in 65% yield. The Piancatelli product is obtained as a *cis/trans* mixture, which is at odds with an electrocyclization process that would lead to the *trans* product exclusively. A multistep ionic alternative mechanism has also been suggested but seems unlikely.[38] Products **13** and **14** are probably formed by post facto equilibration following the standard mechanism discussed in the previous section. Alternatively, an aldol-type mechanism could also explain the low selectivity observed for product **13** (see "Aldol-Type Condensation" below).

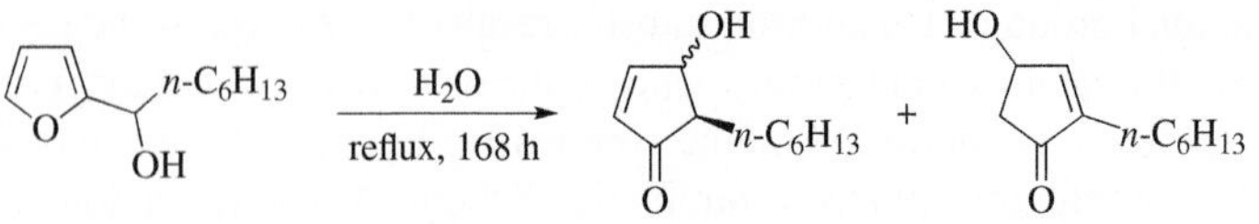

13 (32%)   **14** (65%)

trans/cis = 4:1

Scheme 5

Aldol-Type Condensation. A third type of mechanism was proposed based on the results obtained with 2-furylcarbinols that possess a hydroxyalkyl chain at C5 (Scheme 6).[39] Copper(II) sulfate mediates the formation of spiroketal **16** in high yield from carbinol **15**. In the presence of zinc(II) chloride and water, the spiroketal reacts to generate the fused oxacyclopentenone **17** in 88% yield.[40] This process most likely involves a spiro-cyclopentenone intermediate. The direct transformation of furyl-carbinol **15** into fused oxacyclopentenone **17** can be achieved using zinc(II) chloride in a similar yield of 84%.[41] Importantly, this latter reaction can be performed in anhydrous solvents such as DME or methanol, whereas the transformation of spiroketal **16** into **17** requires at least one equivalent of water, and the addition of more water accelerates the reaction. Spiroketal **16** is most likely an intermediate in the zinc(II) chloride catalyzed direct conversion of **15** into **17**, and because water is generated during the formation of spiroketal **16** from furylcarbinol **15**, an anhydrous solvent can be employed. In contrast, the conversion of ketal **16** into the oxacyclopentenone **17** requires the addition of water. These observations led to the conclusion that water co-catalyzes the isomerization of spiroketal **16** into **17**.

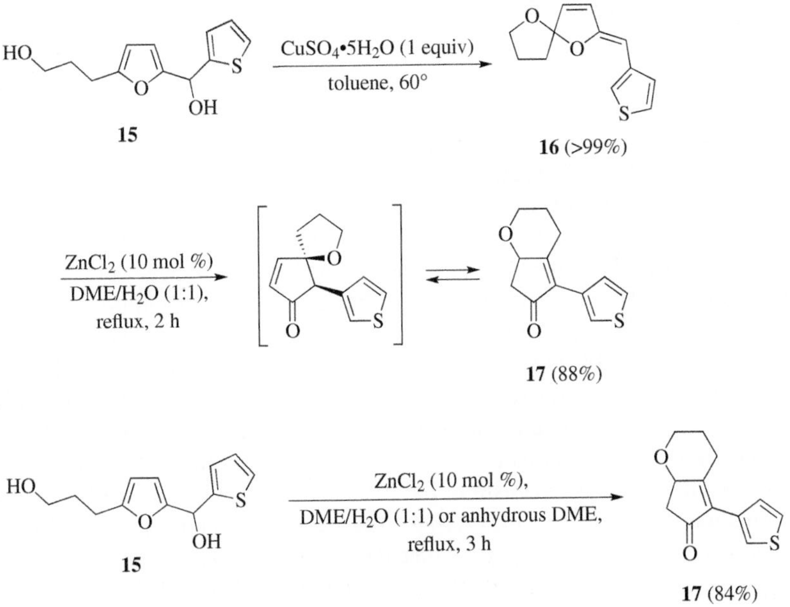

Scheme 6

The mechanism shown in Scheme 7 was proposed for the formation of enone **17**, and more generally, of enone **24**. The process is initiated with the Lewis acid promoted formation of spiroketal **19** from carbinol **18**. Protonation of the latter triggers a ring-opening of the tetrahydrofuran moiety to generate oxocarbenium ion **20**, which is quenched with water to afford the protonated hemiacetal **21** upon proton transfer. Ring-opening of **21** affords the enol **22**, which undergoes an intramolecular aldol reaction. The resulting 4-hydroxycyclopent-2-enone **23** is then transformed into the fused cyclopentenone product **24** via additional steps.

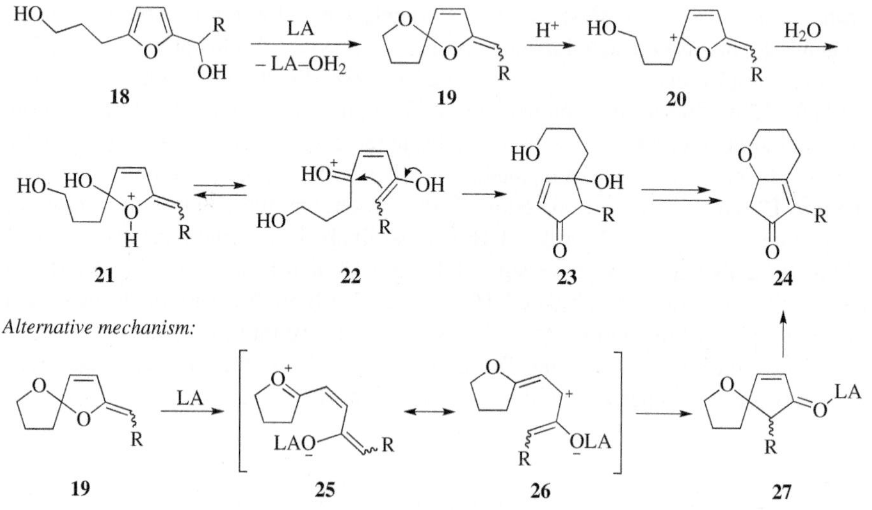

Scheme 7

An alternative route from spiroketal **19** could also account for the formation of cyclopentenone **24**. This sequence involves the cleavage of the spiroketal motif by the Lewis acid to afford the oxocarbenium ion **25**, which is a resonance structure of pentadienyl cation **26**. The cyclization of this intermediate leads to adduct **27**, in which the spirocyclopentenone is coordinated to the Lewis acid. The fused cyclopentenone **24** is then obtained after either an elimination/addition sequence or a 1,2-shift of the tetrahydrofuran oxygen. Since the Lewis acid catalyzed transformation of spiroketal **16** into enone **17** is not possible in an anhydrous solvent, this Lewis acid mediated pathway was not proposed. Instead, a mechanism involving water is preferred, and therefore an aldol-type condensation was suggested over an electrocyclization of the 4π-electron system present in intermediate **22** (Scheme 7). When considering the mechanism of the Piancatelli reaction depicted in Scheme 2, the formation of the five-membered ring could occur from either structure **5** or **4** via a 4π-electrocyclization and an intramolecular aldol-type condensation, respectively. The specific mechanistic pathway has not been definitively decided and may, in fact, be dependent on the specific substrate and reaction conditions.

SCOPE AND LIMITATIONS

Synthesis of Furylcarbinols

2-Furylcarbinols are generally accessible by two simple approaches: (1) addition of a Grignard or organolithium reagent to furfural, and (2) the addition of a C2-lithiated furan to an aldehyde or ketone derivative (Scheme 8).[42]

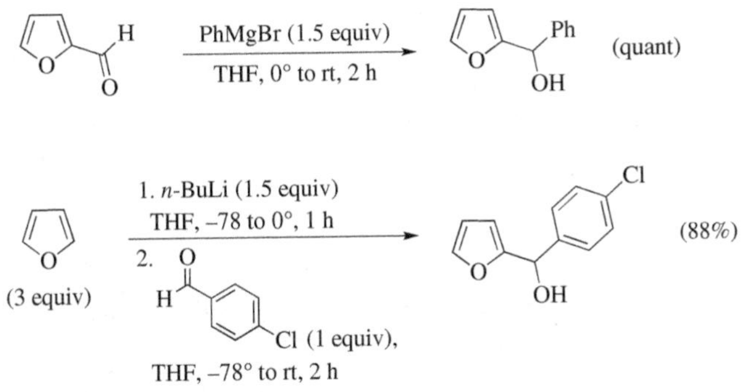

Scheme 8

Additional steps can be employed to access more extensively functionalized 2-furylcarbinols. For instance, 3-bromo-2-furaldehyde can undergo a Sonogashira cross-coupling followed by the addition of phenylmagnesium bromide to generate the corresponding 2-furylcarbinol (Scheme 9).[43]

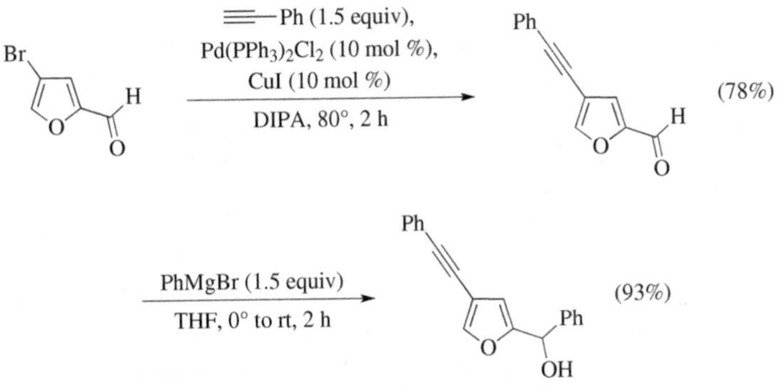

Scheme 9

For the synthesis of precursors used in the preparation of azaspirocycles, the most common starting material is methyl 5-bromo-2-furoate (Scheme 10).[44] For example, a Heck reaction with allyl alcohol followed by reductive amination and concomitant reduction of the ester with lithium aluminum hydride provides the desired 2-furylcarbinol. Notably, most of the starting materials presented in the following sections employ these types of synthetic approaches.

Oxa-Piancatelli Reactions

4-Hydroxycyclopent-2-enones. 2-Furylcarbinols can be transformed into 4-hydroxycyclopent-2-enones using a wide range of Lewis and Brønsted acids. Early examples of this transformation employed strong acids (e.g., formic acid,

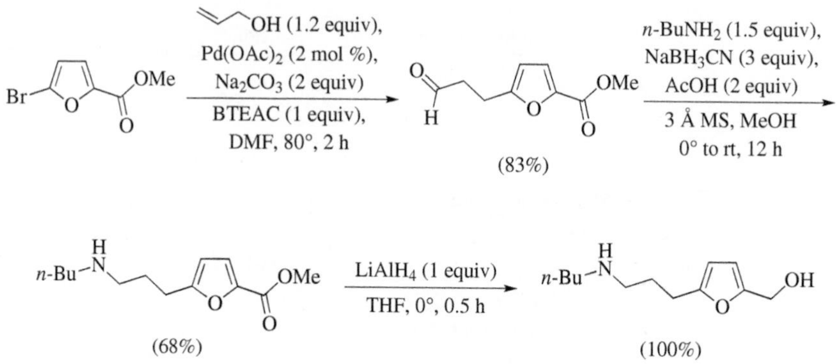

Scheme 10

polyphosphoric acid, and 4-toluenesulfonic acid) in a refluxing acetone/water mixture (Scheme 11).[21] This method affords the desired cyclopentenones in moderate to good yields (up to 70%).

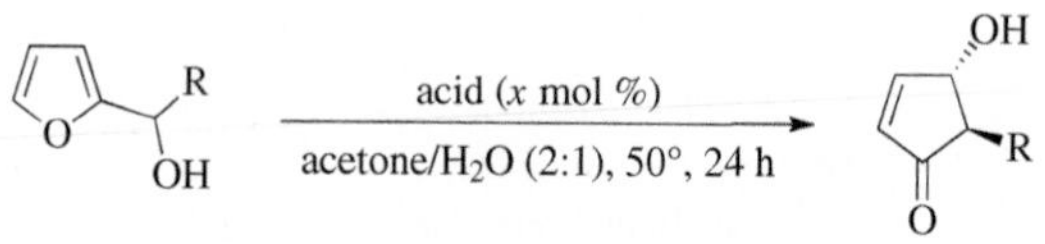

R	Acid	x	Yield (%)
Ph	HCO$_2$H	23	65
n-C$_7$H$_{15}$	H$_3$PO$_4$	33	70
Me	4-TsOH	6	30

Scheme 11

In the case of 5-methyl-2-furylcarbinols, these conditions are too harsh, but the transformation can be achieved in the presence of a mild Lewis acid (zinc(II) chloride) in a mixture of acetone/water at 60° (Scheme 12).[22,45] Nevertheless, a nearly stoichiometric amount of the Lewis acid is required to promote the cyclization. Under these reaction conditions, 2-furylcarbinols bearing aromatic substituents deliver the corresponding products in excellent yields (up to 85%), whereas systems bearing alkyl substituents lead to significantly lower yields owing to the formation

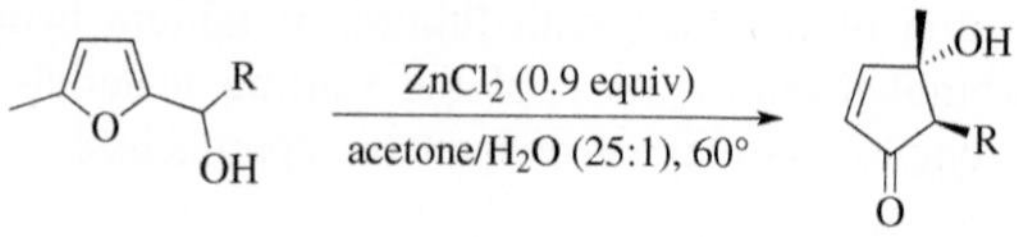

R	Time (h)	Yield (%)
Ph	24	70
2-thienyl	4	85
n-Bu	72	18

Scheme 12

of unidentified side products. The reaction has also been attempted with other 5-substituted-2-furylcarbinols, such as 5-nitro- and 5-methoxy-2-furylcarbinols, but these reactions result in either the recovery of the starting materials or the formation of 4-ylidenebutenolides.[46,47]

The Piancatelli reaction can also be conducted in the absence of an acid catalyst by using microwave irradiation in water. In these cases, the reaction is complete within minutes and furnishes 4-hydroxycyclopent-2-enone products in modest yields and selectivity (Scheme 13).[48] Although this method can avoid degradation of the product that may occur with prolonged heating, the level of diastereocontrol is lower than the corresponding acid-catalyzed reactions. Furthermore, the reaction is not suitable for substrates that contain long nonpolar alkyl chains, which is presumably due to their low solubility in water.

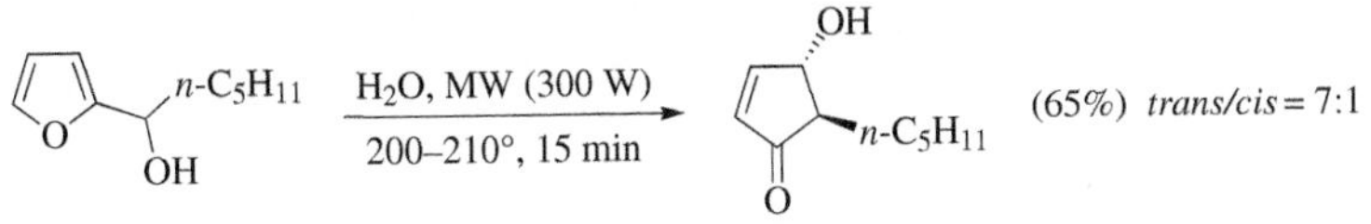

Scheme 13

4-Alkoxyclopent-2-enones. The oxa-Piancatelli reaction has been accomplished using Lewis acids (e.g., dysprosium(III) trifluoromethanesulfonate) as catalysts, and oxaspirocycles can be readily prepared from 5-alkyl-2-furylcarbinols with excellent diastereoselectivity (Scheme 14).[49] However, substrates with alkyl substituents at the carbinol moiety (R^3) are problematic (when R^1, R^2 = H) and lead only to decomposition products. This problem is attenuated by the presence of geminal dimethyl groups adjacent to the alcohol on the alkyl chain (i.e., R^1, R^2 = Me), albeit no explanation is given to account for this phenomenon.

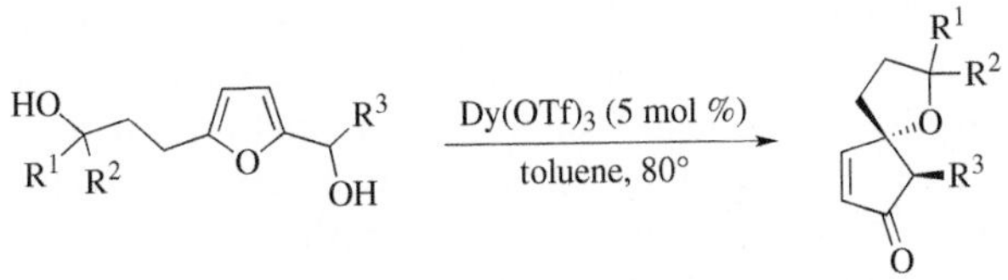

R^1	R^2	R^3	Time (h)	Yield (%)
H	H	Ph	6	91
H	H	n-Bu	6	0
H	Me	n-Bu	1	30
Me	Me	n-Bu	1	98

Scheme 14

Cascade Oxa-Piancatelli Reactions. Depending on the reaction conditions, the oxa-Piancatelli reaction can also be used to generate fused oxabicyclic compounds

(Scheme 15).[41] For instance, 2-furylcarbinols are converted into fused oxabicycles in good-to-excellent yields (up to 90%) when treated with a catalytic amount of zinc(II) chloride in aqueous DME at reflux (cf. Scheme 7). However, this rearrangement is only successful with 2-furylcarbinols bearing unsaturated functionalities (e.g., aryl, thienyl, and styryl) at the carbinol moiety.

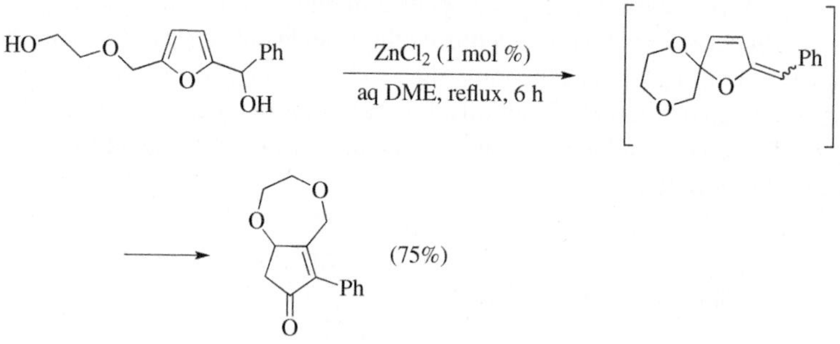

(75%)

Scheme 15

Aza-Piancatelli Reactions

4-Aminocyclopent-2-enones. The extension of the Piancatelli reaction to nitrogen nucleophiles is a significant advance in the development of this transformation. For instance, 4-aminocyclopent-2-enones can be prepared from 2-furylcarbinols using anilines in the presence of dysprosium(III) trifluoromethanesulfonate at 80° in acetonitrile.[23,50] These products are extremely valuable as synthons for accessing aminocyclopentitols,[51] which are present in a variety of bioactive molecules such as peramivir, pactamycin, and trehazolin. The reaction is versatile, tolerating a wide range of functional groups (i.e., both electron-donating and electron-withdrawing groups) at the carbinol moiety and on the aniline, and products are obtained in yields up to 92% (Scheme 16, top reaction).[21] Nevertheless, sterically demanding anilines, such as 2,6-dimethylaniline, preferentially form the corresponding Friedel–Crafts-type adducts as the major products (Scheme 16, bottom reaction).

Intramolecular versions of this reaction furnish the corresponding azaspirocycles, including both 6-azaspiro[4.5]decanes and 1-azaspiro[4.4]nonanes, in good-to-excellent yields (up to 97%) (Scheme 17).[52]

Tertiary 2-furylcarbinols possessing alkyl substituents are prone to dehydration and, therefore, cannot be used to access 4-aminocyclopentenones with a quaternary center adjacent to the carbonyl group. Consequently, the Piancatelli reaction can be performed with a donor–acceptor cyclopropane to circumvent this problem and afford the corresponding 4-aminocyclopent-2-enone under mild conditions and with good diastereoselectivity (Scheme 18).[53] Although this is an interesting approach, it is relatively limited in scope because the cyclopropane moiety must contain an aryl group, and replacing the esters with different functional groups has proven challenging.

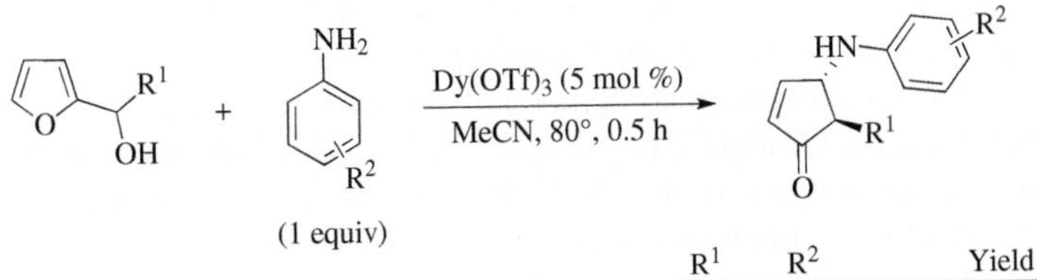

R^1	R^2	Yield (%)
Ph	4-I	92
Ph	3,5-(CF$_3$)$_2$	75
Ph	4-OMe	62
Me	2,4,6-Me$_3$	74
i-Pr	3-Cl	89

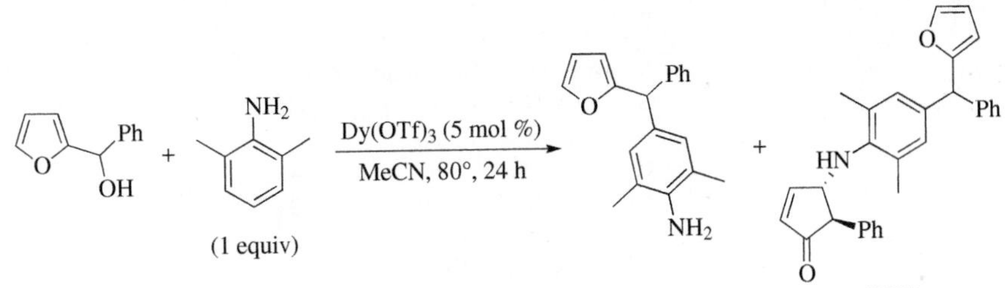

(38%) (15%)

Scheme 16

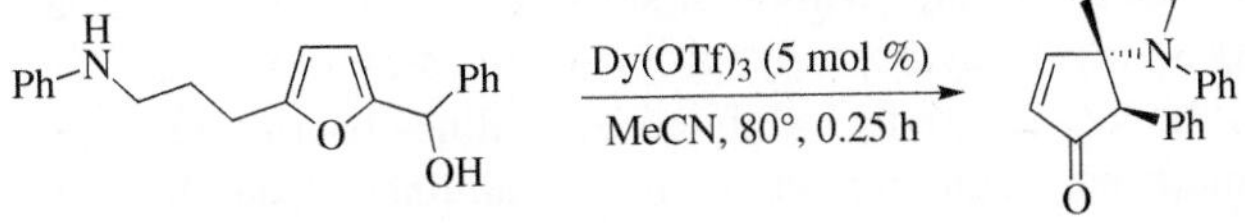

(96%)

Scheme 17

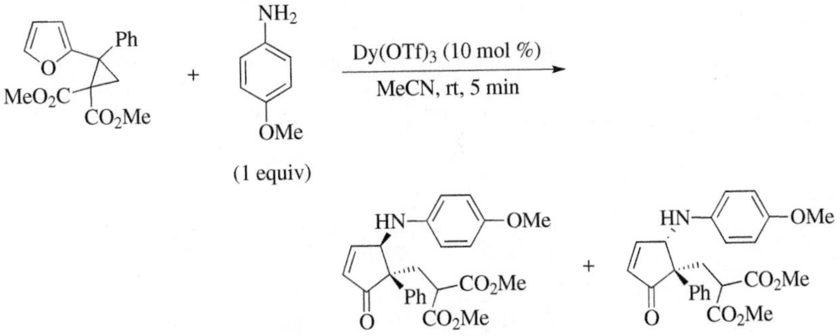

28 29

28 + **29** (57%)
28/29 = 6:1

Scheme 18

Calcium(II) salts can also be used as catalysts for the aza-Piancatelli reaction in nitromethane to give highly functionalized 4-aminocyclopentenones (Scheme 19).[54] These reactions conditions allow the incorporation of various functional groups (e.g., alkyl, aryl, and halide) at the C2 and C3 positions of the 2-furylcarbinols. In contrast, the decomposition of the substrate is observed with 5-alkyl-2-furylcarbinols, which is probably the result of steric hindrance preventing the nucleophilic addition of the aniline to the oxocarbenium ion (structure **2**, Scheme 4).

R^1	R^2	R^3	Time (min)	Yield (%)
Br	H	H	15	93
H	Me	H	30	86
H	H	Me	60	—
Ph	H	H	20	93

Scheme 19

The aza-Piancatelli reaction conditions have been improved by replacing nitromethane with 1,1,1,3,3,3-hexafluoro-2-propanol (HFIP) (Scheme 20).[55] Specifically, the new conditions avoid competitive side reactions such as the Friedel–Crafts alkylation (in the case of anilines without 4-substituents) and deoxyamination (with substrates bearing a vinyl or cyclopropyl group at the carbinol moiety). This solvent also circumvents the low reactivity exhibited by anilines bearing electron-donating groups in nitromethane. Although electron-rich anilines generally suppress the Piancatelli reaction by forming stable complexes with the Lewis acid, HFIP is a strong hydrogen-bond donor that decomplexes the Lewis acid and restores the catalytic activity.

R^1	R^2	R^3	Temp (°)	Time (min)	Yield (%)
vinyl	H	I	20	120	61
cyclopropyl	H	I	20	15	84
Ph	Bn	H	60	60	51
Ph	H	NHAc	60	60	70

Scheme 20

The ability to employ alkyl amines in the aza-Piancatelli has proven particularly challenging, which is likely due to sequestering of the Lewis acid catalyst by the basic alkyl amine (vide supra). Consequently, Mitsunobu conditions have been employed to circumvent this limitation in the presence of pyridinium 4-toluenesulfonate (PPTS) to generate the 1-azaspiro[4.4]nonanes in good-to-excellent yields (Scheme 21).[44] These reaction conditions can also promote the formation of 6-azaspiro[4.5]decanes, albeit the yields are significantly lower than those of 1-azaspiro[4.5]nonanes. Regarding the mechanism, triphenylphosphine reacts with diethyl azodicarboxylate (DEAD) to generate intermediate **30**, which is then protonated by PPTS to provide adduct **31**. The alcohol then attacks phosphonium salt **31** to form cyclization substrate **32**, which loses triphenylphosphine oxide to afford the spiro intermediate **33**. The latter rearranges into enol **34** after proton migration, and the remainder of the mechanism is identical to that described in Scheme 4. This method is limited to primary 2-furylcarbinols, and no examples of intermolecular reactions have been reported.

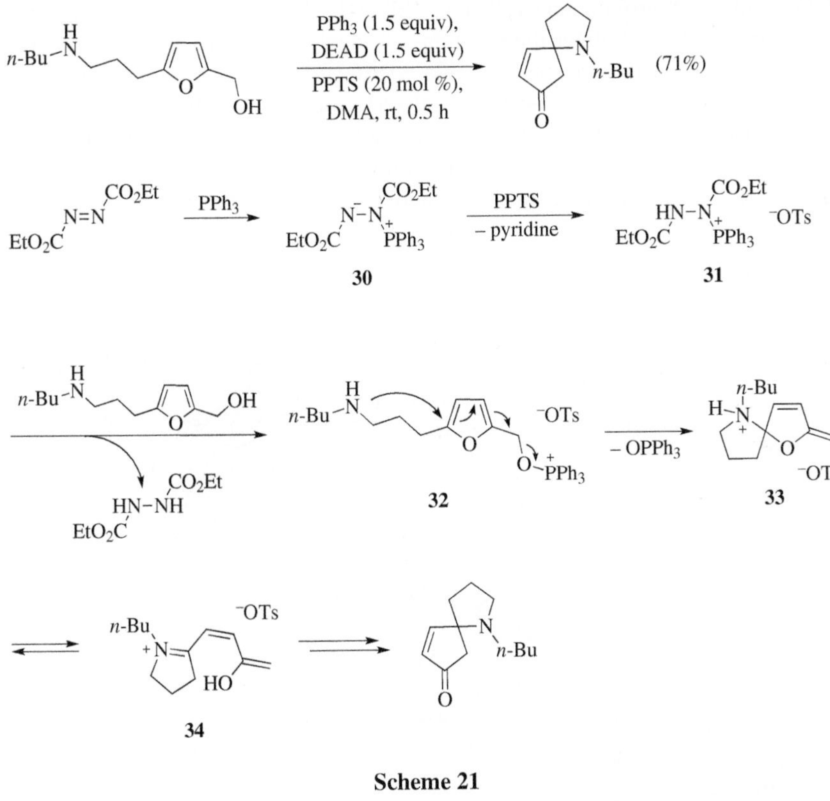

Scheme 21

The reaction can also be applied to other nitrogen derivatives, such as hydroxylamines (Scheme 22),[56] since they have pK_a values similar to anilines and thus exhibit similar reactivity. Nevertheless, only *N*-substituted *O*-protected hydroxylamines afford the corresponding 4-aminocyclopentenones in excellent yields (up to 95%) using dysprosium(III) trifluoromethanesulfonate. In contrast, the 2-furylcarbinol is unreactive with *N*-benzylhydroxylamine and *O*-benzylhydroxylamine as nucleophiles. This lack of reactivity can presumably be ascribed to the complexation of the catalyst by the basic, unhindered hydroxylamine.

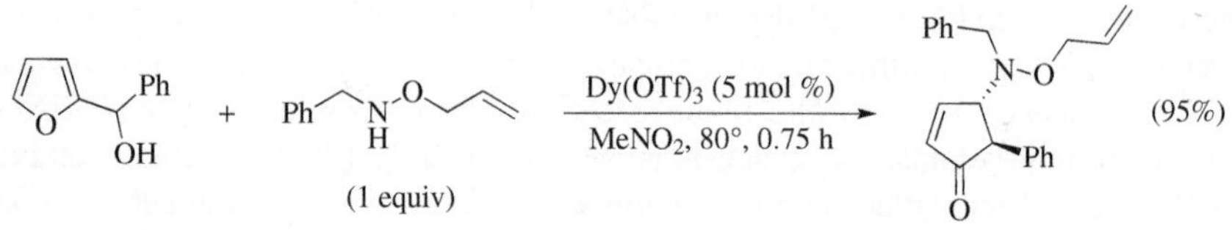

Scheme 22

Enantioselective Aza-Piancatelli Reactions. Enantioselective versions of this reaction have been investigated using chiral phosphoric acids, derived from BINOL and 1,1′-spirobiindane-7,7′-diol (SPINOL) chiral scaffolds, to activate the carbinol moiety,[42,57,58] albeit the reactions are again limited to aniline derivatives. The steric nature of the phosphoric acids is crucial in providing the products with high enantioselectivities, and the best results are obtained with bulky substituents at the 3,3′-positions of the SPINOL backbone (Scheme 23).[42] The reaction is favored in systems having electron-withdrawing groups on the anilines, while electron-rich and electron-neutral anilines appear to deactivate the catalyst.

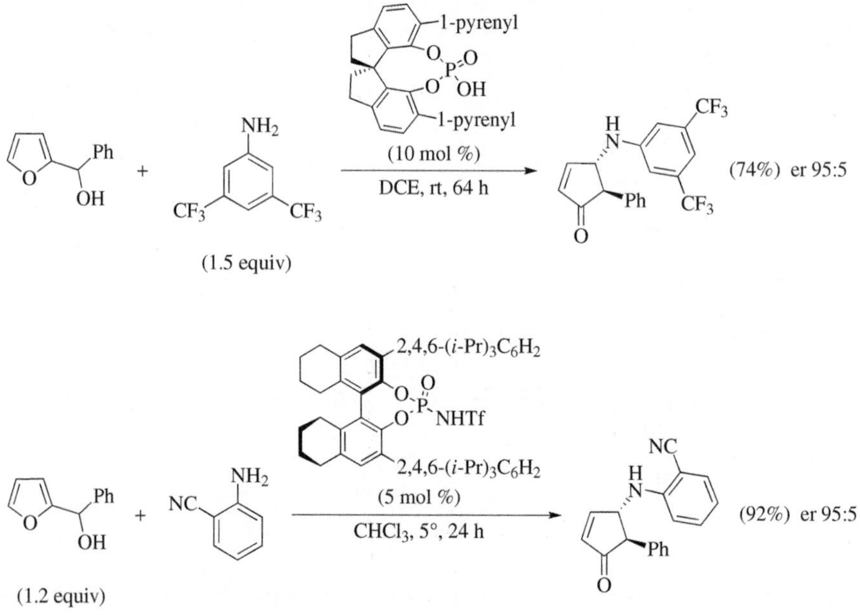

Scheme 23

Cascade Aza-Piancatelli Reactions. The combination of the Piancatelli reaction with other transformations provides an opportunity to generate more complex compounds. Specifically, the presence of a second nucleophile in the 2-position of the aniline derivative promotes a conjugate addition to give a variety of tri- and tetracyclic derivatives using the appropriate Lewis acid (lanthanum(III) triflate or indium(III) triflate) or Brønsted acid (4-TsOH) catalyst. Among the nucleophilic units used for this purpose, heteroatom ($NHTs$, OH, SH and $CONH_2$) and carbon (pyrrole) functionalities successfully provide products in excellent yields as single diastereomers (Schemes 24 and 25).[59,60] In the latter system, however, replacing the pyrrole by an indole or thiophene moiety fails to afford the desired products.

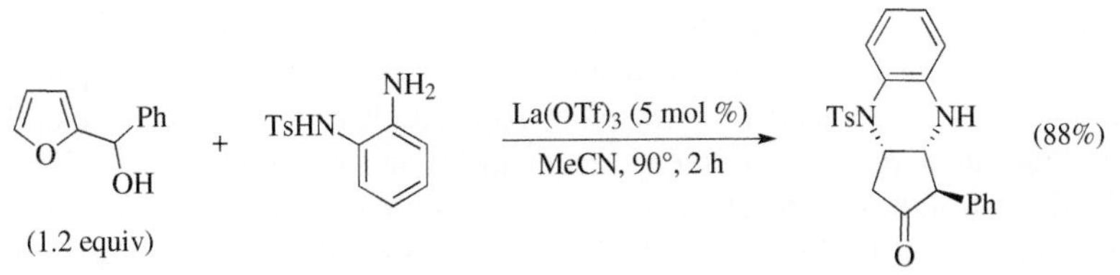

Scheme 24

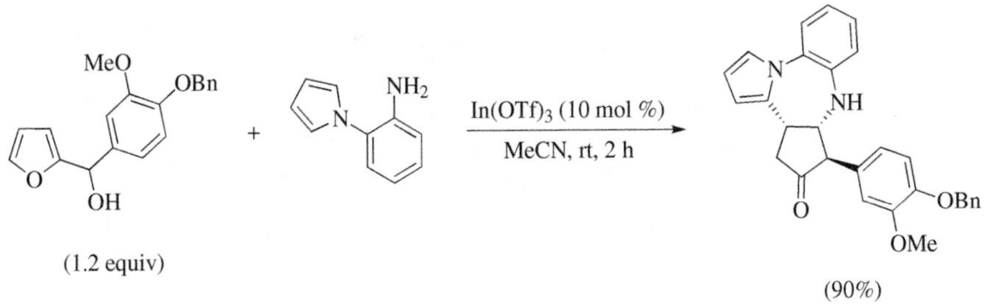

Scheme 25

For this type of reaction sequence, it is possible to control the product of the reaction by selecting the appropriate conditions. For example, in the case of the reaction of amide **35** with furan-2-yl(phenyl)methanol, the usual aza-Piancatelli compound **36** is obtained as the sole product after one hour, whereas extending the reaction time leads exclusively to the corresponding 1,4-benzodiazepin-5-one **37** (Scheme 26).[61] Unfortunately, this type of tandem sequence is not compatible with 2-furylmethanol.

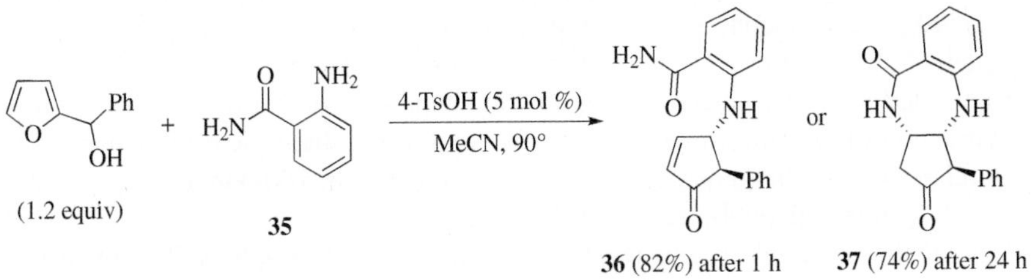

Scheme 26

Synthesis of Pyrroles. Increasing the molecular complexity of the aza-Piancatelli products can be accomplished by the introduction of a second catalyst capable of promoting a subsequent transformation. For example, the combination of calcium(II) and copper(II) salts affords the cyclopenta[b]pyrroles in good-to-excellent yields from the reaction of an array of 4-alkynyl-2-furylcarbinols with anilines. One such example, using 4-iodoaniline, is depicted in Scheme 27.[43] In the proposed mechanism, the calcium-catalyzed step leads to aza-Piancatelli product **38**, which undergoes copper-catalyzed hydroamination to afford the bicyclic intermediate **39**. Isomerization of the cyclopentenone double bond generates pyrrole **40**. The solvent is critical to the success of this transformation since it must be capable of accommodating both reactions. In this respect, only DCE and HFIP give satisfactory yields in comparison with other solvents that are commonly employed in aza-Piancatelli reactions (toluene, acetonitrile, and nitromethane). In the case of HFIP, controlling the reaction time is also a crucial element, since prolonged reaction times lead to the dearomatization of cyclopenta[b]pyrrole **40** to form the corresponding cyclopenta[b]pyrroline, compound **41**.[62]

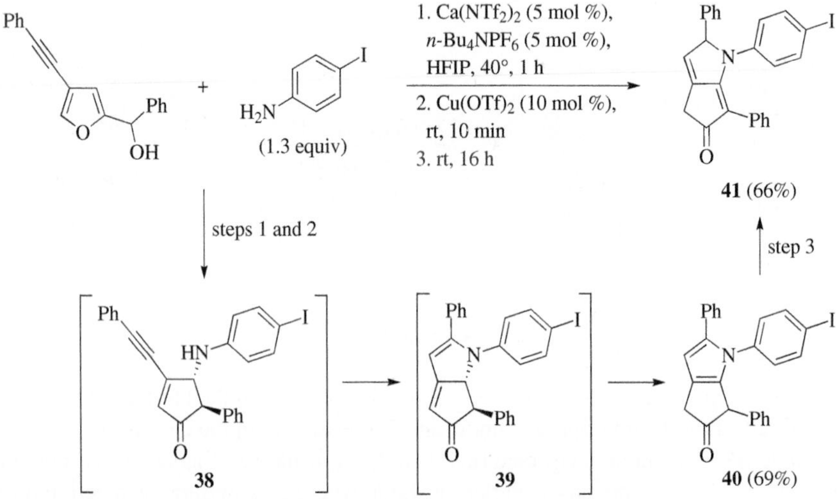

Scheme 27

In another approach to functionalizing the cyclopenta[b]pyrroles (e.g., compound **40**), the nucleophilicity of the pyrrole ring can be exploited in a subsequent Friedel–Crafts reaction. After the first reaction is complete, the introduction of an activated secondary alcohol (e.g., diphenylmethanol derivative **42**) permits the construction of tetrasubstituted pyrroles in good-to-excellent yield (Scheme 28).[43] Although the reaction is compatible with a wide range of benzyl, vinyl, and propargyl alcohols, primary alcohols are not suitable precursors.

In another feature of this reaction sequence, which employs 2'-substituted- and 2',6'-disubstituted-[1,1'-biphenyl]-2-amines, N–C axial chirality is created during the hydroamination step to generate atropisomers, as exemplified in the

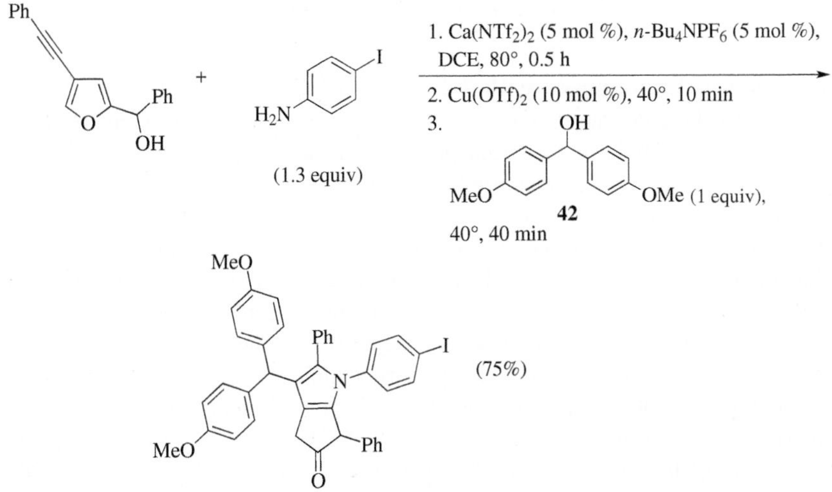

Scheme 28

reaction of **43** to form pyrrole **44** (Scheme 29).[62] The steric hindrance exhibited by the *ortho*-substituent controls the atropodiastereoselectivity of the reaction, and a single atropisomer is formed if the substituent is sufficiently bulky. Nevertheless, if the *ortho*-substituent is too bulky, hydroamination is precluded, and the 4-aminocyclopentenone is obtained as the sole product.

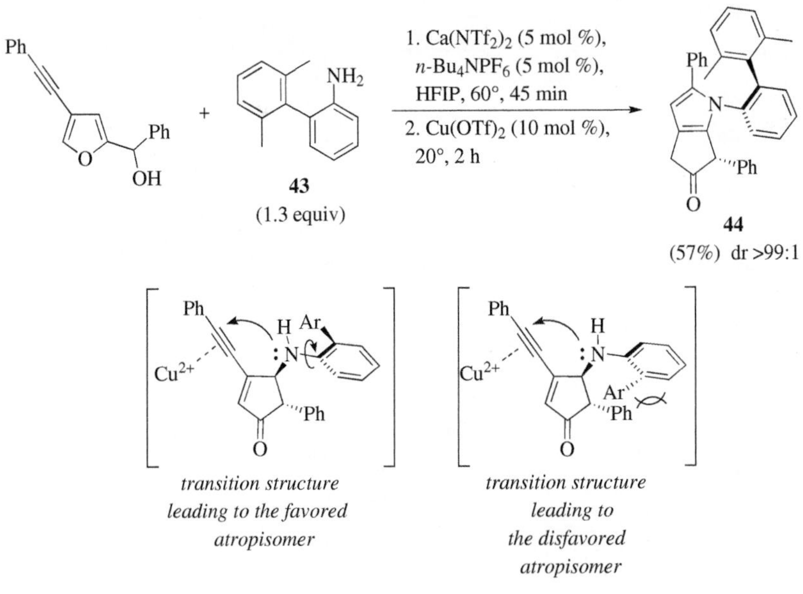

Scheme 29

Carba-Piancatelli Reactions

Reaction of Enaminones with 2-Furylcarbinols. Most of the investigations pertaining to the Piancatelli reaction involve oxygen and nitrogen nucleophiles. Significantly less emphasis has been placed on using carbon nucleophiles, which is presumably because these reactions often suffer from competitive Friedel–Crafts reactions with the carbinol moiety. The first examples of carba-Piancatelli reactions involve the combination of 2-furylcarbinols with enaminones to generate cyclopenta[b]pyrroles in good yields. For example, treatment of 2-furylcarbinol **45** with zinc(II) chloride generates oxocarbenium ion **47**, which undergoes nucleophilic attack by enaminone **46** (Scheme 30).[63] The resulting intermediate, compound **48**, rearranges to pentadienyl carbocation **50** by proton migration and ring-opening of protonated oxacycle **49**. Cyclopent-2-enone **51** is then formed by a 4π-electrocyclization of intermediate **50**, followed by intramolecular conjugate addition of the amino group to furnish the dihydropyrrole ring in **52** with a

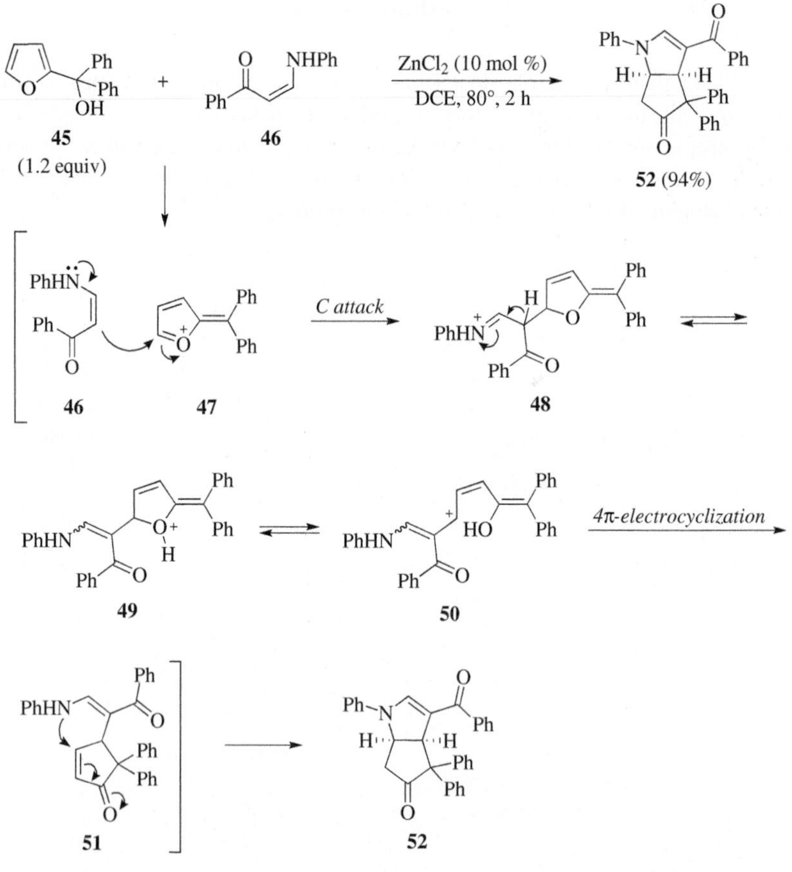

Scheme 30

cis configuration that was confirmed by X-ray crystallography. Unfortunately, the reaction does not tolerate alkyl-substituted tertiary 2-furylcarbinols, because they are prone to dehydration under the reaction conditions.

Reactions of Indoles with 2-Furylcarbinols to Prepare Cyclopent-2-enones. Indole derivatives can be used to access indole-substituted cyclopent-2-enones in the presence of zinc(II) chloride in DCE at 80° (Scheme 31).[64] Nevertheless, this transformation suffers from several drawbacks. The reaction is only compatible with diaryl-2-furylcarbinols, since other substrates result in direct nucleophilic substitution of the alcohol at the C3 position of the indole. Furthermore, the presence of a strong electron-withdrawing group (e.g., a nitro group) at the C5 position on the indole or a Boc protecting group on the indole nitrogen inhibits the reaction because of the reduced nucleophilicity.

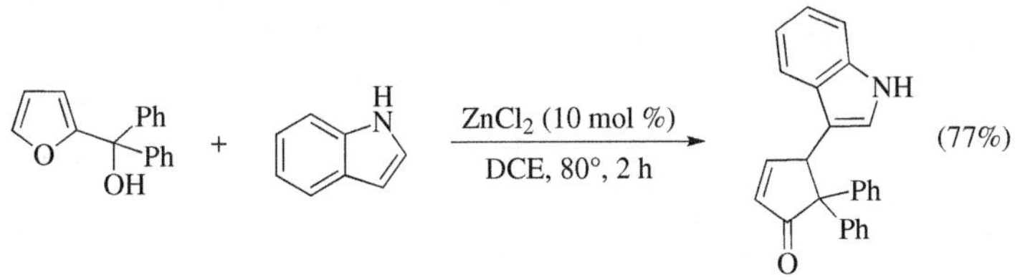

Scheme 31

Cascade Carba-Piancatelli Reactions. Aryl functionalities can also be employed as nucleophiles in the intramolecular variant of the Piancatelli reaction, albeit not in a single operation. For example, in the presence of copper(II) sulfate pentahydrate and acetic acid in toluene at 100°, 2-furylcarbinol **53** affords the stable spirofurooxindole derivative **55** via the intermediacy of the oxocarbenium ion **54** (Scheme 32).[24,65,66] This reaction requires the presence of an electron-donating group at the 3-position of the aromatic ring. Spirofurooxindole **55** is then transformed into a mixture of diastereomeric spiropentenoneoxindoles **57** and **58**, favoring the former, by heating at 130° in DCE. This result is consistent with the formation of the pentadienyl cation **56**, which undergoes a conrotatory 4π-electrocyclization.

APPLICATIONS TO SYNTHESIS

Prostaglandins and Analogues

Prostaglandins are naturally occurring hormones found in humans and other mammals that are biosynthetically derived from arachidonic acid through a cyclooxygenase-enzyme-medicated cascade.[67] Consequently, significant effort has been devoted to developing synthetic methods for their preparation. There are three common strategies used to deliver these derivatives,[68] one of which is based

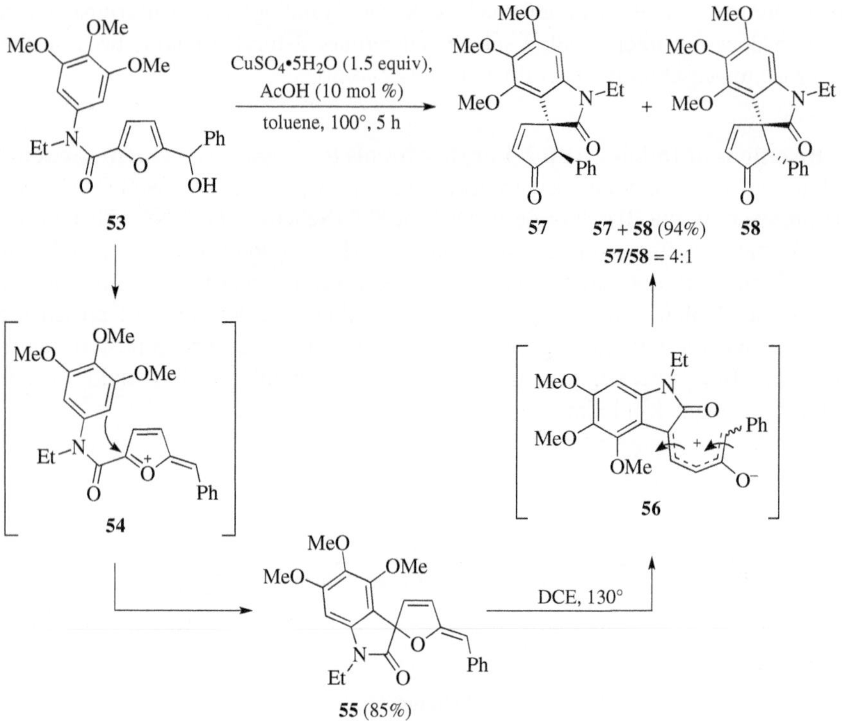

Scheme 32

on the preparation of a key intermediate—the so-called "Corey lactone"[69]—and can be used to prepare prostaglandins on an industrial scale. Several routes have also emerged that utilize the Piancatelli reaction to form the central prostaglandin cyclopentenol skeleton.[68,70–78] The key intermediate is obtained from a suitably functionalized furylmethanol derivative, which is subjected to a two-component coupling strategy to introduce the remaining side chain on the cyclopentyl ring. A related strategy is based on a three-component coupling procedure involving a 4-hydroxy-cyclopent-2-enone derivative that was devised by Noyori.[79]

The synthesis of the prostaglandin derivatives follows the general pathway outlined in Scheme 33. Intermediate **6** is generated by an oxa-Piancatelli reaction using water as the nucleophile in the presence of either a weak Lewis acid (zinc(II) chloride) or a Brønsted acid (4-TsOH). Rearrangement of the α,β-unsaturated cyclopentenone **6** delivers the thermodynamically more stable hydroxyenone **59**.[16,80,81] Although the Piancatelli reaction described in Scheme 33 is not particularly efficient, the operational simplicity of the reaction makes it a practical synthetic route. Subsequent functionalization of cyclopentenone **59** then affords the prostaglandin derivatives as racemic mixtures.

The Piancatelli reaction has been used for the synthesis of the functionalized cyclopentanone **63**, a (3E,5Z)-diene analogue of misoprostol, which is a drug used

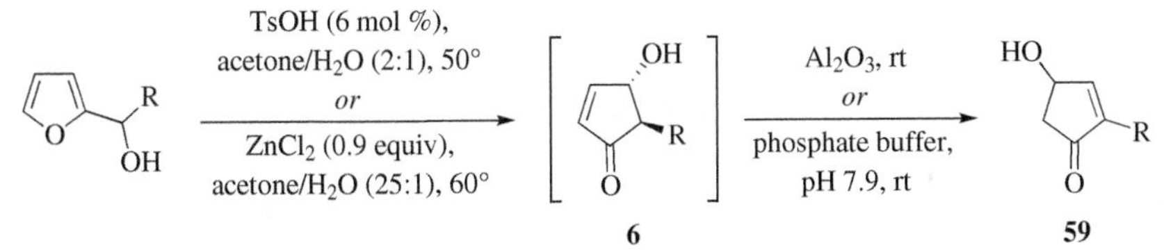

Scheme 33

for treating peptic ulcer disease (Scheme 34).[70] The reaction involves heating 1-furan-2-yl-but-3-yn-1-ol (**60**) in the presence of a catalytic amount of 4-TsOH (0.5 mol %) in a mixture of dioxane/H$_2$O (8:1) at 83–85° over 36 hours. Despite attempts to optimize the reaction, only modest yields (~20–30%) of Piancatelli product **61** are obtained. Trituration of product **61** on alumina facilitates the formation of 4-hydroxy-2-prop-2-ynylcyclopent-2-enone (**62**) by the migration of the alcohol functionality to form the more thermodynamically stable α,β-unsaturated cyclopentenone en route to the misoprostol analogue **63**.

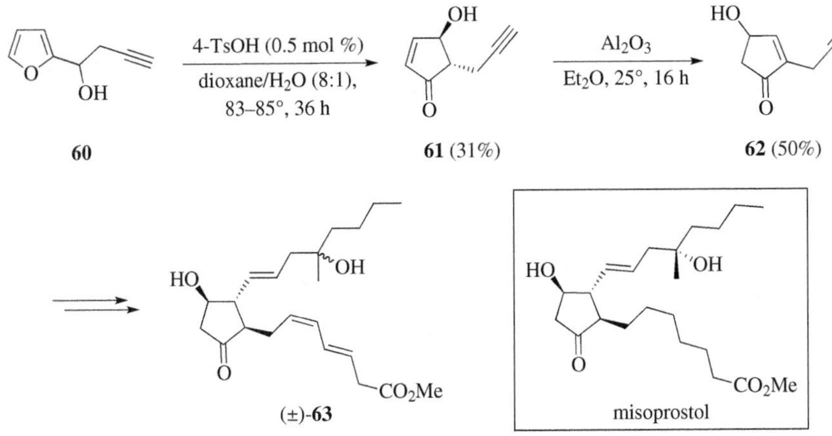

Scheme 34

The Lewis acid catalyzed Piancatelli reaction is used to prepare the 4-fluoro analogue **66** of enisoprost, a compound that possesses gastric antisecretory and mucosal protective activity. Treatment of the furylmethanol derivative **64** with a large excess of zinc(II) chloride in refluxing dioxane/water for 32 hours affords the rearranged Piancatelli product **65**, albeit in only 27% yield (Scheme 35).[71]

Zinc(II) chloride cannot be used for the Piancatelli reaction with furylcarbinol derivatives that contain a methyl ester on the side chain, because these compounds are prone to hydrolysis. Hence, the industrial-scale syntheses of numerous prostaglandin analogues generally employ the common chiral intermediate **69**, which possesses a 4-silyloxycyclopentenone backbone (Scheme 36).[76–78] The isopropyl ester on the side chain is less prone to hydrolysis during the zinc(II) chloride catalyzed Piancatelli

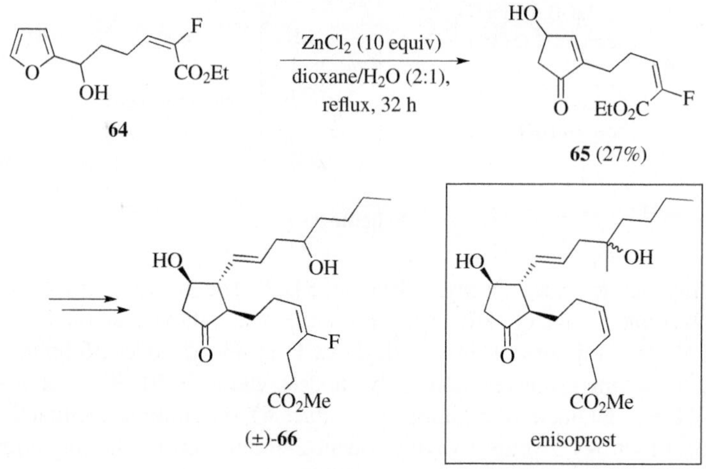

Scheme 35

reaction, which permits this group to be either retained for targeted analogues or removed at a later stage. Although the preparation of compound **68** can be performed on a large scale (15 kg), the reaction would benefit from optimization experiments to improve the 55% overall yield from 2-furylcarbinol **67**.

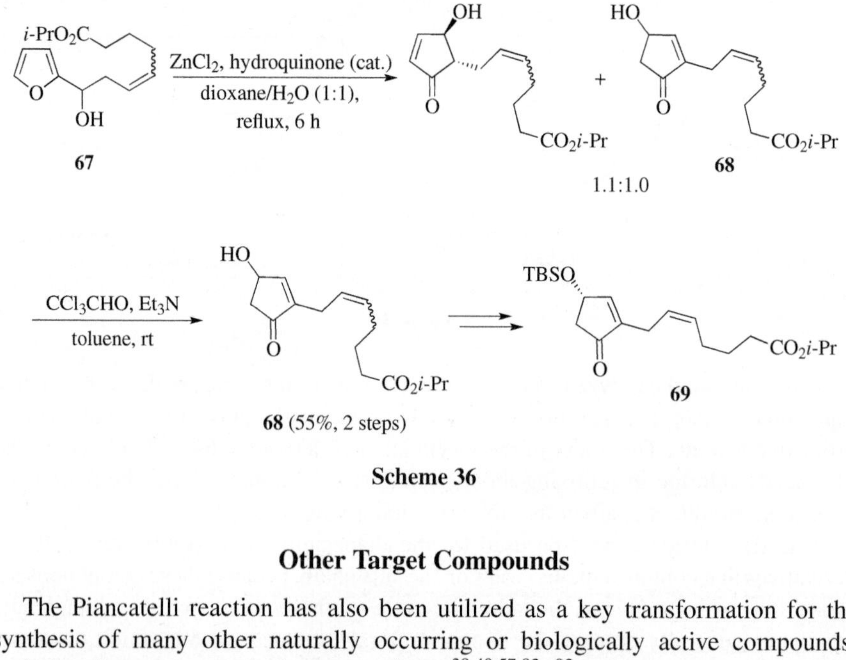

Scheme 36

Other Target Compounds

The Piancatelli reaction has also been utilized as a key transformation for the synthesis of many other naturally occurring or biologically active compounds, including important synthetic intermediates.[39,40,57,82–92] In this context, it is worth mentioning that Piancatelli also discovered the acid-mediated rearrangement of a

2-furylcarbinol moiety into a 3-oxo-5-hydroxycyclopentene unit for the synthesis of a 2-furylsteroid.[82]

(±)-Havellockate. The synthesis of various functionalized molecules that possess a terpenoid skeleton has also been initiated from furfural derivatives by the conversion into the corresponding furylcarbinol for the Piancatelli reaction.[90–92] For example, the synthesis of an advanced synthetic intermediate for (±)-havellockate, a metabolite isolated from the soft coral *Sinularia granosa* collected in the Indian Ocean, involves the preparation of diene **71** from furan-2-ylmethanol (Scheme 37).[88] The treatment of furan-2-ylmethanol with potassium dihydrogen phosphate in refluxing water for 40 hours, followed by *tert*-butyldimethylsilyl protection of the hydroxyl functionality of the cyclopentenone, affords the silyl ether **70**, albeit in only 20% yield over two steps. Iodination and subsequent Stille cross-coupling furnish the vinyl enone **71**, which is converted to the tertiary alcohol **72** by the addition of methyllithium. The functionalized hydrindane **73** is then obtained in 70% yield as a single diastereomer by a hydroxyl-directed Diels–Alder reaction between tertiary alcohol **72** and acrolein in the presence of magnesium bromide ethyl etherate. Twelve additional synthetic steps are required to prepare **74**, which is a potential precursor to (±)-havellockate. Although the total synthesis of this natural product has not yet been reported, intermediate **74** possesses the complete carbon framework in addition to six of the eight stereogenic centers with the correct relative configurations.

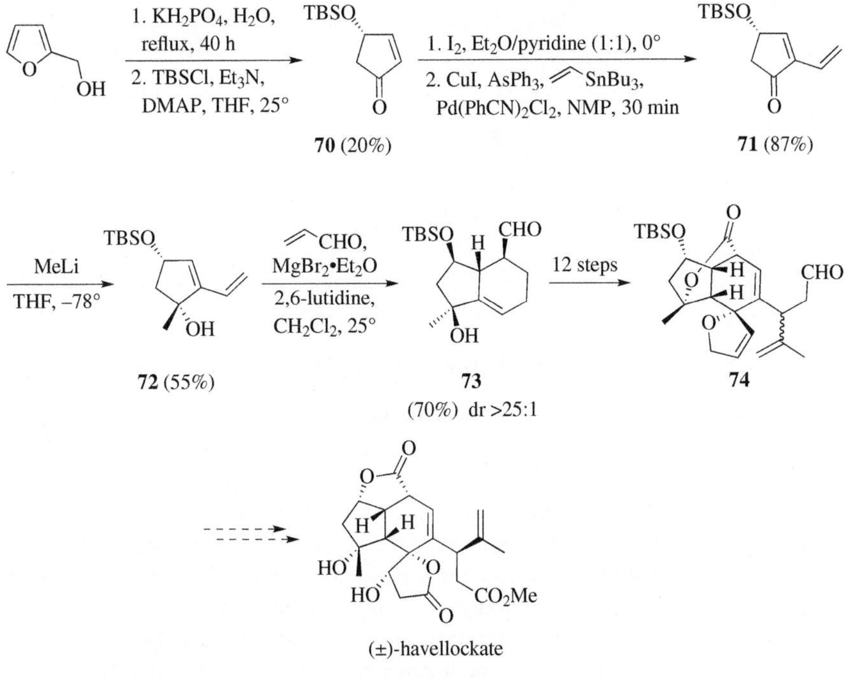

Scheme 37

C8-*epi*-Guanacastepene O. The synthetic utility of the Piancatelli transformation is further illustrated by the preparation of key cyclopentenone derivative **77** (Scheme 38),[86] which is used in the total synthesis of C8-*epi*-guanacastepene O, a member of a family of naturally occurring compounds possessing antibacterial and hemolytic activity. Treatment of furan **75** with zinc(II) chloride affords the target cyclopentenone derivative,[85] which upon the addition of methyllithium delivers the tertiary alcohol **76**. Chemoselective protection of the secondary alcohol and a subsequent PCC-mediated oxidative rearrangement gives cyclopentenone **77**. This compound is then converted to diene **78**, and a key intramolecular Diels–Alder reaction is used to furnish the tricycle **79**, which contains the (5-7-6) tricyclic core of C8-*epi*-guanacastepene O. Notably, this one-pot Piancatelli rearrangement–isomerization sequence has also featured in the enantioselective syntheses of the related terpenes (+)-heptemerone G and (+)-guanacastepene A.[89]

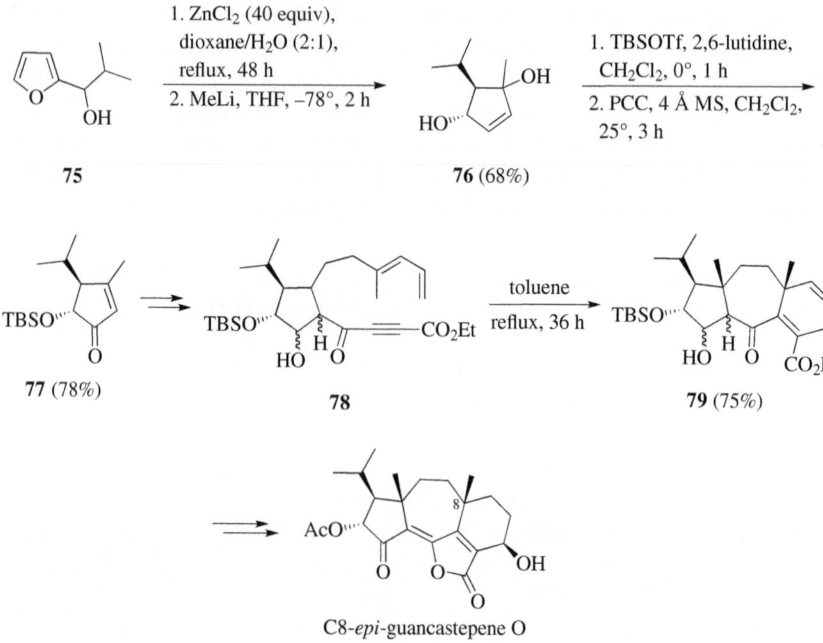

Scheme 38

hNK1 Antagonist Analogue. A catalytic asymmetric aza-Piancatelli reaction[42,57,58] is utilized in the synthesis of the human NK1 antagonist analogue **82**, which is prepared in three steps from the highly enantioenriched cyclopentenone **81** (Scheme 39).[57] Treatment of a mixture of furan-2-yl(phenyl)methanol and methyl 2-aminobenzoate with a catalytic amount of the chiral Brønsted acid **80** in chloroform at low temperature furnishes cyclopentenone **81** in high yield and with excellent diastereo- and enantioselectivities. Luche reduction of the enone followed by the hydrogenation of the alkene and alkylation of the residual hydroxy group deliver the cyclopentane-based hNK1 antagonist **82** in a highly enantioenriched form.

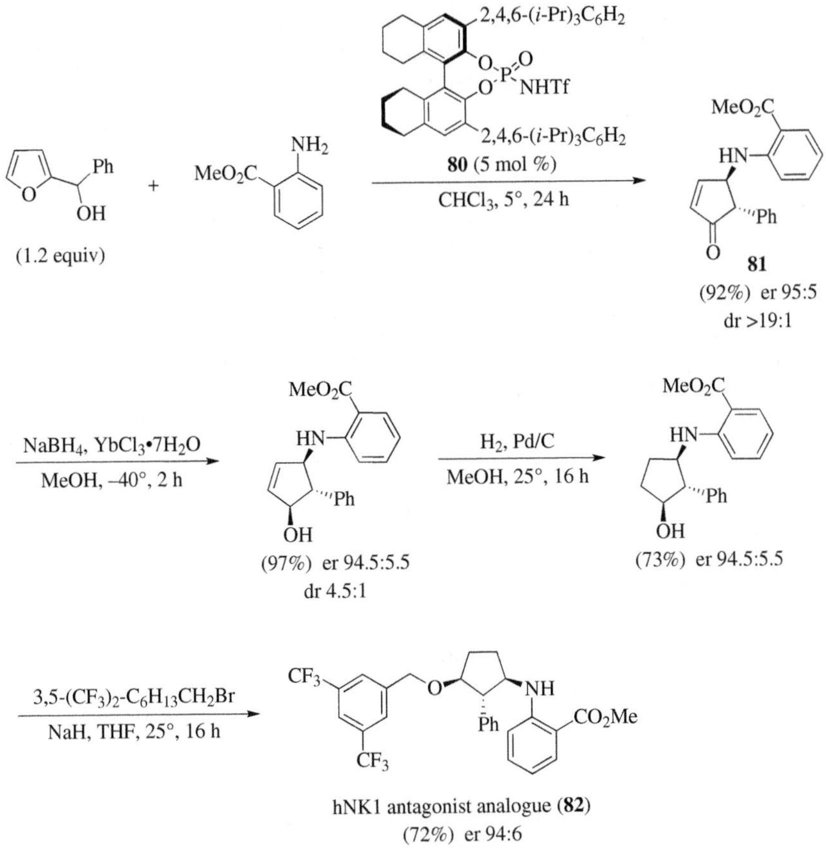

Scheme 39

COMPARISON WITH OTHER METHODS

Oxidation of Cyclopent-4-ene-1,3-diol Derivatives

4-Hydroxycyclopent-2-enone and related derivatives are widely employed in target-directed synthesis, particularly in the preparation of prostaglandins.[68,70−78] Aside from the Piancatelli reaction, these intermediates can also be obtained by the oxidation of *meso*-cyclopent-2-ene-1,3-diol, which is formed by a photochemical-mediated oxidation of cyclopentadiene.[93] Alternatively, 4-hydroxycyclopent-2-enones and their *O*-silyl- and *O*-acetyl-derivatives can be prepared in enantiomerically pure form by lipase-catalyzed mono-hydrolysis of a meso diester, followed by oxidation with either pyridinium dichromate[94] or Dess–Martin periodinane.[95]

The palladium-catalyzed allylic substitution of *meso*-1,4-allylic dibenzoates is an efficient method for the oxidation of allylic esters.[96,97] Consequently, in the presence of chiral ligand **85**, *meso* cyclopentene **83** reacts with the resulting chiral palladium complex to form the π-allylpalladium intermediate **86,** which undergoes selective *O*-alkylation with nitronate **84**. The fragmentation of **87** forms the γ-benzoyloxycyclopentenone **88** in good yield and with excellent enantioselectivity (Scheme 40).

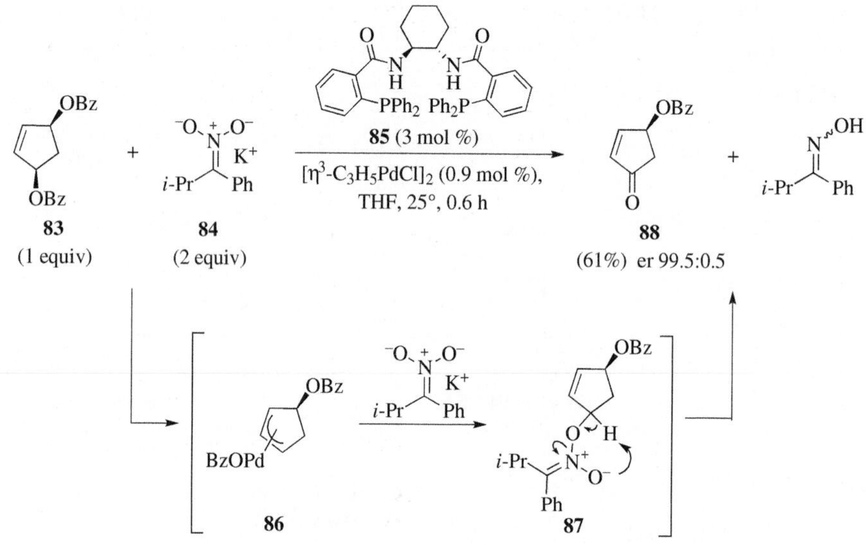

Scheme 40

Organocatalytic Asymmetric Desymmetrization–Fragmentation of Epoxides

Another route to enantiomerically enriched 4-hydroxycyclopentenone derivatives involves the asymmetric desymmetrization of a *meso*-epoxide with an organocatalyst (Scheme 41).[98] Heating 6-oxabicyclo[3.1.0]hexan-3-one (**89**) in the presence of the bifunctional thiourea–*Cinchona* alkaloid catalyst **90** in dichloromethane facilitates enantioselective ring-opening to afford (*R*)-4-hydroxycyclopent-2-enone (**91**) in excellent yield and with high enantiomeric purity.

Ring-Closing Metathesis

A ring-closing metathesis reaction features in the synthesis depicted in Scheme 42. Notably, this route can be performed on a large scale to provide enantiomerically pure 4-hydroxycyclopent-2-enone.[99] The synthesis may be initiated either using D-arabinose or 2-deoxy-D-ribose, which can be converted to dithio intermediate **92** in a few steps (Scheme 42). After a series of functional-group manipulations involving hydroxyl group protection, oxidation cleavage to the aldehyde, and the installation of the vinyl groups, the hydroxyl diene **93** is obtained. The ring-closing

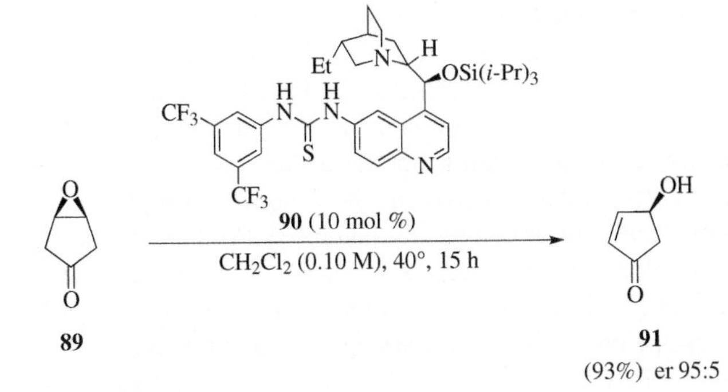

Scheme 41

metathesis reaction with Grubbs' first-generation catalyst delivers the cyclopentene **94** in 89% yield. Dess–Martin oxidation of the allylic alcohol followed by the removal of the silyl group provides (S)-4-hydroxycyclopent-2-enone (**95**) in high enantiomeric purity, albeit via a rather lengthy synthetic sequence.

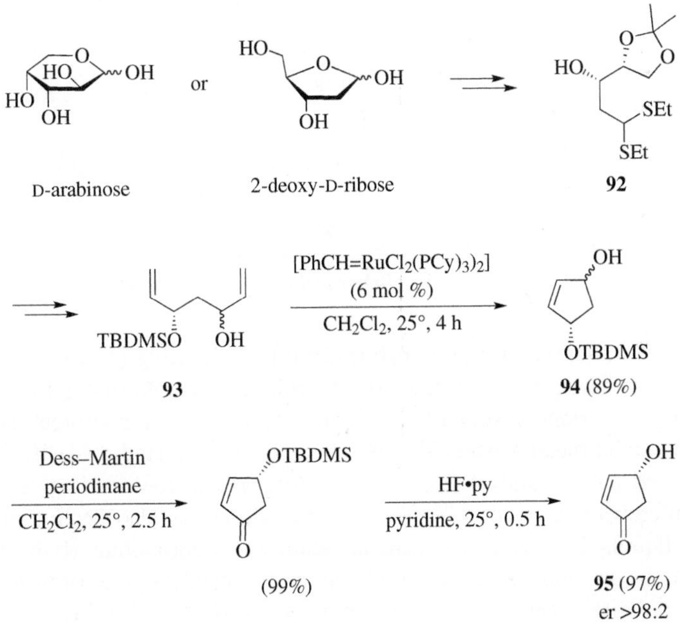

Scheme 42

Condensation of Furaldehydes with Amines

A simple transformation related to the aza-Piancatelli reaction enables the formation of *trans*-4,5-diaminocyclopentenone motifs. The first preparation of a

4,5-diaminocyclopent-2-en-1-one derivative resulted from heating furfural with two equivalents of a primary aniline in an alcohol solution, which afforded the product in low yield.[100] Thirty years later, this reaction has been reinvestigated using Lewis acid catalysis, and the target products are isolated in significantly higher yields.[101−103] More specifically, catalytic dysprosium(III) triflate promotes the reaction in acetonitrile at room temperature, delivering the products in high yields using either secondary aliphatic amines or anilines (Scheme 43). Primary aliphatic amines fail to give the products, albeit primary anilines are successful when scandium(III) triflate is used as the catalyst.[101] The *trans*-4,5-diaminocyclopentenones are believed to arise through the ring-opening of iminium ion intermediate **96** to afford the deprotonated Stenhouse salt[34] **97**, which undergoes a conrotatory 4π-electrocyclization to form cyclopentenone **98**.

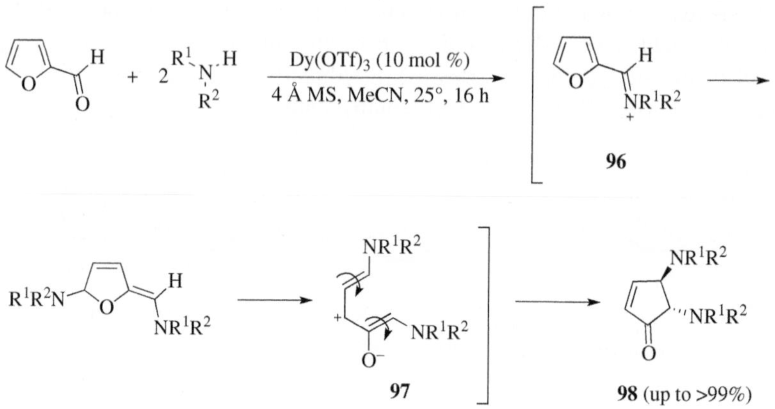

Scheme 43

Additional variations of this transformation include using (1) copper(II) triflate (0.1 mol %) in water;[104] (2) erbium(III) chloride (0.1 mol %) in ethyl lactate as an environmentally friendly solvent;[105] (3) new heteronuclear coordination clusters comprising a 3d-metal (nickel(II) or cobalt(II)) and a lanthanide(III) metal as air-stable and active catalysts that can be employed at low loading levels;[106,107] and (4) microwave activation.[108] The transformation has also been conducted in an ionic liquid, 1-hexyl-3-methylimidazolium tetrafluoroborate ([HMim][BF$_4$]), that functions as both a reusable solvent and a catalyst; the Brønsted acidity and high polarity greatly enhance the rate of reaction.[109] 4-Toluenesulfonamide also serves as an efficient catalyst for this transformation.[110] Nonetheless, these developments have not significantly broadened the scope of substrates that can be employed. The synthetic utility of this procedure is illustrated in the synthesis of the pyrrolo-2-aminoimidazole alkaloid, (±)-agelastatin A, which was isolated from a marine sponge (Scheme 44).[111] The key reaction between furaldehyde and diallylamine affords diaminocyclopentenone **99**, which can then be converted via a five-step sequence to the target alkaloid in 15% overall yield from furfural.

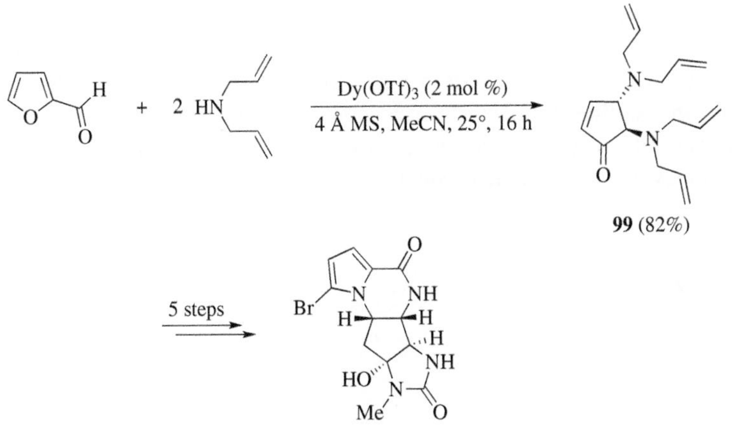

99 (82%)

(±)-agelastatin A
(15%) overall

Scheme 44

The functionalization of the *trans*-4,5-diaminocyclopentenones can be readily accomplished by conjugate addition of nucleophiles (thiols or amines) followed by amine elimination to afford a wide range of 2,4-difunctionalized cyclopentenones.[25] For instance, aluminum chloride catalyzed reaction of furfural with morpholine and subsequent addition of hexanethiol under basic conditions affords sulfide **100** in 79% yield (Scheme 45).

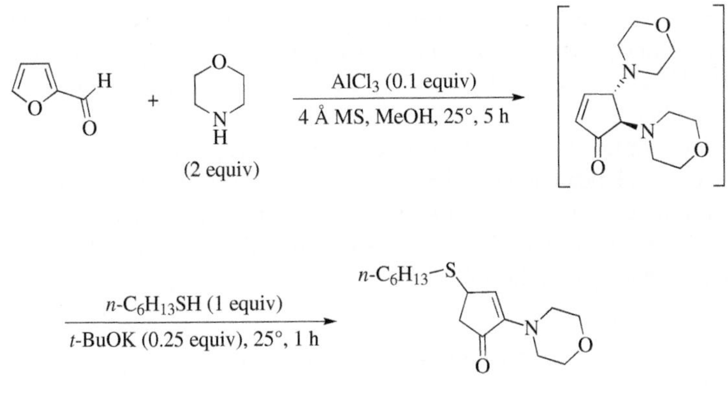

100 (79%)

Scheme 45

Condensation of Glycals with Amines

An interesting alternative to the aza-Piancatelli reaction uses glycals as precursors and indium(III) bromide as a Lewis acid either in dichloromethane at room temperature[112] or in a combination of water and sodium dodecylbenzene sulfonate (SDBS) (a surfactant) at 100° (Scheme 46).[113] This transformation is proposed to occur by the formation of oxocarbenium ion **101** from the glycal. Nucleophilic attack by the aniline, followed by ring-opening and the loss of water, generates intermediate **102**, which undergoes a 4π-conrotatory electrocyclization to furnish the 4-aminocyclopent-2-enone product.

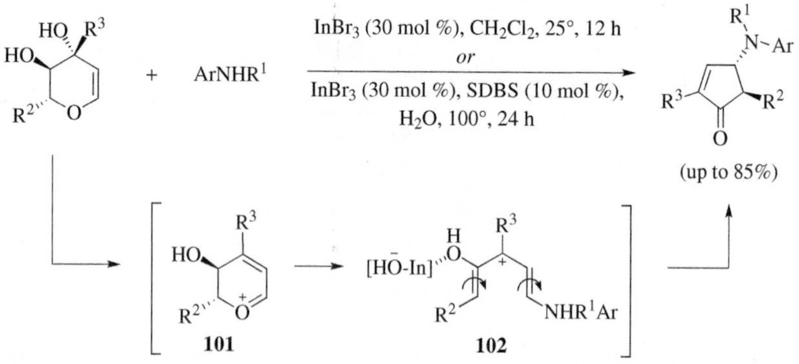

Scheme 46

Cascade Imino-Nazarov Cyclizations

Another method for preparing 4-aminocyclopent-2-enones is based on the formation of 1-aminopentadienyl cations from 4,6-dimethoxyhexa-2,4-dienal and secondary anilines; these cations undergo a Lewis acid catalyzed imino-Nazarov cyclization.[114,115] For example, in the presence of tin(IV) chloride, the starting enal and aniline react to form iminium salt **103**, which undergoes conrotatory 4π-electrocyclization to generate cyclic intermediate **104** that upon hydrolysis furnishes the cyclopentenone derivative (Scheme 47).[114] Interestingly, silver perchlorate promotes a cascade imino-Nazarov cyclization with the same starting materials: the oxyallyl cation is trapped by electrophilic aromatic substitution of the proximal aromatic ring of the aniline.[114]

EXPERIMENTAL CONDITIONS

In general, no special precautions are required for oxa-Piancatelli reactions.[21,116] The aza-Piancatelli reactions carried out in the presence of the Lewis acids Dy(OTf)$_3$ and Ca(NTf$_2$)$_2$ are likewise conducted under an atmosphere of air using reagent-grade solvents.[43,49,54] The aza-Piancatelli reaction run under Mitsunobu conditions with PPTS should be performed under an Ar atmosphere with anhydrous solvents.[44,55] The catalytic asymmetric aza-Piancatelli rearrangement also requires the use of an argon atmosphere and dry reaction tubes,[57] and this is also the case with carba-Piancatelli reactions of enaminones.[63]

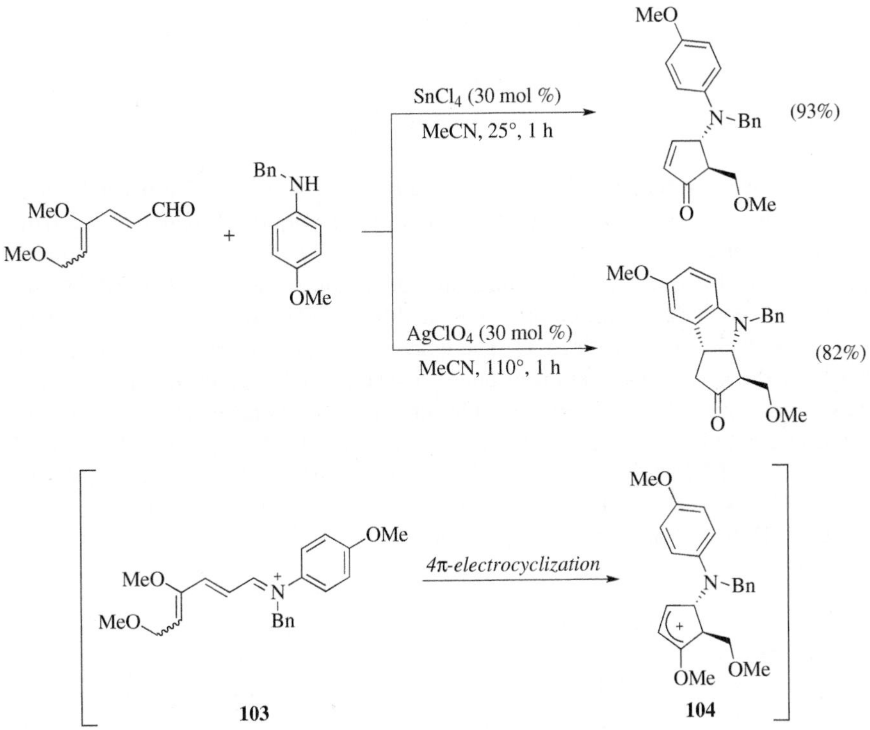

Scheme 47

EXPERIMENTAL PROCEDURES

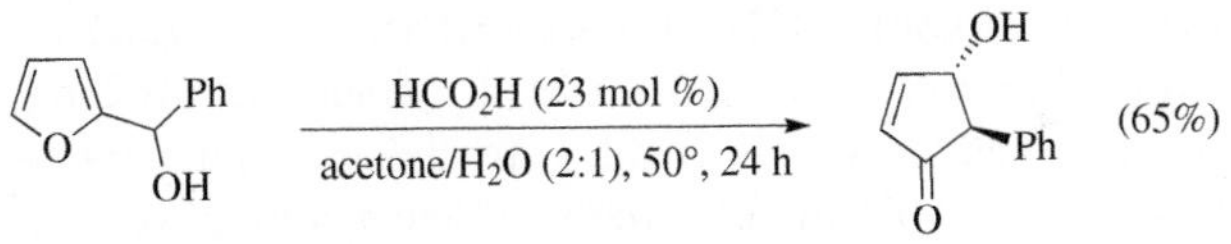

4-Hydroxy-5-phenylcyclopent-2-en-1-one [Oxa-Piancatelli Reaction with a Brønsted Acid].[21] Furan-2-yl(phenyl)methanol (3.00 g, 17.2 mmol) was dissolved in a mixture of acetone/H$_2$O (2:1, 40 mL). Formic acid (0.15 mL, 3.9 mmol, 23 mol %) was added, and the solution was stirred at 50° for 24 h. After being cooled to rt, the mixture was diluted with EtOAc. The organic phase was washed sequentially with NaHCO$_3$ and H$_2$O, dried over anhydrous Na$_2$SO$_4$, and concentrated to yield a reddish oil. This material was purified by column chromatography on silica gel (benzene/Et$_2$O, 2:1) to give the title compound (1.95 g, 65%) as an oil: IR (CHCl$_3$) 1714, 1602 cm^{-1}; ^{1}H NMR (CDCl$_3$) δ 7.38 (dd, J = 3.0, 6.0 Hz, 1H), 6.10 (dd, J = 1.5, 6.0 Hz, 1H), 4.70 (br s, 1H), 3.28 (d, J = 2.5 Hz, 1H).

4-Hydroxy-4-methyl-5-phenylcyclopent-2-en-1-one [Oxa-Piancatelli Reaction with a Lewis Acid].[22] (5-Methylfuran-2-yl)(phenyl)methanol (500 mg, 2.66 mmol) and $ZnCl_2$ (350 mg, 2.57 mmol, 0.9 equiv) were dissolved in a mixture of acetone (20 mL) and H_2O (0.8 mL), and the mixture was stirred at 60° for 24 h. After being cooled to rt, the solution was acidified to pH 3 with 2 N H_2SO_4 and was extracted several times with Et_2O. The combined organic extracts were dried over Na_2SO_4, the solvent was removed under reduced pressure, and the crude product was purified by column chromatography on silica gel (benzene/Et_2O, 1:1) to give the title compound (350 mg, 70%) as a dense oil: IR (CCl_4) 3400, 1700, 1605, 1595 cm^{-1}; ^{1}H NMR (CCl_4) δ 7.25 (d, $J = 6.0$ Hz, 1H), 7.10 (m, 5H), 5.98 (d, $J = 6.0$ Hz, 1H), 4.20 (br s, 1H), 3.57 (s, 1H), 0.85 (s, 3H). Anal. Calcd for $C_{12}H_{12}O_2$: C, 76.57; H, 6.43. Found: C, 76.40; H, 6.62.

4-((4-Iodophenyl)amino)-5-phenylcyclopent-2-en-1-one   [Aza-Piancatelli Reaction with Dysprosium(III) Trifluoromethanesulfonate].[48] Furan-2-yl (phenyl)methanol (128 mg, 0.74 mmol) and 4-iodoaniline (162 mg, 0.74 mmol, 1 equiv) were dissolved in MeCN (6 mL). To the reaction mixture at 23° was added Dy(OTf)$_3$ (23 mg, 0.038 mmol, 5 mol %). The resulting reaction mixture was heated at 80° for 30 min. The reaction was then quenched at rt by the addition of saturated aqueous sodium bicarbonate (5 mL), and the mixture was extracted with EtOAc (3 × 10 mL). The combined organic layers were dried over $MgSO_4$ and concentrated in vacuo. The residue was purified by column chromatography on silica gel to afford the title product (255 mg, 92%) as an oil: IR (thin film) 3776, 3026, 1704, 1588, 1496, 1316 cm^{-1}; ^{1}H NMR (500 MHz, $CDCl_3$) δ 7.72 (dd, $J = 2.4$, 5.7 Hz, 1H), 7.40–7.28 (m, 5H), 7.15–7.09 (m, 2H), 6.40 (dd, $J = 1.7$, 5.7 Hz, 1H), 6.27 (dddd, $J = 2.0, 2.0, 2.0, 9.8$ Hz, 2H), 4.69 (dd, $J = 1.7, 8.2$ Hz, 1H), 4.19 (d, $J = 8.3$ Hz, 1H), 3.35 (d, $J = 2.6$ Hz, 1H); ^{13}C NMR (125 MHz, $CDCl_3$) δ 206.4, 161.5, 145.9, 138.0, 137.9, 135.0, 129.2, 128.0, 127.6, 116.0, 79.6, 63.2, 60.1; HRMS–ESI (m/z): [M + Na]$^+$ calcd for $C_{17}H_{14}INO$, 398.0012; found, 398.0022.

(1.1 equiv)

5-Cyclopropyl-4-((4-iodophenyl)amino)cyclopent-2-en-1-one [Aza-Pianca-telli Reaction with Calcium(II) Bis(trifluoromethanesulfonimide)].[55] To a solution of cyclopropyl(furan-2-yl)methanol (30 mg, 0.217 mmol) and 4-iodoaniline (52 mg, 0.239 mmol, 1.1 equiv) in HFIP (0.65 mL) were added Ca(NTf$_2$)$_2$ (6.2 mg, 0.011 mmol, 5 mol %) and n-Bu$_4$NPF$_6$ (4.2 mg, 0.011 mmol, 5 mol %). The reaction mixture was stirred at 20° for 30 min. The mixture then was transferred to a silica gel chromatography column (pentane/EtOAc, 9:1) to give the title compound (62 mg, 81%) as an orange oil: IR (neat) 3371, 3080, 3004, 1703, 1588, 1496, 1438, 1345, 1317, 1294, 1264, 1183 cm^{-1}; ^{1}H NMR (250 MHz, CDCl$_3$) δ 7.56 (dd, J = 2.3, 5.8 Hz, 1H), 7.46 (d, J = 8.9 Hz, 2H), 7.38–7.27 (br s, 1H), 6.49 (d, J = 8.9 Hz, 2H), 6.24 (dd, J = 1.6, 5.8 Hz, 1H), 4.47 (dd, J = 2.2, 4.0 Hz, 1H), 1.70 (dd, J = 2.6, 8.6 Hz, 1H), 0.96–0.84 (m, 1H), 0.70–0.60 (m, 1H), 0.56–0.40 (m, 2H), 0.30–0.25 (m, 1H); ^{13}C NMR (91 MHz, CDCl$_3$) δ 207.6, 160.4, 146.1, 138.1, 134.5, 115.8, 79.3, 60.3, 56.9, 11.5, 3.5, 2.5; HRMS–ESI (m/z): [M + H]$^+$ calcd for C$_{14}$H$_{15}$INO, 340.0198; found, 340.0190.

1-Butyl-1-azaspiro[4.4]non-8-en-7-one [Aza-Piancatelli Reaction Under Mitsunobu Conditions with PPTS].[23] Under an argon atmosphere, (5-(3-(butyl-amino)propyl)furan-2-yl)methanol (66 mg, 0.313 mmol) was dissolved in DMA (6 mL). Triphenylphosphine (124 mg, 0.47 mmol, 1.5 equiv) and PPTS (16 mg, 0.063 mmol, 20 mol %) were added successively, and the mixture was stirred at 0° for 2 min. Next, DEAD (82 mg, 74 μL, 0.47 mmol, 1.5 equiv) was added to the solution, and the reaction mixture was warmed to rt and stirred at that temperature for 30 min. Dimethylacetamide was removed by vacuum distillation, and the residue was purified by column chromatography on silica gel (EtOAc/petroleum ether/TEA, 1:1:0.01 to 3:1:0.01) to yield the title compound as a colorless liquid (43 mg, 71%): IR (KBr) 2958, 1720, 1461, 1408, 1340, 1218, 1069, 798 cm^{-1}; ^{1}H NMR (400 MHz, CDCl$_3$) δ 7.39 (d, J = 5.6 Hz, 1H), 6.13 (d, J = 5.6 Hz, 1H), 3.08–3.02 (m, 1H), 2.65–2.52 (m, 1H), 2.46 (d, J = 18.5 Hz, 1H), 2.32–1.71 (m, 7H), 1.49–1.33 (m, 2H), 1.34–1.20 (m, 2H), 0.87 (t, J = 7.3 Hz, 3H); ^{13}C NMR (101 MHz, CDCl$_3$) δ 208.0, 168.1, 133.9, 71.6, 51.6, 48.9, 40.7, 37.9, 31.4, 21.6, 20.5, 13.9; HRMS–ESI (m/z): [M + H]$^+$ calcd for C$_{12}$H$_{19}$NO, 193.1467; found, 193.1469.

(4S*,5R*)-4-((3,5-Bis(trifluoromethyl)phenyl)amino)-5-phenylcyclopent-2-en-1-one [Asymmetric Aza-Piancatelli Reaction with a Chiral Brønsted Acid].[57] A dry reaction tube equipped with a stir bar was charged with 2-aminobenzonitrile (5.9 mg, 0.05 mmol) and the catalyst (2.2 mg, 0.0025 mmol, 5 mol %). The tube was capped with a rubber septum, evacuated, and backfilled with argon, and then $CHCl_3$ (2.0 mL) was added. The resulting mixture was cooled to 5°, and furan-2-yl(phenyl)methanol (10.4 mg, 0.06 mmol, 1.2 equiv) was added. The reaction mixture was stirred at 5° for 24 h. After being warmed to rt, the mixture was purified by column chromatography on silica gel (hexane/EtOAc, 5:1 to 3:1) to yield the title compound (12.6 mg, 92%, er 95:5, dr >19:1): HPLC t_R (major) 16.4 min, t_R (minor) 14.7 min (Chiralcel IA, hexane/i-PrOH, 4:1, 1.0 mL/min, 254 nm); $[\alpha]^{rt}_D$ − 239.7 (c 0.6, CH_2Cl_2); IR (ATR) 3351, 3068, 2214, 1708, 1590, 1511, 1456, 1322, 1280, 1166, 1043, 745, 698 cm^{-1}; ^{1}H NMR (400 MHz, $CDCl_3$) δ 7.72 (dd, J = 2.3, 5.7 Hz, 1H), 7.40 (dd, J = 1.5, 7.7 Hz, 1H), 7.37–7.26 (m, 3H), 7.24–7.17 (m, 1H), 7.16–7.08 (m, 2H), 6.70 (t, J = 7.6 Hz, 1H), 6.47 (dd, J = 1.7, 5.7 Hz, 1H), 6.37 (d, J = 8.5 Hz, 1H), 4.94 (d, J = 7.8 Hz, 1H), 4.87–4.73 (m, 1H), 3.42 (d, J = 2.6 Hz, 1H); ^{13}C NMR (101 MHz, $CDCl_3$) δ 205.4, 160.0, 148.4, 137.5, 135.6, 134.3, 133.0, 129.1, 127.9, 127.6, 117.9, 117.4, 111.7, 96.8, 62.6, 60.2; HRMS–ESI (m/z): [M + Na]$^+$ calcd for $C_{18}H_{14}N_2O$, 297.0998; found, 297.0994.

1-(4-Iodophenyl)-2-phenyl-6-(p-tolyl)-4,6-dihydrocyclopenta[b]pyrrol-5(1H)-one [Synthesis of a Pyrrole].[43] To a solution of (4-(phenylethynyl)furan-2-yl)(p-tolyl)methanol (25 mg, 0.087 mmol) and 4-iodoaniline (25 mg, 0.11 mmol, 1.3 equiv) in DCE (0.3 mL) were added $Ca(NTf_2)_2$ (2.6 mg, 0.0043 mmol, 5 mol %) and n-Bu$_4$NPF$_6$ (1.7 mg, 0.0043 mmol, 5 mol %). The reaction mixture was stirred at 80°

for 10 min, and then $Cu(OTf)_2$ (3.2 mg, 0.0087 mmol, 10 mol %) was added. The reaction mixture was stirred at 40° for 10 min. After being cooled to rt, the mixture was directly purified by column chromatography on silica gel (pentane/EtOAc, 97.5:2.5) to give the title product (32 mg, 75%): mp 83–86°; IR (neat) 3056, 3025, 2921, 2857, 1751, 1683, 1595, 1509, 1489, 1463, 1390, 1361, 1264, 1181, 1058 cm^{-1}; ^{1}H NMR (360 MHz, CDCl$_3$) δ 7.41 (d, J = 8.4 Hz, 2H), 7.25–7.10 (m, 5H), 7.03 (d, J = 7.8 Hz, 2H), 6.88 (d, J = 7.9 Hz, 2H), 6.60 (d, J = 8.5 Hz, 2H), 6.47 (s, 1H), 4.43 (s, 1H), 3.54 (d, J = 21.4 Hz, 1H), 3.45 (d, J = 21.4 Hz, 1H), 2.29 (s, 3H); ^{13}C NMR (91 MHz, CDCl$_3$) δ 213.1, 138.8, 138.0, 137.2, 137.0, 136.6, 134.9, 132.6, 129.6, 128.5, 128.3, 127.6, 126.8, 122.0, 106.4, 91.9, 55.7, 39.1, 21.2 (one carbon was not observed); HRMS–ESI (m/z): [M + Na]$^+$ calcd for $C_{26}H_{20}INO$, 512.0467; found, 512.0482.

(1.0 equiv)

(94%)

(3a,6a)-3-Benzoyl-1,4,4-triphenyl-3a,4,6,6a-tetrahydrocyclopenta[*b*]pyrrol-5(1*H*)-one [Carba-Piancatelli Reaction with an Enaminone].[63] To a solution of ZnCl$_2$ (4.1 mg, 0.03 mmol, 10 mol %) in DCE (3 mL) was added furan-2-yldiphenylmethanol (75.1 mg, 0.3 mmol, 1.0 equiv) and (Z)-1-phenyl-3-(phenyl-amino)prop-2-en-1-one (67 mg, 0.3 mmol). The resulting solution was stirred at 80° for 2 h. After cooling to rt, the solvent was evaporated under reduced pressure, and the residue was purified by column chromatography on silica gel (petroleum ether/EtOAc, 3:1) to afford the title compound (129 mg, 94%) as a brown solid: mp 211–214°; ^{1}H NMR (400 MHz, CDCl$_3$) δ 7.86–7.84 (m, 2H), 7.44–7.26 (m, 8H), 7.19–7.15 (m, 3H), 7.10–7.02 (m, 4H), 6.89–6.83 (m, 4H), 5.29 (d, J = 10.0 Hz, 1H), 5.11–5.07 (m, 1H), 3.05 (dd, J = 8.0, 18.8 Hz, 1H), 2.80 (d, J = 18.8 Hz, 1H); ^{13}C NMR (100.6 MHz, CDCl$_3$, Me$_4$Si) δ 215.4, 189.3, 147.9, 141.6, 140.6, 139.5, 139.1, 130.5, 130.3, 129.7, 128.7, 128.5, 127.9, 127.8, 127.5, 127.1, 126.5, 123.1, 120.1, 116.2, 67.2, 60.4, 52.6, 41.7; HRMS–ESI (m/z): [M + H]$^+$ calcd for $C_{32}H_{26}NO_2$, 456.1964; found, 456.1953.

The literature was extensively searched using CAS databases, and the following tables incorporate reactions reported in the literature through April 2020. All examples of Piancatelli reactions, aza-Piancatelli reactions, and carba-Piancatelli reactions are described in the tables. Tandem reactions including Piancatelli rearrangements are also included. An em-dash (—) indicates that the authors did not report a yield, dr, or er for this example. In cases where multiple sets of conditions were reported as part of an optimization study, only the conditions giving the highest yield are shown.

The tables are organized by function of the product framework. Thus, cyclopent-2-enones and cyclopentanones are treated separately and the corresponding tables are defined by the main substituents. All tables are organized by increasing carbon count of the furylcarbinol. Simple alkyl, aryl, and protecting groups on oxygen and nitrogen have been eliminated from the count in order to juxtapose starting materials with similar cores.

The following abbreviations, excluding those found in *"The Journal of Organic Chemistry* Standard Abbreviations and Acronyms"* are used in the text and the Tables.

BTEAC	benzyltriethylammonium chloride
Cy	cyclohexyl
DBE	dibasic ester
DEAD	diethyl azodicarboxylate
DMAc	dimethylacetamide
HFIP	1,1,1,3,3,3-hexafluoro-2-propanol
HMim	1-hexyl-3-methylimidazolium
LA	Lewis Acid
MW	microwave heating
Np	naphthyl
PMA	phosphomolybdic acid
PMP	4-methoxyphenyl
PPTS	pyridinium 4-toluenesulfonate
SDBS	sodium dodecylbenzenesulfonate
SPINOL	1,1'-spirobiindane-7,7'-diol
TBDPS	*tert*-butyldiphenylsilyl

CHART 1. CATALYSTS AND LIGANDS USED IN TEXT AND TABLES

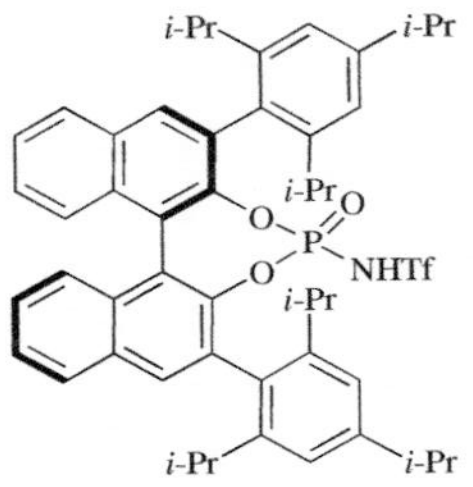

1

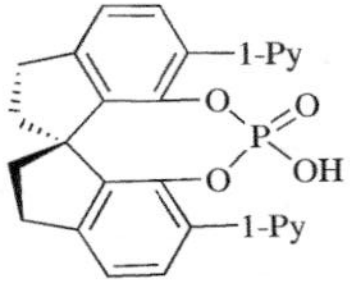

2

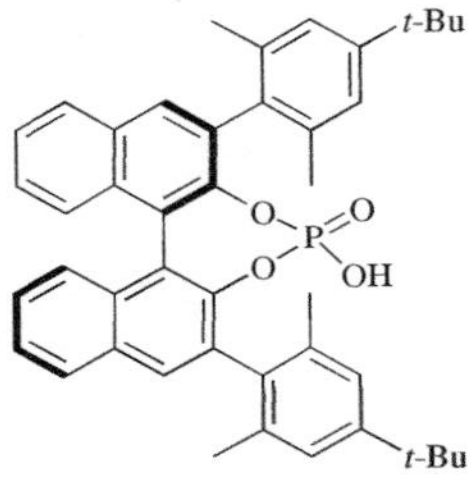

3

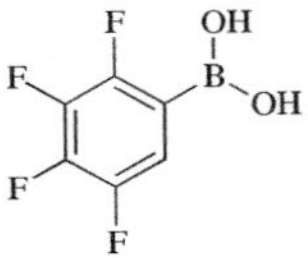

4

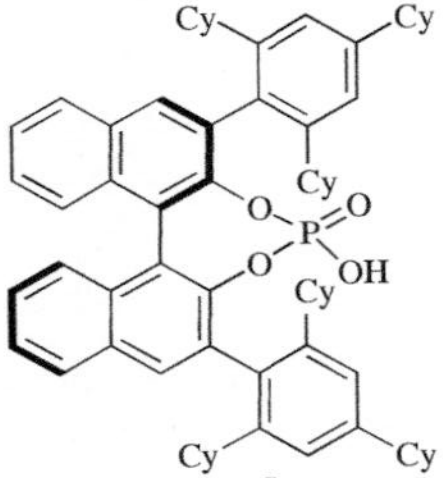

5

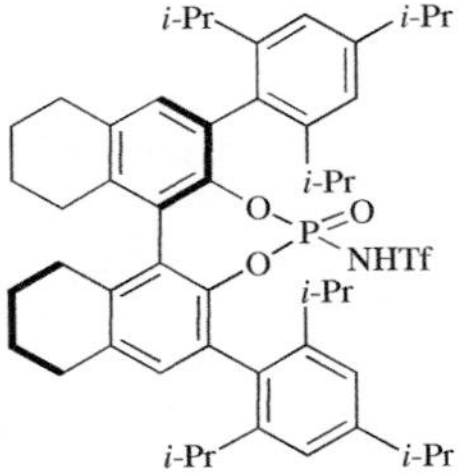

6

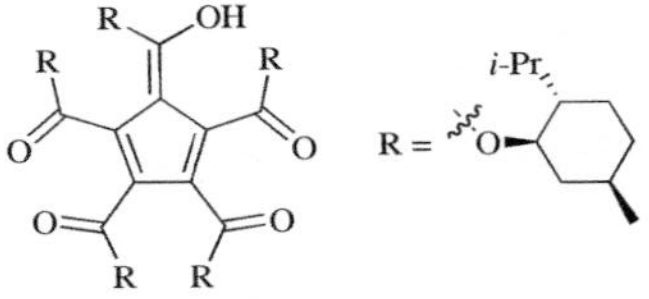

7

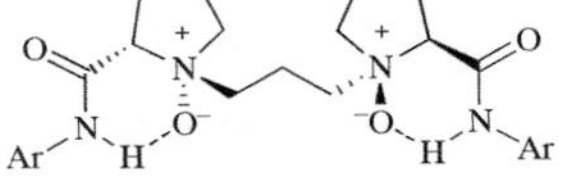

8 Ar = 2,4,6-$Cy_3C_6H_2$

TABLE 1. 4-HYDROXYCYCLOPENT-2-ENONES

Substrate	Conditions	Product(s) and Yield(s) (%)	Refs.
C$_5$			
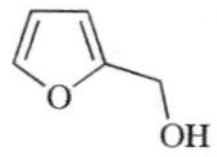	1. KH$_2$PO$_4$ (36.5 eq), H$_2$O, reflux, 40 h 2. TBSCl (0.94 eq), Et$_3$N (1.58 eq), DMAP, THF, rt	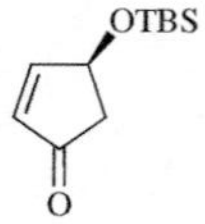 (20)	88
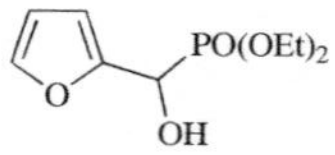	1. Et$_3$N, MeSO$_2$Cl, rt, 18 h 2. DME/H$_2$O (9:1), rt, 48 h	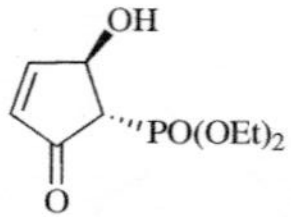 (54)	117
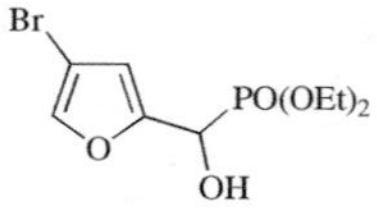	1. Et$_3$N, MeSO$_2$Cl, rt, 4 h 2. DME/H$_2$O (9:1), rt, 96 h	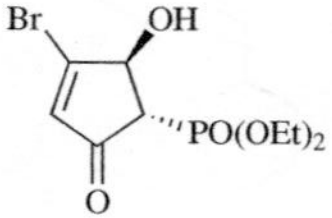 (60)	117
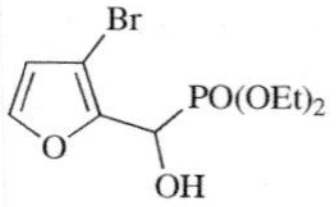	1. Et$_3$N, MeSO$_2$Cl, rt, 1 h 2. SiO$_2$, EtOAc, rt	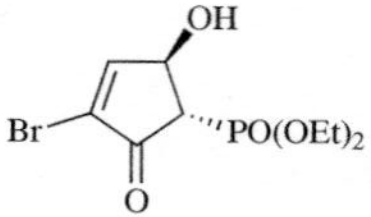 (78)	117
C$_6$			
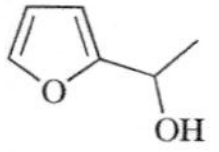	10% aq MgSO$_4$, 160°, 3 h	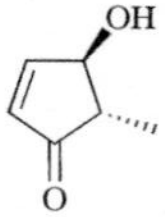 (56)	89, 91

538

21

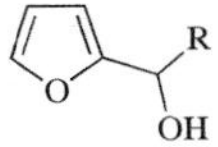

4-TsOH (6 mol %),
acetone/H$_2$O (2:1), 50°, 24 h

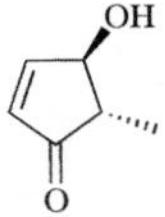

(30)

118

H$_2$SO$_4$ (concd),
acetone/H$_2$O (1.25:1), 50°, 35 h

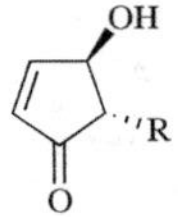

R	
Me	(18)
i-Pr	(39)
Bn	(41)
PhCH$_2$CH$_2$	(45)

116

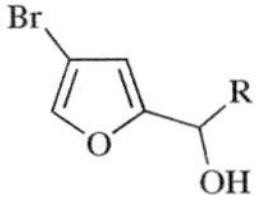

H$_2$SO$_4$ (concd),
DME/H$_2$O (2:1), 85°

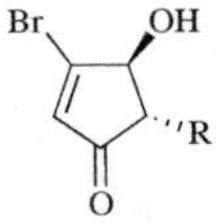

R	Time (h)	
Me	6	(81)
Ph	1	(85)
n-C$_{12}$H$_{25}$	6	(75)

117

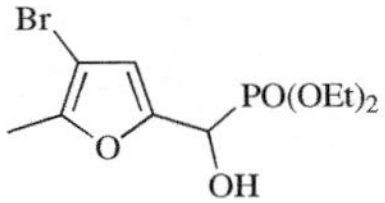

1. Et$_3$N, MeSO$_2$Cl, rt, 1 h
2. DME/H$_2$O (9:1), rt, 24 h

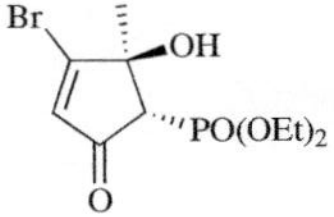

(33)

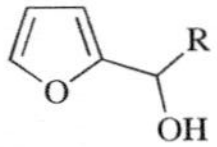

H$_2$O, MW (300 W), 210°

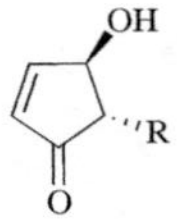

R	Time (min)		dr
Et	15	(54)	7:1
(E)-MeCH=CHCH$_2$	2	(73)	12:1 48
n-C$_5$H$_{11}$	15	(65)	7:1
Ph	2	(96)	1:5

C$_{6-11}$

C$_6$

C$_{7-12}$

TABLE 1. 4-HYDROXYCYCLOPENT-2-ENONES (*Continued*)

Substrate	Conditions	Product(s) and Yield(s) (%)	Refs.

C$_7$

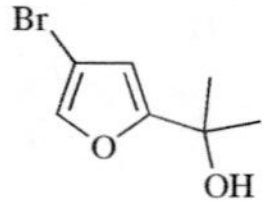

H$_2$SO$_4$ (concd),
DME/H$_2$O (2:1), 85°, 12 h

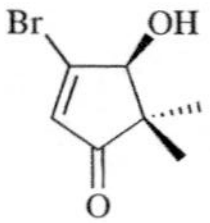

(28)

116

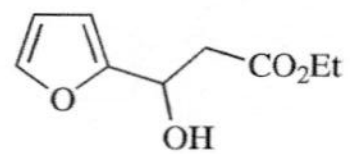

1. ZnCl$_2$ (10 eq),
1,4-dioxane. H$_2$O, reflux, 3 h
2. chloral (0.6 eq). Et$_3$N (1.5 eq),
EtOAc, rt, 12 h

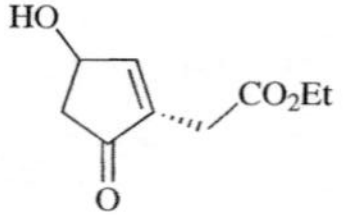

(33)

92

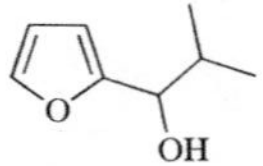

ZnCl$_2$, HCl (pH 6),
1,4-dioxane. H$_2$O, reflux, 48 h

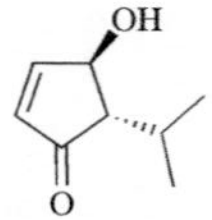

(85)

85

ZnCl$_2$ (4.1 eq), HCl (pH 5.5),
1,4-dioxane/H$_2$O (1.9:1), reflux, 48 h

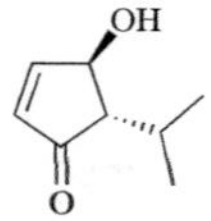

(80)

86

C$_8$

R	Time (h)	
allyl	15	(72)
i-Pr	34	(56)
n-Bu	16	(90)
2-thienyl	1	(72)
Ph	6	(84)

C$_{8-20}$

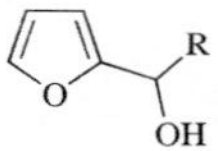

Dy(OTf)$_3$ (10 mol %),
TFA (5 mol %),
t-BuOH/H$_2$O (5:1), 80°

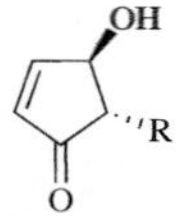

R	Time (h)	
4-NCC$_6$H$_4$	18	(72)
2,4,6-Me$_3$C$_6$H$_2$	3.5	(80)
2-Np	5	(86)
2,4,6-(*i*-Pr)$_3$C$_6$H$_2$	2	(84)

119

			70
			38
			120
			71
			21

C$_8$

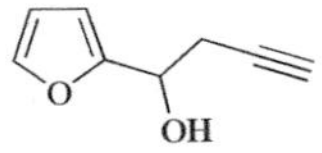

4-TsOH (7.1 mol %),
1,4-dioxane/H$_2$O (8:1), 85°, 36 h

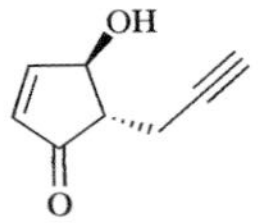

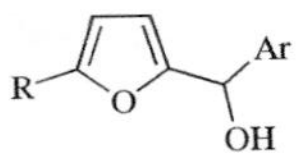

(31)

C$_{9-12}$

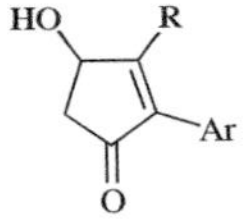

H$_2$O, 100°

Ar	R	Time (h)	
2-thienyl	H	24	(95)
2-thienyl	Me	36	(95)
Ph	H	24	(99)
Ph	Me	36	(96)

C$_{10-12}$

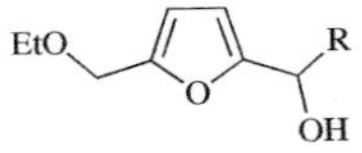

Dy(OTf)$_3$ (10 mol %),
t-BuOH/H$_2$O (5:1), 90°, 24 h

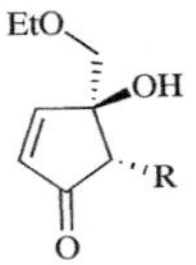

R		dr
n-Bu	(9)	5:1
2-thienyl	(37)	4:1
Ph	(60)	23:1
3-F-4-MeOC$_6$H$_3$	(66)	15:1

C$_{10}$

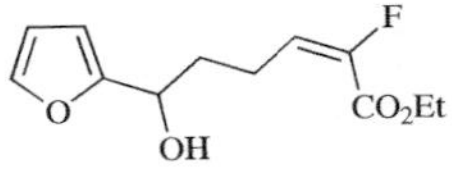

ZnCl$_2$ (10 eq)
dioxane/H$_2$O (2:1), reflux, 32 h

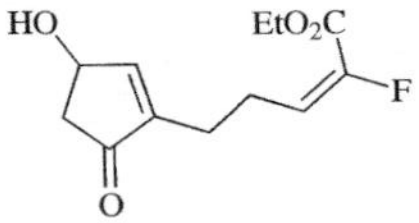

(27)

C$_{11}$

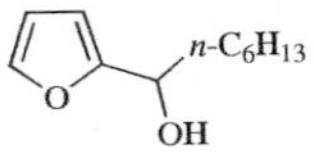

PMA (33 mol %),
acetone/H$_2$O (2:1), 50°, 24 h

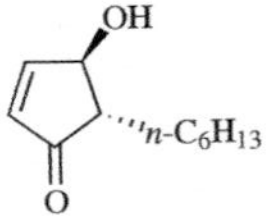

(70)

541

TABLE 1. 4-HYDROXYCYCLOPENT-2-ENONES (Continued)

Substrate	Conditions	Product(s) and Yield(s) (%)	Refs.

C₁₁

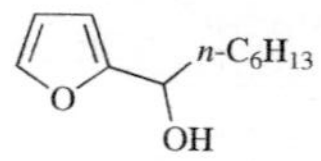

H₂O, 100°, 168 h

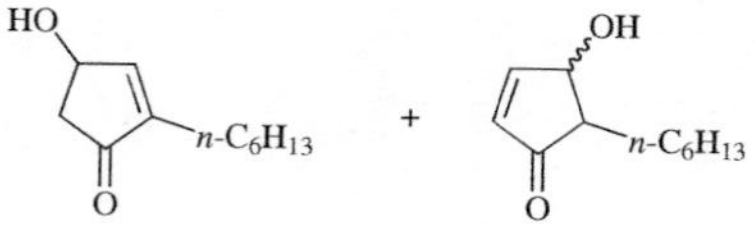

(65) + (32)

38

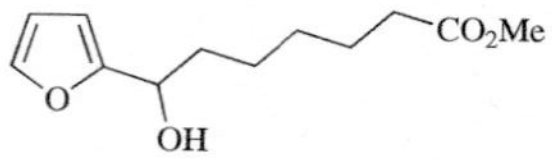

1. ZnCl₂ (5 eq), MeI (5 eq),
dioxane/acetone (1.5:1), reflux, 16 h
2. Chloral (0.5 eq), Et₃N (1.5 eq),
toluene, reflux, 6 h

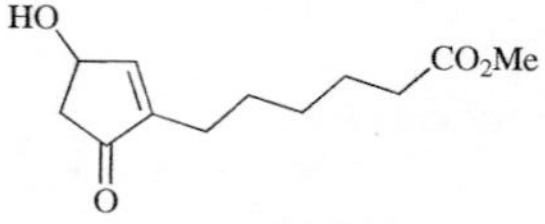

(31)

75

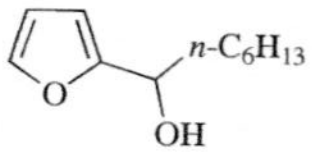

PMA (33 mol %),
acetone/H₂O (2:1), 50°, 24 h

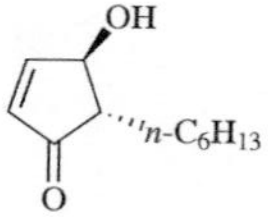

(70)

21

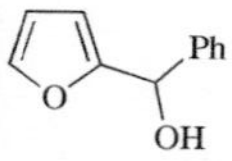

HCO₂H (23 mol %),
acetone/H₂O (2:1), 50°, 24 h

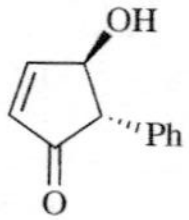

(65)

21

C$_{11-17}$

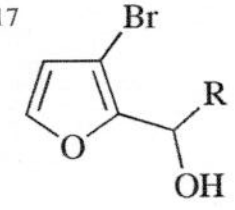

H$_2$SO$_4$ (concd),
DME/H$_2$O (2:1), 90°, 0.3 h

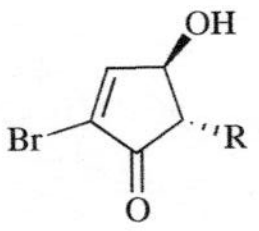

R	
Ph	(82)
n-C$_{12}$H$_{25}$	(85)

116

C$_{11}$

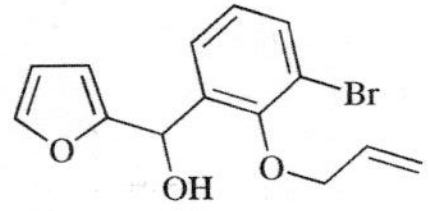

1. 4-TsOH, THF/H$_2$O
2. Chloral, Et$_3$N

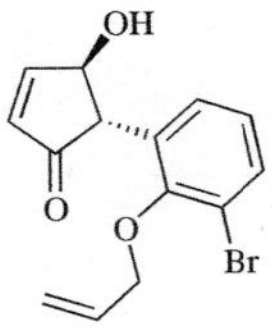

(50)

73

C$_{12}$

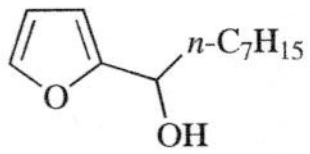

PPA (9.7 mol %),
acetone/H$_2$O (2:1), 65°, 48 h

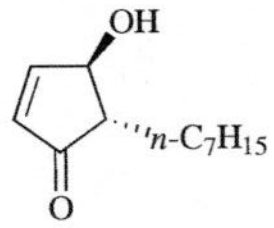

n-C$_7$H$_{15}$

(50)

87

CO$_2$Me

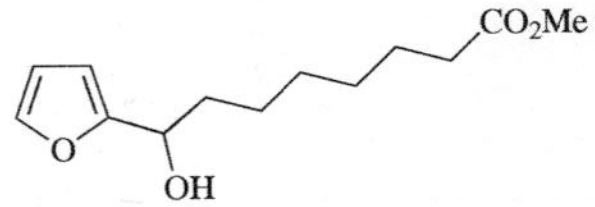

1. ZnCl$_2$ (3.8 eq),
   dioxane/H$_2$O (1.5:1), reflux, 6 h
2. Chloral (0.6 eq), Et$_3$N (0.6 eq),
   toluene, rt

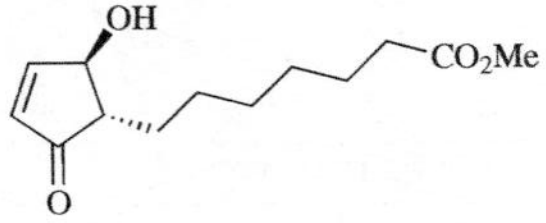

CO$_2$Me

(72)

68

TABLE 1. 4-HYDROXYCYCLOPENT-2-ENONES (Continued)

Substrate	Conditions	Product(s) and Yield(s) (%)	Refs.
C₁₂			
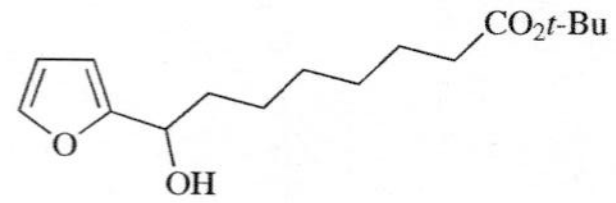	PPA (11 mol %), acetone/H₂O (2:1), 50°, 24 h	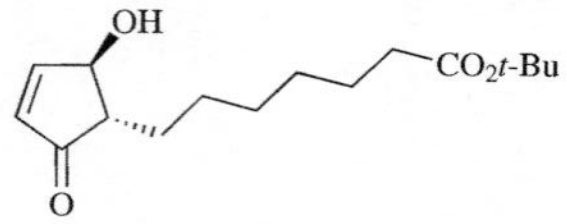 (51)	45
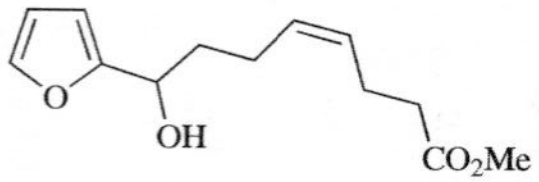	1. ZnCl₂ (3.9 eq), dioxane/H₂O (1.5:1), reflux, 22 h 2. MeI (3.4 eq), K₂CO₃ (3.4 eq), acetone, reflux, 9 h	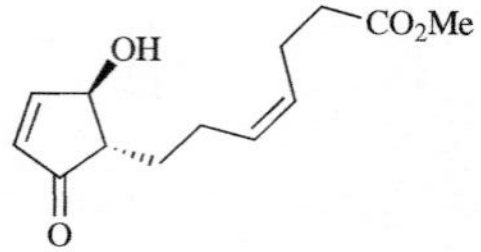 (94)	72
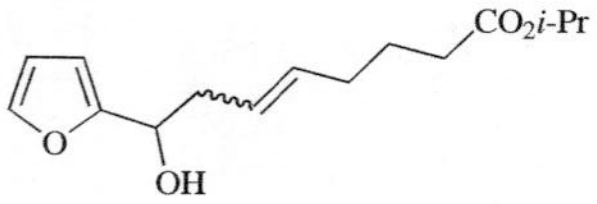	1. ZnCl₂, hydroquinone, dioxane, H₂O, reflux, 12 h 2. Chloral, Et₃N, toluene, rt	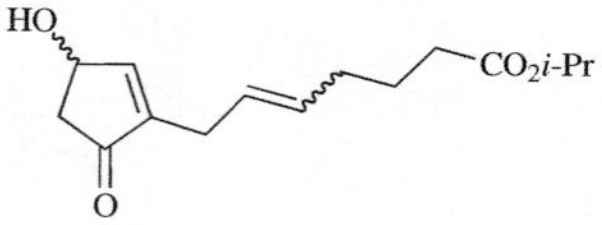 (55)	78
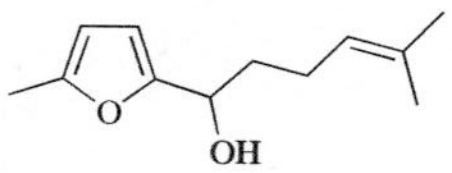	MgCl₂ (4 eq), H₂O/dioxane (1.67:1), reflux, 48 h	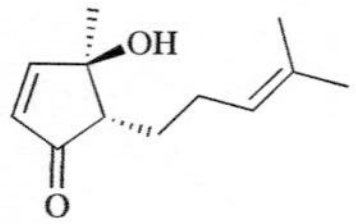 (58)	90
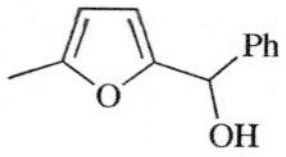	Dy(OTf)₃ (10 mol %), TFA (5 mol %), t-BuOH/H₂O, 80°, 15.5 h	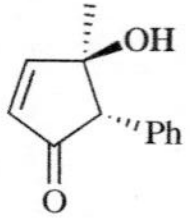 (40) dr 5:1	119

544

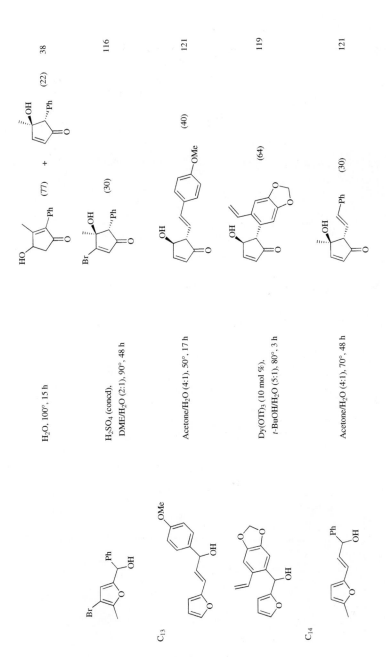

TABLE 1. 4-HYDROXYCYCLOPENT-2-ENONES (*Continued*)

Substrate	Conditions	Product(s) and Yield(s) (%)	Refs.
C_{15–16} 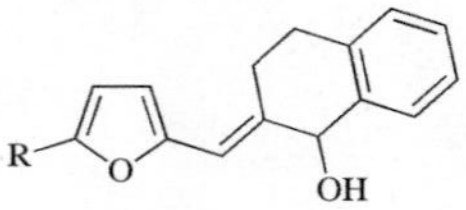	Acetone/H₂O (4:1), 48 h	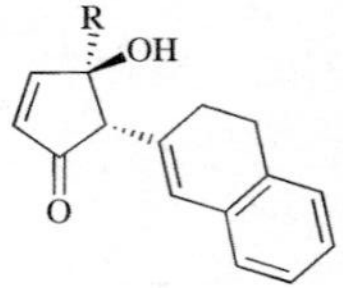 R, Temp (°): H 50 (43); Me 70 (32)	121
C₁₇ 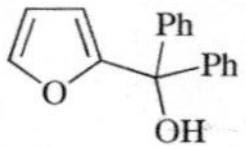	Dy(OTf)₃ (10 mol %), TFA (5 mol %), *t*-BuOH/H₂O (5:1), 80°, 43 h	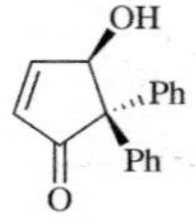 (64)	119
C₂₆ 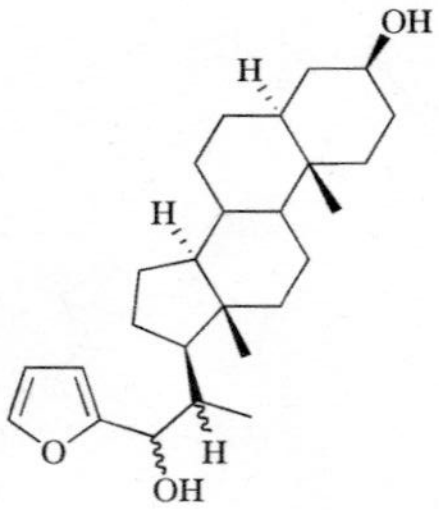	H₂SO₄ (concd), acetone/H₂O (1.25:1), 50°, 30 h	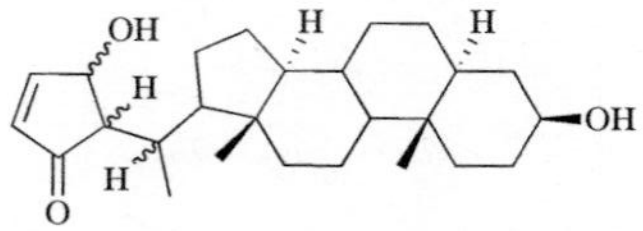 (90)	82

TABLE 2. 4-ALKOXYCYCLOPENT-2-ENONES

Substrate	Conditions	Product(s) and Yield(s) (%)				Refs.

C12-18

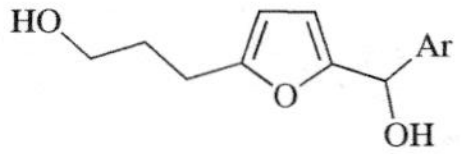

Substrate: furan-derived alcohol with Ar and HO(CH2)3 chain

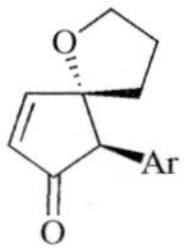

Conditions: Dy(OTf)3 (5 mol %), toluene

Ar	Temp (°)	Time (h)	
2-thienyl	80	4	(90)
Ph	80	6	(91)
4-BrC6H4	80	24	(83)
4-O2NC6H4	100	16	(75)
4-MeOC6H4	80	20	(89)
4-CF3C6H4	100	16	(88)
2,4,6-Me3C6H2	80	8	(78)
2-Np	80	6	(86)

Refs. 49

C12-16

Conditions: Aqueous ethylene glycol, dimethyl ether, ZnCl2 (1 mol %), reflux

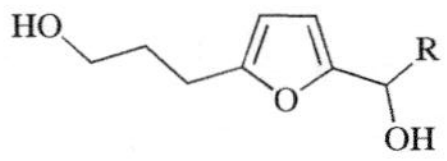

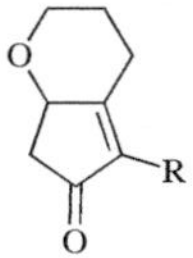

R	Time (h)	
2-thienyl	4	(83)
Ph	5	(81)
4-O2NC6H4	6	(75)
4-MeOC6H4	3.5	(90)
4-MeC6H4	4	(85)
(E)-PhCH=CH	3	(84)

Refs. 41

C13

Conditions: Dy(OTf)3 (5 mol %), toluene, 80°, 1 h

Product: (25)

Refs. 49

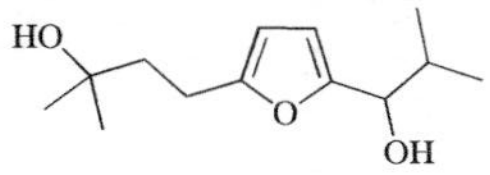

C13

Conditions: Dy(OTf)3 (5 mol %), toluene, 80°, 1 h

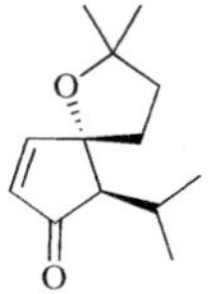

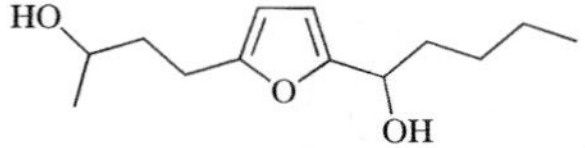

Product: (30) dr 2:1

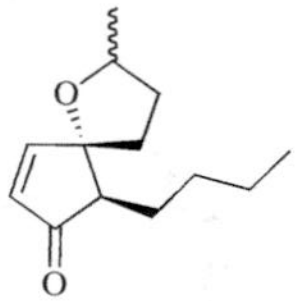

Refs. 49

TABLE 2. 4-ALKOXYCYCLOPENT-2-ENONES (*Continued*)

Substrate	Conditions	Product(s) and Yield(s) (%)	Refs.
C₁₄			
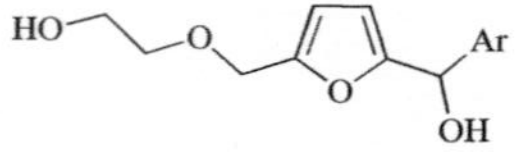	Aqueous ethylene glycol, dimethyl ether, ZnCl₂ (1 mol %), reflux	 	R	Time (h)	
---	---	---
Ph	6	(75)
4-O₂NC₆H₄	8	(70)
4-MeOC₆H₄	5	(86)		41
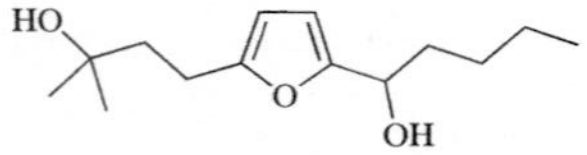	Dy(OTf)₃ (5 mol %), toluene, 80°, 1 h	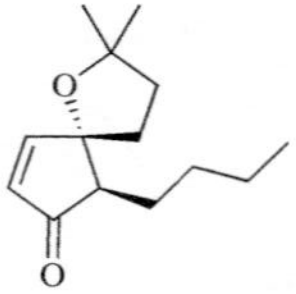 (98)	49
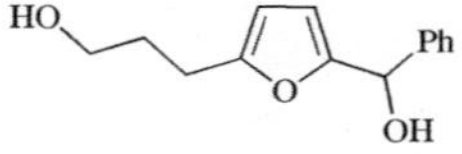	Amberlyst 15, toluene, 60°	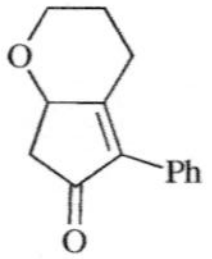 (39)	122
C₁₅₋₂₀			
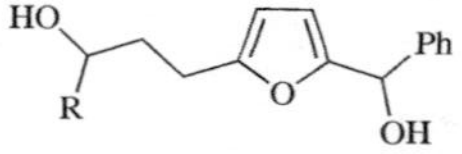	Dy(OTf)₃ (5 mol %), toluene, 80°	 | 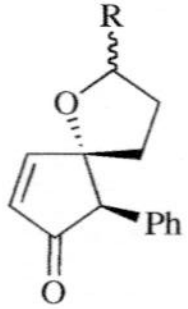 R	Time (h)		dr
Me	4	(73)	2:1
i-Pr	20	(71)	2:1
Ph	72	(90)	2:1		49
C₁₅			
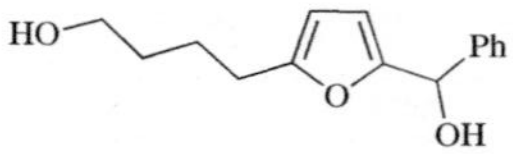	Aqueous ethylene glycol, dimethyl ether, ZnCl₂ (1 mol %), reflux, 6 h	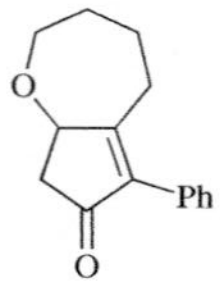 (78)	41

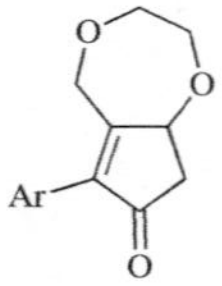

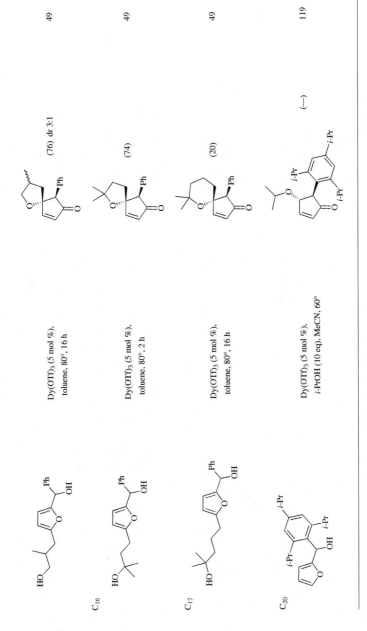

Dy(OTf)$_3$ (5 mol %),
toluene, 80°, 16 h

Dy(OTf)$_3$ (5 mol %),
toluene, 80°, 2 h

Dy(OTf)$_3$ (5 mol %),
toluene, 80°, 16 h

Dy(OTf)$_3$ (5 mol %),
i-PrOH (10 eq), MeCN, 60°

(76) dr 3:1

(74)

(20)

49

49

49

119

TABLE 3. 4-AMINOCYCLOPENT-2-ENONES

Substrate	Nucleophile	Conditions	Product(s) and Yield(s) (%)	Refs.

*Please refer to the chart preceding the tables for the structures indicated by the **bold** numbers.*

C_5

Substrate: furan-CH₂OH, 1.2 eq

Nucleophile: benzamide (H_2N, H_2N-C=O, R)

Conditions: 4-TsOH (5 mol %), MeCN, 80°, 2 h

Product:

R	
H	(32)
Br	(40)

Refs.: 61

C_{5-12}

Substrate: furan with R^1, CHO

Nucleophile: HNR^2R^3 (2.2 eq)

Conditions: $Dy(OTf)_3$ (5 mol %), 4 Å MS, MeCN, rt, o/n

Product: cyclopentenone with NR^2R^3, R^1

Refs.: 102

Nucleophile table:

R^1	R^2	R^3	
Me	$-(CH_2)_2O(CH_2)_2-$		(100)
TMSC≡C	$CH_2CH=CH_2$	$CH_2CH=CH_2$	(41)
2-thienyl	Bn	Bn	(100)
4-pyridyl	$CH_2CH=CH_2$	$CH_2CH=CH_2$	(73)
Ph	$CH_2CH=CH_2$	$CH_2CH=CH_2$	(87)
Ph	Bn	Bn	(100)

Product table:

R^1	R^2	R^3	
$4\text{-}FC_6H_4$	$-(CH_2)_2O(CH_2)_2-$		(84)
$4\text{-}FC_6H_4$	$CH_2CH=CH_2$	$CH_2CH=CH_2$	(85)
$4\text{-}FC_6H_4$	Bn	Bn	(93)
$4\text{-}MeOC_6H_4$	Bn	Bn	(94)
$2\text{-}MeC_6H_4$	$-(CH_2)_2O(CH_2)_2-$		(100)
$4\text{-}MeC_6H_4$	$-(CH_2)_5-$		(97)
$4\text{-}MeC_6H_4$	$CH_2CH=CH_2$	$CH_2CH=CH_2$	(70)

C_6

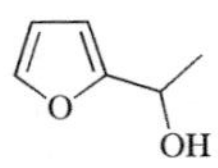

NH_2R

x eq

Catalyst (*y* mol %),
additive (5 mol %),
solvent

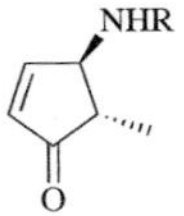

R	*x*	Catalyst	*y*	Additive	Solvent	Temp (°)	Time (h)		
2-ClC_6H_4	3.0	(2,3,4,5-tetrafluorophenyl)boronic acid	20	—	MeCN	60	—	(42)	123
3-ClC_6H_4	1.0	$Dy(OTf)_3$	5	—	MeCN	80	3	(10)	23
4-IC_6H_4	1.0	$Dy(OTf)_3$	5	—	MeCN	80	2.5	(68)	23
4-IC_6H_4	1.1	$Ca(NTf_2)_2$	5	*n*-Bu_4NPF_6	$MeNO_2$	80	1	(80)	54
4-$MeOC_6H_4$	1.1	$Ca(NTf_2)_2$	5	*n*-Bu_4NPF_6	$MeNO_2$	100	6	(62)	55
4-$MeOC_6H_4$	1.1	$Ca(NTf_2)_2$	5	*n*-Bu_4NPF_6	HFIP	60	0.75	(77)	55
2,4,6-$Me_3C_6H_2$	1.0	$Dy(OTf)_3$	5	—	MeCN	80	3.5	(74)	23

C_{6-12}

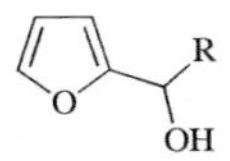

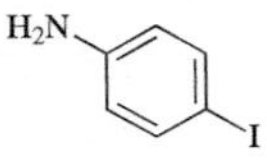

1.1 eq

$Ca(NTf_2)_2$ (5 mol %),
n-Bu_4NPF_6 (5 mol %),
$MeNO_2$

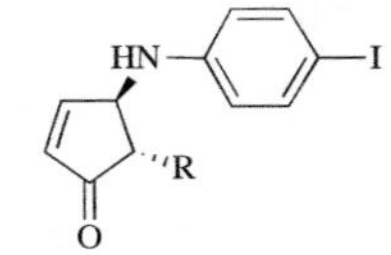

54

R	Temp (°)	Time (min)	
$TBDPSOCH_2$	100	20	(58)
4-$MeOC_6H_4$	40	10	(81)
Bn	100	60	(75)

TABLE 3. 4-AMINOCYCLOPENT-2-ENONES (Continued)

Substrate	Nucleophile	Conditions	Product(s) and Yield(s) (%)	Refs.

*Please refer to the chart preceding the tables for the structures indicated by the **bold** numbers.*

C$_{7-11}$

Nucleophile: H$_2$N–C$_6$H$_4$–I, 1.1 eq

Conditions: Ca(NTf$_2$)$_2$ (5 mol %), n-Bu$_4$NPF$_6$ (5 mol %), solvent

Ref. 55

R	Solvent	Temp (°)	Time		dr
CH$_2$=CH	MeNO$_2$	80	10 min	(tr)	—
CH$_2$=CH	HFIP	20	2 h	(61)	4:1
(E)-TMSCH=CH	HFIP	20	1.5 h	(28)	—
CH$_2$=C(Me)	HFIP	20	—	(0)	—
(E)-MeCH=CH	HFIP	20	2 h	(67)	5:1

R	Solvent	Temp (°)	Time		dr
(E)-c-C$_3$H$_5$CH=CH	HFIP	20	15 min	(74)	—
t-Bu	MeNO$_2$	100	24 h	(59)	—
t-Bu	HFIP	60	3 h	(86)	—
Et$_2$CHCH$_2$	MeNO$_2$	100	16 h	(30)	—
Et$_2$CHCH$_2$	HFIP	60	20 min	(81)	—

C$_8$

Substrate: x eq

Nucleophile: NH$_2$R

Conditions: Catalyst (5 mol %), solvent

x	R	Catalyst	Solvent	Temp (°)	Time (h)		er	
1.0	3-ClC$_6$H$_4$	Dy(OTf)$_3$	MeCN	80	6	(52)	—	23
1.0	4-IC$_6$H$_4$	Dy(OTf)$_3$	MeCN	80	5	(73)	—	23
1.0	2,4,6-Me$_3$C$_6$H$_2$	Dy(OTf)$_3$	MeCN	80	24	(89)	—	23
2.0	2-NCC$_6$H$_4$	**1**	CHCl$_3$	rt	48	(50)	96.0:4.0	57

552

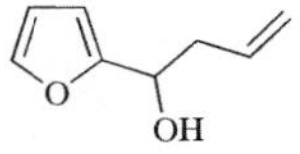

Catalyst (5 mol %), additive (5 mol %), solvent, 80°

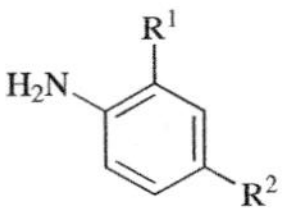

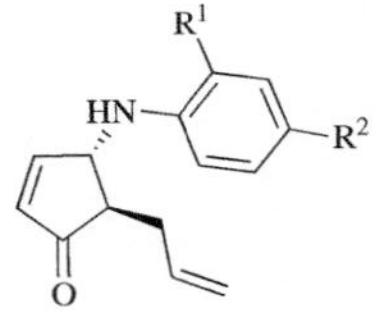

x	R^1	R^2	y	Catalyst	Additive	Solvent	Time (h)		
1.0	H	I	1.1	Ca(NTf$_2$)$_2$	n-Bu$_4$NPF$_6$	MeNO$_2$	0.5	(86)	54
1.2	H$_2$NOC	H	1.0	4-TsOH	—	MeCN	2	(85)	61
1.2	H$_2$NOC	Br	1.0	4-TsOH	—	MeCN	2	(80)	61

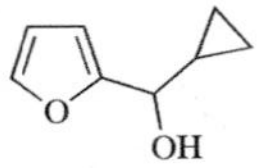

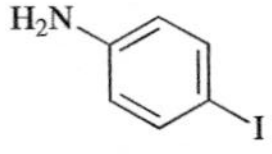

Ca(NTf$_2$)$_2$ (5 mol %), n-Bu$_4$NPF$_6$ (5 mol %), HFIP, 20°, 0.25 h

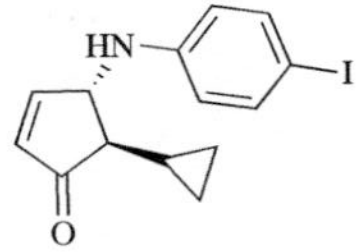

1.1 eq

(84)   55

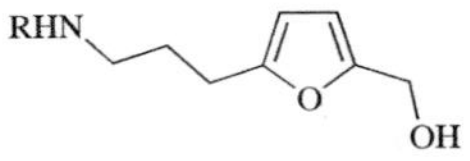

PPh$_3$ (1.5 eq), DEAD (1.5 eq), additive (20 mol %), DMAc, rt, 0.5 h

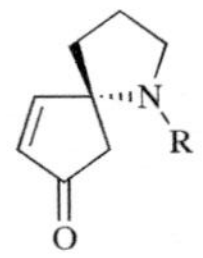

R	Additive		dr
n-Bu	PPTS	(71)	—
Ph	none	(83)	—
4-ClC$_6$H$_4$	none	(93)	—
2-MeOC$_6$H$_4$	none	(88)	—
4-MeOC$_6$H$_4$	none	(89)	—

R	Additive		dr
Bn	PPTS	(72)	—
(R)-CHMePh	PPTS	(73)	1.27:1
PhCH$_2$CH$_2$	PPTS	(72)	—
3,4-(MeO)$_2$C$_6$H$_3$CH$_2$CH$_2$	PPTS	(78)	—

44

TABLE 3. 4-AMINOCYCLOPENT-2-ENONES (*Continued*)

Substrate	Nucleophile	Conditions	Product(s) and Yield(s) (%)	Refs.

*Please refer to the chart preceding the tables for the structures indicated by the **bold** numbers.*

C$_8$

Substrate: structure with RHN chain, furan ring bearing CH_2OH

Conditions: $Dy(OTf)_3$ (x mol %), MeCN, 80°

Product: spiro N–R cyclopentenone

R	x	Time (h)	
Ph	5	15	(90)
4-MeOC$_6$H$_4$	10	48	(57)

Refs: 52

C$_{9-12}$

Substrate: Ar–CH(OH)–(2-furanyl), 1.5 eq

Nucleophile: H_2N–C$_6$H$_3$(CF$_3$)$_2$ (3,5-bis-CF$_3$ aniline)

Conditions: **2** (10 mol %), DCE, rt

Product: HN(3,5-(CF$_3$)$_2$C$_6$H$_3$)-, Ar-substituted cyclopentenone

Ar	Time (h)		dr	er
2-furanyl	43	(62)	13:1	96.0:4.0
2-thienyl	16	(97)	10:1	94.0:6.0
4-MeSC$_6$H$_4$	10	(88)	—	95.5:4.5
3-MeC$_6$H$_4$	49	(90)	—	96.0:4.0

Refs: 42

C$_9$

Substrate: (3-thienyl)–CH(OH)–(2-furanyl), 2.0 eq

Nucleophile: H_2N–C$_6$H$_4$–NC (2-aminobenzonitrile)

Conditions: **1** (5 mol %), CHCl$_3$, rt, 24 h

Product: HN(2-NC-C$_6$H$_4$)-, 3-thienyl-substituted cyclopentenone, (75) er 93.5:6.5

Refs: 57

C_{9-14}

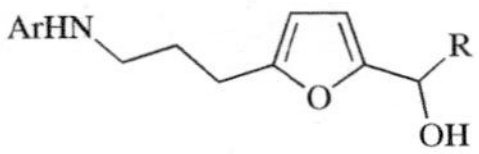

Catalyst (x mol %),
MeCN

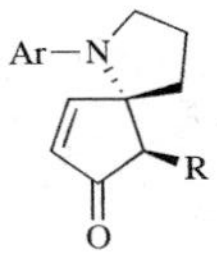

R	Ar	Catalyst	x	Temp (°)	Time		
Me	Ph	4	20	60	—	(86)	123
Me	Ph	Dy(OTf)$_3$	5	80	15 min	(91)	52
Ph	Ph	4	20	60	—	(93)	123
Ph	Ph	Dy(OTf)$_3$	5	80	15 min	(96)	52
Ph	4-ClC$_6$H$_4$	4	20	60	—	(87)	123
Ph	4-ClC$_6$H$_4$	Dy(OTf)$_3$	5	80	10 min	(96)	52
Ph	4-MeOC$_6$H$_4$	4	20	60	—	(72)	123
Ph	4-MeOC$_6$H$_4$	Dy(OTf)$_3$	5	80	2.5 h	(74)	52
4-BrC$_6$H$_4$	Ph	4	20	60	—	(70)	123
4-BrC$_6$H$_4$	Ph	Dy(OTf)$_3$	5	80	5 h	(84)	52
4-MeOC$_6$H$_4$	Ph	4	20	60	—	(84)	123
4-MeOC$_6$H$_4$	Ph	Dy(OTf)$_3$	5	80	15 min	(81)	52

C_9

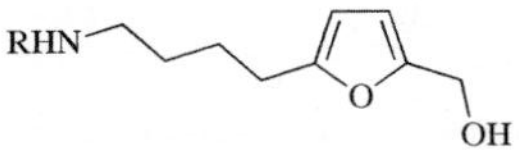

PPh$_3$ (1.5 eq),
DEAD (1.5 eq),
additive (20 mol %),
DMAc, rt, 12 h

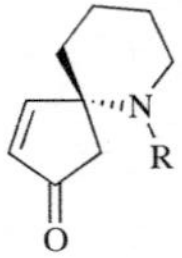

R	Additive		
n-Bu	PPTS	(44)	
Ph	none	(28)	44
4-ClC$_6$H$_4$	none	(42)	
4-MeOC$_6$H$_4$	none	(35)	
Bn	PPTS	(45)	

TABLE 3. 4-AMINOCYCLOPENT-2-ENONES (Continued)

*Please refer to the chart preceding the tables for the structures indicated by the **bold** numbers.*

Substrate	Nucleophile	Conditions	Product(s) and Yield(s) (%)	Refs.
C$_9$				
(furan with ArHN chain, OH)		Dy(OTf)$_3$ (x mol %), MeCN, 80°		52
			Ar x Time (h)	
			Ph 5 48 (70)	
			4-IC$_6$H$_4$ 5 8 (90)	
			4-MeOC$_6$H$_4$ 20 72 (37)	
C$_{10}$				
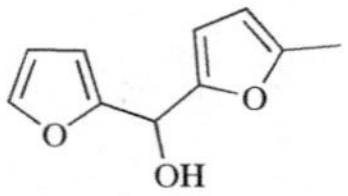 (furan, OH)	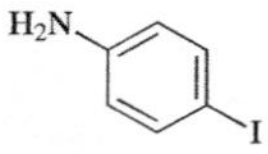 H$_2$N–C$_6$H$_4$–I 1.1 eq	HFIP, 20°, 0.17 h	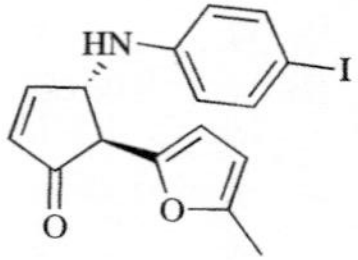 (89) dr 12:1	55
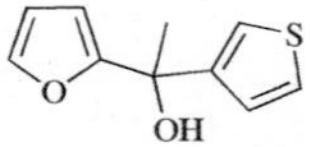 (thiophene, furan, OH) 1.5 eq	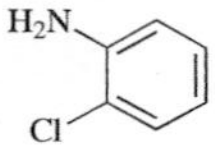 H$_2$N–C$_6$H$_4$–Cl	1 (5 mol %), CHCl$_3$, rt, 24 h	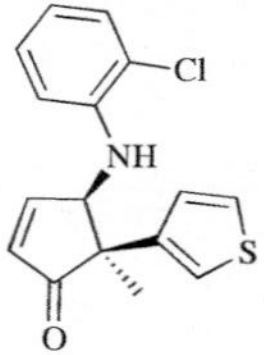 (60) dr 2:1, er 97.5:2.5	57
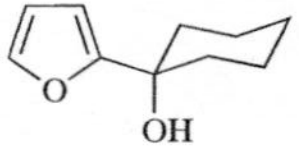 (cyclohexyl, furan, OH)	PhNH$_2$ 1.2 eq	3 (5 mol %), PhF, 50°, 20 h	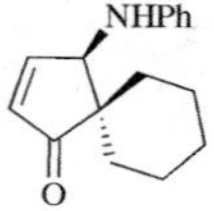 (59) er 80:20	58

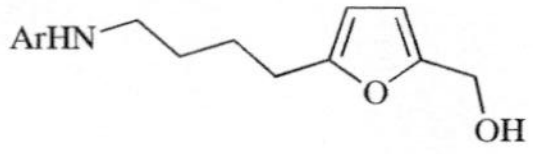

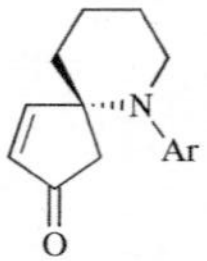

556

C_{10-17}

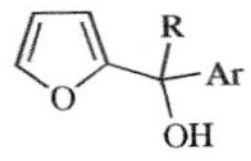

1.5 eq

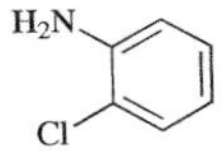

H_2N — C_6H_4Cl (2-Cl)

1 (5 mol %), CHCl$_3$, rt

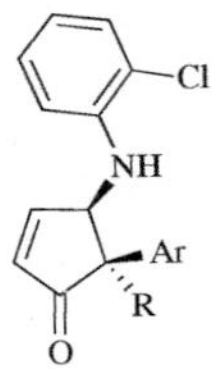

57

R	Ar	Time (h)		dr	er
Me	2-thienyl	24	(60)	2:1	97.5:2.5
Me	4-ClC$_6$H$_4$	36	(70)	5:1	97.5:2.5
Me	4-MeOC$_6$H$_4$	24	(83)	8:1	98.0:2.0
Me	3-MeC$_6$H$_4$	24	(82)	18:1	98.0:2.0
Me	4-MeC$_6$H$_4$	24	(84)	14:1	97.5:2.5

R	Ar	Time (h)		dr	er
Me	4-CF$_3$C$_6$H$_4$	36	(72)	3:1	97.5:2.5
Me	2-Np	24	(85)	11:1	97.0:3.0
Et	Ph	24	(90)	>19:1	97.0:3.0
i-Pr	Ph	24	(95)	>19:1	95.0:5.0
c-C$_6$H$_{11}$	Ph	24	(96)	>19:1	95.0:5.0

C_{10-12}

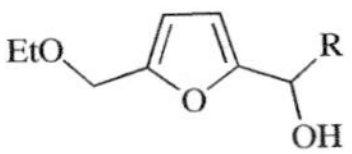

PhNH$_2$

Dy(OTf)$_3$ (10 mol %), t-BuOH, 90°

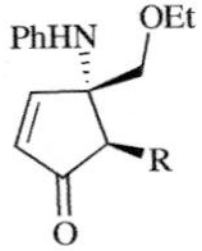

120

R	Time (h)		dr
n-Bu	4	(6)	>95:5
2-thienyl	3	(10)	>95:5
Ph	4	(50)	>95:5
3-F-4-MeOC$_6$H$_3$	3	(28)	>95:5

TABLE 3. 4-AMINOCYCLOPENT-2-ENONES (*Continued*)

Substrate	Nucleophile	Conditions	Product(s) and Yield(s) (%)	Refs.

*Please refer to the chart preceding the tables for the structures indicated by the **bold** numbers.*

C$_{11}$

Substrate:

Nucleophile: ArNH$_2$, 1.1 eq

Conditions: Ca(NTf$_2$)$_2$ (5 mol %), n-Bu$_4$NPF$_6$ (5 mol %), solvent

Product:

Ar	Solvent	Temp (°)	Time		Refs.
Ph	MeNO$_2$	80	30 min	(90)	54
2-IC$_6$H$_4$	MeNO$_2$	80	20 min	(62)	54
3-IC$_6$H$_4$	MeNO$_2$	80	20 min	(92)	54
4-IC$_6$H$_4$	MeNO$_2$	80	10 min	(92)	54
4-HOC$_6$H$_4$	MeNO$_2$	100	6 h	(55)	55
4-HOC$_6$H$_4$	HFIP	60	45 min	(87)	55
4-O$_2$NC$_6$H$_4$	MeNO$_2$	80	30 min	(25)	55
4-O$_2$NC$_6$H$_4$	HFIP	rt	30 min	(59)	55
2-MeOC$_6$H$_4$	MeNO$_2$	90	40 min	(59)	55
2-MeOC$_6$H$_4$	HFIP	20	3 h	(70)	55
4-MeOC$_6$H$_4$	MeNO$_2$	100	2.5 h	(81)	55
4-MeOC$_6$H$_4$	HFIP	20	1.5 h	(90)	55
4-AcHNC$_6$H$_4$	MeNO$_2$	100	16 h	(41)	55
4-AcHNC$_6$H$_4$	HFIP	60	1 h	(70)	55
4-NCC$_6$H$_4$	MeNO$_2$	80	20 min	(74)	54
4-MeC$_6$H$_4$	MeNO$_2$	90	30 min	(75)	54
4-MeO$_2$CC$_6$H$_4$	MeNO$_2$	80	20 min	(91)	54
3-AcC$_6$H$_4$	MeNO$_2$	90	2 h	(92)	54

x eq	ArNH$_2$ y eq		Catalyst (z mol %), solvent								
x	Ar	y	Catalyst	z	Solvent	Temp (°)	Time		dr	er	
1.0	Ph	1.0	Dy(OTf)$_3$	5	MeCN	80	1.5 h	(86)	—	—	23
1.0	Ph	1.2	3	5	C$_6$D$_5$F	50	20 h	(74)	>19:1	90.0:10.0	58
1.0	Ph	1.0	4	20	MeCN	60	—	(72)	—	—	123
1.2	Ph	1.0	PMA	0.03	MeCN	83	55 min	(80)	—	—	124
1.2	4-FC$_6$H$_4$	1.0	PMA	0.03	MeCN	83	50 min	(85)	—	—	124
1.0	2,4,6-F$_3$C$_6$H$_2$	3.0	Dy(OTf)$_3$	5	MeCN	80	3 h	(74)	—	—	23
1.2	2-ClC$_6$H$_4$	1.0	1	5	CHCl$_3$	rt	24 h	(89)	—	—	57
1.0	3-ClC$_6$H$_4$	1.0	Dy(OTf)$_3$	5	MeCN	80	20 min	(82)	—	—	23
1.0	3-ClC$_6$H$_4$	3.0	4	20	MeCN	60	—	(91)	—	—	123
1.2	4-ClC$_6$H$_4$	1.0	PMA	0.03	MeCN	83	45 min	(89)	—	—	124
1.2	2-BrC$_6$H$_4$	1.0	1	5	CHCl$_3$	rt	24 h	(89)	—	—	57
1.0	4-BrC$_6$H$_4$	3.0	4	20	MeCN	60	—	(92)	—	—	123
1.0	4-IC$_6$H$_4$	1.0	Dy(OTf)$_3$	5	MeCN	80	30 min	(92)	—	—	23
1.0	4-IC$_6$H$_4$	3.0	4	20	MeCN	60	—	(65)	—	—	123
1.0	2-O$_2$NC$_6$H$_4$	1.5	2	10	DCE	rt	31 h	(72)	8:1	91.0:9.0	42
1.0	2-F-4-O$_2$NC$_6$H$_3$	1.5	2	10	DCE	rt	42 h	(76)	>20:1	96.0:4.0	42
1.0	4-MeOC$_6$H$_4$	1.0	Dy(OTf)$_3$	20	MeCN	80	18 h	(62)	—	—	23
1.2	4-MeOC$_6$H$_4$	1.0	PMA	0.03	MeCN	83	1 h	(81)	—	—	124
1.0	2-MeO-4-O$_2$NC$_6$H$_3$	15	2	10	DCE	rt	31 h	(79)	>20:1	90.0:10.0	42

TABLE 3. 4-AMINOCYCLOPENT-2-ENONES (Continued)

Please refer to the chart preceding the tables for the structures indicated by the **bold** numbers.
Continued from previous page.

C_{11}

| | Substrate | | Nucleophile | | Conditions | | | Product(s) and Yield(s) (%) | | | Refs. |
| | Ar–CH(OH)–furan, x eq | | NH$_2$Ar, y eq | | Catalyst (z mol %), solvent | | | | | | |
x	Ar	y	Catalyst	z	Solvent	Temp (°)	Time	(yield)	dr	er	
1.2	2-MeC$_6$H$_4$	1.0	PMA	0.03	MeCN	83	55 min	(83)	—	—	124
1.0	4-MeC$_6$H$_4$	3.0	4	20	MeCN	60	—	(56)	—	—	123
1.2	4-MeC$_6$H$_4$	1.0	PMA	0.03	MeCN	83	52 min	(85)	—	—	124
1.0	2-Me-4-O$_2$NC$_6$H$_3$	1.5	2	10	DCE	rt	26 h	(78)	>20:1	92.0:8.0	42
1.0	2,6-Me$_2$C$_6$H$_3$	1.0	Dy(OTf)$_3$	5	MeCN	80	3 h	(33)	—	—	23
1.0	3,5-Me$_2$C$_6$H$_3$	1.0	Dy(OTf)$_3$	5	MeCN	80	3 h	(81)	—	—	23
1.2	2-CF$_3$C$_6$H$_4$	1.0	1	5	CHCl$_3$	5	24 h	(87)	>19:1	95.0:5.0	57
1.2	3-CF$_3$C$_6$H$_4$	1.0	1	5	CHCl$_3$	5	24 h	(82)	>19:1	87.5:12.5	57
1.2	4-CF$_3$C$_6$H$_4$	1.0	1	5	CHCl$_3$	5	24 h	(81)	>19:1	87.0:13.0	57
1.0	2-CF$_3$-4-ClC$_6$H$_3$	1.5	2	10	DCE	rt	34 h	(92)	>20:1	94.5:5.5	42
1.0	2-Br-4-CF$_3$C$_6$H$_3$	1.5	2	10	DCE	rt	45 h	(76)	>20:1	92.5:7.5	42
1.0	2-CF$_3$-4-O$_2$NC$_6$H$_3$	1.5	2	10	DCE	rt	21 h	(92)	>20:1	95.0:5.0	42
1.0	2,5-(CF$_3$)$_2$C$_6$H$_3$	1.5	2	10	DCE	rt	22 h	(83)	12:1	96.5:3.5	42
1.0	3,5-(CF$_3$)$_2$C$_6$H$_3$	1.0	Dy(OTf)$_3$	5	MeCN	80	24 h	(75)	—	—	23
1.5	3,5-(CF$_3$)$_2$C$_6$H$_3$	1.5	2	10	DCE	rt	75 h	(73)	>20:1	94.5:5.5	42
1.0	3,5-(CF$_3$)$_2$C$_6$H$_3$	3.0	4	20	MeCN	60	—	(82)	—	—	123
1.0	2-CF$_3$-4-NCC$_6$H$_3$	1.5	2	10	DCE	rt	48 h	(76)	>20:1	97.0:3.0	42
1.2	2-NCC$_6$H$_5$	1.0	1	5	CHCl$_3$	5	24 h	(92)	>19:1	95.0:5.0	57
1.2	2-H$_2$NOCC$_6$H$_4$	1.0	4-TsOH	5	MeCN	80	1 h	(84)	—	—	61
1.2	2-H$_2$NOC-4-BrC$_6$H$_3$	1.0	4-TsOH	5	MeCN	80	2 h	(—)	—	—	61
1.2	2-MeO$_2$CC$_6$H$_4$	1.0	1	5	CHCl$_3$	5	24 h	(92)	>19:1	93.5:6.5	57
1.0	4-MeO$_2$CC$_6$H$_4$	1.0	Dy(OTf)$_3$	5	MeCN	80	1 h	(86)	—	—	23
1.0	2,4,6-Me$_3$C$_6$H$_2$	1.0	Dy(OTf)$_3$	5	MeCN	80	4 h	(91)	—	—	23

C$_{11-13}$

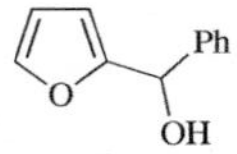

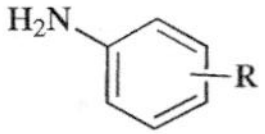

7 (5 mol %), CH$_2$Cl$_2$, 23°, 5 d

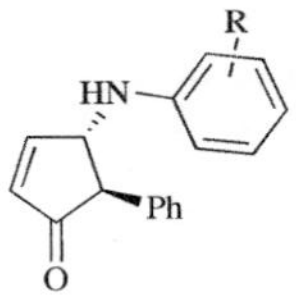

R		er
2,4-F$_2$C$_6$H$_3$	(65)	81.5:18.5
2,4,6-F$_3$C$_6$H$_2$	(53)	80.5:19.5
4-IC$_6$H$_4$	(70)	87.5:12.5
4-CF$_3$C$_6$H$_4$	(64)	88.0:12.0
4-NCC$_6$H$_4$	(50)	85.5:14.5
2-HO$_2$CC$_6$H$_4$	(59)[a]	92.0:8.0

R		er
2-MeO$_2$CC$_6$H$_4$	(44)	90.0:10.0
3-HO$_2$CC$_6$H$_4$	(19)[a]	87.0:13.0
4-MeO$_2$CC$_6$H$_4$	(71)	86.5:13.5
3,5-(CF$_3$)$_2$C$_6$H$_3$	(44)	79.0:21.0
4-AcC$_6$H$_4$	(65)	88.0:12.0

C$_{11-14}$

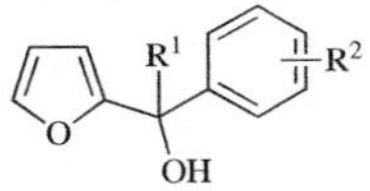

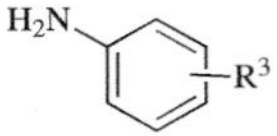

7 (5 mol %), CH$_2$Cl$_2$, 23°, 5 d

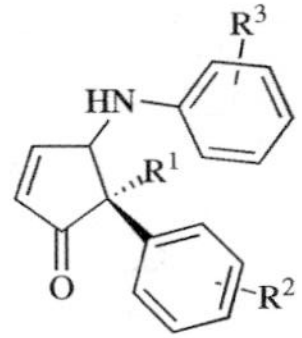

R^1	R^2	R^3		er
H	3-MeO	4-I	(50)	88.0:12.0
H	3-MeO	2-HO$_2$C	(36)[a]	93.5:6.5
H	4-MeO	4-I	(17)	87.5:12.5
H	4-MeO	2-HO$_2$C	(52)[a]	93.0:7.0

R^1	R^2	R^3		er
H	2,4,6-Me$_3$C$_6$H$_2$	4-I	(66)	58.5:41.5
H	2,4,6-Me$_3$C$_6$H$_2$	2-HO$_2$C	(28)[a]	64.0:36.0
Ph	H	4-I	(77)	70.5:29.5
Ph	H	2-HO$_2$C	(26)[a]	86.0:14.0

TABLE 3. 4-AMINOCYCLOPENT-2-ENONES (Continued)

Substrate	Nucleophile	Conditions	Product(s) and Yield(s) (%)	Refs.

*Please refer to the chart preceding the tables for the structures indicated by the **bold** numbers.*

C$_{11}$

Substrate	Nucleophile	Conditions	Product(s) and Yield(s) (%)	Refs.
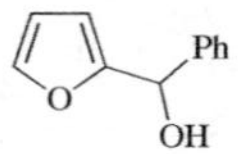 (Ph–CH(OH)–furan) 1.2 eq	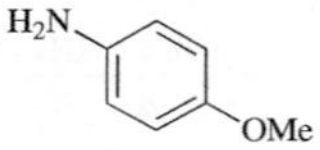 H$_2$N–C$_6$H$_4$–OMe, 1.1 eq	Zn(OTf)$_2$ (5 mol %), HFIP, 20°, 24 h	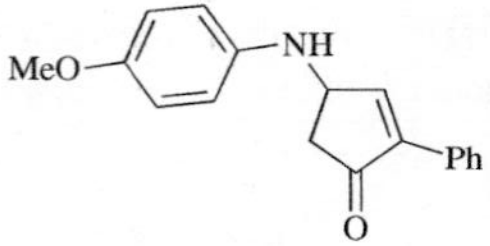 (85)	55
	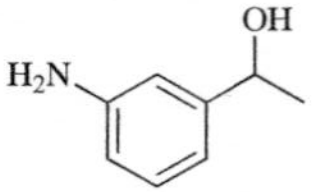 H$_2$N–(aryl)CH(OH)CH$_3$	Dy(OTf)$_3$ (5 mol %), MeCN, 80°, 2 h	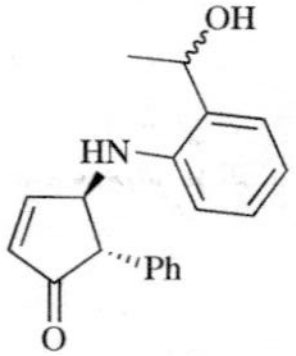 (88) dr 1:1	23
	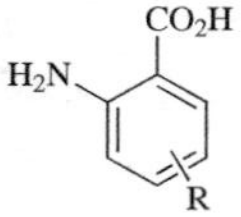 H$_2$N–aryl(CO$_2$H)(R)	1. **1** (5 mol %), CHCl$_3$, 5°, 24 h; 2. TMSCHN$_2$, MeOH, THF, 0°, 3 h	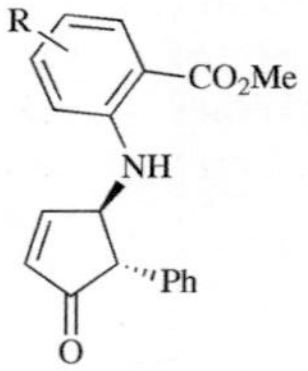	57

Conditions table (row 2):

R		er
H	(83)	95.5:4.5
4-F	(72)	95.0:5.0
5-F	(78)	96.5:3.5

Product table (row 3):

R		er		R		er
4-Cl	(70)	95.0:5.0		4-MeO	(70)	97.0:3.0
5-Cl	(76)	95.0:5.0		5-MeO	(62)	92.5:7.5
6-Cl	(73)	95.0:5.0		4-Me	(70)	91.5:8.5

Dy(OTf)$_3$ (5 mol %),
MeCN, 80°, 24 h

(15) dr 1:1   23

Catalyst (y mol %),
additive (z mol %),
solvent

H$_2$N— (2,6-dimethylphenyl)

HNR^{1}R^2
x eq

R^1	R^2	x	Catalyst	y	Additive	z	Solvent	Temp (°)	Time (h)		
Ph	Me	1.0	Dy(OTf)$_3$	5	—	—	MeCN	80	2.5	(74)	23
Ph	Me	1.1	Ca(NTf$_2$)$_2$	5	n-Bu$_4$NPF$_6$	5	MeNO$_2$	80	1.5	(82)	54
Ph	Me	3.0	4	20	—	—	MeCN	60	—	(70)	123
Ph	CH$_2$=CHCH$_2$	1.1	Ca(NTf$_2$)$_2$	5	n-Bu$_4$NPF$_6$	5	MeNO$_2$	80	1.5	(76)	54
Ph	≡—CH$_2$	1.3	Ca(NTf$_2$)$_2$	5	n-Bu$_4$NPF$_6$	5	MeNO$_2$	90	1	(35)	55
Ph	≡—CH$_2$	1.3	Ca(NTf$_2$)$_2$	5	n-Bu$_4$NPF$_6$	5	HFIP	20	3	(71)	55
Ph	Bn	1.3	Ca(NTf$_2$)$_2$	5	n-Bu$_4$NPF$_6$	5	MeNO$_2$	90	1	(18)	55
Ph	Bn	1.3	Ca(NTf$_2$)$_2$	5	n-Bu$_4$NPF$_6$	5	HFIP	60	1	(51)	55
4-BrC$_6$H$_4$	Me	1.0	Dy(OTf)$_3$	5	—	—	MeCN	80	0.75	(88)	23
4-MeOC$_6$H$_4$	Bn	1.1	Ca(NTf$_2$)$_2$	5	n-Bu$_4$NPF$_6$	5	MeNO$_2$	100	0.5	(88)	55
4-MeOC$_6$H$_4$	Bn	1.1	Ca(NTf$_2$)$_2$	5	n-Bu$_4$NPF$_6$	5	HFIP	20	1	(92)	55

TABLE 3. 4-AMINOCYCLOPENT-2-ENONES (Continued)

Substrate	Nucleophile	Conditions	Product(s) and Yield(s) (%)	Refs.

*Please refer to the chart preceding the tables for the structures indicated by the **bold** numbers.*

C₁₁

Dy(OTf)₃ (5 mol %), MeCN, 80°, 18 h

(67)

23

C₁₁₋₁₇

1.2 eq

1 (5 mol %), CHCl₃, 24 h

(75) er 95.5:4.5

57

Ar	Temp (°)		dr	er
4-FC₆H₄	5	(75)	—	95.5:4.5
3,4-Cl₂C₆H₃	15	(73)	—	96.0:4.0
4-OCH₂CH=CH₂	5	(80)	—	94.5:5.5
3,4-OCH₂OC₆H₃	5	(80)	—	94.5:5.5
4-NCC₆H₄	15	(55)	13:1	96.5:3.5

Ar	Temp (°)		dr	er
4-MeO₂CC₆H₄	15	(70)	—	95.0:5.0
3,5-Me₂C₆H₃	5	(76)	>19:1	94.0:6.0
1-Np	5	(72)	—	94.0:6.0
4-PhC₆H₄	5	(92)	—	95.0:5.0

C₁₁

x eq

ArNH₂
y eq

Catalyst (z mol %), solvent

x	Ar	y	Catalyst	z	Solvent	Temp (°)	Time (h)	er	Refs.
1.0	3,5-(CF₃)₂C₆H₃	1.5	**2**	10	DCE	rt	70	(76) 94.5:5.5	42
1.2	2-NCC₆H₄	1.0	**1**	5	CHCl₃	5	24	(81) 95.5:4.5	57
1.2	2-H₂NOCC₆H₄	1.0	4-TsOH	5	MeCN	80	2	(55) —	61
1.2	2-H₂NOC-4-BrC₆H₃	1.0	4-TsOH	5	MeCN	80	2	(78) —	61

Table (top)

Substrate: 4-ClC6H4 furyl carbinol (x eq) + ArNH2 (y eq), Catalyst (z mol %), solvent → NHAr-cyclopentenone (4-Cl)

x	Ar	y	Catalyst	z	Solvent	Temp (°)	Time (h)		er	
1.0	3,5-(CF$_3$)$_2$C$_6$H$_3$	1.5	2	10	DCE	rt	103	(68)	94.5:5.5	42
1.2	2-NCC$_6$H$_4$	1.0	1	5	CHCl$_3$	5	24	(88)	95.0:5.0	57

Table (second)

Substrate: 4-BrC6H4 furyl carbinol (x eq) + ArNH$_2$ (y eq), Catalyst (5 mol %), solvent → NHAr-cyclopentenone (4-Br)

x	Ar	y	Catalyst	Solvent	Temp (°)	Time (h)	dr		er	
1.0	Ph	1.2	3	PhF	50	20	>19:1	(65)	95.0:5.0	58
1.2	2-NCC$_6$H$_4$	1.0	1	CHCl$_3$	5	24	>19:1	(80)	94.0:6.0	57

C$_{11-15}$

Substrate: 2-NHAc furyl carbinol (1.3 eq) + 4-methoxyaniline (H$_2$N–C$_6$H$_4$–OMe, 1.5 eq)

1. Ca(NTf$_2$)$_2$ (5 mol %), n-Bu$_4$NPF$_6$ (5 mol %), toluene/HFIP (3:1), 60°, 3 h
2. Et$_3$N (2 eq), 20°, 17 h

→ product (81) — 126

Substrate: Ar furyl carbinol + 2,5-bis(CF$_3$)aniline (H$_2$N–C$_6$H$_3$(CF$_3$)$_2$)

2 (10 mol %), DCE, rt → product — 42

Ar	Time (h)		er	
2-MeOC$_6$H$_4$	93	(51)	94.0:6.0
6-MeO-2-naphthyl	6	(92)	95.0:5.0

TABLE 3. 4-AMINOCYCLOPENT-2-ENONES (Continued)

	Substrate	Nucleophile	Conditions	Product(s) and Yield(s) (%)	Refs.

*Please refer to the chart preceding the tables for the structures indicated by the **bold** numbers.*

C$_{11}$

Substrate: (4-methoxyphenyl)(furan-2-yl)methanol, x eq

Nucleophile: $ArNH_2$, y eq

Conditions: Catalyst (z mol %), solvent

Product: cyclopentenone with NHAr and 4-OMe-phenyl substituents

x	Ar	y	Catalyst	z	Solvent	Temp (°)	Time		dr	er	Refs.
1.0	Ph	1.2	**3**	5	PhF	50	20 h	(68)	>19:1	92.0:8.0	58
1.2	Ph	1.0	Dy(OTf)$_3$	0.03	MeCN	83	45 min	(90)	—	—	124
1.0	3-ClC$_6$H$_4$	1.0	**4**	5	MeCN	rt	6.5 h	(82)	—	—	23
1.0	3-ClC$_6$H$_4$	3.0	**4**	20	MeCN	60	—	(85)	—	—	123
1.2	4-ClC$_6$H$_4$	1.0	PMA	0.03	MeCN	83	45 min	(92)	—	—	124
1.2	2-BrC$_6$H$_4$	1.0	PMA	0.03	MeCN	83	55 min	(80)	—	—	124
1.0	4-IC$_6$H$_4$	1.0	Dy(OTf)$_3$	5	MeCN	rt	1.5 h	(68)	—	—	23
1.0	4-IC$_6$H$_4$	3.0	**4**	20	MeCN	60	—	(73)	—	—	123
1.2	4-MeOC$_6$H$_4$	1.0	PMA	0.03	MeCN	83	45 min	(85)	—	—	124
1.2	2-MeC$_6$H$_4$	1.0	PMA	0.03	MeCN	83	50 min	(70)	—	—	124
1.2	4-MeC$_6$H$_4$	1.0	PMA	0.03	MeCN	83	40 min	(88)	—	—	124
1.2	2-NCC$_6$H$_4$	1.0	**1**	5	CHCl$_3$	5	24 h	(70)	—	94.0:6.0	57
1.2	2-H$_2$NOCC$_6$H$_4$	1.0	4-TsOH	5	MeCN	80	1.5 h	(78)	—	—	61
1.2	2-H$_2$NOC-4-BrC$_6$H$_4$	1.0	4-TsOH	5	MeCN	80	1 h	(82)	—	—	61
1.0	2,5-(CF$_3$)$_2$C$_6$H$_3$	3.0	**2**	10	DCE	-20	37 h	(65)	—	94.0:6.0	42
1.0	2-CF$_3$-4-NCC$_6$H$_3$	3.0	**2**	10	DCE	rt	17 h	(72)	—	96.5:3.5	42
1.0	2,4,6-Me$_3$C$_6$H$_2$	1.0	Dy(OTf)$_3$	5	MeCN	rt	5 h	(89)	—	—	23

566

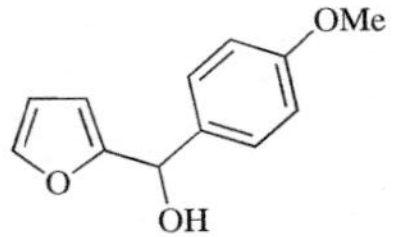

I — 4-methoxyphenyl(furan-2-yl)methanol

1. Co(BF$_4$)$_2$•6H$_2$O (10 mol %); **8** (10 mol %), THF, 35°, 30 min; remove solvent
2. **I, II** (1 eq), o-xylene, 65°, 16 h

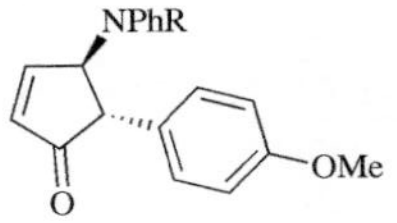

R		er	127
Me	(89)	78.0:22.0	
Bn	(73)	75.0:25.0	

PhNHR **II**

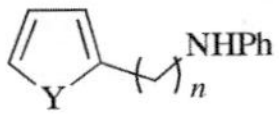

1. Co(BF$_4$)$_2$•6H$_2$O (10 mol %); **8** (10 mol %), THF, 35°, 30 min; remove solvent
2. **I, II** (1 eq), o-xylene, 65°, 16 h

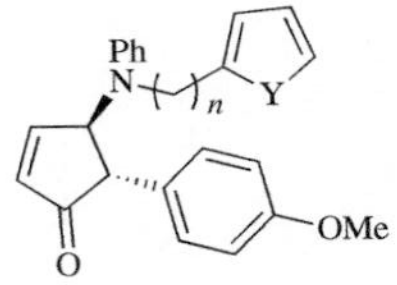

Y	n		er	127
O	3	(83)	71.5:28.5	
S	1	(95)	94.5:5.5	

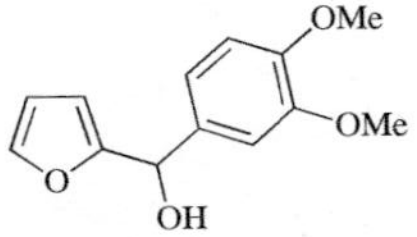

II

ArNH$_2$
y eq

Catalyst (z mol %), solvent

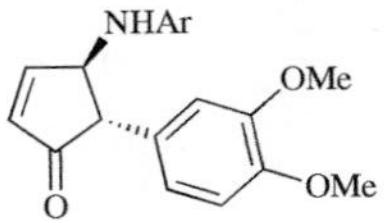

x eq

x	Ar	y	Catalyst	z	Solvent	Temp (°)	Time (h)	er	
1.0	3,5-(CF$_3$)$_2$C$_6$H$_3$	1.5	**2**	10	DCE	rt	27	(94)	96.5:3.5	42
1.2	2-H$_2$NOCC$_6$H$_4$	1.0	4-TsOH	5	MeCN	80	1.5	(70)	—	61
1.2	2-H$_2$NOC-4-BrC$_6$H$_3$	1.0	4-TsOH	5	MeCN	80	2	(60)	—	61

TABLE 3. 4-AMINOCYCLOPENT-2-ENONES (*Continued*)

Substrate	Nucleophile	Conditions	Product(s) and Yield(s) (%)	Refs.

*Please refer to the chart preceding the tables for the structures indicated by the **bold** numbers.*

C_{11}

ArNH$_2$
y eq

Catalyst (5 mol %),
n-Bu$_4$NPF$_6$ (*z* mol %),
solvent

x eq

x	Ar	*y*	Catalyst	*z*	Solvent	Temp (°)	Time	dr	er	
1.0	4-IC$_6$H$_4$	1.1	Ca(NTf$_2$)$_2$	5	MeNO$_2$	90	15 min (93)	—	—	54
1.2	2-NCC$_6$H$_4$	1.0	**1**	0	CHCl$_3$	5	24 h (95)	> 19:1	93.5:6.5	57

ArNH$_2$
y eq

Catalyst (5 mol %),
n-Bu$_4$NPF$_6$ (*z* mol %),
solvent

x eq

x	Ar	*y*	Catalyst	*z*	Solvent	Temp (°)	Time (h)	dr	er	
1.0	4-IC$_6$H$_4$	1.1	Ca(NTf$_2$)$_2$	5	MeNO$_2$	80	0.5 (65)	—	—	54
1.2	2-NCC$_6$H$_4$	1.0	**1**	0	CHCl$_3$	5	24 (72)	> 19:1	65.5:34.5	57

C_{12}

ArNH$_2$

1.2 eq

Catalyst (5 mol %),
solvent

Ar	Catalyst	Solvent	Temp (°)	Time (h)	er	
4-MeC$_6$H$_4$	**1**	CHCl$_3$	5	24 (81)	94.5:5.5	57
2-H$_2$NOCC$_6$H$_4$	4-TsOH	MeCN	80	1 (82)	—	61
2-H$_2$NOC-4-BrC$_6$H$_3$	4-TsOH	MeCN	80	2 (68)	—	61

568

x eq

$ArNH_2$

y eq

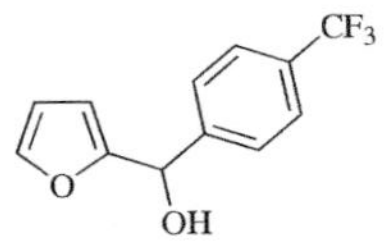

Catalyst (z mol %),
solvent

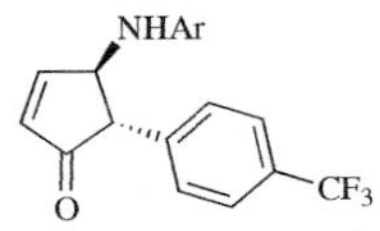

x	Ar	y	Catalyst	z	Solvent	Temp (°)	Time (h)		er	
1.0	3-ClC_6H_4	1.0	$Dy(OTf)_3$	5	MeCN	80	2	(87)	—	23
1.0	3-ClC_6H_4	3.0	**4**	20	MeCN	100	—	(47)	—	123
1.0	4-IC_6H_4	1.0	$Dy(OTf)_3$	5	MeCN	80	4.5	(83)	—	23
1.0	$2,4,6\text{-Me}_3C_6H_2$	1.0	$Dy(OTf)_3$	10	MeCN	80	8	(78)	—	23
1.2	2-NCC_6H_4	1.0	**1**	5	$CHCl_3$	15	24	(56)	95.5:4.5	57

1.5 eq

$ArNH_2$

1 (5 mol %),
$CHCl_3$, rt, 24 h

57

Ar		dr	er
2-FC_6H_4	(88)	15:1	96.0:4.0
2-ClC_6H_4	(90)	15:1	98.0:2.0
2-BrC_6H_4	(88)	15:1	98.0:2.0
2-MeC_6H_4	(79)	5:1	82.0:18.0
$2\text{-CF}_3C_6H_4$	(75)	9:1	98.0:2.0
$3\text{-CF}_3C_6H_4$	(92)	13:1	94.5:5.5
$4\text{-CF}_3C_6H_4$	(87)	19:1	95.0:5.0
2-NCC_6H_4	(82)	>19:1	98.0:2.0
4-NCC_6H_4	(81)	13:1	93.5:6.5
$2\text{-MeO}_2CC_6H_4$	(85)	7:1	96.0:4.0
2-AcC_6H_4	(86)	8:1	96.5:3.5
2-BzC_6H_4	(85)	8:1	97.0:3.0

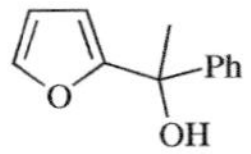

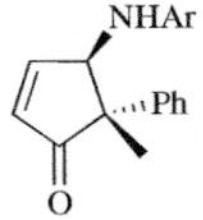

TABLE 3. 4-AMINOCYCLOPENT-2-ENONES (*Continued*)

Substrate	Nucleophile	Conditions	Product(s) and Yield(s) (%)	Refs.

*Please refer to the chart preceding the tables for the structures indicated by the **bold** numbers.*

C_{12}

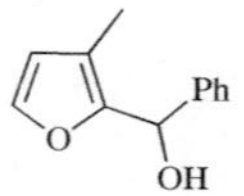	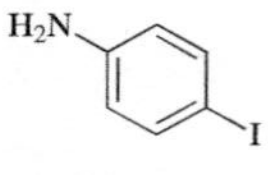 1.1 eq	Ca(NTf$_2$)$_2$ (5 mol %), n-Bu$_4$NPF$_6$ (5 mol %), MeNO$_2$, 80°, 0.33 h	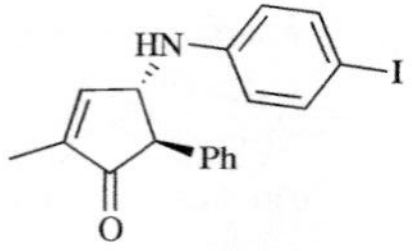 (97)	54
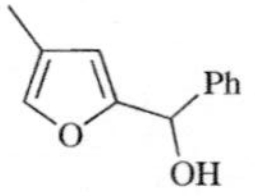	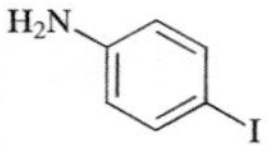 1.1 eq	Ca(NTf$_2$)$_2$ (5 mol %), n-Bu$_4$NPF$_6$ (5 mol %), MeNO$_2$, 90°, 0.5 h	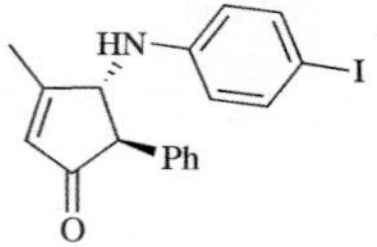 (86)	54

C_{12-21}

| 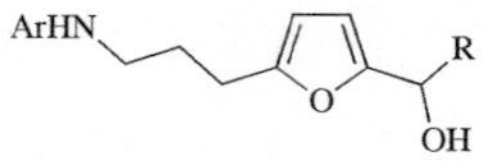 | | Dy(OTf)$_3$ (5 mol %), MeCN, 80° | 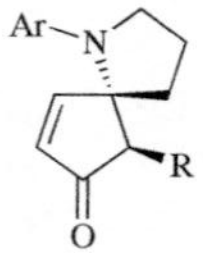 (97) | 52 |

R	Ar	Time	
n-Bu	Ph	15 min	(97)
n-Bu	4-IC$_6$H$_4$	15 min	(79)
Ph	4-BrC$_6$H$_4$	10 min	(78)
Ph	4-IC$_6$H$_4$	5 min	(84)
Ph	4-MeC$_6$H$_4$	30 min	(79)
Ph	4-CF$_3$C$_6$H$_4$	15 min	(82)

R	Ar	Time	
4-O$_2$NC$_6$H$_4$	Ph	5 h	(67)
3,4-OCH$_2$O-6-EtC$_6$H$_2$	4-MeOC$_6$H$_4$	120 h	(96)
3,4-OCH$_2$O-6-H$_2$C=CHC$_6$H$_2$	4-MeOC$_6$H$_4$	1.5 h	(86)
2,4,6-Me$_3$C$_6$H$_2$	Ph	2 h	(74)
Ph$_2$CH	Ph	15 h	(53)

C_{13}

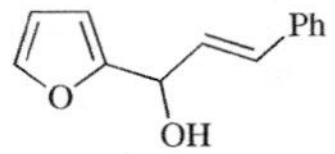

NHR^1R^2

x eq

Catalyst (y mol %),
n-Bu_4NPF_6 (z mol %),
solvent

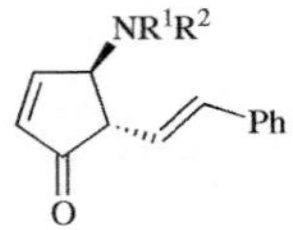

R^1	R^2	x	Catalyst	y	z	Solvent	Temp (°)	Time		dr	er	
$CH_2{=}CHCH_2$	4-$MeOC_6H_4$	1.1	$Ca(NTf_2)_2$	5	5	HFIP	20	30 min	(76)	>20:1	—	55
$\equiv$—CH_2	Ph	1.1	$Ca(NTf_2)_2$	5	5	HFIP	20	30 min	(66)	16:1	—	55
4-IC_6H_4	H	1.1	$Ca(NTf_2)_2$	5	5	HFIP	20	15 min	(75)	6.7:1	—	55
3,5-$(CF_3)_2C_6H_3$	H	1.5	**2**	10	0	DCE	rt	14	(73)	8:1	94.0:6.0	42

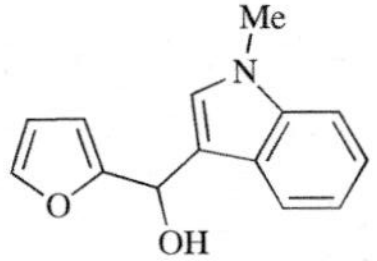

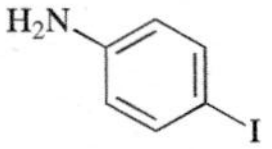

1.1 eq

HFIP, 20°, 0.17 h

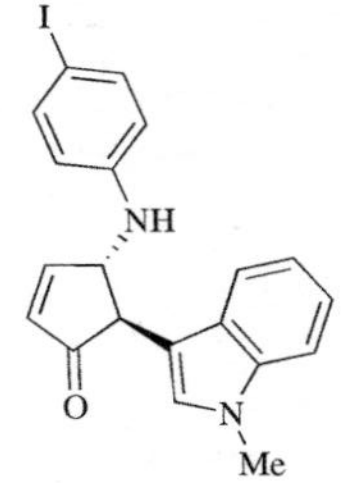

(94) 55

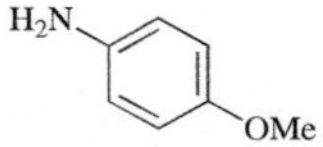

1.1 eq

HFIP, 20°, 0.17 h

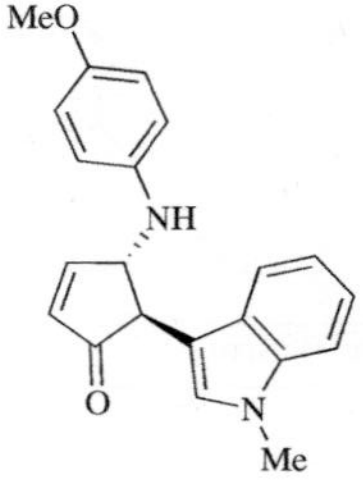

(89) 55

TABLE 3. 4-AMINOCYCLOPENT-2-ENONES (*Continued*)

Substrate	Nucleophile	Conditions	Product(s) and Yield(s) (%)	Refs.

*Please refer to the chart preceding the tables for the structures indicated by the **bold** numbers.*

C$_{13}$

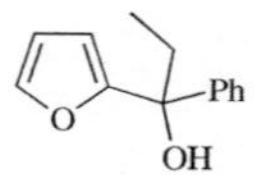

1.5 eq

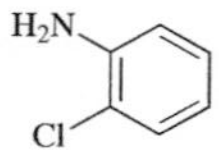

1 (5 mol %),
CHCl$_3$, rt, 24 h

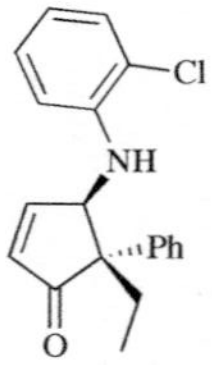

(90) dr >19:1,
er 97.0:3.0

57

C$_{13-15}$

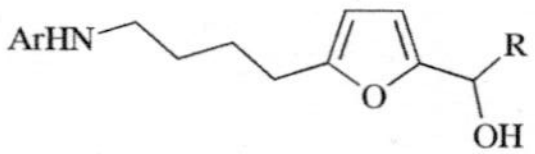

Dy(OTf)$_3$ (5 mol %),
MeCN, 80°

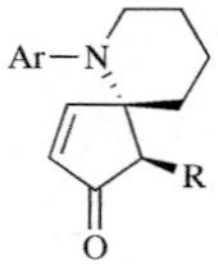

R	Ar	Time	
n-Bu	Ph	15 h	(54)
n-Bu	4-IC$_6$H$_4$	2 h	(65)
Ph	Ph	1 h	(75)
Ph	4-IC$_6$H$_4$	5 min	(69)
Ph	4-MeOC$_6$H$_4$	75 h	(74)

52

C$_{14}$

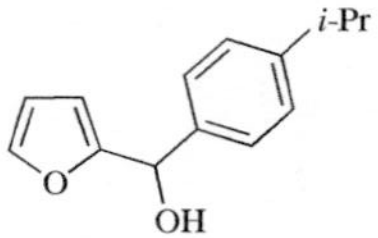

PhNH$_2$

1.2 eq

3 (5 mol %), PhF, 50°, 20 h

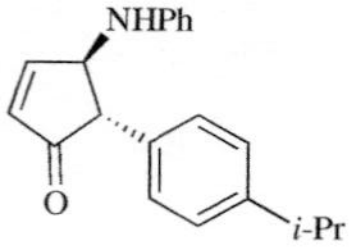

(65) dr >19:1,
er 95.0:5.0

58

572

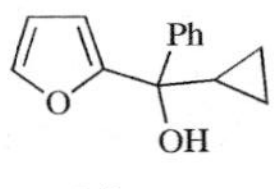

55

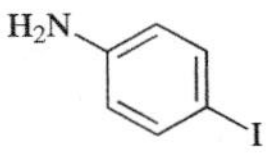

Ca(NTf$_2$)$_2$ (5 mol %),
n-Bu$_4$NPF$_6$ (5 mol %),
HFIP, 60°, 1.5 h

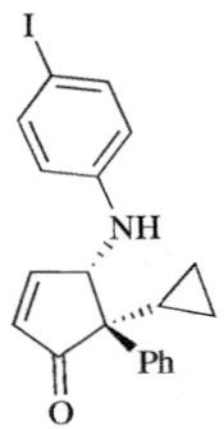

(50)

H$_2$N—⬡—I

2.0 eq

42

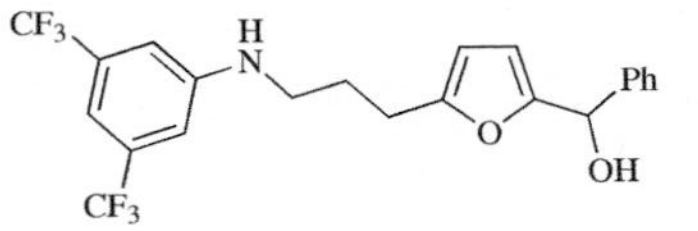

5 (10 mol %),
DBE, rt, 18 h

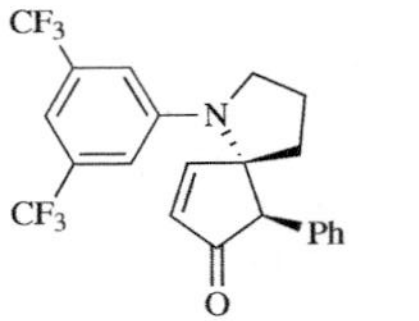

(95) dr >20:1,
er 89.0:11.0

52

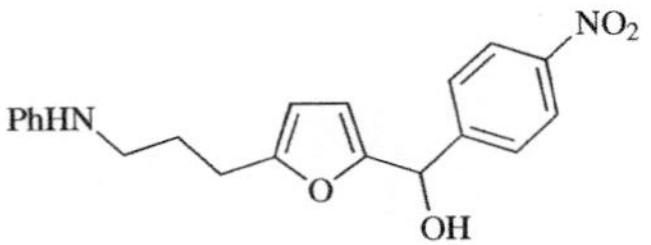

Dy(OTf)$_3$ (5 mol %),
MeCN, 80°, 5 h

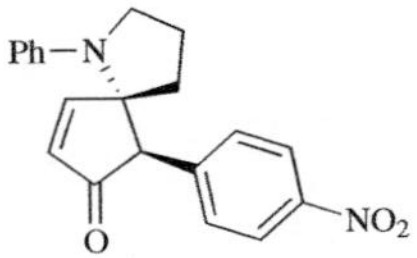

(67)

C$_{15}$

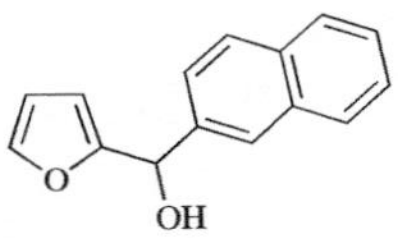

Catalyst (z mol %),
solvent

ArNH$_2$
y eq

x eq

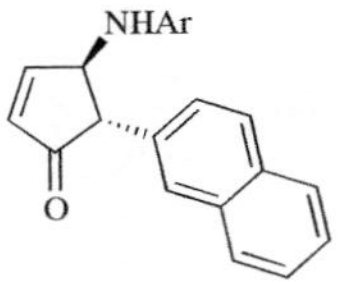

x	Ar	y	Catalyst	z	Solvent	Temp (°)	Time (h)	er	
1.0	3,5-(CF$_3$)$_2$C$_6$H$_3$	1.5	**2**	10	DCE	rt	32 (72)	95.0:5.0	42
1.0	2-CF$_3$-4-NCC$_6$H$_3$	1.5	**2**	10	DCE	rt	29 (65)	97.0:3.0	42
1.2	2-NCC$_6$H$_4$	1.0	**1**	5	CHCl$_3$	5	24 (86)	94.0:6.0	57

TABLE 3. 4-AMINOCYCLOPENT-2-ENONES (*Continued*)

Substrate	Nucleophile	Conditions	Product(s) and Yield(s) (%)	Refs.

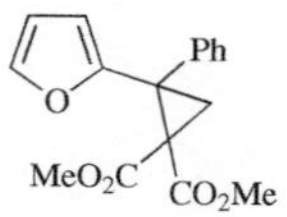

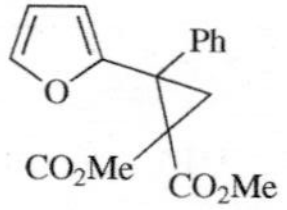

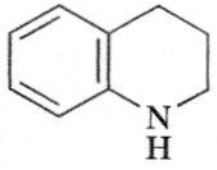

*Please refer to the chart preceding the tables for the structures indicated by the **bold** numbers.*

C₁₅

Dy(OTf)₃ (10 mol %), MeCN

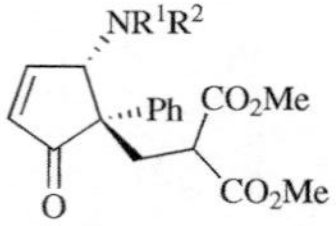

53

R¹	R²	Temp (°)	Time (min)		dr
H	Ph	rt	30	(89)	13:1
H	Ph	80	2	(87)	30:1
H	3-ClC₆H₅	rt	30	(81)	29:1
H	4-IC₆H₅	rt	30	(82)	28:1
H	4-MeOC₆H₅	rt	30	(57)	6:1
H	4-MeOC₆H₅	80	2	(82)	16:1
H	4-MeC₆H₅	rt	30	(80)	10:1
H	4-MeC₆H₅	80	2	(87)	14:1
H	4-CF₃C₆H₅	rt	30	(72)	60:1
Me	Ph	rt	30	(76)	57:1

Dy(OTf)₃ (10 mol %), MeCN, rt, 0.5 h

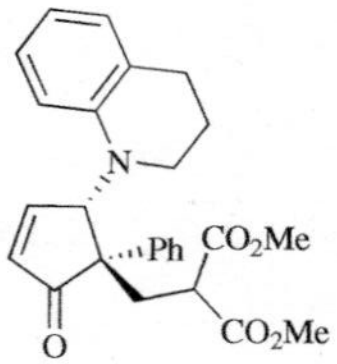

(65) dr 17:1

53

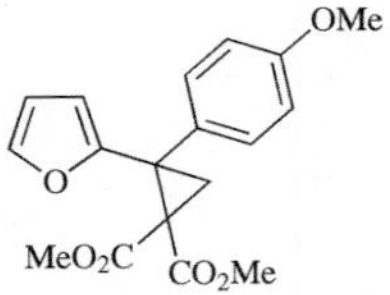

C16

Reaction 1: ArNH2, Dy(OTf)3 (10 mol %), MeCN, rt, 0.5 h

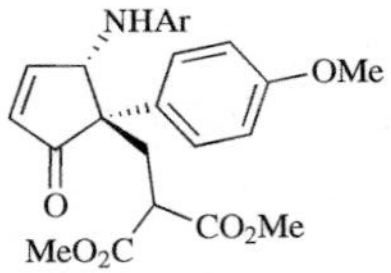

Ar		dr
Ph	(84)	32:1
4-MeOC6H4	(87)	25:1
4-CF3C6H4	(63)	22:1

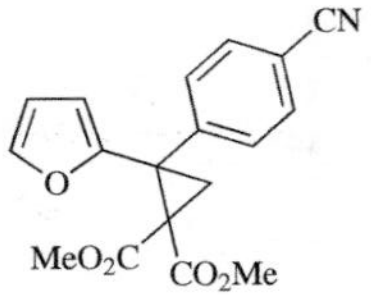

53

Reaction 2: ArNH2, Dy(OTf)3 (10 mol %), MeCN

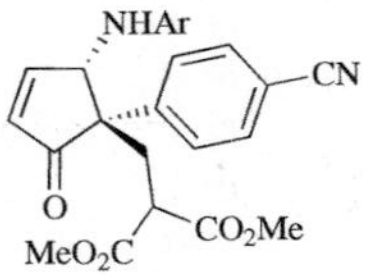

Ar	Temp (°)	Time		dr
Ph	rt	1 h	(76)	1:1
Ph	80	2 min	(77)	3:1
4-MeOC6H4	rt	1 h	(58)	2:1
4-MeOC6H4	80	2 min	(65)	1:1
4-CF3C6H4	rt	1 h	(65)	5:1
4-CF3C6H4	80	2 min	(66)	22:1

53

TABLE 3. 4-AMINOCYCLOPENT-2-ENONES (Continued)

Substrate	Nucleophile	Conditions	Product(s) and Yield(s) (%)	Refs.

Please refer to the chart preceding the tables for the structures indicated by the bold numbers.

C17

Substrate: Ph, Ph, OH, furanyl, x eq

Nucleophile: ArNH2, y eq

Conditions: Catalyst (5 mol %), n-Bu4NPF6 (z mol %), solvent

Product: NHAr, Ph, Ph, O (cyclopentenone)

x	Ar	y	Catalyst	z	Solvent	Temp (°)	Time (h)		er	Refs.
1.0	Ph	1.2	3	0	PhF	50	20	(84)	97.0:3.0	58
1.0	2-IC6H4	1.2	3	0	PhF	50	20	(90)	96.0:4.0	58
1.0	2-BrC6H4	1.2	3	0	PhF	50	20	(86)	98.0:2.0	58
1.0	4-BrC6H4	1.2	3	0	PhF	50	20	(94)	98.0:2.0	58
1.0	4-FC6H4	1.2	3	0	PhF	50	20	(87)	98.0:2.0	58
1.0	3-ClC6H4	1.2	3	0	PhF	50	20	(86)	94.0:6.0	58
1.0	4-MeOC6H4	1.1	Ca(NTf2)2	5	MeNO2	100	0.33	(83)	—	55
1.0	4-MeOC6H4	1.1	Ca(NTf2)2	5	HFIP	20	0.33	(96)	—	55
1.0	2-NCC6H4	1.2	3	0	PhF	50	20	(88)	96.0:4.0	58
1.2	2-NCC6H4	1.0	1	0	CHCl3	5	20	(95)	92.5:7.5	57
1.0	3-MeC6H4	1.2	3	0	PhF	50	20	(81)	93.0:7.0	58
1.0	4-MeC6H4	1.2	3	0	PhF	50	20	(79)	96.0:4.0	58
1.0	4-NCCH2C6H4	1.2	3	0	PhF	50	20	(78)	91.0:9.0	58
1.0	4-MeOCC6H4	1.2	3	0	PhF	50	20	(85)	89.0:11.0	58
1.0	2-i-PrC6H4	1.2	3	0	PhF	50	20	(87)	96.0:4.0	58
1.0	1-Np	1.2	3	0	PhF	50	20	(67)	98.0:2.0	58

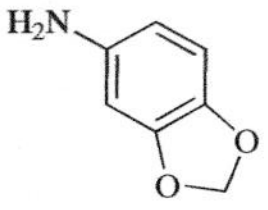

3 (5 mol %), PhF, 50°, 20 h

(70) er 94.0:6.0   58

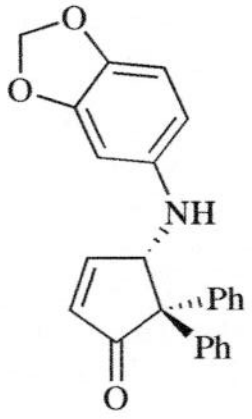

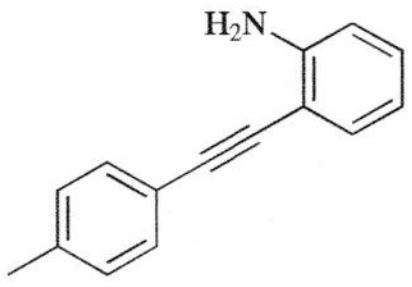

3 (5 mol %), PhF, 50°, 20 h

(84) er 83.0:17.0   58

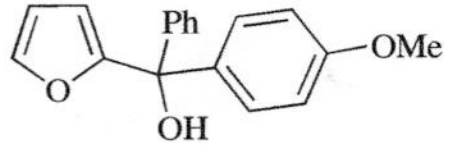

3 (5 mol %), PhF, 50°, 20 h

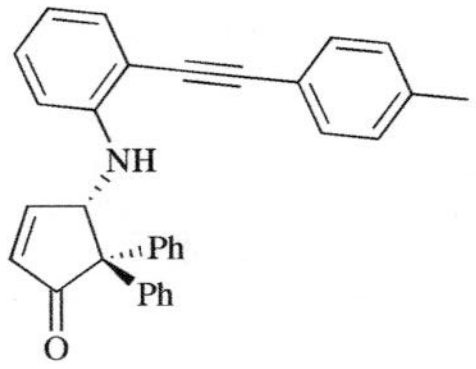

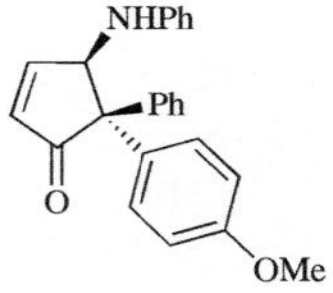

(76) dr 1:1
er 94.0:6.0, 84.0:16.0   58

H₂N (benzodioxole) 1.2 eq

H₂N (alkyne-tolyl) 1.2 eq

PhNH₂ 1.2 eq

TABLE 3. 4-AMINOCYCLOPENT-2-ENONES (*Continued*)

Substrate	Nucleophile	Conditions	Product(s) and Yield(s) (%)	Refs.

*Please refer to the chart preceding the tables for the structures indicated by the **bold** numbers.*

C_{17}

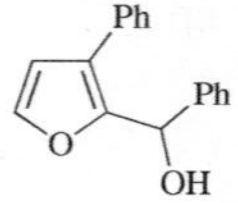

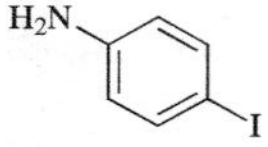

Nucleophile (1.1 eq): H_2N–C$_6$H$_4$–I (4-iodoaniline)

Conditions: Ca(NTf$_2$)$_2$ (5 mol %), n-Bu$_4$NPF$_6$ (5 mol %), MeNO$_2$, 90°, 0.33 h

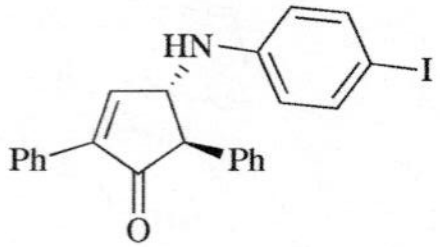

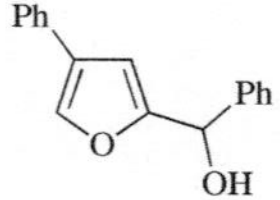

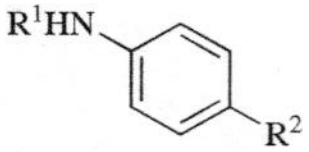

Product: (93) Refs. 54

Nucleophile (1.1 eq): R^{1}HN–C$_6$H$_4$–R^2

Conditions: Ca(NTf$_2$)$_2$ (5 mol %), n-Bu$_4$NPF$_6$ (5 mol %), solvent

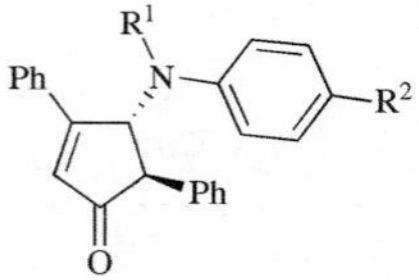

R^1	R^2	Solvent	Temp (°)	Time (h)		Refs.
H	I	MeNO$_2$	80	0.33	(87)	54
Bn	MeO	MeNO$_2$	100	20	(58)	55
Bn	MeO	HFIP	60	12	(80)	55

C_{19}

Nucleophile (1.3 eq): H_2N–C$_6$H$_4$–I (4-iodoaniline)

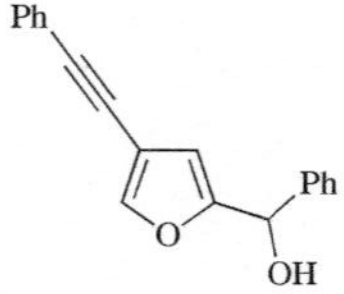

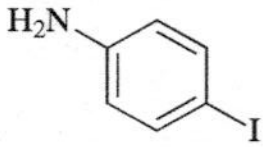

Conditions: Ca(NTf$_2$)$_2$ (5 mol %), n-Bu$_4$NPF$_6$ (5 mol %), HOTf (10 mol %), DCE, 80°, 12 min

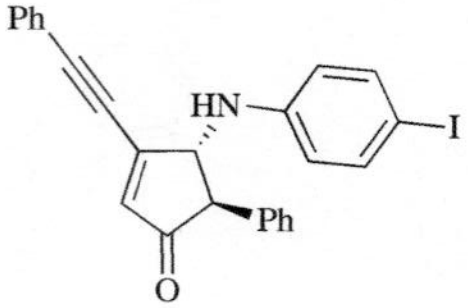

Product: (89) Refs. 43

[a]The yield and er were determined after conversion of the acid to the methyl ester with TMSCH=N$_2$.

TABLE 4. 4-HYDROXYAMINOCYCLOPENT-2-ENONES

Substrate	Nucleophile	Conditions	Product(s) and Yield(s) (%)	Refs.
C_8	H–R–N–OBn	Dy(OTf)$_3$ (10 mol %), MeNO$_2$, 80°		56

R	Time (h)	
t-BuCH$_2$	96	(51)
Bn	24	(50)

Substrate	Nucleophile	Conditions	Product(s) and Yield(s) (%)	Refs.
C_{9-20}	H–Bn–N–OBn	Dy(OTf)$_3$ (10 mol %), MeNO$_2$, 80°		56

R	Time (h)	
n-Bu	48	(53)
2-thienyl	0.5	(92)
4-MeOC$_6$H$_4$	0.5	(81)
4-NCC$_6$H$_4$	17	(84)
2,4,6-(i-Pr)$_3$C$_6$H$_2$	0.67	(80)

Nucleophile: H–R^1–N–OR^2

Conditions: Dy(OTf)$_3$ (x mol %), MeNO$_2$

R^1	R^2	x	Temp (°)	Time (h)	
n-Pr	PhCO$_2$	30	rt	58	(50)
n-C$_5$H$_{11}$	Bn	10	80	1	(76)
t-BuCH$_2$	Bn	5	80	0.75	(81)
Bn	Me	5	80	0.5	(91)
Bn	allyl	5	80	0.75	(95)

Substrate	Product(s) and Yield(s) (%)	Refs.
C_{11}		56

R^1	R^2	x	Temp (°)	Time (h)	
Bn	t-Bu	5	80	1	(76)
Bn	Bn	5	80	0.5	(88)
4-MeOC$_6$H$_4$CH$_2$	Bn	5	80	2	(76)
4-MeO$_2$CC$_6$H$_4$CH$_2$	Bn	5	80	1	(82)
styrylCH$_2$	Bn	5	80	1.5	(86)

TABLE 4. 4-HYDROXYAMINOCYCLOPENT-2-ENONES (Continued)

Substrate	Nucleophile	Conditions	Product(s) and Yield(s) (%)	Refs.
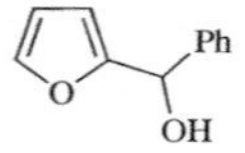 C₁₁	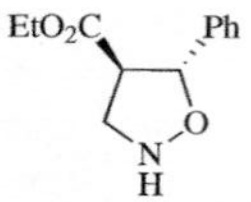	Dy(OTf)₃ (5 mol %), MeNO₂; 80°, 1.25 h	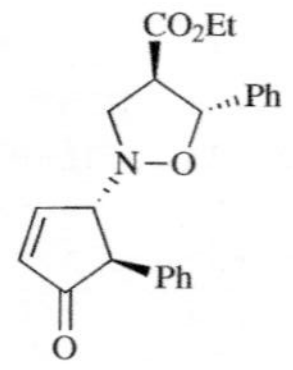 (50) dr 1:1	56
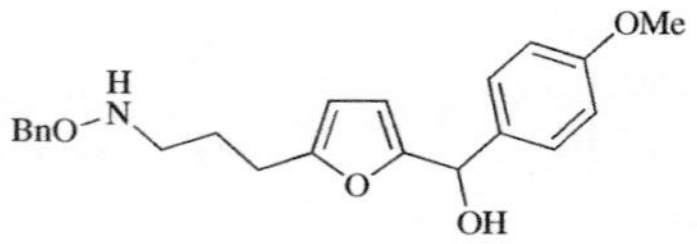 C₁₄		Dy(OTf)₃ (5 mol %), MeNO₂; 80°, 1 h	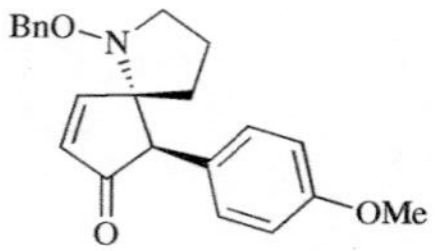 (82)	56
C₁₇		Dy(OTf)₃ (5 mol %), MeNO₂; 80°		56

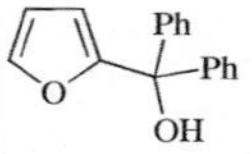

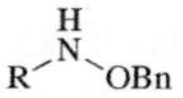

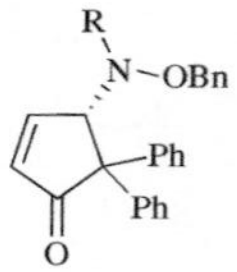

For the C₁₇ product:

R	Time (h)	
t-BuCH₂	18	(83)
Bn	4	(76)

TABLE 5. 3,4-DIAMINOCYCLOPENTANONES

Substrate	Nucleophile	Conditions	Product(s) and Yield(s) (%)	Refs.
C$_5$ furfuryl alcohol (–OH), 1.2 eq	benzene-1,2-(NH$_2$)(NHTs)	La(OTf)$_3$ (5 mol %), MeCN, 90°, 2 h	Ts-N fused tricyclic aminocyclopentanone (tr)	59
C$_{6-13}$ furan-CH(R)OH	ArNH$_2$	1. Ca(NTf$_2$)$_2$ (x mol %), n-Bu$_4$PF$_6$ (x mol %), HFIP, rt, time 1 2. Br–C(Me)$_2$–C(O)NHOBn (2 eq), Na$_2$CO$_3$ (3 eq), temp, time 2	cyclopentanone with R, N–Ar, BnO–N–C(O)C(Me)$_2$	128

R	Ar	x	Time 1 (h)	Temp	Time 2 (h)	Yield
Me	4-IC$_6$H$_4$	5	3	rt	1.5	(69)
allyl	4-IC$_6$H$_4$	5	2	rt	1.5	(52)
c-C$_3$H$_5$	4-IC$_6$H$_4$	5	2	rt	3	(84)
2-thienyl	4-IC$_6$H$_4$	0	2	rt	2.5	(73)
Ph	Ph	5	2	rt	2	(67)
Ph	4-IC$_6$H$_4$	5	0.15	rt	2	(88)
Ph	4-MeOC$_6$H$_4$	5	2	rt	2	(82)
Ph	4-MeC$_6$H$_4$	5	0.5	rt	2	(84)
Ph	4-CF$_3$C$_6$H$_4$	5	0.5	rt	1	(80)
Ph	4-MeO$_2$CC$_6$H$_4$	5	2	rt	1	(72)
Ph	2,4,6-Me$_3$C$_6$H$_2$	5	2	40°	2	(67)
2,4,6-Me$_3$C$_6$H$_2$	4-IC$_6$H$_4$	5	1	rt	1	(77)
(E)-PhCH=CH	4-IC$_6$H$_4$	5	1.5	rt	1	(61)
1-Me-3-indolyl	4-IC$_6$H$_4$	0	0.15	rt	2	(75)

TABLE 5. 3,4-DIAMINOCYCLOPENTANONES (Continued)

Substrate	Nucleophile	Conditions	Product(s) and Yield(s) (%)	Refs.
C$_{6-17}$		1. Ca(NTf$_2$)$_2$ (5 mol %), n-Bu$_4$PF$_6$ (5 mol %), HFIP, temp, time 1 2. (2 eq), Na$_2$CO$_3$ (3 eq), rt, time 2	 | R^1	R^2	Temp	Time 1 (h)	Time 2 (h)
Me	N$_3$	60°	1	1 (81)
Ph	Ph	rt	0.5	3 (78)		128
C$_7$ 1.2 eq		La(OTf)$_3$ (5 mol %), MeCN, 90°, 2 h	 (tr)	59
C$_{11}$ 1.2 eq		La(OTf)$_3$ (5 mol %), MeCN, 90°, 2 h	 | R	
Ms	(83)
Ac	(tr)
Ts	(88)		59

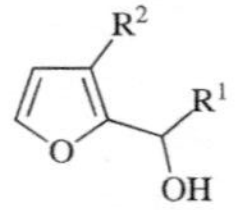

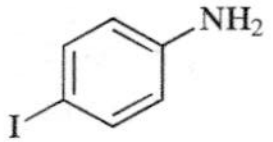

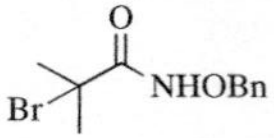

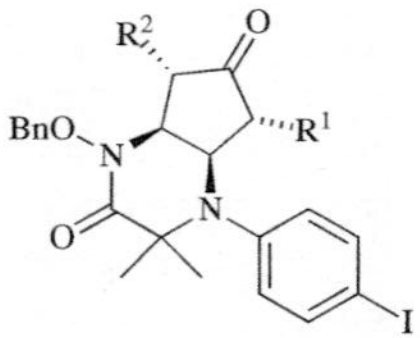

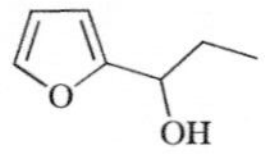

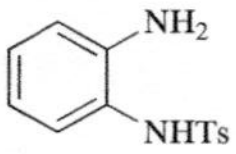

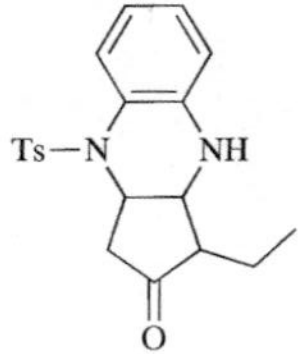

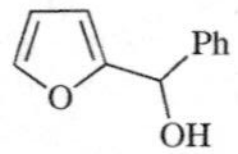

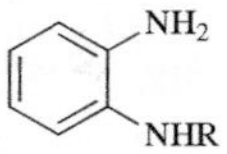

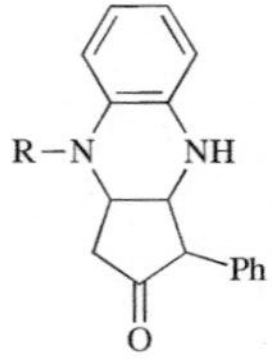

582

C11-12

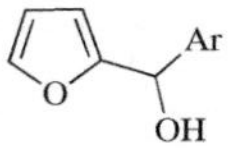

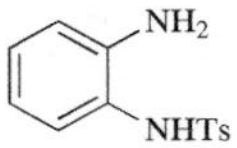

La(OTf)$_3$ (5 mol %),
MeCN, 90°, 2 h

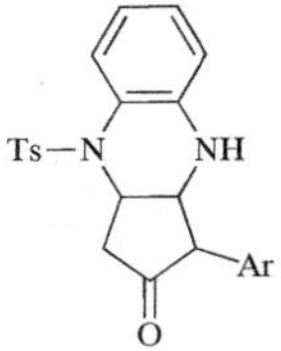

Ar	
Ph	(88)
4-FC$_6$H$_4$	(68)
4-MeOC$_6$H$_4$	(45)
4-CF$_3$C$_6$H$_4$	(tr)

59

C11-17

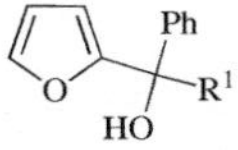

ArNH$_2$

1. Ca(NTf$_2$)$_2$ (5 mol %),
n-Bu$_4$PF$_6$ (5 mol %),
HFIP, temp, time 1

2. 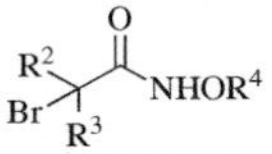 (2 eq), Na$_2$CO$_3$ (3 eq),
rt, time 2

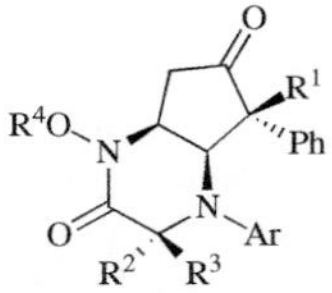

128

R^1	R^2	R^3	R^4	Ar	Temp	Time 1 (h)	Time 2 (h)		dr
H	Me	Me	Me	4-IC$_6$H$_4$	rt	0.15	1	(95)	—
H	Me	Me	t-Bu	4-IC$_6$H$_4$	rt	0.15	1	(88)	—
H	Me	Me	Ph	4-IC$_6$H$_4$	rt	0.15	1	(90)	—
H	Ph	H	Bn	4-CF$_3$C$_6$H$_4$	rt	0.5	1	(53)	4.5:1
H	—(CH$_2$)$_5$—		Bn	4-IC$_6$H$_4$	rt	0.15	4	(86)	—
c-C$_3$H$_5$	Me	Me	Bn	4-IC$_6$H$_4$	60°	1.5	2	(50)	—

TABLE 6. 3-AMINO-4-HYDROXYCYCLOPENTANONES

Substrate	Nucleophile	Conditions	Product(s) and Yield(s) (%)	Refs.
C$_7$ 1.2 eq		La(OTf)$_3$ (5 mol %), MeCN. 80°, 4 h	(19)	59
C$_9$ 1.2 eq		La(OTf)$_3$ (5 mol %), MeCN. 80°, 4 h	R H (24) NO$_2$ (53)	59
C$_{11}$ 1.2 eq		Catalyst (x mol %), MeCN		

584

R¹	R²	Catalyst	x	Temp (°)	Time (h)		
H	H	La(OTf)₃	5	80	4	(81)	59
H	NO₂	La(OTf)₃	5	80	4	(tr)	59
Cl	H	La(OTf)₃	5	80	4	(66)	59
Cl	H	In(OTf)₃	10	rt	3.5	(78)	129
H	Me	In(OTf)₃	10	rt	3.5	(73)	129
Me	H	La(OTf)₃	5	80	4	(48)	59
t-Bu	H	La(OTf)₃	5	80	4	(53)	59

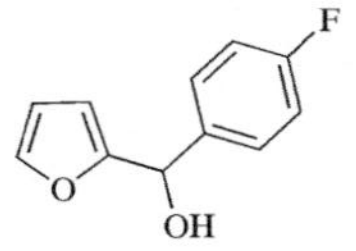

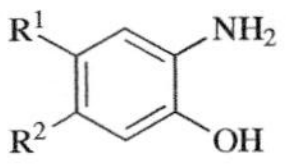

Catalyst (x mol %), MeCN

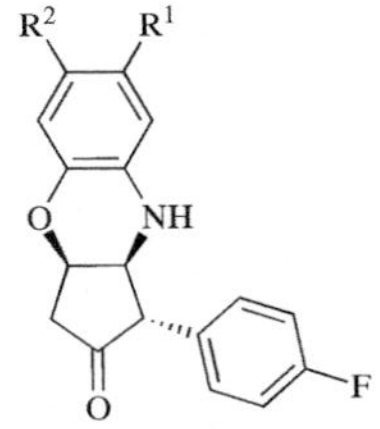

R¹	R²	Catalyst	x	Temp (°)	Time (h)		
H	H	La(OTf)₃	5	80	4	(46)	59
H	H	In(OTf)₃	10	rt	3.5	(81)	129
H	O₂N	La(OTf)₃	5	80	4	(tr)	59
Cl	H	La(OTf)₃	5	80	4	(78)	59
Cl	H	In(OTf)₃	10	rt	3.5	(80)	129
H	Me	In(OTf)₃	10	rt	3.5	(78)	129
Me	H	La(OTf)₃	5	80	4	(48)	59
t-Bu	H	La(OTf)₃	5	80	4	(74)	59

1.2 eq

TABLE 6. 3-AMINO-4-HYDROXYCYCLOPENTANONES (Continued)

Substrate	Nucleophile	Conditions	Product(s) and Yield(s) (%)	Refs.

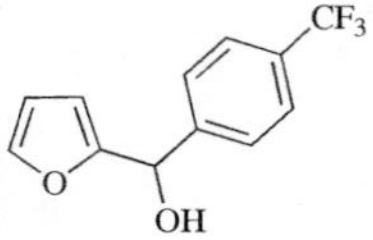

C$_{11}$

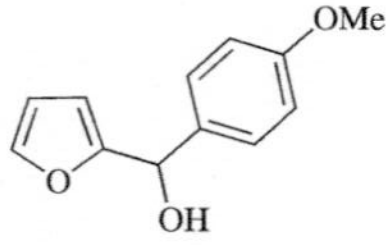

Substrate: 4-MeO-phenyl furyl carbinol, OH, O; 1.2 eq

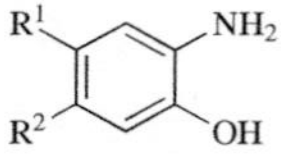

Nucleophile: R^1, R^2 aminophenol (NH$_2$, OH)

Conditions: Catalyst (x mol %), MeCN

Product:

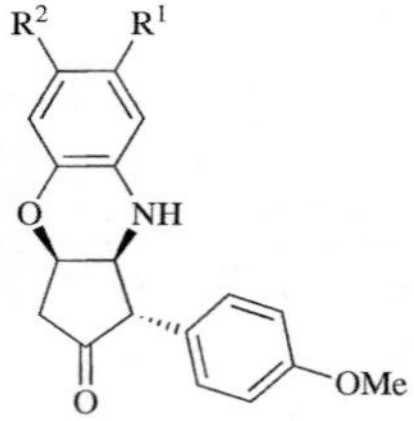

R^1	R^2	Catalyst	x	Temp (°)	Time (h)		Refs.
H	H	La(OTf)$_3$	5	80	4	(46)	59
H	H	In(OTf)$_3$	10	rt	3	(75)	129
H	O$_2$N	La(OTf)$_3$	5	80	4	(tr)	59
Cl	H	La(OTf)$_3$	5	80	4	(51)	59
Cl	H	In(OTf)$_3$	10	rt	2.5	(82)	129
H	Me	In(OTf)$_3$	10	rt	2.5	(80)	129
Me	H	La(OTf)$_3$	5	80	4	(60)	59
t-Bu	H	La(OTf)$_3$	5	80	4	(41)	59

Substrate: 4-CF$_3$-phenyl furyl carbinol, OH, O; 1.2 eq

Nucleophile: R^1, R^2 aminophenol (NH$_2$, OH)

Conditions: La(OTf)$_3$ (5 mol %), MeCN, 80°, 4 h

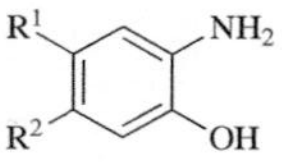

Product (with CF$_3$):

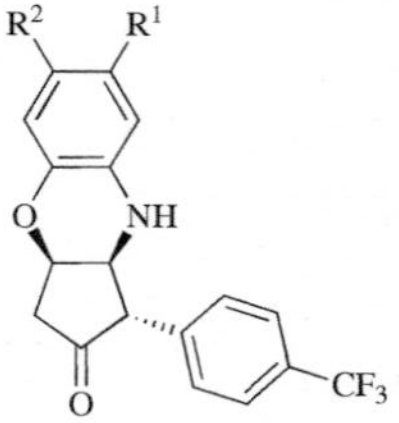

R^1	R^2		Refs.
H	H	(10)	59
H	O$_2$N	(92)	
Cl	H	(82)	
Me	H	(tr)	
t-Bu	H	(tr)	

586

TABLE 7. 3-AMINO-4-MERCAPTOCYCLOPENTANONES

Substrate	Nucleophile	Conditions	Product(s) and Yield(s) (%)	Refs.
C₁₁				

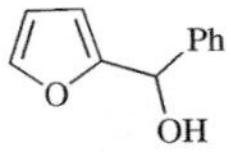

Row 1:

Substrate: Ph–CH(OH)–furan, 1.2 eq

Nucleophile: R, NH₂ / SH (aminothiophenol)

Conditions: In(OTf)₃ (10 mol %), MeCN, rt

Product:

R	Time (h)	
H	2	(86)
Cl	2.5	(85)

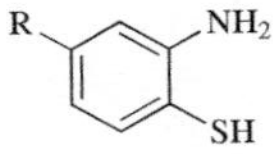

Refs.: 129

Row 2:

Substrate: 4-F-C₆H₄–CH(OH)–furan, 1.2 eq

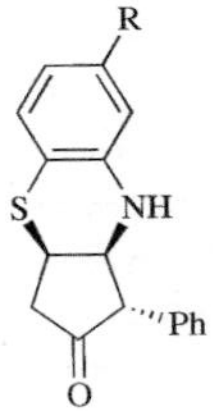

Nucleophile: NH₂ / SH

Conditions: In(OTf)₃ (10 mol %), MeCN, rt, 2.5 h

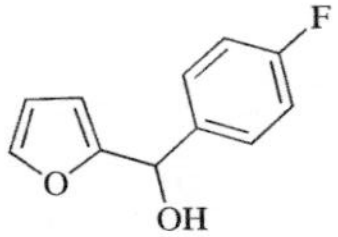

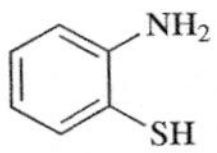

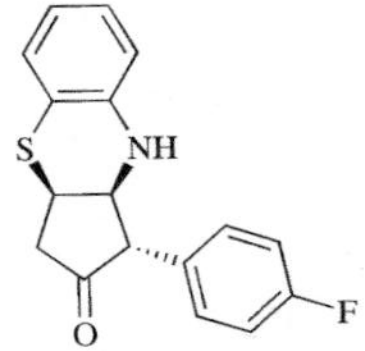

Product: (85)

Refs.: 129

Row 3:

Substrate: 4-OMe-C₆H₄–CH(OH)–furan, 1.2 eq

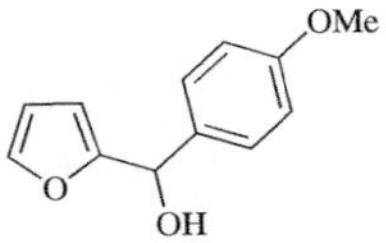

Nucleophile: R, NH₂ / SH

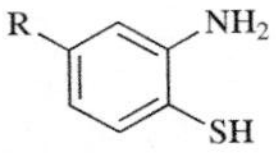

Conditions: In(OTf)₃ (10 mol %), MeCN, rt

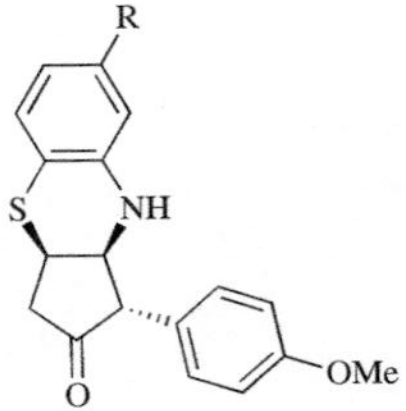

Product:

R	Time (h)	
H	2	(87)
Cl	2.5	(88)

Refs.: 129

TABLE 8. 3-AMINOCYCLOPENTANONES

Substrate	Nucleophile	Conditions	Product(s) and Yield(s) (%)	Refs.

*Please refer to the chart preceding the tables for the structures indicated by the **bold** numbers.*

C$_{8-9}$

Yb(OTf)$_3$ (10 mol %), toluene, 80°

R	Ar	Time (h)	
i-Pr	Ph	9	(64)
i-Pr	4-ClC$_6$H$_4$	9	(68) 130
2-thienyl	Ph	0.25	(64)
2-thienyl	4-ClC$_6$H$_4$	0.25	(58)

130

C$_{9-15}$

I

NHAr^2 furan

II

1. Co(BF$_4$)$_2$•6H$_2$O (10 mol %); **8** (10 mol %), THF, 35°, 30 min; remove solvent
2. **I**, **II** (1 eq), o-xylene, 65°, 16 h

dr >19:1

127

Ar^1	Ar^2		er
2-furanyl	4-ClC$_6$H$_4$	(51)	95.5:4.5
2-thienyl	4-ClC$_6$H$_4$	(61)	93.0:6.0
Ph	Ph	(20)	96.0:4.0
Ph	4-ClC$_6$H$_4$	(42)	95.5:4.5
Ph	4-BrC$_6$H$_4$	(34)	97.0:3.0
4-ClC$_6$H$_4$	4-ClC$_6$H$_4$	(48)	91.5:8.5

Ar^1	Ar^2		er
2-MeOC$_6$H$_4$	4-ClC$_6$H$_4$	(79)	96.5:3.5
3-MeOC$_6$H$_4$	4-ClC$_6$H$_4$	(58)	95.0:5.0
3,4-OCH$_2$OC$_6$H$_3$	4-ClC$_6$H$_4$	(68)	95.5:4.5
3,4-(MeO)$_2$C$_6$H$_3$	4-ClC$_6$H$_4$	(68)	96.5:3.5
4-MeC$_6$H$_4$	4-ClC$_6$H$_4$	(50)	97.0:3.0
2-Np	4-ClC$_6$H$_4$	(31)	97.0:3.0

C$_{11}$

I

1. Co(BF$_4$)$_2$•6H$_2$O
(10 mol %); **8** (10 mol %),
THF; 35°, 30 min;
remove solvent

2. **I**, **II** (1 eq),
o-xylene, 65°, 16 h

II

dr >19:1   127

R	Ar		er
H	4-FC$_6$H$_4$	(74)	97.5:2.5
H	3-ClC$_6$H$_4$	(64)	95.0:5.0
H	4-ClC$_6$H$_4$	(79)	96.5:3.5
H	4-BrC$_6$H$_4$	(71)	96.5:3.5
H	4-MeOC$_6$H$_4$	(62)	91.0:9.0
H	3-MeC$_6$H$_4$	(75)	94.5:5.5

R	Ar		er
H	4-MeC$_6$H$_4$	(80)	96.0:4.0
H	3-CF$_3$C$_6$H$_4$	(77)	96.0:4.0
H	4-t-BuC$_6$H$_4$	(74)	95.5:4.5
H	3,5-(t-Bu)$_2$C$_6$H$_3$	(66)	95.0:5.0
H	2-Np	(56)	95.5:4.5
Me	Ph	(90)	94.5:5.5

C$_{11-15}$

1.2 eq

In(OTf)$_3$ (10 mol %),
MeCN, rt   60

Ar	Time (h)	
Ph	3	(78)
4-FC$_6$H$_4$	5	(70)
2-ClC$_6$H$_4$	4	(72)
4-BrC$_6$H$_4$	3	(75)
4-MeOC$_6$H$_4$	1.5	(87)

Ar	Time (h)	
3,4-OCH$_2$OC$_6$H$_3$	1.5	(85)
3,4-(MeO)$_2$C$_6$H$_3$	2	(87)
3-MeO-4-BnOC$_6$H$_3$	2	(90)
2,3,4-(MeO)$_3$C$_6$H$_2$	1.5	(85)
4-MeC$_6$H$_4$	2	(78)

Ar	Time (h)	
4-i-PrC$_6$H$_4$	2	(80)
4-t-BuC$_6$H$_4$	2.5	(82)
1-Np	3.5	(75)

TABLE 8. 3-AMINOCYCLOPENTANONES (*Continued*)

Substrate	Nucleophile	Conditions	Product(s) and Yield(s) (%)	Refs.

*Please refer to the chart preceding the tables for the structures indicated by the **bold** numbers.*

C_{11}

Substrate: Ar–CH(OH)–(2-furyl), 1.2 eq

Nucleophile: 2-Ph-pyrrole, N-(2-aminophenyl)

Conditions: $In(OTf)_3$ (10 mol %), MeCN, rt

Product:

Ar	Time	
Ph	3	(79)
$2\text{-}ClC_6H_4$	4	(70)

Refs.: 60

C_{11-15}

Substrate: Ar–CH(OH)–(2-furyl), 1.5 eq

Nucleophile: pyrrole, N-(2-aminophenyl), 1 eq

Conditions: **6** (10 mol %), 5 Å MS, $CHCl_3$, rt, 16 h

Product: dr >20:1

Ar		er
3-thienyl	(89)	92.5:7.5
Ph	(92)	94:6
$2\text{-}FC_6H_4$	(94)	93:7
$3\text{-}FC_6H_4$	(73)	93.5:6.5
$3\text{-}ClC_6H_4$	(67)	93:7
$4\text{-}ClC_6H_4$	(68)	88:12

Ar		er
$3\text{-}MeOC_6H_4$	(95)	95.5:4.5
$2,3\text{-}(MeO)_2C_6H_3$	(91)	96:4
$3,5\text{-}(MeO)_2C_6H_3$	(93)	96:4
$3,4,5\text{-}(MeO)_3C_6H_2$	(91)	96.5:3.3
$2\text{-}MeC_6H_4$	(81)	90.5:9.5
$3\text{-}MeC_6H_4$	(95)	92:8

Ar		er
$2\text{-}Me\text{-}4\text{-}FC_6H_3$	(87)	92:8
$3\text{-}Me\text{-}4\text{-}FC_6H_3$	(69)	93:7
1-Np	(91)	92:8
2-Np	(93)	92.5:7.5

Refs.: 131

C$_{11-17}$

Ar^1—CH(OH)—furan + furfuryl-NHAr^2

Yb(OTf)$_3$ (10 mol %),
toluene, 80°, time

Ar^1	Ar^2	Time (min)	
Ph	Ph	9	(87)
Ph	4-ClC$_6$H$_4$	20	(81)
Ph	4-MeOC$_6$H$_4$	9	(80)
Ph	4-EtO$_2$CC$_6$H$_4$	45	(75)
Ph	3,5-Me$_2$C$_6$H$_3$	15	(80)
4-FC$_6$H$_4$	Ph	30	(81)
4-FC$_6$H$_4$	4-ClC$_6$H$_4$	30	(75)

Ar^1	Ar^2	Time (min)	
2-ClC$_6$H$_4$	4-ClC$_6$H$_4$	9	(80)
2-MeOC$_6$H$_4$	4-ClC$_6$H$_4$	9	(84)
4-MeOC$_6$H$_4$	Ph	15	(86)
4-MeOC$_6$H$_4$	4-ClC$_6$H$_4$	15	(83)
4-MeC$_6$H$_4$	Ph	15	(87)
4-MeC$_6$H$_4$	4-ClC$_6$H$_4$	9	(82)
4-t-BuC$_6$H$_4$	4-ClC$_6$H$_4$	15	(74)

Ar^1	Ar^2	Time (min)	
2,4,6-Me$_3$C$_6$H$_2$	Ph	15	(73)
2,4,6-Me$_3$C$_6$H$_2$	4-MeOC$_6$H$_4$	9	(78)
2,4,6-Me$_3$C$_6$H$_2$	4-EtO$_2$CC$_6$H$_4$	60	(70)
2,4,6-Me$_3$C$_6$H$_2$	3,5-Me$_3$C$_6$H$_3$	60	(67)
4-PhC$_6$H$_4$	Ph	15	(80)
4-PhC$_6$H$_4$	4-ClC$_6$H$_4$	15	(77)

C$_{11}$

+ furan-CH(Ar)-OH, 1.2 eq

PMA (10 mol %),
MeCN, reflux

Ar	Time (h)		
Ph	3	(75)	
4-FC$_6$H$_4$	4	(68)	
2-ClC$_6$H$_4$	3	(70)	
4-BrC$_6$H$_4$	3	(72)	132
4-MeOC$_6$H$_4$	1.5	(82)	
2,3,4-(MeO)$_3$C$_6$H$_2$	2	(80)	
4-MeC$_6$H$_4$	2	(76)	
4-i-PrC$_6$H$_4$	2	(78)	
4-t-BuC$_6$H$_4$	3	(80)	
1-Np	3	(69)	

TABLE 8. 3-AMINOCYCLOPENTANONES (*Continued*)

Substrate	Nucleophile	Conditions	Product(s) and Yield(s) (%)	Refs.

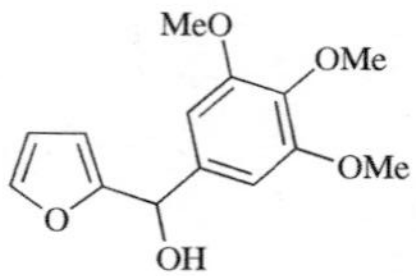

*Please refer to the chart preceding the tables for the structures indicated by the **bold** numbers.*

C₁₁

6 (10 mol %), 5 Å MS, CHCl₃, rt, 16 h

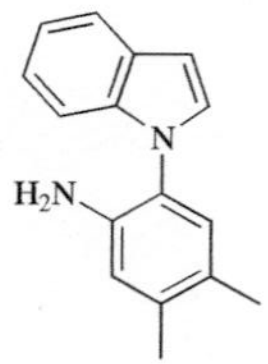

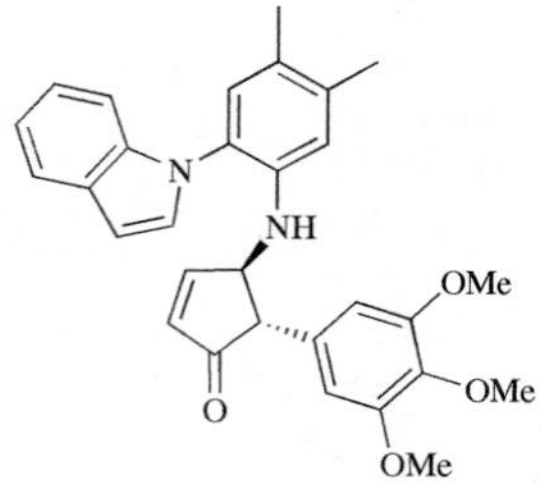

(94) dr >20:1
er 98.5:1.5

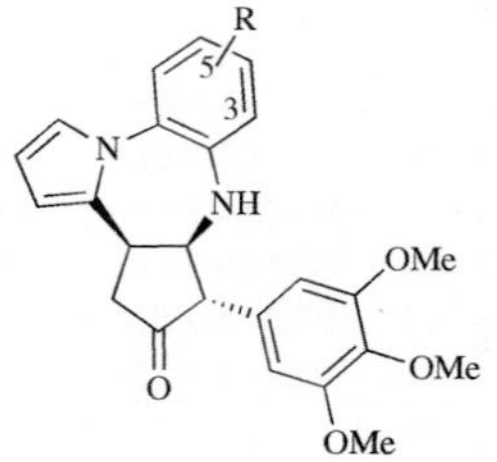

131

R		er
5-Cl	(98)	97.5:2.5
5-Br	(92)	97:3
5-MeO	(90)	98:2
3-Me	(75)	93.5:6.5

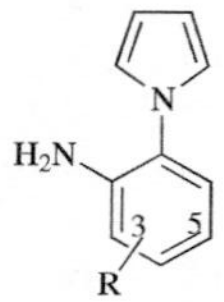

6 (10 mol %), 5 Å MS, CHCl₃, 0°, 16 h

R		er
4-Me	(88)	97:3
5-CF₃	(91)	97.5:2.5
4-Me-5-Cl	(97)	97.5:2.5
4,5-Me₂	(95)	98.5:1.5

131

592

C$_{17}$

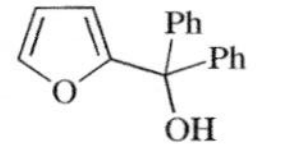

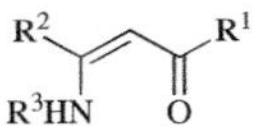

ZnCl$_2$ (10 mol %),
DCE, 80°

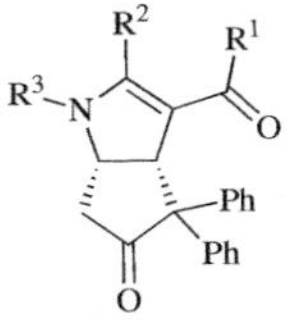

R^1	R^2	R^3	Time (h)	
Cy	H	Ph	4	(99)
Ph	H	Ph	2	(94)
Ph	H	t-Bu	38	(90)
Ph	H	n-C$_6$H$_{13}$	49	(65)
Ph	H	4-ClC$_6$H$_4$	4	(78)
Ph	H	3,4,5-(MeO)$_3$C$_6$H$_2$	5	(93)
Ph	H	Bn	10	(83)

R^1	R^2	R^3	Time (h)	
Ph	H	2,6-Me$_2$C$_6$H$_3$	5	(85)
Ph	H	2,6-i-Pr$_2$C$_6$H$_3$	3	(92)
4-ClC$_6$H$_4$	H	Ph	3	(82)
3,4,5-(MeO)$_3$C$_6$H$_2$	H	Ph	3	(87)
n-Bu	Ph	Ph	14	(91)
1-Np	H	Ph	3	(91)
Ph	Ph	Ph	48	(53)

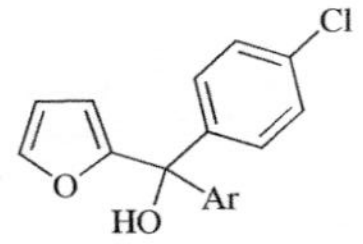

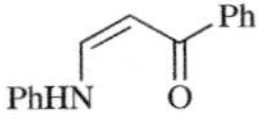

63

ZnCl$_2$ (10 mol %),
DCE, 80°, 7 h

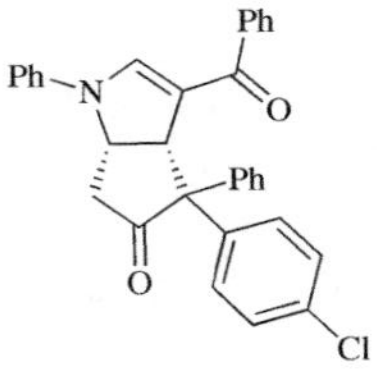

Ar		dr
Ph	(72)	1:1.2
4-ClC$_6$H$_4$	(85)	—

63

TABLE 9. 1,4-BENZODIAZEPIN-5-ONES

Substrate	Nucleophile	Conditions	Product(s) and Yield(s) (%)	Refs.
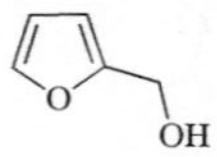 C₅ ... 1.2 eq	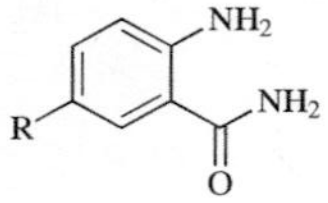 ... NH₂ / NH₂	4-TsOH (5 mol %), MeCN, 90°, 24 h	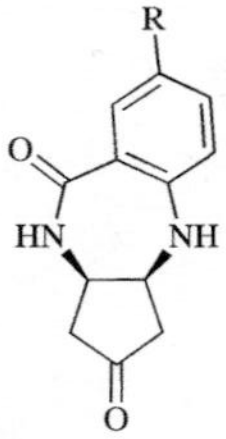 R: H (—); Br (—)	61
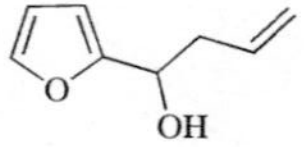 C₈ ... 1.2 eq	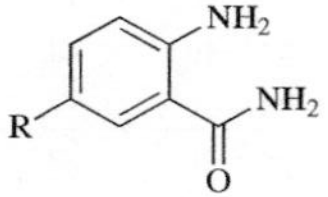 ... NH₂ / NH₂	4-TsOH (5 mol %), MeCN, 90°, 20 h	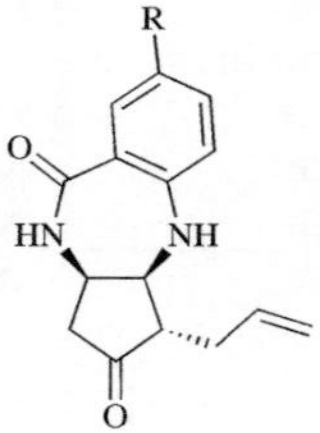 R: H (78); Br (80)	61
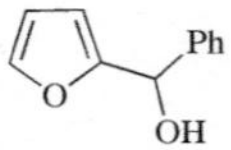 C₁₁ ... Ph ... 1.2 eq	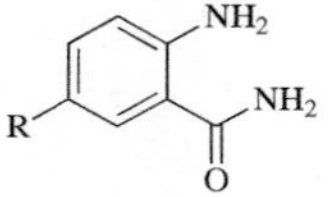 ... NH₂ / NH₂	4-TsOH (5 mol %), MeCN, 90°	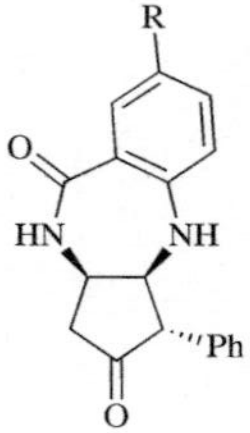 R / Time (h) / — : H, 18, (76); Br, 12, (81)	61

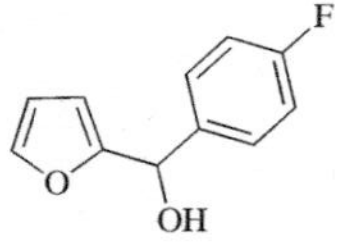

C12

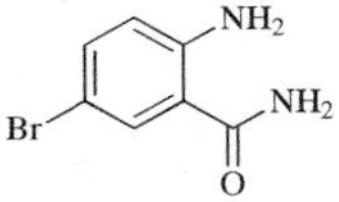

4-TsOH (5 mol %), MeCN, 90°, 16 h — (76) — 61

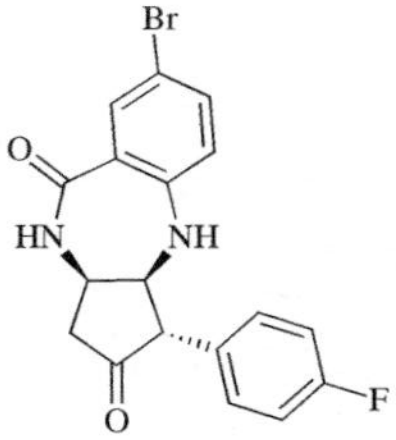

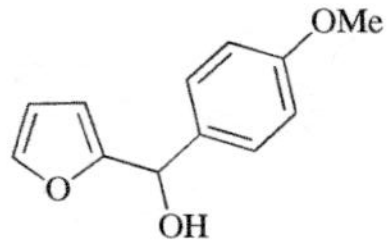

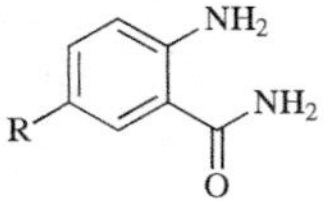

4-TsOH (5 mol %), MeCN, 90°

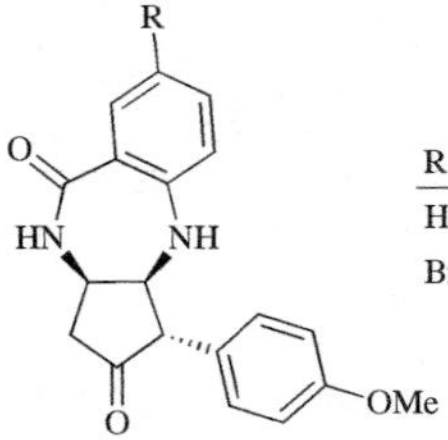

R	Time (h)	
H	18	(69)
Br	16	(74)

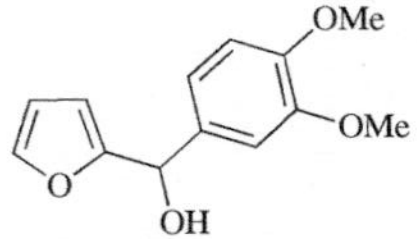

61

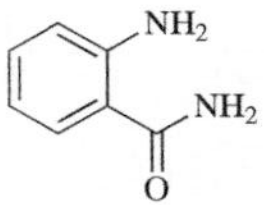

4-TsOH (5 mol %), MeCN, 90°, 18 h — (68) — 61

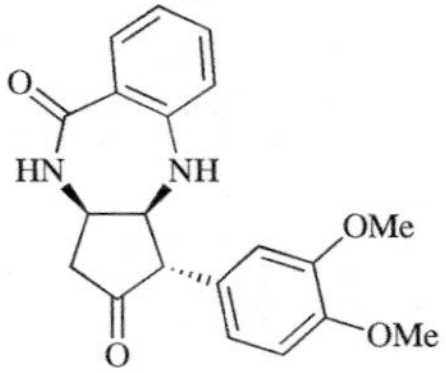

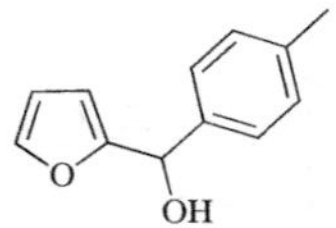

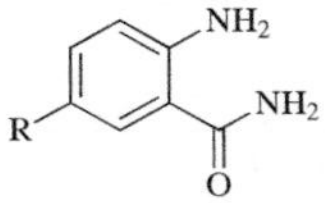

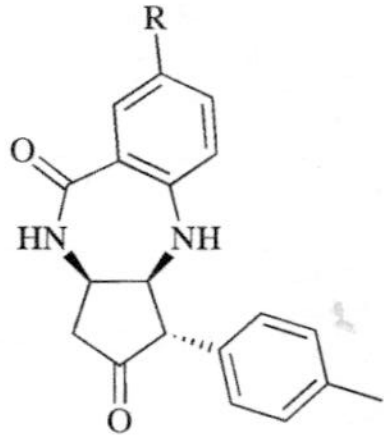

4-TsOH (5 mol %), MeCN, 90°

R	Time (h)	
H	12	(80)
Br	15	(70)

61

1.2 eq

TABLE 10. 4,6-DIHYDROCYCLOPENTA[b]PYRROL-5(1H)-ONES

Substrate	Nucleophile	Conditions	Product(s) and Yield(s) (%)	Refs.
C13-23	1.3 eq	1. Ca(NTf2)2 (5 mol %), n-Bu4NPF6 (5 mol %), DCE, 80°, time 1 2. Cu(OTf)2 (10 mol %), DCE, temp, time 2		43

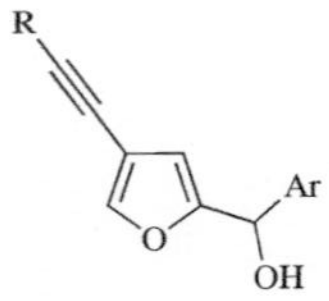

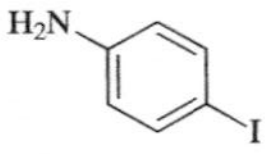

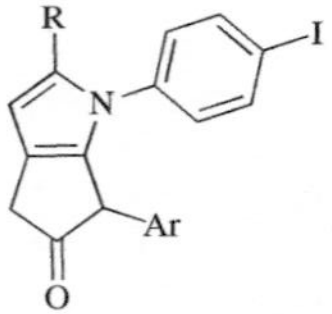

Ar	R	Time 1 (h)	Temp (°)	Time 2 (h)
2-thienyl	Ph	0.15	40	16 (68)
Ph	H	1	80	0.15 (36)
Ph	2-thienyl	1	40	0.15 (64)
Ph	4-MeOC6H4	0.3	40	0.15 (67)
Ph	PhCH2CH2	1.3	40	0.3 (67)

Ar	R	Temp (°)	Time 1 (h)	Time 2 (h)
Ph	1-Np	40	0.75	0.5 (62)
4-MeOC6H4	Ph	40	0.15	0.15 (67)
4-MeC6H4	Ph	40	0.15	0.15 (75)
4-EtO2CC6H4	Ph	40	0.5	0.67 (65)

Substrate	Nucleophile	Conditions	Product(s) and Yield(s) (%)	Refs.
C14-17	1.5 eq	Ca(NTf2)2 (x mol %), n-Bu4NPF6 (x mol %), Cu(OTf)2 (y mol %), MeNO2, 90°		43

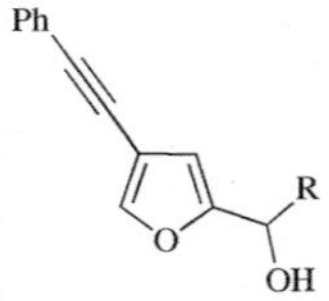

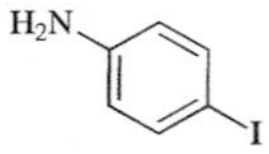

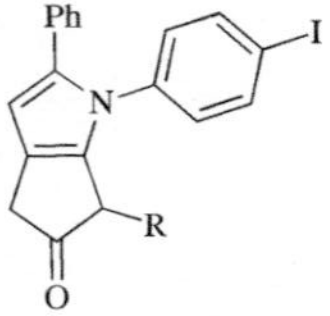

R	x	y	Time (h)	
Me	10	10	0.3	(60)
Et	10	10	0.5	(65)
allyl	5	15	1	(43)
i-Bu	10	10	0.67	(56)

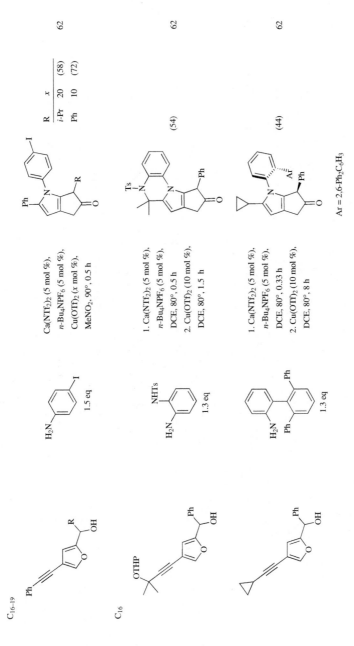

62

62

62

R	x	
i-Pr	20	(58)
Ph	10	(72)

(54)

(44)

Ar = 2,6-Ph$_2$C$_6$H$_3$

Ca(NTf$_2$)$_2$ (5 mol %),
n-Bu$_4$NPF$_6$ (5 mol %),
Cu(OTf)$_2$ (x mol %),
MeNO$_2$, 90°, 0.5 h

1. Ca(NTf$_2$)$_2$ (5 mol %),
n-Bu$_4$NPF$_6$ (5 mol %),
DCE, 80°, 0.5 h
2. Cu(OTf)$_2$ (10 mol %),
DCE, 80°, 1.5 h

1. Ca(NTf$_2$)$_2$ (5 mol %),
n-Bu$_4$NPF$_6$ (5 mol %),
DCE, 80°, 0.33 h
2. Cu(OTf)$_2$ (10 mol %),
DCE, 80°, 8 h

1.5 eq

1.3 eq

1.3 eq

C$_{16-19}$

C$_{16}$

TABLE 10. 4,6-DIHYDROCYCLOPENTA[*b*]PYRROL-5(1*H*)-ONES (*Continued*)

Substrate	Nucleophile	Conditions	Product(s) and Yield(s) (%)	Refs.

C₁₆₋₂₂

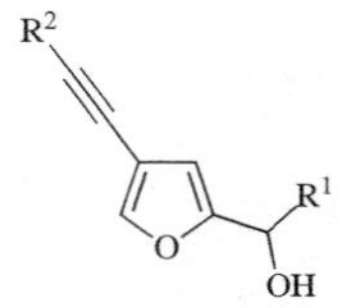

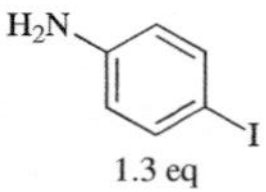

(rotated chemistry table)

Substrate: furan bearing R¹ CH(OH) and R² alkyne, structure with R²–C≡C and OH, R¹

Nucleophile: 4-iodoaniline (H₂N–C₆H₄–I), 1.3 eq

Conditions:
1. Ca(NTf₂)₂ (5 mol %), *n*-Bu₄NPF₆ (5 mol %), HFIP, temp 1, time 1
2. Cu(OTf)₂ (10 mol %), HFIP, temp 2, time 2

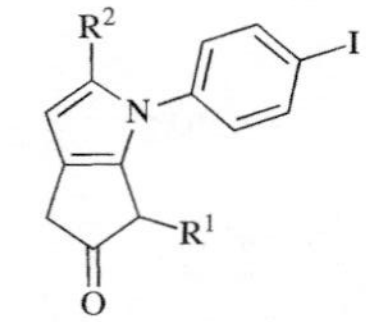

Product:

R¹	R²	Temp 1 (°)	Time 1 (h)	Temp 2 (°)	Time 2 (h)	
c-C₃H₅	Ph	20	1	20	0.15	(61)
Ph	*c*-C₃H₅	80	1.25	40	0.15	(59)
Ph	1-cyclohexenyl	80	0.15	40	0.15	(42)
Ph	2-BrC₆H₄	40	1	20	0.5	(55)
Ph	4-MeO₂CC₆H₄	40	0.75	20	1	(57)
Ph	2,4,6-Me₃C₆H₂	40	0.5	80	1	(34)
2,4,6-Me₃C₆H₂	Ph	60	0.15	20	0.15	(88)

43

C₁₉

Substrate: furan with Ph–C≡C and CH(Ph)(OH)

Nucleophile: H₂N–C₆H₄–R (2-substituted aniline), 1.5 eq

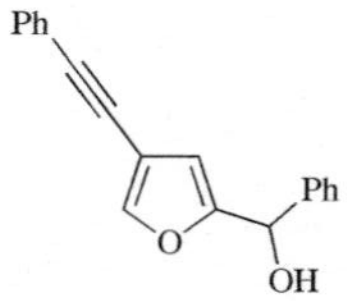

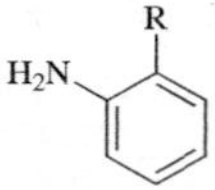

Conditions: Ca(NTf₂)₂ (5 mol %), *n*-Bu₄NPF₆ (5 mol %), Cu(OTf)₂ (10 mol %), MeNO₂, 90°

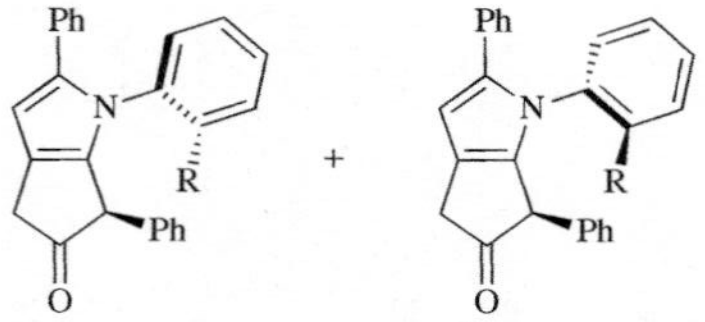

Products:

I + II

R	Time (h)	I + II	dr II
TsHN	0.5	(56)	6:1
i-Pr	1.5	(50)	1.5:1
t-Bu	1.5	(59)	1:0
2,6-(MeO)₂C₆H₃	3	(55)	1:0
2-PhC₆H₄	1	(45)	1:0:0
phenanthren-9-yl	0.5	(52)	2.5:1:0:0

62

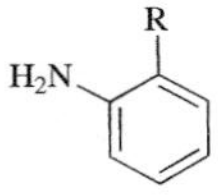

1. Ca(NTf$_2$)$_2$ (5 mol %),
   n-Bu$_4$NPF$_6$ (5 mol %),
   solvent, temp 1, time 1
2. Cu(OTf)$_2$ (10 mol %),
   solvent, temp 2, time 2

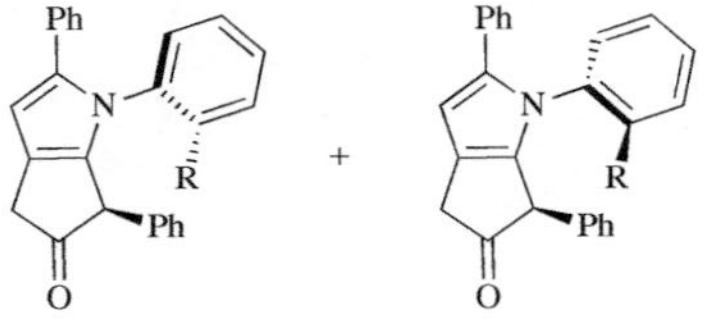

1.3 eq

I + II

62

R	Solvent	Temp 1 (°)	Time 1 (h)	Temp 2 (°)	Time 2 (h)	I + II	dr
Ph	DCE	80	0.15	40	1	(61)	1.3:1
2,6-Cl$_2$C$_6$H$_4$	DCE	80	0.75	40	0.75	(58)	2.1:1
Bz	HFIP	60	0.33	20	7	(48)	1:0
2,6-Me$_2$C$_6$H$_4$	DCE	80	0.2	80	0.2	(60)	4:1
benzofuran-2-yl	HFIP	60	0.75	20	2	(57)	1:0
1-Np	DCE	80	0.75	40	0.5	(88)	1.4:1:0:0
Ph$_2$P(O)	DCE	80	0.5	80	1.5	(81)	1.4:1:0:0
	DCE	80	1.5	80	0.16	(65)	1:0

Ca(NTf$_2$)$_2$ (5 mol %),
n-Bu$_4$NPF$_6$ (5 mol %),
Cu(OTf)$_2$ (10 mol %),
MeNO$_2$, 90°, 1.5 h

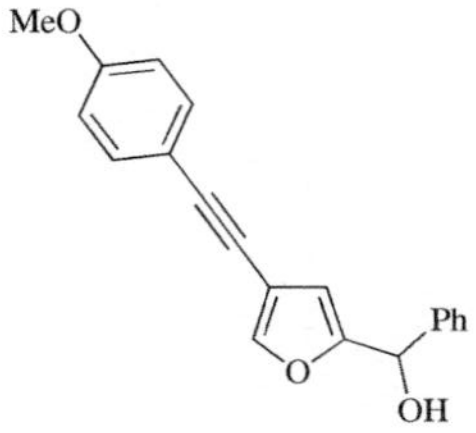

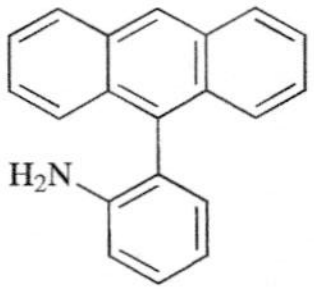

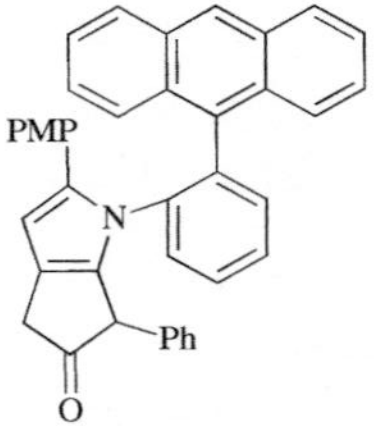

1.5 eq

(56) dr 1:0

62

TABLE 10. 4,6-DIHYDROCYCLOPENTA[b]PYRROL-5(1H)-ONES (Continued)

Substrate	Nucleophile	Conditions	Product(s) and Yield(s) (%)	Refs.

C19

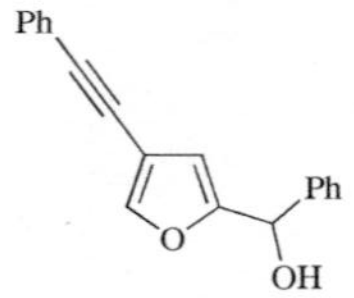

Ph—C≡C— ... furan with CH(Ph)OH substituent

NH₂R
1.3 eq

1. Ca(NTf₂)₂ (5 mol %),
n-Bu₄NPF₆ (5 mol %),
solvent, temp 1, time 1
2. Cu(OTf)₂ (10 mol %),
solvent, temp 2, time 2

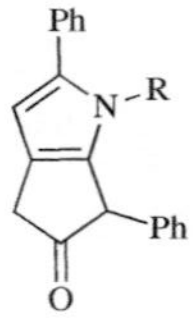

R	Solvent	Temp 1 (°)	Temp 2 (°)	Time 1 (h)	Time 2 (h)		Refs.
4-FC₆H₄	DCE	80	40	0.15	0.5	(75)	43
4-ClC₆H₄	DCE	80	40	0.5	0.15	(68)	43
3-IC₆H₄	DCE	80	40	0.3	0.15	(62)	43
4-IC₆H₄	DCE	80	40	0.5	0.15	(74)	43
4-MeOC₆H₄	HFIP	60	20	2	0.15	(47)	43
3-NCC₆H₄	DCE	80	80	0.15	0.33	(67)	43
4-MeO₂CC₆H₄	DCE	80	40	0.15	1.5	(67)	43
3-AcC₆H₄	DCE	80	40	0.3	0.5	(67)	43
2,4,6-Me₃C₆H₂	HFIP	60	20	0.75	0.15	(78)	43
(E)-4-C₆H₅N₂C₆H₄	HFIP	60	20	2	1	(73)	62

C22

H₂N— ... benzene with Ph substituent
1.3 eq

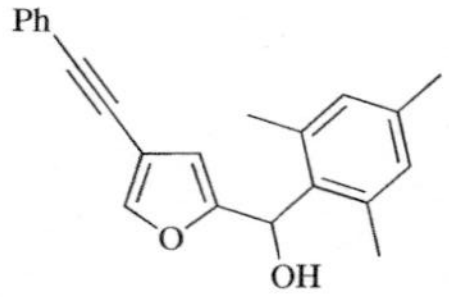

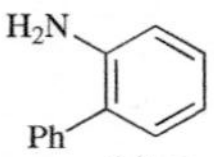

1. Ca(NTf₂)₂ (5 mol %),
n-Bu₄NPF₆ (5 mol %),
HFIP, 60°, 1 h
2. Cu(OTf)₂ (10 mol %),
HFIP, 20°, 0.5 h

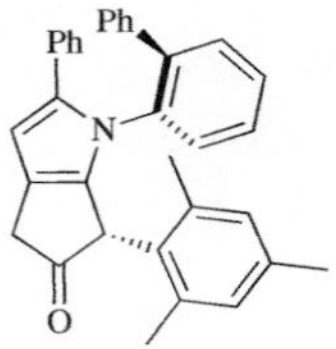

(75) dr 1:0

62

C$_{25}$

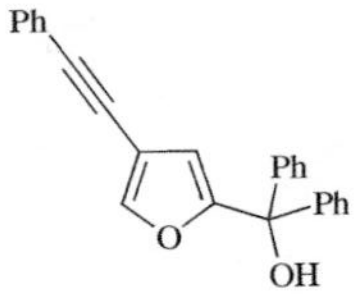

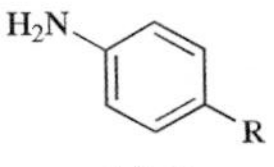

1.3 eq

1. Ca(NTf$_2$)$_2$ (5 mol %),
   n-Bu$_4$NPF$_6$ (5 mol %),
   HFIP, temp 1, time 1
2. Cu(OTf)$_2$ (10 mol %),
   HFIP, temp 2, time 2

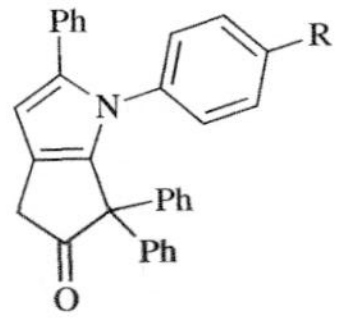

R	Temp 1 (°)	Time 1 (h)	Temp 2 (°)	Time 2 (h)		
H	60	0.3	20	0.15	(60)	43
I	20	0.67	20	0.15	(76)	43
MeO	60	0.5	20	0.15	(98)	43
AcHN	60	0.08	20	0.15	(81)	62
Me	60	0.3	20	0.15	(74)	43

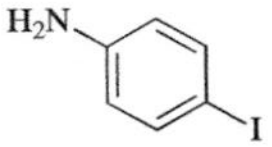

1. Ca(NTf$_2$)$_2$ (6 mol %),
   n-Bu$_4$NPF$_6$ (6 mol %),
   HFIP, 20°, 1 h
2. Cu(OTf)$_2$ (10 mol %),
   HFIP, 20°, 0.15 h
3. Cyclohex-2-en-1-ol,
   DCE, 80°, 0.15 h

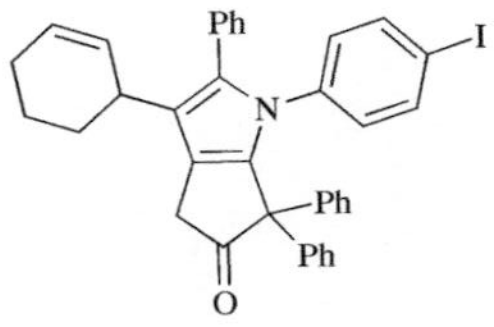

43

TABLE 10. 4,6-DIHYDROCYCLOPENTA[*b*]PYRROL-5(1*H*)-ONES (*Continued*)

Substrate	Nucleophile	Conditions	Product(s) and Yield(s) (%)		Refs.

C$_{25}$

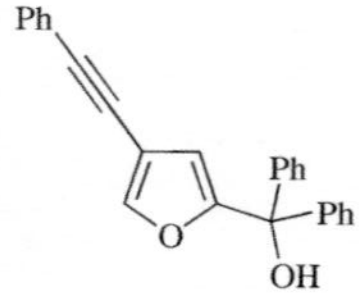

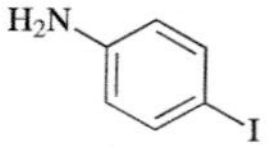

Substrate: (furan with Ph and OH substituents, Ph-alkyne)

Nucleophile: 4-iodoaniline (H_2N—C$_6$H$_4$—I)

Conditions:
1. Ca(NTf$_2$)$_2$ (6 mol %), n-Bu$_4$NPF$_6$ (6 mol %), HFIP, 20°, 1 h
2. Cu(OTf)$_2$ (10 mol %), HFIP, 20°, 0.15 h
3. $\underset{R^1}{\overset{OH}{|}}$R^2, HFIP, temp. time

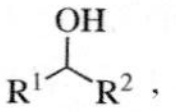

Product: (cyclopenta[b]pyrrolone with N-aryl-I, Ph, Ph, Ph, R^1, R^2, =O)

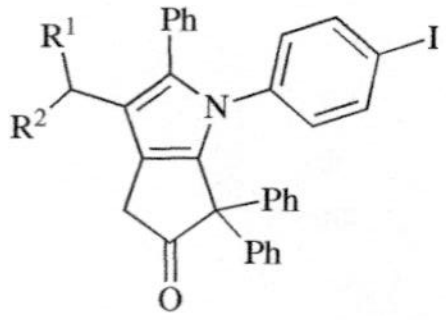

43

R^1	R^2	Temp (°)	Time (h)	
Me	prop-1-en-1-yl	40	1	(79)
Me	styryl	20	0.5	(95)
Ph	cyclopropyl	20	0.15	(65)
Ph	Ph	60	0.67	(74)
Ph	1*H*-indol-3-yl	20	0.15	(63)
Ph	6-methoxynaphthalen-1-yl	40	2	(72)
Ph	styryl	20	0.5	(75)
PMP	Me	20	3	(70)
PMP	PMP	20	0.15	(87)
PMP	phenylethynyl	20	0.15	(73)
PMP	4-methoxystyryl	20	0.5	(58)

TABLE 11. 1,2-DIHYDROCYCLOPENTA[b]PYRROL-5(4H)-ONES

Substrate	Nucleophile	Conditions	Product(s) and Yield(s) (%)	Refs.
C_{16-21}	1.3 eq	1. Ca(NTf_2)_2 (5 mol %), n-Bu_4NPF_6 (5 mol %), HFIP, 40°, time 1 2. Cu(OTf)_2 (10 mol %), HFIP, temp, time 2		62
			R Time 1 (h) Temp (°) Time 2 (h)	
			c-C_3H_5 0.75 20 1.25 (50)	
			2-thienyl 1.25 40 4 (50)	
			1-cyclohexenyl 1 20 0.15 (46)	
			Ph 1 20 16 (66)	
			4-MeOC_6H_4 2 20 16 (44)	
			4-MeO_2CC_6H_4 0.75 20 16 (66)	
			PhCH_2CH_2 1 20 2 (77)	
C_{16}	1.3 eq	Ca(NTf_2)_2 (5 mol %), n-Bu_4NPF_6 (5 mol %), Cu(OTf)_2 (10 mol %), DCE, 80°, 0.2 h	(53)	62

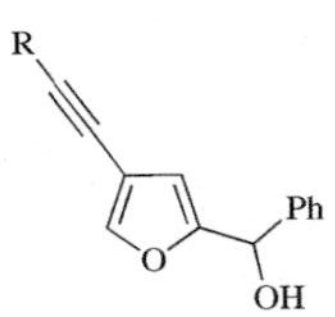

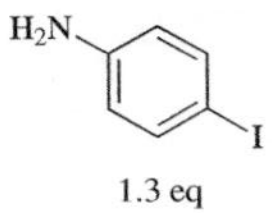

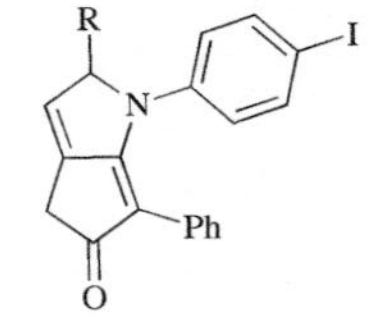

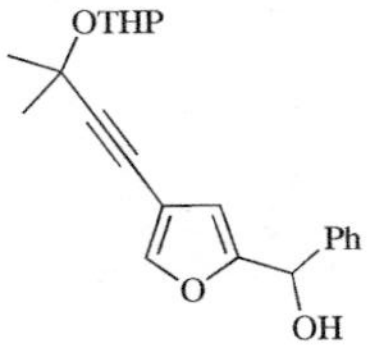

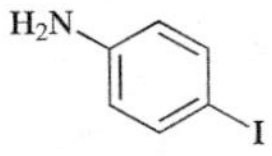

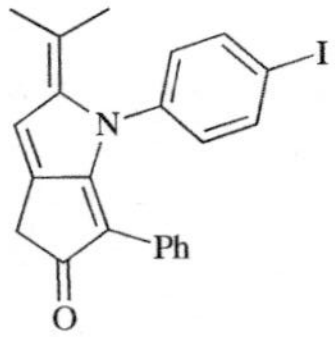

TABLE 11. 1,2-DIHYDROCYCLOPENTA[b]PYRROL-5(4H)-ONES (*Continued*)

Substrate	Nucleophile	Conditions	Product(s) and Yield(s) (%)	Refs.
C19 (Ph–C≡C–furan–CH(Ph)OH)	(2-Ar-aniline) H2N-Ar, 1.3 eq	1. Ca(NTf2)2 (5 mol %), n-Bu4NPF6 (5 mol %), HFIP, temp 1, time 1 2. Cu(OTf)2 (10 mol %), HFIP, temp 2, time 2	(pyrrolone with N-Ar, Ph, Ph)	62

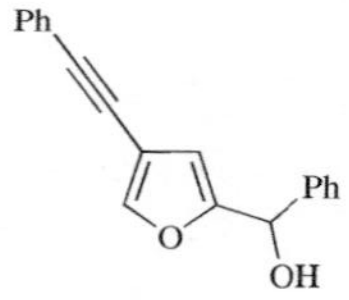

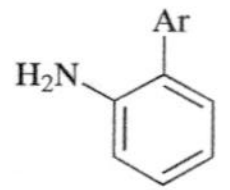

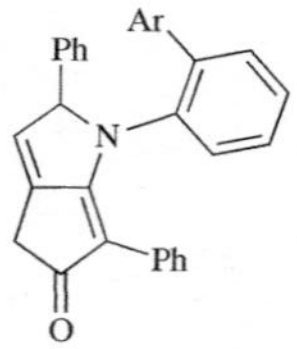

Ar	Temp 1 (°)	Time 1 (h)	Temp 2 (°)	Time 2 (h)
2-benzofuranyl	40	1.25	20	16 (69)
1-Np	60	0.3	20	20 (69)
9-anthracenyl	60	0.3	60/40	2/10 (68)
2-pyranyl	60	0.3	20	16 (73)

| | (4-OMe-aniline) H2N-C6H4-OMe, 1.3 eq | 1. Ca(NTf2)2 (5 mol %), <br> n-Bu4NPF6 (5 mol %),
 HFIP, 60°, 2 h
 2. Cu(OTf)2 (10 mol %),
 HFIP, 20°, 16 h | (pyrrolone, N-C6H4-OMe, Ph, Ph) (51) | 62 |

| | (2-P(O)Ph2-aniline) H2N-C6H4-P(O)Ph2, 1.3 eq | 1. Ca(NTf2)2 (5 mol %), <br> n-Bu4NPF6 (5 mol %),
 HFIP, 40°, 0.5 h
 2. Cu(OTf)2 (10 mol %),
 HFIP, 40°, 1 h | (pyrrolone, N-C6H4-P(O)Ph2, Ph, Ph) (70) | 62 |

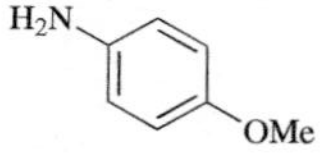

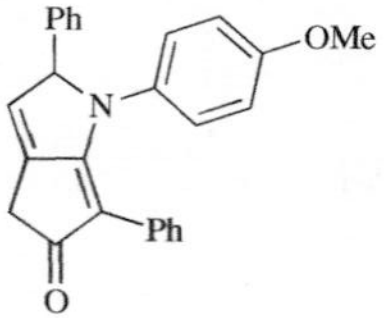

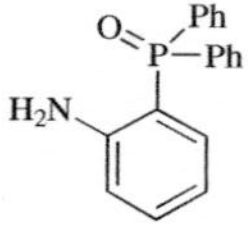

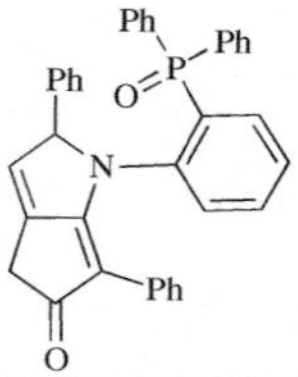

TABLE 12. 3-(1H-INDOL-3-YL)CYCLOPENTEN-2-ONES

Substrate	Nucleophile	Conditions	Product(s) and Yield(s) (%)	Refs.

C_{17}

Substrate: furan–C(Ph)(Ph)OH

Nucleophile: indole (R^1, R^2 on pyrrole ring; R^3, R^4, R^5 on benzene ring)

R^1	R^2	R^3	R^4	R^5	Time (h)	
H	H	H	H	H	2	(79)
H	H	H	H	Br	5	(37)
H	H	H	H	Cl	5	(61)
H	H	H	H	MeO	2	(65)
H	H	H	H	BnO	2	(87)

Conditions: ZnCl$_2$ (10 mol %), DCE, 80°

Product: 3-(indol-3-yl)cyclopentenone (R^1–R^5, Ph, Ph, O)

R^1	R^2	R^3	R^4	R^5	Time (h)	
H	H	H	H	O$_2$N	2	(tr)
H	H	H	Cl	H	4	(48)
H	Boc	H	H	H	2	(tr)
H	H	H	H	MeO$_2$C	2	(tr)
H	H	Me	H	H	2	(79)

R^1	R^2	R^3	R^4	R^5	Time (h)	
H	Me	H	H	H	2	(70)
Me	H	H	H	H	2	(80)
Ph	H	H	H	H	3	(80)

Refs.: 64

C_{17-18}

Substrate: furan–C(Ar^1)(Ar^2)OH

Nucleophile: indole (NH)

Conditions: ZnCl$_2$ (10 mol %), DCE, 80°

Product: 3-(indol-3-yl)cyclopentenone (Ar^1, Ar^2, O)

Ar^1	Ar^2	Time (h)		dr
Ph	4-MeC$_6$H$_4$	2	(79)	1:1
2-FC$_6$H$_4$	2-FC$_6$H$_4$	2	(89)	—
4-ClC$_6$H$_4$	4-ClC$_6$H$_4$	2	(78)	—
2-MeOC$_6$H$_4$	2-MeOC$_6$H$_4$	3	(72)	—
4-t-BuC$_6$H$_4$	4-t-BuC$_6$H$_4$	2	(54)	—

Refs.: 64

TABLE 12. 3-(1*H*-INDOL-3-YL)CYCLOPENTEN-2-ONES (*Continued*)

Substrate	Nucleophile	Conditions	Product(s) and Yield(s) (%)	Refs.
C$_{25}$		ZnCl$_2$ (10 mol %), DCE, 80°, 0.5 h	(67)	64

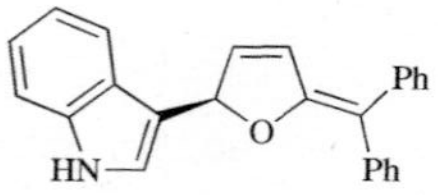

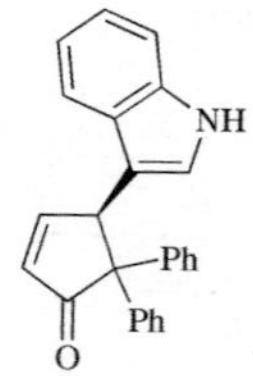

606

TABLE 13. MISCELLANEOUS PRODUCTS

Substrate	Nucleophile	Conditions	Product(s) and Yield(s) (%)	Refs.

C7

Substrate: furan–CH(OH)–CH₂CH₂–NHTs (1.3 eq)

Nucleophile: ArNH₂

Conditions:
1. Ca(NTf₂)₂ (5 mol %), n-Bu₄NPF₆ (5 mol %), toluene/HFIP (3:1), 40°, 48 h
2. Et₃N (2 eq), 20°, 36 h

Product: I (ArHN, H, Ts, R¹, O) + II (H, Ts, NHAr, R¹, O)

Ar	I	II
4-IC₆H₄	(35)	(30)
4-O₂NC₆H₅	(33)	(0)
4-MeO₂CC₆H₄	(45)	(20)

Refs: 126

C11–15

Substrate: furan–CH(OH)–Ar(NHTs) with R (positions 4, 2) (1.3 eq)

Nucleophile: ArNH₂

Conditions:
1. Ca(NTf₂)₂ (5 mol %), n-Bu₄NPF₆ (5 mol %), toluene/HFIP (3:1), temp, time
2. Et₃N (2 eq), 20°, 5 min

Product: ArHN, H, Ts bicyclic (positions 2, 4), O

Ar	R	Temp (°)	Time (h)	
Ph	H	20	9	(90)
4-FC₆H₄	H	20	9	(87)
4-ClC₆H₄	H	20	9	(87)
4-BrC₆H₄	H	20	9	(76)
4-IC₆H₄	H	20	9	(81)
4-IC₆H₄	4-Cl	20	9	(53)
4-IC₆H₄	4-MeO	20	9	(81)

Ar	R	Temp (°)	Time (h)	
4-IC₆H₄	4-Me	20	9	(87)
4-IC₆H₄	3,4-(CH=CH)₂	20	9	(88)
4-O₂NC₆H₄	H	20	17	(68)
4-MeOC₆H₄	H	40	17	(75)
2-MeC₆H₄	H	20	9	(79)
3-MeC₆H₄	H	20	9	(87)
4-MeC₆H₄	H	20	9	(81)

Ar	R	Temp (°)	Time (h)	
4-NCC₆H₄	H	20	9	(78)
4-MeO₂CC₆H₄	H	20	9	(84)
3,5-(CF₃)₂C₆H₃	H	20	9	(73)
4-AcC₆H₄	H	20	9	(77)
2-CH=CHCH₂	H	40	17	(66)
2-Ph	H	20	17	(75)
2-PhC≡C	H	20	17	(76)

Refs: 126

C11

Substrate: furan–CH(OH)–C₆H₄(ortho-NHTs) (1.3 eq)

Nucleophile: (BnO)BnNH

Conditions:
1. Ca(NTf₂)₂ (5 mol %), n-Bu₄NPF₆ (5 mol %), toluene/HFIP (3:1), 20°, 9 h
2. Et₃N (2 eq), 20°, 5 min

Product: Bn, BnO, N, Ts bridged structure, O (81)

Refs: 126

TABLE 13. MISCELLANEOUS PRODUCTS (*Continued*)

Substrate	Nucleophile	Conditions	Product(s) and Yield(s) (%)	Refs.
C$_{11-17}$ 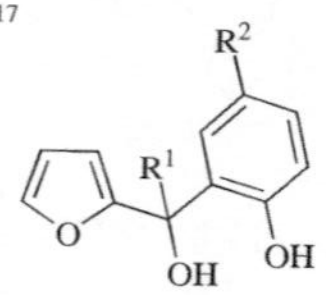 1.3 eq	ArNH$_2$ <table> R^1 R^2 Ar H H 4-IC$_6$H$_4$ (82) H H 4-MeOC$_6$H$_4$ (76) H H 4-MeC$_6$H$_4$ (75) </table>	1. Ca(NTf$_2$)$_2$ (5 mol %), n-Bu$_4$NPF$_6$ (5 mol %), toluene/HFIP (3:1), 20°, 9 h 2. Et$_3$N (2 eq), 20°, 5 min	R^1 R^2 Ar H H 4-MeO$_2$CC$_6$H$_4$ (68) H H 4-ClC$_6$H$_4$ (58) Me H 4-MeOC$_6$H$_4$ (93)	126
C$_{11}$ 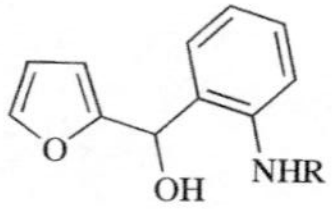 1.3 eq	ArNH$_2$	1. Ca(NTf$_2$)$_2$ (5 mol %), n-Bu$_4$NPF$_6$ (5 mol %), toluene/HFIP (3:1), temp, time 1 2. Et$_3$N (2 eq), 20°, time 2	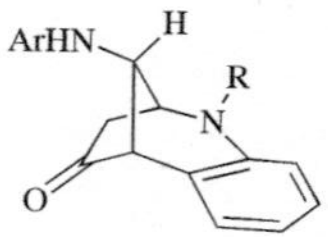 R Ar Temp(°) Time 1(h) Time 2 H 4-IC$_6$H$_4$ 20 — — (0) Ac 4-MeOC$_6$H$_4$ 60 3 17 h (0)a 2-O$_2$NC$_6$H$_4$SO$_2$ 4-BpinC$_6$H$_4$ 20 17 5 min (89) 2-O$_2$NC$_6$H$_4$SO$_2$ 4-IC$_6$H$_4$ 20 8 5 min (87)	126
C$_{12-17}$ 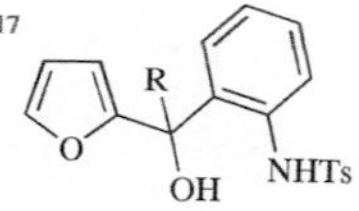 1.3 eq	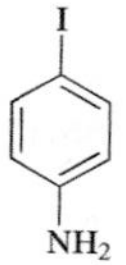	1. Ca(NTf$_2$)$_2$ (5 mol %), n-Bu$_4$NPF$_6$ (5 mol %), toluene/HFIP (3:1), 70°, 17 h 2. Et$_3$N (2 eq), 20°, 17 h	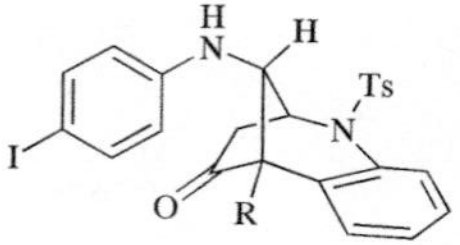 R Me (56) Ph (63)	126

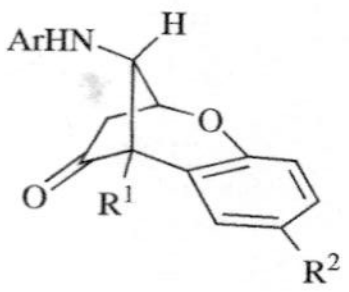

a The product was NH(4-MeOC$_6$H$_4$) formed in 81% yield.

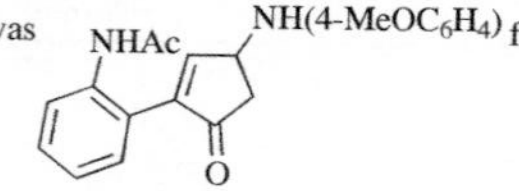

REFERENCES

1 Blanco-Urgoiti, J.; Añorbe, L.; Pérez-Serrano, L.; Domínguez, G.; Pérez-Castells, J. *Chem. Soc. Rev.* **2004**, *33*, 32.
2 Laschat, S.; Becheanu, A.; Bell, T.; Baro, A. *Synlett* **2005**, 2547.
3 Gibson, S. E.; Mainolfi, N. *Angew. Chem., Int. Ed.* **2005**, *44*, 3022.
4 Shibata, T. *Adv. Synth. Catal.* **2006**, *348*, 2328.
5 Lee, H.-W.; Kwong, F.-Y. *Eur. J. Org. Chem.* **2010**, 789.
6 Schore, N. E. *Org. React.* **1991**, *40*, 1.
7 Pellissier, H. *Tetrahedron* **2005**, *61*, 6479.
8 Frontier, A. J.; Collison, C. *Tetrahedron* **2005**, *61*, 7577.
9 Tius, M. A. *Eur. J. Org. Chem.* **2005**, *11*, 2193.
10 Nakanishi, N.; West, F. G. *Curr. Opin. Drug Discovery Dev.* **2009**, *12*, 732.
11 Shimada, N.; Stewart, C.; Tius, M. A. *Tetrahedron* **2011**, *67*, 5851.
12 Vaidya, T.; Eisenberg, R.; Frontier, A. J. *ChemCatChem* **2011**, *3*, 1531.
13 Spencer, W. T., III; Vaidya, T.; Frontier, A. J. *Eur. J. Org. Chem.* **2013**, 3621.
14 Tius, M. A. *Chem. Soc. Rev.* **2014**, *43*, 2979.
15 Wenz, D. R.; Read de Alaniz, J. *Eur. J. Org. Chem.* **2015**, 23.
16 Habermas, K. L.; Denmark, S. E.; Jones, T. K. *Org. React.* **1994**, *45*, 1.
17 Scarpi, D.; Petrović, M.; Fiser, B.; Gómez-Bengoa, E.; Occhiato, E. G. *Org. Lett.* **2016**, *18*, 3922.
18 Shi, X.; Gorin, D. J.; Toste, F. D. *J. Am. Chem. Soc.* **2005**, *127*, 5802.
19 Mézailles, N.; Ricard, L.; Gagosz, F. *Org. Lett.* **2005**, *7*, 4133.
20 Rautenstrauch, V. *J. Org. Chem.* **1984**, *49*, 950.
21 Piancatelli, G.; Scettri, A.; Barbadoro, S. *Tetrahedron Lett.* **1976**, *17*, 3555.
22 Piancatelli, G.; Scettri, A.; David, G.; D'Auria, M. *Tetrahedron* **1978**, *34*, 2775.
23 Veits, G. K.; Wenz, D. R.; Read de Alaniz, J. *Angew. Chem., Int. Ed.* **2010**, *49*, 9484.
24 Yin, B.; Haung, L.; Wang, X.; Liu, J.; Jiang, H. *Adv. Synth. Catal.* **2013**, *355*, 370.
25 Nunes, J. P. M.; Afonso, C. A. M.; Caddick, S. *RSC Adv.* **2013**, *3*, 14975.
26 Piancatelli, G.; D'Auria, M.; D'Onofrio, F. *Synthesis* **1994**, 867.
27 Roche, S. P.; Aitken, D. J. *Eur. J. Org. Chem.* **2010**, 5339.
28 Piutti, C.; Quartieri, F. *Molecules* **2013**, *18*, 12290.
29 Palmer, L. I.; Read de Alaniz J., *Synlett* **2014**, 8.
30 Palframan, M. J.; Pattenden, G. *Chem. Commun.* **2014**, *50*, 7223.
31 Cabrele, C.; Reiser, O. *J. Org. Chem.* **2016**, *81*, 10109.
32 Lebœuf, D.; Gandon, V. *Synthesis* **2017**, *49*, 1500.
33 Verrier, C.; Moebs-Sanchez, S.; Queneau, Y.; Popowycz, F. *Org. Biomol. Chem.* **2018**, *16*, 676.
34 Gomes, R. F. A.; Coelho, J. A. S.; Afonso, C. A. M. *Chem.—Eur. J.* **2018**, *24*, 9170.
35 Faza, O. N.; López, C. S.; Álvarez, R.; de Lera, Á. R. *Chem.—Eur. J.* **2004**, *10*, 4324.
36 Yu, D.; Thai, V. T.; Palmer, L. I.; Veits, G. K.; Cook, J. E.; Read de Alaniz, J.; Hein, J. E. *J. Org. Chem.* **2013**, *78*, 12784.
37 Chung, R.; Yu, D.; Thai, V. T.; Jones, A. F.; Veits, G. K.; Read de Alaniz, J.; Hein, J. E. *ACS Catal.* **2015**, *5*, 4579.
38 D'Auria, M. *Heterocycles* **2000**, *52*, 185.
39 Gao, Y.; Wu, W.-L.; Ye, B.; Zhou, R.; Wu, Y.-L. *Tetrahedron Lett.* **1996**, *37*, 893.
40 Yin, B.-L.; Wu, Y.; Wu, Y.-L. *J. Chem. Soc., Perkin Trans. 1* **2002**, 1746.
41 Yin, B.-L.; Wu, Y.-L.; Lai, J.-Q. *Eur. J. Org. Chem.* **2009**, 2695.
42 Li, H.; Rongbiao, T.; Sun, J. *Angew. Chem., Int. Ed.* **2016**, *55*, 15125.
43 Marin, L.; Gandon, V.; Schulz, E.; Lebœuf, D. *Adv. Synth. Catal.* **2017**, *359*, 1157.
44 Xu, Z.-L.; Xing, P.; Jiang, B. *Org. Lett.* **2017**, *19*, 1028.
45 Piancatelli, G.; Scettri, A. *Tetrahedron Lett.* **1977**, *18*, 1131.
46 D'Auria, M.; Piancatelli, G.; Scettri, A. *Tetrahedron* **1980**, *36*, 1877.
47 D'Auria, M.; Piancatelli, G.; Scettri, A. *Tetrahedron* **1980**, *36*, 3071.
48 Ulbrich, K.; Kreitmeier, P.; Reiser, O. *Synlett* **2010**, 2037.
49 Palmer, L. I.; Read de Alaniz, J. *Org. Lett.* **2013**, *15*, 476.
50 Nichol, M. F.; Limon, L.; de Alaniz, J. R. *Org. Synth.* **2018**, *95*, 46.
51 Berecibar, A.; Grandjean, C.; Siriwardena, A. *Chem. Rev.* **1999**, *99*, 779.
52 Palmer, L. I.; Read de Alaniz, J. *Angew. Chem., Int. Ed.* **2011**, *50*, 7167.
53 Wenz, D. R.; Read de Alaniz, J. *Org. Lett.* **2013**, *13*, 3250.
54 Lebœuf, D.; Schulz, E.; Gandon, V. *Org. Lett.* **2014**, *16*, 6464.

[55] Lebœuf, D.; Marin, L.; Michelet, B.; Perez-Luna, A.; Guillot, R.; Schulz, E.; Gandon, V. *Chem.—Eur. J.* **2016**, *22*, 16165.

[56] Veits, G. K.; Wenz, D. R.; Palmer, L. I.; St. Amant, A. H.; Hein, J. E.; Read de Alaniz, J. *Org. Biomol. Chem.* **2015**, *13*, 8465.

[57] Cai, Y.; Tang, Y.; Atodiresei, I.; Rueping, M. *Angew. Chem., Int. Ed.* **2016**, *55*, 14126.

[58] Gade, A. B.; Patil, N. T. *Synlett* **2017**, *28*, 1096.

[59] Liu, J.; Shen, Q.; Yu, J.; Zhu, M.; Han, J.; Wang, L. *Eur. J. Org. Chem.* **2012**, 6933.

[60] Reddy, B. V. S.; Reddy, Y. V.; Singarapu, K. *Org. Biomol. Chem.* **2016**, *14*, 1111.

[61] Nayani, K.; Cinsani, R.; Hussaini SD, A.; Mainkar, P. S.; Chandrasekhar, S. *Eur. J. Org. Chem.* **2017**, 5671.

[62] Marin, L.; Guillot, R.; Gandon, V.; Schulz, E.; Lebœuf, D. *Org. Chem. Front.* **2018**, *5*, 640.

[63] Wang, C.; Dong, C.; Kong, L.; Li, Y.; Li, Y. *Chem. Commun.* **2014**, *50*, 2164.

[64] Xu, J.; Luo, Y.; Xu, H.; Chen, Z.; Miao, M.; Ren, H. *J. Org. Chem.* **2017**, *82*, 3561.

[65] Yin, B.-L.; Lai, J.-Q.; Zhang, Z.-R.; Jiang, H.-F. *Adv. Synth. Catal.* **2011**, *353*, 1961.

[66] Huang, L.; Zhang, X.-H.; Li, J.; Ding, K.; Li, X.; Zheng, W.; Yin, B. *Eur. J. Org. Chem.* **2014**, 338.

[67] Collins, P. W.; Djuric, S. W. *Chem. Rev.* **1993**, *93*, 1533.

[68] Rodríguez, A.; Nomen, M.; Spur, B. W.; Godfroid, J.-J. *Eur. J. Org. Chem.* **1999**, 2655.

[69] Corey, E. J.; Weinshenker, N. M.; Schaaf, T. K.; Huber, W. *J. Am. Chem. Soc.* **1969**, *91*, 5675.

[70] Collins, P. W.; Kramer, S. W.; Gasiecki, A. F.; Weier, R. M.; Jones, P. H.; Gullikson, G. W.; Bianchi, R. G.; Bauer, R. F. *J. Med. Chem.* **1987**, *30*, 193.

[71] Collins, P. W.; Kramer, S. W.; Gullikson, G. W. *J. Med. Chem.* **1987**, *30*, 1952.

[72] Dygos, J. H.; Adamek, J. P.; Babiak, K. A.; Behling, J. R.; Medich, J. R.; Ng, J. S.; Wieczorek, J. J. *J. Org. Chem.* **1991**, *56*, 2549.

[73] Yoshida, Y.; Sato, Y.; Okamoto, S.; Sato, F. *J. Chem. Soc., Chem. Commun.* **1995**, 811.

[74] Rodríguez, A.; Spur, B. W. *Tetrahedron Lett.* **2003**, *44*, 7411.

[75] Harikrishna, M.; Mohan, H. R.; Dubey, P. K.; Subbaraju, G. V. *Synth. Commun.* **2009**, *39*, 2763.

[76] Henschke, J. P.; Liu, Y.; Chen, Y.-F.; Meng, D.; Sun, T. (ScinoPharm Taiwan Ltd.) U.S. Patent 2009/0259058, October 15, 2009.

[77] Henschke, J. P.; Liu, Y.; Chen, Y.-F.; Meng, D.; Sun, T. (Scinopharm Taiwan Ltd.) U.S. Patent 7897795 B2, March 1, 2011.

[78] Henschke, J. P.; Liu, Y.; Huang, X.; Chen, Y.; Meng, D.; Xia, L.; Wei, X.; Xie, A.; Li, D.; Huang, Q.; Sun, T.; Wang, J.; Gu, X.; Haung, X.; Wang, L.; Xiao, J.; Qiu, S. *Org. Process Res. Dev.* **2012**, *16*, 1905.

[79] Suzuki, M.; Yanagisawa, A.; Noyori, R. *J. Am. Chem. Soc.* **1988**, *110*, 4718.

[80] Scettri, A.; Piancatelli, G.; D'Auria, M.; David, G. *Tetrahedron* **1979**, *35*, 135.

[81] Csákÿ, A. G.; Mba, M.; Plumet, J. *J. Org. Chem.* **2001**, *66*, 9026.

[82] Piancatelli, G.; Scettri, A. *Tetrahedron* **1977**, *33*, 69.

[83] Gao, Y.; Wu, W.-L.; Wu, Y.-L.; Ye, B.; Zhou, R. *Tetrahedron* **1998**, *54*, 12523.

[84] Yin, B.-L.; Yang, Z.-M.; Hu, T.-S.; Wu, Y.-L. *Synthesis* **2003**, 1995.

[85] Csákÿ, A. G.; Mba, M.; Plumet, J. *Synlett* **2003**, 2092.

[86] Li, C.-C.; Wang, C.-H.; Liang, B.; Zhang, X.-H.; Deng, L.-J.; Liang, S.; Chen, J.-H.; Wu, Y.-D.; Yang, Z. *J. Org. Chem.* **2006**, *71*, 6892.

[87] Otero, M. P.; Santín, E. P.; Rodríguez-Barrios, F.; Vaz, B.; de Lera, Á. R. *Bioorg. Med. Chem. Lett.* **2009**, *19*, 1883.

[88] Beingessner, R. L.; Farand, J. A.; Barriault, L. *J. Org. Chem.* **2010**, *75*, 6337.

[89] Michalak, K.; Wicha, J. *Synlett* **2013**, 1387.

[90] Katsuta, R.; Aoki, K.; Yajima, A.; Nakada, T. *Tetrahedron Lett.* **2013**, *54*, 347.

[91] Michalak, K.; Wicha, J. *Tetrahedron* **2014**, *70*, 5073.

[92] Oh, M. H.; Lee, J. W.; Lee, D. H. *Bull. Korean Chem. Soc.* **2016**, *37*, 1169.

[93] Crandall, J. K.; Banks, D. B.; Coyler, R. A.; Watkins, R. J.; Watkins, J. P. *J. Org. Chem.* **1968**, *33*, 423.

[94] Le Liepvre, M.; Ollivier, J.; Aitken, D. J. *Eur. J. Org. Chem.* **2009**, 5953.

[95] Holec, C.; Sandkuhl, D.; Rother, D.; Kroutil, W.; Pietruszka, J. *ChemCatChem* **2015**, *7*, 3125.

[96] Trost, B. M.; Richardson, J.; Yong, K. *J. Am. Chem. Soc.* **2006**, *128*, 2540.

[97] Trost, B. M.; Masters, J. T.; Lumb, J.-P.; Fateen, D. *Chem. Sci.* **2014**, *5*, 1354.

[98] Dickmeiss, G.; De Sio, V.; Udmark, J.; Poulsen, T. B.; Marcos, V.; Jørgensen, K. A. *Angew. Chem., Int. Ed.* **2009**, *48*, 6650.

[99] Chang, C.-T.; Jacobo, S. H.; Powell, W. S.; Lawson, J. A.; FitzGerald, G. A.; Praticoc, D.; Rokacha, J. *Tetrahedron Lett.* **2005**, *46*, 6325.

[100] Lewis, K. G.; Mulquiney, C. E. *Tetrahedron* **1977**, *33*, 463.

101 Li, S.-W.; Batey, R. A. *Chem. Commun.* **2007**, 3759.
102 Hiscox, A.; Ribeiro, K.; Batey, R. A. *Org. Lett.* **2018**, *20*, 6668.
103 Estevão, M. S.; Afonso, C. A. M. *Tetrahedron Lett.* **2017**, *58*, 302.
104 Gomes, R. F. A.; Esteves, N. R.; Coelho, J. A. S.; Afonso, C. A. M. *J. Org. Chem.* **2018**, *83*, 7509.
105 Procopio, A.; Costanzo, P.; Curini, M.; Nardi, M.; Oliveiro, M.; Sindona, G. *ACS Sustainable Chem. Eng.* **2013**, *1*, 541.
106 Griffiths, K.; Gallop, C. W. D.; Abdul-Sada, A.; Vargas, A.; Navarro, O.; Kostakis, G. E. *Chem.—Eur. J.* **2015**, *21*, 6358.
107 Griffiths, K.; Kumar, P.; Mattock, J. D.; Abdul-Sada, A.; Pitak, M. B.; Coles, S. J.; Navarro, O.; Vargas, A.; Kostakis, G. E. *Inorg. Chem.* **2016**, *55*, 6988.
108 Nardi, M.; Costanzo, P.; De Nino, A.; Di Gioia, M. L.; Olivito, F.; Sindona, G.; Procopio, A. *Green Chem.* **2017**, *19*, 5403.
109 Ramesh, D.; Srikanth, R. T.; Narasimhulu, M.; Rajaram, S.; Suryakiran, N.; Chinni, M. K.; Venkateswarlu, Y. *Chem. Lett.* **2009**, *38*, 586.
110 Liu, J.; Zhu, M.; Li, J.; Zheng, X.; Wang, L. *Synlett* **2013**, 2165.
111 Duspara, P. A.; Batey, R. A. *Angew. Chem., Int. Ed.* **2013**, *52*, 10862.
112 Li, F.; Ding, C.; Wang, M.; Yao, Q.; Zhang, A. *J. Org. Chem.* **2011**, *76*, 2820.
113 Wang, S.; William, R.; Seah, K. K. G. E.; Liu, X.-W. *Green Chem.* **2013**, *15*, 3180.
114 William, R.; Wang, S.; Ding, F.; Arviana, E. N.; Liu, X.-W. *Angew. Chem., Int. Ed.* **2014**, *53*, 10742.
115 William, R.; Leng, W. L.; Wang, S.; Liu, X.-W. *Chem. Sci.* **2016**, *7*, 1100.
116 D'Auria, M.; D'Onofrio, F.; Piancatelli, G.; Scettri, A. *Gazz. Chim. Ital.* **1986**, *116*, 173.
117 Castagnino, E.; D'Auria, M.; De Mico, A.; D'Onofrio, F.; Piancatelli, G. *J. Chem. Soc., Chem. Commun.* **1987**, 907.
118 West, F. G.; Gunawardena, G. U. *J. Org. Chem.* **1993**, *58*, 2402.
119 Fisher, D.; Palmer, L. I.; Cook, J. E.; Davis, J. E.; Read de Alaniz, J. *Tetrahedron* **2014**, *70*, 4105.
120 Bredihhin, A.; Luiga, S.; Vares, L. *Synthesis* **2016**, *48*, 4181.
121 Antonioletti, R.; De Mico, A.; Piancatelli, G.; Scettri, A. *Gazz. Chim. Ital.* **1986**, *116*, 745.
122 Palmer, L. I.; Veits, G. K.; Read de Alaniz J., *Eur. J. Org. Chem.* **2013**, 6237.
123 Tang, W.-B.; Cao, K.-S.; Meng, S. S.; Zheng, W.-H. *Synthesis* **2017**, *49*, 3670.
124 Reddy, B. V. S.; Narasimhulu, G.; Lakshumma, P. S.; Reddy, Y. V.; Yadav, J. S. *Tetrahedron Lett.* **2012**, *53*, 1776.
125 Hammersley, G. R.; Nichol, M. F.; Steffens, H. C.; Delgado, J. M.; Veits, G. K.; Read de Alaniz, J. *Beilstein J. Org. Chem.* **2019**, *15*, 1569.
126 Wang, S.; Guillot, R.; Carpentier, J. F.; Sarazin, Y.; Bour, C.; Gandon, V.; Lebœuf, D. *Angew. Chem., Int. Ed.* **2020**, *59*, 1134.
127 Shen, B.; He, Q.; Dong, S.; Liu, X.; Feng, X. *Chem. Sci.* **2020**, *11*, 3862.
128 Baldé, B.; Force, G.; Marin, L.; Guillot, R.; Schulz, E.; Gandon, V.; Lebœuf, D. *Org. Lett.* **2018**, *20*, 7405.
129 Reddy, B. V. S.; Reddy, Y. V.; Lakshumma, P. S.; Narasimhulu, G.; Yadav, J. S.; Sridhar, B.; Reddy, P. P.; Kunwar, A. C. *RSC Adv.* **2012**, *2*, 10661.
130 Gouse, S.; Reddy, N. R.; Baskaran, S. *Org. Lett.* **2019**, *21*, 3822.
131 Wei, Z.; Zhang, J.; Yang, H.; Jiang, G. *Org. Lett.* **2019**, *21*, 2790.
132 Sarnikar, Y. P.; Biradar, D. O.; Mane, Y. D.; Khaded, B. C. *J. Heterocyclic Chem.* **2019**, *56*, 1111.

ORGANIC REACTIONS

CHAPTER 3

TRANSITION-METAL-MEDIATED AND TRANSITION-METAL-CATALYZED CARBON–FLUORINE BOND FORMATION

CONSTANZE N. NEUMANN AND TOBIAS RITTER

Max Planck Institute für Kohlenforschung, Kaiser-Wilhelm-Platz 1, D-45470 Mülheim an der Ruhr, Germany

Edited by P. ANDREW EVANS

CONTENTS

ritter@kofo.mpg.de
How to cite: Neumann C.N.; Ritter T. Transition-Metal-Mediated and Transition-Metal-Catalyzed Carbon–Fluorine Bond Formation. *Org. React.* **2020**, *104*, 613–870.
© 2020 Organic Reactions, Inc. Published in 2020 by John Wiley & Sons, Inc.
Doi:10.1002/0471264180.or104.03

INTRODUCTION

The importance of fluorinated molecules in the pharmaceutical and agrochemical industries, as well as in new materials, has led to an increase in research efforts on fluorination chemistry.[1−3] The advancement of positron emission tomography (PET) as a viable clinical diagnostic method provides an additional impetus for the development of fluorination reactions and thus sets a higher standard for operational convenience and late-stage modification of complex molecules.[4−7] Progress has been made in the development of inexpensive and efficient fluorination reagents with limited environmental impact for use in larger-scale applications, as well as in the targeted late-stage modification of complex molecules, where functional-group tolerance, mild reaction conditions, and the scope outweigh the price of reagents and amount of waste generated as primary considerations behind reaction development. Fluorination reactions are an area of chemistry in which comparatively little inspiration can be drawn from nature. For instance, while haloperoxidase enzymes are known to introduce other halides, none have been identified for the introduction of the fluoride anion, which is highly electronegative and difficult to oxidize. Hence, only a handful of fluorinated molecules that are made through biosynthesis are known.[8,9]

Milder fluorination reagents and the use of catalysts that lower the barrier for carbon–fluorine bond formation have decreased the need for high-energy fluorinating reagents. Carbon–fluorine bond formation is a very productive and active field for catalysis, because the formation of a carbon–fluorine bond is often favorable but restricted by high kinetic barriers.[10,11] This chapter highlights recently developed transition-metal-catalyzed and transition-metal-mediated reactions but does not aspire to be a comprehensive review of the field. Discoveries in the area of trifluoromethylation, perfluorination, and reactions in which a carbon–carbon or carbon–heteroatom bond is formed alongside carbon–fluorine bonds are not discussed herein, nor is fluorination that coincides with a major rearrangement of the carbon skeleton. Several recent reviews have addressed these subjects.[7,12–16]

MECHANISM AND STEROCHEMISTRY

Fluorination reactions can proceed via a number of different reaction mechanisms. A common mechanistic differentiation is made between fluorination reactions that employ a nucleophilic source of fluoride, such as an alkali metal fluoride (nucleophilic fluorination reactions), and reactions in which the fluorinating reagent concomitantly serves as an oxidant, such as Selectfluor or N-fluoro-benzenesulfonimide (NFSI) (electrophilic fluorination reactions). In this chapter, transition-metal-catalyzed and transition-metal-mediated reactions are further classified by the role the transition metal plays in the transformation, such as withdrawing electron density (Lewis acid catalysis) or mediating redox reactions (redox catalysis). Within these mechanistic scenarios, we highlight whether the transition metal is directly bonded to fluoride, to the substrate to be fluorinated, or to both in the carbon–fluorine bond-forming step (Figure 1). Transition metals mediate and catalyze carbon–fluorine bond formation reactions either by redox catalysis, where the oxidation state of the transition metal changes during each turnover, or via Lewis acid/base catalysis, where the transition metal lowers the activation barrier for the fluorination by complexation to either the substrate or the fluorination reagent. Distinctions are drawn between mechanisms in which either the substrate, fluoride, or both reaction components are bound to the transition metal during the carbon–fluorine bond forming step by adding either 'sub' (to indicate a metal–substrate bond) or 'F' (to indicate a metal–fluoride bond) to the classifier for the fluorination mechanism. The product stereochemical outcome that results from a specific mechanism is discussed in general terms here and in "Scope and Limitations" for specific transformations.

Nucleophilic Fluoride Sources

Many nucleophilic fluoride sources are commercially available, including simple alkali fluorides as well as organic-solvent-soluble sources of fluoride, such as tetrabutylammonium fluoride (TBAF) and tris(dimethylamino)sulfonium difluoro-trimethylsilicate (TASF). Nucleophilic sources of fluoride often suffer from low reactivity in the presence of water because of the high solvation energy of the fluoride ion. As a consequence, anhydrous reaction conditions must be employed for many nucleophilic fluorination reactions.[17–19] The presence of nucleophilic functional groups (e.g., hydroxyl groups, amines, or carboxylic acids) can interfere with nucleophilic fluorination reactions, whereas easily oxidized functional groups, such as thiols, are more readily tolerated.[20–23]

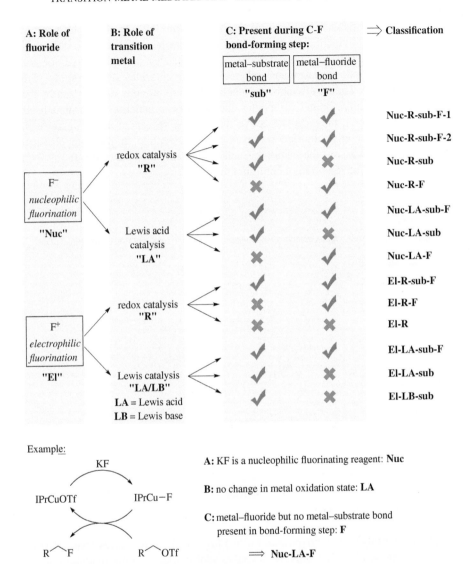

Figure 1. The classification system used in this chapter arranges the mechanisms by (A) the type of fluorination reagent, (B) the type of transition-metal catalysis, and (C) the presence of a bond between the metal and substrate, and between metal and fluoride in the C-F bond-forming reaction step.

Redox Catalysis via Organometallic and Metal Fluoride Intermediates (Nuc-R-sub-F-2). For redox transformations in which the substrate and fluoride are metal-bound, two scenarios are possible: the fluoride ion and the substrate can be bound to the same metal center or to different ones. The palladium-catalyzed fluorination of allylic chlorides proceeds by a homobimetallic mechanism, which involves the intermediacy of a neutral π-allylpalladium fluoride as the nucleophile and a cationic π-allylpalladium complex as the electrophile (Scheme 1).[24] Typical allylic leaving groups such as carbonates and acetates are ineffective for the palladium-catalyzed displacement, whereas allylic chlorides and bromides are readily displaced. The requirement for a halide leaving group is rationalized by the inability of palladium-bound carbonates and acetates to undergo anion metathesis with fluoride, which prevents the active fluorinating reagent from being formed. Hence, the metal allyl fluoride serves as the nucleophile to promote classical outer-sphere attack on the cationic metal allyl intermediate (Nuc-R-sub-F-2) (Scheme 1) with the retention of configuration.

Nuc-R-sub-F-2

nucleophilic fluoride - redox catalysis - metal–substrate bond - metal–fluoride bond - bimetallic

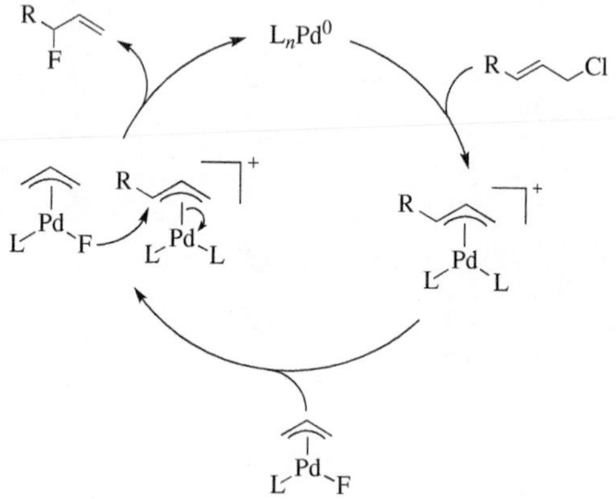

Scheme 1

Redox Catalysis via Organometallic Fluoride Intermediates (Nuc-R-sub-F-1). Activation of C–X bonds by oxidative addition leads to the oxidation of the transition-metal center and generation of a reactive organometallic intermediate. Anion metathesis with a nucleophilic fluoride source furnishes an intermediate in which fluoride is bound to the transition-metal center as an X-type ligand, and reductive elimination leads to the formation of a carbon–fluorine bond. A classic example of the Nuc-R-sub-F-1 mechanistic pathway is the Pd(0)/Pd(II)-catalyzed fluorination of aryl (pseudo)halides (Scheme 2).[25]

Much of the seminal work in aromatic fluorination chemistry centers on the Nuc-R-sub-F-1 mechanism, in which carbon–fluorine bond formation occurs via a mechanism analogous to the well-developed carbon–carbon bond-forming cross-coupling reaction manifold. Early efforts focused on successfully forming the transition-metal–fluoride bond while avoiding the creation of fluoride-bridged dimer

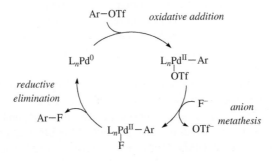

Nuc-R-sub-F-1

nucleophilic fluoride - redox catalysis - metal–substrate bond - metal–fluoride bond - monometallic

Scheme 2

structures, which tend to be resistant to reductive elimination (Figure 2).[26] When palladium(II) complexes bearing both an aryl and a fluoride ligand are subjected to high temperature, phosphorus–fluorine bond formation and other side reactions commonly occur instead of carbon–fluorine bond formation. DFT studies indicate that carbon–fluorine reductive elimination through a three-coordinate palladium(II) environment is thermodynamically feasible and kinetically competitive (Figure 2).[26]

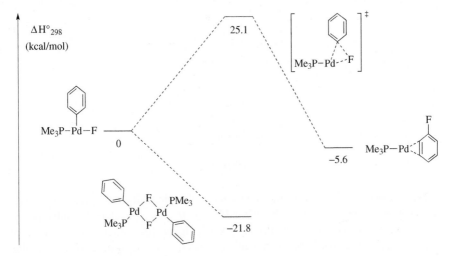

Figure 2. Activation energies for C–F reductive elimination from 3-coordinate Pd(Ar)(F) complexes calculated using DFT.

4-Fluoronitrobenzene is obtained in 10% yield in the palladium-mediated flu-orination using 2-di-*tert*-butylphosphino-2′,4′,6′-triisopropylbiphenyl (*t*-BuXPhos) as the ligand.[26] Although both the formation of the fluoride-bridged aryl palladium dimer and C–H activation of the ligand are disfavored processes, this transformation could not be extended to other substrates.[26-28] Moreover, facile carbon–fluorine bond formation in the absence of palladium by an S_NAr mechanism on arenes that have strongly electron-withdrawing substituents casts doubt on whether the carbon–fluorine bond in 4-nitrofluorobenzene is formed by reductive elimination.[29,30] Palladium fluoride complexes can liberate highly reactive desolvated fluoride ions, and a competitive carbon–phosphine reductive elimination/S_NAr mechanism is postulated to account for the 'net Ar–F reductive elimination' from the ligand.[10,31]

Bulky, monodentate-phosphine-ligated palladium complexes have been formed from aryl halides, and the resulting complexes have been studied in the course of work on C–N cross-coupling. These complexes are monomeric, and the analogous monomeric aryl fluoride complexes can also be obtained.[25] Efficient carbon–fluorine reductive elimination from (**L1**)PdArF takes place in the presence of Ar–X substrates that are capable of trapping the palladium(0) species (Scheme 3). Generally, less than two percent of ArH is formed alongside the aryl fluoride product; however, a number of substrates provide the desired fluorinated products as mixtures of constitutional isomers.

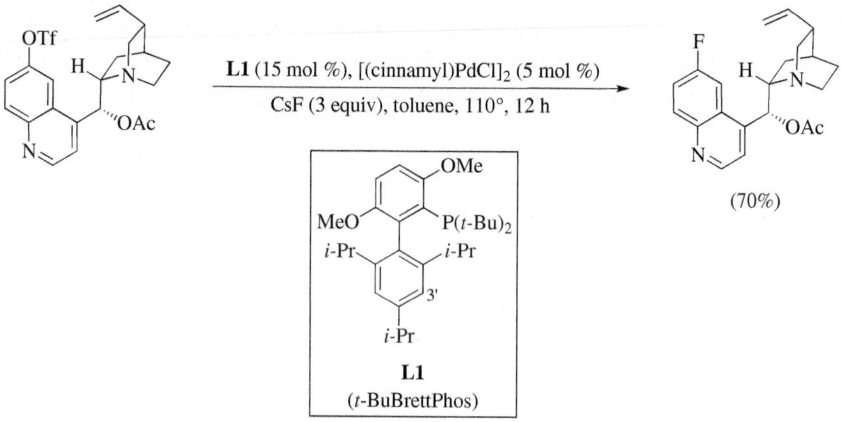

Scheme 3

Mechanistic work reveals that a modified phosphine ligand, generated in situ under the reaction conditions by the addition of one equivalent of substrate to the phosphine ligand, serves as the supporting ligand for reductive elimination of electron-rich substrates.[32] Since not all the substrates can form a ligand suitable for efficient carbon–fluorine bond formation, the substrate scope is limited by the necessity to permit in situ ligand derivatization. Consequently, a new generation of 3'-arylated ligands was developed that extends the scope of palladium-catalyzed aromatic fluorination to electron-rich arenes and heteroarenes.[33] Constitutional isomer formation appears to be independent of the 3'-arylation of the phosphine ligand (to form the modified ligand) and is particularly problematic for substrates with electron-donating groups in the 4-position and for substrates that contain only a

3-substituent.[34] Mechanistic work using the palladium complex with the 3′-arylated ligand indicates that oxidative insertion of the substrate and subsequent deprotonation with basic fluoride (present in excess under catalytic conditions) affords the palladium–aryne complex **1** (Scheme 4).[30,35] From this intermediate, competing pathways provide the two constitutional isomer products. The reaction temperature can be lowered by employing the phosphine ligand di-1-adamantyl(4″-butyl-2″,3″,5″,6″-tetrafluoro-2′,4′,6′-triisopropyl-2-methoxy-*meta*-terphenyl)phosphine (AlPhos), which renders reductive elimination more facile compared to previous ligand generations and also decreases the yield of the isomeric fluorinated products.[30,36] Additional modifications of the reaction conditions have made aryl halides and heteroaryl halides suitable substrates for the palladium-catalyzed fluorination reaction.[33,36,37]

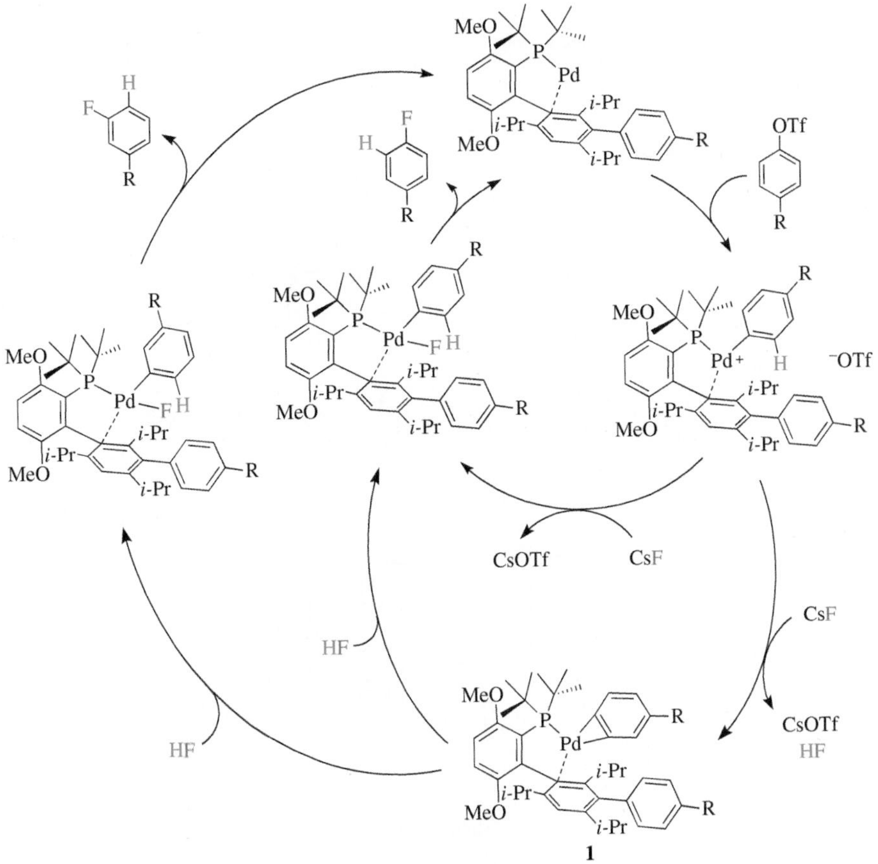

Scheme 4

Redox Catalysis via Organometallic Intermediates and Exogenous Fluoride (Nuc-R-sub). Stereochemical considerations are particularly relevant for allylic fluorination reactions, which can generate two stereoisomers of both the linear and branched fluorinated products. Catalysts that selectively deliver the linear[38,39] and branched[39] products by allylic substitution have been reported, and the enantioselective construction of branched products is also known.[40,41] In transformations occurring via the intermediacy of metal η^3-allyl complexes, fluoride substitution can occur with inversion or retention of configuration at the displacement site, depending on whether the fluoride nucleophile is bound to the transition metal center (inner-sphere, Nuc-R-sub-F) or free in solution (outer-sphere, Nuc-R-sub). In the case of the copper-catalyzed allylic fluorination reaction of allylic halides, both mechanisms are possible (Scheme 5), and neither pathway can be ruled out based on the available data.[42]

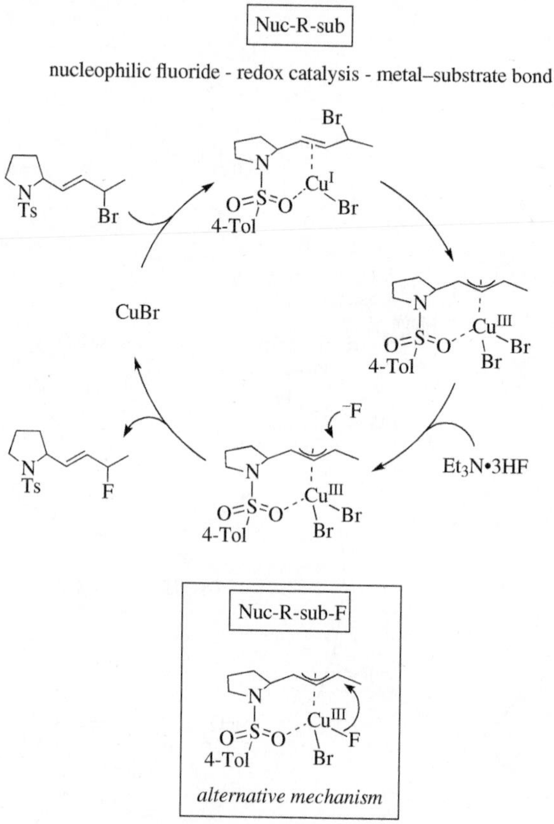

Scheme 5

α-Fluorination of benzylic aldehydes with a nucleophilic fluoride source can be accomplished in the presence of a ruthenium complex, even though enolic aldehydes are nucleophilic at the α-position (Scheme 6).[43] The substrate umpolung likely results from the oxidation of the ruthenium enolate by silver(I), which are the only fluoride salts effective for this transformation. The proposed mechanism involves deprotonation of the aldehyde by fluoride to form a ruthenium oxyallyl complex. Two-electron oxidation of the ruthenium oxyallyl complex is inferred from the necessity to employ two equivalents of silver(I) fluoride and the appearance of a grey precipitate of silver. Outer-sphere attack of the fluoride affords the α-fluorination product (Nuc-R-sub); however, the available data does not preclude the coordination of fluoride to the ruthenium center before carbon–fluorine bond formation (Nuc-R-sub-F). Overall, the occurrence of a Nuc-R-sub mechanism in transition-metal-catalyzed fluorination reactions is much more speculative than the more common Nuc-R-sub-F process, most likely resulting from the affinity of fluoride for high-valent transition metals.

Nuc-R-sub

nucleophilic fluoride - redox catalysis - metal–substrate bond

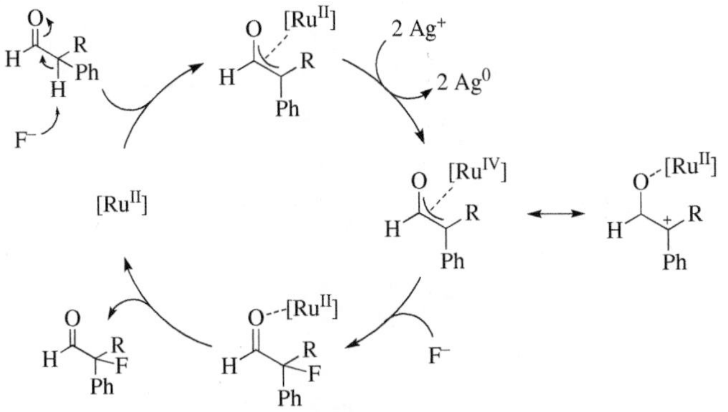

Scheme 6

Redox Catalysis via Metal Fluoride Intermediates (Nuc-R-F). Transition-metals can catalyze or mediate C–F bond formation between alkyl or benzyl radicals and nucleophilic fluorinating reagents to furnish the corresponding alkyl and benzyl fluorides by a Nuc-R-F mechanism. If the radical intermediate can achieve a planar conformation, then a racemic product is formed in the absence of a chiral catalyst that can promote the formation of one enantiomer. Site-selectivity can be determined through chelation, whereby a functional group in the substrate binds to the catalyst to ensure that the reaction occurs at a nearby carbon–hydrogen bond. Alternatively, electronic and steric features of the substrate or catalyst can determine the site of reaction, as exemplified by the cases in which (1) a bulky catalyst preferentially interacts with a terminal carbon–hydrogen bond or (2) carbon–hydrogen bonds adjacent to electron-withdrawing substituents react more slowly than other carbon–hydrogen bonds. In the manganese-porphyrin-catalyzed, aliphatic C–H fluorination reaction, a high-valent catalyst reacts with alkanes to generate alkyl radicals, which undergo carbon–fluorine bond formation with manganese-bound fluorides (Scheme 7).[44] The metal-bound fluoride is generated by anion metathesis at the manganese(IV) center, followed by substrate hydrogen-atom abstraction. The lowest unoccupied molecular orbitals of the low-spin d^2-oxomanganese(V) complex are expected to be the two orthogonal Mn–O π^* orbitals, which would direct the scissile carbon–hydrogen bond into a bent π^* approach trajectory.[44]

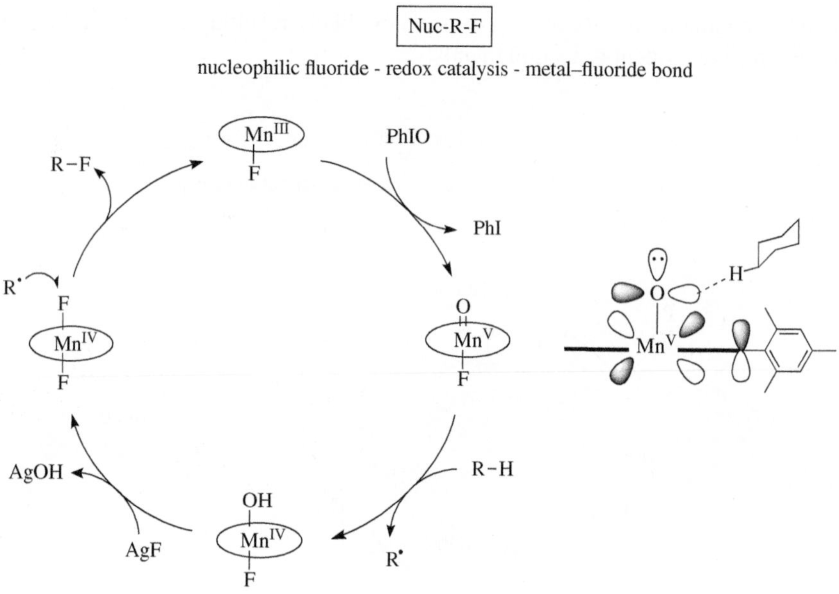

Scheme 7

Lewis Acid Catalysis via Organometallic Fluoride Intermediates (Nuc-LA-sub-F). The enantioselective ring-opening of aziridines and epoxides has been reported using both manganese–salen and cobalt–salen complexes. In the case of the cobalt-catalyzed fluorination reaction, the transition-metal complex functions as a Lewis acid catalyst to activate both the epoxide and the fluoride nucleophile (Scheme 8).[45] Mechanistic work using a monometallic catalyst suggested a bimetallic mechanism, in which the analogous tethered bimetallic system exhibits increased reaction rates, thereby supporting the proposed mechanism.[45,46]

Lewis Acid Catalysis via Organometallic Intermediates and Exogenous Fluoride (Nuc-LA-sub). Large, highly polarizable gold complexes catalyze alkyne fluorinations, and in these reactions, the carbon–fluorine bond-forming step does not involve a metal–fluoride bond. Compared with most transition metals, gold forms weak bonds to fluoride. Coordination of the triple bond to the cationic gold complex

Nuc-L-sub-F

nucleophilic fluoride - Lewis acid catalysis - metal-substrate bond - metal-fluoride bond

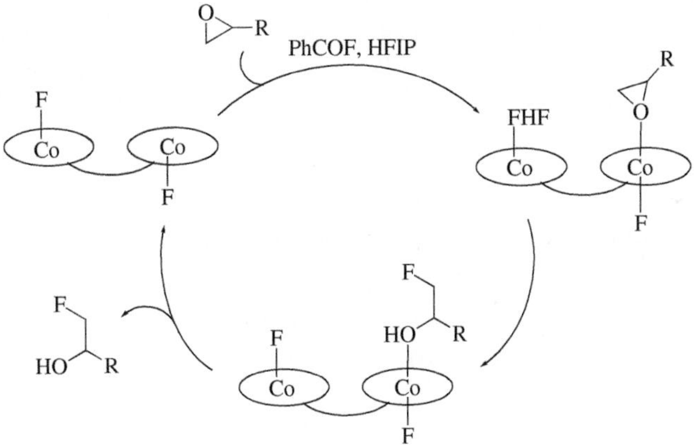

Scheme 8

activates the alkyne π-system toward attack by a fluoride ion (Scheme 9).[47] In the highlighted example, the presence of an amide group adjacent to the alkyne results in improved regioselectivity compared to other gold-catalyzed alkyne fluorination reactions, since the directing group imparts a bias on the regioselectivity of fluoride attack to furnish a gold chelate with favorable ring size.[47,48]

Nuc-LA-sub

nucleophilic fluoride - Lewis acid catalysis - metal–substrate bond

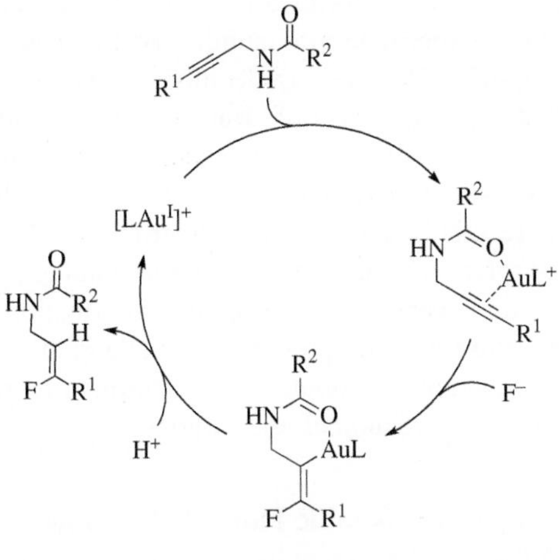

Scheme 9

Lewis Acid Catalysis via Metal Fluoride Intermediates (Nuc-LA-F).
Fluorination at aliphatic and benzylic sites proceeds by an S_N2 displacement with transition-metal-bound fluoride nucleophiles, resulting in inversion of configuration at the displacement site (Scheme 10).[49] In addition to modulating the reactivity of fluoride, ligation of fluoride to the transition-metal center renders the fluoride soluble in organic solvents. Hence, it enables the use of simple alkali metal halides rather than more expensive sources of soluble fluorides, such as TBAF and TASF.[49-51]

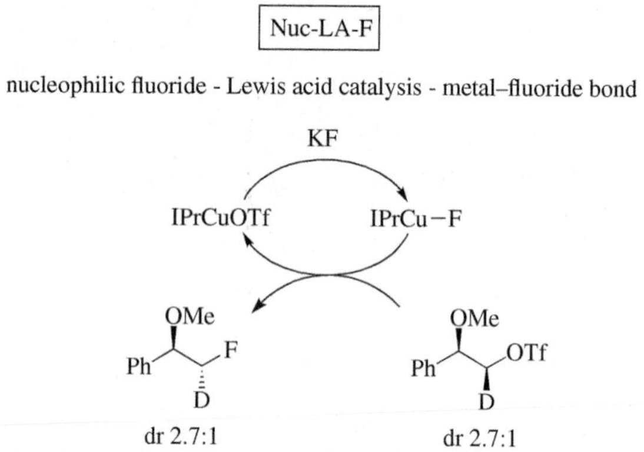

Scheme 10

Electrophilic Fluoride Sources

Early examples of electrophilic fluorinating reagents were notorious for their high reactivity, which prompted the development of milder 'F+' sources with more controlled reactivity that has led to a rapid expansion of electrophilic fluorination reactions. Since fluorine is the most electronegative element, it carries a partial negative charge whenever it forms a bond to other elements. Nevertheless, reagents such as N-fluorodibenzenesulfonamide (NFSI) and 1-chloromethyl-4-fluoro-1,4-diazoniabicyclo[2.2.2]octane (Selectfluor), which contain a weak nitrogen–fluorine bond, are capable of transferring fluoride oxidatively to transition-metal centers and can also serve as a source of fluorine radical. Because NFSI, Selectfluor, and N-fluoropyridinium salts can furnish formal equivalents of 'F+' and 'F', their reactivity differs markedly from simple fluoride salts, and thus, these reagents are referred to as electrophilic fluorinating agents.[52] Compared to nucleophilic fluorination reactions, electrophilic fluorination reactions tend to be less sensitive to moisture and the presence of nucleophilic functional groups, but the considerable oxidation potential of the fluorinating reagents can limit functional-group tolerance toward amines and sulfides.

Redox Catalysis via Organometallic Fluoride Intermediates (El-R-sub-F).

The fluorination of pyridines can be accomplished using silver(II) fluoride. The mechanism is thought to involve the coordination of silver fluoride to a nitrogen atom in the heteroarene, which results in hydrogen-atom abstraction (Scheme 11).[53] Silver(II) activates the substrate by withdrawing electron density from the nitrogen center of the heteroarene. Notably, since the activation does not occur at the reactive carbon atom, this silver-mediated fluorination of heteroarenes represents an unusual type of El-R-sub-F mechanism.

El-R-sub-F

electrophilic fluorine source - redox catalysis - metal–substrate bond - metal–fluoride bond

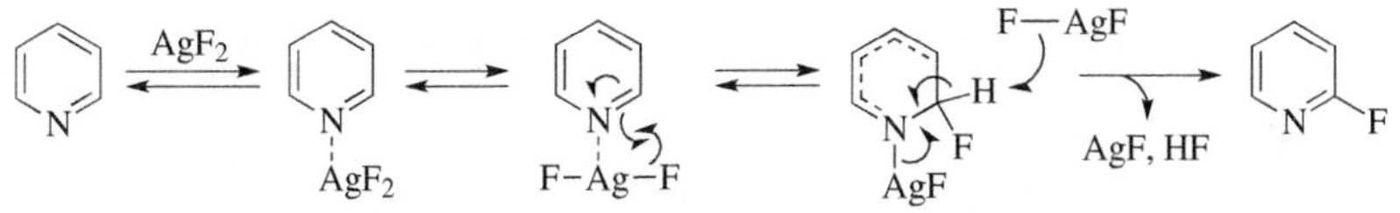

Scheme 11

Copper-mediated fluorination of aryl boronates proceeds by an El-R-sub-F mechanism wherein a copper(I) catalyst is oxidized to a copper(III) fluoride complex, followed by rate-determining transmetalation (Scheme 12).[54] Detailed mechanistic studies reveal that this reaction involves two stable copper intermediates: oxidative fluorine transfer forms a copper(III) fluoride complex, and aryl transmetalation generates a second intermediate that is proposed to be a neutral arylboronate copper(III) fluoride based on ^{1}H, ^{19}F, and ^{11}B NMR spectroscopic evidence.

El-R-sub-F

electrophilic fluorine source - redox catalysis - metal–substrate bond - metal–fluoride bond

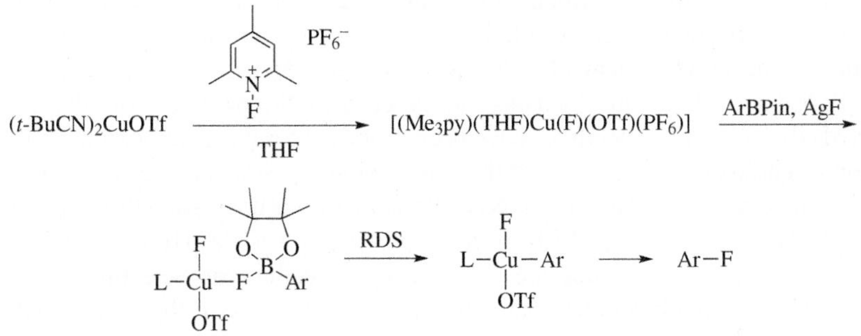

Scheme 12

Redox Catalysis via Metal Fluoride Intermediates (El-R-F). The conventional approach for transition-metal-catalyzed C–H functionalization of aryl rings involves metalation of the arene, which is often facilitated by a directing group or by the use of a superstoichiometric amount of the arene substrate, relative to the oxidant. After ligand exchange on the metal center, reductive elimination furnishes the desired product. Alternatively, C–H fluorination can occur by an El-R-F mechanism without the need for superstoichiometric amounts of arene (Scheme 13).[55] In this reaction, a specially designed ligand enables the oxidation of the palladium(II) catalyst to a palladium(IV) fluoride complex using Selectfluor. Following the direct transfer of a fluorine radical from the high-valent metal fluoride to the substrate, a single-electron transfer (SET) and proton loss generate the aryl fluoride and regenerate the palladium(II) catalyst.

El-R-F

electrophilic fluorine source - redox catalysis - metal–fluoride bond

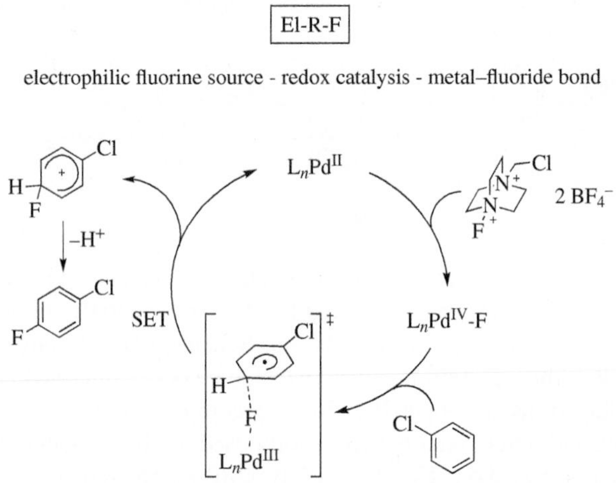

Scheme 13

The silver-catalyzed fluorination of alkylboronates is also proposed to occur by an El-R-F mechanism: Selectfluor oxidizes the silver(I) catalyst (silver nitrate) to a silver(III) fluoride complex, which then promotes an SET with the alkylboronate to furnish a silver(II) fluoride and an alkyl radical.[56] Fluorine abstraction from silver(II) fluoride by the alkyl radical affords the alkyl fluoride product and regenerates silver(I). Related El-R-F mechanisms that proceed by the oxidation of silver(I) to silver(III) fluoride with Selectfluor have been proposed for the decarboxylative fluorination of carboxylic acids[57,58] and the ring-opening of strained cyclic alcohols.[59] However, a detailed mechanistic study suggests that decarboxylative fluorination of carboxylic acids occurs via an El-R mechanism with an Ag(I)/Ag(II) catalytic cycle, in which silver(I) carboxylates are oxidized by Seleefluor or Selectfluor radical cation to silver(II) carboxylates that can decarboxylate to form the corresponding alkyl radicals.[60] The low probability of silver(III) formation in the absence of strongly stabilizing ligands is also discussed, which calls into question mechanistic proposals that feature silver(III) fluoride complexes in the absence of ligands other than the solvent and Selectfluor.[56–59]

Redox Catalysis with Exogenous Fluoride (El-R). The fluorination of aryl trifluoroboronates is proposed to occur by redox catalysis using a palladium complex. In contrast to the redox processes discussed above, this mechanism involves neither a palladium–fluoride nor a palladium–substrate bond (Scheme 14).[61] Single-electron transfer from the palladium catalyst leads to a one-electron reduction of the electrophilic fluorinating reagent, which then transfers a fluorine atom to the aryl trifluoroboronate substrate. A second SET step promotes a one-electron oxidation of the aryl radical intermediate, and loss of boron trifluoride generates the fluorinated product. The interaction of the transition-metal catalyst with the coupling partners is reminiscent of photoredox catalysis, except that the energy required to drive the reaction is provided by the high energy Selectfluor reagent rather than by irradiation.

El-R

electrophilic fluorine source - redox catalysis

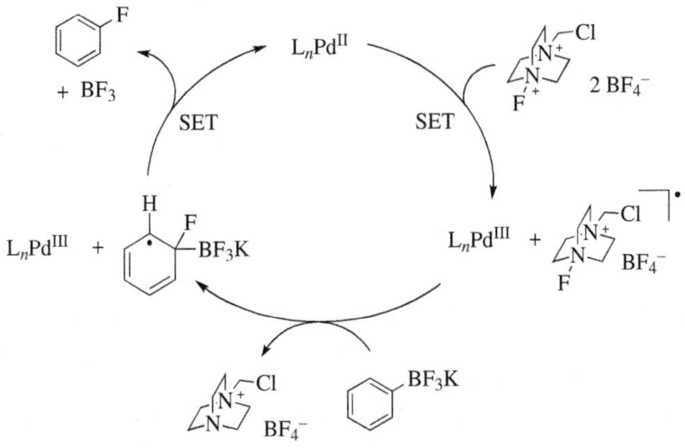

Scheme 14

The fluorination of unactivated aliphatic carbon–hydrogen bonds can also be accomplished using a photoexcited decatungstate anion, which generates an alkyl radical from an alkane by C–H abstraction.[62] Formation of the alkyl fluoride product is proposed to occur either by fluorine atom abstraction from NFSI or by SET to form a radical cation followed by fluoride transfer (El-R).

Lewis Acid Catalysis via Organometallic Fluoride Intermediates (El-LA-sub-F). The titanium-mediated α-fluorination of acyloxazolidinones is proposed to proceed via the coordination of the metal to both the substrate and NFSI during the carbon–fluorine bond formation step (Scheme 15).[63] In the El-R-sub-F and Nuc-R-sub-F mechanisms, the transition metal interacts directly with fluorine, but in the El-LA-sub-F mechanism, the reactivity of NFSI is enhanced by coordination of the Lewis acid to the sulfonamide oxygen atom. However, because the reaction requires superstoichiometric amounts of titanium(IV) chloride (1.5–1.75 equivalents), an alternative mechanism is also plausible, wherein the substrate undergoes soft enolization by titanium(IV) chloride, and the excess Lewis acid activates NFSI. The importance of Lewis acid activation of NFSI for the α-fluorination of (2-arylacetyl)thia- and oxazolidin-2-ones is illustrated by a nickel-catalyzed fluorination reaction, in which the presence of a second Lewis acid (triethylsilyl trifluoromethanesulfonate) is required to effect fluorination.[64]

El-LA-sub-F

electrophilic fluorine source - Lewis acid catalysis - metal–substrate bond - metal–fluorine bond

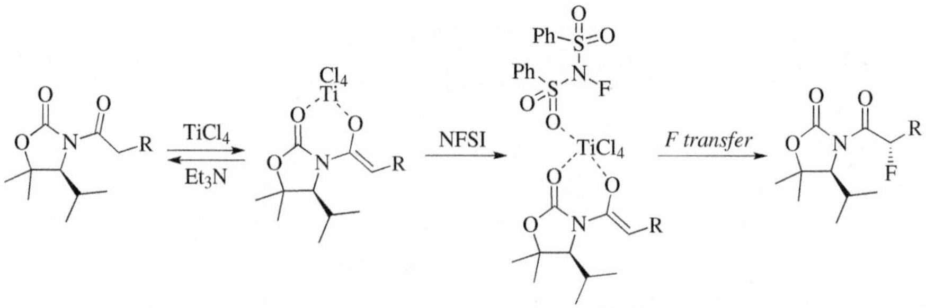

Scheme 15

Lewis Acid Catalysis via Organometallic Intermediates and Exogenous Fluoride (El-LA-sub). Fluorination adjacent to carbonyl groups is dominated by the reaction of metal enolates with electrophilic fluorination reagents. The fluorination of many carbonyl derivatives can occur in the absence of a catalyst, and thus, the main challenge addressed with catalysis is enantioinduction at the site of fluorination. Fluorination of a disubstituted rather than a trisubstituted carbon atom introduces the additional challenges of avoiding difluorination and racemization because the monofluorinated product is more readily enolized than the starting material.[65] In contrast, the fluorinations of ketenes, epoxides, and alkynes to furnish the α-fluorinated carbonyl derivatives proceed by alternative and more varied mechanisms. For instance, titanium-catalyzed carbonyl α-fluorination proceeds by Lewis acid catalysis: soft enolization of the starting material is followed by nucleophilic attack of the electrophilic fluorinating reagent on the less-hindered *Si*-face of the substrate–catalyst complex (Scheme 16).[66–68] Palladium-,[65,69–81] titanium-,[63,66,67,82] iron-,[83,84] gold-,[85–87] cobalt-,[88] ruthenium-,[43,89] scandium-,[90] iridium-,[91,92] nickel-,[64,76,93–98] and copper-based[99–104] catalysts have been explored for the (enantioselective) α-fluorination of carbonyl derivatives.

Lewis Base Catalysis via Organometallic Intermediates and Exogenous Fluoride (El-LB-sub). Alkyl ketenes can be used to generate α-fluoro esters in a catalytic and enantioselective fashion using a chiral organometallic Lewis base (Scheme 17).[84] Kinetic data is inconsistent with initial adduct formation between the

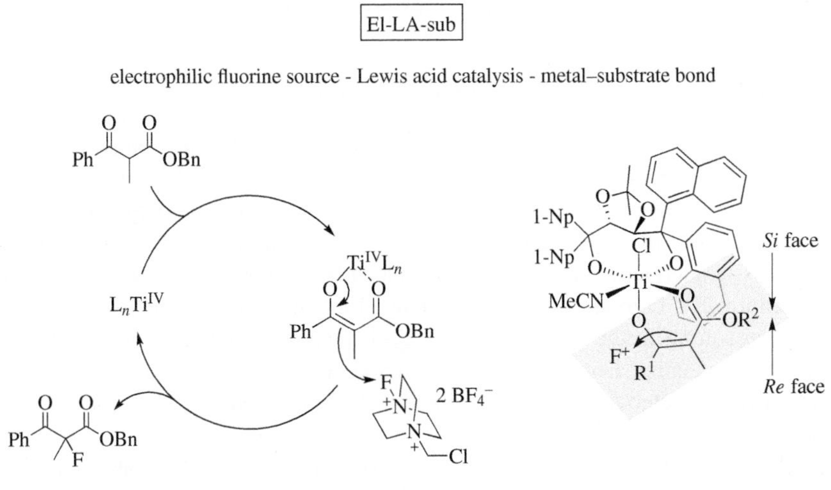

El-LA-sub

electrophilic fluorine source - Lewis acid catalysis - metal–substrate bond

Scheme 16

chiral catalyst and NFSI but does support an El-LB-sub-mechanism, in which nucleo-philic addition of (–)-PPY ((*R*)-(+)-4-pyrrolidinopyrindinyl(pentamethylcyclo-pentadienyl)iron) to the ketene generates a chiral enolate. In the absence of an external nucleophile, the fluorinated *N*-acylated intermediate was isolated, and a crystal structure was obtained for an anion-exchanged derivative. The judicious choice of nucleophile is crucial for efficient turnover. Notably, the nucleophile must be sufficiently reactive to release the catalyst from the *N*-acylated intermediate (thereby forming the ester), but sufficiently unreactive to suppress the competitive attack of the ketene substrate, which would result in the formation of an achiral enolate and, thus, a racemic product.

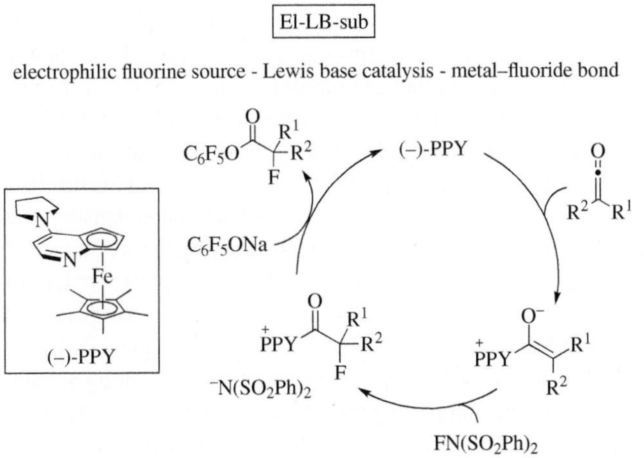

El-LB-sub

electrophilic fluorine source - Lewis base catalysis - metal–fluoride bond

Scheme 17

Aryl Fluoride Synthesis

Early attempts at transition-metal-catalyzed carbon–fluorine bond formation were foiled by the formation of fluoride-bridged dimeric palladium complexes, which are unreactive toward reductive elimination and tend to undergo competitive coupling between fluoride and the supporting ligands on the metal.[10] Given the difficulties associated with the development of a carbon–fluorine reductive elimination at palladium(II), carbon–fluorine reductive elimination from palladium(IV) was pursued, resulting in the first catalytic C_{aryl}–F bond forming reaction.[105] The first reductive elimination from a well-defined palladium complex followed suit, along with X-ray characterization of the palladium(IV) intermediate.[106,107] Advances in C–X cross-coupling reactions involving reductive elimination from palladium(II) ultimately informed the development of analogous reductive eliminations to form carbon–fluorine bonds.[25] Generally, palladium(II)-catalyzed fluorination relies on phosphine ligands, whereas palladium(IV) catalysis utilizes nitrogen-based ligands.

Aryl fluorides have also been prepared from isolated transition-metal complexes in the course of mechanistic studies that focus on the reductive elimination step and in the synthesis of ^{18}F-aryl fluorides.[106–115] Although many carbon–fluorine bond forming reactions proceed via organometallic intermediates, transition-metal catalysis can also lower fluorination barriers without leading to the formation of intermediates bearing carbon–metal bonds (El-R mechanism).[61]

Aryl Fluoride Synthesis from Aryl Halides and Aryl Pseudohalides. The formation of carbon–fluorine bonds via the reductive elimination from palladium(II) has proven elusive; however, the bulky biaryl phosphine ligand 2-(di-*tert*-butylphosphino)-2′,4′,6′- triisopropyl-3,6-dimethoxy-1,1′-biphenyl (**L1**, *t*-BuBrettPhos) mediates reductive elimination from (**L1**)PdArF and also serves as the supporting ligand in the palladium-catalyzed fluorination of aryl triflates in the presence of cesium fluoride. These reactions generate a mixture of constitutional isomers and a chlorinated side product formed by the competitive addition of a chloride ion that is liberated from the [(cinnamyl)PdCl]$_2$ precatalyst. This limitation prompted the development of a precatalyst that is activated without releasing nucleophiles or byproducts that inhibit this process, thereby eliminating the formation of chlorinated side products.[37] Interestingly, using **L2** (HGPhos) in place of **L1** affords higher fluorination yields for challenging substrates such as electron-rich aromatics and heteroarenes. In addition, the scope of the reaction can be extended to aryl bromides by using silver(I) fluoride to facilitate the transmetalation step (Scheme 18).[33] The C3′-substituted birarylphosphine ligand AlPhos enables the palladium-catalyzed fluorination of a variety of activated (hetero)aryl triflates at room temperature,[36] and it also provides a high degree of site-selectivity for substrates that afford mixtures of fluorinated isomers using **L1**.[35,36]

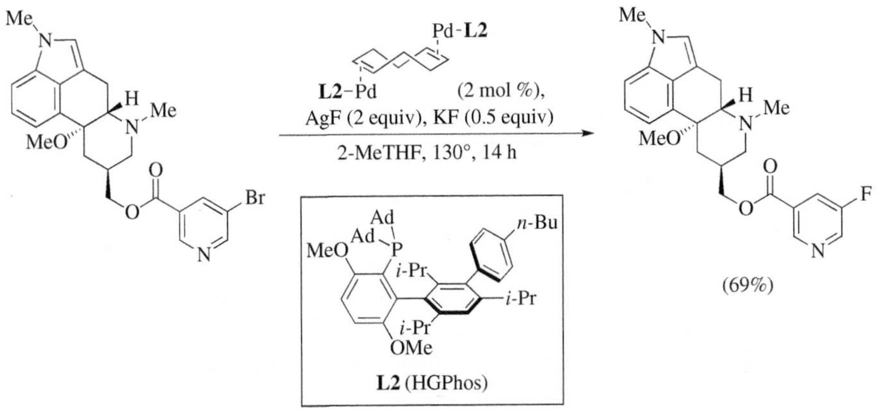

Scheme 18

Aryl fluorides can be prepared from phenols in a one-pot procedure by generating aryl nonafluorobutylsulfonate intermediates in situ and using them directly in a palladium-catalyzed fluorination.[116] Interestingly, a competition experiment using cesium fluoride, cesium chloride, and cesium bromide in combination with the aryl nonaflate substrate affords the aryl fluoride product almost exclusively. This result is rationalized by DFT calculations, which indicate that the barriers for aryl fluoride and aryl chloride reductive elimination are 22.7 kcal•mol^{-1} and 24.1 kcal•mol^{-1}, respectively, and that the former process is the more exothermic reaction.

An aromatic halide exchange reaction in a macrocyclic system employs a Cu(I)/Cu(III) catalytic cycle, in which nitrogen chelation in the macrocycle lowers the barriers for oxidative addition (Scheme 19).[117] Halide exchange establishes an equilibrium between the substrate and product aryl halide, favoring the product with the stronger C–X bond. Interestingly, if the halide salt MY precipitates while MX remains soluble, the direction of the reaction can be reversed because the product

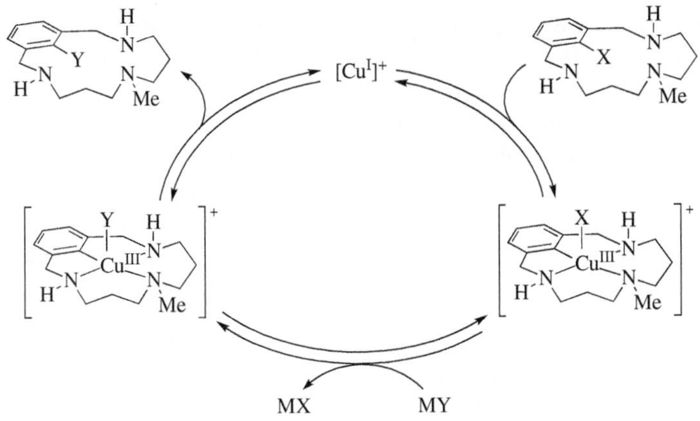

Scheme 19

halide salt is removed from the reaction equilibrium, similar to a copper-catalyzed, aromatic Finkelstein reaction. Halide exchange is the rate-limiting step of the reaction, and DFT analysis suggests that the barrier for carbon–chlorine reductive elimination (26.9 kcal•mol^{-1}) exceeds that for carbon–fluorine reductive elimination (16.2 kcal•mol^{-1}) by a considerable margin.

A pyridyl directing group is essential for stabilizing the copper(I) catalyst and accelerating the oxidative addition in the copper-catalyzed fluorination reaction of aryl iodides and aryl bromides (Scheme 20).[118] Although reduction products (Ar–H) are not formed, the diaryl ether adducts are formed in <5% yields from the incorporation of serendipitous water. Copper(II) fluoride displays some activity for the fluorination of aryl iodides,[119] but the oxidation of copper(I) by silver(I) fluoride can be prevented by the presence of the pyridine in this system. X-ray absorption near-edge structure/extended X-ray absorption fine structure (XANES/EXAFS) data indicates that while the copper(I) catalyst is fully oxidized to copper(II) when iodobenzene is subjected to the copper-catalyzed fluorination, copper(I) is still present at completion in the reaction of the corresponding 2-pyridyl-substituted iodobenzene.

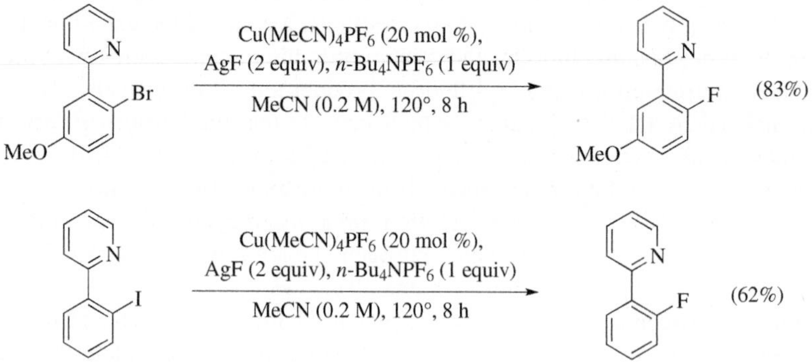

Scheme 20

The development of efficient copper catalysis for aryl halide fluorination in the absence of a directing group is based on the discovery of a ligand that inhibits rapid decomposition of the copper(I) fluoride species (Scheme 21).[120] The basic hypothesis presumes that weakly donating ligands and a non-coordinating counter ion would favor carbon–fluorine reductive elimination from high-valent copper complexes. Excess copper relative to silver fluoride is critical for the fluorination of aryl iodides, possibly because the silver mediates unproductive redox processes that deplete the amount of copper(I) present. Deuterium-labeling studies suggest

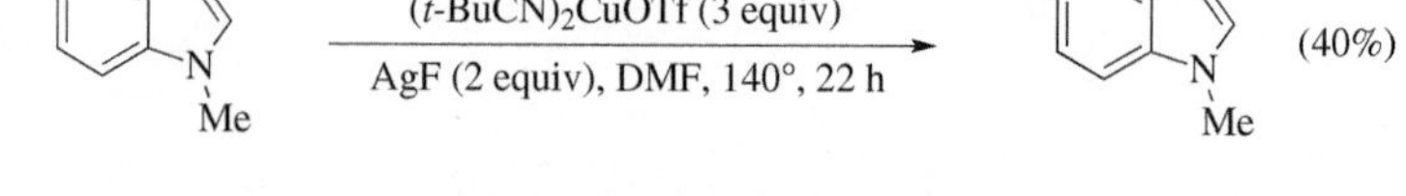

Scheme 21

that either water or the acetonitrile ligand provides the proton incorporated in the protodehalogenated side product, which is difficult to separate from the aryl fluoride product.

Aryl Fluoride Synthesis from Arenes. The direct synthesis of carbon–fluorine bonds from carbon–hydrogen bonds using hydrogen fluoride is thermodynamically unfavorable. However, in the presence of molecular oxygen, the formation of water provides the required thermodynamic driving force.[121] The development of an environmentally acceptable process with water as the only byproduct requires a metal–fluorine bond that (1) is capable of oxidizing a carbon–hydrogen bond and (2) can be formed by oxidation of the metal using only oxygen and hydrogen fluoride. Copper(II), silver(I), tellurium(IV), and mercury(II) metal salts were identified as promising candidates because the reduction potentials of these salts are between 1.0 and 0.0 V versus a standard hydrogen electrode (SHE). The selection of copper(II) is primarily based on the cost and environmental impact.[121] At 450–550°, a stream of benzene in nitrogen is passed over a reactor bed charged with copper(II) fluoride to furnish fluorobenzene with 95% selectivity, albeit with only 30% conversion (Scheme 22). Nevertheless, the spent catalyst can be regenerated by exposure to an HF/O_2 stream at 400°, making this a relatively attractive process.

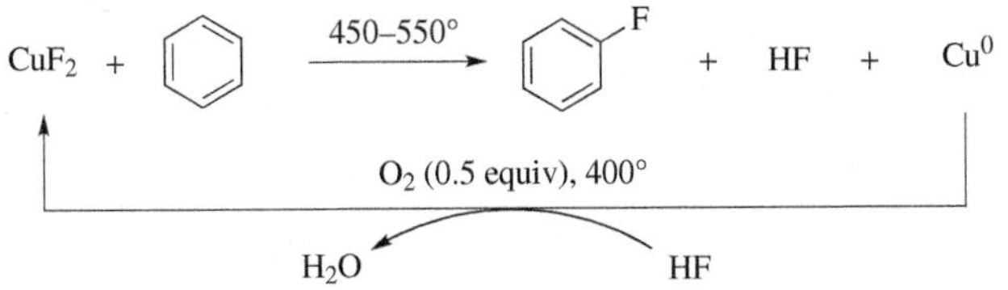

Scheme 22

The palladium-catalyzed C–H fluorination of aryl derivatives requires an ancillary ligand that facilitates the Pd(II)/Pd(IV) oxidation with the electrophilic fluorinating agent before any interaction with the arene substrate (Scheme 23).[55] Fluoride transfer and substrate oxidation are proposed to occur in a concerted fashion, forming the aryl fluoride product via a single transition state. A derivative of the palladium(IV) fluoride intermediate, in which 2-chlorophenanthroline is replaced by phenanthroline, can be isolated and characterized, and is an active C–H fluorination reagent. An 'oxidation first' mechanism is well precedented in aliphatic C–H functionalization chemistry, but this report constitutes the first example of an aromatic C–H functionalization reaction, in which the interaction between a high-valent catalytic intermediate and the arene substrate is directly investigated.[55] Late-stage fluorination of complex arenes proceeds with synthetically useful yields in the presence of a number of functional groups, including ketones and free hydroxyl groups. In most cases, the reaction affords a mixture of two constitutionally isomeric aryl fluoride products, which have to be separated by HPLC.

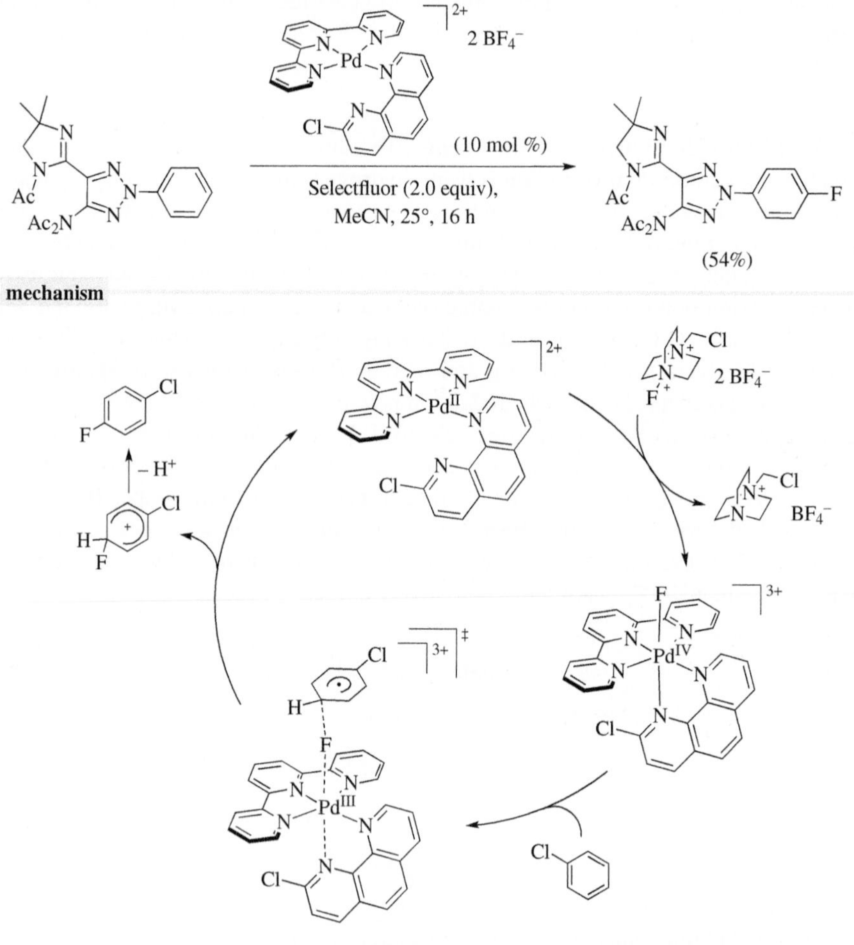

Scheme 23

Directing groups are often employed in C–H functionalization reactions to facilitate arene metalation; however, the directed palladation of C–H bonds can be inhibited by electrophilic fluorinating reagents. A supporting pyridine ligand present in the fluorinating reagent (e.g., N-fluoropyridinium tetrafluoroborate) can compete with the directing group and thus competitively bind the palladium(II). In addition, the electrophilic fluorine atom can form a complex with the Lewis basic directing group, which lowers the ability of the directing group to bind to the metal center.[122] Nevertheless, both benzylic and aromatic C–H bonds in proximity to pyridine or quinoline directing groups are susceptible to a palladium-catalyzed C–H fluorination in the presence of N-fluoropyridinium tetrafluoroborate under microwave irradiation (Scheme 24).[105] Microwave heating leads to a significant rate acceleration in the fluorination and increases the selectivity for the formation of aryl fluorides over competing carbon–carbon and carbon–oxygen bond formation pathways.

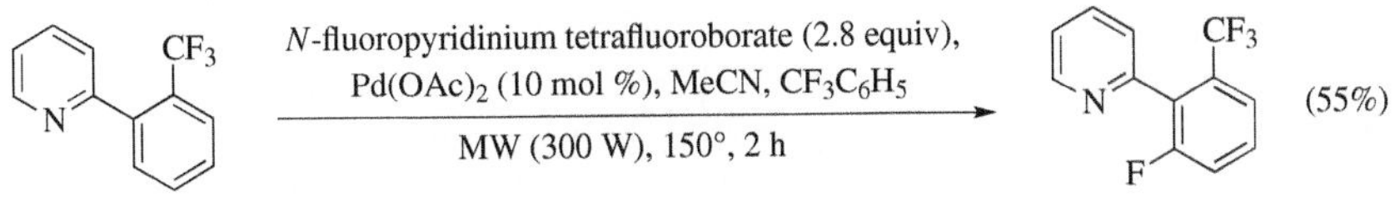

N-fluoropyridinium tetrafluoroborate (2.8 equiv),
Pd(OAc)$_2$ (10 mol %), MeCN, CF$_3$C$_6$H$_5$

MW (300 W), 150°, 2 h

(55%)

Scheme 24

Strongly coordinating X-type ligands (e.g., triflamides), which donate one electron to the transition-metal center upon binding, can function as directing groups in C–H fluorination reactions but lead to difluorination owing to the slow displacement of the monofluorinated substrate from the palladium center. A weakly coordinating acidic amide directing group often permits facile ligand displacement at the palladium center and thus favors the formation of the monofluorinated adducts.[122] Employing acetonitrile as the solvent is crucial for attaining the monofluorinated product with good selectivity, which is consistent with the solvent competitively displacing the product by behaving as a weakly coordinated L-type ligand for the metal center.

Depending on the reaction conditions, an oxalyl amide directing group enables the selective formation of either monofluorination or difluorination products in a palladium-catalyzed C–H fluorination reaction.[123] Fluoro-, chloro-, bromo-, and iodo-substituted aromatic compounds undergo highly site-selective monofluorination at the less sterically encumbered of the two available *ortho* positions, and basic hydrolysis of the oxalyl amide liberates a primary amino group. Directed *ortho*-fluorination of arenes substituted with quinoxaline, pyrazole, benzo[*d*]oxazole, and pyrazine groups can be achieved with a combination of Pd(OAc)$_2$, NFSI, and TFA.[124]

A Pd(II)/Pd(IV) catalytic cycle involving an in situ generated cationic Pd(NO$_3$)$^+$ species has been proposed in an oxime-directed C–H fluorination reaction (Scheme 25).[125] Although the monofluorinated product dominates when the reaction is performed at 25°, complete difluorination occurs upon heating the reaction mixture to 110°.

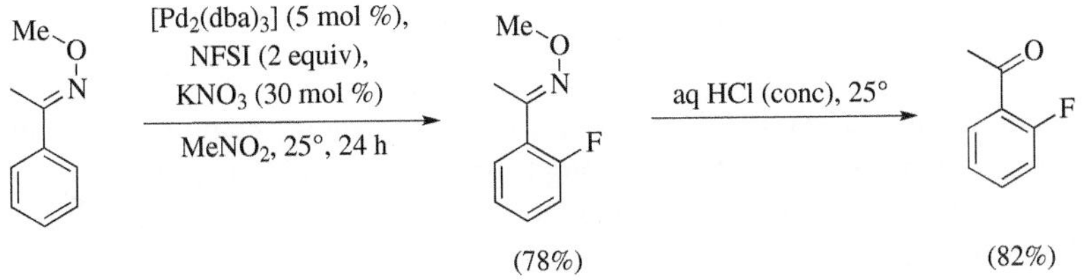

[Pd$_2$(dba)$_3$] (5 mol %),
NFSI (2 equiv),
KNO$_3$ (30 mol %)

MeNO$_2$, 25°, 24 h

aq HCl (conc), 25°

(78%)   (82%)

Scheme 25

Selective C–H mono- and difluorination of aminoquinolinyl benzamides can be achieved with a copper catalyst in the presence of silver(I) fluoride.[126] Difluorination requires higher catalyst loadings, longer reaction times, and a pyridine additive, which slows the decomposition of the starting material and reduces the rate of fluorination. Palladium-catalyzed C–H fluorination using NFSI provides access to 2-functionalized aryl tetrazines.[127] The limited stability of the fluorinated aryl tetrazines on silica generally leads to isolated yields below 50%. Copper mediates the aerobic C–H activation of azacalix[1]arene[3]pyridines to afford a well-defined arylcopper(III) intermediate **2** (Scheme 26).[128] The fluorination of substrate **2** is more efficient using potassium fluoride than using cesium fluoride, and there is no reaction using lithium fluoride or sodium fluoride.

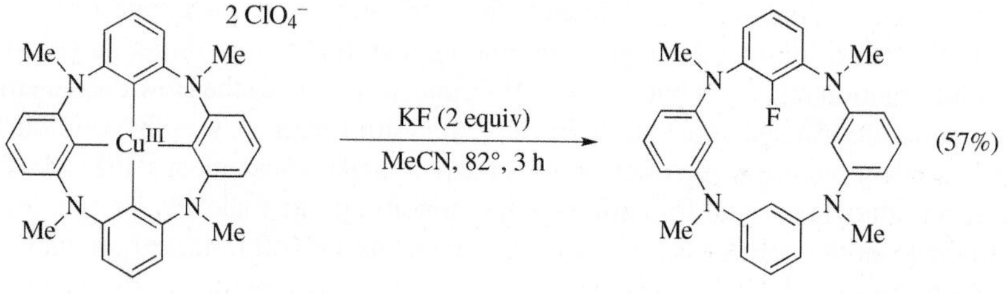

Scheme 26

The combination of Pd(OTf)$_2$•2H$_2$O and NMP catalyzes the triflimide-directed C–H fluorination of arenes (Scheme 27).[129] Two competing side reactions (acetoxylation and carbonylative lactamization) are avoided by replacing Pd(OAc)$_2$ with Pd(OTf)$_2$•2H$_2$O, which avoids the presence of acetate in the reaction system. Interestingly, the solvent impacts the selectivity: NMP leads predominantly to difluorinated products, and DMF affords the monofluorinated arenes (along with ~7% of the difluorinated arene).

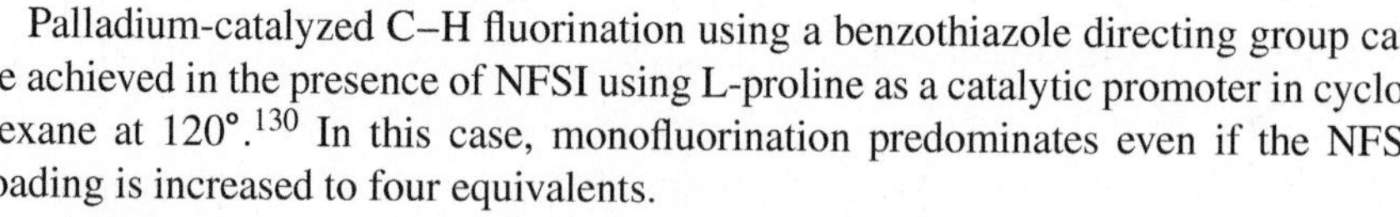

Scheme 27

Palladium-catalyzed C–H fluorination using a benzothiazole directing group can be achieved in the presence of NFSI using L-proline as a catalytic promoter in cyclohexane at 120°.[130] In this case, monofluorination predominates even if the NFSI loading is increased to four equivalents.

Aryl Fluoride Synthesis from Aryl Boronic Acid Derivatives, Aryl Silanes, and Aryl Stannanes. Electrophilic fluorination of highly reactive main-group organometallics can proceed by the direct fluorination of the carbon–metal σ-bond with electrophilic fluorinating reagents. Catalysts are required for the fluorination of aryl groups bound to main-group metals with lower-lying σ-orbitals, such as boron and tin, with fluorinating agents that are less reactive than fluorine gas or acetyl hypofluorite. Silver(I) salts catalyze the fluorination of arylstannanes with Selectfluor, and higher yields are obtained when the counterion is hexafluorophosphate rather than tetrafluoroborate.[131] Reductive elimination is thought to occur via a bimetallic intermediate comprising two silver atoms, each of which undergoes a one-electron redox process. The development of catalytic conditions derived from the silver-mediated fluorination of arylstannanes is complicated by protodestannylation, a significant side reaction when substoichiometric amounts of the silver(I) salts are employed.[23] Protodestannylation likely occurs when tributyltin triflate (formed in the transmetalation step) is hydrolyzed in situ to generate triflic acid. Formation of the Ar–H bond can be suppressed by addition of sodium bicarbonate, which also leads to the precipitation of the tin residue, thus facilitating purification (Scheme 28).[23] That fluorination proceeds via transmetalation to silver and reductive elimination rather than direct fluorination of the carbon-tin bond is evident from the formation of phenol products when water is added to the reaction mixture. Phenol formation in the presence of water is consistent with the exchange of hydroxide for fluoride at a high-valent silver center, but it would not be expected if an aryl–silver bond was not also formed. Thioethers and electron-rich amines that possess β-hydrogens are outside the scope of the transformation owing to undesirable side reactions with Selectfluor.

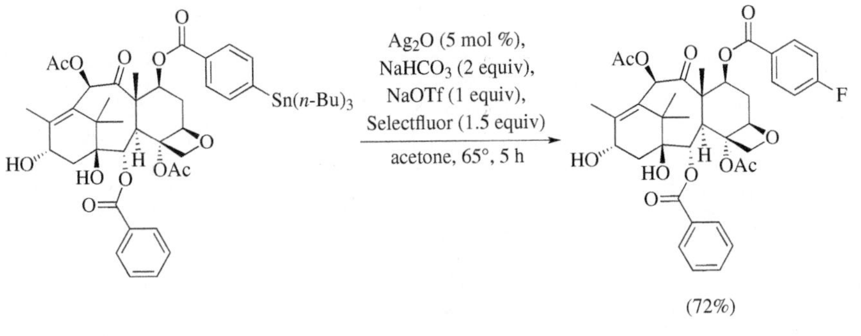

Scheme 28

The silver-mediated fluorination of arylstannanes commonly results in a 10–20% yield of the reduced product in addition to the desired aryl fluoride product.[131] In contrast, the reduction adduct (Ar–H) cannot be detected in the analogous fluorination of arylboronic acids.[132] These reaction conditions are compatible with protic, nucleophilic, and electrophilic functional groups, as well as with 5- and 6-membered nitrogenous heterocycles. Methanol participation leads to the formation of the aryl methyl ether rather than the aryl fluoride, and therefore, a solvent exchange must be performed after the boron-to-silver transmetalation step.

The first metal-catalyzed fluorination of arylboronic acid derivatives is illustrated in Scheme 29.[61] A wide range of aryl trifluoroborates is compatible with this reaction, except for heteroarene substrates. The formation of aryltrifluoroborates in situ through the addition of sodium fluoride and potassium hydrogenfluoride enables the fluorination of arylboronic acids and esters. The reduced side product (Ar–H) is not observed, which is attributed to the lack of organopalladium intermediates (cf. Scheme 14). The oxidation of [(terpy)Pd(MeCN)](BF$_4$)$_2$ (terpy = 2,2′:6′,2″-terpyridine) (**3**) is turnover-limiting, and the unusual palladium(III) intermediate, compound **4**, can be isolated and characterized. Carbon–fluorine bond formation is thought to occur by a fluorine radical transfer from the Selectfluor radical cation to the aryltrifluoroborate substrate. Electron transfer from the substrate–fluorine adduct to complex **4** and loss of BF$_3$, which is sequestered by the sodium fluoride additive, completes the proposed catalytic cycle.[61]

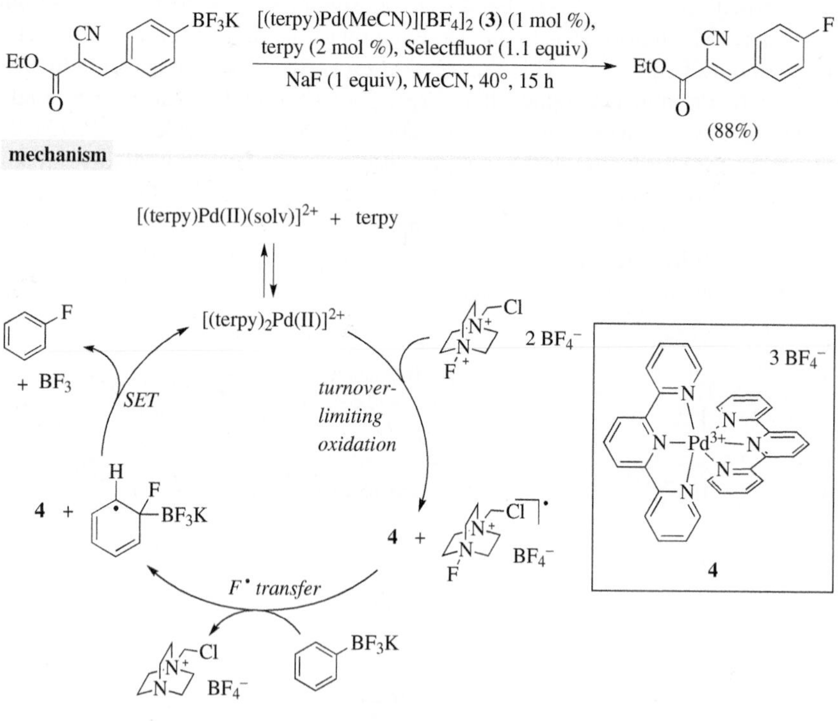

mechanism

Scheme 29

Arylstannanes and aryltrifluoroborates can be fluorinated in a two-step, one-pot reaction sequence, wherein the arene substrate is added to a pre-combined mixture of the copper(I) complex and electrophilic fluorinating reagent.[133] A subsequently developed copper-mediated fluorination of aryltrifluoroborates and arylboronic esters employs a nucleophilic source of fluoride.[134] Although halide-containing substrates are susceptible to a competing halodeboronation, keto and formyl groups are compatible with the reaction conditions, and only trace quantities (<3%) of protodeboronation is observed.[134] A copper-mediated fluorination of arylboronic esters that relies on an electrophilic fluorinating reagent (Scheme 30)[54] also tolerates formyl and NH-amido functionalities. Mechanistic studies indicate that a facile oxidation of copper(I) to copper(III) is followed by a rate-determining transmetalation of the aryl boronate substrate to copper(III) and then fast reductive elimination from copper(III).

Silver(I) fluoride activates the aryl boronic ester to transfer the aryl substrate from boron to copper by forming a fluoroboronate. Notably, the low nucleophilicity of silver(I) fluoride prevents the activation of the ArBPin from inducing transmetalation of the aryl substrate to copper(I), but transmetalation to the more electrophilic copper(III) species proceeds readily. The use of silver(I) fluoride thus prevents the formation of unstable copper(I) fluoride (cf. Scheme 12). Unfortunately, the fluorinated product is formed as a mixture with the reduced product (>10%), and separation of the two adducts has not been attempted for most reported reactions.

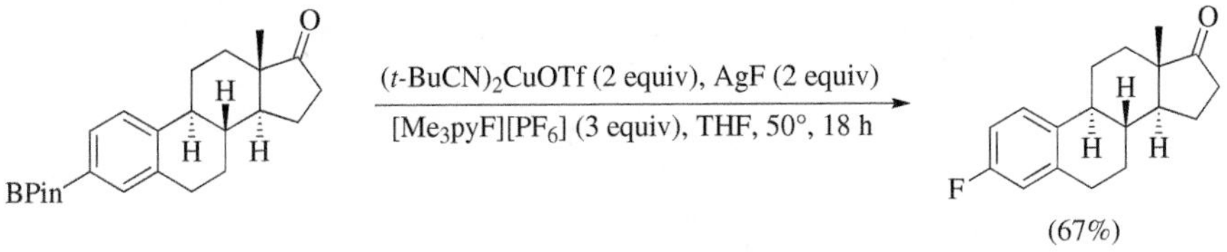

Scheme 30

Silver salts can be used to mediate the fluorination of aryltrifluoroborates, albeit the formation of the desired aryl fluoride is accompanied by varying amounts of fluorinated isomers.[135] Fluoride salts, particularly potassium fluoride, promote an iron-mediated, electrophilic fluorination of aryltrifluoroborates (Scheme 31).[136] Moderate-to-good yields are observed for electron-rich and electron-neutral substrates, but protodeboronation is a problematic side reaction. Furthermore, fluorinated isomers can be formed in substantial amounts (up to 25% yield for electron-rich aryl trifluoroborates) along with the desired product.

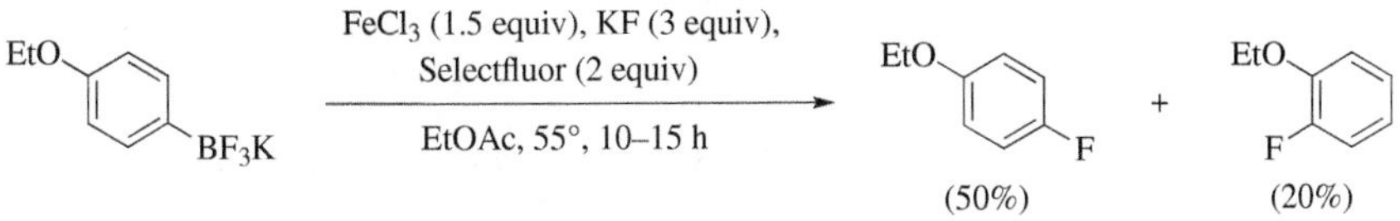

Scheme 31

Stoichiometric amounts of silver salts are required for the fluorination of aryltrihydroxysilanes and aryltrialkoxysilanes, and the highest yields are obtained using two equivalents of silver(I) oxide.[137] The addition of barium oxide suppresses the formation of protodeboronation side products (Ar–H) to trace amounts (≤5%), a result that is attributed to the sequestration of silicon-based Lewis acids. Furthermore, the addition of 2,6-lutidine reduces the amount of difluorination in electron-rich substrates.

Aryl Fluoride Synthesis from Diaryliodonium Salts. In fluorination reactions of unsymmetrical diaryliodonium salts, the less electron-rich aryl substituent on iodine is preferentially fluorinated. The uncatalyzed fluorination of diaryliodonium salts containing a mesityl group and a second, different aryl substituent often exhibits a preference for the fluorination of the mesityl substituent, which is attributed to the "ortho effect" (Scheme 32).[138] Copper catalysis of diaryliodonium fluorination enhances the yield of the electron-poor aryl fluoride product with a concomitant decrease in the yield of mesityl fluoride (Scheme 32, top reaction). A more significant effect is observed in the fluorination of iodonium salts that contain both mesityl and electron-rich aryl substituents: while mesityl fluoride is formed almost exclusively in the absence of copper, the electron-rich aryl fluoride is formed in high yield in the presence of catalytic amounts of copper(II) triflate (Scheme 32, bottom reaction).[139] The starting iodonium salts can be conveniently accessed by treating the corresponding arylboronic acids with MesI(OAc)$_2$.

Additive	Yield (%)	Product Ratio	
—	64	50	50
Cu(OTf)$_2$	73	94	6

Additive	Yield (%)	Product Ratio	
—	77	1	99
Cu(OTf)$_2$	83	96	4

Scheme 32

The energy barrier for carbon–fluorine reductive elimination from the copper(III) intermediate is 4.4 kcal·mol^{-1} (calculated by DFT), which indicates that aryl transfer to copper is the rate-limiting step of the transformation ($\Delta G^{\ddagger} = 8.0$ kcal·mol^{-1})[140] Employing Cu(MeCN)$_4$OTf instead of Cu(OTf)$_2$ affords higher and more reproducible yields of the aryl fluoride product in the copper-mediated ^{18}F-fluorination of diaryliodonium salts.[141] While the synthetic utility of copper-catalyzed fluorination of diaryliodonium salts suffers somewhat from the challenges associated with the preparation of the starting materials, the low barriers for carbon–fluorine bond formation are a great asset when working with the radioactive isotope ^{18}F.

Heteroaromatic Fluorination Reactions

Many aliphatic and aromatic fluorination reactions are compatible with hetero-cyclic substituents present on the substrates. However, the direct fluorination of the heteroaromatic ring itself is challenging, and as a result, only a handful of reactions that have been reported for the fluorination of carbocyclic arenes can be applied to heteroaromatic fluorination. Advances in ligand design for the palladium-catalyzed fluorination of aryltriflates and arylbromides minimize the formation of constitutionally isomeric arylfluorides and render the reaction suitable for the fluorination of heteroaromatic triflates and bromides.[37] Heteroaromatic nonaflates are also fluorinated under palladium catalysis,[116] and the silver-catalyzed fluorination of a pyridyltrifluoroborate substrate has been reported.[135] Interestingly, the palladium-catalyzed fluorination of five-membered heteroaromatic rings is only possible for specialized substrate classes, such as 3-bromothiophenes that contain both an electron-withdrawing substituent and phenyl groups in the ortho positions, or ortho-substituted benzo[b]furans.[142] However, in the presence of a directing group, palladium catalysis is successful with a variety of five-membered heteroaromatic rings and occurs in a site-specific manner (Scheme 33).[126,130]

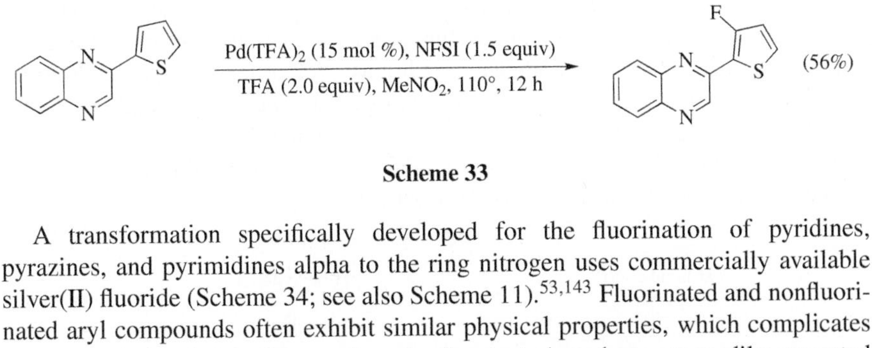

$$Pd(TFA)_2 \text{ (15 mol \%), NFSI (1.5 equiv)}$$
$$TFA \text{ (2.0 equiv), } MeNO_2, 110°, 12 \text{ h}$$

(56%)

Scheme 33

A transformation specifically developed for the fluorination of pyridines, pyrazines, and pyrimidines alpha to the ring nitrogen uses commercially available silver(II) fluoride (Scheme 34; see also Scheme 11).[53,143] Fluorinated and nonfluori-nated aryl compounds often exhibit similar physical properties, which complicates product isolation, but in this reaction, the fluorinated products are readily separated from starting materials by silica gel column chromatography or acid/base extraction.

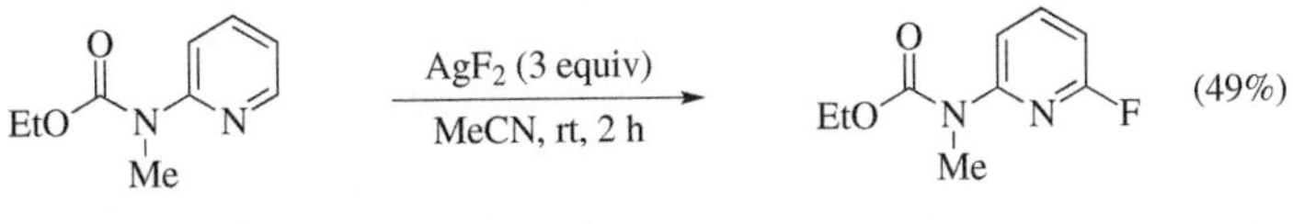

$$AgF_2 \text{ (3 equiv)}$$
$$MeCN, \text{ rt, 2 h}$$

(49%)

Scheme 34

Isolated palladium complexes of syndnones are fluorinated with Selectfluor to furnish the precursors for a cyclization reaction with functionalized alkynes (Scheme 35).[144] Although the need for a stoichiometric amount of palladium limits the synthetic utility of the transformation, the scarcity of methods available for the introduction of fluorine substituents into five-membered heteroaromatic rings and the excellent functional-group tolerance make this an important and interesting transformation.

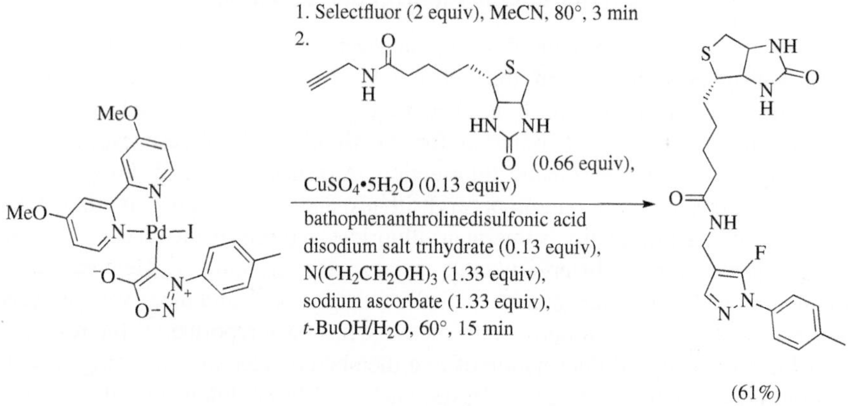

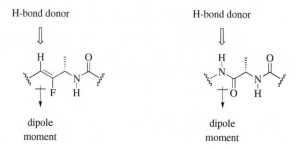

1. Selectfluor (2 equiv), MeCN, 80°, 3 min

CuSO$_4$•5H$_2$O (0.13 equiv)

bathophenanthrolinedisulfonic acid
disodium salt trihydrate (0.13 equiv),
N(CH$_2$CH$_2$OH)$_3$ (1.33 equiv),
sodium ascorbate (1.33 equiv),
t-BuOH/H$_2$O, 60°, 15 min

(61%)

Scheme 35

Alkenyl Fluoride Synthesis

The development of synthetic methods for accessing fluoroalkenes has gathered momentum because these compounds function as bioisosteres of amides and are difficult to cleave under physiological conditions.[145] Fluoroalkenes are conformationally locked and can either mimic an *s-trans* or an *s-cis* amide bond conformation. Moreover, fluoroalkenes and amides possess similar dipole moments, and the hydrogen-bond-donating ability of a C(sp^2)–H bond adjacent to a fluorine substituent mimics the donating ability of an amide (Figure 3).

H-bond donor   H-bond donor

dipole   dipole
moment   moment

Figure 3. Alkenyl fluorides as hydrolytically stable bioisosteres of amides.

Synthesis of Alkenyl Fluorides from Alkynes. Gold(I)–fluorine complexes undergo reversible carbon–fluorine bond formation in the presence of excess alkyne, and protodeauration with triflic acid liberates the alkenyl fluoride product.[48] Gold-catalyzed *trans*-hydrofluorination of alkynes can be achieved with triethylamine trihydrofluoride (as a source of hydrogen fluoride) and PhNMe$_2$•HOTf (as a dichloromethane-soluble acid co-catalyst). For unsymmetrical alkyne substrates, synthetically useful regioselectivities ($\geq 5:1$) can be obtained with alkynes containing both alkyl and aryl substituents. Directing groups dramatically improve the regio-selectivities for unsymmetrically substituted alkynes (Scheme 36).[47] In this case, the formation of a favorable gold chelate determines the regiochemical outcome of the hydrofluorination reaction (see Scheme 9).

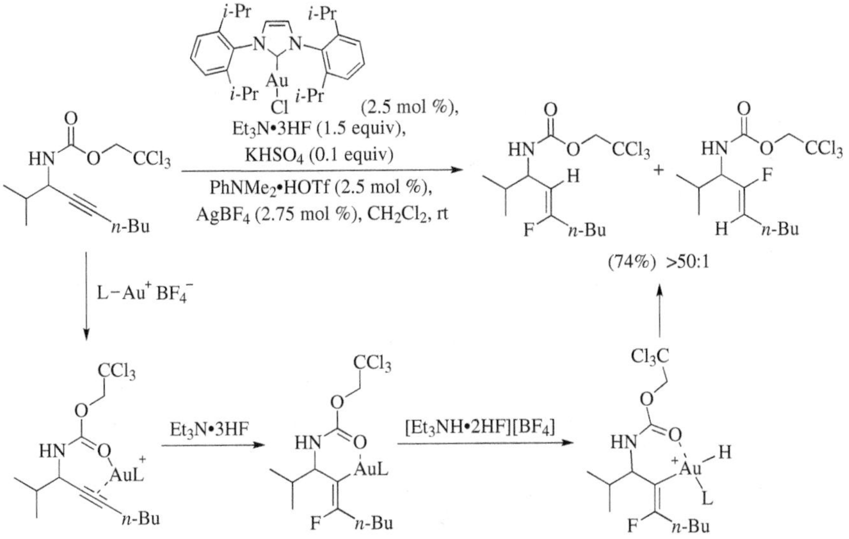

Scheme 36

In addition to gold(I)–fluorine complexes, gold(I) bifluoride complexes have been investigated as catalysts for the hydrofluorination of alkynes (Scheme 37).[146] Several ligated gold(I) hydroxides furnish gold(I) bifluorides upon treatment with triethylamine trihydrofluoride, wherein the liberation of water is proposed to favor the formation of the latter. Gold(I) bifluorides are air- and moisture-stable and can be considered anhydrous sources of fluoride. The hydrofluorination of alkynyl sulfides has also been reported, and the resulting products are promising biomimetic surrogates for enol(ates) of thioesters.[146]

The use of DMPU•HF instead of triethylamine trihydrofluoride leads to fewer side reactions with cationic metal complexes because of the increased acidity and the lower metal-coordinating ability of the base. The reagent is readily prepared by mixing DMPU with an aqueous solution of hydrofluoric acid (65 wt %) in an 11.8:1 molar ratio.[147] DMPU is a weaker base but a stronger hydrogen-bond acceptor than the more commonly used bases, triethylamine and pyridine, which makes it a more versatile reagent. For example, DMPU•HF enables the gold(I)-catalyzed hydrofluo-rination of terminal alkynes and removes the necessity for an acid additive.

Synthesis of Alkenyl Fluorides from Alkenylstannanes and Alkenylboronic Acids.
The regioselective and stereospecific fluorination of alkenylstannanes with a

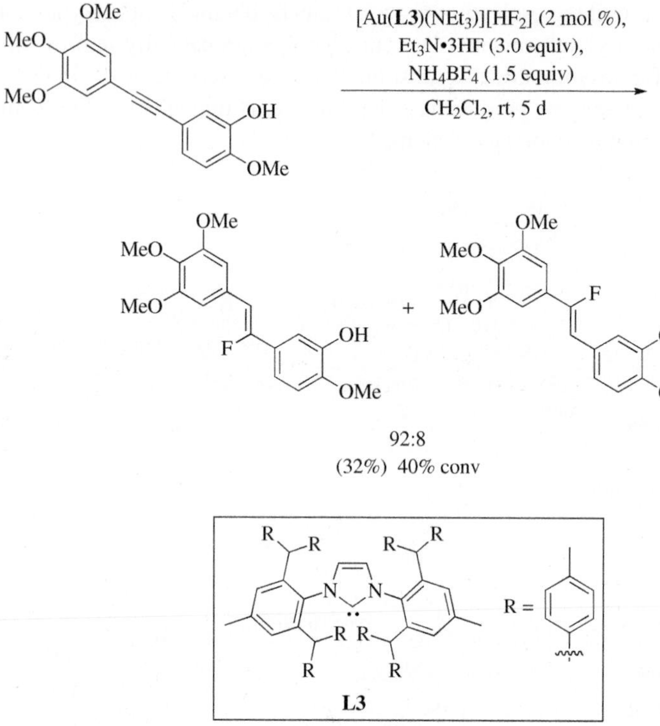

Scheme 37

silver catalyst was initially performed using a combination of xenon difluoride and silver hexafluorophosphate.[148-150] Later studies demonstrated that silver triflate was superior by promoting faster transmetalation (Scheme 38).[150] Interestingly, silver triflate is thought to react with xenon difluoride to form silver(I) fluoride and a more reactive source of fluorine, either trifluoromethanesulfonyl hypofluorite or fluoroxenon triflate.

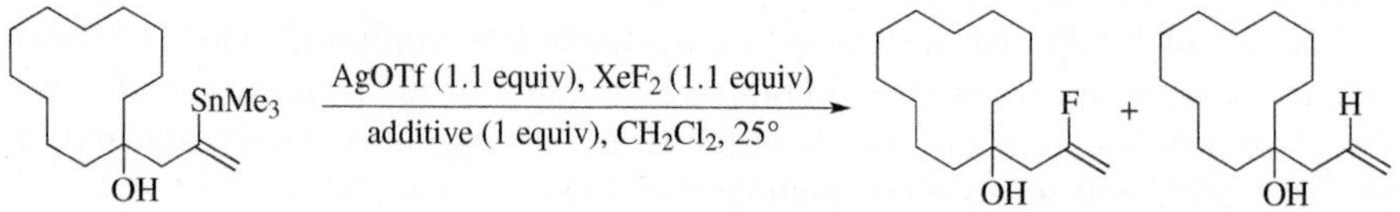

Additive	Yield (%)	
—	55	45
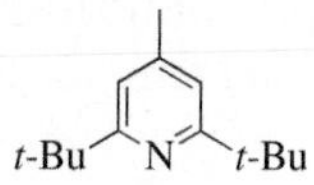 *t*-Bu pyridine *t*-Bu	74	19

Scheme 38

Silver-mediated fluorination of alkenylstannanes in the presence of the milder electrophilic fluorinating reagent F-TEDA-PF$_6$ proceeds at room temperature (Scheme 39).[145] With the exception of an alkenylstannane derived from a symmetrical alkyne, all substrates feature an allylic alcohol or tosyl-protected allylic amine group to promote high regioselectivity in the initial hydrostannylation reaction used to prepare the starting materials.

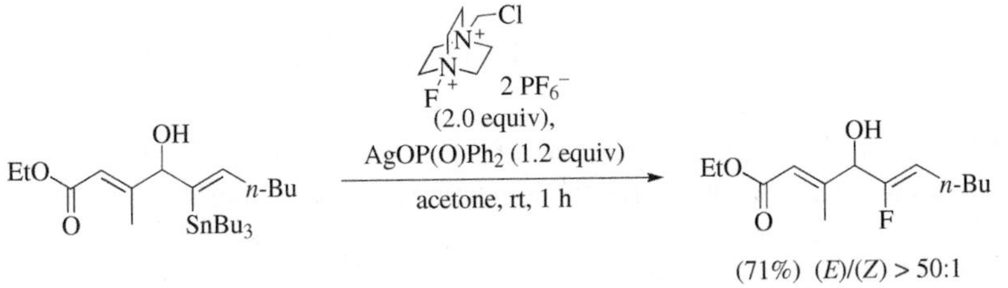

Scheme 39

Treatment of an alkenyl trifluoroborate with Selectfluor affords the corresponding fluoroalkene, typically as a 1:1 mixture of (*E*)- and (*Z*)-isomers.[151] Notably, the addition of a silver triflate to the two-step, one-pot reaction sequence promotes fluorination of alkenylboronic acids with retention of the alkene geometry (Scheme 40).[132] Selective formation of the (*E*)-isomer from the (*E*)-alkenylboronic acid is consistent with transmetalation to silver, followed by oxidation of silver and stereospecific carbon–fluorine reductive elimination.

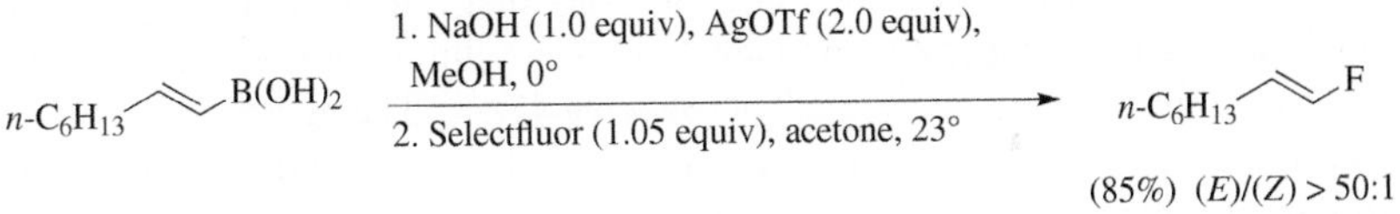

Scheme 40

Synthesis of Alkenyl Fluorides from Alkenes. Aromatic and olefinic C–H fluorination reactions can be accomplished with *O*-methyl oxime ether directing groups (Scheme 41).[125] Based on mechanistic studies, a Pd(II)/Pd(IV) catalytic cycle is proposed in which a nitrate promoter plays a crucial role. Similar results

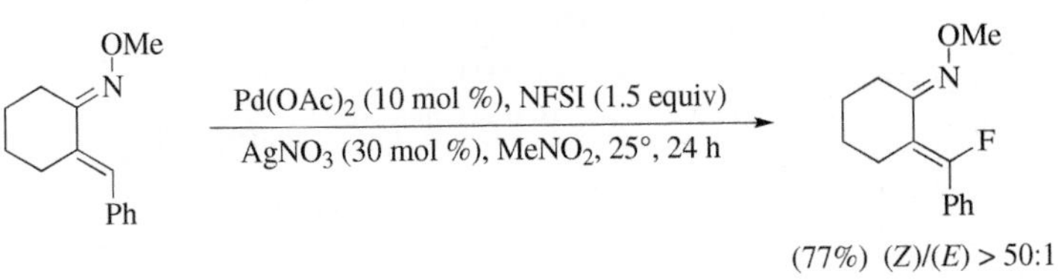

Scheme 41

are obtained when palladium(II) nitrate is used instead of a combination of the palladium pre-catalyst and a nitrate salt (silver nitrate or potassium nitrate), which is consistent with a cationic $[Pd(NO_3)]^+$ species effecting the C–H activation step. The nitrate-promoted transformation constitutes the first example of chelation-assisted olefinic $C(sp^2)$–H bond fluorination. This method permits the fluorination of acyclic oxime esters, including five-, six-, seven-, and eight-membered cyclic O-methyl oxime ethers, which are fluorinated in 65–78% yields.

Aliphatic Fluorination

The use of transition metals in the synthesis of aliphatic fluorides includes metal catalysis of traditional reaction pathways such as S_N2 reactions of alkyl halides and pseudohalides. Many methods for the preparation of alkyl fluorides rely on radical intermediates, and electrophilic fluorinating reagents can act as a source of the fluorine radical (e.g., Selectfluor, NFSI, and transition-metal fluorides) to facilitate the development of novel transformations. Methods that form carbon–carbon or carbon–heteroatom bonds in conjunction with the synthesis of an aliphatic fluoride or in which fluorination terminates or initiates a cyclization or rearrangement are beyond the scope of this chapter.[14,15] The synthesis of alkyl fluorides from isolated transition-metal compounds has often been studied in mechanistic investigations into carbon–fluorine reductive elimination from transition metals.[113,152–155]

Alkyl Fluoride Synthesis from Alkenes. Free-radical hydrofluorination reactions of unactivated alkenes result in exclusive Markovnikov addition and typically rely on either HF/pyridine or HF/amine as the fluorinating agent. However, these transformations require harsh reaction conditions that are deleterious to substrate stability and reaction scope. Alternatively, an electrophilic fluorinating reagent can be used as a source of the fluorine radical (Scheme 42).[156] The hydrogen and fluorine atoms originate from sodium borohydride and Selectfluor, respectively, and $Fe_2(ox)_3$ is proposed to mediate the transformation via the formation of Fe-H intermediates. The hydrofluorination is relatively insensitive to the nature of the iron(III) source, but a large excess of reagents is required, as is high dilution. Hydrofluorination by non-diastereospecific addition across the double bond is observed for many substrates, albeit styrenes furnish benzylic alcohols and dimers.

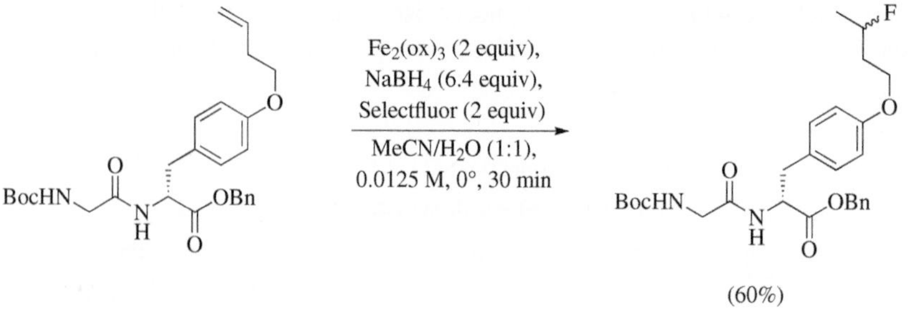

$Fe_2(ox)_3$ (2 equiv),
$NaBH_4$ (6.4 equiv),
Selectfluor (2 equiv)
────────────────
MeCN/H_2O (1:1),
0.0125 M, 0°, 30 min

(60%)

Scheme 42

Exclusive Markovnikov selectivity is obtained in a cobalt-catalyzed hydrofluorination reaction (Scheme 43),[157] albeit hydration is a problematic side reaction if the reaction mixture is not degassed. A variety of functional groups (aryl bromides, nitro groups, thiophenes, and hydroxyl groups) are well tolerated, whereas amino groups, carboxylic acids, and alkynes are not suitable for this process. Alkene isomerization reduces efficiency in some cases, although *gem*-disubstituted alkenes afford reasonable yields. The proposed mechanism proceeds by cobalt–hydrogen addition across the alkene bond, followed by alkyl radical release from the cobalt–alkyl intermediate and subsequent fluorination of the radical by the electrophilic fluorinating reagent.

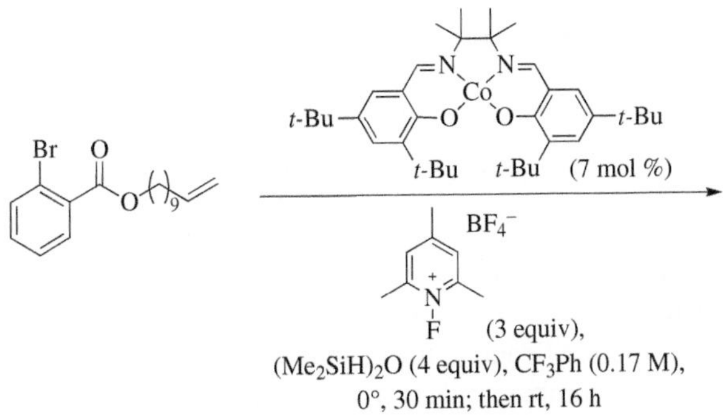

Scheme 43

Alkyl Fluoride Synthesis from Alkyl Bromides and Alkyl Triflates. Alkyl fluorides can be accessed from alkyl halide and pseudohalide precursors in the absence of a catalyst. Displacement reactions using simple halide salts often result in low conversion, which can be improved by including alcohols,[158] or iconic liquids[159,160] as (co)solvents. Elimination often competes with nucleophilic displacement when activated substrates, soluble fluoride sources, and forcing reaction conditions are employed, leading to mixtures of alkenes and alkyl fluorides that are challenging to separate.[161,162] However, the copper-catalyzed fluorination of alkyl triflates is complete within an hour (Scheme 44).[49] Common side reactions, such as elimination, are circumvented by employing mild reaction conditions and the low basicity fluorinating reagent, IPrCuF (IPr = 1,3-bis(2,6-diisopropylphenyl)imidazol-2-ylidene), which is prepared in situ by treatment of IPrCuOTf with potassium fluoride (see Scheme 10). The σ-donating ability of the IPr ligand contributes significantly to the nucleophilicity of the copper–fluoride bond.

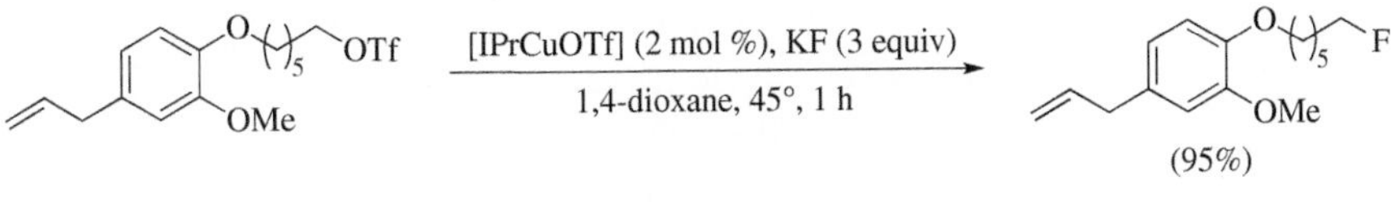

Scheme 44

Superstoichiometric amounts of copper fluorination reagents are generally required for efficient syntheses of alkyl fluorides from alkyl bromides (Scheme 45).[50] Notably, the S_N2-type displacement of alkyl bromides may also be accomplished in the presence of unprotected alcohols and aromatic bromides. Alkene byproducts are observed with secondary alkyl bromide substrates, but primary benzyl bromides are readily converted to the corresponding fluorides.

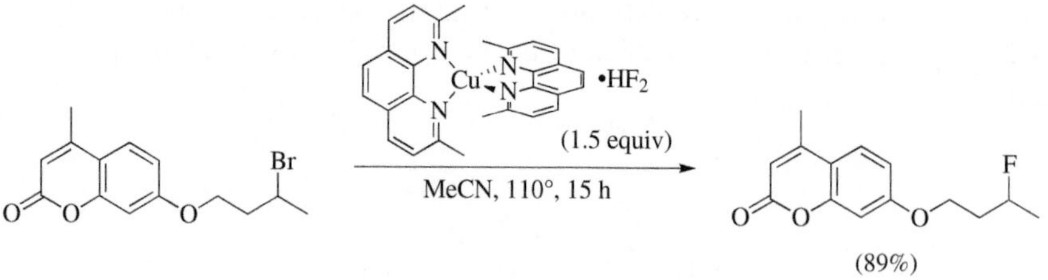

Scheme 45

Alkyl Fluoride Synthesis from Alkanes. Hydroxylation of C–H bonds can be achieved with high-valent metal oxo complexes via a radical-rebound mechanism. Rapid displacement of the hydroxide ligand in the manganese(IV) hydroxyl intermediate by fluoride can redirect the biomimetic C–H hydroxylation mechanism to facilitate the coupling of alkyl radicals with metal-bound fluorides (Scheme 46; see also Scheme 7).[44,163] Conversions are typically around 70%, but the reaction also affords alcohols and ketones (15–20%), so that alkyl fluoride products are commonly isolated in ~50% yield. A 2:1 ratio of oxygenated and fluorinated products is formed in the absence of TBAF, whereas no fluorination is observed in the absence of silver(I) fluoride. Positions adjacent to electron-withdrawing substituents are unreactive towards fluorination. Interestingly, the reaction preferentially fluorinates methylenes in preference to methines because steric repulsion between the substrate and the porphyrin in the transition state renders the cleavage of the weaker tertiary carbon–hydrogen bonds unfavorable. The efficient overlap between the degenerate Mn–O π^* LUMO

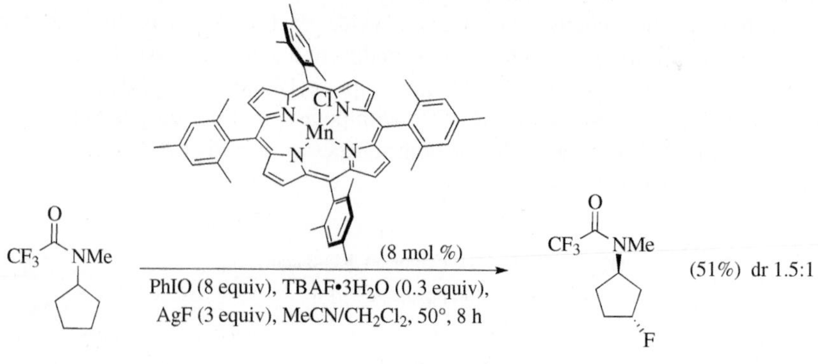

Scheme 46

orbitals and the substrate C–H bond σ-orbital requires a bent π* approach trajectory, which necessitates a close approach of the substrate to the porphyrin ligand (see Scheme 7).[44]

Copper(I) iodide catalyzes the fluorination of adamantane with Selectfluor, and the efficiency is improved by the addition of $KB(C_6F_5)_4$ and a diimine copper ligand to render the copper salt more soluble. The substrate scope of the transformation can be expanded by using the co-catalyst **5**, which forms an *O*-centered radical in the presence of redox-active metals, and potassium iodide, which promotes the formation of a cuprate complex (Scheme 47).[164] Longer reaction times decrease the yield of the fluorinated product, and the observation of acetamide products (from incorporation of acetonitrile) suggests that the fluorinated product undergoes S_N1 solvolysis. Solvolysis of the alkyl fluoride is particularly problematic with tertiary substrates, owing to the stability of the carbocation.[164,165] Allylic and benzylic fluorinations can be performed with the same catalytic system.[164] A radical mechanism is postulated for the copper-catalyzed fluorination because (1) higher yields are observed in the absence of molecular oxygen, (2) there is minimal interference by acetonitrile, and (3) only trace amounts of product are observed in the presence of TEMPO (Scheme 47).[165]

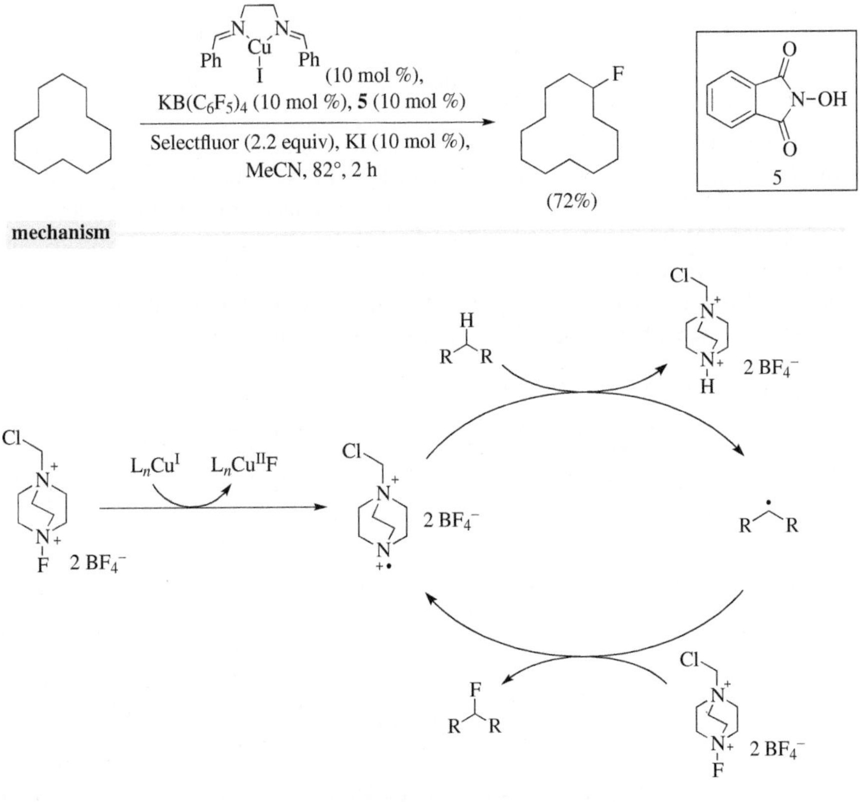

mechanism

Scheme 47

The surprising preference for monofluorination, however, is attributed to the partial ionic character of the transition state, which is also consistent with the deactivation conferred by adjacent acetate groups.

Uranyl nitrate is a hydrogen abstraction catalyst that can mediate the fluorination of certain alkanes upon irradiation with visible light (Scheme 48).[166] The highly oxidizing excited state $[UO_2]^{2+*}$ of depleted uranium-238 can be accessed using blue light ($\lambda = 450–495$ nm); the +2.6 V reduction potential of $[UO_2]^{2+*}$ renders it sufficiently reactive to abstract a hydrogen atom from unactivated carbon–hydrogen bonds.[166,167] While cyclooctane is almost quantitatively fluorinated, cyclohexane and cyclopentane furnish yields of 42% and 32%, respectively, and straight-chain alkanes afford a mixture of isomers that are monofluorinated in non-terminal positions (55–60% yields). Carbonyl groups quench the uranyl excited state by a reversible, inner-sphere electron-transfer mechanism (exciplex decay). In an interesting competition experiment, when an equimolar mixture of cyclopentanone and cyclooctane is subjected to the reaction conditions, fluorocyclooctane is formed in 74% yield, and only traces of fluorinated cyclopentanone are observed.[166]

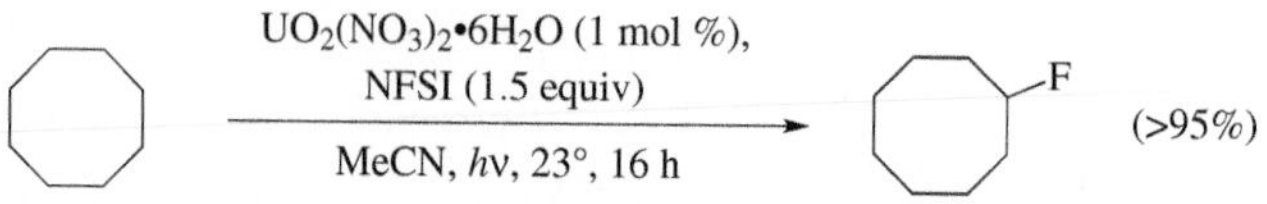

$$UO_2(NO_3)_2 \cdot 6H_2O \ (1 \ mol \ \%),$$
$$NFSI \ (1.5 \ equiv)$$
$$MeCN, \ h\nu, \ 23°, \ 16 \ h$$

(>95%)

Scheme 48

Vanadium(III) oxide catalyzes the fluorination of aliphatic C–H bonds, generally affording a mixture of mono- and difluorinated products (~10:1). Notably, the fluorination of 1-adamantanol occurs only at tertiary positions in good-to-moderate yields (Scheme 49),[168] whereas there is no fluorination alpha to carbonyl substituents, and benzylic positions are fluorinated in only moderate yields. Although the reaction is highly sensitive to molecular oxygen, the product purification is relatively simple: catalyst and reagent byproducts can be removed by filtration.

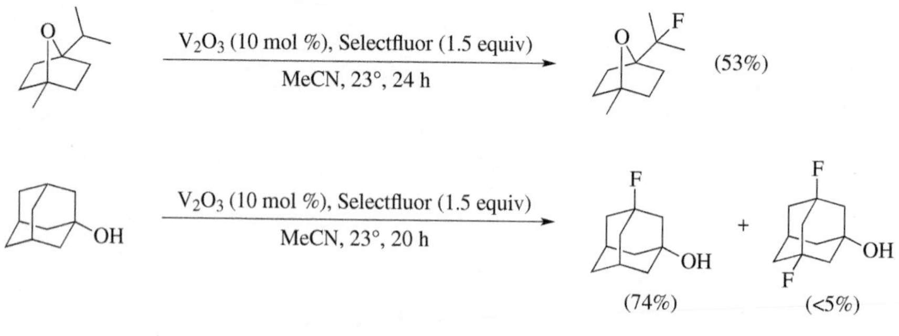

V_2O_3 (10 mol %), Selectfluor (1.5 equiv)

MeCN, 23°, 24 h

(53%)

V_2O_3 (10 mol %), Selectfluor (1.5 equiv)

MeCN, 23°, 20 h

(74%)   (<5%)

Scheme 49

The palladium-catalyzed, directed aliphatic and benzylic C–H fluorination reaction is proposed to proceed via a palladium(IV) intermediate in which both a fluoride and sulfonamide fragment (originating from the fluorinating agent NFSI) are bound to the palladium center (Scheme 50; see also Scheme 74).[169] The addition of one equivalent of pivalic acid increases the yield of the desired product by favoring carbon–fluorine (rather than carbon–nitrogen) bond reductive elimination from the palladium(IV) intermediate.[152,169] Secondary aliphatic and benzylic positions undergo successful fluorination, whereas primary aliphatic carbon–hydrogen bonds are not good substrates. For instance, there is only a single reported example of the fluorination at a primary position (33% yield). The presence of a directing group is required for efficient fluorination to occur under the conditions depicted in Scheme 50, albeit aqueous hydrobromic acid can be used to readily remove the 8-aminoquinoline group to furnish the fluorinated carboxylic acid.

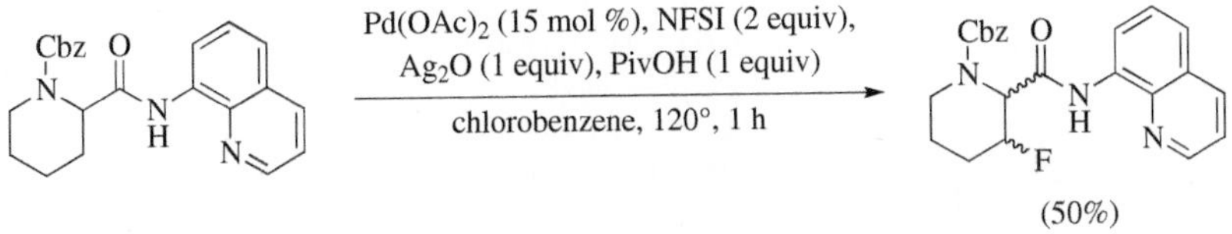

Scheme 50

Several unnatural, enantiopure, fluorinated α-amino acids can be obtained by directed C(sp³)–H activation followed by stereoselective fluorination (Scheme 51), which is proposed to proceed via an El-R-sub-F mechanism (see Scheme 12).[170] Selectfluor is employed because of its lower propensity to interfere with the C–H palladation step compared with pyridine-containing fluorinating reagents. A stereogenic center is constructed during the C(sp³)–H activation step and the diastereoselectivity of this reaction relies on the facile formation of *trans*-substituted five-membered palladacycle intermediates.

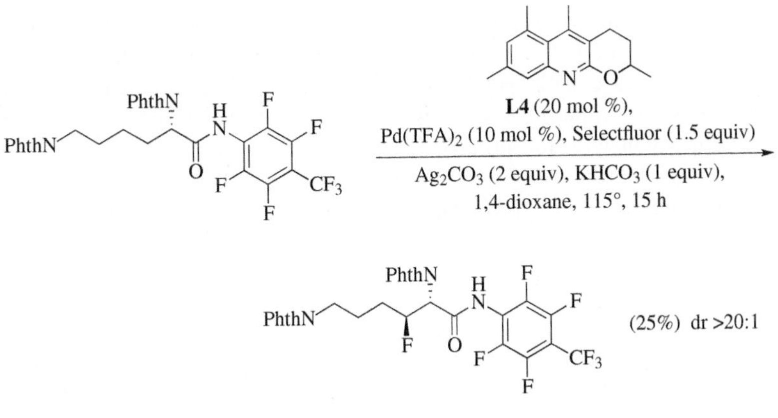

Scheme 51

The palladium-catalyzed fluorination of aliphatic or benzylic C–H bonds located beta to an amide carbonyl group by Selectfluor gives significantly higher yields in the presence of either a manganese(II) acetate or an iron(II) acetate cocatalyst.[171] The cocatalyst is proposed to aid in releasing the fluorinated product from the palladium(II) catalyst by providing an additional Lewis acidic binding site. No epimerization of α-stereogenic centers is observed, and a preference for the formation of a five-membered-ring palladacycle ensures high site-selectivity. In a concurrent study, the analogous palladium-catalyzed, directed C–H fluorination is accomplished using a different cocatalyst at lower reaction temperatures (Scheme 52).[172] Removal of the 2-(pyridine-2-yl)isopropyl amine (PIP) directing group is readily accomplished by an in situ esterification of the highly electrophilic pyridinium triflate intermediate, which forms upon treatment of the fluorination product with triflic anhydride.

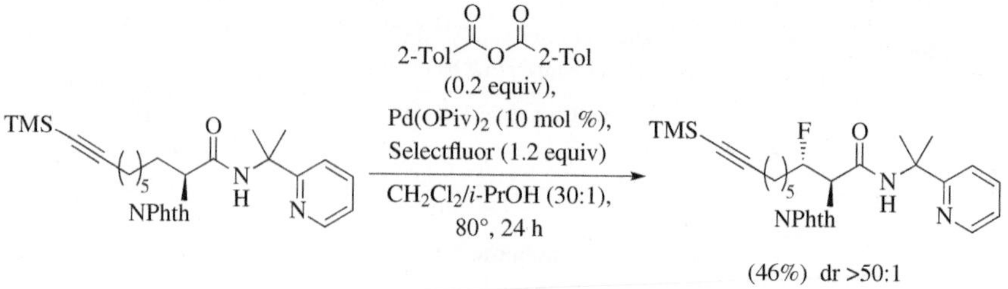

Scheme 52

Alkyl Fluoride Synthesis from Alkyl Carboxylic Acids. The first catalytic, decarboxylative fluorination that employed a nucleophilic fluoride source (triethylamine trihydrofluoride) utilized a manganese catalyst and the iodine(III) carboxylate that is formed in situ from iodosylbenzene and the carboxylic acid substrate (Scheme 53).[173] Alkenes and carbon–hydrogen bonds that participate in common metalloporphyrin-catalyzed oxidative reactions are unreactive under these conditions. In contrast to Selectfluor-based methods, substrates that contain electron-rich aromatic groups are fluorinated without competing aryl-ring fluorination. The ease with which carboxylic acids undergo manganese-catalyzed decarboxylative fluorination scales with the C–COOH bond dissociation energies, which is consistent with the proposed radical mechanism. Namely, the process is proposed to involve the transfer of a manganese-bound fluorine to a carbon-based radical intermediate. This protocol can also be used for ^{18}F-radiofluorination and affords radiochemical conversions above 50%.

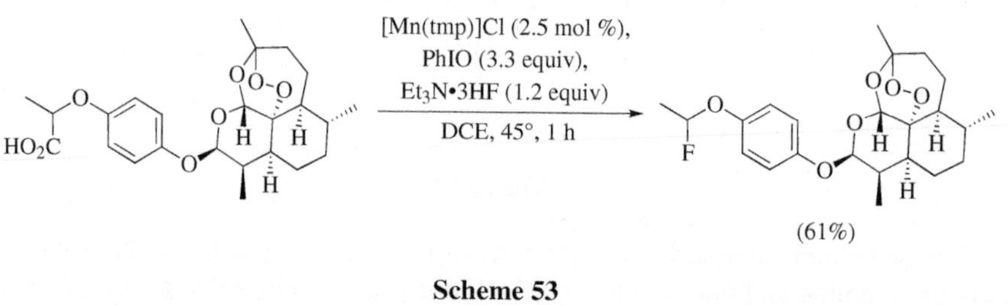

Scheme 53

A photoexcited iridium complex can be used in conjunction with Select-fluor to mediate the fluorodecarboxylation of a carboxylic acid (Scheme 54).[174] Single-electron transfer from the transiently formed iridium(IV) species to an alkyl carboxylate occurs with the extrusion of carbon dioxide to afford an alkyl radical. The reaction with Selectfluor furnishes the product along with Selectfluor radical cation, which is capable of oxidizing Ir(III)* to complete the catalytic cycle. Primary, secondary, and tertiary carboxylic acids are efficiently transformed into alkyl fluorides, and importantly, β-fluoro carbonyl derivatives do not undergo fluoride elimination under the basic reaction conditions.

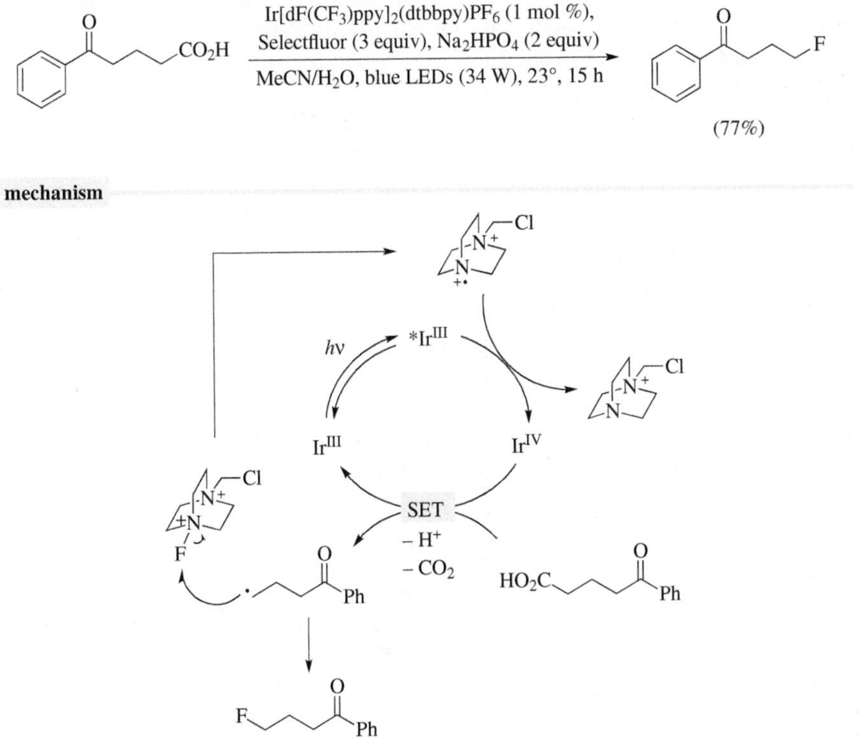

mechanism

Scheme 54

Common silver(I) salts, namely $AgBF_4$, AgOAc, AgOTf, and $AgNO_3$, have comparable efficiency in the silver-catalyzed fluorodecarboxylation reaction of alkyl carboxylic acids with Selectfluor (Scheme 55).[58] Water appears to be essential to the reaction, as no conversion is observed in several anhydrous solvents. The reactivity of the carboxylic acids is directly proportional to the stability of the resulting radical, with tertiary carboxylates being the most reactive. The reaction can be conducted in water for substrates prone to polyfluorination, such as adamantane carboxylate: the low solubility of the alkyl fluoride products reduces any further fluorination.

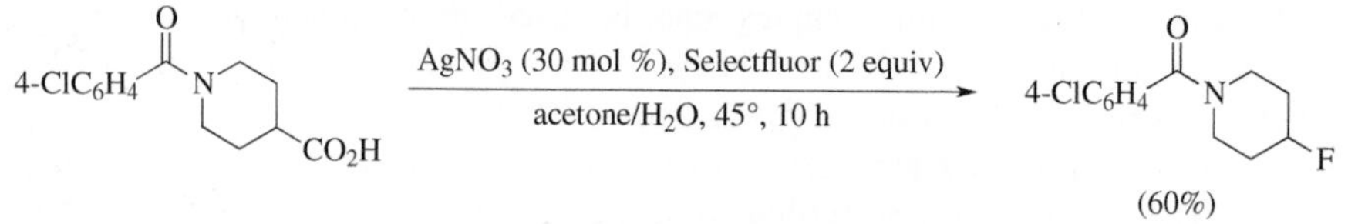

Scheme 55

Shorter reaction times can be achieved for the silver-catalyzed fluorination of carboxylic acids by the addition of sodium persulfate because the rate-determining oxidation of silver(I) carboxylate to the silver(II)–carboxylate complex proceeds more efficiently with sodium persulfate than with Selectfluor.[60] A specialized set of reaction conditions has been developed to improve the yields for the silver-catalyzed fluoro–decarboxylation to make fluorinated γ-butyrolactones, which are present in many biologically active building blocks.[57]

The first example of photoredox catalysis for the formation of carbon–fluorine bonds was reported in 2014 (Scheme 56).[175] Consequently, the key to success is selecting substrates that have an oxidation potential sufficiently high that Selectfluor does not directly oxidize them, but low enough to be susceptible to SET from the triplet metal-to-ligand charge transfer (^{3}MLCT) state of Ru(bpy)$_3{}^{2+}$. The reduction of Selectfluor by SET from the excited photocatalyst has been established by transition absorption spectroscopy, which indicates the growth of a new absorption band centered at 450 nm in the presence of Selectfluor with a concomitant decrease in the intensity in the band attributed to the photocatalyst. However, no change in the excited state difference spectrum between the photocatalyst and a mixture of the substrate and photocatalyst is observed. The authors suggest that Selectfluor radical cation (formed via photoexcitation) oxidizes the substrate, and then decarboxylation and fluorination occur.

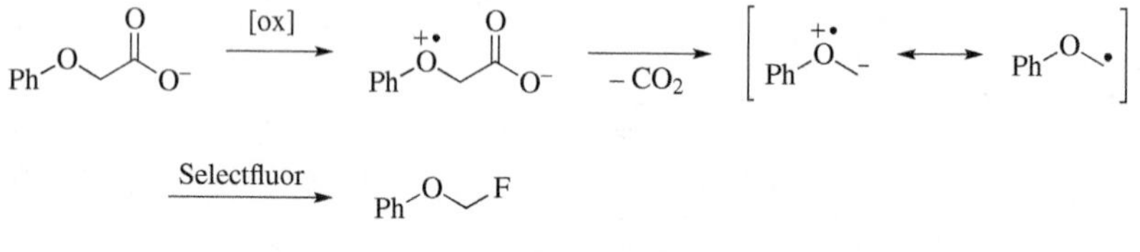

Scheme 56

Alkyl Fluoride Synthesis through Ring-Opening of Epoxides and Aziridines. The enantioselective ring-opening of *meso* epoxides and terminal epoxides with fluoride can be achieved using a two-catalyst system comprising an amine (e.g., **L5**) and a cobalt Lewis acid (Scheme 57).[46] The combination of hydrofluoric acid with the chiral Lewis acids leads to competitive background reactions and catalyst inhibition. Hence, these conditions employ a latent hydrofluoric acid source. Namely, benzoyl fluoride and 1,1,1,3,3,3-hexafluoro-2-propanol (HFIP) are used in conjunction with the chiral amine **L5** to generate hydrogen fluoride in situ and mitigate the aforementioned problems. Nevertheless, the reaction is slow and requires 72–120 hours to reach completion, despite high catalyst loadings.

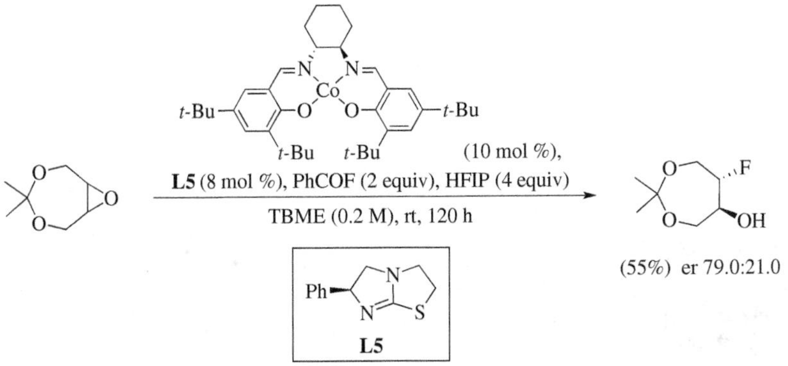

Scheme 57

Hydrolytic kinetic resolution of terminal epoxides can be performed with the same cobalt catalyst with DBN, and under these conditions, the reaction exhibits a k_{rel} of up to 300.[45,46] Kinetic studies of the ring-opening fluorination of *meso* epoxides reveal an apparent first-order dependence on the cobalt catalyst. However, substituent effects, non-linear effects, and studies using a catalyst consisting of two molecules of cobalt–salen linked by a connector provides evidence for a rate-limiting bimetallic ring-opening step (see Scheme 8).[45] The linked cobalt–salen complex affords faster reaction rates, higher turnover numbers, and improved substrate scope than the monomeric catalyst.

The combination of two Lewis acids—a chiral cobalt–salen complex and an achiral titanium(IV) co-catalyst—catalyzes the enantioselective ring-opening of aziridines to furnish *trans*-β-fluoro amines with a latent source of hydrofluoric acid (Scheme 58).[176] A chelating aziridine protecting group is critical for obtaining excellent enantioselectivity (92:8 er) in the ring-opening fluorination of *meso* aziridines. Kinetic resolution of piperidine-derived aziridines can be performed with $k_{rel} = 6.6$. One enantiomer of the β-fluoroamine can thus be obtained from an unsymmetrically substituted racemic aziridine in 41% yield with 81.0:19.0 er, while the unconsumed starting aziridine is left enantiomerically enriched. The reaction proceeds by a bimetallic mechanism, in which the chiral cobalt–salen catalyst binds the nucleophile, and the titanium co-catalyst activates the aziridine.

Alkyl Fluoride Synthesis from Alkylstannanes and Alkylboronic Acid Derivatives. Primary, secondary, and tertiary alkylboronic acid derivatives can be fluorinated with Selectfluor in a site-specific manner by a radical mechanism (Scheme 59).[56] No reaction occurs in the absence of silver(I) nitrate or if arylboronic acids are used as substrates. The reaction is chemoselective, in that carboxylic acid functional groups are tolerated[56] despite being substrates for the previously reported silver-catalyzed fluorodecarboxylation reactions.[57,58,60,173]

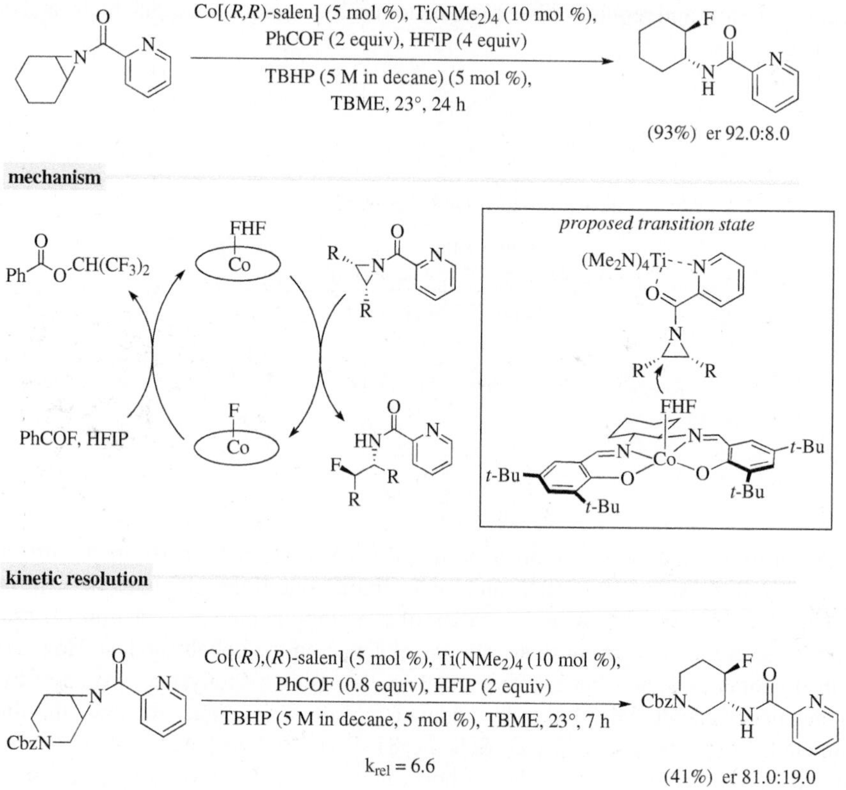

(93%) er 92.0:8.0

mechanism

kinetic resolution

$k_{rel} = 6.6$

(41%) er 81.0:19.0

Scheme 58

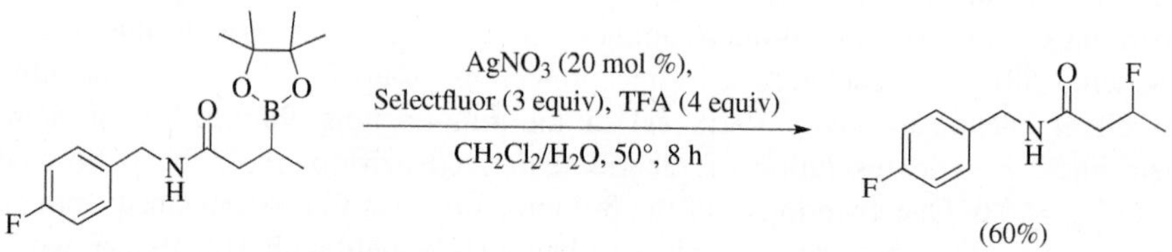

(60%)

Scheme 59

Allylic Fluorination

The deployment of fluoride as a nucleophile in allylic displacements is complicated by (1) the low nucleophilicity of fluoride in coordinating solvents, (2) the high basicity of the anion in non-coordinating solvents, and (3) the potential reversibility of the allylic displacement. The rate of leaving-group displacement for palladium- and iridium-catalyzed allylic substitutions follows the trend: methyl carbonate > fluoride > acetate, whereas the fluoride is more reactive than a methyl carbonate in the presence of platinum catalysts.[38,39]

Synthesis of Allylic Fluorides from Allylic Halides. Attempts to prepare allylic fluorides by nucleophilic attack on π-allyl palladium intermediates predominantly lead to elimination, generating *trans*-diene products.[177,178] Interestingly, the inclusion of silver(I) fluoride facilitates the enantioselective palladium-catalyzed allylic fluorination (Scheme 60), which is driven by the precipitation of silver(I) chloride to promote carbon–fluorine bond formation.[177] Notably, under these conditions, the diene product is formed in less than 10% yield. Overall retention of configuration is observed for the displacement of chloride by fluoride, indicating that the reaction occurs by an outer-sphere nucleophilic attack of the fluoride ion on the π-allyl intermediate.

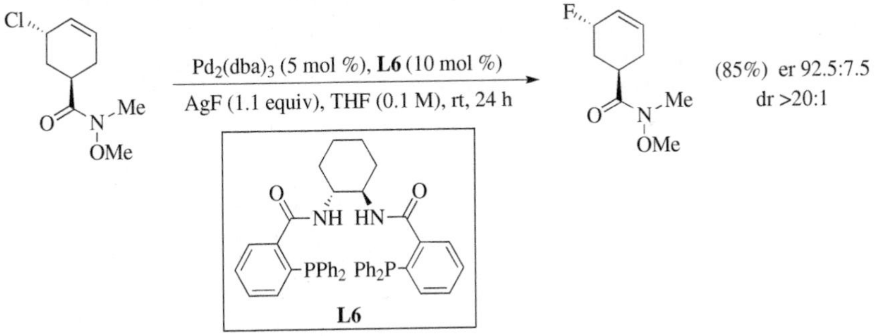

Scheme 60

The synthesis of enantioenriched branched allylic fluorides from linear allylic chlorides and bromides can be achieved using ligand **L7** with a palladium catalyst (Scheme 61).[41] Higher branched/linear (b/l) selectivities are obtained for phosphine ligands with large bite angles, which is commonly the case for allylic displacement reactions that proceed by an outer-sphere attack.[41,179] Mechanistic studies suggest that this fluorination proceeds by a homobimetallic process, wherein a palladium-bound fluoride attacks a π-allyl palladium complex[24] (see Scheme 1). Nonpolar solvents decrease the rate of the unselective background reactions and favor hydrogen-bonding between ligand **L7** and the nucleophile. Free hydroxyl

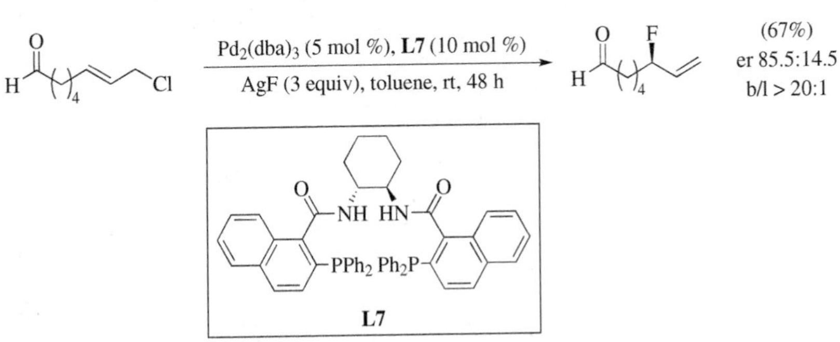

Scheme 61

groups are tolerated, but competitive etherification is observed, and secondary alkyl amines are not compatible because of facile N-alkylation.

Copper catalysis can improve the yield, diastereomeric ratio, and regioselectivity of the fluorination of allyl chlorides and bromides with silver(I) fluoride.[42] However, the reaction requires a heteroatom-containing substituent in the allylic or homoallylic position capable of ligating the copper catalyst (see Scheme 5).

Synthesis of Allylic Fluorides from Allylic Alcohol and Allylic Thiol Derivatives. Preliminary attempts to facilitate the palladium-catalyzed displacement of activated esters using TBAF led to the formation of allylic alcohols, whereas TBAF•(t-BuOH)$_4$ generates the desired allylic fluoride product (Scheme 62).[38] Improved yields are obtained using an activated ester rather than a methyl carbonate leaving group. Palladium-catalyzed allylic fluorination has been successfully adapted for the use of ^{18}F-labeled allylic fluorides using ^{18}F-TBAF in acetonitrile.

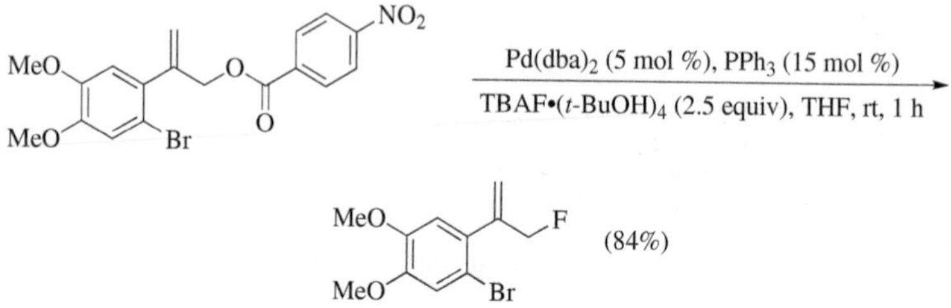

Scheme 62

An unusual conservation of regio- and stereochemistry is observed in the preparation of branched and linear (E)- and (Z)-allyl fluorides from the corresponding allyl carbonates using an iridium catalyst (Scheme 63).[39] The fluoride

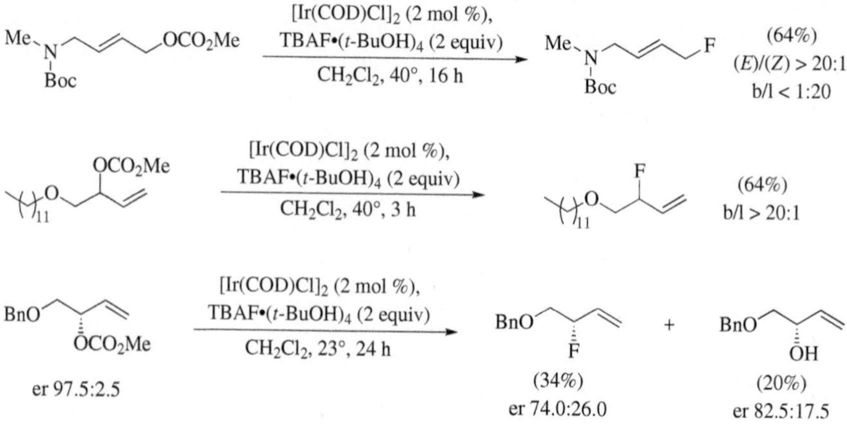

Scheme 63

substituent is introduced in a regiospecific manner, so that branched products are formed from branched starting materials, and linear carbonates furnish linear allyl fluorides. ^{18}O-Labeling studies reveal the intermediacy of an unsymmetrical (η^1-allyl)iridium rather than a fully developed (η^3-allyl)iridium. The rates of displacement for the carbonate starting material increase in the following order: branched > (Z)-linear ≈ (E)-linear. Adaptation to ^{18}F chemistry is made possible using either K[18]F•4,7,13,16,21,24-hexaoxa-1,10-diazabicyclo[8.8.8]hexacosane (K[18]F•Cryptand 222) or ^{18}F-Et$_4$NF, with the latter resulting in more reproducible results.

The iridium-catalyzed allylic substitution of trichloroacetimidates with fluoride affords high yields of branched allylic fluorides (Scheme 64).[180] No linear product is detected, and both silyl ether groups and terminal alkynes are compatible with these reaction conditions. Tertiary acetimidates react to form the corresponding tertiary allylic fluorides, which are unstable. In such cases, the alkene must be reduced in situ prior to isolating the product. The reaction can be adapted for the synthesis of ^{18}F-allylic fluorides using ^{18}F-KF and Cryptand 222 in the presence of camphorsulfonic acid (CSA).

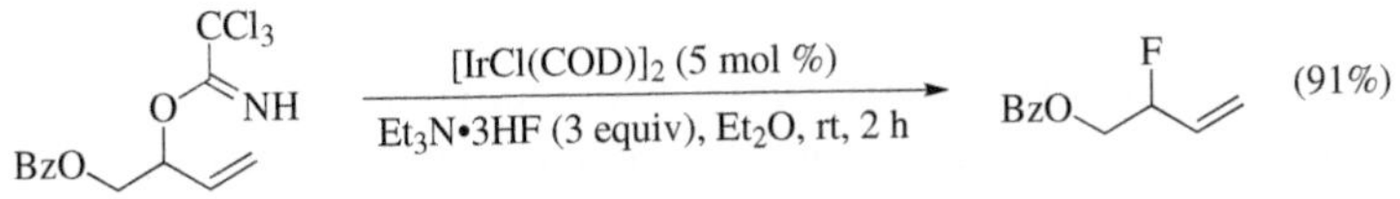

Scheme 64

The enantioselective fluorination of racemic, branched trichloroacetimidates can be performed with an iridium complex and chiral diene ligand (S,S)-**L8** (Scheme 65).[40] The bulky diene ligand decelerates the rate of fluoride attack, thereby rendering it slow compared to the rate of equilibration of the two diastereomeric π-allyl iridium complexes, which facilitates a dynamic, kinetic transformation that is both efficient and stereoselective (>99:1 dr).

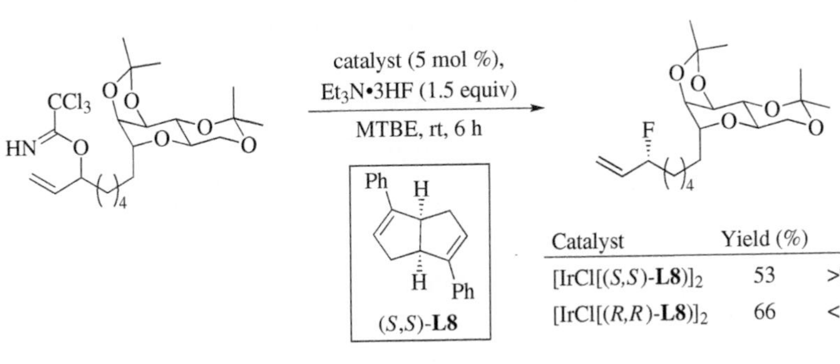

Catalyst	Yield (%)	dr
[IrCl[(S,S)-**L8**]]$_2$	53	>99:1
[IrCl[(R,R)-**L8**]]$_2$	66	<1:99

Scheme 65

The S$_N$2′ nucleophilic displacement of bridgehead leaving groups takes place in the presence of a Josiphos–rhodium catalyst to afford 1,2-*trans*-fluorohydrin

products as single constitutional and diastereoisomers (Scheme 66).[181] Nevertheless, bridgehead-substituted substrates, azabicyclic alkenes, and non-benzofused
substrates are unreactive under these conditions. A site-selective, rhodium-catalyzed
ring-opening of vinyl epoxides can be used to prepare branched allylic fluorohydrins
at room temperature.[182]

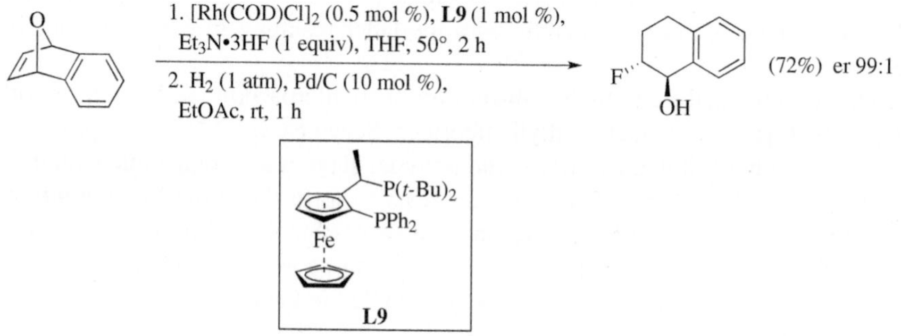

Scheme 66

In contrast to some allylic halides, allylic phosphorothioates are stable to moisture,
air, and silica gel chromatography, and can be readily transformed into allylic fluorides in the presence of a palladium catalyst and silver(I) fluoride (Scheme 67).[183]
Rapid π-σ-π rearrangement results in the same allylic fluoride being formed from
two constitutionally isomeric starting materials (wherein the leaving group can be
located at either allylic site). An interaction between the silver salt and substrate can
be observed by ^{19}F and ^{31}P NMR, and may facilitate the ionization of the phospho-
rothioate ester.

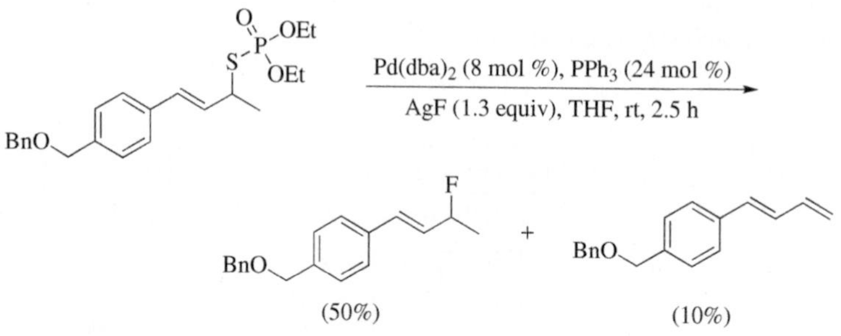

Scheme 67

Allylic Fluoride Synthesis from Alkenes. Allylic C–H fluorination can be
accomplished with good selectivity for the formation of branched allylic fluorides
(Scheme 68).[184] The bimetallic palladium–chromium catalyst system is derived
from a palladium(II)–sulfoxide complex that was developed for allylic C–H

functionalization.[185] Since nucleophiles that contain acidic X–H bonds are the most successful for allylic functionalization with the palladium(II)–sulfoxide catalyst, triethylamine trihydrofluoride is used as the fluorination reagent. In contrast to many other nucleophilic fluorinating reagents, triethylamine trihydrofluoride is not a strong Lewis base, making it less likely to interfere with C–H activation and lead to the formation of undesired elimination products. This fluorination reaction can be conducted in air using non-anhydrous solvents, and pyridines, phenols, and secondary amides are suitable substrates, resulting in complete conversion and minimal diene formation.

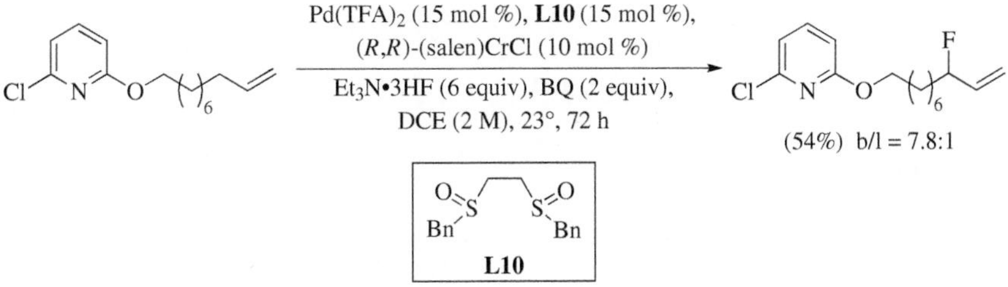

Scheme 68

Benzylic Fluorination

The motivation for the development of benzylic fluorination is the ongoing desire to increase in vivo drug lifetimes by generating compounds that are less vulnerable to benzylic oxidation by P450 enzymes.[186,187] An early example of transition-metal-catalyzed fluorination employs a 16-electron ruthenium complex to permit the fluorination of benzylic bromides and iodides using thallium fluoride.[51] Stringent exclusion of water and Teflon-lined reaction vessels are required, and the reaction exhibits poor functional-group tolerance. Interestingly, the iridium-based fluorination reagent Cp*(PMe₃)Ir(Ph)F facilitates the formation of benzyl fluoride and difluoromethane from benzyl bromide and dichloromethane, respectively.[188] The first, catalytic, enantioselective hydrofluorination reaction utilizes a chiral palladium catalyst with xenon difluoride (Scheme 69).[189] The formation of the (S)-enantiomer is consistent with the intermediacy of an (η^3-benzyl)palladium(II) complex undergoing oxidation to afford a palladium(IV) fluoride intermediate, which then generates the enantioenriched carbon–fluorine bond (with retention of configuration) via reductive elimination.[153]

The preparation of racemic benzylic fluorides by the hydrofluorination of styrenes is readily accomplished with catalytic tetrakis(triphenylphosphine)palladium(0), triethylsilane, and Selectfluor, which is a milder fluorinating reagent compared to xenon difluoride (Scheme 70).[189] The reaction proceeds by a *syn*-specific sequential addition of hydride and fluorine; for example, (E)- and (Z)-2-benzylidene-3-methylbutan-1-ol afford the *anti* and *syn* benzylic fluoride products, respectively.

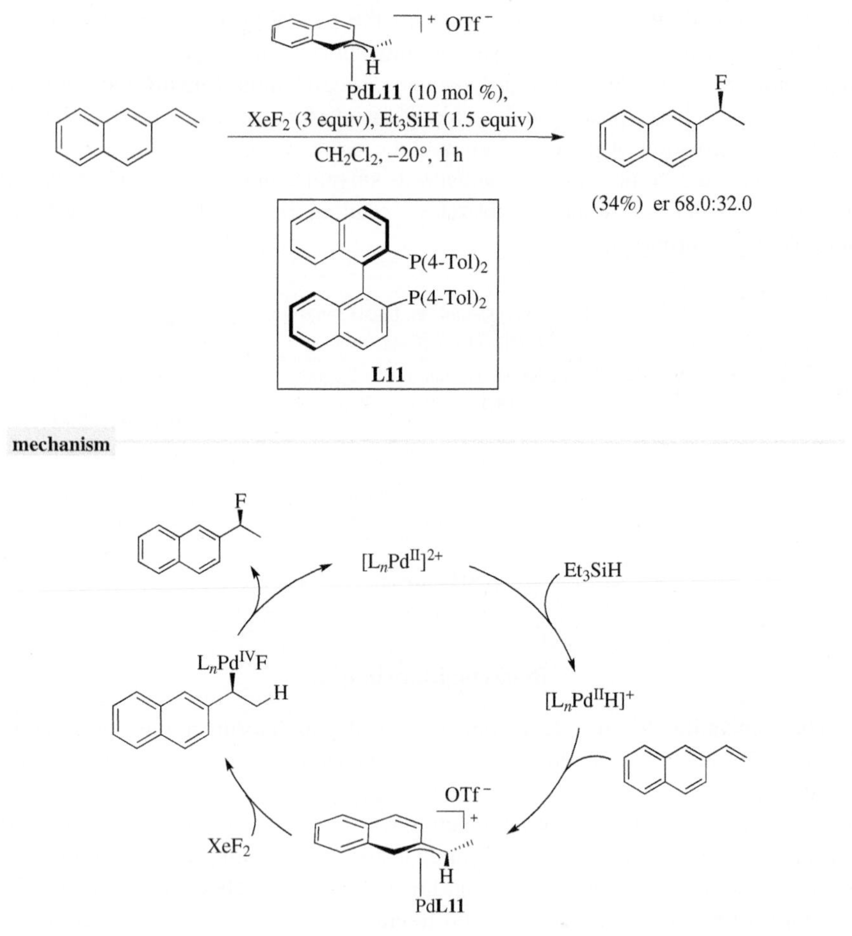

(34%) er 68.0:32.0

mechanism

Scheme 69

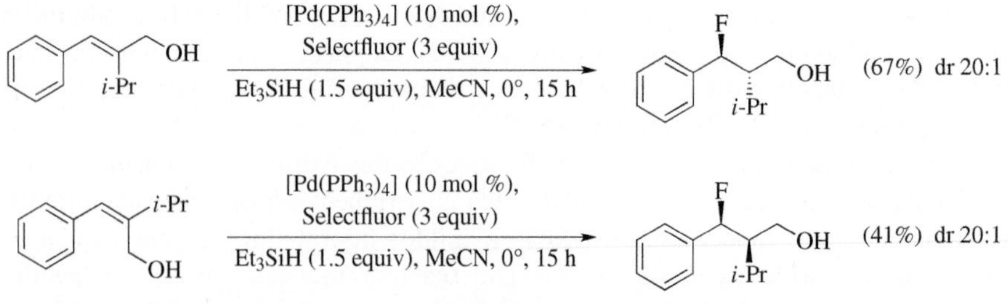

(67%) dr 20:1

(41%) dr 20:1

Scheme 70

Pyridine and quinoline directing groups facilitate the palladium-catalyzed fluorination of methyl substituents on (adjacent) aromatic systems with N-fluoropyridinium reagents. The reaction requires forcing reaction conditions and is only successful with relatively simple pyridine- and quinoline-based aryl methyl groups (Scheme 71).[105]

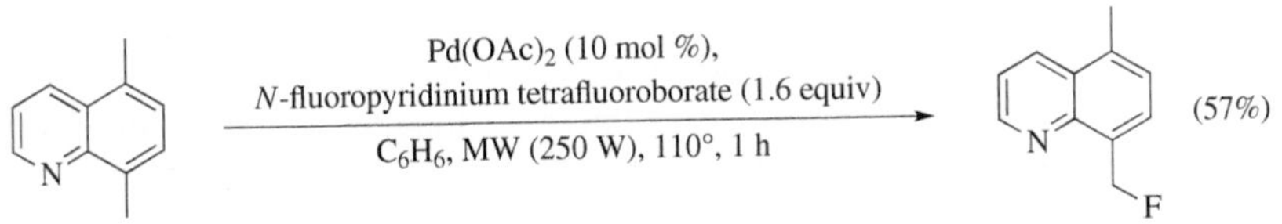

Pd(OAc)$_2$ (10 mol %),
N-fluoropyridinium tetrafluoroborate (1.6 equiv)
C$_6$H$_6$, MW (250 W), 110°, 1 h

(57%)

Scheme 71

The N-fluoropyridinium reagent serves as the source of fluorine, but it is also critical for oxidizing palladium to a higher oxidation state, which enables carbon–fluorine bond formation via a more facile reductive elimination. A later variant of the directed, palladium-catalyzed, benzylic C–H fluorination instead utilizes a combination of PhI(OPiv)$_2$, which oxidizes palladium, and silver(I) fluoride, which serves as a nucleophilic fluorinating agent.[190]

Iron(II) acetylacetonate provides high yields in iron-catalyzed benzylic fluorinations with Selectfluor.[186] Notably, the hard, polydentate O-donor ligands facilitate access to the higher oxidation states of iron, and electron-donating groups on aromatic rings in the substrates lead to competitive aryl and benzylic fluorinations. Although treatment with Selectfluor generally results in fluorination alpha to the carbonyl group, the addition of the iron catalyst results in fluorination at the benzylic position (Scheme 72).[191] The iron- and copper-catalyzed benzylic fluorination reactions are complementary; steric constraints have a more significant influence on selectivity in the former, and radical stability is more important for the latter. The fluorination of cymene demonstrates these divergent outcomes: iron catalysis permits fluorination of the benzylic methyl group, and copper catalysis affords the tertiary benzylic fluoride.[164,186]

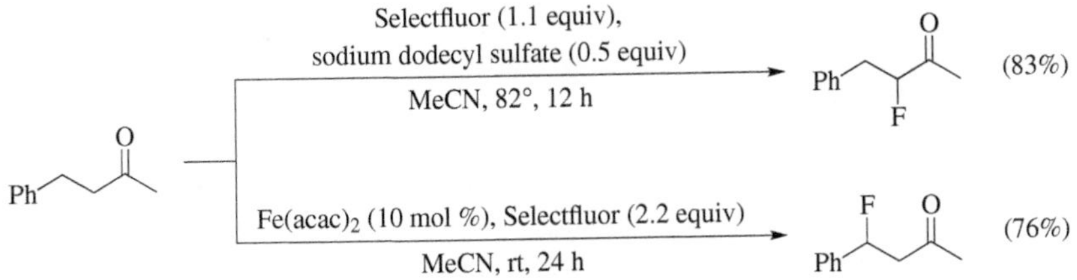

Selectfluor (1.1 equiv),
sodium dodecyl sulfate (0.5 equiv)
MeCN, 82°, 12 h

(83%)

Fe(acac)$_2$ (10 mol %), Selectfluor (2.2 equiv)
MeCN, rt, 24 h

(76%)

Scheme 72

Oxomanganese(V) intermediates, formed from a chiral manganese–salen catalyst and iodosobenzene, can generate benzylic radicals by C–H abstraction.[192] In analogy to the manganese-catalyzed fluorination of aliphatic C–H bonds (see Schemes 7 and 46), radical rebound occurs after anion metathesis (from hydroxide to fluoride) at the manganese(IV) center to furnish benzylic fluorides. Low levels of enantioinduction are observed, which is likely because the transition state for the fluorine transfer step is very early, and it occurs with a linear Mn–F–C geometry. The high enantiomeric ratios observed for alkene epoxidation by manganese–salen complexes are attributed to a side-on approach of the substrate π-bond to the manganyl oxo group of the catalyst, thereby creating steric constraints during oxygen atom transfer.[193] In the fluorination reaction, the reduced steric hindrance inherent to the linear transition-state geometry is thought to reduce the interaction with the stereogenic centers on the ligand environment, resulting in poor enantioinduction.[192]

Benzylic C–H fluorination in the presence of decatungstanate ion **6** is proposed to involve C–H abstraction by a photoexcited decatungstanate followed by fluorine transfer from NFSI (Scheme 73).[187] The formation of an acetamide side product from the acid-catalyzed displacement of the benzylic fluoride by acetonitrile can be suppressed by the addition of either sodium bicarbonate or lithium carbonate. The low solubility of NFSI at room temperature leads to the equilibration of the benzylic radicals in substrates with more than one benzylic position, and different selectivity is observed for specific substrates using AIBN as the radical initiator at 75°.

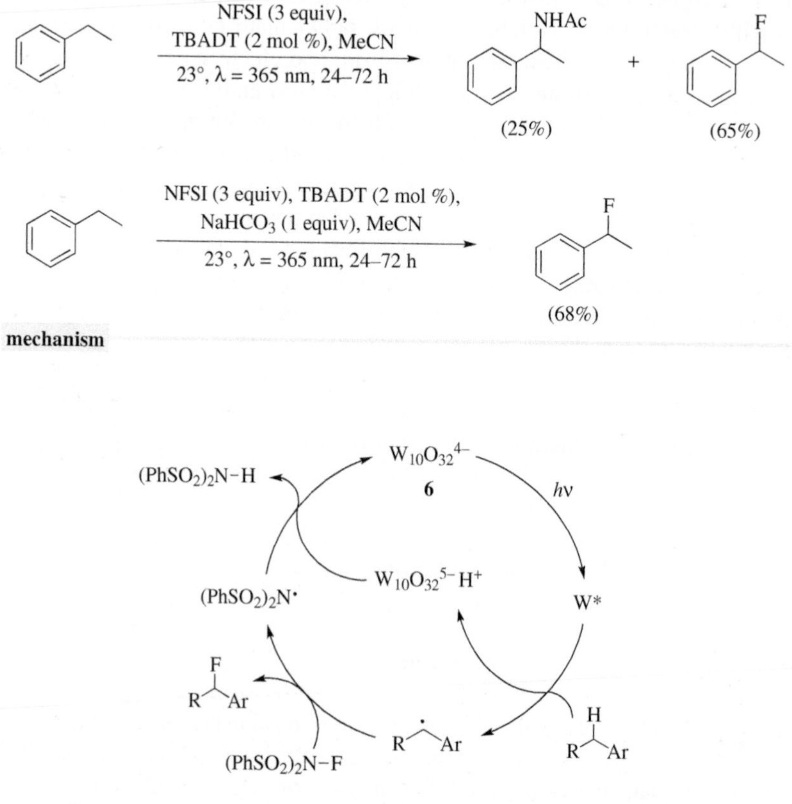

mechanism

Scheme 73

For example, 4-ethyltoluene undergoes selective fluorination of the benzylic ethyl and methyl groups with decatungstanate and AIBN, respectively.[187]

Several aliphatic fluorination methods are also useful for the preparation of benzylic fluorides; for example, secondary and tertiary benzylic fluorides can be prepared from the corresponding carboxylic acids in the presence of Selectfluor and a photocatalyst.[174] Manganese-catalyzed decarboxylation can also be used to promote a range of reactions, including benzylic fluorinations.[173]

Four mechanistically related palladium-catalyzed reactions can be employed for directed C–H benzylic fluorination (Scheme 74).[169–172] The nitrogen-containing directing group is proposed to facilitate C–H activation with the palladium(II) catalyst via a five-membered palladacycle. Oxidation of palladium(II) by either Selectfluor or NFSI affords a palladium(IV) fluoride intermediate, which undergoes reductive elimination to form a carbon–fluorine bond. With the exception of substrates relying on an 8-aminoquinoline-based directing group, the examples outlined in Scheme 74 all contain a phthalimide-protected nitrogen substituent alpha to the site of fluorination that directs the stereochemical outcome of the fluorination reaction. Products are obtained with modest-to-excellent diastereocontrol (up to >20:1 dr).[170–172] Notably, the new stereocenter is established during the C–H activation step and is controlled by the preferred formation of the *trans*-substituted palladacycle intermediate.

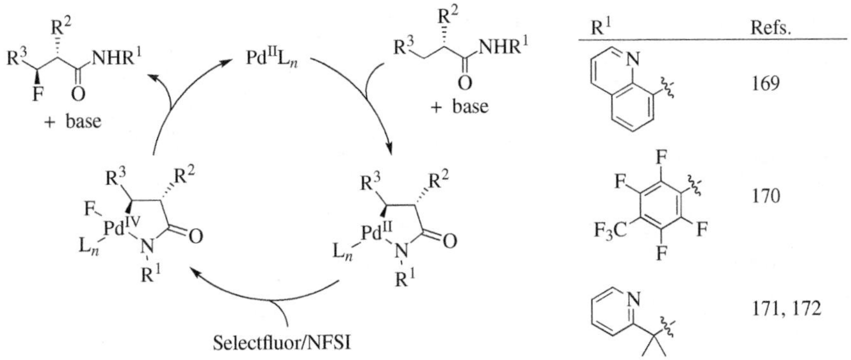

Scheme 74

Synthesis of α-Fluorinated Carbonyl Compounds

Although uncatalyzed fluorination alpha to an electron-withdrawing functional group can occur with electrophilic fluorinating reagents, transition metals are often employed in the development of regio-, diastereo-, and enantioselective fluorination reactions.

Synthesis of α-Fluorinated Ketones and Aldehydes. Ketones bearing similar substituents on either side of the carbonyl group can be challenging to fluorinate selectively. One of the few examples of a site-selective synthesis of α-fluoro ketones proceeds by the isomerization of allylic alcohols into enolates in the presence of [Cp*IrCl$_2$]$_2$, followed by electrophilic fluorination with Selectfluor (Scheme 75).[91,92] The allylic alcohol precursors allow base-sensitive electrophiles to be utilized and ensures site-selective enolate formation.

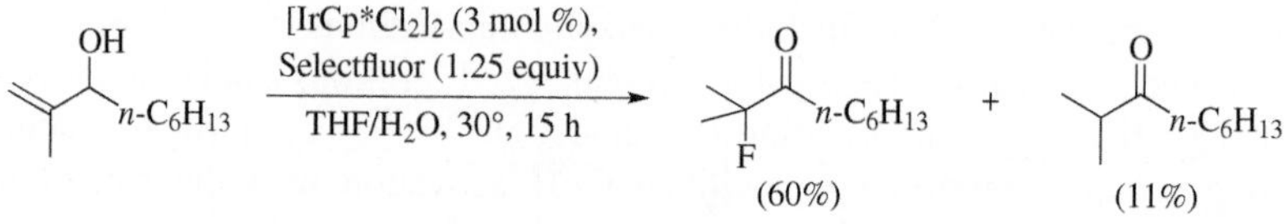

Scheme 75

Gold complexes permit the synthesis of α-fluoro acetals and α-fluoro ketones from both internal and terminal alkynes with a high degree of chemo- and site-selectivity (Scheme 76).[85] In line with the Dewar–Chatt–Duncanson model, the σ-bonding contribution is larger than the π-back-bonding contribution in the metal–alkyne bond of complexes formed from late transition metals with alkynes, thereby rendering the metal–alkyne bond susceptible to nucleophilic attack.[85,194] Unsymmetrical disubstituted alkynes are fluorinated in a highly site-selective manner without competitive fluorination of the ketone present in the product. Notably, ketones have been excluded

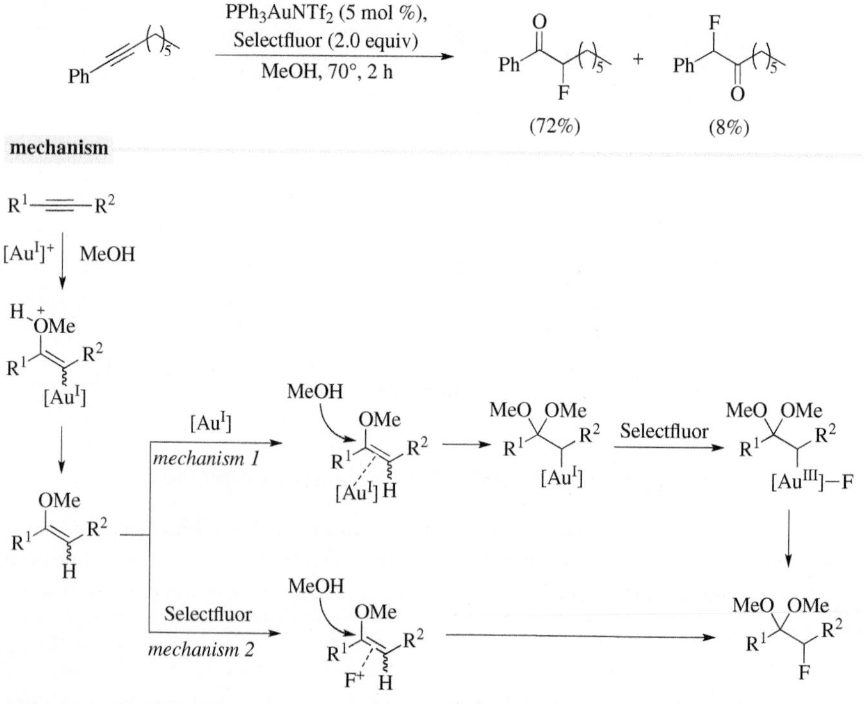

Scheme 76

as productive intermediates in the formation of α-fluoro acetals and α-fluoro ketones from alkynes, which has been rationalized by two mechanisms that involve either direct fluorination of the enol ether intermediate by Selectfluor or re-activation of the same intermediate by the gold catalyst and fluorination of the gold to facilitate reductive elimination.[85]

Allenyl carbinol esters can also serve as starting materials for the synthesis of α-fluoro ketones. The isomerization of the starting material under the influence of gold catalysis delivers (E)/(Z)-mixtures of dienes, which undergo fluorination with Selectfluor to furnish monofluoro-α,β-unsaturated ketones with exclusive (E)-selectivity.[86]

α-Fluorinated ketones can be formed by hydrogen fluoride insertion into a gold carbene intermediate generated by the gold-catalyzed addition of N-oxide **L12** to an alkyne (Scheme 77).[87] Internal alkynes react sluggishly in this reaction, and alkynes with nucleophilic substituents furnish side products from competitive addition of these substituents to the alkyne.

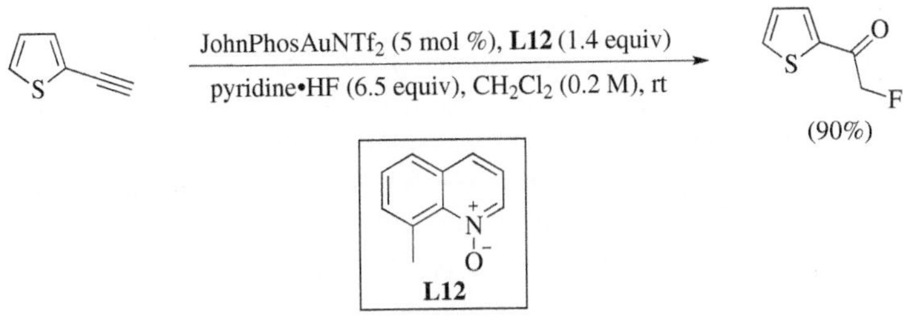

Scheme 77

The asymmetric construction of tertiary α-fluoro esters can be accomplished from ketenes using a chiral iron catalyst with enantiomeric ratios of up to 99.5:0.5 er (see Scheme 17).[84] Several aryl alkyl ketenes and dialkyl ketenes furnish good levels of enantioinduction, and generate a tertiary stereocenter adjacent to only one side of the activating carbonyl group. Ruthenium catalysis promotes a Meinwald rearrangement in phenyl-substituted epoxides to furnish 2-alkylphenylacetaldehydes, which are fluorinated at the α-position with AgHF$_2$.[43] Aldehydes bearing two different α-substituents can be directly transformed into the chiral α-fluorinated aldehydes (see Scheme 6), albeit with moderate enantioselectivities.

Synthesis of α-Fluorinated Amides. Nickel-catalyzed enantioselective α-fluorination proceeds by a 'dual-activation' strategy, in which a chiral nucleophile is combined with an achiral transition-metal-based Lewis acid catalyst to access a metal-coordinated, chiral ammonium enolate (Scheme 78).[77] The addition of a Lewis acidic lithium salt (e.g., lithium perchlorate) activates the fluorinating reagent and thereby improves the yields for aliphatic substrates. Different carbonyl derivatives are readily accessible depending on the nucleophile employed to quench the reaction.[76] Enantioinduction is controlled by the chiral Lewis base, benzoylquinidine

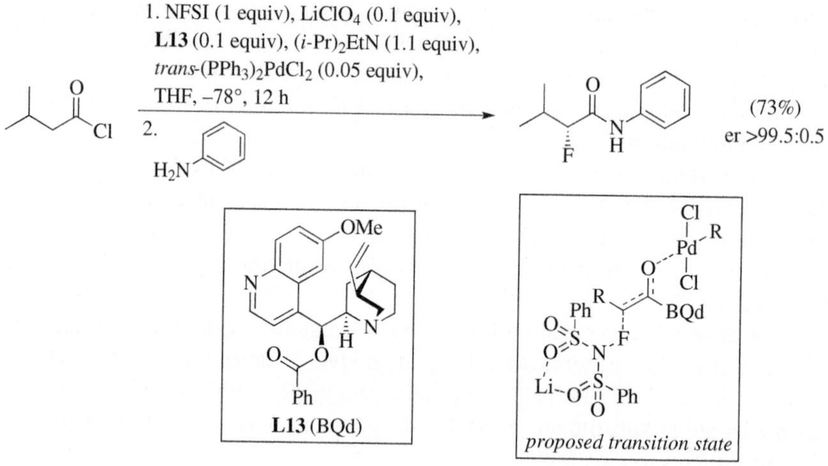

Scheme 78

(**L13**, BQd), and the other diastereoisomer of the product can be obtained using benzoylquinine, which is the pseudoenantiomer of **L13**.

The direct α-fluorination of *N*-acyloxazolidinones can be achieved using titanium enolates generated by soft enolization (see Scheme 15).[63] Alkenes are compatible with the reaction conditions, but substrates with terminal alkynes are less efficient. The competing α-chlorination observed for certain substrates is attributed to the reaction of NFSI with titanium(IV) chloride. This unproductive pathway can be mitigated by using $TiCl_2(Oi\text{-}Pr)_2 \cdot 2Et_3N \cdot HCl$.

The addition of catalytic quantities of HFIP is crucial for obtaining high levels of enantioinduction in the nickel(II)-catalyzed α-fluorination of arylacetate derivatives (Scheme 79).[95] HFIP is proposed to protonate the metal–oxygen bond, which releases the products from the catalytic center. Consequently, the reaction temperature can be reduced from 0° to –60°, which improves the enantiomeric ratios without impacting the overall efficiency.

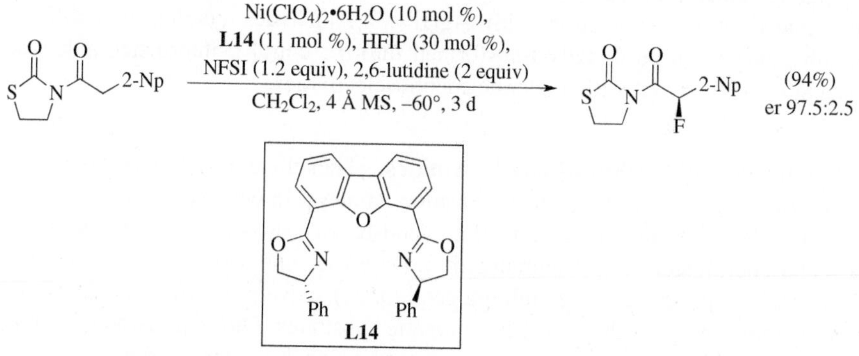

Scheme 79

Synthesis of α-Fluorinated β-Keto Esters. The first catalytic enantio-selective fluorination reaction was based on the observation that activated methylene groups react with electrophilic fluorinating reagents, and that sub-stoichiometric amounts of a chiral Lewis acid accelerate the enolization process.[67] Experimentally, titanium-based Lewis acids are the most effective catalysts, and enantiopure TADDOLate (TADDOL = α,α,α′,α′-tetraaryl-2,2-disubstituted 1,3-dioxolane-4,5-dimethanol) ligands furnish fluorinated products with enan-tiomeric purities that are comparable with stoichiometric methods.[67] The titanium–TADDOLate catalyst system facilitates the asymmetric α-fluorination of several different types of activated carbonyl substrates.[82] β-Keto esters with large benzyl ester groups undergo α-fluorination to provide products with enantiomeric ratios of 80.0:20.0 to 95.0:5.0, whereas phenyl ester substituents lead to enantiomeric ratios between 83.5:16.5 and 94.0:6.0. Double stereochemical differentiation of β-keto esters bearing chiral ester substituents increases diastereoselection and affords the products with diastereomeric ratios of up to 96.5:3.5. Crude reaction mixtures often contain a chlorinated side product, which is proposed to be formed by a radical mechanism. Fluorinating agents less reactive than Selectfluor increase the enantiomeric ratio for carbonyl substrates with large contributions of the enol tautomer. Notably, thiocarbonyl groups decompose under the reaction conditions, and β-keto amides are poor substrates, because of the resistance to enolization resulting from A^{1,3}-strain in the enol form. A transition-state model has been proposed, based on X-ray crystallographic studies and computational modeling, to rationalize the stereochemical outcome for this process. The shielding of the *Re*-face of the metal-ligated substrate complex by the aryl group of the catalyst directs the electrophilic fluorinating agent to the *Si*-face (see Scheme 16).[66,68,195]

The catalytic activity (and in some cases, the enantioselectivity) of the ruthenium-catalyzed, electrophilic fluorination of 1,3-dicarbonyl compounds is improved by the inclusion of a cosolvent that can accept a hydrogen bond, thereby facilitating the enolization process (Scheme 80).[89] Since the ruthenium-catalyzed reaction uses a softer Lewis acid and a milder electrophilic fluorinating agent, the reaction rates are lower than those of titanium-catalyzed fluorinations, albeit bulky substituents in the 2-position are better tolerated in the ruthenium-catalyzed process.

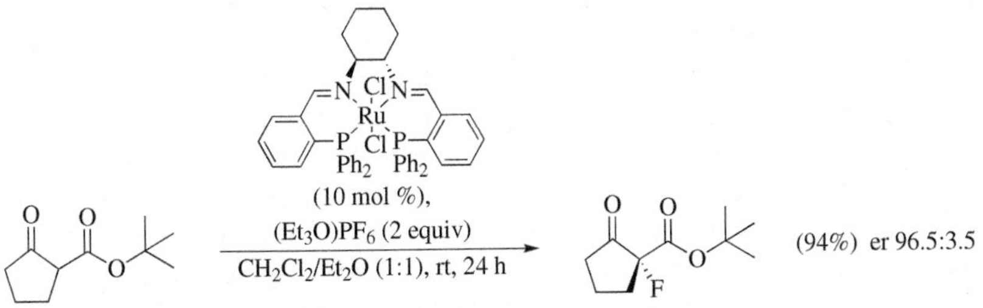

Scheme 80

Modest enantiomeric ratios are obtained when carbohydrate-substituted bipyridines are employed as chiral ligands in the copper(II)-catalyzed asymmetric fluorination of β-keto esters.[100] The low enantioselection with tartrate-derived bisoxazolines in the copper-catalyzed fluorination of aliphatic β-keto esters can be improved using a ligand that has an additional stereogenic center.[102] The fluorination of 2-substituted β-keto esters with elemental fluorine is catalyzed by Ni(NO₃)₂•6H₂O and racemic-BINAP.[90] Enantiopure BINAP does not provide any asymmetric induction, because elemental fluorine is likely too reactive to permit differentiation of the enantiotopic faces of the substrate. In the fluorination of β-keto esters, nickel halide salts with spirocyclic diamine ligands furnish enantiomeric ratios of up to 81.5:18.5 er.[98] Reactions in the absence of nickel are slow but proceed with the same level of enantioinduction, suggesting that the chiral ligand can act as an organocatalyst. Nickel-catalyzed enantioselective α-fluorination of α-chloro-β-keto esters in the presence of a different chiral diamine ligand affords α-chloro-α-fluoro-β-keto esters with enantioselectivities of up to 99.5:0.5 er.[96]

Using a cobalt catalyst to facilitate the α-fluorination of a β-keto ester (Scheme 81), both the yield and the enantiomeric excess increase at lower temperature.[88] Cyclic substrates bearing methyl, ethyl, and *tert*-butyl esters react efficiently, whereas the results with acyclic substrates are modest (64% yield and er 85.5:14.5).

Temp	Yield (%)	er
rt	41	70.5:29.5
0°	68	92.5:7.5
−20°	84	94.5:5.5

Scheme 81

The copper-catalyzed enantioselective fluorination of cyclic (67.5:32.5 to 92.5:7.5 er) and acyclic (up to 76.0:24.0 er) β-keto esters is readily accomplished by the combination of the chiral Cu(OTf)₂(bisoxazoline) with NFSI as the electrophilic fluorine source.[99] The enantiomeric excess is highly solvent-dependent, and the addition of one equivalent of HFIP is crucial for the release of the fluorinated product from the catalyst. Several metal salts have been examined for the α-fluorination of β-keto esters in the presence of a linked bis-thiazoline ligand. For instance, enantioselectivities of up to >99.5:0.5 er are observed using copper(II) triflate (Scheme 82).[104] Although acyclic substrates afford racemic products, tetralone and indanone carboxylates with fused aromatic rings react efficiently and provide products with excellent enantioselectivities. The transformation is proposed to proceed by bifunctional catalysis, with coordination to the copper cation activating the β-keto ester, and the N–H–O hydrogen bond and the π–π interaction directing NFSI attack to the *Si*-face of the substrate.

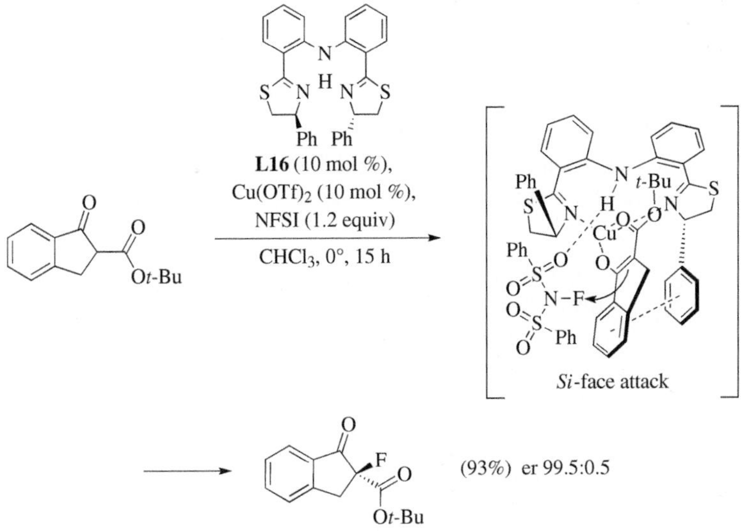

Scheme 82

A partially enantiomerically enriched nickel catalyst facilitates the asymmetric fluorination of β-keto esters with enantioselectivities of up to 99.5:0.5 er. This reaction is the first significant example of asymmetric amplification in enantioselective fluorination reactions, in which a catalyst with 72.5:17.5 er furnishes products with >95.0:5.0 er (Scheme 83).[93]

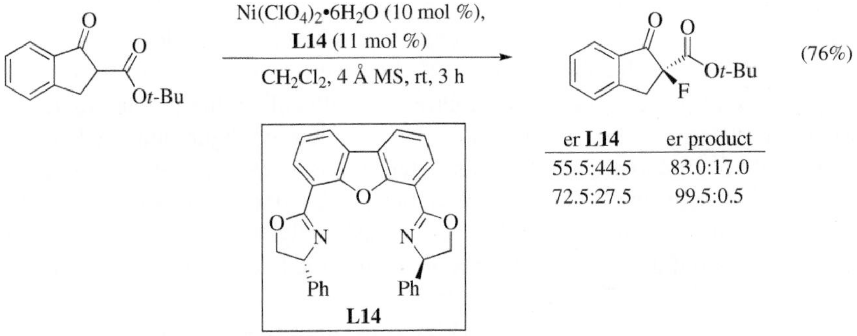

er **L14**	er product
55.5:44.5	83.0:17.0
72.5:27.5	99.5:0.5

Scheme 83

The enantioselective fluorination of α-keto esters can also be achieved with a mildly basic palladium–diphosphine catalyst system.[79] Interestingly, this process does not furnish difluorinated products and minimizes the amount of racemization, which is attributed to the mildly basic reaction conditions that disfavor the enolization of the monofluorinated products. Excellent enantioinduction is also obtained using a similar palladium–phosphine catalytic system in the fluorination of β-keto esters,[72] and a modified version of this catalyst effects the fluorination of the less acidic *tert*-butoxycarbonyl lactones and lactams.[69]

Synthesis of α-Fluorinated Oxindoles. The *N*-donor pincer ligand **L17** is used in nickel-catalyzed, enantioselective fluorinations of oxindoles and β-keto esters, furnishing products with enantiomeric ratios above 96.0:4.0 (Scheme 84).[97]

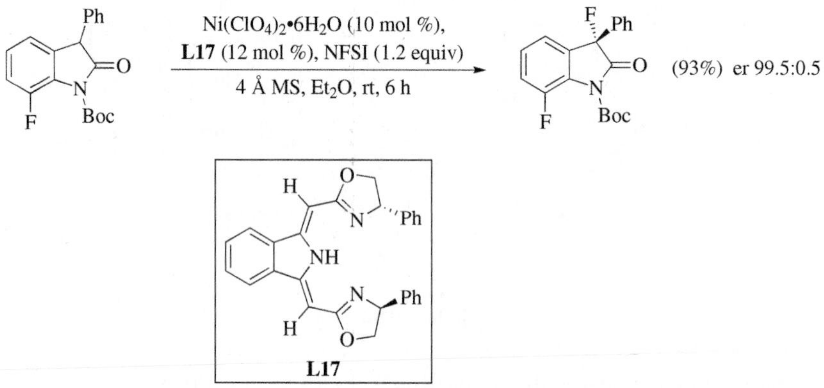

Scheme 84

The enantioselective α-fluorination of β-keto esters and *N*-Boc-oxindoles with a chiral iron(II) salen complex provides products with high enantiomeric purities.[83] In this case, molecular sieves and a bulky ester substituent are critical for obtaining high asymmetric induction.

The enantioselective fluorination of unprotected oxindoles is relatively slow with the (*S*)-DM-BINAP palladium complex **7**;[65] however, the protection of the oxindole nitrogen with a Boc group provides improved results, albeit the product undergoes partial deprotection during the reaction. The absolute configuration predicted for a chiral palladium enolate model is consistent with the experimentally determined product stereochemistry.[65] The highly enantioselective fluorination of the secondary benzylic site in an oxindole requires rapid methanolysis in order to prevent the racemization and difluorination of the more enolizable fluorinated oxindole product (Scheme 85).[65,75]

Moderate enantiomeric ratios are achieved for the palladium-catalyzed fluorination of oxindoles using Selectfluor, provided a Boc-protecting group is present on the oxindole nitrogen atom.[80]

Synthesis of α-Fluorinated β-Keto Phosphonates. Mono- and difluorinated phosphonates are used in drug design as phosphate mimics, which can be attributed to the second pK_a value (~6.5) being similar to that of biological phosphates. Enantioselective fluorination of β-keto phosphonates is catalyzed by a chiral palladium complex and affords enantiomeric ratios between 97.0:3.0 and 99.0:1.0 er for

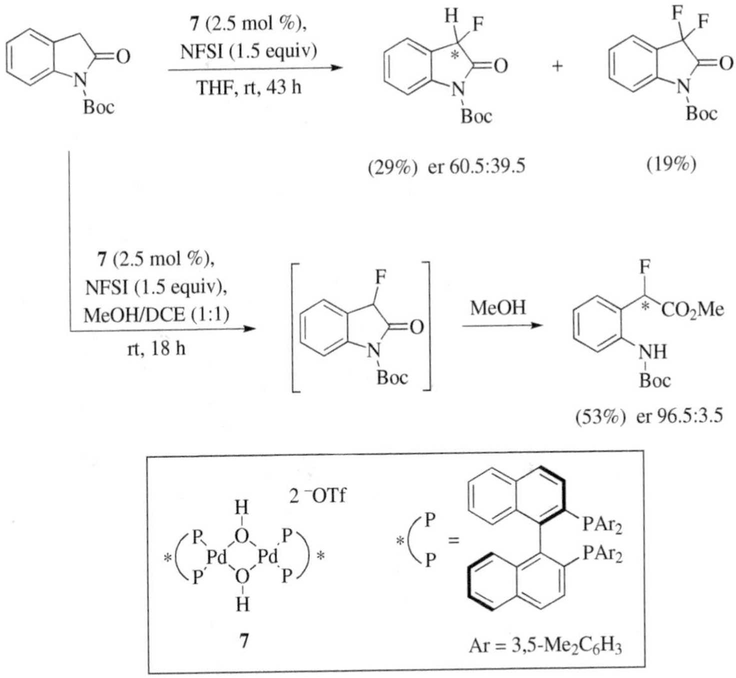

Scheme 85

both cyclic and acyclic substrates.[71] Acyclic substrates generally react more slowly and result in moderate yields of products with high enantiomeric ratios. Alcoholic solvents, acetone, and THF afford optimal results in a different palladium-catalyzed α-fluorination of β-keto phosphonates: cyclic phosphonates are fluorinated in methanol (>70% yields, er >96.5:3.5) and acyclic substrates are fluorinated in THF to provide products in 50–79% yield and with enantiomeric ratios greater than 93.5:6.5 er.[74]

The sequential chlorination–fluorination of β-keto esters and β-keto phosphonates relies on the enantioselective fluorination with the complex derived from copper(II) triflate and the chiral ligand SPYMOX (**L18**) (Scheme 86).[101]

Aliphatic, aromatic, and heterocyclic β-keto esters are compatible with the one-pot chlorination–fluorination procedure. Interestingly, the displacement of the α-chloro-β-keto ester with azide is proposed to occur via an S_N2 mechanism because it affords the corresponding α-azido-α-fluoro-β-keto ester with similar enantiomeric excess.[101] The copper–SPYMOX complex also facilitates the fluorination of α-alkyl-β-keto esters (Scheme 87) and alkylmalonates.[103]

The enantioselective fluorination of α-chloro-β-keto phosphonates can also be accomplished using a low loading of an air- and moisture-stable palladium catalyst.[81] A critical component to the success of this process is identifying a bulky base that does not coordinate to the transition metal and thereby accelerates the transformation without affecting the enantioselectivity.

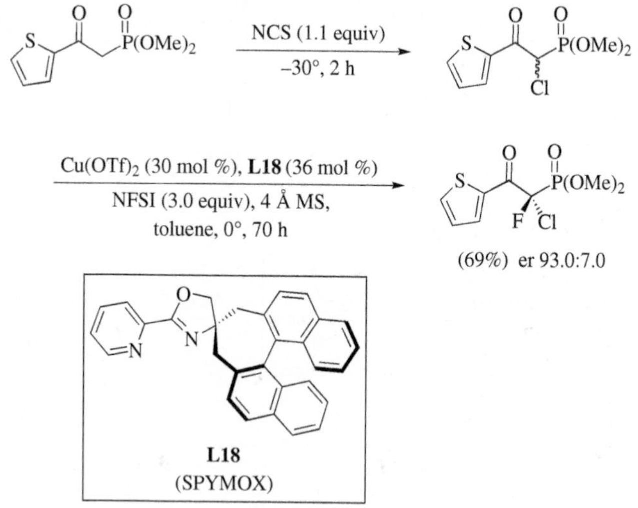

Scheme 86

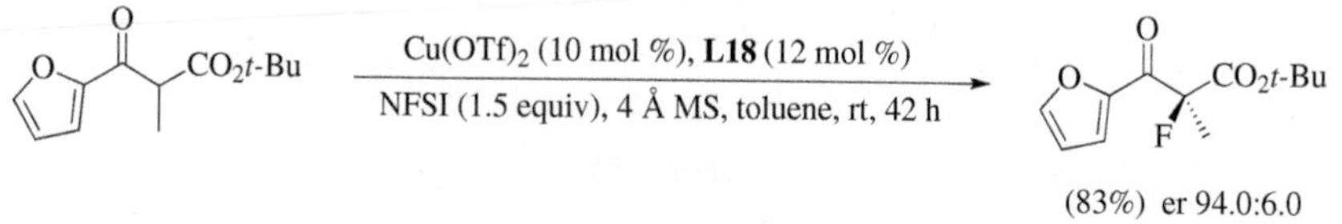

Scheme 87

Synthesis of α-Fluorinated α-Cyano Acetates. 8,8′-Bis(diphenylphosphino)-
3,3′,4,4′-tetrahydro-4,4,4′,4′,6,6′-hexamethyl-2,2′-spirobi[2H-1-benzopyran] [SPAN
(PPh₂)₂] (L19), a diphosphane with a wide bite-angle, facilitates the palladium-
catalyzed enantioselective fluorination of α-cyanoacetates with enantioselectivities
of up to 96.5:3.5 er (Scheme 88).[78]

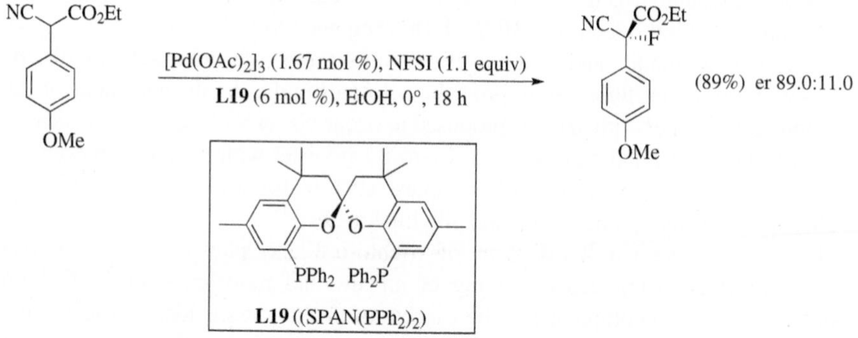

Scheme 88

The preparation of α-fluoro-α-cyanoacetates with enantiomeric ratios between 92.5:7.5 and 99.5:0.5 er can be achieved using a different palladium catalyst and NFSI.[73] Interestingly, α-alkyl substituted α-cyanoacetates are unreactive under these conditions, which limits the scope of this process to α-aryl derivatives.

The scope of most practical fluorination reactions is still quite narrow. Moreover, many of the useful fluorination reactions require expensive fluorinating reagents and result in the formation of mixtures of products that are challenging to separate. Transition-metal-catalyzed reactions have improved the accessibility of fluorinated molecules, maybe more so than the analogous advances with the other halides, but so far, relatively little attention has focused on the development of practical, cost-efficient fluorination reactions. We now have a better appreciation of fluorine's sophisticated reactivity patterns and can access more sustainable and less expensive fluorine sources, which provides exciting opportunities to develop fluorination reactions that can rival the utility of other halogenation reactions.

APPLICATIONS TO SYNTHESIS

In view of the harsh conditions and the limited functional-group tolerance of many traditional fluorination reactions, such as the Balz–Schiemann reaction and S_N2 displacements using alkali halides, a 'building-block approach' has traditionally been employed to incorporate fluorine into complex and densely functionalized molecules. For instance, fluorine is incorporated either by attaching a simple fragment that contains the requisite carbon–fluorine bond or by using a precursor at the outset of the synthesis that contains the desired carbon–fluorine bond. Given the stability of the carbon–fluorine bond to a wide range of reaction conditions, the early incorporation of the fluorine substituent is often preferable to performing a fluorination reaction at a later stage in the synthesis, even now that several fluorination reactions are compatible with complex molecules. The ability to readily access a large number of fluorinated fragments is therefore beneficial for the construction of complex fluorinated molecules. For example, redox-active phenyltetrazole sulfone groups can serve as a convenient alternative to traditional leaving groups, such as halides, in radical cross-coupling reactions.[196] Elaboration of the sulfones via α-fluorination prior to radical cross-coupling provides a convenient avenue for the installation of fluoride into coupling fragments. Sulfone reagents for monofluoromethylation and difluoromethylation can be prepared on >200-gram scales and undergo radical cross-coupling with a variety of aryl zinc reagents.[196] For example, radical cross-coupling with a fluoroalkyl phenyltetrazole (Scheme 89) gives access to a fluorinated pharmaceutical building block in three steps from commercially available starting materials,[196] whereas a previous synthesis required eight or nine steps.[197]

A strong motivation for the late-stage construction of carbon–fluorine bonds exists in cases where the de novo synthesis of the entire carbon skeleton with the fluorine present is not desirable, and when specific modifications to an existing structure are required, as in diversity-oriented synthesis and the establishment of structure–activity relationships.[198] The synthesis of corticosteroid **8**, for example, is

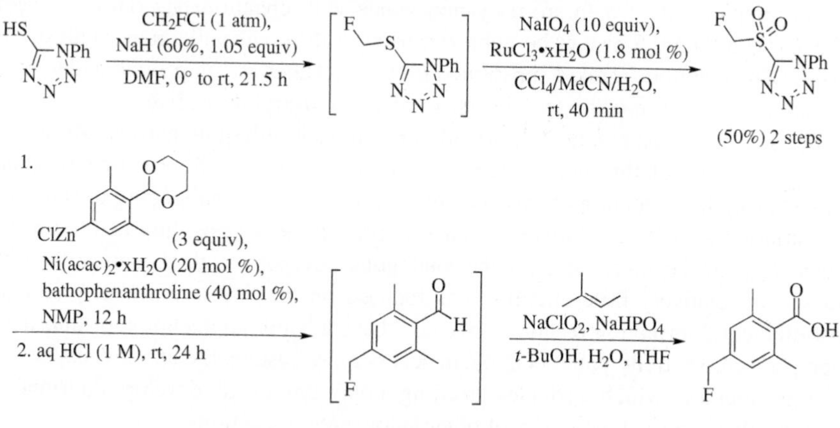

Scheme 89

streamlined by a late-stage fluorination reaction (Scheme 90).[199] A decarboxylative fluorination approach obviates the need for expensive and toxic reagents, which are used in other synthetic approaches towards corticosteroid **8**.[199] Interestingly, the formation of the side-product **9** is thought to occur via the hydrolysis of thioester **8**, followed by decarboxylative fluorination.

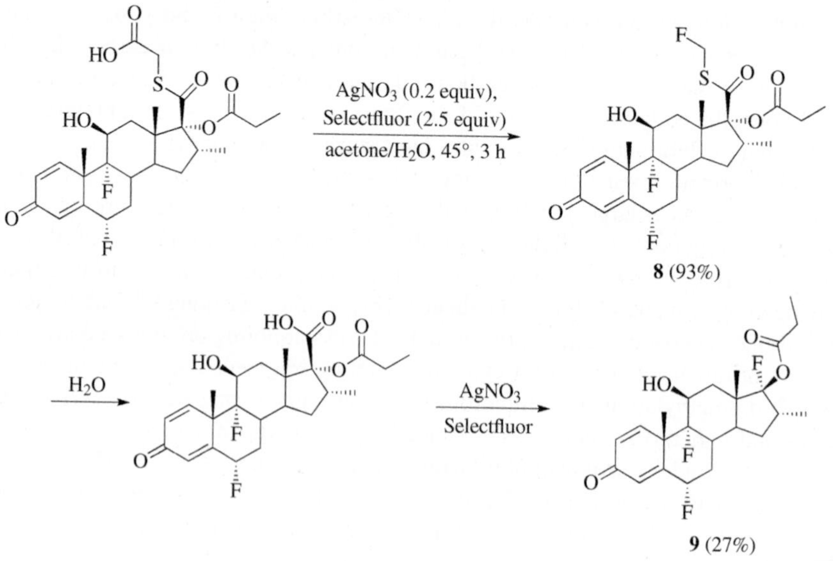

Scheme 90

One of the primary goals for the development of late-stage fluorination chemistry has been potential applications in fluorine-18 positron-emission tomography (^{18}F-PET) tracer synthesis. The unnatural ^{18}F isotope has a half-life of 109 minutes and is the preferred isotope for many applications of noninvasive positron-emission medical imaging.[200] Given the short lifetime of fluorine-18, the preparation of the C–[18]F bond in the synthesis of PET tracers must occur at a late stage, preferably as the last step. In addition to the operational difficulties of working with radioactive reagents on a micromolar scale, ^{18}F-fluorination chemistry is rendered challenging by limited access to electrophilic fluorinating reagents and the practical challenges of ensuring anhydrous working conditions. The scope of electrophilic and nucleophilic fluorination reactions is nearly orthogonal, and heavy reliance on nucleophilic fluorination reactions has made particular substrate classes, and functional groups difficult to radiolabel.[111,201–203] Until recently, very few nucleophilic fluorination reactions could be successfully conducted in the presence of moisture,[110] and the adaptation of these methods to ^{18}F-radiochemistry necessitated either the addition of [19]F[–204] or laborious azeotropic drying steps to render [18]F[–] sufficiently reactive.[205] Complex ^{18}F-labeled molecules can now be readily prepared from either phenol (Scheme 91)[206–208] or boronate ester precursors (Scheme 92).[209–211] Both transformations rely on a nucleophilic fluoride source; however, both reactions proceed efficiently with substrates containing protic functional groups, and more importantly, tolerate the adventitious moisture present under radiochemical conditions.

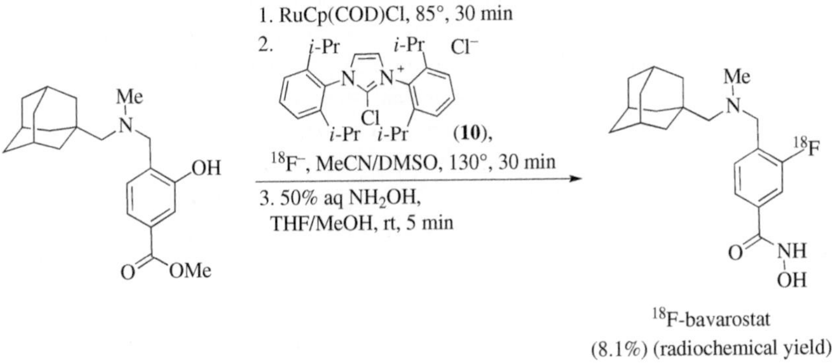

Scheme 91

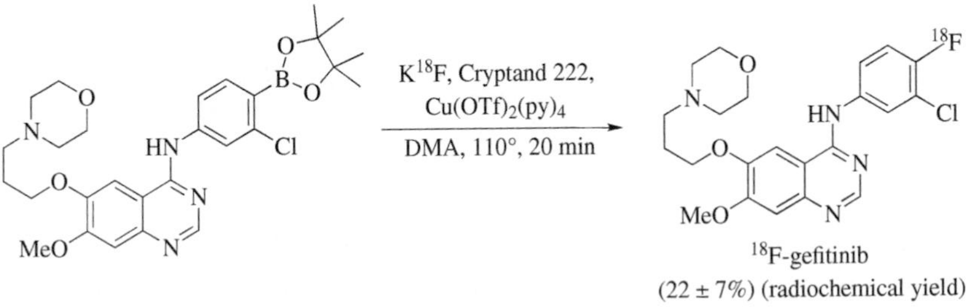

Scheme 92

[18]F-Labeled bavarostat (Scheme 91) can be used to map histone deacetylase 6 (HDAC6) in the living brain, which has been demonstrated in PET imaging

experiments in rodents and non-human primates.[212] The HDAC6 enzyme is a pharmaceutical target of interest in connection with major depressive disorder, Alzheimer's disease, and Parkinson's disease, among others, but the development of both drugs and biological reporter molecules selective for a particular zinc-dependent HDAC among the 11 existing paralogues is challenging. Ruthenium-mediated, late-stage radiofluorination permits efficient manual radiolabeling and automated radiosynthesis of bavarostat, and thus facilitates imaging studies of a class of ^{18}F-labeled hydroxamic acids that is difficult to access by other approaches.[212]

An ^{18}F-labeled version of the drug gefitinib (Scheme 92), an inhibitor of the epidermal growth factor receptor-tyrosine kinase (EGFR-TK) used in the treatment of nonsmall-cell lung cancer, can be prepared using copper-mediated radiolabeling of the corresponding boronate ester derivative.[209] Because the fluorination reaction tolerates a wide range of heteroarenes and other functional groups, the radioactive ^{18}F-substituent can be introduced in the last synthetic step. A previously disclosed synthesis of ^{18}F-gefitinib, which is used to study the intratumoral concentration of the drug using non-invasive positron emission imaging, requires a three-step radiosynthesis starting with the ^{18}F-fluorination of 3-chloro-4-trimethylammonium-nitrobenzene.[209,213]

The development of new and even more powerful methods for installing fluorine-18 is important, as are thorough studies to determine the limitations of these methods; the information gleaned will minimize potential problems and assess the reliability of using such methods as a final step in a synthetic sequence.[209,212] Radiofluorination (^{18}F) has also been achieved in the enantiomorphic, regioselective ring-opening of epoxides using a cobalt catalyst (Scheme 93).[214,215] ^{18}F-THK-5105 is a promising PET tracer candidate for the imaging of tau pathology, an important biomarker for Alzheimer's disease.[214] A previously reported synthesis of racemic

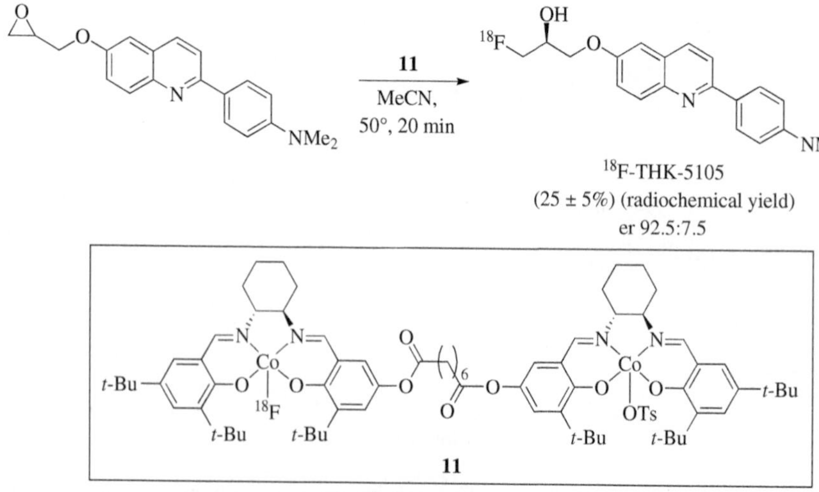

Scheme 93

^{18}F-THK-5105 involves an S_N2 substitution with [18]F$^-$ on a differentially protected diol substrate, followed by deprotection.[216] In contrast, cobalt-mediated radiofluorination furnishes the desired radioligand in 92.5:7.5 er, and the radioisotope is introduced in the final synthetic step.

Late-stage fluorination is a powerful tool in diversity-oriented synthesis, in radiochemistry, and for structure-activity relationship studies. Nevertheless, installing the required synthetic handles for fluorination (e.g., halides, boronic esters, carboxylic acids, or alkenes) can be problematic. Particularly enabling synthetic strategies rely on either an undirected C–H functionalization (Scheme 94),[55] or the combination of an undirected C–H functionalization reaction with the requisite fluorination of the new functional group. Unfortunately, the undirected C–H fluorination of complex substrates at room temperature with a palladium complex often leads to the formation of two fluorinated constitutional isomer products. Although separation of the mixture of isomers is required, two fluorinated derivatives of the pesticide procymidone can thus be accessed in a single transformation starting from the commercially available compound (Scheme 94). The ability to replace aromatic C–H bonds with a fluorine substituent in the presence of many pre-existing functional groups makes possible the rapid generation and evaluation of fluorinated analogues of pharmaceutical lead compounds or agrochemicals.

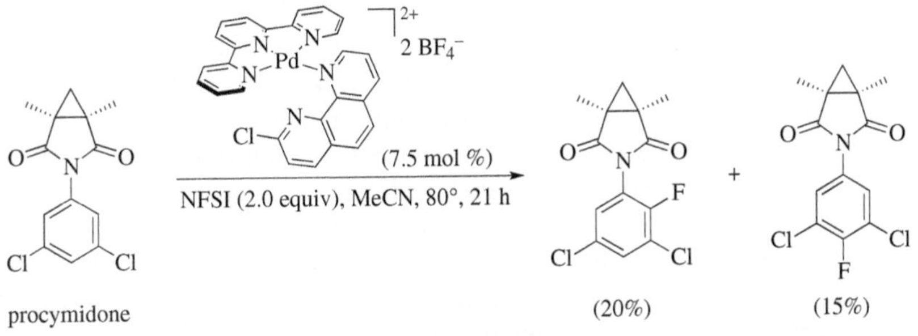

Scheme 94

Powerful strategies for directed C–H fluorination are available with several directing groups; the usefulness of this strategy for controlling the positional selectivity in the fluorination step is offset by the ability to either incorporate the directing group or readily convert it to the desired motif in the target molecule. The readiness with which certain transformations are used at a late stage in a synthesis relies mainly on their reliability and predictability. A thorough study of the optimal deoxyfluorination reaction conditions for certain alcohols has been conducted (see Scheme 100 below), and a predictive computational model has been generated.[217] Investigations on the behavior of fluorination reactions with different types of substrates combined with the studies of fluorination reactions in complex, densely functionalized substrates lend credibility to the notion that fluoride no longer needs to be installed at the outset of a synthetic sequence.

In certain contexts, fluorination has become a sufficiently predictable and functional-group tolerant transformation that fluorine atoms are installed for synthetic convenience even though no fluorine atom is present in the target structure. For example, a fluorination–nucleophilic aromatic substitution sequence has been reported for the functionalization of pyridines (Scheme 95) and diazines.[143]

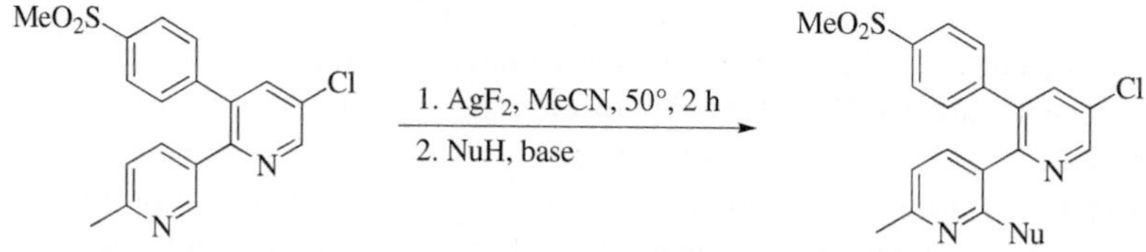

Nu	Yield (%)
MeO	55
NC–	40
pyrazolyl	48
$n\text{-}C_8H_{17}NH$	51

Scheme 95

COMPARISON WITH OTHER METHODS

Improvements to traditional methods for the synthesis of aryl fluorides have been reported, and among the most notable improvements is the ability to employ anhydrous TBAF for nucleophilic aromatic substitution.[218] Tetrabutylammonium fluoride is notoriously hygroscopic and decomposes upon heating, which precludes using harsh drying procedures. A novel route to the reagent from less hygroscopic precursors permits the isolation of anhydrous and highly reactive TBAF. Consequently, with this formulation, the desolvation of fluoride ions no longer contributes to the activation barrier for nucleophilic displacement (Scheme 96).[218,219] Similarly, anhydrous tetramethylammonium fluoride (TMAF)[220] or an anhydrous acyl azolium fluoride reagent[221] may be used to facilitate nucleophilic aromatic substitution.

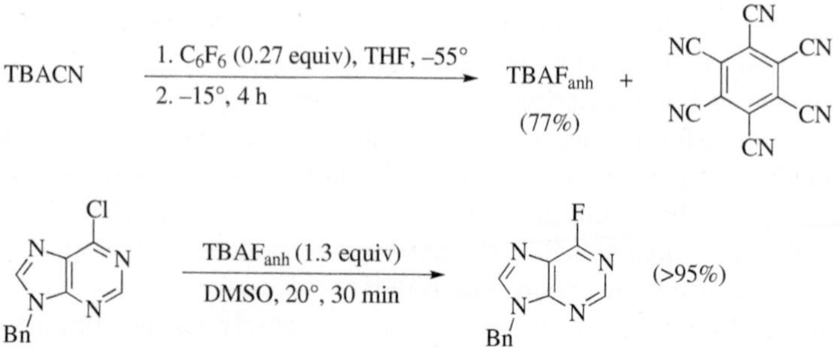

Scheme 96

Although transition-metal-based methods for the fluorination of arenes rely largely on aryl triflates, aryl halides, arylmetaloids, and arenes as precursors, several metal-free methods have been reported for the conversion of phenols to aryl fluorides. The imidazolium-based PhenoFluor reagent effects deoxyfluorination of phenols[206,222,223] and complex alcohols[224] (Scheme 97).[222] An unusual, concerted, nucleophilic, aromatic substitution mechanism permits the introduction of fluoride into arenes substituted with electron-withdrawing and electron-donating substituents.[206] The development of the PhenoFluorMix reagent[225] (for phenols) and the AlkylFluor reagent[226] (for alcohols) obviates the necessity of handling a moisture-sensitive reagent while maintaining substrate scope and selectivity for the PhenoFluor deoxyfluorination (Scheme 98).

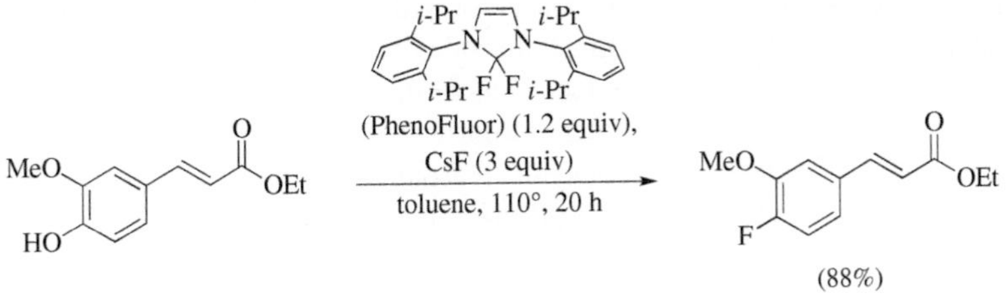

Scheme 97

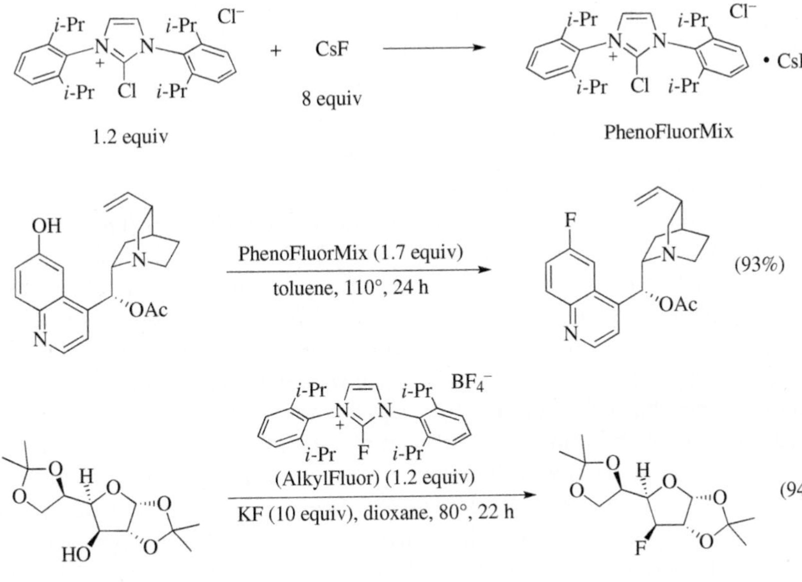

Scheme 98

Although more limited in reaction scope, the PyFluor deoxyfluorination reagent is less expensive than PhenoFluor and demonstrates high selectivity for fluorination over elimination in aliphatic and benzylic deoxyfluorination reactions (Scheme 99).[227]

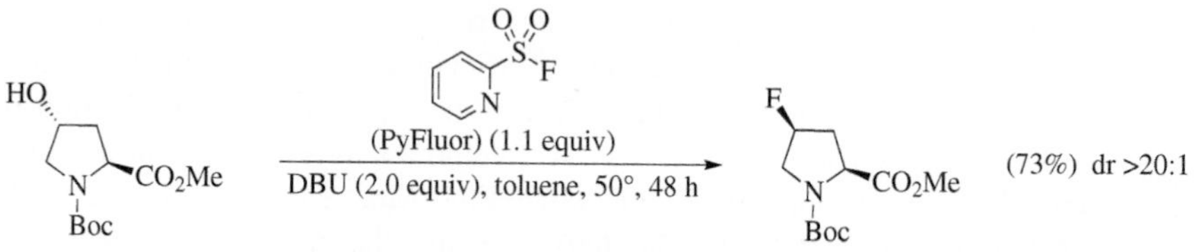

Scheme 99

The deoxyfluorinations of electron-deficient and electron-neutral aryl and heteroaryl fluorosulfonate intermediates, formed in situ from the reaction of the corresponding phenols with sulfuryl fluoride, is accomplished using TMAF.[220] High-yielding reaction conditions (selected from several sulfonyl fluoride reagents and nitrogenous bases) for the deoxyfluorination of alcohol substrates can be predicted with high fidelity using machine learning.[217] Scheme 100 depicts one such deoxyfluorination reaction that was optimized by this method.

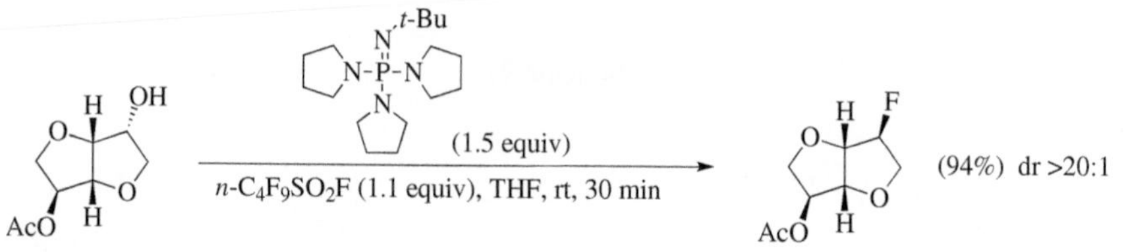

Scheme 100

The irradiation of an organic sensitizer in the presence of either Selectfluor or NFSI results in the conversion of alkanes to alkyl fluorides (Scheme 101). Notably, the process is selective for methines and benzylic C–H bonds. Efficient sensitizers include acetophenone,[228] anthraquinone,[229] and 1,2,4,5-tetracyanobenzene.[230]

A fluorinated derivative of the anticancer drug candidate salinosporamide A is prepared by the introduction of the fluorinase gene from *S. cattleya* into a heterologous host.[231,232] Other complex fluorinated molecules have also been prepared using synthetic biology, relying on a biochemical pathway that consists of two polyketide synthase systems that are engineered to accept fluoroacetate as a substrate.[233]

High-throughput screening of a library of chiral amines can be used to identify catalysts for enantioselective α-fluorination of ketones by enamine activation, an example of which is depicted in Scheme 102.[234] High yields and enantiomeric ratios for the α-fluorination with pyrrolidine- and imidazolinone-based organocatalysts had previously only been possible with aldehyde substrates since ketones undergo slow enamine formation that limits the equilibrium concentration of the critical enamine intermediate. While the concentration of the two most reactive *N*-alkene rotational isomers can differ quite significantly in the case of aldehyde-derived enamines, ketone-derived enamines often have a much less pronounced difference in

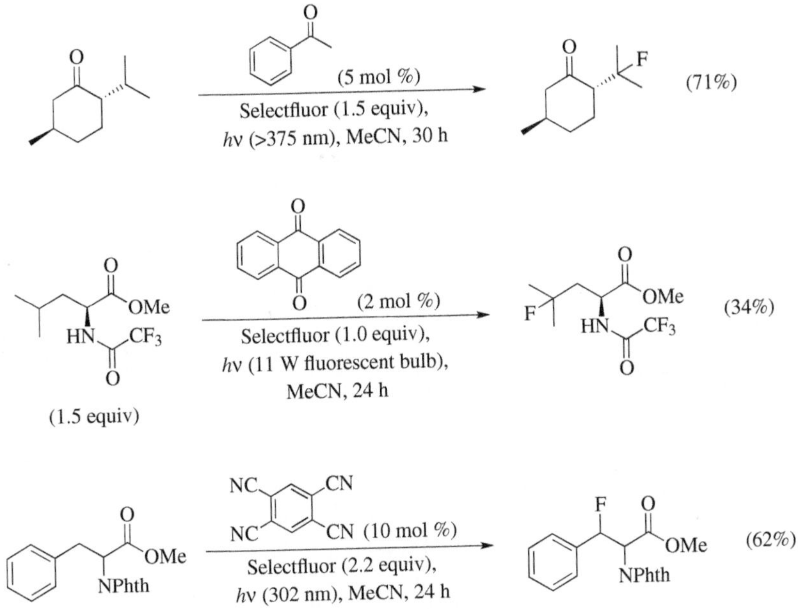

Scheme 101

the concentration of rotational isomers, which results in low enantioinduction.[234,235] Computational evidence suggests that high enantiofacial selectivity in ketone α-fluorination with cinchona-based alkaloid catalysts is a result of the preference for the seven-membered ring to adopt a chair-like conformation during the fluorine transfer step, coupled with the size of the C9-quinoline substituent in the organocatalyst (Scheme 102).[235] The mild reaction conditions prevent isomerization and minimize difluorination for a wide range of cyclic ketones, but acyclic substrates afford products in low yields and with poor enantiomeric ratios.

Selectfluor can be used for enantioselective fluorination by pairing it with catalytic quantities of a chiral anion, which both solubilizes Selectfluor in the nonpolar reaction medium and provides a chiral environment for the reaction.[236,237] The insolubility of Selectfluor in the reaction medium in the absence of the chiral catalyst effectively eliminates the racemic background reaction, and the slow release of Selectfluor results in improved functional-group tolerance. A nonlinear effect supports the exchange of both tetrafluoroborate counteranions in Selectfluor for chiral phosphates in the fluorocyclization of alkenes.[237] In the enantioselective fluorination of enamines (Scheme 103),[238] it is proposed that one oxygen atom of the phosphate anion is ion-paired with the Selectfluor reagent, while the other oxygen atom is hydrogen-bonded to the enamide to activate the substrate towards fluorination.[236,238] Asymmetric-counteranion-directed catalysis is also applicable to the fluorinative dearomatization of phenols,[239] directed fluorination of alkenes,[240,241] and α-fluorination of ketones[242] using a protected amino acid organocatalyst.[243]

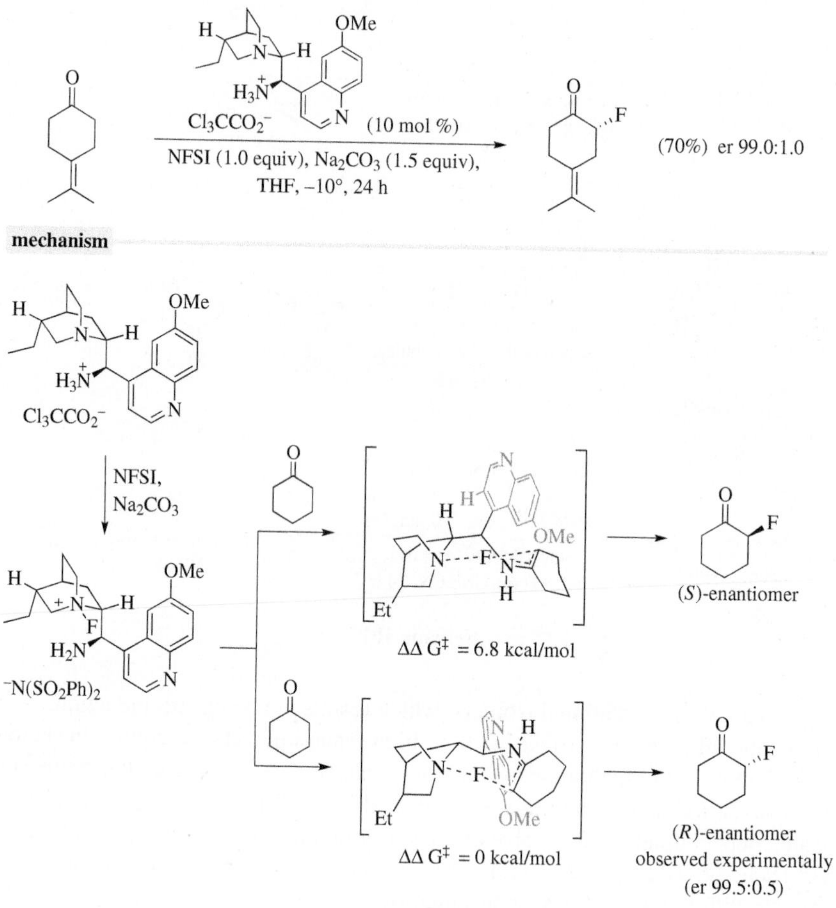

mechanism

Scheme 102

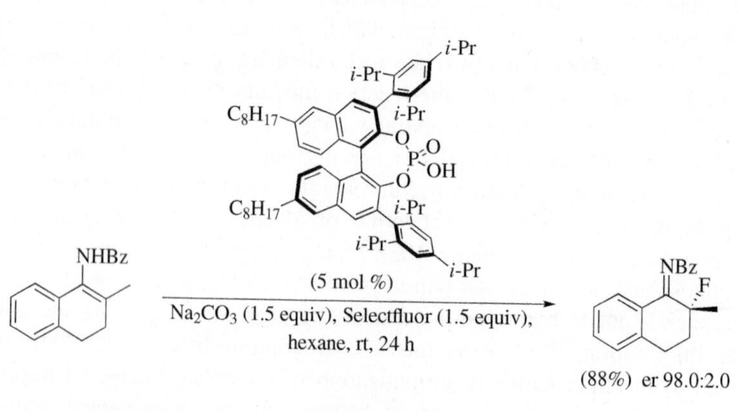

(88%) er 98.0:2.0

Scheme 103

The combination of hydrogen-bonding and phase-transfer catalysis for the development of enantioselective fluorination can be further extended to simple alkali metal fluoride salts (Scheme 104).[244] The hydrogen-bond network between the fluoride nucleophile, the substrate, and the catalyst is reminiscent of the only enzymatically-catalyzed fluorination reaction in nature, which involves a sulfonium leaving group and leads to the formation of a primary alkyl fluoride.[245]

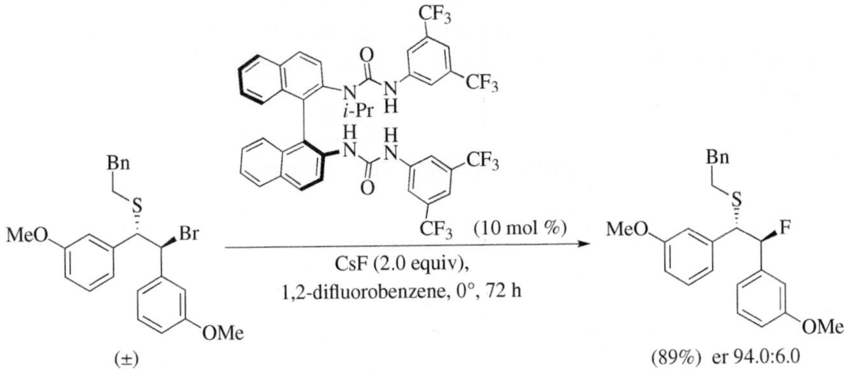

Scheme 104

The synthesis of alkenyl fluorides can be accomplished by constructing the carbon skeleton from fluorinated building blocks. Terminal alkenyl fluorides can be installed by the cross-metathesis of terminal alkenes with (Z)-bromofluoroethene using a molybdenum catalyst.[246] Wittig and Horner–Wadsworth–Emmons reactions can also be used to access alkenyl fluorides, but these reactions often result in variable (Z)/(E) selectivity.[247]

EXPERIMENTAL CONDITIONS

Many sources of nucleophilic fluoride are hygroscopic and should be stored in a glovebox, glove bag, or desiccator because hydration decreases the nucleophilicity of fluoride. The CsF used as a fluoride source in palladium-catalyzed fluorination reactions of aryl triflates and aryl bromides is dried at 180° under vacuum for 48 h.[25,33,36,37,248] The dried CsF is then transferred to a nitrogen-filled glovebox where it is thoroughly ground using an oven-dried mortar and pestle and then passed through a sieve to reduce the size of the CsF particles. Because AgF is light sensitive, weighing of AgF should be carried out in a darkened area, and reactions should be performed in amber vials or glassware wrapped in aluminum foil.[163]

Caution: Hydrogen fluoride is highly toxic and difficult to handle safely in both its gaseous form and as a concentrated (48%) solution in water. Complexes of HF with amines such as pyridine (Pyr•HF, Olah's reagent) and triethylamine (Et₃N•3HF, TREAT-HF) provide increased operational convenience and abate the risk of exposure to HF because the vapor pressure of HF is reduced. Olah's reagent contains a small equilibrium concentration of HF, and exposure to the reagent should be treated in the same way as exposure to HF.[249,250] As a source of HF, $Et_3N{\cdot}3HF$ can corrode glass, and depending on the reaction conditions under which the reagent is employed, polyethylene, polypropylene, or PTFE vials may be necessary.[48,177] Selectfluor (F-TEDA) is a free-flowing, virtually nonhygroscopic powder that is stable in water and many organic solvents, but it decomposes in dilute NaOH. The reagent reacts slowly with DMF upon heating and reacts rapidly and exothermically with DMSO.[251,252] Selectfluor can be stored under air at room temperature, but exothermic decomposition can occur at temperatures above 80°.[252,253] In comparison with Selectfluor, NFSI is a weaker oxidant and exhibits higher solubility in less polar solvents.[254]

EXPERIMENTAL PROCEDURES

F-**Tolvaptan [Fluorination of an Arene].**[55] A mixture of palladium complex **3** (10.4 mg, 18.8 µmol, 7.50 mol %) and 2-chlorophenanthroline (4.0 mg, 18.8 µmol, 7.5 mol %) was dissolved in MeCN (1.0 mL) and added to a 4-mL vial containing a solution of NFSI (158 mg, 500 µmol, 2.00 equiv) and tolvaptan (**12**) (112 mg, 250 µmol) in MeCN (1.5 mL, final concentration of tolvaptan (**12**) = 0.10 M). The reaction mixture was then heated to 50°. After stirring at 50° for 20 h, the reaction mixture was allowed to cool to rt and then was filtered through a pad of Celite, eluting with CH_2Cl_2 (3 × 4 mL). The filtrate was concentrated in vacuo, and the residue was purified by column chromatography on silica gel (hexane/EtOAc, 1.5:1) to afford a mixture of products **13** and **14** as a white solid. Further purification by HPLC (YMC-Pack Pro C18, 10 × 150 mm, 5 µm, MeOH/H_2O, 70:30, 4.5 mL/min, 35°) provided products **13** (34 mg, 73 µmol, 29%) and **14** (25 mg, 54 µmol, 21%).

13: ^{1}H NMR (500 MHz, DMSO-d_6, major rotamer) δ 9.97 (s, 1H), 7.56 (m, 1H), 7.52 (d, J = 2.4 Hz, 1H), 7.43 (d, J = 7.3 Hz, 1H), 7.36 (dd, J = 7.5, 1.0 Hz, 1H), 7.28

(m, 1H), 7.26 (m, 1H), 7.10 (dd, $J = 8.3, 2.5$ Hz, 1H), 6.84 (d, $J = 8.3$ Hz, 1H), 6.79 (d, $J = 10.3$ Hz, 1H), 5.74 (d, $J = 4.6$ Hz, 1H), 4.93 (m, 1H), 4.62 (ddd, $J = 12.8, 3.4, 3.4$ Hz, 1H), 2.71 (m, 1H), 2.36 (s, 3H), 2.32 (s, 3H), 2.10 (m, 1H), 1.98 (m, 1H), 1.74 (m, 1H), 1.49 (m, 1H); ^{13}C NMR (126 MHz, DMSO-d_6, major rotamer) δ 167.8, 166.9, 152.4 (d, $J = 245.9$ Hz), 145.1, 138.1, 136.2, 135.5, 133.8 (d, $J = 5.9$ Hz), 131.9, 131.1 (d, $J = 3.3$ Hz), 130.5, 129.8, 129.5, 127.4, 127.2, 126.8, 125.7 (d, $J = 13.1$ Hz), 125.6, 125.1, 113.2 (d, $J = 19.4$ Hz), 69.3, 46.0, 35.3, 25.5, 19.4, 18.7; ^{19}F NMR (471 MHz, DMSO-d_6) δ −125.7 (br s), −126.6 (m), −127.0 (m); HRMS–ESI (*m/z*): [M + Na]$^+$ calcd for $C_{26}H_{24}ClFN_2O_3$, 489.1352; found, 489.1357.

14: ^{1}H NMR (500 MHz, DMSO-d_6, major rotamer) δ 9.99 (s, 1H), 7.51 (d, $J = 2.3$ Hz, 1H), 7.44 (d, $J = 7.3$ Hz, 1H), 7.38 (m, 1H), 7.36 (m, 1H), 7.28 (d, $J = 7.5$ Hz, 1H), 7.26 (dd, $J = 7.5, 7.5$ Hz, 1H), 7.09 (dd, $J = 8.3, 2.6$ Hz, 1H), 6.81 (d, $J = 8.3$ Hz, 1H), 6.70 (d, $J = 8.4$ Hz, 1H), 5.71 (d, $J = 4.6$ Hz, 1H), 4.90 (ddd, $J = 10.4, 3.2, 3.2$ Hz, 1H), 4.64 (ddd, $J = 13.3, 3.6, 3.6$ Hz, 1H), 2.72 (m, 1H), 2.36 (s, 3H), 2.28 (d, $J = 1.5$ Hz, 3H), 2.12 (m, 1H), 1.98 (m, 1H), 1.76 (m, 1H), 1.50 (m, 1H); ^{13}C NMR (126 MHz, DMSO-d_6, major rotamer) δ 167.9, 166.9 (d, $J = 2.1$ Hz), 153.1 (d, $J = 246.5$ Hz), 145.0, 138.3, 136.3, 135.5, 134.6 (d, $J = 3.2$ Hz), 131.8, 130.5, 129.8, 129.4, 127.4, 126.8, 125.8 (d, $J = 13.1$ Hz), 125.6, 125.1, 122.9 (d, $J = 17.1$ Hz), 122.1, 121.3, 69.4, 45.9, 35.3, 25.6, 19.4, 11.8 (d, $J = 4.7$ Hz); ^{19}F NMR (471 MHz, DMSO-d_6) δ −124.4 (br s), −124.6 (s), −125.3 (s); HRMS–ESI (*m/z*): [M + Na]$^+$ calcd for $C_{26}H_{24}ClFN_2O_3$, 489.1352; found, 489.1358.

1-(4-Fluorobenzenesulfonyl)-4-methylpiperazine [Fluorination of an Aryl Triflate].[36]

In a nitrogen-filled glovebox, an oven-dried reaction tube (20 × 150 mm) equipped with a stir bar was charged sequentially with **L20** (49 mg, 25 μmol, 5 mol %), CsF (228 mg, 1.50 mmol, 3.00 equiv), aryl triflate **15** (194 mg, 0.50 mmol), and anhydrous toluene (5 mL). The reaction tube was sealed with a screw cap containing a Teflon septum and was removed from the glovebox. After being stirred vigorously at rt for 96 h, the reaction mixture was filtered through a pad of Celite, eluting with Et_2O, and the filtrate was concentrated in vacuo. The crude material was purified by column chromatography on silica gel (MeOH/CH$_2$Cl$_2$, 0:100 to 1:99 to 2:98) to provide product **16** as a pale-yellow, crystalline solid. The reaction was performed in duplicate; 109 mg product (85% yield) was obtained in the first run and 115 mg product (89% yield) in the second run: mp 84°; IR (neat) 1348, 1149, 1121, 1092, 937, 838, 837, 819, 735, 610, 547 cm^{-1}; ^{1}H NMR (500 MHz, CDCl$_3$) δ 7.81–7.71 (m, 2H), 7.20 (t, $J = 8.5$ Hz, 2H), 3.02 (s, 4H), 2.47 (s, 4H), 2.26 (s, 3H); ^{13}C NMR (126 MHz, CDCl$_3$) δ 165.4 (d, $J = 255.1$ Hz), 131.6, 130.6 (d, $J = 9.3$ Hz), 116.4 (d, $J = 22.5$ Hz), 54.1, 46.1, 45.8; ^{19}F NMR (282 MHz, CDCl$_3$) δ –105.21. Anal. Calcd for C$_{11}$H$_{15}$FN$_2$O$_2$S: C, 51.15; H, 5.85. Found: C, 51.42; H, 5.95.

(E)-Ethyl 2-Cyano-3-(4-fluorophenyl)acrylate [Fluorination of an Aryl Trifluoroborate].[61] Palladium precatalyst **3** (4.4 mg, 8.0 μmol, 0.020 equiv), terpy (3.7 mg, 16 μmol, 0.040 equiv), aryl trifluoroborate **17** (123 mg, 400 μmol), Selectfluor (170 mg, 480 μmol, 1.20 equiv), and NaF (16.8 mg, 400 μmol, 1.00 equiv) were sequentially added to a round-bottomed flask (200 mL), followed by MeCN (4.0 mL, 0.1 M) at 23°. An air-cooled reflux condenser was attached, and the reaction mixture was heated to 40°. After being stirred at 40° for 15 h open to air, the reaction mixture was allowed to cool to 23° and was transferred to a separatory funnel. The reaction vial was rinsed with additional MeCN (2 × 4 mL), and these washes were added to the separatory funnel as well. Pentane (20 mL) was added, and the organic layer was washed with water (20 mL). The aqueous layer was extracted with pentane (5 × 20 mL). The combined organic layers were dried over sodium sulfate, filtered, and concentrated in vacuo to afford a colorless solid. The solid was purified by column chromatography on silica gel (pentane/Et$_2$O, 17:3), to afford product **18** (84.5 mg, 385 μmol, 96%) as a colorless crystalline solid: ^{1}H NMR (500 MHz, CDCl$_3$) δ 8.21 (s, 1H), 8.05–8.01 (m, 2H), 7.22–7.17 (m, 2H), 4.39 (q, $J = 7.1$ Hz, 2H), 1.40 (t, $J = 7.2$ Hz, 3H); ^{13}C NMR (125 MHz, CDCl$_3$) δ 165.2 (d, $J = 256$ Hz), 162.2, 153.2, 133.4 (d, $J = 9.1$ Hz), 127.7 (d, $J = 3.6$ Hz), 116.5 (d, $J = 21.9$ Hz), 115.3, 102.4 (d, $J = 2.8$ Hz), 62.6, 14.0; ^{19}F NMR (375 MHz, CDCl$_3$) δ –106.0; HRMS–FIA (m/z): [M + H]$^+$ calcd for C$_{12}$H$_{10}$FNO$_2$, 220.0768; found, 220.0769.

4-Acetyl-6-*tert*-butyl-3-fluoro-1,1-dimethylindane (*F*-Celestolide) [Directed Aliphatic C–H Fluorination].[163] To a 25-mL Schlenk flask containing a Teflon-coated magnetic stir bar were added 4-acetyl-6-*tert*-butyl-1,1-dimethylindane (Celestolide, **19**) (500 mg, 2.05 mmol), Mn(salen)Cl (260 mg, 0.41 mmol, 20 mol %), and AgF (780 mg, 6.15 mmol, 3.00 equiv). The flask was capped with a rubber septum, connected to a Schlenk line, evacuated, and backfilled with nitrogen; this evacuation–backfill process was performed three times. Triethylamine trihydrofluoride (130 μL, 0.80 mmol, 0.38 equiv) was diluted with dry and degassed MeCN (1 mL) in a 4-mL vial fitted with a septum, and the resulting solution was flushed with N_2 for 4 min. The solution of Et₃N•3HF was transferred into the Schlenk flask using a syringe; the vial was rinsed with MeCN (0.5 mL), and this solution was also transferred into the reaction flask. The Schlenk flask was placed in a preheated water bath at 50°, and the speed of stirring was set to ~600 rpm. Iodosobenzene (2.25 g, 10.2 mmol, 5.0 equiv) was then added in small portions over the course of 5 h, and MeCN (0.5 mL) was added to the reaction mixture for every 2 equiv of PhIO added. The reaction mixture was diluted with CH_2Cl_2 (10 mL) and filtered through a pad of silica gel. The filtrate was concentrated under reduced pressure at 25°. Column chromatography on silica gel (hexanes/EtOAc, 100:0 to 85:15) provided product **20** as a yellow solid (295 mg, 55%): ¹H NMR (500 MHz, CDCl₃) δ 7.77 (d, J = 1.7 Hz, 1H), 7.43 (t, J = 1.5 Hz, 1H), 6.44 (ddd, J = 53.9, 5.9, 1.5 Hz, 1H), 2.65 (s, 3H), 2.39–2.06 (m, 2H), 1.37 (s, 12H), 1.34 (s, 3H); ¹³C NMR (125 MHz, CDCl₃) δ 199.9, 155.9, 154.2, 135.2, 134.8, 125.7, 123.6, 93.9, 48.4, 42.7, 35.2, 31.5, 31.4, 29.0, 28.6; ¹⁹F NMR δ –158.6; LRMS–EI (*m/z*): [M]⁺ 262.2.

3-Fluoropentadecane [Aliphatic Fluorodecarboxylation].[58] 2-Ethyltetradecanoic acid (**21**) (51.2 mg, 0.20 mmol), AgNO₃ (6.8 mg, 0.04 mmol, 20 mol %), and Selectfluor (141.6 mg, 0.40 mmol, 2.00 equiv) were placed in a Schlenk tube. The reaction vessel was evacuated and filled with nitrogen. Acetone (2 mL) and water (2 mL) were added, and the reaction mixture was heated to 60° and stirred at that temperature for 10 h. The resulting mixture was cooled to rt and was extracted with CH_2Cl_2 (3 × 15 mL). The combined organic phases were dried over anhydrous Na_2SO_4, filtered, and concentrated in vacuo to afford the crude product. Purification by column chromatography on silica gel (hexanes) furnished product **22** as a colorless oil (42.7 mg, 93%): FTIR (neat) 2926, 2855, 1465 cm^{-1}; ^{1}H NMR (400 MHz, CDCl$_3$) δ 4.49–4.29 (m, 1H), 1.69–1.26 (m, 24H), 0.96 (t, J = 7.6 Hz, 3H), 0.88 (t, J = 7.2 Hz, 3H); ^{13}C NMR (100 MHz, CDCl$_3$) δ 95.7 (d, J = 166.2 Hz), 34.7 (d, J = 21.2 Hz), 31.9, 29.8, 29.7, 29.6, 29.56, 29.55, 29.43, 29.37, 28.1 (d, J = 21.2 Hz), 25.2 (d, J = 4.4 Hz), 22.7, 14.1, 9.4 (d, J = 5.9 Hz); ^{19}F NMR (282 MHz, CDCl$_3$) δ –181.6 (m); HRMS (*m/z*): calcd for $C_{15}H_{31}F$, 230.2410; found, 230.2414.

23

Pd(dmdba)$_2$ (9 mol %), **L7** (9 mol %)

AgF (2.75 equiv), toluene, rt, 48 h

24
(62%) er 98.5:1.5
b/l > 20:1

L7

(*S*)-4-(1-Fluoroallyl)tetrahydro-2*H*-pyran [**Fluorination of an Allylic Chloride**].[41] In a nitrogen-filled glovebox, a flask was charged with **L7** (79.1 mg, 0.10 mmol, 0.09 equiv), AgF (381 mg, 3.0 mmol, 2.75 equiv), and bis(3,5,3,5-dimethoxydibenzylideneacetone)palladium (81.5 mg, 0.10 mmol, 0.09 equiv). The flask was then sealed with a rubber septum and removed from the glovebox. A separate flame-dried flask was charged with (*E*)-4-(3-chloroprop-1-en-1-yl)tetrahydro-2*H*-pyran (**23**) (175.6 mg, 1.09 mmol), and then evacuated and backfilled with nitrogen; this evacuation–backfill procedure was performed three times. The flask was charged with toluene (22 mL), and then 20 mL of this solution was transferred under nitrogen to the flask containing the other reagents. The reaction flask was wrapped in aluminum foil, and the reaction mixture was stirred at rt for 48 h. At this point, an aliquot was removed, filtered through a short plug of silica gel, and analyzed by GC to determine site-selectivity (branched/linear > 20:1; GC t_R(S) = 51.4 min, t_R(R) = 46.2 min (Macherey–Nagel Hydrodex β-TBDAc, 25 m × 0.25 mm, 70°, 1 mL/min). This aliquot was combined with the reaction mixture, and the mixture thus obtained was loaded directly onto a column of silica gel (pentanes/Et$_2$O, 100:0 to 85:15 with 1% triethylamine) to furnish product **24** (90 mg, 0.624 mmol, 62%, er 98.5:1.5) as a colorless oil: [α]$^{21}_D$ +4.4 (*c* 1.6, CHCl$_3$); FTIR (thin film) 2950, 2844, 1446, 1430, 1389, 1269, 1239, 1148, 1092, 1016, 988, 974, 933, 878, 825 cm^{-1}; ^{1}H NMR (500 MHz, CDCl$_3$) δ 5.82 (dddd, J = 17.3, 14.0, 10.6, 6.7 Hz, 1H), 5.37–5.27

(m, 2H), 4.61 (ddd, J = 48.1, 6.7, 6.6 Hz, 1H), 4.05–3.95 (m, 2H), 3.37 (dt, J = 11.8, 2.2 Hz, 2H), 1.88–1.72 (m, 2H), 1.53–1.37 (m, 3H); ^{13}C NMR (125 MHz, CDCl$_3$) δ 134.52 (d, J = 20.0 Hz), 118.82 (d, J = 12.3 Hz), 96.96 (d, J = 169.9 Hz), 67.72, 67.62, 39.79 (d, J = 21.9 Hz), 28.14 (d, J = 5.4 Hz), 28.08 (d, J = 4.9 Hz); ^{19}F NMR (376 MHz, CDCl$_3$) δ –184.23 (m), –209.69* (m, peak corresponding to linear isomer); HRMS–ESI (*m/z*): [M – F]$^+$ calcd for C$_8$H$_{12}$O, 125.0966; found, 125.0963.

TABULAR SURVEY

Transition-metal-catalyzed and transition-metal-mediated fluorination reactions are surveyed. Reactions are grouped into aromatic, heteroaromatic, alkenyl, aliphatic, benzylic, and allylic fluorination reactions as well as fluorination reactions occurring alpha to a carbonyl group. The fluorination reactions are further subdivided by the type of starting material used in the reaction. The tables contain selected examples of the title reaction that were found in the open literature through May 2019.

Examples within each table are arranged by increasing carbon count of the title substrate undergoing fluorination. To place like examples proximal to one another, the carbon count is based on the main skeleton of the substrate, and excludes protecting groups and small groups attached to heteroatoms. All reported yields are provided in parentheses. An em-dash indicates that no data is provided in the primary reference. Enantioselectivity units have been converted to enantiomeric ratios (er).

The following abbreviations, in addition to those included in *The Journal of Organic Chemistry* Standard Abbreviations and Acronyms, are used in the tables:

AlPhos	di-1-adamantyl(4″-butyl-2″,3″,5″,6″-tetrafluoro-2′,4′,6′-triisopropyl-2-methoxy-*meta*-terphenyl)phosphine
t-Amyl-OH	2-methylbutanol-2-ol
b	branched
BAr$_F$	tetrakis[3,5-bis(trifluoromethyl)phenyl]borate
BINAP	2,2-bis(diphenylphosphino)-1,1′-binaphthyl
BMIDA	*N*-methyliminodiacetic acid boronate
BOX	bis(oxazoline)
BPin	pinacolboronate
bpz	2,2′-bipyrazine
BQ	1,4-benzoquinone
t-BuBrettPhos	2-(di-*tert*-butylphosphino)-2′,4′,6′-triisopropyl-3,6-dimethoxy-1,1′-biphenyl
t-BuXPhos	2-di-*tert*-butylphosphino-2′,4′,6′-triisopropylbiphenyl
COD	cyclooctadiene
Cp*	pentamethylcyclopentadienyl
18-crown-6	1,4,7,10,13,16-hexaoxacyclooctadecane
Cryptand-222	4,7,13,16,21,24-hexaoxa-1,10-diazabicyclo[8.8.8]hexacosane
dba	dibenzylideneacetone
Dbfox-Ph	phenyl-bis(oxazoline)
dF(CF$_3$)ppy	2-(2,4-difluorophenyl)-5-(trifluoromethyl)pyridine
dmdba	3,5,3,5-dimethoxydibenzylideneacetone
dtbbpy	4,4′-di-*tert*-butyl-2,2′-dipyridyl
dtbyp	4,4′-di-*tert*-butyl-2,2′-bipyridine
F-TEDA	Selectfluor; 1-chloromethyl-4-fluoro-1,4-diazoniabicyclo[2.2.2]octane bis(tetrafluoroborate)
HFIP	1,1,1,3,3,3-hexafluoro-2-propanol

HGPhos	(diadamantan-1-yl)(4″-butyl-2′,4′,6′-triisopropyl-3,6-dimethoxy-[1,1′:3′,1″-terphenyl]-2-yl)phosphane
IPr	1,3-bis(2,6-diisopropylphenyl)imidazol-2-ylidene
l	linear
mCPBA	3-chloroperoxybenzoic acid
[Me$_3$pyF]BF$_4$	1-fluoro-2,4,6-trimethylpyridinium tetrafluoroborate
[Me$_3$pyF]PF$_6$	1-fluoro-2,4,6-trimethylpyridinium hexafluorophosphate
2-Me-THF	2-methyltetrahydrofuran
MS	molecular sieves
MW	microwave heating
NfF	nonafluorobutanesulfonyl fluoride
NHPI	*N*-hydroxyphthalimide
NFSI	*N*-fluorobenzenesulfonimide
NFTPT	1-fluoro-2,4,6-trimethylpyridinium triflate
NHPI	*N*-hydroxyphthalimide
Np	naphthyl
ox	oxalate
Pd/C	palladium on carbon
Phen	phenanthroline
PhI(OPiv)$_2$	bis(*tert*-butylcarbonyloxy)iodobenzene
Phth	phthalyl
PIP	2-(pyridin-2-yl)isopropyl amine
(–)-PPY	(*R*)-(+)-4-pyrrolidinopyrindinyl (pentamethylcyclopentadienyl)iron
RDS	rate-determining step
salen	*N,N′*-bis(salicylidene)ethylenediamine
Selectfluor	F-TEDA; 1-chloromethyl-4-fluoro-1,4-diazoniabicyclo[2.2.2]octane bis(tetrafluoroborate)
SIPr	1,3-bis(2,6-diisopropylphenyl)dihydroimidazolidene
SPAN(PPh$_2$)$_2$	8,8′-bis(diphenylphosphino)-3,3′,4,4′-tetrahydro-4,4,4′,4′,6,6′-hexamethyl-2,2′-spirobi[2*H*-1-benzopyran]
TADDOL	α,α,α′,α′-tetraaryl-2,2-disubstituted 1,3-dioxolane-4,5-dimethanol
TASF	tris(dimethylamino)sulfonium difluorotrimethylsilicate
TBADT	tetra-*n*-butylammonium decatungstate
TBDPS	*tert*-butyldiphenylsilyl
terpy	2,2′:6′,2″-terpyridine
TMAF	tetramethylammonium fluoride
tmp	tetramesitylporphyrin
Tol	tolyl, methyphenyl

CHART 1. LIGANDS USED IN THE TEXT AND TABLES

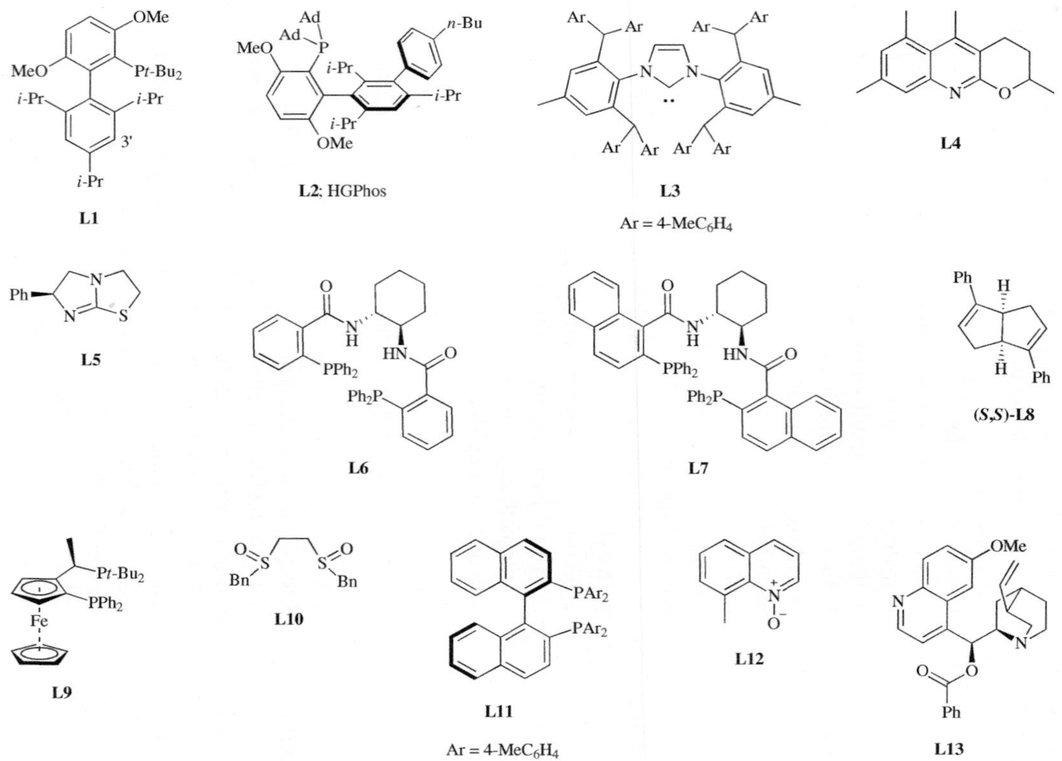

L4

(S,S)-L8

L13

L3

Ar = 4-MeC$_6$H$_4$

L7

L12

L2: HGPhos

L6

L11

Ar = 4-MeC$_6$H$_4$

L10

L1

L5

L9

L14, L15, L16, L17, L18, L19, L20, L21, L22, L23, L24

CHART 1. LIGANDS USED IN THE TEXT AND TABLES (*Continued*)

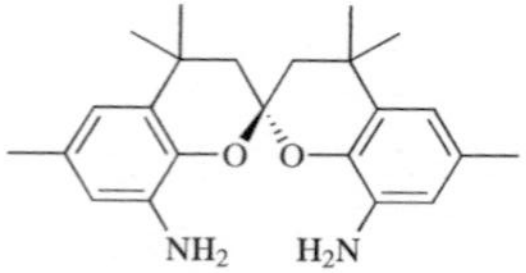

L25

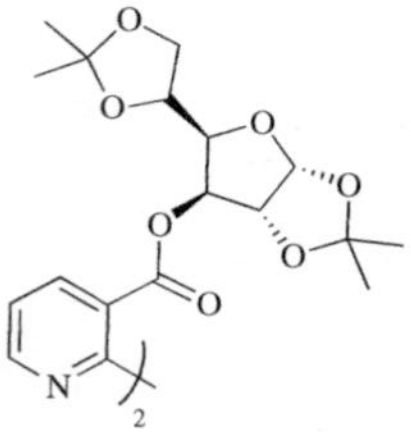

L26

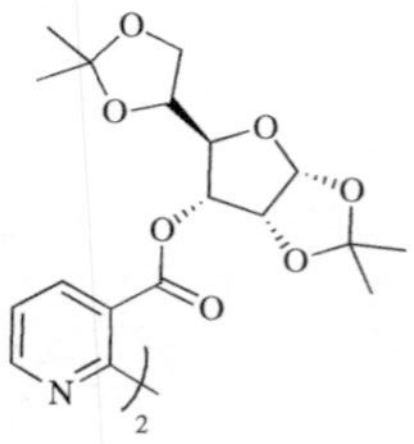

L27

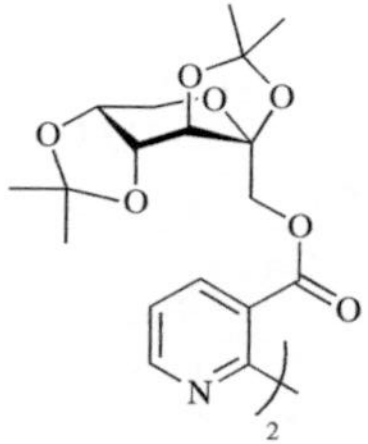

L28

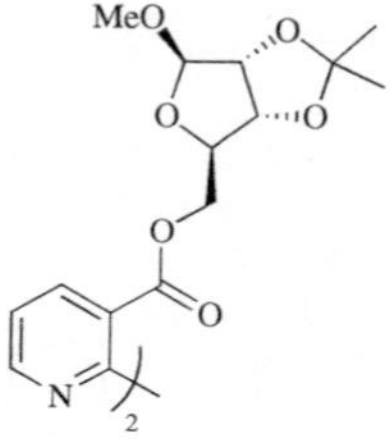

L29

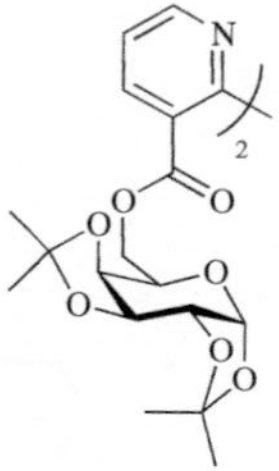

L30

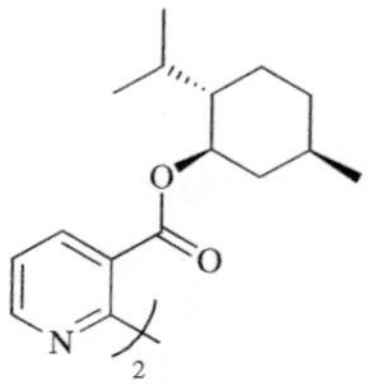

L31

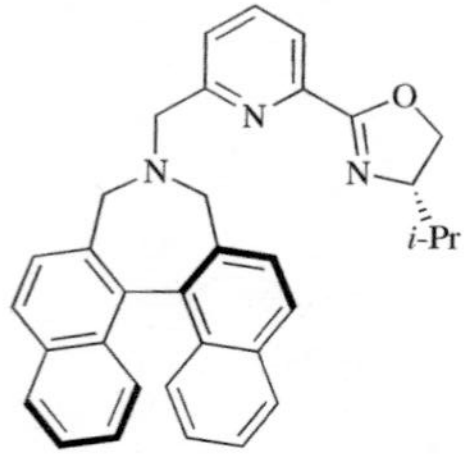

L32

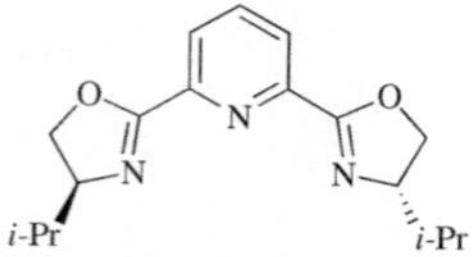

L33

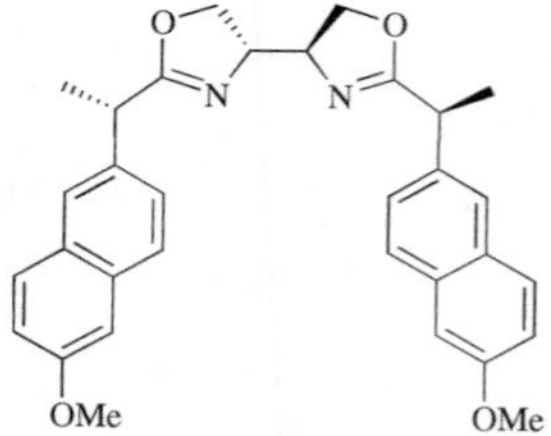

L34

CHART 2. CATALYSTS USED IN THE TEXT AND TABLES

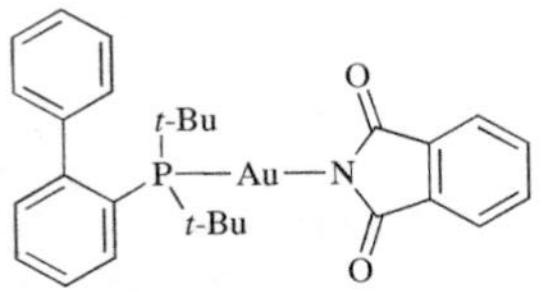

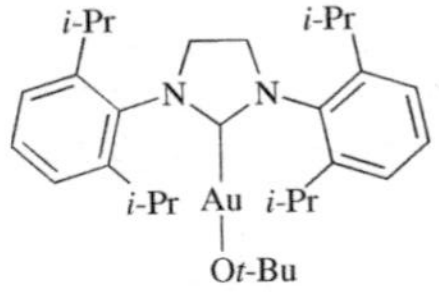

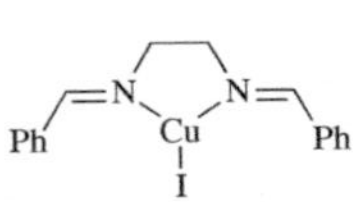

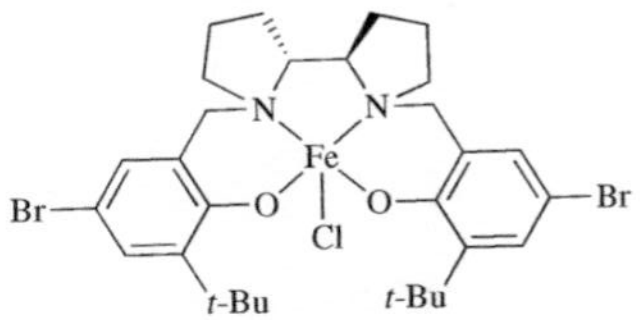

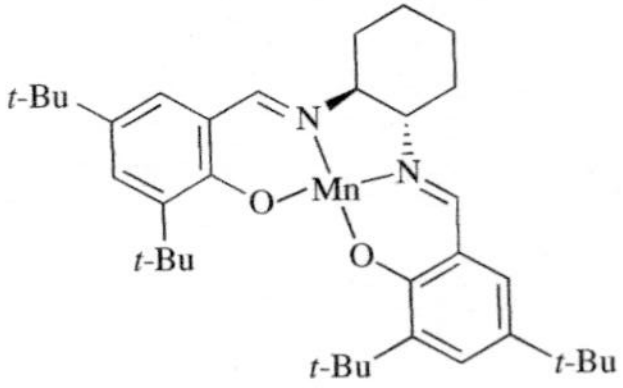

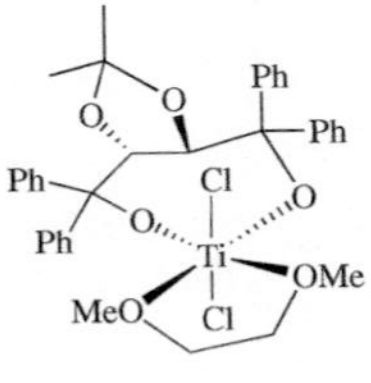

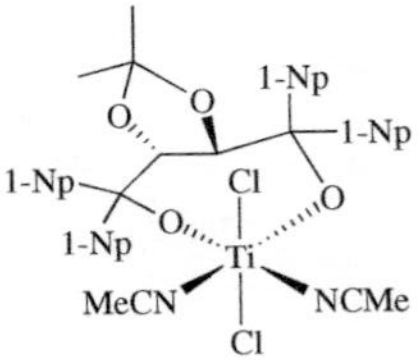

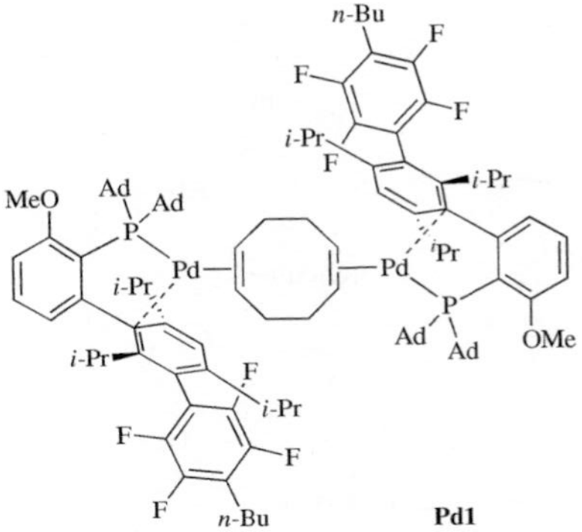

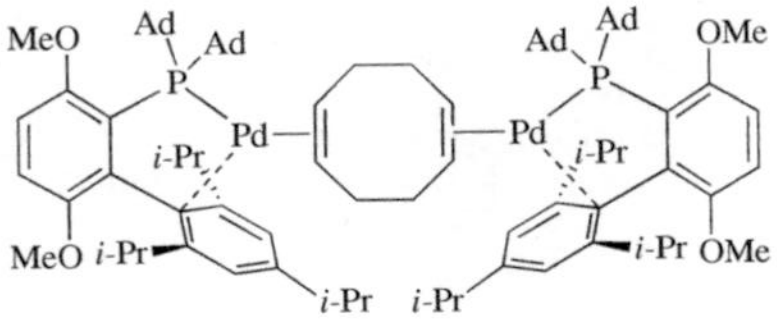

CHART 2. CATALYSTS USED IN THE TEXT AND TABLES (*Continued*)

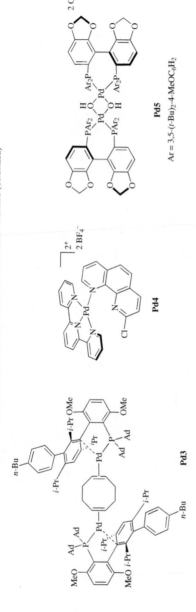

Pd3

Pd4

Pd5

Ar = 3,5-(*i*-Bu)₂-4-MeOC₆H₂

	Ar
Pd6	3,5-Me₂C₆H₃
Pd7	3,5-(*i*-Bu)₂-4-MeOC₆H₂

	Ar
Pd8	3,5-Me₂C₆H₃
Pd9	Ph

Pd10

Ar = 4-MeC₆H₄

Pd11

Ar = 3,5-Me₂C₆H₃

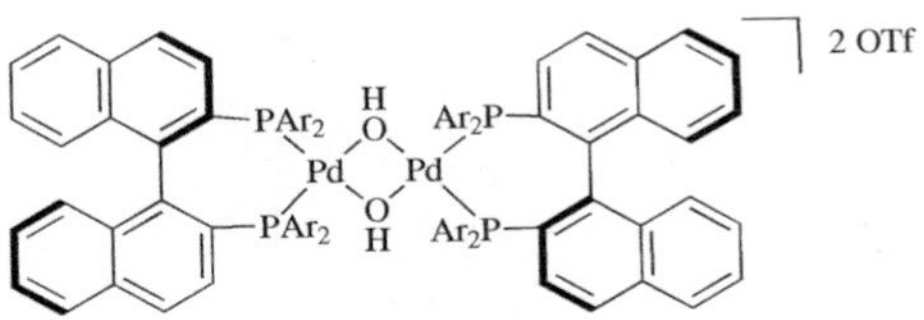

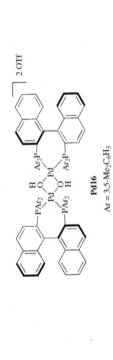

Pd13

2 OTf

Pd12

Ar = 3,5-Me₂C₆H₃

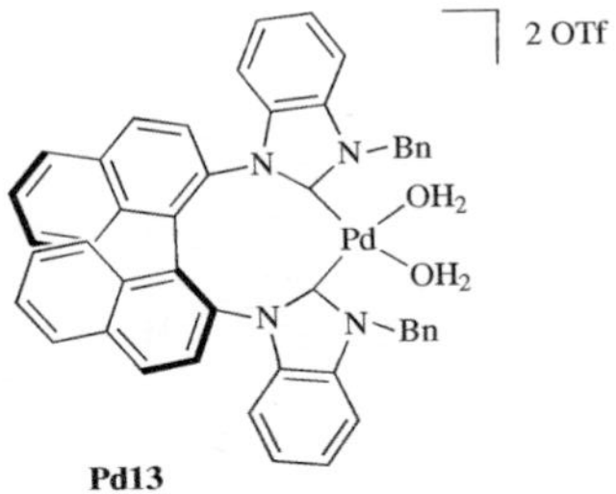

2 OTf

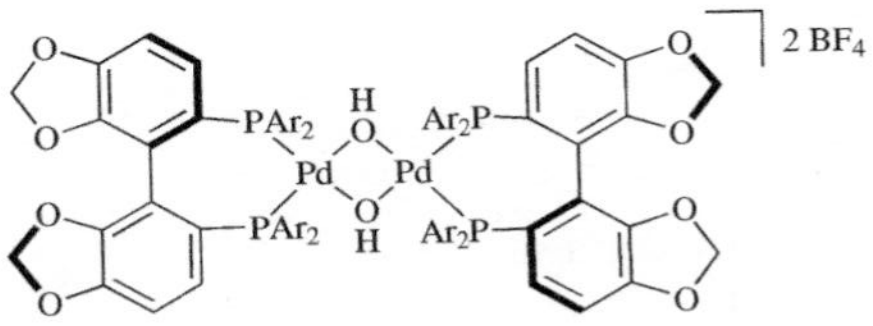

Pd14

Ar = 3,5-(t-Bu)₂-4-MeOC₆H₂

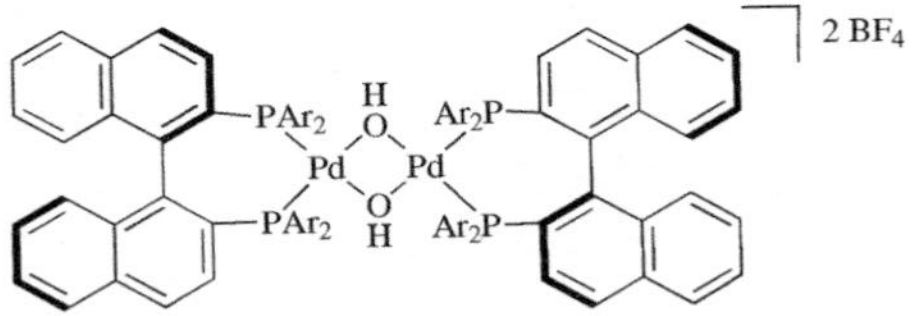

2 BF₄

Pd15

Ar = 3,5-Me₂C₆H₃

2 BF₄

Pd16

Ar = 3,5-Me₂C₆H₃

2 OTf

TABLE 1A. ARYL FLUORIDE SYNTHESIS FROM ARYL HALIDES AND ARYL PSEUDOHALIDES

Substrate	Conditions	Product(s) and Yield(s) (%)	Refs.

*Please refer to the charts preceding the tables for the structures indicated by the **bold** numbers.*

C$_6$

Substrate	Conditions	Product(s) and Yield(s) (%)	Refs.
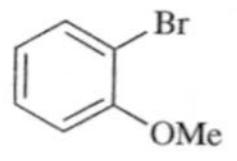 2-bromoanisole (Br, OMe)	**Pd2** (2 mol %), AgF (2 eq), KF (0.5 eq), cyclohexane, 130°, 14 h	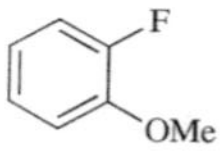 (F, OMe) (73)	33
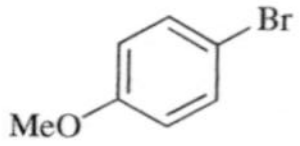 4-bromoanisole (MeO, Br)	**Pd2** (2 mol %), AgF (2 eq), KF (0.5 eq), cyclohexane, 130°, 14 h	 **I** (3-F, MeO) + **II** (4-F, MeO) I + II (70), I/II = 2.7:1	33

C$_{6-13}$

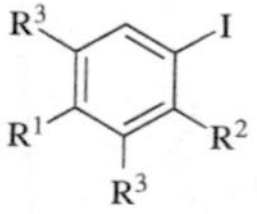

(t-BuCN)$_2$CuOTf (3 eq), AgF (2 eq), DMF, 140°, 22 h   120

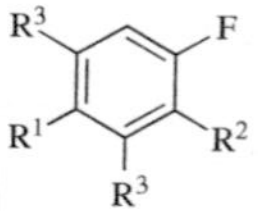

R^1	R^2	R^3	
PhO	H	H	(68)
BnO	H	H	(48)
t-BuCOHN	H	H	(40)
t-BuCOMeN	H	H	(62)
H	Me	H	(87)
CF$_3$	H	H	(40)
OHC	H	H	(43)
EtO$_2$C	H	H	(46)
H	H	Me	(68)
n-Bu	H	H	(62)
t-Bu	H	H	(78)
Ph	H	H	(57)
Bz	H	H	(56)

C₆

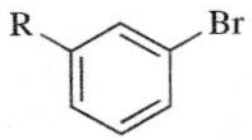

		R	x	
		BnO	1.5	(88)
		MeS	2	(69)

Pd2 (x mol %), AgF (2 eq), KF (0.5 eq), cyclohexane, 110°, 14 h — 33

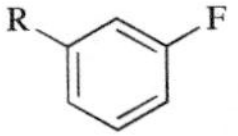

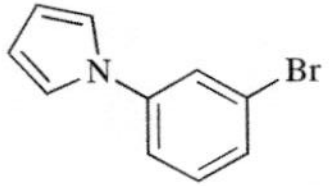

Pd2 (1 mol %), AgF (2 eq), KF (0.5 eq), cyclohexane, 120°, 14 h — 33

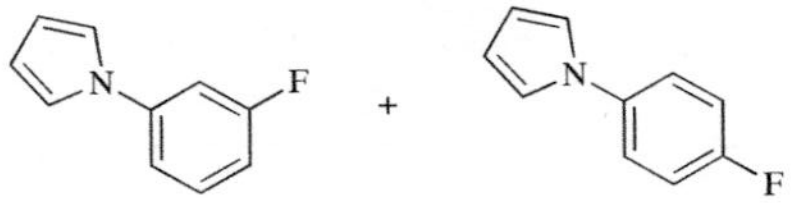

I + **II** (86), **I/II** > 50:1   **II** — 33

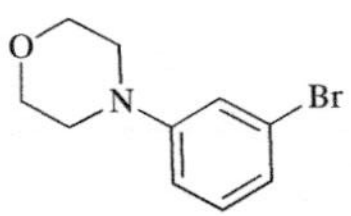

Pd2 (1.5 mol %), AgF (2 eq), KF (0.5 eq), cyclohexane, 120°, 14 h   (86) — 33

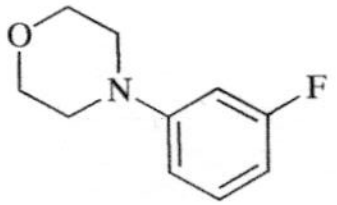

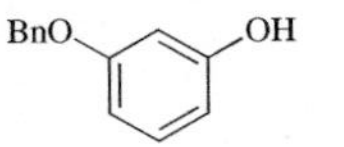

1. NfF (1.5 eq), CsF (2.0 eq), toluene, 50°, 1 h
2. Pd₂(dba)₃ (2 mol %), **L1** (6 mol %), MW, toluene, 180°, 0.5 h   (56) — 116

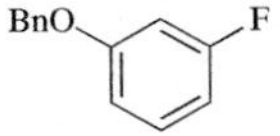

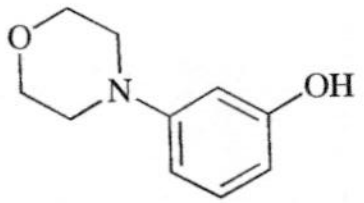

1. NfF (1.5 eq), CsF (2.0 eq), toluene, 50°, 24 h
2. Pd₂(dba)₃ (2 mol %), **L1** (6 mol %), MW, toluene, 180°, 0.5 h   (41) — 116

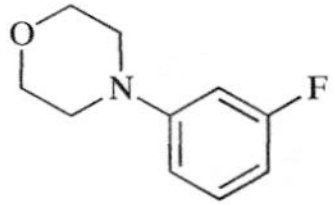

TABLE 1A. ARYL FLUORIDE SYNTHESIS FROM ARYL HALIDES AND ARYL PSEUDOHALIDES (*Continued*)

Substrate	Conditions	Product(s) and Yield(s) (%)	Refs.

*Please refer to the charts preceding the tables for the structures indicated by the **bold** numbers.*

C$_6$

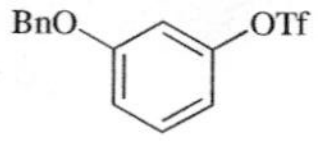

Pd$_2$(dba)$_3$ (2 mol %), **L1** (6 mol %), CsF (2 eq), MW, toluene, 180°, 0.5 h

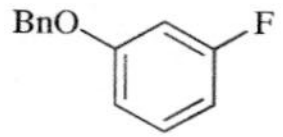

(63)

116

C$_{6-7}$

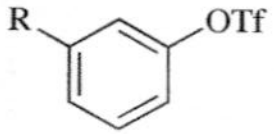

[(cinnamyl)PdCl]$_2$ (*x* mol %), **L1** (*y* mol %), CsF (2 eq), cyclohexane, 12 h

R	x	y	Temp (°)	I	II
BnO	5	15	130	(57)	(2)
Me$_2$N	2	6	130	(84)	(1)
PhOCH$_2$	2.5	7.5	110	(90)	(<1)

25

C$_6$

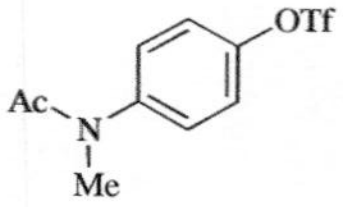

Pd2 (4 mol %), CsF (3 eq), cyclohexane, 120°, 14 h

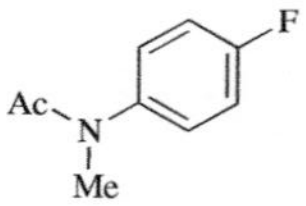

(71)

37

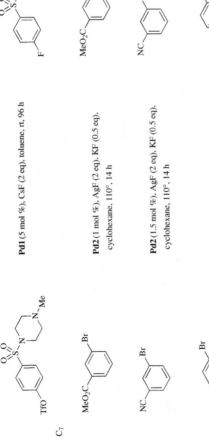

Substrate	Conditions	Product (Yield)	Ref.
C$_7$ — piperazinyl sulfonyl arene, TfO	**Pd1** (5 mol %), CsF (2 eq), toluene, rt, 96 h	piperazinyl sulfonyl arene, F (87)	36
MeO$_2$C-aryl-Br	**Pd2** (1 mol %), AgF (2 eq), KF (0.5 eq), cyclohexane, 110°, 14 h	MeO$_2$C-aryl-F (93)	33
NC-aryl-Br	**Pd2** (1.5 mol %), AgF (2 eq), KF (0.5 eq), cyclohexane, 110°, 14 h	NC-aryl-F (73)	33
Br-aryl-C(O)NMe$_2$	**Pd2** (1.5 mol %), AgF (2 eq), KF (0.5 eq), cyclohexane, 90°, 14 h	F-aryl-C(O)NMe$_2$ (92)	33
Br-aryl (CF$_3$, SO$_2$Me)	**Pd2** (1 mol %), AgF (2 eq), KF (0.5 eq), cyclohexane, 110°, 14 h	F-aryl (CF$_3$, SO$_2$Me) (75)	33

TABLE 1A. ARYL FLUORIDE SYNTHESIS FROM ARYL HALIDES AND ARYL PSEUDOHALIDES (Continued)

*Please refer to the charts preceding the tables for the structures indicated by the **bold** numbers.*

C7

Substrate	Conditions	Product(s) and Yield(s) (%)	Refs.
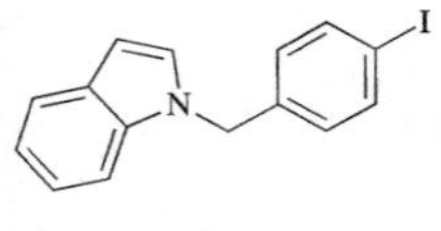	(t-BuCN)₂CuOTf (3.0 eq), AgF (2.0 eq). DMF, 140°, 22 h.	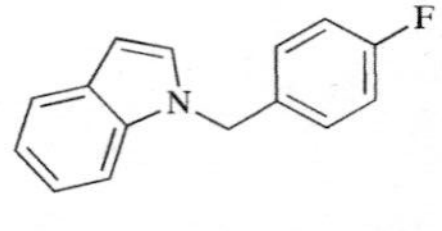 (59)	120
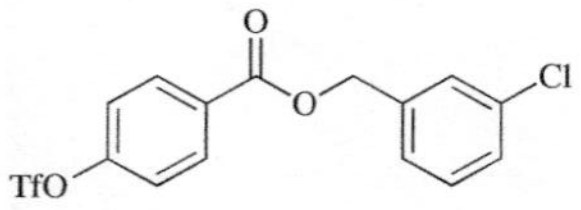	**Pd1** (5 mol %), CsF (3 eq), toluene, rt, 96 h	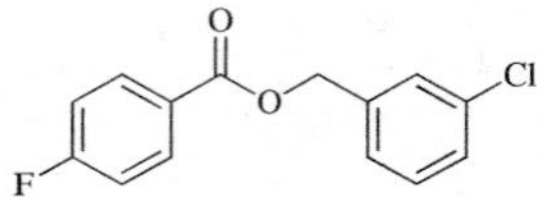 (90)	36
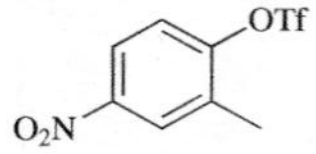	[(Cinnamyl)PdCl]₂ (2 mol %), **L1** (6 mol %), CsF (2 eq), toluene, 110°, 12 h	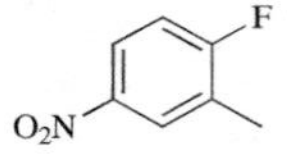 (80)	25
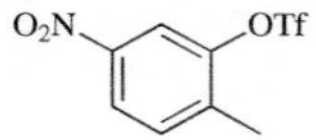	[(Cinnamyl)PdCl]₂ (2 mol %), **L1** (6 mol %), CsF (2 eq), toluene, 110°, 12 h	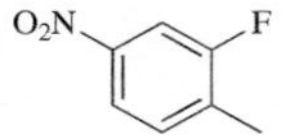 (83)	25
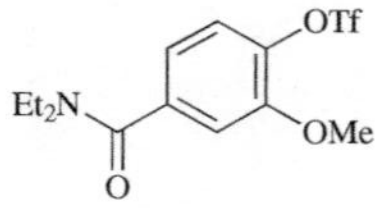	**Pd2** (2 mol %), CsF (3 eq), toluene, 100°, 14 h	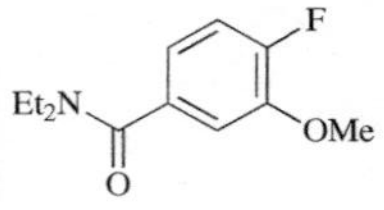 (91)	37

Substrate	Conditions	Product (yield)	Ref
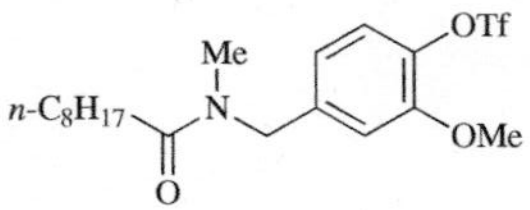 C_8 — n-C_8H_{17}, Me-N, C=O, OMe, OTf benzyl derivative	**Pd2** (1.5 mol %), CsF (3 eq), toluene, 130°, 14 h 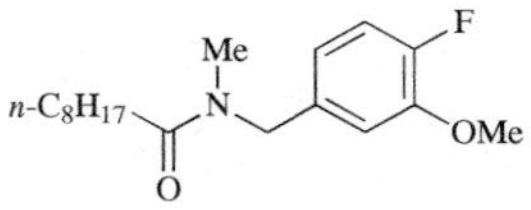	n-C_8H_{17}, Me-N, C=O, OMe, F benzyl derivative (85)	37
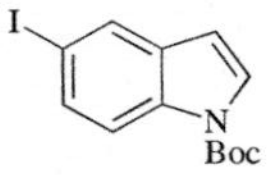 5-iodo-indole, N-Boc	(t-BuCN)$_2$CuOTf (3.0 eq), AgF (2.0 eq), DMF, 140°, 22 h 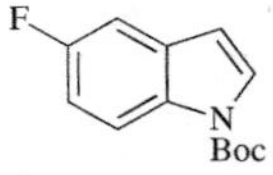	5-fluoro-indole, N-Boc (40)	120
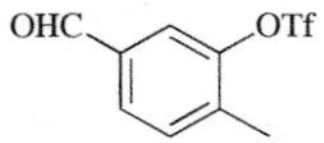 OHC, OTf, methylbenzene	**Pd1** (1 mol %), CsF (3 eq), 2-Me-THF, rt, 96 h 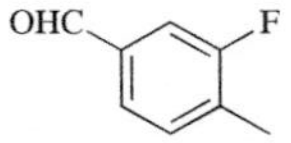	OHC, F, methylbenzene (82)	36
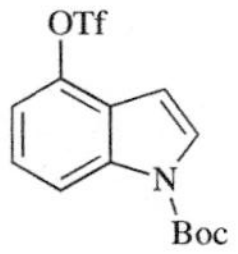 4-OTf-indole, N-Boc	[(Cinnamyl)PdCl]$_2$ (2 mol %), **L1** (6 mol %), CsF (3 eq), toluene, 110°, 12 h 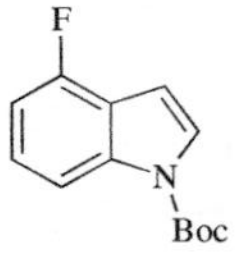	4-fluoro-indole, N-Boc (73)	25

TABLE 1A. ARYL FLUORIDE SYNTHESIS FROM ARYL HALIDES AND ARYL PSEUDOHALIDES (Continued)

Substrate	Conditions	Product(s) and Yield(s) (%)	Refs.

*Please refer to the charts preceding the tables for the structures indicated by the **bold** numbers.*

C₈

Substrate	Conditions	Product(s) and Yield(s) (%)	Refs.
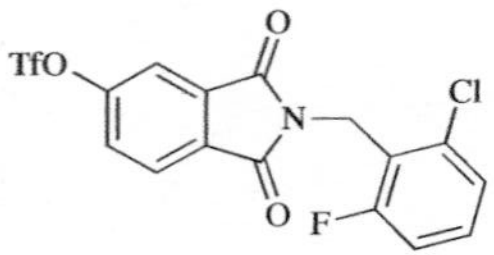	**Pd1** (2 mol %), CsF (3 eq), toluene, rt, 96 h	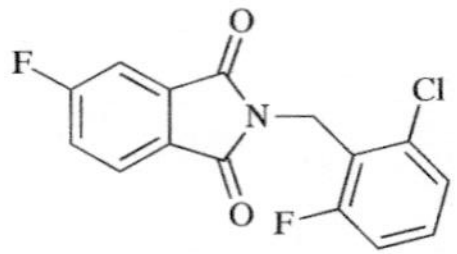 (87)	36

C₉

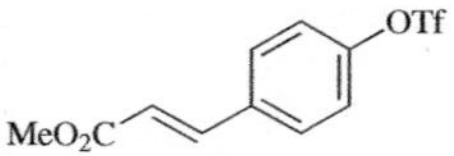	**Pd2** (2 mol %), CsF (3 eq), toluene, 80°, 14 h	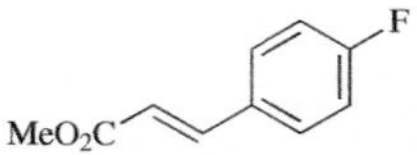 (92)	37
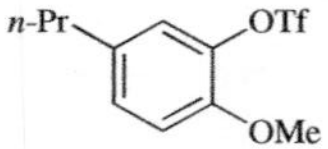	[(Cinnamyl)PdCl]₂ (2.5 mol %), **L1** (7.5 mol %), CsF (2 eq), cyclohexane, 130°, 12 h	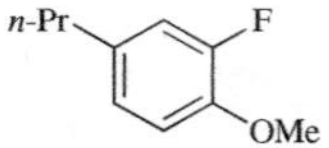 (60) 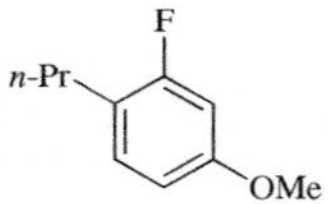 + (1)	25
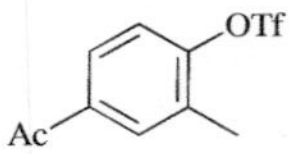	[(Cinnamyl)PdCl]₂ (0.75 mol %), **L1** (1.1 mol %), toluene, 120°, 20 min residence time (35 μL min⁻¹)	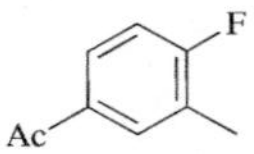 (85)[a]	248
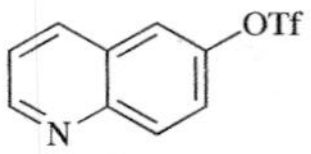	[(Cinnamyl)PdCl]₂ (2 mol %), **L1** (6 mol %), CsF (2 eq), toluene, 120°, 12 h	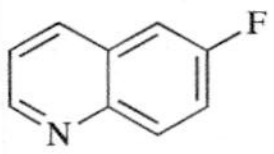 (78)	25

C_{10}

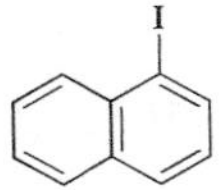

(t-BuCN)$_2$CuOTf (3 eq), AgF (2 eq),
DMF, 140°, 22 h

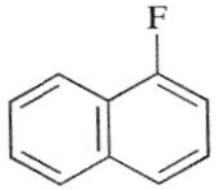

(50)

120

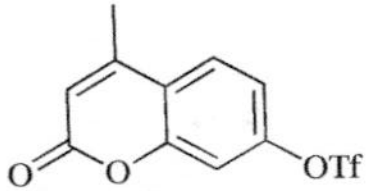

C_{11-14}

Pd2 (2 mol %), CsF (3 eq),
toluene, 110°, 14 h

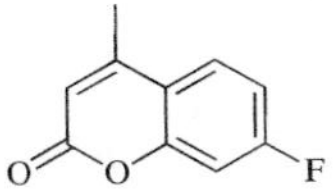

(92)

37

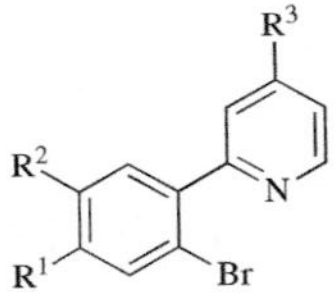

[Cu(MeCN)$_4$]PF$_6$ (20 mol %), AgF (2.0 eq),
n-Bu$_4$NPF$_6$ (1.0 eq), MeCN, 120°, 8 h

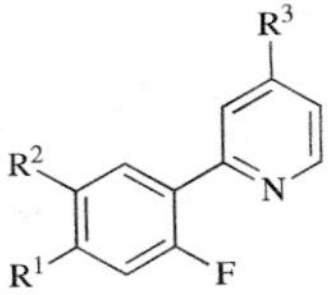

255

R^1	R^2	R^3	
H	H	H	(66)
H	H	Cl	(28)
H	Br	Me	(53)
H	Cl	Me	(67)
H	MeO	Me	(83)
H	CF$_3$O	Me	(42)
H	O$_2$N	Me	(45)
Cl	Cl	Me	(50)
MeO	MeO	Me	(63)
H	Me	Me	(81)
H	CF$_3$	Me	(56)
H	MeO$_2$C	Me	(76)
H	Ac	Me	(76)
Me	Me	Me	(68)

TABLE 1A. ARYL FLUORIDE SYNTHESIS FROM ARYL HALIDES AND ARYL PSEUDOHALIDES (*Continued*)

Substrate	Conditions	Product(s) and Yield(s) (%)	Refs.

*Please refer to the charts preceding the tables for the structures indicated by the **bold** numbers.*

C₁₁

Substrate	Conditions	Product(s) and Yield(s) (%)	Refs.
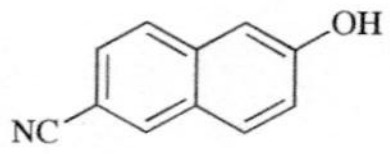 (naphthalene, NC–, –OH)	1. NfF (1.5 eq), CsF (2.0 eq), toluene, 50°, 48 h 2. Pd₂(dba)₃ (2 mol %), **L1** (6 mol %), MW, toluene, 180°, 0.5 h	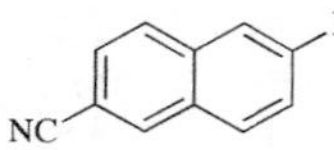 (naphthalene, NC–, –F) (64)	116

C₁₂

Substrate	Conditions	Product(s) and Yield(s) (%)	Refs.
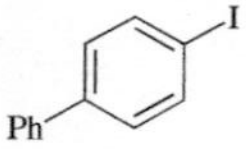 (biphenyl, I, Ph)	(*t*-BuCN)₂CuOTf (3 eq), AgF (2 eq), DMF, 140°, 22 h	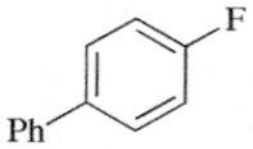 (biphenyl, F, Ph) (57)	120
(pyridine with R¹, R², Br)	[Cu(MeCN)₄]PF₆ (20 mol %), AgF (2.0 eq), *n*-Bu₄NPF₆ (1.0 eq), MeCN, 120°, 8 h	(pyridine with R¹, R², F) R¹ / R²: H / Me (73); Me / H (76)	255
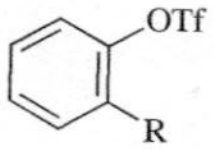 (OTf, R)	[(Cinnamyl)PdCl]₂ (*x* mol %), **L1** (*y* mol %), toluene, 120°, 20 min residence time (35 mL min⁻¹)	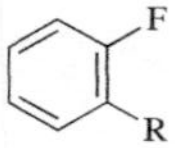 (F, R) R / x / y: *c*-C₆H₁₁ / 2 / 6 (85)ᵃ; Ph / 2.5 / 7.5 (79)ᵃ	248
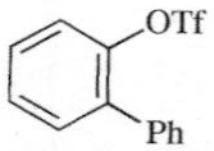 (OTf, Ph)	[(Cinnamyl)PdCl]₂ (2 mol %), **L1** (6 mol %), CsF (2 eq), toluene, 110°, 12 h	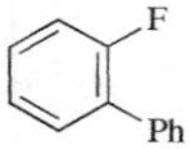 (F, Ph) (82)	25

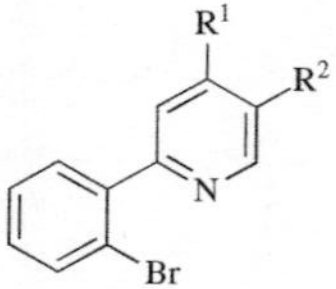

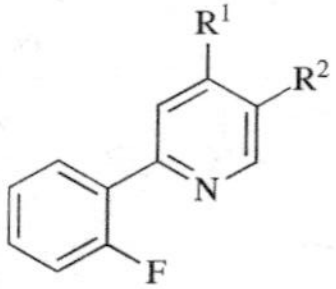

710

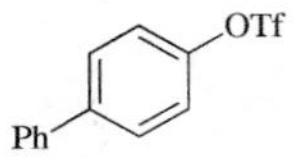

25

(3)   +

(78)   +

[(Cinnamyl)PdCl]$_2$ (2.5 mol %), **L1** (7.5 mol %), CsF (2 eq), cyclohexane, 100°, 12 h

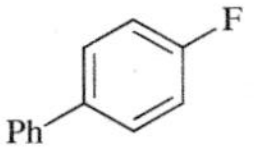

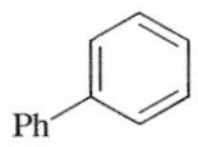

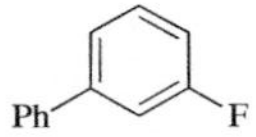

37

(2)   +   (82)

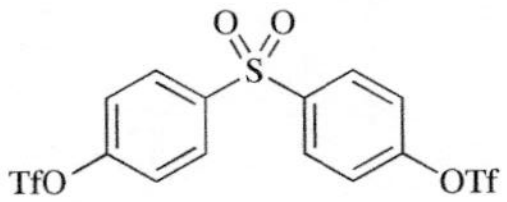

Pd2 (6 mol %), CsF (3 eq), toluene, 130°

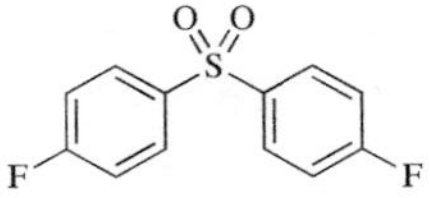

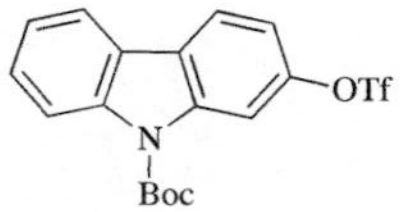

25

(1)

(76)   +

[(Cinnamyl)PdCl]$_2$ (2.5 mol %), **L1** (7.5 mol %), CsF (2 eq), cyclohexane, 110°, 12 h

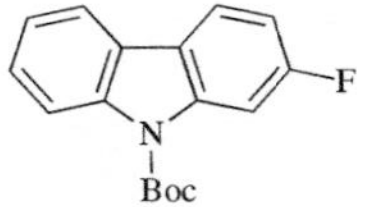

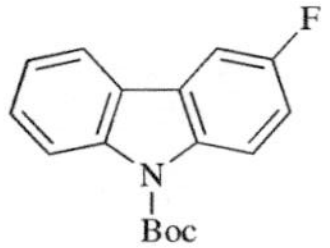

33

(96)

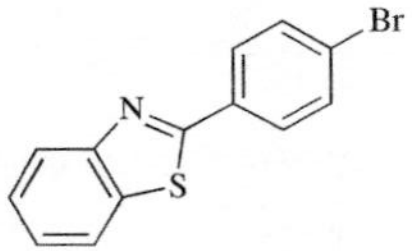

Pd2 (1 mol %), AgF (2 eq), KF (0.5 eq), cyclohexane, 90°, 14 h

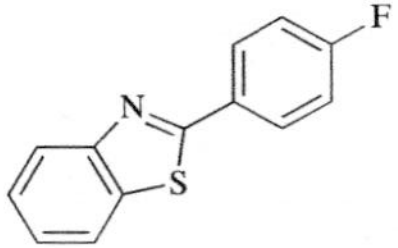

C$_{13}$

TABLE 1A. ARYL FLUORIDE SYNTHESIS FROM ARYL HALIDES AND ARYL PSEUDOHALIDES (*Continued*)

Substrate	Conditions	Product(s) and Yield(s) (%)	Refs.

*Please refer to the charts preceding the tables for the structures indicated by the **bold** numbers.*

C13

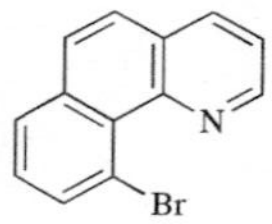

	[Cu(MeCN)4]PF6 (20 mol %), n-Bu4NPF6 (1 eq), AgF (2 eq), MeCN, 120°, 8 h	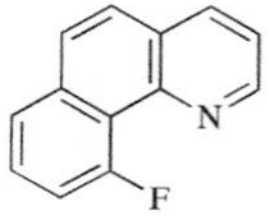 (58)	255
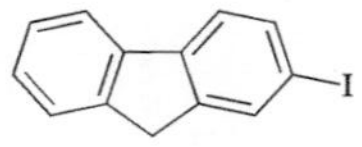	**Pd2** (2 mol %), AgF (2 eq), KF (0.5 eq), cyclohexane, 130°, 14 h	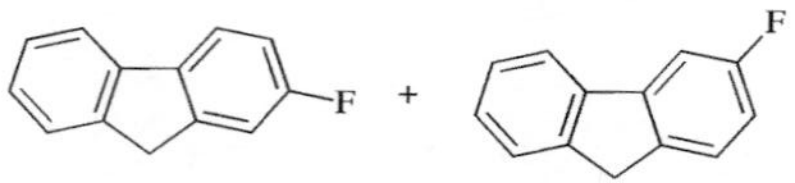 **I** + **II** (76), **I/II** = 17:1	33
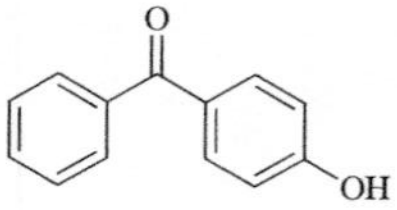	1. NfF (1.5 eq), CsF (2 eq), toluene, 50°, 16 h 2. Pd2(dba)3 (2 mol %), **L1** (6 mol %), MW, toluene, 180°, 0.5 h	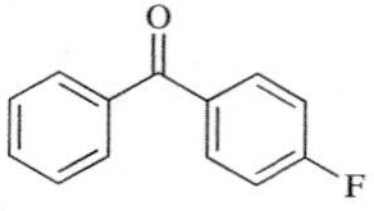 (66)	116
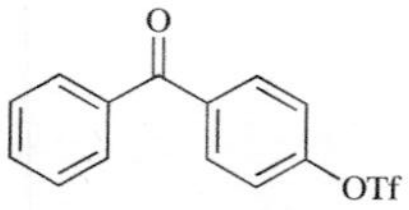	Pd2(dba)3 (2 mol %), **L1** (6 mol %), CsF (2 eq), MW, toluene, 180°, 0.5 h	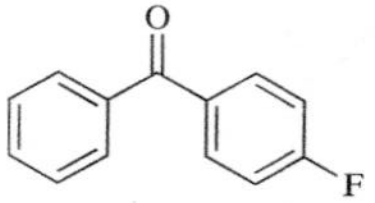 (82)	116

C14

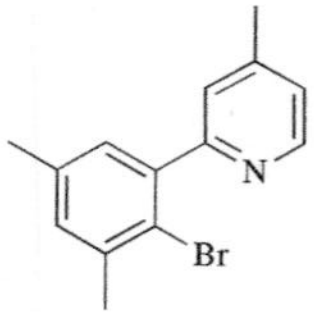

| | [Cu(MeCN)4]PF6 (20 mol %), AgF (2.0 eq), <br> n-Bu4NPF6 (1.0 eq), MeCN, 120°, 8 h | 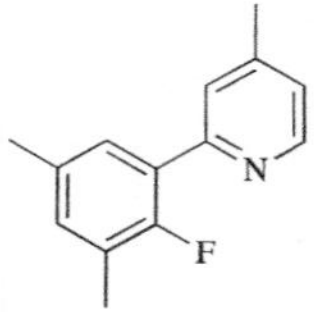 (51) | 255 |

712

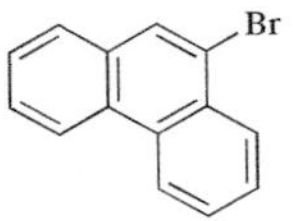

33

(94)

Pd2 (2 mol%), AgF (2 eq), KF (0.5 eq).
cyclohexane, 90°, 14 h

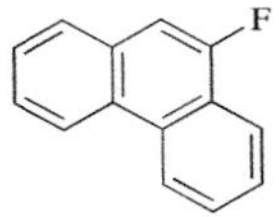

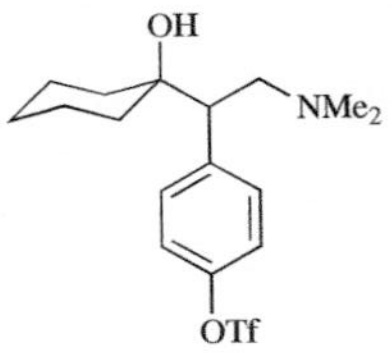

Pd2 (3 mol %), CsF (3 eq).
toluene, 130°, 14 h

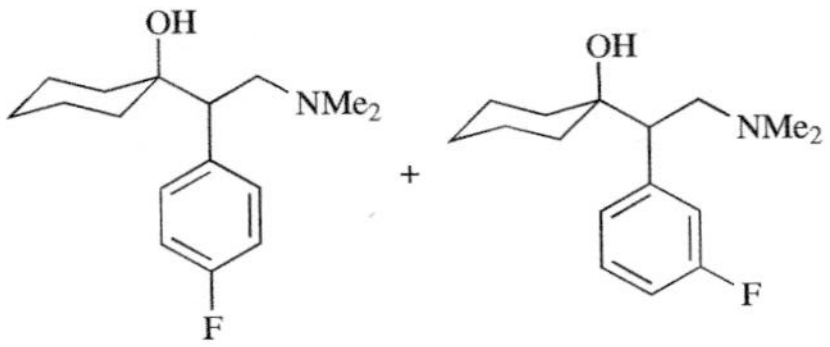

I + II (67), I/II = 3:1   **II**

37

I

36

(95)

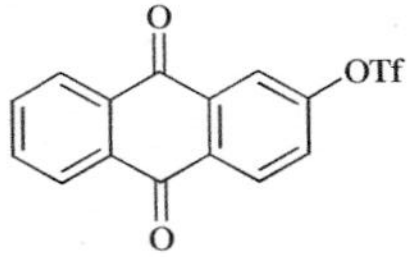

Pd1 (2.5 mol %), CsF (3 eq). toluene. rt, 96 h

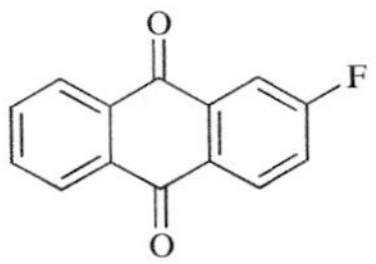

255

(69)

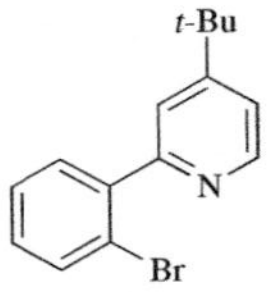

[Cu(MeCN)₄]PF₆ (20 mol %), AgF (2.0 eq),
n-Bu₄NPF₆ (1.0 eq), MeCN, 120°, 8 h

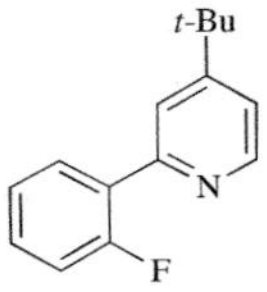

C₁₅

713

Substrate	Conditions	Product(s) and Yield(s) (%)	Refs.

*Please refer to the charts preceding the tables for the structures indicated by the **bold** numbers.*

C₁₅

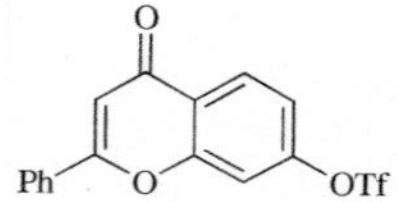

[(Cinnamyl)PdCl]₂ (2 mol %), **L1** (6 mol %), CsF (3 eq), toluene, 80°, 12 h

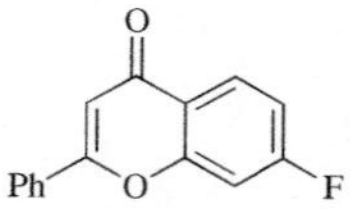

(63)   (2)

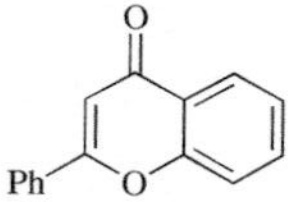

25

Pd1 (x mol %), CsF (3 eq), toluene, 110°, 24 h

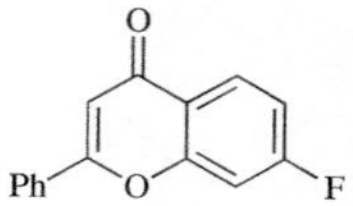

x	
0.25	(94)
6	(99)

36

C₁₆

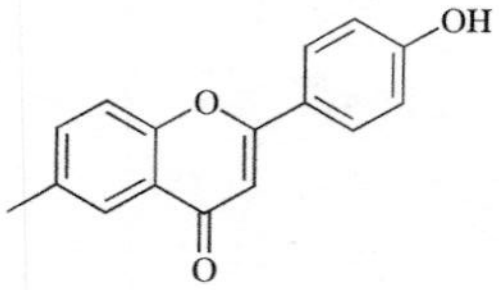

1. NfF (1.5 eq), CsF (2.0 eq), toluene, 50°, 20 h
2. Pd₂(dba)₃ (2 mol %), **L1** (6 mol %), MW, toluene, 180°, 0.5 h

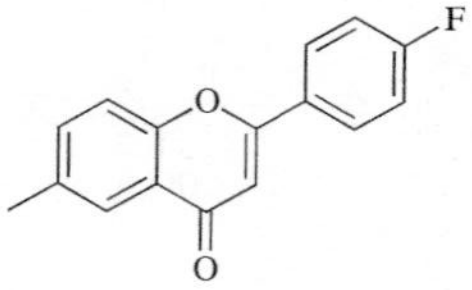

(69)

116

C₁₈

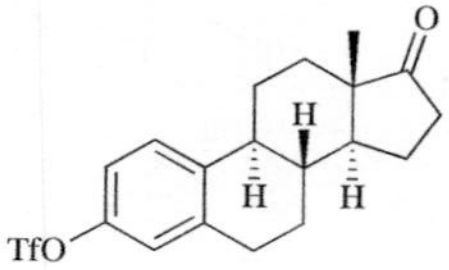

Pd2 (1 mol %), CsF (3 eq), cyclohexane, 120°, 14 h

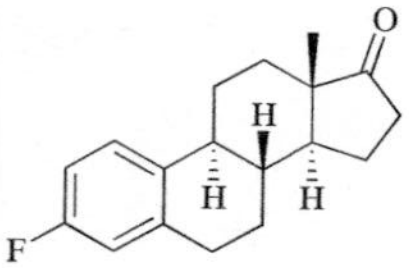

(74)

25

C$_{19}$

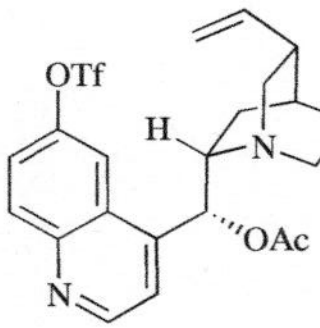

[(Cinnamyl)PdCl]$_2$ (5 mol %), **L1** (7.5 mol %), CsF (2 eq), toluene, 110°, 12 h

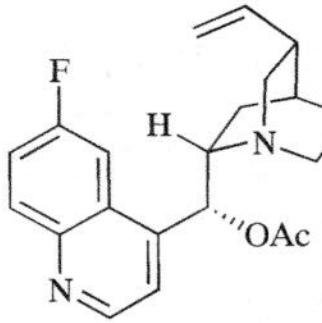

(70)

37

C$_{20}$

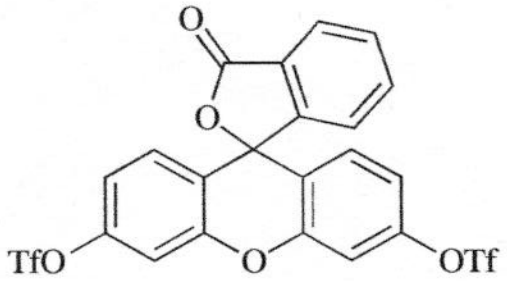

[(Cinnamyl)PdCl]$_2$ (4 mol %), **L1** (6 mol %), CsF (6 eq), toluene, 110°, 12 h

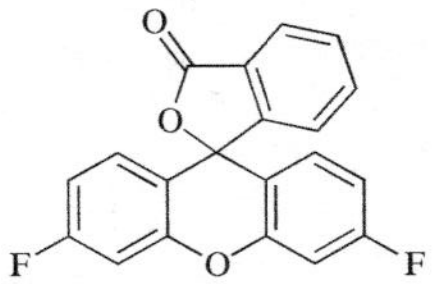

(73)

25

C$_{27}$

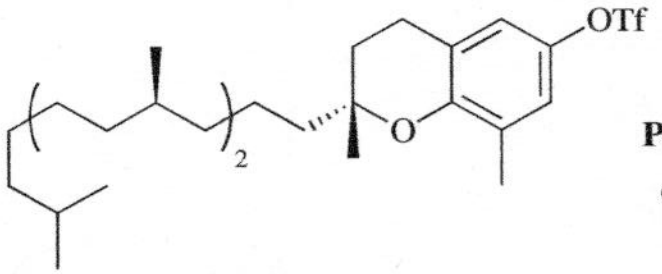

Pd2 (1.5 mol %), CsF (3 eq), cyclohexane, 130°

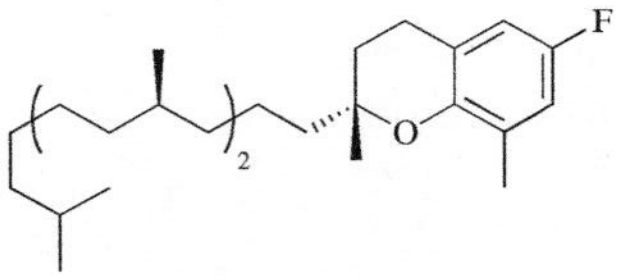

(88)

37

^{a} The reactions were conducted in flow in a CsF-packed-bed reactor.

TABLE 1B. ARYL FLUORIDE SYNTHESIS FROM ARYL BORONIC ACIDS AND THEIR DERIVATIVES

Boronic Substrate	Conditions	Product(s) and Yield(s) (%)	Refs.

*Please refer to the charts preceding the tables for the structures indicated by the **bold** numbers.*

C$_{6-9}$

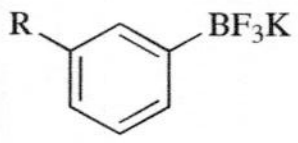

R—⟨benzene ring⟩—BF$_3$K

AgOTf (3 eq), Selectfluor (1.2 eq),
LiOH•H$_2$O (1.2 eq), EtOAc, 55°, 5–15 h

R—⟨benzene ring⟩—F

R	
MeO	(42)
EtO	(42)
BnO	(52)
Me	(63)
CF$_3$	(66)
MeO$_2$C	(67)
Ac	(65)
i-Pr	(68)

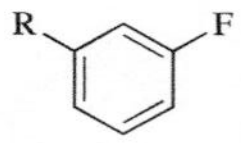

135

C$_{6-12}$

R—⟨benzene ring⟩—BF$_3$K

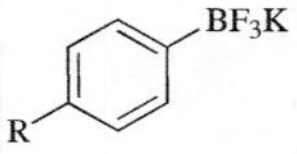

AgOTf (3 eq), Selectfluor (1.2 eq),
LiOH•H$_2$O (1.2 eq), EtOAc, 55°, 5–15 h

R—⟨benzene ring⟩—F

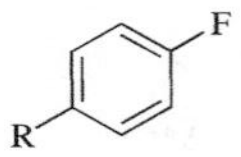

R	
H	(51)
F	(65)
Cl	(60)
Br	(63)
MeO	(65)
EtO	(65)
BuO	(75)
BnO	(55)
Me	(63)
CF$_3$	(68)
MeO$_2$C	(82)
NC–	(40)
Ac	(75)
t-Bu	(70)
n-C$_6$H$_{13}$	(78)
Ph	(85)

135

136

C_{6-18}

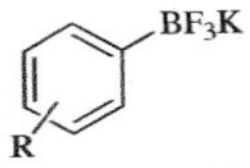

FeCl$_3$ (1.5 eq), Selectfluor (2.0 eq),
KF (3.0 eq), EtOAc, 55°, 10–15 h

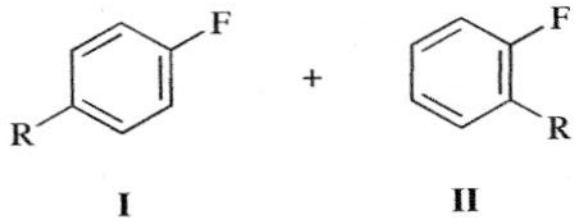

R	I	II
H	I + II (40)	
2-MeO	(2)	(40)
3-MeO	(25)	(15)
3-EtO	(40)	(11)
4-F	(60)	(0)
4-MeO	(55)	(15)
4-EtO	(50)	(20)
4-n-BuO	(50)	(25)
4-BnO	(60)	(5)
2-Me	(0)	(60)
3-Me	(39)	(21)
4-Me	(40)	(0)
4-CF$_3$	(0)	(0)
4-NC	(10)	(0)
3-i-Pr	(29)	(36)
3,4-t-Bu$_2$	(40)	(0)
3,4-Ph$_2$	(54)	(0)

TABLE 1B. ARYL FLUORIDE SYNTHESIS FROM ARYL BORONIC ACIDS AND THEIR DERIVATIVES (*Continued*)

Boronic Substrate	Conditions	Product(s) and Yield(s) (%)	Refs.
*Please refer to the charts preceding the tables for the structures indicated by the **bold** numbers.*			
C_6			
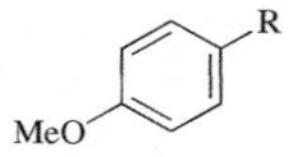	AgOTf (3 eq), Selectfluor (1.2 eq), $LiOH{\bullet}H_2O$ (1.2 eq), EtOAc, 55°, 5–15 h	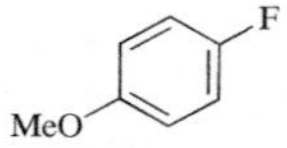 R: $B(OH)_2$ (16); 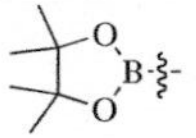 (19); 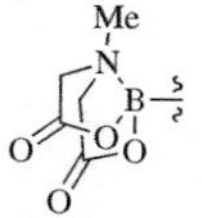 (4); BF_3K (70)	135
C_{6-9}			
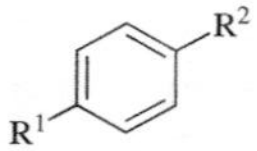	[(Terpy)Pd(MeCN)]$(BF_4)_2$ (*x* mol %), terpy (*y* mol %), NaF (1 eq), Selectfluor (*z* eq), solvent, 15 h	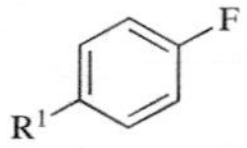	61

R^1	R^2	*x*	*y*	*z*	Solvent	Temp (°)	
Br	BF_3K	2	4	1.2	MeCN	40	(96)
PhO	BF_3K	2	4	1.1	DMF	23	(99)
PhO	BMIDA	2	4	1.5	DMF	23	(70)
H_2NOC	BF_3K	2	4	1.2	MeCN	40	(81)
HO_2CCH_2	BF_3K	5	10	1.2	DMF	4	(74)
$AcCH_2$	BF_3K	2	4	1.2	DMF	23	(70)

C_{6-7}

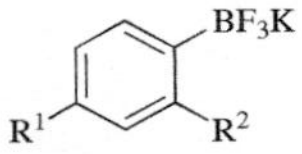

NFTPT (2 eq), (t-BuCN)$_2$CuOTf (x eq),
EtOAc, 80°, 12 h

I + **II**

133

R¹	R²	x	I	II
H	PhO	2.0	(47)	(4)
PhO	H	2.0	(56)	(0)
AcHN	H	1.2	(42)	(0)
MeO$_2$C	H	1.2	(53)	(4)

Cu(OTf)$_2$ (4 eq), KF (4 eq), MeCN, 60°, 20 h

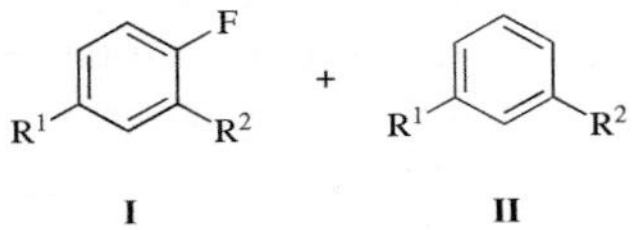

I + **II**

134

R¹	R²	I	II
H	PhO	(66)	(2)
PhO	H	(45)	(1)
O$_2$N	H	(65)	(2)
OHC	H	(57)	(2)
BnO$_2$C	H	(68)	(2)
NC–	H	(63)	(2)

TABLE 1B. ARYL FLUORIDE SYNTHESIS FROM ARYL BORONIC ACIDS AND THEIR DERIVATIVES (*Continued*)

Boronic Substrate	Conditions	Product(s) and Yield(s) (%)	Refs.

*Please refer to the charts preceding the tables for the structures indicated by the **bold** numbers.*

C_{6-8}

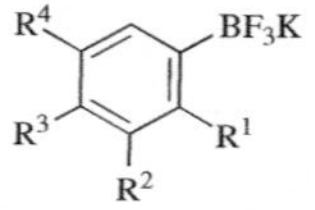

AgOTf (3 eq), Selectfluor (1.2 eq),
LiOH•H_2O (1.2 eq), EtOAc, 55°, 5–15 h

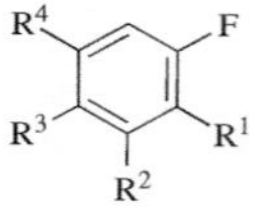

R^1	R^2	R^3	R^4	
H	Cl	Cl	H	(75)
H	Cl	MeO	H	(77)
H	MeO	MeO	MeO	(35)
F	H	H	H	(65)
F	H	MeO	H	(60)
MeO	H	H	H	(40)
Me	H	H	H	(71)
Me	Me	H	H	(90)

135

C_{6-9}

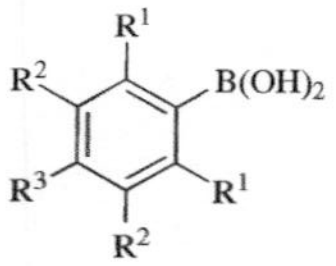

1. NaOH (x eq), AgOTf (y eq),
   MeOH, 0°, 0.5 h
2. Selectfluor (1.05 eq),
   3 Å MS, acetone, 23°, 1 h

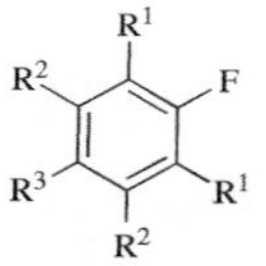

R^1	R^2	R^3	x	y	
H	H	Br	1.0	2	(73)
H	H	HO	1.2	3	(70)
H	H	MeO	1.0	2	(84)
H	H	AcHN	1.0	2	(77)

132

H	H	CF$_3$	1.2	3	(86)
H	H	OHC	1.0	2	(71)
H	H	MeO$_2$C	1.0	2	(76)
H	H	NC–	1.2	3	(77)
H	Me	H	1.0	2	(78)
Me	H	Me	1.0	2	(73)

C$_6$

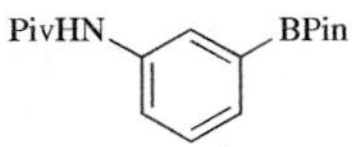

PivHN—BPin

(t-BuCN)$_2$CuOTf (2 eq),
[Me$_3$pyF]PF$_6$ (3 eq),
AgF (2 eq), THF, 50°, 18 h

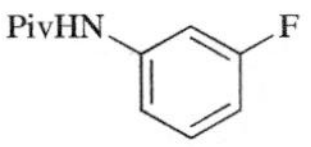

PivHN—F   (64)

54

C$_7$

(HO)$_2$B— indazole-N-Boc

1. NaOH (1.0 eq), AgOTf (2.0 eq),
MeOH, 0°, 0.5 h
2. Selectfluor (1.05 eq),
3 Å MS, acetone, 23°, 1 h

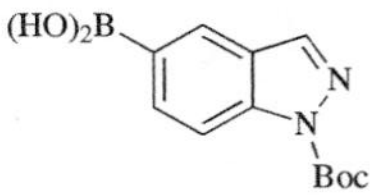

F—indazole-N-Boc   (75)

132

C$_8$

BocO / BocO — BF$_3$K, succinimide

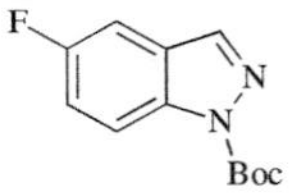

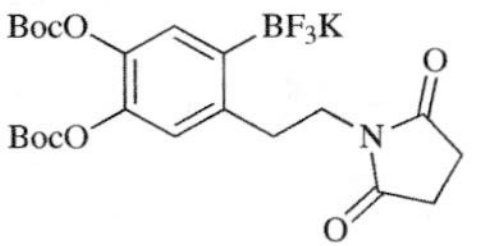

[(Terpy)Pd(MeCN)](BF$_4$)$_2$ (2 mol %),
terpy (4 mol %), NaF (1 eq),
Selectfluor (1.2 eq), MeCN, 40°, 15 h

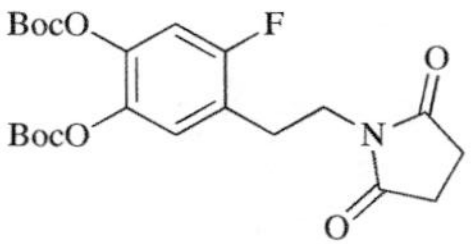

BocO / BocO — F, succinimide   (74)

61

(HO)$_2$B— indole-N-Boc

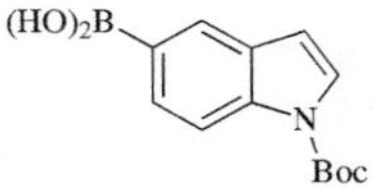

1. NaOH (1.0 eq), AgOTf (2.0 eq),
MeOH, 0°, 0.5 h
2. Selectfluor (1.05 eq),
3 Å MS, acetone, 23°, 1 h

F—indole-N-Boc   (75)

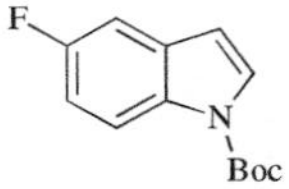

132

TABLE 1B. ARYL FLUORIDE SYNTHESIS FROM ARYL BORONIC ACIDS AND THEIR DERIVATIVES (*Continued*)

*Please refer to the charts preceding the tables for the structures indicated by the **bold** numbers.*

Boronic Substrate	Conditions	Product(s) and Yield(s) (%)	Refs.
C₈			
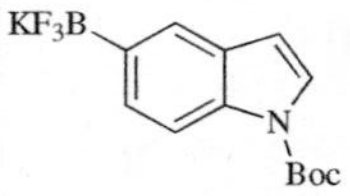	AgOTf (3 eq), Selectfluor (1.2 eq), LiOH•H₂O (1.2 eq), EtOAc, 55°, 5–15 h	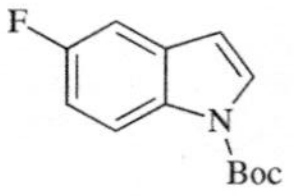 (42)	135
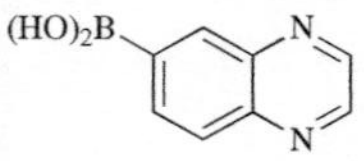	1. NaOH (1.2 eq), AgOTf (3.0 eq), MeOH, 0°, 0.5 h 2. Selectfluor (1.05 eq), 3 Å MS, acetone, 23°, 1 h	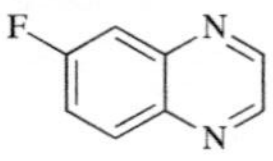 (71)	132
C₉			
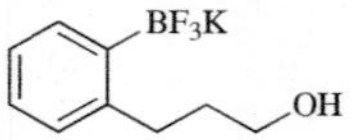	[(Terpy)Pd(MeCN)](BF₄)₂ (2 mol %), terpy (4 mol %), NaF (1 eq), Selectfluor (1.2 eq), DMF, 23°, 15 h	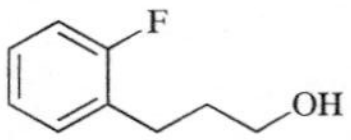 (71)	61
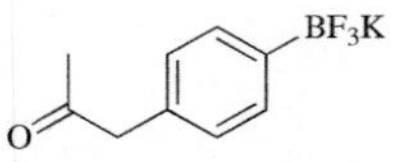	[(Terpy)Pd(MeCN)](BF₄)₂ (2 mol %), terpy (4 mol %), NaF (1 eq), Selectfluor (1.1 eq), DMF, 23°, 15 h	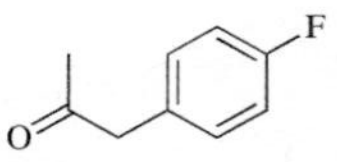 (70)	61
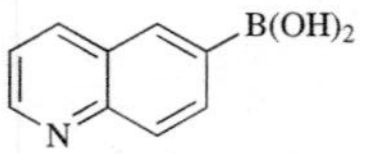	1. NaOH (1.2 eq), AgOTf (3.0 eq), MeOH, 0°, 0.5 h 2. Selectfluor (1.05 eq), 3 Å MS, acetone, 23°, 1 h	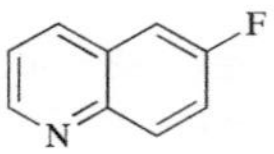 (75)	132
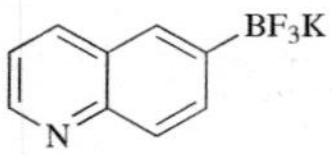	Cu(OTf)₂ (4 eq), KF (4 eq), MeCN, 60°, 20 h	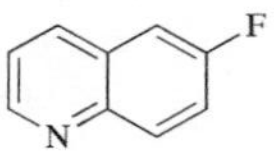 (36)	133

Substrate	Conditions	Product	Refs.
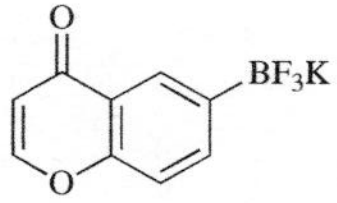 C₁₀ (BF₃K-chromone)	[(Terpy)Pd(MeCN)](BF₄)₂ (2 mol %), terpy (4 mol %), NaF (1 eq), Selectfluor (1.2 eq), MeCN, 23°, 15 h	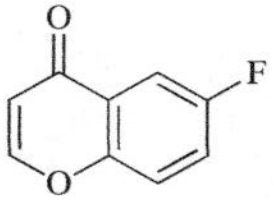 (83)	61
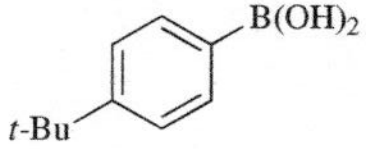 B(OH)₂, t-Bu	1. NaOH (1.0 eq), AgOTf (2.0 eq), MeOH, 0°, 0.5 h 2. Selectfluor (1.05 eq), 3 Å MS, acetone, 23°, 1 h	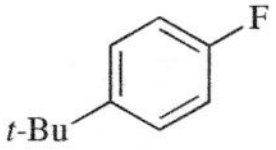 (82)	132
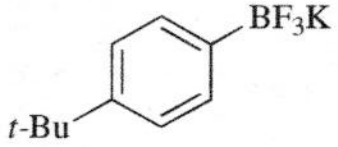 BF₃K, t-Bu	[(Terpy)Pd(MeCN)](BF₄)₂ (2 mol %), terpy (4 mol %), NaF (1 eq), Selectfluor (1.2 eq), DMF, 23°, 15 h	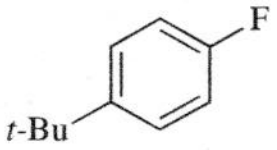 (98)	61
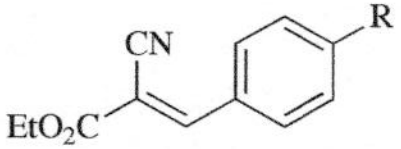 (EtO₂C, CN, R-cinnamonitrile)	[(Terpy)Pd(MeCN)](BF₄)₂ (2 mol %), terpy (4 mol %), NaF (1 eq), Selectfluor (1.2 eq), additive (2 eq), MeCN, 40°, 15 h	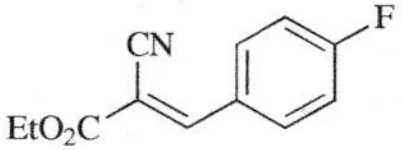 R — Additive (HO)₂B — KHF₂ — (86) PinB — KHF₂ — (95) KF₃B — — — (96)	61
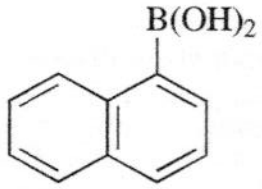 B(OH)₂-naphthalene	1. NaOH (1.0 eq), AgOTf (2.0 eq), MeOH, 0°, 0.5 h 2. Selectfluor (1.05 eq), 3 Å MS, acetone, 3°, 1 h	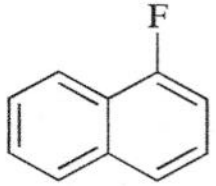 (82)	132

TABLE 1B. ARYL FLUORIDE SYNTHESIS FROM ARYL BORONIC ACIDS AND THEIR DERIVATIVES (*Continued*)

Please refer to the charts preceding the tables for the structures indicated by the **bold** numbers.

Boronic Substrate	Conditions	Product(s) and Yield(s) (%)	Refs.
C$_{10}$			
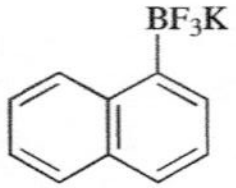	NFTPT (2 eq), Cu(MeCN)$_4$BF$_4$ (1.2 eq), t-BuCN (10 eq), EtOAc, 80°, 12 h	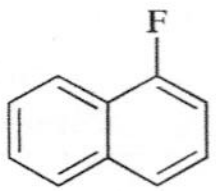 (56)	133
	AgOTf (3 eq), Selectfluor (1.2 eq), LiOH•H$_2$O (1.2 eq), EtOAc. 55°, 5–15 h	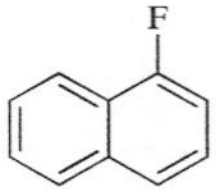 (63)	135
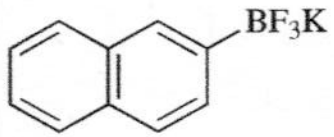	AgOTf (3 eq), Selectfluor (1.2 eq), LiOH•H$_2$O (1.2 eq), EtOAc. 55°, 5–15 h	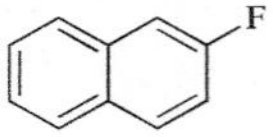 (60)	135
	NFTPT (2 eq), Cu(MeCN)$_4$BF$_4$ (1 eq), EtOAc. 80°, 12 h	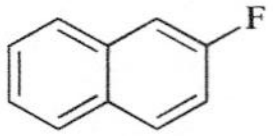 (64)	133
C$_{11}$			
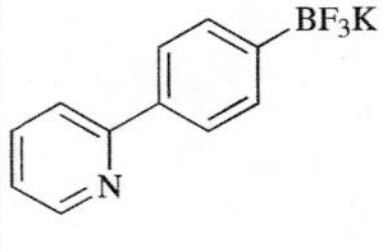	[(Terpy)Pd(MeCN)](BF$_4$)$_2$ (2 mol %), terpy (4 mol %), NaF (1 eq), Selectfluor (1.2 eq), DMF, 23°, 15 h	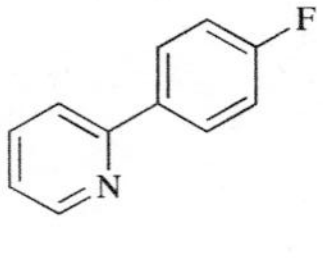 (86)	61

724

C₁₂

Substrate	Conditions	Product(s) (yield %)	Refs.

C_{12}

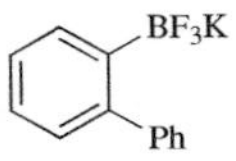 BF_3K / Ph (2-substituted)	[(Terpy)Pd(MeCN)](BF_4)_2 (2 mol %), terpy (4 mol %), NaF (1 eq), Selectfluor (1.2 eq), DMF, 23°, 15 h	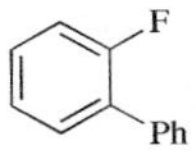 F, Ph (85) 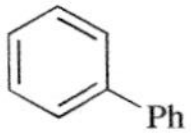 + Ph (4)	61
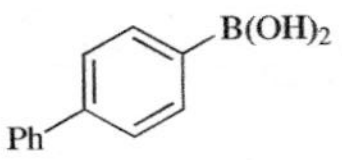 B(OH)_2 / Ph (4-substituted)	1. NaOH (1.0 eq), AgOTf (2.0 eq), MeOH, 0°, 0.5 h 2. Selectfluor (1.05 eq), 3 Å MS, acetone, 23°, 1 h	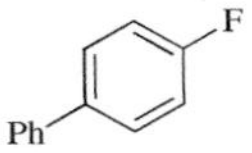 F, Ph (48)	132
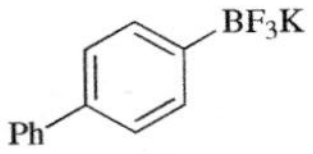	NFTPT (2 eq), Cu(MeCN)_4BF_4 (1.2 eq), EtOAc, 80°, 12 h	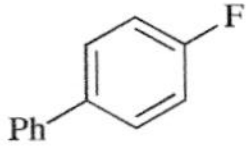 F, Ph (65) 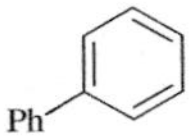 + Ph (6)	133
	Cu(OTf)_2 (4 eq), KF (4 eq), MeCN, 60°, 20 h	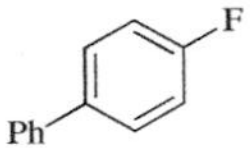 F, Ph (71) 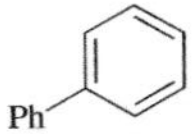 + Ph (4)	134
BF_3K / Ph (4-substituted)	[(Terpy)Pd(MeCN)](BF_4)_2 (2 mol %), terpy (4 mol %), Selectfluor (1.2 eq), DMF, 23°, 15 h	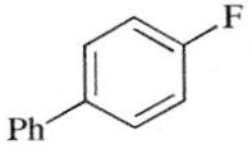 F, Ph (73)	61

725

TABLE 1B. ARYL FLUORIDE SYNTHESIS FROM ARYL BORONIC ACIDS AND THEIR DERIVATIVES (*Continued*)

Boronic Substrate	Conditions	Product(s) and Yield(s) (%)	Refs.

*Please refer to the charts preceding the tables for the structures indicated by the **bold** numbers.*

C$_{12}$

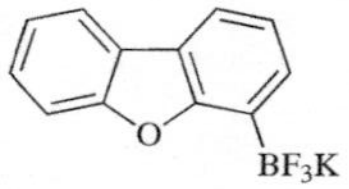

Cu(OTf)$_2$ (4 eq), KF (4 eq), MeCN, 60°, 20 h

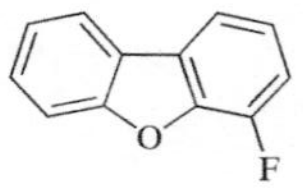

(67)

134

C$_{13}$

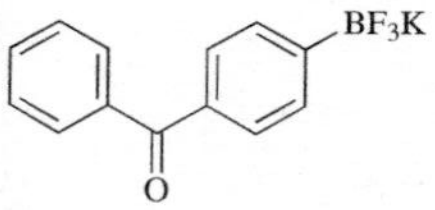

Cu(OTf)$_2$ (4 eq), KF (4 eq), MeCN, 60°, 20 h

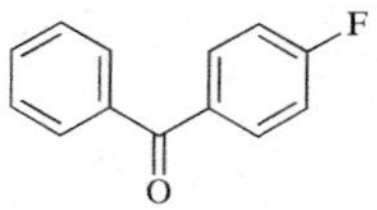

(66)

134

C$_{15}$

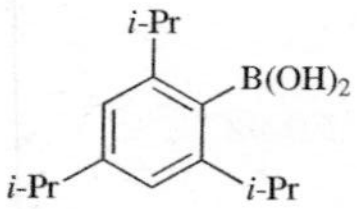

[(Terpy)Pd(MeCN)](BF$_4$)$_2$ (2 mol %),
terpy (4 mol %), NaF (1 eq), KHF$_2$ (6 eq),
Selectfluor (1.2 eq), DMF, 4°, 15 h

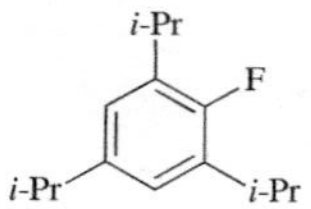

(63)

61

TABLE 1C. ARYL FLUORIDE SYNTHESIS FROM ARYL STANNANES

Stannane	Conditions	Product(s) and Yield(s) (%)	Refs.
C_{6-7}	F-TEDA•2PF$_6$ (1.2 eq), AgOTf (2 eq), acetone, 23°, 20 min	 R H (82) F (73) HO (72) MeO (76) OHC (77) NC– (76) Me$_2$(O)NCH$_2$ (63)	131
C_8	F-TEDA•2PF$_6$ (1.2 eq), AgOTf (2 eq), acetone, 23°, 20 min	(81)	131
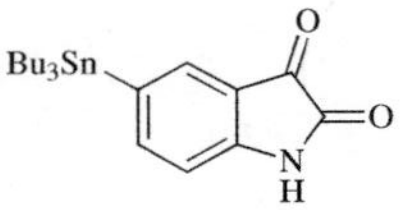	F-TEDA•2PF$_6$ (1.2 eq), AgOTf (2 eq), acetone, 23°, 20 min	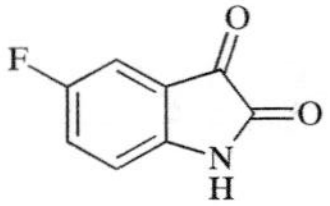 (72)	131
C_9	Ag$_2$O (5 mol %), NaHCO$_3$ (2.0 eq), NaOTf (1.0 eq), F-TEDA•2PF$_6$ (1.5 eq), acetone, 65°, 5 h	(85)	23

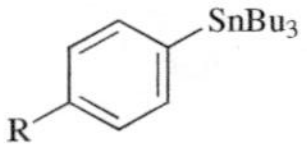

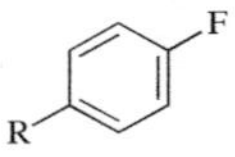

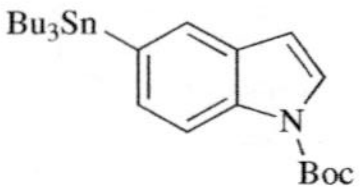

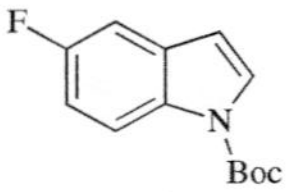

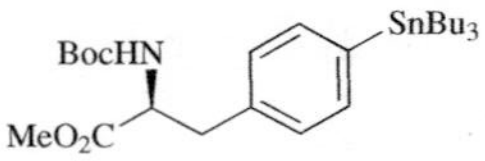

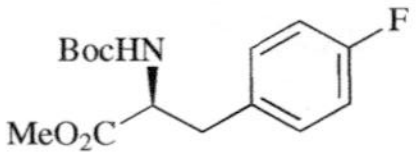

TABLE 1C. ARYL FLUORIDE SYNTHESIS FROM ARYL STANNANES (*Continued*)

Stannane	Conditions	Product(s) and Yield(s) (%)	Refs.
C₉			
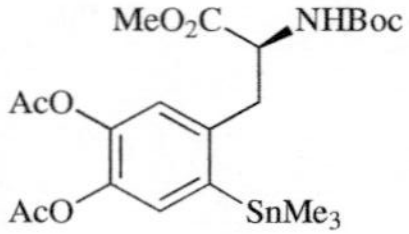 MeO₂C–NHBoc; AcO, AcO, SnMe₃	Ag₂O (5 mol %), NaHCO₃ (2.0 eq), NaOTf (1.0 eq), F⁻TEDA•2PF₆ (1.5 eq), acetone, 65°, 5 h	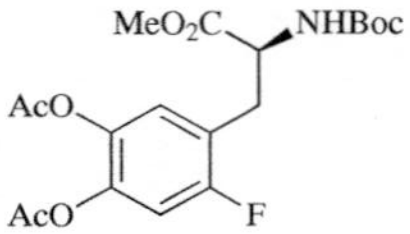 MeO₂C–NHBoc; AcO, AcO, F (78)	23
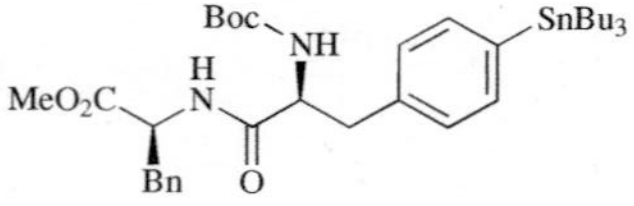 Boc–NH, SnBu₃, MeO₂C, Bn	Ag₂O (5 mol %), NaHCO₃ (2.0 eq), NaOTf (1.0 eq), F⁻TEDA•2PF₆ (1.5 eq), acetone, 65°, 5 h	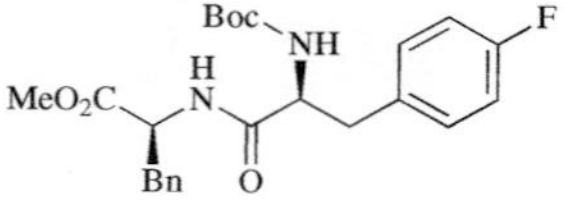 Boc–NH, F, MeO₂C, Bn (92)	23
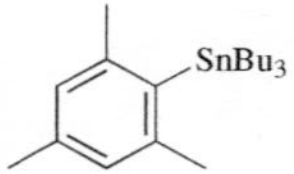 mesityl–SnBu₃	Selectfluor (1.2 eq), AgOTf (2 eq), acetone, 23°, 20 min	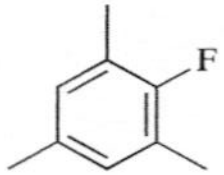 F (73)	131
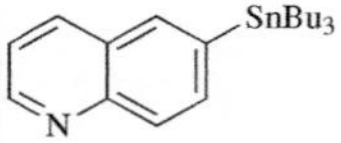 quinoline–SnBu₃	F⁻TEDA•2PF₆ (1.2 eq), AgOTf (2 eq), acetone, 23°, 20 min	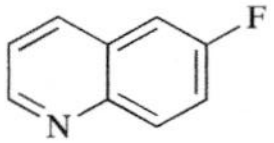 F (79)	131
C₁₀			
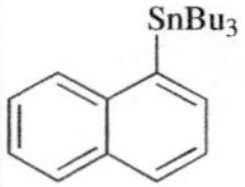 naphthalene–SnBu₃	NFTPT (2 eq), (t-BuCN)₂CuOTf (1.2 eq), EtOAc, rt, 12 h	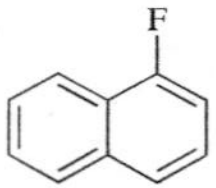 F (60)	133

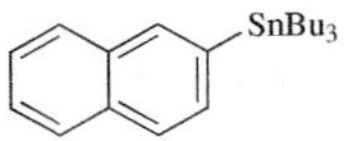

C$_{12}$			

SnBu$_3$ (naphthalene)

NFTPT (2 eq), (t-BuCN)$_2$CuOTf (1 eq),
EtOAc, rt, 12 h

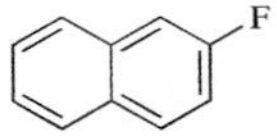

F-naphthalene (71)

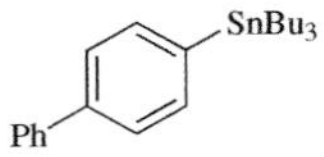

133

SnBu$_3$ (biphenyl)

NFTPT (2 eq), (t-BuCN)$_2$CuOTf (1 eq),
EtOAc, rt, 12 h

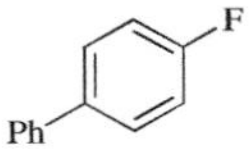

F-biphenyl (67) + Ph-phenyl (8)

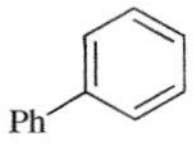

133

F-TEDA•2PF$_6$ (1.2 eq), AgOTf (2 eq),
acetone, 23°, 20 min

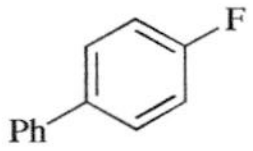

F-biphenyl (83)

131

C$_{14}$

SnBu$_3$ (cyclic peptide)

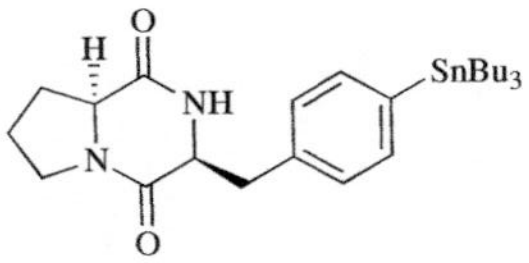

Ag$_2$O (5 mol %), NaHCO$_3$ (2.0 eq),
NaOTf (1.0 eq), F-TEDA•2PF$_6$ (1.5 eq),
acetone, 65°, 5 h

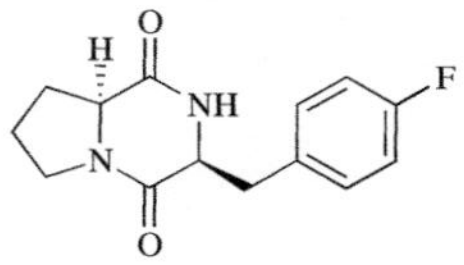

F-product (78)

23

729

TABLE 1C. ARYL FLUORIDE SYNTHESIS FROM ARYL STANNANES (*Continued*)

Stannane	Conditions	Product(s) and Yield(s) (%)	Refs.
C₁₅	Ag₂O (5 mol %), NaHCO₃ (2.0 eq), NaOTf (1.0 eq), F-TEDA•2PF₆ (1.5 eq), acetone, 65°, 5 h	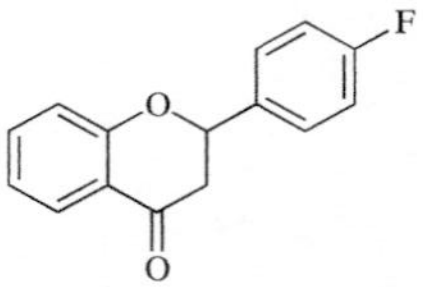 (90)	23
C₁₈	Ag₂O (5 mol %), NaHCO₃ (2.0 eq), NaOTf (1.0 eq), F-TEDA•2PF₆ (1.5 eq), acetone, 65°, 5 h	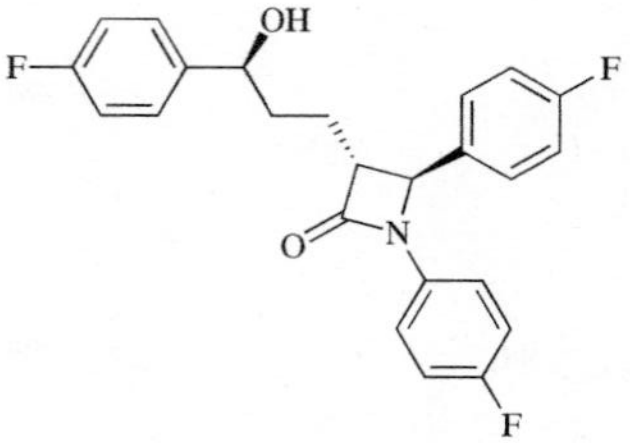 (90)	23
	Ag₂O (5 mol %), NaHCO₃ (2.0 eq), NaOTf (1.0 eq), F-TEDA•2PF₆ (1.5 eq), acetone, 65°, 5 h	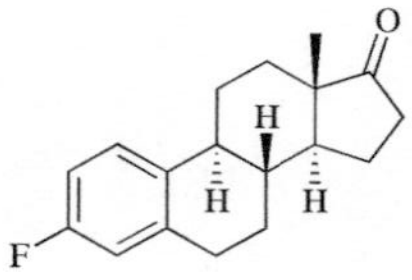 (81)	23

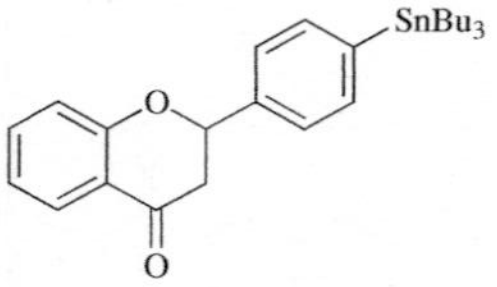

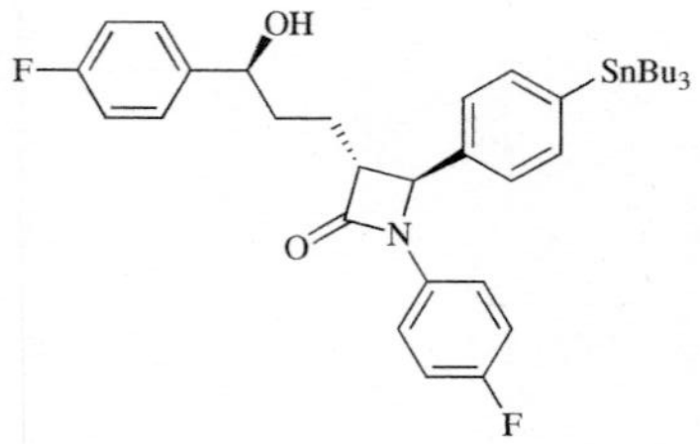

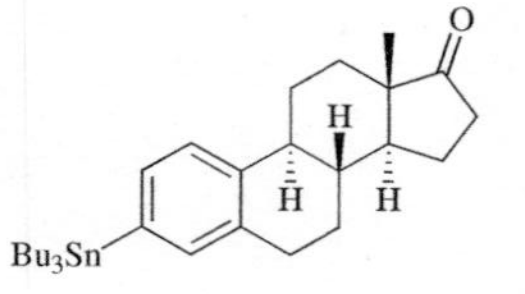

C$_{19}$

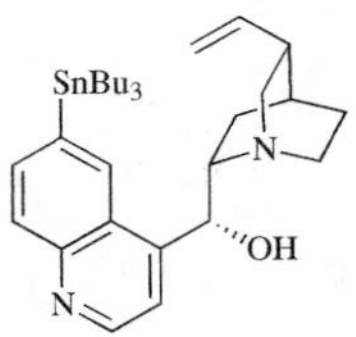

Ag$_2$O (20 mol %), NaHCO$_3$ (2.0 eq), NaOTf (2.0 eq), F-TEDA•2PF$_6$ (2.0 eq), acetone, 90°, 2 h

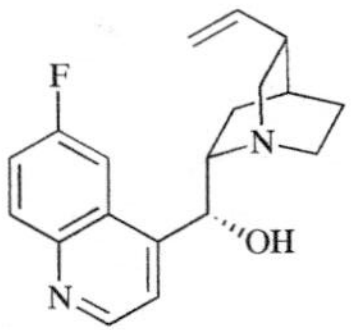

(75)

23

C$_{21}$

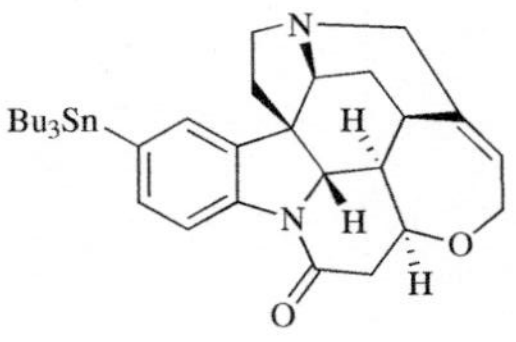

1. BnBr (1.0 eq), acetone, 23°, 12 h
2. AgOTf (1.0 eq), acetone, –78°, 2 min
3. Ag$_2$O (5 mol %), NaHCO$_3$ (2.0 eq), NaOTf (1.0 eq), F-TEDA•2PF$_6$ (1.5 eq), acetone, 65°, 2 h
4. 1,4-Cyclohexadiene, Pd/C (10%), MeOH, 40°, 4 h

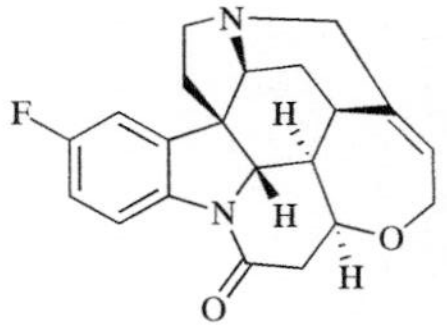

(60)

23

C$_{28}$

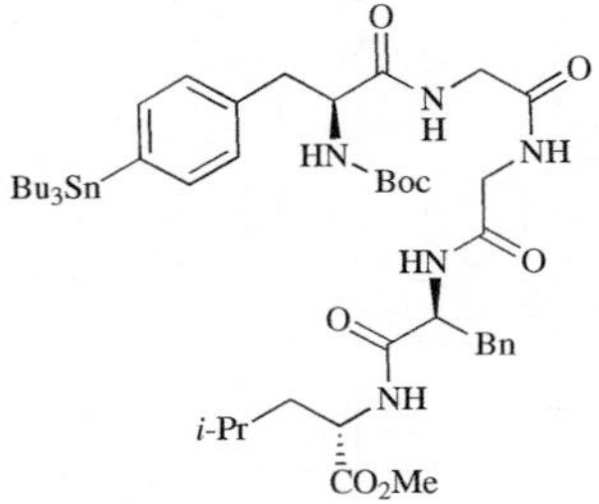

Ag$_2$O (5 mol %), NaHCO$_3$ (2.0 eq), NaOTf (1.0 eq), F-TEDA•2PF$_6$ (1.5 eq), acetone, 65°, 5 h

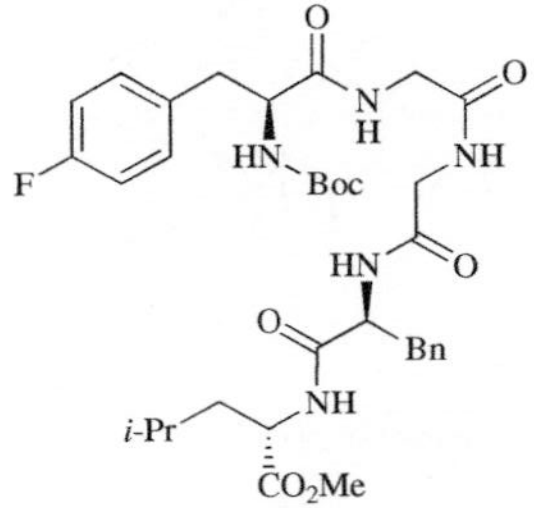

(81)

23

TABLE 1C. ARYL FLUORIDE SYNTHESIS FROM ARYL STANNANES (*Continued*)

Stannane	Conditions	Product(s) and Yield(s) (%)	Refs.
C_{28}	Ag$_2$O (5 mol %), NaHCO$_3$ (2.0 eq), NaOTf (1.0 eq), F-TEDA•2PF$_6$ (1.5 eq), acetone, 65°, 5 h	(72)	23
C_{41}	Ag$_2$O (20 mol %), NaHCO$_3$ (2.0 eq), NaOTf (2.0 eq), F-TEDA•2PF$_6$ (1.5 eq), MeOH (5 eq), acetone, 65°, 3 h	(65)	23

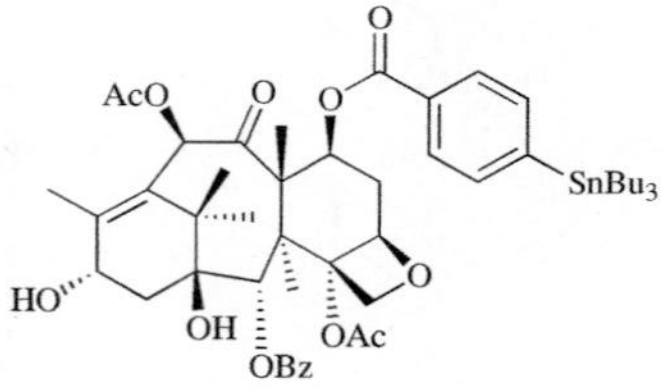

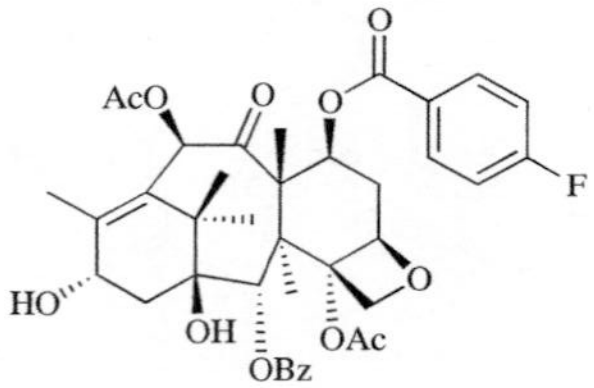

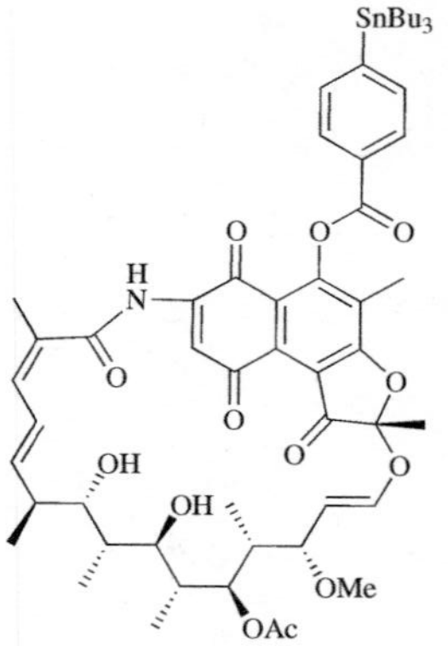

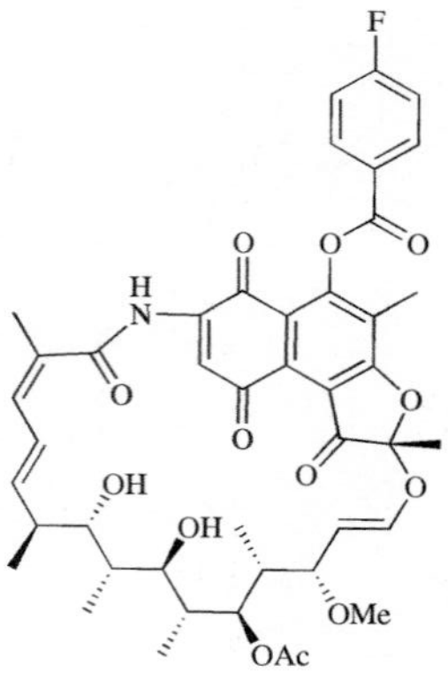

TABLE 1D. ARYL FLUORIDE SYNTHESIS FROM ARENES

Arene	Conditions	Product(s) and Yield(s) (%)	Refs.

*Please refer to the charts preceding the tables for the structures indicated by the **bold** numbers.*

C$_6$

Arene	Conditions	Product(s) and Yield(s) (%)	Refs.
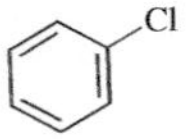 (chlorobenzene)	**Pd4** (5 mol %), 2-chlorophenanthroline (5 mol %), Selectfluor (2 eq), MeCN, 50°, 12 h	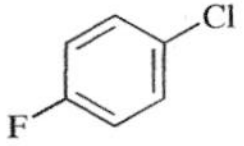 (30) 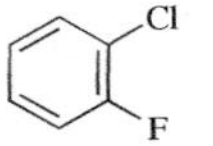 + (21)	55
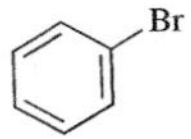 (bromobenzene)	**Pd4** (5 mol %), 2-chlorophenanthroline (5 mol %), Selectfluor (2 eq), MeCN, 50°, 12 h	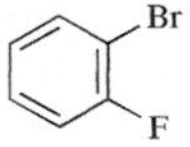 (25) 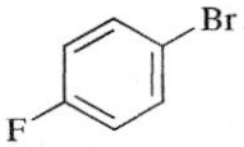 + (24)	55
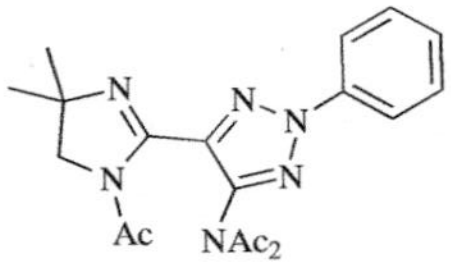 (triazole–imidazoline NAc$_2$ compound)	**Pd4** (10 mol %), 2-chlorophenanthroline (10 mol %), Selectfluor (2 eq), MeCN, 25°, 16 h	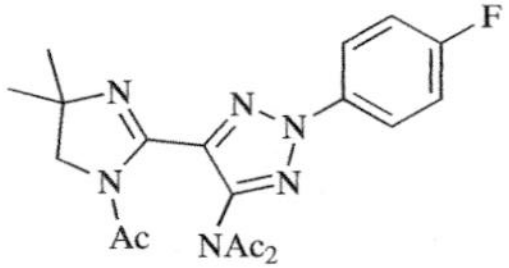 (54)	55
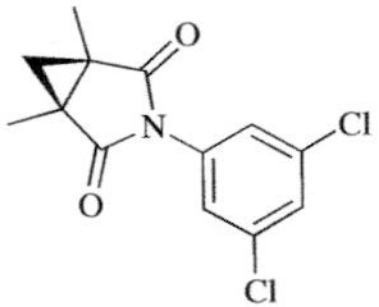 (3,5-dichlorophenyl cyclopropane-succinimide compound)	**Pd4** (7.5 mol %), 2-chlorophenanthroline (7.5 mol %), NFSI (2 eq), MeCN, 80°, 21 h	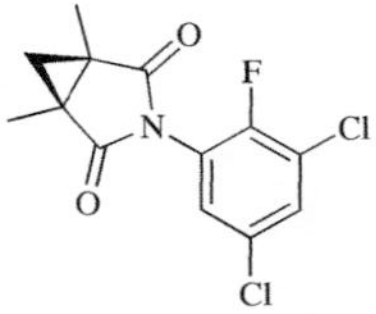 (17) 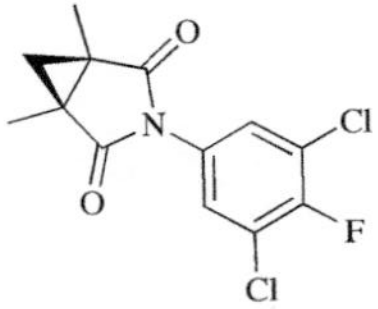 + (15)	55

TABLE 1D. ARYL FLUORIDE SYNTHESIS FROM ARENES (*Continued*)

*Please refer to the charts preceding the tables for the structures indicated by the **bold** numbers.*

Arene	Conditions	Product(s) and Yield(s) (%)	Refs.
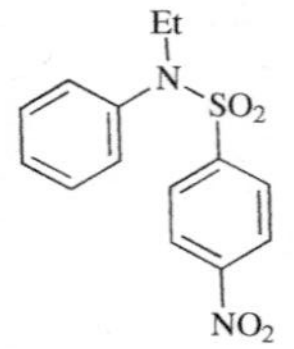 **C₆**	**Pd4** (5 mol %), 2-chlorophenanthroline (5 mol %), Selectfluor (2 eq), MeCN, 50°, 14 h	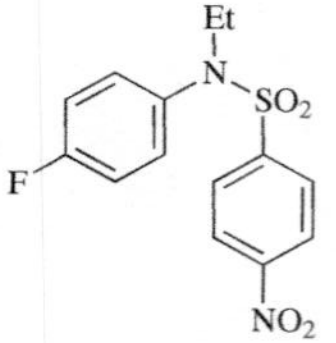 (31) 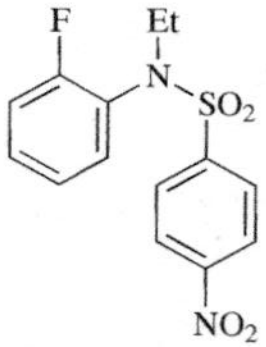 + (29)	55
C₆₋₇	Pd(OAc)₂ (10 mol %), NFSI (1.5 eq), TFA (2.0 eq), CH₃NO₂, 110°, 12 h	R: H (75), Cl (62), Me (74)	124
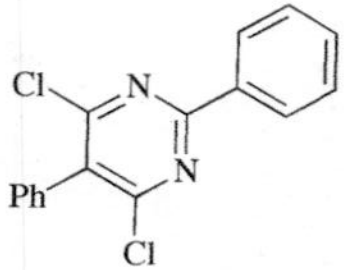 **C₇**	**Pd4** (5 mol %), 2-chlorophenanthroline (5 mol %), Selectfluor (2 eq), MeCN, 50°, 14 h	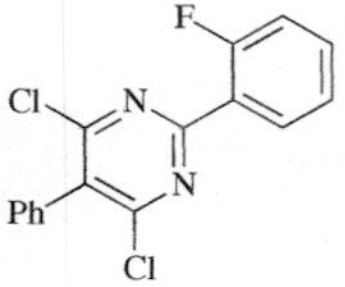 (31) 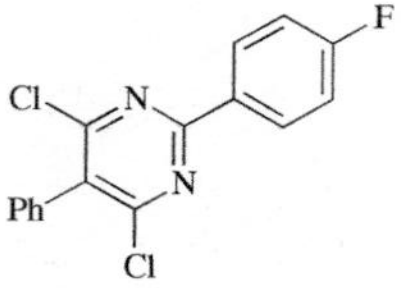 + (14)	55
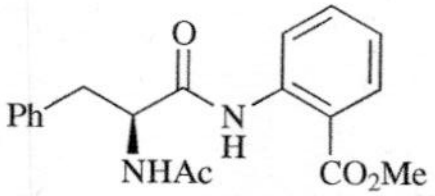	**Pd4** (5 mol %), 2-chlorophenanthroline (5 mol %), NFSI (2 eq), MeCN, 25°, 14 h	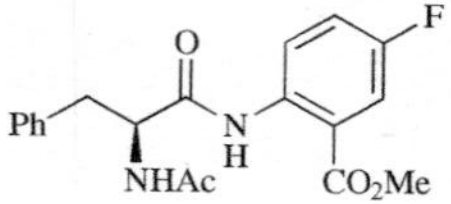 (45) + 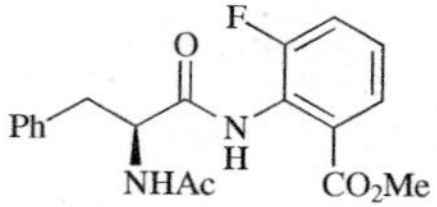 (7)	55

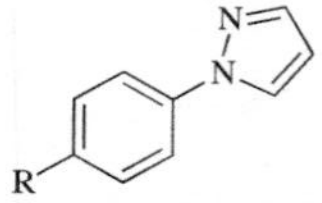

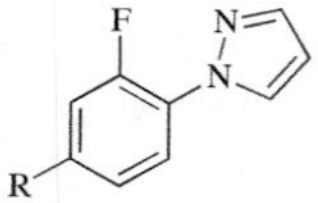

734

C$_{7-11}$

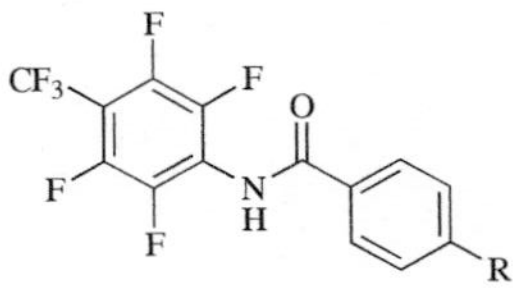

Pd(OTf)$_2$(MeCN)$_4$ (10 mol %),
NMP (50 mol %),
NFTPT (3 eq), PhCF$_3$, 120°, 2 h

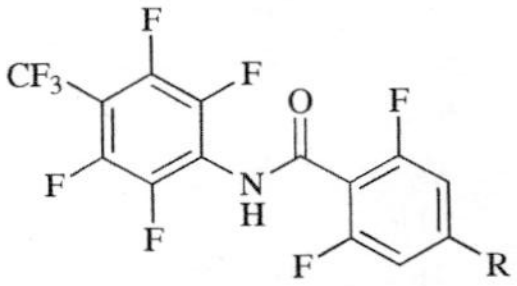

R	
H	(74)
Cl	(66)
t-Bu	(88)

122

C$_{7-8}$

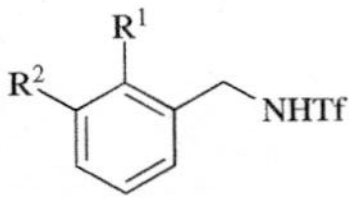

Pd(OTf)$_2$•6H$_2$O (x mol %),
additive (0.5 eq),
NFTPT (y eq), DCE, 120°

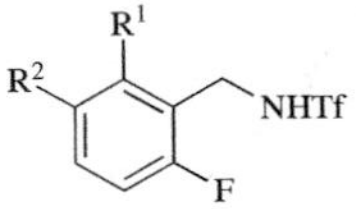

129

R^1	R^2	x	Additive	y	Time (h)	
H	H	10	DMF	2.0	4	(41)
H	Cl	10	NMP	2.0	2	(70)
H	Br	10	NMP	2.0	2	(70)
F	H	10	NMP	2.0	4	(84)
Cl	H	5	NMP	1.5	2	(82)
Br	H	10	NMP	1.5	2	(83)
OMe	H	10	NMP	1.5	1	(60)
H	Me	10	DMF	2.0	1	(58)
H	CF$_3$	10	NMP	2.0	8	(81)
Me	H	5	NMP	1.5	0.5	(82)
Me	Cl	10	NMP	2.0	1	(80)
CF$_3$	H	10	NMP	1.5	4	(88)

TABLE 1D. ARYL FLUORIDE SYNTHESIS FROM ARENES (Continued)

Arene	Conditions	Product(s) and Yield(s) (%)	Refs.

*Please refer to the charts preceding the tables for the structures indicated by the **bold** numbers.*

C$_{7-8}$

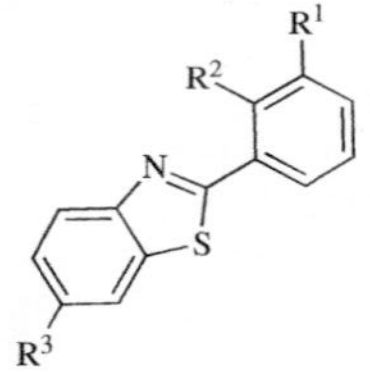

Conditions: Pd(PPh$_3$)$_4$ (15 mol %), L-proline (40 mol %), NFSI (3 eq), cyclohexane, 120°, 5 h

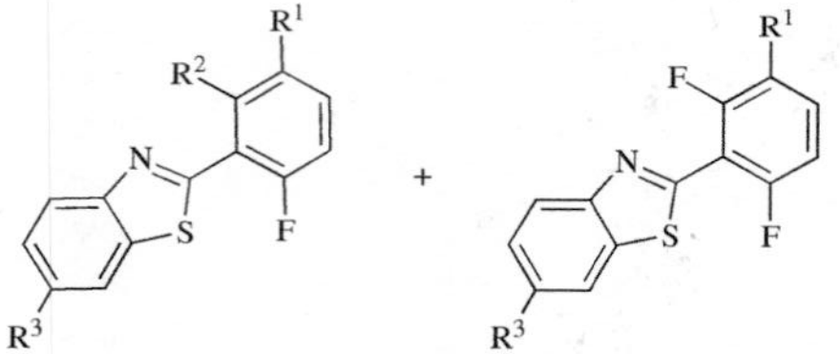

Product structures **I** and **II**

Ref. 130

R^1	R^2	R^3	I	II
H	H	H	(79)	(—)
H	H	Cl	(53)	(—)
H	F	H	(69)	(—)
H	F	F	(68)	(—)
H	F	Cl	(63)	(—)
H	Cl	H	(68)	(—)
H	Cl	F	(82)	(—)
H	MeO	H	(68)	(tr)
H	O$_2$N	H	(tr)	(—)
Cl	H	H	(63)	(4)
H	H	Me	(56)	(37)
H	F	Me	(35)	(8)
H	Cl	Me	(54)	(15)
H	Me	H	(74)	(—)
H	Me	F	(73)	(—)
H	Me	Cl	(65)	(—)
Cl	H	Me	(34)	(33)
Me	H	H	(60)	(—)
Me	H	F	(44)	(17)
Me	H	Cl	(47)	(16)

C7-14

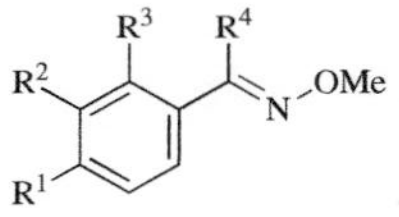

[Pd2(dba)3] (5 mol %),
KNO3 (30 mol %), NFSI (2.0 eq),
CH3NO2, 12–24 h

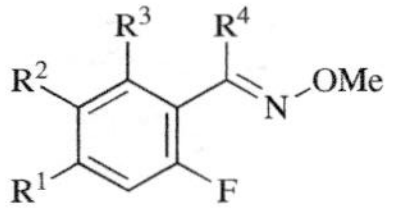

R^1	R^2	R^3	R^4	Temp (°)		(E)/(Z)
H	H	H	H	70	(71)	—
H	H	H	Me	25	(78)	—
H	H	Cl	Me	90	(76)	—
H	Cl	H	Me	40	(72)	2:1
F	H	H	Me	40	(83)	—
Cl	H	H	Me	40	(82)	—
Br	H	H	Me	40	(75)	—
I	H	H	Me	40	(81)	—
MeO	H	H	Me	25	(86)	—
BnO	H	H	Me	25	(80)	—
O2N	H	H	Me	90	(85)	—
MeO2S	H	H	Me	50	(79)	—
H	H	H	Et	25	(82)	—
H	H	Me	Me	90	(90)	—
H	CF3	H	Me	70	(80)	—
Me	H	H	Me	25	(80)	—
MeO2C	H	H	Me	40	(82)	—
NC–	H	H	Me	70	(82)	3:1
H	H	H	c-Pr	40	(73)	—
Br	H	H	n-Bu	40	(73)	—
H	H	H	Ph	25	(86)	—
Ph	H	H	Me	25	(79)	—

TABLE 1D. ARYL FLUORIDE SYNTHESIS FROM ARENES (*Continued*)

Arene	Conditions	Product(s) and Yield(s) (%)	Refs.
(structure)	Pd(OAc)$_2$ (5 mol %), NFSI (3 eq), solvent, 80°	(structure)	123

*Please refer to the charts preceding the tables for the structures indicated by the **bold** numbers.*

C$_{7-9}$

R^1	R^2	R^3	Solvent	Time (h)	
H	H	H	1,4-dioxane/t-amyl-OH (3:1)	4	(56)
H	H	F	1,4-dioxane/t-amyl-OH (3:1)	4	(78)
H	H	Cl	1,4-dioxane/t-amyl-OH (3:1)	4	(83)
H	H	Br	1,4-dioxane/t-amyl-OH (3:1)	6	(81)
H	H	MeO	t-amyl-OH	12	(85)
H	Cl	H	1,4-dioxane/t-amyl-OH (3:1)	8	(81)
H	Br	H	1,4-dioxane/t-amyl-OH (3:1)	4	(80)
H	I	H	1,4-dioxane/t-amyl-OH (3:1)	8	(72)
H	MeO	H	dioxane	10	(77)
Br	H	H	1,4-dioxane/t-amyl-OH (3:1)	6	(54)
MeO	H	H	DCE/t-amyl-OH (1:1)	4	(61)
MeO	H	MeO	DCE	6	(72)
H	H	Me	t-amyl-OH	6	(94)
H	H	Me	t-amyl-OH	12	(32)
H	H	CF$_3$	t-amyl-OH	24	(82)
H	Me	H	t-amyl-OH	6	(93)
H	Me	Me	t-amyl-OH	6	(95)
Me	Me	H	t-amyl-OH	6	(95)
CH$_2$=CH	H	H	1,4-dioxane/t-amyl-OH (3:1)	4	(55)

C_{7–11}

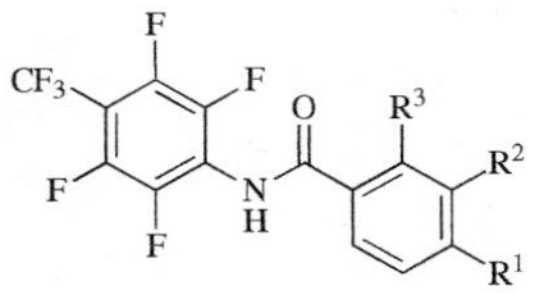

Pd(OTf)$_2$(MeCN)$_4$ (10 mol %),
NMP (20 mol %),
NFTPT (1.5 eq), MeCN, 120°

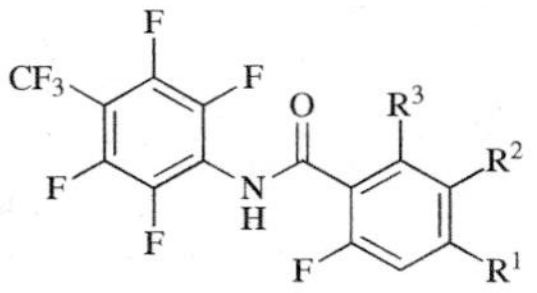

R^1	R^2	R^3	Time (h)	
H	H	H	8–12	(65)
H	H	F	8–12	(69)
H	H	Cl	8–12	(68)
H	Cl	H	8–12	(73)
H	Br	H	8–12	(78)
H	AcO	H	8–12	(37)
Cl	H	H	8–12	(66)
Br	H	H	8–12	(69)
MeO	H	H	8–12	(54)
AcO	H	H	8–12	(44)
BnO	H	H	2–3	(78)
H	H	Me	2–3	(78)
H	H	CF$_3$	2	(70)
H	Me	H	2–3	(75)
NC–	H	H	8–12	(62)
H	Ac	H	8–12	(36)
t-Bu	H	H	8–12	(76)

TABLE 1D. ARYL FLUORIDE SYNTHESIS FROM ARENES (Continued)

Arene	Conditions	Product(s) and Yield(s) (%)	Refs.

*Please refer to the charts preceding the tables for the structures indicated by the **bold** numbers.*

C$_{7-8}$

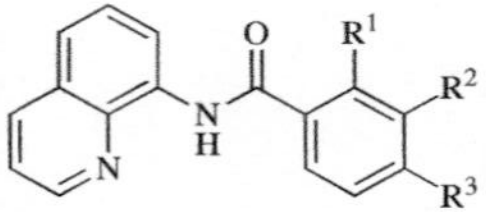

CuI (x mol %), AgF (y eq), NMO (z eq), DMF

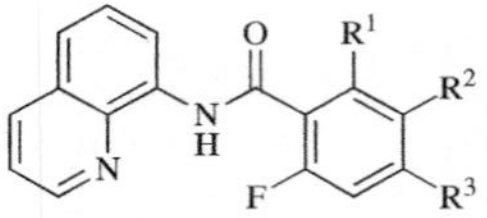

R^1	R^2	R^3	x	y	z	Temp (°)	Time (min)	
H	H	O$_2$N	12	4	5	60	120	(60)
H	H	Me	25	3.5	4.5	105	75	(75)
H	H	CF$_3$	15	3	4	80	30	(60)
H	H	MeO$_2$C	10	4	5	90	60	(56)
H	H	NC–	12	3.5	4.5	65	75	(62)
H	CF$_3$	H	12	4	5	100	45	(71)
Me	H	H	25	4	5	120	20	(63)
CF$_3$	H	H	15	4	5	120	90	(80)

Refs. 126

C$_8$

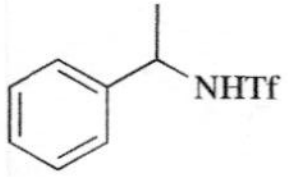

Pd(OTf)$_2$·6H$_2$O (10 mol%), NMP (0.5 eq), NFTPT (3.0 eq), PhCF$_3$, 120°, 4 h

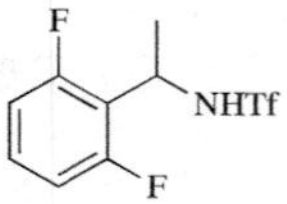

(70)

Refs. 129

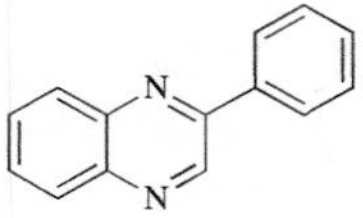

Pd(OAc)$_2$ (15 mol %), NFSI (3 eq), TFA (4 eq), CH$_3$NO$_2$, 110°, 12 h

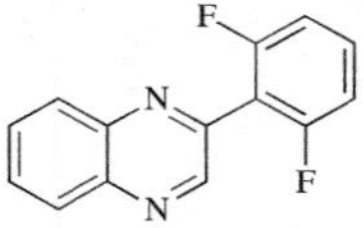

(84)

Refs. 124

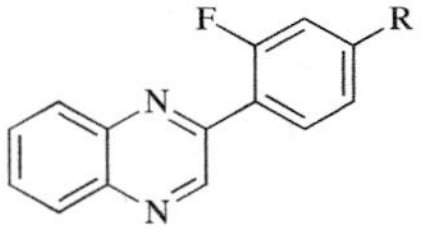

R	x	
H	1	(74)
F	10	(67)
Cl	10	(68)
Br	10	(65)
MeO	1	(78)
Ph	1	(76)

129

(53)

124

55

(29)

(21)

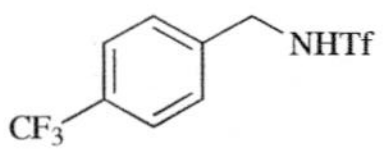

Pd(OTf)$_2$•6H$_2$O (20 mol%), NMP (0.5 eq), NFTPT (3.0 eq), MW, PhCF$_3$, 150°, 2 h

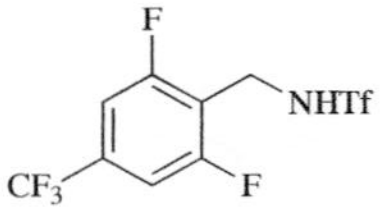

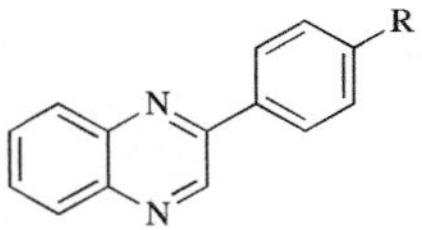

Pd(OAc)$_2$ (10 mol %), NFSI (1.5 eq), TFA (2 eq), CH$_3$NO$_2$/MeCN (x/1, v/v), 110°, 12 h

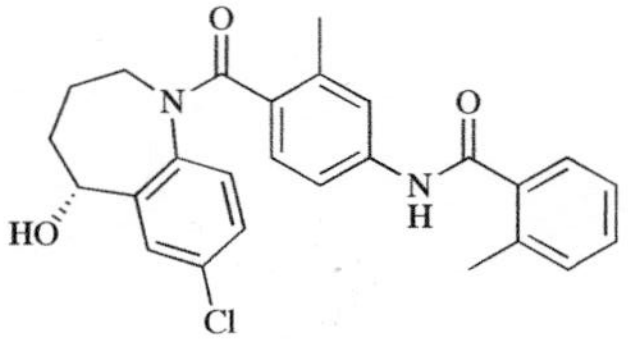

Pd4 (7.5 mol %), 2-chlorophenanthroline (7.5 mol %), NFSI (2 eq), MeCN, 50°, 20 h

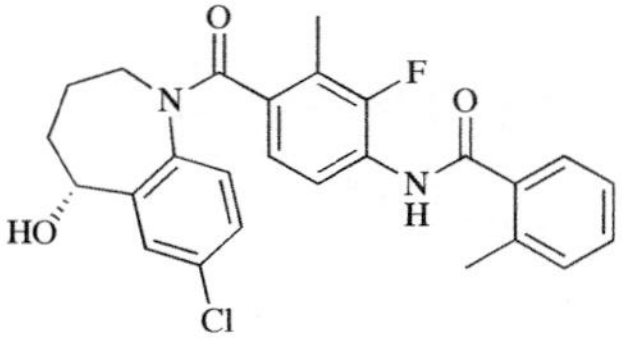

C$_{8-14}$

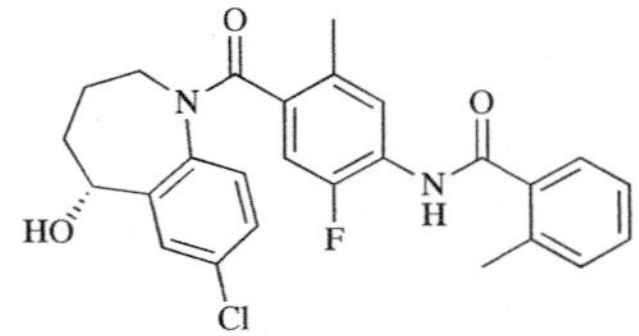

C$_8$

TABLE 1D. ARYL FLUORIDE SYNTHESIS FROM ARENES (Continued)

Arene	Conditions	Product(s) and Yield(s) (%)	Refs.

*Please refer to the charts preceding the tables for the structures indicated by the **bold** numbers.*

C₉

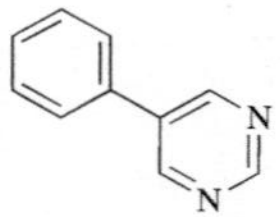	**Pd4** (5 mol %), 2-chlorophenanthroline (5 mol %), Selectfluor (2 eq), MeCN, 50°, 14 h	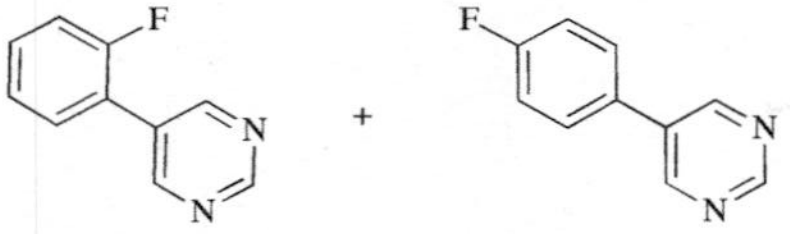 (31) + (15)	55
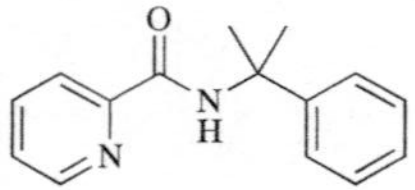	CuI (50 mol %), AgF (7.0 eq), NMO (7.0 eq), DMPU, 110°, 12 h	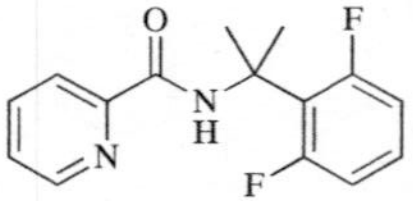 (52)	126
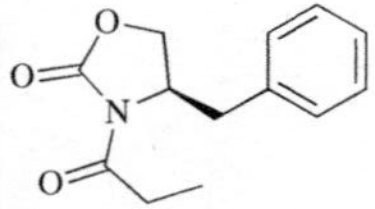	**Pd4** (5 mol %), 2-chlorophenanthroline (5 mol %), Selectfluor (2 eq), MeCN, 0°, 36 h	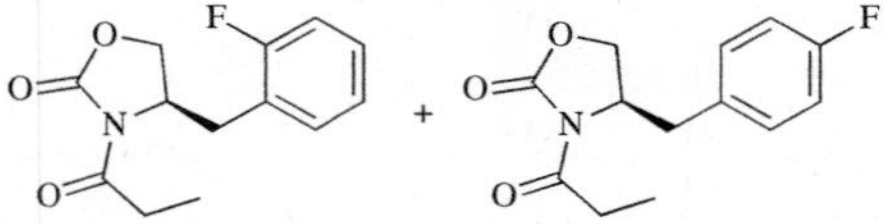 (51) + (20)	55
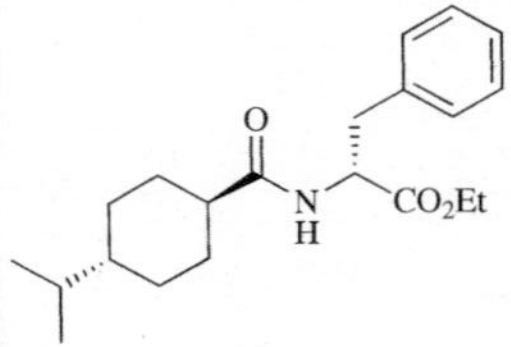	**Pd4** (5 mol %), 2-chlorophenanthroline (5 mol %), Selectfluor (2 eq), MeCN, 25°, 24 h	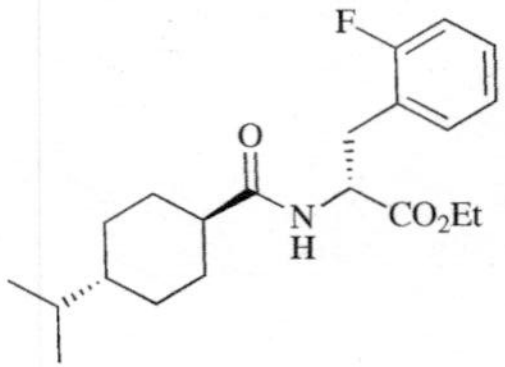 (25) + 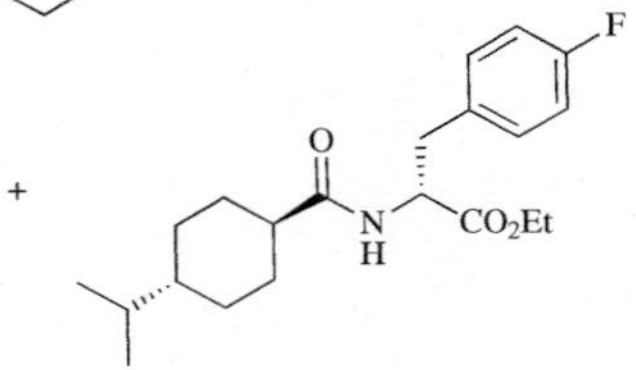 (14)	55

742

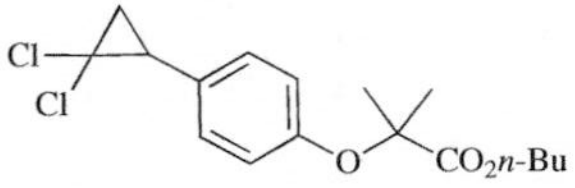

Pd4 (7.5 mol %),
2-chlorophenanthroline (7.5 mol %),
NFSI (2 eq), MeCN, 25°, 8 h

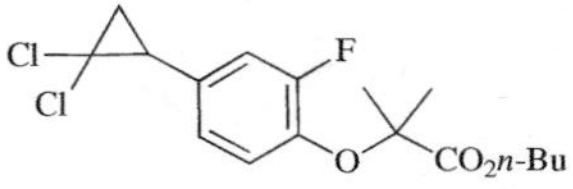

(30)

55

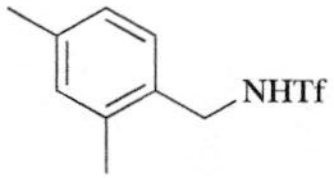

Pd(OTf)$_2$•6H$_2$O (10 mol %),
NMP (0.5 eq), NFTPT (1.5 eq),
DCE, 120°, 0.5 h

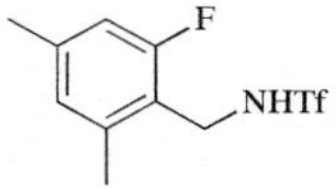

(85)

129

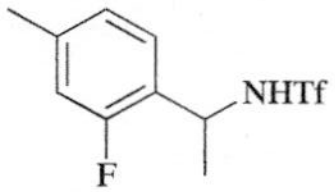

Pd(OTf)$_2$•6H$_2$O (10 mol %),
NMP (0.5 eq), NFTPT,
PhCF$_3$, 120°, 4 h

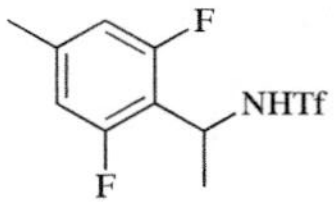

(71)

129

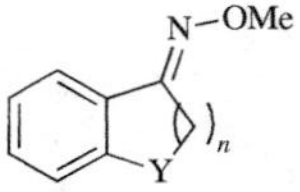

[Pd$_2$(dba)$_3$] (5 mol %),
KNO$_3$ (30 mol %), NFSI (2.0 eq),
CH$_3$NO$_2$, 12–24 h

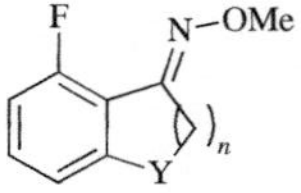

n	Y	Temp (°)	
1	CH$_2$	25	(—)
2	O	40	(83)
2	CH$_2$	25	(87)
3	CH$_2$	40	(82)

55

C$_{9-11}$

TABLE 1D. ARYL FLUORIDE SYNTHESIS FROM ARENES (*Continued*)

Arene	Conditions	Product(s) and Yield(s) (%)	Refs.

*Please refer to the charts preceding the tables for the structures indicated by the **bold** numbers.*

C_{11}

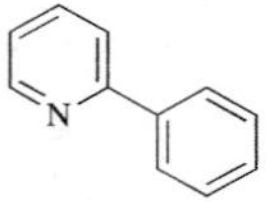	**Pd4** (5 mol %), 2-chlorophenanthroline (5 mol %), Selectfluor (2 eq), MeCN, 50°, 14 h	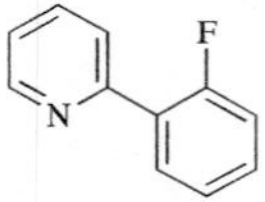 (42) + 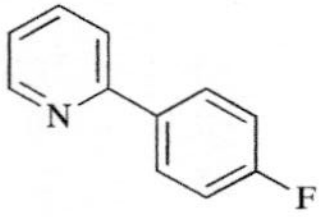 (13)	55
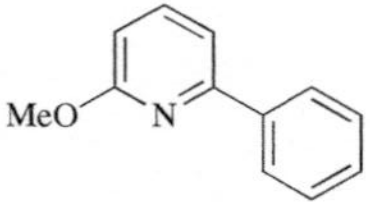	Pd(OAc)$_2$ (10 mol %), N-fluoropyridinium BF$_4$ (4.5 eq), PhCF$_3$, MW (300 W), MeCN, 150°, 2 h	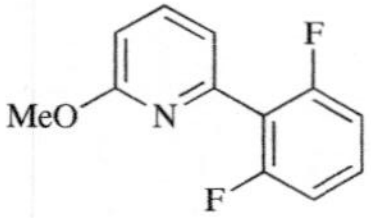 (60)	105
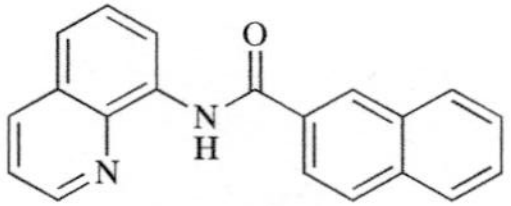	CuI (23 mol %), AgF (6.0 eq), NMO (8.0 eq), pyridine (8.0 eq), DMF, 100°, 1.5 h	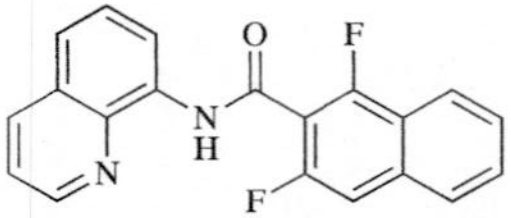 (70)	126

C_{12}

| 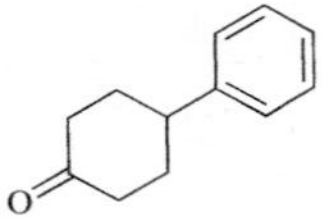 | **Pd4** (5 mol %), 2-chlorophenanthroline (5 mol %), Selectfluor (2 eq), MeCN, 0°, 24 h | 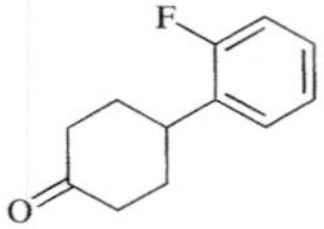 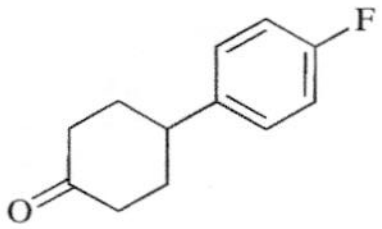 (26) + (30) | 55 |

744

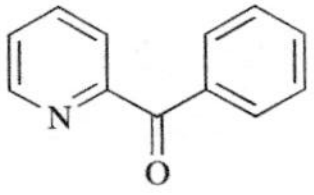

(48)

Pd(OAc)$_2$ (10 mol %),
N-fluoropyridinium BF$_4$ (3.3 eq),
PhCF$_3$, MW (300 W),
MeCN, 150°, 2 h

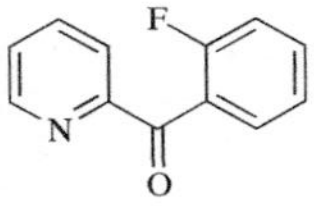

105

(41)

(18)

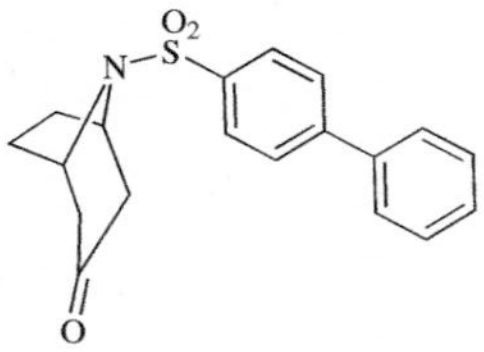

Pd4 (5 mol %),
2-chlorophenanthroline (5 mol %),
NFSI (2 eq), MeCN, DCE, 25°, 25 h

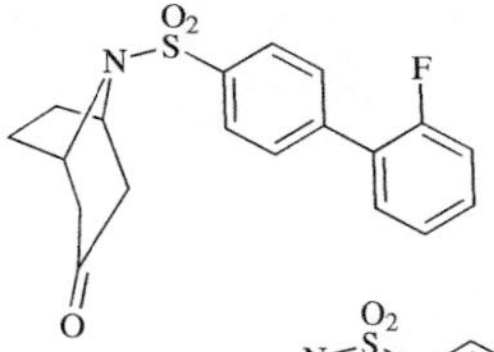

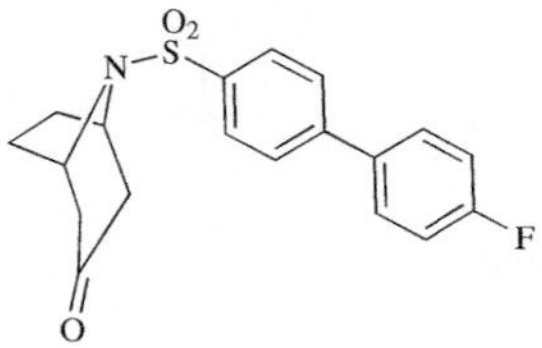

55

+

(81)

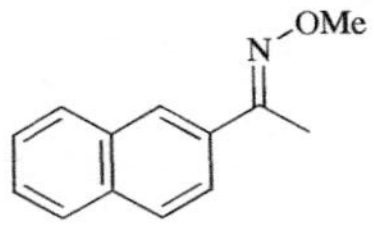

[Pd$_2$(dba)$_3$] (5 mol %),
KNO$_3$ (30 mol %), NFSI (2.0 eq),
CH$_3$NO$_2$, 25°, 12–24 h

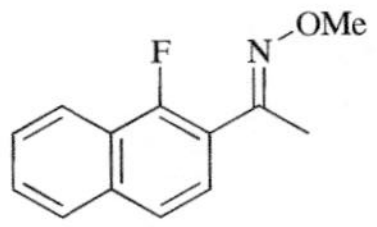

125

TABLE 1D. ARYL FLUORIDE SYNTHESIS FROM ARENES (Continued)

Arene	Conditions	Product(s) and Yield(s) (%)	Refs.

*Please refer to the charts preceding the tables for the structures indicated by the **bold** numbers.*

C₁₃

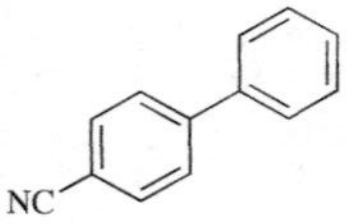

Pd4 (5 mol %),
2-chlorophenanthroline (5 mol %),
NFSI (2 eq), MeCN, 25°, 20 h

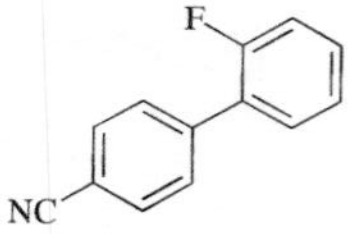

(42)   (19)

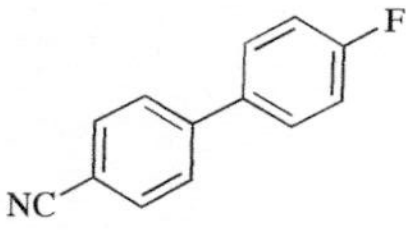

55

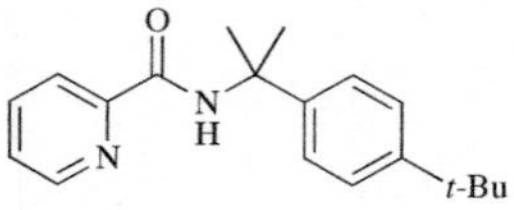

Cul (50 mol %), AgF (7.0 eq),
NMO (7.0 eq), DMPU, 110°, 4 h

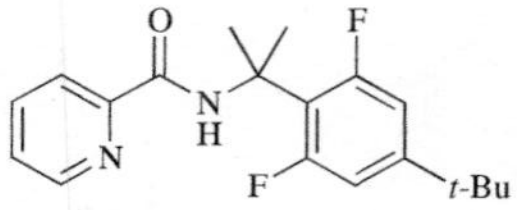

(43)

126

C₁₄

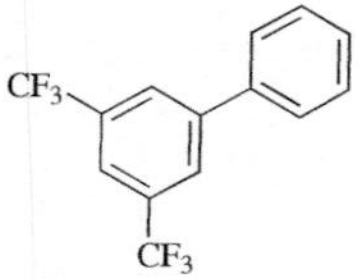

Pd4 (5 mol %),
2-chlorophenanthroline (5 mol %),
Selectfluor (2 eq). MeCN, 50°, 16 h

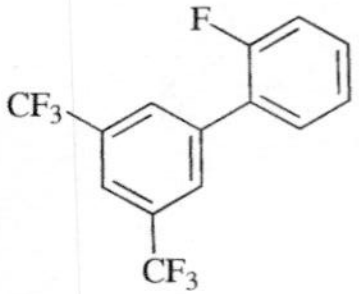

(59)   (26)

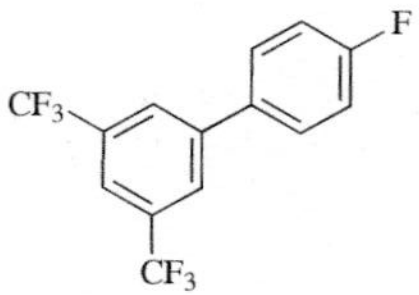

55

746

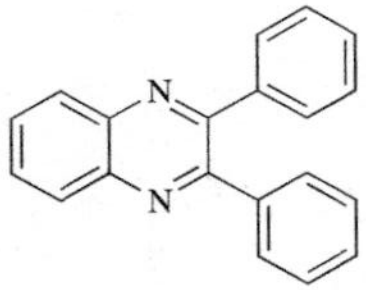

Pd(OAc)$_2$ (15 mol %),
NFSI (3 eq), TFA (2 eq),
CH$_3$NO$_2$/MeCN (10:1, v/v),
110°, 12 h

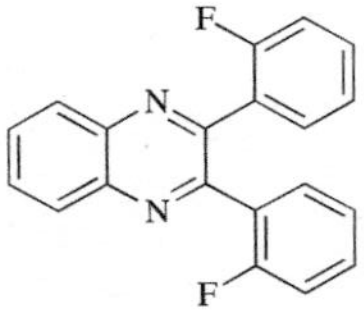

(55)

124

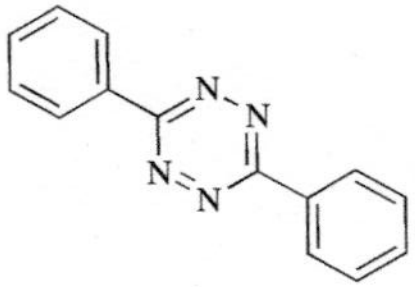

[Pd(dba)$_2$] (20 mol %),
NFSI (12 eq), MW (200 W),
PhCF$_3$, 110°, 0.5 h

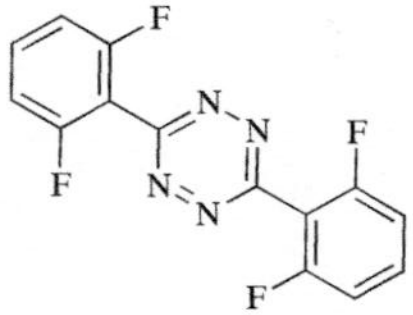

(36)

127

TABLE 1E. ARYL FLUORIDE SYNTHESIS FROM ARYL SILANES

Silane	Conditions	Product(s) and Yield(s) (%)	Refs.
C$_{6-13}$	Selectfluor (2.0 eq), Ag$_2$O (2.0 eq), BaO (1.1 eq), acetone, 90°, 2 h	R: H (90), Cl (86), Br (86), MeO (76), BzO (78), AcHN (70), Me (79), CF$_3$ (90), EtO$_2$C (85), Ac (82), Ph (83), Bz (85)	137
C$_7$	F-TEDA (2.0 eq), Ag$_2$O (2.0 eq), BaO (1.1 eq), acetone, 90°, 2 h	(74)	137
C$_9$	F-TEDA (2.0 eq), Ag$_2$O (3.0 eq), BaO (1.1 eq), acetone, 90°, 2 h	(60)	137

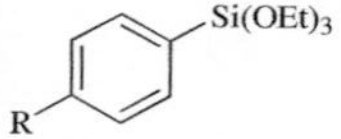

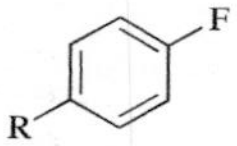

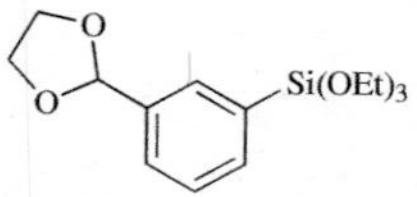

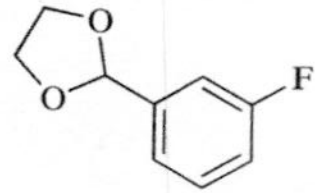

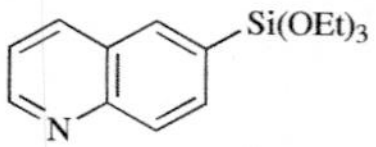

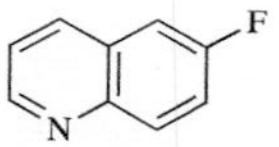

748

C$_{10}$

| 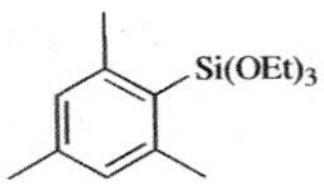 F-TEDA (2.0 eq), Ag$_2$O (3.0 eq), 2,6-lutidine (1.1 eq), acetone, 90°, 2 h | 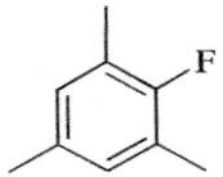 (73) | 137 |

| 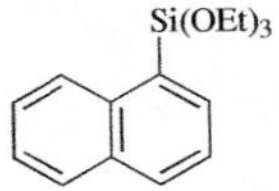 F-TEDA (2.0 eq), Ag$_2$O (2.0 eq), 2,6-lutidine (1.1 eq), acetone, 90°, 2 h | 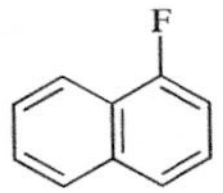 (75) | 137 |

TABLE 1F. ARYL FLUORIDE SYNTHESIS FROM IODONIUM SALTS

Iodonium Salt	Conditions	Product(s) and Yield(s) (%)	Refs.
C_{6-13}	Cu(OTf)$_2$ (x eq), 18-crown-6 (0.4 eq), KF (1.1 eq), DMF, 60°, 18 h	 R x MeO 0.2 (84) PhO 0.5 (97) BnO 0.5 (81) Ph 0.2 (82) Bz 0.2 (72)	139
C_6	Cu(OTf)$_2$ (20 mol %), 18-crown-6 (40 mol %), KF (1.1 eq), DMF, 60°, 18 h	(74)	139
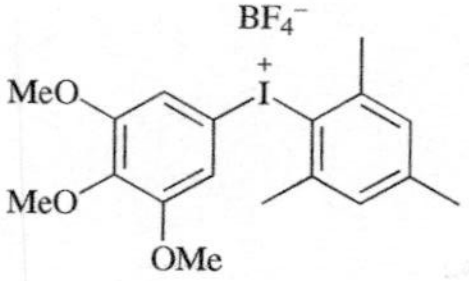	Cu(OTf)$_2$ (20 mol %), 18-crown-6 (40 mol %), KF (1.1 eq), DMF, 60°, 18 h	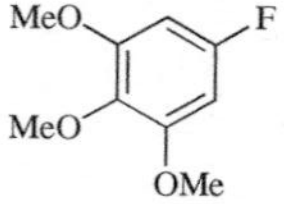 (53)	139
C_7	Cu(OTf)$_2$ (20 mol %), 18-crown-6 (40 mol %), KF (1.1 eq), DMF, 60°, 18 h	(67)	139
C_{10}	Cu(OTf)$_2$ (0.5 eq), 18-crown-6 (0.4 eq), KF (1.1 eq), DMF, 60°, 18 h	(66)	139

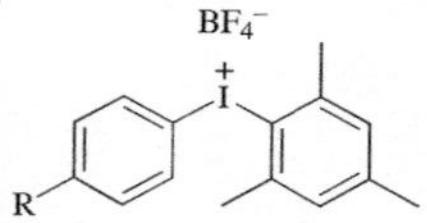

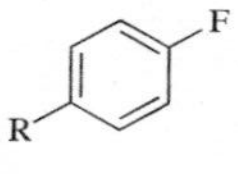

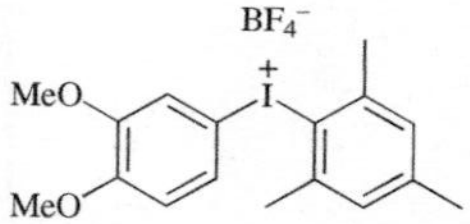

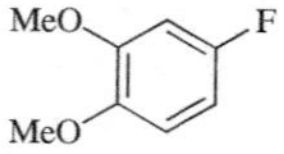

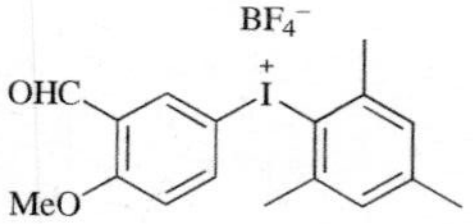

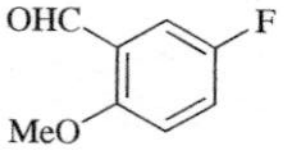

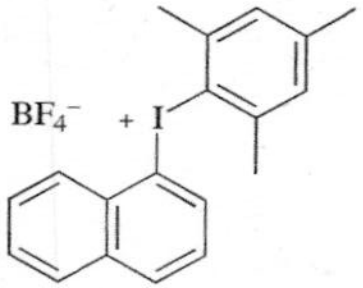

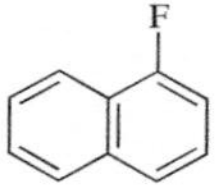

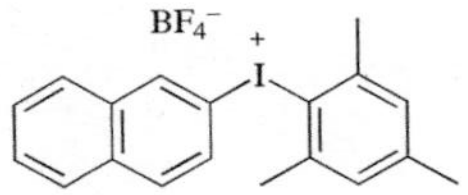

C$_{12}$

BF$_4^-$ Cu(OTf)$_2$ (0.5 eq),
18-crown-6 (0.4 eq),
KF (1.1 eq), DMF, 60°, 18 h   (78)   139

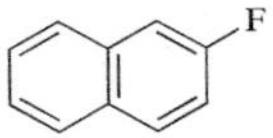

BF$_4^-$ Cu(OTf)$_2$ (0.5 eq),
18-crown-6 (0.4 eq),
KF (1.1 eq), DMF, 60°, 18 h   (61)   139

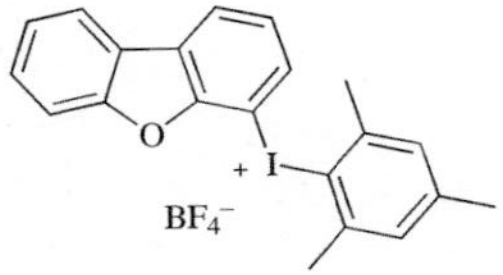

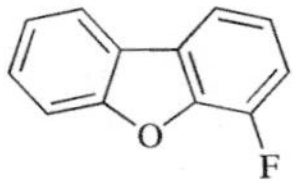

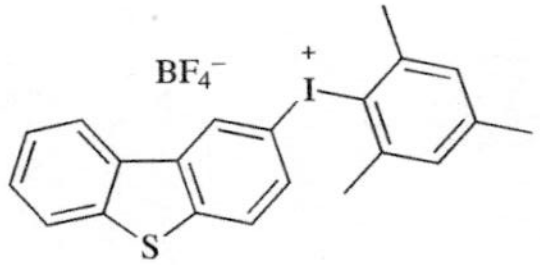

BF$_4^-$ Cu(OTf)$_2$ (0.5 eq),
18-crown-6 (0.4 eq),
KF (1.1 eq), DMF, 60°, 18 h   (56)   139

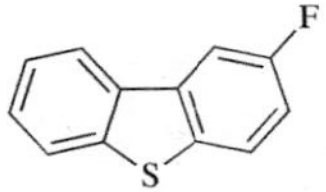

C$_{18}$

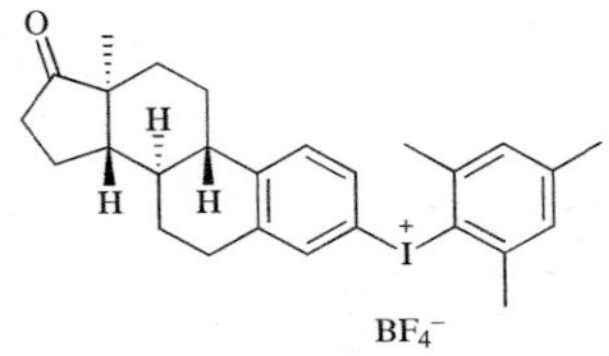

BF$_4^-$ Cu(OTf)$_2$ (0.5 eq),
18-crown-6 (0.4 eq),
KF (1.1 eq), DMF, 60°, 18 h   (83)   139

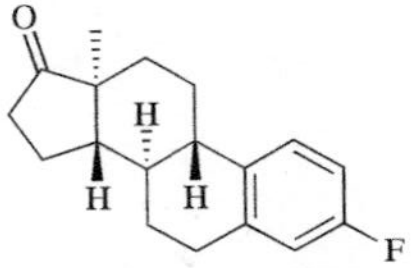

TABLE 2. SYNTHESIS OF FLUORINATED HETEROARENES

Heteroaromatic	Conditions	Product(s) and Yield(s) (%)	Refs.

*Please refer to the charts preceding the tables for the structures indicated by the **bold** numbers.*

C₄

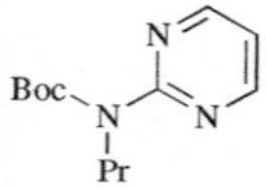

AgF₂ (3 eq), MeCN, rt, 1 h

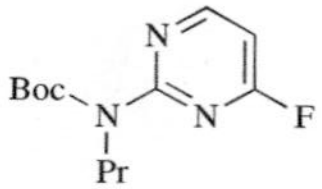

(77)

53

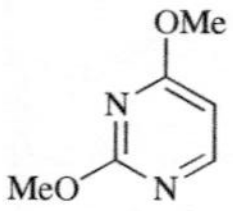

AgF₂ (3 eq), MeCN, rt, 1 h

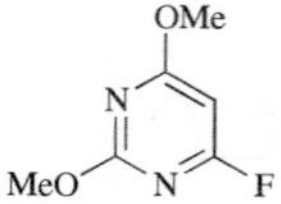

(76)

143

C₅

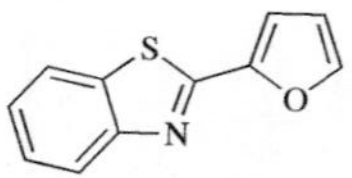

Pd(PPh₃)₄ (15 mol %), L-proline (40 mol %), NFSI (3 eq), cyclohexane, 120°, 5 h

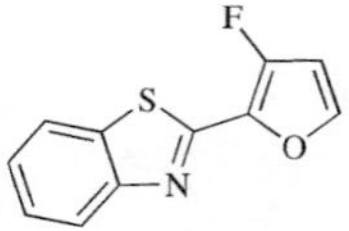

(tr)

130

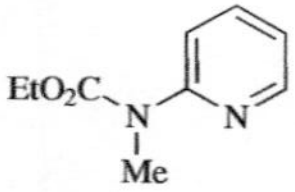

AgF₂ (3 eq), MeCN, rt, 2 h

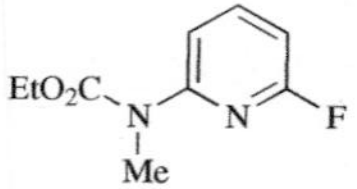

(49)

53

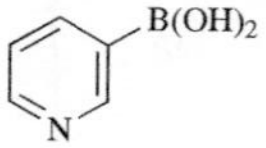

1. NaOH (1.2 eq), MeOH, AgOTf (3.0 eq), 0°
2. Selectfluor (1.05 eq), 3 Å MS, acetone, 23°

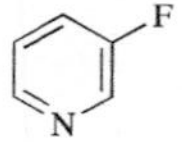

(72)

132

752

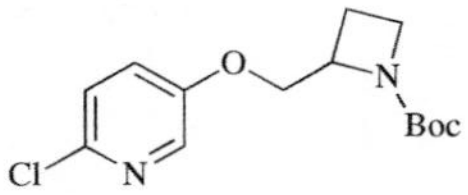

143

53, 143

143

124

(65)

(69)

(75)

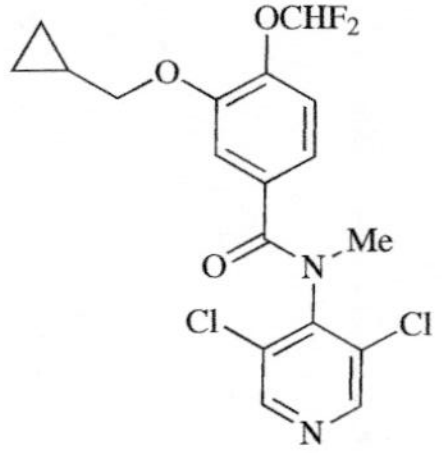

(56)

AgF$_2$ (3 eq), MeCN, rt, 1 h

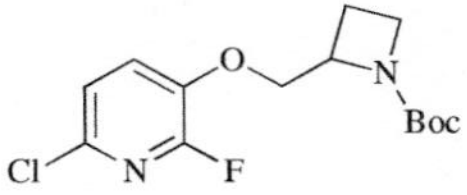

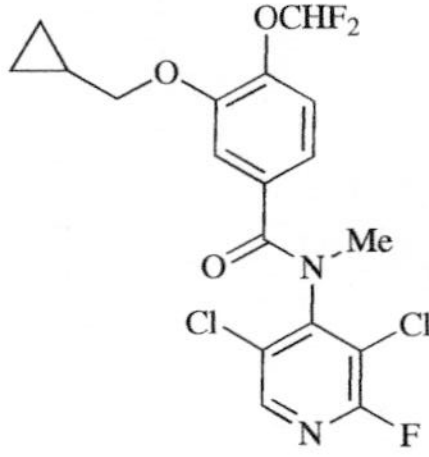

AgF$_2$ (3 eq), MeCN, rt, 1 h

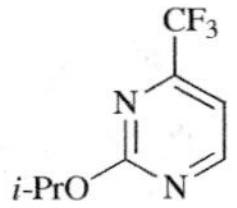

AgF$_2$ (3 eq), MeCN, rt, 1 h

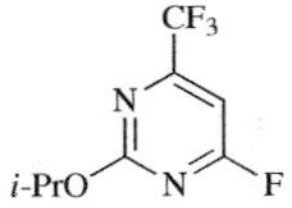

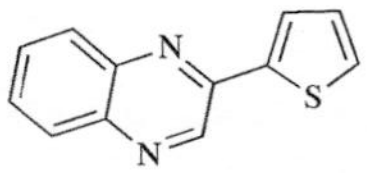

Pd(CF$_3$CO$_2$)$_2$ (15 mol %), NFSI (2 eq),
TFA (2 eq), CH$_3$NO$_2$, 110°, 12 h

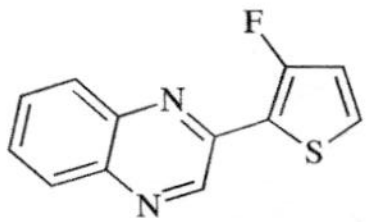

C$_6$

TABLE 2. SYNTHESIS OF FLUORINATED HETEROARENES (*Continued*)

Heteroaromatic	Conditions	Product(s) and Yield(s) (%)	Refs.

Please refer to the charts preceding the tables for the structures indicated by the bold numbers.

C_6

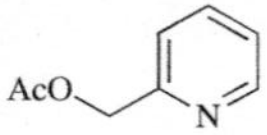

AgF$_2$ (3 eq), MeCN, rt, 1 h

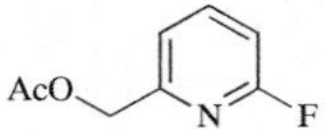

(71)

53

C_{6-12}

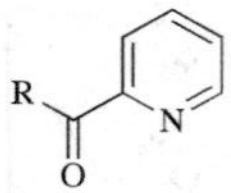

AgF$_2$ (3 eq), MeCN, rt, 1 h

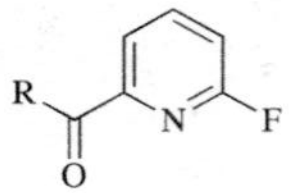

R	
EtO	(67)
Et$_2$N	(75)
Ph	(54)

53

C_6

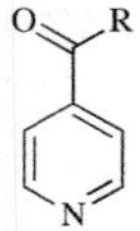

AgF$_2$ (3 eq), MeCN, 1 h

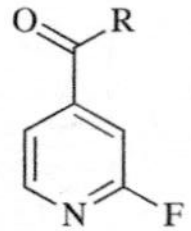

R	Temp (°)	
EtO	50	(44)
Et$_2$N	23	(66)

53

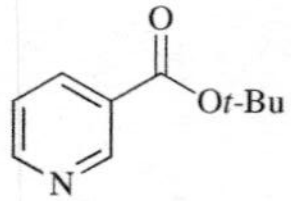

AgF$_2$ (3 eq), MeCN, rt, 1 h

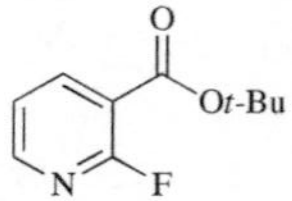

(42) +   (21)

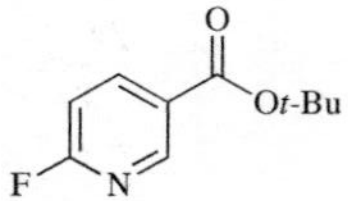

53

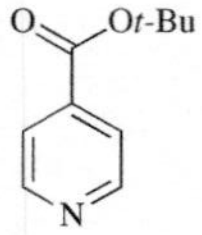

AgF$_2$ (3 eq), MeCN, rt, 1 h

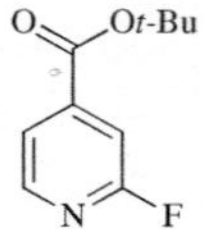

(64)

143

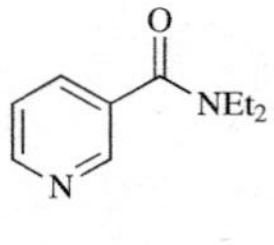

AgF$_2$ (3 eq), MeCN, rt, 1 h

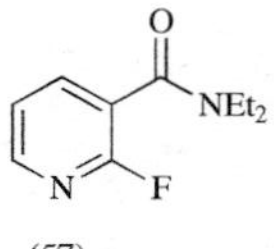

(13) + (57)

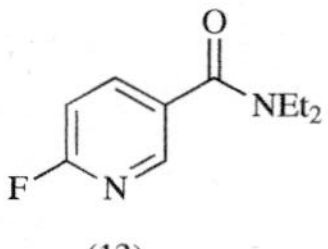

53

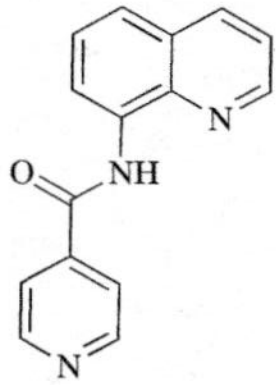

CuI (20 mol %), AgF (4 eq),
NMO (5 eq),
pyridine, 65°, 2 h

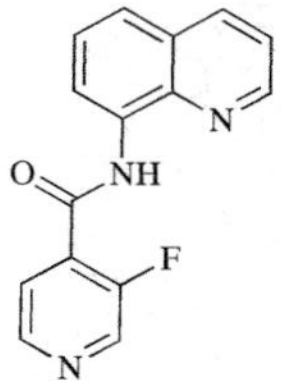

(62)

126

CuI (25 mol %), AgF (6.0 eq),
NMO (8.0 eq),
pyridine, 85°, 1.5 h

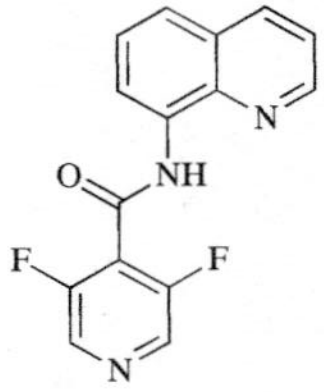

(61)

126

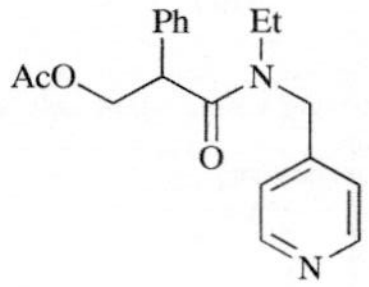

AgF$_2$ (3 eq), MeCN, 50°, 1 h

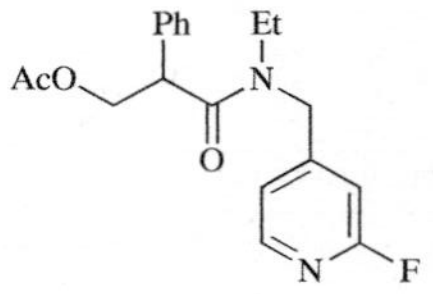

(74)

53

TABLE 2. SYNTHESIS OF FLUORINATED HETEROARENES (*Continued*)

Heteroaromatic	Conditions	Product(s) and Yield(s) (%)	Refs.

*Please refer to the charts preceding the tables for the structures indicated by the **bold** numbers.*

C₆

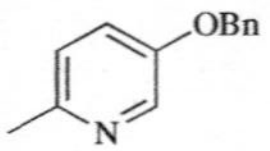	AgF_2 (3 eq), MeCN, rt, 1 h	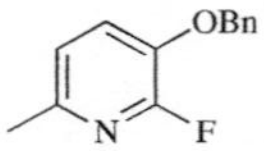 (99)	53, 143
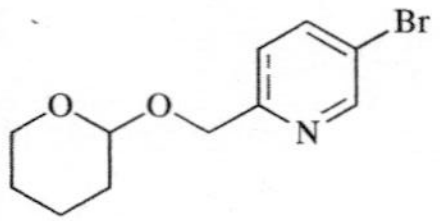	AgF_2 (3 eq), MeCN, rt, 1 h	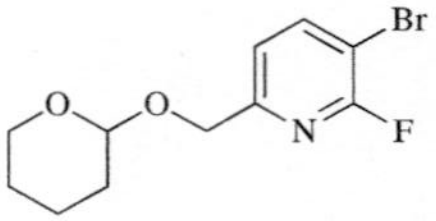 (87)	143
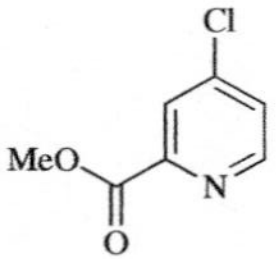	AgF_2 (3 eq), MeCN, rt, 1 h	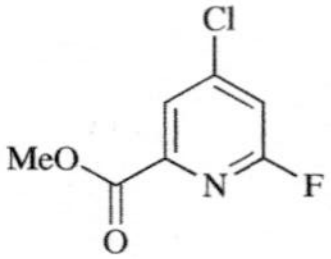 (72)	53
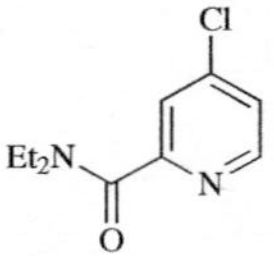	AgF_2 (3 eq), MeCN, rt, 1 h	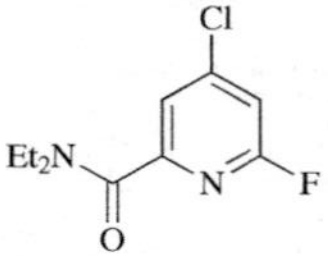 (82)	143
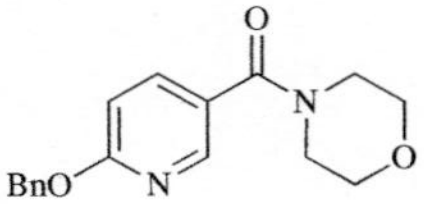	AgF_2 (3 eq), MeCN, rt, 1 h	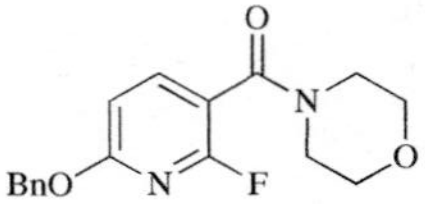 (73)	53

C₇

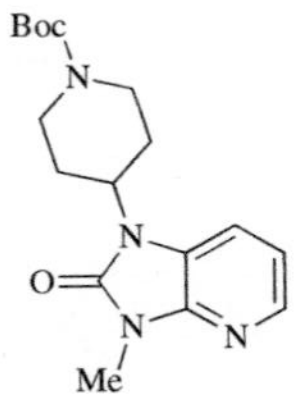

AgF₂ (3 eq), MeCN, rt, 1 h

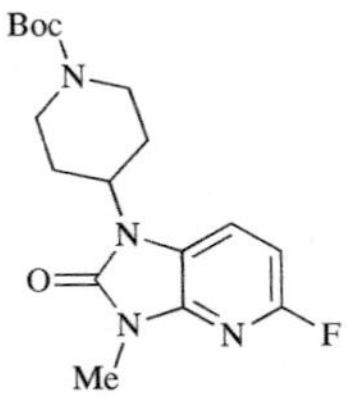

AgF₂ (3 eq), MeCN, rt, 1 h

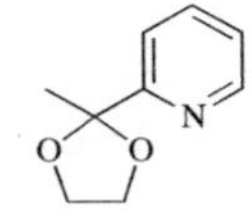

AgF₂ (3 eq), MeCN, rt, 1 h

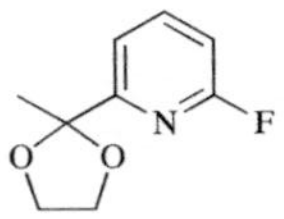

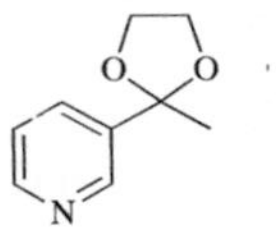

AgF₂ (3 eq), MeCN, rt, 1 h

(75)

(91)

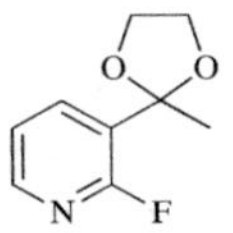

+

(50)

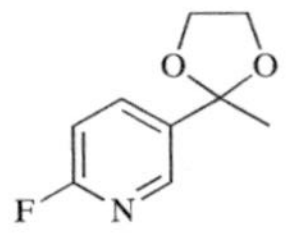

(98)

(22)

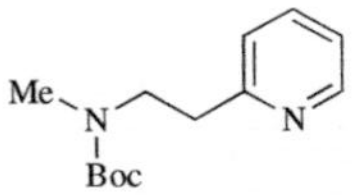

53

53

53

143

Boc—N ... N—Me ... F

Me—N—Boc

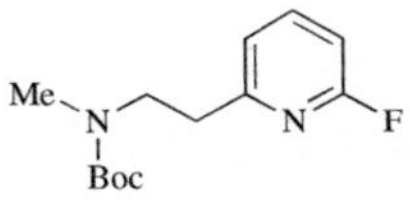

TABLE 2. SYNTHESIS OF FLUORINATED HETEROARENES (*Continued*)

Heteroaromatic	Conditions	Product(s) and Yield(s) (%)	Refs.

*Please refer to the charts preceding the tables for the structures indicated by the **bold** numbers.*

C₇

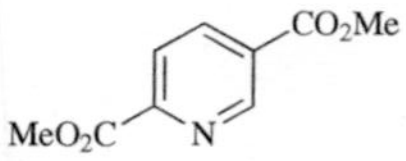

AgF₂ (3 eq), MeCN, rt, 1 h — (59) — 53

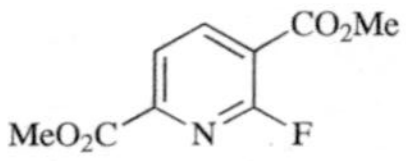

C₈

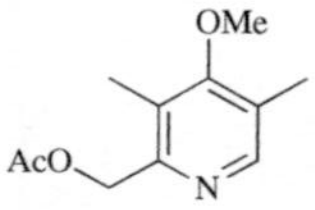

AgF₂ (3 eq), MeCN, rt, 1 h — (81) — 53

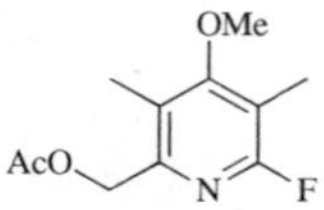

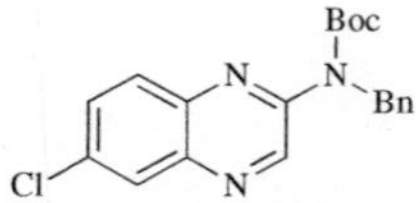

AgF₂ (3 eq), MeCN, rt, 1 h — (99) — 143

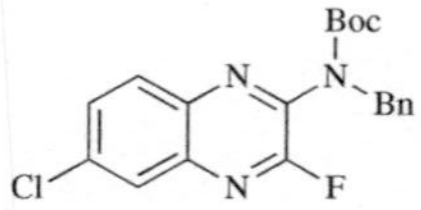

C₉

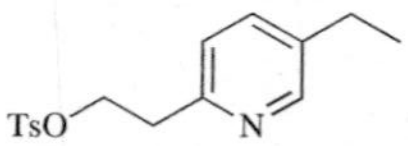

AgF₂ (3 eq), MeCN, rt, 1 h — (83) — 53

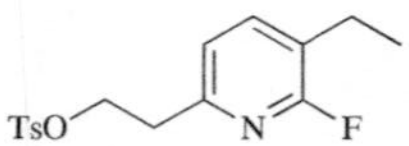

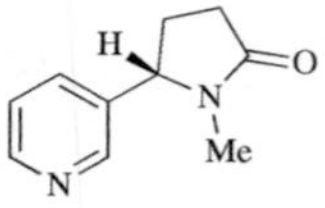

AgF₂ (3 eq), MeCN, 50°, 2 h — (50) + (10) — 53

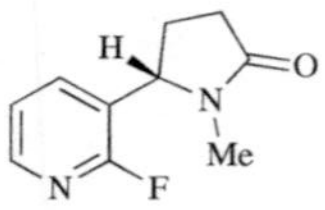

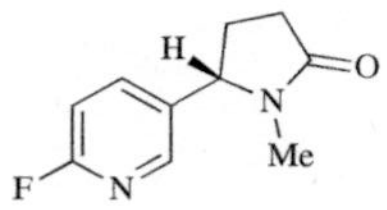

758

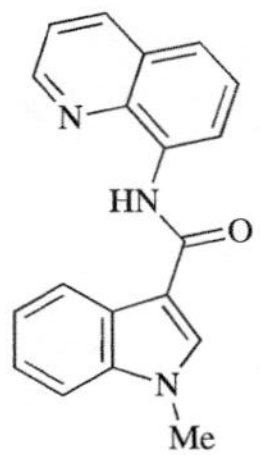

CuI (10 mol %), AgF (4 eq),
NMO (5 eq),
DMF, 50°, 1 h

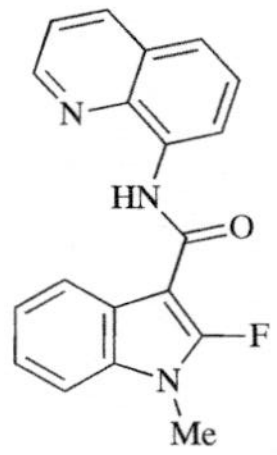

(54)

126

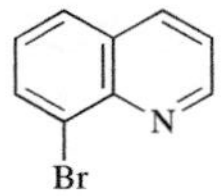

AgF$_2$ (3 eq). MeCN, 50°, 1 h

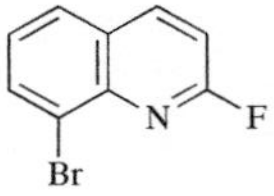

(61)

53

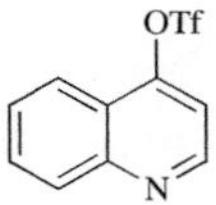

[(Cinnamyl)PdCl]$_2$ (2 mol %),
L1 (6 mol %), toluene, 120°,
20 min residence time (35 μL min^{-1})[a]

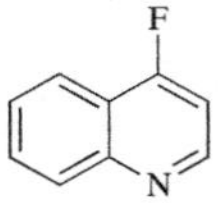

(71)

248

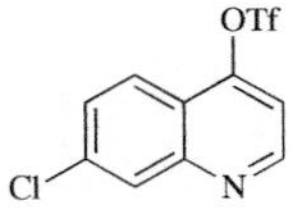

Pd1 (2 mol %), CsF (3 eq), toluene, rt, 96 h

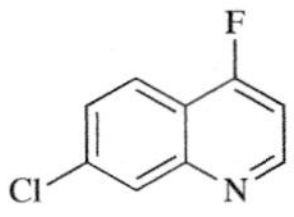

(85)

36

TABLE 2. Synthesis of Fluorinated Heteroarenes (*Continued*)

Heteroaromatic	Conditions	Product(s) and Yield(s) (%)	Refs.

*Please refer to the charts preceding the tables for the structures indicated by the **bold** numbers.*

C₉

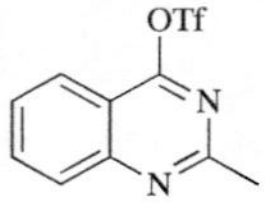

Pd1 (2.5 mol %), CsF (3 eq),
2-Me-THF, rt, 96 h

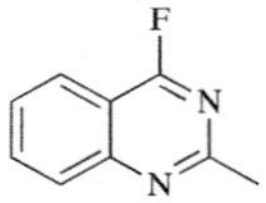

(83)

36

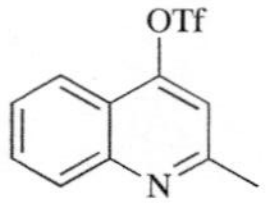

[(Cinnamy)PdCl]₂ (2 mol %),
L1 (6 mol %), toluene, 120°,
20 min residence time (35 µL min⁻¹)ᵃ

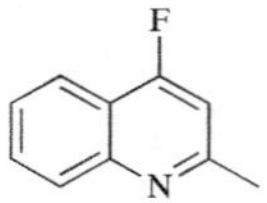

(85)

248

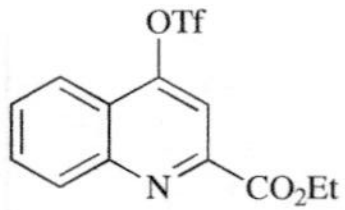

Pd1 (1 mol %), CsF (3 eq),
2-Me-THF, 60°, 24 h

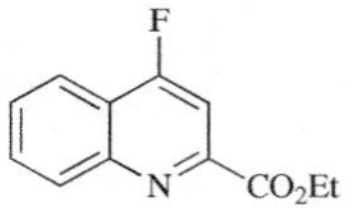

(95)

36

[(Cinnamy)PdCl]₂ (2.5 mol %),
L1 (7.5 mol %), toluene, 120°,
20 min residence time (35 µL min⁻¹)ᵃ

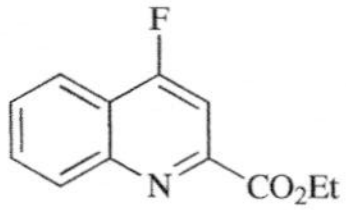

(60)

248

C₁₁

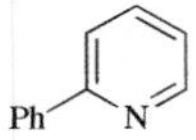

AgF₂ (3 eq), MeCN, rt, 1 h

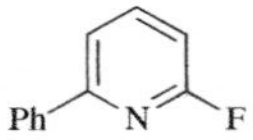

(88)

53, 143

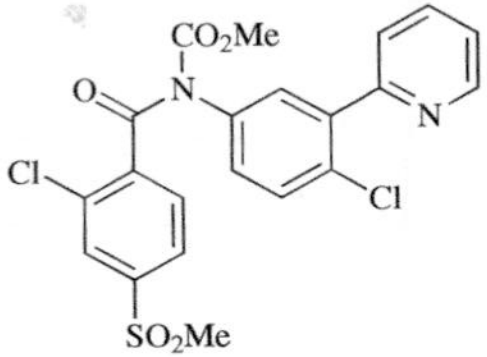

AgF$_2$ (3 eq), MeCN, rt, 2 h   (99)   143

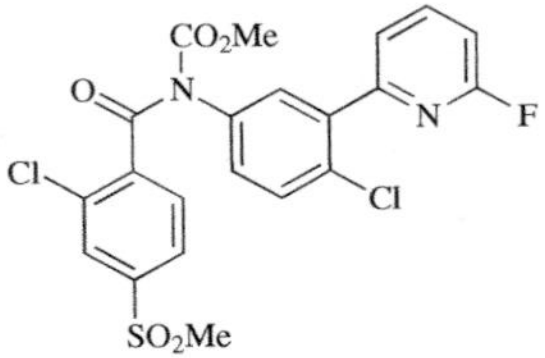

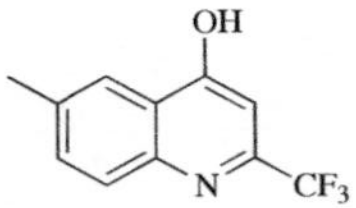

1. NfF (1.5 eq), CsF (2.0 eq), toluene, 50°, 31 h;
2. Pd$_2$(dba)$_3$ (2 mol %), L1 (7.5 mol %),
CsF (3 eq), MW, toluene, 180°, 0.5 h   (85)   116

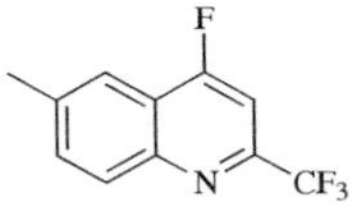

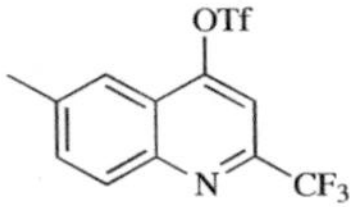

[(Cinnamyl)PdCl]$_2$ (2 mol %),
L1 (6 mol %), toluene, 120°
20 min residence time (35 μL min^{-1})^a (64)   248

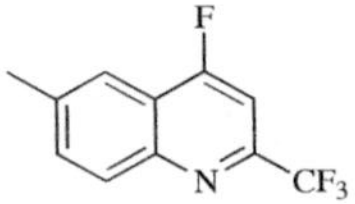

Pd1 (2 mol %), CsF (3 eq), toluene, rt, 96 h   (84)   36

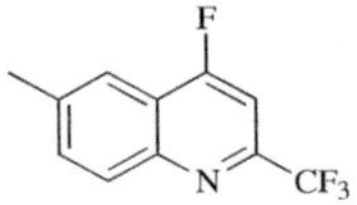

TABLE 2. SYNTHESIS OF FLUORINATED HETEROARENES (Continued)

Heteroaromatic	Conditions	Product(s) and Yield(s) (%)	Refs.

Please refer to the charts preceding the tables for the structures indicated by the **bold** numbers.

C$_{12}$

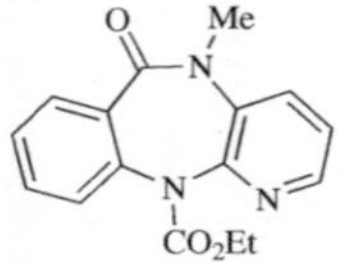

AgF$_2$ (3 eq), MeCN, rt, 1 h

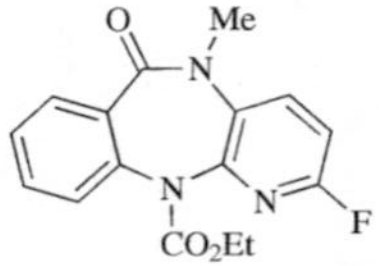

(57)

143

C$_{13}$

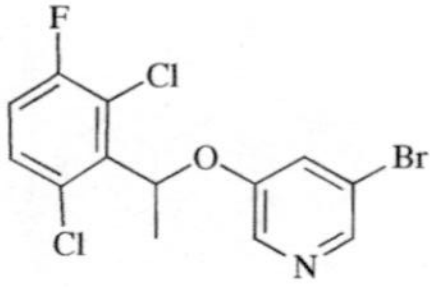

AgF$_2$ (3 eq), MeCN, rt, 1 h

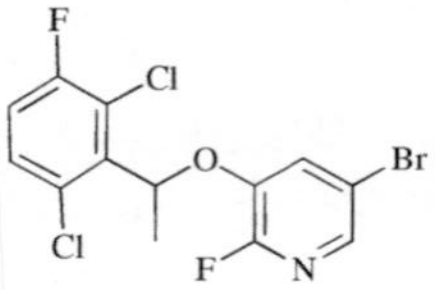

I

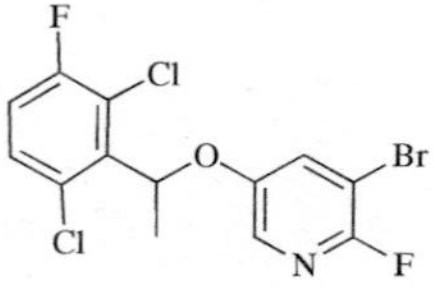

+

II

I + II (90), **I/II** = 7.2:1

143

C$_{14}$

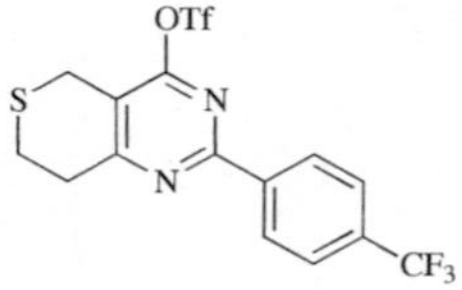

PdI (4 mol %), CsF (3 eq),
2-Me-THF, rt, 96 h

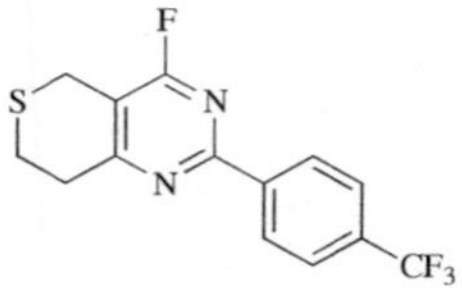

(89)

36

C$_{15}$

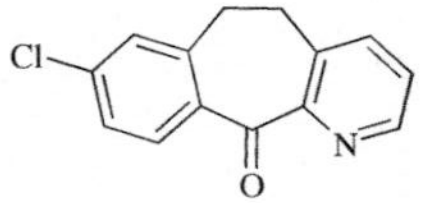

AgF$_2$ (3 eq), MeCN, rt, 1 h

(53)

143

C$_{17}$

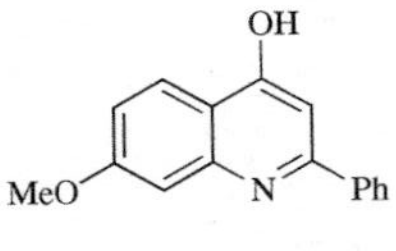

1. NfF (1.5 eq), CsF (2.0 eq), toluene, 50°, 7 d
2. Pd$_2$(dba)$_3$ (2 mol %), **L1** (6 mol %),
CsF (3 eq), MW, toluene, 180°, 0.5 h

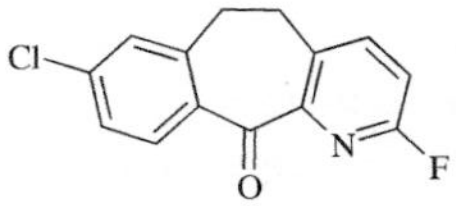

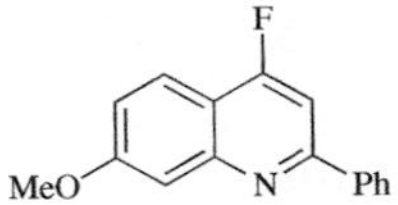

(73)

116

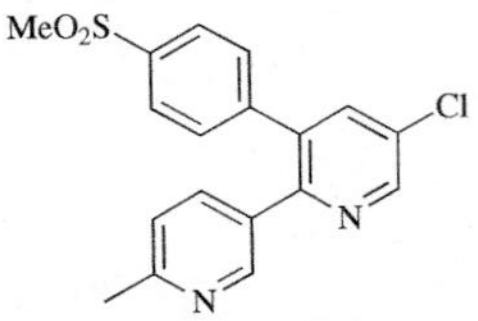

AgF$_2$ (3 eq), MeCN, 50°, 2 h

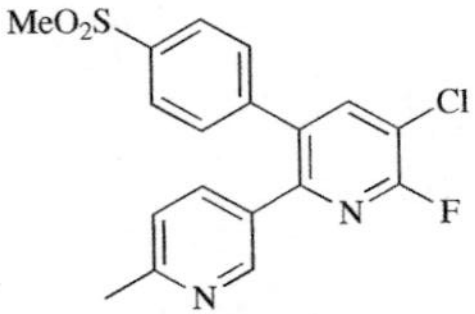

(69)

143

[a]The reactions were conducted in flow in a CsF-packed-bed reactor.

TABLE 3. ALKENYL FLUORIDE SYNTHESIS

*Please refer to the charts preceding the tables for the structures indicated by the **bold** numbers.*

Alkyne	Conditions	Product(s) and Yield(s) (%)	Refs.
C₄			
(Et₂N—OAc)	IPrAuCl (2.5 mol %), PhNMe₂•HOTf (2.5 mol %), AgBF₄ (2.75 mol %), Et₃N•3HF (1.5 eq), KHSO₄ (0.1 eq). CH₂Cl₂, rt, 15 h	N/A (0)	47
	IPrAuCl (2.5 mol %), PhNMe₂•HOTf (2.5 mol %), AgBF₄ (2.75 mol %), Et₃N•3HF (1.5 eq), KHSO₄ (0.1 eq). CH₂Cl₂, rt, 15 h	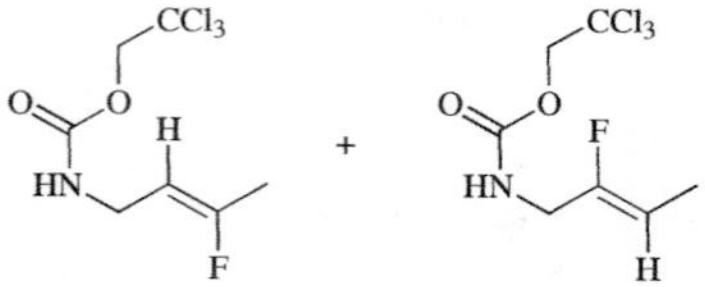 **I** (Z)/(E) = 83:17 + **II** **I + II** (57), **I/II** > 50:1	47
	IPrAuCl (2.5 mol %), PhNMe₂•HOTf (2.5 mol %), AgBF₄ (2.75 mol %), Et₃N•3HF (1.5 eq), KHSO₄ (0.1 eq). CH₂Cl₂, rt, 15 h	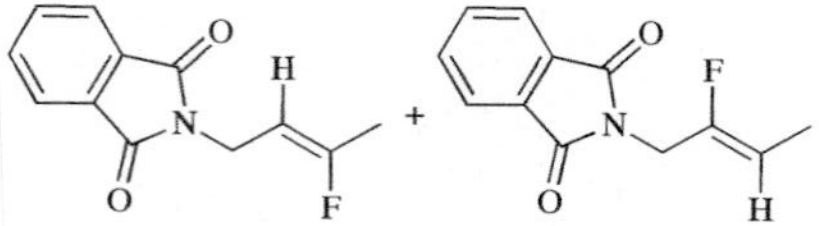 **I** (Z)/(E) = 84:14 + **II** **I + II** (53), **I/II** > 50:1	47
C₆			
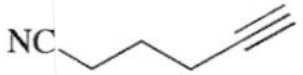	**Au1** (2 mol %), DMPU/HF (65 wt %, 1.2 eq), DCE, 55°, 3 h	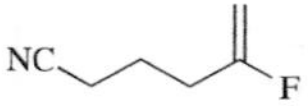 (89)	147

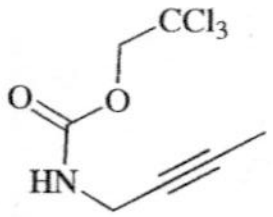

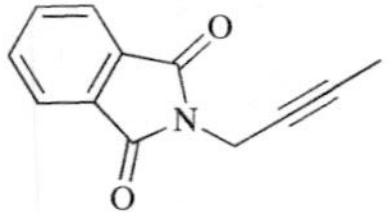

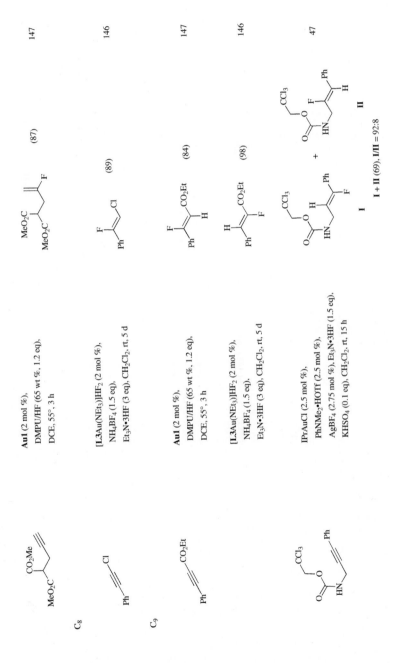

TABLE 3. ALKENYL FLUORIDE SYNTHESIS (*Continued*)

Alkyne	Conditions	Product(s) and Yield(s) (%)	Refs.
	Please refer to the charts preceding the tables for the structures indicated by the **bold** numbers.		
C$_{10}$			
	**Au1** (2 mol %), DMPU/HF (65 wt %, 1.2 eq), DCE, 55°, 3 h	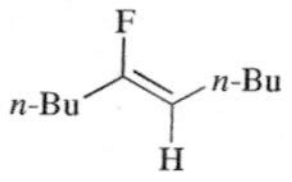 (84)	147
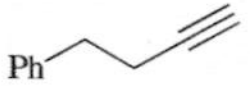	**Au1** (2 mol %), DMPU/HF (65 wt %, 1.2 eq), DCE, 55°, 3 h	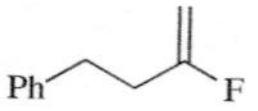 (81)	147
	[L3Au(NEt$_3$)]HF$_2$ (2 mol %), NH$_4$BF$_4$ (1.5 eq), Et$_3$N•3HF (3 eq), CH$_2$Cl$_2$, rt, 5 d	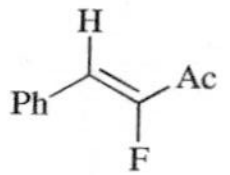 (86) 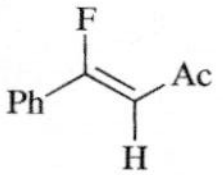 + (10)	146
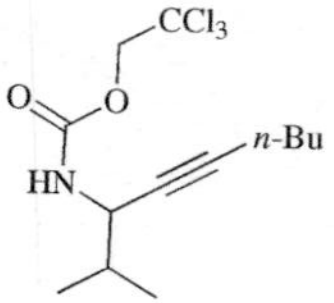	IPrAuCl (2.5 mol %), PhNMe$_2$•HOTf (2.5 mol %), AgBF$_4$ (2.75 mol %), Et$_3$N•3HF (1.5 eq), KHSO$_4$ (0.1 eq), CH$_2$Cl$_2$, rt, 15 h	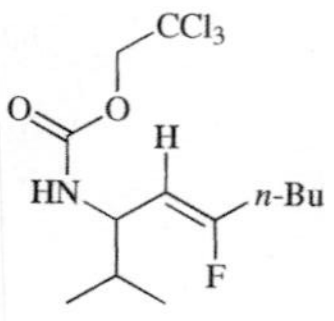 **I** 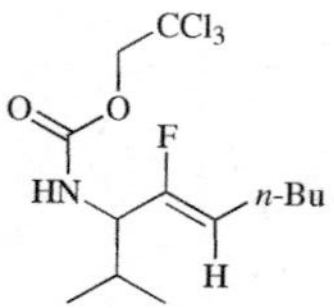 + **II** I + II (74), I/II > 50:1	47
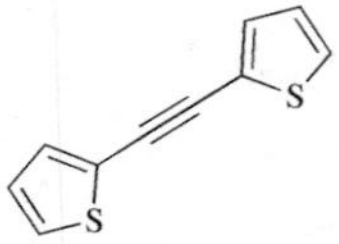	[L3Au(NEt$_3$)]HF$_2$ (2 mol %), NH$_4$BF$_4$ (1.5 eq), Et$_3$N•3HF (3 eq), CH$_2$Cl$_2$, 50°, 24 h	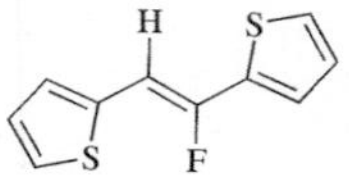 (94)	146

Substrate	Conditions	Product(s) (%)	Ref.
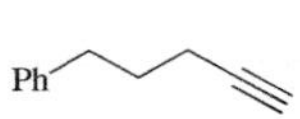 C_{11} Ph(CH$_2$)$_3$C≡CH (2-fluoro-5-phenyl-1-pentene)	**Au1** (2 mol %), DMPU/HF (65 wt %, 1.2 eq), DCE, 55°, 3 h	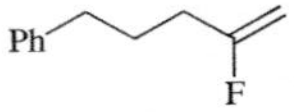 vinyl fluoride product (79)	147
C_{12} n-C$_5$H$_{11}$C≡CC≡Cn-C$_5$H$_{11}$	**L21**AuCl (2.5 mol %), AgBF$_4$ (2.5 mol %), PhNMe$_2$•HOTf (10 mol %), Et$_3$N•3HF (1.5 eq), KHSO$_4$ (1 eq), CH$_2$Cl$_2$, rt, 30 h	n-C$_5$H$_{11}$, F vinyl fluoride (81)	48
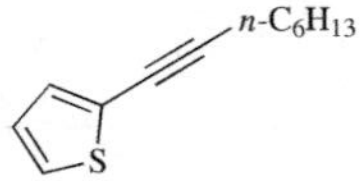 n-C$_6$H$_{13}$C≡C–(2-thienyl)	**Au2** (2.5 mol %), PhNMe$_2$•HOTf (10 mol %), Et$_3$N•3HF (1.5 eq), KHSO$_4$ (1 eq), CH$_2$Cl$_2$, rt, 30 h	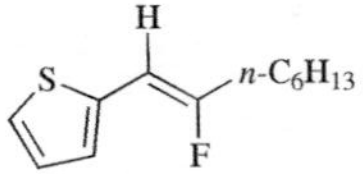 n-C$_6$H$_{13}$, F thienyl vinyl fluoride (74)	48
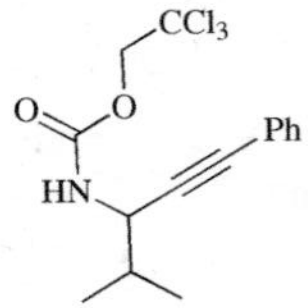 Cl$_3$CCH$_2$OC(O)NH–CH(iPr)–C≡C–Ph	IPrAuCl (2.5 mol %), AgBF$_4$ (2.75 mol %), PhNMe$_2$•HOTf (2.5 mol %), Et$_3$N•3HF (1.5 eq), KHSO$_4$ (0.1 eq), CH$_2$Cl$_2$, rt, 15 h	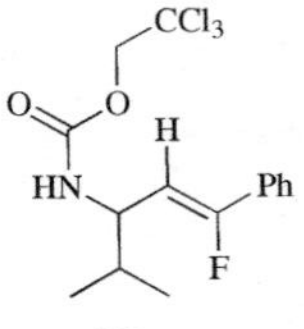 (60) 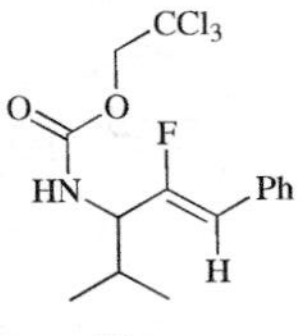 + (5) isomeric vinyl fluorides	47

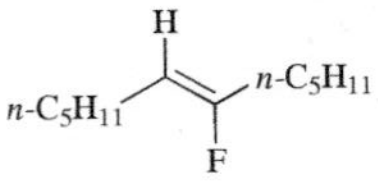

TABLE 3. ALKENYL FLUORIDE SYNTHESIS (*Continued*)

Alkyne	Conditions	Product(s) and Yield(s) (%)	Refs.

*Please refer to the charts preceding the tables for the structures indicated by the **bold** numbers.*

C13

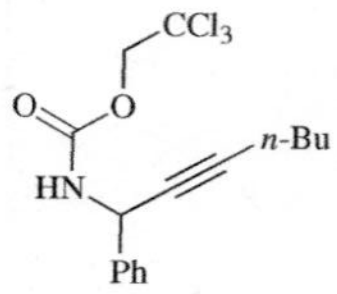

IPrAuCl (2.5 mol %), AgBF$_4$ (2.75 mol %),
PhNMe$_2$•HOTf (2.5 mol %),
Et$_3$N•3HF (1.5 eq),
KHSO$_4$ (0.1 eq), CH$_2$Cl$_2$, rt, 15 h

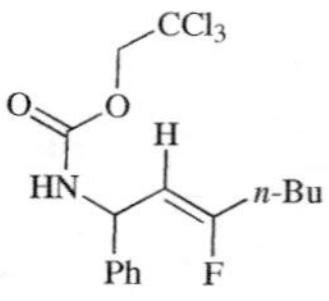

I + **II** (68), **I/II** > 50:1

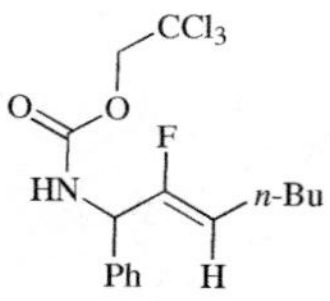

47

C14

Au2 (2.5 mol %),
PhNMe$_2$•HOTf (10 mol %),
Et$_3$N•3HF (1.5 eq),
KHSO$_4$ (1 eq), CH$_2$Cl$_2$, rt, 24 h

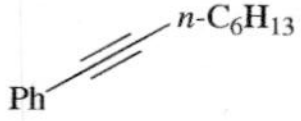

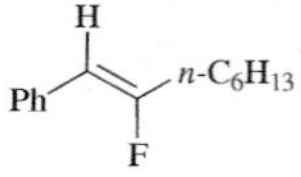

(72) + (6)

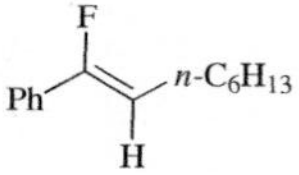

48

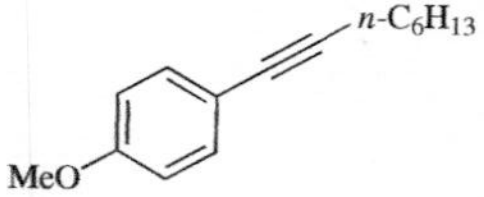

Au2 (2.5 mol %),
PhNMe$_2$•HOTf (10 mol %),
Et$_3$N•3HF (1.5 eq),
KHSO$_4$ (1 eq), CH$_2$Cl$_2$, rt, 24 h

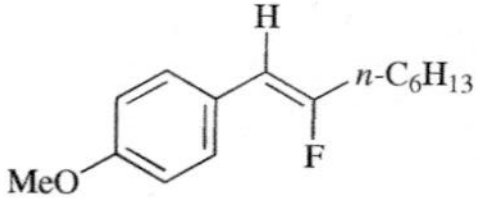

(52) + (11)

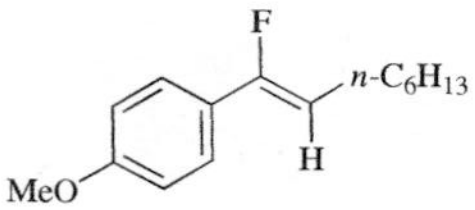

48

[L3Au(NEt$_3$)]HF$_2$ (2 mol %),
NH$_4$BF$_4$ (1.5 eq),
Et$_3$N•3HF (3 eq), CH$_2$Cl$_2$, 50°, 24 h

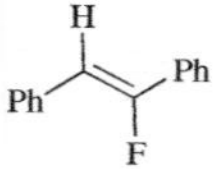

(99)

146

	L2IAuCl (2.5 mol %), AgBF$_4$ (2.5 mol %), PhNMe$_2$•HOTf (10 mol %), Et$_3$N•3HF (1.5 eq), KHSO$_4$ (1 eq), CH$_2$Cl$_2$, rt, 30 h	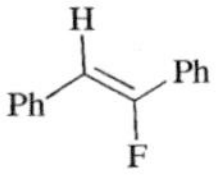 (86) 48
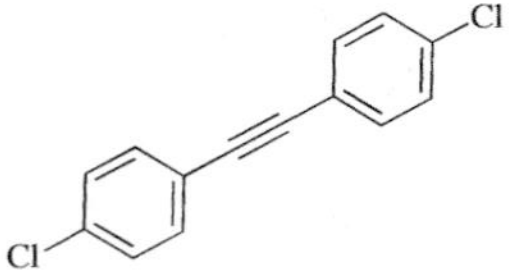	**[L3Au(NEt$_3$)]HF$_2$** (2 mol %), NH$_4$BF$_4$ (1.5 eq), Et$_3$N•3HF (3 eq), CH$_2$Cl$_2$, 50°, 24 h	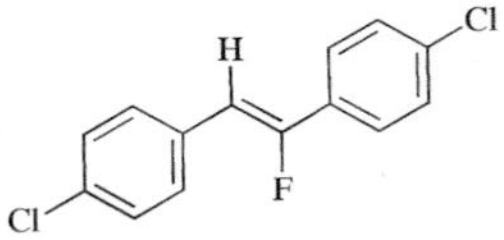 (95) 146
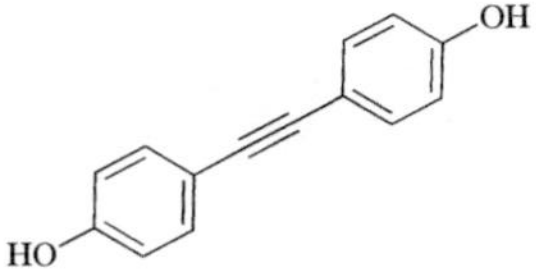	**[L3Au(NEt$_3$)]HF$_2$** (2 mol %), NH$_4$BF$_4$ (1.5 eq), Et$_3$N•3HF (3 eq), CH$_2$Cl$_2$, 50°, 48 h	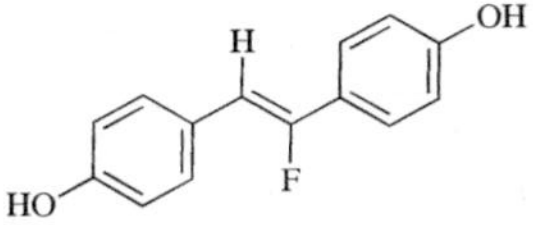 (99) 146
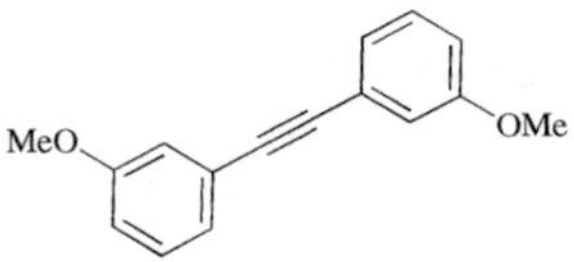	**[L3Au(NEt$_3$)]HF$_2$** (2 mol %), NH$_4$BF$_4$ (1.5 eq), Et$_3$N•3HF (3 eq), CH$_2$Cl$_2$, 50°, 24 h	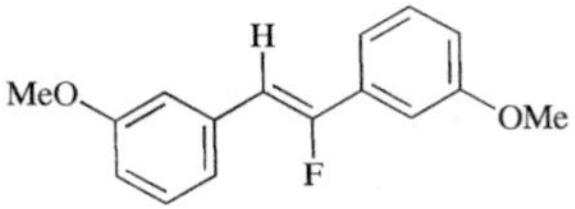 (94) 146

TABLE 3. ALKENYL FLUORIDE SYNTHESIS (Continued)

Alkyne	Conditions	Product(s) and Yield(s) (%)	Refs.

*Please refer to the charts preceding the tables for the structures indicated by the **bold** numbers.*

C14

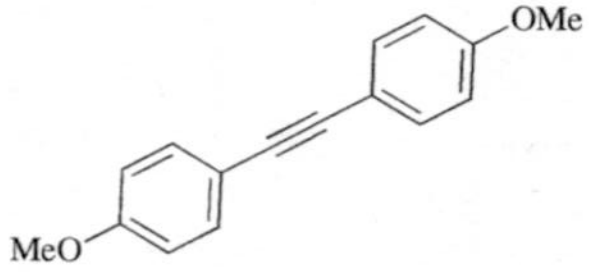

[L3Au(NEt₃)]HF₂ (2 mol %),
NH₄BF₄ (1.5 eq),
Et₃N•3HF (3 eq), CH₂Cl₂, 50°, 24 h

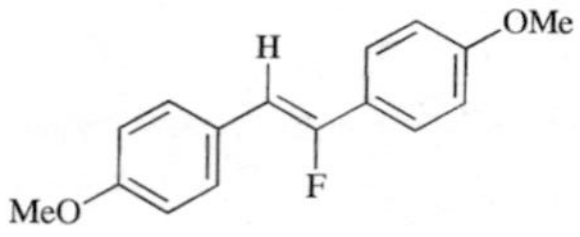

(99)   146

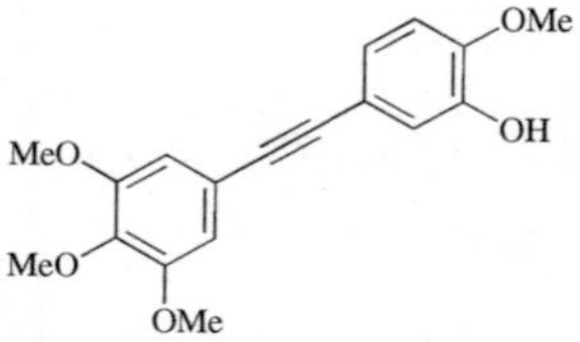

[L3Au(NEt₃)]HF₂ (2 mol %),
NH₄BF₄ (1.5 eq),
Et₃N•3HF (3 eq), CH₂Cl₂, rt, 5 d

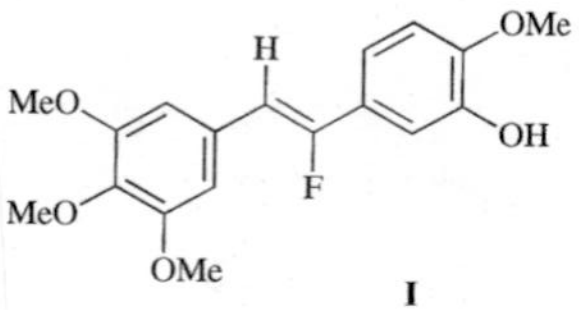

I

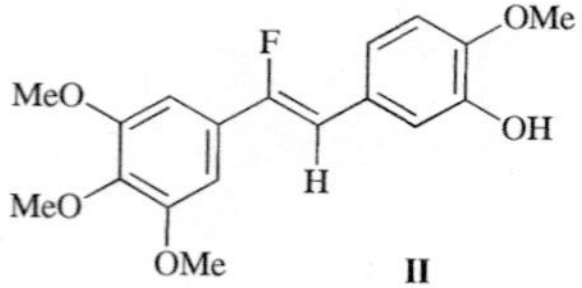

II

I + II (32), **I/II** = 92:8

146

C15

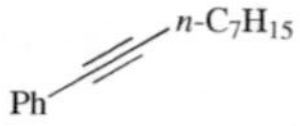

[L3Au(NEt₃)]HF₂ (2 mol %),
NH₄BF₄ (1.5 eq),
Et₃N•3HF (3 eq), CH₂Cl₂, rt, 5 d

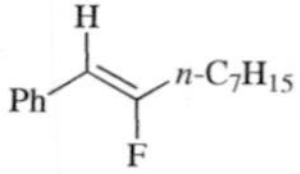

(94)   146

C_16

| | | | |

[L3Au(NEt₃)]HF₂ (2 mol %), NH₄BF₄ (1.5 eq), Et₃N•3HF (3 eq), CH₂Cl₂, rt, 5 d

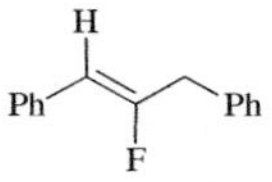

(91)

146

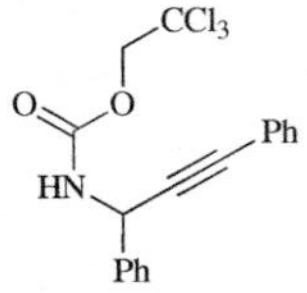

IPrAuCl (2.5 mol %), PhNMe₂•HOTf (2.5 mol %), AgBF₄ (2.75 mol %), Et₃N•3HF (1.5 eq), KHSO₄ (0.1 eq), CH₂Cl₂, rt, 15 h

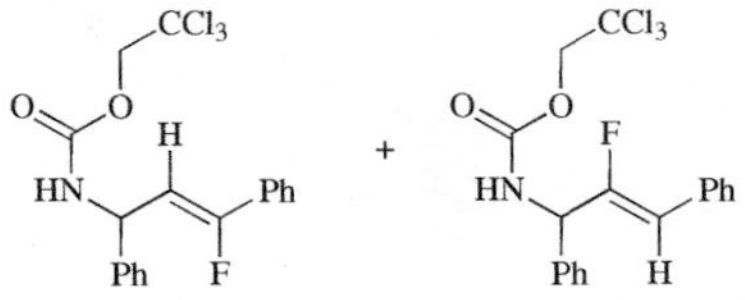

I + II (65), **I/II** = 98:2

I   +   **II**

47

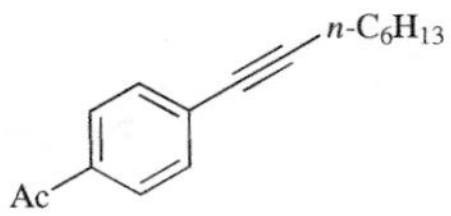

(SIPr)AuOt-Bu (2.5 mol %), AgBF₄, PhNMe₂•HOTf (10 mol %), Et₃N•3HF (1.5 eq), KHSO₄ (1 eq), CH₂Cl₂, rt, 30 h

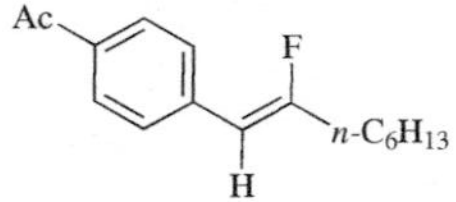

(82)

48

TABLE 3. ALKENYL FLUORIDE SYNTHESIS (*Continued*)

Alkyne	Conditions	Product(s) and Yield(s) (%)	Refs.

*Please refer to the charts preceding the tables for the structures indicated by the **bold** numbers.*

C$_{16}$

[L3Au(NEt$_3$)]HF$_2$ (2 mol %),
NH$_4$BF$_4$ (1.5 eq),
Et$_3$N•3HF (3 eq), CH$_2$Cl$_2$, 50°, 24 h

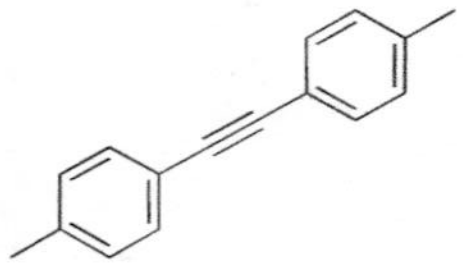

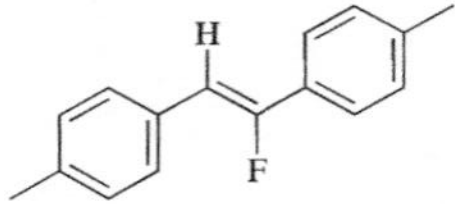

(90)

146

C$_{18}$

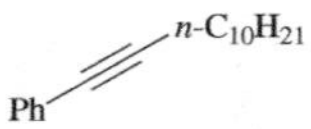

[L3Au(NEt$_3$)]HF$_2$ (2 mol %),
NH$_4$BF$_4$ (1.5 eq),
Et$_3$N•3HF (3 eq), CH$_2$Cl$_2$, rt, 5 d

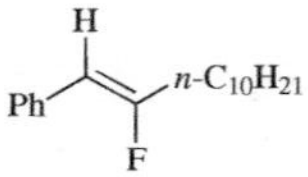

(98)

146

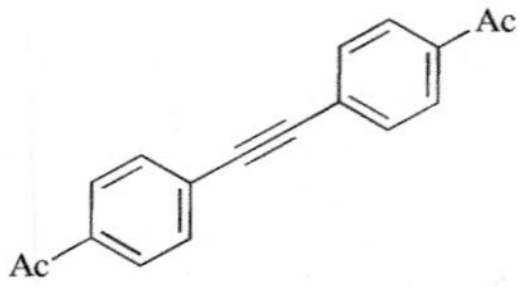

[L3Au(NEt$_3$)]HF$_2$ (4 mol %),
NH$_4$BF$_4$ (1.5 eq),
Et$_3$N•3HF (3 eq), CH$_2$Cl$_2$, 50°, 24 h

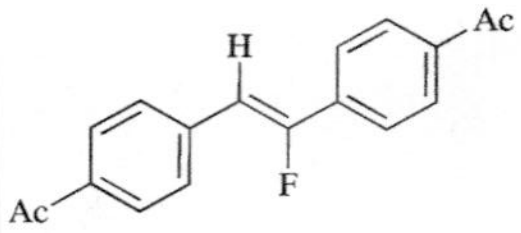

(86)

146

C$_{22}$

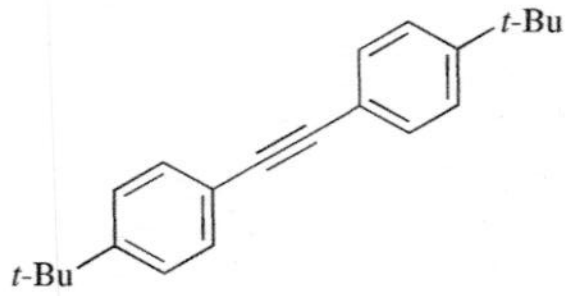

[L3Au(NEt$_3$)]HF$_2$ (2 mol %),
NH$_4$BF$_4$ (1.5 eq),
Et$_3$N•3HF (3 eq), CH$_2$Cl$_2$, 50°, 24 h

(91)

146

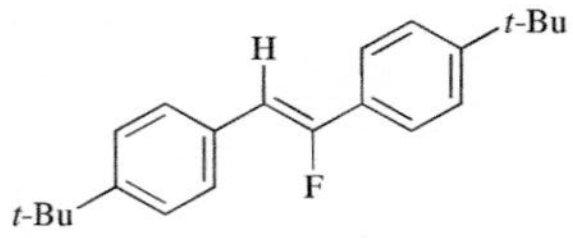

772

TABLE 4A. ALKYL FLUORIDE SYNTHESIS FROM ALKANES

*Please refer to the charts preceding the tables for the structures indicated by the **bold** numbers.*

Alkane	Conditions	Product(s) and Yield(s) (%)	Refs.
C$_6$			
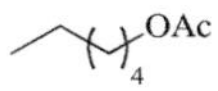 ![OAc structure]	**CuI** (10 mol %), KB(C$_6$F$_5$)$_4$ (10 mol %), NHPI (10 mol %), Selectfluor (2.2 eq.), KI (1.2 eq.), MeCN, 81°, 1.5 h	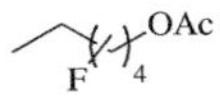 ![F OAc structure] (56)	164
C$_7$			
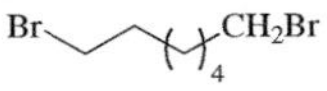 Br⌇CH$_2$Br	**CuI** (10 mol %), KB(C$_6$F$_5$)$_4$ (10 mol %), NHPI (10 mol %), Selectfluor (2.2 eq.), KI (1.2 eq.), MeCN, 81°, 3 h	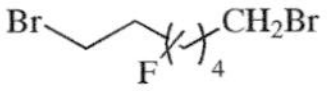 Br⌇CH$_2$Br, F (55)	164
 ![cycloheptane]	**CuI** (10 mol %), KB(C$_6$F$_5$)$_4$ (10 mol %), NHPI (10 mol %), Selectfluor (2.2 eq.), KI (1.2 eq.), MeCN, 81°, 1 h	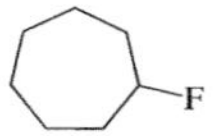 ![fluorocycloheptane] (66)	164
C$_8$			
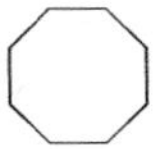 ![cyclooctane]	**CuI** (10 mol %), KB(C$_6$F$_5$)$_4$ (10 mol %), NHPI (10 mol %), Selectfluor (2.2 eq.), KI (1.2 eq.), MeCN, 81°, 0.5 h	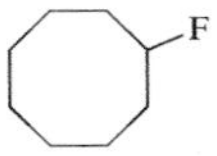 ![fluorocyclooctane] (41)	164
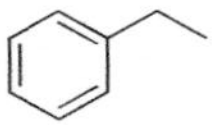 ![ethylbenzene]	**CuI** (10 mol %), KB(C$_6$F$_5$)$_4$ (10 mol %), NHPI (10 mol %), Selectfluor (2.2 eq.), MeCN, 24 h, 25°	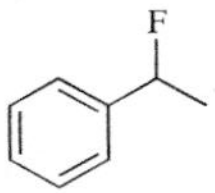 ![1-fluoroethylbenzene] (28)	164

773

TABLE 4A. ALKYL FLUORIDE SYNTHESIS FROM ALKANES (*Continued*)

*Please refer to the charts preceding the tables for the structures indicated by the **bold** numbers.*

Alkane	Conditions	Product(s) and Yield(s) (%)	Refs.
C₉			
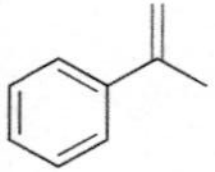	**CuI** (10 mol %), KB(C$_6$F$_5$)$_4$ (10 mol %), NHPI (10 mol %), Selectfluor (2.2 eq), MeCN, 24 h, 25°	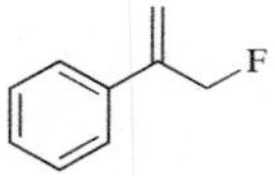 (53)	164
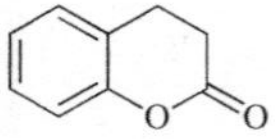	**CuI** (10 mol %), KB(C$_6$F$_5$)$_4$ (10 mol %), NHPI (10 mol %), Selectfluor (2.2 eq), KI (1.2 eq), MeCN, 3 h, 25°	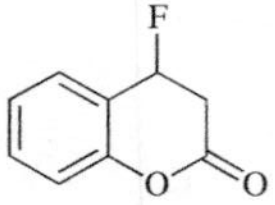 (47)	164
C₁₀			
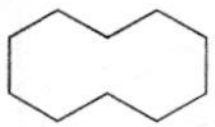	**CuI** (10 mol %), KB(C$_6$F$_5$)$_4$ (10 mol %), NHPI (10 mol %), Selectfluor (2.2 eq), KI (1.2 eq), MeCN, 0.5 h, 81°	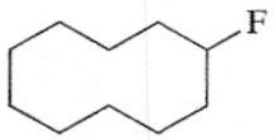 (47)	164
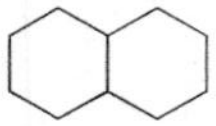	**CuI** (10 mol %), KB(C$_6$F$_5$)$_4$ (10 mol %), NHPI (10 mol %), Selectfluor (2.2 eq), KI (1.2 eq), MeCN, 2 h, 25°	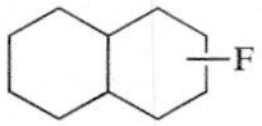 (52)	164
	Mn(salen)Cl (6 mol %), PhIO (6 eq), TBAF (0.3 eq), AgF (3 eq), MeCN/DCM (3:1), 50°, 6–15 h	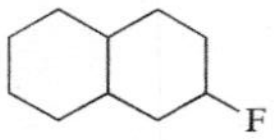 (40) + 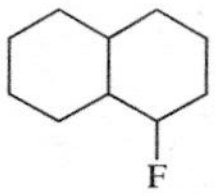 (11)	44

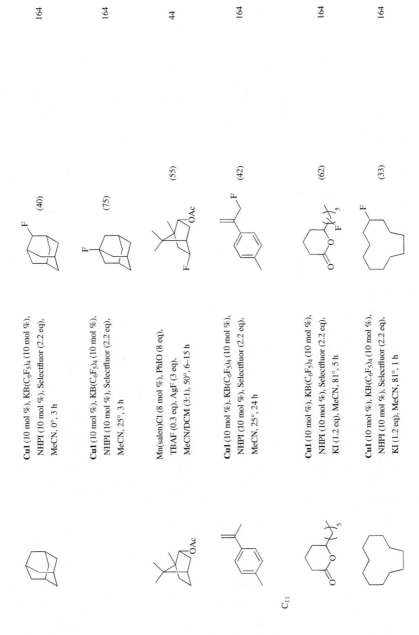

Substrate	Conditions	Product	Ref.
adamantane	**CuI** (10 mol %), KB(C$_6$F$_5$)$_4$ (10 mol %), NHPI (10 mol %), Selectfluor (2.2 eq), MeCN, 0°, 3 h	(40)	164
adamantane	**CuI** (10 mol %), KB(C$_6$F$_5$)$_4$ (10 mol %), NHPI (10 mol %), Selectfluor (2.2 eq), MeCN, 25°, 3 h	(75)	164
OAc	Mn(salen)Cl (8 mol %), PhIO (8 eq), TBAF (0.3 eq), AgF (3 eq), MeCN/DCM (3:1), 50°, 6–15 h	(55)	44
	CuI (10 mol %), KB(C$_6$F$_5$)$_4$ (10 mol %), NHPI (10 mol %), Selectfluor (2.2 eq), MeCN, 25°, 24 h	(42)	164
	CuI (10 mol %), KB(C$_6$F$_5$)$_4$ (10 mol %), NHPI (10 mol %), Selectfluor (2.2 eq), KI (1.2 eq), MeCN, 81°, 5 h	(62)	164
	CuI (10 mol %), KB(C$_6$F$_5$)$_4$ (10 mol %), NHPI (10 mol %), Selectfluor (2.2 eq), KI (1.2 eq), MeCN, 81°, 1 h	(33)	164

C$_{11}$

TABLE 4A. ALKYL FLUORIDE SYNTHESIS FROM ALKANES (*Continued*)

*Please refer to the charts preceding the tables for the structures indicated by the **bold** numbers.*

Alkane	Conditions	Product(s) and Yield(s) (%)	Refs.
C_{12}	**CuI** (10 mol %), KB(C_{6}F_{5})_{4} (10 mol %), NHPI (10 mol %), Selectfluor (2.2 eq), KI (1.2 eq), MeCN, 81°, 2 h	(63)	164
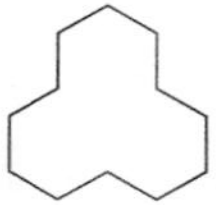 C_{14}	**CuI** (10 mol %), KB(C_{6}F_{5})_{4} (10 mol %), NHPI (10 mol %), Selectfluor (2.2 eq), KI (10 mol %), MeCN, 81°, 2 h	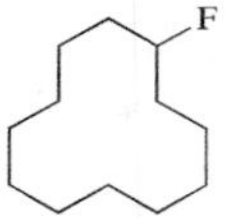 (72)	164
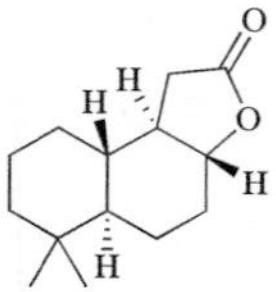	Mn(TMP)Cl (12 mol %), PhIO (10 eq), TBAF (0.3 eq), AgF (3 eq), MeCN/DCM (3:1), 50°, 6–15 h	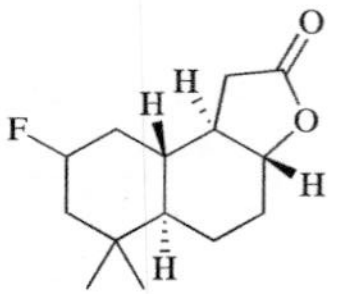 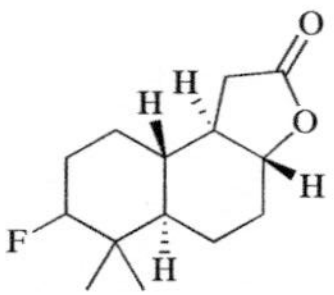 (16), α/β = 7.8 + (42), α/β = 3.1	44
C_{19}	Mn(salen)Cl (8 mol %), PhIO (8 eq), TBAF (0.3 eq), AgF (3 eq), MeCN/DCM (3:1), 50°, 6–15 h	(23), α/β = 6.2 + (32), α/β = 4.5	44

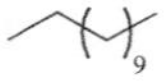

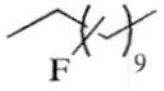

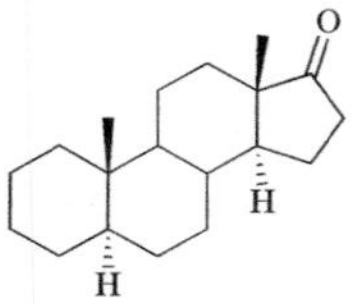

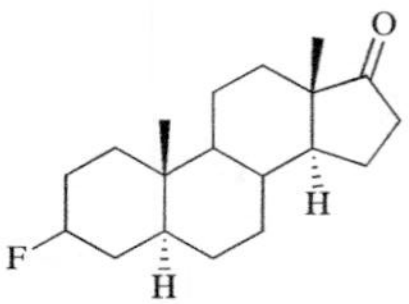

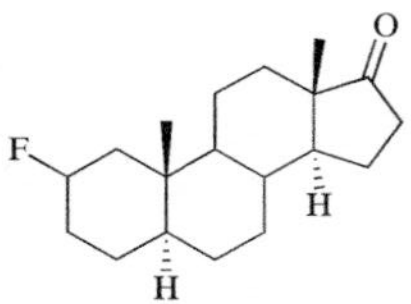

776

TABLE 4B. ALKYL FLUORIDE SYNTHESIS FROM ALKYL CARBOXYLIC ACIDS

Carboxylic Acid	Conditions	Product(s) and Yield(s) (%)	Refs.
C₅	Ir{dF(CF₃)ppy}₂(dtbbpy)PF₆ (1 mol %), Na₂HPO₄ (2.0 eq), Selectfluor (2.0 eq), MeCN/H₂O (3:1), 34 W blue LEDs, 23°, 1 h	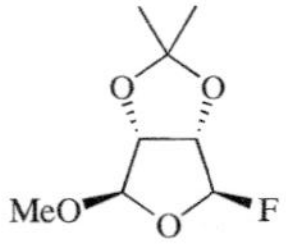 (92), (E)/(Z) = 2.5:1	174
C₆	Ir{dF(CF₃)ppy}₂(dtbbpy)PF₆ (1 mol %), Na₂HPO₄ (2.0 eq), Selectfluor (3.0 eq), MeCN/H₂O (1:1), 34 W blue LEDs, 23°, 12 h	 (79)	174
	Ir{dF(CF₃)ppy}₂(dtbbpy)PF₆ (1 mol %), Na₂HPO₄ (2.0 eq), Selectfluor (3.0 eq), MeCN/H₂O (1:1), 34 W blue LEDs, 23°, 15 h	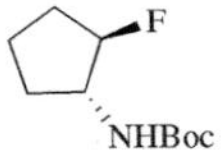 (82)	174
	Ir{dF(CF₃)ppy}₂(dtbbpy)PF₆ (1 mol %), Na₂HPO₄ (2.0 eq), Selectfluor (3.0 eq), MeCN/H₂O (1:1), 34 W blue LEDs, 23°, 1 h	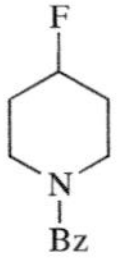 (90)	174
	Ru(bpz)₃(PF₆)₂ (1 mol %), Na₂HPO₄ (2.0 eq), Selectfluor (2.0 eq), MeCN/H₂O (1:1), 34 W blue LEDs, 23°, 3 h	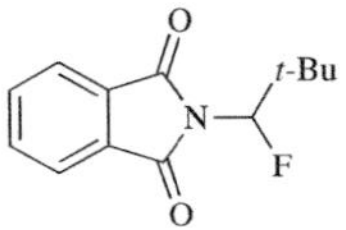 (90)	174

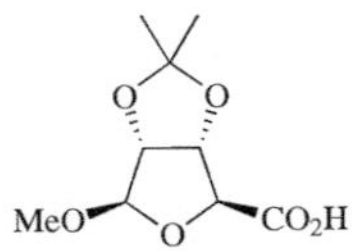

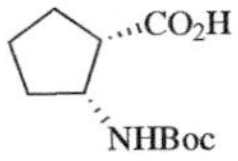

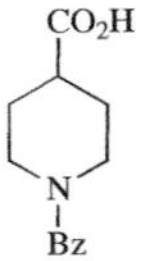

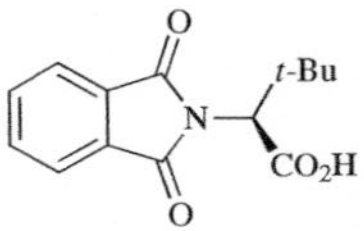

777

TABLE 4B. ALKYL FLUORIDE SYNTHESIS FROM ALKYL CARBOXYLIC ACIDS (*Continued*)

Carboxylic Acid	Conditions	Product(s) and Yield(s) (%)	Refs.
C₇	Ir[dF(CF₃)ppy]₂(dtbbpy)PF₆ (1 mol %), Na₂HPO₄ (2.0 eq), Selectfluor (3.0 eq), MeCN/H₂O (1:1), 34 W blue LEDs, 23°, 15 h	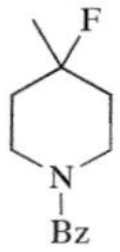 (76)	174
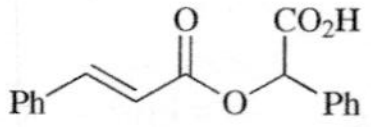 C₈	[Mn(tmp)]Cl (2.5 mol %), PhIO (3.3 eq), Et₃N·HCl (1.2 eq), DCE, 45°, 0.75–1.5 h	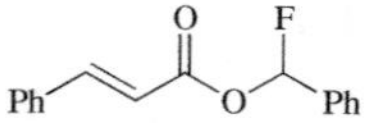 (56)	173
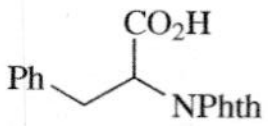 C₉	[Mn(tmp)]Cl (2.5 mol %), PhIO (3.3 eq), Et₃N·HCl (1.2 eq), DCE, 45°, 0.75–1.5 h	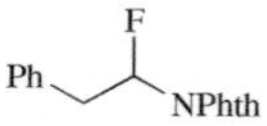 (36)	173
	Ir[dF(CF₃)ppy]₂(dtbbpy)PF₆ (1 mol %), Na₂HPO₄ (2.0 eq), Selectfluor (2.0 eq), MeCN/H₂O (1:1), 34 W blue LEDs, 23°, 1 h	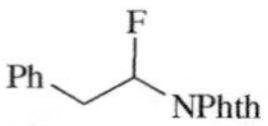 (90)	174
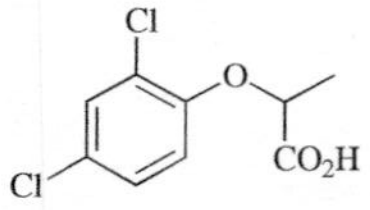	[Mn(tmp)]Cl (2.5 mol %), PhIO (3.3 eq), Et₃N·HCl (1.2 eq), DCE, 45°, 0.75–1.5 h	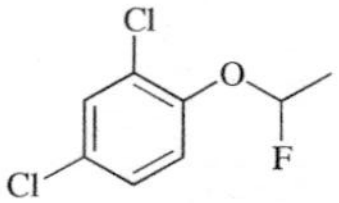 (70)	173

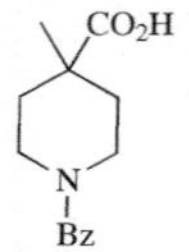

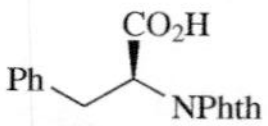

Substrate	Conditions	Product(s) (%)	Refs.
C10 indan-2-yl–CO$_2$H	Ru(bpz)$_3$(PF$_6$)$_2$ (1 mol %), Na$_2$HPO$_4$ (2.0 eq), Selectfluor (2.0 eq), MeCN/H$_2$O (1:1), 34 W blue LEDs, 23°, 1 h	2-fluoroindane (92)	174
C11 4-t-Bu-cyclohexyl–CO$_2$H	Ir[dF(CF$_3$)ppy]$_2$(dtbbpy)PF$_6$ (1 mol %), Na$_2$HPO$_4$ (2.0 eq), Selectfluor (3.0 eq), MeCN/H$_2$O (3:1), 34 W blue LEDs, 23°, 15 h	cis-1-F-4-t-Bu-cyclohexane (50) + trans-1-F-4-t-Bu-cyclohexane (20)	174
PhC(O)(CH$_2$)$_3$CO$_2$H	Ir[dF(CF$_3$)ppy]$_2$(dtbbpy)PF$_6$ (1 mol %), Na$_2$HPO$_4$ (2.0 eq), Selectfluor (3.0 eq), MeCN/H$_2$O (1:1), 34 W blue LEDs, 23°, 15 h	PhC(O)(CH$_2$)$_3$F (77)	174
PhC(O)CH$_2$CH(CH$_3$)CO$_2$H	Ru(bpz)$_3$(PF$_6$)$_2$ (1 mol %), Na$_2$HPO$_4$ (2.0 eq), Selectfluor (2.0 eq), MeCN/H$_2$O (1:1), 34 W blue LEDs, 23°, 1 h	PhC(O)CH$_2$CH(CH$_3$)F (96)	174
1,2,3,4-tetrahydronaphthalen-2-yl–CO$_2$H	Ru(bpz)$_3$(PF$_6$)$_2$ (1 mol %), Na$_2$HPO$_4$ (2.0 eq), Selectfluor (3.0 eq), MeCN/H$_2$O (1:1), 34 W blue LEDs, 23°, 15 h	2-F-1,2,3,4-tetrahydronaphthalene (71)	174
C12 n-Bu–CH(n-C$_6$H$_{13}$)–CO$_2$H	Ir[dF(CF$_3$)ppy]$_2$(dtbbpy)PF$_6$ (1 mol %), Na$_2$HPO$_4$ (2.0 eq), Selectfluor (3.0 eq), MeCN/H$_2$O (1:1), 34 W blue LEDs, 23°, 1 h	n-Bu–CHF–n-C$_6$H$_{13}$ (83)	174

TABLE 4B. ALKYL FLUORIDE SYNTHESIS FROM ALKYL CARBOXYLIC ACIDS (*Continued*)

Carboxylic Acid	Conditions	Product(s) and Yield(s) (%)	Refs.
C$_{12}$			
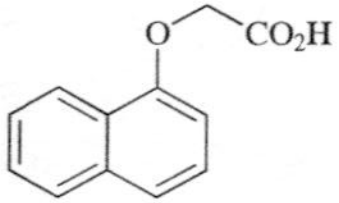	[Mn(tmp)]Cl (2.5 mol %), PhIO (3.3 eq), Et$_3$N•HCl (1.2 eq), DCE, 45°, 0.75–1.5 h	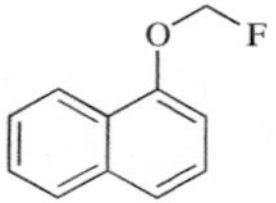 (58)	173
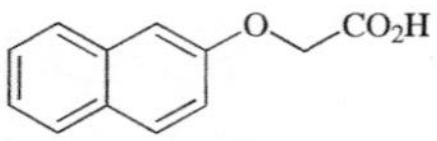	[Mn(tmp)]Cl (2.5 mol %), PhIO (3.3 eq), Et$_3$N•HCl (1.2 eq), DCE, 45°, 0.75–1.5 h	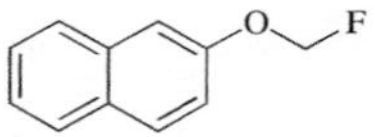 (46)	173
C$_{13}$			
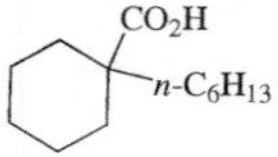	Ir[dF(CF$_3$)ppy]$_2$(dtbbpy)PF$_6$ (1 mol %), Na$_2$HPO$_4$ (2.0 eq), Selectfluor (3.0 eq), MeCN/H$_2$O (1:1), 34 W blue LEDs, 23°, 15 h	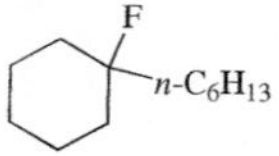 (79)	174
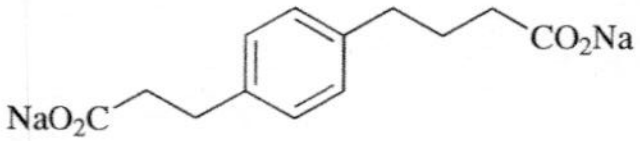	Ir[dF(CF$_3$)ppy]$_2$(dtbbpy)PF$_6$ (1 mol %), Na$_2$HPO$_4$ (2.0 eq), Selectfluor (3.0 eq), MeCN/H$_2$O (1:1), 34 W blue LEDs, 23°, 1 h	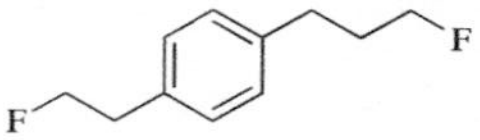 (71)	174
C$_{14}$			
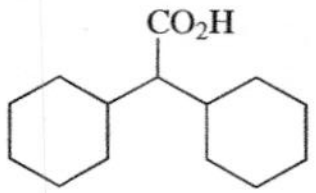	[Mn(tmp)]Cl (2.5 mol %), PhIO (3.3 eq), Et$_3$N•HCl (1.2 eq), DCE, 45°, 0.75–1.5 h	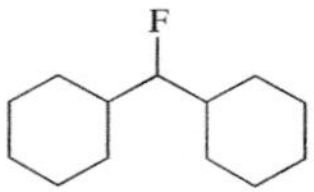 (15)	173

Substrate	Conditions	Product	Yield	Refs.
C₁₅ (4-phenylphenylacetic acid, CO₂H)	Ir[dF(CF₃)ppy]₂(dtbbpy)PF₆ (1 mol %), Na₂HPO₄ (2.0 eq), Selectfluor (2.0 eq), MeCN/H₂O (1:1), 34 W blue LEDs. 23°, 1 h	(benzyl fluoride, Ph)	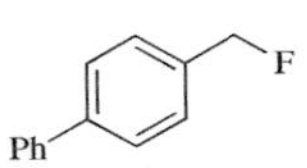 (87)	174
	Ir[dF(CF₃)ppy]₂(dtbbpy)PF₆ (1 mol %), Na₂HPO₄ (2.0 eq), Selectfluor (3.0 eq), MeCN/H₂O (1:1), 34 W blue LEDs, 23°, 1 h	(F, OH, Ph)	(80)	174
C₁₆	[Mn(tmp)]Cl (2.5 mol %), PhIO (3.3 eq), Et₃N·HCl (1.2 eq), DCE, 45°, 0.75-1.5 h	(CF₃ pyridine ether)	(74)	173
C₁₇	Ir[dF(CF₃)ppy]₂(dtbbpy)PF₆ (1 mol %), Na₂HPO₄ (2.0 eq), Selectfluor (2.0 eq), MeCN/H₂O (1:1), 34 W blue LEDs, 23°, 1 h	(Ph)	(90)	174
	Ir[dF(CF₃)ppy]₂(dtbbpy)PF₆ (1 mol %), Na₂HPO₄ (2.0 eq), Selectfluor (2.0 eq), MeCN/H₂O (1:1), 34 W blue LEDs, 23°, 1 h	(Ph)	(88)	174

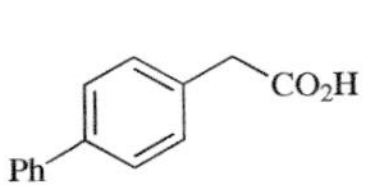

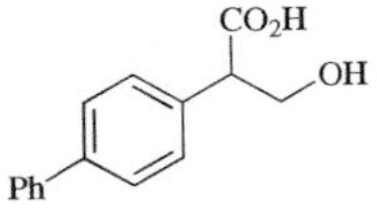

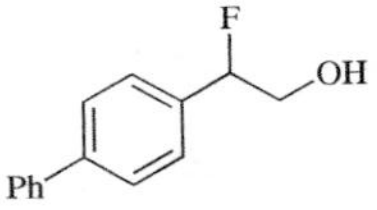

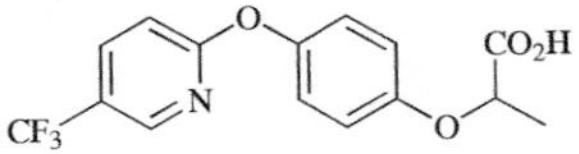

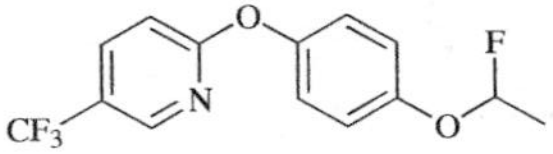

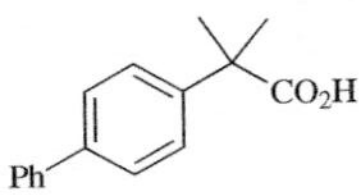

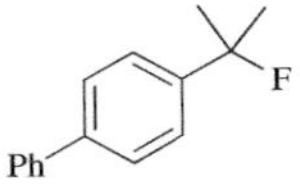

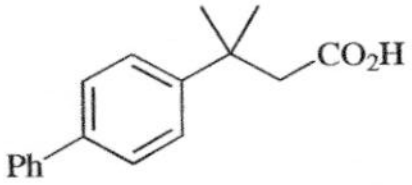

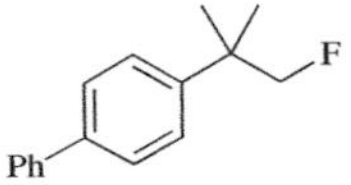

Carboxylic Acid	Conditions	Product(s) and Yield(s) (%)	Refs.
C$_{21}$			
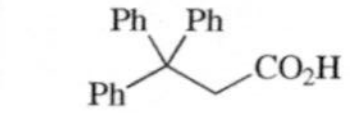	Ir[dF(CF$_3$)ppy]$_2$(dtbbpy)PF$_6$ (1 mol %), Na$_2$HPO$_4$ (2.0 eq), Selectfluor (3.0 eq), MeCN/H$_2$O (1:1), 34 W blue LEDs, 23°, 1 h	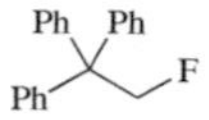 (82)	174
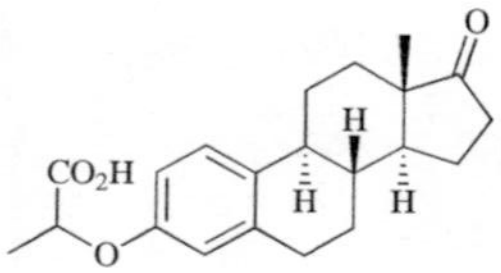	[Mn(tmp)]Cl (2.5 mol %), PhIO (3.3 eq), Et$_3$N•HCl (1.2 eq), DCE, 45°, 0.75 h	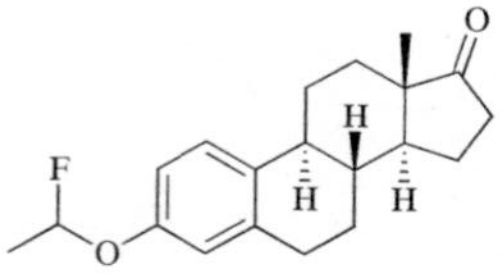 (65)	173
C$_{24}$			
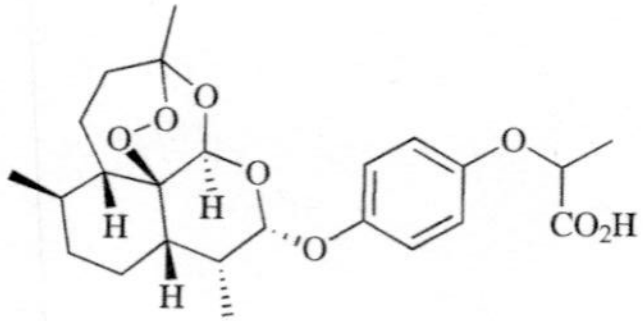	[Mn(tmp)]Cl (2.5 mol %), PhIO (3.3 eq), Et$_3$N•HCl (1.2 eq), DCE, 45°, 1 h	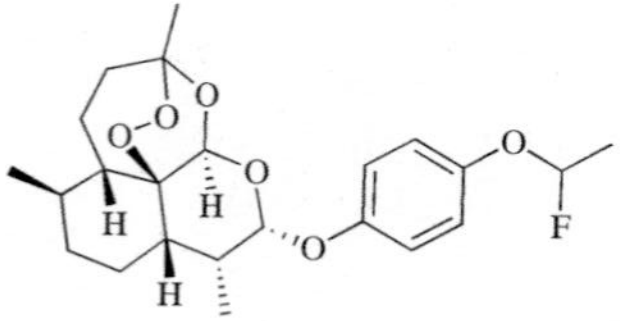 (61)	173

TABLE 4C. ALKYL FLUORIDE SYNTHESIS FROM ALKENES

Alkene	Conditions	Product(s) and Yield(s) (%)	Refs.
C₃ 	Fe₂(ox)₃ (2 eq), NaBH₄ (6.4 eq), Selectfluor (2 eq), MeCN/H₂O (1:1), 0°, 0.5 h	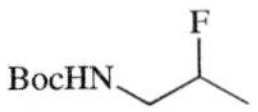 (61)	156
	Fe₂(ox)₃ (2 eq), NaBH₄ (6.4 eq), Selectfluor (2 eq), MeCN/H₂O (1:1), 0°, 0.5 h	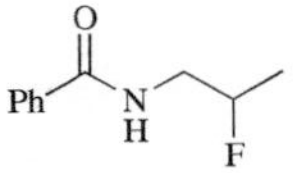 (67)	156
C₅₋₆ 	Fe₂(ox)₃ (2 eq), NaBH₄ (6.4 eq), Selectfluor (2 eq), MeCN/H₂O (1:1), 0°, 0.5 h	 R HO₂C (66) HO₂CCH₂ (73) BnO₂CCH₂ (79) BnHNCOCH₂ (75)	156
C₈ 	Fe₂(ox)₃ (2 eq), NaBH₄ (6.4 eq), Selectfluor (2 eq), MeCN/H₂O (1:1), 0°, 0.5 h	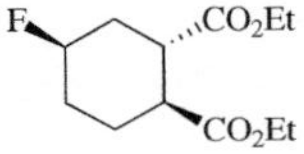 (52) dr 5:1	156

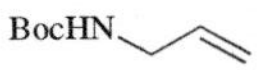

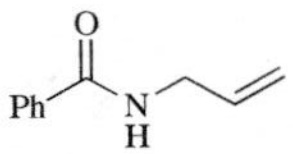

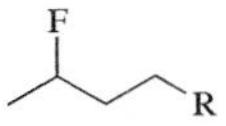

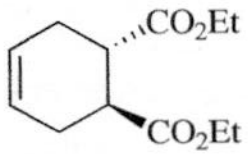

TABLE 4C. ALKYL FLUORIDE SYNTHESIS FROM ALKENES (*Continued*)

Alkene	Conditions	Product(s) and Yield(s) (%)			Refs.

C$_{8-14}$

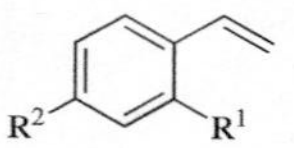

Pd(PPh$_3$)$_4$ (10 mol %),
Selectfluor (3 eq),
Et$_3$SiH (1.5 eq), MeCN, 0°, 2 h

R^1	R^2	
H	H	(58)
H	F	(55)
H	Br	(78)
H	AcO	(99)
H	O$_2$N	(61)
MeO	H	(30)
H	CF$_3$	(72)
H	OHC	(76)
H	EtO$_2$C	(64)
H	NC–	(70)
H	PhHNCO	(78)
H	PhthHNCO	(87)
EtO$_2$C	H	(58)
H	t-Bu	(43)
H	Ph	(69)

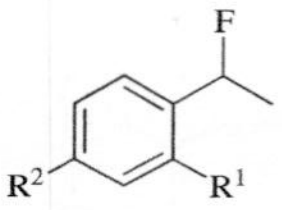

189

C$_{9-16}$

Pd(PPh$_3$)$_4$ (10 mol %),
Selectfluor (3 eq),
Et$_3$SiH (1.5 eq), MeCN, 0°, 2 h

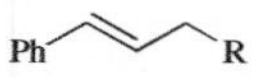

R	
AcO	(54)
PhthN	(66)
MeO$_2$C	(51)
4-MeC$_6$H$_4$	(73)

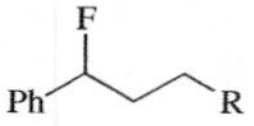

189

784

C$_9$

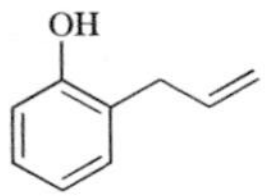

Fe$_2$(ox)$_3$ (2 eq), NaBH$_4$ (6.4 eq), Selectfluor (2 eq), MeCN/H$_2$O (1:1), 0°, 0.5 h

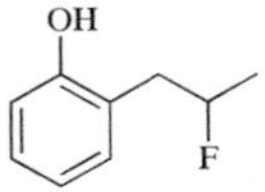

(60)

156

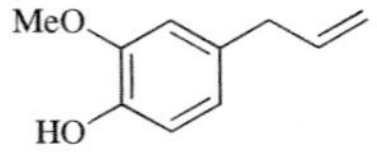

Fe$_2$(ox)$_3$ (2 eq), NaBH$_4$ (6.4 eq), Selectfluor (2 eq), MeCN/H$_2$O (1:1), 0°, 0.5 h

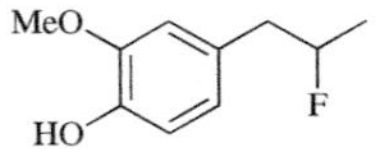

(54)

156

C$_{10}$

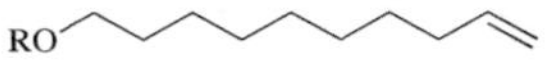

Fe$_2$(ox)$_3$ (2 eq), NaBH$_4$ (6.4 eq), Selectfluor (2 eq), MeCN/H$_2$O (1:1), 0°, 0.5 h

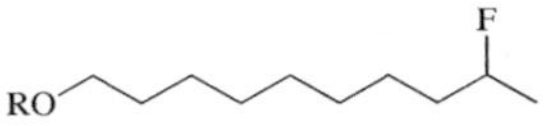

156

R	
H	(75)
PMB	(70)

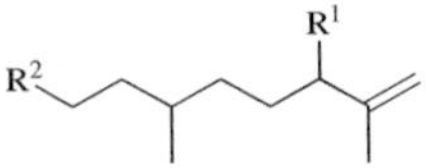

Fe$_2$(ox)$_3$ (2 eq), reductant (6.4 eq), Selectfluor (2 eq), MeCN/H$_2$O (1:1), 0°, 0.5 h

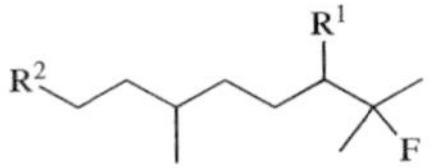

156

R^1	R^2	Reductant	
H	HO	NaBH$_4$	(70)
H	H$_2$N	NaBH$_4$	(52)
D	HO	NaBD$_4$	(68)

TABLE 4C. ALKYL FLUORIDE SYNTHESIS FROM ALKENES (*Continued*)

Alkene	Conditions	Product(s) and Yield(s) (%)	Refs.
C$_{10}$			
	Fe$_2$(ox)$_3$ (2 eq), NaBH$_4$ (6.4 eq), Selectfluor (2 eq), MeCN/H$_2$O (1:1), 0°, 0.5 h	(56)	156
	Fe$_2$(ox)$_3$ (2 eq), NaBH$_4$ (6.4 eq), Selectfluor (2 eq), MeCN/H$_2$O (1:1), 0°, 0.5 h	(55)	156
	Pd(PPh$_3$)$_4$ (10 mol %), Selectfluor (3 eq), Et$_3$SiH (1.5 eq), MeCN, 0°, 2 h	(65)	189
C$_{12}$			
	Pd(PPh$_3$)$_4$ (10 mol %), Selectfluor (3 eq), Et$_3$SiH (1.5 eq), MeCN, 0°, 15 h	(67) dr > 20:1	189
	Pd(PPh$_3$)$_4$ (10 mol %), Selectfluor (3 eq), Et$_3$SiH (1.5 eq), MeCN, 0°, 2 h	(41) dr > 20:1	189

786

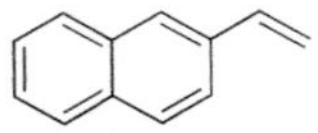

C$_{14}$

Pd(PPh$_3$)$_4$ (10 mol %),
Selectfluor (3 eq),
Et$_3$SiH (1.5 eq), MeCN, 0°, 2 h

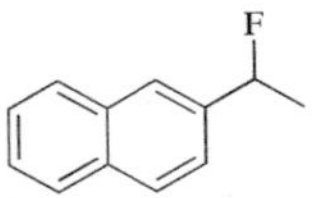

(46)

189

Pd(PPh$_3$)$_4$ (10 mol %),
Selectfluor (3 eq),
Et$_3$SiH (1.5 eq), MeCN, 0°, 2 h

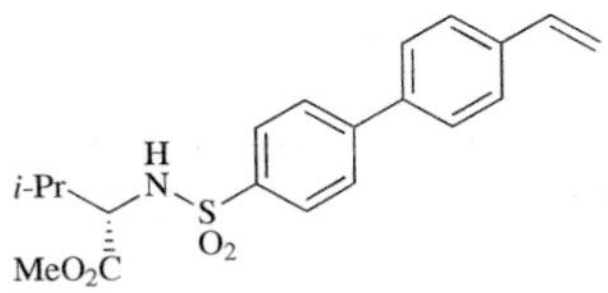

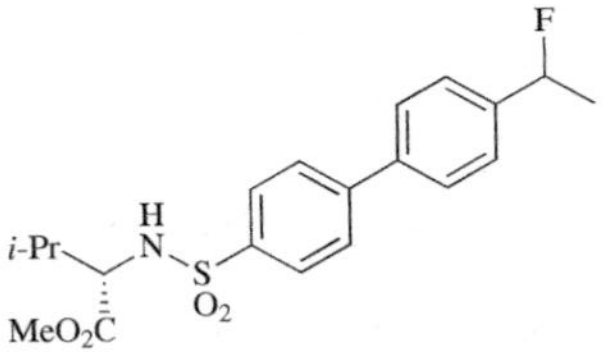

(80) dr 1:1

189

C$_{16}$

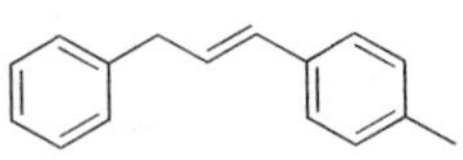

Pd(PPh$_3$)$_4$ (10 mol %),
Selectfluor (3 eq),
Et$_3$SiH (1.5 eq), MeCN, 0°, 2 h

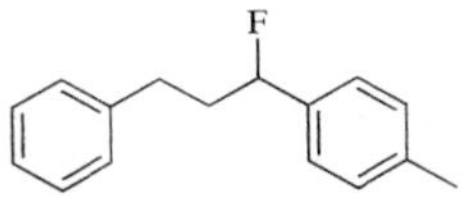

(65)

189

C$_{28}$

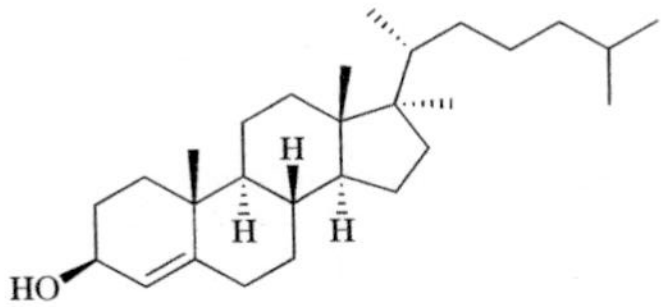

Fe$_2$(ox)$_3$ (4 eq), NaBH$_4$ (12.8 eq),
Selectfluor (4 eq),
THF/MeCN/H$_2$O (4:2:3), 0°, 0.5 h

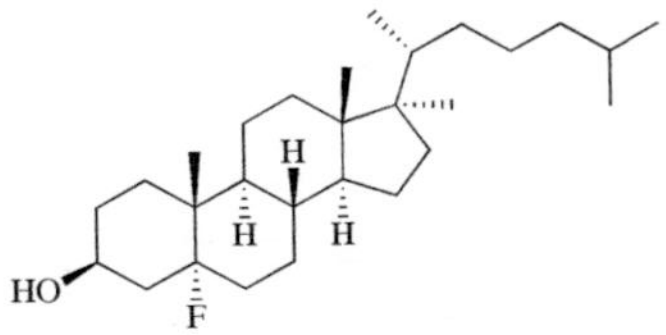

(41)

156

787

TABLE 5. SYNTHESIS OF BENZYLIC FLUORIDES

Substrate	Conditions	Product(s) and Yield(s) (%)	Refs.

*Please refer to the charts preceding the tables for the structures indicated by the **bold** numbers.*

C₈

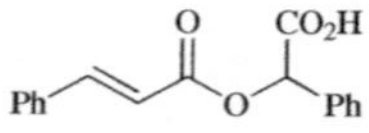

[Mn(tmp)]Cl (2.5 mol %), PhIO (3.3 eq),
Et₃N•HF (1.2 eq), DCE, 45°, 0.75–1.5 h

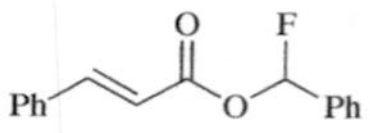

(56)

173

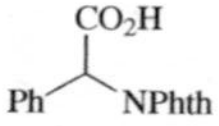

[Mn(tmp)]Cl (2.5 mol %), PhIO (3.3 eq),
Et₃N•HF (1.2 eq), DCE, 45°, 0.75–1.5 h

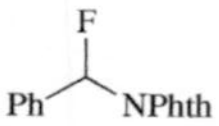

(65)

173

C₈₋₁₄

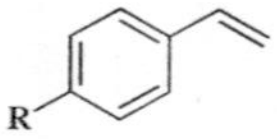

[Pd(PPh₃)₄] (10 mol %),
Selectfluor (3.0 eq),
Et₃SiH (1.5 eq), MeCN, 0°, 2 h

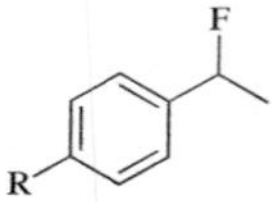

R	
H	(58)
F	(55)
Br	(78)
AcO	(99)
O₂N	(61)
CF₃	(72)
OHC	(76)
PhthNCH₂	(87)
NC–	(70)
PhHN(O)C	(78)
t-Bu	(43)
Ph	(69)

189

C8

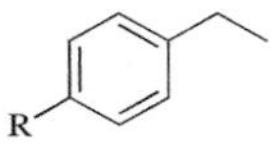

Mn1 (20 mol %), PhIO (6–8 eq),
Et$_3$•HF (x eq), AgF (y eq),
MeCN, 50°, 6–8 h

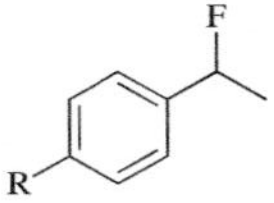

R	x	y	
Cl	0.5	3	(52)
Br	0.5	3	(44)
I	0.5	3	(53)
AcO	1.5	0	(58)

192

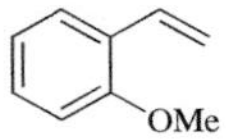

[Pd(PPh$_3$)$_4$] (10 mol %),
Selectfluor (3.0 eq),
Et$_3$SiH (1.5 eq), MeCN, 0°, 2 h

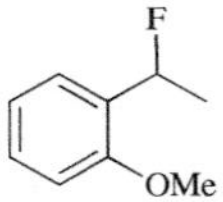

(30)

189

C9

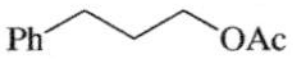

Fe(acac)$_2$ (10 mol %),
Selectfluor (2.2 eq),
MeCN, rt, 24 h

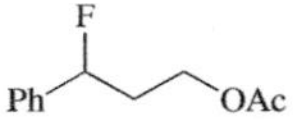

(71)

186

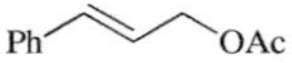

[Pd(PPh$_3$)$_4$] (10 mol %),
Selectfluor (3.0 eq),
Et$_3$SiH (1.5 eq), MeCN, 0°, 2 h

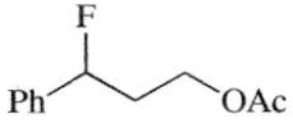

(54)

189

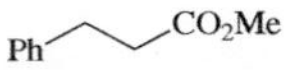

Fe(acac)$_2$ (10 mol %),
Selectfluor (2.2 eq),
MeCN, rt, 24 h

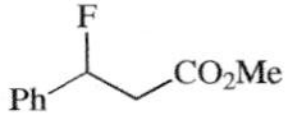

(52)

186

TABLE 5. SYNTHESIS OF BENZYLIC FLUORIDES (Continued)

*Please refer to the charts preceding the tables for the structures indicated by the **bold** numbers.*

C9

Substrate	Conditions	Product(s) and Yield(s) (%)	Refs.
 Ph～NPhth	**Mn1** (20 mol %), PhIO (6.0–9.0 eq), Et$_3$N•HF (0.5 eq), AgF (3.0 eq), MeCN, 50°, 6–9 h	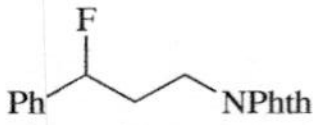 F, Ph, NPhth (47)	192
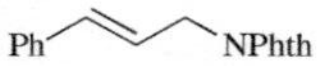 Ph～NPhth	[Pd(PPh$_3$)$_4$] (10 mol %), Selectfluor (3.0 eq), Et$_3$SiH (1.5 eq), MeCN, 0°, 2 h	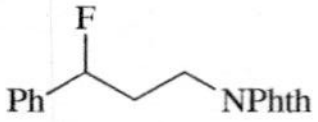 F, Ph, NPhth (66)	189
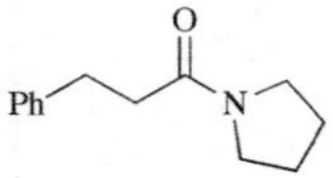 Ph～C(O)–N(pyrrolidine)	Fe(acac)$_2$ (10 mol %), Selectfluor (2.2 eq), MeCN, rt, 24 h	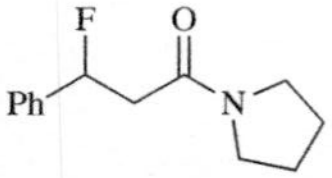 F, Ph, O, N(pyrrolidine) (41)	186

C9–10

Substrate	Conditions	Product(s) and Yield(s) (%)	Refs.
R^1/R^2/R^3-aryl–CH$_2$–CH(NPhth)–C(O)NHAr Ar = 4-CF$_3$C$_6$F$_4$	Selectfluor (1.5 eq), Pd(TFA)$_2$ (10 mol %), **L4** (10 mol %), Ag$_2$CO$_3$ (2.0 eq), 1,4-dioxane, 115°, 15 h	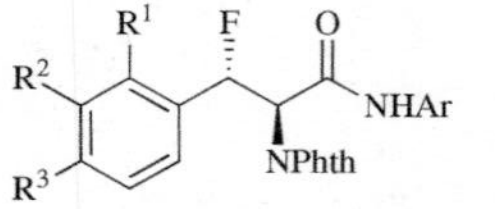 (see table below) dr >20:1	170

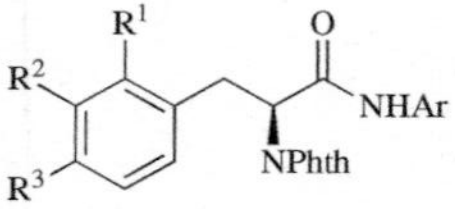

R^1	R^2	R^3	
H	H	H	(79)
H	H	F	(70)
H	H	Cl	(67)
H	H	Br	(68)
H	H	O$_2$N	(52)

C$_9$

H	H	H	CF₃	(76)
H	H	H	MeO₂C	(77)
H	H	H	NC–	(87)
H	H	MeO₂C	H	(78)
H	MeO₂C	H	H	(77)

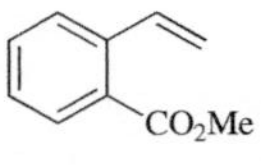

[Pd(PPh₃)₄] (10 mol %),
Selectfluor (3.0 eq),
Et₃SiH (1.5 eq), MeCN, 0°, 2 h

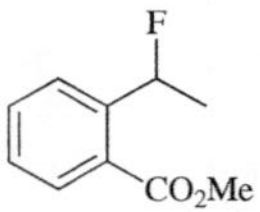

(58)

189

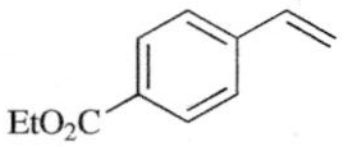

[Pd(PPh₃)₄] (10 mol %),
Selectfluor (3.0 eq),
Et₃SiH (1.5 eq), MeCN, 0°, 2 h

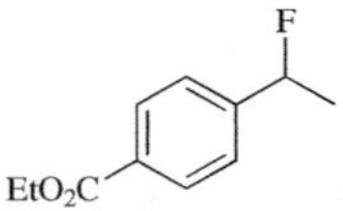

(64)

189

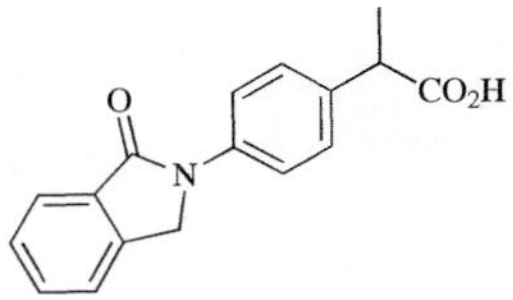

[Mn(tmp)]Cl (2.5 mol %),
PhIO (3.0 eq), Et₃N·HF (1.2 eq),
DCE, 45°, 0.75 h

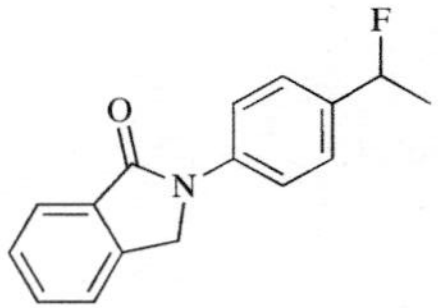

(66)

173

TABLE 5. SYNTHESIS OF BENZYLIC FLUORIDES (*Continued*)

*Please refer to the charts preceding the tables for the structures indicated by the **bold** numbers.*

Substrate	Conditions	Product(s) and Yield(s) (%)	Refs.
C₉₋₁₁			

C_{9-11}

Substrate	Conditions	Product(s) and Yield(s) (%)	Refs.
(structure: R^1, R^2-substituted arene with O, NHPIP, NPhth)	Pd(OAc)$_2$ (6 mol %), Selectfluor (1.05 eq), DCM/i-PrCN (30:1), 80°, 24 h	(structure: F, R^1, R^2, O, NHPIP, NPhth)	256

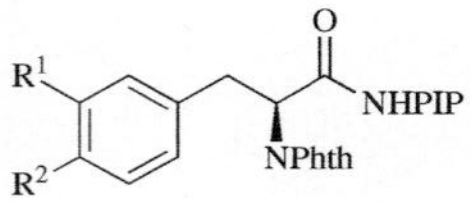

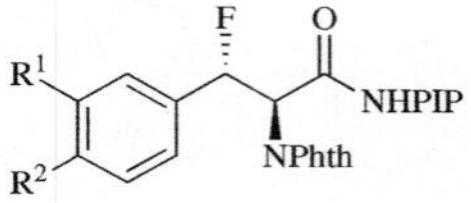

R^1	R^2	
H	F	(66)
H	Cl	(70)
H	Br	(73)
H	MeO	(33)
H	O$_2$N	(44)
H	BPin	(53)
F	H	(66)
F	F	(66)
F	Cl	(67)
F	O$_2$N	(44)
MeO	H	(58)
O$_2$N	MeO	(73)
H	Ac	(70)

C_9

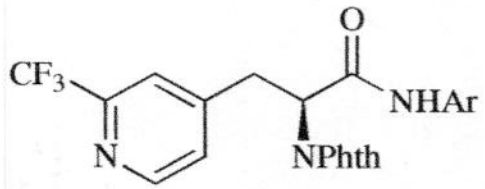

Substrate	Conditions	Product(s) and Yield(s) (%)	Refs.
(structure: CF$_3$-pyridine with O, NHAr, NPhth) Ar = 4-CF$_3$C$_6$F$_4$	Selectfluor (1.5 eq), Pd(TFA)$_2$ (10 mol %), **L4** (10 mol %), Ag$_2$CO$_3$ (2.0 eq), 1,4-dioxane, 115°, 15 h	(structure: F, CF$_3$-pyridine, O, NHAr, NPhth) (43) dr >20:1	170

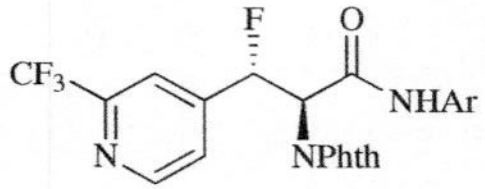

Substrate	Conditions	Product	Ref.
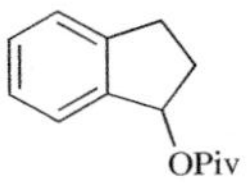	**Mn1** (20 mol %), PhIO (6.0–9.0 eq), Et₃N•HF (0.5 eq), AgF (3.0 eq), MeCN, 50°, 6–9 h	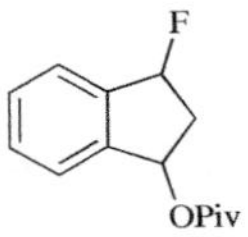 (51) dr 8.5:1	192
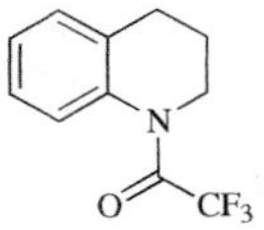	**Mn1** (20 mol %), PhIO (6.0–9.0 eq), Et₃N•HF (1.5 eq), MeCN, 50°, 6–9 h	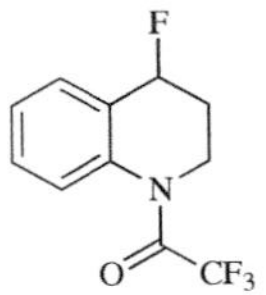 (55)	192
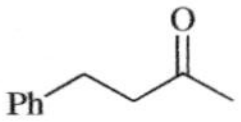 C₁₀	Fe(acac)₂ (10 mol %), SelectFluor (2.2 eq), MeCN, rt, 24 h	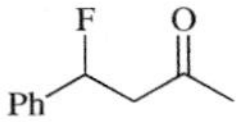 (76)	186
	**Mn1** (20 mol %), PhIO (6.0–9.0 eq), Et₃N•HF (0.5 eq), AgF (3.0 eq), MeCN, 50°, 6–9 h	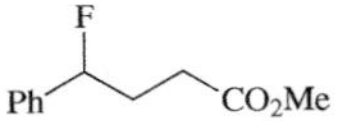 (52)	192
	[Pd(PPh₃)₄] (10 mol %), Selectfluor (3.0 eq), Et₃SiH (1.5 eq), MeCN, 0°, 2 h	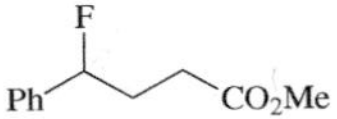 (51)	189

TABLE 5. SYNTHESIS OF BENZYLIC FLUORIDES (Continued)

Substrate	Conditions	Product(s) and Yield(s) (%)	Refs.

*Please refer to the charts preceding the tables for the structures indicated by the **bold** numbers.*

C$_{10}$

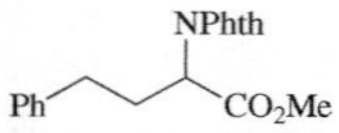	**Mn1** (20 mol %), PhIO (6.0–9.0 eq), Et$_3$N•HF (1.5 eq), MeCN, 50°, 6–9 h	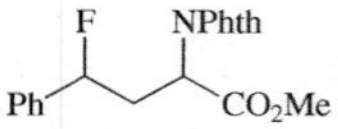 (48) dr 5:1	192
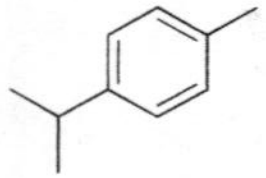	Fe(acac)$_2$ (10 mol %), Selectfluor (2.2 eq), MeCN, rt, 24 h	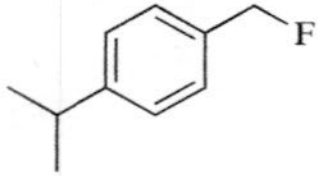 (66)	186
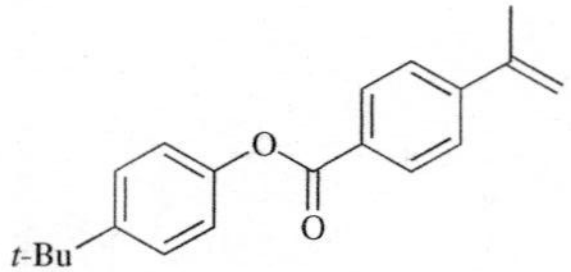	[Pd(PPh$_3$)$_4$] (10 mol %), Selectfluor (3.0 eq), Et$_3$SiH (1.5 eq), MeCN, 0°, 2 h	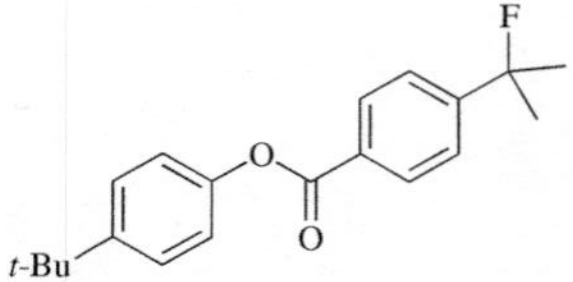 (65)	189
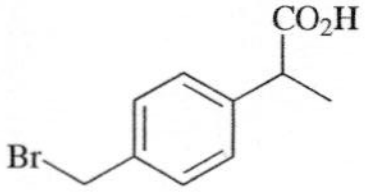	[Mn(tmp)]Cl (2.5 mol %), PhIO (3.0 eq), Et$_3$N•HF (1.2 eq), DCE, 45°, 0.75–1.5 h	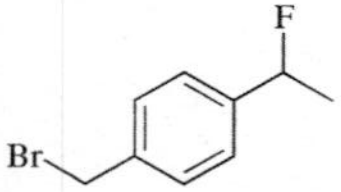 (61)	173

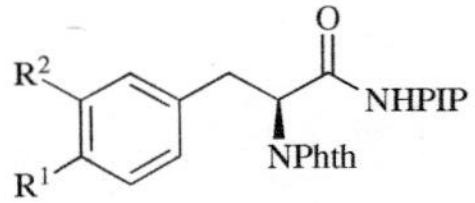

Pd(OAc)₂ (6 mol %),
Selectfluor (1.05 eq),
DCM/i-PrCN (30:1), 80°, 24 h

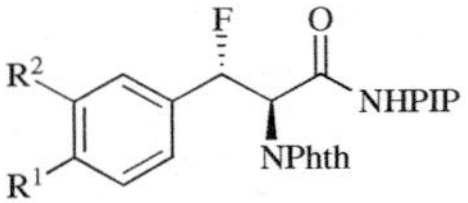

R¹	R²	
H	NC–	(56)
Me	H	(60)
CF₃	H	(57)
MeO₂C	H	(64)
NC–	H	(58)

256

Mn1 (20 mol %), PhIO (6.0–9.0 eq),
Et₃N•HF (1.5 eq), MeCN, 50°, 6–9 h

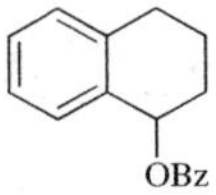

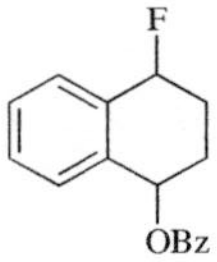

(70) dr 3:1

192

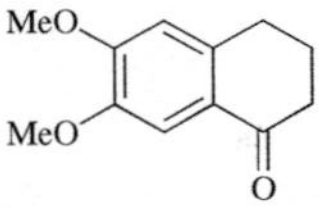

Mn1 (20 mol %), PhIO (6.0–9.0 eq),
Et₃N•HF (0.5 eq), AgF (3.0 eq).
MeCN, 50°, 6–9 h

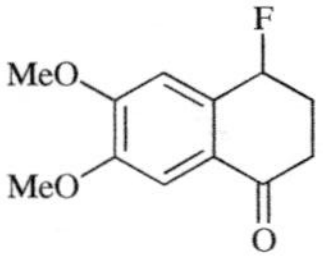

(53)

192

TABLE 5. SYNTHESIS OF BENZYLIC FLUORIDES (*Continued*)

Substrate	Conditions	Product(s) and Yield(s) (%)	Refs.

*Please refer to the charts preceding the tables for the structures indicated by the **bold** numbers.*

C₁₀

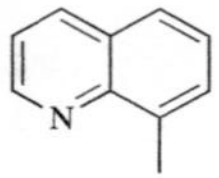	Pd(OAc)$_2$ (10 mol %), PhI(OPiv)$_2$ (2.0 eq), AgF (5.0 eq), MgSO$_4$ (2.0 eq), DCM, 60°, 16 h	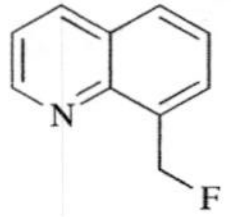 (49)	190
	[Me$_3$pyF]BF$_4$ (1.5 eq), Pd(OAc)$_2$ (10 mol %), MW (250W), benzene, 110°, 1 h	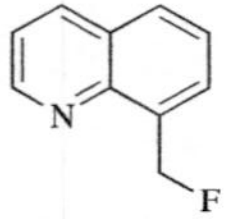 (59)	105
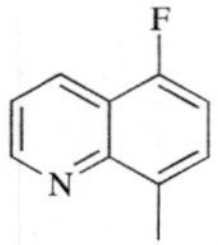	Pd(OAc)$_2$ (7 mol %), [Me$_3$pyF]BF$_4$ (1.5 eq), MW (250W), benzene, 100°, 4 h	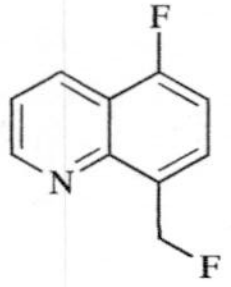 (49)	105
	Pd(OAc)$_2$ (10 mol %), PhI(OPiv)$_2$ (2.0 eq), AgF (5.0 eq), MgSO$_4$ (2.0 eq), DCM, 60°, 16 h	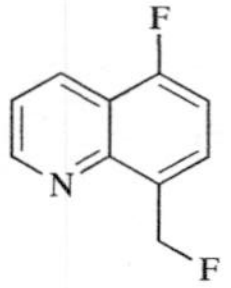 (30)	190

Substrate	Conditions	Product	Ref
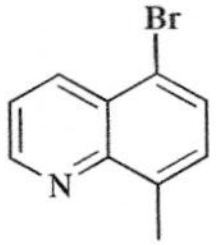	Pd(OAc)$_2$ (10 mol %), PhI(OPiv)$_2$ (2.0 eq), AgF (5.0 eq), MgSO$_4$ (2.0 eq), DCM, 60°, 16 h	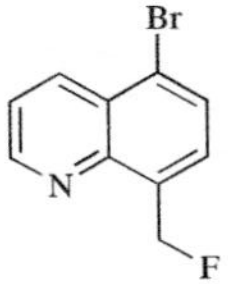 (44)	190
	[Me$_3$pyF]BF$_4$ (2.0 eq), Pd(OAc)$_2$ (10 mol %), MW (250W), benzene, 110°, 4 h	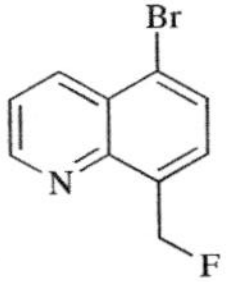 (53)	105
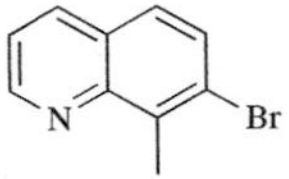	Pd(OAc)$_2$ (10 mol %), PhI(OPiv)$_2$ (2.0 eq), AgF (5.0 eq), MgSO$_4$ (2.0 eq), DCM, 60°, 16 h	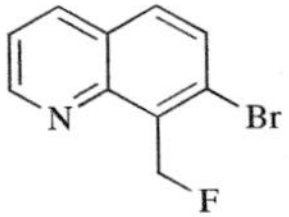 (42)	190
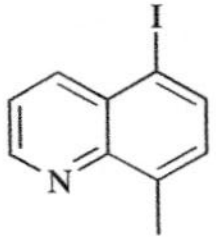	Pd(OAc)$_2$ (10 mol %), PhI(OPiv)$_2$ (2.0 eq), AgF (5.0 eq), MgSO$_4$ (2.0 eq), DCM, 60°, 24 h	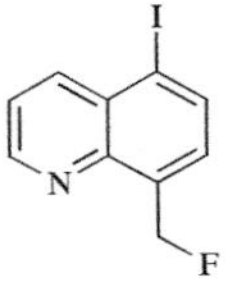 (55)	190

TABLE 5. SYNTHESIS OF BENZYLIC FLUORIDES (*Continued*)

Substrate	Conditions	Product(s) and Yield(s) (%)	Refs.
C$_{10}$	Pd(OAc)$_2$ (10 mol %), PhI(OPiv)$_2$ (2.0 eq), AgF (5.0 eq), MgSO$_4$ (2.0 eq), DCM, 60°, 24 h	(41)	190
C$_{11}$ Ar = 4-CF$_3$C$_6$F$_4$	Pd(TFA)$_2$ (10 mol %), Selectfluor (1.5 eq), L4 (10 mol %), Ag$_2$CO$_3$ (2.0 eq), 1,4-dioxane, 115°, 15 h	(66) dr >20:1	170
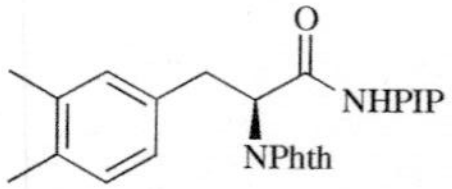	Pd(OAc)$_2$ (6 mol %), Selectfluor (1.05 eq), DCM/i-PrCN (30:1), 80°, 24 h	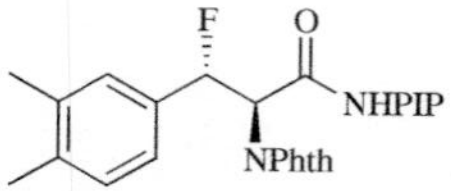 (54)	256
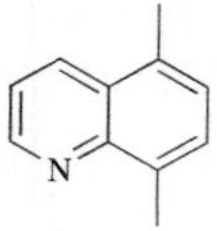	Pd(OAc)$_2$ (10 mol %), PhI(OPiv)$_2$ (2.0 eq), AgF (5.0 eq), MgSO$_4$ (2.0 eq), fluorobenzene, 60°, 16 h	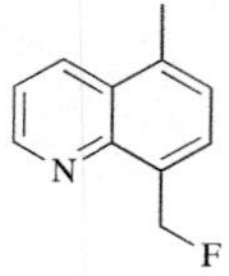 (39)	190
	[Me$_3$pyF]BF$_4$ (1.6 eq), Pd(OAc)$_2$ (10 mol %), MW (250W), benzene, 110°, 1 h	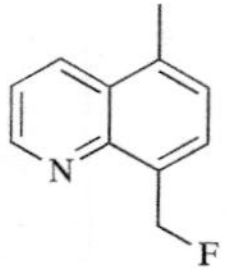 (57)	105

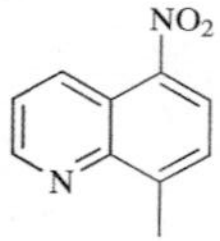

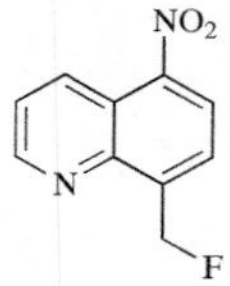

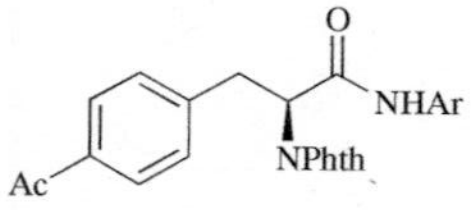

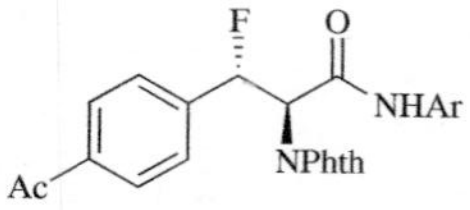

*Please refer to the charts preceding the tables for the structures indicated by the **bold** numbers.*

C_{12}

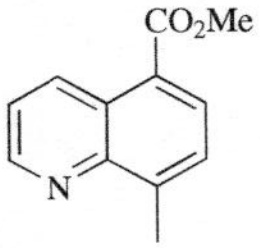

Substrate	Conditions	Product	Yield	Ref.
(quinoline, CO_2Me, Me)	Pd(OAc)_2 (10 mol %), PhI(OPiv)_2 (2.0 eq), AgF (5.0 eq), MgSO_4 (2.0 eq), DCM, 60°, 24 h	(quinoline, CO_2Me, CH_2F)	(59)	190
(quinoline, CN, Me)	Pd(OAc)_2 (10 mol %), PhI(OPiv)_2 (2.0 eq), AgF (5.0 eq), MgSO_4 (2.0 eq), DCM, 60°, 16 h	(quinoline, CN, CH_2F)	(70)	190
(Ph, i-Pr, CH_2OH alkene)	[Pd(PPh_3)_4] (10 mol %), Selectfluor (3.0 eq), Et_3SiH (1.5 eq), MeCN, 0°, 2 h	(F, Ph, i-Pr, OH)	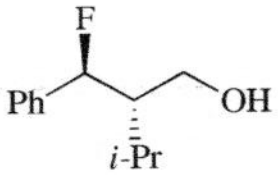 (67) dr >20:1	189
(Ph, i-Pr, HO alkene)	[Pd(PPh_3)_4] (10 mol %), Selectfluor (3.0 eq), Et_3SiH (1.5 eq), MeCN, 0°, 2 h	(F, Ph, i-Pr, OH)	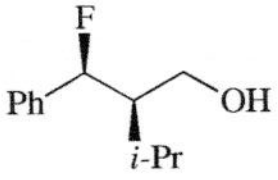 (41) dr >20:1	189
(naphthyl vinyl)	[Pd(PPh_3)_4] (10 mol %), Selectfluor (3.0 eq), Et_3SiH (1.5 eq), MeCN, 0°, 2 h	(naphthyl, F, CH_3)	(46)	189

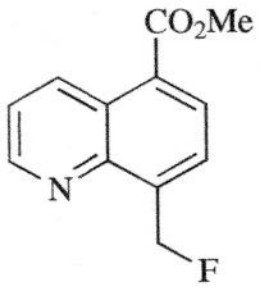

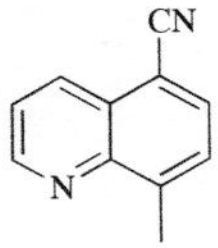

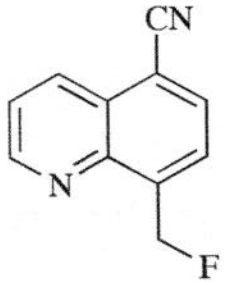

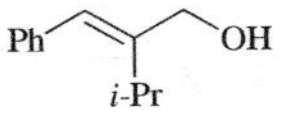

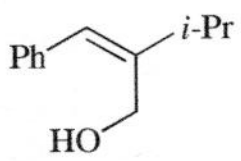

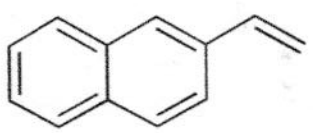

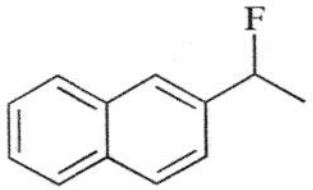

TABLE 5. SYNTHESIS OF BENZYLIC FLUORIDES (Continued)

*Please refer to the charts preceding the tables for the structures indicated by the **bold** numbers.*

C$_{13}$

Substrate	Conditions	Product(s) and Yield(s) (%)	Refs.
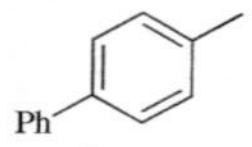	Fe(acac)$_2$ (10 mol %), Selectfluor (2.2 eq), MeCN, rt, 24 h	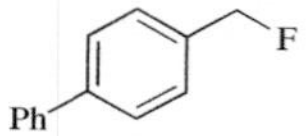 (59)	186
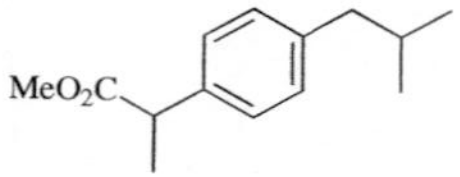	**Mn1** (20 mol %), PhIO (6.0–9.0 eq), Et$_3$N•HF (1.5 eq), AgF (3.0 eq), MeCN, 50°, 6–9 h	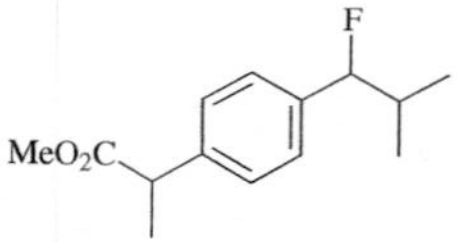 (55)	192
	Mn1 (20 mol %), PhIO (4.0 eq), Et$_3$N•3HF (0.5 eq), AgF (3.0 eq), MeCN, 50°, 4 h	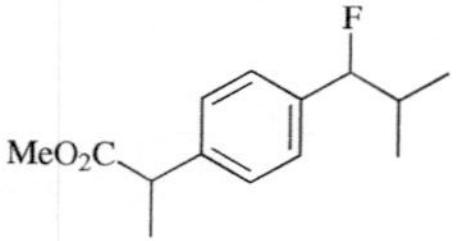 (55)	163
	Fe(acac)$_2$ (10 mol %), Selectfluor (2.2 eq), MeCN, rt, 24 h	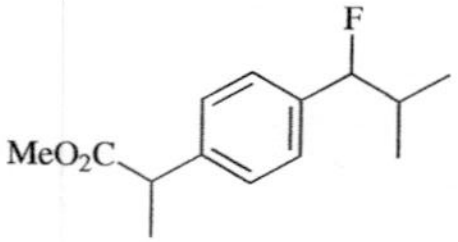 (64)	186
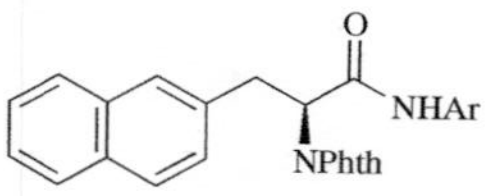	Pd(TFA)$_2$ (10 mol %), Selectfluor (1.5 eq), **L4** (10 mol %), Ag$_2$CO$_3$ (2.0 eq), 1,4-dioxane, 115°, 15 h	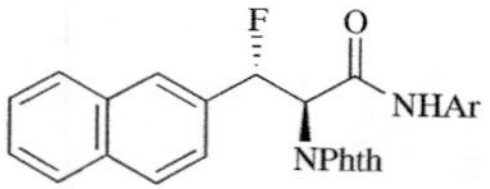 (53) dr 7:1	170

Ar = 4-CF$_3$C$_6$F$_4$

C_{14}				
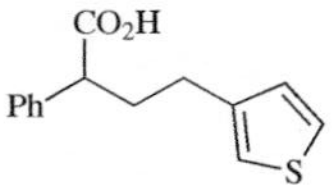	[Mn(tmp)]Cl (2.5 mol %), MesIO (3.3 eq), Et_3N•HF (1.2 eq), DCE, 45°, 0.75–1.5 h	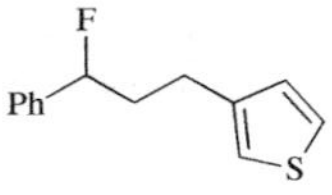	(56)	173
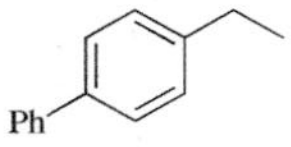	**Mn1** (20 mol %), PhIO (6.0–9.0 eq), Et_3N•HF (1.5 eq), MeCN, 50°, 6–9 h	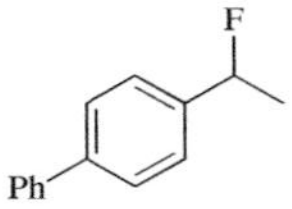	(58)	192
	[Mn(tmp)]Cl (2.5 mol %), PhIO (3.3 eq), Et_3N•HF (1.2 eq), DCE, 45°, 0.75–1.5 h	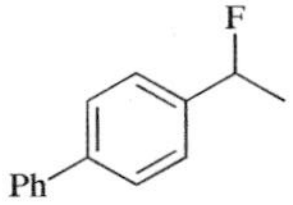	(33)	173
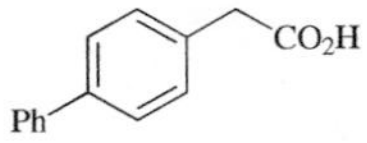	Ir[dF(CF_3)ppy]$_2$(dtbbpy)PF_6 (1 mol %), Na_2HPO_4 (2.0 eq), Selectfluor (2.0 eq), MeCN/H_2O (1:1), 34 W blue LEDs, 23°, 1 h	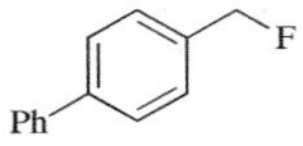	(87)	174
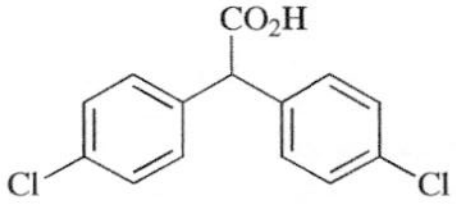	[Mn(tmp)]Cl (2.5 mol %), PhIO (3.3 eq), Et_3N•HF (1.2 eq), DCE, 45°, 0.75–1.5 h	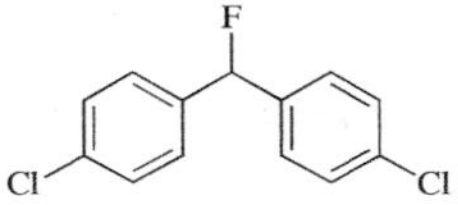	(58)	173

TABLE 5. SYNTHESIS OF BENZYLIC FLUORIDES (*Continued*)

Please refer to the charts preceding the tables for the structures indicated by the **bold** numbers.

Substrate	Conditions	Product(s) and Yield(s) (%)	Refs.
C₁₄			
	[Pd(PPh₃)₄] (10 mol %), Selectfluor (3.0 eq), Et₃SiH (1.5 eq), MeCN, 0°, 2 h	(80) dr 1:1	189
C₁₅			
	Fe(acac)₂ (10 mol %), Selectfluor (2.2 eq), MeCN, rt, 24 h	(61)	186
	[Mn(tmp)]Cl (2.5 mol %), PhIO (3.3 eq). Et₃N•HF (1.2 eq), DCE, 45°, 0.75–1.5 h	(60)	173
	[Ir{dF(CF₃)ppy}₂(dtbbpy)]PF₆ (1 mol %), Na₂HPO₄ (2.0 eq), Selectfluor (2.0 eq), MeCN/H₂O (3:1), 34 W blue LEDs, 23°, 1 h	(85)	174
	[Mn(tmp)]Cl (2.5 mol %), PhIO (3.3 eq), Et₃N•HF (1.2 eq), DCE, 45°, 0.75–1.5 h	(48)	173

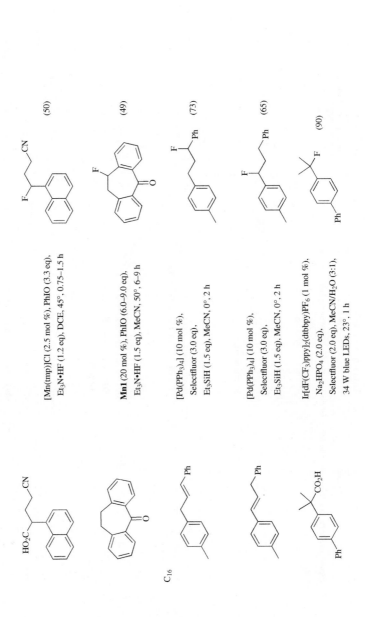

[Mn(tmp)]Cl (2.5 mol %), PhIO (3.3 eq), Et₃N•HF (1.2 eq), DCE, 45°, 0.75–1.5 h	(50)	173
Mn1 (20 mol %), PhIO (6.0–9.0 eq), Et₃N•HF (1.5 eq), MeCN, 50°, 6–9 h	(49)	192
[Pd(PPh₃)₄] (10 mol %), Selectfluor (3.0 eq), Et₃SiH (1.5 eq), MeCN, 0°, 2 h	(73)	189
[Pd(PPh₃)₄] (10 mol %), Selectfluor (3.0 eq), Et₃SiH (1.5 eq), MeCN, 0°, 2 h	(65)	189
Ir[dF(CF₃)ppy]₂(dtbbpy)PF₆ (1 mol %), Na₂HPO₄ (2.0 eq), Selectfluor (2.0 eq), MeCN/H₂O (3:1), 34 W blue LEDs, 23°, 1 h	(90)	174

C₁₆

803

TABLE 5. SYNTHESIS OF BENZYLIC FLUORIDES (*Continued*)

*Please refer to the charts preceding the tables for the structures indicated by the **bold** numbers.*

Substrate	Conditions	Product(s) and Yield(s) (%)	Refs.
C$_{16}$	Pd(OAc)$_2$ (10 mol %), PhI(OPiv)$_2$ (2.0 eq), AgF (5.0 eq), MgSO$_4$ (2.0 eq), fluorobenzene, 60°, 28 h	(39)	190
C$_{17}$	**Mn1** (20 mol %), PhIO (6.0–9.0 eq), Et$_3$N•HF (1.5 eq), MeCN, 50°, 6–9 h	(67)	192
	Mn1 (20 mol %), PhIO (5.0 eq), AgF (3.0 eq), Et$_3$N•HF (0.5 eq), MeCN, 50°, 6–8 h	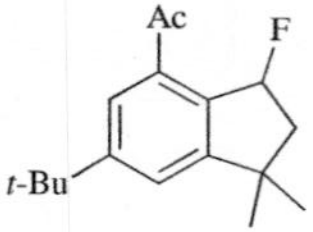 (67)	163
C$_{27}$	**Mn1** (20 mol %), PhIO (6.0–9.0 eq), Et$_3$N•HF (1.5 eq), MeCN, 50°, 6–9 h	(53) dr 2.2:1	192

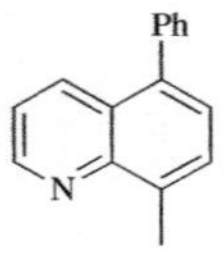

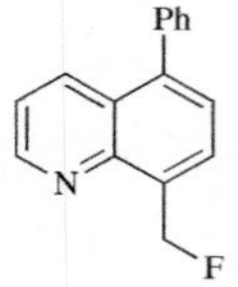

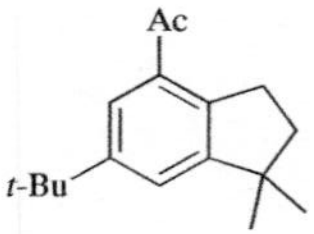

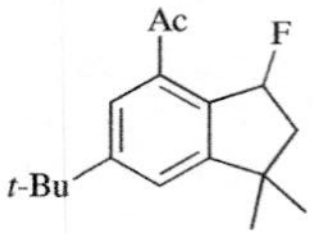

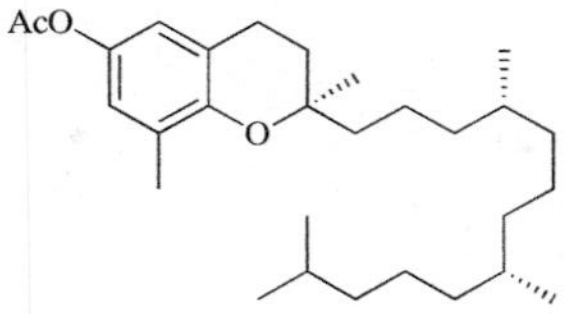

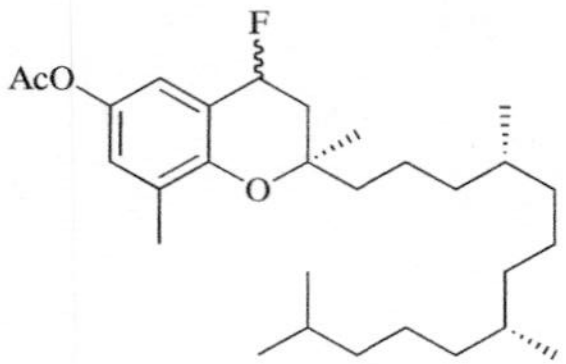

TABLE 6. SYNTHESIS OF ALLYLIC FLUORIDES

*Please refer to the charts preceding the tables for the structures indicated by the **bold** numbers.*

Allylic Substrate	Conditions	Product(s) and Yield(s) (%)	Refs.

C$_4$

Substrate	Conditions	Product	Refs.
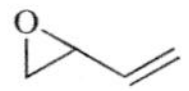 (epoxide with vinyl)	1. Rh(COD)$_2$BF$_4$ (5 mol %), Et$_3$N•3HF (3 eq), Et$_2$O, 25°, 0.5 h 2. 4-Fluorobenzoylchloride, pyridine, DMAP, rt, 18 h	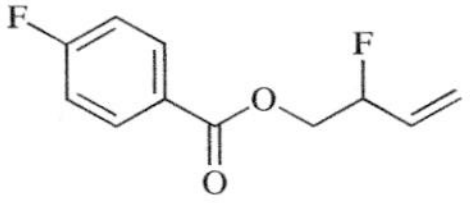 (49)	182
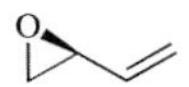 (epoxide with vinyl)	1. Rh(COD)$_2$BF$_4$ (5 mol %), Et$_3$N•3HF (3 eq), Et$_2$O, 25°, 0.5 h 2. 4-Fluorobenzoylchloride, pyridine, DMAP, rt, 18 h	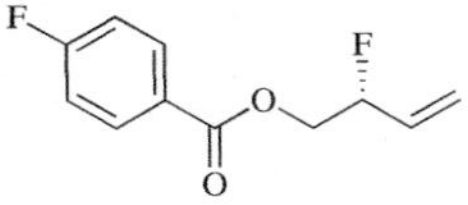 (43) er 84.5:15.5	182
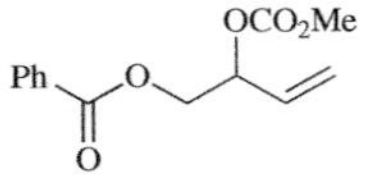 (OCO$_2$Me substrate, PhCO$_2$)	[Ir(COD)Cl]$_2$ (4 mol %), TBAF•4-t-BuOH (2 eq), DCM, 40°, 24 h	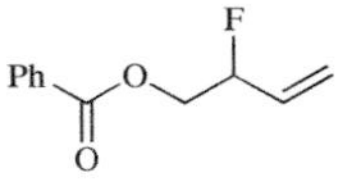 (32)	39
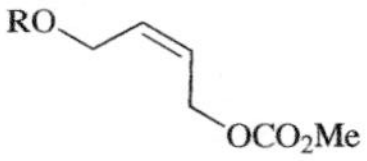 (RO OCO$_2$Me substrate)	[Ir(COD)Cl]$_2$ (x mol %), TBAF•4-t-BuOH (2 eq), DCM, 40°	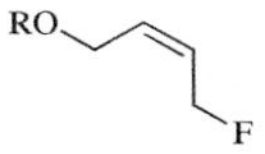 (see table below)	39

R	x	Time (h)		(Z)/(E)
4-BrC$_6$H$_4$CH$_2$	2	16	(52)	—
Bz	2	24	(65)	—
C$_{12}$H$_{25}$	2	3	(65)	—
Ph$_3$C	4	5	(50)	18:1

805

TABLE 6. SYNTHESIS OF ALLYLIC FLUORIDES (*Continued*)

Allylic Substrate	Conditions	Product(s) and Yield(s) (%)	Refs.

*Please refer to the charts preceding the tables for the structures indicated by the **bold** numbers.*

C₄

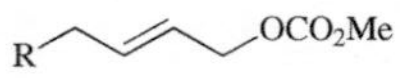

[Ir(COD)Cl]₂ (2 mol %),
TBAF•4-*t*-BuOH (2 eq),
DCM, 40°, 16 h

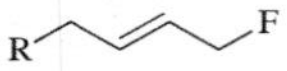

R			Linear/Branched
Boc(Me)N	(64)		>20:1
PhthHN	(48)		15:1

39

C₄₋₁₁

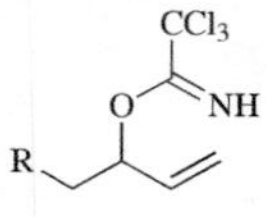

[IrCl(COD)]₂ (5 mol %),
Et₃N·3HF (3 eq), Et₂O, rt

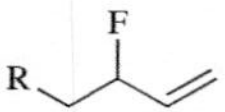

R	Time (h)	
	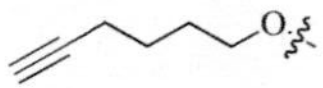 1.5	(68)
	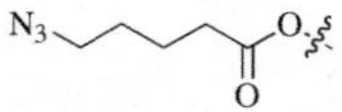 1	(64)
4-FC₆H₄O	1.5	(90)
BnO	2	(85)
BzO	2	(91)
PhthN	1.5	(89)
4-BrC₆H₄CO	1.5	(83)

180

C₄

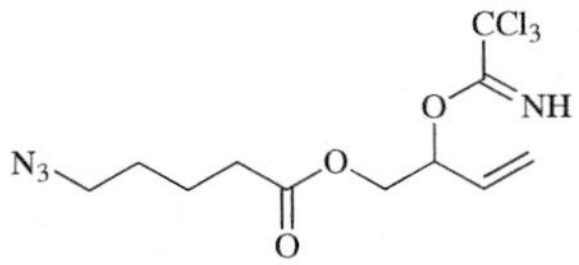

[IrCl(**L22**)]₂ (2.5 mol %), Et₃N•3HF (1.5 eq), MTBE, 25°, 3 h

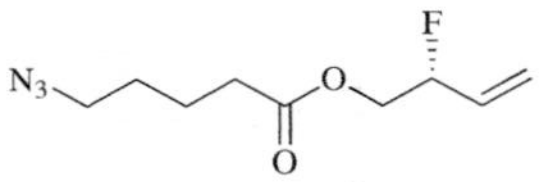

(61) er 97.5:2.5   40

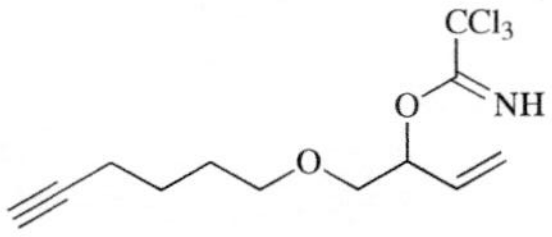

C₄₋₁₀

[IrCl(**L22**)]₂ (2.5 mol %), Et₃N•3HF (1.5 eq), MTBE, 25°, 0.5 h

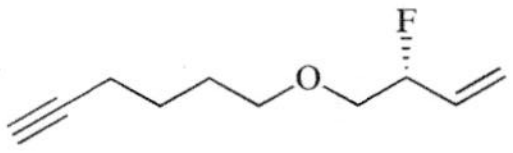

(74) er 99.5:0.5   40

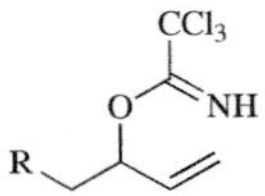

[IrCl(**L22**)]₂ (2.5 mol %), Et₃N•3HF (1.5 eq), MTBE, 25°

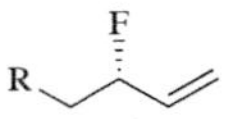

40

R	Time (h)		er
4-FC₆H₄O	1	(82)	96.5:3.5
2-BrC₆H₄O	1	(70)	96.0:4.0
4-MeOC₆H₄O	1.5	(77)	97.5:2.5
BnO	1	(90)	96.0:4.0
PhCO₂	1	(75)	96.5:3.5
TBDPSO	1	(69)	97.5:2.5
PhthN	1	(73)	91.0:9.0
TBDPS	1.5	(70)	98.5:1.5
Ph	1	(82)	95.0:5.0

807

TABLE 6. SYNTHESIS OF ALLYLIC FLUORIDES (*Continued*)

Allylic Substrate	Conditions	Product(s) and Yield(s) (%)	Refs.

*Please refer to the charts preceding the tables for the structures indicated by the **bold** numbers.*

C₄

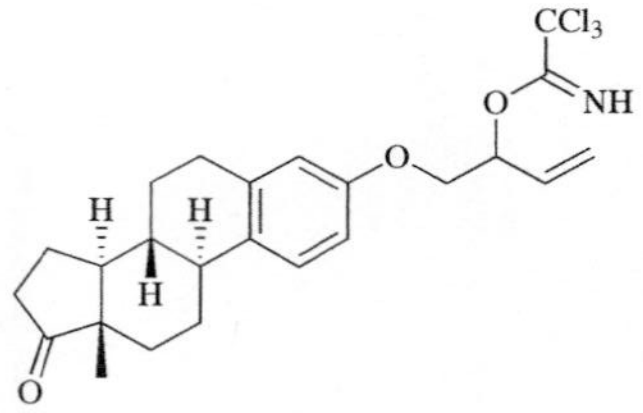

Catalyst (5 mol %),
Et₃N·3HF (1.5 eq), **MTBE**, rt, 6 h

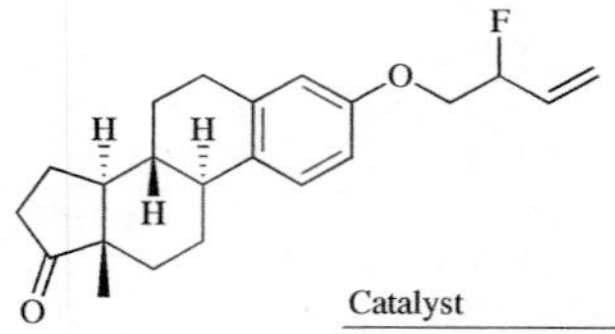

Catalyst		dr
[IrCl(*S,S*)-**L8**]₂	(54)	1:3
[IrCl(*R,R*)-**L8**]₂	(52)	24:1

40

C₅

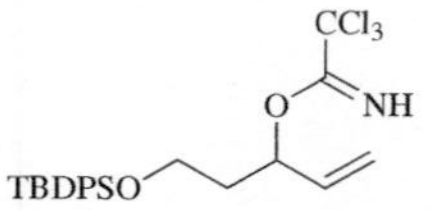

[IrCl(COD)]₂ (5 mol %),
Et₃N·3HF (3.0 eq), Et₂O, rt, 1 h

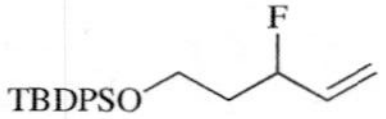

(78)

180

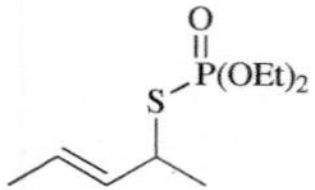

Pd(dba)₂ (8 mol %),
PPh₃ (24 mol %), AgF (1.3 eq),
THF, rt, 2.5 h

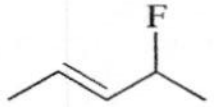

(50)

183

C$_{5-8}$

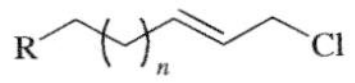

Pd$_2$(dba)$_3$ (5 mol %),
ligand (x mol %), AgF (3.0 eq),
solvent, rt, 48 h

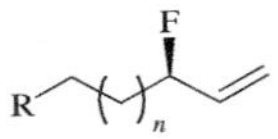

41

n	R	Ligand	x	Solvent		Branched/Linear	er
1	BnO	L7	10	toluene	(83)	>20:1	60.5:39.5
3	BnO	L7	10	toluene	(78)	>20:1	79.0:21.0
4	Br	L7	10	toluene	(66)	>20:1	—
4	HO	PPh$_3$	30	THF	(52)	6:1	—
4	TBSO	L7	10	toluene	(84)	>20:1	79.0:21.0

C$_{5-6}$

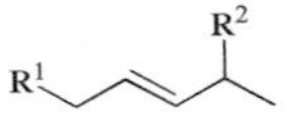

CuBr (30 mol %),
Et$_3$N•3HF (6 eq), MeCN, 24 h

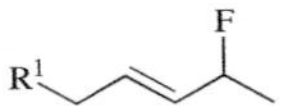

42

R^1	R^2	Temp (°)	
BzO	Br	35	(18)
Ts(Me)N	Br	35	(70)
Ts(Me)N	Cl	50	(65)
Ts(Ph)N	Br	35	(83)
Ts(Bn)N	Br	35	(78)
EtO$_2$C	Br	35	(49)
EtO$_2$C	Cl	50	(51)^a

TABLE 6. SYNTHESIS OF ALLYLIC FLUORIDES (Continued)

Allylic Substrate	Conditions	Product(s) and Yield(s) (%)	Refs.

*Please refer to the charts preceding the tables for the structures indicated by the **bold** numbers.*

C_{5-6}

| 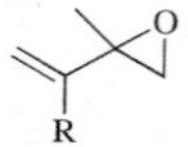 | 1. Rh(COD)$_2$BF$_4$ (5 mol %),
Et$_3$N•3HF (3 eq), Et$_2$O, 25°, 0.5 h
2. 4-Fluorobenzoylchloride,
pyridine, DMAP, rt, 18 h | 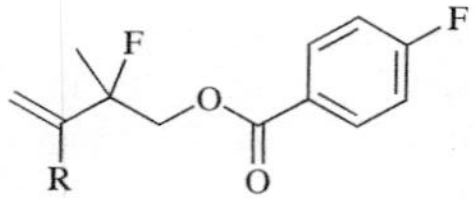 $\dfrac{\text{R}}{\begin{array}{l}\text{H} \quad (70)\\ \text{Me} \quad (56)\end{array}}$ | 182 |

C_{5-12}

| 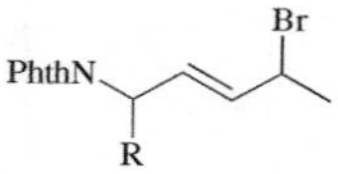 | CuBr (20 mol %),
Et$_3$N•3HF (6 eq),
MeCN, 35°, 24 h | 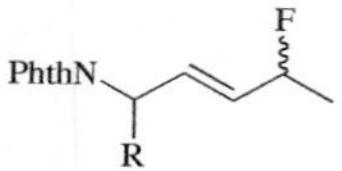 $\dfrac{\text{R}}{\begin{array}{l}\text{H} \quad (92)\\ \text{Me} \quad (81)\\ \text{Bn} \quad (85)\end{array}}$ | 42 |

C$_5$

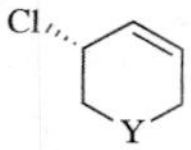

| | Pd$_2$dba$_3$ (5 mol %), **L7** (10 mol %),
AgF (x eq), solvent, rt, 24 h | | 177 |

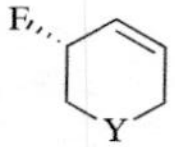

Y	x	Solvent		er
O	1.1	THF	(62)	95.0:5.0
TsN	2.0	toluene	(74)	98.0:2.0

C$_6$

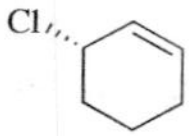

| | Pd$_2$dba$_3$ (5 mol %), **L6** (10 mol %),
AgF (1.1 eq), THF, rt, 24 h | (85) er 94.0:6.0 | 177 |

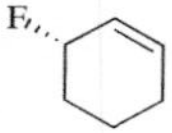

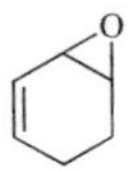

182

184

184

1. Rh(COD)$_2$BF$_4$ (5 mol %), Et$_3$N•3HF (3 eq), Et$_2$O, 25°, 0.5 h
2. 4-Fluorobenzoylchloride, pyridine, DMAP, rt, 18 h

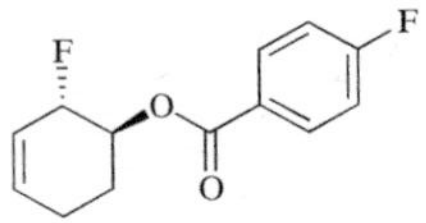

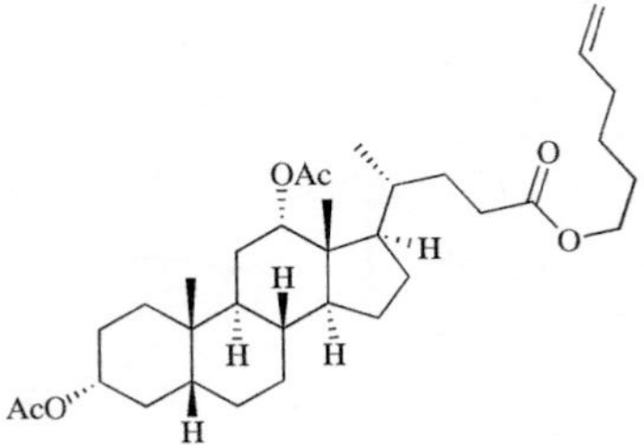

(79)

Pd(TFA)$_2$ (15 mol %), **L10** (15 mol %), (*R,R*)-(salen)CrCl (10 mol %), Et$_3$N•3HF (6.0 eq), BQ (2.0 eq), DCE, 23°, 72 h

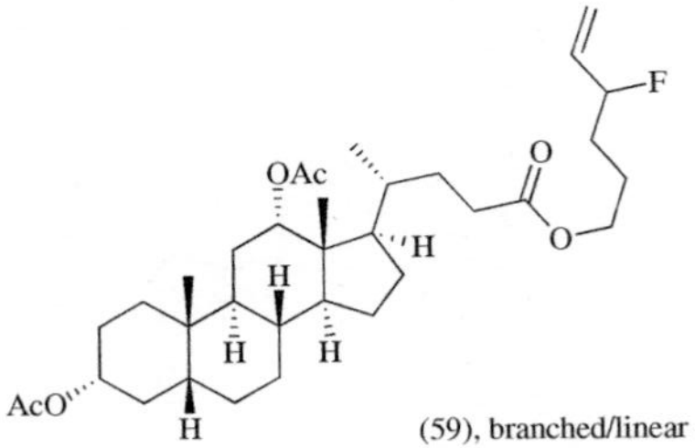

(59), branched/linear = 8:1

Pd(TFA)$_2$ (15 mol %), **L10** (15 mol %), (*R,R*)-(salen)CrCl (10 mol %), Et$_3$N•3HF (6.0 eq), BQ (2.0 eq), DCE, 23°, 72 h

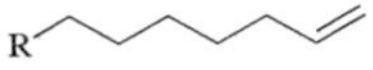

R		Branched/Linear
2-HOC$_6$H$_4$CO$_2$	(68)	7.4:1
PhHNCO$_2$	(67)	6.5:1
PhthN	(51)	6.7:1
BrCH$_2$	(67)	7:1
BnOCH$_2$	(59)	6.9:1

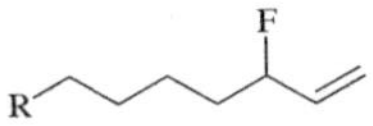

C$_{7-8}$

811

TABLE 6. SYNTHESIS OF ALLYLIC FLUORIDES (Continued)

Allylic Substrate	Conditions	Product(s) and Yield(s) (%)	Refs.

*Please refer to the charts preceding the tables for the structures indicated by the **bold** numbers.*

C7

BnO⟶OCO₂Me

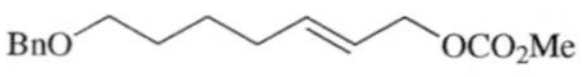

[Ir(COD)Cl]₂ (2 mol %),
TBAf•4t-BuOH (2 eq),
DCM, 40°, 16 h

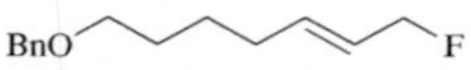

F (68)

39

BnO⟶ (epoxide)

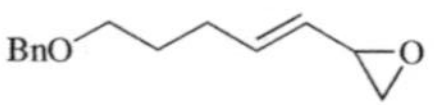

1. Rh(COD)₂BF₄ (5 mol %),
Et₃N•3HF (3 eq), Et₂O, 25°, 3 h
2. 4-Fluorobenzoylchloride,
pyridine, DMAP, rt, 18 h

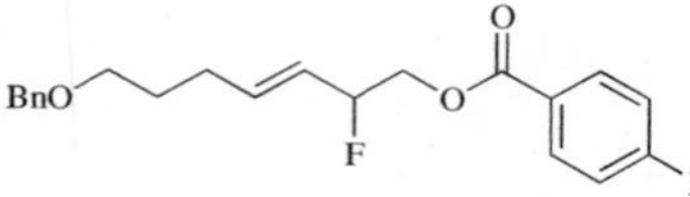

4-fluorobenzoate ester product (48)

182

(trichloroacetimidate sugar substrate, dr 1:1)

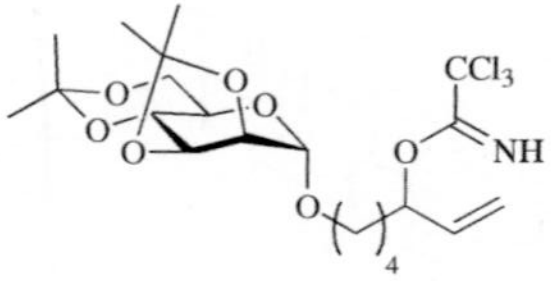

Catalyst (5 mol %),
Et₃N•3HF (1.5 eq), MTBE, rt, 6 h

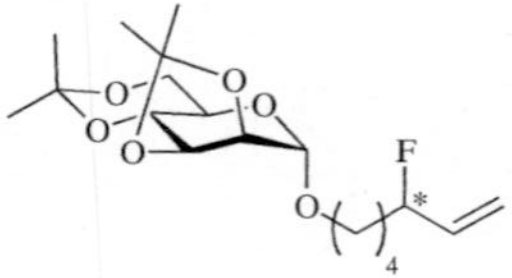

(sugar fluoride product)

Catalyst	dr	*
[IrCl(S,S)-**L8**]₂ (53)	1:99	(R)
[IrCl(R,R)-**L8**]₂ (66)	99:1	(S)

40

C7–8

R^1⟶Br pyrrolidine (N-R^2), dr 1:1

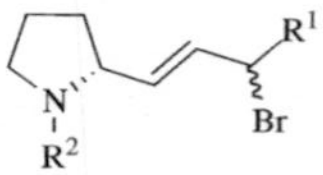

CuBr (20 mol %),
Et₃N•3HF (6 eq),
MeCN, 35°, 24 h

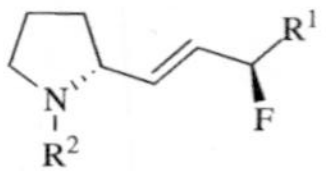

R^1⟶F pyrrolidine product

R^1	R^2		dr
H	Ts	(tr)	—
Me	Ns	(89)	6:1
Me	Ts	(82)	4:1

42

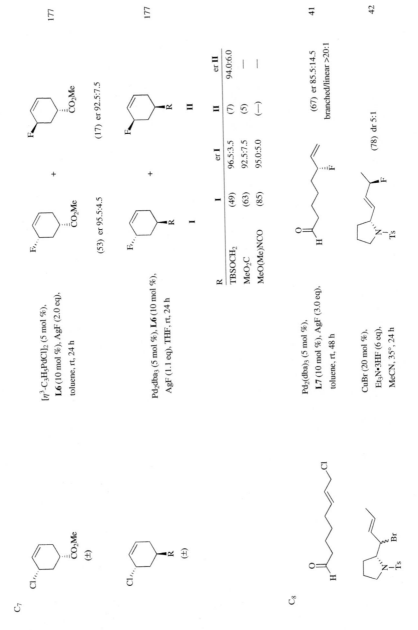

R	I	er I	II	er II
TBSOCH₂	(49)	96.5:3.5	(7)	94.0:6.0
MeO₂C	(63)	92.5:7.5	(5)	—
MeO(Me)NCO	(85)	95.0:5.0	(—)	—

177

177

41

42

TABLE 6. SYNTHESIS OF ALLYLIC FLUORIDES (Continued)

*Please refer to the charts preceding the tables for the structures indicated by the **bold** numbers.*

C₈

Allylic Substrate	Conditions	Product(s) and Yield(s) (%)	Refs.
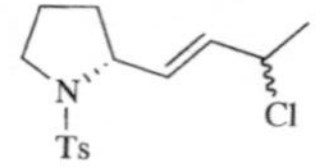	CuBr (30 mol %), Et$_3$N·3HF (6 eq), MeCN, 50°, 24 h	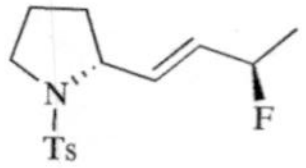 (64) dr 4:1	42
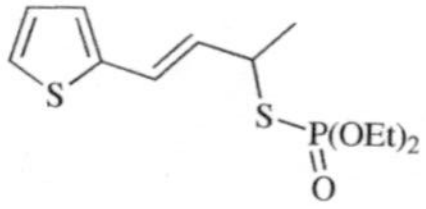	Pd(dba)$_2$ (8 mol %), PPh$_3$ (24 mol %), AgF (1.3 eq), THF, rt, 2.5 h	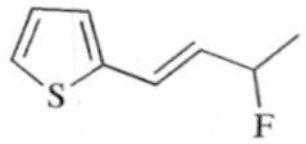 (55) 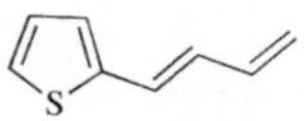 + (10)	183
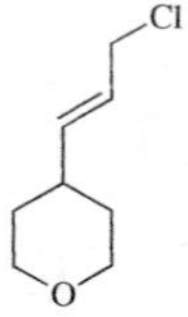	Pd$_2$(dba)$_3$ (5 mol %), **L7** (10 mol %), AgF (3.0 eq), toluene, rt, 48 h	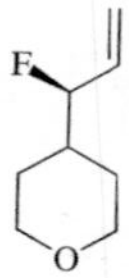 (62) er 98.5:1.5 branched/linear >20:1	41
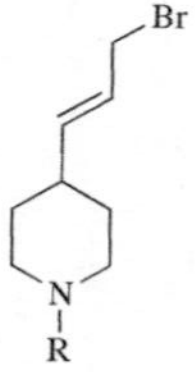	Pd$_2$(dba)$_3$ (5 mol %), **L7** (10 mol %), AgF (3.0 eq), toluene, rt, 48 h	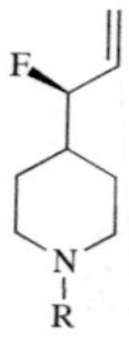 R Boc (85) Cbz (88) er 96.5:3.5 branched/linear = 10:1	41

C_9

Reactant	Conditions	Product(s) and Yield(s) (%)	Refs.
	$Pd(TFA)_2$ (15 mol %), **L10** (15 mol %), (*R*,*R*)-(salen)CrCl (10 mol %), $Et_3N•3HF$ (6.0 eq), BQ (2.0 eq), DCE, 23°, 72 h	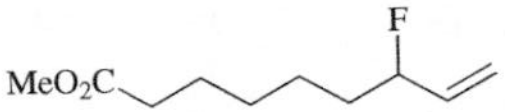 (53), branched/linear = 7.5:1	184
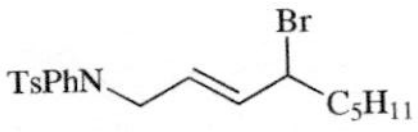	CuBr (20 mol %), $Et_3N•3HF$ (6 eq), MeCN, 35°, 24 h	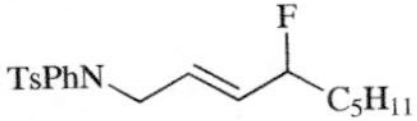 (50)	42
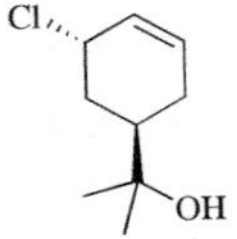 dr 5:1	Pd_2dba_3 (5 mol %), **L6** (10 mol %), AgF (1.1 eq), THF, rt, 24 h	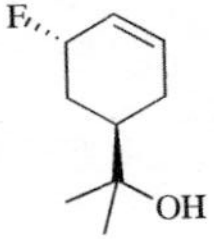 (59) dr 20:1, er 93.5:6.5	177
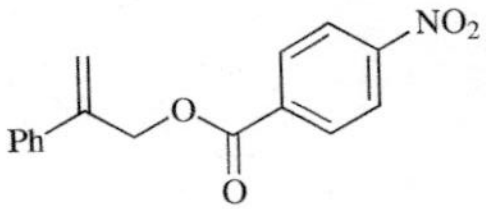	$Pd(dba)_2$ (5 mol %), PPh_3 (15 mol %), TBAF•(*t*-BuOH)$_4$ (2.5 eq), THF, rt, 1 h	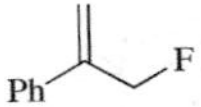 (66)	38
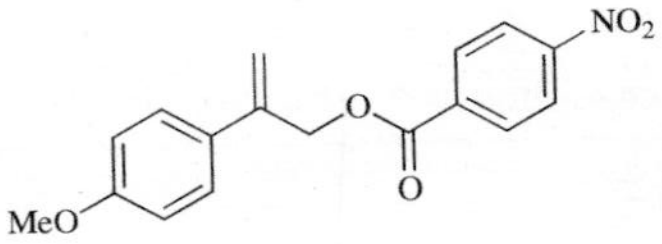	$Pd(dba)_2$ (5 mol %), PPh_3 (15 mol %), TBAF•(*t*-BuOH)$_4$ (2.5 eq), THF, rt, 1 h	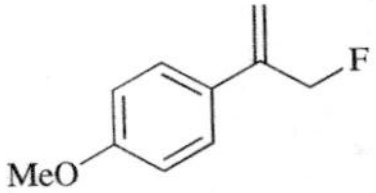 (85)	38

TABLE 6. SYNTHESIS OF ALLYLIC FLUORIDES (Continued)

Allylic Substrate	Conditions	Product(s) and Yield(s) (%)	Refs.

*Please refer to the charts preceding the tables for the structures indicated by the **bold** numbers.*

C₉

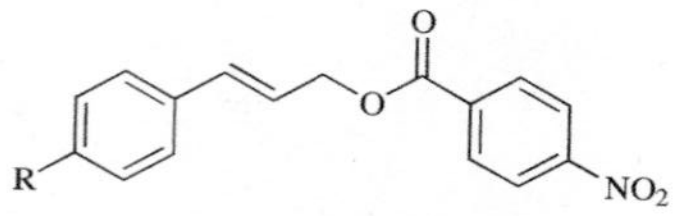

Pd(dba)₂ (5 mol %),
PPh₃ (15 mol %),
TBAF•(t-BuOH)₄ (2.5 eq), THF

R	Temp (°)	Time (min)	
H	23	60	(>95)
Br	23	60	(53)
MeO	23; then 40	60; then 20	(39)

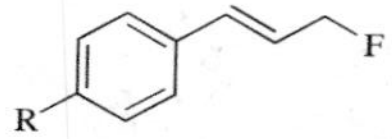

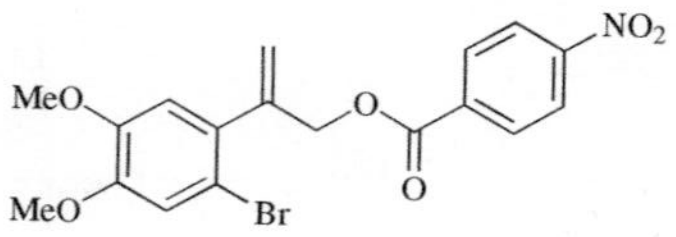

38

Pd(dba)₂ (5 mol %),
PPh₃ (15 mol %),
TBAF•(t-BuOH)₄ (2.5 eq),
THF, rt, 1 h

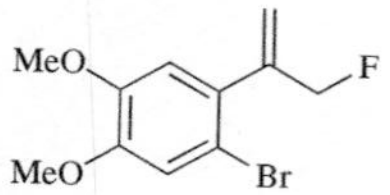

(84)

38

C₁₀

Pd(TFA)₂ (15 mol %),
L10 (15 mol %),
(R,R)-(salen)CrCl (10 mol %),
Et₃N•3HF (6.0 eq), BQ (2.0 eq),
DCE, 23°, 72 h

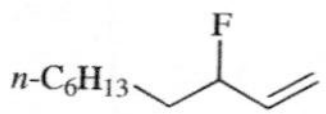

(56), branched/linear = 7:1

184

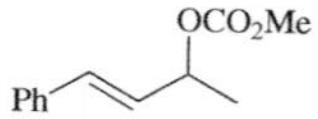

[Ir(COD)Cl]₂ (4 mol %),
TBAF•(t-BuOH)₄ (2 eq),
DCM, 40°, 16 h

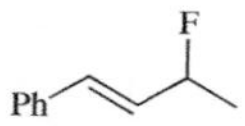

(66)

39

Substrate	Conditions	Product	Yield	Refs.
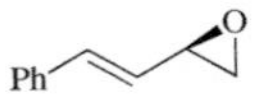 (epoxide, Ph)	1. Rh(COD)$_2$BF$_4$ (5 mol %), Et$_3$N•3HF (3 eq), Et$_2$O, 25°, 0.5 h 2. 4-Fluorobenzoylchloride, pyridine, DMAP, rt, 18 h	(4-fluorobenzoate ester)	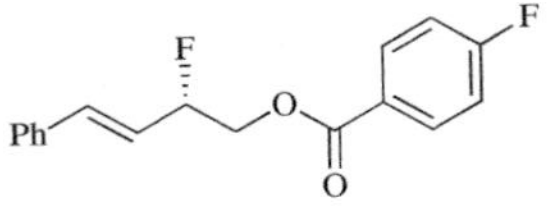 (51) er 55.0:45.0	182
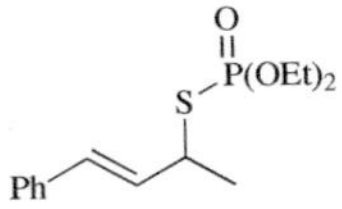 (S–P(OEt)$_2$ substrate, Ph)	Pd(dba)$_2$ (8 mol %), PPh$_3$ (24 mol %), AgF (1.3 eq), THF, rt, 2.5 h	 (66) + (9)		183
(S–P(OEt)$_2$ substrate, Ph)	Pd(dba)$_2$ (2 mol %), PPh$_3$ (6 mol %), AgF (1.3 eq), THF, rt	see table below		183
(4-nitrobenzoate ester, CH$_2$Cl)	Pd(dba)$_2$ (5 mol %), PPh$_3$ (15 mol %), TBAF•(t-BuOH)$_4$ (2.5 eq), THF, rt, 1 h	(benzylic chloride, F)	(60)	38

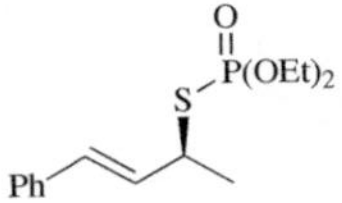

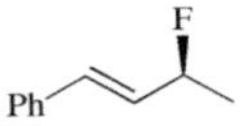

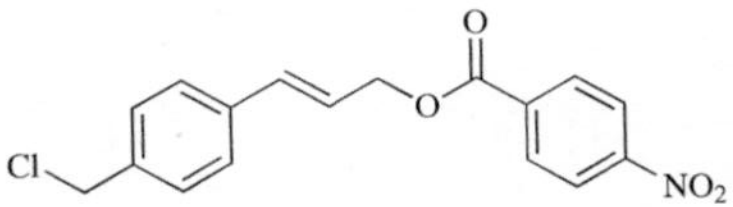

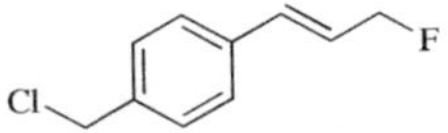

Time (h)		er
1	(37)	96.0:4.0
5.5	(65)	80.0:20.0
12	(—)	79.0:21.0

TABLE 6. SYNTHESIS OF ALLYLIC FLUORIDES (*Continued*)

*Please refer to the charts preceding the tables for the structures indicated by the **bold** numbers.*

Allylic Substrate	Conditions	Product(s) and Yield(s) (%)	Refs.
C$_{10}$	Pd(dba)$_2$ (5 mol %), PPh$_3$ (15 mol %), TBAF•(t-BuOH)$_4$ (2.5 eq), THF, 40°, 4 h	(35)	38
C$_{10-11}$	1. Rh(COD)$_2$BF$_4$ (5 mol %), Et$_3$N•3HF (3 eq), Et$_2$O, 25°, 0.5 h 2. 4-Fluorobenzoylchloride, pyridine, DMAP, rt, 18 h		182

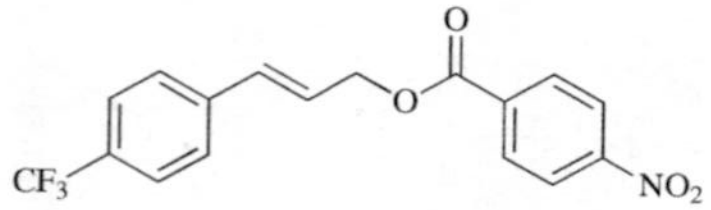

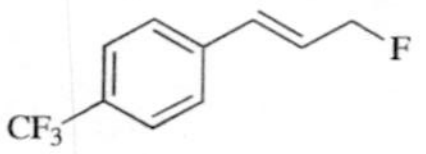

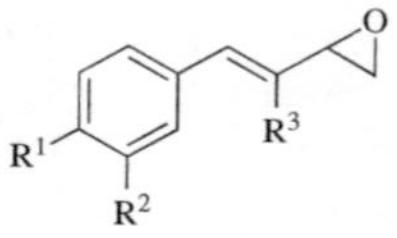

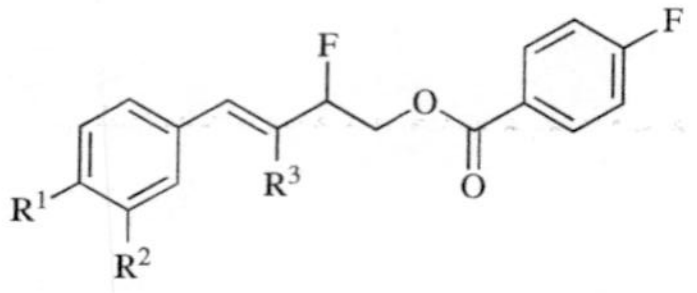

R^1	R^2	R^3	
H	H	H	(58)
H	Br	H	(43)
Cl	H	H	(46)
H	H	Me	(60)
Cl	H	Me	(62)
CF$_3$	H	H	(46)

183

$$I + II$$

R^1	R^2	I+II	I	II
F	H	(66)	(83)	(17)
MeO	H	(65)	(81)	(19)
BnOCH$_2$	H	(60)	(84)	(16)
t-BuO$_2$C	H	(53)	(84)	(16)
H	CF$_3$	(66)	(82)	(18)
TIPSO(Me)CH	H	(79)	(84)	(16)

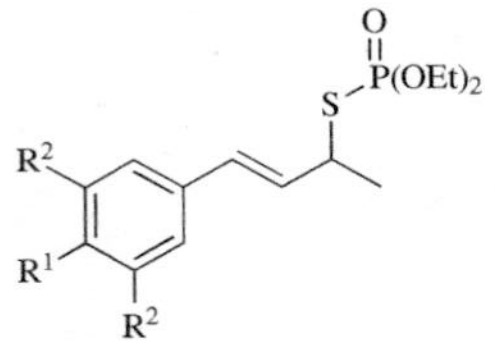

Pd(dba)$_2$ (8 mol %),
PPh$_3$ (24 mol %),
AgF (1.3 eq), THF, rt, 2.5 h

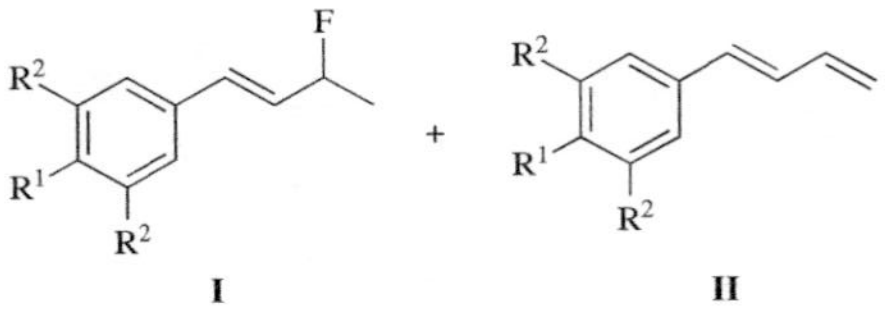

C$_{10-12}$

184

(54), branched/linear = 7:1

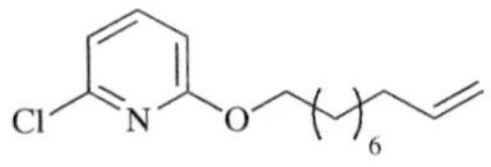

Pd(TFA)$_2$ (15 mol %),
L10 (15 mol %),
(R,R)-(salen)CrCl (10 mol %),
Et$_3$N·3HF (6.0 eq), BQ (2.0 eq),
DCE, 23°, 72 h

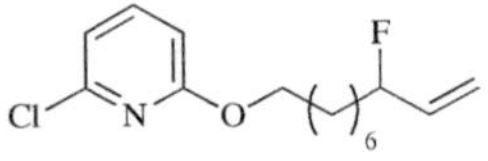

C$_{10}$

819

TABLE 6. SYNTHESIS OF ALLYLIC FLUORIDES (Continued)

Allylic Substrate	Conditions	Product(s) and Yield(s) (%)	Refs.

Please refer to the charts preceding the tables for the structures indicated by the bold numbers.

C₁₁

Allylic Substrate	Conditions	Product(s) and Yield(s) (%)	Refs.
(CCl₃–C(O)–NH–O– substrate, Ph chain)	[IrCl(COD)]₂ (5 mol %), Et₃N·3HF (3.0 eq), Et₂O, rt, 1 h	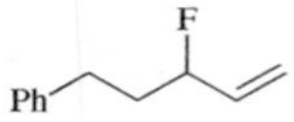 (68)	180
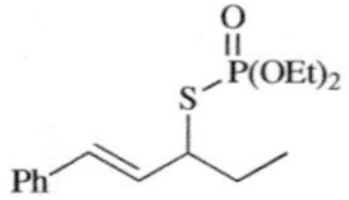 (S–P(OEt)₂ (O) substrate, Ph chain)	Pd(dba)₂ (8 mol %), PPh₃ (24 mol %), AgF (1.3 eq), THF, rt, 2.5 h	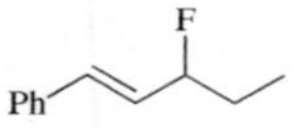 (53) 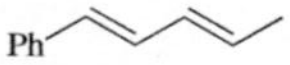 + (10)	183
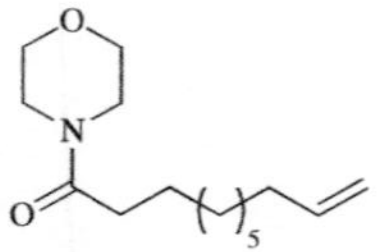 (morpholine amide substrate with terminal alkene)	Pd(TFA)₂ (15 mol %), **L10** (15 mol %), (R,R)-(salen)CrCl (10 mol %), Et₃N·3HF (6.0 eq), BQ (2.0 eq), DCE, 23°, 72 h	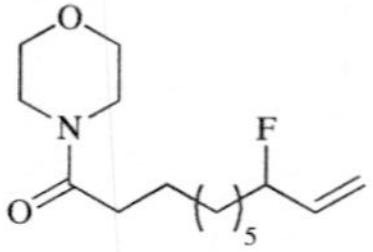 F-substituted morpholine amide (53), branched/linear = 7:1	184
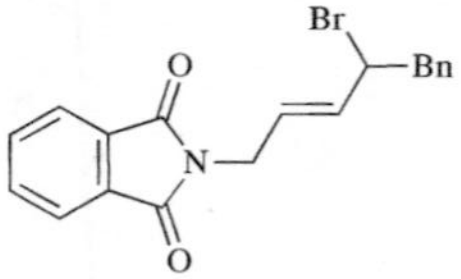 (Br, Bn, phthalimide substrate)	CuBr (20 mol %), Et₃N·3HF (6 eq), MeCN, 35°, 24 h	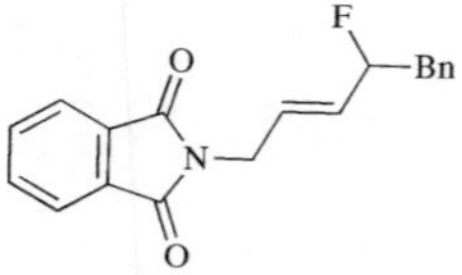 F, Bn, phthalimide product (45)	42

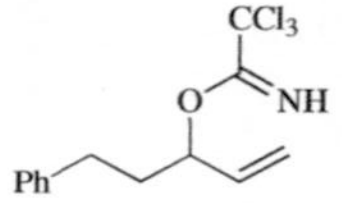

C_{12}

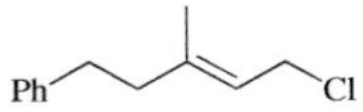 Pd$_2$(dba)$_3$ (5 mol %), PPh$_3$ (30 mol %), AgF (3 eq), toluene, rt, 48 h	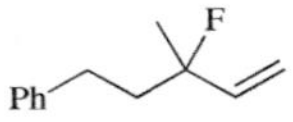 (42), branched/linear >20:1	41
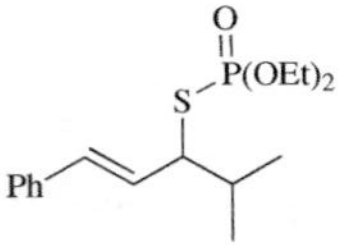 Pd(dba)$_2$ (8 mol %), PPh$_3$ (24 mol %), AgF (1.3 eq), THF, rt, 2.5 h	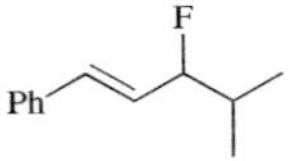 (57) 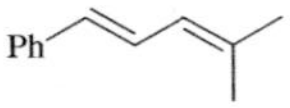 + (11)	183
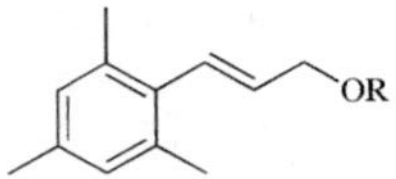 Pd(dba)$_2$ (5 mol %), PPh$_3$ (15 mol %), TBAF•(t-BuOH)$_4$ (2.5 eq), THF, rt, 1 h	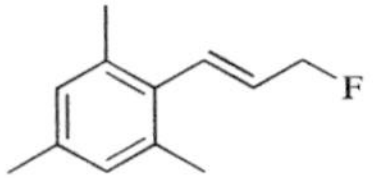 (65)	38

C_{13}

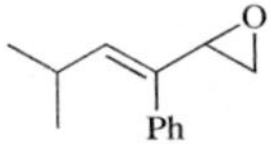 1. Rh(COD)$_2$BF$_4$ (5 mol %), Et$_3$N•3HF (3 eq), Et$_2$O, 25°, 0.5 h 2. 4-Fluorobenzoylchloride, pyridine, DMAP, rt, 18 h	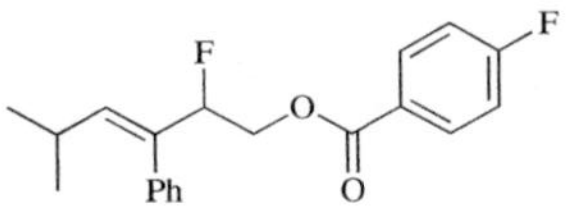 (59)	182

TABLE 6. SYNTHESIS OF ALLYLIC FLUORIDES (Continued)

Allylic Substrate	Conditions	Product(s) and Yield(s) (%)	Refs.

*Please refer to the charts preceding the tables for the structures indicated by the **bold** numbers.*

C13

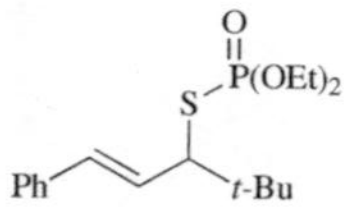

Pd(dba)₂ (8 mol %),
PPh₃ (24 mol %),
AgF (1.3 eq), THF, rt, 2.5 h — (74) — 183

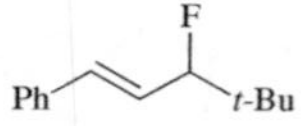

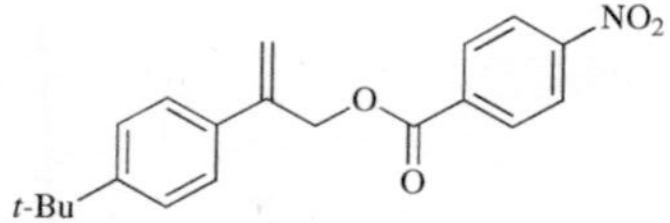

Pd(dba)₂ (5 mol %),
PPh₃ (15 mol %),
TBAF•(t-BuOH)₄ (2.5 eq),
THF, rt, 1 h — (>95) — 38

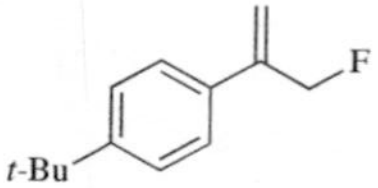

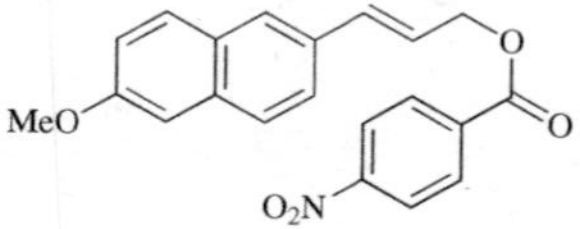

Pd(dba)₂ (5 mol %),
PPh₃ (15 mol %),
TBAF•(t-BuOH)₄ (2.5 eq),
THF, rt, 1 h — (46) — 38

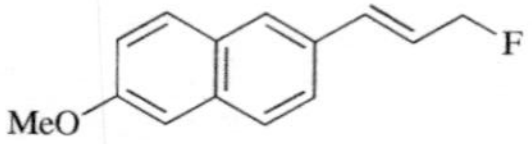

C14

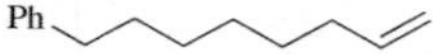

Pd(TFA)₂ (15 mol %),
L10 (15 mol %),
(R,R)-(salen)CrCl (10 mol %),
Et₃N•3HF (6.0 eq),
BQ (2.0 eq), DCE, 23°, 72 h — (64), branched/linear = 7:1 — 184

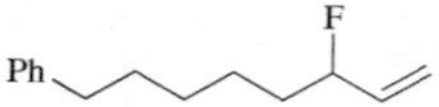

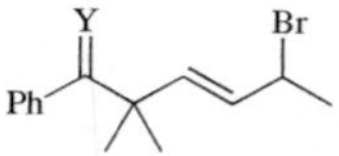

CuBr (20 mol %),
Et₃N•3HF (6 eq),
MeCN, 35°, 24 h

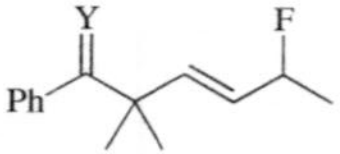

Y	
O	(54)
MeON	(59)

— 42

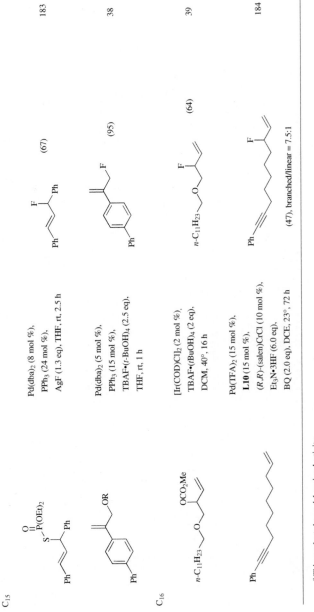

C15

Pd(dba)₂ (8 mol %),
PPh₃ (24 mol %),
AgF (1.3 eq), THF, rt, 2.5 h

(67)

183

Pd(dba)₂ (5 mol %),
PPh₃ (15 mol %),
TBAF•(t-BuOH)₄ (2.5 eq),
THF, rt, 1 h

(95)

38

C16

[Ir(COD)Cl]₂ (2 mol %),
TBAF•(BuOH)₄ (2 eq),
DCM, 40°, 16 h

(64)

39

Pd(TFA)₂ (15 mol %),
L10 (15 mol %),
(R,R)-(salen)CrCl (10 mol %),
Et₃N•3HF (6.0 eq),
BQ (2.0 eq), DCE, 23°, 72 h

(47), branched/linear = 7.5:1

184

ᵃThis product shows 4:1 regioselectivity.

TABLE 7A. FLUORINATION OF β-KETO ESTERS

*Please refer to the charts preceding the tables for the structures indicated by the **bold** numbers.*

β-Keto Ester	Conditions	Product(s) and Yield(s) (%)	Refs.
C_4			
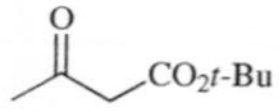 (structure: O=, CO₂*t*-Bu)	1. NCS (1.1 eq), rt, 2 h 2. **L18** (12 mol %), Cu(OTf)₂ (10 mol %), NFSI (3 eq), 4 Å MS, benzene, 40°, 8 h	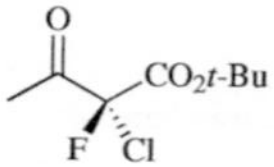 (structure: CO₂*t*-Bu, F, Cl) (70) er 95.0:5.0	101
C_5			
(structure: O=, CO₂Et)	[RuCl₂(**L23**)] (10 mol %), NFSI (1.08 eq), (Et₃O)PF₆ (0.21 eq), solvent	(structure: CO₂Et, F)	89

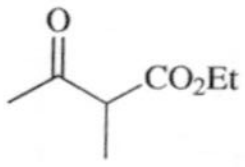

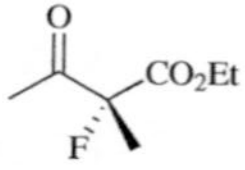

Solvent	Temp	Time (h)	er
DCM	rt	24	(91) 79.5:20.5
DCM/Et₂O (1:1)	—	4	(96) 88.5:11.5

β-Keto Ester	Conditions	Product(s) and Yield(s) (%)	Refs.
	Cu(ClO₄)₂·6H₂O (5 mol %), **L24** (7.5 mol %), NFSI (1.1 eq), HFIP (10 mol %), toluene, 0°, 8 h	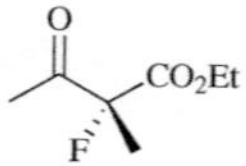 (structure: CO₂Et, F) (84) er 85.0:15.0	102
	Co(acac)₂ (10 mol %), **L15** (10 mol %), NFSI (1.4 eq), Et₂O, rt, 12 h	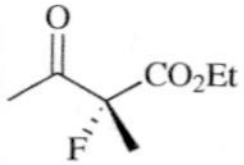 (structure: CO₂Et, F) (64) er 85.5:14.5	88
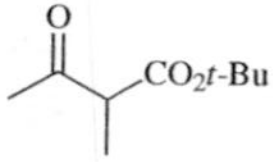 (structure: O=, CO₂*t*-Bu)	**Pd5** (2.5 mol %), NFSI (1.5 eq), EtOH, 20°, 72 h	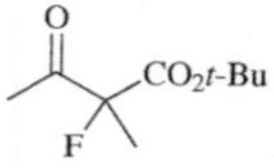 (structure: CO₂*t*-Bu, F) (49) er 95.5:4.5	70

824

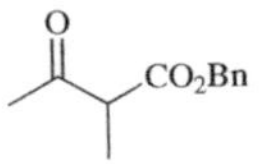

[RuCl$_2$(**L23**)] (10 mol %), NFSI (1.08 eq), (Et$_3$O)PF$_6$ (0.21 eq), DCM, rt, 24 h

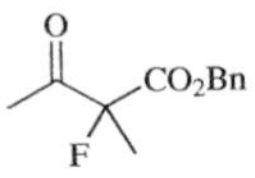

(82) er 92.0:8.0   89

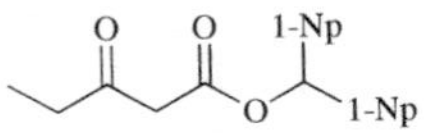

1. NCS (1.1 eq), rt, 2 h

2. **L18** (12 mol %), Cu(OTf)$_2$ (10 mol %), NFSI (3 eq), 4 Å MS, benzene, 40°, 17 h

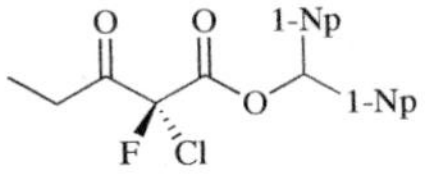

(67) er 93.5:6.5   101

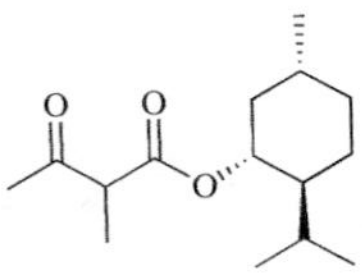

Catalyst (5 mol %), Selectfluor (1.15 eq), MeCN, rt, 1 h

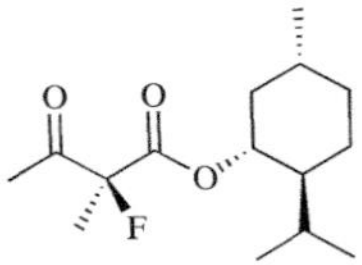

(59)   82

Catalyst	dr
CpTiCl$_3$	37:63
Ti1	51:49
Ti2	75:25

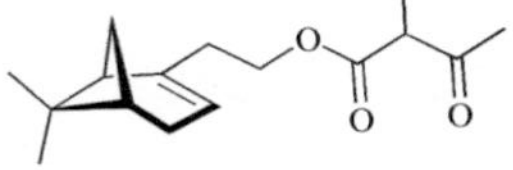

Catalyst (5 mol %), Selectfluor (1.15 eq), MeCN, rt, 1 h

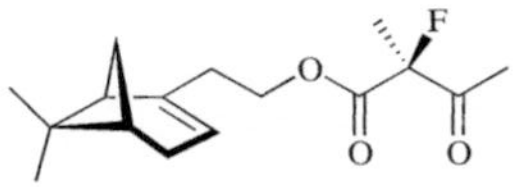

(—)   82

Catalyst	dr
CpTiCl$_3$	50:50
Ti1	59:41
Ti2	75:25

TABLE 7A. FLUORINATION OF β-KETO ESTERS (Continued)

β-Keto Ester	Conditions	Product(s) and Yield(s) (%)	Refs.

*Please refer to the charts preceding the tables for the structures indicated by the **bold** numbers.*

C₅

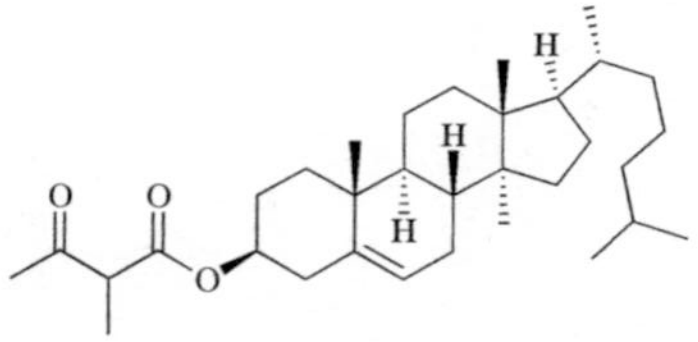

Tf2 (5 mol %), Selectfluor (1.15 eq),
MeCN, rt, 1 h

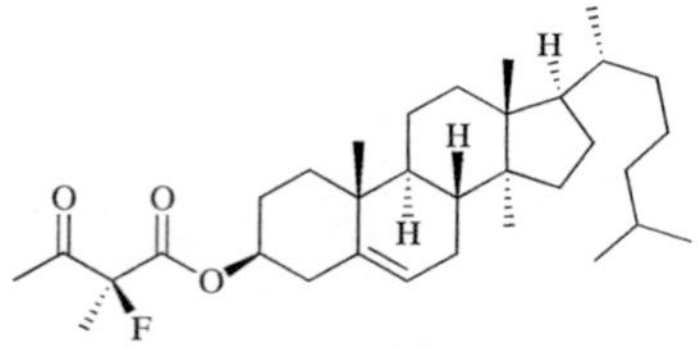

(**56**) dr 80:20

82

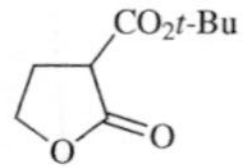

Catalyst (*x* mol %), NFSI (1.5 eq),
solvent, rt

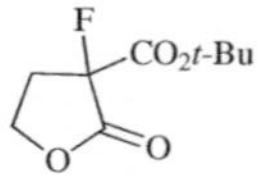

69

Catalyst	*x*	Solvent	Time (h)		er
Pd5	2.5	*t*-BuOH	24	(75)	99.0:1.0
Pd6	5	*i*-PrOH	6	(78)	93.5:6.5
Pd7	5	*i*-PrOH	24	(74)	98.5:1.5
Pd8	5	*i*-PrOH	6	(54)	91.0:9.0
Pd8	5	*i*-PrOH	6	(89)	90.0:10.0
Pd8	5	*t*-BuOH	6	(96)	89.5:10.5
Pd8	5	CH₂Cl₂	24	(tr)	—
Pd8	5	THF	24	(10)	75.5:24.5
Pd8	5	EtOH	24	(49)	87.5:12.5
Pd9	5	*i*-PrOH	6	(79)	88.5:11.5

C_{6-18}

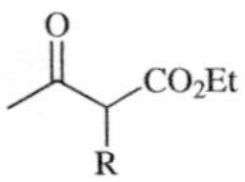

Catalyst (5 mol %), F-TEDA (1.15 eq).
MeCN, rt, 1 h

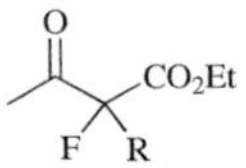

R	Catalyst		er
Et	Tf1	(44)	60.5:39.5
Et	Tf2	(44)	72.5:27.5
Ph	Tf1	(55)	62.0:38.0
Ph	Tf2	(55)	90.5:9.5
Bn	Tf1	(96)	65.0:35.0–68.5:31.5
Bn	Tf2	(96)	82.5:17.5
$C_6F_5CH_2$	Tf1	(88)	77.5:22.5
$C_6F_5CH_2$	Tf2	(88)	78.5:21.5
$4\text{-}O_2NC_6H_4CH_2$	Tf1	(—)	73.0:27.0
$4\text{-}O_2NC_6H_4CH_2$	Tf2	(—)	80.5:19.5
$3,5\text{-}Me_2C_6H_3CH_2$	Tf1	(57)	62.0:38.0
$3,5\text{-}Me_2C_6H_3CH_2$	Tf2	(57)	87.0:13.0
$2,4\text{-}Me_2C_6H_3CH_2$	Tf1	(83)	68.5:31.5
$2,4\text{-}Me_2C_6H_3CH_2$	Tf2	(83)	86.5:13.5
$2,4,6\text{-}Me_3C_6H_2CH_2$	Tf1	(92)	74.5:25.5
$2,4,6\text{-}Me_3C_6H_2CH_2$	Tf2	(92)	83.5:17.5
$1\text{-}NpCH_2$	Tf1	(80)	75.5:24.5
$1\text{-}NpCH_2$	Tf2	(80)	84.5:15.5
Ph_2CHCH_2	Tf1	(90)	67.5:32.5–72.0:28.0
Ph_2CHCH_2	Tf2	(90)	68.5:31.5

TABLE 7A. FLUORINATION OF β-KETO ESTERS (Continued)

β-Keto Ester	Conditions	Product(s) and Yield(s) (%)	Refs.

*Please refer to the charts preceding the tables for the structures indicated by the **bold** numbers.*

C_6

| 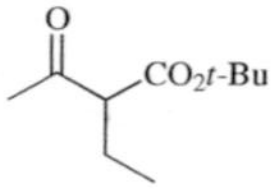 | **Pd8** (2.5 mol %), NFSI (1.5 eq), EtOH, 20°, 42 h | 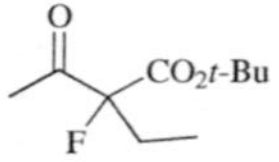 CO_2t-Bu (88) er 93.5:6.5 | 75 |
| 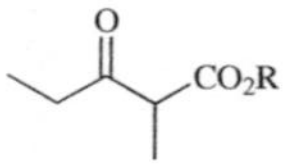 | Catalyst (5 mol %), Selectfluor (1.15 eq), MeCN, rt, 1 h | 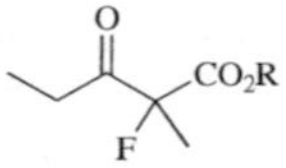 CO_2R | 82 |

R	Catalyst		er
Et	Ti1	(—)	65.0:35.0
Et	Ti2	(—)	81.0:19.0
Ph	Ti1	(50)	72.5:17.5
Ph	Ti2	(50)	94.0:6.0
4-MeOC$_6$H$_4$	Ti1	(35)	64.0:36.0
4-MeOC$_6$H$_4$	Ti2	(35)	93.0:7.0
Bn	Ti1	(82)	74.0:26.0
Bn	Ti2	(82)	85.5:14.5
Ph$_2$CH	Ti1	(87)	79.0:21.0
Ph$_2$CH	Ti2	(87)	90.5:9.5
2,6-t-Bu$_2$-4-MeC$_6$H$_2$	Ti1	(55)	57.5:42.5
2,6-t-Bu$_2$-4-MeC$_6$H$_2$	Ti2	(55)	65.0:35.0
2,4,6-i-Pr$_3$C$_6$H$_2$CH$_2$	Ti1	(89)	77.0:23.0
2,4,6-i-Pr$_3$C$_6$H$_2$CH$_2$	Ti2	(89)	94.5:5.5

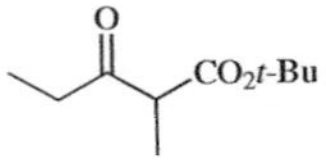

Pd12 (2.5 mol %), NFSI (1.5 eq), EtOH, 20°, 72 h — (47) er 84.5:15.5 — 75

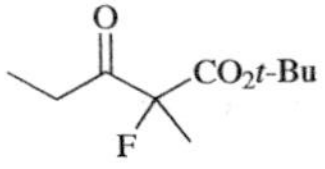

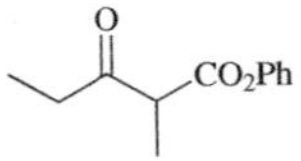

[RuCl$_2$(**L23**)] (10 mol %), NFSI (1.08 eq), (Et$_3$O)PF$_6$ (0.21 eq), DCM, rt, 24 h — (34) er 76.5:23.5 — 89

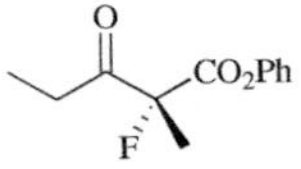

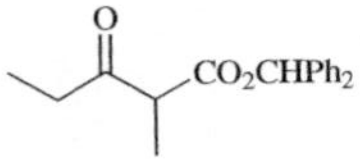

Ni(ClO$_4$)$_2$•6H$_2$O (10 mol %), **L14** (11 mol %), NFSI (1.2 eq), 4 Å MS, CH$_2$Cl$_2$, 18 h — (75) er 91.5:8.5 — 93

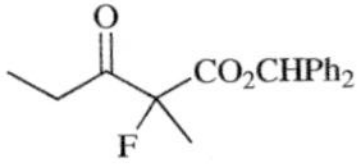

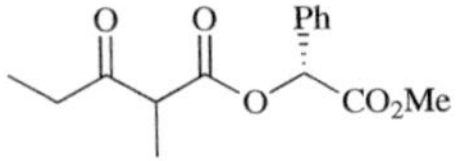

Catalyst (5 mol %), Selectfluor (1.15 eq), MeCN, rt, 1 h — (59) — 82

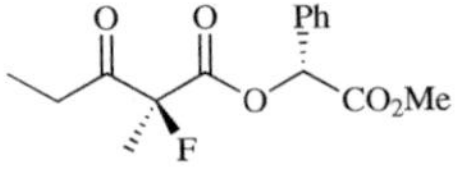

Catalyst	dr
CpTiCl$_3$	66:34
Ti1	88:12
Ti2	96.5:3.5

β-Keto Ester	Conditions	Product(s) and Yield(s) (%)	Refs.

*Please refer to the charts preceding the tables for the structures indicated by the **bold** numbers.*

C_6

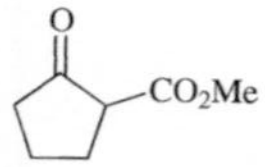

Cu(ClO₄)₂·6H₂O (5 mol %),
L24 (7.5 mol %), NFSI (1.1 eq),
HFIP (10 mol %), toluene, 0°, 6 h

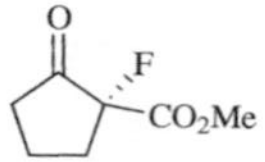

(96) er 86.0:14.0

102

C_{6-8}

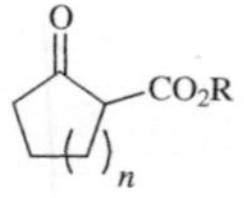

Co(acac)₂ (10 mol %),
L15 (10 mol %),
NFSI (1.4 eq), Et₂O, 12 h

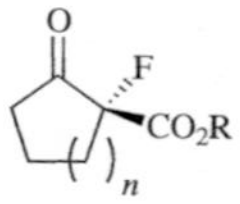

n	R	Temp (°)		er
1	Me	−20	(74)	95.0:5.0
1	t-Bu	−20	(65)	93.0:7.0
2	Me	0	(65)	89.5:10.5
2	Et	0	(65)	87.5:12.5
3	Me	0	(75)	89.5:10.5

88

C_6

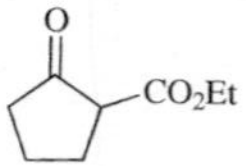

[RuCl₂(**L23**)] (10 mol %),
NFSI (1.08 eq),
(Et₃O)PF₆ (0.21 eq), solvent

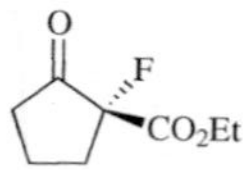

Solvent	Temp	Time (h)		er
DCM	rt	24	(82)	79.0:21.0
DCM/Et₂O (1:1)	—	4	(84)	82.5:17.5

89

10% F_2/N_2 (3.0 eq),
Ni(NO$_3$)$_2$·6H$_2$O (10 mol %),
rac-BINAP (10 mol %), CH$_3$CN, 0°

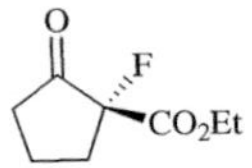

(81)   90

Co(ClO$_4$)$_2$·6H$_2$O (10 mol %),
ligand (10 mol %),
NFSI (1.4 eq), THF, rt, 12 h

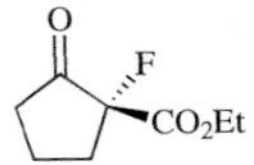

Ligand		er
L15	(50)	54.0:46.0
L33	(55)	62.5:37.5

88

Co(acac)$_2$ (10 mol %),
ligand (10 mol %),
NFSI (1.4 eq), solvent, 12 h

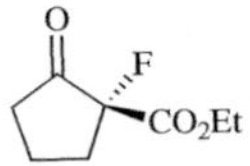

88

Ligand	Solvent	Temp (°)		er
L15	Et$_2$O	–20	(84)	94.5:5.5
L15	Et$_2$O	0	(68)	92.5:7.5
L15	Et$_2$O	rt	(41)	86.5:13.5
L15	THF	rt	(60)	80.0:20.0
L33	THF	rt	(86)	67.0:33.0

NiCl$_2$ (10 mol %), **L25** (12 mol %),
NFSI (1.1 eq), CH$_2$Cl$_2$, rt, 3 h

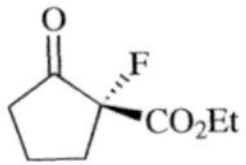

(66) er 75.5:24.5   98

TABLE 7A. FLUORINATION OF β-KETO ESTERS (*Continued*)

β-Keto Ester	Conditions	Product(s) and Yield(s) (%)	Refs.

*Please refer to the charts preceding the tables for the structures indicated by the **bold** numbers.*

C₆

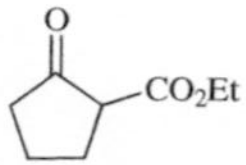

L25 (12 mol %), NFSI (1.1 eq),
CH₂Cl₂, rt, 15 h

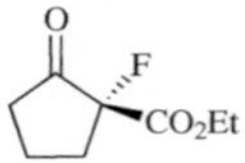

(96) er 71.5:28.5

98

L34 (7.5 mol %), NFSI (1.1 eq),
additive (5 mol %), DCM

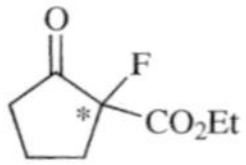

102

Additive	Temp (°)	Time (h)		er	*
Cu(OTf)₂	0	10	(89)	63.0:37.0	(*S*)
Cu(OTf)₂	25	4	(97)	64.0:36.0	(*S*)
Cu(ClO₄)₂•6H₂O	25	3	(91)	67.0:33.0	(*S*)
Zn(OTf)₂	25	6	(73)	61.0:39.0	(*R*)

Cu(ClO₄)₂•6H₂O (5 mol %),
L24 (7.5 mol %), NFSI (1.1 eq),
additive (x mol %), toluene

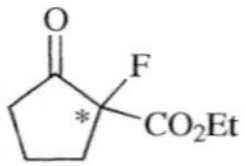

102

Additive	x	Temp (°)	Time (h)		er	*
4 Å MS	—	25	3	(93)	91.0:9.0	(*R*)
NaBAr_F	10	25	3	(78)	65.5:34.5	(*S*)
HFIP	10	25	4	(94)	90.0:10.0	(*S*)
HFIP	10	0	8	(93)	93.0:7.0	(*R*)
HFIP	10	−20	12	(88)	93.0:7.0	(*R*)

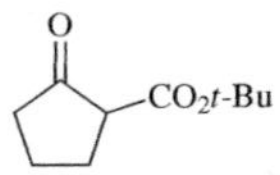

Ni(ClO$_4$)$_2$•6H$_2$O (10 mol %),
L14 (11 mol %), NFSI (1.2 eq),
4 Å MS, CH$_2$Cl$_2$, rt, 2 h

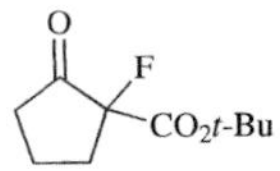

93

L25 (11 mol %), NFSI (1.1 eq),
additive (10 mol %), CH$_2$Cl$_2$, rt

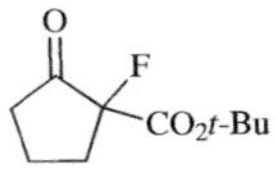

98

Additive	Time (h)		er
none	15	(48)	80.5:19.5
NiCl$_2$	3	(87)	81.5:18.5

Cu(OTf)$_2$-(R)-Ph-BOX (1 mol %),
NFSI (1.5 eq), HFIP (1 eq),
Et$_2$O, 20°, 0.5 h

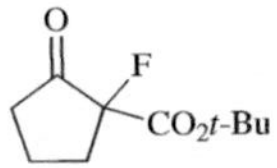

(96) er 92.5:7.5

99

Ni(ClO$_4$)$_2$ (10 mol %), L32 (5 mol %),
NFSI (1.1 eq), 4 Å MS, CH$_2$Cl$_2$, rt

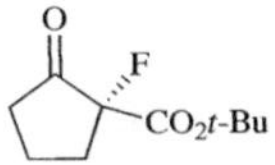

94

Time (h)		er
19	(93)	82.0:18.0
24	(90)[a]	96.0:4.0
24	(65)[b]	67.5:32.5

TABLE 7A. FLUORINATION OF β-KETO ESTERS (*Continued*)

β-Keto Ester	Conditions	Product(s) and Yield(s) (%)	Refs.

*Please refer to the charts preceding the tables for the structures indicated by the **bold** numbers.*

C$_6$

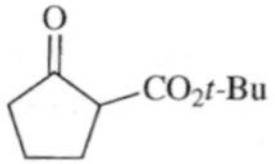

[RuCl$_2$(**L23**)] (x mol %), NFSI (1.08 eq), additive (0.21 eq), solvent

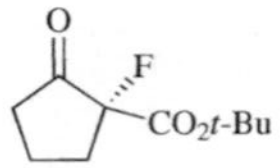

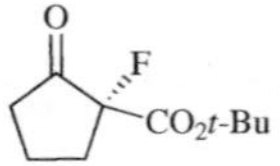

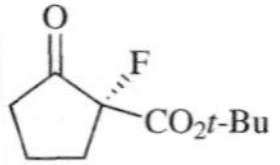

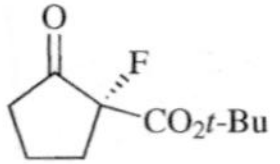

x	Additive	Solvent	Temp	Time (h)		er
2	—	DCM/Et$_2$O (1:1)	—	4	(96)	94.5:5.5
10	(Et$_3$O)PF$_6$	DCM	rt	24	(91)	94.0:6.0
10	(Et$_3$O)PF$_6$	DCM/Et$_2$O (1:1)	—	4	(94)	96.5:3.5

89

Cu(ClO$_4$)$_2$•6H$_2$O (5 mol %), **L24** (7.5 mol %), NFSI (1.1 eq), HFIP (10 mol %), toluene, 0°, 10 h

(97) er 91.5:8.5

102

10% F$_2$/N$_2$ (3.0 eq). Ni(NO$_3$)$_2$•6H$_2$O (10 mol %), *rac*-BINAP (10 mol %), MeCN, 0°

(88)

90

Fe1 (2 mol %), AgClO$_4$ (2 mol %), NFSI (1.2 eq), 4 Å MS, MeCN, 0°, 10 h

(88) er 97.5:2.5

83

Pd5 (2.5 mol %), NFSI (1.5 eq), solvent

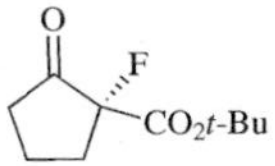

(cyclopentanone with F, CO$_2$t-Bu)

Solvent	Temp (°)	Time (h)		er
THF	10	48	(83)	96.0:4.0
i-PrOH	20	24	(90)	96.0:4.0

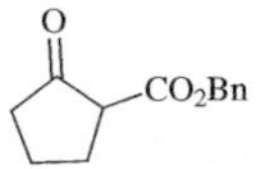

100

Cu(OTf)$_2$ (10 mol %), ligand (12 mol %), NFSI (1.5 eq), DCM, rt

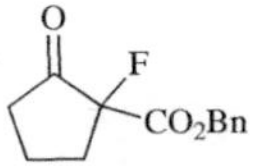

(cyclopentanone with F, CO$_2$Bn)

Ligand		er
L26	(77)	63.5:26.5
L27	(57)	57.0:43.0
L28	(68)	52.5:47.5
L29	(82)	52.5:47.5
L30	(59)	52.5:47.5
L31	(64)	52.5:47.5

82

Catalyst (5 mol %), F-TEDA (1.15 eq), MeCN, rt, 1 h

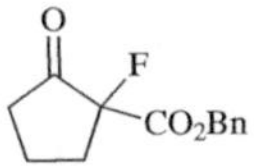

(cyclopentanone with F, CO$_2$Bn)

Catalyst		er
Ti1	(74)	65.5:34.5
Ti2	(74)	78.5:21.5

69

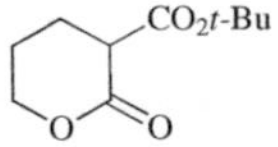

1. **Pd5** (5 mol %), NFSI (1.5 eq), t-BuOH, rt, 27 h
2. BnNH$_2$ (2 eq), rt, 4 h

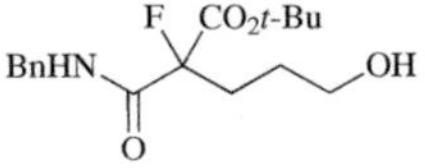

(BnHN, F, CO$_2$t-Bu, OH structure)

(35) er 98.5:0.5

(cyclopentanone with CO$_2$Bn)

(lactone with CO$_2$t-Bu)

TABLE 7A. FLUORINATION OF β-KETO ESTERS (Continued)

β-Keto Ester	Conditions	Product(s) and Yield(s) (%)	Refs.

*Please refer to the charts preceding the tables for the structures indicated by the **bold** numbers.*

C₇

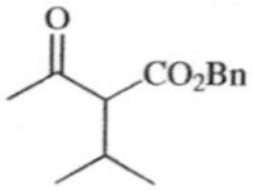	**T2** (5 mol %), Selectfluor (1.15 eq), MeCN, rt, 1 h	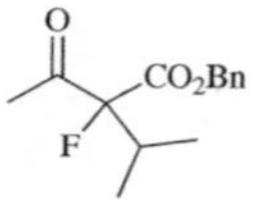 (40) er 62.0:38.0	66
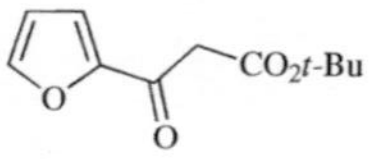	1. NCS (1.1 eq), rt, 2 h 2. **L18** (36 mol %), Cu(OTf)₂ (30 mol %), NFSI (3 eq), 4 Å MS, benzene, 40°, 24 h	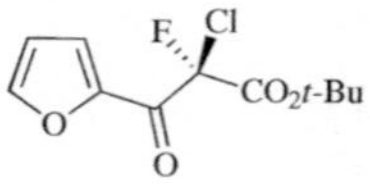 (73) er 91.0:9.0	101
	Cu(ClO₄)₂•6H₂O (5 mol %), **L24** (7.5 mol %), NFSI (1.1 eq). HFIP (10 mol %), toluene, 0°, 8 h	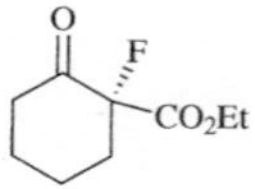 (90) er 76.0:24.0	102
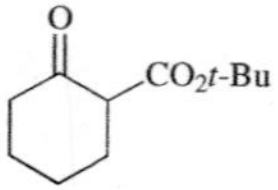	Cu(OTf)₂–(R)-Ph-BOX (1 mol %), NFSI (1.5 eq). HFIP (1.0 eq). Et₂O, 20°, 3 h	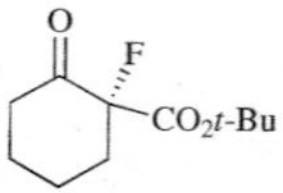 (92) er 81.5:18.5	99
	Ni(ClO₄)₂•6H₂O (10 mol %), **L14** (11 mol %), NFSI (1.2 eq). 4 Å MS, DCM, 2 h	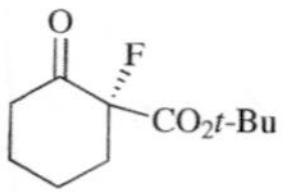 (86) er 99.5:0.5	93

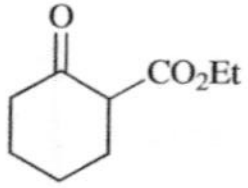

836

C9

			Temp (°)	er		
[RuCl$_2$(**L23**)] (10 mol %), NFSI (1.08 eq), (Et$_3$O)PF$_6$ (0.21 eq), DCM, rt, 24 h					(61) er 82.5:17.5	89

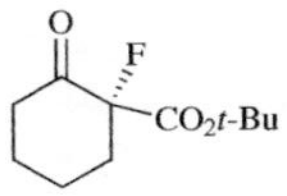

Pd14 (x mol %), NFSI (1.5 eq), EtOH, 20 h

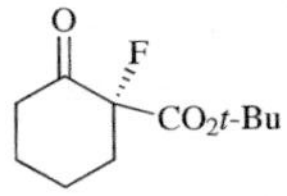

x	Temp (°)	er		
1	0	95.5:4.5	(82)	75
2.5	−10	97.0:3.0	(91)	70, 75

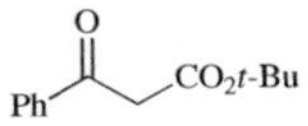

1. NCS (1.1 eq), rt, 2 h
2. **L18** (12 mol %), Cu(OTf)$_2$ (10 mol %), NFSI (3 eq), 4 Å MS, benzene, 80°, 4 h

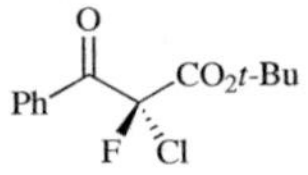

(67) er 96.0:4.0 101

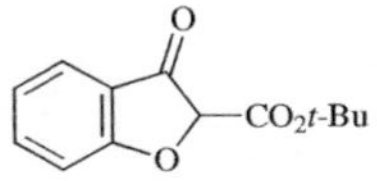

Fe1 (2 mol %), AgClO$_4$ (2 mol %), NFSI (1.2 eq), 4 Å MS, MeCN, 0°, 10 h

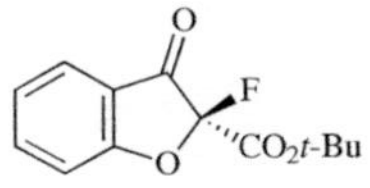

(91) er 91.5:8.5 83

β-Keto Ester	Conditions	Product(s) and Yield(s) (%)	Refs.

*Please refer to the charts preceding the tables for the structures indicated by the **bold** numbers.*

C$_{10}$

Ph–C(=O)–CH(CH$_3$)–CO$_2$R

Catalyst (5 mol %), Selectfluor (1.15 eq),
MeCN, rt, 4–16 h

Ph–C(=O)–C(F)(CH$_3$)–CO$_2$R

82

R	Catalyst		er
Me	**T1**	(76)	67.5:32.5
Me	**T2**	(76)	83.5:16.5
Et	**T1**	(71)	64.0:36.0
Et	**T2**	(71)	81.0:19.0
EtMe$_2$C	**T1**	(69)	71.0:29.0
EtMe$_2$C	**T2**	(69)	86.5:13.5
Ph	**T1**	(53)	—
Ph	**T2**	(53)	83.5:16.5
4-MeOC$_6$H$_4$	**T1**	(34)	58.5:41.5
4-MeOC$_6$H$_4$	**T2**	(34)	82.5:17.5
Bn	**T1**	(85)	69.0:21.0
Bn	**T2**	(85)	85.0:15.0
C$_6$F$_5$CH$_2$	**T1**	(44)	78.5:21.5
C$_6$F$_5$CH$_2$	**T2**	(44)	81.0:19.0
4-CF$_3$-C$_6$H$_4$CH$_2$	**T1**	(—)	77.5:22.5
4-CF$_3$-C$_6$H$_4$CH$_2$	**T2**	(—)	84.5:15.5
Ph$_2$CH	**T1**	(59)	79.5:20.5
Ph$_2$CH	**T2**	(59)	91.0:9.0

Ph–C(=O)–CH(CH$_3$)–CO$_2$Et

[RuCl$_2$(**L23**)] (10 mol %),
NFSI (1.08 eq), (Et$_3$O)PF$_6$ (0.21 eq),
DCM, rt, 24 h

Ph–C(=O)–C(F)(CH$_3$)–CO$_2$Et (65) er 62.0:38.0

89

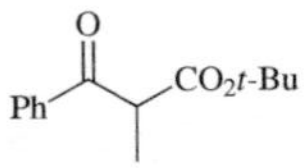

Catalyst (2.5 mol %),
NFSI (1.5 eq), solvent, 20°

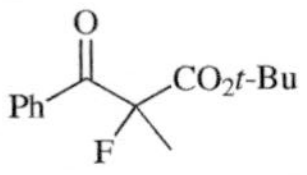

Catalyst	Solvent	Time (h)		er
Pd12	H_2O	75	(76)	94.5:5.5
Pd12	acetone/H_2O (9:1)	96	(73)	94.0:6.0
Pd12	EtOH	40	(54)	95.5:4.5
Pd12	EtOH/H_2O (1:10)	66	(57)	95.5:4.5
Pd12	EtOH/H_2O (4:1)	120	(53)	92.5:7.5
Pd15	EtOH	40	(92)	95.5:4.5

70

Cu(OTf)$_2$-(R)-Ph-BOX (1 mol %),
NFSI (1.5 eq), HFIP (1.0 eq),
Et$_2$O, 20°, 96 h

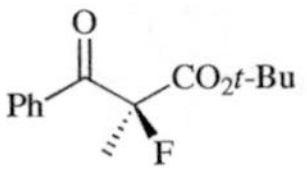

(56) er 71.5:28.5

99

Fe1 (2 mol %), AgClO$_4$ (2 mol %),
NFSI (1.2 eq), 4 Å MS, MeCN, 0°, 10 h

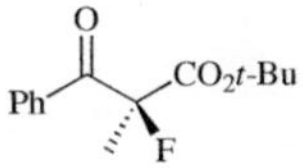

(87) er 97.0:3.0

83

839

TABLE 7A. FLUORINATION OF β-KETO ESTERS (Continued)

*Please refer to the charts preceding the tables for the structures indicated by the **bold** numbers.*

β-Keto Ester	Conditions	Product(s) and Yield(s) (%)	Refs.
C$_{10}$			
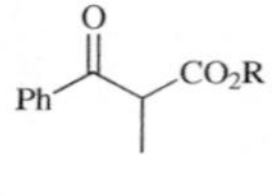	Cu(OTf)$_2$-(R)-Ph-BOX (1 mol %), NFSI (1.5 eq), HFIP (1.0 eq), Et$_2$O, 20°	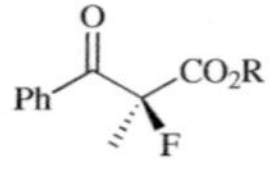	99

R	Time (h)		er
1-NpCH$_2$	2	(88)	70.0:30.0
Ph$_2$CH	48	(72)	76.0:24.0

β-Keto Ester	Conditions	Product(s) and Yield(s) (%)	Refs.
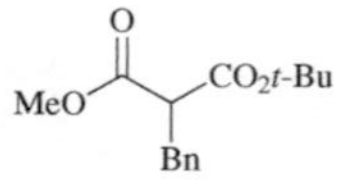	**Fe1** (2 mol %), AgClO$_4$ (2 mol %), NFSI (1.2 eq), 4 Å MS, MeCN, 0°, 10 h	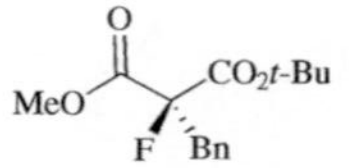 (72) er 75.5:24.5	83
	Ti1 (2 mol %), AgClO$_4$ (2 mol %), NFSI (1.2 eq), 4 Å MS, MeCN, 0°, 10 h	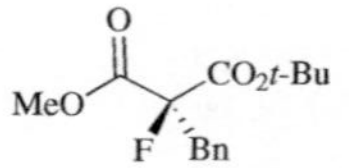 (72) er 75.5:24.5	83
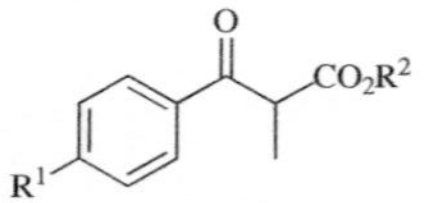	Catalyst (5 mol %), Selectfluor (1.15 eq), MeCN, rt, 4–16 h	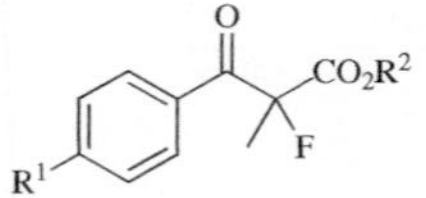	82

R^1	R^2	Catalyst		er
MeO	Me	**Ti1**	(94)	63.0:37.0
MeO	Me	**Ti2**	(94)	80.5:19.5
MeO	Et	**Ti1**	(73)	64.0:36.0
MeO	Et	**Ti2**	(73)	78.0:22.0
O$_2$N	Et	**Ti1**	(76)	60.5:39.5
O$_2$N	Et	**Ti2**	(76)	77.5:22.5

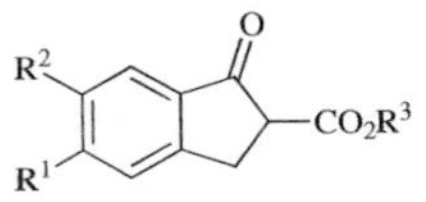

Fe1 (2 mol %), AgClO$_4$ (2 mol %),
NFSI (1.2 eq), 4 Å MS, MeCN, 0°, 10 h

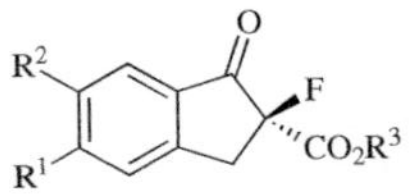

R^1	R^2	R^3		er
H	H	Me	(99)	73.0:27.0
H	H	Et	(99)	79.5:20.5
H	H	i-Pr	(98)	97.0:3.0
H	H	t-Bu	(96)	97.5:2.5
H	H	Ad	(99)	96.0:4.0
H	F	t-Bu	(95)	97.5:2.5
H	Cl	t-Bu	(98)	97.0:3.0
H	MeO	t-Bu	(99)	97.5:2.5
Cl	H	t-Bu	(93)	97.0:3.0
MeO	MeO	t-Bu	(99)	96.0:4.0

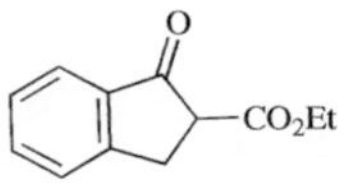

L25 (12 mol %), additive (10 mol %),
NFSI (1.1 eq), CH$_2$Cl$_2$, rt

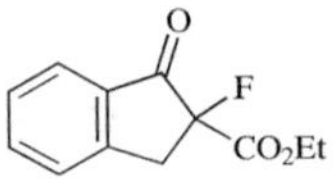

Additive	Time (h)		er
none	15	(<5)	—
NiCl$_2$	3	(100)	54.5:45.5

TABLE 7A. FLUORINATION OF β-KETO ESTERS (*Continued*)

β-Keto Ester	Conditions	Product(s) and Yield(s) (%)	Refs.

*Please refer to the charts preceding the tables for the structures indicated by the **bold** numbers.*

C$_{10}$

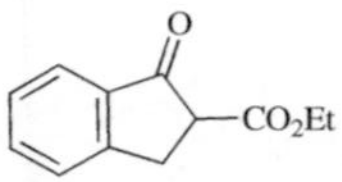

Cu(ClO$_4$)$_2$•6H$_2$O (5 mol %),
L24 (7.5 mol %), NFSI (1.1 eq),
HFIP (10 mol %), toluene, 0°, 6 h

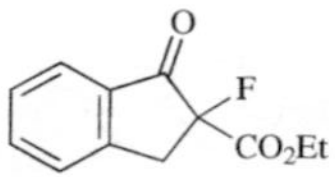

(98) er 67.0:33.0 — 102

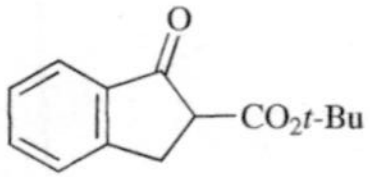

Pd12 (2.5 mol %), NFSI (1.5 eq).
EtOH, –20°, 36 h

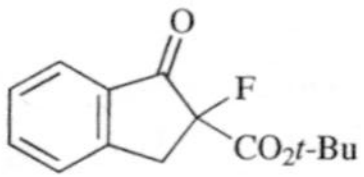

(85) er 91.5:8.5 — 75

L32 (5 mol %), additive (5 mol %),
NFSI (1.1 eq). 4 Å MS, CH$_2$Cl$_2$, rt, 2 h

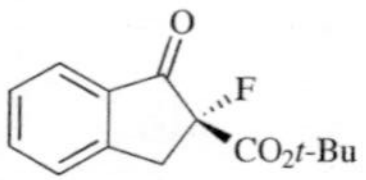

— 94

Additive		er
Ni(ClO$_4$)$_2$	(99)	97.0:3.0
Ni(ClO$_4$)$_2$	(99)^a	99.5:0.5
Ni(ClO$_4$)$_2$	(94)^b	72.0:28.0
Mg(ClO$_4$)$_2$	(99)	95.5:4.5
Mg(ClO$_4$)$_2$	(99)^a	99.5:0.5
Mg(ClO$_4$)$_2$	(86)^b	63.5:36.5

L25 (12 mol %), additive (10 mol %),
NFSI (1.1 eq), CH$_2$Cl$_2$, rt, 3 h

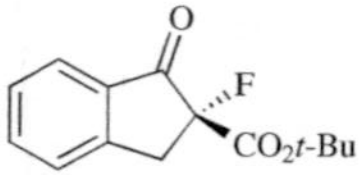

— 98

Additive	Time (h)		er
none	15	(<5)	—
NiCl$_2$	3	(100)	63.5:36.5

C_{10-11}

C_{10}

Ni(ClO$_4$)$_2$·6H$_2$O (2 mol %),
L14 (2 mol %), NFSI (1.2 eq),
4 Å MS, CH$_2$Cl$_2$, 6 h

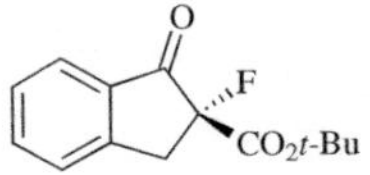

(93) er 99.5:0.5   93

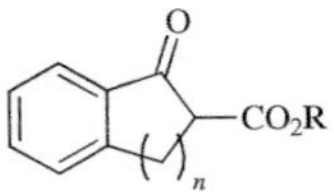

Ni(ClO$_4$)$_2$·6H$_2$O (10 mol %),
L14 (11 mol %), NFSI (1.2 eq),
4 Å MS, DCM

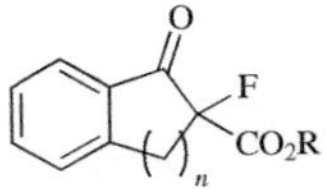

93

n	R	Time (h)		er
1	t-Bu	3	(76)	99.5:0.5
1	Ad	2	(71)	99.5:0.5
1	L-menthol	2	(66)	99.5:0.5
2	Ad	3	(88)	97.5:2.5

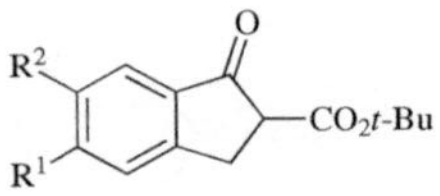

Ni(ClO$_4$)$_2$·6H$_2$O (10 mol %),
L17 (12 mol %), NFSI (1.2 eq),
4 Å MS, Et$_2$O, rt, 6 h

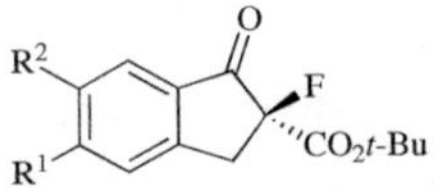

97

R^1	R^2		er
H	H	(92)	98.5:1.5
Br	H	(93)	96.0:4.0
MeO	H	(93)	98.5:1.5
MeO	MeO	(92)	97.5:2.5

TABLE 7A. FLUORINATION OF β-KETO ESTERS (*Continued*)

*Please refer to the charts preceding the tables for the structures indicated by the **bold** numbers.*

β-Keto Ester	Conditions	Product(s) and Yield(s) (%)	Refs.
C_{10}			
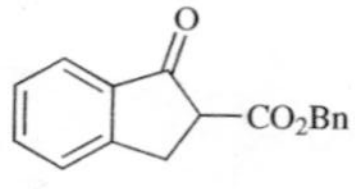	**L32** (5 mol %), $Ni(ClO_4)_2$ (5 mol %), NFSI (1.1 eq), 4 Å MS, DCM, rt, 2 h	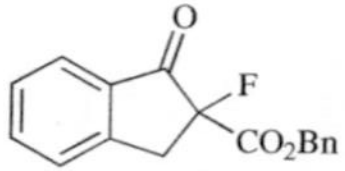 (99)[a] er 95.5:4.5	94
	$Cu(OTf)_2$-(*R*)-Ph-BOX (1 mol %), NFSI (1.5 eq), HFIP (1 eq), Et_2O, 20°, 0.5 h	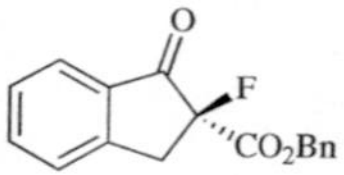 (94) er 67.5:32.5	99
C_{11}			
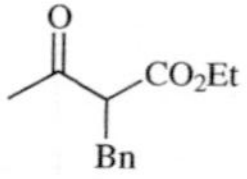	$[RuCl_2($**L23**$)]$ (10 mol %), NFSI (1.08 eq), $(Et_3O)PF_6$ (0.21 eq), DCM, rt, 24 h	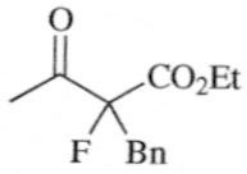 (79) er 86.5:13.5	89
	Ti2 (5 mol %), Selectfluor (1.15 eq), MeCN, rt, 4–16 h	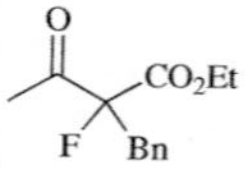 (83) er 53.0:47.0	82
	$Cu(OTf)_2$ (10 mol %), **L26** (12 mol %), NFSI (1.5 eq), DCM, rt	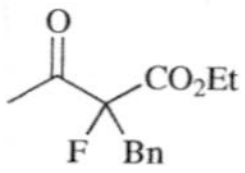 (75) er 54.5:45.5	100
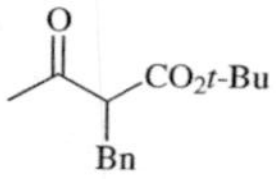	**Fe1** (2 mol %), $AgClO_4$ (2 mol %), NFSI (1.2 eq), 4 Å MS, MeCN, 0°, 10 h	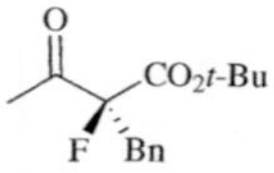 (96) er 93.5:6.5	83

Substrate	Conditions	Product
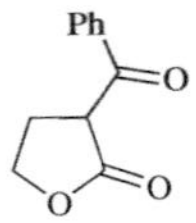	[RuCl$_2$(**L23**)] (10 mol %), NFSI (1.08 eq), (Et$_3$O)PF$_6$ (0.21 eq), DCM, rt, 24 h	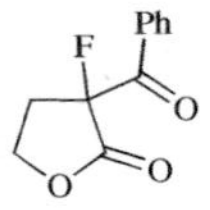 (60) er 85.5:14.5 89
	Tf2 (5 mol %), Selectfluor (1.15 eq), MeCN, rt	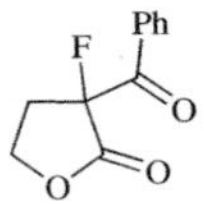 (86) er 64.5:35.5 82
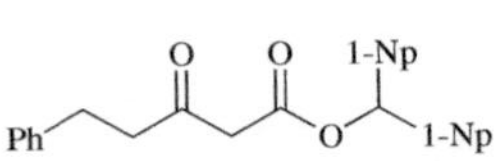	1. NCS (1.1 eq), rt, 2 h 2. **L18** (12 mol %), Cu(OTf)$_2$ (10 mol %), NFSI (3 eq), 4 Å MS, benzene, 40°, 19 h	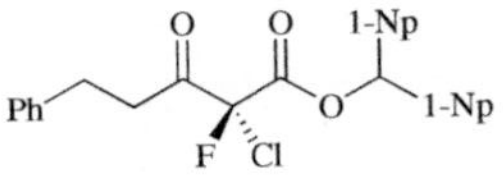 (63) er 89.5:10.5 101
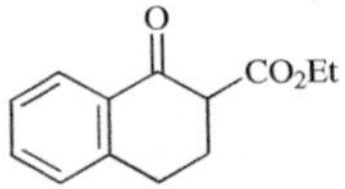	Cu(ClO$_4$)$_2$•6H$_2$O (5 mol %), **L24** (7.5 mol %), NFSI (1.1 eq), HFIP (10 mol %), toluene, 0°, 6 h	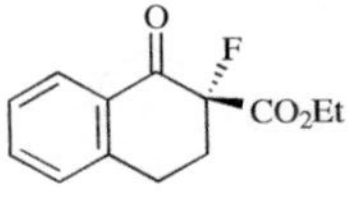 (90) er 58.0:42.0 100, 102

TABLE 7A. FLUORINATION OF β-KETO ESTERS (Continued)

β-Keto Ester	Conditions	Product(s) and Yield(s) (%)	Refs.

*Please refer to the charts preceding the tables for the structures indicated by the **bold** numbers.*

C₁₁

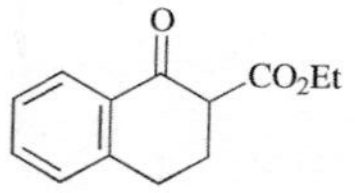

Tf2 (5 mol %), Selectfluor (1.15 eq), MeCN, rt, 4–16 h

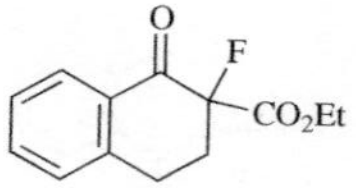

(93) er 60.0:40.0

82

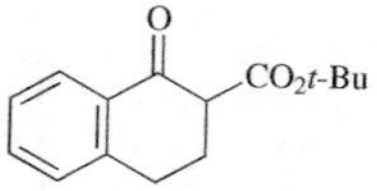

L18 (5 mol %), Ni(ClO₄)₂ (5 mol %), NFSI (1.1 eq), 4 Å MS, CH₂Cl₂, rt

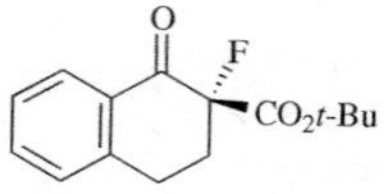

94

Time (h)		er
4	(99)	67.0:33.0
12	(99)^a	96.0:4.0
12	(75)^b	56.0:44.0

Cu(ClO₄)₂•6H₂O (5 mol %), **L24** (7.5 mol %), NFSI (1.1 eq), HFIP (10 mol %), toluene, 0°, 6 h

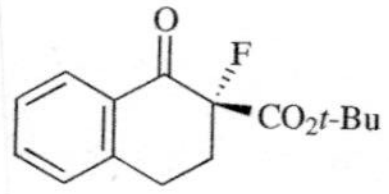

(90) er 58.0:42.0

102

Ni(ClO₄)₂•6H₂O (10 mol %), **L17** (12 mol %), NFSI (1.2 eq), 4 Å MS, Et₂O, rt, 6 h

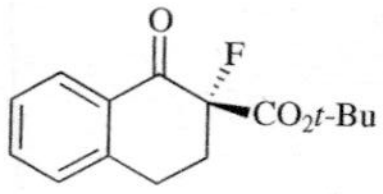

(92) 96.5:3.5

97

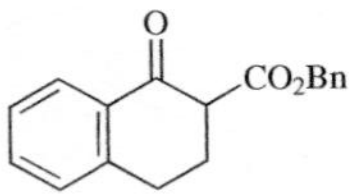

C₁₂

Fe1 (2 mol %), AgClO₄ (2 mol %),
NFSI (1.2 eq), 4 Å MS, MeCN, 0°, 10 h

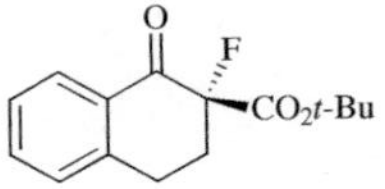

(96) er 84.5:15.5   83

Cu(OTf)₂–(R)-Ph-BOX (1 mol %),
NFSI (1.5 eq), HFIP (1 eq),
Et₂O, 20°, 0.5 h

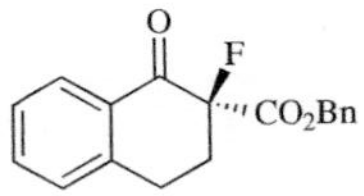

(92) er 69.0:31.0   99

Cu(OTf)₂ (10 mol %), ligand (12 mol %),
NFSI (1.5 eq), DCM, rt, 12 h

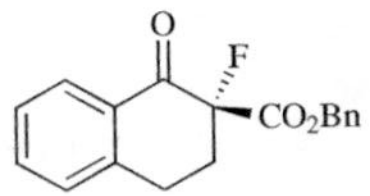

100

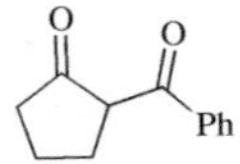

Ligand		er
L26	(54)	59.0:41.0
L27	(62)	54.5:45.5
L28	(47)	52.5:47.5
L29	(70)	52.5:47.5
L30	(67)	52.5:47.5

Tf2 (5 mol %), Selectfluor (1.15 eq),
MeCN, rt, 1 h

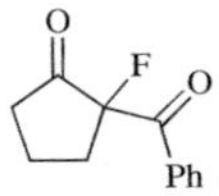

(90) er 50.0:50.0   82

TABLE 7A. FLUORINATION OF β-KETO ESTERS (Continued)

β-Keto Ester	Conditions	Product(s) and Yield(s) (%)	Refs.

*Please refer to the charts preceding the tables for the structures indicated by the **bold** numbers.*

C$_{12}$

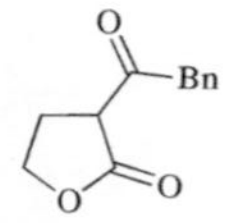

T2 (5 mol %), Selectfluor (1.15 eq),
MeCN, rt, 1 h

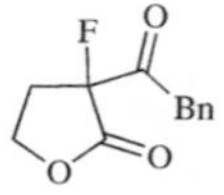

(63) er 75.5:24.5

82

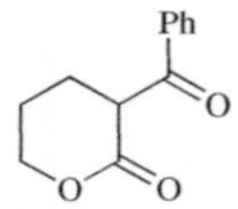

T2 (5 mol %),
additive (1.15 eq), MeCN, rt

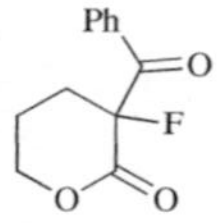

Additive	Time (h)		er
Selectfluor	1	(70)	58.0:42.0
NFSI	24	(99)	62.0:38.0
NFPy•BF$_4$	120	(75)	69.5:30.5

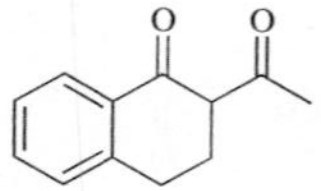

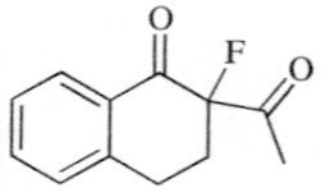

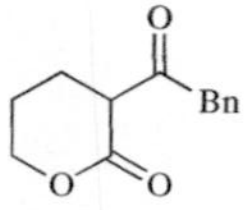

82

C$_{13}$

T2 (5 mol %), Selectfluor (1.15 eq),
MeCN, rt, 1 h

(67) er 50.0:50.0

82

T2 (5 mol %),
additive (x eq), MeCN, rt

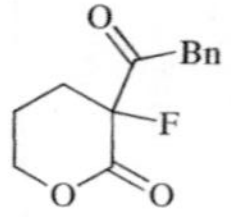

Additive	x	Time (h)		er
Selectfluor	1.15	1	(85)	56.5:43.5
NFPy•BF$_4$	—	120	(75)	69.5:30.5

82

C$_{14}$

(structure: 2-naphthyl C(=O)CH$_2$CO$_2$t-Bu)

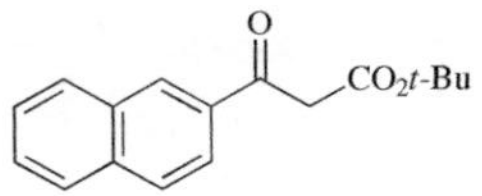

1. NCS (1.1 eq), rt, 2 h
2. **L18** (12 mol %), Cu(OTf)$_2$ (10 mol %), NFSI (3 eq), 4 Å MS, benzene, 40°, 6 h

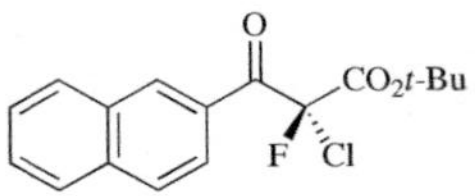

(product: 2-naphthyl C(=O)C(F)(Cl)CO$_2$t-Bu)

(71) er 96.0:4.0   101

(structure: 2-naphthyl C(=O)CH(CH$_3$)CO$_2$Et)

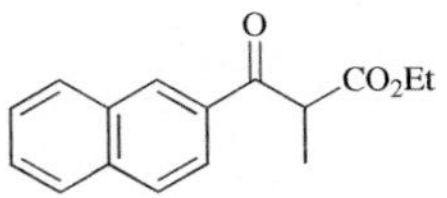

Catalyst (5 mol %),
Selectfluor (1.15 eq), MeCN, rt, 1 h

(product: 2-naphthyl C(=O)C(CH$_3$)(F)CO$_2$Et)

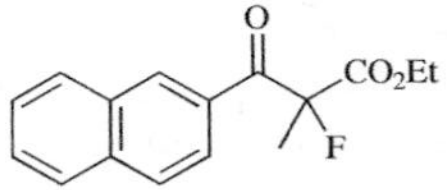

(89)

Catalyst	er
Ti1	67.0:33.0
Ti2	80.5:19.5

82

C$_{18}$

(structure: lactone with PhCH(Ph)C(=O)— substituent on γ-butyrolactone)

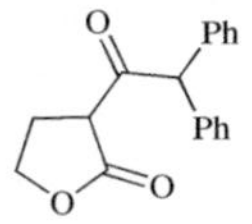

Ti2 (5 mol %), Selectfluor (1.15 eq),
MeCN, rt, 1 h

(product: fluorinated lactone with PhCH(Ph)C(=O)— substituent)

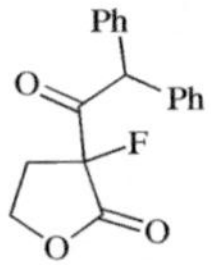

(89) er 52.5:47.5   82

---

[a] NFSI was slowly added to the reaction mixture.
[b] The β-keto ester was slowly added to the reaction mixture.

TABLE 7B. FLUORINATION OF β-KETO PHOSPHONATES

β-Keto Phosphonate	Conditions	Product(s) and Yield(s) (%)	Refs.

*Please refer to the charts preceding the tables for the structures indicated by the **bold** numbers.*

C_4

β-Keto Phosphonate structure with $P(O)(OEt)_2$

Pd8 (10 mol %), NFSI (1.5 eq), EtOH, 40°, 48 h

Product: $P(O)(OEt)_2$ with F — (57) er 97.0:3.0

Ref. 71

C_{5-15}

β-Keto Phosphonate structure R^1, R^2, $P(O)(OEt)_2$

Pd10 (5 mol %), NFSI (1.5 eq), THF, rt

Product: R^1, R^2, F, $P(O)(OEt)_2$

R^1	R^2	Time (h)		er
Et	Me	90	(65)	93.5:6.5
MeCH=CH	Me	58	(79)	96.5:3.5
Ph	Me	94	(62)	95.5:4.5
4-ClC$_6$H$_4$	Me	94	(68)	95.5:4.5
4-MeOC$_6$H$_4$	Me	78	(61)	95.5:4.5
4-O$_2$NC$_6$H$_4$	Me	90	(78)	93.5:6.5
Ph	Bn	86	(50)	95.5:4.5

Ref. 74

C_5

β-Keto Phosphonate cyclopentanone structure with $P(O)(OEt)_2$

Catalyst (5 mol %), NFSI (1.5 eq), solvent, rt

Product: cyclopentanone with F and $P(O)(OEt)_2$

Catalyst	Solvent	Time (h)		er	Refs.
Pd8	EtOH	2	(91)	97.5:2.5	71
Pd10	MeOH	45	(67)	97.5:2.5	74

C_6–12

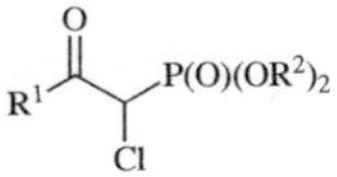

Pd11 (0.5 mol %),
2,6-di-*tert*-butyl-4-methylpyridine (2.0 eq),
NFSI (1.1 eq), MeOH, rt, 1 d

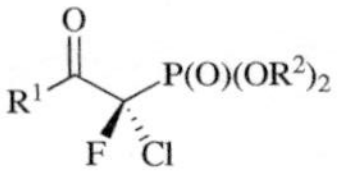

81

R^1	R^2		er
2-thienyl	Et	(40)	92.5:7.5
n-C$_5$H$_{11}$	Et	(56)	96.5:3.5
Ph	Et	(80)	95.5:4.5
Ph	Me	(76)	94.5:5.5
4-FC$_6$H$_4$	Et	(88)	96.5:3.5
4-ClC$_6$H$_4$	Et	(64)	95.5:4.5
4-BrC$_6$H$_4$	Et	(83)	95.5:4.5
4-MeOC$_6$H$_4$	Et	(79)	95.5:4.5
4-O$_2$NC$_6$H$_4$	Et	(78)	91.5:8.5
4-MeC$_6$H$_4$	Et	(75)	97.5:2.5
2-naphthyl	Et	(85)	97.5:2.5

C$_6$

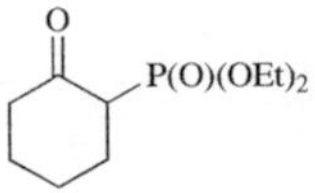

Catalyst (5 mol %), NFSI (1.5 eq), solvent, rt

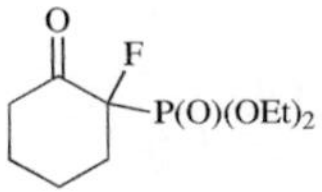

Catalyst	Solvent	Time (h)		er
Pd8	EtOH	8	(93)	98.0:2.0	71
Pd10	MeOH	86	(73)	97.5:2.5	74

TABLE 7B. FLUORINATION OF β-KETO PHOSPHONATES (Continued)

*Please refer to the charts preceding the tables for the structures indicated by the **bold** numbers.*

β-Keto Phosphonate	Conditions	Product(s) and Yield(s) (%)	Refs.
C₆			
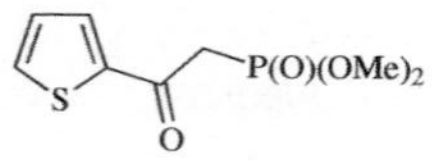	1. NCS (1.1 eq), –30°, 2 h 2. NFSI (3.0 eq), Cu(OTf)₂ (30 mol %), **L18** (36 mol %), 4 Å MS, toluene, 0°, 70 h	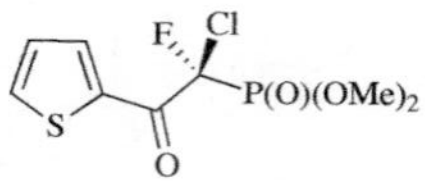 (69) er 93.0:7.0	101
C₈			
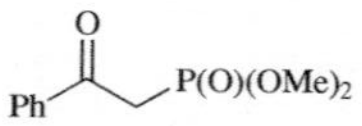	1. NCS (1.1 eq), –30°, 2 h 2. NFSI (3.0 eq), Cu(OTf)₂ (10 mol %), **L18** (12 mol %), 4 Å MS, toluene, 0°, 24 h	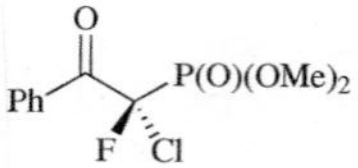 (73) er 96.0:4.0	101
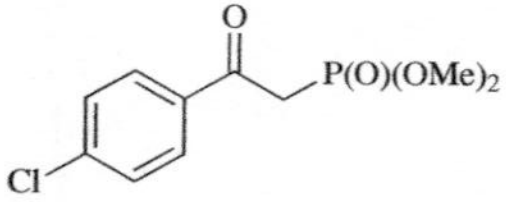	1. NCS (1.1 eq), –30°, 2 h 2. NFSI (3.0 eq), Cu(OTf)₂ (10 mol %), **L18** (12 mol %), 4 Å MS, toluene, 0°, 24 h	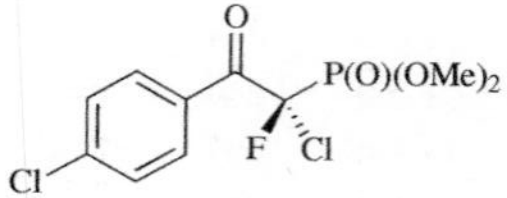 (55) er 95.0:5.0	101
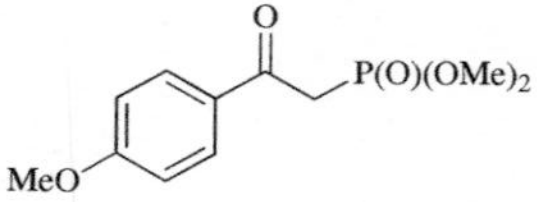	1. NCS (1.1 eq), –30°, 2 h 2. NFSI (3.0 eq), Cu(OTf)₂ (10 mol %), **L18** (12 mol %), 4 Å MS, toluene, 0°, 50 h	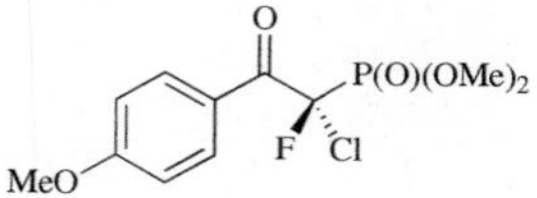 (52) er 96.0:4.0	101
C₉			
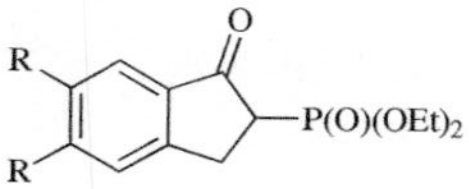	**Pd11** (5 mol %), NFSI (1.5 eq), MeOH, rt 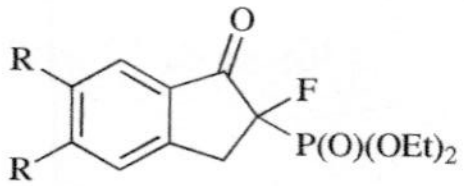

R	Time (h)		er
H	3	(91)	98.5:1.5
MeO	11	(86)	97.5:2.5		74

C_{10}

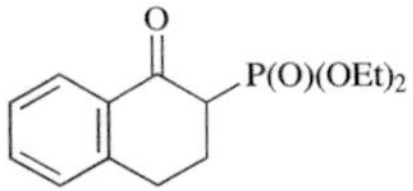

Pd8 (5 mol %), EtOH, rt, 3 h

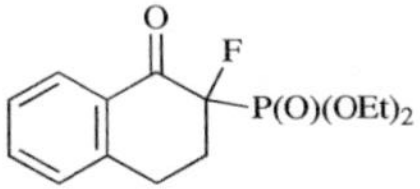

(97) er 97.0:3.0   71

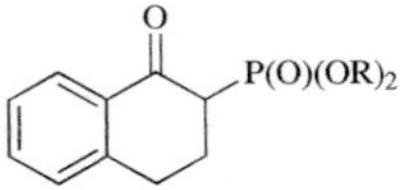

C_{10}

Pd11 (5 mol %), NFSI (1.5 eq), MeOH, rt

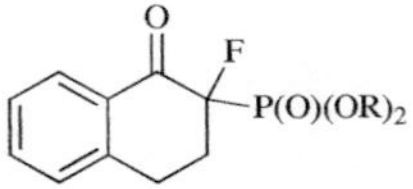

R	Time (h)		er
Et	8	(93)	98.5:1.5
Me	6	(89)	96.5:3.5
i-Pr	23	(91)	97.5:2.5

74

$C_{10–12}$

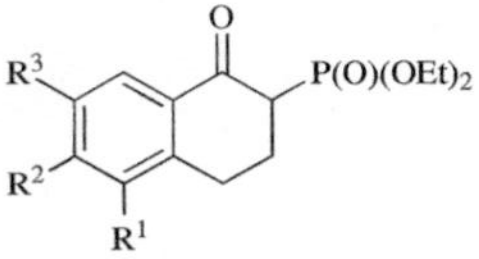

Pd11 (5 mol %), NFSI (1.5 eq), MeOH, rt

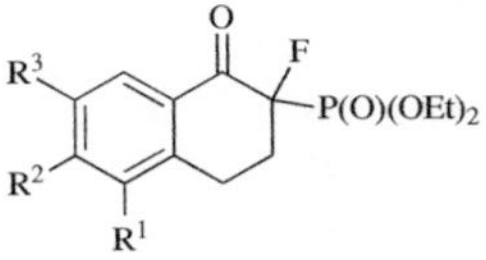

R^1	R^2	R^3	Time (h)		er
H	MeO	H	10	(92)	97.5:2.5
Me	H	Me	12	(84)	97.5:2.5

74

C_{12}

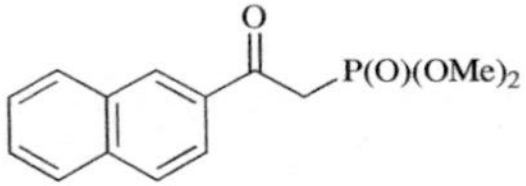

1. NCS (1.1 eq), –30°, 2 h
2. NFSI (3.0 eq), Cu(OTf)$_2$ (10 mol %),
   L18 (12 mol %), 4 Å MS, toluene, 0°, 65 h

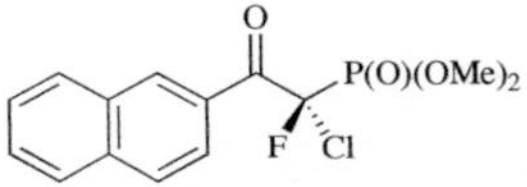

(64) er 95.0:5.0   101

TABLE 7C. FLUORINATION OF β-CYANO ESTERS

β-Cyano Ester	Conditions	Product(s) and Yield(s) (%)	Refs.

Please refer to the charts preceding the tables for the structures indicated by the bold numbers.

C_{9-10}

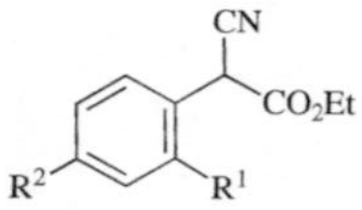

Pd(OAc)$_2$ (1.67 mol %), NFSI (1.1 eq),
L19 (6 mol %), EtOH, 18 h

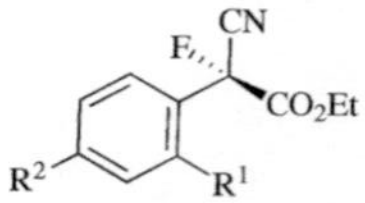

R^1	R^2	Temp		er
H	H	0°	(97)	96.5:3.5
H	H	rt	(97)	89.0:11.0
H	MeO	0°	(89)	89.0:11.0
H	MeO	rt	(81)	85.0:15.0
Cl	H	0°	(35)	54.5:45.5
Cl	H	rt	(55)	52.5:47.5
Br	H	0°	(35)	54.5:45.5
Br	H	rt	(47)	53.0:47.0
H	Me	0°	(92)	85.0:15.0
H	Me	rt	(95)	64.0:36.0
H	EtO$_2$C	0°	(83)	72.5:27.5
H	EtO$_2$C	rt	(88)	61.0:39.0

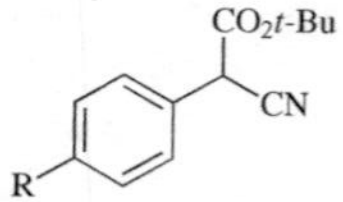

78

[(R)-BINAP]Pd(OH$_2$)(MeCN)PF$_6$ (5 mol %),
NFSI (1.0 eq), MeOH, 0°

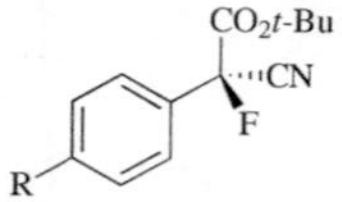

R	Time (h)		er
H	60	(83)	99.5:0.5
Cl	17	(94)	92.5:7.5
MeO	72	(85)	99.5:0.5
Me	60	(85)	96.5:3.5

73

C_{13}

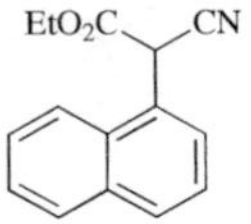

Pd(OAc)$_2$ (1.67 mol %),
L19 (6 mol %), EtOH, 18 h

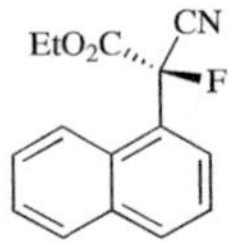

	Temp		er
	0°	(75)	59.0:41.0
	rt	(95)	55.0:45.0

78

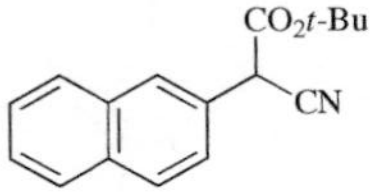

[(R)-BINAP]Pd(OH$_2$)(MeCN)PF$_6$ (5 mol %),
NFSI (1.0 eq), MeOH, 0°, 60 h

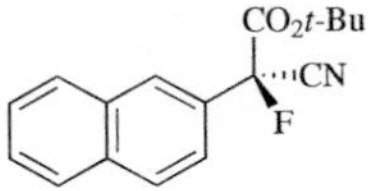

(88) er 96.5:3.5

73

C_{17}

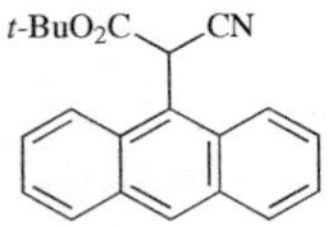

[(R)-BINAP]Pd(OH$_2$)(MeCN)PF$_6$ (5 mol %),
NFSI (1.0 eq), MeOH/THF (1:1), rt, 52 h

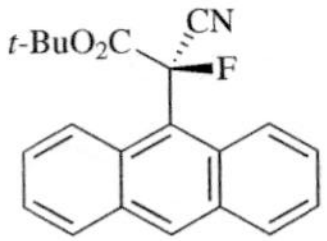

(42) er 95.5:4.5

73

TABLE 7D. α-FLUORINATION OF CARBOXAMIDES

Carboxamide	Conditions	Product(s) and Yield(s) (%)	Refs.

Please refer to the charts preceding the tables for the structures indicated by the **bold** numbers.

C₂₋₈

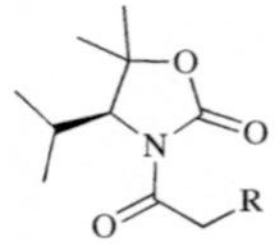

TiCl₄ (1.5 eq), Et₃N (2 eq), NFSI (2 eq), CH₂Cl₂, rt, 1 h

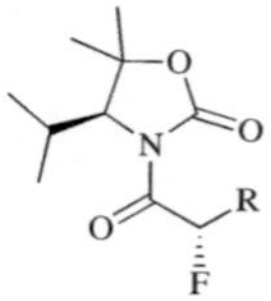

R		dr
Cl	(67)	96:4
Me	(91)	98:2
i-Pr	(77)	98:2
n-Bu	(94)	98:2
Ph	(92)	96:4

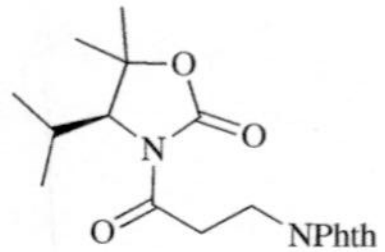

63

C₃

TiCl₄ (1.5 eq), Et₃N (2 eq), NFSI (2 eq), CH₂Cl₂, rt, 1 h

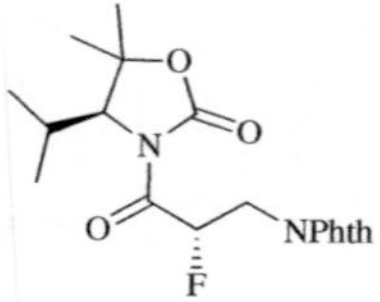

(70) dr 94:6

63

C₄

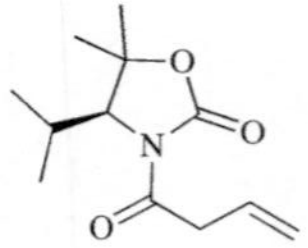

TiCl₄ (1.5 eq), Et₃N (2 eq), NFSI (2 eq), CH₂Cl₂, rt, 1 h

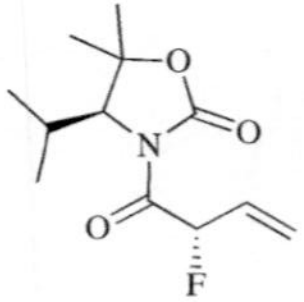

(46) dr 98:2

63

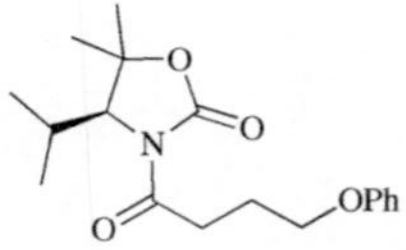

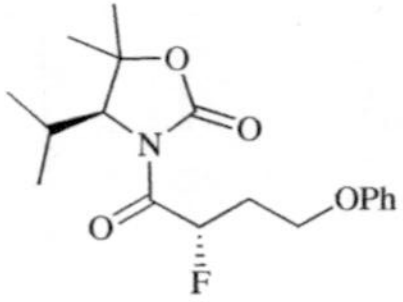

TiCl₄ (1.5 eq), Et₃N (2 eq), NFSI (2 eq), CH₂Cl₂, rt, 1 h

(90) dr 98:2

63

C_5

Substrate	Conditions	Product	Yield
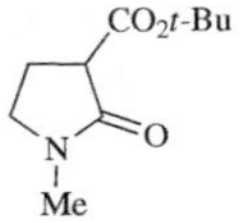 pyrrolidinone, CO₂t-Bu, N–Me	**Pd5** (5 mol %), NFSI (1.5 eq), 2,6-lutidine (0.5 eq), EtOH, rt, 48 h	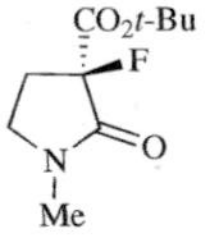 (77) er 99.5:0.5	75
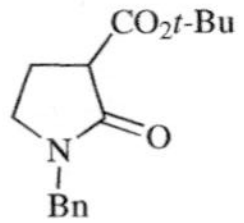 pyrrolidinone, CO₂t-Bu, N–Bn	**Pd5** (5 mol %), NFSI (1.5 eq), 2,6-lutidine (0.5 eq), EtOH, rt, 48 h	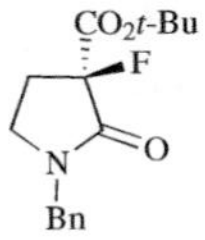 (89) er 99.0:1.0	75
oxazolidinone, OBn	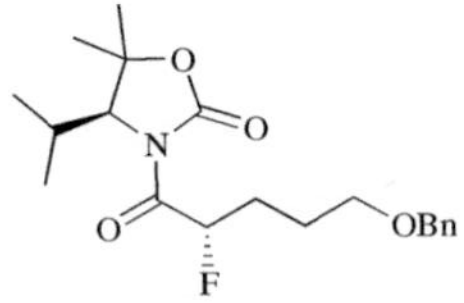 ZrCl₄ (1.5 eq), Et₃N (2 eq), NFSI (2 eq), CH₂Cl₂, rt, 1 h	(68) dr 96:4	63
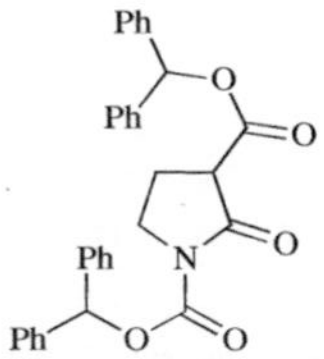 pyrrolidinone, dibenzhydryl ester	**Pd6** (5 mol %), NFSI (1.5 eq), 2,6-lutidine (0.5 eq), EtOH, rt, 48 h	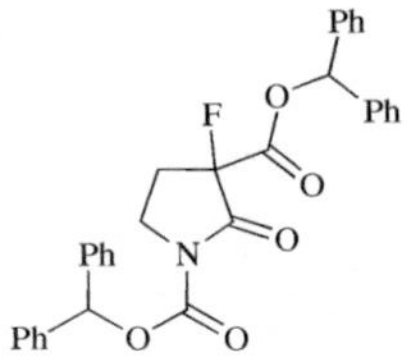 (45) er 99.5:0.5	69

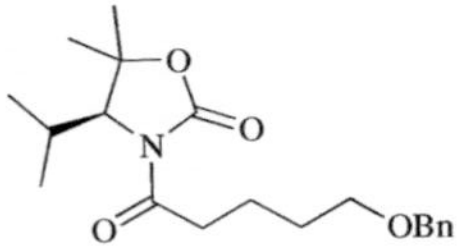

TABLE 7D. α-FLUORINATION OF CARBOXAMIDES (*Continued*)

Carboxamide	Conditions	Product(s) and Yield(s) (%)	Refs.

*Please refer to the charts preceding the tables for the structures indicated by the **bold** numbers.*

C₆

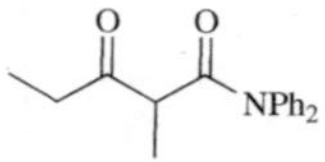

Catalyst (5 mol %),
Selectfluor (1.15 eq),
MeCN, rt, 15 h

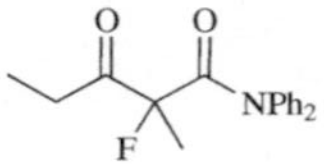

Catalyst	er
Tl1	(75) 68.0:32.0
Tl2	(75) 77.5:22.5

82

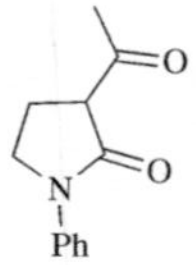

[RuCl₂(**L23**)] (10 mol %),
NFSI (1.08 eq), (Et₃O)PF₆ (0.21 eq),
DCM, rt, 24 h

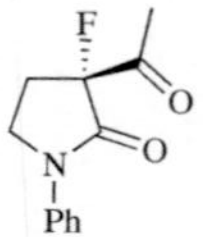

(90) er 54.0:46.0

89

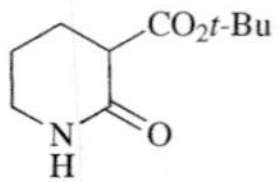

Pd5 (5 mol %), NFSI (1.5 eq),
2,6-lutidine (0.5 eq), EtOH, rt, 48 h

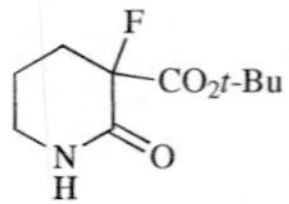

(74) er 97.0:3.0

75

C₇

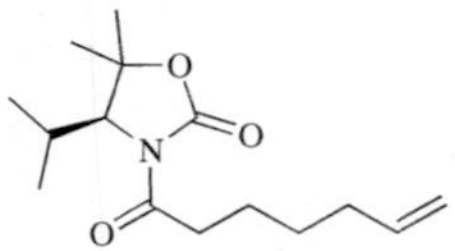

TiCl₄ (1.5 eq), Et₃N (2 eq),
NFSI (2 eq), CH₂Cl₂, rt, 1 h

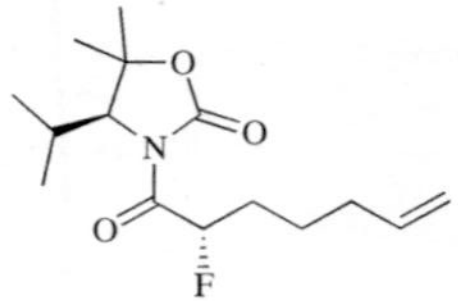

(80) dr 98:2

63

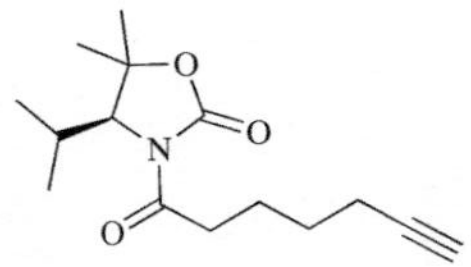

63

(56) dr 5:1

89

(69) er 55.5:44.5

63

(92) dr 96:4

TiCl$_4$ (1.5 eq), Et$_3$N (2 eq), NFSI (2 eq), CH$_2$Cl$_2$, rt, 1 h

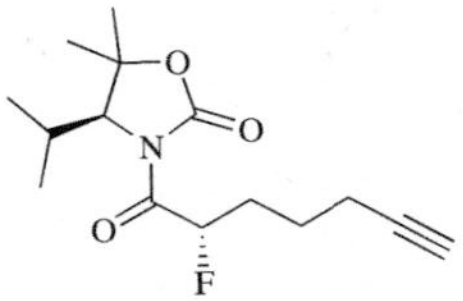

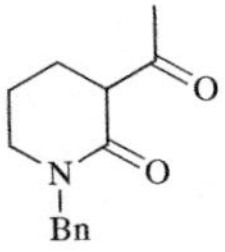

[RuCl$_2$(**L25**)] (10 mol %), NFSI (1.08 eq), (Et$_3$O)PF$_6$ (0.21 eq), DCM, rt, 24 h

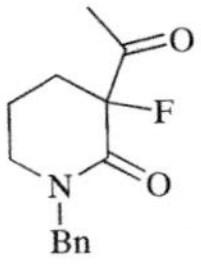

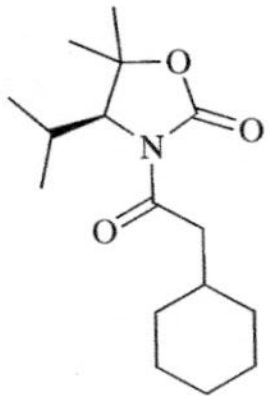

TiCl$_4$ (1.5 eq), Et$_3$N (2 eq), NFSI (2 eq), CH$_2$Cl$_2$, rt, 1 h

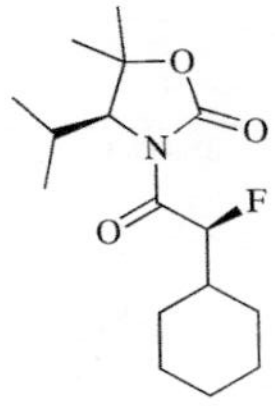

C$_8$

TABLE 7D. α-FLUORINATION OF CARBOXAMIDES (Continued)

Carboxamide	Conditions	Product(s) and Yield(s) (%)	Refs.

Please refer to the charts preceding the tables for the structures indicated by the bold numbers.

C$_{8-12}$

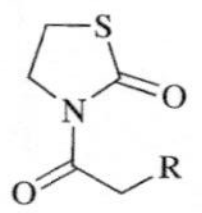

Ni(ClO$_4$)$_2$·6H$_2$O (10 mol %), **L14** (11 mol %),
HFIP (30 mol %), NFSI (1.2 eq),
2,6-lutidine (2.0 eq), 4 Å MS, CH$_2$Cl$_2$

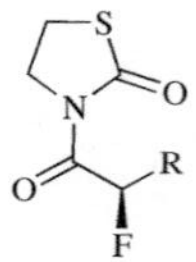

95

R	Temp (°)	Time (d)		er
Ph	–60	2	(91)	99.0:1.0
4-FC$_6$H$_4$	–60	3	(90)	97.0:3.0
4-BrC$_6$H$_4$	–60	3	(96)	96.5:3.5
4-BrC$_6$H$_4$	–80	7	(87)	99.5:0.5
3-MeOC$_6$H$_4$	–60	3	(93)	98.0:2.0
4-MeOC$_6$H$_4$	–60	2	(93)	98.0:2.0
3-MeC$_6$H$_4$	–60	4	(93)	99.0:1.0
3-MeC$_6$H$_4$	–80	7	(80)	99.5:0.5
4-MeC$_6$H$_4$	–60	3	(90)	98.0:2.0
4-CF$_3$C$_6$H$_4$	–60	4	(94)	97.0:3.0
1-naphthyl	–60	7	(87)	96.0:4.0
2-naphthyl	–60	3	(94)	97.5:2.5
2-naphthyl	–80	7	(70)	99.5:0.5

C$_8$

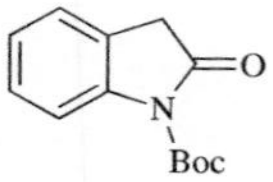

Pd16 (2.5 mol %), NFSI (1.5 eq),
ClCH$_2$CH$_2$Cl/MeOH (1:1), rt, 18 h

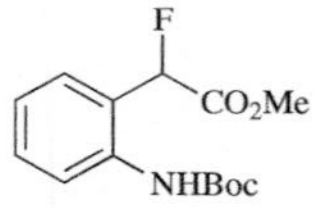

(53) er 96.5:3.5

65

C$_{9-10}$ starting material (oxazolidinone with aryl group bearing R^1, R^2)

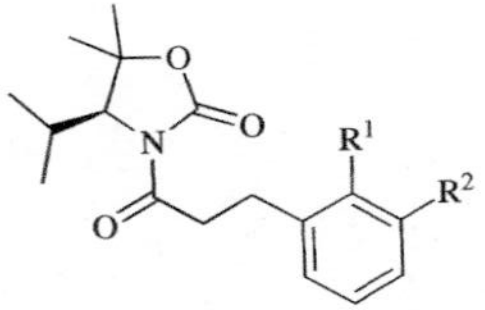

TiCl$_4$ (1.5 eq), Et$_3$N (2 eq), NFSI (2 eq), CH$_2$Cl$_2$, rt, 1 h

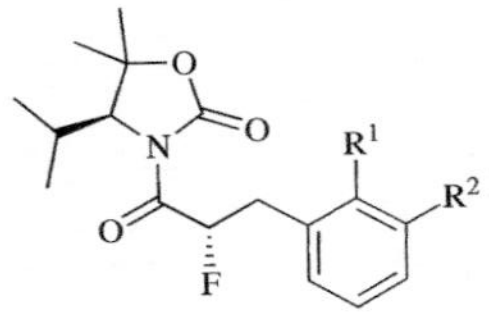

R^1	R^2		dr
H	H	(88)	98:2
Cl	H	(76)	96:4
H	CF$_3$	(61)	98:2

63

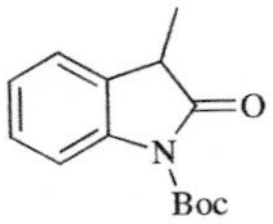

Ni(ClO$_4$)$_2$•6H$_2$O (10 mol %), L17 (12 mol %), NFSI (1.2 eq), 4 Å MS, Et$_2$O, rt, 6 h

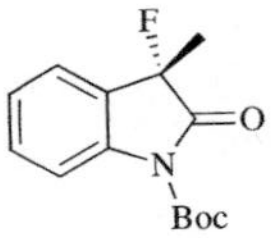

(91) er 97.5:2.5

97

Ni(OAc)$_2$•4H$_2$O (10 mol %), L14 (11 mol %), NFSI (1.2 eq), 4 Å MS, CH$_2$Cl$_2$, rt, 35 h

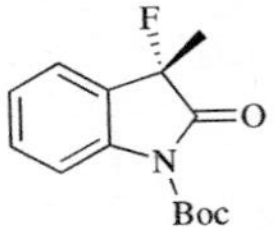

(73) er 96.5:3.5

93

TABLE 7D. α-FLUORINATION OF CARBOXAMIDES (Continued)

Carboxamide	Conditions	Product(s) and Yield(s) (%)	Refs.

*Please refer to the charts preceding the tables for the structures indicated by the **bold** numbers.*

C$_{10-16}$

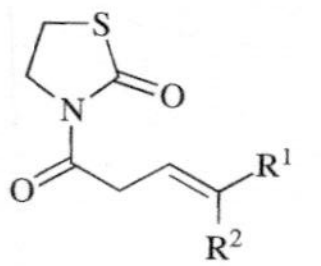

Ni(ClO$_4$)$_2$•6H$_2$O (10 mol %), **L14** (11 mol %),
HFIP (30 mol %), NFSI (1.2 eq),
2,6-lutidine (2.0 eq), 4 Å MS, CH$_2$Cl$_2$

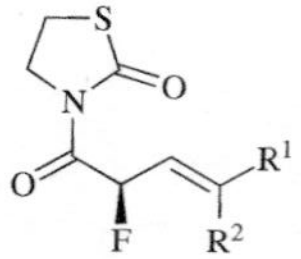

R^1	R^2	Temp (°)	Time (h)		er
Ph	H	–70	24	(58)	91.5:8.5
Ph	H	–60	24	(95)	89.0:11.0
Ph	Me	–80	92	(50)	95.5:4.5
Ph	Me	–70	26	(99)	93.5:6.5
Ph	Me	–60	24	(99)	92.5:7.5
4-ClC$_6$H$_4$	Me	–70	21	(99)	93.0:7.0
4-BrC$_6$H$_4$	Me	–60	21	(99)	90.0:10.0
4-MeC$_6$H$_4$	Me	–60	21	(99)	92.5:7.5
2-naphthyl	Me	–60	24	(97)	92.5:7.5
Ph	Ph	–60	8	(99)	93.0:7.0

95

C$_{10-11}$

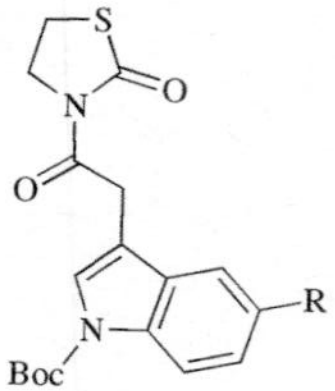

Ni(ClO$_4$)$_2$•6H$_2$O (10 mol %), **L14** (11 mol%),
HFIP (30 mol %), NFSI (1.2 eq),
2,6-lutidine (2.0 eq), 4 Å MS, CH$_2$Cl$_2$, –60°

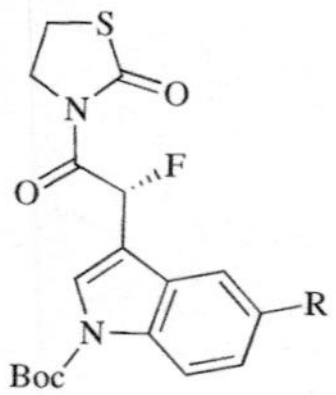

R	Time (d)		er
H	3	(88)	99.0:1.0
Br	3	(99)	98.5:1.5
MeO	4	(64)	99.5:0.5
Me	2	(92)	99.5:0.5

95

C$_{14}$

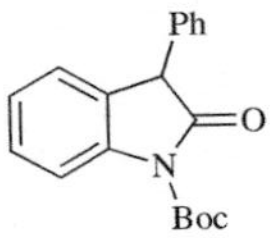

Ni(OAc)$_2$•4H$_2$O (10 mol %), **L14** (11 mol %),
NFSI (1.2 eq), 4 Å MS, CH$_2$Cl$_2$, 5 h

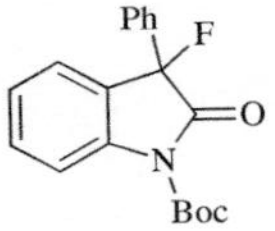

(72) er 98.0:2.0   93

C$_{14-15}$

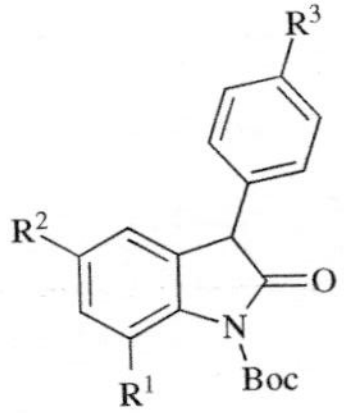

Ni(ClO$_4$)$_2$•6H$_2$O (10 mol %), **L17** (12 mol %),
NFSI (1.2 eq), 4 Å MS, Et$_2$O, rt, 6 h

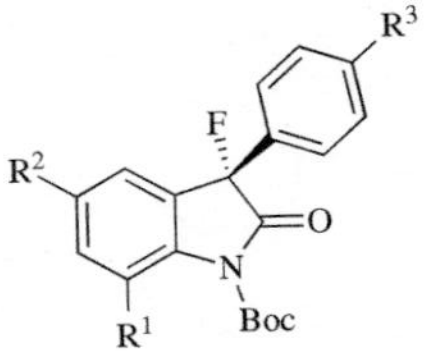

97

R^1	R^2	R^3		er
H	H	H	(95)	99.5:0.5
H	H	F	(95)	99.5:0.5
H	H	OMe	(89)	99.5:0.5
H	F	H	(91)	99.5:0.5
H	MeO	H	(91)	99.5:0.5
H	MeO	F	(92)	99.5:0.5
F	H	H	(93)	99.5:0.5
H	H	Me	(94)	99.0:1.0
H	MeO	Me	(91)	99.5:0.5
H	Me	F	(92)	99.0:1.0

TABLE 7D. α-FLUORINATION OF CARBOXAMIDES (*Continued*)

Carboxamide	Conditions	Product(s) and Yield(s) (%)	Refs.

*Please refer to the charts preceding the tables for the structures indicated by the **bold** numbers.*

C_{15}

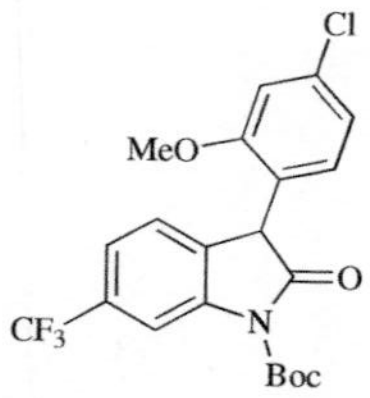	Ni(OAc)$_2$•4H$_2$O (10 mol %), **L14** (11 mol %), NFSI (1.2 eq), 4 Å MS. CH$_2$Cl$_2$, 14 h	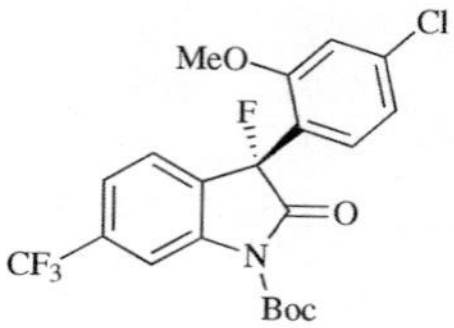 (71) er 96.5:3.5	93
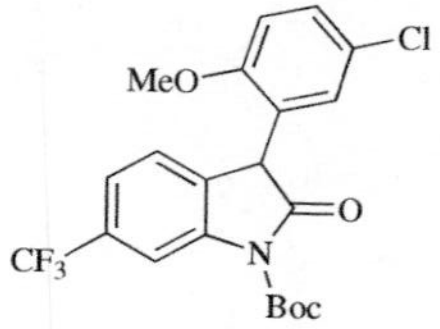	Ni(ClO$_4$)$_2$•6H$_2$O (10 mol %), **L17** (12 mol %), NFSI (1.2 eq), 4 Å MS, Et$_2$O, rt, 6 h	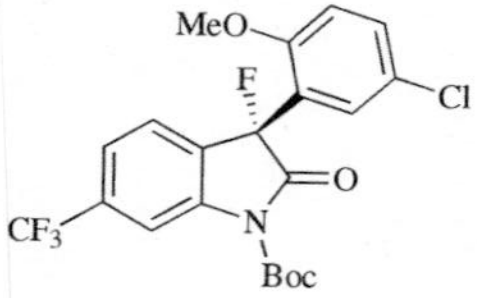 (90) er 99.5:0.5	97
	Pd16 (2.5 mol %), NFSI (1.5 eq), acetone, 0°, 18 h	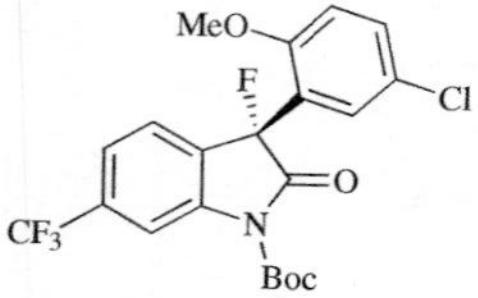 (90) er 85.5:14.5	65

REFERENCES

1 Brown, J. M.; Gouverneur, V. *Angew. Chem., Int. Ed.* **2009**, *48*, 8610.
2 Furuya, T.; Kamlet, A. S.; Ritter, T. *Nature* **2011**, *473*, 470.
3 Balz, G.; Schiemann, G. *Ber. Dtsch. Chem. Ges.* **1927**, *60*, 1186.
4 Hollingworth, C.; Gouverneur, V. *Chem. Commun.* **2012**, *48*, 2929.
5 Campbell, M. G.; Ritter, T. *Org. Process Res. Dev.* **2014**, *18*, 474.
6 Liang, S. H.; Vasdev, N. *Angew. Chem., Int. Ed.* **2014**, *53*, 11416.
7 Preshlock, S.; Tredwell, M.; Gouverneur, V. *Chem. Rev.* **2016**, *116*, 719.
8 O'Hagan, D.; Harper, D. B. In *Asymmetric Fluoroorganic Chemistry: Synthesis, Applications, and Future Directions*; Ramachandran, P. V., Ed.; ACS Symposium Series 746; American Chemical Society: Washington, DC, 2000; pp 210–224.
9 Carvalho, M. F.; Oliveira, R. S. *Crit. Rev. Biotechnol.* **2017**, *37*, 880.
10 Grushin, V. V. *Acc. Chem. Res.* **2010**, *43*, 160.
11 Neumann, C. N.; Ritter, T. *Acc. Chem. Res.* **2017**, *50*, 2822.
12 Campbell, M. G.; Mercier, J.; Genicot, C.; Gouverneur, V.; Hooker, J. M.; Ritter, T. *Nat. Chem.* **2016**, *9*, 1.
13 Ma, J.-A.; Cahard, D. *Chem. Rev.* **2008**, *108*, PR1.
14 Champagne, P. A.; Desroches, J.; Hamel, J.-D.; Vandamme, M.; Paquin, J.-F. *Chem. Rev.* **2015**, *115*, 9073.
15 Liang, T.; Neumann, C. N.; Ritter, T. *Angew. Chem., Int. Ed.* **2013**, *52*, 8214.
16 Yang, X.; Wu, T.; Phipps, R. J.; Toste, F. D. *Chem. Rev.* **2015**, *115*, 826.
17 Zhan, C.-G.; Dixon, D. A. *J. Phys. Chem. A* **2004**, *108*, 2020.
18 Bathgate, R. H.; Moelwyn-Hughes, E. A. *J. Chem. Soc.* **1959**, 2642.
19 Vincent, M. A.; Hillier, I. H. *Chem. Commun.* **2005**, 5902.
20 Kim, D. W.; Ahn, D.-S.; Oh, Y.-H.; Lee, S.; Kil, H. S.; Oh, S. J.; Lee, S. J.; Kim, J. S.; Ryu, J. S.; Moon, D. H.; Chi, D. Y. *J. Am. Chem. Soc.* **2006**, *128*, 16394.
21 Neumann, C. N.; Ritter, T. *Angew. Chem., Int. Ed.* **2015**, *54*, 3216.
22 Stavber, S. *Molecules* **2011**, *16*, 6432.
23 Tang, P.; Furuya, T.; Ritter, T. *J. Am. Chem. Soc.* **2010**, *132*, 12150.
24 Katcher, M. H.; Norrby, P.-O.; Doyle, A. G. *Organometallics* **2014**, *33*, 2121.
25 Watson, D. A.; Su, M.; Teverovskiy, G.; Zhang, Y.; García-Fortanet, J.; Kinzel, T.; Buchwald, S. L. *Science* **2009**, *325*, 1661.
26 Yandulov, D. V.; Tran, N. T. *J. Am. Chem. Soc.* **2007**, *129*, 1342.
27 Grushin, V. V. *Organometallics* **2000**, *19*, 1888.
28 Grushin, V. V. *Chem.—Eur. J.* **2002**, *8*, 1006.
29 Grushin, V. V.; Marshall, W. J. *Organometallics* **2007**, *26*, 4997.
30 Sather, A. C.; Buchwald, S. L. *Acc. Chem. Res.* **2016**, *49*, 2146.
31 Grushin, V. V.; Marshall, W. J. *Organometallics* **2008**, *27*, 4825.
32 Maimone, T. J.; Milner, P. J.; Kinzel, T.; Zhang, Y.; Takase, M. K.; Buchwald, S. L. *J. Am. Chem. Soc.* **2011**, *133*, 18106.
33 Lee, H. G.; Milner, P. J.; Buchwald, S. L. *J. Am. Chem. Soc.* **2014**, *136*, 3792.
34 Milner, P. J.; Maimone, T. J.; Su, M.; Chen, J.; Müller, P.; Buchwald, S. L. *J. Am. Chem. Soc.* **2012**, *134*, 19922.
35 Milner, P. J.; Kinzel, T.; Zhang, Y.; Buchwald, S. L. *J. Am. Chem. Soc.* **2014**, *136*, 15757.
36 Sather, A. C.; Lee, H. G.; De La Rosa, V. Y.; Yang, Y.; Müller, P.; Buchwald, S. L. *J. Am. Chem. Soc.* **2015**, *137*, 13433.
37 Lee, H. G.; Milner, P. J.; Buchwald, S. L. *Org. Lett.* **2013**, *15*, 5602.
38 Hollingworth, C.; Hazari, A.; Hopkinson, M. N.; Tredwell, M.; Benedetto, E.; Huiban, M.; Gee, A. D.; Brown, J. M.; Gouverneur, V. *Angew. Chem., Int. Ed.* **2011**, *50*, 2613.
39 Benedetto, E.; Tredwell, M.; Hollingworth, C.; Khotavivattana, T.; Brown, J. M.; Gouverneur, V. *Chem. Sci.* **2013**, *4*, 89.
40 Zhang, Q.; Stockdale, D. P.; Mixdorf, J. C.; Topczewski, J. J.; Nguyen, H. M. *J. Am. Chem. Soc.* **2015**, *137*, 11912.
41 Katcher, M. H.; Sha, A.; Doyle, A. G. *J. Am. Chem. Soc.* **2011**, *133*, 15902.
42 Zhang, Z.; Wang, F.; Mu, X.; Chen, P.; Liu, G. *Angew. Chem., Int. Ed.* **2013**, *52*, 7549.
43 Althaus, M.; Togni, A.; Mezzetti, A. *J. Fluorine Chem.* **2009**, *130*, 702.
44 Liu, W.; Huang, X.; Cheng, M.-J.; Nielsen, R. J.; Goddard, W. A.; Groves, J. T. *Science* **2012**, *337*, 1322.
45 Kalow, J. A.; Doyle, A. G. *J. Am. Chem. Soc.* **2011**, *133*, 16001.

[46] Kalow, J. A.; Doyle, A. G. *J. Am. Chem. Soc.* **2010**, *132*, 3268.

[47] Gorske, B. C.; Mbofana, C. T.; Miller, S. J. *Org. Lett.* **2009**, *11*, 4318.

[48] Akana, J. A.; Bhattacharyya, K. X.; Müller, P.; Sadighi, J. P. *J. Am. Chem. Soc.* **2007**, *129*, 7736.

[49] Dang, H.; Mailig, M.; Lalic, G. *Angew. Chem., Int. Ed.* **2014**, *53*, 6473.

[50] Liu, Y.; Chen, C.; Li, H.; Huang, K.-W.; Tan, J.; Weng, Z. *Organometallics* **2013**, *32*, 6587.

[51] Barthazy, P.; Togni, A.; Mezzetti, A. *Organometallics* **2001**, *20*, 3472.

[52] Timofeeva, D. S.; Ofial, A. R.; Mayr, H. *J. Am. Chem. Soc.* **2018**, *140*, 11474.

[53] Fier, P. S.; Hartwig, J. F. *Science* **2013**, *342*, 956.

[54] Fier, P. S.; Luo, J.; Hartwig, J. F. *J. Am. Chem. Soc.* **2013**, *135*, 2552.

[55] Yamamoto, K.; Li, J.; Garber, J. A. O.; Rolfes, J. D.; Boursalian, G. B.; Borghs, J. C.; Genicot, C.; Jacq, J.; van Gastel, M.; Neese, F.; Ritter, T. *Nature* **2018**, *554*, 511.

[56] Li, Z.; Wang, Z.; Zhu, L.; Tan, X.; Li, C. *J. Am. Chem. Soc.* **2014**, *136*, 16439.

[57] Phae-nok, S.; Soorukram, D.; Kuhakarn, C.; Reutrakul, V.; Pohmakotr, M. *Eur. J. Org. Chem.* **2015**, 2879.

[58] Yin, F.; Wang, Z.; Li, Z.; Li, C. *J. Am. Chem. Soc.* **2012**, *134*, 10401.

[59] Zhao, H.; Fan, X.; Yu, J.; Zhu, C. *J. Am. Chem. Soc.* **2015**, *137*, 3490.

[60] Patel, N. R.; Flowers, R. A. *J. Org. Chem.* **2015**, *80*, 5834.

[61] Mazzotti, A. R.; Campbell, M. G.; Tang, P.; Murphy, J. M.; Ritter, T. *J. Am. Chem. Soc.* **2013**, *135*, 14012.

[62] Halperin, S. D.; Fan, H.; Chang, S.; Martin, R. E.; Britton, R. *Angew. Chem., Int. Ed.* **2014**, *53*, 4690.

[63] Alvarado, J.; Herrmann, A. T.; Zakarian, A. *J. Org. Chem.* **2014**, *79*, 6206.

[64] Suzuki, T.; Hamashima, Y.; Sodeoka, M. *Angew. Chem., Int. Ed.* **2007**, *46*, 5435.

[65] Hamashima, Y.; Suzuki, T.; Takano, H.; Shimura, Y.; Sodeoka, M. *J. Am. Chem. Soc.* **2005**, *127*, 10164.

[66] Hintermann, L.; Perseghini, M.; Togni, A. *Beilstein J. Org. Chem.* **2011**, *7*, 1421.

[67] Hintermann, L.; Togni, A. *Angew. Chem., Int. Ed.* **2000**, *39*, 4359.

[68] Piana, S.; Devillers, I.; Togni, A.; Rothlisberger, U. *Angew. Chem., Int. Ed.* **2002**, *41*, 979.

[69] Suzuki, T.; Goto, T.; Hamashima, Y.; Sodeoka, M. *J. Org. Chem.* **2007**, *72*, 246.

[70] Hamashima, Y.; Suzuki, T.; Takano, H.; Shimura, Y.; Tsuchiya, Y.; Moriya, K.-i.; Goto, T.; Sodeoka, M. *Tetrahedron* **2006**, *62*, 7168.

[71] Hamashima, Y.; Suzuki, T.; Shimura, Y.; Shimizu, T.; Umebayashi, N.; Tamura, T.; Sasamoto, N.; Sodeoka, M. *Tetrahedron Lett.* **2005**, *46*, 1447.

[72] Hamashima, Y.; Yagi, K.; Takano, H.; Tamás, L.; Sodeoka, M. *J. Am. Chem. Soc.* **2002**, *124*, 14530.

[73] Kim, H. R.; Kim, D. Y. *Tetrahedron Lett.* **2005**, *46*, 3115.

[74] Kim, S. M.; Kim, H. R.; Kim, D. Y. *Org. Lett.* **2005**, *7*, 2309.

[75] Hamashima, Y.; Sodeoka, M. *J. Synth. Org. Chem., Jpn.* **2007**, *65*, 1099.

[76] Paull, D. H.; Scerba, M. T.; Alden-Danforth, E.; Widger, L. R.; Lectka, T. *J. Am. Chem. Soc.* **2008**, *130*, 17260.

[77] Erb, J.; Paull, D. H.; Dudding, T.; Belding, L.; Lectka, T. *J. Am. Chem. Soc.* **2011**, *133*, 7536.

[78] Jacquet, O.; Clément, N. D.; Blanco, C.; Belmonte, M. M.; Benet-Buchholz, J.; van Leeuwen, P. W. N. M. *Eur. J. Org. Chem.* **2012**, 4844.

[79] Suzuki, S.; Kitamura, Y.; Lectard, S.; Hamashima, Y.; Sodeoka, M. *Angew. Chem., Int. Ed.* **2012**, *51*, 4581.

[80] Zhang, R.; Wang, D.; Xu, Q.; Jiang, J.; Shi, M. *Chin. J. Chem.* **2012**, *30*, 1295.

[81] Woo, S. B.; Suh, C. W.; Koh, K. O.; Kim, D. Y. *Tetrahedron Lett.* **2013**, *54*, 3359.

[82] Bertogg, A.; Hintermann, L.; Huber, D. P.; Perseghini, M.; Sanna, M.; Togni, A. *Helv. Chim. Acta* **2012**, *95*, 353.

[83] Gu, X.; Zhang, Y.; Xu, Z.-J.; Che, C.-M. *Chem. Commun.* **2014**, *50*, 7870.

[84] Lee, S. Y.; Neufeind, S.; Fu, G. C. *J. Am. Chem. Soc.* **2014**, *136*, 8899.

[85] de Haro, T.; Nevado, C. *Adv. Synth. Catal.* **2010**, *352*, 2767.

[86] Jin, Z.; Hidinger, R. S.; Xu, B.; Hammond, G. B. *J. Org. Chem.* **2012**, *77*, 7725.

[87] Zeng, X.; Liu, S.; Shi, Z.; Liu, G.; Xu, B. *Angew. Chem., Int. Ed.* **2016**, *55*, 10032.

[88] Kawatsura, M.; Hayashi, S.; Komatsu, Y.; Hayase, S.; Itoh, T. *Chem. Lett.* **2010**, *39*, 466.

[89] Althaus, M.; Becker, C.; Togni, A.; Mezzetti, A. *Organometallics* **2007**, *26*, 5902.

[90] Chambers, R. D.; Nakano, T.; Okazoe, T.; Sandford, G. *J. Fluorine Chem.* **2009**, *130*, 792.

[91] Ahlsten, N.; Martin-Matute, B. *Chem. Commun.* **2011**, *47*, 8331.

[92] Ahlsten, N.; Bartoszewicz, A.; Agrawal, S.; Martín-Matute, B. *Synthesis* **2011**, 2600.

[93] Shibata, N.; Kohno, J.; Takai, K.; Ishimaru, T.; Nakamura, S.; Toru, T.; Kanemasa, S. *Angew. Chem., Int. Ed.* **2005**, *44*, 4204.

[94] Shibatomi, K.; Tsuzuki, Y.; Iwasa, S. *Chem. Lett.* **2008**, *37*, 1098.

95 Reddy, D. S.; Shibata, N.; Horikawa, T.; Suzuki, S.; Nakamura, S.; Toru, T.; Shiro, M. *Chem.—Asian J.* **2009**, *4*, 1411.

96 Kang, S. H.; Kim, D. Y. *Adv. Synth. Catal.* **2010**, *352*, 2783.

97 Deng, Q.-H.; Wadepohl, H.; Gade, L. H. *Chem.—Eur. J.* **2011**, *17*, 14922.

98 Jacquet, O.; Clément, N. D.; Freixa, Z.; Ruiz, A.; Claver, C.; van Leeuwen, P. W. N. M. *Tetrahedron: Asymmetry* **2011**, *22*, 1490.

99 Ma, J.-A.; Cahard, D. *Tetrahedron: Asymmetry* **2004**, *15*, 1007.

100 Assalit, A.; Billard, T.; Chambert, S.; Langlois, B. R.; Queneau, Y.; Coe, D. *Tetrahedron: Asymmetry* **2009**, *20*, 593.

101 Shibatomi, K.; Narayama, A.; Soga, Y.; Muto, T.; Iwasa, S. *Org. Lett.* **2011**, *13*, 2944.

102 Balaraman, K.; Vasanthan, R.; Kesavan, V. *Tetrahedron: Asymmetry* **2013**, *24*, 919.

103 Narayama, A.; Shibatomi, K.; Soga, Y.; Muto, T.; Iwasa, S. *Synlett* **2013**, *24*, 375.

104 Peng, J.; Du, D.-M. *RSC Adv.* **2014**, *4*, 2061.

105 Hull, K. L.; Anani, W. Q.; Sanford, M. S. *J. Am. Chem. Soc.* **2006**, *128*, 7134.

106 Furuya, T.; Benitez, D.; Tkatchouk, E.; Strom, A. E.; Tang, P.; Goddard, W. A.; Ritter, T. *J. Am. Chem. Soc.* **2010**, *132*, 3793.

107 Furuya, T.; Ritter, T. *J. Am. Chem. Soc.* **2008**, *130*, 10060.

108 Dubinsky-Davidchik, I. S.; Potash, S.; Goldberg, I.; Vigalok, A.; Vedernikov, A. N. *J. Am. Chem. Soc.* **2012**, *134*, 14027.

109 Brandt, J. R.; Lee, E.; Boursalian, G. B.; Ritter, T. *Chem. Sci.* **2014**, *5*, 169.

110 Lee, E.; Hooker, J. M.; Ritter, T. *J. Am. Chem. Soc.* **2012**, *134*, 17456.

111 Lee, E.; Kamlet, A. S.; Powers, D. C.; Neumann, C. N.; Boursalian, G. B.; Furuya, T.; Choi, D. C.; Hooker, J. M.; Ritter, T. *Science* **2011**, *334*, 639.

112 Hoover, A. J.; Lazari, M.; Ren, H.; Narayanam, M. K.; Murphy, J. M.; van Dam, R. M.; Hooker, J. M.; Ritter, T. *Organometallics* **2016**, *35*, 1008.

113 Zhao, S.-B.; Wang, R.-Y.; Nguyen, H.; Becker, J. J.; Gagne, M. R. *Chem. Commun.* **2012**, *48*, 443.

114 Ball, N. D.; Sanford, M. S. *J. Am. Chem. Soc.* **2009**, *131*, 3796.

115 Kaspi, A. W.; Yahav-Levi, A.; Goldberg, I.; Vigalok, A. *Inorg. Chem.* **2008**, *47*, 5.

116 Wannberg, J.; Wallinder, C.; Ünlüsoy, M.; Sköld, C.; Larhed, M. *J. Org. Chem.* **2013**, *78*, 4184.

117 Casitas, A.; Canta, M.; Solà, M.; Costas, M.; Ribas, X. *J. Am. Chem. Soc.* **2011**, *133*, 19386.

118 Mu, X.; Liu, G. *Org. Chem. Front.* **2014**, *1*, 430.

119 Grushin, V.; U.S. Patent 2006/0074261, April 6, 2006.

120 Fier, P. S.; Hartwig, J. F. *J. Am. Chem. Soc.* **2012**, *134*, 10795.

121 Subramanian, M. A.; Manzer, L. E. *Science* **2002**, *297*, 1665.

122 Chan, K. S. L.; Wasa, M.; Wang, X.; Yu, J.-Q. *Angew. Chem., Int. Ed.* **2011**, *50*, 9081.

123 Chen, C.; Wang, C.; Zhang, J.; Zhao, Y. *J. Org. Chem.* **2015**, *80*, 942.

124 Lou, S.-J.; Xu, D.-Q.; Xia, A.-B.; Wang, Y.-F.; Liu, Y.-K.; Du, X.-H.; Xu, Z.-Y. *Chem. Commun.* **2013**, *49*, 6218.

125 Lou, S.-J.; Xu, D.-Q.; Xu, Z.-Y. *Angew. Chem., Int. Ed.* **2014**, *53*, 10330.

126 Truong, T.; Klimovica, K.; Daugulis, O. *J. Am. Chem. Soc.* **2013**, *135*, 9342.

127 Testa, C.; Gigot, É.; Genc, S.; Decréau, R.; Roger, J.; Hierso, J.-C. *Angew. Chem., Int. Ed.* **2016**, *55*, 5555.

128 Yao, B.; Wang, Z.-L.; Zhang, H.; Wang, D.-X.; Zhao, L.; Wang, M.-X. *J. Org. Chem.* **2012**, *77*, 3336.

129 Wang, X.; Mei, T.-S.; Yu, J.-Q. *J. Am. Chem. Soc.* **2009**, *131*, 7520.

130 Ding, Q.; Ye, C.; Pu, S.; Cao, B. *Tetrahedron* **2014**, *70*, 409.

131 Furuya, T.; Strom, A. E.; Ritter, T. *J. Am. Chem. Soc.* **2009**, *131*, 1662.

132 Furuya, T.; Ritter, T. *Org. Lett.* **2009**, *11*, 2860.

133 Ye, Y.; Sanford, M. S. *J. Am. Chem. Soc.* **2013**, *135*, 4648.

134 Ye, Y.; Schimler, S. D.; Hanley, P. S.; Sanford, M. S. *J. Am. Chem. Soc.* **2013**, *135*, 16292.

135 Dubbaka, S. R.; Narreddula, V. R.; Gadde, S.; Mathew, T. *Tetrahedron* **2014**, *70*, 9676.

136 Dubbaka, S. R.; Gadde, S.; Narreddula, V. R. *Synthesis* **2015**, *47*, 854.

137 Tang, P.; Ritter, T. *Tetrahedron* **2011**, *67*, 4449.

138 Ross, T. L.; Ermert, J.; Hocke, C.; Coenen, H. H. *J. Am. Chem. Soc.* **2007**, *129*, 8018.

139 Ichiishi, N.; Canty, A. J.; Yates, B. F.; Sanford, M. S. *Org. Lett.* **2013**, *15*, 5134.

140 Ichiishi, N.; Canty, A. J.; Yates, B. F.; Sanford, M. S. *Organometallics* **2014**, *33*, 5525.

141 Ichiishi, N.; Brooks, A. F.; Topczewski, J. J.; Rodnick, M. E.; Sanford, M. S.; Scott, P. J. H. *Org. Lett.* **2014**, *16*, 3224.

142 Milner, P. J.; Yang, Y.; Buchwald, S. L. *Organometallics* **2015**, *34*, 4775.

143 Fier, P. S.; Hartwig, J. F. *J. Am. Chem. Soc.* **2014**, *136*, 10139.

[144] Liu, H.; Audisio, D.; Plougastel, L.; Decuypere, E.; Buisson, D.-A.; Koniev, O.; Kolodych, S.; Wagner, A.; Elhabiri, M.; Krzyczmonik, A.; Forsback, S.; Solin, O.; Gouverneur, V.; Taran, F. *Angew. Chem., Int. Ed.* **2016**, *55*, 12073.

[145] Sommer, H.; Fürstner, A. *Chem.—Eur. J.* **2017**, *23*, 558.

[146] Nahra, F.; Patrick, S. R.; Bello, D.; Brill, M.; Obled, A.; Cordes, D. B.; Slawin, A. M. Z.; O'Hagan, D.; Nolan, S. P. *ChemCatChem* **2015**, *7*, 240.

[147] Okoromoba, O. E.; Han, J.; Hammond, G. B.; Xu, B. *J. Am. Chem. Soc.* **2014**, *136*, 14381.

[148] Tius, M. A.; Kawakami, J. K. *Synth. Commun.* **1992**, *22*, 1461.

[149] Tius, M. A.; Kawakami, J. K. *Synlett* **1993**, 207.

[150] Tius, M. A.; Kawakami, J. K. *Tetrahedron* **1995**, *51*, 3997.

[151] Petasis, N. A.; Yudin, A. K.; Zavialov, I. A.; Prakash, G. K. S.; Olah, G. A. *Synlett* **1997**, 606.

[152] Pérez-Temprano, M. H.; Racowski, J. M.; Kampf, J. W.; Sanford, M. S. *J. Am. Chem. Soc.* **2014**, *136*, 4097.

[153] Racowski, J. M.; Gary, J. B.; Sanford, M. S. *Angew. Chem., Int. Ed.* **2012**, *51*, 3414.

[154] Zhao, S.-B.; Becker, J. J.; Gagné, M. R. *Organometallics* **2011**, *30*, 3926.

[155] Mankad, N. P.; Toste, F. D. *Chem. Sci.* **2012**, *3*, 72.

[156] Barker, T. J.; Boger, D. L. *J. Am. Chem. Soc.* **2012**, *134*, 13588.

[157] Shigehisa, H.; Nishi, E.; Fujisawa, M.; Hiroya, K. *Org. Lett.* **2013**, *15*, 5158.

[158] Kim, D. W.; Jeong, H.-J.; Lim, S. T.; Sohn, M.-H.; Katzenellenbogen, J. A.; Chi, D. Y. *J. Org. Chem.* **2008**, *73*, 957.

[159] Kim, D. W.; Song, C. E.; Chi, D. Y. *J. Am. Chem. Soc.* **2002**, *124*, 10278.

[160] Shinde, S. S.; Lee, B. S.; Chi, D. Y. *Org. Lett.* **2008**, *10*, 733.

[161] Cox, D. P.; Terpinski, J.; Lawrynowicz, W. *J. Org. Chem.* **1984**, *49*, 3216.

[162] Pilcher, A. S.; Ammon, H. L.; DeShong, P. *J. Am. Chem. Soc.* **1995**, *117*, 5166.

[163] Liu, W.; Huang, X.; Groves, J. T. *Nat. Protoc.* **2013**, *8*, 2348.

[164] Bloom, S.; Pitts, C. R.; Miller, D. C.; Haselton, N.; Holl, M. G.; Urheim, E.; Lectka, T. *Angew. Chem., Int. Ed.* **2012**, *51*, 10580.

[165] Pitts, C. R.; Bloom, S.; Woltornist, R.; Auvenshine, D. J.; Ryzhkov, L. R.; Siegler, M. A.; Lectka, T. *J. Am. Chem. Soc.* **2014**, *136*, 9780.

[166] West, J. G.; Bedell, T. A.; Sorensen, E. J. *Angew. Chem., Int. Ed.* **2016**, *55*, 8923.

[167] Neumann, C. N.; Ritter, T. *Nat. Chem.* **2016**, *8*, 822.

[168] Xia, J.-B.; Ma, Y.; Chen, C. *Org. Chem. Front.* **2014**, *1*, 468.

[169] Zhu, Q.; Ji, D.; Liang, T.; Wang, X.; Xu, Y. *Org. Lett.* **2015**, *17*, 3798.

[170] Zhu, R.-Y.; Tanaka, K.; Li, G.-C.; He, J.; Fu, H.-Y.; Li, S.-H.; Yu, J.-Q. *J. Am. Chem. Soc.* **2015**, *137*, 7067.

[171] Miao, J.; Yang, K.; Kurek, M.; Ge, H. *Org. Lett.* **2015**, *17*, 3738.

[172] Zhang, Q.; Yin, X.-S.; Chen, K.; Zhang, S.-Q.; Shi, B.-F. *J. Am. Chem. Soc.* **2015**, *137*, 8219.

[173] Huang, X.; Liu, W.; Hooker, J. M.; Groves, J. T. *Angew. Chem., Int. Ed.* **2015**, *54*, 5241.

[174] Ventre, S.; Petronijevic, F. R.; MacMillan, D. W. C. *J. Am. Chem. Soc.* **2015**, *137*, 5654.

[175] Rueda-Becerril, M.; Mahé, O.; Drouin, M.; Majewski, M. B.; West, J. G.; Wolf, M. O.; Sammis, G. M.; Paquin, J.-F. *J. Am. Chem. Soc.* **2014**, *136*, 2637.

[176] Kalow, J. A.; Doyle, A. G. *Tetrahedron* **2013**, *69*, 5702.

[177] Katcher, M. H.; Doyle, A. G. *J. Am. Chem. Soc.* **2010**, *132*, 17402.

[178] Hintermann, L.; Läng, F.; Maire, P.; Togni, A. *Eur. J. Inorg. Chem.* **2006**, 1397.

[179] van Haaren, R. J.; Goubitz, K.; Fraanje, J.; van Strijdonck, G. P. F.; Oevering, H.; Coussens, B.; Reek, J. N. H.; Kamer, P. C. J.; van Leeuwen, P. W. N. M. *Inorg. Chem.* **2001**, *40*, 3363.

[180] Topczewski, J. J.; Tewson, T. J.; Nguyen, H. M. *J. Am. Chem. Soc.* **2011**, *133*, 19318.

[181] Zhu, J.; Tsui, G. C.; Lautens, M. *Angew. Chem., Int. Ed. Engl.* **2012**, *51*, 12353.

[182] Zhang, Q.; Nguyen, H. M. *Chem. Sci.* **2014**, *5*, 291.

[183] Lauer, A. M.; Wu, J. *Org. Lett.* **2012**, *14*, 5138.

[184] Braun, M.-G.; Doyle, A. G. *J. Am. Chem. Soc.* **2013**, *135*, 12990.

[185] Chen, M. S.; White, M. C. *J. Am. Chem. Soc.* **2004**, *126*, 1346.

[186] Bloom, S.; Pitts, C. R.; Woltornist, R.; Griswold, A.; Holl, M. G.; Lectka, T. *Org. Lett.* **2013**, *15*, 1722.

[187] Nodwell, M. B.; Bagai, A.; Halperin, S. D.; Martin, R. E.; Knust, H.; Britton, R. *Chem. Commun.* **2015**, *51*, 11783.

[188] Veltheer, J. E.; Burger, P.; Bergman, R. G. *J. Am. Chem. Soc.* **1995**, *117*, 12478.

[189] Emer, E.; Pfeifer, L.; Brown, J. M.; Gouverneur, V. *Angew. Chem., Int. Ed.* **2014**, *53*, 4181.

[190] McMurtrey, K. B.; Racowski, J. M.; Sanford, M. S. *Org. Lett.* **2012**, *14*, 4094.

[191] Stavber, G.; Zupan, M.; Stavber, S. *Synlett* **2009**, 589.

192 Liu, W.; Groves, J. T. *Angew. Chem., Int. Ed.* **2013**, *52*, 6024.
193 Zhang, W.; Loebach, J. L.; Wilson, S. R.; Jacobsen, E. N. *J. Am. Chem. Soc.* **1990**, *112*, 2801.
194 Bistoni, G.; Belanzoni, P.; Belpassi, L.; Tarantelli, F. *J. Phys. Chem. A* **2016**, *120*, 5239.
195 Perseghini, M.; Massaccesi, M.; Liu, Y.; Togni, A. *Tetrahedron* **2006**, *62*, 7180.
196 Merchant, R. R.; Edwards, J. T.; Qin, T.; Kruszyk, M. M.; Bi, C.; Che, G.; Bao, D.-H.; Qiao, W.; Sun, L.; Collins, M. R.; Fadeyi, O. O.; Gallego, G. M.; Mousseau, J. J.; Nuhant, P.; Baran, P. S. *Science* **2018**, *360*, 75.
197 Baroudy, B. M.; Clader, J. W.; Josien, H. B.; McCombie, S. W.; McKittrick, B. A.; Miller, M. W.; Neustadt, B. R.; Palani, A.; Smith, E. M.; Steensma, R.; Tagat, J. R.; Vice, S. F.; Gilbert, E.; Labroli, M. A. (Schering Corporation) U.S. Patent 6391865 B1, May 21, 2002.
198 Keita, T.; Kazunori, F.; Takafumi, I.; Hiroyuki, N.; Yuko, U.; Kazutake, T.; Takashi, I.; Takehiko, Y.; Shuji, A. *Eur. J. Org. Chem.* **2016**, 1562.
199 Zhou, J.; Jin, C.; Su, W. *Org. Process Res. Dev.* **2014**, *18*, 928.
200 Ametamey, S. M.; Honer, M.; Schubiger, P. A. *Chem. Rev.* **2008**, *108*, 1501.
201 Stenhagen, I. S. R.; Kirjavainen, A. K.; Forsback, S. J.; Jorgensen, C. G.; Robins, E. G.; Luthra, S. K.; Solin, O.; Gouverneur, V. *Chem. Commun.* **2013**, *49*, 1386.
202 Shah, A.; Pike, V. W.; Widdowson, D. A. *J. Chem. Soc., Perkin Trans. 1* **1998**, 2043.
203 Rotstein, B. H.; Stephenson, N. A.; Vasdev, N.; Liang, S. H. *Nat. Commun.* **2014**, *5*, 4365.
204 Cardinale, J.; Ermert, J.; Kügler, F.; Helfer, A.; Brandt, M. R.; Coenen, H. H. *J. Labelled Compd. Radiopharm.* **2012**, *55*, 450.
205 Miller, P. W.; Long, N. J.; Vilar, R.; Gee, A. D. *Angew. Chem., Int. Ed.* **2008**, *47*, 8998.
206 Neumann, C. N.; Hooker, J. M.; Ritter, T. *Nature* **2016**, *534*, 369.
207 Beyzavi, M. H.; Mandal, D.; Strebl, M. G.; Neumann, C. N.; D'Amato, E. M.; Chen, J.; Hooker, J. M.; Ritter, T. *ACS Cent. Sci.* **2017**, *3*, 944.
208 Rickmeier, J.; Ritter, T. *Angew. Chem., Int. Ed.* **2018**, *57*, 14207.
209 Taylor, N. J.; Emer, E.; Preshlock, S.; Schedler, M.; Tredwell, M.; Verhoog, S.; Mercier, J.; Genicot, C.; Gouverneur, V. *J. Am. Chem. Soc.* **2017**, *139*, 8267.
210 Preshlock, S.; Calderwood, S.; Verhoog, S.; Tredwell, M.; Huiban, M.; Hienzsch, A.; Gruber, S.; Wilson, T. C.; Taylor, N. J.; Cailly, T.; Schedler, M.; Collier, T. L.; Passchier, J.; Smits, R.; Mollitor, J.; Hoepping, A.; Mueller, M.; Genicot, C.; Mercier, J.; Gouverneur, V. *Chem. Commun.* **2016**, *52*, 8361.
211 Tredwell, M.; Preshlock, S. M.; Taylor, N. J.; Gruber, S.; Huiban, M.; Passchier, J.; Mercier, J.; Génicot, C.; Gouverneur, V. *Angew. Chem., Int. Ed.* **2014**, *53*, 7751.
212 Strebl, M. G.; Campbell, A. J.; Zhao, W.-N.; Schroeder, F. A.; Riley, M. M.; Chindavong, P. S.; Morin, T. M.; Haggarty, S. J.; Wagner, F. F.; Ritter, T.; Hooker, J. M. *ACS Cent. Sci.* **2017**, *3*, 1006.
213 Seimbille, Y.; Phelps, M. E.; Czernin, J.; Silverman, D. H. S. *J. Labelled Compd. Radiopharm.* **2005**, *48*, 829.
214 Graham, T. J. A.; Lambert, R. F.; Ploessl, K.; Kung, H. F.; Doyle, A. G. *J. Am. Chem. Soc.* **2014**, *136*, 5291.
215 Buckingham, F.; Gouverneur, V. *Chem. Sci.* **2016**, *7*, 1645.
216 Okamura, N.; Furumoto, S.; Harada, R.; Tago, T.; Yoshikawa, T.; Fodero-Tavoletti, M.; Mulligan, R. S.; Villemagne, V. L.; Akatsu, H.; Yamamoto, T.; Arai, H.; Iwata, R.; Yanai, K.; Kudo, Y. *J. Nucl. Med.* **2013**, *54*, 1420.
217 Nielsen, M. K.; Ahneman, D. T.; Riera, O.; Doyle, A. G. *J. Am. Chem. Soc.* **2018**, *140*, 5004.
218 Sun, H.; DiMagno, S. G. *J. Am. Chem. Soc.* **2005**, *127*, 2050.
219 Sun, H.; DiMagno, S. G. *Angew. Chem., Int. Ed.* **2006**, *45*, 2720.
220 Schimler, S. D.; Cismesia, M. A.; Hanley, P. S.; Froese, R. D. J.; Jansma, M. J.; Bland, D. C.; Sanford, M. S. *J. Am. Chem. Soc.* **2017**, *139*, 1452.
221 Ryan, S. J.; Schimler, S. D.; Bland, D. C.; Sanford, M. S. *Org. Lett.* **2015**, *17*, 1866.
222 Tang, P.; Wang, W.; Ritter, T. *J. Am. Chem. Soc.* **2011**, *133*, 11482.
223 Fujimoto, T.; Becker, F.; Ritter, T. *Org. Process Res. Dev.* **2014**, *18*, 1041.
224 Sladojevich, F.; Arlow, S. I.; Tang, P.; Ritter, T. *J. Am. Chem. Soc.* **2013**, *135*, 2470.
225 Fujimoto, T.; Ritter, T. *Org. Lett.* **2015**, *17*, 544.
226 Goldberg, N. W.; Shen, X.; Li, J.; Ritter, T. *Org. Lett.* **2016**, *18*, 6102.
227 Nielsen, M. K.; Ugaz, C. R.; Li, W.; Doyle, A. G. *J. Am. Chem. Soc.* **2015**, *137*, 9571.
228 Xia, J.-B.; Zhu, C.; Chen, C. *Chem. Commun.* **2014**, *50*, 11701.
229 Kee, C. W.; Chin, K. F.; Wong, M. W.; Tan, C.-H. *Chem. Commun.* **2014**, *50*, 8211.
230 Bloom, S.; Knippel, J. L.; Lectka, T. *Chem. Sci.* **2014**, *5*, 1175.
231 Dong, C.; Huang, F.; Deng, H.; Schaffrath, C.; Spencer, J. B.; O'Hagan, D.; Naismith, J. H. *Nature* **2004**, *427*, 561.

[232] Eustáquio, A. S.; O'Hagan, D.; Moore, B. S. *J. Nat. Prod.* **2010**, *73*, 378.
[233] Walker, M. C.; Thuronyi, B. W.; Charkoudian, L. K.; Lowry, B.; Khosla, C.; Chang, M. C. Y. *Science* **2013**, *341*, 1089.
[234] Kwiatkowski, P.; Beeson, T. D.; Conrad, J. C.; MacMillan, D. W. C. *J. Am. Chem. Soc.* **2011**, *133*, 1738.
[235] Lam, Y.-h.; Houk, K. N. *J. Am. Chem. Soc.* **2014**, *136*, 9556.
[236] Phipps, R. J.; Hamilton, G. L.; Toste, F. D. *Nat. Chem.* **2012**, *4*, 603.
[237] Rauniyar, V.; Lackner, A. D.; Hamilton, G. L.; Toste, F. D. *Science* **2011**, *334*, 1681.
[238] Phipps, R. J.; Hiramatsu, K.; Toste, F. D. *J. Am. Chem. Soc.* **2012**, *134*, 8376.
[239] Phipps, R. J.; Toste, F. D. *J. Am. Chem. Soc.* **2013**, *135*, 1268.
[240] Wu, J.; Wang, Y.-M.; Drljevic, A.; Rauniyar, V.; Phipps, R. J.; Toste, F. D. *Proc. Natl. Acad. Sci. U.S.A.* **2013**, *110*, 13729.
[241] Zi, W.; Wang, Y.-M.; Toste, F. D. *J. Am. Chem. Soc.* **2014**, *136*, 12864.
[242] Yang, X.; Phipps, R. J.; Toste, F. D. *J. Am. Chem. Soc.* **2014**, *136*, 5225.
[243] Lin, J.-H.; Xiao, J.-C. *Tetrahedron Lett.* **2014**, *55*, 6147.
[244] Pupo, G.; Ibba, F.; Ascough, D. M. H.; Vicini, A. C.; Ricci, P.; Christensen, K. E.; Pfeifer, L.; Morphy, J. R.; Brown, J. M.; Paton, R. S.; Gouverneur, V. *Science* **2018**, *360*, 638.
[245] Zhu, X.; Robinson, D. A.; McEwan, A. R.; O'Hagan, D.; Naismith, J. H. *J. Am. Chem. Soc.* **2007**, *129*, 14597.
[246] Koh, M. J.; Nguyen, T. T.; Zhang, H.; Schrock, R. R.; Hoveyda, A. H. *Nature* **2016**, *531*, 459.
[247] Sano, S.; Matsumoto, T.; Nanataki, H.; Tempaku, S.; Nakao, M. *Tetrahedron Lett.* **2014**, *55*, 6248.
[248] Noël, T.; Maimone, T. J.; Buchwald, S. L. *Angew. Chem., Int. Ed.* **2011**, *50*, 8900.
[249] Kotun, S. P.; Prakash, G. K. S.; Hu, J. Pyridinium Poly(hydrogen fluoride). *Encyclopedia of Reagents for Organic Synthesis* [Online]; Wiley, Posted March 15, 2007.
[250] Davenport, K. G. Hydrogen Fluoride. *Encyclopedia of Reagents for Organic Synthesis* [Online]; Wiley, Posted April 15, 2001.
[251] Banks, R. E.; Mohialdin-Khaffaf, S. N.; Lal, G. S.; Sharif, I.; Syvret, R. G. *J. Chem. Soc., Chem. Commun.* **1992**, 595.
[252] Banks, R. E.; Murtagh, V.; An, I.; Maleczka, R. E. 1-(Chloromethyl)-4-fluoro-1,4-diazoniabicyclo [2.2.2]octane Bis(tetrafluoroborate). *Encyclopedia of Reagents for Organic Synthesis* [Online]; Wiley, Posted September 17, 2007.
[253] Nyffeler, P. T.; Durón, S. G.; Burkart, M. D.; Vincent, S. P.; Wong, C. H. *Angew. Chem., Int. Ed.* **2005**, *44*, 192.
[254] Poss, A. J. 4-Hydroxy-1-fluoro-1,4-diazoniabicyclo[2.2.2]octane Bis(tetrafluoroborate). *Encyclopedia of Reagents for Organic Synthesis* [Online]; Wiley, Posted October 15, 2003.
[255] Mu, X.; Zhang, H.; Chen, P.; Liu, G. *Chem. Sci.* **2014**, *5*, 275.
[256] Zhang, Q.; Yin, X. S.; Chen, K.; Zhang, S. Q.; Shi, B. F. *J. Am. Chem. Soc.* **2015**, *137*, 8219.

CUMULATIVE CHAPTER TITLES BY VOLUME

Volume 1 (1942)

1. **The Reformatsky Reaction:** Ralph L. Shriner

2. **The Arndt-Eistert Reaction:** W. E. Bachmann and W. S. Struve

3. **Chloromethylation of Aromatic Compounds:** Reynold C. Fuson and C. H. McKeever

4. **The Amination of Heterocyclic Bases by Alkali Amides:** Marlin T. Leffler

5. **The Bucherer Reaction:** Nathan L. Drake

6. **The Elbs Reaction:** Louis F. Fieser

7. **The Clemmensen Reduction:** Elmore L. Martin

8. **The Perkin Reaction and Related Reactions:** John R. Johnson

9. **The Acetoacetic Ester Condensation and Certain Related Reactions:** Charles R. Hauser and Boyd E. Hudson, Jr.

10. **The Mannich Reaction:** F. F. Blicke

11. **The Fries Reaction:** A. H. Blatt

12. **The Jacobson Reaction:** Lee Irvin Smith

Volume 2 (1944)

1. **The Claisen Rearrangement:** D. Stanley Tarbell

2. **The Preparation of Aliphatic Fluorine Compounds:** Albert L. Henne

3. **The Cannizzaro Reaction:** T. A. Geissman

4. **The Formation of Cyclic Ketones by Intramolecular Acylation:** William S. Johnson

5. **Reduction with Aluminum Alkoxides (The Meerwein-Ponndorf-Verley Reduction):** A. L. Wilds

6. **The Preparation of Unsymmetrical Biaryls by the Diazo Reaction and the Nitrosoacetylamine Reaction:** Werner E. Bachmann and Roger A. Hoffman

7. **Replacement of the Aromatic Primary Amino Group by Hydrogen:** Nathan Kornblum

8. **Periodic Acid Oxidation:** Ernest L. Jackson

9. **The Resolution of Alcohols:** A. W. Ingersoll

10. **The Preparation of Aromatic Arsonic and Arsinic Acids by the Bart, Béchamp, and Rosenmund Reactions:** Cliff S. Hamilton and Jack F. Morgan

Volume 3 (1946)

1. **The Alkylation of Aromatic Compounds by the Friedel-Crafts Method:** Charles C. Price

2. **The Willgerodt Reaction:** Marvin Carmack and M. A. Spielman

3. **Preparation of Ketenes and Ketene Dimers:** W. E. Hanford and John C. Sauer

4. **Direct Sulfonation of Aromatic Hydrocarbons and Their Halogen Derivatives:** C. M. Suter and Arthur W. Weston

5. **Azlactones:** H. E. Carter

6. **Substitution and Addition Reactions of Thiocyanogen:** John L. Wood

7. **The Hofmann Reaction:** Everett L. Wallis and John F. Lane

8. **The Schmidt Reaction:** Hans Wolff

9. **The Curtius Reaction:** Peter A. S. Smith

Volume 4 (1948)

1. **The Diels-Alder Reaction with Maleic Anhydride:** Milton C. Kloetzel

2. **The Diels-Alder Reaction: Ethylenic and Acetylenic Dienophiles:** H. L. Holmes

3. **The Preparation of Amines by Reductive Alkylation:** William S. Emerson

4. **The Acyloins:** S. M. McElvain

5. **The Synthesis of Benzoins:** Walter S. Ide and Johannes S. Buck

6. **Synthesis of Benzoquinones by Oxidation:** James Cason

7. **The Rosenmund Reduction of Acid Chlorides to Aldehydes:** Erich Mosettig and Ralph Mozingo

8. **The Wolff-Kishner Reduction:** David Todd

Volume 5 (1949)

1. **The Synthesis of Acetylenes:** Thomas L. Jacobs

2. **Cyanoethylation:** Herman L. Bruson

Volume 15 (1967)

1. **The Dieckmann Condensation:**   John P. Schaefer and Jordan J. Bloomfield

2. **The Knoevenagel Condensation:**   G. Jones

Volume 16 (1968)

1. **The Aldol Condensation:**   Arnold T. Nielsen and William J. Houlihan

Volume 17 (1969)

1. **The Synthesis of Substituted Ferrocenes and Other π-Cyclopentadienyl-Transition Metal Compounds:**   Donald E. Bublitz and Kenneth L. Rinehart, Jr.

2. **The γ-Alkylation and γ-Arylation of Dianions of β-Dicarbonyl Compounds:**   Thomas M. Harris and Constance M. Harris

3. **The Ritter Reaction:**   L. I. Krimen and Donald J. Cota

Volume 18 (1970)

1. **Preparation of Ketones from the Reaction of Organolithium Reagents with Carboxylic Acids:**   Margaret J. Jorgenson

2. **The Smiles and Related Rearrangements of Aromatic Systems:**   W. E. Truce, Eunice M. Kreider, and William W. Brand

3. **The Reactions of Diazoacetic Esters with Alkenes, Alkynes, Heterocyclic, and Aromatic Compounds:**   Vinod Dave and E. W. Warnhoff

4. **The Base-Promoted Rearrangements of Quaternary Ammonium Salts:**   Stanley H. Pine

Volume 19 (1972)

1. **Conjugate Addition Reactions of Organocopper Reagents:**   Gary H. Posner

2. **Formation of Carbon–Carbon Bonds via π-Allylnickel Compounds:**   Martin F. Semmelhack

3. **The Thiele-Winter Acetoxylation of Quinones:**   J. F. W. McOmie and J. M. Blatchly

4. **Oxidative Decarboxylation of Acids by Lead Tetraacetate:**   Roger A. Sheldon and Jay K. Kochi

Volume 20 (1973)

1. **Cyclopropanes from Unsaturated Compounds, Methylene Iodide, and Zinc-Copper Couple:**   H. E. Simmons, T. L. Cairns, Susan A. Vladuchick, and Connie M. Hoiness

2. **Sensitized Photooxygenation of Olefins:**   R. W. Denny and A. Nickon

3. **The Synthesis of 5-Hydroxyindoles by the Nenitzescu Reaction:**   George R. Allen, Jr.

4. **The Zinin Reaction of Nitroarenes:**   H. K. Porter

Volume 21 (1974)

Volume 22 (1975)

Volume 23 (1976)

Volume 24 (1976)

Volume 25 (1977)

Volume 26 (1979)

Volume 40 (1991)

1. **The Pauson-Khand Cycloaddition Reaction for Synthesis of Cyclopentenones:**
   Neil E. Schore

2. **Reduction with Diimide:**   Daniel J. Pasto and Richard T. Taylor

3. **The Pummerer Reaction of Sulfinyl Compounds:**   Ottorino DeLucchi, Umberto Miotti, and
   Giorgio Modena

4. **The Catalyzed Nucleophilic Addition of Aldehydes to Electrophilic Double Bonds:**
   Hermann Stetter and Heinrich Kuhlmann

Volume 41 (1992)

1. **Divinylcyclopropane-Cycloheptadiene Rearrangement:**   Tomáš Hudlický, Rulin Fan,
   Josephine W. Reed, and Kumar G. Gadamasetti

2. **Organocopper Reagents: Substitution, Conjugate Addition, Carbo/Metallo-cupration, and
   Other Reactions:**   Bruce H. Lipshutz and Saumitra Sengupta

Volume 42 (1992)

1. **The Birch Reduction of Aromatic Compounds:**   Peter W. Rabideau and Zbigniew
   Marcinow

2. **The Mitsunobu Reaction:**   David L. Hughes

Volume 43 (1993)

1. **Carbonyl Methylenation and Alkylidenation Using Titanium-Based Reagents:**
   Stanley H. Pine

2. **Anion-Assisted Sigmatropic Rearrangements:**   Stephen R. Wilson

3. **The Baeyer-Villiger Oxidation of Ketones and Aldehydes:**   Grant R. Krow

Volume 44 (1993)

1. **Preparation of α,β-Unsaturated Carbonyl Compounds and Nitriles by Selenoxide Elimination:**   Hans J. Reich and Susan Wollowitz

2. **Enone Olefin [2 + 2] Photochemical Cyclizations:**   Michael T. Crimmins and
   Tracy L. Reinhold

Volume 45 (1994)

1. **The Nazarov Cyclization:**   Karl L. Habermas, Scott E. Denmark, and Todd K. Jones

2. **Ketene Cycloadditions:**   John Hyatt and Peter W. Raynolds

Volume 46 (1994)

1. **Tin(II) Enolates in the Aldol, Michael, and Related Reactions:**   Teruaki Mukaiyama and
   Shū Kobayashi

Volume 97 (2018)

1. **[2+2+2] Cycloadditions of Alkynes with Heterocumulenes and Nitriles:**   Nicholas D. Staudaher, Ryan M. Stolley, and Janis Louis

2. **Amide-Forming Ligation Reactions:**   Vijaya R. Pattabiraman, Ayodele O. Ogunkoya, and Jeffrey W. Bode

Volume 98 (2019)

1. **The Saegusa Oxidation and Related Procedures:**   Jean LeBras and Jacques Muzart

2. **The Asymmetric Vinylogous Mukaiyama Aldol Reaction:**   Martin H. C. Cordes and Markus Kalesse

Volume 99 (2019)

1. **Addition of Non-Stabilized Carbon-Based Nucleophilic Reagents to Chiral *N*-Sulfinyl Imines:**   Melissa A. Herbage, Jolaine Savoie, Joshua D. Sieber, Jean-Nicolas Desrosiers, Yongda Zhang, Maurice A. Marsini, Keith R. Fandrick, Daniel Rivalti, and Chris H. Senanayake

2. **Iridium-Catalyzed, Enantioselective, Allylic Alkylations With Carbon Nucleophiles:**   Jian-Ping Qu, Günter Helmchen, Ze-Peng Yang, Wei Zhang, and Shu-Li You

Volume 100 (2020)

1. **The Negishi Cross-Coupling Reaction:**   Colin Diner and Michael G. Organ

2. **Generation and Trapping of Functionalized Aryl- and Heteroarylmagnesium and -Zinc Compounds:**   Paul Knochel, Ferdinand H. Lutter, Maximilian S. Hofmayer, Jeffrey M. Hammann, and Vladimir Malakhov

3. **Copper-Catalyzed, Enantioselective Hydrofunctionalization of Alkenes:**   Haoxuan Wang and Stephen L. Buchwald

4. **The Catalytic, Enantioselective Favorskii Reaction: in Situ Formation of Metal Alkynylides and Their Addition to Aldehydes:**   Yeshua Sempere and Erick M. Carreira

5. **Enantioselective Lithiation-Substitution of Nitrogen-Containing Heterocycles:**   Kevin Kasten, Nico Seling, and Peter O'Brien

6. **Catalytic, Enantioselective, Transfer Hydrogenation:**   Masahiro Yoshimura and Masato Kitamura

7. **Hydrofunctionalization of Alkenes by Hydrogen-Atom Transfer:**   Ryan A. Shenvi, Jeishla L. M. Matos, and Samantha A. Green

8. **Carbon-Carbon Bond Formation by Metallaphotoredox Catalysis:**   Eric D. Nacsa and David W. C. MacMillan

AUTHOR INDEX, VOLUMES 1–104

Volume number only is designated in this index

Organic Reactions, Vol. 104, Edited by P. Andrew Evans et al.
© 2020 Organic Reactions, Inc. Published in 2020 by John Wiley & Sons, Inc.

CHAPTER AND TOPIC INDEX, VOLUMES 1–104

Many chapters contain brief discussions of reactions and comparisons of alternative synthetic methods related to the reaction that is the subject of the chapter. These related reactions and alternative methods are not usually listed in this index. In this index, the volume number is in **boldface**, the chapter number is in ordinary type.

Organic Reactions, Vol. 104, Edited by P. Andrew Evans et al.
© 2020 Organic Reactions, Inc. Published in 2020 by John Wiley & Sons, Inc.